THE CAMBRIDGE ENCYCLOPEDIA OF DARWIN AND EVOLUTIONARY THOUGHT

This volume is a comprehensive reference work on the life, labors, and influence of the great evolutionist Charles Darwin. With more than sixty essays written by an international group representing the leading scholars in the field, this is the definitive work on Darwin. It covers the background to Darwin's discovery of the theory of evolution through natural selection, the work he produced and his contemporaries' reactions to it, and his influence on science in the 150 years since the publication of *Origin of Species*. It also explores the implications of Darwin's discoveries in religion, politics, gender, literature, culture, philosophy, and medicine, critically evaluating Darwin's legacy. Fully illustrated and clearly written, it is suitable for scholars and students as well as the general reader. The wealth of information it provides about the history of evolutionary thought makes it a crucial resource for understanding the controversies that surround evolution today.

MICHAEL RUSE is Lucyle T. Werkmeister Professor of Philosophy and Director of the Program in the History and Philosophy of Science at Florida State University. He is the author of twenty books and the founding editor of *Biology and Philosophy*.

A portrait of Charles Darwin by the noted watercolorist George Richmond, painted in 1839 in celebration of his marriage to his first cousin Emma Wedgwood. Permission: English Heritage

THE CAMBRIDGE ENCYCLOPEDIA OF
DARWIN AND EVOLUTIONARY THOUGHT

Edited by

MICHAEL RUSE

Florida State University

CAMBRIDGE UNIVERSITY PRESS
Cambridge, New York, Melbourne, Madrid, Cape Town,
Singapore, São Paulo, Delhi, Mexico City

Cambridge University Press
32 Avenue of the Americas, New York, NY 10013-2473, USA

www.cambridge.org
Information on this title: www.cambridge.org/9780521195317

© Cambridge University Press 2013

This publication is in copyright. Subject to statutory exception
and to the provisions of relevant collective licensing agreements,
no reproduction of any part may take place without the written
permission of Cambridge University Press.

First published 2013

Printed in the United States of America

A catalog record for this publication is available from the British Library.

Library of Congress Cataloging in Publication Data
The Cambridge encyclopedia of Darwin and evolutionary thought /
[edited by] Michael Ruse.
p. cm.
Includes bibliographical references and index.
ISBN 978-0-521-19531-7 (hardback)
1. Darwin, Charles, 1809–1882 – Encyclopedias. 2. Evolution
(Biology) – Encyclopedias. I. Ruse, Michael.
QH360.2.C36 2012
576.8'203–dc23   2012010226

ISBN 978-0-521-19531-7 Hardback

Cambridge University Press has no responsibility for the persistence or
accuracy of URLs for external or third-party Internet Web sites referred
to in this publication and does not guarantee that any content on such
Web sites is, or will remain, accurate or appropriate.

Contents

Contributors	*page* vii
Preface	xi
Acknowledgments	xv

    Introduction 1

1 Origins and the Greeks 32
    Jeremy Kirby

2 Evolution before Darwin 39
    Michael Ruse

3 Charles Darwin's Geology: The Root of His Philosophy of the Earth 46
    David Norman

4 Looking Back with "Great Satisfaction" on Charles Darwin's Vertebrate Paleontology 56
    Paul D. Brinkman

5 The Origins of the *Origin*: Darwin's First Thoughts about the Tree of Life and Natural Selection, 1837–1839 64
    Jonathan Hodge

6 Darwin and Taxonomy 72
    Mary Pickard Winsor

7 Darwin and the Barnacles 80
    Marsha L. Richmond

8 The Analogy between Artificial and Natural Selection 88
    Bert Theunissen

9 The *Origin of Species* 95
    Michael Ruse

10 Sexual Selection 103
    Richard A. Richards

11 Darwin and Species 109
    James Mallet

12 Darwin and Heredity 116
    Robert Olby

13 Darwin and Time 124
    Keith Bennett

14 Darwin's Evolutionary Botany 131
    Richard Bellon

15 Mimicry and Camouflage 139
    William Kimler and Michael Ruse

16 Chance and Design 146
    John Beatty

17 Darwin and Teleology 152
    James G. Lennox

18 The Evolution of the *Origin* (1859–1872) 158
    Thierry Hoquet

19 Alfred Russel Wallace 165
    John van Wyhe

20 Darwin and Humans 173
    Gregory Radick

21 Darwin and Language 182
    Stephen G. Alter

22 Darwin and Ethics 188
    Eric Charmetant

23 Social Darwinism 195
    Naomi Beck

24 Darwin and the Levels of Selection 202
    Daniel Deen, Brian Hollis, and Chris Zarpentine

25 Darwin and Religion 211
    Mark Pallen and Alison Pearn

26 Darwinism in Britain 218
    Peter Bowler

27 Darwinism in the United States, 1859–1930 226
   Mark A. Largent

28 The German Reception of Darwin's Theory,
   1860–1945 235
   Robert J. Richards

29 Darwin and Darwinism in France before 1900 243
   Jean Gayon

30 Encountering Darwin and Creating Darwinism
   in China 250
   Yang Haiyan

31 Darwinism in Latin America 258
   Thomas F. Glick

32 Botany: 1880s to 1920s 264
   Dawn Mooney Digrius

33 Population Genetics 273
   Michael Ruse

34 Synthesis Period in Evolutionary Studies 282
   Joe Cain

35 Ecological Genetics 293
   David W. Rudge

36 Darwin and Darwinism in France after 1900 300
   Jean Gayon

37 Botany and the Evolutionary Synthesis, 1920–1950. . . 313
   Vassiliki Betty Smocovitis

38 The Emergence of Life on Earth and the
   Darwinian Revolution 322
   Iris Fry

39 The Evolution of the Testing of Evolution 330
   Steven Hecht Orzack

40 Mimicry and Camouflage: Part Two 336
   Joseph Travis

41 The Tree of Life 340
   Joel D. Velasco

42 Sociobiology. 346
   Mark E. Borrello

43 Evolutionary Paleontology. 353
   David Sepkoski

44 Darwin and Geography 361
   David N. Livingstone

45 Darwin and the Finches 368
   Frederick Rowe Davis

46 Developmental Evolution. 375
   Manfred D. Laubichler and Jane Maienschein

47 Darwin's Evolutionary Ecology 383
   James Justus

48 Darwin and the Environment 391
   David Steffes

49 Molecular Biology: Darwin's Precious Gift 397
   Francisco J. Ayala

50 Challenging Darwinism: Expanding, Extending,
   Replacing 405
   David J. Depew and Bruce H. Weber

51 Human Evolution after Darwin 412
   Jesse Richmond

52 Language Evolution since Darwin 420
   Barbara J. King

53 Cultural Evolution 428
   Kenneth Reisman

54 Literature 436
   Gowan Dawson

55 Darwin and Gender 443
   Georgina M. Montgomery

56 Evolutionary Epistemology 451
   Tim Lewens

57 Ethics after Darwin. 461
   Richard Joyce

58 Darwin and Protestantism 468
   Diarmid A. Finnegan

59 Creationism 476
   Ronald L. Numbers

60 Darwin and Catholicism 485
   John F. Haught

61 Judaism, Jews, and Evolution. 493
   Marc Swetlitz

62 Religion: Islam 499
   Martin Riexinger

63 From Evolution and Medicine to Evolutionary
   Medicine. 505
   Tatjana Buklijas and Peter Gluckman

Bibliography 515
Index 551
Color illustrations follow pages 94, 130, and 382

Contributors

STEPHEN G. ALTER is an associate professor of history at Gordon College, Wenham, Massachusetts. He is the author of *William Dwight Whitney and the Science of Language* (2005). During 2010 he held a sabbatical year fellowship from the American Philosophical Society.

FRANCISCO J. AYALA is University Professor and Donald Bren Professor of Biological Sciences at the University of California, Irvine. In 2001 he received the U.S. National Medal of Science and in 2010, the Templeton Prize. His research focuses on molecular evolution, as well as on the philosophy of biology. For further information, see http://www.faculty.uci.edu/profile.cfm?faculty_id=2134.

JOHN BEATTY teaches history and philosophy of science, and social and political philosophy at the University of British Columbia. His research focuses on the theoretical foundations, methodology, and sociopolitical dimensions of evolutionary biology.

NAOMI BECK is a Research Fellow at the Max Planck Institute for Economics in Jena, Germany. Her recent research examines Nobel Laureate Friedrich A. von Hayek's use of evolutionary concepts and theories in defense of free-market economics.

RICHARD BELLON teaches the history of science and science policy at Michigan State University, where he holds a joint appointment in the Lyman Briggs College and the Department of History. His most recent research examines the role of botany in the reception of the *Origin of Species* and the moral status of the inductive method in Victorian British science.

KEITH BENNETT is professor of Late Quaternary environmental change at Queen's University Belfast. He held a Royal Society–Wolfson Merit Award (2007–12) and has recently been elected to the Royal Irish Academy.

MARK E. BORRELLO is an associate professor in the Graduate Program for the History of Science, Technology and Medicine and the Department of Ecology, Evolution and Behavior at the University of Minnesota. He is the author of *Evolutionary Restraints: The Contentious History of Group Selection* (2010).

PETER BOWLER is professor emeritus of the history of science at Queen's University, Belfast. He is a Fellow of the British Academy and a member of the Royal Irish Academy.

PAUL D. BRINKMAN is assistant director of the Paleontology and Geology Research Lab at the North Carolina Museum of Natural Sciences. His most recent book is entitled *The Second Jurassic Dinosaur Rush: Museums and Paleontology in America at the Turn of the Twentieth Century* (2010).

TATJANA BUKLIJAS is a historian of science and medicine and Research Fellow of the Liggins Institute at the University of Auckland. Her key research interests are history of anatomy and embryology, development and disease, medicine and evolution, and science and medicine in Central Europe.

JOE CAIN is professor of history and philosophy of biology at University College London. He edited a new edition of Darwin's *The Expression of the Emotions in Man and Animals* (2nd ed., 2009).

ERIC CHARMETANT is associate professor of philosophy at Centre Sèvres–Jesuit Faculties of Paris. He recently published "Darwin et l'éthique – Une rencontre précoce, un chantier toujours ouvert" (2010).

FREDERICK ROWE DAVIS is associate professor of history and the history and philosophy of science at Florida State University. His first book is *The Man Who Saved Sea Turtles: Archie Carr and the Origins of Conservation Biology* (2007).

GOWAN DAWSON is senior lecturer in Victorian studies at the University of Leicester. He is the author of *Darwin, Literature and Victorian Respectability* (Cambridge, 2007), and coauthor of *Science in the Nineteenth-Century Periodical: Reading the Magazine of Nature* (Cambridge, 2004).

DANIEL DEEN is an assistant professor of philosophy at Concordia University, Irvine, California. His interests are in the philosophy of religion and the philosophy of science.

Contributors

DAVID J. DEPEW is professor emeritus of communication studies and rhetoric of inquiry, University of Iowa. He is coauthor with Bruce Weber of *Darwinism Evolving: Systems Dynamics and the Genealogy of Natural Selection* (1995) and coauthor with Marjorie Grene of *Philosophy of Biology: An Episodic History* (2004).

DAWN MOONEY DIGRIUS is assistant professor of history at Stevens Institute of Technology. Her chapter on Gregor Mendel and his contributions to evolution is included in Greenwood Press's *Icons of Evolution*, edited by Brian Regal (2007).

DIARMID A. FINNEGAN is lecturer in human geography at the Queen's University of Belfast. He is the author of *Natural History Societies and Civic Culture in Victorian Scotland* (2009) and has published widely on the historical geographies of scientific knowledge in the nineteenth century. He is currently working on the rhetorical geography of science and religion debates in mid-Victorian Britain and Ireland.

IRIS FRY teaches in the Department of Humanities and Arts at the Technion–Israel Institute of Technology. Her book *The Emergence of Life on Earth: A Historical and Scientific Overview* (2000) was published by Rutgers. She has published articles on the science, history, and philosophy of the origin-of-life problem and on the interaction between science and religion.

JEAN GAYON is professor of philosophy and history of science at the University of Paris–Panthéon Sorbonne. He is author of *Darwin's Struggle for Survival: Heredity and the Natural Selection Hypothesis* (Cambridge, 1998).

THOMAS F. GLICK is professor of history at Boston University. He received an honorary doctorate from the University of Valencia in 2010.

PETER GLUCKMAN, FRS, is University Distinguished Professor, professor of pediatric and perinatal biology, and head of the Centre for Human Evolution, Adaptation and Disease of the Liggins Institute at the University of Auckland, specializing in fetal and developmental origins of disease. He has written extensively on evolutionary medicine.

JOHN F. HAUGHT, PhD, is a Senior Fellow at Woodstock Theological Center at Georgetown University and author of *Making Sense of Evolution: Darwin, God, and the Drama of Life* (2010).

JONATHAN HODGE, at the University of Leeds, researches the history and philosophy of theories of creation and evolution. He is coeditor, with Gregory Radick, of *The Cambridge Companion to Darwin* (2nd ed., 2009). Two volumes reprint many of his papers: *Before and After Darwin* (2008) and *Darwin Studies* (2009).

BRIAN HOLLIS is a postdoctoral researcher at the University of Lausanne. He studies sexual selection and social systems using experimental evolution.

THIERRY HOQUET teaches history and philosophy of biology at the Université Jean Moulin Lyon 3, where he is a professor in the Faculty of Philosophy. He is the scientific editor of the Web site http://www.buffon.cnrs.fr/ and in 2009 published *Darwin contre Darwin*.

RICHARD JOYCE is professor of philosophy at Victoria University of Wellington, New Zealand. He previously worked in Australia and the United Kingdom after gaining his PhD from Princeton. He is author of *The Myth of Morality* (Cambridge, 2001) and *The Evolution of Morality* (2006) as well as numerous journal articles and book chapters, mostly in the field of metaethics.

JAMES JUSTUS is an assistant professor in the Departments of Philosophy and History and Philosophy of Science at Florida State University, and he was recently a postdoctoral Research Fellow at the Sydney Centre for the Foundations of Science. His research focuses on philosophy of science (biology in particular), environmental philosophy, formal epistemology, and logical empiricism.

WILLIAM KIMLER is associate professor of History and Alumni Distinguished Undergraduate Professor at North Carolina State University. Recent public lectures include Darwin Day events and an American Scientist podcast on Darwin.

BARBARA J. KING, a biological anthropologist, teaches at the College of William and Mary. Her latest book, *How Animals Grieve* (2013), and her contributions to NPR's 13.7 *Cosmos and Culture* blog (see www.barbarajking.com) focus on animal cognition/emotion and the animal-human bond.

JEREMY KIRBY is associate professor of philosophy at Albion College. He is the author of *Aristotle's Metaphysics: Form, Matter, and Identity* (2011) and of several articles concerning epistemology in antiquity.

MARK A. LARGENT is an associate professor in James Madison College at Michigan State University, where he teaches courses in the history of science and in science policy. He is the author of *Breeding Contempt: The History of Coerced Sterilization in the United States* (2011) and *Vaccine: The Debate in Modern America* (2012).

MANFRED D. LAUBICHLER is President's Professor at Arizona State University, where he directs the Center for Social Dynamics and Complexity and is associate director of the Origins Project; adjunct professor at the Marine Biology Laboratory in Woods Hole, Massachusetts; external professor at the Santa Fe Institute; external faculty member at the Konrad Lorenz Institute in Altenberg, Austria; and visiting scholar at the Max Planck Institute for the History of Science in Berlin.

JAMES G. LENNOX is professor of history and philosophy of science at the University of Pittsburgh. Recent publications include "Darwinism and Neo-Darwinsim" in the *Blackwell Companion to the Philosophy of Biology* (2008) and "The

Darwin/Gray Correspondence, 1857–1869: An Intelligent Discussion about Chance and Design" (*Perspectives on Science*, 2010). He is author of *Aristotle's Philosophy of Biology* (Cambridge, 2001) and coeditor of *Being, Nature, and Life in Aristotle* (Cambridge, 2010), and he has held fellowships at Clare Hall, Cambridge (1986–87), and the Istituto di Studi Avanzati, University of Bologna (2006).

TIM LEWENS is reader in philosophy of the sciences at the University of Cambridge, where he is also a Fellow of Clare College. His publications include *Darwin* (2007).

DAVID N. LIVINGSTONE is professor of geography and intellectual history at Queen's University Belfast and a Fellow of the British Academy. His most recent book is *Adam's Ancestors: Race, Religion and the Politics of Human Origins* (2008). He was awarded the Founder's Medal of the Royal Geographical Society in 2011 and has been invited to deliver the Gifford Lectures in 2014.

JANE MAIENSCHEIN is Regents' Professor, President's Professor, and Parents Association Professor at Arizona State University, where she directs the Center for Biology and Society, and adjunct professor and director of the HPS Program at the Marine Biological Laboratory in Woods Hole, Massachusetts.

JAMES MALLET is professor of biological diversity at University College London. His research is mainly on evolutionary genetics, particularly on color pattern mimicry and origins of species in butterflies of the Amazon basin. He has also published a number of historical papers on Darwin, Wallace, and their species concept.

GEORGINA M. MONTGOMERY is an assistant professor in Lyman Briggs College and the Department of History at Michigan State University. She recently coedited *Making Animal Meaning* (2011), a collection of ten original essays, three that capture some of the most compelling theoretical underpinnings of animal meaning and seven that illustrate meaning making through studies of specific spaces, species, and human-animal relations.

DAVID NORMAN is director of the Sedgwick Museum of Earth Sciences, reader in Vertebrate Palaeobiology at the University of Cambridge, and Odell Fellow in the Natural Sciences at Christ's College, Cambridge. His most recent book is *Dinosaurs* (2005) in the prestigious Very Short Introductions series published by Oxford University Press, and in 2011 he published a lengthy scientific description of the skull of the very early ornithischian dinosaur *Heterodontosaurus* in the *Zoological Journal of the Linnean Society of London*.

RONALD L. NUMBERS is the Hilldale Professor of the History of Science and Medicine at the University of Wisconsin, Madison. Among his books are *The Creationists: From Scientific Creationism to Intelligent Design* (2006) and *Darwinism Comes to America* (1998), both published by Harvard University Press.

ROBERT OLBY is a historian of the biological sciences. Trained in the United Kingdom, he has taught at Leeds and Pittsburgh. His most recent book, *Francis Crick: Hunter of Life's Secrets*, was published in 2009: Now he has returned to work on the history of genetics.

STEVEN HECHT ORZACK is a senior research scientist at the Fresh Pond Research Institute. His current research concerns conservation biology, demography, human genetics, evolutionary biology, and the history and philosophy of biology.

MARK PALLEN is a medical microbiologist at the University of Birmingham with wide-ranging interests in evolution, Darwin, and the history of ideas. He is author of the *Rough Guide to Evolution* (2009) and lives in Malvern a few hundred yards from where Darwin stayed in 1849 and 1851.

ALISON PEARN is the associate director of the Darwin Correspondence Project at the University of Cambridge. She has jointly edited nine of the first nineteen volumes of the definitive edition of the *Correspondence of Charles Darwin* (1985–).

GREGORY RADICK is professor of history and philosophy of science at the University of Leeds. His book *The Simian Tongue: The Long Debate about Animal Language* (2007) was awarded the 2010 Suzanne J. Levinson Prize of the History of Science Society.

KENNETH REISMAN holds a PhD from Stanford University, where he studied philosophy of science and evolutionary biology. He is now a consultant with McKinsey & Company.

RICHARD A. RICHARDS, professor of philosophy at the University of Alabama, writes on Darwin and the philosophy and history of biology. He recently published *The Species Problem*, with Cambridge University Press (2010).

ROBERT J. RICHARDS is the Morris Fishbein Professor of the History of Science and Medicine at the University of Chicago. He has written several books on the impact of evolutionary theory and on German intellectual thought. Currently he is coauthoring a book with Michael Ruse entitled *Debating Darwin*.

JESSE RICHMOND earned his PhD in History and Science Studies from the University of California, San Diego, in 2009. He now teaches courses in the history of science at Union College in Schenectady, New York.

MARSHA L. RICHMOND is an associate professor in the Department of History, Wayne State University, Detroit, Michigan. A former editor of the *Correspondence of Charles Darwin* (1985–) and winner of the 2010 Margaret W. Rossiter History of Women in Science Prize of the History of Science Society, her research focuses on heredity studies from Darwin through classical genetics.

MARTIN RIEXINGER, MA Tübingen, PhD Freiburg, Habilitation Göttingen, is associate professor at the Unit for Arabic and Islamic Studies at Aarhus University (Denmark).

Contributors

DAVID W. RUDGE is an associate professor at Western Michigan University. He has published extensively on H. B. D. Kettlewell's classic investigations of the phenomenon of industrial melanism from historical, philosophical, and science education perspectives.

MICHAEL RUSE is the author of *The Darwinian Revolution: Science Red in Tooth and Claw* (1979, 2nd ed., 1999). He is the coeditor with Robert J. Richards of *The Cambridge Companion to the* Origin of Species (2009).

DAVID SEPKOSKI is a Research Scholar at the Max Planck Institute for the History of Science in Berlin. His most recent book is *Rereading the Fossil Record: The Growth of Paleobiology as an Evolutionary Discipline* (2012).

VASSILIKI BETTY SMOCOVITIS is Distinguished Alumni Professor 2009–11 at the University of Florida and teaches the history of science in the Departments of Biology and History. She is the author of *Unifying Biology: The Evolutionary Synthesis and Evolutionary Biology* (1996) and is completing a scientific biography of G. Ledyard Stebbins Jr.

DAVID STEFFES received his PhD in the history of science from the University of Oklahoma in 2008. He has recently held postdoctoral fellowships from the National Science Foundation at Arizona State University (2009–11) and Florida State University (2008–9), where he has researched the historical intersection of ecology, evolutionary biology, and environmental ideology in the twentieth century.

MARC SWETLITZ holds a PhD from the University of Chicago and taught at the Massachusetts Institute of Technology and the University of Oklahoma. Swetlitz coedited, along with Geoffrey Cantor, and contributed to *Jewish Tradition and the Challenge of Darwinism* (2006).

BERT THEUNISSEN is professor of the history of science at the Descartes Centre for the History of the Sciences and the Humanities at Utrecht University, the Netherlands. His current work focuses on the history of animal breeding.

JOSEPH TRAVIS is the Robert O. Lawton Distinguished Professor of Biological Science and former dean of the College of Arts and Sciences at Florida State University. With Michael Ruse, he is the editor of *Evolution: The First Four Billion Years*, published by Harvard University Press in 2009.

JOHN VAN WYHE is a senior lecturer at the National University of Singapore who specializes in Darwin and Wallace. He has edited both *The Complete Work of Charles Darwin Online* (http://darwin-online.org.uk/) and *Wallace Online* (http://wallace-online.org/). He is the author or editor of a number of books, including *Charles Darwin's Notebooks from the Voyage of the* Beagle (2009), edited with John Chancellor, and *Charles Darwin's Shorter Publications, 1829–1883* (2009). He is completing a book on Wallace's voyage in the Malay archipelago and the independent path to evolution by natural selection.

JOEL D. VELASCO received his PhD from the University of Wisconsin, Madison. He has been an Andrew W. Mellon Fellow in the Humanities at Stanford University and a visiting professor at Cornell. He is currently the Ahmanson Postdoctoral Instructor in Philosophy at the California Institute of Technology.

BRUCE H. WEBER is Professor of Biochemistry Emeritus at California State University, Fullerton, as well as Robert H. Woodworth Chair Emeritus in Science and Natural Philosophy at Bennington College. He is the coauthor with David Depew of *Darwinism Evolving: Systems Dynamics and the Genealogy of Natural Selection* (1995).

MARY PICKARD WINSOR (known as Polly) retired in 2004 from teaching history of biology, especially systematics and evolution, at the Institute for History and Philosophy of Science and Technology (IHPST), University of Toronto. Since that time she has spoken at conferences and published several articles, including "Creation of the Essentialism Story: An Exercise in Metahistory" in *History and Philosophy of the Life Sciences*, and "Taxonomy Was the Foundation of Darwin's Evolution" in *Taxon*.

YANG HAIYAN is an associate professor in the Department of Medical Humanities at Peking University and was a visiting scholar (2008–9) in the Department of History and Philosophy of Science at the University of Cambridge. With a PhD in the history and philosophy of science from the Department of Philosophy, Peking University, her research focuses on cultural history of biology and medical humanities.

CHRIS ZARPENTINE recently received his PhD from Florida State University. He currently teaches at the University of Utah.

Preface

CHARLES ROBERT DARWIN, the fifth child (of six) and second son of Dr. Robert Darwin and his wife Susannah (formerly Wedgwood) of Shrewsbury (pronounced Shrowsberry), a town in the English Midlands next to the Welsh border, was born on 12 February 1809 (the same day as Abraham Lincoln across the Atlantic) (Fig. Preface.1). He was sent to one of England's famous public (in reality private) schools and then at a young age was directed north, to Edinburgh, the capital of Scotland, to study medicine. After two years he realized that medicine was not for him, and so he moved south to Cambridge, to work for a degree and prepare for the life of a clergyman in the Church of England. He graduated in 1831.

Through connections he had made as a student, Darwin was offered the chance to join the British warship HMS *Beagle*, as it set off for South America to map the coastline (Plate I). The voyage took five years, eventually going all the way around the world, returning to England in 1836. By this time, all thoughts of a clerical life had vanished, and Darwin, supported by family money, settled into full-time work as a scientist. He became an evolutionist shortly after the *Beagle* voyage and discovered the mechanism for which he is famous, natural selection, in 1838. He married his first cousin Emma Wedgwood early in 1839 (Fig. Preface.2), and by that time, starting to show the signs of a still-unknown illness that plagued him for the rest of his life, he settled into the role of a somewhat reclusive invalid. The couple moved to a house in Kent and in all had ten children, seven of whom lived to maturity.

Darwin did not publish for twenty years, and then did so only because a young naturalist, Alfred Russel Wallace, sent him an essay containing virtually the same ideas that he had discovered in the late 1830s. Rapidly Darwin wrote up his theory, and *On the Origin of Species by Means of Natural Selection, or the Preservation of Favoured Races in the Struggle for Life* appeared late in 1859 (Plate II). The work caused great controversy and was attacked by many, including leading churchmen (notably Samuel Wilberforce, bishop of Oxford), and defended by many others, including leading scientists (notably Thomas Henry Huxley, morphologist, paleontologist, college teacher and administrator, and general man of letters). Twelve years after *Origin*, in 1871, Darwin followed with a work on our own species, *The Descent of Man and Selection in Relation to Sex*. He died at home, his heart exhausted, in April

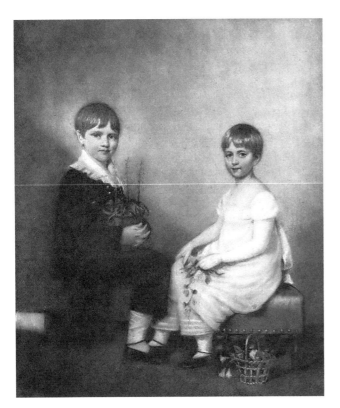

FIGURE PREFACE.1. Charles Darwin at about age six with his younger sister at about age four. From H. E. Litchfield, *Emma Darwin, Wife of Charles Darwin: A Century of Family Letters* (Cambridge: privately printed by Cambridge University Press, 1904)

FIGURE PREFACE.2. An etching of the partner picture of Emma Wedgwood, by George Richmond, painted to mark her marriage to Charles Darwin. From H. E. Litchfield, *Emma Darwin, Wife of Charles Darwin: A Century of Family Letters* (Cambridge: privately printed by Cambridge University Press, 1904)

1882. By general acclaim he was buried in Westminster Abby, where he lies today, next to the great Isaac Newton.

Darwin is famous. Darwin is controversial. In England to this day his memory is cherished, and his name honored. His face, nearly covered by his full beard, is on the back of the ten-pound note, a successor to that other great Victorian, Charles Dickens. Elsewhere also the name of Darwin is held in high esteem. Yet in many parts of America, and increasingly in other areas of the world, he is taken to be the apotheosis of all that is wrong with modern society: parents and teachers, church leaders and politicians, do all that they can to exclude him from the classroom. The Darwinian Revolution ranks up there in the history of science with the Copernican Revolution. No one today doubts that the Earth goes around the sun. Many today doubt that we humans are modified monkeys.

This is an encyclopedia about Darwin and his influence, written by a team of experts, drawn from the ranks of practicing scientists as well as from those on the side of the humanities. We are united in the conviction that Charles Darwin and his work were and are very important – in science and in many other fields of human activity and inquiry, including philosophy, theology, linguistics, and literature. If we do not infect you with our enthusiasm and leave you sharing our conviction, we have failed in our task. What we are not trying to do is convince you that Darwin was always right. He was not. Nor, conversely, are we trying to show you that Darwin was basically wrong. He was not. And we are certainly not trying to show that, right or wrong, overall Darwin's influence is either malicious or overrated. This is not true, although we agree fully that things are far more complex than simple good or ill. Finally, thank goodness, we are not trying to show that everything is known and that everyone agrees. You will see in these pages that often we differ among ourselves about some very important points. This is a good part of what makes it all so exciting. Charles Darwin was one of the towering figures in Western civilization, and his legacy is with us still today. We want to share with you our knowledge and our thrill at great ideas. Some monkeys! Some modification!

The volume is intended to be entire unto itself, but you might want to flesh out your reading by turning to some original sources. The *Origin of Species* is a remarkably readable book for a classic. All references in this volume, unless it is explicitly stated otherwise, are to the first edition, and this is the one that you should read. (You can tell if you have the first edition because the work does not contain the alternative name for natural selection, "survival of the fittest," added to some of the later editions.) There is a facsimile (paperback) edition, with a short introduction by the eminent, twentieth-century evolutionist Ernst Mayr, published by Harvard University Press. All of Darwin's published material is now online (*The Complete Work of Charles Darwin Online*, http://darwin-online.org.uk/).

Thanks to immense labors by John van Wyhe, it is a wonderful resource with much supplementary material, including reviews and the like. It includes the invaluable bibliography *Darwin: A Reader's Guide*, by Michael T. Ghiselin.

For almost three decades now a dedicated team of researchers has been producing the definitive edition of Darwin's voluminous correspondence. In this *Encyclopedia*, published letters are referred to by volume and page, and also identify the sender and recipient. Thus: Darwin 1985–, 14:423, letter to M. E. Boole, 14 December 1866. There is much work still to be done. It is possible to access online almost all of the letters thus far edited, with synopses of letters yet to be edited and published. Consult the "Darwin Correspondence Project," http://www.darwinproject.ac.uk/. Unpublished letters, part of the Correspondence Project, are referred to by catalog number: DCP, 9105, letter to G. H. Darwin, 21 October [1873]. The notebooks in which Darwin worked out his ideas about evolution were transcribed, edited, and published in 1987 by a team headed by Paul Barrett. References are to this edition, with name and page of the particular notebook. Thus: Barrett et al. 1987, D2. Unpublished manuscripts in the Darwin Archives at the University Library, Cambridge, are referenced by catalog number. Thus: CUL DAR 210.8: 42.

Acknowledgments

"I am rather sorry that you are Editor, as I have always heard that an Editor's life is one of ceaseless trouble & anxiety." Sharing with Jane Austen and Michael Ruse a lifelong inability to master the spelling of the English language, this is from a letter written by Charles Darwin, on 24 December 1866, to Benjamin Dann Walsh, an American-residing entomologist who had been at Cambridge at the same time as Darwin and who was the associate editor of the *Practical Entomologist*. Forget the orthography and focus on the sentiment. Never were truer words said! Walsh had no illusions. After the failure of the *Practical Entomologist*, he took up the editorship of the *American Entomologist*. On 29 August 1868, quoting Proverbs, he wrote to Darwin: "I have recently returned like a dog to his vomit, & again become Editor of a Monthly Periodical (of which I enclose a Prospectus) devoted to Economic Entomology."

So one asks oneself why one takes time and effort to edit a volume such as this. The most obvious reason does not apply. There can be no pretense that this volume is designed to introduce readers to a new and growing field. To use a metaphor, Darwin Studies is a field very well plowed indeed. But that in a way gives us the reason. It became clear after 2009, the year of many conferences and publications celebrating the 200th anniversary of the birth of Charles Darwin and the 150th anniversary of the publication of his great book, *On the Origin of Species*, that there was huge interest in his ideas and their consequences. Much work was ongoing and many new and interesting facts were coming to light and equally new and interesting interpretations being offered. It was the genius of my editor at Cambridge University Press, Beatrice Rehl, to see that this was so and that there was now need of a volume that gathered all together in one place to give people a full sense of scholarship today on Darwin and his importance. This was a need obviously intensified by the fact that Darwin's work is still highly controversial in the United States and increasingly elsewhere and that how we think about the topic has immediate consequences for education and much more.

I agreed with her, and after some hesitation – hesitation that would have been much greater had I realized the work involved – I committed myself to editing such a volume. So let me start by thanking all of my contributors (and some who in the end did not join the collection) for sharing with me my enthusiasm for such a project and for putting up with my constant prodding and pushing. Ultimately this volume stands or falls by the essays that they have written. I should say that it has been very gratifying to have many very senior people join in the project when the last thing they needed was one more invitation to write a piece for a collection. It has been no less gratifying to find younger people working on all aspects of Darwin and his influence, and to them also I extend my thanks for taking this project so seriously. And it must be said that I am very much in the debt of my scientist authors who saw that we all have much to gain when scholars from different sides of the campus come together to share in understanding the work of a great man and his ever-widening influence of so many aspects of life today. One of the real joys of a job such as this is bringing into the public light work that you have long thought terrific, albeit regrettably unknown. Let me embarrass William Kimler by saying that I was determined to have his wonderful work on mimicry in this volume, and now I do.

Contributors are vital. So also is the editor. Such a volume as this cannot be just a collection of pieces, however good. The editor has to have an underlying and unifying vision. That sounds a little pretentious, but it is true. My vision starts with the fact that ultimately and fundamentally Charles Darwin was a scientist – not a theologian, not a philosopher, not a literary critic, whatever his influence in those fields. This is the thread, the backbone, of this book, and if you do not see this, then I have failed myself and I have failed you. On this linking theme I have tried to build, to add and extend, in many different directions – those connected to the areas just mentioned like religion, philosophy, literature, as well as to other fields like politics, ecology (both in the more technical scientific sense and in the more popular value-laden sense), feminist theory, and more. I wanted also to show that although Darwin may have been the quintessential Englishman, his influence spread out across the world, changing and yet being changed by the cultures into which it entered. I am particularly pleased that I was able to conclude the volume with an essay on one of the newest areas of evolutionary thinking, its application to medicine. I am very grateful to one of today's most eminent scientist-physicians for having agreed to coauthor this piece.

My thinking about Darwin and his importance was formed by and has persisted from a year (1972–73) that I spent at Cambridge University attached to the Department of History and Philosophy of Science. I like to joke that I rarely agree with the opinions of the Marxist scholar Robert M. Young, and he never agrees with mine, but I still think that his was the most original mind that turned to the study of Darwin. His influence was reinforced by contact with the great historian of geology Martin Rudwick, as well as by Roy Porter, then the equivalent of a postdoctoral student, and the future historian of medicine William Bynum. Across in the University Library, in charge of the Darwin Archives, was the ever-knowledgeable and helpful Peter Gautrey. Always available and willing to talk and share ideas was Sydney Smith of St. Catharine's College, who concealed a keen intellect and immense background understanding behind the facade of being the archetypal, old English buffer.

In the four decades since that tremendous year I have come to know many scientists who worked and continue to work in the tradition of Darwin. Above all, my life has been enriched by my friendship with the great student of social behavior, Edward O. Wilson of Harvard University. If I had not already made the joke about Bob Young, I would make the same joke about Ed Wilson. There isn't much we agree on, in science or in philosophy. But we are bound by deep feelings of friendship and on my side by the realization that it is the science of men such as he that enrich our understanding of Darwin himself and of his great influence and importance. The same can be said of the many other evolutionists that I have read and met and argued with through the years: these include Ernst Mayr, George C. Williams, John Maynard Smith, William D. Hamilton, Nicholas Davies, Steven Jay Gould, John J. Sepkoski, Francisco J. Ayala, Richard Dawkins, and Randolph Nesse. Deduce who would be most likely to be doing mathematics while having flasks of home-made beer behind him fermenting on the windowsill. Then check your answer.

Throughout my career I enjoyed the friendship and support and advice of my fellow philosopher and historian of science, the late David Hull. I miss his presence every day and my greatest regret is that he was unable to contribute to this volume. He knew about this project and thought I was crazy to do it. Very much alive and contributors are my fellow-born Englishman Jonathan Hodge, whose incredible generosity to all scholars is deservedly legendary; Robert J. Richards, who has convinced me of the great importance to our story of the polymath Herbert Spencer, something of which Bob approves heartily and I do not; Jean Gayon, whose piece was way over-length but so interesting and ground breaking that I broke all of my rules and divided it into two so I had an excuse to use all of it; and Ronald L. Numbers, whose mistaken obsession with American college football is more than balanced by tremendous sensitivity to the history of the relationship between science and religion, especially since the coming of evolutionary thinking. More immediately, during those dark nights that afflict any editor, I have turned for support and advice to Bob Richards, Joe Cain, Greg Radick, David Sepkoski, Jane Maienschein, John Beatty, and many times to my colleague Fritz Davis.

Beatrice Rehl has been my friend and supporter and backbone through this whole task. She has been aided by her assistants and more recently by my production editor Brian MacDonald and indexer Lin Maria Riotto. Martin Young, my illustrator, worked and reworked the material I gave to him. Eric Rogers, my research assistant, found sources and contacts that neither he nor I dreamed really existed. Mary Tudor complained that when she died they would find "Calais" (the last British possession in France) engraved on her heart. They will find the word "Permissions" on mine. As always my family was there for support. Words cannot tell of my love for my wife Lizzie, who always knows when I need criticism and when, emphatically, I do not. This debt should not be taken as an excuse to get yet another dog. Finally, I am very much obliged to my home institution, Florida State University, not just for giving me the time and support to do such a job as this, but for financial aid through the Werkmeister funds attached to my professorship and through other sources from the Program in the History and Philosophy of Science and the College of Arts and Sciences. I am proud that my former dean and present friend Joe Travis agreed to contribute to this encyclopedia.

Introduction

THE ANCIENT GREEKS were not evolutionists (Essay 1, "Origins and the Greeks"). It was not that they had an a priori prejudice against a gradual developmental origin for organisms (including humans) but that they saw no real evidence for it. More importantly, they could not see how blind law – that is to say, natural law without a guiding intelligence – could lead to the intricate complexity of the world, complexity serving the ends of things, particularly organisms. This need to think in terms of consequences or purposes, what Aristotle called "final causes," was taken to speak definitively against natural origins.

It was not until the seventeenth century – what is known as the Age of the Enlightenment – that we get the beginnings of evolutionary thinking (Essay 2, "Evolution before Darwin"). This could have happened only if there was something, an ideology, sufficiently strong to overcome the worry about ends. Such an ideology did appear, that of progress: the belief that through unaided effort humans could themselves improve society and culture. It was natural for many to move straight from progress in the social world to progress in the biological world, and so we find people arguing for a full-scale climb upward from primitive forms, all the way up to the finest and fullest form of being, *Homo sapiens*: from "monad to man," as the saying went (Fig. Introduction.1). It was not generally an atheistic doctrine, being more one in line with "deism," the belief that God works through unbroken law. But it did increasingly challenge any biblical reading of the past, and it went against evangelical claims about Providence, the belief that we humans unaided can do nothing except for the sacrifice of Jesus on the cross.

Radical claims like these did not go unchallenged. Critics, notably the German philosopher Immanuel Kant and his French champion, the comparative anatomist Georges Cuvier, continued to argue that final causes stand in the way of all such speculations. Moreover, particularly after the French Revolution, many thought the idea of progress to be both false and dangerous. For this reason, evolution was hardly a respectable notion. It had all of the markings of a "pseudoscience," like mesmerism (the belief in bodily magnetism) or phrenology (the belief that bumps on the skull give clues to psychological traits). It existed as an epiphenomenon of a cultural ideology; it was valued because it was value laden through and through. This is not to say

FIGURE INTRODUCTION.1. Particularly popular in medieval times were sketches of the "chain of being," showing the structural order of things, from the simplest of nonliving things (like stones) up to the ultimately important, God. This is from the *Ladder of Ascent and Descent of the Mind* (1305) by the Catalan philosopher Ramon Lull (1232–1315), first printed edition 1512. Although not in itself dynamic, it resonated in the eighteenth century with thoughts of progress and was surely an influencing factor in the thinking of early evolutionists. From M. Ruse, *Monad to Man* (Cambridge, Mass.: Harvard University Press, 1996)

FIGURE INTRODUCTION.2. The anatomist Robert Grant (1793–1874) was an ardent evolutionist and a close acquaintance of Darwin when the latter spent two years in Edinburgh training to be a physician. Darwin was lucky in his teachers and mentors, but clearly he had a nose for picking out those who could instruct and help. Permission: Wellcome

that it was an unpopular idea. As we see in our own day, manifested by such pseudosciences as homeopathy (the belief in the curative power of small doses of the poison that in quantity kills), pseudosciences can be very popular. But enthusiasm lay generally with the public and not with the professional community.

The *Origin of Species* (1859) set out to change all of this. It is important therefore, from the beginning, to get Charles Darwin right. And as a start on this, we must recognize that the autobiography that he penned toward the end of his life, although captivating and very informative, is in many respects highly misleading. Darwin characterizes himself as a charming young man, not terribly directed or motivated, keenest of all on the country sports of shooting and the like, who almost by chance backed into one of the greatest discoveries of all time. This is simply not true. We must keep balance and perspective and not let the English penchant for self-deprecating modesty cloud the story. As an individual, Darwin was genuinely warm and friendly, loyal to family and friends, a good master to his servants, and for all that he was very careful with his money, good at managing it, and generous to those in need. He was loved and with good reason. He was also hard working, even to the point of obsession. He did not have the kind of mind that is good at doing things that impress schoolteachers. He was not that gifted at mathematics, nor was he a brilliant success with languages, dead or living. That put him at a disadvantage, given that back then these were precisely the talents needed for formal academic success. But he was clearly very intelligent; moreover, older people (especially when he went to Cambridge) saw this and almost rushed to be his friends and mentors (see Fig. Introduction.2 and Plate III). Above all, Darwin had an oversized, inventive and discerning eye for a good theory or hypothesis. Added to this is the fact that he was ruthless in his pursuit of an idea and the supporting facts, using others (particularly by courtesy of the penny post introduced in 1840) to gather information for his speculations. He was indeed sick – possibly a psychological sickness but even more possibly purely physical – but he used this sickness to avoid distractions and other commitments. One of his biographers has written of Darwin as having a sliver of ice through his heart, and never were truer words written.[1]

[1] The comment is made by Janet Browne in the introduction to her two-volume biography of Darwin: *Charles Darwin: Voyaging* (1995) and *Charles Darwin: The Power of Place* (2002). In this Introduction, I have relied heavily on this biography for details of Darwin's life and work. I have also used my own earlier writings, including *The Darwinian Revolution: Science Red in Tooth and Claw* (1999a); *Taking*

Introduction

FIGURE INTRODUCTION.3. A cartoon by one of Darwin's fellow Cambridge students (Albert Way) making fun both of Darwin's love of horse riding and of his passion for beetle collecting. Permission: Cambridge University Library

That "Darwin of the *Beagle*" became "Darwin of the *Origin*" was no mere chance. The abilities and drive meshed smoothly with Darwin's background and training. There was a great deal of money in the Darwin-Wedgwood family, and it was kept that way by the frequent intermarriages of which Charles Darwin and his cousin Emma Wedgwood were but one instance. Father Robert was a physician and also a very shrewd businessman, arranging mortgages between those with money to lend (generally industrialists) and those with need of money (often aristocrats with land to provide security). Maternal grandfather Josiah Wedgwood was the founder of the great pottery works, one of the biggest successes in the Industrial Revolution (see Plate IV). Charles inherited the cash, and one immediate payoff was that he never had to work formally to make a living. Not for him the boring jobs of marking papers and sitting on departmental committees. Darwin also inherited much that led to the making of the cash. He was no country bumpkin, nor was he (for all that he had been intended for the church) an ethereal scholar with thoughts fixed only on abstruse points of logic or theology. Science and technology lay behind the revolution, and it was this that grasped Charles Darwin from the beginning. From their earliest days, he and his older brother Erasmus were junior chemists with their own garden-shed laboratory. Then both at Edinburgh and increasingly at Cambridge, Darwin immersed himself in the biological sciences of the day – collecting, reading, listening to others, and attending courses pertinent to these interests (Fig. Introduction.3).

The earth sciences he also pursued, an area of inquiry that was growing and thriving by leaps and bounds. Industry demands fuel, coal now that the trees were vanishing, and materials, iron, copper, and the like. It also has need of transportation, initially waterways, including man-made canals, and then in the nineteenth century the highly successful railway system. All of this demands knowledge of the rocks. No serious businessman wants to invest in a mine that might come up dry after vast expenditures. Equally, no serious businessman wants great effort made to drill tunnels through solid granite when a system of locks going up or around would be much cheaper. Geology holds the key to understanding what exists beyond direct sight, and by the time that Darwin was an undergraduate at Cambridge, the science was a ferment of action and discovery and controversy. That there was a frisson of worry about the time demands of the earth sciences, and the time restrictions of scripture read conservatively, added to its interest – especially given that, almost to a man, the Cambridge professors had to be ordained members of the Church of England.

It was entirely natural that when Darwin set off on the *Beagle* voyage – itself an opportunity to naturalize in new and strange parts of the world – geology should have been something foremost in his mind (Essay 3, "Charles Darwin's Geology: The Root of His Philosophy of the Earth"). It was

Darwin Seriously: A Naturalistic Approach to Philosophy (1986); *Monad to Man: The Concept of Progress in Evolutionary Biology* (1996); *Mystery of Mysteries: Is Evolution a Social Construction?* (1999b); *Can a Darwinian Be a Christian? The Relationship between Science and Religion* (2001); *Darwin and Design: Does Evolution Have a Purpose?* (2003a); *The Evolution-Creation Struggle* (2005); *Darwinism and Its Discontents* (2006); *Charles Darwin* (2008); *Philosophy after Darwin: Classic and Contemporary Readings* (2009c); and *The Philosophy of Human Evolution* (2012).

Introduction

an exciting time to take up the subject, for opinion (in Britain) was starkly divided, between those (the "catastrophists" represented by one of Darwin's Cambridge mentors, Adam Sedgwick, professor of geology) who thought that every now and then the earth is shaken up by huge earthquakes and the like (after which organisms are created, miraculously, anew) and those (the "uniformitarians" represented by Scottish lawyer-turned-geologist Charles Lyell) who thought that ongoing regular processes, like rain and snow and deposition and erosion, suffice to create the earth's geological history. Lyell had just started publishing his *Principles of Geology* (1830–33), and Darwin devoured it and believed. It was ever the basis for his thinking about earth history and was the foundation of the three books on geology that Darwin published in the ten years after the *Beagle* voyage. No doubt time alone on the ship and the independence forced upon him by the distance from the British scientific community was significant, both in his thinking about geology and also on his mind frame as he now started to work toward the problem of organic origins.

That Darwin, in the mid-1830s – always remember that it was in this decade that Darwin did his creative work, not the future decade of the 1850s when he finally published – was interested in organic origins is no surprise at all. The Cambridge professors loathed and detested evolution, thinking it would subvert both science and religion – they were themselves treading a rather fine, delicate line with their fondness for science and so had to insist to the orthodox that religiously they were purer than pure. Like Mr. Dick in *David Copperfield*, evolution was their King Charles's Head. They could not stay away from the topic. A bright young entrant like Darwin had to sense that there was something of interest here – a sense that would be confirmed when (in 1836) the leading astronomer and philosopher of science John F. W. Herschel wrote to Lyell (in a letter that became public) that origins is the "mystery of mysteries" (Cannon 1961). That it was Charles Darwin of all people who became an evolutionist (the usual word was "transmutation," and "evolution" became generally used for organic origins only in the 1850s and 1860s) is less of a surprise than it might have been. His father's father, Erasmus Darwin – physician, inventor, friend of business – was an ardent evolutionist, and as a youth Charles Darwin had read his grandfather's major work, *Zoonomia*. (Volume 1 was published in 1794 and Volume 2 in 1796. It is in the first volume that the evolutionary speculations occur.) (Fig. Introduction.4). Then, when at Edinburgh, Darwin had been close to one of the very few open evolutionists in Britain at that time, the anatomist Robert Grant. Finally, thanks to Lyell – who gave a detailed exposition in the second volume of his *Principles* – Darwin knew in detail about the evolutionary theory of the Frenchman Jean Baptiste de Lamarck. (Lyell introduced the theory to criticize it. More than one, including Darwin's contemporary and fellow evolutionist Herbert Spencer, read Lyell and was converted to evolution!)

It is always nice and romantic to suppose that new ideas demand a Road to Damascus experience. Probably for Darwin,

FIGURE INTRODUCTION.4. Erasmus Darwin (1731–1802) was one of the early evolutionists. His *Zoonomia* was widely read, including by his grandson Charles. This is a copy of a painting from 1770 by Joseph Wright of Derby. Permission: Wellcome

becoming an evolutionist was a bit more gradual. There is no question but that major influences, along with the geology that was making him think about the operation of laws in nature and implications for such things as time and place, were the fossils that he was collecting on the *Beagle* trip. His finds were almost forcing him to think about origins and changes and causes, and Darwin said as much in his autobiography. We must not exaggerate. Again we see that the young Darwin was, from the first, right in the heart of science in a full-time and professional way. Yet, Darwin was not as skilled and knowledgeable a paleontologist as he was geologist (Essay 4, "Looking Back with 'Great Satisfaction' on Darwin's Vertebrate Paleontology"). It is a field that demanded more biological knowledge than he had in those early years. But equally he was no mere tyro, and certainly, when he returned to England, he was keen to get the best authorities to study his findings – an ambition speaking not just to his own knowledge and abilities but also to his rapidly rising status in the scientific community as one who could expect and get the leaders in the field to work with or for him. Richard Owen, anatomist and paleontologist, was the obvious choice, and (given the quality and freshness of the fossils) it was clearly in the interests of both when Owen did work on Darwin's collection. There is a poignant paradox here, for later it was Owen who became the outstanding opponent of the Darwinians and their theorizing. At first, however, Darwin and Owen were friendly, and although Owen always had yearnings for more metaphysical, German-influenced readings of life's history, one suspects that the two may well have discussed origins and transmutation, not necessarily in an entirely hostile fashion (Rupke 1994). One thing always to be kept in mind is that Owen never had Darwin's privileged start in life or financial independence. He was in the thrall of men who hated evolution. Later, when he himself moved to a public evolutionary stance, one has trouble seeing if his big complaint with the Darwinians is that they are wrong or that they have stolen ideas that he (Owen) had all along.

Along with the fossils, Darwin was certainly set on the path to evolution by the distributions of the organisms – birds and reptiles particularly – that he saw when the *Beagle* in 1835 visited the Galapagos Archipelago in the mid-Pacific. Even more certainly, his thinking solidified early in 1837 when the taxonomist studying his bird collection confirmed that from island to island there are genuinely different species. It was at this point Darwin opened a series of private notebooks (the key species notebooks are B through E, and the key human notebooks are M and N) and jotted down thoughts on evolution. And its causes! Darwin was a graduate of the University of Cambridge, the home two hundred years previously of the great Isaac Newton. Again and again Darwin's mentors stressed that Newton's overriding achievement was to provide causal understanding of the major advances in physics in the Scientific Revolution. Kant, in his *Critique of Judgement* (1790), had denied that there could be a "Newton of the blade of grass." Darwin, determined to show him wrong, set out deliberately to find the cause of evolutionary change, the biological equivalent of Newton's law of gravitational attraction.

The key insight leading to the discovery of the mechanism of natural selection, the systematic differential reproduction of organisms brought on by the limited supplies of food and space, came late in September 1838. It was then that Darwin read the *Essay on a Principle of Population* (1826) by the Reverend Thomas Robert Malthus, who argued the population pressures in humans lead to inevitable struggles for existence. Darwin generalized to all species – actually Malthus mentioned that he got his inspiration from a more general discussion by, of all people, Benjamin Franklin – and then argued that success in the struggle will (on average) be a function of the different variations of the competitors and that this will lead to ongoing change – change moreover of a particular kind, namely in the direction of features or characteristics (like the hand and the eye) that aid their possessors. In other words, this process of natural selection (the term is not used for another two or three years) produces contrivances or adaptations, things that seem as if designed for the ends they serve. That is to say, the process or mechanism gives a natural (in the sense of working according to blind, unguided law) explanation of Aristotelian final causes. There is no need to suppose outside, divine intervention.

Thanks to the notebooks, we can map in some detail the exact route to discovery of the mechanism and the thinking that came thereafter (Essay 5, "The Origins of the *Origin*: Darwin's First Thoughts about the Tree of Life and Natural Selection, 1837–1839"). In a sense, though, we do have somewhat of an embarrassment of riches, especially when you add in our possession of many of the pertinent works that Darwin read (and annotated extensively) at that time. This has led to some controversy about what the later Darwin said, especially in his autobiography, about his discovery and what the jottings seem to reveal. Particularly there are questions about the exact role played in the discovery by the analogy with artificial selection, the ways in which agriculturalists and fanciers choose the specimens they favor and use as breeding stock. Darwin claimed that it was this that led directly to natural selection, but the notebooks (a reading endorsed by the essay given) suggest otherwise. Perhaps the answer is somewhere in the middle. Darwin was certainly conscious of artificial selection and its importance – an industrial revolution demands an agricultural revolution, to feed the workers, and Shrewsbury is in the heart of rural Britain (and the Wedgwoods particularly were interested in breeding) – but whether it played quite the direct role in discovery might be doubted. What is certainly the case – pointed out in no uncertain fashion to Darwin after the *Origin* was published – is that others had also hit on the notion of natural selection. Darwin at this time even read a pamphlet toying with the idea and noted it. He read: "A severe winter, or scarcity of food, by destroying the weak and unhealthy, has all the good effects of the most skilful selection." About this (in the margin), showing that he sees that something pertinent is at work here although he still doesn't quite get the full analogy, Darwin wrote: "In plants man presents mixtures, varies conditions and destroys, the unfavourable kind – could he do this last effectively and keep the same exact conditions for many generations he would make species, which would be infertile with other species." What does seem to be true is that only Darwin was exploring the possibility that selection could lead to full-blown, permanent change. Others deserve a footnote and little more. (The pamphlet is by Sir John Sebright, a noted breeder mentioned in the first chapter of the *Origin*. See Ruse 1975b.)

A mechanism is not a theory. The public Darwin was getting married and starting a family, falling sick, and working and publishing frenetically on geology (Fig. Introduction.5). The private Darwin was thinking furiously and by 1842 felt sufficiently confident to put his ideas on paper in a 35-page preliminary essay (usually known as the "Sketch"), and then some two years later in 1844 he expanded his ideas to a much longer, 230-page essay (usually known as the "Essay.") We know that he did show material to a young botanist, Joseph Hooker (to become one of Darwin's lifelong friends and a source of much material, physical and intellectual), and he left a note to his wife arranging for publication were he to die prematurely – something he thought quite possible. But that was it, and now the flat-out activity rather slowed as Darwin – the professional, public Darwin – turned increasingly away from geology and toward the life sciences. Obviously, they had always been part of his work and life: the fossils, the Galapagos (and many South American) specimens, both animal and plant, and more. Classification, what biologists call "taxonomy," was both a vital tool and (certainly for the private evolutionist) a great font of inspiration. In the century previously, the great Swedish biologist Linnaeus had formulated the basic principles of classification (the "Linnaean system"),

FIGURE INTRODUCTION.5. In 1842 Charles and Emma Darwin moved to Down House, which Darwin's father bought for the young couple for £2,200. They immediately set about making renovations and additions. Darwin lived here for the rest of his life. From H. E. Litchfield, *Emma Darwin, Wife of Charles Darwin: A Century of Family Letters* (Cambridge: privately printed by Cambridge University Press, 1904)

where organisms are assigned hierarchically to nested sets of ever-greater power and generality – from species at the lowest basic level to kingdoms at the highest. For Darwin, especially for a Darwin whose thinking about evolution was ever influenced by those Galapagos organisms hopping from island to island and changing as they went and thus bringing a treelike history to life (very unlike Lamarck's parallel upward progressions), it was almost a truism that his developmental thinking was the explanation of the fanlike, distributive pattern that epitomized Linnaeus's system (Essay 6, "Darwin and Taxonomy").

It is very probable that it was taxonomic thinking that pushed Darwin to what he considered the major conceptual addition to his theory – the "principle of divergence" – that occurred in the years from the "Essay of 1844" to the *Origin*. Why should there be the range of different forms that we find? Is it just accidental, or is there a deeper reason? In the notebooks, things seem to happen almost by default. "The enormous number of animals in the world depends on their varied structure and complexity; hence as the forms became complicated, they opened fresh means of adding to their complexity; but yet there is no necessary tendency in the simple animals to become complicated although all perhaps will have done so from the new relations caused by the advancing complexity of others" (Barrett et al. 1987, 422–3, E, 95). Then,

Darwin saw how this all comes about by selection, because it is advantageous to organisms to differ from potential competitors and thus occupy different niches reducing conflict. "The same spot will support more life if occupied by very diverse forms.... Each new variety or species, when formed will generally take the place of and so exterminate its less well-fitted parent. This, I believe, to be the origin of the classification or arrangement of all organic beings at all times. These always *seem* to branch and sub-branch like a tree from a common trunk; the flourishing twigs destroying the less vigorous, – the dead and lost branches rudely representing extinct genera and families" (Darwin 1985–, 6:448–49, letter to Asa Gray, 5 September 1857) (see Fig. Introduction.6).

Publicly taxonomy was now at the fore, as Darwin plunged into what was going to be an eight-year-long study of barnacles, marine invertebrates that had first captured his fancy when on board the *Beagle* (Essay 7, "Darwin and the Barnacles"). This took him right into the next decade and apparently in some quarters made him a bit of a figure of fun, as the archetypal scientist-scholar who devotes his whole life to the study of something that to the layperson seems of unbelievably trivial importance. But why did Darwin, the ambitious Darwin, go off at this tangent? Why barnacles indeed? Although there are comments and moves made that make for fascinating significance, given our knowledge that Darwin was now an

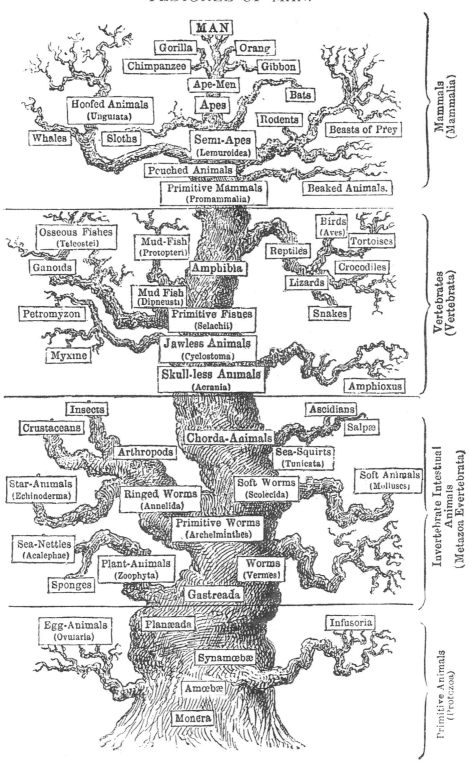

FIGURE INTRODUCTION.6. The tree of life as drawn later in the nineteenth century by Darwin's great German supporter Ernst Haeckel. Note how thoroughly progressionist it is, with simple forms at the bottom (monads) and humans at the top (man). Haeckel used the term "monera," referring to prokaryotes, single-celled organisms without a nucleus. From E. Haeckel, *The Evolution of Man* (New York: Appleton, 1897)

evolutionist, he could not – he certainly did not – come out and profess the convictions that he thought made causal sense of his work. Why did Darwin delay? Why did he not publish the "Essay of 1844"? The note to his wife made it clear that Darwin wanted his thinking made public at some point. Like his sickness, there are as many answers as people who ask the question. Probably various factors were involved. He was sick and felt unable to fight vigorously for his ideas. He never really expected the delay to be so long – twenty-plus years from the Malthus moment to the appearance of the *Origin*. The barnacle studies just stretched and stretched, and the years went by. Most importantly, the public work of the 1830s had paid off. His mentors who had pushed his career were seeing their efforts rewarded. By the mid-1840s Darwin was established as a serious and important scientist. He was cherished by the community, especially by the Cambridge professors and their set who had helped him launch his career. And here's the rub. They went on hating evolution – Cuvier was their scientific hero – and someone going that way would be criticized and ostracized. Added to this, 1844 was the year that the Scottish publisher Robert Chambers published (anonymously) his *Vestiges of the Natural History of Creation*, a pro-evolutionary work that was anathematized by the scientific establishment (as it was equally lauded by the uninformed and ignorant). Darwin, whose great public success was now being reinforced by the general and enthusiastic reception of a book (the *Voyage of the Beagle*) based on his travel years, had no desire to put all in jeopardy.

Finally, however, particularly at the urging of friends who gradually were being let into the secret – after Hooker came Lyell and then in England the young anatomist Thomas Henry Huxley (grandfather of the novelist Aldous Huxley), reinforced in America by the Harvard botanist Asa Gray – Darwin started work on a massive volume, intended to overwhelm with fact and footnote. Huxley always praised Darwin for the delay, arguing that the barnacle work gave him invaluable understanding and experience of the organic world. There may be some truth in this, although one cannot honestly say, despite the principle of divergence, that the differences between the "Essay of 1844" and the *Origin* seem worth quite such a wait and effort. What was important was the growing status and the new network surrounding Darwin, a network that was going to be much more inclined than the older Cambridge set to accept and promote his ideas. But also Darwin did work hard in the 1850s on the empirical evidence for his evolutionary thinking, doing, for instance, careful experiments on the survivability of seeds in salt water, a crucial piece of information for his claims about how organisms could spread around the world, given the barriers of the oceans. (Remember, we are a hundred years too early for plate tectonics.) And it is clear that, whatever may have been the truth back in the late 1830s, by the 1850s the analogy with artificial selection was growing increasingly in his mind. He was delving carefully into the successes of breeders and judging the relevance to his concerns. What does seem probable, and perhaps we should not really be that surprised, is that Darwin was himself fairly selective in this direction, picking out precisely those results that were favorable to his thinking and glossing over those that were not (Essay 8, "The Analogy between Artificial and Natural Selection").

Then came the thunderbolt. In the summer of 1858, Alfred Russel Wallace, a young naturalist and professional collector, formerly in Brazil and now in the Far East, someone with whom Darwin had been corresponding, sent to Darwin (of all people) a short essay with exactly the same ideas that had been fermenting for nigh twenty years (Fig. Introduction.7). Friends, Lyell and Hooker, came to the rescue. Wallace had to be acknowledged but there must be no nonsense about Darwin's priority and so, along with Wallace's essay, pertinent extracts from the "Essay of 1844" and the already-quoted, informative letter sent to Asa Gray (about the principle of divergence) were published in the *Proceedings of the Linnaean Society of London*. Then Darwin sat down to write an overview of his theory. Thus it was that, in the late fall of 1859, the *Origin of Species* arrived on the scene.

Read the essay on the *Origin* in the light of what it is trying to do (Essay 9, "The *Origin of Species*"). It is taking seriously Darwin's own comment that the book contains "one long argument" and is setting out to show the nature of that argument. Because it is exposing the conceptual skeleton of the *Origin* rather than trying to give a full synopsis of the work, one should use the essay as a map to more detailed discussions in later essays, for instance about species or sexual selection or heredity. Note how Darwin runs together the argument for evolution (and the tree of life) and the argument for the mechanism of natural selection. One point of interest will be the extent to which readers separated out these two aims. Darwin never talks explicitly in the *Origin* about those whom he is opposing, those who argue for some kind of non-natural creation of life. Although there were biblical literalists (like today's American creationists) back then, these are not his target. He has in mind real, respectable scientists, like his old friend Adam Sedgwick, professor of geology at Cambridge and, perhaps reaching even further back, the great French anatomist Georges Cuvier. More immediately, the Swiss-born, American-transplant, ichthyologist and geologist (expert on glaciers and their effects) Louis Agassiz would have been in his sights – particularly in light of his neo-Cuvierian *Essay on Classification* published in 1857. Agassiz sent Darwin a copy. In a letter of 13 March 1859, Darwin wrote to Huxley, who admittedly liked to hear these sorts of things, that it was "utterly impracticable rubbish" (Darwin 1985–, 7:262).

Given the central importance of the *Origin*, we must turn and consider in some detail aspects of the argumentation given in the work. The obvious place to start is with the mechanisms of change. Darwin always thought that, although natural selection is by far the most important mechanism of evolutionary change, it is by no means the only one. The major alternative was always a secondary form of selection, so-called sexual selection (Essay 10, "Sexual Selection") This appears even in the "Sketch of 1842," so it is not some late "add on," although it is not until he comes to write his major work on

FIGURE INTRODUCTION.7. Alfred Russel Wallace (1823–1913), the co-discoverer of natural selection, in 1853. He was already an ardent evolutionist. From A. R. Wallace, *My Life* (London: Chapman and Hall, 1905)

our species, *The Descent of Man and Selection in Relation to Sex*, that Darwin gives the mechanism extended treatment. Whereas natural selection involves a struggle against the elements and other organisms for space and food and the like, leading to reproduction, sexual selection occurs only within species and is a function of competition for mates.

Given, whatever the exact relationship, the central importance in Darwin's thinking of the analogy between artificial and natural selection, it is surely plausible to think that Darwin founded his distinction between the two kinds of selection on the distinction one finds in the world of the breeders, between those selecting for profit – fatter pigs, shaggier sheep – and those selection for pleasure – more tuneful birds and fiercer dogs. This supposition gains further strength when one finds that Darwin divided sexual selection into two kinds: selection between males through conflicts for females ("male combat") and selection by females for more desirable males ("female choice") – thus the magnificent antlers of the stag and the gorgeous feathers of the peacock, respectively. These correspond – and Darwin points out the correspondence – to breeders selecting for fighting spirits in their dogs and cocks and breeders selecting for prettier feathers on their budgerigars and like pets.

What is particularly interesting is the fate of sexual selection over the years. Initially, most people inclined to think with Alfred Russel Wallace that truly the distinction is not that significant – certainly not sufficiently significant to overcome worries that the whole process seems fatally anthropomorphic. Why should one suppose that peahens have the same standards of beauty as humans? Starting in the 1960s, however, particularly with the rise of sociobiology (of which more later), sexual selection has come to play a larger and larger role in the thinking of evolutionists. It is thought to be a really significant aspect of the biological world. Darwin, as we shall see, thought it very important in the context of humans, an assumption as controversial then as it is now. Remember that selection (of whatever kind) leads not just to change but to change of a particular kind, namely adaptive change. Put this in the context of the sexual selection of human beings, and you are plunged right into discussions about male-female differences and whether they are natural (meaning biological) or cultural (meaning more environmental). But whether sexual selection is accepted or whether it is rejected, it is realized that it cannot be ignored, and for this reason, if for no other, demands careful and explicit scrutiny.

A lot of not-always-tremendously-helpful things are said about the *Origin*, at the head of which list is the claim that the work is mislabeled because it is not about the origin of species at all. It is true that the work is basically on evolution and its major mechanism of natural selection, but there is much on species, their nature and their causes. What else is the principle of divergence but an attempt to show why the world comes cut up at the joints, to use a phrase of Plato? It is obvious that Darwin is going to have some tricky discussion about the nature of species. On the one hand, he wants them to be things that are real enough to merit discussion about natures and causes. On the other hand, he wants them not to be so fixed that they cannot change and evolve. Some or all are in constant motion and change. So there is a paradox of a kind here, but it is not mysterious and not in Darwin's opinion beyond understanding. What is surely true is that often discussion of the topic has been clouded by later proposals about species, not to mention enthusiasts' eagerness to claim Darwin as one of their precursors – or conversely, to promote their own importance by contrasting their successes with Darwin's supposed failures (Essay 11, "Darwin and Species").

This much we can say, that Darwin surely thought that species are real in some sense. There may be many borderline cases – one hopes that there are borderline cases! – but species are real. We can also say that Darwin was keenly aware that reproductive isolation is an important part of the story. Cabbages and humans don't share offspring. However, there is little doubt that Darwin was unwilling (unlike many taxonomists in the twentieth century) to put the entire burden on reproductive isolation. He thought it broke down too often to be reliable. Also, he was worried about the role of selection in reproductive isolation. Or, rather, he was not so worried about its role – he didn't think it was there when it came to producing hybrid sterility – but about the consequences for such issues as the reality of species. As we shall see shortly, factors like these take us to the heart of some of the most difficult and contentious issues surrounding natural selection, so there is hardly any surprise that Darwin's thinking on the species issue generally causes differences of opinion. These started as soon as he published and continue to this day. If ever proof was needed that scientific understanding is more than simply determining matters of brute fact, demanding also philosophical and like (including historical) judgments, the species problem provides it.

The most (deservedly) influential work in the twentieth century about scientific change was Thomas Kuhn's *The Structure of Scientific Revolutions*. Well known is Kuhn's notion of a "paradigm," a kind of way of thinking within which scientists do all of their work ("normal science") almost all of the time. Equally well known is the claim that sometimes paradigms break down and there is a switch to a new one, a switch not entirely rational and much akin to a political or religious conversion, after which science resumes its normal state and work proceeds now in the new paradigm. I don't think anyone would deny that something of this nature went on in the Darwinian Revolution. Darwin's teachers and elders, men like Adam Sedgwick and William Whewell, really did see the world in one way, and Darwin's followers like Joseph Hooker and Thomas Henry Huxley really did see the world in another way. It is comforting to say that one side is wrong and the other side is right, and in a way this is certainly true. But it is not quite all of the truth. Sedgwick and Whewell were as bright and informed as Hooker and Huxley. A kind of conversion experience had occurred.

Having said this, it is clear that Kuhn often tells only part of the story, and this is certainly true in the Darwinian case. The impression certainly is that everything happens once and for all in one decisive stroke. Now you believe in miraculous creation of organisms. Now you believe that organisms are made by natural selection. In fact, this was not – or at least only rarely – true. As we shall see, although generally speaking evolution was a terrific success – by about 1870 it was becoming the standard view in much of the Western world (the American South, of course, excepted) – natural selection was far less successful. It was not until the past century was into its fourth and fifth decades that selection really started to catch fire. So certainly, whatever it was in the way of a paradigm that Darwin provided, it was not something within which the scientific community from thenceforth happily worked.

There were various reasons for the caution about natural selection, including a couple of scientific reasons that were very important. The first was the problem of heredity (Essay 12, "Darwin and Heredity"). Natural selection demands a constant supply of new variations, but then it is vital that these variations stay around and be passed on. There is little point in being a winner in the struggle for existence if your offspring don't have the very features that made you a success. Darwin spent a lot of time struggling with these issues, even inventing a hypothesis – "pangenesis" – to explain matters. Basically his problem was that although he could see that sometimes features do persist from generation to generation, he could not get away from the belief that often features, however admirable, get blended away in breeding – half in the next generation, then a quarter, and so forth. Before long even the best new variation is lost. And his critics seized on this point and used it as a refutation of the effectiveness of natural selection. The problem was not to be solved until the beginning of the new century. In Darwin's defense, let us say that no one else had much idea about what to do, except to criticize. It is true, of course, that across Europe in his monastery garden in Brno, in the Austro-Hungarian Empire, the monk Gregor Mendel was doing work on pea plants that was to be recognized as the foundational inquiry that led to modern theories of heredity, genetics. But before one immediately concludes that it was a tragedy that Darwin and Mendel never worked together or symbiotically, one must recognize that Mendel was working on technical issues of plant breeding and not setting out to fill a gap opened by the *Origin*. Indeed, although Darwin never read Mendel – he could have done, had he been searching in that direction – Mendel read the *Origin*. But (as we can tell from his marginalia), Mendel did not see that he had the solution to the problem. To be honest, it does not seem that Mendel was either bowled over by Darwin or horrified. He was interested in evolution, but it was not really his problem (Fairbanks and Rytting 2001).

The other big scientific problem that Darwin faced was that of time (Essay 13, "Darwin and Time"). Today we know that, although evolution can be slow, it can and does at times go really quite quickly. Natural selection can make for major changes in short periods if need be. In any case, there is plenty of time for evolution, fast or slow – life on earth started nearly four billion years ago. Darwin always thought that selection would be slow, probably too slow for us to record in our lifetimes (a point of significance to be noted shortly), and he had little idea about the available time. He made some calculations, suggesting that the earth is pretty old, but these were derided by the geologists. Then, as the physicists started to get involved in the problem, increasingly it seemed as if the time available for change was very short and restricted. People did not take this as a blow against evolution, but they thought it told against natural selection. Darwin – who almost amusingly was made very much aware of the problem by his son George, a brilliant mathematician who was working with William Thomson (later Lord Kelvin), the chief critic of a long-age earth – did what he could to cover up or avoid the problem, but basically he had to hope that some solution would eventually appear. It did, of course, when the physicists discovered that radioactive decay generates heat and that, with this factor acknowledged, the earth is plenty old enough for selection, at whatever intensity or speed.

One should keep a sense of balance. Natural selection had its many critics. Yet not all was gloom and doom. Darwin himself was convinced that his mechanism mattered. Botany was a long-established passion, going back to Darwin's attending lectures on the subject when an undergraduate at Cambridge (Essay 14, "Darwin's Evolutionary Botany"). But it was after the *Origin* that the interest and work really increased – a good strategy for a sick man, living in the countryside, with money to indulge his interests with greenhouses and gardeners, and with well-connected botanical chums like Hooker ever ready to send him specimens. There emerged a string of papers and books on domesticated varieties, on climbing plants, on insectivorous plants, on methods of fertilization, and much more. Making a value judgment, the really delightful studies came early in the 1860s on orchids. It seems clear not only that Darwin was looking for something that would be a relief and welcome change from the strain of writing the *Origin* but that he was also after something where he could show that natural selection really does produce adaptations and that this is something of which the whole biological world should take note. As an example of the new world into which Darwin led us, even to this day there is no better introduction than *On the Various Contrivances by Which British and Foreign Orchids Are Fertilised by Insects, and on the Good Effects of Intercrossing*. It is fascinating, for all that it is in competition for the world's most technical and boring title.

Even better than the flowers were the insects. If you think about it, it is obvious that if natural selection is going to find any supporters, it will be with insect biologists. They are dealing with fast-breeding organisms, where strong and effective adaptations – for survival, for food, for reproduction, above all for avoiding predators – are going to be absolutely crucial. That predator avoidance is fundamental was well known before the *Origin*, and as soon as natural selection appeared on the scene, it was being used to explain the techniques of such avoidance, the adaptive strategies taken by insects (Essay 15, "Mimicry and Camouflage"). Much successful effort was put into explaining mimicry as a product of a differential reproduction brought on by the struggle for existence. It is nice to be able to report that Darwin was very appreciative of this work, even to the extent of finding the major player (Henry Walter Bates) a good job (albeit one that rather took him away from his science). What is rather puzzling is that, although the work found its way into later editions of the *Origin*, Darwin never moved it quite as much up front and center as one might have expected. This possibly could be related to the point just noted: Darwin always had doubts about testing natural selection, simply because of (what he thought was) its slow-acting nature and ability. Today, as we shall learn later, evolutionists do not have such doubts and qualms, and mimicry and camouflage continue to have an important role in evolutionary studies.

One thing that must be recognized is that, although the problem of final cause was certainly shifted and changed by the *Origin*, it was not obviously expelled or anathematized (Essay 16, "Chance and Design; Essay 17, Darwin and Teleology"). The Greeks had not been able to see how blind, unguided law was able to create objects, organisms particularly, that seem made with ends in view – entities that seem as if designed. The eye exists for the purpose or end of seeing, even if now it is not actually seeing (or if, for some reason, it never did see). Natural selection is supposed to speak to this, because as we know it does not just bring about change but change after a certain fashion, namely in the direction of adaptation or contrivance (to use the term in Darwin's title) or design-like features. The point has been made already that Darwin's creative work occurred in the 1830s, not 1850s. In the former period, the thinking of someone like Cuvier held sway, namely the thinking of someone who agreed with the Greeks that final cause is all important. By the 1850s, anatomists like Owen and Huxley – in their day-to-day science they were often a lot closer to each other than either liked to admit – were focusing much more on homologies, similarities of structure, than on adaptation. After all, the specimens with which they dealt were usually dead, often very long dead, and so the needs of organisms were not pressing issues, and adaptation was downplayed. Structure persists, and finding links and connections was taken to be the major task at hand.

So Darwin was solving a problem that, by the time he published, many did not find pressing. This is obviously another (major) reason why natural selection was not hugely successful. It was solving a problem that many did not really see as needing solving. Although conversely there were others, like the botanist Asa Gray, who not only saw the need but were not sure that selection was quite up to the job. They wanted more, including special shoves in the right direction, from God through the medium of new variations. Darwin had some trouble expressing himself on this issue – in part because of his own evolving thinking about the deity – but he was very clear that, whether or not God exists, he must be kept out of science. In this wise, Darwin certainly looked toward the secular science of today rather than backward to the god-impregnated inquiries that were, for instance, the staple of his Cambridge mentors and teachers.

And finally, before moving on from looking at the *Origin* directly, mention must be made of the fact that the book itself was an exercise in evolution (Essay 18: " The Evolution of the *Origin* (1859–1872)"). It went through six editions and involved a huge amount of rewriting and often expansion. It used to be that it was always the sixth and final edition that was reprinted. But scholarly opinion today has swung against this and toward the first edition. For a start, the first edition is certainly easier to read, not having been torn apart and reworked so often that it really does resemble something produced by a committee. For a second, there are issues about whether the corrections introduced by Darwin were always for the best. At the linguistic level, Darwin added Herbert Spencer's alternative term for natural selection, the "survival of the fittest." This has led to endless mistaken claims that natural selection now reduces to the uninformative "those that survive are those that survive." At the conceptual level, Darwin messed endlessly with his discussion of heredity, digging ever deeper pits into which to jump. He tried to speed things up to account for the (mistaken) constraints on time. And more.

This said, there are some very interesting changes, perhaps most of all on the topic of progress. Before Darwin, it was progress that fueled evolutionary speculation and acceptance. Evolution was a pseudoscience. Darwin changed that. Evolution (at the least) was now accepted fact; it was common sense. But what about progress and what about the status of evolutionary thinking? In a passage quoted earlier, Darwin made it clear that he did not accept an inevitable upward charge, a kind of teleological force producing humankind. On the other hand, he was a good Victorian, living off the wealth of the Industrial Revolution, so he was not about to turn his back on progress in society or progress in biology. The latter, however, had to be done in terms of selection – no god, no special forces, no nothing like that. But Darwin certainly implied that, by defining progress in terms of division of labor (a kind of functioning complexity) and then invoking what today's biologists call arms races – evolving lines compete and the adaptations get better – we get not only comparative improvement but also a kind of absolute improvement leading to human brains and thinking.

> If we look at the differentiation and specialisation of the several organs of each being when adult (and this will include the advancement of the brain for intellectual purposes) as the best standard of highness of organisation, natural selection clearly leads towards highness; for all physiologists admit that the specialisation of organs, inasmuch as they perform in this state their functions better, is an advantage to each being; and hence the accumulation of variations tending towards specialisation is within the scope of natural selection. (Darwin 1861, 134)

What is interesting is that this passage does not come in the first edition of the *Origin* but has to wait until the third, admittedly appearing only two years later in 1861. Perhaps this tells us something about both the status Darwin hoped to achieve for his theory and the success he had in his efforts. Darwin wanted to hit the jackpot; he wanted to elevate the status of evolutionary thought from that of a pseudoscience to what we might call a "professional science," that is, something done by full-time researchers in the laboratory or the university or the like – the status that something like physics, and physiology for that matter, now held. For this reason, although even in the first edition of the *Origin* there are passages that betray a progressivist commitment, generally Darwin tried to stay away from such frank ideology. This is not the stuff of professional science. The lack of enthusiasm for natural selection suggests that Darwin was not fully successful in his aim, and the (rapid) consequent bringing in of explicit discussion of progress – the very thing that made for pseudoscience status – suggests that Darwin realized that he had not achieved all he wanted. The fact is that when the *Origin* was published, most people immediately read it as a peon to progress, often mixing it up with the thinking of Spencer, a fanatical progressionist. And one suspects that Darwin, who was himself in favor of progress of all kinds, simply decided to go with the flow and get what he could. What this all means for the actual status of evolution after the *Origin* is something to which we will have to return.

Given that Darwin was a biological progressionist, one infers that he thought our species, *Homo sapiens*, is relatively important. This is indeed true, although from the first he was convinced that we are completely and utterly part of the animal world. No special divine interventions are needed to explain our origins. Most probably the *Beagle* voyage, especially the encounter with the Tierra del Fuegians, the denizens of the land at the foot of South America, convinced the ship's naturalist of this (Fig. Introduction.8). Even the highest form of human (aka the English) is but a step from the savage state, a point made heavily by the rapid reversion to the norm of three natives who had been brought to England on a previous voyage, who had been turned into presentable Europeans, and who were now being returned to lift up the general moral and cultural level of their fellows at the foot of the continent. Darwin never had doubts about human evolutionary origins, and indeed the first explicit discussion of natural selection that we find in the notebooks (about a month after Darwin read Malthus) is applying the mechanism not only to humankind but to our brains and intellectual abilities. "An habitual action must some way affect the brain in a manner which can be transmitted.–this is analogous to a blacksmith having children with strong arms. The other principle of those children. which *chance?* produced with strong arms, outliving the weaker one, may be applicable to the formation of instincts, independently of habits.–" (Barrett et al. 1987, N, 42).

In the *Origin*, Darwin did not want to conceal his views about humans, but neither did he want the discussion swamped before he could get the main points of the theory out on the table. He therefore contented himself with the greatest understatement of the nineteenth century: "Light will be thrown on the origin of man and his history" (Darwin 1859, 488). But no one was deceived. As soon as Darwin published, the world started talking about the "monkey theory,"

FIGURE INTRODUCTION.8. Darwin's encounters with the Tierra del Fuegians in their natural habitat really disturbed the young naturalist. From then on, he had no doubt but that even the most civilized of human being is but a short step from the "savage." Permission: Wellcome

and no one had any doubt that the real battle was going to be over our species. Thus, for instance, Huxley and Owen battled over the human brain, with the former making us part of the primate world and the latter making us distinct from all other living forms. Book after book started to pour forth on the origins of *Homo sapiens*. And the popular press picked up the idea and ran with it. The comic magazine *Punch* made much of the controversy. For everyone, even if there had been doubt about whether the English are right at the top, there was absolutely none that the Irish are right at the bottom (Fig. Introduction.9). Labored jokes about the doings of Mr. G. O'Rilla became commonplace. For this reason, if for no other, given his extreme reluctance to break from his isolation, one suspects that (whatever his personal views), all other things being equal, Darwin would have stayed out of the debate about our species.

FIGURE INTRODUCTION.9. Darwinism became part of general Victorian culture and was used to support various beliefs and prejudices, as in this cartoon (from the magazine *Puck* in 1882) showing the Irishman as being apelike. It is titled "The King of A-Shantee," thus also bringing in prejudice against Africans, with the pun on shanty (meaning run-down house) and Ashanti (an African tribe from Ghana).

All other things were not equal. Alfred Russel Wallace is interesting not only in his own right but as a contrast with Charles Darwin (Essay 19, "Alfred Russel Wallace"). Coming from a segment of society much down the scale from Darwin – the lowest level of the middle classes rather than the highest – Wallace was an autodidact with respect to everything including science, teaching himself the basics both through reading and then through observations of nature as he pursued a profession as a collector, visiting first South America and then the Malay Peninsula as he sought exotic specimens. Whereas Darwin was completely and utterly the professional scientist, always working within the system, Wallace was anything but. To speak of him as a "maverick" is a kind way of avoiding words like "flake." He was brilliant and capable of serious and lasting science. He discovered natural selection independently, and his success, let there be no mistake, was the end point of a long effort to pin down the origins of organisms. He thought creatively about such issues as mimicry and also about biogeography. But he was always (quite fearlessly for consequences) adopting strange and unconventional ideas, starting with evolution itself in the mid-1840s when, thanks to *Vestiges* (which turned Wallace into an evolutionist), evolution was the epitome of a pseudoscience. This was followed later by enthusiasms for socialism, vegetarianism, land reformism,

and more – including, to the horrified amazement of the scientific establishment, total and utter commitment to spiritualism (a cozy belief that he shared, incidentally, with Chambers). In the mid-1860s, Wallace became convinced that only by invoking unseen spirit forces could one explain the evolution of humans. We have features like our hairlessness and our large brains that simply cannot have been produced by natural selection. There must have been SOMETHING MORE.

Darwin was appalled. This would destroy their joint child. He was spurred to action, and in 1871 he produced his own work on our species, *The Descent of Man and Selection in Relation to Sex* (Essay 20, "Darwin and Humans"). It is, as any reader can vouch, a very oddly balanced work, for a full three-fifths is on the secondary mechanism of sexual selection. Interesting as this is, one feels that it is somewhat out of place in a book ostensibly on human evolution. Until, that is, one realizes that it is all a response to Wallace. Darwin agreed with Wallace that there are aspects of human nature that it is hard to put down causally to natural selection. Why, for instance, do we find the racial variations that we do? And it was here that sexual selection came to the rescue, for Darwin argued that human evolution is deeply indebted to the ways in which humans (males in great extent) are into the business of choosing mates. He included some prime nineteenth-century anthropological speculations. Quoting the explorer Richard Burton on the large bottoms of some African women, we learn that the men "are said to choose their wives by ranging them in a line, and by picking her out who projects farthest *a tergo*, Nothing can be more hateful to a negro than the opposite form" (Darwin 1871a, 2:346) (Plate V). Recent scholarship has made it very clear that Darwin's thinking about humankind was greatly influenced by his family's detestation of slavery, a major issue in British circles in the early part of the nineteenth century. The truth of this, however, should not conceal that Darwin is a child of his time in other respects also, and that even a liberal Victorian would have views of non-Europeans that make us blanch today. The question is how easy it is, or if it is truly possible, when faced with someone like Darwin from a culture so different from ours, to distinguish between the objective scientific findings and the subjective cultural prejudices.

Whatever the merits of Darwin's argumentation about humans, no one can deny that one thing that he did was open up and inspire much more fully a host of related inquiries about human nature – inquiries that were gathering steam as researchers spread out across the globe (in the wake of the empire building of the Victorians) gathering comparative information on very diverse societies. One area that received attention from Darwin himself and that was of keen interest to many more generally was that of language or linguistics (Essay 21, "Darwin and Language"). In a way, it was almost natural that this topic would be of such interest, particularly given the exposure not only to the range of European languages, but also now to the languages of the East. Whatever the moral merits, governing a subcontinent demanded knowledge of the local tongues – an urge that was being felt strongly by the time the *Descent* was published, because after the Indian Mutiny (of 1857) the crown took over the governing of the country and started to introduce significantly more professionalism in its running. With this increasing exposure to and understanding of language, although not all were immediately enthusiastic, the time was ripe for an evolutionary analysis, trying (as did Darwin in the *Descent*) to show how it might have come into being and how it might have diverged as societies themselves diverged and moved apart. One point of some interest was how one was to explain causally the spread and divergence of languages. Although, as we shall see later, the analogical invocation of the pressure of the struggle with consequent selection is still somewhat controversial, it was something that appealed to Darwin. "As [Oxford Sanskritist] Max Müller has well remarked: – 'A struggle for life is constantly going on amongst the words and grammatical forms in each language. The better, the shorter, the easier forms are constantly gaining the upper hand, and they owe their success to their own inherent virtue'" (Darwin 1871a, 1:60).

Morality also was something of great interest to Darwin, and there is an extensive discussion of the topic in the *Descent of Man* (Essay 22, "Darwin and Ethics"). Darwin thought carefully about how a mechanism with a struggle for existence at its heart could nevertheless produce beings that are genuinely thoughtful and caring for the well-being of others. It is worth pointing out that there is a major difference between Darwin's treatment of the topic and that of others, most particularly that of his contemporary Herbert Spencer (Essay 23, "Social Darwinism"). For the latter, the aim is essentially one of justification: how do we ground moral claims, what makes them right? With what many Victorians perceived as the fall of religion and its failure to provide a firm backing for the morality needed in an industrial society, evolution for Spencer and his many followers seemed like an attractive modern, secular alternative. And it was here that progress came into play, for it was taken to be the ground of right action. In his 1857 essay, "Progress: Its Law and Cause," Spencer staked his banner even before the *Origin*: "Now, we propose in the first place to show, that this law of organic progress is the law of all progress. Whether it be in the development of the Earth, in the development of Life upon its surface, in the development of Society, of Government, of Manufactures, of Commerce, of Language, Literature, Science, Art, this same evolution of the simple into the complex, through successive differentiations, holds throughout" (Spencer 1857, 244). In Spencer's mind, and those of his many followers, doing good means cherishing and aiding progress. Doing bad means ignoring or hurting progress. For all that in the *Descent* and elsewhere his own personal moral convictions often shone through, Darwin was not really into this sort of enterprise. He was more working in the role of a scientist, trying to show the nature of morality and how it is that it has come about and stays in action. Having said this, philosophically one does see Darwin in the tradition of British empiricism, where morality is ultimately a matter of emotion rather than correspondence to some disinterested objective truth. What else can one say about a person who actually contemplates a situation where, were selection to dictate such an action, the highest moral imperative might be to kill one's brothers?

Note incidentally just how misleading it is to lump together all who took seriously the possible worth of biology for ethical behavior. Herbert Spencer was drawn to laissez-faire capitalism, thinking that it was this that leads to an upwardly rising society; although, after visiting America, he inclined to think that all work and no play certainly does make Jack a rather dull boy. Wallace, with his inclinations to socialism, was more into group explanations and the eventual emergence of good feelings toward all. Somewhat paradoxically, given that he was not at all keen on the idea of sexual selection through female choice in the animal world, he rather thought it might be effective in the human world. Society will be upgraded by young women choosing only the best young men as breeding partners. If he was basing this on personal family observation, one can only conclude that the Wallace children must have been as odd as their father. Huxley, although he was dedicated to progress, to improving the lot of his fellow countrymen, almost from the first had grave doubts about sunny optimistic readings of the evolutionary process. He saw the necessity of a lot more struggle against our animal nature than did someone like Spencer. He referred to himself as a "Calvinist" and when one thinks of his frequently gloomy take on humankind, there is much truth in this. No doubt the fact that he himself was subject to crushing depressions fed into this philosophy.

In this context, there has been much debate about the category in which we should place Darwin himself. Was Darwin a Social Darwinian? The answer is mixed. If you are thinking about a harsh master, of the kind often found in the novels of Charles Dickens, then clearly not. But he was very much a child of his time, particularly of his manufacturing, capitalist class. As he made clear in a letter to a correspondent (Swiss law professor Heinrich Fick, on July 26, 1872), he had little or no time for working men's unions, writing that "the rule insisted on by all our Trades-Unions, that all workmen, – the good and bad, the strong and weak, – shd all work for the same number of hours and receive the same wages. The unions are also opposed to piece-work, – in short to all competition. I fear that Cooperative Societies, which many look at as the main hope for the future, likewise exclude competition. This seems to me a great evil for the future progress of mankind." More through hope than conviction, he added: "Nevertheless under any system, temperate and frugal workmen will have an advantage and leave more offspring than the drunken and reckless. – " (DCP, 8427f).

One issue that lay behind natural selection from its first introduction and that becomes a matter of real, pressing importance by the time of the discussion of morality in the *Descent* is that of the level at which natural selection might be expected to operate (Essay 24, "Darwin and the Levels of Selection"). When Darwin introduces the struggle in the *Origin*, he makes it clear that it is every individual for itself – that "as more individuals are produced than can possibly survive, there must in every case be a struggle for existence, either one individual with another of the same species, or with the individuals of distinct species, or with the physical conditions of life" (Darwin 1859, 63). But how then can he explain what is now known as "altruism," where one organism gives to another even at the cost of its own reproduction? The social insects were particularly troublesome, because here you find sterile female workers, who give their all to the nest and apparently do nothing for themselves. Darwin did not have the insights of modern genetics, so any solution he offered was bound to be at best partial. But he did sense that relatedness was the key – somehow, even the greatest altruist is helping relatives and not mere strangers. Today, we use the term "group selection" (as opposed to "individual selection") to denote selection producing (at cost to oneself) features that help others. The question is whether the term should be restricted to those others who are nonrelatives or whether it can be extended to all, related or not. Most biologists today would restrict the term, in which case Darwin is not a group selectionist in the *Origin*. Call him, if you will, a "family selectionist" or some such thing; but recognize that individual selectionists would claim that as one of their own. Certainly this seems in line with the years after the *Origin*, when Darwin and Wallace thrashed out the topic with Wallace (an ardent socialist) always inclining toward group selection (and incidentally iffy about aspects of sexual selection, something firmly individualist). The *Descent* was and is a matter of great controversy, with even those inclined not to think there is group selection earlier agreeing that, when it comes to morality, Darwin finally softens and allows group selection (involving nonrelatives). The weasel word in the discussion is "tribe." If this includes nonrelatives, then Darwin is truly a group selectionist (as the essay on language in this encyclopedia claims him to be). But don't overlook the letter Darwin wrote later to a son, where explicitly he likened a tribe to a hive of bees or a nest of ants. This suggests that he was consistent to the end, never wanting to go beyond family selection, something more on the individual-selection end of the scale than the group-selection one.

And so we come to the topic of ongoing fascination: Charles Darwin and religion (Essay 25, "Darwin and Religion"). There has been much disagreement, but usually this reflects the diverse interests of those asking and discussing, for actually there is a lot of pertinent material, and the main points are pretty clear. Darwin's religious life fell into three phases. The first from childhood up to the time on the *Beagle* was when he was a fairly conventional and committed Christian, secure in the beliefs of the Church of England. Then his formal commitment started to fade (quite quickly), and he became what should be described as a "deist," that is, one who believes in a kind of god who is an unmoved mover. This is a god who set everything in motion and now sits back and lets events unfold, as by clockwork. Fairly obviously, evolution is a testament to the power and magnificence of such a god, for no miracles are needed. It was a powerful and natural vision for a child of the Industrial Revolution – a god who works through machine rather than by hand – and it was backed by the arguments of the protocomputer inventor Charles Babbage, a good friend of Darwin's brother Erasmus, who (in his *Ninth Bridgewater Treatise*) showed how miraculous-type exceptions could be programmed in and occur occasionally entirely by virtue of unbroken law (Fig. Introduction.10). The deism lasted right through the writing of the *Origin*, but then this too started to fade and vanish. Darwin never became an atheist, in the sense of total denial of any kind of god, but he was certainly happy to adopt Huxley's new term of "agnostic." It should be added that, like many nonbelieving Victorians, it was not science that turned Darwin from religious conviction but theology – he could not stomach the eternal damnation of nonbelievers and that sort of thing.

As the essay on religion points out truly, what does come across very strongly when studying Darwin and his life is just how nonemotionally involved he was in religion. He had to think about it quite a bit, both as he was growing up and then when he had his theory, one that so clearly did impinge on religious belief – but he never seems to agonize over it, nor is it an obsession. In this he contrasts strongly not only with his Cambridge teachers, clergymen down the line, but also with his friends. Lyell, who worshiped with the Unitarians for a while, obsessed about the status of humankind. Huxley, who was the arch nonbeliever, nevertheless kept picking away at religion like a scab that never heals. Darwin just assumes that humans are part of the selective landscape without a hint of a worry. That's just not his fight. And the same is true of the discussion of religion in the *Descent*. Although he covers himself by saying that discussion of origins does not tell you about truth value, his neo-Humean theorizing about the rise of religion – likening it to the antics of his dog on a windy day when the parasol flaps around – suggests that he thinks it all pretty much superstition. Compared to morality, the treatment of religion is brief. For Darwin, that is as it should be. Morality matters. Religion does not.

Having said this, one should never underestimate the extent to which the religion of his early years left its mark on Darwin's thinking. Most obviously there is the obsession with adaptation, a direct result of the heavy influence of British natural theology with its great regard for the argument from design. To this day, ultra-adaptationists tend to have grown up immersed in this theology, and critics tend to be those for whom the tradition is quite alien. Then there is the tree of life, something lifted (metaphorically) right out of Genesis. Even that wonderful concluding passage of the *Origin* may well be a modification of a natural theological peon of praise to the Creator. Compare the earliest version that we have (from the "Sketch of 1842") with a passage, written by the Scottish physicist David Brewster, something read by Darwin just before he discovered natural selection. First Brewster:

> In considering our own globe as having its origin in a gaseous zone, thrown off by the rapidity of the solar rotation, and as consolidated by cooling from the chaos of its

FIGURE INTRODUCTION.10. Charles Babbage, the inventor of a protocomputer, was a good friend of Charles Darwin's brother Erasmus. This book was an unofficial addition to a series of works on natural theology sponsored by the Earl of Bridgewater. Babbage showed how he could set his computing machine to produce the natural numbers in regular order up to a million and one, and then without interference but according to initial conditions the succession would change to produce all sorts of unexpected numbers (in Babbage's example, 100,010, 002 rather than 100,000,002). The conclusion was drawn that miracles, meaning unexpected occurrences, can be part of the natural order without need of divine intervention. Title page of C. Babbage, *Ninth Bridgewater Treatise* (London: John Murray, 1838)

elements, we confirm rather than oppose the Mosaic cosmogony, whether allegorically or literally interpreted....

In the grandeur and universality of these views, we forget the insignificant beings which occupy and disturb the planetary domains. Life in all its forms, in all its restlessness, and in all its pageantry, disappears in the magnitude and remoteness of the perspective. The excited mind sees only the gorgeous fabric of the universe, recognises only its Divine architect, and ponders but on its cycle and desolation. (Brewster 1838, 301)

Then Darwin:

There is a simple *grandeur* in the *view* of *life* with its powers of growth, assimilation and reproduction, being *originally* breathed into matter under one or a few *forms*, and that whilst this our *planet* has gone circling on according to fixed laws, and land and water, in a *cycle* of *change*, have gone on replacing each other, and from so simple an *origin*, through the process of gradual selection of infinitesimal *changes*, endless *forms* most beautiful and most wonderful have been evolved. (Darwin 1909, 52; italicized words are those echoed)

Darwin died in 1882. Even before that, though, the world was starting to pick up and move on, taking his ideas, using them, modifying them, and sometimes rejecting them. Looking at the reception of Darwinism in every country would be a huge task, quite swamping all else. Fortunately, there are now many good surveys, freeing us here to focus more on specific countries and examples. Let us start, as we must, with the two chief Anglophone countries, Britain and America. The former is the home of Darwin and his ideas, and the latter is, by any measure, the country that has done most in working on and developing evolutionary ideas, notwithstanding the paradox that it is also the country where opposition has been highest. Then let us move to the two countries that have the greatest in-depth history of evolutionary theorizing, Germany and France. What happened in those two lands and how did they handle the fact that it was an Englishman, thinking in a very English fashion, that made the major evolutionary moves? After that, the choice becomes more open, and many countries (like Russia) have good reason to be considered and discussed. Included here is an essay on China, illustrating how evolutionary thinking moved right across the world and how it was received in a culture that, although modernizing rapidly, was still (from Western perspectives) alien in the extreme. And concluding is a discussion of Darwin's fate in South America, something fairly deserving attention because it was after all in that part of the world that the young Darwin traveled and began his evolutionary speculations. Throughout the aim is not so much to emphasize specific issues but to give general assessments and to see how social and cultural factors affected the story of Darwin's science.

Already we have been primed for the story of the reception of Darwin's thinking in his home country (Essay 26, "Darwinism in Britain"). Evolution is accepted. In large measure natural selection is not. What does seem clear is that Darwinism, meaning the ideas inspired by his thinking, is – part cause and part effect – a major element in the overall cultural and metaphysical shift that we see in Britain in the second half of the nineteenth century. As already intimated, by midcentury it was becoming very clear that the old norms and ways were simply not adequate for a country that had industrialized and become (almost overnight) an urban-based rather than rural society. For many, religion was increasingly being seen as not just false but irrelevant; new, more professional methods of running the country (and empire) were needed; science itself was becoming more university based and, although we know it would be very misleading to refer to Darwin as an amateur, the kind of gentleman-researcher that he represented was increasingly becoming rare and out of place; and there was much more, including the arrival of universal male suffrage and state-supported elementary education for all – after the Reform Act of 1867 it was quipped that "we must educate our masters." In an important way, the move was from a spiritually based, oligarchic society to a materialist-based, democratic society, and science and technology had central roles in this new system. There is a reason why the debate (at the British Association in 1860) between the bishop of Oxford Samuel Wilberforce and the professor of mines Thomas Henry Huxley took on such mythic proportions, for the clash was between the old ideology – decked out in Elizabethan clerical robes – and the new – dressed in a modern business suit (see Plates VI and VII).

For Huxley and his supporters, Darwinism was much more than a science – it was a secular substitute for the old religion, a metaphysical foundation for the new order of things. In good reason, this was why the actual mechanism was of less importance. It was what Darwinism represented, blind law working endlessly, to bring on change, that really counted. Although remember what has been noted already. It was not blind law working to no purpose. It was law bringing on progress, in society mirroring progress in biology. Darwin, we know, endorsed this vision, but – a point to which we are already sensitized – above all it was the philosophy of Herbert Spencer. As it happens, by century's end, troubles in the empire, poverty and depressions at home, and military arms races with Germany were making hopes of progress seem empty and shallow. In the light of what was just said, it is little wonder that Huxley was led to write his great essay, "Evolution and Ethics," denying that we see such an upward process. But the underlying vision of a material world, unaided by spirit forces from without, persisted. This is not to say that there was no evolutionary science, but it tended not to be very causal and significantly increasingly it was something to be found more in the museums, places that existed for display and education, than in universities, places for research and the advance of knowledge.

The story of the reception of Darwin in America is a fascinating tale of how preexisting culture and needs affect and condition the reception of new ideas (Essay 27, "Darwinism in the United States, 1859–1930"). The major and well-known clash at the time of Darwin himself was between his champion, Asa Gray, professor of botany at Harvard, and the likewise Harvard-based Louis Agassiz, a strong antievolutionist. Because evolution eventually won the day, the usual assumption is that Gray would have been the major influence in the New World. This was not true, even though it raises the massive paradox that the very person whose ideas were the focus of attack in the *Origin* was he who had the real influence. Agassiz had the students. After the Civil War, in the North, as in Britain, industry and urban society grew exponentially

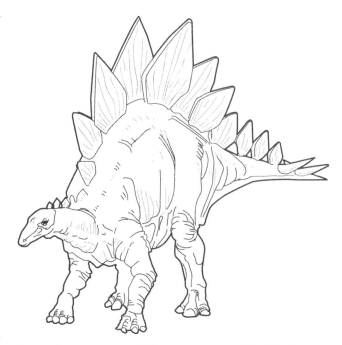

FIGURE INTRODUCTION.11. The stegosaurus, one of the giant dinosaurs discovered in the American West toward the end of the nineteenth century

and, as in Britain, a science- and technology-based world picture grew to dominate. Evolution was at the heart of this vision. It is therefore not surprising that virtually all of those students, including Agassiz's own son, became evolutionists. Expectedly, however, the form of their evolution owed far more to morphology- and homology-exhibiting archetypes – precisely those things cherished by the nonevolutionary Agassiz, student of the *Naturphilosophen* philosopher Friedrich Schelling and anatomist Lorenz Oken – than to natural selection and its explanations of British adaptationism. Naturally, Herbert Spencer, with his message of progress, was deeply appreciated.

Of course, particularly with a country as big and diverse as America was then becoming, one should be careful about sweeping generalizations. Given the demands of agriculture in that country, intensified after the war with the building of the railroads and the opening of the prairies and the routes to the West, there was much interest in methods of breeding, and this certainly spilled over to an appreciation of the merits of natural selection. As in Britain, however, one senses that much that occurred was less than fully focused causally, or invoked causes more liked for the metaphysical (often progressivist) implications than for their scientific merits. The magnificent fossil discoveries in the West of the United States and Canada bolstered the beliefs in evolution as such, but also they contributed to what (as in Britain) was becoming a pattern, where museums became very much the homes of the evolutionist, places of display and education and less of ground-breaking research (Fig. Introduction.11). This was reinforced in the United States particularly with a turn by biological investigators from broad historical studies to much more reductionistic laboratory studies. A bright student went for Germanic-type training to one of the new universities like Johns Hopkins, spending summers at research institutes like the one at Woods Hole in Massachusetts, rather than roaming the West for fossils or the fields and forests for butterflies.

What of Germany itself? (Essay 28, "The German Reception of Darwin's Theory, 1860–1945") From at least the end of the eighteenth century there were thinkers who accepted some kind of evolutionary perspective or another. It was usually if not always mixed up with analogies with individual development and thus led to a kind of progressivist reading of life's history, the kind that made Darwin so uncomfortable when he separated himself from views about inevitable, upward change. This continued after the *Origin*, especially at the hands of the great morphologist Ernst Haeckel – he who popularized the individual-group connection with his so-called biogenetic law, "ontogeny recapitulates phylogeny." To this day, there are debates about just how much of a Darwinian we should consider Haeckel. Undoubtedly he was an enthusiast for the *Origin*, and nigh hero worshiped Darwin himself. But his writings show strong evidence of his own intellectual heritage, with a taste for tracing trees – that he himself illustrated memorably – rather than working on the ways in which a mechanism like natural selection could produce organisms and their adaptations. This was the same for others too. Perhaps directly as a result of Haeckel's own urges to make a full-blown metaphysical picture of his science – at times, he even gave tremors to Huxley – evolution in Germany in the later years of the nineteenth century was rarely quiet or unchallenged. It figured in debates about society and religion and more. How long-lasting were these effects and what their ultimate outcome is still contested today. American biblical literalists, fundamentalist or creationists (of which more later), combine their critiques of Darwinian evolution as science with the claim that it is morally pernicious, having led in a fairly direct line to the vile doctrines of the National Socialists. As you will learn from the essay given here, the truth is very different. Something had to lead to Hitler and his vile minions, and no one would deny the racism of the nineteenth century – shared pretty much by everyone including Darwin – must have had some input. But to pick out Darwin and his follower Haeckel for special condemnation is to make a politically motivated moral charge on the back of a historical falsehood.

General opinion among English-speaking historians of evolution is that after the *Origin* the French went into a century-plus sulk, from which they are only just now emerging, if that. They did not discover natural selection and, as the country that had done most in the century and a half before to put evolution on the map, the failure and the perceived disgrace was too much to bear. They wanted nothing to do with Darwin or anything connected with him. In fact, as is so often the case with oft-told tales, there is some truth in all of this, but the real story is much more complex and interesting, so much so that there are two essays covering the period from the *Origin* to the present. Certainly today Darwin is genuinely acknowledged and respected for his work. The magnificent Muséum National d'Histoire Naturelle, in the botanic gardens on the bank of the Seine, has an exhibit on evolution that gives Charles Darwin all of the credit that he deserves. But it is true that it was a long time coming. After the *Origin*, Darwin as a scientist was respected, Darwin as a support for all sorts of speculations about human nature and society was eagerly turned to good use – generally, much to the chagrin of Darwin himself – but Darwin as an evolutionist among professional biological circles was a nonstarter (Essay 29: "Darwin and Darwinism in France before 1900"). The great French biological scientists of the day, notably Claude Bernard and Louis Pasteur, set the pace and the standards, and their kind of hard-nosed, bench-based, experimental science was not welcoming toward the kind of naturalist-inspired speculation of the *Origin*. (We shall see the same story with Germany and botany.) And, of course, there was the home-grown Lamarckism ever-ready to provide answers for those who asked the pertinent questions. So overall, we should probably see French reactions as part of a general type of reaction to the *Origin* – eagerness to co-opt for ideological ends and a sense that Darwin's style was out of kilter with the direction of professional biology – and not necessarily as something specific to that particular country.

At the burial of Karl Marx – somewhat amusingly he lies in Highgate Cemetery London, literally facing the remains of Herbert Spencer – his great supporter Friedrich Engels praised Darwin for having done in the biological world what Marx had done in the social world. In fact, Marx's reaction to Darwin was interestingly nuanced. He devoured the *Origin* as soon as it appeared, writing a couple of years later to Engels: "It is remarkable how Darwin rediscovers, among the beasts and plants, the society of England with its division of labour, competition, opening up of new markets, 'inventions' and Malthusian 'struggle for existence'" (Marx and Engels 1975–2005, 41:380; letter from Marx to Engels, June 18, 1862). He did think sufficiently highly of Darwin that he sent a copy of *Das Kapital* to Darwin. (It remained in Darwin's library uncut!) Because of this, in those countries taken over by groups ostensibly following in the footsteps of Marx, Darwin got high praise, even when, judged objectively, the science of the land was being perverted by politically influenced factors, referring especially to the Soviet Union and the disastrous effects of the charlatan agronomist Trofim Lysenko. Expectedly, the praise is usually directed toward the ends of the speakers and their patrons. In Communist China, we find that Darwin is lauded as much for his materialism-atheism as for anything strictly scientific (Essay 30, "Encountering Darwin and Creating Darwinism in China"). You should not think that this use of Darwin for political and social ends was something new. Long before the communists, Chinese intellectuals were using Darwin's ideas – and, as often as not, Herbert Spencer's ideas flying under the colors of Darwinism – in the cause of deserved cultural changes. After the devastating war with Japan at the end of the nineteenth century, Darwin's claims about the struggle for existence found favorable readers, as did various thoughts of progress and of the need to strive for success. One reason why Darwin was praised was because he showed the kind of reverence for ancestors that the Chinese appreciate. Was he not following in the footsteps of Grandfather Erasmus? Unfortunately the first war with Japan was followed by a second starting in the 1930s, which morphed into the general worldwide conflict ending only in 1945, at which point a civil war took over. Science generally in China suffered, and this affected evolutionary studies in particular. Today, as is well known, particularly thanks to fabulous fossil discoveries, Chinese evolutionary studies are thriving, and it will be interesting to see if they challenge the overall dominance of the West as the country seems to be doing in the economic field.

Finally, there is South America (Essay 31, "Darwinism in Latin America"). There are many different countries in the region with many different challenges, so it is hard to make firm generalizations. Positivism in some version was a major influence on the thinking of scientists and others, including politicians. Here as elsewhere, when one speaks of evolution, it is usually better to think first of Herbert Spencer (and Haeckel to a certain extent) and only secondarily of Darwin, although it is the latter who usually gets the great praise and respect. Some kind of evolutionary positivism or naturalism seems to have been the mark of the forward-looking thinker. Sometimes, perhaps expectedly but regrettably, the ideas of evolution were used to rationalize beliefs and practices that would have shocked the old scientist in his greenhouses down in Kent. This applies particularly to the extermination of the natives in Argentina, something a troubled Darwin wrote about in the *Voyage of the Beagle*. For all of his Victorian views about race, in the *Descent* Darwin made it very clear that his sense of the struggle between races (and the consequent fitter elements) was that the real focus should not be on violence and who beats whom but on the immunity of Western races to diseases that wipe out native populations. What should never be forgotten, however, is that, though the countries of the continent often used evolutionary ideas more for political and social ends than for strict science, Brazil was the home of the German-born Fritz Müller. Given that Bates and Wallace both worked in Brazil, there must be something overwhelmingly inspiring about the insect life in that region, for it was Müller who (following Bates) made significant and lasting contributions to our understanding of mimicry. That part of the world will always have a special place in the hearts of Darwinians.

And now, as we move from Darwin's nineteenth century into the recently finished twentieth century, let us pause again to take the temperature of the times, or rather to assess the status of evolutionary thinking. For the first 150 years of its life, evolution was a pseudoscience, riding on the back of the ideology of progress. Charles Darwin set out to change things. He put together the evidence for evolution so that it became common sense. He provided a mechanism of change, one that spoke to the big problem of final cause. Darwin himself wanted to create a mature science of evolutionary studies, what we can call a professional science – or if you like, normal science working within an established paradigm. We must conclude that he was only partially successful. Obviously there were professional scientists doing evolutionary studies. Ernst Haeckel is a case in point, and it would be wrong to deny that his attempt to work out relationships and histories was professional science. But note how often the work being done was either noncausal or all over the place with respect to what made things work and change. Again and again, people were far more interested in the social implications of evolution than in working on technical problems about the nature of living beings. There was some work using natural selection, but it was very much the exception rather than the norm. And by century's end, evolution was truly much more the science of the museum than it was of the laboratory. Historians of the period talk of the "revolt from morphology," meaning that around the beginning of the new century there was a whole new breed of biologists – people like the geneticist William Bateson in England, the cytologist Edmund B. Wilson, and the future geneticist Thomas Hunt Morgan in America, who were turning to bench studies, highly reductionist in outlook, determined to make of biology a science to stand with any other (namely the physical sciences). Evolutionary studies were out of the loop.

By and large, evolution became what one might call a "popular science" – respectable (more or less) but not cutting-edge science, more philosophical and background than anything else. In some hands, it became virtually a secular religion, an alternative suited for the industrial, urban world, to compensate for the perceived failure of the more conventional religions of the past. It is amusing how often the palaces of evolution, otherwise known as natural history museums, now being built in major city after major city, were so often modeled on medieval cathedrals (see Plate VIII). Instead of going to the Church of Christ on a Sunday morning, the family could go to the Church of Darwin on a Sunday afternoon. Who was responsible for all of this? As we have seen, there were many factors, from the problems of the science to the need of alternative philosophies. Darwin himself was perhaps a major culprit. A rich man who could afford to do as he pleased, he did rather shut himself away, pursuing his own interests, leaving it to lieutenants like Huxley to go out and do the hard work of proselytizing. Had he been prepared to pour some of his considerable fortune into a research institute of selection studies, perhaps things might have been a little different. But it was not to be, and, to be fair, remember, apart from the real handicap of the ongoing sickness, Darwin probably did not think that such an effort would really pay dividends.

And yet, the story did not end there. Today, if anything is a professional science, a paradigm supporting normal science, it is evolutionary biology, and Darwin's contributions are right at the center. His ideas matter. So let us pick up the thread and see what happened next, starting with the rediscovery of the work of Gregor Mendel. It is satisfying to begin the story with botany (Essay 32, "Botany: 1880s to 1920s") – satisfying both because so often botany gets pushed aside in favor of animal studies and because botany, in fact, has always played a vital role in evolutionary studies. Mendel, after all was working on pea plants, not fruit flies, and the same focus on plants is true of many who followed him, including some of the key figures of the twentieth century such as Ronald A. Fisher. What is important is the way in which we see the beginnings of the move from a rather low-grade science to one that is much more rigorous and professionally acceptable. It did not happen in an easy, straight line because we have Darwin himself using selection and yet, with some good reason, criticized for his rather old-fashioned experimental methods, and then we have leading German researchers like Julius von Sachs, who for all his sophistication did not embrace much by way of evolutionary causation and certainly not selection. But all of this was about to change, and plant studies as much as animal studies were part of the work and evidence.

The rediscovery of Mendel's ideas at the beginning of the twentieth century was the crucial event, moving evolution from its past toward its future. How much Mendel himself truly realized what he had done, and how much later thinkers read back into his work what they wanted to find, are still matters of historical debate. The point is that now the way was being opened for an adequate theory of heredity, something so lacking and so needed by the theory of the *Origin* (Essay 33, "Population Genetics"). What was necessary was that the genetics be extended from individual organisms to factors of heredity working in populations. Unlike Lamarckism, to take an example, natural selection is something that acts not on the individual but is meaningful only in groups. Thanks to some very mathematically gifted biologists, this work was done, and so by around 1930, the framework of a full theory or paradigm of evolutionary change was starting to emerge.

But even with the mathematics done, this, to use an analogy, was just the skeleton. Now, the task turned to the naturalists and the experimentalists to supply the empirical flesh. What was needed was not simply people committed to evolution and trained in the pertinent science, but people with vision, the Thomas Henry Huxleys of their days, able to build groups and find funding and attract students and do all of the things needed to get an area of science functioning as mature work – as normal science, to use Kuhn's phrase, or what has been termed as professional science. In the United States, the key figure was the Russian-born geneticist Theodosius Dobzhansky (Essay 34, "Synthesis Period in Evolutionary Studies"). He took a proposal by the American geneticist Sewall Wright, the "shifting balance theory" – at least, he took the version that used the pictorial metaphor of an "adaptive landscape" (Dobzhansky, to be candid, was never very strong on mathematics) – and used it to pursue studies in the wild and in the laboratory. Following his teacher Morgan in taking the little fruit fly as the model organism, Dobzhansky and his associates and students followed in detail the physical and chromosomal changes over generations, trying to work out how forces of selection and of drift bring on changes. His work and that of those in his orbit (particularly the taxonomist Ernst Mayr, the paleontologist George Gaylord Simpson, and the botanist G. Ledyard Stebbins) did much to establish Darwinian selection as a major force in nature,

FIGURE INTRODUCTION.12. Herbert Spencer (1820–1903), Darwin's contemporary, was an ardent evolutionist. Wildly popular in his day, at his death his reputation sank like a stone. Nevertheless, his fingerprints are all over twentieth-century evolutionary thinking. From David Duncan, *Herbert Spencer* (London: Williams and Norgate, 1911)

although there was often a non-Darwinian flavor to the work, especially when their thinking was influenced by (what modern scholars are now seeing as the) deep roots that Wright's thinking had in Herbert Spencer as much as Charles Darwin (Fig. Introduction.12).

Socially, what was crucially important was the way in which Dobzhansky and his fellows worked hard to bring evolutionary studies into the universities, making them part of the biological curriculum. Dobzhansky went to Columbia, Mayr left the American Museum of Natural History for Harvard and a year or two later Simpson followed, Stebbins went west and worked at the University of California at Berkeley and then at the new campus at Davis. An evolution society was founded; funds were sought and found to start a journal (*Evolution*), one dedicated to the kinds of causal studies now being effected; grants were awarded (thanks, especially after the Second World War, to the great rise in available federal money through the National Science Foundation); and in Dobzhansky's laboratory especially there was a flow of new graduate students and post-docs. Above all, there was a conscious awareness that evolutionary studies had had low-grade status as a science, and a major factor was the way in which it had acted as a vehicle for nonscientific cultural hopes and aspirations, especially about social progress (being reflected in claims about biological progress). All of these new professional evolutionists were deeply committed to both biological and social progress. All knew that such professions in their science would be fatal to their professionalizing ends. So thoughts of progress were suppressed and kept out of the university science, reserved for the popular books that poured forth from their pens – as such popular books about evolution continue to pour forth today.

Something very much parallel happened in England also (Essay 35, "Ecological Genetics"). The key figure there was E. B. Ford, universally known as "Henry." He allied himself with Fisher in much the way that Dobzhansky allied himself with Wright, and one immediate consequence was that non-Darwinian notions like genetic drift got short shrift. Working in the British tradition of Bates and Wallace, Ford and his students, including Philip Sheppard, Arthur Cain, and Bernard Kettlewell, did highly influential studies of fast-breeding organisms showing the workings of natural selection in bringing on subtle adaptations. Sheppard and Cain did seminal studies of shell color and banding of snails, showing how the colors and patterns adjust according to the backgrounds – hedges, ditches, forests, and the like – and Kettlewell continued the studies of industrial melanism that had so excited nineteenth-century lepidopterists. No less adept than Dobzhansky at finding funds, Ford convinced one of Britain's largest private research foundations – the Nuffield Foundation, started by England's counterpart to the real Henry Ford of Detroit – that insects are great models for humans. For instance, the studies of his group could tell much about the spread and retentions of various genes, information that could be very important when studying genetic factors in humankind. Also, as was the case with the Dobzhansky group, it is interesting that as biology felt the huge effects of the molecular revolution – epitomized by the discovery in 1953 of the structure of the DNA molecule – it filtered almost seamlessly into evolutionary studies. Fears that molecular studies might replace whole-organism studies entirely were soon followed by the realization that molecular biology could be a very powerful tool for throwing light on hitherto-intractable evolutionary problems.

Let us return to France for a sense of how these ideas started to spread out to other countries. We should not expect to find much action until around 1930 or later, and we do not (Essay 36, "Darwin and Darwinism in France after 1900"). For instance, although to a person the paleontologists were evolutionists, that was about as far as they would go, being even reluctant to speculate on phylogenies. Given the harsh criticism that greeted Teilhard de Chardin's attempts to reconcile science and religion (in his *Phenomenon of Man*, published posthumously in 1955), it is worth noting explicitly that, judged as a paleontologist, Teilhard's brilliant reconstructions stood out as significant exceptions. Where real change did come – as in America and certainly influenced by America – was with respect to population studies of the actions and effects of selection and of how these play out for overall evolutionary changes. As soon as the theoreticians had done their work, eager young French researchers (significantly, with good mathematical strengths) were picking up the ideas and putting them to the test. Indeed, one of the most important experimental innovations – population cages – came from that country. And before long, important work was being done on key issues such as the ways in which selection pressures can vary. It cannot be said that the ideas of neo-Darwinism were universally and immediately welcomed in France – Lamarckism had great staying power – but a beachhead was established, pointing to the universal acceptance of today.

Finally, as part of the story of the making of modern evolutionary biology, botany must again get full mention (Essay 37: "Botany and the Evolutionary Synthesis, 1920–1950"). The importance of getting the right subject to study can never be overestimated. As intimated, the little fruit fly *Drosophila* showed itself a perfect organism for genetic studies – it breeds easily and quickly, requires minimal maintenance, has no odd sexual system, has giant chromosomes that are easy to study, and can be found readily in the wild in accessible places. Mendel got the right plant (the pea) when he sought the principles of heredity – even to the point of having different features of study controlled from different chromosomes, so that there were no immediate complicating factors. The early geneticists of the twentieth century were not so lucky in their choice of the evening primrose, because it proved to have a very complex system that led, among other things, to the belief that changes are large and sudden – saltationism. But by the 1920s, things had righted themselves and then for the next thirty years botanists – notably the Carnegie group at Stanford and others at Berkeley – did path-breaking studies to work out principles of speciation and the like. Animal studies tend perhaps to be more glamorous. But from Darwin on, the plants have provided more than their share of information about the evolutionary process. That Stebbins was a key figure in the making of the evolutionary synthesis was no anomaly.

Somewhat artificially, let us position ourselves now in 1959. It is the 150th anniversary of the birth of Darwin and the 100th anniversary of the publication of the *Origin*. Evolution, as an area of science, is still somewhat tentative in respects and threatened from without by various forces, not the least being the way that molecular studies (for all that they were on the verge of being seized upon as tools by evolutionists) were exploding in size and threatening to take all students and grants of the life sciences. But notwithstanding the worries and insecurities, we have now a functioning, professional science. What then were evolutionists able to do in the half century following? Staying now with the science, it is to this question that we turn next, starting with the problem of the origin of life (Essay 38, " The Emergence of Life on Earth and the Darwinian Revolution").

In a way, this problem reminds one of the problem faced by Sherlock Holmes in the story about Silver Blaze, the missing racehorse. Asked if there was anything to which he wanted to draw attention, Holmes replied that he was puzzled by the dog that barked in the night. But the dog did not bark in the night, came the reply. Exactly! It should have done, and because it didn't Holmes inferred (correctly) that it was an inside job. The same is true of the *Origin of Species* and the origin-of-life question. What does Darwin have to say on the topic? Nothing! And now the question is why, because the omission had to be deliberate. Before Darwin, people like Lamarck and Chambers assumed automatically that one must discuss life's origins, and the same was true of people like Haeckel after. Darwin realized fully that talking about origins, especially at a time when (over in France) Pasteur was showing that much thinking on the topic was simply wrong, would only lead to trouble. So he spoke simply of life "having been originally breathed into a few forms or into one" and left it at that. Basically, although he had some private thoughts, it was not really his problem, and he pushed it to one side. But of course it could not be sidelined indefinitely, and the past century saw much interesting and fruitful, if far from definitive, work on the problem. The coming of the molecular age obviously transformed things, and today there are many exciting areas of inquiry. Is this Darwinian science? Well, in one sense, perhaps not. In another sense, obviously at some point the evolutionist has to face the topic, and moreover it is clear that even at the earliest molecular stages, when one would hardly want to speak of things as "living," something very much like natural selection is going to be active and important.

The traditional philosophical view of scientific theories sees everything happily integrated into one massive system, generally thought to be an axiom system with high-powered principles or laws at the top, and then everything seen to be deductively connected on the way down to lower-level empirical claims. This view of theories is not entirely wrong – it is almost certainly the one held by Darwin himself – but most today realize that actual science tends to be far messier, with small areas of theory or modeling connected loosely together with others, sharing some ideas and theory but not necessarily entirely consistently throughout. As it has grown, covering as it does so many areas, this lack of systematic purpose has often plagued evolutionary studies (Essay 39, "The Evolution of the Testing of Evolution"). It does not mean that all is lost, for in various areas there is much serious and important work. For instance, the number of studies demonstrating the action of selection in experiment and in the wild, building on the work of the mid-twentieth century, has grown exponentially. But it is clear that researchers are not always as meticulous as they might be in distinguishing their aims. Is the claim, for instance, that everything is adaptive, or only in part? These are points particularly to be kept in mind as we move now through work being done today across the spectrum of topics falling within the Darwinian consilience.

Mimicry and camouflage were important for Darwin studies back in the years immediately following the *Origin*. They continue to be so today (Essay 40, "Mimicry and Camouflage: Part Two"). What is fascinating about this area is how often researchers are working not just in the Darwinian mode but actually with hypotheses that Darwin himself formulated. A good example is the question of sexual dimorphism, where female butterflies mimic other species, whereas the males do not. Darwin suggested that natural selection is the factor in making the female mimics but that sexual selection is the factor making for males to stay with the original species colors and patterns. Recent studies have confirmed the truth of his hypothesis, underlining not just the importance of adaptation as a central biological concept but also that selection does so often work on and for the individual and not the group. Sauce for the goose is not always sauce for the gander. Notice however that, in the tradition of the very best science, solving one problem is not the end of the story. There are always new problems to be solved. Molecular techniques for instance show that complex adaptations are created again and again rather than simply inherited, and now the race is to find the reason why. Critics of evolutionary biology, especially those with religious axes to grind, often point with glee to the unsolved problems of the science. They quite miss the point – something stressed strongly by Kuhn's philosophy of science – that good science throws up new problems constantly. It is always forward moving rather than resting on its laurels.

Darwin always had a somewhat ambiguous attitude toward the actual history of life. It was he after all who established beyond doubt that there is a history of life, one produced by evolution. And if you look at some of his writings, there are heavy hints about what he thinks the course of life truly was. The barnacle work is a case in point. In the *Descent*, he opted explicitly for an African ancestry for humankind. But although he gave a stylized-tree picture – his only diagram – in the *Origin*, he was not much into providing actual histories or phylogenies (see Plate IX). This is perhaps what one might have expected because, ultimately, a great deal of phylogeny tracing is not very Darwinian, if one means doing something using natural selection. Indeed, with reason, natural selection is often thought something of a handicap because it covers up true relationships with superficial adaptations. One must dig beneath, to find homologies, to trace paths. Within bounds, this is much the same today, although the methods of inquiry have become far more sophisticated and reliable, especially in this molecular age (Essay 41: "The Tree of Life"). Moreover, thanks to such new devices such as the "molecular clock" – based on the rate at which mutations occur and change accumulates – we can put some absolute dates on events, hitherto unknown. But is it simply a matter of things meshing, with non-Darwinian work fitting nicely with Darwinian selection studies? One fascinating new finding is about how, thanks to viruses, genes can be passed between very different branches of the tree of life. Does "lateral gene transfer" show that Darwin was wrong? Two comments are in order. First, although it may be a major factor with simple-celled organisms (prokaryotes), it is unlikely to be so great a factor with complex-celled organisms (eukaryotes). Second, even if it did mess up the tree of life significantly, it is not obvious that the importance of Darwinian factors are downgraded. The adaptive values of lateral transfer are not obvious, so no one is saying that natural selection suddenly becomes unimportant.

The study of the evolution of instinct and social behavior, brought together under the name "sociobiology," has been one of the most fertile and controversial areas of evolutionary biology in the past fifty years (Essay 42, "Sociobiology"). After years of ignoring issues to do with the level of selection, finally in the 1960s biologists started to face the question squarely and (in major reaction to work by the English-born Vero Wynne-Edwards) a thoroughly neo-Darwinian individualist stance was taken. Huge amounts of very profitable work have been done right through the animal kingdom, from the social insects to the primates. New ideas such as "kin selection" (where genes are passed on by proxy, as it were, through close relatives) and "local mate competition" (where sex ratios are skewed because of the waste when siblings compete for the same reproductive opportunities) have been devised and used highly effectively in order to understand the workings of organisms in groups. However, there has always been a minority that has group-selection yearnings, and recently their ranks have been joined by the man who wrote the bible of the whole movement, Edward O. Wilson, author of *Sociobiology: The New Synthesis*. He argues now that a more integrated, "holistic" approach must be taken to animal behavior. Perhaps significantly, Wilson stands in direct intellectual line to an earlier, Harvard ant specialist, William Morton Wheeler – who was in turn much influenced by Herbert Spencer, especially by analogies that the earlier evolutionist drew between the individual and the group. It could be that we are hearing echoes of divisions between evolutionary visions that go back to the middle years of the nineteenth century.

Paleontology also has been vibrant in the past fifty years (Essay 43, "Evolutionary Paleontology"). What is fascinating is the gap between the professional and the public. Most people, if asked why evolution is true, would say "because of the fossils" (the same reply that would be given by those asked to defend their view that evolution is not true). Yet Darwin expended much effort in the *Origin* to saying why the fossil record does not deny his evolutionary thinking and for years afterward paleontologists either ignored the whole question of evolution or went off in search of non-Darwinian mechanisms. The action was within the reduction-happy sciences like genetics, and paleontology was mainly a source of nice fossils for the museums. G. G. Simpson, Dobzhansky's associate, started to change all of that, and since then – particularly with the rise of "paleobiology" – much effort has been made to give paleontology full status within the evolutionary family. Some, if not much of the work, both theoretical and empirical, would bring delight to Darwin. One of the biggest problems he faced in the *Origin* was the total absence of pre-Cambrian fossils, leading to the invention of remarkable ad hoc hypotheses to explain away this worrying phenomenon. Now we have a remarkably detailed record back to the earliest forms of life nearly four billion years ago. More than this, we have lots of very sophisticated adaptationist studies. A classic analysis is of the plates on the backs of the dinosaur stegosaurus, showing how they were almost certainly used for heat regulation.

However, sometimes the thinking of paleontologists is at best neutrally Darwinian, and sometimes verges on the unfriendly. The well-known theory of "punctuated equilibrium" of Niles Eldredge and Stephen Jay Gould, suggesting that the course of evolution is not smooth but goes in fits and starts, went through various incarnations, but in Gould's hands was not particularly selection friendly. The very name of the theory had echoes of a theory, "dynamic equilibrium," from another tradition. In a like vein, John J. Sepkoski Jr. (student of both Gould and Wilson) did sterling work in mapping the major events in life's history, producing neo-Spencerian pictures of the repeated upward spurts of complexity, followed by subsequent balance. It was work that could be given Darwinian underpinnings, but not work starting with Darwinism.

One encounters some of the same sorts of issues when one turns from time to space, from paleontology to biogeography (Essay 44, "Darwin and Geography"). One thing that cannot be overemphasized is just how important the experience of new lands and new flora and fauna were to Darwin. It is hotly contested as to how far one should think of Darwin as being influenced by the romantic movement, but it cannot be denied that his early writings, when he writes of his experiences of nature in its many varieties around the world, show a rapture worthy of Goethe or (closer to home) Wordsworth. Expectedly however, we find that, although this enthusiasm for nature and its variety around our globe found its way firmly into the *Origin*, those who followed in his footsteps as often as not reflected their national trends rather than anything strictly Darwinian. This was notably so when it came to human themes. And as always, the specter of Herbert Spencer loomed in the background. This was especially true in America in the twentieth century as biologists moved to more ecological studies trying to map the differences in organisms in different climates and lands. Mention must also be made of the great effects on biogeographical studies brought by the geological theory of plate tectonics. Many of Darwin's own anomalies – for instance, the similarities between plants in the Southern Hemisphere on lands often separated by vast expanses of ocean – are now seen as the direct result of the slow but steady movement of continents around our planet.

The Galapagos Archipelago has always had a special place in the hearts of Darwinian evolutionists, for it was from his visit and his later reflections that Darwin's move to evolution really started to gather steam (Fig. Introduction.13). There has been some controversy about the exact organisms that really excited and prompted Darwin, but it was not long before the drab little finches of the island started to play a significant role in Darwin's thinking, and these tiny birds continue in that role down to this day (Essay 45, "Darwin and the Finches"). In many respects, the story of the finches is the story of Darwinian evolutionary theory in miniature. Darwin was excited about the finches. Those who came after him thought them interesting and perhaps significant, but in no wise did people want to use them as evidence for natural selection. This continued true even after the population geneticists had brought selection back into style. It was thought that the non-Darwinian genetic drift was the real cause of change. Then the tide changed, and increasingly the finches were seen as paradigmatic end results of a struggle for existence resulting in many different adaptations, for living in the face of both the environment and competing finch forms. Today, thanks particularly to the stunning,

Introduction

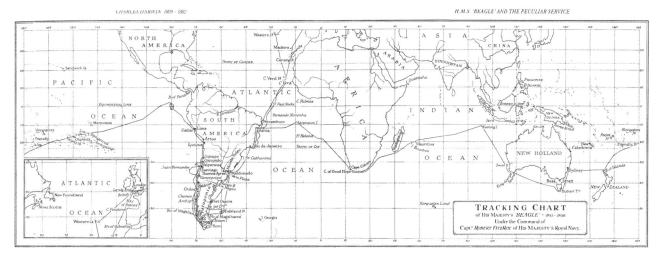

Figure Introduction.13. The voyage of the *Beagle* (1831–36). Note the visit to the Galapagos Archipelago in 1835. From C. Darwin, *Journal of Researches* (London: John Murray, 1845)

long-term studies of the husband and wife team of Peter and Rosemary Grant, the finches are at the top of the list of well-defined selection studies. The results, moreover, are of the best kind of science. They show that Darwin was right in his basic theory but that there is far more to the story than he did or could have dreamed of – about speciation, about adaptation, and (very excitingly) about rates of evolution. Darwin was, for example, completely wrong about the inevitable slow working of natural selection. One likes to think that no one would be more excited than he about this discovery.

Embryology has always had an intricate but somewhat uneasy relationship with Darwinian thinking (Essay 46, "Developmental Evolution"). For Darwin himself in the *Origin*, the discussion of embryology was a triumph of selectionist thinking, of which he was very proud. But although embryology continued for the rest of the nineteenth and into the twentieth century as one of the most important areas of the life sciences, one that was completely bound up with evolutionary thinking, it tended not to be very Darwinian. It was rather used to work out relationships between organisms, in which work (as noted above) adaptations were generally a nuisance taken (with reason) to conceal true homologies, and of course – thanks to the biogenetic law – it was much involved in the tracing of phylogenies, something else that paid scant attention to natural selection. It was perhaps understandable that, when Darwinian selection and Mendelian genetics were synthesized, there was something of a tendency to regard organisms as black boxes with genes making the input and fully grown organisms emerging and not much interest in what happened in between. Things have changed dramatically in the past three or four decades, thanks particularly to the coming of the molecules, and the tracing of development from an evolutionary perspective (evo-devo, so called) is a big business. And some of the findings have been truly astonishing. For instance, we now know that there are significant molecular homologies between the genes controlling development in fruit flies and in humans. It turns out that organisms are built on the Lego principle, with the same building blocks put together in different ways and ratios. Whether this is now all that Darwinian is a different matter. No one denies selection outright, but it must be allowed that sometimes the impression given is that development is where the real evolutionary action occurs and then selection comes along to clean things up, tweaking advances and removing failures. No doubt this is a debate that will continue.

Ecology is the study of living organisms and their relationships to each other and to the environment. Although the term was not invented until the decade after the *Origin* (by Haeckel) and although the concept does not get the separate treatment of, say, paleontology, it is obviously something that is threaded right through Darwin's great work (Essay 47, "Darwin's Evolutionary Ecology"). That Darwin did have very important insights no one would ever deny. He showed in great detail how the welfare of any group of organisms is intricately bound up with the welfares of others, and he made significant contributions to our understanding of key ecological notions like niches. However, care must be taken not to confuse surface similarities with deep differences (Essay 48, "Darwin and the Environment"). The concept of a balance of nature is one deeply embedded in first Greek and then Christian thought. In a way, it almost follows from the biblical story for one would expect God to have ordered things so that the world would continue in happy equilibrium for the benefit of all, especially humans. Darwin certainly made some efforts to capture the notion, arguing that selection would often act to balance things out. But truly the balance does not fit tremendously comfortably with evolution through natural selection. On the one hand, one is always expecting some change, at some point or another. On the other hand, Darwinian selection is (in the opinion of most) never working directly for the good of the whole but for the individual. This means that the balance is never an end in itself, but always a consequence. Once again one must recognize that there were other sources for the enthusiasm about equilibrium positions, and once again the influence of Spencer cannot be ignored. Today, with the threats of global warming and the like, it is realized how selection can work for the short-term gain rather than the long-term harmony. A point to be noted is how often in discussions about these topics one gets an uneasy mix of the professional and the more popular. The professional evolutionary ecologist is trying to understand the workings of nature, whereas the ecologist in the popular realm invariably has moral or social issues foremost. Referring to Darwin is rarely neutral at these times, as he is alternatively praised as the first person to understand properly the issues at stake or he is condemned as the progenitor of ultimate selfishness in the face of upcoming environmental catastrophe.

Is molecular biology truly no more than a handmaiden to the evolutionary biologist, or does it carry within it deeper threats (Essay 49: "Molecular Biology: Darwin's Precious Gift")? One of the most exciting ideas to emerge from the new approach was that of the neutral theory of evolution. Could it be that down at the molecular level a great deal goes on beneath the reach of selection? Could it therefore be that, at this level, random forces – drift – were the chief causes of change? This idea was seized on with enthusiasm, for at once it seemed that one had a very accurate way of determining relationships between different organisms, perhaps with some real time estimates. One simply works out the rate at which change is occurring, steady change that is occurring, and then one can generate real and accurate phylogenies. As it happens, it now seems that the initial enthusiasm was a little too high; although no one doubts that there is some real truth and value to the idea. This is often spoken of as "non-Darwinian" evolution – the "neutral theory of evolution" – and in a way of course it is. But note that it is not really "anti-Darwinian" evolution. No one is saying that the hand and the eye were produced by drift. Rather that there are dimensions of the biological world where selection does not reach, or (probably) does not reach as readily as it does others.

What is certainly the case is that the molecular revolution is not going to vanish and that the face of evolutionary studies is changed forever. The Human Genome Project is still only on the verge of being fully exploited, as biologists study the vast amount of information that has been revealed about our genetic makeup – and the makeup of many other organisms, also. Obviously many surprises lie ahead. Whether these will finally convince people that Darwinism is now outmoded, on a par with Newtonian theory or (worse) phlogiston theory, is something the future will tell. Most Darwinians today would argue not. But there are, as there always have been, those who beg to differ (Essay 50, "Challenging Darwinism: Expanding, Extending, Replacing"). Part of the time – too much of the time – the differences are more linguistic than substantive. Is individual selection now on its way out? So much depends on how you define your terms. If individual selection can encompass the family, then not obviously; if it cannot, then probably. But there are genuine differences, and one suspects that these are as much philosophical as scientific. From the *Naturphilosophen* on, there have always been those who find

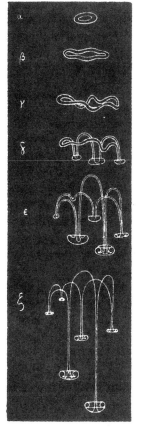

FIGURE INTRODUCTION.14. The Scottish morphologist D'Arcy Wentworth Thompson (1860–1948) believed that much organic form is simply a function of the laws of physics. Here he is trying to show that the shape of the jellyfish is the result of the same laws of physics that determine the descent of more-dense liquids through less-dense ones. From Thompson, *Growth and Form* (Cambridge: Cambridge University Press, 1917)

offensive the kind of blind, reductionistic approach epitomized by Darwinian selection. Today's representatives have seized on the notion (found in Schelling) of "self organization" – there is something inherent in matter itself that makes for organic form no need for selection. We get "order for free." Perhaps its plausibility is best left as an exercise for the reader (Fig. Introduction.14).

If humans were not part of the story, would anyone care very much about evolution? Well, there are people who care about organic chemistry, so probably some, but one much doubts that there would be the intense interest that there has been from the eighteenth century down to the present. It is we who make the subject so fascinating and so fraught with tension for so many. One suspects that even (perhaps especially) the just-mentioned critics of Darwinism have humans somewhere in their minds. Could we just be the product of blind, random force? What is true is that the 150 years since the *Origin* have seen huge effort put into discovering our evolutionary past and, despite setbacks and prejudices and outright fraud, the effort has paid immense dividends (Essay 51, "Human Evolution after Darwin"). There is much fossil evidence, and it is backed by findings from other areas, notably in recent years from molecular biology. Many hitherto unanswered questions are now settled. Humans came from Africa, not Asia. Humans split from the apes around five or six million years ago, not earlier. Humans got up on their hind legs, and then their brains exploded up in size, rather than conversely. There are still mysteries, including the crucial one about precisely why we came down from the trees in the jungle and walked out over the plains. And always puzzling new finds emerge, most recently the little being (the "hobbit") in Indonesia. Is the work in this area recognizably Darwinian? Obviously, in an overall sense it is, for selection is thought to have played (and still plays) a vital role in human change, although, just as obviously, the science has moved on dramatically from the speculations in the *Descent of Man*. And everyone today is keenly aware of the fallacy that plagued the field long after Darwin, namely the assumption that the closer something is to being European, the more it is favored by natural selection.

In his *Meditations*, published around 1637, the great French philosopher René Descartes held forth on the significance of language: "It is a very remarkable thing that there are no men, not even the insane, so dull and stupid that they cannot put words together in a manner to convey their thoughts. On the contrary, there is no other animal however perfect and fortunately situated it may be, that can do the same." He pointed out that this does not seem to be necessarily an anatomical matter. Magpies can say words as well as we. And deaf and dumb humans find other physical ways to communicate. He concluded that this all "proves not merely animals have less reason than men but that they have none at all, for we see that very little is needed to talk." Like many generations of English dog lovers, Darwin thought this absurd, and he had the theory to back up his beliefs. Language was fascinating to Darwin, and, for evolutionists it has continued to be down to the present (Essay 52, "Language Evolution since Darwin"). It poses major challenges, obviously, because words do not fossilize, but as with other elusive features, ways are devised to overcome this issue. One is comparative studies with other animals, particularly primates, and even more particularly the great apes. Another is by looking for related fossil evidence – for instance, the parts of the brain and of the vocal organs. For a while a (now-refuted) hypothesis was floated that the Neanderthals could not talk properly because they lacked the necessary anatomy. Artifacts are also suggestive. Sophisticated technology implies the ability to communicate efficiently. Famously, one of the most important moves to filling out the story came from the American linguist Noam Chomsky, who argued that all languages share the same innate deep structure. Famously, Chomsky himself denied that this was an evolutionary hypothesis – he was almost with Descartes on the separation of human and beast – although he has now recanted, and his students and collaborators have done major work in showing how the innate structure relates to biology. New hypotheses are still being produced about the nature and origin of language, and today it is one of the most exciting areas of evolutionary study.

The evolution of language slides easily into the more comprehensive topic of the evolution of culture generally. It is an underexaggeration to say that it has been a happy home of many and varied hypotheses (Essay 53, "Cultural Evolution"). Roughly, these can be divided into two camps. First, there are those who argue that culture can be divided into units and that these units function like genes or organisms, struggling for survival and reproduction – that is, being passed on to other thinkers – and knowledge and culture is an outcome of this selective process. Richard Dawkins's (1976) theory of memes is a prime example. Note how readily these views soak up wishes and prejudices – for instance, that memes are parasitic and hence prone to produce (supposed) corruptions of human well-being like religion. Second, there are those who argue that culture is in some sense informed by innate, selection-produced beliefs or traits. Darwin subscribed to something along these lines, writing in his early notebooks: "Plato says in Phaedo that our '*necessary ideas*' arise from the preexistence of the soul, are not derivable from experience. – read monkeys for preexistence – " (Barrett et al. 1987, 551, M, 128, 4 September 1838). Those of our would-be ancestors who took logical and mathematical reasoning seriously survived and reproduced, and those who did not did not. The devil of course is in the details, and much effort today is being put into finding how learning and like abilities are involved in the overall picture.

Creative artists are an important part of culture, and particularly in literature evolution generally and Darwinian ideas more specifically have been picked up and used and transformed and presented favorably or unfavorably as the writer or the times declared (Essay 54, "Literature"). Before the *Origin*, poets and novelists were using evolutionary themes – Tennyson in *In Memoriam* using Robert Chambers's progressivist vision of life's history to suggest that his dead friend Arthur Hallam was a superior specimen who had come too early, Dickens in *Bleak House* using dinosaur examples to suggest that industrial London was the kind of primitive world that would contain such brutes (see Plate X) – and after the *Origin* they continued with such themes, sometimes directly Darwinian and sometimes less so and more Spencerian and indebted to other evolutionists – Samuel Butler, for instance, using recapitulatory ideas in his late novel *The Way of All Flesh*. The worries and hopes of society can be depicted vividly through fiction. H. G. Wells shared the fears of his countrymen at the end of the nineteenth century that progress was over and only decline lay in the future. These worries come starkly in his novel *The Time Machine*, where in the future our race has divided into two, equally unsatisfactory groups: the Eloi above ground, beautiful but childlike; and the Morlocks below ground, intelligent and hardworking, but vile and ugly. In America then and later, we see the themes of struggle and competition being worked out in fiction, by Jack London and others. And this continues to the present, for instance in the work of the English novelist Ian McEwen, who tries to use Darwinian psychology to show the motivations of his characters. No doubt as we extend our understanding of evolution in different cultures, we shall see more and more evidence of how creative thinkers have used evolutionary ideas to the particular ends and causes that drive them in their writings.

Increasingly, we have become aware of the extent to which gender issues permeate culture, and the contribution of Darwin – both as part of culture and as an aid to explaining culture – has been a topic of much debate (Essay 55, "Darwin and Gender"). The obvious analysis is that Darwin was a Victorian sexist, especially in his discussions in the *Descent of Man*, and that his thinking has been used to legitimate such sexism, from then until the present, as is shown by discussions to be found in such works as Edward O. Wilson's *Sociobiology: The New Synthesis*. That there is truth in this can hardly be denied. Women are simply portrayed as childlike, obviously lower down the evolutionary scale of being. The Darwin family life was much the same with all being focused on the father and then the sons. Of course, things are never that simple. In the family, it is pretty clear that Emma was in charge, and Charles knew and approved. And the whole point about both natural and sexual selection is that, unless in some fundamental way the sexes are equal, things are out of balance. Fisher was good on this. If it is better to be a boy, then parents are going to have boys, and conversely. This is why it cannot be permanent in those societies today (like India and China) where boys are prized over girls and there is real sex selection. Before long girls are going to be such a rare commodity that parents will strive to get them. Some Darwinian evolutionists have made much of points like these, arguing that in fact we see that evolution has compensated for the features males have. These and like ideas are controversial, but right or wrong, reactionary or visionary, Darwin's thinking is still very much part of the debate.

Fifty years ago, the suggestion that Darwinism might make some contribution to philosophical understanding would have been greeted somewhat like a bad smell at a vicarage tea party. This was not always so. The American pragmatists were very keen on Darwinian evolution, thinking it gave keen insights into the nature of knowledge, its acquisition, and its status. No doubt this enthusiasm was a factor in the decline of appreciation of evolution for philosophy. The founders of analytic philosophy like Bertrand Russell rather thought that pragmatism was not just wrong but positively immoral. But when you think about it, this is surely a wrongheaded attitude. That we are the product of a long, slow, natural, nondirected process of change from probably inorganic material rather than the cherished climax of a Good God's week of creative activity has to matter for both the theory of knowledge (epistemology) and the theory of morality (ethics). And increasingly in the past fifty years philosophers have started to agree.

In respects, especially in epistemology, many of the points made about culture generally (especially about the different possible approaches) apply directly (Essay 56, "Evolutionary Epistemology"). One major question has been whether in some sense a Darwinian account of knowledge implies that one is getting ever closer to a true description of an objectively existing world. One might think so. After all, if fire doesn't really burn, why should we think that it does? However, as has been pointed out, ultimately selection does not really care about truth or objectivity. Being successful in the struggle is what really counts. If we are deceived part of the time or even all of the time, so long as we reproduce, that is what matters. Some critics, notably the well-known Calvinist philosopher Alvin Plantinga, have seized upon this to argue against the possibility of any kind of naturalistic approach – one that depends on blind law – to the world and its understanding. Others doubt one need go that far. From Kant on, it has been appreciated that knowledge is never pure and simple – at the least, we structure experience according to our psychology. Perhaps Darwinism simply takes us further down this path, and we must recognize that while we can certainly distinguish good knowledge from bad – Darwin was right and Sedgwick and Agassiz were wrong – there is necessarily an evolutionary input to all understanding.

Evolutionary morality was very heavily criticized in the years after the *Descent* (Essay 57: "Ethics after Darwin"). One should recognize, however, that the main object of attack was not Darwin, who was mainly concerned to show the origins of morality rather than its justification. The focus of fire was Spencer, who used his belief in the nature of evolution to argue that morally we should promote the evolutionary process because that is the way in which value is kept and increased. The philosophers, first Henry Sidgwick and then G. E. Moore (who introduced his famous "naturalistic fallacy"), argued that claims about matter of empirical fact could not support moral claims. The scientists, notably Thomas Henry Huxley, denied that things are all that progressive and pointed out in any case doing the right thing often means going against our animal nature. The past four decades however have seen a great rise in interest in and enthusiasm for an evolutionary approach to morality. Great credit goes to Edward O. Wilson, who in his writings (especially his *On Human Nature*) has argued that evolution is the key to moral understanding and justification. The general public took up the cause with enthusiasm, arguing in a way that would have excited the Spencer of metaphysical excess and appalled the Spencer of lifelong bachelordom, who lived in a drab boardinghouse that he not get too excited and distracted from his life mission, that now we have justification for even our mortal sins. "Do men *need* to cheat on their women?" asked the *Playboy* cover of August 1978. "A new science says yes," it assured its readers. Most of a philosophical vein, however, deny the neo-Spencerian approach taken by Wilson and make other connections. Great controversy has surrounded the claims of some thinkers that a Darwinian approach points to some kind of moral nonrealism, where morality is simply (to use a phrase) "an illusion put in place by our genes to make us social." In fairness, it should be pointed out that the illusion is not morality itself – modern Darwinians are not into unrestricted rape and pillage – but the belief that morality has an objective foundation. The claim is simply that morality has no base beyond human emotions.

We come to religion. It is appropriate to start with Protestant Christianity, for it was within that version of faith

FIGURE INTRODUCTION.15. In 1925 William Jennings Bryan, three times candidate for the presidency, led the prosecution of John Thomas Scopes for teaching evolution in the public schools of Tennessee. Here, Bryan – having offered himself as an expert witness on the Bible – is being examined by the leader of the defense, notorious freethinker Clarence Darrow. Permission: Smithsonian Institution Archives

that Darwin worked and that so influenced the form of his theory. The problem obviously when it comes to discussing reactions to Darwin is that there is as much variation among Protestants as there is variation among animals in the natural world (Essay 58, "Darwin and Protestantism"). And, expectedly, this is reflected in the reactions to Darwin's theorizing. Some were very comfortable with his ideas, starting with the Reverend Baden Powell (father of the scout master), who endorsed Darwin in 1860 in his contribution to the notorious Anglican, iconoclastic volume *Essays and Reviews*. Some liked Darwin's ideas but wanted to supplement them, as did Asa Gray in America, seeking to give some kind of nonnatural direction to new variations. And some rejected the whole message, as did the doyen of American Presbyterians, Charles Hodge at Princeton Theological Seminary. *What is Darwinism?* asked one of his books. "It is atheism" came the stern reply. What does seem to be the case overall is that simplistic pictures of science at warfare with religion are just wrong, and even those most critical often find points where agreement is possible. What also seems to be the case as we come into the twentieth century is that Darwin did continue to fascinate and disturb Protestant thinkers, and this continues to this day. This is hardly surprising given the far-reaching, Darwinian implications for such key Christian notions as miracles and morality and original sin. In 2011 an eminent theologian at Calvin College, a leading American liberal arts college, lost his job because he suggested that perhaps modern evolutionary theory is incompatible with a literally existing Adam and Eve. (It is!) (See Plate XI.)

In discussions about the religious implications of Darwin's ideas, much to the dismay of conventional Protestants, most people have in mind the opposition by a large branch of the American evangelical movement to any and all kinds of evolution (Essay 59, "Creationism"). It is important to note, therefore, that so-called creationists (using this in the modern sense and not of the people whom Darwin was countering in the *Origin*) accept a somewhat idiosyncratic form of Protestantism coming out of America in the middle of the nineteenth century. What is surprising, and probably would be to most of today's creationists, is the historical significance of the Seventh-day Adventist movement, with its emphasis on a literal six days of creation, about six thousand years ago. As a widespread phenomenon, this Young Earth Creationism (YEC) is fairly new. Three-time presidential candidate William Jennings Bryan, prosecuting attorney in the Scopes Monkey Trial of 1925, believed in an old earth, where the six days of creation are to be interpreted as six long ages (Fig. Introduction.15). YEC really caught fire only in the 1960s with the publication of *Genesis Flood* by biblical scholar John C. Whitcomb and hydraulic engineer Henry M. Morris. The fondest hope of its advocates is that it be introduced into publicly supported schools (in the United States) alongside teaching about evolutionary origins. Thus far, the First Amendment separation of church and state has been effective in denying fulfillment of this hope, both for YEC and for its somewhat milder successor "intelligent design theory." Whether this will continue to be the case remains to be seen, as also the extent to which these various views will be able to establish themselves beyond the American borders.

Having found themselves in hot water over the Galileo affair, Catholics generally have been happy to sit back and let Protestants fight the battle over Darwin (Essay 60, "Darwin and Catholicism"). This is not to say that the ideas were universally accepted, even though there were some early Catholic converts to some kind of (generally guided) evolution. There was a condemnation of the materialism thought inherent in all evolutionary theories, especially Darwin's – although in fairness, it should be noted that a lot of the early opposition was primarily a function of general issues facing the church and was part of the general move to a much more conservative position after the unification of Italy. The twentieth century saw a slow but increasing acceptance of evolutionary ideas. Almost paradoxically, the thinker who did most in the twentieth century to bring evolutionary science and Christian religion together fruitfully was the Jesuit priest and paleontologist Pierre Teilhard de Chardin. He was forbidden to publish (while he was alive) by his church, and when his ideas did appear in print (around the time of the hundredth anniversary of the *Origin*), he was scorned by the scientific community. But in the years following, the spirit if not always the details of his thinking has gained more and more respect. Not that this means that, even now, the church thinks all that creatively about evolution, looking for ways in which the science might enrich the theology rather than threaten it. However, the way is open for such thinking, thanks especially to the fact that, almost at the end of the century, Pope John Paul II gave a surprisingly strong endorsement not only of evolution but also of Darwinian ideas. As always, things were somewhat qualified, by insisting that the creation of souls (as nonscientific notions) stands beyond science. But the principle of creative interaction is acknowledged and expected.

Christianity and Darwinism are like parent and child. You think they are different, and then suddenly when least expected, in the half light, you see a staggering similarity – origins, trees, design, humans, and more. Focusing on that relationship is not prejudice against other faiths but simple recognition of history. But, of course, the story does not stand still, it is not finished, and it certainly is not isolated. By the middle of the nineteenth century, Jews in Western Europe particularly were integrating more and more into general society, and this continued and intensified with the great migration across the Atlantic (Essay 61, "Judaism, Jews, and Evolution"). Naturally prizing education and learning, Jewish teachers and intellectuals encountered and starting contributing to science in major ways, and this has continued to the present – in Israel too since the founding of that country. In some significant respects, one sees a mirroring of the Christian response to evolution. The more liberal parts of the group, particularly the Reform branch of Judaism, accepted evolutionary ideas reasonably readily. More conservative parts of the group, particularly the conservative and Orthodox branches, had more trouble, and especially in ultra-Orthodox branches the opposition to evolution continues strongly to this day. But, as with conservative Christians, it is dangerous to generalize too facilely or readily, for there are Orthodox Jews who are very comfortable with evolution, with Darwinism even. What is still far from fully understood or researched are the implications of the differences between Judaism and Christianity. For instance, given that the design argument has not played the role for Judaism that it has for Christianity, does this mean that selection has less hold on the Jewish evolutionist than it has for the Christian (or Christian-cultured) evolutionist? The late Stephen Jay Gould used to claim that, because he had no Christian upbringing, he was no ultra-adaptationist; although in his case perhaps an argument can be made for the secular, very left-wing milieu of his childhood rather than anything specifically Jewish.

The interaction of Darwin and Islam is a rather different story, for here we truly do have the meeting of alien world pictures – a meeting that was bound to be slow at first, because of the widespread illiteracy in Muslim countries and the lack of interest in science generally (Essay 62, "Religion: Islam"). The *Origin* was not translated into Arabic until well into the twentieth century. As was the case of the spread of evolutionary ideas in countries like China and those of South America, often the interest in evolution was less in its virtues as science and more for its supposed ideological components, materialism and so forth. It is not surprising that evolution was popular early in the past century among the reforming "Young Turks" opposed to the status quo in the Ottoman Empire. Moving down toward the present, one finds that (as one might expect, given that Islam is so widespread a religion over many lands and cultures) there are all shades of acceptance of evolution, although (as is still the case with many Christians) often even when positive it is some kind of theistic evolution that is most favored. But, again perhaps a function of ignorance and illiteracy, one finds that most people in Muslim lands either reject evolution or are indifferent to or ignorant of the whole idea. A form of creationism, not entirely unlike its American counterpart, is spreading (especially in places like Turkey). As always, one suspects that underlying the motives are factors more from the cultural and moral or social realm than from pure science. Darwinism is caught up in more general debates about Western culture and its worth.

And so finally we come to evolution and medicine (Essay 63, "From Evolution and Medicine to Evolutionary Medicine"). This has not been left to last by default. Anything but! Rather, it is one of (if not the) newest branches of the Darwinian family, really only just starting to develop and gain ground. It seemed appropriate to end the volume on something that is so very definitely looking forward and not back – if that is not too much of a paradox for a field that derives its very being from history. Actually, from Darwin on, there was concern about human health and whether evolution can throw light on its problems. Are we breeding the wrong kind of people, the weak and the sick and the profligate? This led to many years of theorizing and of proposing solutions for its amelioration – so called eugenics. Primarily because of the appalling events in Germany under the Third Reich, outright calls for the biological alteration of humankind are now less common, although vestiges of eugenics still persist under such more friendly names as "genetic counseling." But now we have a rather different approach to human health, one that plunges right into questions about sickness and disease and tries to uncover pertinent evolutionary facts and implications. For instance, why do we have fevers and what should we do about them? The usual advice is to take a painkiller and reduce the temperature. But what if the high temperature has some real biological value in fighting infection – a fever is an adaptation? At a more complex level, how should we understand serious problems like high blood pressure in pregnancy? Could it be that it is a result of mother and fetus fighting it out for supremacy, the mother having one set of biological interests and the infant having others?

No one would pretend that we have now a fully fledged area of medical science, and expectedly often the ideas have to fight to be taken seriously. Back in the years after the *Origin*, Huxley was much involved in reforming medical education and working to see that basic biology became part of the training. But although he made anatomy and physiology required subjects for would-be doctors, he never thought to push evolutionary studies as part of the curriculum. He (no big friend of natural selection) could see no good reason for this in the program. Such thinking continues to this day. But evolutionary medicine is growing and gathering more and more supporters, significantly among younger researchers, and the hope is that one day it too will take a full place at the table of Darwinian evolutionary studies.

This is for the future. Now the time is to turn to the individual essays of the *Encyclopedia*. Their broad range and their exciting content speak without need of further proof of the importance of Charles Darwin and his theory of evolution.

Essay 1

Origins and the Greeks

Jeremy Kirby

Thomas Henry Huxley's reaction to Darwin's idea is understandable: "How extremely stupid of me not to have thought of that!" Darwin's argument is not arcane. Why did we have to wait so long for an idea as simple and attractive as Darwin's? An attempt to answer this question by means of a narrow set of influences is likely to produce an account that is, at best, impressionistic. However, that teleology played a leading role is widely accepted, as Darwin himself always recognized that the appearance of design is distinctive of the organic world, having been raised on the teleological argument of Archdeacon William Paley. And an important part of the conceptual network wherein Darwin found himself developed in antiquity. Classical thinkers erected much of the scaffolding with which evolutionists have had to work, framing the debate over teleology in important ways. Early cosmologists thought the idea of the world's coming to be from nothing as unintelligible. Early teleologists thought getting order out of chaos, equally unsettling, akin to getting something from nothing.

THE EARLY COSMOLOGISTS

Natural philosophy before Socrates can be said to concern the origin of the cosmos. The problem of origin, as one might call it, can be characterized in terms of the following three claims:

(1) Coming to exist involves a transition from nonbeing to being.
(2) Things come to exist.
(3) Only that which has being can undergo a transition of any kind.

Suppose, in accordance with (2), that the cosmos comes to be. By (1), the cosmos underwent a transition from nonbeing to being. According to (3), however, this means that the cosmos *existed prior to itself* in order to undergo the transition in question. One might, of course, attempt to avoid this unwelcome conclusion by maintaining that the transition in question involves the alteration of a substance that is at one time not the cosmos and at another time is. To what substrate might one appeal?

One might, instead, give up on the very idea of providing an origin. And this is precisely what a number of philosophers from Elea (ca. 490 B.C.E.) recommended.

Parmenides and his student Zeno went so far as to reject the second proposition. If the cosmos came to exist, they argued, it did so from either what is or what is not. If the former, then it already is, and something cannot come to exist when it already exists. If the latter, then something will come to exist out of nothing. As the saying was later popularized, however, *ex nihilo nihil fit*. Thus, the cosmos did not come to exist – nor did anything else for that matter.

On the other end of the spectrum, a few philosophers, followers of Heraclitus (fl. 500 B.C.E.), sought to disarm the problem of origin by maintaining that it is the very use of the term "being" which has issued in the difficulty. Likening all things to a river, these philosophers maintained that all is in flux, and that whatever is in flux is not in a state of being. It is, rather, in a state of becoming. The first and third claims of our problem of cosmogony, on this view, are considered unintelligible.

Under the entry *cosmos*, in a standard Greek lexicon, one will find that the term expresses arrangement and order. One might, therefore, take the project of origin to be that of explaining how it is that a rather inchoate existence might evolve into a cosmos with a level of organization deserving of the name. Philosophers attempted to explain this origin by discussing the ways and means whereby organization may be imposed upon a rather inchoate and insufficiently formed material substrate. Rejecting the first proposition in the problem, these natural philosophers maintained that coming to exist involves a transition from not being such and such to being such and such. The statue, for example, comes into being not *ex nihilo* but from the clay upon which the artisan imposes the form.

Among this group of materialists were the monists, who attempted to trace our origin to one substance. These natural philosophers, perhaps more than any others, sought explanation of the differentiated in the undifferentiated, taking cosmogony to be a move from the homogeneous to the heterogeneous. Thales (ca. 600 B.C.E.), living as he did on the coast of Asia Minor, thought that the substance out of which everything was born was water. For this view he offered various reasons (many of which one might consider empirical). He faced the difficulty of explaining, ultimately, how everything could come to be from water, when some elements, fire, for example, seem to be eliminated by water. His successor, Anaximander (ca. 610–546 B.C.E.), sought to circumnavigate this problem by making the substrate *apeiron*, which is to say, indefinite. Lacking any natural characteristic whatsoever, this substance would not, as it were, be incompatible with any natural substance. Of course, any substance that lacks natural characteristics altogether seems itself unnatural. In this way, *apeiron* seems a little like the playwright's *deus ex machina*. Perhaps with this in mind, Anaximenes (585–528 B.C.E.) maintained that air is the better candidate. Air is something observable in the form of breath and wind, or so it was thought, and compatible with both fire and water. Anaximenes seems to have had the intuition of Galileo – that the qualitative should be explained in terms of the quantitative – as he took fire to be the result of a reduction in air density, whereas water and then earth were the result of an increase in terms of the density of air molecules.

The atomists, most notably Leucippus (fl. 440–435 B.C.E.) and Democritus (fl. 435 B.C.E.) may have been the first of the Presocratic philosophers to attempt to eliminate intelligent design at the primary causal level. Rather than taking intelligence as an irreducible feature of matter, early Greek atomism countenanced atoms and void as the primary realities, giving intelligence a status akin to the epiphenomenal:

> Democritus sometimes does away with what appears to the senses, and says that none of these appears according to truth but only according to opinion: the truth in real things is that there are atoms and void. "By convention sweet," he says, "by convention bitter, by convention hot, by convention cold, by convention color: but in reality atoms and void." (Kirk, Raven, and Schofield 1983, 410)

In a way similar to that of the monists, these atomists rejected the first of the propositions composing the problem of origin. Composite entities come into being as atoms configure together and diminish as the atoms part ways. In contrast to the monists, the fundamental realities, though they are eternal and infinite in number, are not divine. This fairly mundane status of the ultimate realities, however, issues in the question of how the order or arrangement of the cosmos would come about. The atomists answered this question by maintaining, on the one hand, that inanimate particles may self-organize, as the pebbles on a beach may be said to collect in virtue of their size and shape in a certain order. On the other hand, the atomists thought the supply of atoms infinite (with infinite void), composing an infinite number of worlds throughout the universe. The advocate of intelligent design, for example, might consider the chances of our world coming about through blind chance – with its structured ecology – akin to the likelihood of Hamlet being produced by the random striking of keys on a keyboard. On the atomist's strategy, however, the analogy is misguided. Were one to countenance an infinite number of keyboards, as the atomists recognized an infinite number of worlds, all being struck randomly throughout the universe, for an infinite amount of time, the suggestion that Hamlet could thus be produced is less difficult to entertain.

While some philosophers sought to reduce the elements to one substance and others to atoms, still others favored a pluralistic approach, countenancing four basic nondiscrete substances: air, water, earth, and fire. Empedocles (490–439 B.C.E.), for example, held this view (see Plate XII). He thought that these four elements, which he called "roots," are divine and eternal and are organized by the divine struggle between Love, which provides mixture, and Strife, which brings about separation. When Strife has fully gained the upper hand, the four elements are stratified with earth at the bottom, water thereafter, air the penultimate, and fire (the lightest) at the top. When Love gets her revenge, however, portions of the elements are blended and living things occupy the world. As Love's strength begins to wane, and as separation results in the perishing of one generation of living things, another generation of living things is brought about by Strife.

> A twofold tale I shall tell: at one time they [i.e., the roots] grew to be one alone out of many, at another time they grew apart to be many out of one. Double is the birth of mortal things and double their failing; for the one is brought to birth and destroyed by the coming together of all things, the other is nurtured and flies apart as they grow apart again. And these never cease their continual interchange, now through Love all coming together into one, now again each carried apart by the hatred of Strife. (Kirk, Raven, and Schofield 1983, 287)

Empedocles' double zoogony is interesting for a number of reasons. Primary among these is that the second stage of the zoogony is thought to contain an idea that bears some similarity to the idea that survival goes to the fittest.

> Empedocles says that…next came together those ox-headed man-progeny, i.e., made of an ox and a human. And all the parts that were fitted together in a manner which enabled them to be preserved became animals and remained because they fulfilled each other's needs – the teeth cutting and softening the food, the stomach digesting it, the liver turning it into blood. And when the head of a human came together with a human body, it caused the whole to be preserved, but it does not fit together with the body of an ox, and so it is destroyed. For whatever did not come together according to the appropriate formula perished. (McKirahan 1994, 279)

The image suggested by this fragment of text is that of body parts being randomly thrown together into various combinations with the resultants being fit or otherwise for survival. And Empedocles, it bears mentioning, takes present-day species to be, in turn, the resultants of earlier generations:

> Empedocles held that the first generations of animals and plants were not complete but consisted of separate limbs not joined together; the second, arising from the joining of these limbs, were like creatures in dreams; the third was the generation of whole natured forms; and the fourth no longer from the homogeneous substances such as the earth or water, but by intermingling…in other cases because feminine beauty excited sexual urge. (Kirk, Raven, and Schofield 1983, 303)

There are, to be sure, interesting similarities that one may identify concerning Empedocles' outlook and Darwin's. At the same time, it is equally clear that to say that Empedocles anticipated Darwin's theory is very wide of the mark. We can discern key differences, for example, from the passages we have here considered. Reproduction and death are part and parcel of Darwin's theory. For Empedocles, on what I take to be the natural reading, reproduction does not seem to play a significant role until the fourth generation. And by the time the fourth generation has arrived, it would seem that the recombination of somatic parts is a fait accompli. With Empedocles, we have modification, and we have descent, but it is not clear that we have descent *with* modification. Furthermore, on Empedocles' view the cosmos is brought into existence by the work of two deities, adversaries though they may be, Love and Strife. Darwin, in stark contrast, sought to explain phenomena without appeal to the supernatural (Ruse 2006, 13).

THE SOCRATIC PARADIGM

So radical is the break that Socrates makes with the early cosmologists that it is common for historians to arrange their accounts in terms of before and after Socrates (Cornford 1932). Many of the individuals we have discussed thus far sought to ground explanation in the nature of the material. And, as we have seen, on the program of the atomists, the mind – insofar as the mental is epiphenomenal – is fairly inefficacious. There is little room for design or intention on such a view as this. Socrates seems to think that the accounts that attempt to reduce all to the physical promise little in terms of explanatory return.

There is tragic irony in the fact that Socrates was sentenced to death for impiety and atheism. In Plato's *Phaedo*, as Socrates awaits the effects of the hemlock, the reader is treated to arguments supporting the claim that the soul is immortal. Toward this end, Socrates maintains that a material description of the events we care most about will, of necessity, be incomplete (J. M. Cooper 1997, 98c2–99b2). In the course of the discussion, it is suggested that Socrates can successfully escape from prison – an alternative that he declines (see Plato's *Crito*). With this is mind, he argues in the following manner:

(1) If a materialist description is adequate, then a description of the bones, blood, tendons, et cetera will explain why Socrates is sitting in prison.
(2) But bones, blood, and tendons, are just as good for escaping as sitting.
(3) So bones, blood, and tendons cannot be singled out to explain why Socrates is sitting in prison.
(4) Ergo, the description of the materialist is inadequate.

Socrates, of course, takes the psychological explanation to be the important one, as it will concern the purposes and intentions that underwrite his conviction that staying in prison is the right course to take. To explain his intentions by appeal to that which lacks intention is to get something from nothing. Indeed, Socrates does not restrict this kind of teleological explanation to human action, as he is willing to extend design, purpose, and intention to the craftsperson responsible for living things in general.

In Xenophon's *Memorabilia*, furthermore, Socrates is said to have offered an argument to an individual – an argument akin to the teleological argument – with the aim of convincing his interlocutor to have a proper appreciation for the beneficial effects of providence:

> Don't you feel that there are other things too that look like effects of providence? For example, because our

eyes are delicate, they have been shuttered with eyelids which open when we have occasion to use them, and close in sleep; and to protect them from injury by the wind, eyelashes have been made to grow as a screen; and our foreheads have been fringed with eyebrows to prevent damage even from the sweat of our head. Then our hearing takes on all sounds, yet never gets blocked up by them. And the front teeth of all animals are adapted for cutting, whereas the molars are adapted for masticating what is passed on to them. (Waterfield and Tredennick 1990, 90)

As David Sedley (2007, 82) has suggested, the "creative power of accident," as it was presented by the atomists, had emerged as a mode of explanation aspiring to compete with intelligent causation. We find in Xenophon's Socrates an advocate for the teleological argument, or something near enough. Socrates, it bears mentioning, does not seem concerned with a return to the status quo ante. For he seems to take materialists – those who claim to locate intelligence within nature – to be paying mere lip service to the role of intelligence:

> I do not any longer persuade myself that I know why ... anything ... comes to be or exists by the old method of investigation, but I have a confused method of my own. One day I heard someone reading ... from a book of Anaxagoras, and saying that it is Mind that directs and is the cause of everything. I was delighted with this cause and it seemed to me good, in a way, that Mind should be the cause of all. I thought that if this were so, the directing mind would direct everything and arrange each thing in the way that was best. If then one wished to know the cause of each thing, why it comes to be or perishes or exists, one had to find what was the best way for it to be.... On these premises it befitted a man to investigate only ... what is best. As I reflected on this subject I was glad to think that I had found in Anaxagoras a teacher about the cause of things after my own heart, and that he would tell me, first, whether the earth was flat or round, and then explain why it is so of necessity, saying which is better, and that it was better to be so ... This wonderful hope was dashed as I went on reading and saw that the man made no use of Mind, nor gave it any responsibility for the management of things, but mentioned as causes air, and ether, and water, and many other strange things. (J. M. Cooper 1997, 97b4–98c)

Explanation is radically teleological for Socrates. He believes it strange that one might appeal merely to material elements in explaining natural phenomena, in the same way that it would be strange to attempt to explain his staying in prison by reference to blood and sinews. That it is best, in a describable way, that he remains in prison is the proper explanation for his thus remaining. The description of why it is best for natural phenomena to be as they are, in the same way, may not be omitted from explanation.

THE BIOLOGY OF COSMOLOGY

Plato's greatest influence was Socrates (Plate XIII). However, Plato was also influenced by the followers of Heraclitus, who thought that all that is perceptible is in flux. That which is in continual flux, as the reasoning goes, is not, because of its continual change, describable in terms of the universal. Definition, a necessary condition for knowledge, is, however, of the universal. Knowledge, so he reasoned, is not of the sensible world. Plato countenanced, therefore, a world beyond the perceptual, which, as the Eleatics might favorably consider, does not admit of change. Geometry was among the most exciting sciences of the day. (Rumor has it that geometry was a prerequisite in Plato's Academy.) And, in this way, Plato posited a world of realities akin to those of Euclid for the other branches of inquiry, so that philosophers might press their minds up and against such realities in their various modes of research.

It has been argued that Plato's influence and his countenance of immutable forms, or *eidē*, made it difficult for later thinkers to take seriously the idea that one species might evolve into another. A fixed number of forms for species cannot, so it is argued, accommodate the unfixed number of species recognized by the evolutionist:

> Any ... commitment to an unchanging *eidos* precludes belief in descent with modification. The concept of evolution rejects the *eidos*, replacing it with the variable population. Gradual evolution and natural selection ... are inconceivable except through population thinking. (Mayr 1964, XX)

Of course, a major part of the work accomplished by the introduction of the forms involved explaining why it is that the world appears to be organized. The namesake for Plato's *Timaeus* describes therein a Demiurge who – making use of the forms – designs a living world with its own soul. Intelligence is imposed upon the inchoate Heraclitean flux by the design of the Demiurge. And if Plato's account here is to be taken literally, design becomes something of a first principle. When one is doing natural science, one is discovering the formal realities that inspired the Demiurge to produce the world:

> Now surely it's clear to all that it was the eternal model that he looked at, for, of all things that have come to be, our universe is the most beautiful, and of causes the Demiurge is the most excellent. This then is how it came to be: it is a work of craft, modeled after that which is changeless and is grasped by a rational account, that is, wisdom. (J. M. Cooper 1997, 29a)

The debate over whether Plato's account is to be taken literally extends nearly to the time of Plato himself, with members of Plato's own Academy taking opposing positions. But the literal reading was favored by Aristotle, the Epicureans, the Stoics, Galen, and at least two Platonists of the second century C.E., Plutarch and Atticus (Zeyl 2000). On this reading,

Plato is prepared to introduce intelligent design into the curriculum at the Academy:

> Now it wasn't permitted (nor is it now) that one who is supremely good should do anything but what is best. Accordingly, the god reasoned and concluded that in the realm of things that are naturally visible no unintelligent thing could be as a whole better than anything which does possess intelligence as a whole.... When the maker made our world, what living thing did he make it resemble? Let us not stoop to think it was any of those that have the natural character of a part, for nothing that is a likeness of anything incomplete could ever turn out beautiful. Rather, let us lay it down that the universe resembles more closely than anything else the Living of which all other living things are parts.... Since the god wanted nothing more than to make the world like the best of the intelligible things, complete in every way, he made it a single visible living thing, which contains within itself all the living things whose nature it is to share. (J. M. Cooper 1997, 30a–31a)

As one commentator has put the matter, Anaxagorean Mind, in Plato's later thought, becomes a divine craftsmen (Lennox 2001, 287). The decree of Socrates in the *Phaedo* – that explanation is teleological explanation – is upheld a fortiori in Plato's mature work. For Plato, the world is a living organism. The whole is not to be understood in terms of its parts, but the parts in terms of their contribution to what is best. And to expect what is best to come to be from what is not is to expect a surplus to come to be from what is less.

THE MASTER OF THOSE WHO KNOW

Plato's best student is no creationist. Linking time inextricably to motion, Aristotle believed that the variety of creation found in the *Timaeus* commits one to the idea that time came into being (Plate XIV). The idea that time came into being issues in the idea that the period erstwhile involves – what appears to be a *contradictio in adjectivo* – a time before time. Aristotle is not an atheist. In the twelfth book of his *Metaphysics*, however, he argues that contemplation is the best activity and the only activity befitting God. Creation, as it were, would make God's existence less contemplative. And the idea that God has an existence less than ideal is unseemly. Nevertheless, insofar as God's existence is a paradigm at which all things aim, God is in a certain sense a first mover.

With respect to the problem of origin, on the one hand, Aristotle believed with the materialists that the world need not come to be *ex nihilo* and that change requires a substrate. On the other hand, Aristotle believed that those materialists who find order generated from the random accept something akin to the idea that one might get something from nothing. In *Physics* II.3, Aristotle develops an account that delineates four aspects of explanation. In one way, he says, the thing out of which a thing comes to be is called a cause, for example, the bronze of the statue. The scholastics referred to this part of an explanatory account as *causa materialis*, or the material cause. In another way, there is the *causa formalis*, the form or archetype of the explanandum, the shape imposed upon the bronze by the sculptor. There is also the primary source of the *explanandum* to consider, or that which makes what is made, as Aristotle puts it. This is the *causa efficiens* or the efficient cause, for example, the sculptor who brings about the artifact. And, finally, there is the *causa finalis*, the end or that for the sake of which the thing made is made. Aristotle takes his materialist predecessors to have focused upon the material and efficient variety of causes at the neglect of the final and formal. An enumeration of the raw materials that compose a house does not present a sufficient reason for the existence thereof. That the materials are arranged in a certain way for a certain end is also required. So, if we neglect form and function, we appear to think that things can come about without sufficient reason.

Aristotle develops his account of teleological causation by means of an analogy with crafts, but he thinks the same goes *mutatis mutandis* for nature:

> If then it is both by nature and for an end that the swallow makes its nest, and the spider its web, and plants grow leaves for the sake of fruit and send their roots down for the sake of nourishment, it is plain that this kind of cause is operative in things which come to be and are by nature. (Barnes 1995, 199a25–30)

But what reason is there for thinking that organs and organisms are analogous to artifacts? The latter are brought about by means of deliberation. Aristotle cannot, like Plato, appeal to the deliberation of the Demiurge in accounting for the former. So he attempts to minimize the difference by maintaining that, ultimately, craft and deliberation part ways:

> It is ridiculous for people not to believe that something is coming about for a purpose if they do not see the moving cause has deliberated. Yet craft too does not deliberate. (199b26–28)

The claim is astonishing. His point seems to be that the production of an artifact does not rest ultimately on the deliberation of the craftsperson. The deliberation on the part of the builder is simply the means of getting to the craft itself, which we need not think itself deliberates:

> In investigating the cause of each thing it is always necessary to seek what is most precise (as also in other things): thus a man builds because he is a builder, and a builder builds in virtue of his art of building. This last cause then is prior. (195b21–25)

Even the blueprint is based upon something. And the forms of reality we find in nature, and only derivatively in art, make for the essences within the *scala naturae*. Moreover, in Aristotle's view, if we accept the alternative, the view for example of the atomists, we may avoid the commitment that things come to be out of nothing. But we are, nevertheless, considering ourselves entitled to another kind of free lunch (195a25–35).

While a fair amount of Aristotle's thinking on biology is influenced by his thoughts on theology, to conclude that Aristotle was content to pronounce *ex cathedra* on nature would be a mistake. The following is taken from his *Historia Animalium*, where Aristotle traces the embryonic developments of chicks for twenty days, by fracturing, successively, twenty eggs:

> With the common hen after three days and three nights there is the first indication of the embryo; with larger birds the interval being longer, with smaller birds shorter. Meanwhile, the yolk comes into being, rising toward the sharp end, where the primal element of the egg is situated, and where the egg gets hatched; and the heart appears, like a speck of blood, in the white of the egg. This point beats and moves as though endowed with life, and from it, as it grows, two vein-ducts with blood in them trend in a convoluted course towards each of the two circumjacent integuments; and a membrane carrying bloody fibers now envelops the white, leading off from the vein ducts. (561a5-16)

As it is thus unfair to describe Aristotle's approach toward nature as simply stipulative or a priori, it is also nearsighted to criticize Aristotle's work as simply panglossian. Final causes are often described as picturesque but scientifically barren. But Aristotle's appreciation for teleological explanation has been fruitful. The following is a locus classicus for the concept of homology.

> There are some animals whose parts are neither identical in form nor differing in the way of excess or defect, but they are the same only in the way of analogy, as for instance, bone is only analogous to fish-bone, nail to hoof, hand to claw, and scale to feather; for what the feather is in a bird, the scale is in a fish. (486b17-22)

The by now familiar insistence that explanation should indicate why it is best that things be as they are, the countenance of *causa finalis*, leads Aristotle to individuate and understand organs in terms of their function. And the recognition that organs among different species are homologous is a concept that one can hardly do without. Common descent – it hardly bears mentioning – was Darwin's way of explaining homology. Of course, Aristotle's theoretical work concerning homology issues the following question. If Aristotle countenanced homology, why did he not consider common descent a potential explanation thereof? As the following passage reinforces, Aristotle's well-fortified commitment to teleology could not be displaced by a theory that putatively made chance or spontaneity prior to intelligence.

> Why then should it not be the same with the parts in nature, e.g., that our teeth should come up of necessity – the front teeth sharp, fitted for tearing, the molars broad and useful for grounding down the food – since they did not arise for this end, but it was merely a coincident result; and so with all the other parts in which we suppose there is a purpose? Wherever then all the parts came about just what they would have been if they came to be for an end, such things survived, being organized spontaneously in a fitting way; whereas those which grew otherwise, perished and continue to perish, as Empedocles says his "man-faced oxprogeny" did. Such are the arguments (and others of the kind) that may cause difficulty on this point. Yet it is impossible that this should be the true view. For teeth and all other natural things either invariably or for the most part come about in a given way; but of not one of the results of chance or spontaneity is this true. We do not ascribe to chance or mere coincidence the frequency of rain in winter, but frequent rain in summer we do; nor heat in summer but only if we have it in winter. If then, it is agreed that things are either the result of coincidence or for the sake of something, and these cannot be the result of coincidence or spontaneity, it follows that they are for the sake of something; and that such things are all due to nature even the champions of the theory which is before us would agree. Therefore action for an end is present in things that come to be and are by nature. (198b24-199a8)

Aristotle attributes a theory to Empedocles where survival continues to belong to the fittest. As I had suggested earlier, Empedocles' view seems to have the fittest surviving at only one ancestral stage. And that Aristotle would use the locution "continue to perish" in describing the view he has in mind might mean that he is entertaining something like an application of Empedocles' view to successive generations. He argues that organs cannot ultimately be the product of chance, because regularity is not the product of chance, and organs are formed in a regular way. Of course, when he argued that chance could not account for the regularity with which we find incisors in the front and molars in the back of the mouth, he seems to overlook the fact that the explanation in question concerned not the question of how teeth are regularly formed in this way but, rather, the question of how the organs originated (Sedley 2007, 191). And, given Darwin's systematic use and articulation of natural selection, this is a distinction that is not so easy to overlook.

THE PHILOSOPHER'S LEGACY

Aristotle's influence can hardly be exaggerated. With the rediscovery of Aristotle's works in the twelfth and thirteenth centuries comes Aristotle's deification, as Aristotle is often referred to simply as The Philosopher. All of the five ways provided by Saint Thomas Aquinas are unmistakably Aristotelian, and the fifth is a locus classicus for the argument from design. In the fourteenth century, Dante described Aristotle as "the master of those who know." While some modern philosophers sought to eliminate *causa finalis* from explanation – René Descartes, for example, thought that there was an element of hubris in trying to discern God's intentions – for others (e.g., Leibniz), it is clear that final cause played an important role in their thinking. The Scottish skeptic and empiricist David Hume, in his *Dialogues Concerning Natural Religion*, provides a compelling criticism of the use of teleology in the argument from design. Even here, however, it has been suggested that Hume believed the appearance of design so manifest that – much akin to belief in the external world, or the efficacy of induction – it was immune from serious doubt (Butler 1960). In just this way, in the *Critique of Judgment*, Immanuel Kant argued that plants and animals must be considered natural ends, as we can conceive of their possibility only on the assumption that they came about in accordance with design. Darwin's contemporaries held similar convictions. The French anatomist George Cuvier, for example, who was influential in Darwin's day, held teleological commitments that were explicitly Aristotelian:

> Natural history nevertheless has a rational principle that is exclusive to it and which it employs with great advantage on many occasions; it is the conditions of existence or, popularly, final causes. As nothing may exist which does not include the conditions which made it possible, the different parts of each creature must be coordinated in such a way as to make possible the whole organism, not only in itself but in relationship to those which surround it, and the analysis of these conditions often leads to general laws as well founded as those of calculation or experiment. (Cuvier 1817, 1:6)

Darwin himself had a very high opinion of Aristotle. In a letter to William Ogle, an English translator of Aristotle's *De Partibus Animalium*, Charles Darwin expresses respect and admiration for Aristotle:

> You must let me thank you for the pleasure which the Introduction to the Aristotle book has given me. I have rarely read anything which has interested me more.... From quotations which I had seen I had a high notion of Aristotle's merits, but I had not the most remote notion what a wonderful man he was. Linnaeus and Cuvier have been my two gods, though in very different ways, but they are mere school-boys to old Aristotle.... I never realized before reading your book to what an enormous summation of labour we owe even our common knowledge. (DCP, 13697, letter from Darwin to Ogle, February 22, 1882, quoted in Gotthelf 1999, 15)

Of course, Darwin always took final cause very seriously. And it was to final cause that natural selection ultimately spoke (Ruse 2008, 16). Finding a cause, as we have seen, is very important. Things really do not come about *ex nihilo*. Darwin's mechanism of natural selection filled the gap. Having done so, he was able to explain instinct, paleontology, geographical distribution, morphology, and embryology, with the fact of evolution (39). It is understandable that the most influential philosophers of antiquity, who could not avert their gaze from this appearance of design – so too their heirs – put so much energy into defending it. Teleology became, as a result, a centerpiece around which a great deal of thought was built. Their view taken as a whole was not unattractive or simpleminded. By making natural selection a centerpiece, however, Darwin was able to show that a great deal of important scientific thought could be constructed without the commitment to *causa finalis*.

❧ Essay 2 ❦

Evolution before Darwin

Michael Ruse

EVOLUTION IS A child of the Enlightenment. For the Greeks, ongoing organic change was simply ruled out by the design-like nature of organisms (Sedley 2008). They must be understood in terms of what Aristotle called "final causes." The eye is for seeing, the hand is for grasping. The Greeks could not see how this intricate, purposeful complexity could have come about by blind law, and so they denied the possibility of what Charles Darwin called "descent with modification." This fixity was reinforced when Christianity appeared and when the new religion decided to take on board what Thomas Carlyle called "Jewish old clothes." It was by no means obvious to the early members of the church that they had to adopt the Old Testament as canonical – it was after all the record of a group who had brought on the death of the Christian Savior. However, particularly under the influence of Saint Augustine (ca. 400 c.e.), it was appreciated that making sense of Christian drama demanded the background of the Jewish history, and so the early chapters of Genesis became an integral foundation of the Christian religion.

This happy fusion of Greek philosophy and Jewish religion lasted for more than fifteen hundred years. The earth was young, organisms were created by divine fiat, humans (just one pair) came last, and shortly thereafter almost everything was wiped out by a worldwide flood. As we move from the seventeenth to the eighteenth century, various factors acted to break the stranglehold of this vision. First there was the rise of science: in the physical sciences there arose thoughts that perhaps origins could be explained naturally; in the earth sciences, suspicions that so complex an entity as the globe could not be quite as young as once suspected. Then there were the pressures from philosophy and religion. The great French philosopher René Descartes lived and died a practicing Catholic. But his *Meditations* introduced the idea of a malignant being that deceives us all, and although Descartes himself thought he had contained this being, others were not quite so sure. Even more corrosive was both the Reformation (with at least two variants of Western Christianity, why should one be true and not the other?) and the discovery of non-Christian faiths in the East (who is to say that Buddhists or Confucians are completely wrong in their ignorance of the Christian story?).

FIGURE 2.1. During the French Revolution, Jean-Baptiste Pierre Antoine de Monet, Chevalier de la Marck (1744–1829), throve happily through the troubles and upheavals because he was deeply committed to its ideals of change and progress. Today he is best known as the father of "Lamarckism," the inheritance of acquired characteristics, but in fact he was not the first to propose this mechanism, and in his eyes it was never the main force for change. This picture dates from 1802, just around the time when he became an evolutionist. From *Archives du Muséum d'Histoire Naturelle*, 6th ser., 6 (1930)

Most importantly of all, a new ideology that appeared on the scene did not completely remove worries about final causes (in biology that is); but, for many it did trump the total opposition to organic change, or "transmutation" as it was generally known (Ruse 1996, 2010). This was the idea of progress, the belief that it is possible through human effort to improve our lot – we can learn more about the world through science and then, applying this knowledge, can improve our living conditions, our health care, our education, and much more that makes for a full and satisfying life (Bury 1920). Almost immediately, enthusiasts for cultural progress picked it up and applied it to the world of life, arguing that where there is progress there is also transmutation, something that was then generally used in a good circular fashion to argue for cultural progress.

Denis Diderot, the French encyclopedist, was a pioneer. "Just as in the animal and vegetable kingdoms, an individual begins, so to speak, grows, subsists, decays and passes away, could it not be the same with the whole species?" (Diderot 1943, 48, quoting *On the Interpretation of Nature*, 1754) He made no bones about seeing a link between his social views and his scientific speculations. "The Tahitian is at a primary stage in the development of the world, the European is at its old age. The interval separating us is greater than that between the new-born child and the decrepit old man" (Diderot 1943, 152, quoting *Supplement to Bougainville's Voyage*, 1772). A couple of points should be made, applying not only to Diderot but also to subsequent thinkers. First, one should not automatically assume that embracing transmutation meant that one was an atheist. Generally, most transmutationists moved to some form of deism, meaning a belief in a god who is an unmoved mover – a god who has created and set the universe in motion and now lets everything unfold through unbroken law. For such a god, transmutation is, if anything, proof of his existence and power rather than a refutation. This does not mean that there was now no clash between science and religion. But usually there was no obsessive worry about the exact text of scripture; although this was always in the background, Saint Augustine had given us the tools to tackle this problem (McMullin 1985). He himself took Genesis literally, but he argued that if science conflicts with scripture, then it might be necessary to interpret scripture allegorically or metaphorically. Biblical literalism today comes out of nineteenth-century America and is not really to be found in traditional Christianity. The big concern then was the clash between progress and Providence. Providence means that we are sinners and can do nothing except through the redeeming grace of Jesus' sacrifice. Progress challenges this viewpoint, arguing that we humans can indeed, unaided, improve things. In the eyes of many Christians this is unjustified, prideful hubris.

The other point to be made is that generally no one was under major illusions that transmutationism was more than an epiphenomenon of cultural value-laden notions of progress. It was not something grounded in fact and theory. Until well into the nineteenth century, the empirical evidence was just not there. No one in the eighteenth century had a good grasp of the fossil record, or of the geographical distributions of organisms. There was some suggestive evidence – Aristotle had recognized what we today called "homologies," the isomorphisms between animals of different species (like the bones of the forelimbs of humans, horses, bats, birds, and porpoises). But this was hardly overwhelming. The actual term "pseudoscience" is an invention of the 1820s, but the notion was fully articulated in the eighteenth century, and basically this was the category into which transmutation was deposited.

One can say that, whatever its status, once planted ideas of transmutation proved to be a hardy plant or thriving weed, depending on one's perspective. To the chagrin of professional scientists, pseudosciences can be very popular. Over in Britain, toward the end of the eighteenth century, a full-blooded Enlightenment figure, keen on industry and medicine and science – not to mention political revolutions (first in America and then, until things got out of hand, in France) – was the physician and future grandfather of Charles Darwin, Erasmus Darwin. Much given to writing in verse, the merits of which are not entirely obvious to posterity, Erasmus Darwin gushed forth on the topic of organic change.

> *Organic Life beneath the shoreless waves*
> *Was born and nurs'd in Ocean's pearly caves;*
> *First forms minute, unseen by spheric glass,*
> *Move on the mud, or pierce the watery mass;*
> *These, as successive generations bloom,*
> *New powers acquire, and larger limbs assume;*

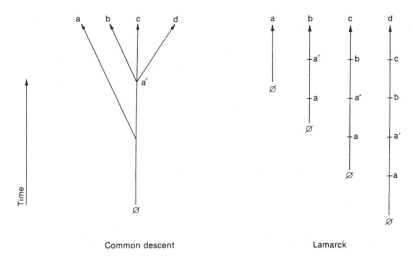

FIGURE 2.2. This diagram shows the big difference between the thinking about the history of Charles Darwin (*left*) and Lamarck (*right*). For Darwin, common ancestry was always part of the picture, whereas for Lamarck it was ever a climb up the Chain of Being, with new life always starting the climb again. For Darwin, extinction is forever, whereas for Lamarck, if a species goes extinct in one period, it will reappear in a later period. Permission: Ruse drawing

Whence countless groups of vegetation spring,
And breathing realms of fin, and feet, and wing.

Thus the tall Oak, the giant of the wood,
Which bears Britannia's thunders on the flood;
The Whale, unmeasured monster of the main,
The lordly Lion, monarch of the plain,
The Eagle soaring in the realms of air,
Whose eye undazzled drinks the solar glare,
Imperious man, who rules the bestial crowd,
Of language, reason, and reflection proud,
With brow erect who scorns this earthy sod,
And styles himself the image of his God;
Arose from rudiments of form and sense,
An embryon point, or microscopic ens!

(E. Darwin 1803, 1, lines 295–314)

This vision was all bound up with the cultural idea of progress: "This idea [that the organic world had a natural origin] is analogous to the improving excellence observable in every part of the creation; ... such as in the progressive increase of the wisdom and happiness of its inhabitants" (E. Darwin 1794, 509).

Moving back to France and entering the nineteenth century, just as Erasmus Darwin's writings were coming to an end the torch was picked up by the best known of all of the pre–Charles Darwin evolutionists (Fig. 2.1). Jean-Baptiste Pierre Antoine de Monet, Chevalier de la Marck, was a minor aristocrat who (having wisely changed his name to Citoyen J-B Lamarck) nevertheless rose up through the ranks of science during and after the Revolution – leading one to suspect, what is indeed true, that he was a supporter of change and upward cultural progress. The actual spur to his becoming a transmutationist was most likely that he was classifying marine invertebrate fossils and could not account readily for the absence today of some earlier forms. Lamarck thought their (watery) living conditions were so comfortable that they would not have gone extinct – and must therefore be around still in different forms (Burkhardt 1977).

In his major work, *Philosophie Zoologique* (published in 1809, the year of Charles Darwin's birth), Lamarck presented his evolutionary theory illustrating it with a tree of life – an initial trunk and then different forms branching off. Actually, however, this was a little misleading, for although there were by then people illustrating life's history using trees (not necessarily interpreting the past in an evolutionary fashion but thinking more in terms of an ongoing nonnatural creation), and although Darwin in the *Origin* rather stitched evolution to the tree for all time, Lamarck did not really have this view of history. He thought that life is constantly being "spontaneously generated," with worms and the like being formed by electricity striking little ponds. Then organisms start moving up the ladder of life – a ladder that was a combination of medieval beliefs about organisms taking their places in an upward-rising Chain of Being with eighteenth-century beliefs about progress. This means therefore that, at any point in time, the cross section of living things comes from a set of unequal-length parallel lines of development. Unlike a tree of life, today's lions and today's humans are not related. At some point in the distant future, lions will evolve into humans. Extinction is never forever because more primitive forms will eventually evolve into those forms which have gone extinct (Ruse 1999a) (Fig. 2.2).

Lamarck is best known today for the mechanism that bears his name, the inheritance of acquired characteristics. The blacksmith's arms get strong through work at the forge, and the blacksmith's children are born with stronger arms. The giraffe stretches its neck to get to the branches, and its offspring are born with longer necks (Fig. 2.3). This mechanism was not a discovery of Lamarck. It is to be found, for instance, in the writings of Erasmus Darwin. More famously, after natural selection it was always a secondary mechanism for Charles Darwin. For Lamarck himself, it too was a secondary mechanism after the main force for change, a kind of directed power forcing organisms ever upward. Lamarckism, the inheritance of acquired characteristics, was laid on top of this basic picture, rather spoiling the smooth upward flow. It was this that Lamarck was trying to illustrate with his tree of life.

Lamarck was a professional scientist, well respected for his work on botanical and invertebrate classification. He had, however, something of a reputation for wild speculation, costing the government a great deal of money because of his fantasies about meteorology. His thinking about transmutation was likewise regarded skeptically by many, simply because of what it was. Lamarck's evolutionary ideas had their admirers and supporters (Corsi 1988). Yet one somewhat unfortunate fact

FIGURE 2.3. Giraffes arrived at the Muséum d'Histoire Naturelle around 1830. They seemed to confirm the claim of Lamarck's *Philosophie Zoologique* (1809) that much stretching and stretching had resulted in permanent neck lengths. Nineteenth-century print

of timing for all such thinking was that, with the horrors of the French Revolution still raw in people's minds, the philosophies that led to it – beliefs in progress being at the forefront – were regarded with much political suspicion and distaste. Erasmus Darwin had likewise felt this wave of disapproval. The most devastating rebuttal of his ideas came not as reasoned objections to his ideas but, at the hands of conservative politicians, through parodies making cruel fun of his poetic speculations (Canning, Frere, and Ellis 1798).

A more measured critique of evolutionary thinking came from the great German philosopher Immanuel Kant. His objection was that of the Greeks. He could see no way to reconcile creation through blind law with the design-like nature of organisms. He wrote that an organism, although governed by the natural laws of physics, is not "a mere machine, for that has merely *moving* power, but it possesses in itself *formative* power of a self-propagating kind which it communicates to its materials though they have it not of themselves; it organises them, in fact, and this cannot be explained by the mere mechanical faculty of motion." You have to invoke some principle of organization, of final cause, as organisms, and their parts are seen to be working toward ends. This led to the notorious claim that "it is alike certain that it is absurd for men to make any such attempt or to hope that another *Newton* will arise in the future, who shall make comprehensible by us the production of a blade of grass according to natural laws which no design has ordered. We must absolutely deny this insight to men" (Kant 1790, 54).

Kant did not think that the idea of organic evolution is silly. Indeed, like some of the evolutionists of his day (e.g., Erasmus Darwin), Kant thought that the isomorphisms between the parts of different organisms rather point in the way of evolution. "This analogy of forms, which in all their differences seem to be produced in accordance with a common type, strengthens the suspicion that they have an actual kinship due to descent from a common parent." Kant even went on to spell things out, speaking of our ability to "trace in the gradual approximation of one animal species to another, from that in which the principle of ends seems best authenticated, namely from man, back to the polyp, and from this back even to mosses and lichens, and finally to the least perceivable stage of nature" (Kant 1790, 78–79). But ultimately it appears that these connections are all ideal, connections in theory, and not in actuality. There is no common descent. Evolution is untrue.

Georges Cuvier, the great French comparative anatomist at the beginning of the nineteenth century, was almost certainly influenced by Kant, not least in his principle of understanding that he believed governed all thinking in the biological sciences (Fig. 2.4).

> Natural history nevertheless has a rational principle that is exclusive to it and which it employs with great advantage on many occasions; it is the *conditions of existence* or, popularly, *final causes*. As nothing may exist which does not include the conditions which made its existence possible, the different parts of each creature must be coordinated in such a way as to make possible the whole organism, not only in itself but in its relationship to those which surround it, and the analysis of these conditions often leads to general laws as well founded as those of calculation or experiment. (Cuvier 1817, 1:6)

For Cuvier, like Kant, all of this made evolution impossible. Apart from anything else, an organism midway between two established forms would be literally neither fish nor fowl. It would be out of adaptive or final-cause focus and could not survive. Not that this theoretical impossibility deterred Cuvier from offering detailed empirical reasons why evolution cannot be true. Almost paradoxically, it was he who first started to see the fossil record in a roughly progressive fashion, but for him the gap between forms was definitive – as was the fact that the mummified forms (of cats and birds and the like) being brought back from Egypt (thanks to Napoleon's excursions into Africa) were exactly the same species as exist today. So many years, so little change.

Unlike most of his countrymen, Cuvier was a Protestant. But it seems that this had little significance in his opposition to evolution. More likely underlying his distaste was a conservative nature, one that served him well as a servant of the state (for he held important bureaucratic roles both under Napoleon and in the years after) and that included a deep hostility to philosophies of progress that he saw as having led to nothing but disruption and confusion. However, it is hard to keep a good idea down, and as the century moved along, increasingly ideas of progress started again to flourish.

Germany, distinctively, reflected the philosophical idealism popular in that country. The group of thinkers known as *Naturphilosophen* and their sympathizers all subscribed to some kind of upward progression of ideas and possibly all reality (R. J. Richards 2002b). Sometimes, as for instance in the case of the philosopher Hegel, this was kept firmly in the realm of the ideal.

> Nature is to be regarded as a *system of stages*, one arising necessarily from the other and being the proximate truth of the stage from which it results: but it is not generated *naturally* out of the other but only in the inner Idea which constitutes the ground of Nature. *Metamorphosis* pertains only to the Notion as such, since only its alteration is development. But in Nature, the Notion is partly only something inward, partly existent only as a living individual: *existent* metamorphosis, therefore, is limited to this individual alone. (Hegel 1817, 21)

Others, like the biologist Lorenz Oken and most probably the poet Goethe by the end of his long life (1832), were seeing this as the actual course of history. Note, however, that, whereas for Charles Darwin, much influenced by British natural theology (above all by Archdeacon William Paley) and almost certainly (through his Cambridge mentors) by Cuvier's thinking on conditions of existence, the final-cause-like nature of organisms was the overwhelming item in need of explanation, for the German thinkers it was always isomorphisms – homologies – that were central to their vision. It was in and because of these similarities, reflecting ideal archetypes or *Baupläne*, that their Neoplatonic vision of the ultimate connection or oneness of all being was grounded.

Perhaps independently, this kind of formalistic thinking – as opposed to the Kant-Cuvier-Paley-Darwin functionalist

FIGURE 2.4. The French biologist Georges Cuvier (1769–1832) is rightly known as the father of comparative zoology. Deeply influenced by both Aristotle and Kant (he was educated in Germany), Cuvier made final causes – what he called "conditions of existence" – the linchpin of his scientific thought. For him, evolution was not just false but theoretically impossible, for it would mean that organisms would have to travel from one integrated functioning form to another, passing over a space of nonfunctionality. Nineteenth-century lithograph

thinking – also found a happy home in France. Cuvier's sometime friend and then great biological rival Etienne Geoffroy Saint-Hilaire endorsed a form of Lamarckian evolution, spiced with the notion of archetypes on which groups of organisms were modeled (Appel 1987). Starting with the mammalian ear – "Strictly, it will suffice for you to consider man, a ruminant, a bird, and a bony fish. Dare to compare them directly and you will reach in one stroke all that anatomy can furnish you of the most general and philosophical nature"– Geoffroy was soon generalizing out to other bones in the vertebrate body – "An organ is sooner altered, atrophied, or annihilated than transposed" – and before long (to Cuvier's great ire) he was seeing links between invertebrates and vertebrates, and nothing could stop an enthusiasm for total, organic change (Geoffroy 1818, xxxviii).

These ideas crossed over to Britain. Once the Napoleonic Wars were firmly finished and memories receding, and industry and commerce again starting to thrive and grow – particularly the incredible explosion of the railway system – ideas of progress started to find renewed support and expression. The Scottish anatomist Robert Grant was one who absorbed the message of archetype-based transmutationism (Desmond 1989). Among those who were exposed to his enthusiasm in the 1820s was the young Edinburgh University medical student Charles Darwin (who had apparently already read some of his grandfather's works). Grant moved to London, but his career was never successful. The very opposite is true of the English anatomist Richard Owen, who by the 1840s was by far the country's most important biologist (Rupke 1994). Because he was always dependent on conservative patrons, he had to conceal

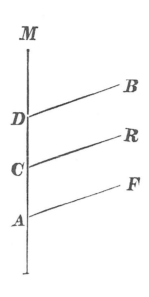

FIGURE 2.5. Publisher Robert Chambers (1802–71) was an enthusiast for many unorthodox ideas including phrenology, evolution, and spiritualism. (Later, he and Alfred Russel Wallace had comforting exchanges on the last-named phenomenon.) Like other early evolutionists, his enthusiasm for evolution was a function of his enthusiasm for cultural progress. His mechanism for change, based on Germanic ideas about development, saw embryos as going through stages, and, if birth is delayed, then they develop to the next higher form on the chain of being. *Left*: Nineteenth-century lithograph. *Right*: F stands for fish; R for reptiles; B for birds; and M for mammals. A, C, D, simply mark points of divergence. From R. Chambers, *Vestiges* (London: Churchill, 1844)

his more radical thinking. But Owen always had a strong liking for Germanic-type thinking. In the Darwinian story he is, with justification, labeled the anti-Christ because of his opposition to the *Origin*; but jealousy at not being able to express his own evolutionary yearnings was a major factor there. Certainly, later in life, Owen was explicitly an idealist evolutionist, but even in the 1840s he was giving hints of such thinking. He made much of the vertebrate archetype, and although ambiguous, a sympathetic thinker could interpret in a fairly concrete way his understanding of the upward rise of organic life (Owen 1849).

There was no such ambiguity in the anonymously published *Vestiges of the Natural History of Creation*. Appearing in 1844, and now known to have been written by the Scottish publisher Robert Chambers, it was an interesting mishmash of idiosyncratic understandings of physics, amateur paleontological gleanings, up-from-the-cutting-floor ruminations about embryology, speculative hypothesizing of a sociopolitical nature, and much more (Fig. 2.5). It preached a doctrine of upward change from the inorganic – the frost patterns on windows in winter were thought particularly suggestive – through the major classes of organisms, where every now and then a developing embryo stays a little longer in the womb and thus goes on to become a new species. And through and through the message was one of progress.

> A progression resembling development may be traced in human nature, both in the individual and in large groups of men.... Now all of this is in conformity with what we have seen of the progress of organic creation. It seems but the minute hand of a watch, of which the hour hand is the transition from species to species. Knowing what we do of that latter transition, the possibility of a decided and general retrogression of the highest species towards a meaner type is scarce admissible, but a forward movement seems anything but unlikely. (Chambers 1846, 400–2)

This really was the fast food of science, and like fast food always, it was loathed and condemned by the establishment and loved and gobbled up by the general population (Secord 2000). Whatever its status as a work of science – and all agreed that later editions (where Chambers sought professional scientific collaborative help) were much improved over the first edition – it was written by a man who knew how to make a case and present it to the public and, moreover, by a man who could see a good point if it was there, for by now the fossil record was getting better known and the embryological evidence was (to put it mildly) highly suggestive. By midcentury in Britain, evolution was an idea known by all, hated by most professional scientists, and loved by altogether-too-many people in the more gullible parts of the population – as the professionals noted gloomily, in the more gullible, distaff parts of the population. The smash-hit success of the age was a tribute to a friend who died young, *In Memoriam*, which was written over twenty years and published in 1850 by Alfred Tennyson. Picking up on the optimistic theme of *Vestiges*, the poem ended echoing the evolutionary tract, suggesting that the dead friend was a higher type who had arrived too early.

> *A soul shall strike from out the vast*
> *And strike his being into bounds,*
>
> *And moved thro' life of lower phase,*
> *Result in man, be born and think,*
> *And act and love, a closer link*
> *Betwixt us and the crowning race*
> . . . .
> *Whereof the man, that with me trod*
> *This planet, was a noble type*
> *Appearing ere the times were ripe,*
> *That friend of mine who lives in God.*

The dear Queen (Victoria) found this a great comfort when Prince Albert died.

When Darwin published – and by midcentury he had already been sitting on his ideas for a decade – there was no shock to evolution as such. The shock was more that now all must accept it, or at least take it seriously, given Darwin's status as a very professional scientist. Even had *Vestiges* started to recede in memory, a new authority on the science was now making certain that evolution – and it was he who really popularized the word (which hitherto had been applied more to individual than to group development) – was an idea that all must acknowledge if not accept. Herbert Spencer, from England's middle classes and an enthusiastic sponge for all radical ideas – be they extreme laissez-faire economics or a thoroughly naturalized philosophy of knowledge and morality – preached (and that is not an inappropriate term) evolution right through the 1850s, an evolution whose backbone was progress (Richards 1987): from the simple to the complex, from what Spencer called – and he acknowledged explicitly his debt to German thinking including Goethe and (especially) the philosopher Friedrich Schelling – the homogeneous to the heterogeneous.

> Now, we propose in the first place to show, that this law of organic progress is the law of all progress. Whether it be in the development of the Earth, in the development of Life upon its surface, in the development of Society, of Government, of Manufactures, of Commerce, of Language, Literature, Science, Art, this same evolution of the simple into the complex, through successive differentiations, holds throughout. (H. Spencer 1857, 2–3)

Scholars are now realizing just how influential Spencer's thought has been, especially with American evolutionists in the twentieth century, but if we look at it first in cold daylight it comes across as very odd. In its fullest form, it seems to involve a kind of stability, disrupted on occasion by external forces and which then strives to reachieve stability at a higher level (H. Spencer 1862). This vision of "dynamic equilibrium," as it is called, is part metaphysical, part based on an eclectic reading of then-contemporary physics, and part a half-baked understanding of German morphology and philosophy. It is thoroughly non-Darwinian and, inasmuch as there are physical causes (as opposed to metaphysical destiny), they are firmly Lamarckian – the inheritance of acquired characteristics. In this Spencer was at one with most pre-Darwinian evolutionists, who seem generally to have put the burden of change more on upward-reaching metaphysical impulses than on real physical causes.

In Spencer's case this is almost paradoxical. It has long been realized – it was firmly drawn to Darwin's attention – that many people had hit on the notion of natural selection before he did. The physician William Wells (1820) had floated the idea when Darwin was a child. Patrick Mathew (1831), a writer on timber, had had the idea. Richard Owen always claimed that the idea was his. Darwin had probably read an article mentioning the idea, and he had certainly read a pamphlet by an animal breeder who suggested the mechanism in passing (Ruse 1975b). None of this really means much, because no one was making it the basis of a theory of evolutionary change – when Darwin read the pamphlet, he actually noted the passage but (this was some months before he had the big breakthrough that did lead to his seeing the significance of natural selection) did not make that much of it, and certainly did not put it into a full evolutionary context. Spencer too was one who hit on the idea of natural selection: even back at the beginning of the 1850s, he suggested that it is working among humans. Taking note of the dreadful story of the Irish – remember Spencer was writing a year or two after the potato famine – we learn: "For as those prematurely carried off must, in the average of cases, be those in whom the power of self-preservation is the least, it unavoidably follows, that those left behind to continue the race are those in whom the power of self-preservation is the greatest – are the select of their generation" (H. Spencer 1852, 500). But almost typically, Spencer never thought this insight significant, incorporating it into a very non-Darwinian context. Like Darwin, Spencer was impressed by the Reverend Thomas Robert Malthus's gloomy calculations that population numbers outstrip food supplies and thus lead to an inevitable struggle for existence. But whereas Darwin used this idea to fuel his selective mechanism of change, Spencer rather argued that, as you go up the evolutionary scale and intelligence rises, reproductive abilities and inclinations decline – the dumb herring reproduces much and the clever elephant a little – and that hence eventually the Malthusian pressure falls away. It is indeed true that, in the *Descent*, Darwin likewise worried about the large-familied Irish and the small-familied Scots (eventually deciding that the Scots win because they look after their children better), but overall the distance between Spencer and Darwin could not be starker.

We should not read the history of pre-Darwinian evolution too much one way or the other. Darwin was not the first evolutionist. By the time he published, the idea was well known, and he certainly did not have to fight to bring it to people's attention. In Britain – and elsewhere – there were already many accepting or at least favorable to some version of evolution. And even before Darwin, it is clear that religious reactions would be at least mixed and not necessarily universally unfavorable. Having said this, even by the time of the *Origin* the status of evolution was in many respects that of a pseudoscience, something existing primarily on the back of an enthusiasm for various notions of social or cultural progress. And even if one goes so far as to say that there were hints in the wind about natural selection, until Alfred Russel Wallace made his genuine independent discovery in 1858 no one else sensed that here was something that could fuel evolutionary change and speak to the worry about design, about final causes. Putting Charles Darwin in historical context in no way detracts from the significance of what he did to further our understanding of origins.

❦ Essay 3 ❦

Charles Darwin's Geology: The Root of His Philosophy of the Earth

David Norman

Charles Robert Darwin was the epitome of the nineteenth-century natural philosopher by temperament and by training. Nevertheless, the ambit of his researches, which had roots that were firmly planted in the interwoven fields of chemistry, mineralogy, and geology, is quite extraordinary in the way that it came to encompass the physical and biological world that he inhabited (A. Geikie 1909; Judd 1909; Browne 1995; Herbert 2005).

ORIGINS AND INFLUENCES

Darwin was born into a comfortable and well-respected family in the county town of Shrewsbury. His father, Robert, a noted physician, astute businessman, and financier (Browne 1995), was not scholarly in an academic sense; however, Darwin's grandfather Erasmus was a renowned intellect. Among Erasmus's published works were geological as well as what we might now refer to as evolutionary interpretations. The *Botanic Garden* (E. Darwin 1791) reveals, especially in its "Philosophical Notes XV–XXIV" and "Geological Recapitulation," that Erasmus was well versed in contemporary debates about minerals, rocks, and earth processes. He used this to make reasoned proposals about the formation of granites, lavas, coal, limestone, clays, and ironstone and envisaged a dynamic structure to the earth that was driven by a hot fluid interior (Herbert 1991).[1] This "dynamic" perspective also resonated with his understanding of animal life: he understood (as did his approximate contemporary in Paris, Jean-Baptiste Lamarck) that change pervades the living world. For example, Erasmus used the term "evolution" but used it as a descriptor of growth (ontogeny): the changes in structure and appearance that occur during the lifetime of any individual as it develops on a trajectory from fertilized egg to adult. However, it is also clear that Erasmus perceived the possibility of plasticity of animal form and appearance over much longer periods of time. He used the example of the existence of purposeless or rudimentary features, such as the accessory toes seen in

---

[1] It is interesting to note that he particularly mentions the rocks of Arthur's Seat in Edinburgh as demonstrating clear signs of their having formed originally under conditions of extreme heat.

the feet of cattle and pigs, as suggestive that such animals formerly possessed fully functional toes but that they had become vestigial with the passage of time; and he proposed more forthright views on "transmutation" in his book *Zoonomia* (E. Darwin 1796).

The extent to which Erasmus's works influenced Darwin remain a matter of debate (Browne 1995). Darwin acknowledged much later (notably in relation to Erasmus's quasi-evolutionary speculations) that he was aware of his grandfather's writings but did not ascribe any strong influence to them in relation to the development of his theory of natural selection. As we shall see, however, such important Darwin-family books imposed themselves upon Darwin cumulatively during his pre-*Beagle* years. And while Erasmus's zoological and botanical writings have a "transcendent" quality to them, laced as they are with poetical musings, the geological observations are closely argued and seem very relevant to Darwin's early intellectual development.

CHILDHOOD

Darwin's formal education started at the age of eight, at Rev. Case's Unitarian chapel school in Shrewsbury. During this time, Darwin's fascination with the natural world was fostered; his father spent time with him in his garden, which was well stocked with a variety of plants, and no doubt helped him to identify and name them, perhaps by reference to Erasmus's verse compendium *The Botanic Garden*. Darwin assisted his father with entries in the "Garden Book" that recorded seasonal changes, in a manner reminiscent of the curate of Selborne (G. White 1789). In Darwin's autobiographical sketch (written in August 1838, at the time of his engagement to his cousin Emma Wedgwood), he recalled this time in his life:

> I ... formed a strong taste for collecting ... pebbles & minerals ... when about 9 or 10 I distinctly recollect the desire I had of being able to know something about every pebble in front of the hall door. (Darwin 1985–, 2:439)

His father, recognizing Darwin's enthusiasms, presented him with two illustrated reference books on natural history (Brookes 1763a, 1763b); these had been owned by Robert's elder brother, Charles (after whom Darwin had been named). Uncle Charles died at medical school, and these were Dr. Darwin's last reminders of this brother (Browne 1995).

A year or so later, Darwin's mother Susanna died and, as had earlier been decided, young Charles was sent away to board at Dr. Butler's school in Shrewsbury. Though a mere fifteen minutes from home, such an abrupt change to his life – losing his mother and being simultaneously wrenched from a warm and supportive home – must have been traumatic for the young boy. Furthermore, the school was tough and austere, and his memories of school in later life were clearly jaundiced:

> Nothing could have been worse for the development of my mind than Dr. Butler's school. (Darwin 1958b, 27)

Charles's natural interests and enthusiasms were, however, encouraged by increasingly close emotional and intellectual ties to his elder brother Erasmus ("Eras"). During free time throughout Charles's formative years Eras was a companion and inveterate experimenter: he set up a "chemistry lab" in a garden outhouse, with Charles recruited as his assistant. No doubt making "stinks and bangs," they also analyzed the composition of minerals (using chemical textbooks) and even purchased a goniometer to measure the angles between crystal faces. When Eras left to train for a medical degree in Cambridge in 1822, Charles continued with experiments and became particularly proficient in the use of the blowpipe to assist with the analysis of the chemical composition of a variety of materials; this required him to blow air through the pipe into the flame of a gas light to create very high temperatures to melt or fuse materials under study and naturally enough earned him the nickname "Gas" among friends at school.

Charles's intellectual stagnation at school was noted, and it is not surprising that Charles was withdrawn from school and sent, in 1825, to join Eras, who had by then moved to Edinburgh to augment his medical training before becoming a practicing physician. This suited the sixteen-year-old Charles and the Darwin family admirably because Eras could act as a mentor, guide, and companion; they lodged together, and it seems that his father saw, in Charles's demeanor and constitution, promise as a future physician (Browne 1995).

EDINBURGH

Edinburgh at the time of Darwin's arrival was an academic crucible: radical, dynamic, and well connected with the European (French, German, and Italian) powerhouses of intellectual progress. Darwin attended medical classes in his first year, but his enthusiasm for medicine waned (F. Darwin 1892). Other interests were, however, nourished (in particular) by extracurricular courses given by Thomas Hope and Robert Jameson (Rudwick 2008). Hope taught chemistry and included mineralogy, crystallography, meteorology, and theories of the earth, among other topics; Jameson offered natural history and encouraged debate and practical research by his students (Secord 1991b). Jameson, in particular, used his museum to demonstrate his lecture material, and Darwin became very familiar with it and its curator William Macgillivray. As Secord (1991b) noted, often Darwin's annotations in his copy of Jameson's *Manual of Mineralogy* (1821) match the museum displays case by case. He also learned how to discriminate between mammoth and mastodon remains, which would stand him in good stead during his *Beagle* years.

Hope and Jameson were also theatrical antagonists: at heart, Jameson was a "Wernerian" (having been taught by Abraham Werner in Freiberg) and advocated (somewhat anachronistically) the view that the geological structure of the earth (its layers or strata) had settled out in succession from a former universal ocean. In stark contrast, Hope was a "plutonist" following Edinburgh-based geologist James Hutton's view that the earth had been continuously modified by internal (volcanic) heat (Rudwick 2008). One particularly apposite example that embodies the disagreement

FIGURE 3.1. Adam Sedgwick (1785–1873), professor of geology at the University of Cambridge, taught Darwin the basic methods of geology. He was always opposed to evolution and wrote bitterly against the *Origin*. From J. W. Clark and T. M. Hughes, *Life and Letters of Adam Sedgwick* (Cambridge: Cambridge University Press, 1890)

between these two men was the occasion of a field trip led by Jameson to Salisbury Crags. As Darwin recalled later, while demonstrating an outcrop displaying a trap dyke, Jameson said "with a sneer that there were men who maintained that it had been injected from beneath in a molten condition" (F. Darwin 1892).

Darwin was by now proficient in the rudiments of geology and knew that Jameson was mistaken (incidentally siding with his grandfather). Janet Browne (1995) revealed that during 1826–27 Darwin purchased and read some important books, including Erasmus Darwin's *Botanic Garden* (1791) and *Zoonomia* (1796). Darwin also met, toward the close of 1826, the zoologist Robert Grant, who taught Darwin zoological dissection and the complexity of the life cycles of marine creatures collected from the Firth of Forth; through Grant's involvement with the Plinian Society, Darwin developed a taste for the presentation of novel research observations. The complex nature of animal life cycles and the variation seen in fossils prompted Grant to suggest that it was reasonable to suspect that animal species may likewise have adapted and changed over time; and he expressed admiration for similar views held by Darwin's grandfather and Lamarck (prompting Darwin to read both authors: [Browne 1995]).

Darwin's exposure to aspects of mineralogy and general geology and the intellectual debates surrounding these subjects were remarkably timely (Porter 1977, 1978; Laudan 1987; Rudwick 2008) but might not have been *pivotal* had it not been for parental intervention when it became apparent that Darwin had not been attending to his medical studies in Edinburgh.

CAMBRIDGE

A frank reappraisal by father and son of lack of progress on the medical course followed in the summer of 1827. It was decided that Charles would transfer to Cambridge in order to study for an "ordinary degree" as a necessary prelude to taking Holy Orders and becoming a clergyman. Dr. Darwin was anxious to avoid the risk of Charles becoming (as Eras had) an "idle man" so a safe, established career path and one that would leave Darwin with time to indulge his passion for natural history appealed to father and son.

Christ's College Cambridge proved to be a comfortable base for work, sport, and hobbies for Darwin. The ordinary degree appears not to have been overly demanding – Darwin graduated 10th in a class of 178. But of far greater importance was the fact that Darwin's time at Cambridge (1828–31) overlapped that of a number of young, extremely gifted and influential natural philosophers ("men of science"). Leading among these were John Henslow, who had been a professor of mineralogy from 1822 to 1827 but then switched his attention to botany. Henslow occupied a pivotal position in Cambridge; he and Adam Sedgwick (the Woodwardian Professor of Geology) founded the Philosophical Society for the purpose of debating and publishing articles on mathematics and the sciences. Among the group of like-minded academics were future stars of nineteenth-century science: William Whewell (who succeeded Henslow to the chair of mineralogy), George Airy (the Lucasian Professor of Mathematics), and other Cambridge luminaries such as Charles Babbage, John Herschel, and George Peacock.

Henslow held open house once a week; these convivial occasions encouraged philosophical discussion and could be attended by undergraduates who professed an interest

FIGURE 3.2. Charles Lyell's *Principles of Geology* (1830–33) was probably the single greatest scientific influence on Darwin, for all that it denied evolution. From K. Lyell, *Life and Letters of Charles Lyell* (London: John Murray, 1881)

in natural philosophy. Darwin gained an invitation to one of these soirées and fitted so well socially and intellectually that he developed a strong friendship with Henslow and his family. He also became Henslow's regular classroom assistant and, when Darwin's final examinations were over in January 1831, was adopted as Henslow's personal tutee for the remainder of the academic year becoming "the man who walks with Henslow" (F. Darwin 1892). Under Henslow's instruction, Darwin read Alexander von Humboldt's remarkable narrative of his expedition to South America (Humboldt 1814–29) and John Herschel's equally inspirational *Discourse* on natural philosophy (Herschel 1830); both books demonstrated in practice and in theory, respectively, the importance of geology as an exciting observation-based science.

Under this spell, Darwin planned a small-scale expedition of his own to the Canary Islands (aping Humboldt). However, while his skills in zoology, entomology, botany, mineralogy, and chemistry were adequate to the task, his geological field skills were entirely theoretical (as he discovered to his shame when attempting to make his own geological map of the area around Shrewsbury [Herbert 2005]). So, on Henslow's bidding, Darwin became Adam Sedgwick's field assistant on a geological excursion to North Wales in the summer of 1831. Their task was a comparatively simple one: to confirm the distribution and structure of rocks that had been described in that area in George Greenough's (1820) geological map (Fig. 3.1). Sedgwick was skeptical of the map, but proof would be necessary and fieldwork was the only solution. It is probably a tribute to Sedgwick, as a teacher, that Darwin converted theoretical knowledge into a suite of practical skills in observation, collection, note taking, identification, measurement of dip and strike, and mapping so effectively during their excursion (Herbert 2005). Darwin was an ideal assistant, fired, as he was, by the need to perfect these skills in preparation for his own expedition, and he returned knowing that his own observations and notes had contributed to scientific knowledge by correcting the authority of Greenough (Secord 1991a).

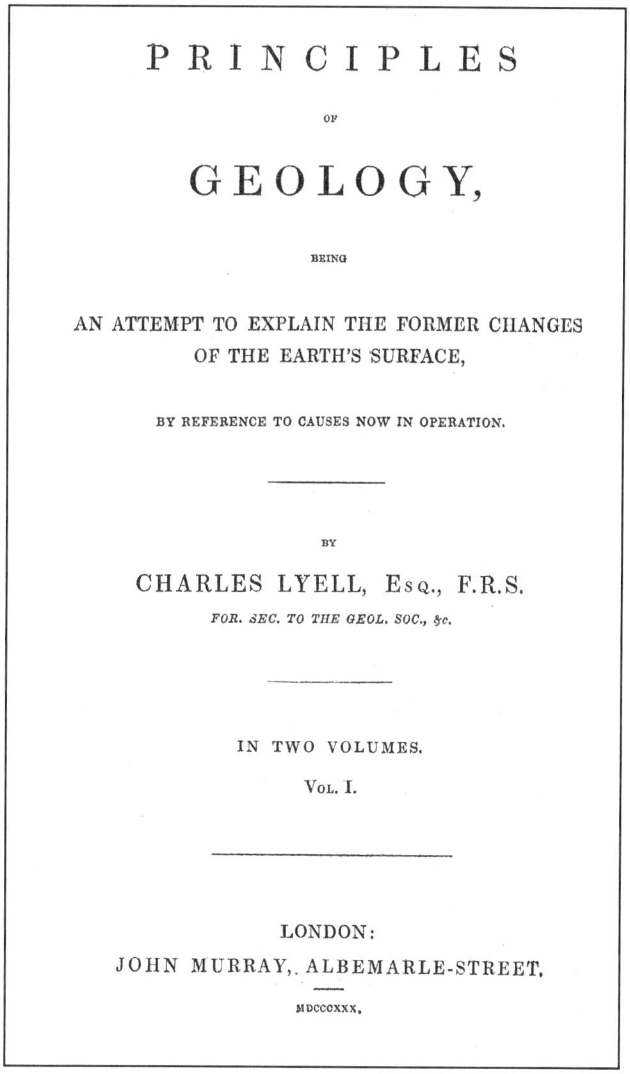
PRINCIPLES

OF

GEOLOGY,

BEING

AN ATTEMPT TO EXPLAIN THE FORMER CHANGES
OF THE EARTH'S SURFACE,

BY REFERENCE TO CAUSES NOW IN OPERATION.

BY

CHARLES LYELL, ESQ., F.R.S.

FOR. SEC. TO THE GEOL. SOC., &c.

IN TWO VOLUMES.

VOL. I.

LONDON:
JOHN MURRAY, ALBEMARLE-STREET.

MDCCCXXX.

FIGURE 3.3. The *Principles of Geology* by Charles Lyell was the definitive statement of what William Whewell called the "uniformitarian" position on life's history. As the subtitle stresses, all is explained by law, of an intensity and type working today. From C. Lyell, *Principles of Geology* (London: John Murray, 1830)

THE *BEAGLE* EXPERIENCES

Darwin's plan to visit the Canary Islands was overturned upon his return from North Wales by the offer (facilitated by Henslow) of a place aboard HMS *Beagle* as the ship's naturalist. Captain FitzRoy wanted a geologist for the voyage, and the circumstances of Darwin's training in Cambridge had prepared him for the task. As an innately skilled observer, he was prepared; as one who was conversant with many aspects of applications of geological understanding to an interpretation of landscapes, he was prepared; and as one who was familiar with the ambit of the geological sciences, especially in terms of their potential to inquire into major causal questions as advocated so clearly by his grandfather, Hope, Herschel, Humboldt, and Sedgwick, he was primed and ready. And in a slightly more subtle way, Sedgwick had inculcated into Darwin a mathematical (geometric) component to the exploration and understanding of the earth. In Wales he had learned that rocks had been bent, cleaved, upended, folded, elevated, or depressed in response to ancient forces and that such outcomes, which gave landscapes their form, could be observed, measured, and interpreted freely and thoughtfully.

Volcanoes and Their Effects

One of many extraordinary pieces of good timing that so influenced Darwin's career was the publication of the first volume of Charles Lyell's *Principles of Geology* (1830–33) (Figs. 3.2 and 3.3). Henslow recommended that Darwin take the book with him on the voyage (but that he should read it with due skepticism!), and FitzRoy bought a copy as a personal gift for "his" naturalist and future shipboard companion. Though now regarded universally as one of the most important geological books ever published, it aroused considerable suspicion and antipathy among leading geologists (Buckland, Sedgwick, Greenough, Conybeare, and Murchison) when it first appeared (Porter 1978). Aboard the *Beagle*, and away from what might have been the insidious influence of others, Darwin was able to read Lyell and explore his views dispassionately. As described by Darwin, the effect was immediate:

> The very first place which I examined, namely St. Jago in the Cape Verde islands, showed me clearly the wonderful superiority of Lyell's manner of treating geology. (Darwin 1958b, 77)

Careful measurement of the raised beaches around the volcanic island of St. Jago (São Tiago) demonstrated very clearly two facts: that *elevation* of the land, rather than depression of sea level, had occurred (Herbert 1991); and that *subsidence* had also occurred after the period of volcanically driven uplift (Secord 1991a). This combination of observations spanned the range of theoretical geological models that had been generated by Humboldt, von Buch, Scrope, and Lyell (Dean 1980). The evidence suggested that, just as Humboldt and von Buch (von Buch 1820) had argued, a volcanic cone was the product of the pressure of molten lava bulging upward below the surface, rather than, as Lyell and Scrope (Scrope 1825) supposed, the spewing out and piling up of lava around a crack in the surface of the earth; however, it also showed that once the lava had been ejected the elevatory pressure had been relieved and the volcanic cone began to subside.

At his very first landfall, Darwin made original observations and generated novel explanations. As Secord (1991a) and Herbert (1991) have shown, this event made Darwin consider himself a geologist and contemplate writing a book on the subject. As a consequence his principal efforts of collecting, note taking, and theorizing became geological throughout the voyage (Rhodes 1991; Pearson and Nicholas 2007). Thus galvanized, Darwin's voyage of exploration presented him with approximately three and a half years on the geologically unstable continent of South America (while FitzRoy charted its southern coastline). From landfall in Brazil, Darwin was able to record raised beaches or terraces similar to those he had seen in the Cape Verde islands and the terraces became more prominently marked as he followed the coastline south. With assistance from the ship's crew Darwin compiled detailed records of the height and extent of the terraces in Patagonia (Herbert 1991).

This was a remarkably expansive project because Darwin was assembling data that would support the idea that the entire continent of South America had been progressively elevated and tilted over a substantial period of time. He attempted to date the periods of elevation of the terraces using the marine fossil shells that he was able to collect, and because some fossils on the lower terraces also retained their original color, he supposed that they had been elevated comparatively recently. Darwin's general model of elevation

FIGURE 3.4. The frontispiece of the first volume of Lyell's *Principles*, showing an Italian ruin. Note the weathering on the columns starts about eight feet up, suggesting that the land had first sunk (and so the bases of the columns were submerged) and then later risen again. This is confirmation of Lyell's "grand theory of climate," suggesting that, rather like a waterbed, as one area of the earth's surface is subsiding, another part is rising. Geographical distributions are vital evidence for the theory, and it was this that stimulated Lyell's follower Charles Darwin to take seriously the distributions of the animals on the Galapagos Archipelago. From C. Lyell, *Principles of Geology* (London: John Murray, 1830)

of the land was that it had occurred relatively steadily with occasional interruption (a Lyellian uniformitarian stance) in the sense that the scarp faces of the terraces represent periods of stasis during which marine erosion cut the terraces back. He was anxious to continue the measurement on the west coast in order to establish whether the entire continent had been elevated; this he did successfully during the first half of 1835 (Fig. 3.4).

While reflecting on the extraordinary motion of South America, Darwin became witness (20 February 1835) to elevation at first hand. While ashore at Valdivia (Chile), he felt the shock of the great earthquake that wrecked Concepcion (Moorehead 1969). A few days later the *Beagle* entered the harbor at Concepcion and, amid the devastation of the town noted, with a guilty geological relish, that the vibrations created by the earthquake had a *direction* (in accordance with the views of John Michell [1760]) manifested in the fact that buildings with specific orientations were unaffected while others were destroyed; and that the land and adjacent shoreline had been *elevated* permanently by several feet. Not for nothing was Darwin led to remark:

> Daily it is forced home on the mind of the geologist that nothing, not even the wind that blows, is so unstable as the level of the crust of the earth. (1845, 321)

And this physical demonstration supported his conclusion (Darwin 1838, 659) that "thousands of miles of both coasts of South America have been upraised within the recent period by a slow, long-continued, intermittent, movement."

Darwin was also able to confirm that, coincident with the earthquake, two large volcanoes on the Cordillera had erupted violently (Darwin 1840a). On this basis, he concluded that vulcanism and earthquakes were causally linked and associated with elevation of the land. He became convinced that all these movements were linked to the hot and fluid nature of the earth's interior that was "subject to some change, – its cause completely unknown, – its action slow, intermittent, but irresistible" (Darwin 1840a, 631).

Traverses that Darwin undertook across the Cordillera while in Chile allowed him to map, in cross section the distribution of beds on either side of their central axis (Darwin 1846). On this basis he was able to confirm that the mountains had been forced upward by the injection of igneous rock from below; this allowed him to explain the greatly elevated position of fossil-bearing marine rocks and even fossil forests that he discovered on these excursions.

Galapagos: A Volcanic Province

Continuing the theme of vulcanism and general earth processes, during the summer of 1835 Darwin visited the Galapagos Islands. Knowing them to be of relatively recent origin and almost entirely volcanic, Darwin was well prepared to investigate these islands firsthand. As Herbert and her coauthors (2009) show, Darwin spent the majority of her time ashore on James Island and undertook a systematic study and collection of its volcanic rocks. What emerged (Darwin 1844) was a series of remarkable insights into igneous rock formation. He confirmed that volcanoes are capable of generating different varieties ("species") of igneous rock during phases of eruptive activity, which were then crudely separated into "trachytic" and "basaltic" components (Harker 1909; Herbert et al. 2009; Gibson 2010). Darwin was also able to show from direct observation that molten rock (lava) is a complex mixture of chemicals and that, over a range of temperatures, some components aggregate into crystals while the remainder stays fluid and that such crystallized components may settle within the fluid lava according to their density (Gibson 2009; Herbert et al. 2009).

Darwin used this evidence to propose that the *diversity* of igneous rock types and mineral aggregations on earth was not time based (appearing one after the other as the earth aged) but rather a product of dynamic processes operating within volcanoes all the time (Gibson 2010); and that this diversity was augmented during the cooling phase of fluid lava flows on the flanks of volcanoes (and, by implication, during magmatic intrusion within the earth's crust). Alfred Harker (1909) fully recognized Darwin's important contributions in his very aptly titled *The Natural History of Igneous Rocks*.

South American Fossils

The South American sojourn also allowed Darwin the freedom to explore its geology more intimately. The Patagonian coast yielded a rich harvest of fossils. Some, such as those of the giant ground sloth (*Megatherium*) he recognized, but he was also able to collect a range of previously undescribed material (Herbert and Norman 2009); among these, Darwin was able to recognize fossil animals that mimicked the living fauna typical of South America but were of much greater size (Rachootin 1985). But along with these, two particular discoveries that he made stood out dramatically. One was represented by the highly distinctive teeth of a mastodon (an entirely extinct elephant-like creature) that had come to prominence through the researches of Thomas Jefferson and Benjamin Franklin (K. S. Thomson 2008). This suggested that mastodons had become extinct in South America in the past. Even more surprisingly, he discovered the distinctive teeth of a fossil horse, which showed that the horse had been a native of South America, long before the arrival of the conquistadors.

Many of these discoveries offered insights into the geological past of South America and prompted pertinent questions concerning extinctions, taxonomic identity, systematics, ecological change, ancient environments, and their influence on fossil preservation and successional changes in faunal composition over time; all became extremely pertinent to Darwin's later theoretical work on evolutionary change in the organic world (Rachootin 1985).

Coral Islands

The structure and variety of coral islands attracted Darwin's attention during the post-Galapagos phase (1835–36) of the voyage (Darwin 1842c) (Fig. 3.5). Their origin and a consistent explanation of their diversity of form were matters of controversy, with which Lyell was involved. Darwin's observations and explanation provided a wonderfully simple resolution. He recognized that the ocean floor on occasion threw up an eruptive volcanic cone that breached the ocean surface to form a new island. Once emerged, the island would attract floating marine organisms, such as coral polyps that would grow their calcitic skeletons in the shallows around the margins of the islands. The coral polyps would grow only in clear, warm waters that were sunlit (hence their tropical distribution); sunlight was essential because the coral

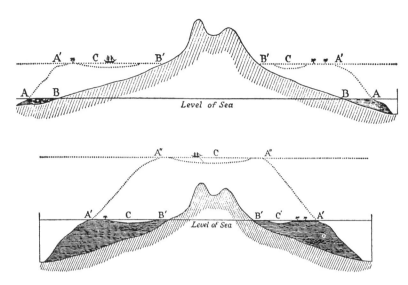

FIGURE 3.5. Darwin's theory of coral reefs. Note that this depends on the earth gradually subsiding, as suggested by Lyell's climate theory. From C. Darwin, *The Structure and Distribution of Coral Reefs* (London: Smith, Elder, 1842)

organisms (polyps) coexisted with minute photosynthetic algae (plantlike organisms) embedded in their bodies that were able to generate vital sugars to sustain the life of each polyp.

Once the newly emerged volcanic island had stabilized, it would, according to Darwin's observations in the Cape Verde islands, begin to gradually subside. As it did so, the geometry of the essentially conical island implied that the coral fringe would not only sink but gradually migrate away from the land surface, creating a potential fringing coral reef separated from the land by a shallow lagoon. All that was required was that the subsidence of the cone did not exceed the rate at which the polyps could grow fresh coral skeletons before they sank too deep for light to penetrate the sea water. If this process is allowed to continue, the volcanic cone will eventually subside beneath the ocean surface leaving the familiar ring-like coral atoll structure with its central shallow lagoon (hiding the crater of the original volcanic cone).

In addition, from a "physical equilibrium" perspective, this would have been very appealing to the larger view of Darwin, having demonstrated that elevation of the land could involve entire continents, such as South America: if the crust of the earth was being raised over huge, continent-sized areas, it should, by the application of simple logic, be undergoing depression elsewhere – and where better than the oceanic floor? Darwin knew perfectly well of the debate concerning the loading of the seafloor adjacent to continental areas, caused by the erosion of huge quantities of sediment; the symmetry of this model created a global vision of the dynamic earth.

AFTER THE *BEAGLE* VOYAGE

With the exception of brief excursions to Scotland and North Wales, Darwin's geological fieldwork came to an end once he was back in England. Despite this, three geologically based topics consumed his time in the period up to the mid-1850s and heralded the onset of his much more broadly focused species work.

Earthworms and Landscapes

Darwin provided a remarkable insight into the action of earthworms and their effect upon geomorphology. The idea was first introduced in a short paper on the formation of "vegetable mould" presented to the Geological Society in 1837 (Darwin 1840b). Darwin eventually produced a monograph on the topic as his last major contribution (Darwin 1881). The realization, prompted by a conversation with his uncle Josiah Wedgwood, that earthworms play a major role in the recycling and restructuring of soil and that this effect could be measured in less than a decade by cutting simple soil profiles, was remarkable. In an echo of the Charles Lyell's uniformitarianism, it became clear that small and comparatively insignificant earthworms, given sufficient time (by implication the millions of years available within the geological time scale), were capable of playing a major role in shaping the landscape of the earth (Gould 1982).

The Parallel Roads of Glen Roy

Encouraged by his success in building his own model to explain the dynamics of the earth based upon his observations in South America (Darwin 1840a) and, in particular, the prevalence of elevation of land, Darwin's attention became focused closer to home by a field trip to explore the parallel roads of Glen Roy (Rudwick 1974; Herbert 2005) (Fig. 3.6). These remarkable geomorphological features, the source of renown and much discussion of their cause, comprise two parallel ridges (the "roads") that mark sharply defined changes in slope that follow the contours around the bases of the hills that enclose the glen. Darwin's general conclusion was that the parallel roads represented another example of elevation of the land, the roads representing ancient marine terraces or strandlines that had been subsequently abandoned as the area had been uplifted.

At the outset, Darwin's interpretations seemed perfectly plausible, driven as they were by his observations and experience elsewhere; however, within the year he was shown to be entirely wrong by Agassiz's new glacial action model (Herbert 2005). Though chastened by the experience, this, no doubt, taught Darwin an extremely timely lesson about the need for extreme caution in drawing interpretations from observational data. It is indeed likely that his later work benefited considerably from this personal setback (Rudwick 1974).

FIGURE 3.6. Darwin thought that the parallel roads of Glen Roy were caused by the sea, which had since run out given the rise of the land. Louis Agassiz showed that they were caused by a lake, trapped by ice. Note that Darwin's false hypothesis is, like the true hypothesis about coral reefs, a consequence of Lyell's theory of climate. From C. R. Darwin, Observations on the parallel roads of Glen Roy, and of other parts of Lochaber in Scotland, with an attempt to prove that they are of marine origin, *Philosophical Transactions of the Royal Society* 129 (1839): 39–81

Barnacles

The late 1840s and early 1850s were also dominated by his work on living and fossil barnacles. This culminated in four notable monographic studies (Darwin 1852, 1854; 1851, 1855; see also Darwin 1985–, 4: Appendix). These cross the intellectual divide between the geological and biological worlds of science; they established his competence as a paleontologist and the place of fossils in the history of life on earth, while the work on living forms underpinned his species work by giving him a fundamental grasp of the principles of taxonomy and systematics.

SUMMARY

Darwin's innate ability to observe, consider, and suggest causal mechanisms for natural phenomena was remarkable and is widely appreciated within the biological natural sciences. His contributions to the physical natural sciences have been obscured by his *Origin of Species*. Frank Rhodes (1991) began to redress this absence of balance by focusing on Darwin's first major geological paper (Darwin 1840a), which the president of the Royal Society, Sir Archibald Geikie (1909, 29), described as

> one of the most brilliant and suggestive essays which th[e Geological] Society [of London] ever published....
> It was the first attempt to treat this subject not as a mere matter of idle speculation, but on a basis of personal observation in the field.

Darwin's broad conclusions can be applied to his geologically based work as follows:

Continental Earth
1. Mountains form by the accumulation of small, intermittent vertical movements.
2. Mountains and mountain chains were built by a gradual pumping mechanism involving repeated intrusions of molten rock, followed by periods of cooling.
3. Slow, gradual continental elevation was more probable than occasional "catastrophic" global paroxysms.
4. A common subcrustal mechanism linked earthquakes, volcanoes, mountains, and continental elevation.
5. The cross-sectional structure of mountain chains was described, and Darwin noted that their axes were formed by igneous intrusion; he even suggested, on the basis of differential elevation of islands close to the coast near Concepcion, that the axis of elevation of the Cordillera was located a few tens of miles off the coast of Chile – where we now know that the oceanic trench is located.
6. The region of the earth beneath the crust was fluid and hot.

7. He accurately characterized the principal physical effects and motions created during earthquakes.
8. He also acknowledged that earthquakes can occur in areas of *subsidence* and suggested that these needed to be studied in detail.
9. That the mechanisms that he had demonstrated had acted on the entire South American continent, causing it to be uplifted and tilted, implied that the entire globe consisted of continents that floated upon a sea of molten rock that were subject to the similar forces.
10. Geomorphology, and the landscapes that are so familiar to us, are being constantly remodeled by the action of earthworms.

Oceanic Earth

1. Depression and elevation of the ocean floor and oceanic islands are seen as ongoing and dictated by loading of the oceanic crust (owing to the formation of volcanoes and sediment runoff from the continents) and the logical requirement for the earth to maintain a global equilibrium of the crust (overall elevation must logically balance subsidence). That is to say Charles Lyell was correct.
2. Coral islands present an integrated demonstration of inorganic (vulcanism, elevation, and subsidence) and organic (the growth of coral polyps) proof of the continued vertical displacement of the earth's crust.

Mechanisms for generating petrological diversity

1. Volcanoes behave like chemical factories whose products (igneous rock "species") vary depending upon its relative state and maturity.
2. Differential crystallization and density-dependent fractionation within fluid lava flows also generate a diverse array of rock (petrological) "species."

The history of life

1. Distinct geographical regions of the earth tend to show distinctive faunas, often with gigantic fossil ancestors of the living fauna.
2. Extinctions may be global in the case of some species and much more local for others.
3. The principles of taxonomy and systematics appear to apply equally to fossil and living species, provided that their skeletal remains are well preserved.

As Geikie rightly observed, the words "brilliant" and "suggestive" characterize nearly all of these contributions. The majority of these proposals *contradicted* (and a few reinforced on the basis of new observations) what was considered to be a consensus of the time. They demonstrate the arrival of a leading, insightful, and original geologist, during what has often been referred to as the "Heroic Age of Geology" (Porter 1977). As pointed out by Rhodes (1991), Darwin knew intuitively that the geology of the earth would turn out to be simple, and he would have reveled in the underlying simplicity of plate tectonics when it emerged a little over a century after Darwin's truly insightful work.

But it is also clear that alongside the purely geological aspects of Darwin's discoveries and interpretations, which he was attempting to develop into a personal "Theory of the Earth" (Rhodes 1991), it is possible to draw out some remarkably cross-disciplinary insights as well.

The integration of the physical world and biological processes that underpins his work on coral islands is one obvious example. But also, in the manner of his approach to the petrology and mineralogy of the earth, his approach (and perhaps his underlying philosophy) was redolent of that which he would deploy much later as he developed his theory of evolution by means of natural selection. P. N. Pearson (1996) drew together some of the intellectual threads with respect to Darwin's approach to the origin and creation of diversity in igneous rocks. Darwin's general thesis was that molten rock is of a generally uniform consistency, yet comprises a multitude of chemical ingredients in a form of molten rock "soup." The processes that occur inside the volcano (the equivalent of a chemical retort) and within the lava, as it cascades from the volcanic edifice and cools, generate petrological diversity.

Taking, for example, the lava "soup," the crystallization and removal of one mineral "species" from the body of molten rock changes the overall composition of the remaining melt. Subtraction, removal, or "extinction" is one of the essences of Darwin's theory of natural selection: a disadvantageous trait among members of a species may be removed or subtracted from a breeding population because such individuals are "unfit." Evolutionary change is thus directed away from the genetic composition of the "unfit" organism: it has been selected against. In a pure metaphorical sense, the direction of magmatic differentiation is away from (against) the composition of the crystallizing mineral: those particular "individuals" having been removed from the "parental" population of minerals left in the melt.

Diversity generation in igneous rocks follows from the existence of semi-molten rocks, chemical and density differences between minerals, and the influence of gravity; these factors constitute a natural "algorithm" (Dennett 1995). Crystal segregation does not however amount to "evolution" in an accepted biological sense – it is the *variation* introduced during reproduction, as well as the *hereditary principle*, that creates the variation between organic being from which traits may be selected in the living world. Nevertheless, density-dependent segregation of crystals in melts serves to sift, sort, and impart some degree of order to inorganic systems and represents a limited (that is to say *invariant*) form of selection.

Is it possible that Darwin's mechanism for generating diversity in the inorganic world – the "petrological kingdom" – represents an intellectual staging post for his mechanism for generating diversity in the biological kingdom? Perhaps this is taking things too far; he had already started developing his theory of natural selection in 1837. However, if a little credence is given to this suggestion of an element of continuity of philosophy (exploring nature's ability to generate variety by deducible mechanisms), it might demonstrate – at least to my way of thinking – how transparent to Darwin were the walls that tend today to separate the biological and physical natural sciences. And that he was indeed such a "clever, clever man" (Gould 1983).

ACKNOWLEDGEMENTS

This essay is utterly dependent on the work of a number of genuine Darwin historians: Sandra Herbert, Janet Browne, Jim Secord, and Frank Rhodes. I am deeply grateful to JS, SH, and Michael Ruse for taking the time to read and comment upon the shortcomings of an earlier version of this manuscript. All errors that remain are my own and carry an implicit apology.

Essay 4

Looking Back with "Great Satisfaction" on Charles Darwin's Vertebrate Paleontology

Paul D. Brinkman

Charles Darwin, with the cooperation of shipmates and a local network of landowners, merchants, and guides, made an important collection of fossil vertebrates from South America during the second expedition of HMS *Beagle*, from 1831 to 1836. Many of the particulars of Darwin's fossil collecting have been confused or omitted in previous accounts of his voyage. The present account, drawing on several previously underutilized resources, adds interesting details to the story and corrects a few misconceptions. It also explores the nationalistic aspects of Darwin's science – the network of expatriated Englishmen who helped him and their loyalist motives. Finally, it examines the study and description of Darwin's fossils by Richard Owen (Fig. 4.1). A review of Owen's results shows that, despite claims to the contrary, Darwin's field identifications were remarkably good.

Darwin's shipmates were not uniformly friendly to paleontology. His collection of vertebrate fossils attracted heaps of abuse, some good-natured, some hostile. He endured "sundry sneers about Seal & Whale bones" from the crew. Worse, First Lieutenant John C. Wickham, who was "always growling about [Darwin] bringing more dirt on board than any ten men," referred to his deck cluttering specimens as "damned beastly devilment" (Barlow 1945, 103). And FitzRoy (1839, 107) called them "cargoes of apparent rubbish." Even Darwin himself was plagued by doubts about the usefulness of his fossils. He confessed to his Cambridge mentor John Stevens Henslow that he was "not feeling quite sure of the value of such bones as I before sent you" (Barlow 1967, 81). In time, however, the fossil vertebrates would prove to be the most personally satisfying, as well as one of the most scientifically significant, collections Darwin made during the voyage.

THE GALLOPING *NATURALISTA*

Darwin found vertebrate fossils in South America for the first time at Punta Alta, a modest outcrop on the coast southwest of Buenos Aires (Fig. Introduction. 13). On a bright and calm 22 September 1832, Darwin rowed ashore with FitzRoy and Lieutenant Bartholomew J. Sulivan. A Spanish major at Bahía Blanca, a nearby fort, was skeptical about the peaceful intentions of the British survey ship, and especially suspicious of Darwin, who had been introduced as a *naturalista*, or "a man that knows every thing" (FitzRoy 1839, 104). The major sent a nervous troop of gaucho soldiers who stood watch while the party landed to examine some conspicuous rocks. Darwin described a bed of fossiliferous gravel and a fifteen-foot-thick layer of "red earthy clay containing ... small pebbles & ... shells," adopting the local term *tosca* for this layer (see Darwin's geology notes, CUL). He also found a few fossils exposed in the bedrock and broken fragments on the beach. The *Beagle* remained in the area for several weeks, so Darwin returned to collect as often as possible. The ship's fiddler and odd-job man Syms Covington probably worked at times as Darwin's assistant. FitzRoy (1839, 107) wrote that Darwin and Covington, who was later released from duty to become Darwin's personal servant, "used their pick-axes in earnest" to acquire the bones. Darwin was overjoyed to find a skull, which he tentatively attributed to a rhinoceros – he wrestled for three hours to extract it from bedrock, then dragged it aboard the *Beagle* after nightfall. He also collected a disassociated mandible, bearing a lone tooth, and fragments of a bony shell (Darwin 1988). It was probably here that Darwin despaired at having "had to break off the projecting end of a huge, partly excavated, bone, when the boat waiting for him would wait no longer" – a memory that Darwin recalled with much regret in later years (Judd 1911, 9).

Punta Alta was the most productive fossil vertebrate locality Darwin ever found (Fig. 4.2). He spent all or part of at least five days collecting fossils at Punta Alta, 22–23 and 25 September and 7 and 16 October 1832, and returned the following August. Darwin's own account makes no mention of Covington's participation in 1832. He does credit Covington and another assistant for helping make collections there in August 1833, however. FitzRoy may have confused the two visits in his narrative, something that Darwin was also prone to do in his publications. Covington's journal (http://www.asap.unimelb.edu.au/bsparcs/covingto/chap_3.htm), chapter 3, contains the following entry: "AT Bahia Blanca, near Johnsons Point [Punta Alta], WE ALSO found the remains of bones of *Megatherium*." Whether Covington participated personally in the excavation is not clear from the context, but it seems likely that he did.

Weeks later, on 2 October, FitzRoy ordered a shore party to build a cairn atop the sea cliffs at Monte Hermoso twenty miles east of Punta Alta. Fierce weather and breaking waves drove the *Beagle* off, stranding the men with scant provisions. Darwin and seventeen others spent two "sufficiently miserable" nights under a shelter improvised from the sails of their whaleboat, shivering against the wind and rain. When food ran low, the party scavenged a dead hawk and a fish found floating in the tide. On the evening of the second day, FitzRoy sent a boat in close enough to toss a cask with provisions into the surf. Some sailors swam out to retrieve it. Darwin and the shore party suffered through a cold, sleepless, second night, and were finally rescued the following day (Darwin 1988). Meanwhile, a member of the shore party found "many curious fossils" in the cliffs. On

FIGURE 4.1. Richard Owen (1804–92) was by far the most important British biologist (anatomist and paleontologist) of the first half of the nineteenth century. This is a portrait from about 1850 when he was at the height of his powers and before he had started his controversies with the Darwinians. Permission: Wellcome

his friend's sacrifices for science, FitzRoy (1839, 112) wrote: "Mr. Darwin was also on shore, having been searching for fossils, and he found this trial of hunger quite long enough to satisfy even his love of adventure." The *Beagle* made sail for Buenos Aires to resupply on 19 October, but FitzRoy landed for half an hour at Monte Hermoso to make some final observations. Darwin accompanied the captain to "Starvation Point" – as he called it – and had the good fortune to collect "some well preserved fossil[s] of two or three sorts of Gnawing animals [rodents]. – One of them must have much resembled the Agouti but it is smaller." He also noted the geology (Darwin 1988, 110–11).

Darwin was excited about his fossils. In a letter to his sister Caroline Darwin dated 24 October–24 November [1832], he reported that he had been very fortunate in finding numerous specimens, boasting, "I am almost sure that many of them are quite new; this is always pleasant, but with the antidiluvian animals it is doubly so." He also reported finding "the curious osseous coat, which is attributed to the Megatherium; as the only specimens in Europe are at Madrid ... this alone is enough to repay some wearisome minutes" (Darwin 1985–, 1:276). *Megatherium*, a giant ground sloth described by Georges Cuvier from a skeleton collected near Buenos Aires and mounted in Madrid, was all the rage in

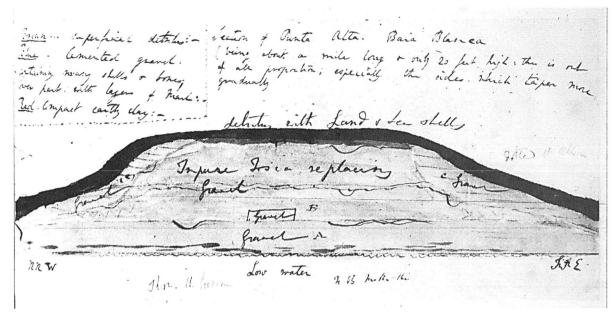

FIGURE 4.2. Darwin drew this rough cross section of Punta Alta, the first and most productive fossil vertebrate locality he found during the *Beagle* voyage. Permission: Cambridge University Library

England. Cuvier had mistakenly attributed some osseous plates to *Megatherium*. When Darwin found similar fossils, he suspected the error. Darwin read an English newspaper concerning additional fossils collected near Buenos Aires and exhibited at the Geological Society of London by Woodbine Parish, a repatriated diplomat. Was Darwin disappointed at being scooped? If so, he made no record of these feelings. Instead, the news that such importance had been accorded to the Parish collection inspired Darwin with the hope that his own fossils might be similarly received. Darwin expected that the fossils would provide him with an entrée into the elite circle of British science.

In November, Darwin wrote to Henslow, fishing for approval and guidance, and summarizing his discoveries at Punta Alta and Monte Hermoso:

> I have been very lucky with fossil bones; I have fragments of at least 6 distinct animals.... – 1st. the Tarsi & Metatarsi very perfect of a Cavia: 2nd the upper jaw & head of some very large animal, with 4 square hollow molars. – & the head greatly produced in front. – I at first thought it belonged either to the Megalonyx or Megatherium. – In confirmation, of this, in the same formation I found a large surface of the osseous polygonal plates, which "late observations" (what are they?) show belong to the Megatherium. – Immediately I saw them I thought they must belong to an enormous Armadillo, living species of which genus are so abundant here: 3d the lower jaw of some large animal, which from the molar teeth, I should think belonged to the Edentata: 4th. some large molar teeth, which in some respects would seem to belong to an enormous Rodentia; 5th, also some smaller teeth belonging to the same order: &c &c. – If it interests you sufficiently to unpack them, I shall be *very curious* to hear something about them[.] (Darwin 1985–, 1:280, letter to Henslow, 24 November 1832)

In his geology notes, Darwin elaborated on the osseous plates:

> At Punta Alta the only organic remain I found in the Tosca ... was a most singular one: it consisted in an extent of about 3 feet by 2 covered with thick osseous polygonal plates ... it resembles the case of Armadillo on a grand scale[.] (Darwin's geology notes, CUL DAR; see also Herbert 2005)

But with no word yet from Henslow on the status of his fossils, some of which had already been shipped home, Darwin was digging blindly. Even so, he tenaciously followed up every fossil lead he learned about from helpful locals. In his next letter to Henslow, for example, Darwin related that he had found and interviewed Parish's agent, a North American named Mr. Oakley. The interview convinced him that the Parish specimen came from the same formation as his own specimens from Punta Alta (Darwin 1985–, 1:308, letter, 11 April 1833). In August 1833, Henslow at last responded to Darwin's request for information on the Parish fossils: "The fossil portions of Megatherium turned out to be extremely interesting as serving to illustrate certain parts of the animal which the specimens formerly received in this country & in France had failed to do." Henslow reported that William Clift, the Hunterian Museum conservator who had restored the Parish specimens, was interested in cleaning, drafting, and describing Darwin's fossils with the object of finding out "how far they serve to illustrate ... the Great Beast." Henslow entreated Darwin to "Send home every scrap of Megatherium skull you can set your eyes upon. – & *all* fossils" (Darwin 1985–, 1:327–28, letter, 31 August 1833). By March 1834, Darwin finally received

Henslow's August letter, which featured precisely the praise and encouragement Darwin needed to reinvigorate his investigations. "I was delighted at receiving your letter..." he replied. "Nothing for a long time has given me so much pleasure.... I am quite astonished that such miserable fragments of the Megatherium should have been worth all the trouble.... It is a most flattering encouragement to find Men, like Mr Clift, who will take such interest, in what I send home" (Darwin 1985–, 1:368–69, letter, March 1834).

In the last days of August 1833, Darwin made a long, overland trip from Patagones to Buenos Aires, stopping in Bahía Blanca to rendezvous with the *Beagle*. After a few idle days waiting, he hired a guide and set out for Punta Alta to watch for the ship. Arriving late in the afternoon of the 22nd, he spent a pleasant evening hunting and marking fossils. Rain set in, so the pair returned to the fort empty handed (Darwin 1988). In his pocket notebook entry for this date, Darwin reasoned that the fossils from Punta Alta and Monte Hermoso must be older than "present shells." How much older he was reluctant to speculate. (See Darwin's Falkland field notebook, p. 138a, CUL.) Once the ship arrived, Darwin joined Sulivan on the 29th. Sulivan's shore party encamped near Punta Alta was far better prepared than the one that suffered at Monte Hermoso. The travelers' four-gallon boiler provided five and a half pints of tea per man per day. They ate salt pork, fresh beef, venison, and biscuit and drank a quarter pint of rum each day. The following morning, Darwin waited for low tide to search the beach for fossils, delaying the entire group. Several important discoveries resulted, including one magnificent specimen, tolerably complete, entombed in bedrock just at the low water mark. Sulivan provided a man to help Covington collect it, while Darwin and the others left for Bahia Blanca. Covington labored for several days with the fossils, while Darwin was content to "superintend." Punta Alta "is a quiet retired spot & the weather beautiful," Darwin (1988, 178) wrote, noting that "the very quietness is almost sublime." Sulivan (1896) remembered that the shore party spent one night huddled together on the yawl, moored in the soft mud just offshore. The men laughed and swapped sea stories under an awning filled with tobacco smoke while thunder and lightning roared around them. It was one of the happiest evenings of his life. Darwin "passed the night pleasantly." (See Darwin's B Blanca field notebook, CUL. Covington spent 29 September to 3 October working at Punta Alta [Darwin 1988].)

Back in Buenos Aires, Darwin made preparations for another overland trip in search of fossils. He had his dentures mended and stocked collecting equipment and provisions, including snuff and cigars. He set out on the afternoon of 27 September 1833, heading north toward Santa Fe. He spent the night near the town of Lujan. The famous Madrid *Megatherium* had been found along the banks of the Rio Lujan, which Darwin crossed the following morning (Darwin 1988). (Darwin was apparently unaware of the river's paleontological significance. Lujan was also the home of Francisco Javier Muniz, Argentina's first naturalist, and first fossil vertebrate collector. See also Darwin's St. Fe field notebook, CUL.)

Traveling by moonlight and arriving at dawn on 1 October, Darwin spent the day scouring the bluffs of the Rio Carcarvana for fossils, netting only one "curious & large cutting tooth." On the Rio Parana, Darwin hired a canoe to pursue some "immense" *Mastodon* bones jutting from the bank. Unfortunately, these were so fragile that he was only able to collect some teeth fragments. Finally, Darwin exhumed a fossil horse's tooth "well buried" in the "Tosca." Darwin took ill with a fever during his travels and returned to Buenos Aires to rendezvous with the *Beagle* in late October (Darwin 1988, 193; Barlow 1945, 210). (See also Darwin's St. Fe field notebook, CUL; and Darwin's geology notes, CUL.)

Finding that city in revolutionary turmoil, Darwin retreated to Montevideo, where he learned that the *Beagle* would not be sailing for another month. The delay afforded him the opportunity to make another "gallop" north to the Rio Negro in late November. There he collected a few broken fragments of "megatherium," and purchased a large skull from a local estancia owner, complaining: "When found the head was quite perfect; but [gauchos] knocked the teeth out with stones, and then set up the head as a mark to throw at" (Darwin 1839a, 181).

From Montevideo, Darwin sailed south along the Patagonian coast. After a lengthy passage, he spent Christmas at Puerto Deseado but found no fossils. The *Beagle* reached Puerto San Julian early in 1834. There, in a relatively young deposit of earthy matter on a terrace above the cliff, Darwin reported collecting "some very perfect bones of some large animal, I fancy a Mastodon. -the bones of one hind extremity are very perfect & solid" (Darwin 1985–, 1:369, letter to Henslow, March 1834). Shortly thereafter the *Beagle* departed Patagonia, and Darwin never collected another fossil mammal, although he did play an important role in promoting further paleontological exploration in South America (see Brinkman 2003).

THE IMPORTANCE OF BEING ENGLISH

An invaluable aid to Darwin's researches was the network of expatriated Englishmen then living in South America. As a gentleman and a guest of the Royal Navy, Darwin had privileged access to the cream of British society in what is now Argentina, including the charge d'affaires. This, in turn, led to contacts with English merchants and landholders and their local network of associates. It also garnered Darwin an invaluable passport as a *naturalista* from none other than General Juan Manuel de Rosas, future dictator of Argentina. Darwin used these privileges to his advantage. For example, Charles Hughes, a childhood acquaintance residing in Buenos Aires, provided information for the overland trip from Montevideo. "Nothing could be more obliging than he was," Darwin wrote in a letter to Caroline, "he obtained a great deal of information for me & has undertaken several troublesome commissions, which otherwise I never could have managed.... I think I have infected him with a slight geological Mania, which I hope he will encourage" (Darwin 1985–, 1:277, letter, 24 November 1832).

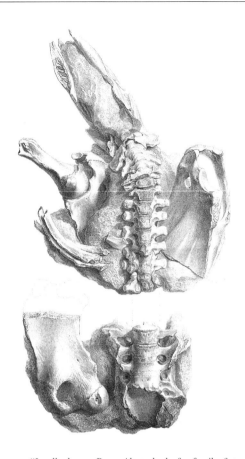

FIGURE 4.3. "I walked on to Punta Alta to look after fossils; & to my great joy I found the head of some large animal, imbedded in a soft rock.... It took me nearly 3 hours to get it out: As far as I am able to judge, it is allied to the Rhinoceros.... I did not get it on board till some hours after it was dark" (from Darwin's diary, 23 September 1832). *Scelidotherium* was one of the largest and most complete fossil mammals Darwin collected in South America. From Richard Owen, Fossil Mammalia, Part 1, in *The Zoology of the Voyage of H.M.S.* Beagle, edited and superintended by Charles Darwin (London: Henry Colburn 1840)

Darwin's most important contact was Edward Lumb, a prosperous English merchant, who graciously placed himself and his property at Darwin's disposal. Lumb and his wife hosted the vagabond naturalist at their estancia on the Pampas near Ensenado and at their Buenos Aires home. Darwin's fossil collecting trip near Santa Fe was outfitted from Lumb's home. There he made the acquaintance of an unnamed Spanish gentleman living near the Rio Tercero, a friend of Lumb's, who promised to collect fossils and forward them to Buenos Aires. When Darwin went to Montevideo and arranged a second overland trip bound for the Rio Negro, Lumb provided him with a letter of introduction to Mr. Keen, an English estancia owner living nearby. Keen accompanied Darwin to the place where the bull's-eye skull was acquired and arranged to ship it to Buenos Aires. Another English landowner, Mr. Hooker, promised to procure fossils from his property and forward them to Lumb also. When Darwin finally learned from Henslow that his fossils had been well received in England, his anxiety about the specimens in Lumb's care became acute. Darwin wrote and implored him to take the greatest possible care (Winslow 1975). Henslow's encouragement had prodded Darwin and his local Anglo-Argentine network to greater efforts on behalf of paleontology.

Traveling naturalists from Europe had a very nationalistic idea of science. Darwin, for instance, was consciously working to serve British science, even to the extent of expressing regret that rival naturalist Alcide d'Orbigny might have skimmed the cream of South America for the benefit of French science (see, e.g., Darwin's letter to Henslow dated [ca. 26 October –]24 November [1832] in Darwin 1985–, 1:280). Parish (1839, xvii), the retired diplomat, was also motivated by nationalism: "I regret that I lost ... the opportunity of making what too late I learnt would have been very acceptable additions to our zoological collections; but I never imagined that our public museums were so entirely destitute.... The collections of some of the museums on the Continent are, I believe, much more complete; especially those of Paris." Lumb expressed his nationalistic view of science in a letter to Darwin dated 13 November 1833: "I do not consider I have done more than what any Englishman should do for the promotion of any scientific end which may tend to the aggrandisement of his Country (Darwin 1985–, 1:355)." And in a letter to Henslow dated 2 May 1834 Lumb wrote: "Permit me this opportunity of offering my Services to you & to assure you that I shall feel highly gratified if by any Information, or Specimens I can obtain in this Country I can contribute to the advancement of Science in my native land" (Darwin 1985–, 1:386).

All of Darwin's fossils were destined for England, where the infrastructure for science was comparatively well developed. There was never any intention to leave anything behind in the Museo Publico de Buenos Aires – the first institution of its kind in South America. Darwin (1988, 114) was not at all impressed by his visit to this "very poor" museum, a struggling assortment of relics housed on the second floor of the convent of Santo Domingo. (He was likewise disparaging of the "Kings collection at Madrid where for all purposes of science [*Megatherium* bones] are nearly as much hidden as if in their primeval rock" [Darwin 1988, 109]). Apparently, there were no fossils in the collection at the time of his visit. Nor was there another facility for paleontology elsewhere in Buenos Aires. In fact, there was virtually no local interest in fossil vertebrates at this time. Darwin suspected and Henslow confirmed that his fossils would be important for science. Had he left them in South America, they would have served no immediate purpose. The idea that a nation should control its own natural resources, as a kind of national scientific patrimony, was uncommon in Darwin's day.

MUSEUM MATTERS

In an August 1834 letter to Caroline, Darwin made a policy statement regarding the disposition of his specimens, writing that "the ultimate destination of *all* my collections will of course be to wherever they may be of most service to Natural History" (Darwin 1985–, 1:404). Loyalty to British science and self-interest were two additional considerations

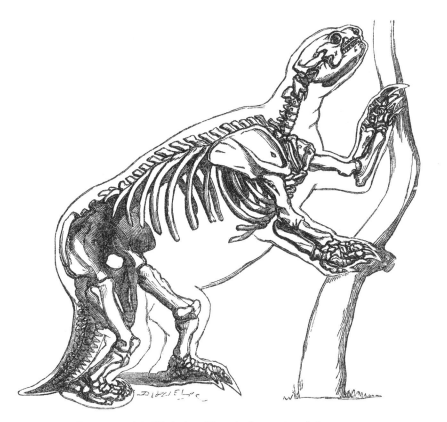

FIGURE 4.4. Reconstructed *Scelidotherium*. The animal was a giant sloth.

he weighed when choosing a repository. Darwin was ill-suited for describing his fossils, so these needed to be relegated to a specialist. On Henslow's recommendation, they were placed in the temporary custody of Clift at the Hunterian Museum. Clift had worked on Parish's "megatherium" and was the best-qualified person in England to prepare Darwin's specimens. And he was eager for the opportunity: Caroline wrote that "you never saw a little man so delighted" (Darwin 1985–, 1:373, letter, 9–28 March 1834). Darwin was pleased and flattered that his fossils were attracting attention, but he was reluctant to commit his specimens permanently to the Hunterian. He was initially inclined to favor the British Museum because of the many favors he had received from His Majesty's Service. But Darwin's generous feelings toward that institution gave way to certain misgivings about the state of its management. Henslow encouraged his protégé to dole out his collections to any interested and qualified naturalists, which Darwin resolved to do.

As many of the fossils were completely new forms, Clift was unfortunately not quite up to snuff. The best comparative anatomy collection in the world, at that time, was at the Museum d'Histoire Naturelle in Paris – Darwin could have sent his fossils there. But grateful nationalism compelled him to remain loyal to Britain. In the end, Darwin decided to donate all the fossils to the Hunterian, with the stipulation that they provide a set of casts for himself and for the British Museum, the Geological Society, Cambridge, and Oxford. "I ought to make up my mind to give my own set [of casts] to Paris,"

Darwin confided in a 19 December 1836 letter to Richard Owen, "but I confess I should be grieved to lose my trophies. I should feel like a knight who had lost his armorial bearings" (Darwin 1985–, 1:527).

Darwin's fossils were then studied and described by Clift's son-in-law, Richard Owen (see Figs. 4.3 and 4.4). An ambitious comparative anatomist and a future dean of British science, Owen had visited Cuvier at his museum in Paris and was eager to establish the Hunterian along similar, scientific lines. He suspected that writing a good, scientific description of Darwin's fossils based on Cuvier's techniques would enhance his reputation. He was right – the work consolidated his reputation as the "British Cuvier" (Rupke 1994). Darwin pitched in by providing a geological introduction for Owen's publication. He also presented a paper on the subject to the Geological Society in May 1837 (Darwin 1838).

Darwin collected fossils primarily for their value as geological specimens: "All the interest which I individually feel about these fossils is their connection with the Geology of the Pampas" (Darwin 1985–, 1:404, letter to Caroline D., 9–12 August 1834). He knew that most of the fossil vertebrates he collected were about the same age. On the basis of the similarities between the living mollusks along the Atlantic coast of South America and the fossil mollusks buried in a bed just below the extinct fossil vertebrates he collected, Darwin reasoned that the coast had been uplifted in relatively recent geological time.

Some historians have unjustly emphasized Darwin's so-called mistakes and misidentifications with his vertebrate fossils (e.g., Sulloway 1982b; Desmond and Moore 1991; Herbert 2005). But with fragmentary fossils still encased in matrix, a limited reference library, and virtually no comparative materials, Darwin's working conditions on the voyage were far from ideal, and field identifications are tentative even under the best of circumstances. Nearly everything Darwin collected was new to science, a fact that he recognized in the field. One cannot fault him for failing to identify taxa that had never been described and named. By 1840, only two extinct fossil genera known to occur in South America, *Megatherium* and *Mastodon*, had appeared in the scientific literature (Owen 1840) (fig. 4.5). Consequently, Darwin referred many of his fossils to these two taxa. Darwin acknowledged his limits as a vertebrate zoologist and his "ignorance of comparative Anatomy" (Darwin 1985–1:368, letter to Henslow, March 1834). He treated his identifications as tentative and routinely expressed doubts about them in notes and letters. He was

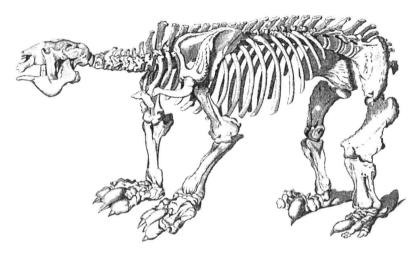

FIGURE 4.5. This reconstruction of *Megatherium* appeared in a book in the *Beagle* library. *Megatherium* was one of only two fossil vertebrates known to occur in South America before Darwin's voyage. From Edward Pidgeon, *The Fossil Remains of the Animal Kingdom* (London: Whittaker, Treacher, & Co., 1830), 132

often reduced to guesswork. Yet his record of fossil identifications is remarkably good.

Owen's *Fossil Mammalia* described ten large quadrupeds and two rodents, most of which were new to science. For example, Darwin's collection included two unusual new genera: a complete skull of *Toxodon*; and some postcranial bones of *Macrauchenia*. In the field, he mistook them for *Megatherium* and *Mastodon*. But the former specimen had only a few broken and badly worn teeth on its lower jaw, and the latter had no skull material at all. Worse, these were the first specimens ever collected of two orders of mammals unique to South America (later called Notoungulata and Litopterna). Darwin's mistakes with these specimens are perfectly understandable when one considers the extenuating circumstances. He identified his other fossils more or less correctly. He referred a number of specimens to *Megatherium*, *Megalonyx*, or to an unspecified "edentate," including: a skull fragment of *Glossotherium*; a jaw and teeth of *Mylodon*; a reasonably complete skeleton of *Scelidotherium*; a jaw of *Megalonyx*; and a skull of *Megatherium*. These five genera are all giant ground sloths of the order Edentata (Xenarthra). The first three were new to science, the fourth was very poorly known, and only the last was relatively well known. Darwin could not distinguish these genera in the field, but he recognized their similarities. He also collected some dermal armor and two small bones, which Owen identified as "Large Edentata" (referred to as *Hoplophorus* in a figure). Darwin often followed European scientific opinion by referring these remains to *Megatherium*, but he did recognize the resemblance of the dermal armor to armadillos, and he often privately referred to them as such. He also correctly identified the molar of an extinct horse, and the teeth and skeletal elements of a *Mastodon*. Other fossils recovered at Monte Hermoso included a jaw and hind foot of an extinct member of *Ctenomys*, an extant rodent endemic to South America, and a molar and some bone fragments of an unnamed animal resembling the capybara, another large extant rodent. Darwin identified these remains as agoutis and unspecified rodents (Owen 1840; Brinkman 2010).

Meanwhile, Darwin had begun secretly working on the subject of transmutation, and the fossil vertebrates of South America provided a key line of evidence: "In July opened first note book on 'Transmutation of Species' – had been greatly struck ... on character of S. American fossils" (Darwin's journal, 1837, CUL). During the voyage, Darwin tried to identify his fossils by comparing them to descriptions and figures of vertebrate fossils in the *Beagle* literature and – more importantly – to certain representatives of the living, endemic fauna of South America. In several cases he recognized a similarity, including fossil "Gnawing animals" with agoutis, and "osseous polygonal plates" with armadillos. It was this curious pattern of resemblance between fossil and living fauna of South America that first inspired Darwin's contemplation of the origin of species.

CONCLUSION

With their privileged access to its fossil resources, British naturalists were ideally positioned to establish the basic body of knowledge of vertebrate paleontology in South America. Darwin's specimens added six genera to the fossil fauna of South America, including five entirely new forms, *Scelidotherium*, *Glossotherium*, *Mylodon*, *Macrauchenia*, and *Toxodon*, and one dubious North American genus, *Megalonyx*. His researches established the preliminary stratigraphic relations for these forms. Darwin and other British naturalists established the research agenda that would form the basis for future work in South American paleontology. Darwin's theory of evolution by natural selection would inspire future fossil explorers in Argentina to search for transitional forms linking the continent's unique taxa to a global phylogeny of mammals.

Darwin recognized the resemblance between some of the fossils he collected and the extant rodents and armadillos of South America during the voyage. He later dubbed this phenomenon the "law of the succession of types" (Darwin 1839a, 210), and claimed that it was one line of evidence that led him to contemplate the origin of species (Darwin 1859). I have argued elsewhere that Darwin became a convinced transmutationist before the end of the voyage largely because of his shipboard contemplation of fossil vertebrate succession

(Brinkman 2010). But whenever his "conversion" happened, the fossil evidence played a crucial role in convincing Darwin personally of the fact of transmutation.

Darwin once wrote with exaggerated modesty that "I, at one time, began to think that the fossil bones would be as troublesome to me, and as of little service, as some other branches of my collection are likely to be. – But now I look back to the trouble I took in procuring them with great satisfaction" (Darwin 1985–, 1:527, letter to Owen, 19 December 1836).

Essay 5

The Origins of the *Origin*: Darwin's First Thoughts about the Tree of Life and Natural Selection, 1837–1839

Jonathan Hodge

Darwin's *Origin of Species* (1859) argues for two big ideas, both metaphorically expressed: the tree of life and natural selection. New species descend from earlier, ancestral species; and these lines of descent with divergent modifications branch and rebranch, like the branches on a tree. So, if every line traces to one first species, all life forms one tree. Natural selection has been the main cause of these changes. By selective breeding, humans make, in a domesticated species, varieties fitted for different ends: heavy horses for plowing, fast ones for racing. In the wild, over eons, natural selective breeding due to the struggle for existence works unlimited changes in branching lines of adaptive, divergent descents, from fish ancestors fitted for swimming to bird descendants fitted for flying and mammals for running.

Darwin first had these ideas more than twenty years before publishing them in the *Origin*. In October 1836, the *Beagle* voyage ended. In July 1837, he opened his private Notebook B with a comprehensive account of the course and causes of life's changes, including a first version of his tree of life. He has the idea of natural selection late in 1838, in Notebook E. The ideas may look like instant insights; but the story is not so simple. Any short telling of the origins of the *Origin* commits misleading omissions and condensations. However, even this very short one can counter two contrasting demands: from rationalists hoping for an edifying tale of universal methodological principles consistently yielding successful solutions to certain given problems specifiable in advance; and from romantics yearning for an epic saga of individual genius bringing imagination and intuition to transcendent reconfigurations of experience, man, nature, and so the whole world. (For documentation of what is said here about Darwin's early theorizing, and for references to the secondary literature, see M. J. S. Hodge 2009b; for the texts of the notebooks, see Barrett et al. 1987. Becquemont 2009 is an important recent study.)

A THEORIST COMES OF AGE

To start seeing why the story cannot satisfy those demands, consider Darwin's earliest ambitious theorizing about the earth and life in the middle years, 1834 and 1835, of the voyage. As a maturing theorist, his main debts were to two mentors: Robert Grant, his informal instructor in invertebrate zoology at Edinburgh, and Charles Lyell, author of *Principles of Geology* in three volumes (1830–33). Grant and Lyell, both Scottish not English, were respectively a doctor and a lawyer, not clerics or indeed Christians; and both sided with French opponents of Georges Cuvier, the Protestant Parisian savant most admired by the Anglican churchmen teaching geology and natural history at Oxford and Cambridge. Grant had given Darwin a preoccupation with all kinds of generation from ordinary growth to sexual reproduction, a preoccupation prominent in Darwin's voyage studies of lower animals. But Grant had presented no system of theory for Darwin to agree and disagree with; Lyell alone had done that for Darwin. There was always more to Darwin, body and soul, than his scientific theorizing: he had his family, his fieldwork, his politics, and so on; but it was Lyell's system of theory that provided the immediate intellectual context for his inaugural practices as an innovative, prospectively publishable theorist.

Lyell taught that ever since the oldest known fossil-bearing rocks were laid down the same causes have acted with the same intensities in the same circumstances, and so produced the same sorts and sizes of effects. The leveling actions of aqueous causes are balanced by unleveling elevations and subsidences due to igneous agencies; and species extinctions and origins occur throughout the past and on into the future. These controversial Lyellian doctrines Darwin will always accept.

In 1835, however, Darwin disagreed with Lyell about the causes of extinctions and about coral island formation. His alternatives to Lyell's theories drew on his Grantian concerns with generation in plants and lower animals. Lyell ascribed extinctions to competitive upsets or defeating invasions caused by changes in climatic and other local circumstances. Darwin, disagreeing, adopted a theory respectfully rejected by Lyell: that a species will eventually die because like an individual animal, and like a graft succession of apple trees, it has, generationally transmitted, an intrinsically limited lifetime.

Lyell did not say what naturalists would see if witnessing a species originating. But he denied that species could arise by transmutations of other species. Each species, he taught, is created separately and at one place and time, determined entirely by adaptational considerations, so that the resemblances among any group of congeneric species are due to their common providential fitting to similar conditions. Darwin may have disagreed, around mid-1836, thinking that this explanation fails for congeneric species original to places with very different conditions and that their common characters are due instead to common descent from a single ancestral species.

If Darwin did accept transmutation on the voyage, he probably did so tentatively and limitedly. What did most to move him to confident and comprehensive transmutationist theorizing were his reflections on the ornithologist John Gould's judgments on the Galapagos land birds – reflections made

FIGURE 5.1. Erasmus Alvey Darwin, Charles Darwin's older brother. He lived as a man about town with many intellectual friends. Intelligent and sweet tempered, although somewhat melancholic, he was deeply loved by all who knew him. He was a major influence on Charles, in introducing him both to science and then (after the *Beagle* voyage) to London society. From H. E. Litchfield, *Emma Darwin, Wife of Charles Darwin: A Century of Family Letters* (Cambridge: privately printed by Cambridge University Press, 1904)

in early March 1837 when taking lodgings in London near his best bachelor buddy, a frail, clubbable, bookish charmer who was likely, uniquely, privy to Charles's covert notebook theorizing: his elder brother, named after their grandfather, Erasmus (Fig. 5.1).

Gould judged what Darwin himself had not even suspected: that many of the land birds collected on the Galapagos were of species peculiar to the islands but very similar to distinct species living on the South American mainland. For Darwin, this generalization raised a decisive geological-geographical issue: these species had originated on these young arid volcanic islands, and yet were closely similar to species living on the nearest lush, forested older continental land, rather than resembling species original to other arid volcanic islands around the world. So, Darwin thought, these similarities could not be explained, Lyell-style, as adaptations to common conditions, but could be ascribed to descent from common ancestors.

Lyell had represented all transmutationist views as an unjustified extrapolation to supraspecific groups – genera, orders, or classes – of the descents from common ancestors taken by most naturalists and ethnologists to explain any intervarietal resemblances within any one species. Darwin's new comprehensive and confident commitment to species transmutations was initially made as just such a common ancestral extrapolation for supraspecific groups. In early

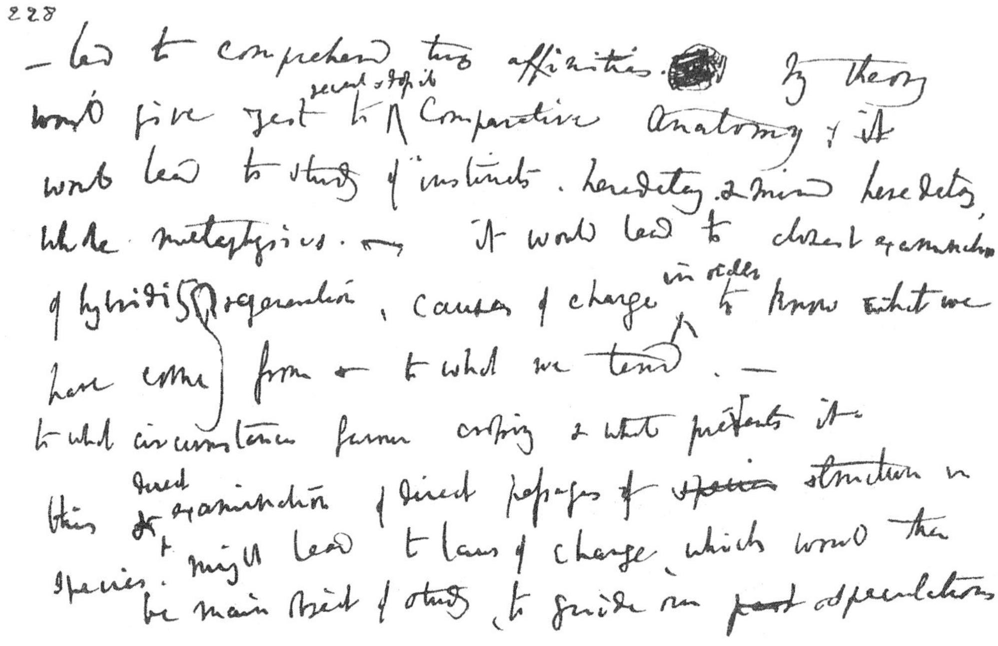

FIGURE 5.2. Some months after returning from the *Beagle* voyage, Darwin opened a series of notebooks in which he began to jot down evolutionary ideas. The key notebooks are B through E, where he speculates on evolution, and M and N, more concerned with human-related issues. This passage is from Notebook B, kept from around the middle of 1837. The transcription reads as follows: "– led to comprehend true affinities. My theory would give zest to recent & Fossil Comparative Anatomy, & it would lead to study of instincts, heredetary. & mind heredetary, whole metaphysics. – it would lead to closest examination of hybridity [to what circumstances favour crossing & what prevents it] & generation, causes of change [in order] to know what we have come from & to what we tend. – this & [direct] examination of direct passages of species structures in species, might lead to laws of change, which would then be main object of study, to guide our past speculations" (bracketed text indicates a passage that has been inserted later by Darwin). Note that, apart from Darwin's inability to spell, he already has the idea of evolution (the discovery of natural selection was a year off) and is excitedly thinking of how powerful an explanatory power it will be. From F. Darwin, *Life and Letters of Charles Darwin* (London: John Murray, 1887), vol. 2

March he integrated this new commitment with his species mortality theory, and with reflections on new judgments made by Richard Owen on his South American fossil mammal specimens.

Lyell warned anyone inclining to the transmutation of species about all the other theses – continued spontaneous generations of the simplest micro-organisms, progressive escalations over eons from these monads to the highest animals, and an ape ancestry for man – comprising the most sustained transmutationist theorizing: what Lyell called Jean-Baptiste Lamarck's system. Darwin took this warning as a challenge no grandson of Erasmus Darwin should evade. By July, at the opening of his Notebook B, he had taken the most consequential decision of his life as a theorist: to elaborate an improved system of zoonomical theory with the scope and structure of Lyell's version of Lamarck's system (Fig. 5.2). Under the heading *Zoonomia*, the laws of life, the title of his grandfather's best-known work, the first two dozen pages of Charles Darwin's notebook sketch just such a system. From now on he would be agreeing and disagreeing with all manner of authors, on all kinds of subjects within and beyond the sciences, but often most critically with himself as author of this, his most ambitious and controversial system of theory.

TREES OF LIVES AND DEATHS

Lamarck himself had the actions of fluids within all living bodies producing over eons recurrent escalations of organization up a series of classes and large families from monads to mammals. Adaptive responses to changing external circumstances, with the inheritance of acquired characters, caused ramifying departures, within classes, from this serial progression. Species mutability made possible both linear progress and arboriform diversification. By contrast, Lyell's version of Lamarck's system opened with an unlimited mutability of species adapting to changing conditions allowing a ramifying common descent, not merely for any family or order of species but, ultimately, for all life from a single, common ancestral origin. Lyell then presented the progress from monads to mammals, its internal causes, and eventual outcome in the ascent of man.

Darwin's systemic sketch matches this bipartite structure. He first has the powers peculiar to sexual generation ensuring adaptive changes in altering circumstances, and hence the formation of new species from old – so explaining how divergent reiterations of such species formations entail over eons a common descent for families and classes, and explaining those geographical and paleontological generalizations about species that remain inexplicable if species originate in independent creations at places and times determined solely adaptationally. Then Darwin's second part introduces the progressive tendencies raising life from monadic, infusorian beginnings up to mammalian perfection. Darwin here invokes no additional causes, internal or external, assuming rather that these progressive tendencies arise as adaptive changes due to the same powers of sexual generation invoked in opening his sketch.

Darwin there ascribes these powers to two features distinguishing all sexual from any asexual generations: maturation in the offspring produced and the mating, crossing, of two parents. The first enables new adaptive, hereditary variations to be acquired in altered circumstances; the second is counterinnovative when offspring are intermediate in character between their two parents. Migration with isolation of a few individuals inbreeding in new circumstances can circumvent this counterinnovative action of crossing, and so allow a new variety to form, and then diverge enough to become intersterile with the parent stock and so become no mere variety but a new species. The ramifying reiterations of such species formations make possible the adaptive diversification of a family or class from its common ancestral species.

Here ends the first part, which has gone from individual sexual reproductions all the way to interfamilial divergences. In Darwin's second part, moving from monadic simplicity to mammalian perfection, all change is not only adaptive but also progressive, and here the tree-of-life metaphor becomes explicit and analogically elaborated. Some lowly species living in constant conditions may not change at all, while other species do so only slowly. There is no necessitation of an invariable rate of change or then of progress. Within any group, high extinct species produced by fast-changing lines of descent can be succeeded by lower species branching out from old, slow, low lines. If ramified and varied in rate according to circumstances, a tendency for progress in all adaptive species formations is reconcilable with regressions – lower fishes, say, coming after higher ones – in the paleontological successions of supraspecific groups. Following Lyell's version of Lamarck, Darwin has progress initiated by monads produced all the time in spontaneous generations; but Darwin supposes that the lifetime of any monad's entire issue, although vast, is limited. Those lines of life that have changed and therefore progressed most must, then, have changed most quickly; hence mammal species have, as Lyell emphasized, shorter species lifetimes than mollusks do; hence, too, among extant species of higher animals there are more gaps of character from more extinctions, and affinities are more circular than linear. Because species deaths by extinctions are compensated by species births in splittings and branchings, the total number of species is, as in Lyell, constant on a long-run average. The branchings of the tree of life are dependent on contingent geographical circumstances, and so numerically irregular, with those branchings making genera being more branched. There is however a tendency toward threefold diversifications into aquatic, aerial, and terrestrial ways of life; and, if a dominant one of these, the terrestrial say, often has further aquatic and aerial issue, then this explains any tendency for groups to have five subgroups, as in the regular arrangements of quinarian taxonomists (Figs. 5.3 and 5.4).

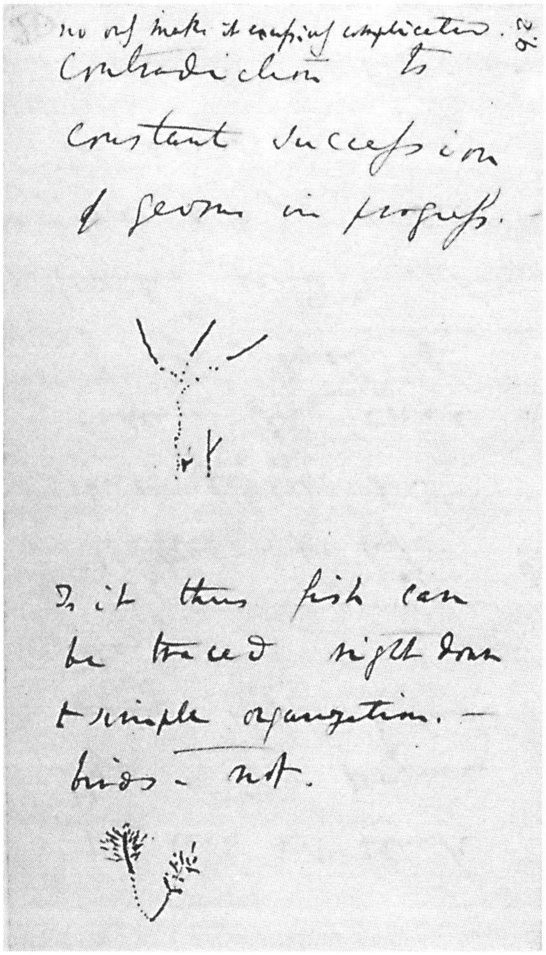

FIGURE 5.3. The first trees of life that Darwin drew (around July 1837). He wrote: "The tree of life should perhaps be called the coral of life, base of branches dead, so that passages cannot be seen." B 26. Permission: Cambridge University Library

Here ends Darwin's sketch of an inaugural zoonomical system. Strikingly, he soon revises not the first part but the second. Rejecting the limited monad lifetime as entailing falsely the eventual simultaneous extinctions of all the species within one family or order, he needs another account of the correlation between greater character gaps, less linear affinities, shorter species lifetimes, and higher grades of organization. He now thinks that gaps within and between groups correlate not with the organizational perfection of those groups but with their taxonomic width. For, in the tree of species branchings, with the total species number constant, when one ancestral species has a dozen descendant species, there must be a dozen lines ending without splitting in extinctions; and in the greater multiplying of species in the diversifying descent of a large group, a class say rather than a genus, there will be vastly more extinctions and more gaps in character, within and between such groups.

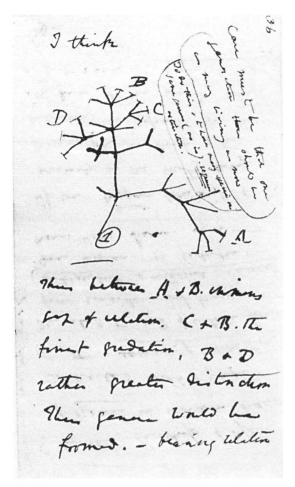

FIGURE 5.4. The famous branching tree, drawn shortly after the first attempts. B 36. Permission: Cambridge University Library

With this new version of the tree of life, any special properties the monads have are explanatorily redundant and no longer invoked. What remain, for all times since the earliest life on earth, are the multiplicative and diversifying splittings and divergings of some species and extinctions of others. In this arboriform process, any species but no supraspecific group has an intrinsic mortality; and any species as a quasi individual is born, lives, and dies just once; and, likewise, any supraspecific group issuing from its single, ancestral species. Moreover, only one line of species in an ancestral group has had descendants in any particular offspring group, so there is no general tendency for fish species, say, to have mammalian descendants. One line of fish species did so once, due presumably to exceptional circumstances, as all the rest have not.

Darwin's new tree of life with its treatment, at once Lyellian and Grantian in its resources, of species as generating, dividing, and multiplying quasi individuals, has now departed fundamentally from any scheme, such as Lamarck's, of recurrent escalations of life through a given array of particular organizational types; and so, indeed, will Darwin understand this tree for the rest of his own life. For he has now, in the summer of 1837, an abstract, referentially anonymous scheme: as in the single illustration in the *Origin*, which has no names of particular groups, fishes or finches, but only letters and numbers representing abstractly the splittings, divergings, and extinctions of any varieties and species – and so of any genera, families, and the rest – in the indefinitely long run of times past, present, and future.

Darwin's tree, in 1837, is not a tree of taxonomic divisions, but of species propagations. In a taxonomic tree, differentiae divide a genus to distinguish its species; in Darwin's tree, an ancestral species divides in making several descendant species of a new genus. Darwin invokes not just the branching structure of a tree but the branching growth producing that structure, instructed by his view that a tree grows as an association or colony of individual buds, some propagating successor buds, others dying without issue. Likewise, then, with species: some divide before dying of old age and produce new species – with their new, intrinsically limited, leases of life – while others die in extinctions without doing so. Darwin has the maturing offspring from a sexual reproduction recapitulating all the long changes undergone in the lines of its entire ancestry: in later jargon, ontogenies recapitulate phylogenies. But these phylogenies do not constitute or conform to any vast mega-ontogeny for the whole tree of life; the branching growth is not fulfilling any determinate, seminal, developmental destiny.

Throughout his zoonomical sketch and its first revisions, Darwin was drawing on Lyell's historical geography of species and their births and deaths as quasi individuals, and on his own Grantian preoccupations – shared with his grandfather whose *Zoonomia* Grant may have directed him to at Edinburgh – with sexual and asexual reproduction, individual and colonial lives, limited and unlimited life. No one before Darwin had formulated such a system of arboriform species propagations and extinctions; but then no one had put such Grantian preoccupations to work in agreeing and disagreeing, as Darwin was, with Lyell, with Lyell's version of Lamarck, and with himself.

AFTER THE TREE, BEFORE NATURAL SELECTION

Within a year after that zoonomical sketch of July 1837, Darwin's voracious reading and uninhibited reflections on myriad topics within and beyond the sciences – from all the divisions of natural history to religion, ethics, aesthetics, and more – reached peaks of energy and ambition never excelled in later decades. Intellectually, he was on a roll. Notebook A, on geology, opened at the same time as Notebook B, on biology, even conjectures about an early nebular earth, a taboo topic for a Lyellian geologist never addressed publicly by Darwin. In July 1838, he started a Notebook M devoted to metaphysics, meaning not the ancient science of being, but – as it had often done for a century now – the study of mind including morality and sociality. Early in September, a section at the back of Notebook D is assigned to generation as a distinct but not separate subject.

Darwin had a general and a specific reason for reading and thinking about mind: his general account of adaptive structural changes in all species, plant or animal, had these changes often initiated by habit, a faculty of mind even in lowly plants. Again, always including man in the tree of life, he now seeks natural, gradual causes for any capacities commonly deemed distinctive of humans, especially language and the moral sense. On the life, mind, and animal ancestry of humans, Darwin comes, within a year, to almost every view published in the early 1870s in *The Descent of Man* and *The Expression of the Emotions*.

Concerning generation, he concludes that in ontogeny and phylogeny hermaphroditic sex precedes the separation of sexes found in higher animals; and that any unfertilized egg or ovule in a female may be like an asexual bud and so incapable of maturing and of acquiring novel hereditary variations from pre- and postnatal influences – these acquisitions being the very purpose of sexual generation and essential for species changes and progress over eons.

From autumn 1837 to the following summer, Darwin's species formation theorizing – as inaugurated in the first part of the zoonomical sketch opening Notebook B – is developed in explaining two permanent changes: adaptive divergence in structures and instincts, and loss of fertility in crossings with the ancestral stock. Cases of nonblending of parental characters, especially in human interracial crosses, Darwin took as signs of incipient constitutional incompatibility between the races. An instinctive aversion to interracial pairing suggested, moreover, that greater constitutional divergence would lead to a consistent disinclination to interbreeding, eventually allowing a divergence entailing intersterility. Racial divergence would have become species divergence, with all the usual criteria for specific distinction met. Amateur ornithologist William Yarrell told Darwin that if two breeds of domestic animals are crossed, the offspring have the characters of the older breed. Elaborating many corollaries from this generalization, Darwin soon took it to show that over successive generations any hereditarily perpetuated characters become so firmly and powerfully embedded in the hereditary constitution that a blending constitutional compromise between two very old breeds becomes impossible, and, through a natural coordination of mind and body, they would be instinctively averse to interbreeding.

This reflection gave him a new way of comparing and contrasting species formation in the wild and race formation in domesticated species. Some domestic breeds, although markedly different in bodies and habits, interbreed readily and successfully, whereas wild species differing that much do not. Darwin took it that domestication itself, this unnatural condition, vitiated the instinctive aversion to interbreeding naturally accompanying in the wild such degrees of bodily and habitual divergence. So, conspecific domestic races provide analogical support for the theory of species formation in the wild, by indicating how character divergences between varieties could arise over a long succession of generations, divergences wider than many wild congeneric species showed; and, on the vitiation of instincts under domestication premise, they confirmed that in the wild such varieties would not interbreed and so would not be counted by naturalists as varieties but as good species. The very absence of very distinct varieties in wild species is, then, evidence that varieties in the wild, unlike races under domestication, do become species by first ceasing to interbreed and later becoming incapable of interbreeding.

From the early months of 1838, Darwin persistently drew a contrast between two sorts of domestic races: natural races or varieties and artificial ones. The natural varieties, due to natural causes rather than to human artifice, are local varieties, isolated so as not to be interbreeding with others, and diverging as they adapt slowly over many generations to local conditions of soil, climate, and so on. By contrast, artificial varieties are often monstrous, distinguished by variations arising as rare, maturational accidents – variations persisting only thanks to the human art of picking, selective breeding, that has made races, often in a few generations, that could never be formed and flourish without benefit of that human art. As Darwin read about the art of selective breeding, he became convinced at this time that species formation in the wild was to be compared with natural variety formation in domesticated species and contrasted with the making of artificial varieties.

Darwin's view of species formation was always that it was an adaptive achievement. Rather than becoming extinct, dying without issue, a species may succeed in adapting sufficiently to new circumstances to give rise to one or more offspring species. Adaptive variations in individuals Darwin came to contrast with monstrous variations. When a puppy moves to a cold climate and grows thicker fur than its parents, that is an adaptation. The variation is induced by the surrounding conditions and is advantageous. By contrast, a puppy growing thicker fur in a warmer country is a monstrous variation: it is a response, even an adaptive response, to rare, unhealthy conditions within the womb. Both adaptive and monstrous variations are made possible by sexual generation; but only the adaptive variations contribute to species formation; rare, monstrous variations are blended out in crossing and are less able to survive and procreate anyway. Darwin thinks adaptive structural variations are often initiated by changes in habits and so in the use of organs. If all the jaguars in a region swim for fish prey on their country becoming flooded, then a new variety with webbed feet could arise through the inheritance of this acquired character. Such webbed-foot exemplars, instantiating Darwin's threefold diversifications into aerial, aquatic, and terrestrial ways of life, were prominent in Lyell's epitome of Lamarck. For Darwin, initiations of structural change by habit changes complemented instinctive aversions to interbreeding as initiating eventual species formation.

Darwin remains throughout the notebook years and beyond seriously committed to progress in the history of life. Here, he worries about challenging Lyell, who opposed all claims that fossils evidenced a progression in the creation of the main types of life. Darwin could avoid a direct challenge by taking his tree growth as a representation only of those changes since the time – whenever that might have been – when the earth was first stocked with all those main types. However, in accepting that individual embryonic maturations recapitulate all past ancestral changes, he had to contemplate an earth when the fish ancestors of today's mammals had not yet had any mammal descendants, an earth that was, moreover, contra Lyell, not fit perhaps for mammals from too little cooling from an original molten state. Again, although reluctant to assume that the eventual formation of man with his distinctive moral life was the sole purpose of all the prior, prehuman progress of life, he did think it was one purpose of the institution by God of those laws of generation that make progress not just possible but inevitable if not invariable.

A decision Darwin was taking in the summer of 1838 served to segregate these commitments concerning progress from the formulation of his theory of species propagation itself. He knew that ideally a causal theory offered to explain certain kinds of facts should be supported in two ways: independently of those facts it is being used to explain, and by showing how well it does explain them. In conformity with this ideal and so too with structural precedents in his July 1837 sketch and in Lyell's version of Lamarck, Darwin resolved to argue for his species propagation theory in two ways. First, he would argue for it by citing the peculiar powers of sexual generation, including Yarrellian constitutional embedding, and by citing the diversification of domesticated species into natural varieties. Here he would be establishing the existence in nature of these causes and their adequacy, their competence, to bring about adaptive species formations in any long run of time, so as to yield such species propagations and diversifications as the tree of life represented. Then, in a second body of argumentation, he would show how this theory could explain, could connect and make intelligible, many different kinds of facts about species: biogeographical facts, paleontological facts, comparative embryological facts, and so on.

This twofold structure and strategy of argumentation is very much what he will adopt in arguing for his theory of natural selection in his unpublished "Sketch of 1842" and "Essay of 1844," and in their published sequel, the *Origin*; and Darwin was committed to it many months before he had first formulated that theory. One consequence, in the summer of 1838, of designing his argumentational case in this way was that those issues – concerning the first forms of life, the subsequent progress in life's ascent, and any correlation that ascent may have had with any cooling and calming of an earth originally nebular and molten – would appear not in the presentation of the species propagation theory itself, or in presenting its evidential credentials independently of its explanatory virtues, but later on in the exposition, when those virtues were elaborated in biogeography, paleontology, embryology, and so on. In the summer of 1838, Darwin was only resolving to write in this way on his theory's behalf; his notebooks contain no sustained acting upon that resolution. What they do show is that he was, even more than before, seeing his various conclusions on diverse topics as being eventually, potentially publishable, public science.

FIGURE 5.5. The Reverend Thomas Robert Malthus (1766–1834). It was his *Essay on the Principle of Population* that sparked the discovery of natural selection for both Darwin and Wallace. Permission: Wellcome

NATURE'S SELECTION

From the middle of September 1838 Darwin's pace slows. The theory of natural selection emerges gradually, from late that month to mid-March 1839, in successive modifications to the earlier theory of adaptive species formations.

A first modification adds to the earlier theory without subtraction or amendment. Near September's end, reflecting on Malthus on population, Darwin dwells initially on how Malthusian superfecundity makes species liable to extinction in even very slightly changing conditions, before considering the implications for the species surviving such changes (Fig. 5.5). Invoking Malthus on some human populations doubling in twenty-five years and on the checks to all populations, Darwin argues that, with all species pressing so hard on others, there is everywhere a fragile competitive balance that slight changes in conditions can upset, causing in some species total population loss. This reflection allows Darwin to return to Lyell's view of extinctions and to abandon his own view, going back to 1835, of some extinctions coming from expiry of a limited vital duration rather than from external contingencies. His generational theory of species extinctions is now replaced with Lyell's ecological one. So much for the losing species then, but what of the winners? In one further sentence Darwin ponders the final cause, the divinely intended benefit, of all this populational pressing, arguing that it is to sort out, to retain, fitting structure and so adapt structure to these changes in conditions. Structure is then adaptively improved in animals and plants, just as, he reflects, Malthus shows how the energy of victorious ancient peoples was providentially enhanced by life and death struggles as excessive fertility forced their tribal migrations and imperial invasions onto contested, occupied ground. Here Darwin responds to Malthus as one theist extending another's teleology and theodicy for superfecundity and empire.

This Malthusian sorting goes on both within and between species; but Darwin draws no analogy with the picking or selecting practiced by human breeders. Nor is there any shift here on how sexual generation ensures adaptive change in altered conditions. What Darwin emphasizes over the next two months is what this sorting entails for advantageous variations acquired in individual maturations, and so for his geological preoccupation with the exchanges of species in changing conditions over vast periods of time. Only a structural variation that is adaptive for the whole lifetime of an individual will, he concludes, be retained in the Malthusian crush of population over many generations; variations adaptive to fetal circumstances alone will not be; and retained variations, eventually becoming strongly hereditary, can be accumulated in prolonged progressive changes. Thus do his new Malthusian insights fit with earlier views on both adaptation and progress.

In late November, in his Notebook N (sequel to M on metaphysics), Darwin illuminates long-run adaptation and progress through his first explicit contrast between two principles explaining adaptive change in structures and habits in the short run. One principle is familiar enough: an adult father blacksmith, thanks to the inherited effects of his habits, has sons with strong arms. The other principle has no exact precedent: any children whom chance has produced with strong arms outlive others. The contrast is direct. Chance production means here, as it has all along for Darwin, production by small, hidden, and rare causes effective prenatally, so that the opposite of chance is postnatal habits. What is new, then, is the conviction that those products of chance with the same benefits as the effects of habits can contribute to adaptive change; because, although rare, individuals with such beneficial variant structures will survive over future generations at others' expense. However, Darwin acknowledges a difficulty in deciding which adaptive structures – and instincts, because these principles apply, he notes, to brain changes – have been due to which of the two principles. A few days later he is, in Notebook E, again considering principles. This time there are three principles, and they can, he says, account for everything. Strikingly, none of the three is new to him: that grandchildren resemble grandfathers; that there is variation in changing circumstances; and that fertility exceeds what food can support. Darwin may well have wanted these three principles to subsume the earlier pair, while circumventing the unresolved difficulty of deciding which adaptive changes to ascribe to which one of that pair.

A further innovation soon comes, seemingly, from Darwin's comparing wild predatory canine species with sporting breeds among domesticated dogs, including, significantly, any webbed-footed breeds. Strikingly reversing himself, Darwin now decides that there is in wild species a selective breeding just as in man's making of artificial varieties of domestic species. Nature's Malthusian sorting is henceforth interpreted as nature's picking or selection. He is soon arguing that because nature's selective breeding is so vastly more prolonged, more discriminating, and more comprehensive than man's, a causal analogy conforms to the traditional proportionality: the greater cause, selection by nature, is adequate to proportionally greater effects than the intraspecific adaptive divergence produced by the much lesser cause, man's selection. These greater effects could include, then, the unlimited interspecific adaptive divergences in the tree of life. Species formations are now compared, by Darwin, not as before with local, natural varieties in domestic species, but with varieties made by the human art that has its natural analog in the selective breeding entailed by the struggle for existence. By March 1839, he is resolving to argue publicly that his theory ascribes species formations to a natural process of selection analogous to man's. The transformations of his older theory making this analogy essential to its very formulation have now given it the structure and content it will have twenty years later in the *Origin*. What these transformations have not done is to resolve the indecision over the two principles of late November. Both artificial and natural selective breeding, Darwin will always accept, work sometimes with chance variations, sometimes with the heritable effects of habits; and sometimes the heritable effects of habits work without selection, whereas chance variations contribute only to sustained, cumulative change with selection. The selective breeding analogy, like the three principles, will always subsume those two principles.

The efficiencies of man's and of nature's selective breeding depend equally on the special powers of sexual as opposed to asexual generation. Comparing and contrasting the two kinds of selective breeding do not make redundant comparisons and contrasts between those two kinds of generation. But the theory of natural selection, as an ecological – economy of nature – theory now constituted by the breeding analogy, will have its argumentation developed separately from any theorizing about all generations. As a theory of the main cause of changes in the tree of life, natural selection, with its Lyellian and Malthusian struggles among and within species, can then be detached from any theorizing about individual sexual and asexual generations, theorizing pursuing Darwin's Grantian preoccupations; and both enterprises will continue to draw inspiration from the grandpaternal precedents set by Erasmus Darwin. The generation theorizing will be fundamentally transformed in the early 1840s when Darwin very probably conceives pangenesis much as published in 1868. His theorizing about natural selection, first fully expounded in his manuscript "Sketch of 1842" and "Essay of 1844," will be reformed, but not fundamentally altered, on being supplemented by new thoughts, in the 1840s and 1850s, about sexual selection and about structural differentiation and functional specialization in progress and adaptive divergence; and these thoughts are articulated as such supplementary argumentation when Darwin finally publishes the *Origin* in 1859.

Essay 6

Darwin and Taxonomy

Mary Pickard Winsor

CHARLES DARWIN WAS born into a world in which taxonomy was already the established scientific language for expressing the diversity of life. In Europe and its colonies around the world, a growing community of museum workers, wealthy collectors, and avid hobbyists named and classified kinds of plants and animals numbering in the tens of thousands, a number that was increasing at a dizzying rate (Farber 2000). As a boy, Darwin absorbed samples of this community's output, using taxonomists' names for flowers in his father's garden and for birds shot for sport; in his university years, he began to interact with taxonomists. Years later, he recollected those carefree days:

> But no pursuit at Cambridge was followed with nearly so much eagerness or gave me so much pleasure as collecting beetles. It was the mere passion for collecting, for I did not dissect them and rarely compared their external characters with published descriptions, but got them named anyhow. I will give a proof of my zeal: one day, on tearing off some old bark, I saw two rare beetles and seized one in each hand; then I saw a third and new kind, which I could not bear to lose, so that I popped the one which I held in my right hand into my mouth. Alas it ejected some intensely acrid fluid, which burnt my tongue so that I was forced to spit the beetle out, which was lost, as well as the third one.
>
> … No poet ever felt more delight at seeing his first poem published than I did at seeing in Stephen's *Illustrations of British Insects* the magic words, "captured by C. Darwin, Esq." (Darwin 1958a, 62–63)

Childish though Darwin's collecting hobby seemed to him later, it meant that as an undergraduate he was familiar with current taxonomic ideas and practices; without this familiarity, Professor Henslow would not have recommended him for the *Beagle* voyage.

THE NATURAL SYSTEM

The importance of the achievements of taxonomy for the discovery and proof of evolution is hard to exaggerate, as Darwin (1859, 128) recognized in the *Origin*:

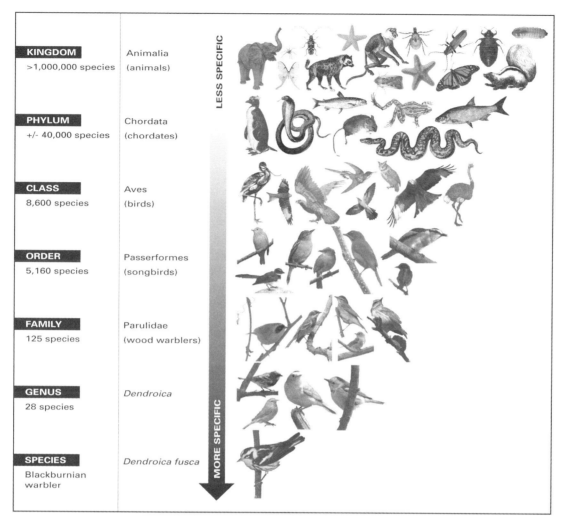

FIGURE 6.1. The classificatory system of Carl Linnaeus (1707–78). It is a series of nested sets (taxa), getting ever-more inclusive as one moves up the hierarchy (categories).

It is a truly wonderful fact – the wonder of which we are apt to overlook from familiarity – that all animals and all plants throughout all time and space should be related to each other in group subordinate to group, in the manner which we everywhere behold – namely, varieties of the same species most closely related together, species of the same genus less closely and unequally related together, forming sections and sub-genera, species of distinct genera much less closely related, and genera related in different degrees, forming sub-families, families, orders, sub-classes, and classes.

One of the virtues of his theory, he claimed, is that it can explain this "wonderful fact," which otherwise is mysterious (Fig. 6.1). That claim has two elements: first, the hierarchical structure of classification inherited from Linnaeus expresses essential relationships among organisms; and, second, the alternative theory gives no explanation. Both claims are roughly, but not exactly, true.

Because it is roughly true that the Linnaean hierarchy corresponds to what we may call the shape of nature, Darwin concluded the *Origin* by assuring taxonomists that after they have accepted his theory, they can carry on as usual – "pursue their labours as at present" (484) – and to a large extent that is what happened. Skillful taxonomists could continue to make sound contributions in their areas of expertise while ignoring his theory, and other biologists could ignore taxonomy as no longer of interest because its central question had been solved. Because the only other explanation for the shape of nature was the plan of God, to which Darwin's explanation was so clearly superior, the gap between religion and science widened into a chasm. These were tragic misunderstandings.

In the mid-eighteenth century, Linnaeus provided the broad and deep foundation for modern taxonomy. While his artificial classes and orders of flowering plants made botany accessible for beginners, he had insisted that skillful naming is the job of an expert, someone with wide enough experience and sound enough understanding to recognize natural kinds of organisms, the entities now called taxa. The word "taxon" was introduced in the twentieth century to clarify the distinction between "species" as a concept or category and particular species like the Magellanic penguin (*Spheniscus magellanicus*) or the dandelion (*Taraxacum officianale*), exactly like the grammatical distinction between common nouns and proper nouns ("river" vs. "Nile"). Species are not the only kinds of organisms people recognize, though; pine trees and birds are also taxa, exemplifying the categories "genus" and "class." Linnaeus's system of naming taxa at the species level is justly famous, but equally important was his consistent giving of proper names to taxa at the higher ranks, like Pinus (white pine, Scots pine, Ponderosa, and other pines) and Aves (all the birds). After the death of Linnaeus, botanists and zoologists building on his framework gradually replaced, just as he had hoped they would, the classes and orders that he had created on the basis of one or two characters with more natural ones. Darwin was not saying there was anything wonderful about the fact that naturalists liked to classify with ranked categories; rather, he was saying that the recognition of taxa by experts was revealing a hierarchical structure that really is the shape of nature. Victorians called this the natural system.

BRANCHING TRANSMUTATION WITH EXTINCTION, 1837

It was not aboard the *Beagle* but in 1837, after expert taxonomists in London classified his specimens, that the meaning of the natural system became blindingly clear to Darwin. The giant fossil bones he had collected in Argentina turned into evidence for evolution only after Darwin heard the anatomist Richard Owen assess the relationship of these extinct mammals to the armadillos, sloths, llamas, and capybaras that live in the same region today. Darwin's many specimens of plain brown birds from the Galapagos took on meaning only when ornithologist John Gould declared that the little ones were new species in a new genus in the finch family, and the larger ones were new species of the mockingbird genus; though the species were all restricted to these geologically young islands, they were taxonomically related to finches and mockingbirds on the nearest mainland.

When Darwin opened his first notebook on transmutation in March of 1837, he was already in a radically different position compared to others who had imagined that species might change. His experiences had given him space, and time, and the shape of nature. Riding across the plains of Argentina and sailing long stretches of ocean gave him a vivid sense of space. Digging out shells from mountaintops and working out a theory to explain coral atolls gave him an understanding of the vastness of geological time. And his assiduous collecting, combined with the taxonomists' practice of naming, gave him the conviction that taxa were real.

Was that not a strange advantage, for a man who was destined to alter forever the fixity of species and quash their status as exemplary natural kinds? In 1859 he would write that "we shall have to treat species in the same manner as those naturalists treat genera, who admit that genera are merely artificial combinations made for convenience. This may not be a cheering prospect; but we shall at least be freed from the vain search for the undiscovered and undiscoverable essence of the term species" (Darwin 1859, 485). Categories were indeed human inventions, and their uniform application sometimes required making arbitrary distinctions, but he certainly did not believe they were nothing more than that. Quite the contrary, taxa large and small are phenomena of the world around us, real phenomena. Their reality had not been so clear to naturalists a hundred years previously, though Linnaeus was fiercely certain that species really do belong each to its proper genus. People who expected taxonomists to gradually fill in the gaps of the great chain of being, because nature makes no leaps, complained that the Linnaean hierarchy breaks apart what is continuous. But as hundreds of naturalists paid closer attention, lumps and clumps of taxa emerged from the fog, and the conviction grew that organisms were truly related, essentially similar, linked by something called affinity. Any naturalist who maintained that genera were nothing more than human inventions would never see in the living world any evidence for evolution. Darwin's final theory said no such thing.

Darwin's explanation for the natural system was not natural selection, which he figured out a year later; his explanation for the natural system was a novel kind of transmutation featuring branching and extinction. Quite early in the notebook, he used the metaphor of branches, noting that "organized beings represent a tree. *irregularly branched* some branches far more branched, – Hence Genera" (Barrett et al. 1987, B121). These short notes show that his idea differed fundamentally from anyone else's, including Lamarck's, for previous theorists had used only living forms. Other naturalists disputed whether nature was continuous or discontinuous, but Darwin's great idea was that it could be both, if the links existed in the past but were now extinct. "We need not think that fish and penguins really pass into each other. ... The tree of life should perhaps be called the coral of life, base of branches dead; so that passages cannot be seen" (B25).

At this point Darwin sketched his first two treelike diagrams (Fig. 5.3). The connection between the class of fish and the class of birds would not be through a bird that swims but through an extinct common ancestor of both classes. Soon after this he made another diagram, exploring his new idea from a different angle (Fig. 5.4). This one expresses how branching evolution and extinction could generate taxonomic groups. "Thus between A and B immense gap of relation, C and B the finest gradation, B and D rather greater distinction. Thus genera would be formed ..." The branches seemed to show a world filling up with new forms, but he then added as a side note that the multiplication of species must be counterbalanced by many deaths of species if the total number of species is to remain roughly constant (Barrett et al. 1987, B25–B39). Extinction is not the sole cause of the great distance between the species marked A and B on his diagram, however; their separation is mostly due to the eastward and southward directions A's ancestors took while B was heading north. Those opposite directions look like what

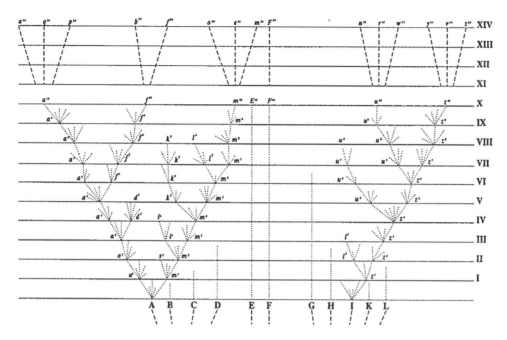

FIGURE 6.2. Darwin's tree of life, the only diagram in the *Origin of Species*. Note that in the context, Darwin is using the tree to explain branching (through his "principle of divergence"), which is why it is so spread out. From C. Darwin, *On the Origin of Species by Means of Natural Selection, or the Preservation of Favoured Races in the Struggle for Life* (London: John Murray, 1859)

Darwin would later call divergence, but that important part of his analysis he set aside and did not return to until ten years later.

Capturing the idea of a "coral of life" on a piece of paper was hard to do. In Darwin's first two diagrams (Fig. 5.3), one dimension of the paper roughly means time, the bottom as the past, moving upward to the present, with extinction indicated by dots. But in his third diagram (Fig. 5.4), the surface of the page gives no time dimension. It is as though we are taking a bird's-eye view, peering down at the branches of an oak tree. All the branches signify the past; the living forms are indicated by short lines tacked on at right angles to the tip of some branches. Omitting time gives him both dimensions of the paper to represent amount of difference, or diversity. In another notebook years later, Darwin pictured the diversity of living forms as dots along a horizontal line (Ospovat 1981, 173), their many differences indicated only by spaces of various lengths. This is also how diversity in the present appears in the *Origin*'s famous tree diagram, where today's taxa are the labeled points along lines X and XIV, a thin portrayal indeed (Fig. 6.2).

Other naturalists who were thinking about similarities among living organisms had experimented with other ways to spread diversity across a page (O'Hara 1991; P. F. Stevens 1994). William Sharp Macleay had proposed in 1819 that members of each natural group sit on the circumference of a circle (Fig. 6.3). Although enthusiasm for his ideas rapidly waned because his demand for numerical symmetry (five per circle) did not mesh with their experience, many London naturalists, Darwin among them, still felt the quinarian system contained intriguing half-truths. Immediately following his first two diagrams, Darwin wrote that "the bottom of branches deaden, *so that* in Mammalia <<birds>> it would only appear like circles" (Barrett et al. 1987, B27). In other words, if the trunk of the bird tree had subdivided into branches diverging away from each other, the similarities between the existing orders of birds would not be linear but somewhat circular.

CHARACTERS, ESSENTIAL AND ANALOGICAL

In the Linnaean hierarchy, the only characters that counted were ones that uniquely located a taxon within its own stack of nested taxa. The quinarian debate enlarged naturalists' attention to include two other kinds of resemblance, both reaching sideways across from one stack to another. Blendings were expected where one circle touches its neighbor. When barnacles, formerly classed among the Mollusca, were found to have embryos like the embryos of crabs, Macleay placed them between Crustacea and Mollusca. The idea of transitional forms was an old one, and continuity was still a good reason to complain that hierarchy could not well capture the shape of nature. But Macleay's other idea, analogies, was something quite new. He would compare two forms, each of which belonged solidly within its own circle of affinities, and find some striking feature they shared.

Late in 1838 Darwin noticed a paper by Owen that seemed to contain a rationale for evaluating characters. Within the Mammalia, the sea cows (manatees and dugongs,

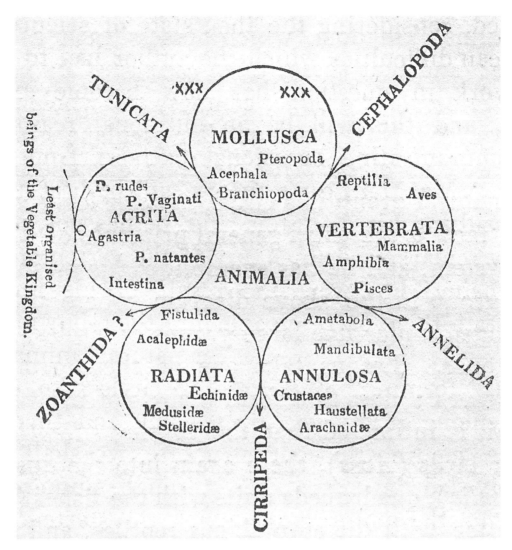

FIGURE 6.3. W. S. Macleay's quinary system of classification, consisting of osculating (touching) circles, as reproduced by his follower William Swainson. From W. Swainson, *Geography and Classification of Animals* (London: Longman, 1835)

Plate XVI) had been grouped with dolphins and whales, but Owen (1838, 39–40) agreed with a proposal connecting them to elephants:

> The generative organs being those which are most remotely related to the habits and food of an animal, I have always regarded as affording very clear indication of its true affinities. We are least likely, in the modifications of these organs, to mistake a merely *adaptive* for an *essential* character.

Darwin scribbled to himself,

> How little *clear* meaning has this to what it might have. – What is the difference between an *essential* character & an *adaptive* one. – are not the essential ones eminently adaptive. – Does it not mean *lately* adapted or transformed & hence not indicative of true affinity. (Barrett et al. 1987, E92e)

Darwin shared with Owen the concept of "true affinity," and they also shared the belief, basic both to Cuvier's principles and to natural theology, that organisms are perfectly adapted, that is, well formed for their role. Owen was using the term "essential character" in a sense that harks back to Linnaeus, as indicative of a taxon's correct place in a natural system (Müller-Wille 2007). At the same time he was beginning to develop a challenge to the old argument from design, which privileged function, in favor of a higher order of design, an abstract morphological type. When Owen spoke of the "modification" of an organ, he did not mean a physical transformation but one taking place now, in our imagination, and originally, in the mind of the Creator. Darwin already realized that his theory promised a more meaningful distinction than this problematical contrast, but then he put the issue aside.

In 1840 John Obadiah Westwood, an entomologist whose work Darwin greatly admired, proposed that the analogies so loved by quinarians were "essential characters," except that affinity meant having many characters in common and analogy meant sharing only one (Di Gregorio 1987). The ornithologist Hugh Strickland (1840, 222) fiercely contradicted Westwood's idea. Affinity is "*the relation which subsists between two or more members of a natural group*, or in other words, *an agreement in essential characters*" (emphasis in original). "Analogy, in short, is nothing more than *an agreement in non-essential characters*, or a resemblance which does not constitute affinity." As a result of exchanges like this, Darwin (1859, 427) could declare in 1859 that naturalists had agreed upon "the very important distinction between real affinities and analogical or adaptive resemblances." But what had made Strickland's definition more acceptable than Westwood's? The claim that taxonomy was once simply the reading of pattern out of data ignores the process by which the community of experts chose which data to value and which to discard. One factor involved in the shift of opinion from the 1820s to the 1850s can only be called theological. Because they felt sure that correctly recognized taxa were real, Strickland, Westwood, and Owen all believed that to improve the natural system was to discover God's plan. Who can doubt, said Strickland (1840, 221), "that such groups as Vertebrata, Insecta, Mammalia, Pisces, Coleoptera, &c., are not merely human generalizations, but real apartments in the edifice of the Divine Architect?" Ignoring Westwood, Strickland (1840, 224) cited early quinarians who maintained that God had inserted symbolic representations at strategic points in the natural system to help naturalists read the Book of Nature.

> This has always appeared to me one of the most unsound and unphilosophical of the doctrines maintained by the advocates of the circular system. It seems derogatory to Creative Power to suppose that the principle of *representation* had any place in the scheme of creation, or that certain organs were given to species, not with a view to the discharge of certain destined functions, but for the apparently useless object of *imitating* or *representing* other species in a distant part of the system ... that the long tail of the horse was given it, not for the purpose of brushing off flies, but in order to represent the long "tail" (train) of the peacock.... Without wasting words upon the serious discussion of such puerilities, I will merely repeat my deliberate conviction, that relations of analogy are not to be regarded as affording any evidence of προαίρεσις or *intention*, in the scheme of creation, but are mere coincidences of structure incidental to the grand design....

Owen too explained the homology of vertebrate bones by reference to "some archetypal exemplar on which it has pleased the Creator to frame certain of his living creatures" (Rupke 2009, 170). Darwin (1859, 413) would later declare that the problem with naturalists' belief that the natural system "reveals the plan of the Creator" is that "nothing is thus added to our knowledge," but in Strickland's formulation, a mature, respectful view of God added something important. It tells us that organisms were designed for their own sake and not for ours.

RANKING AND DIVERGENCE

Darwin had no wish to become a taxonomist; he was glad to leave the drudgery of formal description of his collections to others. One of the reasons it was drudgery was that the exploding quantity of material and workers meant that the same thing could easily get renamed by mistake – indeed, this was happening all the time, in several languages. Disentangling nomenclature required the tools of a scholar and the tact of a diplomat, because some taxonomists had an emotional attachment to their own published opinions. Need for regulation was widely felt, and Strickland was the man with the political skills to make it happen. In 1842 the British Association for the Advancement of Science authorized a committee, which Strickland chaired and on which Darwin and Westwood served, to draft the new rules. The Strickland rules, which were soon widely adopted, owed their success to the strategy of focusing on practices already in wide use and leaving contentious questions, like how to define the category "species," unanswered. Very likely this committee work encouraged Darwin not to attempt a facile definition of any taxonomic category in the *Origin* (McOuat 1996).

In the summer of 1843, Darwin's good friend George Waterhouse asked him for advice about the ranking of some odd mammals. It was already understood that the native Australian mammals resemble opossums by giving birth to tiny embryos that grew in a pouch rather than in a placenta. Two other hairy quadrupeds from that distant continent were even more strange. The platypus had a bill like a duck, and the spiny anteaters (echidna) definitely laid eggs (Plate XV). Owen had published an impressive set of dissections that accorded these two creatures a separate new taxon, the monotremes. But Waterhouse was reluctant to allow three species (one platypus and two echidnas) to have the same rank as dense orders like rodents with hundreds of species, so he wanted to alter the definition of marsupials to include the monotremes.

Darwin, in a warm and frank tone, replied in effect: Your problem is that you have no idea at all what your goal is. If you are just making a catalog, you might allow criteria like numerical balance, but if your goal is to save ink ("conveying much information through single words"), you may count only differences. He reminded Waterhouse that a few odd flies called stylopids (parasites of bees) are given ordinal rank, equivalent to the massive order Coleoptera (as they still are). Naturalists seeking the natural system were aiming higher, Darwin knew; they wanted "to discover the laws according to which the Creator has willed to produce organized beings," but he cautioned Waterhouse that such an "empty high-sounding" expression "means just nothing" (Darwin 1985–, 2:375). My own view, said Darwin (1985–, 2:378), is that relationship means consanguinity.

> I believe ... that if every organism, which ever had lived or does live, were collected together (which is impossible as only a few *can* have been preserved in a fossil state) a perfect series would be presented, linking all, say

the Mammals, into one great, quite indivisible group – and I believe all the orders, families & genera amongst the Mammals are merely artificial terms highly useful to show the relationship of those members of the series, *which have not become extinct–* (letter to Waterhouse, 31 July 1843)

Waterhouse merely seemed amused to be pressured to clarify his own principles.

> Naturalists say one animal may have a relationship of affinity with another, or it may have a relationship of analogy without there being any true affinity – I am very much puzzled about this matter.... When ... I say one animal is nearly related to another, I mean that the two agree in several important points, & the relationship is more distant when there are few points of resemblance and those comparatively unimportant. (Darwin 1985–, 2:381–82, letter to Darwin, 9 August 1843)

Waterhouse's cool reaction to Darwin's ideas is understandable, because Darwin admitted that "the difficulty of ascertaining true relationship ie a natural classification remains just the same" for Darwin, "though I know what I am looking for" (1985–, 2:376). Darwin's monumental work on barnacles bears out what he had told Waterhouse, that belief in branching evolution would not make the job of finding true affinities any easier, because the characters already employed by taxonomists were the only ones available (Padian 1999).

Looking back years later, Darwin recalled that it was long after sketching out his theory of natural selection that he finally realized he had overlooked a key problem, namely, why branches tend to diverge from one another. He began to think over this problem around 1847, perhaps because other naturalists were exploring the idea of branching relationships, so the fact that he offered no explanation was a weakness in his own theory (Ospovat 1981). He wanted natural selection to supply the answer and was greatly pleased in 1857 when he saw how it could. He explained his "principle of divergence" in his chapter on natural selection in the *Origin* and illustrated it in his famous diagram, in which variation is represented by lines fanning out (Fig. 6.2). Whatever makes a wide-ranging and variable species vary, he said, will also make its descendant species variable, giving rise to genera. Their repeated divergence is explained because forms can have a selective advantage purely because they differ the most. They will avoid competing with each other and can exploit the environment in new ways. But if a species does not vary, it need not go extinct, but might persist for millions of years, while its cousins were splitting and diverging so richly that taxonomists must create families or orders to contain the new taxa. Thus Darwin was satisfied he had dealt with the irregularity of the natural system. From our perspective, he had underestimated the effect of catastrophes, for paleontologists later documented mass extinctions.

ARGUMENT IN THE *ORIGIN*

After Wallace's stunning paper reached him in 1858, Darwin worked intensely to condense his big book. In mid-March (15) 1859, wrapping up the penultimate chapter, which includes classification, he wrote to Hooker that "the facts seem to me to come out very strong for mutability of species" (Darwin 1985–, 7:265). He closed that chapter with this sentence:

> Finally, the several classes of fact which have been considered in this chapter seem to me to proclaim so plainly, that the inumerable species, genera, and families of organic beings, with which this world is peopled, have all descended, each within its own class or group, from common parents, and have all been modified in the course of descent, that I should without hesitation adopt this view, even if it were unsupported by other facts or arguments. (Darwin 1859, 457–58)

This is a strong claim, for natural selection is not mentioned in this discussion of classification and morphology (though it does play a role in his explanation of embryology), nor are the factors so important in his own conversion in 1837, geography and paleontology. Overlooked during the modern synthesis, when speciation and natural selection were paramount, these "several classes of fact" probably explain why many of Darwin's peers who finished reading the *Origin* still unconvinced by natural selection did accept that branching evolution must have occurred. The momentum had long been building in a favorable direction. For example, he could say in 1859:

> No one regards the external resemblance of a mouse to a shrew, of a dugong to a whale, of a whale to a fish, as of any importance. These resemblances, though so intimately connected to the whole life of the being, are ranked as merely "adaptive or analogical characters;".... (Darwin 1859, 414)

Linnaeus had known in 1758, contrary to common sense, that whales are mammals and shrews are not rodents, but the correct classification of dugongs was more recent (Plate XVI). When Darwin wrote "no one," he meant no one whose opinion mattered, because he felt that taxonomic progress was moving in the direction of uncovering reality.

Darwin, who had great respect for those with the stamina to do careful taxonomy, wrote the classification portion of this chapter treading delicately, because the gist of his argument was that taxonomists did not understand what they were doing. His theory would explain why they had rules of thumb that seem illogical, and why none of their rules apply without exception. He said in private that his theory "will clear away an immense amount of rubbish about the value of characters" (Darwin 1985–, 6:456, letter to T. H. Huxley, 26 September 1857). An "important" character did not mean a feature important to the organism itself but merely one that was useful to the taxonomist. Any feature, such as the shape of a plant's leaf, of an insect's antenna, or a bird's beak, can be "important" to one taxon and useless in another. Darwin (1859, 417) quoted with approval a botanist pointing out certain plants that lack all the characters supposedly defining their group "and thus laugh at our classification." Darwin (1859, 420) offered his solution:

> All the foregoing rules and aids and difficulties in classification are explained, if I do not greatly deceive myself, on the view that the natural system is founded on descent with modification; that the characters which naturalists consider as showing true affinity between any two or more species, are those which have been inherited from a common parent, and, in so far, all true classification is genealogical....

This solution is transformed in his final chapter into a vision of a golden future, once naturalists have accepted evolution: "Our classifications will come to be, as far as they can be so made, genealogies ..." (Darwin 1859, 486). But how far can they be so made, in fact? Actual genealogies of human families do not have a hierarchical structure. Darwin's irregular branching clearly dispensed with any notion that the ranks of categories had absolute meaning, endorsing taxonomists' practice of raising or lowering a taxon's rank at will. An essential character no longer carries the implication that God intended it, but sorting out the recent, from the ancient, from the analogous, still requires close study by experts.

Essay 7

Darwin and the Barnacles

Marsha L. Richmond

BETWEEN 1851 AND 1855, Darwin published a series of four monographs on the cirripedes (barnacles), two on the living Cirripedia (1852, 1854) and two on fossil Cirripedia (1851, 1855).[1] This study consumed eight years of his life, from 1846 to 1854. Sandwiched between penning early drafts of his species theory in 1842 and 1844 (Darwin 1909) and the publication of *Origin of Species* in 1859, the barnacle monographs have been interpreted as delaying Darwin's work on his theory of evolution. It is clear, however, that the study of the barnacles complemented Darwin's earlier preoccupation with invertebrate biology and served to bolster his confidence in his species theory (Sloan 1985; A. C. Love 2002). He gained from this study clarity on points important to his evolutionary theory (the significance of variation, homology, and embryology as keys to affinity, change of function, and the evolution of novelty), as well as significant empirical support that would feature in *Origin*. In addition, his receipt of the Royal Medal of the Royal Society of London in 1853, largely on the basis of his first barnacle volume, gained him widespread recognition as a naturalist of high standing. Coupled with his earlier geological studies, the barnacle monographs reinforced Darwin's scientific reputation. His credentials were thus impeccable prior to publishing a theory of evolution that would attract protracted and vociferous criticism among naturalists and laymen alike.

To be sure, Darwin's barnacle monograph was a significant achievement in its own right. Indeed, despite some errors in interpretation, it remains an important work in cirripede morphology and systematics to this day (Southward 1987; Newman 1987, 4). Certainly it was solidly within the tradition of mid-nineteenth-century natural history. Taxonomy – the grouping and classification of organisms – was a major preoccupation of nineteenth-century naturalists. Botanists and zoologists, following the model set by Carl von Linné (Linnaeus) (1707–78), had long strived to catalog nature's vast array of species, exponentially expanded in the early

---

[1] Although the title pages of *A Monograph on the Sub-class Cirripedia, with Figures of All the Species. The Lepadidae, or Pedunculated Cirripedes* (1852) and *A Monograph on the Fossil Balanidae and Verrucidae of Great Britain* (1855) read "1851" and "1854," respectively, they were delayed in publication and, because taxonomic works establish priority, should be cited according to their actual date of publication (Newman 1993).

nineteenth century by the voyages of exploration such as that of the *Beagle*. Naturalists thus employed a hierarchical approach of dividing plant and animal kingdoms into rational categories – orders, classes, genera, and species – using the Linnaean binomial system of naming the genus and species (R. A. Richards 2009). But the aim was not simply to turn the natural world into a museum. Taxonomy formed a prominent pillar within the particularly British tradition of natural theology – a way to illustrate in the "Great Chain of Being" the order of nature reflecting the design of the Creator.

Yet Darwin's study of the barnacles deviated in significant ways from previous taxonomy studies. It was a highly theoretical approach to investigating animal form and function in the context of systematics. Moreover, his application of new cutting-edge methodological approaches, along with an "adaptationist approach to taxonomy" (Innes 2009), not only resulted in new discoveries but set his work apart from the majority of systematic works. Certainly, *A Monograph on the Sub-class Cirrepedia* was no mere exercise in natural theology. The qualities that were singled out in the citation awarding him the Royal Medal identified prominent characteristics of Darwin's particular approach to understanding nature that were prevalent throughout his career, and which became a model for future biologists imbued with the new evolutionary view of the world.

That Darwin would become engrossed in such an enterprise was not surprising. He had long been fascinated by marine invertebrates, which opened up for him a particularly interesting window into life's mysteries (Winsor 1976; Sloan 1985). This curiosity was especially fostered during his years as a student in Edinburgh through his contact with Robert Edmond Grant (1793–1874), whom he often accompanied "to collect animals in the tidal pools, which I dissected as well as I could" (Darwin 1958a, 49–50). References to dissections of various invertebrates, as well as his particular fascination with their larvae and ova and their modes of generation generally, can be found in his notebook from the period and his first scientific paper (Barrett 1977, 2:285–91; Barrett et al. 1987). This interest was certainly evident throughout the *Beagle* voyage, with both geology and marine zoology capturing his attention (A. C. Love 2002). As Jonathan Hodge (1985) noted, Darwin was truly a "life-long generation theorist."

Darwin's decision to undertake such a major taxonomic work was certainly not a calculated one. In the years following his return from the *Beagle* voyage (1831–36), he had published a series of books and articles on various aspects of the geology and natural history of South America. On 1 October 1846, the day he sent off proofs of his final geological publication, he took down from the shelf his collection of curious invertebrates that remained undescribed and on which he intended to write brief notices (Darwin 1985–, 3:344; Browne 1995, 471, letter to R. FitzRoy, 1 October 1846). One in particular captured his attention. In January 1835, while combing the beaches of the Chonos Archipelago just off the coast of Chile, Darwin picked up a conch shell that was "completely drilled by the cavities formed by this animal. –" Back on board ship, he examined the little boring creature under the microscope, identifying it,

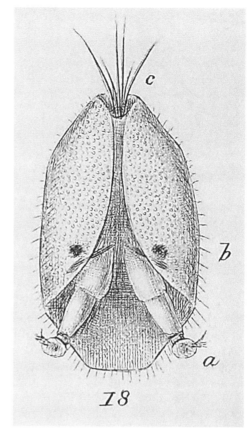

FIGURE 7.1. *Cryptophialus minutus* ("Mr Arthrobalanus"), the barnacle that set Darwin off on his eight-year study of the animals. Depicted is a larva in the last (or pupal) stage. From C. Darwin, *A Monograph on the Sub-class Cirripedia, with Figures of All the Species. The Balanidae (or Sessile Cirripedes; the Verrucidae, etc.)* (London: Ray Society, 1854), pl. XXIV, fig. 18. Redrawn from Darwin's original sketch of January 1835 included in his *Beagle* voyage Zoology Notes (*Charles Darwin's Zoology Notes and Specimen Lists from H.M.S. Beagle*, ed. R. Keynes [Cambridge: Cambridge University Press.2000], 276)

in the detailed description included in his Zoology Notes, as a member of the Balanidae. However, this minuscule parasite was like no other barnacle he had ever seen. It burrowed into its host rather than attaching to the surface and lacked a shell, unlike all other forms; he easily concluded, "It is manifest this curious little animal forms new genus. –" (Darwin 2000, 274, 276; see also Stott 2003, 62–63). Given his interest in generation, it was perhaps his further discovery of developing eggs within the base of the barnacle that heightened his interest. He recorded seeing four different stages in the larval development of this "Balanus" and at the time remarked on the resemblance of one stage to that observed in the metamorphosis of Crustacea (Darwin 2000, 275). This observation was notable, for in 1835 the presence of larval stages of cirripedes was still a matter of dispute among naturalists. It was a decade later, however, before he was finally able to return to this specimen to determine its taxonomic position.

Initially Darwin intended simply to write a paper on this and the other marine invertebrates from the voyage, but his plans were derailed. He became completely engrossed in dissecting "a little animal about the size of a pin's head" (Darwin 1985–, 3:359, letter to R. FitzRoy, 28 October [1846]), and thus this "ill formed little monster" – first humorously called "Mr Arthrobalanus" and later officially christened *Cryptophialus minutus* – led to his eight-year-long work on the barnacle monographs (Darwin 2000, 274) (Fig. 7.1).

THE QUEST TO IDENTIFY "MR ARTHROBALANUS"

Although barnacles were best known for being a nuisance to seagoing vessels, they were of particular theoretical interest to mid-nineteenth-century naturalists. For those seeking to uncover the order of nature, they presented an enigma, for they did not well fit into accepted taxonomic schemes. Linnaeus had placed them along with mollusks in the class "Worms," further divided into the order "testacea" on the basis of their external cases. The French invertebrate specialist Jean-Baptiste Lamarck (1744–1829) noticed their internal appendages or cirri and renamed them "Cirrhipeda," further dividing them into stalked and sessile forms and placing them in between the Annelides and Conchifera (Innes 2009). Georges Cuvier (1769–1832), however, soon returned them to the Mollusca (Winsor 1969).

This was where things stood when John Vaughan Thompson (1779–1847) published a developmental history of cirripedes in 1830. Thompson, living near the Irish seacoast, was able to observe the sequential stages of the metamorphosis of nauplius and cypris larvae into adult barnacles and thereby point out their resemblance to crustacean larvae (J. V. Thompson 1830). This was not at all clear in 1846 when Darwin began reviewing the anomalous Mr Arthrobalanus. Having suddenly been shifted from one branch of the animal kingdom to another – from the Mollusca to the Articulata – the Cirripedia were in dire need of revaluation, in terms of their systematic relationships as well as their anatomical and physiological features in comparison with other Crustacea.

As Darwin explained in the preface to the first volume of *A Monograph on the Sub-class Cirrepedia*, he had not set out to engage in a taxonomic study of the entire subclass (Darwin 1852, vii). In trying to place *Cryptophialus* within barnacle taxonomy, he spent fourteen months undertaking an anatomical study of pedunculated and sessile cirripedes. As he confided to Joseph Dalton Hooker (1817–1911), who aided him immensely in his early study of barnacle anatomy, "I hope to Heaven I am right in spending such a time over one object –" (Darwin 1985–, 4:11, letter to J. D. Hooker, 8 [February 1847]). It soon became obvious, however, that the taxonomy of the group was in a profound state of disarray. Having already made such a sound start, Darwin was sorely tempted by the suggestion of John Edward Gray (1800–75), keeper of the zoological collections at the British Museum and himself a cirripede expert, that he prepare a comprehensive monograph.

In undertaking this massive project, Darwin first read all the historical accounts and contemporary literature available on the topic. He also drew on the expertise and assistance of individuals from around the world, who provided him specimens as well as valuable information (Anderson and Lowe 2010). His considerable skills as a microscopist, honed during the *Beagle* voyage, were put to a supreme test in dissecting minute creatures. The practical aspect of classifying hundreds of species of barnacles was overwhelming, particularly given the tremendous amount of variation in individual forms that astounded even one whose evolutionary theory postulated varieties as incipient species.

He was particularly concerned about the philosophy of classification, not simply its practice. This topic was not, of course, new to him. Darwin had thought much about classification after developing his ideas about transmutation of species soon after the return of the *Beagle* (Desmond and Moore 1991; Browne 1995). Early on he was attracted by the quinarian system of William Sharpe Macleay (1792–1865). Although idealistic in its assumption that all taxa were divisible into five groups, the system nonetheless emphasized the use of analogy and affinity in grouping organisms and potentially well accommodated an anomalous group like the cirripedes that shared properties of two different classes. This approach was particularly attractive to someone like Darwin, who believed that taxonomic relationships were in essence a reflection of genealogical descent (S. Smith 1965; Ospovat 1981, 108). These nascent ideas became sharpened in the 1840s through his frequent discussions with Richard Owen (1804–92), whose expertise in comparative anatomy was complemented by new philosophical ideas coming from the Continent (E. Richards 1987; Sloan 1992; Rupke 1993). Influenced by the philosophical anatomy of Étienne Geoffroy Saint-Hilaire (1772–1844), Owen was at the time formulating new guidelines for taxonomy, including a precise definition of homology to denote parts in different organisms that shared "structural correspondences" with other closely related forms (Fig. 7.2). While for Owen this notion was an ideal – representing a common design among members of a group – within Darwin's maturing evolutionary perspective the archetype became an ancestor, and the principles of natural classification thus began to assume a new meaning (Ospovat 1981; Desmond 1982). As Ghiselin (1969, 83) noted, they "ceased to be merely descriptive and became explanatory" (Fig. 7.3).

Darwin's ideas on classification were well developed prior to beginning the barnacle monograph. His notebooks from the late 1830s and early 1840s indicate that he was thinking deeply about developing a theory of classification based on descent with modification. In corresponding with the well-respected taxonomist George Robert Waterhouse (1810–88) on the topic in 1843, Darwin provided a clear description of his particular understanding of the natural system of classification based on points of resemblance between organisms. "Natural" did not mean for Darwin a reflection of the order of creation, as understood by most naturalists, but rather an arrangement of members of a group that best identified true genealogical relationships (Darwin 1985–, 2:377–78, letter to George Waterhouse, [26 July 1843]). This genealogical understanding of the aims of classification explains why Darwin was

FIGURE 7.2. Étienne Geoffroy Saint-Hilaire (1772–1844), French philosophical naturalist who championed the principle of "unity of composition," that is to say homology as opposed to Cuvier's adaptation. Permission: Wellcome

particularly receptive to Geoffroy Saint-Hilaire's philosophical anatomy, which was based on a transcendental view of the underlying unity in the design of organisms. He easily incorporated the concepts of analogy and homology within his theory of classification but endowed them with new meaning. Within the context of his evolutionary interpretation, homological relations became more than simply tools for description. For Darwin, homology did not depict an ideal plan but actual phylogenetic relationships. Rather than simply being a guide, homology for Darwin was thus an essential tool for identifying evolutionary relationships that linked members of a group. It was also the touchstone for venturing hypotheses about the possible line or lines of descent connecting one species to previously existing forms. Homology, in short, was the foundation of Darwin's theory of classification, and the cirripedes offered him the ideal group on which to test his views.

Darwin's evolutionary interpretation of the meaning of classification also explains why he readily adopted embryology as a methodological tool for revealing homologies. The use of embryological development to reveal systematic relationships emerged from Karl Ernst von Baer (1792–1876) and his magisterial text, *Die Entwickelungsgeschichte der Thiere* (1828–37) (Fig. 7.4). The import of this work for classification was impressed on European naturalists in the late 1830s and early 1840s (Ospovat 1976; Appel 1987; R. J. Richards

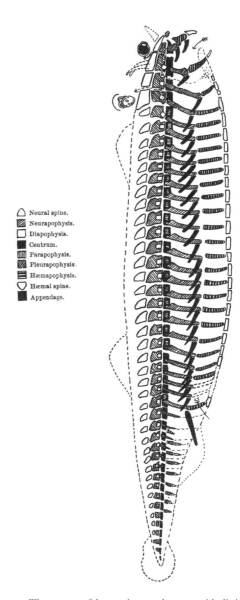

FIGURE 7.3. The concept of the vertebrate archetype, an idealistic Platonic form of the vertebrate, as depicted by Richard Owen. Darwin, who postulated archetypes for both barnacles and orchids (organisms on which he worked), interpreted this as an ancestor. In later years, certainly, this was true also of Owen. From R. Owen, *On the Nature of Limbs: A Discourse* (London: Voorst, 1849)

1992; Rupke 1993). Darwin first came into contact with these ideas as a student in Edinburgh through his association with Grant (Desmond 1984; Sloan 1985). Darwin's own understanding of embryological development, as outlined in his evolutionary "Essay of 1844" (Darwin 1909, 57–255), accorded well with such views. After he began work on the *Beagle* invertebrates in 1846, he was reintroduced to embryological considerations in classification through reading the influential essay on classification (1844) by Henri Milne-Edwards (1800–85).

Like von Baer, Milne-Edwards recognized that comparative embryogenesis could be used to yield information about systematic relationships. Within members of the same branch of the animal kingdom, developmental

FIGURE 7.4. Karl Ernst von Baer (1792–1876), eminent Estonian embryologist and morphologist, whose classic text, *Entwickelungsgeschichte der Thiere* (The Developmental History of Animals), published in two parts (1828, 1837), established principles that were fundamental to the development of comparative embryology. Permission: Wellcome

FIGURE 7.5. Henri Milne-Edwards (1800–85), French invertebrate zoologist whose work *Histoire naturelle des crustacés* (1837–41) was long a standard reference work on crustacea. Milne-Edwards's 1844 essay "Considérations sur quelques principes relatifs à la clasification naturelle des animaux" (Considerations on Some Principles Relative to the Natural Classification of Animals) greatly influenced Darwin's approach to classifying the barnacles. Milne-Edwards's application of Adam Smith's principle of the division of labor to the organic world was also very important for Darwin. Permission: Wellcome

stages generally illustrated an increasing divergence from an early resemblance shared by all members of a class to later features that were characteristic of a particular order, family, genus, and individual species. Milne-Edwards drew from this generalization several principles that became the foundation for many mid-nineteenth-century taxonomists: (1) the most general structures of a class appear earliest in development and can thus be used to establish higher taxonomic affinities; (2) characters shared by organisms reflect the degree of zoological parentage; (3) some organisms, in contrast to the general phenomenon of "progressive" development, exhibit arrested or retrograde development; (4) increasing specialization in embryogenesis illustrates the tendency in higher organisms toward a "division of physiological labor" (a concept he became noted for), and this principle could be used to determine "lowness" and "highness" in particular groups; and (5) embryology, by revealing homological relationships in development, provided an empirical guide on which to base classification ([Richmond] 1989). Darwin was particularly struck by Milne-Edwards's classificatory principles. Indeed, his lengthy abstract of this essay, dated December 1846, opens with the statement: "– This is the most profound paper I have ever seen on Affinities" (CUL DAR 72:117; Ospovat 1981, 174–75). In the monograph on the Cirripedia, begun in 1847, many of the points discussed by Milne-Edwards, combined with Owen's ideas of homology and the archetype, merged with Darwin's own transformist understanding of classification in his treatment of the natural history and systematics of the barnacles (Fig. 7.5).

DARWIN'S SUCCESSES AND "BLUNDERS"

An understanding of the theoretical principles upon which Darwin drew provides the context for assessing the monograph on the Cirripedia. His evolutionary understanding of classification is clearly evident (R. J. Richards 1992, 136–43). For example, his belief that a classification based on homologies established through embryology as well as anatomy would best reveal possible genetic relationships justified his decision to rank the Cirripedia as a separate subclass of the Crustacea. Applying the "embryological criterion of homology," Darwin concluded that the resemblances in larval metamorphosis shared by Crustacea and Cirripedia indicated their community of descent.

This also explains why in both volumes of *A Monograph on the Sub-class Cirripedia*, (1852, 1854) Darwin devoted introductory sections to describing the metamorphosis of cirripedes. As he noted, this was necessary "on account of the great importance of arriving at a correct homological interpretation of the different parts of the mature animal" (Darwin 1852, 25). He based his determination of the archetypal cirripede on Milne-Edwards's model of an archetypal crustacean consisting of twenty-one segments, variously divided in different groups between cephalic, thoracic, and abdominal somites (Milne-Edwards 1844; Appel 1987, 218–19). In barnacles, Darwin identified seventeen of these twenty-one segments, assuming that the four terminal crustacean segments were missing. This archetypal cirripede formed the basis for his making out the anatomical organization of all his barnacle specimens and guided his assessment of taxonomic rank (R. J. Richards 1992) (Fig. 7.6).

In providing a comprehensive phylogenetic treatment of cirripedes, Darwin relied on larval homologies as well as adult morphology, and particularly the number and forms of the "valves" (plaques) forming the shell and the muscles attached to them. In formulating standard nomenclature of these parts, he established terminology that continues to guide systematists (Deutsch 2010). As Newman (1987, 4) has noted, Darwin's monographs "contain a prodigious body of information on the diversity, anatomy, reproduction, geologic chronology and age, and established morphological and taxonomic standards that for the most part have been retained and elaborated upon to the present."

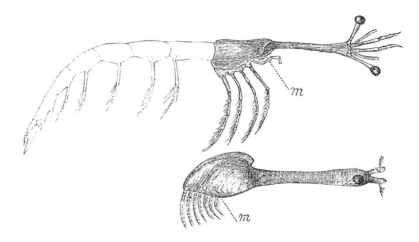

FIGURE 7.6. Woodcut showing the external homologies of a stomatopod crustacean (*above*, taken from Milne-Edwards) with a mature *Lepas* (*below*). As Darwin noted, externally a cirripede consists of "the three anterior segments of the head of a Crustacean, with its anterior end permanently cemented to a surface of attachment, and with its posterior end projecting vertically from it." From C. Darwin, *A Monograph on the Sub-class Cirripedia, with Figures of All the Species. The Lepadidae, or Pedunculated Cirripedes* (London: Ray Society, 1852), 28

The citation of the Royal Medal Darwin received in 1853 pointed to the significant accomplishments of the barnacle monograph, including his description of the metamorphosis of cirripedes and the anatomy of larvae, the use of development to explain homological relations, and his discovery of "new facts" and "promulgation of original views" ([Richmond] 1989, 406-7). Several specific discoveries were singled out. Among this list, two suffice to illustrate the approach Darwin employed. First, his discovery of "complemental" males was not only new but supported his views about the origin of novelty in evolution – in this case, the evolution of sexual dimorphism from a hermaphrodite. Second, his discovery of the cementing apparatus of barnacles indicates how his evolutionary views could generate new knowledge but also sometimes lead him astray.

Darwin's discussion of the sexual systems of cirripedes provides one of the clearest examples of how his transformist views influenced his taxonomic decisions. One of the major characters distinguishing cirripedes from other crustaceans is hermaphroditism. Darwin's discovery of rudimentary males parasitic on the female in the genus *Ibla* provided a clear case of sexual dimorphism in barnacles. But his further finding of minute complemental males attached to a hermaphrodite in both *Ibla* and *Scalpellum* provided evidence of incipient stages in the evolution of separate sexes. As Darwin (1854, 29) noted, "In the series of facts now given, we have one curious illustration more to the many already known, how gradually nature changes from one condition to the other, – in this case from bisexuality to unisexuality." In private, he boasted about this discovery to Hooker, "I never sh^d have made this out, had not my species theory convinced me, that an hermaphrodite species must pass into a bisexual species by insensibly small stages" (Darwin 1985-, 4:140, letter to J. D. Hooker, 10 May 1848). Bolstered by this finding, Darwin looked for similar relations in closely allied genera, and he was duly rewarded by discovering separate sexes in *Alcippe*. This influenced his decision to include this genus within the same family as *Ibla* and *Scalpellum*. Hence, his evolutionary views led him to solve the problem of sexuality in cirripedes (Fig. 7.7).

Darwin was not as fortunate in the interpretation he gave to the discovery of the organs that served to transform a previously mobile barnacle larva into a state of permanent attachment as an adult. Such a system was of particular interest to him from a theoretical standpoint. Thomas Henry Huxley (1825-95) noted that "a Barnacle is, in reality, a Crustacean fixed by its head, and kicking the food into its mouth with its legs" (1857, 238). Dissecting cirripede larvae in the last stage of development prior to attachment to a host, Darwin (1852, 20) observed "two long, rather thick, gut-formed masses, into the anterior ends of which the cement-ducts running from the prehensile antennae could be traced." He came to believe that these were the incipient ovaria and cement glands of the organism, and that the cementing apparatus was homologically equivalent to the ovarian tube. The case appeared to be a striking instance of an organ that had been transformed to perform a new function. Darwin (1854, 151-52) later ventured to suggest the possible "evolution" of this organ system from the ancestral crustacean:

> To conclude with an hypothesis, – those naturalists who believe that all gaps in the chain of nature would be filled up, if the structure of every extinct and existing creature were known, will readily admit, that Cirripedes were once separated by scarcely sensible intervals from some other, now unknown, Crustaceans. Should these intervening forms ever be discovered, I imagine they would prove to be Crustaceans, of not very low rank, with their

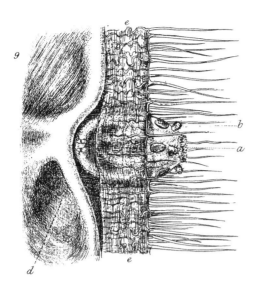

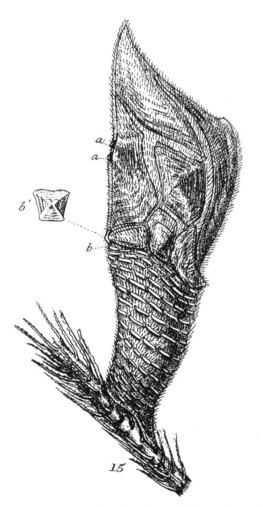

FIGURE 7.7. *Scalpellum vulgare.* Greatly magnified complemental male attached to a hermaphrodite (*top*). Adult magnified, with complemental males attached at "a" (*bottom*). From C. Darwin, *A Monograph on the Sub-class Cirripedia, with Figures of All the Species. The Lepadidae, or Pedunculated Cirripedes* (London: Ray Society 1852), pl. V, figs. 9 and 15

oviducts opening at or near their second pair of antennae, and that their ova escaped, at a period of exuviation, invested with an adhesive substance or tissue, which served to cement them, together, probably, with the exuviae of the parent, to a supporting surface. In Cirripedes, we may suppose the cementing apparatus to have been retained; the parent herself, instead of the exuviae, being cemented down, whereas the ova have come to escape by a new and anomalous course.

Crisp (1983) pointed to Darwin's interpretation of the cement glands as an example in which evolutionary views erroneously influenced his understanding of the phenomena. Certainly he found it difficult to abandon this homology when it was challenged by August David Krohn (1803–91) in 1859 (Krohn 1859). Although Darwin told Charles Lyell, "It is chiefly the interpretation which I put on parts that is so wrong; & not the part which I describe" (Darwin 1985–, 8:396, letter to C. Lyell, 28 [September 1860]), more than interpretation was involved. This case provided his best evidence for descent with modification and was crucial to his picture of how the archetypal cirripede had evolved from the ancestral crustacean.

This was not Darwin's only blunder. The crustacean he chose (from Milne-Edwards) turned out to be an inaccurate representation of the homologies shared with adult cirripedes, and he also included a form, *Proteolepas*, that was later removed from the subclass (Newman 1987, 6; 1993). Deutsch (2010) attributes Darwin's errors to his reliance on determining homology by reference to an ideal archetype. Yet this classificatory system, along with his evolutionary mode of reasoning, was, as we have seen, the foundation for the success Darwin achieved with the barnacle monograph (Innes 2009, 76). Certainly, any errors in interpretation did not detract from the ultimate value of his venture into cirripede systematics and biology.

In addition to its intrinsic value, the barnacle monograph also provided key support for Darwin's theory of evolution, to which he returned on the very day the final volume went to press. "The Cirripedes form a highly varying and difficult group of species to class," he noted in his *Autobiography*, "and my work was of considerable use to me, when I had to discuss in the *Origin of Species* the principles of a natural classification" (Darwin 1958, 118). Indeed, he frequently drew upon cirripedes to illustrate points apart from the link between systematics and descent with modification. One example may suffice. Darwin had long regarded barnacles as a challenge to his view of the value of cross-fertilization. As hermaphrodites with their sexual organs sealed away in a shell, barnacles appeared to be perpetually self-fertilizing. Evidence to the contrary was slim, based solely on his having "scrupulously examined a Balanus, which had had its penis cut off & was imperforate, but in which the ova were impregnated" (Darwin 1985–, 4:179, letter to J. L. R. Agassiz, 22 October 1848). After the barnacle monographs were published, Darwin learned about a naturalist "who watching some shells, saw one protrude its long probosciformed penis, & insert it in the shell of an adjoining individual! So here is a load off my mind. –" (Darwin 1985–, 5:492, 496–97, letter to A. Gray, 29 November [1857]). He mentioned this point in *Origin* to support his view that intercrossing between individuals "gives vigour and fertility to the offspring" (Darwin 1859, 98, 101).

Interestingly, Darwin's description of the sex lives of barnacles not only was noted among naturalists but also entered mid-nineteenth-century popular culture. The barnacle monographs encouraged the middle classes to visit the seaside as a source of edification as well as pleasure. And the "cultural life of barnacles" entered Victorian literature, most notably in Charles Dickens's (1812–70) Barnacle family in *Little Dorrit* (J. Smith 2000). Sex in barnacles indeed continues to excite attention today, part of a genre called "green pornography" (Prairie Starfish Video Productions 2008; Sundance Channel 2009). And fascination with how Darwin's seemingly obsessive quest to unravel the mystery surrounding one "ill-formed little monster" discovered on a South American beach in 1835 could lead to a single-minded eight-year-long study of barnacles continues to beguile scientists and writers to this day (Quammen 1998; Stott 2003; Zelnio 2010).

Essay 8

The Analogy between Artificial and Natural Selection

Bert Theunissen

The "principles of domestication are important for us," Charles Darwin (1868b, 3) wrote in his *Variation of Animals and Plants under Domestication*, to illustrate "that the principle of Selection is all important" in producing evolutionary change (Fig. 8.1). The work of breeders, he explained, might be seen as "an experiment on a gigantic scale" that provided empirical support for his claims about analogous processes in nature. For instance, centuries of artificial selection of small heritable differences (variations) among domestic dogs had produced breeds as different as the bulldog, the greyhound, and the spaniel, each of them specialized to perform a specific task in the human household. In similar fashion, natural selection, by acting on the variations of wild animals and plants, had created the stunning diversity of the living world, in which every species was characterized by adaptations enabling it to survive and reproduce under the circumstances given by its natural surroundings.

Historians and philosophers of science agree that the analogy between artificial and natural selection was a vital element of Darwin's argument in the *Origin of Species* (1859). Philosophers have argued that he deployed the analogy to show that natural selection was a *vera causa*, a true cause, in nature. Darwin proceeded by arguing, first, that domestic races can be produced by sustained selection of individual variations. He then claimed that both these elements, the variations as well as selection, are also present in nature and can in an analogous way, on a much longer time scale, produce new species (Waters 2003). There is some debate on how essential the analogy really was (Ruse 1975a; Gayon 1998). Darwin (1859, 457–59) himself claimed that, even without it, the available evidence spoke convincingly in favor of descent with modification. Nevertheless, he made good use of the analogy in his effort to structure the *Origin* as "one long argument," as he called it.

Besides providing support for his evolutionary views, Darwin repeatedly professed that his study of the stockbreeding literature had also been instrumental in his *discovery* of the principle of natural selection. In 1859, in a letter to Wallace, he wrote: "I came to conclusion that Selection was the principle of change from study of domesticated productions; & then reading Malthus I saw at once how to apply this principle" (Darwin 1985–, 7:279, 6 April 1859). His notebooks and correspondence

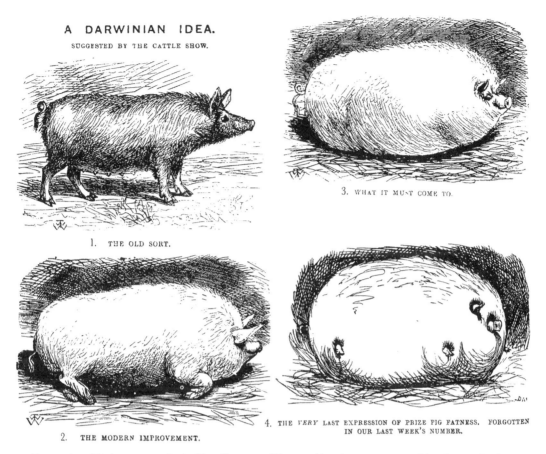

FIGURE 8.1. The importance of animal breeding was well known, although not everyone took it quite as seriously as did Darwin. Nineteenth-century lithograph

confirm that Darwin, after his return from the *Beagle* voyage in 1836, began to read widely in the breeding literature. In early 1838, for instance, he read two pamphlets written by the breeding experts Sir John Sebright and John Wilkinson, and in one of his notebooks he commented: "Whole art of making varieties may be inferred from facts stated" (Barrett et al. 1987, C, 133).

Historians differ, however, as to whether it was domestic breeding or Malthus's essay on population that provided the crucial source of inspiration from which Darwin derived the principle of natural selection. Although it has been argued that it was precisely his reading of Malthus's essay, in September 1838, that triggered Darwin's full appreciation of the mechanism of artificial selection (Herbert 1971), it seems undeniable that Darwin's knowledge of the breeding literature must have prepared him for seeing the evolutionary implications that might be derived from Malthus's claim that more organisms are always being produced than can survive, resulting in a constant struggle for existence (Ruse 1975b).

Whatever the chronology of Darwin's thinking, the idea that his theory of evolution by natural selection sprung to life one evening in September 1838 is, of course, a simplification (Largent 2009a). Darwin may have conceived the basic idea at that time, but he would continue to work on his theory

for some twenty years before he published it, and historians have shown that his understanding of some of the theory's ramifications changed considerably during this period. Yet little has been written until now about Darwin's post-1838 thoughts on domestic breeding. The assumption seems to be that, for Darwin, the matter was by and large settled after he had grasped the analogy between artificial and natural selection.

The situation was more complicated than this, however. Kenneth Waters (2003, 127) wrote that "Darwin lured readers into his new ways of reasoning by introducing this type of reasoning in the uncontroversial setting of breeding techniques." I aim to show that the setting was anything but uncontroversial. It was only with considerable effort that Darwin arrived at the interpretation of how domestic varieties were produced that he presented in the *Origin*. Moreover, he succeeded in establishing the analogy with natural selection only by downplaying the importance of two other breeding techniques – crossing of varieties and inbreeding – that many breeders deemed essential to obtain new breeds. These techniques are still routinely used today, and this calls for a reconsideration of the widespread pedagogic use of the domestic analogy in popular expositions of Darwin's theory.

FIGURE 8.2. Sir John Saunders Sebright, seventh Baronet (1767–1846), was one of the leading breeders of Darwin's day. He discussed the workings of selection in domestic breeding and in natural populations in a pamphlet that Darwin read and annotated. Permission: David Spain

DOMESTIC BREEDING

An instructive starting point for exploring the development of Darwin's views on domestic breeding is provided by the two pamphlets by cattle- and fowl-breeding experts Sebright and Wilkinson that Darwin read in the spring of 1838 (Fig. 8.2). A closer look at these works provides an overview of the main issues that engaged breeders of domestic animals in the first half of the nineteenth century.

The central theme of Sebright's *The Art of Improving the Breeds of Domestic Animals* (1809) was that sustained selection of small heritable differences from generation to generation was essential for successful breeding. First, selection was indispensable to counteract the tendency of domestic varieties to "degenerate." Without selection, Sebright contended, domesticates returned to their "unimproved" natural state or developed more and more defects. Second, selection could be used to enhance a breed's desirable characteristics, such as "the propensity to fatten in cattle, and the fine wool in sheep" (5–6). It is not clear whether Sebright believed that completely new breeds might be created in this way; all he said was that breeds could be greatly improved by selection. A second important tool for the breeder was inbreeding. Here Sebright referred to the impressive results achieved by Robert Bakewell, the Leicestershire breeder whose pioneering experiments in the late eighteenth century had provided the foundation for a new approach to breeding. Sebright praised Bakewell for having shown that inbreeding, provided it was combined with sharp selection, did not necessarily lead to degeneration, as had long been believed. Another important innovation that Bakewell introduced was progeny testing, based on the idea that the quality of breeding stock should ultimately be judged on the basis of the performance of its offspring.

Without inbreeding, Sebright noted, "no one could have been said to be possessed of a particular breed, good or bad" (10). Because desirable new properties always appeared in just one or a few animals, these individuals had to be bred among themselves for some generations in order to "fix" the property and thus to create a new breed. Sebright, however, warned against very close inbreeding, as bad characteristics were as effectively passed on by inbreeding as good ones. It was sustained inbreeding that explained the tendency of domestic varieties to degenerate, he believed. Therefore, constant culling of animals with defects was needed, and even then an occasional outcross with unrelated animals was needed for a breed to retain its vigor. In nature it was through constant outcrossing that degeneration was prevented.

Sebright sounded a cautionary note with respect to a third breeding method: crossing of varieties. He did not object to crossing per se. Breeders of English sheep breeds might safely put Spanish Merino rams to their ewes in order to improve fleece quality, for instance. The difficulties began when two very different breeds were crossed with the objective of combining the good properties of both in a new breed. The first generation offspring of such a cross often looked "tolerable," yet it was "a breed that cannot be continued," as the "mongrels" that were bred from the first generation reverted to the parent breeds or were endowed with "the faults of both" (17–19). Here Sebright was of course describing the phenomena that would only in the early twentieth century be explained as the result of Mendelian dominant-recessive relations and recombination: the variability present in the parent stocks was masked, to a certain extent, by dominance-recessive relations in the first generation, while recombination brought this variability to full light in the second.

In his *Remarks on the Improvement of Cattle* (1820), Wilkinson agreed that selection of heritable variations was a powerful tool for the improvement of domestic breeds. He even came close to stating that new breeds might be made in this way: "The distinction indeed between some [animals improved by selection] and their own particular variety, has scarcely been less, than the distinction between that variety and the whole species" (4–5).

Wilkinson did not share Sebright's pessimistic view of crossing though. He concurred that it was impossible to combine the best properties of two breeds through crossing without some unwanted ones also creeping in. Perfection was unattainable, yet it did not follow that crossing was not useful. As an example, Wilkinson mentioned the widespread use of Shorthorn beef cattle for crossing purposes. Often, in implementing such crosses, farmers had not wanted to lose all the characteristics of their own breed and had created new, intermediary ones. For instance, breeders of Alderney cattle, a breed that produced extremely rich milk in small quantities, had used Shorthorns to obtain a new variety that produced more (slightly less fat) milk and better meat.

Sebright and Wilkinson's pamphlets thus convey the main principles of the art of breeding as they spread rapidly among well-informed breeders in the early 1800s. There was still ample room for discussion though, as their writings also make clear. Furthermore, successful breeding required experience and patience, and attempts to improve a breed often ended in failure. Such experiences continued to fuel the discussion over breeding techniques.

In Britain, for instance, crossing with Merino rams was tried on a wide scale around 1800, but most farmers soon became disappointed with the results. Meat production was important for them, and the Merino's carcass quality was poor. It proved impossible to create a crossed animal that combined excellent meat production with superior fleece quality. Yet many other domestic breeds were successfully crossed with improved varieties. Besides Shorthorn cattle, Bakewell's renowned New Leicester sheep, bred for the production of meat and fat, provides an example. The breed was used widely for crossing, resulting not only in the improvement of local strains but also in new breeds.

Inbreeding as a method for stock improvement remained controversial too. Some breeders saw it as unnatural, and the delicacy and impaired fertility of heavily inbred animals were known to all farmers. Breeding expert William Youatt (1834, 525) wrote that while inbreeding had produced Bakewell's new breeds of cattle and sheep, continued inbreeding had been the cause of their subsequent deterioration.

Darwin was aware of these discussions. In 1839, for instance, he distributed a list of "Questions about the Breeding of Animals" among breeding experts, and the two known respondents to this questionnaire pointed out the ins and outs of crossing and inbreeding in detail to him.

DARWIN ON DOMESTIC BREEDING

It will be clear by now that breeding practices were more diverse and complicated than can be gleaned from Darwin's *Origin*, which presents selection of heritable variations as the predominant method used by stockbreeders. In an earlier attempt at committing his evolutionary thoughts to paper, however, Darwin acknowledged the role of crossing in breeding practices. In a manuscript known as the "Essay of 1844," he wrote:

> When once two or more races are formed ... their crossing becomes a most copious source of new races. When two well-marked races are crossed the offspring in the first generation take more or less after either parent or are quite intermediate between them, or rarely assume characters in some degree new. In the second and several succeeding generations, the offspring are generally found to vary exceedingly.... Much careful selection is requisite to make intermediate or new permanent races: nevertheless crossing has been a most powerful engine, especially with plants. (Darwin 1909, 68–69).

Where animals were concerned, Darwin added a reservation, echoing Sebright's view that "the most skilful agriculturalists now greatly prefer careful selection from a well-established breed, rather than from uncertain cross-bred stocks" (69).

The "Essay of 1844" was published after Darwin's death by his son Francis, and in a footnote to the passages just quoted, the latter remarked: "The effects of crossing is much more strongly stated here than in the *Origin* ... where indeed the opposite point of view is given." The passage in the *Origin* Francis referred to reads:

> Moreover, the possibility of making distinct races by crossing has been greatly exaggerated. There can be no doubt that a race may be modified by occasional crosses, if aided by the careful selection ... but that a race could be obtained nearly intermediate between two extremely different races or species, I can hardly believe.... The offspring from the first cross between two pure breeds is tolerably and sometimes (as I have found with pigeons) extremely uniform ... ; but when these mongrels are crossed one with another for several generations, hardly two of them will be alike, and then the extreme difficulty, or rather utter hopelessness, of the task becomes apparent. Certainly, a breed intermediate between *two very distinct* breeds could not be got without extreme care and long-continued selection; nor can I find a single case on record of a permanent race having been thus formed. (Darwin 1859, 20)

Further on, Darwin added that "all the best breeders are strongly opposed to [crossing], except sometimes amongst closely allied sub-breeds" (31–32).

Thus, in the *Origin*, Darwin downplayed the role of crossing. He did not mention the many crosses involving cattle, horses, and sheep, which according to Wilkinson, Youatt, and other authors had resulted in the creation of new, intermediary breeds. Concluding the chapter on variation under domestication in the *Origin*, Darwin (1859, 43) repeated that "the importance of the crossing of varieties has, I believe, been greatly exaggerated."

Whereas he still saw a significant role for crossing in the "Essay of 1844," Darwin presented inbreeding as a purely detrimental technique in the essay as well as in the *Origin*. In the essay he stated that "injurious consequences follow from long-continued close interbreeding in the same family" (1909, 70–71), and in the *Origin* he repeated this verdict in various formulations. There was no mention of Bakewell and other breeders' positive evaluations of the method as a tool for fixing varieties. Darwin (1859, 43) was adamant that it was selection that was "by far the predominant Power" in creating varieties.

Commenting upon his father's changing appreciation of crossing, Francis Darwin speculated: "His change of opinion may be due to his work on pigeons" (Darwin 1909, 68). He was right.

FIGURE 8.3. Darwin began to study fancy pigeons in 1855, and they would figure strongly in his discussion of artificial selection in the *Origin*. Nineteenth-century lithograph

FANCY PIGEONS

In England, pigeon clubs arose in the eighteenth century (Secord 1981, 1985). By 1850, the fancy had become part of an excited movement for poultry improvement. Darwin took up pigeon breeding as a case study of domestication in 1855 and had a pigeon house built in his garden. He became a member of two pigeon clubs and attended poultry and pigeon shows (Fig. 8.3).

The shows and breed competitions organized by the societies enabled the fanciers to test their breeding skills. The main pigeon breeds had been in existence since the early eighteenth century at the latest. A century later their conformation and characteristic properties had been set down in detail in standards of excellence that were employed by show judges to assess an animal's merit. Understandably, breed constancy or "purity" was of the utmost importance to the breeders. A fancier who bought an expensive bird bred from prize-winning stock expected it to breed true, that is, to beget offspring that approached the breed standard as closely as possible. As poultry journalist William Tegetmeier (1854, 32), Darwin's main adviser on pigeons, explained in his *Profitable Poultry*, crossbreds were "worthless for stock purposes, as they do not breed true to any particular character."

Darwin himself provided a splendid example of the indignation that the surreptitious use of crossing aroused among fanciers. In a letter to Huxley dated 27 November 1859, he wrote:

> For instance I sat one evening in a gin-palace in the Borough amongst a set of Pigeon-fanciers, – when it was hinted that Mr Bult had crossed his Powters with Runts to gain size; & if you had seen the solemn, the mysterious & awful shakes of the head which all the fanciers gave at this scandalous proceeding, you would have

> recognised how little crossing has had to do with improving breeds, & how dangerous for endless generations the process was. (Darwin 1985–, 7:404)

Evidently Darwin accepted the fanciers' protestations of their abhorrence of crossing as truthful. With respect to inbreeding, he probably took Tegetmeier (1854, 18, 24) as his guide, who warned his readers that close inbreeding for more than a few generations, while it might help to preserve special characteristics, resulted in "diseased and weakly offspring."

Darwin's willingness to attach credit to these breeding experts rather than others is understandable: pigeons were his prime example of the power of artificial selection in the *Origin* – he devoted nine of the chapter on domestication's thirty-six pages to them – and if crossing and inbreeding had been unimportant, it was selection alone that had been responsible for the creation of such spectacularly different varieties as the Tumbler, the Pouter, the Jacobin, and the Runt, out of a single wild ancestor, the rock pigeon. (The fact that the rock pigeon could plausibly be argued to have been the only wild ancestor had induced Darwin to choose domestic pigeons for his special study.)

Evidence that crossing of varieties had played a role in the original creation of the main pigeon breeds was not entirely lacking though. Several examples were given in *A Treatise on Domestic Pigeons* published in 1765, a book that Darwin had read. It should be added, however, that nothing conclusive was known about the origin of the main pigeon varieties. In this respect, pigeons were no different from most other domestic breeds, because public record keeping by means of stud books and pedigrees did not develop until the late eighteenth century. Being buried in mystery, the origin of most domesticates could not plead against Darwin's view that crossing had been unimportant (Alter 2007c). In Darwin's own words:

> All that we know, and, in a still stronger degree, all that we do not know, of the history of the great majority of our breeds, even of our more modern breeds, agrees with the view that their production, through the action of unconscious and methodical selection, has been almost insensibly slow. (1868b, 2:244)

The reason was that "the chance will be infinitely small of any record having been preserved of such slow, varying, and insensible changes" (1859, 40).

For a proper understanding of Darwin's perception of breeding practices, it is also important to realize that he was not an experienced breeder. He studied pigeons for some three years, from 1855 until 1858, whereas becoming an experienced practical breeder takes the better part of a lifetime. Obviously Darwin's incursion into pigeon breeding was too short to demonstrate the power of selection or to establish whether artificial selection could create new varieties. Nor did Darwin have the same intentions as the regular pigeon fancier, whose aim it was to breed animals that approached the ideal standard. Darwin was interested in different questions, such as the interfertility of the breeds and the appearance of reversions to the ancestral rock pigeon in crossed animals (which supported his claim of their common descent). For information on breeding methods, he could not but rely on the specialist literature and the breeders' testimonies, meaning that he constantly had to weigh the often contradictory evidence.

He demonstrably struggled with the information thus obtained. For instance, in the letter to Huxley mentioned earlier he wrote:

> I have picked up most by reading really numberless special treatises & *all* Agricultural & Horticultural Journals; but it is work of long years. *The difficulty is to know what to trust.* No one or two statements are worth a farthing, – the facts are so complicated.

The matter of "what to trust" was a recurrent issue in his correspondence. For instance, in a letter to Tegetmeier, he asked: "Can you tell me what sort of man Ferguson the author of a Poultry Book is? Has he had much experience? Is he honest?" And to Hooker he wrote: "Thanks about Beaton.... I can plainly see that he is not to be trusted. He does not well know his own subject of crossing" (Darwin 1985–, 9:38, Darwin to Tegetmeier, 25 February 1861; 9:127, Darwin to Hooker, 14 May 1861). What Darwin presented in the *Origin* reflects the decision he had made with respect to "what to trust": he had decided to trust the pigeon fanciers, whose personal testimonies were fresh in his mind and fully supported the analogy between artificial and natural selection.

Darwin did not leave it at this, however; his grappling with the evidence continued. After 1859 he came across new evidence for successful variety crossing, and in his *Variation* (1868) he felt compelled to slightly shift his position. He conceded that breeding from mongrels was not as impracticable as he had suggested in the *Origin*. And he now also accepted that domestic races had often been intentionally modified by one or two crosses. Yet Darwin did not fundamentally change his mind on the importance of crossing. He remained convinced that only a small number of breeds owed their origin to crosses. It was a technique, moreover, that breeders had mastered only recently. Not until some three-quarters of a century ago had they begun consciously and "methodically," with a specific goal in mind, to modify their breeds, Darwin believed. Before that time they had not worked methodically but "unconsciously," merely selecting what seemed to be the best animals for the propagation of their breeds, without aiming to change them in any particular way (1859, 33–37).

In the *Variation*, Darwin suggested that the main breeds of domesticates had a long history and might have been in existence for thousands of years. He acknowledged that inbreeding might help preserve desirable characters, yet he could not believe that a procedure that affected the fertility and vigor of breeding stock would have played a significant part in the production of the immense variety of domestic breeds. At the end of the book, selection was again presented as the principal method, working slowly over thousands of generations.

DARWIN'S "ENDURING ANALOGY"

Darwin was familiar with the written sources on breeding discussed so far, and he used or could have used them to shore up his rendering of breeding practices in the *Origin*. Historians of animal husbandry and domestic breeding have in recent decades gathered much more information on breeding methods in the eighteenth and nineteenth centuries that Darwin may not have been familiar with (e.g., Trow-Smith 1957, 1959; N. Russell 1986; Wood and Orel 2001; Derry 2003). Their studies confirm that crossing and inbreeding were very much part of the new breeding practices that were developed in the late eighteenth century. Experienced breeders knew that the judicious combination of crossing (which produced new combinations of properties), inbreeding (which helped to fix desirable ones), and selection provided the key to success. An example is the creation of the Thoroughbred in the eighteenth century. This horse breed, which now represents the epitome of a purebred race, was actually a product of prolonged cross-breeding, combined with inbreeding and selection, of Arabians with British breeds.

What was new here was the intention to create breeds that conformed to well-defined standards. There were very few such breeds before the nineteenth century. Contrary to what Darwin believed, most breeds were not of ancient origin. Horses, cattle, sheep, pigs, and fowl were kept in highly variable local strains, which their owners saw no need to breed according to accepted standards. There were no standards; utility was all that mattered. For the same reason, breeders felt no qualms about mixing strains, and they introduced animals from other regions or from abroad whenever they saw fit.

For the pioneers of the breed improvement movement that started in the late eighteenth century, finding ways to reduce the variability of local strains and the concomitant unpredictability of their performance was the main challenge. Thus it was to obtain more uniformity that Bakewell used inbreeding and selection. He probably started his experiments with a highly variable group of animals, possibly different strains, which he crossed until some individuals appeared that combined the characteristics he was looking for. By breeding these in and in, in combination with scrupulous selection, he obtained animals in which all the desired characters were fixed.

As the ideal of uniformity gained prominence in the course of the nineteenth century, breeders became increasingly reluctant to acknowledge their use of crossing, as it might raise the suspicion that their seemingly uniform animals were of mixed origin and would therefore produce variable offspring. The pigeon breeders whom Darwin met at the gin-palace provide an example of such secretiveness. When shows became popular in the mid-nineteenth century, the purity of established breeds became sacrosanct. Purity implied quality, and compromising it became an offense. Standards and stud books had to provide the guarantees that buyers now demanded of the purity of their acquisitions.

Against this background we can better understand the difficulties of interpretation that Darwin faced. Crossing and inbreeding were recommended by some experts and condemned by others, while still others played down or denied their use. It is also clear what made the fancy pigeons different from sheep, pigs, and other utility breeds. They were exceptional in that the main breeds had been in existence and well defined for a comparatively long time. When Darwin entered the fancy, the cult of purity and uniformity had been in place for about a century. While breeders of utility stock were still improving their local strains by means of crosses, Darwin dwelled in circles of fancy pigeon breeders for whom cross-breeding was anathema – or at least couldn't bear the light of day – and who professed that their breeds had a long history of pure descent. Assuming that what was true for pigeons was true for all major breeds, and projecting his reading of contemporary pigeon breeding back onto the past, Darwin concluded that it was sustained selection of small variations that must have created them.

Darwin's deepest thoughts on breeding methods remain inaccessible to us, yet it seems safe to conclude that he devised an interpretation of breeding practices that tried to make the best of the evidence at his disposal while it suited his purposes at the same time. Whereas Darwin professed that he derived the idea of natural section from his study of domesticates, my analysis suggests that the reverse was also true. It was his understanding of natural selection that guided the interpretation of breeding practices that Darwin would ultimately present in the *Origin*. The years he associated with pigeon breeders were of crucial importance for this interpretation.

To this day, Darwin's analogy figures prominently in textbooks and popular works. Even biologists still seem to think that Darwin was basically correct in thinking that it is artificial selection that produces domestic breeds. Yet the practical realities of breeding domesticates do not merely revolve around selection. Breeders still use inbreeding to preserve and enhance desirable properties, and some of them ignore the dangers of taking this too far – the dire consequences of unrestrained inbreeding in popular dog breeds immediately spring to mind. New breeds, fancy varieties as well as utility breeds, are still routinely made by means of the techniques of crossing, inbreeding, and selection. The sheer number and diversity of new domestic varieties that have been created over the past two hundred years suffice to realize that breeders must possess much faster means of producing novel races than mere selection of small chance variations. Crossing produces variation by recombination, and inbreeding and selection help to curb it, so that only the desirable properties remain. Thus a new breed can be created in a restricted number of generations.

Domestic breeding can still be said to illustrate the power of selection, for artificial selection is one of its essential tools. Yet it is not the only one. Darwin's analogy between the production of domestic varieties and species formation in nature belongs to the past and should not be used by modern teachers and popularizers to explain the workings of evolutionary theory.

ACKNOWLEDGMENT

This essay is based on Theunissen 2012.

PLATE I. HMS *Beagle* off the Galapagos. Painted by John Chancellor. Used by permission of his son Gordon Chancellor

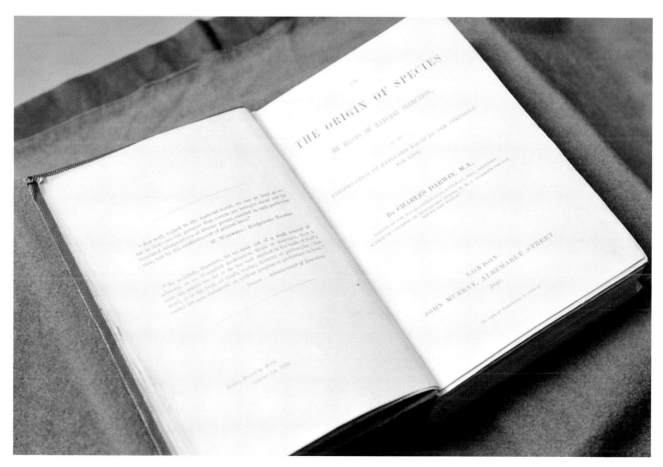

PLATE II. The first edition of the *Origin of Species*, published in 1859. Permission: Christopher Kohler

PLATE III. Christ's College, Cambridge. Charles Darwin was an undergraduate from 1828 to 1831. Tradition has it that Archdeacon William Paley had occupied those very rooms in which Darwin lived. From Rudolf Ackermann, *A History of the University of Cambridge* (London: Harrison and Leigh, 1815)

PLATE IV. The Wedgwood family painted by George Stubbs in 1780. The founder of the pottery works, the original Josiah Wedgwood, is on the bench at the far right. The young man on the horse to the right is his son, also Josiah Wedgwood, the future uncle and father-in-law of Charles Darwin. The young woman in the middle (on the horse) is Darwin's mother, Susannah Wedgwood. From H. E. Litchfield, *Emma Darwin, Wife of Charles Darwin: A Century of Family Letters* (Cambridge: privately printed by Cambridge University Press, 1904)

PLATE V. The "Hottentot Venus" (Sarah Baartman) fascinated Europeans. In the *Descent of Man*, Darwin assured the reader that big bottoms were much admired by certain African tribes. Permission: Wellcome

PLATE VI. Thomas Henry Huxley was Darwin's great supporter. From nineteenth-century *Vanity Fair* cartoon

PLATE VII. Samuel Wilberforce, bishop of Oxford and leader of the High Church faction of the Church of England. (He was the son of William Wilberforce of slavery abolition fame.) From nineteenth-century *Vanity Fair* cartoon

PLATE VIII. The science reformers at the end of the nineteenth century set out deliberately to replace what they regarded as the outmoded ideology of Christianity with a new, naturalistic ideology, based on a progressionist view of life history. Part of the campaign involved building new secular cathedrals, "natural history museums," that mimicked deliberately the edifices of the past. Here we see the Natural History Museum in London at the bottom, very consciously modeled on buildings like the medieval cathedral at Laon in France at the top. In fact Richard Owen, responsible for its construction, probably wanted at least some odor of Christianity to permeate the structure; but it did not take long for the users to appropriate it entirely for their own social ends. Permission: Martin Young

PLATE IX. The *Origin* and almost all of Darwin's other books went through many editions. Permission: Christopher Kohler

PLATE X. "London. Michaelmas term lately over, and the Lord Chancellor sitting in Lincoln's Inn Hall. Implacable November weather. As much mud in the streets as if the waters had but newly retired from the face of the earth, and it would not be wonderful to meet a Megalosaurus, forty feet long or so, waddling like an elephantine lizard up Holborn Hill." The opening lines of *Bleak House* by Charles Dickens. Shown here is a model of Megalosaurus constructed by Benjamin Waterhouse Hawkins for the Crystal Palace and Park in Sydenham, London, which opened in 1854. Today we think the dinosaurs far more agile than the brutes conceived by the Victorians. *Bleak House* appeared first in 1852, seven years before the *Origin*. No wonder Darwin's book was the sensation of the season. Permission: Joe Cain photo

PLATE XI. John Schneider, a theology professor at Calvin College in Michigan, argued that Adam and Eve did not exist literally, and that hence original sin cannot be (as Saint Augustine argued) the consequence of an actual act of disobedience but must be more symbolic of a general human fallibility. For this, he lost his job. Permission: John Schneider

PLATE XII. Empedocles (ca. 490–430 B.C.E.) believed that the four elements – earth, water, air, fire – are constantly mixed by the two divine forces of Love and Strife. This led him to propose a kind of proto-evolutionary theory, with fragmentary parts of bodies appearing naturally and then sometimes these cohering into functional organisms. From the *Nuremberg Chronicle*, printed in the city of that name in 1493 (and hence qualifying as an incunabulum, that is a pre-1500 printed book)

PLATE XIII. Plato (423–347 B.C.E.), the student of Socrates, argued that that blind law and change cannot lead to the functioning of the world as we see it, especially the functioning of organisms. Painting by Raphael, Sistine Chapel, the Vatican

PLATE XIV. Aristotle (384–322 B.C.E.) opposed both the thinking of Plato and of Presocratic materialism. His *Parts of Animals* and *Generation of Animals* originate the systematic, comparative investigation of animals and their development. He defended a natural teleology – development is naturally directed toward producing a well-adapted animal. Painting by Raphael, Sistine Chapel, the Vatican

PLATE XV. The platypus, a mammal that lays eggs, was a constant source of interest and puzzlement in the nineteenth century. Drawn by John Gould, who convinced Darwin that his birds from the Galapagos are of different species. From Gould, *Mammals of Australia* (privately printed, 1845–63)

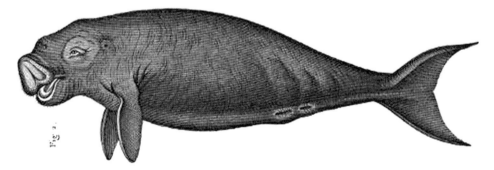

PLATE XVI. Sea cows used to graze vegetation in many shallow coasts around the world, the dugongs, like this one, in the Eastern Hemisphere, the manatees in the Western Hemisphere; all are now endangered. Cuvier classified them with whales, but Linnaeus had correctly placed them next to elephants. The *Beagle* carried a copy of the seventeen-volume *Dictionnaire classique d'histoire naturelle* (Paris, 1822–31) that is the source of this picture. By permission of the Thomas Fisher Rare Book Library, University of Toronto

PLATE XVII. "Dr. Livingstone, I presume!" This imagined drawing of the famous meeting in 1871, in Ujiji, a village on the shores of Lake Tanganyika, between the explorer Henry Stanley and Dr. Livingstone, physician and missionary, shows several different racial types. Darwin thought that such types were a result of sexual selection rather than natural selection. Permission: Wellcome

PLATE XVIII. Ernst Mayr (1904–2005), one of the twentieth century's leading systematists and evolutionists, on his ninetieth birthday. Permission: Frank Sulloway

❧ Essay 9 ❦

The *Origin of Species*

Michael Ruse

In his *Autobiography*, written toward the end of his life, Darwin (1958a, 140) wrote that the *Origin* consists of "one long argument." Let us start there. The argument came in three main parts. In a letter written a year or two after the *Origin* was first published, Darwin outlined his strategy (Fig. 9.1).

> In fact the belief in natural selection must at present be grounded entirely on general considerations. (1) on its being a vera causa, from the struggle for existence; & the certain geological fact that species do somehow change (2) from the analogy of change under domestication by man's selection. (3) & chiefly from this view connecting under an intelligible point of view a host of facts. (Darwin 1985–, 11:433, letter to George Bentham, 22 May 1863)

Note Darwin's use of the term "vera causa." *Verae causae*, or "true causes," were things insisted upon by Isaac Newton, a demand endorsed by those writing on science in Britain in the 1830s. This was just the time when Darwin was thinking creatively about evolution, and it is clear that the young scientist took the exhortation to heart. He wanted to produce an evolutionary theory that would live up to the standards of the best science, meaning the best Newtonian science.

Darwin's authorities, notably the astronomer John F. W. Herschel and the historian and philosopher of science William Whewell, agreed that the best kind of science is based on nature's laws, and that these laws must be shown to be interconnected in an axiom system – premises (which in the case of science make reference to causes) and deduced theorems (Figs. 9.2 and 9.3). In the Newtonian system, we start with the laws of motion and the law incorporating the *vera causa* of gravitational attraction, and then from these follow other laws, about the motions of objects down here on earth and up there in the heavens. But how exactly does one know that one has a true cause? Herschel, who was somewhat of an empiricist, inclined to think that the best evidence is analogical. One experiences a cause oneself – his example was the pull on a piece of string, as one whirls a stone around one's finger – and so one has analogical evidence of the forces of nature – the force pulling the moon in toward the earth as it goes in circles. Whewell was more of a rationalist, inclined to think that the best evidence is incorporated in what he called a "consilience of inductions."

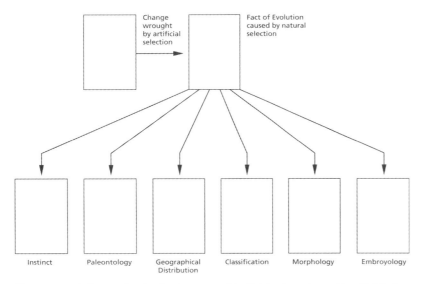

FIGURE 9.1. A diagram showing the three parts of the *Origin*: the analogy with artificial selection, the arguments to natural selection, and the consilience

As in a court of law, one appeals to a wide range of evidence or clues, explaining it through some central cause ("the butler did it"), and then conversely the cause is supported by the evidence. It is a true cause.

We see all three of these points in the letter to Bentham. There is the demand that one put one's cause into a law network; there is the call for a consilience; and there is the demand for an analogy. It is through these requirements that we can understand the argument of the *Origin*.

ARTIFICIAL SELECTION (CHAPTER 1)

The *Origin* opens with the analogy from artificial selection. Whether or not Darwin actually used this analogy in his route to discovery, he intended it in the *Origin* to open the way to natural selection. He wanted to make what goes on in the natural world plausible from the successes of the human world, the world of breeders of animals and plants. It was by no means obvious that Darwin should have used such an analogy. Most people around the middle of the nineteenth century thought that, if anything, breeding and artificial selection disprove evolution. Supposedly, you never get lasting changes in the human world. Alfred Russel Wallace believed this, and the first part of Wallace's essay (the one he sent to Darwin in 1858) argued that the human world has no relevance to the natural world. Darwin disagreed. Partly on the basis of his superior knowledge of the successes of breeders and partly driven by his need to satisfy the empiricist requirement for a *vera causa*, he made much of the world of breeding. Above all, he argued that, from the lowly rock pigeon, breeders have been able to produce the wide variety of fancy pigeon forms that we see today. There is virtually no feature left untouched, no possible form not created.

Darwin did not introduce breeding merely to suggest the possibility of change. It was change of a particular kind, namely toward the production of features that humans find desirable. In other words, artificial selection produces organisms that are adapted to our needs or our fancies. They are explicitly designed to our interests. Of sheep: "It would seem as if they had chalked out upon a wall a form perfect in itself, and then had given it existence" (1859, 31). Also important is the fact that selection can, as it were, go on under the radar. "Methodological selection" is selection done consciously with respect to some desired end. However, selection can and does produce changes that we do not necessarily intend. Continued change (or even trying to keep things stable) makes for inadvertent differences, so that later forms are different from earlier forms, and groups separated simply verge away from each other. "Unconscious selection" thus changes the forms of organisms quite without our knowledge or desire. Obviously, by pointing this out, Darwin intended to prepare the way yet more strongly for a natural form of selection. Differential reproduction can have cumulative effects even without intelligent forethought.

NATURAL SELECTION (CHAPTERS 2-5)

The way prepared, Darwin was now ready to start the argument for the main mechanism of natural selection. First he had to convince the reader that there is widespread variation in the natural world. Without this, obviously no sustained change would be possible. He therefore ranged widely over the world of animals and plants showing that, whenever organisms are looked at in any detail, they exhibit a great deal of variation. Darwin always believed this, but no doubt his extended study of barnacles confirmed his conviction that no two forms are ever exactly identical.

Now came the two crucial chapters. First, Darwin argued that there is always an ongoing struggle for existence. Population pressures put everything under a strain.

> A struggle for existence inevitably follows from the high rate at which all organic beings tend to increase. Every being, which during its natural lifetime produces several eggs or seeds, must suffer destruction during some period of its life, and during some season or occasional year, otherwise, on the principle of geometrical increase, its numbers would quickly become so inordinately great that no country could support the product. Hence, as more individuals are produced than can possibly survive, there must in every case be a struggle for existence, either one individual with another of the same species, or with the individuals of distinct species, or with the physical conditions of life. It is the doctrine of Malthus applied with manifold force to the whole animal and vegetable kingdoms; for in this case there can be no artificial increase of food, and no prudential restraint from

marriage. Although some species may be increasing, more or less rapidly, in numbers, all cannot do so, for the world would not hold them. (63–64)

And so to natural selection. With the struggle, with variations, a differential survival and reproduction follow automatically.

> How will the struggle for existence ... act in regard to variation? Can the principle of selection, which we have seen is so potent in the hands of man, apply in nature? I think we shall see that it can act most effectually. Let it be borne in mind in what an endless number of strange peculiarities our domestic productions, and, in a lesser degree, those under nature, vary; and how strong the hereditary tendency is.... Can it, then, be thought improbable, seeing that variations useful to man have undoubtedly occurred, that other variations useful in some way to each being in the great and complex battle of life, should sometimes occur in the course of thousands of generations? If such do occur, can we doubt (remembering that many more individuals are born than can possibly survive) that individuals having any advantage, however slight, over others, would have the best chance of surviving and of procreating their kind? On the other hand, we may feel sure that any variation in the least degree injurious would be rigidly destroyed. This preservation of favourable variations and the rejection of injurious variations, I call Natural Selection. (80–81)

FIGURE 9.3. William Whewell (1794–1866), Victorian polymath, was instrumental in launching Darwin on his career as a scientist and provided the key methodological principle of a "consilience of inductions." From Mrs. Stair Douglas, *Life and Selections from the Correspondence of William Whewell* (London: Kegan Paul, 1881)

One can hardly say that anything here is particularly formal. However, Darwin was trying as much as possible to offer a law network as demanded by his philosophical mentors. Natural selection is to have the same role as Newtonian gravitation. It is the true cause from which all else stems. Moreover, as with artificial selection, it does not simply bring on change; it works in a particular direction. It makes for design-like features: adaptations.

Along with natural selection, Darwin introduced his subsidiary mechanism of sexual selection. He divided this into two kinds: sexual selection brought about by male combat and sexual selection brought about by female choice. The former produces such things as the antlers of deer and the latter such things as the remarkable tail feathers of some species of bird. It was made clear that the division between natural selection and sexual selection was based on the different intents of human breeders. Some breed for profit, for such things as fleshier cattle and shaggier sheep, and others breed for pleasure, for such things as more vicious fighting cocks and more beautiful birds.

With natural and sexual selection introduced, one might have thought that Darwin would have entered into an extended exposition of these mechanisms in action. You would be disappointed. The treatment is brief and almost entirely hypothetical, with but one quick, casual reference to the possibility of selection working on wolves in parts of the United States. Why was this most crucial discussion virtually nonexistent? Almost certainly because Darwin thought that no direct evidence could be given. "That natural selection will always act

FIGURE 9.2. John F. W. Herschel (1792–1871), astronomer and philosopher of science, inspired Darwin toward a life of science. Permission: Wellcome

with extreme slowness, I fully admit" (108). You simply cannot and do not see selection in action.

More positively, there was another important piece of the picture to be colored in: the principle of divergence. Why do we have so many different forms of organism? Why the range and variety? As is his wont throughout the *Origin*, Darwin introduced the topic by reference to human activity. We aim for distinct forms because they speak to our different needs and whims. The same is true in nature: "[T]he more diversified the descendants from any one species become in structure, constitution, and habits, by so much will they be better enabled to seize on many and widely diversified places in the polity of nature, and so be enabled to increase in numbers" (112). Darwin admitted fully that he saw this divergence explained thanks to one of his favorite metaphors, the division of labor. As appreciated fully by the grandson of one of the greatest heroes of the Industrial Revolution, you get more for your money if different people or things do different tasks. "The advantage of diversification in the inhabitants of the same region is, in fact, the same as that of the physiological division of labour in the organs of the same individual body" (115). Different forms, each with its specialized adaptations, occupy different niches, each with its specialized needs. Thus, thanks to natural selection, the world bears more than it would if every organism were fitted (less efficiently) for every niche. And so, given time, we get the incredible range of forms found on this earth, past and present: "As buds give rise by growth to fresh buds, and these, if vigorous, branch out and overtop on all sides many a feebler branch, so by generation I believe it has been with the great Tree of Life, which fills with its dead and broken branches the crust of the earth, and covers the surface with its ever branching and beautiful ramifications" (130).

Obviously selection cannot work unless there is some way in which advantageous features can be passed on down through the generations. Completing this part of the argument, therefore, Darwin turned naturally to a discussion of heredity and the laws that govern it. He admitted that he was somewhat at sea on this issue because neither he nor anyone else had any real understanding of the underlying principles. We therefore get something of a hodgepodge of possible causes of new variations and the ways in which they get passed on from one generation to the next. It is worth noting that Darwin always subscribed to a form of what is now known as Lamarckism, that is to say the inheritance of acquired characteristics. Although it was always very much secondary to natural selection, there is evidence to suggest that, in later editions of the *Origin*, this supposed mechanism became more important. Overall, from today's perspective, we see a mix of ideas (like Lamarckism) that today we would reject, ideas that today seem important (such as the way in which features can vanish for several generations and then reappear), ideas that are in the thick of discussion today (like correlations of growth), and more. Darwin did not take the issue of heredity casually, but – as he himself would have been the first to admit – what he offered were more problems for the future than solutions for today.

DIFFICULTIES (CHAPTER 6)

Darwin now took some time out from the main argument of the book to talk about some of the difficulties the might have occurred to the reader thus far. Why, for instance, do we rarely see in the fossil record transitional forms between organisms? The imperfection of the record (to be discussed shortly) is one factor, but another is that transitional forms tend to be short-lived, being quite literally neither fish nor fowl. They had to be adapted – otherwise they would have been wiped out completely – but it does not follow that they were particularly well adapted and long lasting. Probably Darwin had Cuvier in mind here with the insistence that transitional forms are impossible. Darwin thought them possible, but he agreed they would not be very stable. Note incidentally that although Darwin necessarily believed that evolution is smooth, in the sense that you go imperceptibly from one form to another (else adaptive focus is lost), he recognized that it could go in fits and spurts.

What about very complex and sophisticated organs like the eye? Could selection possibly have produced them? Darwin took a strategy that is still favored by evolutionists today. He argued that even if the fossil record does not show gradation from simple to complex, among living organisms we find a gradation from the simplest to the most complex. Why therefore should one not have had a similar gradation through time?

What about characteristics that seem to have little purpose or function? Darwin made it clear that there may indeed be many features with little purpose or function, the by-products of growth and so forth, or the legacy of the past. But we should be wary of saying that anything absolutely has no function, because later we could be shown wrong. What was emphasized was that nothing occurs except for the good of the individual. Members of other species might take advantage of features, but selection does not produce features for other species. Implicit here obviously was the counter to a nonnaturalistic account of origins, which might well have the good of the whole foremost.

Finally, Darwin took advantage of this part of his discussion to give his opinion on a matter that has divided biologists from the time of Aristotle down to the present: form versus function.

> It is generally acknowledged that all organic beings have been formed on two great laws – Unity of Type, and the Conditions of Existence. By unity of type is meant that fundamental agreement in structure, which we see in organic beings of the same class, and which is quite independent of their habits of life. On my theory, unity of type is explained by unity of descent. The expression of conditions of existence, so often insisted on by the illustrious Cuvier, is fully embraced by the principle of natural selection. For natural selection acts by either now adapting the varying parts of each being to its organic and inorganic conditions of life; or by having adapted them during

long-past periods of time: the adaptations being aided in some cases by use and disuse, being slightly affected by the direct action of the external conditions of life, and being in all cases subjected to the several laws of growth. Hence, in fact, the law of the Conditions of Existence is the higher law; as it includes, through the inheritance of former adaptations, that of Unity of Type. (206)

INSTINCT AND HYBRIDISM (CHAPTERS 7, 8)

Darwin himself considered the discussion of instinct and hybridism to be more about the theory's difficulties, but already he was starting to segue into the third part of his argument, the consilience of the whole of the life sciences. This is especially so of the discussion of instinct, which illustrates well Darwin's method of discussion – detailed reference to the work of others, discussion of his own work and findings, and all wrapped up under evolution through selection. It is clear that instinct is not some add-on topic but something that Darwin saw as a vital part of the animal world and demanding explanation. Instincts can be as important in survival and reproduction as physical characteristics. They must be the product of natural selection. Ranging across organisms, Darwin looked briefly at cuckoos laying eggs in the nests of others, at the slave-making instincts of ants, and at the cell-making abilities of the honey bee. Darwin accepted fully the idea of what Richard Dawkins has labeled an "extended phenotype," namely that an adaptation does not necessarily have to be part of the organism itself but can be something produced by the organism – as the honeycomb – of benefit to the organism. In the case of the comb, the technique of explanation followed that offered earlier of the eye – it is indeed very complex (Darwin went to some effort to show that it is a marvelous adaptation using the wax very economically), but comb building could readily have come in some stages, as is shown by the less efficient abilities of other bees (the humble bee, for instance) existing today.

Darwin was much interested in the sterility that one finds in insect nests. He did not think the actual production of sterility was a major problem, but obviously it was a challenge as to how it could come through natural selection. One could hardly say that the more sterile a worker, the more offspring it has in the struggle. Drawing as so often on the human-world analogy, Darwin pointed out that breeders can produce features in organisms that never themselves breed, by going back to the family and selecting at that level: "[A] well-flavoured vegetable is cooked, and the individual is destroyed; but the horticulturist sows seeds of the same stock, and confidently expects to get nearly the same variety; breeders of cattle wish the flesh and fat to be well marbled together; the animal has been slaughtered, but the breeder goes with confidence to the same family" (237-38). Likewise in nature. A nest of related individuals in some sense functions as an individual, and selection can have its way: "[A] slight modification of structure, or instinct, correlated with the sterile condition of certain members of the community, has been advantageous to the community: consequently the fertile males and females of the same community flourished, and transmitted to their fertile offspring a tendency to produce sterile members having the same modification" (238).

Following on instinct comes a chapter on hybridization, where again the matter of sterility is a central issue. It is interesting to note that Darwin was reluctant to say that the sterility of hybrids is a direct function of natural selection. This was the assumption of Wallace. He argued that it was better for species that hybrids be sterile because hybrids are probably not that efficient and would be taking up resources needed by full species members. Darwin differed, arguing that the results of hybridization are all over the place, from fertility to sterility, and so no general rule could be pronounced. He offered a simple physiological explanation, namely that sterility simply comes from the breakdown of the unification of two separate systems. But (as became clear in later correspondence with Wallace) underlying his surface argument was Darwin's belief that it could not be of value to individual organisms to produce sterile offspring, even though it might be of benefit to the groups to which they belong. Selection of benefit to a family, a group of related individuals, was one thing. Selection of benefit to a species was another.

THE CONSILIENCE (CHAPTERS 9-13)

Darwin now started to move right into the overall sweep through the life sciences. He turned first to the fossil record. His discussion here was divided into two, with the first part overlapping somewhat with the immediately previous discussion, in that it was still dealing with problems. This was almost forced upon him, because one of the biggest arguments at that time used against evolution was that there seemed to be no transitional forms in the record. Picking up on and continuing earlier argument, Darwin therefore devoted some effort to showing that the record is highly incomplete and that the absence of intermediates is almost to be expected rather than otherwise. Darwin did also take the opportunity to do some rather innovative thinking about absolute time. He argued that from the rate of the denudation of the Weald (the area between the North and South Downs south of London), one can calculate the time since it first started to be eroded away. He put the figure at around 300 million years. As it happens, he was severely criticized by the geologists for this calculation and removed it from later editions of the *Origin*. But it does give some indication of the very large time scale Darwin thought needed for the evolution of organisms here on earth.

Darwin then turned more positively to the fossil record and why it supports a theory such as he was proposing. Most significantly, the further down the record (hence the older) the more difference there is from modern forms. And the older forms are frequently intermediate in some sense between quite different modern forms. "It is a common belief that the more ancient a form is, by so much the more it tends to connect by

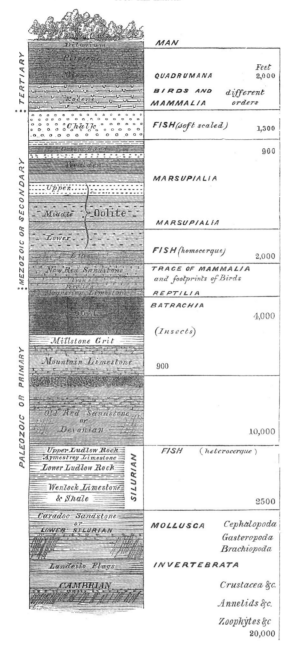

FIGURE 9.4. A depiction of the fossil record as known at the time of the *Origin*. From R. Owen, *Paleontology*, 2nd ed. (Edinburgh: Black, 1861)

than the more ancient; for each new species is formed by having had some advantage in the struggle for life over other and preceding forms" (336–37) (see fig. 9.4).

There is also the case of the embryos. Anticipating years of phylogeny tracing using Haeckel's "biogenetic law," Darwin agreed that ancient forms often look like the embryos of modern forms. This is readily explicable by evolution through natural selection, if we suppose that evolution through time often involves adding on new stages to animal development – the older forms are unchanged and hence (because adult is like embryo) seem embryonic by today's standards.

Chapters on the geographical distribution of organisms came next. As with geology, time was spent showing how seeming difficulties can be explained away. Through the 1850s Darwin had experimented. In order to bolster his belief that much can be understood as the result of life floating to new lands on driftwood or on the feet of migrant birds – or in their bellies! – he ran little experiments, seeing how long seeds can survive in saltwater and the like. He also pointed out that some anomalies can be readily explained. Plants can be more easily transported than animals and that is why they tend to have wider distributions. Other anomalies remain so, waiting explanation – the similarities between plants in New Zealand and the bottom of South America, for instance. But overall, as is fitting for a field that was so intimately connected with Darwin's becoming an evolutionist, the facts of distribution are triumphantly presented as among the strongest pieces of evidence for his theory.

Naturally the denizens of the Galapagos got happy mention (Fig. 9.5). Why should we find such similar (but different) organisms from island to island? Predictably, he notes, "this is just what might have been expected on my view, for the islands are situated so near each other that they would almost certainly receive immigrants from the same original source, or from each other" (400). And most striking of all. How do you explain the fact that the inhabitants of the Galapagos are like the inhabitants of the nearby South American mainland and not like those of Africa, whereas in the case of the Cape Verde islands in the Atlantic off the coast of Africa, the similarities are reversed? "I believe this grand fact can receive no sort of explanation on the ordinary view of independent creation; whereas on the view here maintained, it is obvious that the Galapagos Islands would be likely to receive colonists, whether by occasional means of transport or by formerly continuous land, from America; and the Cape Verde islands from Africa; and that such colonists would be liable to modification; – the principle of inheritance still betraying their original birthplace" (398–99).

Moving on quickly to wrap up his argument, Darwin then dealt in order with a number of topics that had been the focus of much interest by researchers in the half century before the *Origin* was published. What Darwin called "mutual affinities of organic beings" – the basis for classification – were readily explained by his theory. We classify organisms according

some of its characters groups now widely separated from each other"(330). Having given some examples, Darwin agreed that "there is some truth in the remark" (340). Obviously, it follows given natural selection and is "wholly inexplicable on any other view" (342). What about progress, from simple to complex? There seems to be something to this, and again selection explains it. It is a tricky topic: "But in one particular sense the more recent forms must, on my theory, be higher

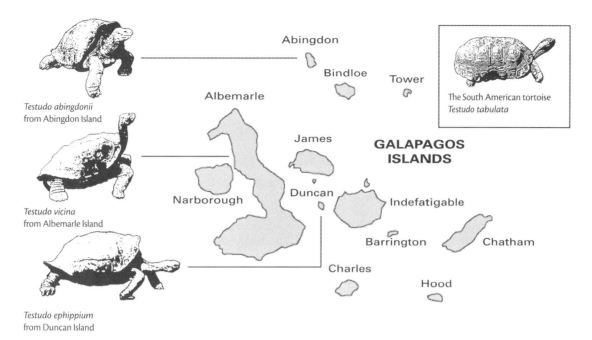

FIGURE 9.5. The distribution of the tortoises on the Galapagos – very similar to each other (and to the mainland form), but slightly different

to their similarities, but the similarities in turn reflect the history of the organisms being so classified. "On the principle of the multiplication and gradual divergence in character of the species descended from a common parent, together with their retention by inheritance of some characters in common, we can understand the excessively complex and radiating affinities by which all the members of the same family or higher group are connected together" (430–31).

Morphology was likewise readily explained on Darwinian principles. "What can be more curious than that the hand of a man, formed for grasping, that of a mole for digging, the leg of the horse, the paddle of the porpoise, and the wing of the bat, should all be constructed on the same pattern, and should include the same bones, in the same relative positions?" (434) Obviously, what we have here is the legacy of evolution from a shared ancestor. Natural selection takes the bones and molds them according to the different needs of their possessors (Fig. 9.6).

Darwin was particularly pleased with his discussion of embryology (Fig. 9.7). Why is it that the embryos of organisms very different as adults are so similar? It is simply because they have a shared ancestor, and Darwin's theory explains how it all comes about. Embryos tend to be protected, and hence, without good reason, selection will not work on them, changing them. Adults, however, have to find their own ways, and so they felt the full force of selection, which causes changes and differences. In a clever move, Darwin swung back to his analogy with artificial selection. He hypothesized that because animal breeders are interested only in the adults, we should find that the young of varieties of domestic animals are considerably more like each other than are the adults. To his delight, this was denied by the breeders themselves. However, on checking dogs and horses, Darwin found that his hypothesis was true. Measurement showed that "puppies had not nearly acquired their full amount of proportional difference." Likewise, "the colts have by no means acquired their full amount of proportional difference" (445).

Finally, Darwin turned to rudimentary organs, for instance the nipples of males and the rudimentary limbs of snakes. This is readily explained on the hypothesis of evolution through selection. However, if you believe in some kind of special creation, then rudimentary organs really make no

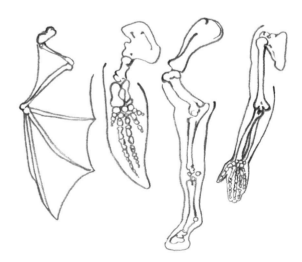

FIGURE 9.6. Homologies: the nonfunctional similarities between the forelimbs of vertebrates

sense at all. Phrases like "for the sake of symmetry" are no genuine explanation.

CONCLUSION (CHAPTER 14)

The argument was now complete. Analogy, law network, consilience. In concluding, because he did not want to be accused of dodging the crucial issue, Darwin made brief reference to our own species. "In the distant future I see open fields for far more important researches. Psychology will be based on a new foundation, that of the necessary acquirement of each mental power and capacity by gradation. Light will be thrown on the origin of man and his history" (488).

And so to the final, famous paragraph:

> It is interesting to contemplate an entangled bank, clothed with many plants of many kinds, with birds singing on the bushes, with various insects flitting about, and with worms crawling through the damp earth, and to reflect that these elaborately constructed forms, so different from each other, and dependent on each other in so complex a manner, have all been produced by laws acting around us. These laws, taken in the largest sense, being Growth with Reproduction; Inheritance which is almost implied by reproduction; Variability from the indirect and direct action of the external conditions of life, and from use and disuse; a Ratio of Increase so high as to lead to a Struggle for Life, and as a consequence to Natural Selection, entailing Divergence of Character and the Extinction of less-improved forms. Thus, from the war of nature, from famine and death, the most exalted object which we are capable of conceiving, namely, the production of the higher animals, directly follows. There is grandeur in this view of life, with its several powers, having been originally breathed into a few forms or into one; and that, whilst this planet has gone cycling on according to the fixed law of gravity, from so simple a beginning endless forms most beautiful and most wonderful have been, and are being, evolved. (489–90)

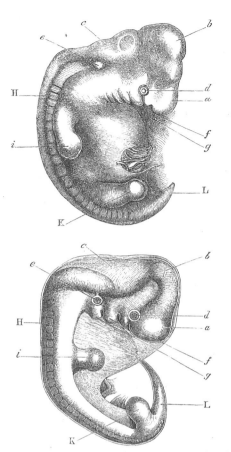

FIGURE 9.7. Thanks to the work of people like Karl Ernst von Baer early in the nineteenth century, the similarities between the embryos of organisms very different as adults were a commonplace by the time Charles Darwin set to work. He used these similarities as strong evidence of shared evolutionary descent. From C. Darwin, *Descent of Man* (London: John Murray, 1871)

Essay 10

Sexual Selection

Richard A. Richards

In his *On the Origin of Species*, Darwin proclaimed that natural selection was the main, but not exclusive mechanism of change. Alongside natural selection, based on the struggle to survive, was sexual selection, based on the struggle to reproduce. Twelve years later, in his two-volume *Descent of Man, and Selection in Relation to Sex,* Darwin focused on sexual selection, devoting part of the first volume and the entire second volume to sexual selection, not just in humans but across biodiversity. He used sexual selection to explain traits not easily explained by natural selection. How, for instance, could natural selection form the peacock's extravagant tail when that tail seemed to be a liability in avoiding predators in the struggle for existence? Perhaps the peacock's tail was instead a way to charm female peahens in the struggle for a mate. The evolutionists who came after Darwin were less inclined to give female choice such an important role, but in the past fifty years there has been a renaissance, and sexual selection now enjoys the enthusiastic support of many who work in the biological and human sciences. It is currently used to explain even more than Darwin had imagined.

Darwin's interest in sexual selection long predated his *Origin*. In his "Sketch of 1842," he included a passage on sexual selection contrasting it with natural selection (Darwin 1909, 10). This passage reappeared in a slightly modified form in his "Essay of 1844":

> Besides this natural means of selection, by which those individuals are preserved, whether in their egg or seed or in their mature state, which are best adapted to the place they fill in nature, there is a second agency at work in most bisexual animals tending to produce the same effect, namely the struggle of the males for the females. These struggles are generally decided by the law of battle; but in the case of birds, apparently, by the charms of their song, by their beauty or their power of courtship. (Darwin 1909, 92–93)

A similar passage was included in the brief and first public statement of his views – the extract of his "big species book" read before the Linnaean Society along with Alfred R. Wallace's paper in 1858 (Darwin and Wallace 1858, 50).

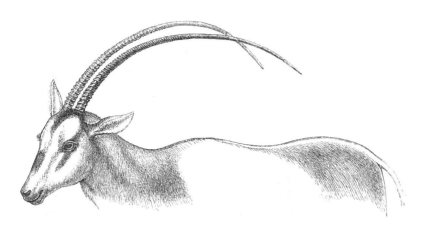

FIGURE 10.1. The exaggerated horns of the antelope is a result of sexual selection through male combat. From C. Darwin, *Descent of Man* (London: John Murray, 1871)

FIGURE 10.2. The beautiful male bird of paradise is a result of sexual selection through female choice. From C. Darwin, *Descent of Man* (London: John Murray, 1871)

A year later Darwin laid out the ideas in more detail in his *Origin*, first in chapter IV on "Natural Selection," devoting a paragraph on each of the main kinds of sexual selection (echoing the preceding passage from the "Essay of 1844"). The first kind was based on the *intra*sexual struggle among males for the "possession of the females." Victory in this battle depends on both the general vigor of the male and the possession of "special weapons" (Fig. 10.1). The second kind was *inter*sexual, where males compete not through battle for dominance, but through their charms to attract the attention of females, who then choose partners on that basis (Darwin 1859, 88–89). This second form of sexual selection was notable and distinct from the first, because competition here depended not just on the capabilities of the males but also on the choices and preferences of females (Fig. 10.2). These two forms of sexual selection could, according to Darwin, explain how males and females who have the same general habits of life, come to "differ in structure, colour, or ornament." They do so because some "individual males have had, in successive generations, some slight advantage over other males, in their weapons, means of defence, or charms; and have transmitted these advantages to their male offspring" (89–90).

Darwin briefly referred to sexual selection in later chapters of the *Origin*, first in "The Laws of Variation"; second in "Difficulties on Theory," in a section on "organs of little apparent importance"; and finally in "Instinct." In his chapter on the laws of variation, he noted that the characters that vary most among the species of a genus are also those parts that vary most between the sexes. He concluded that the parts that become most variable will be modified by *both* natural selection and sexual selection (157–58). Then later, in his discussion of "organs of little apparent importance," he cryptically suggested that "some little light can apparently be thrown on the origin" of the differences among the human races but declined to elaborate on the grounds that, without the details, his reasoning would appear frivolous (199).

THE DESCENT OF MAN, AND SELECTION IN RELATION TO SEX

Darwin followed up on this last suggestion in 1871 in his over 800-page, two-volume *Descent of Man, and Selection in Relation to Sex*. The first 250 pages of volume 1 were devoted to the "descent of man," with the remaining 140 pages devoted to the principles of sexual selection, and the secondary sexual characters of insects. In volume 2, Darwin began with the sexual characters of fishes, amphibians, and reptiles, then turned to birds and mammals, until finally returning to humans in just the last 80 pages.

In light of Darwin's claim in the general introduction that his focus was on humans, all this attention to sexual selection in insects, fishes, amphibians, reptiles, birds, and mammals is puzzling. It was, in part, Darwin's response to the increasing tendency of other evolutionists, Alfred R. Wallace in particular, to use supernatural explanations for human evolution. In the late 1860s, after a turn to spiritualism, Wallace (1869b, 391–94) had increasingly come to doubt the adequacy of natural selection for explaining many distinctive features of the human race: the high intellect, the moral sense, the organs of speech, the hand, the hairless skin, and beauty of the human form. Darwin (1871a, 1:249) agreed with Wallace that natural selection could not explain everything about human evolution. It did not, for instance, seem able to explain the differences among races. But Darwin believed that sexual selection, rather than a supernatural guiding force, could explain racial differences and many of the distinctly human features (see Plate XVII). To show this, a more general and complete treatment of sexual selection would be required.

As we might expect, Darwin's discussion of sexual selection in the *Descent of Man* began with the inability of natural selection to explain certain classes of characters. In his chapter on "the Principles of Sexual Selection," where he first worked out the theoretical basis for sexual selection, Darwin began with the different kinds of sexual dimorphism and the appropriate explanations of each. One kind of sexually dimorphic character was associated with different "habits of life" and was explained by natural selection. Darwin (1871a, 1:254–55) cited species of flies where the females are "blood-suckers" but the males live on flowers, and the barnacle species with the "complemental males" that lack mouths and live like plants on the females. Because the males and females here have different demands placed on them by their habits of life, different characters will be advantageous in the struggle for existence and will be formed by natural selection.

Then there were the dimorphic characters associated with reproduction. Darwin distinguished primary sexual characters – those directly related to reproduction, from secondary sexual characters – those not directly connected (1:253). Some secondary sexual characters graduate into the primary: the organs of sense and locomotion that allow males to find and reach females, and the organs of prehension that allow the male to hold the female securely in copulation. Some females have secondary characters that relate to the nourishment and protection of the young, such as mammary glands and, in marsupials, abdominal sacks (1:254). Darwin thought that these primary and secondary sexually dimorphic characters, like those associated with different habits of life, could be explained by natural selection (1:256). Some dimorphic characters, however, may have been formed by natural selection, such as the organs of sense and locomotion, but perfected by sexual selection and the competition for a mate (1:256–57).

Some sex differences were more "disconnected" from the primary sexual organs: "the greater size, strength and pugnacity of the male, his weapons of offence or means of defense against rivals, his gaudy colouring and various ornaments, his power of song, and other such characters" (1:254). These characters were not required at all for reproduction, but were nonetheless to be explained by sexual selection.

> There are many other structures and instincts which must have been developed through sexual selection – such as the weapons of offence and the means of defence possessed by the males for fighting with and driving away their rivals – their courage and pugnacity – their ornaments of many kinds – their organs for producing vocal or instrumental music – and their glands for emitting odours; most of these latter structure serving only to allure or excite the female. That these characters are the result of sexual and not of ordinary selection is clear, as unarmed, unornamented, or unattractive males would succeed equally well in the battle for life and in leaving a numerous progeny, if better endowed males were not present (1:257–58).

For Darwin then, natural selection explained the characters associated with different habits of life, the primary sex organs, and the secondary sex organs that were most closely connected with reproduction. Sexual selection explained the secondary sex organs and traits least directly connected with reproduction – the ornamentation and special weapons of the males, as well as the perfection of other characters formed by natural selection – organs of sense and locomotion, that gave some males an advantage in the struggle for a mate.

In this chapter on the "Principles of Sexual Selection," Darwin also engaged in four additional projects. First was an analysis of the conditions that lead to sexual selection. If there were an excess of males – more males than females, we should expect a struggle for a mate among the males. Darwin argued that there was *some* excess of males across biodiversity and at various times, but concluded this was no universal law (1:265). The same effect could be achieved in polygamous systems though, where a few males would have great reproductive success, leaving many males – the weaker and less attractive ones – with little or no success. Darwin argued that this would likely explain some, but not all sexual dimorphism. Many monogamous animals also have the relevant kinds of sexual dimorphism (1:265–66). Darwin thought the comparison of closely related species that were polygamous and monogamous was nonetheless informative. The sexes of the polygamous peacock differed greatly, while the sexes of the monogamous guinea fowl differed little. He concluded: "Hence it appears that with birds there often exists a close relation between polygamy and the development of strongly marked sexual differences" (1:270).

Here Darwin also turned to the predictions of sexual selection – what we would expect if sexual selection were operating. The first thing we might expect is that the male would be more modified than the female. He concluded that this was in fact so, because females tended to more closely resemble the young (1:272). Second, the male would likely have stronger passions than the female and be more "eager" in pursuit of potential mates. This Darwin thought to be well confirmed in the observations of fish, alligators, and insects and most strikingly in birds (1:272). Darwin also recognized the occasional development of sex reversal, where the male and female roles are reversed.

> In various classes of animals a few exceptional cases occur, in which the female instead of the male has acquired well pronounced secondary sexual characters, such as brighter colours, greater size, strength, or

pugnacity. With birds ... there has sometimes been a complete transposition of the ordinary characters proper to each sex; the females having become the more eager in courtship, the males remaining comparatively passive, but apparently selecting, as we may infer from the results, the more attractive females. Certain female birds have thus been rendered more highly coloured or otherwise ornamented, as well as more powerful and pugnacious than the males. (1:276)

These apparent exceptions proved the rule: when the sex roles are reversed, and the males become more selective, the patterns of sexual dimorphism are also reversed. For Darwin this confirmed the relation between selection and sexual dimorphism, even though it seemed to contradict broad generalizations about all males and females having certain roles. (Darwin also considered the possibility of simultaneous "double-selection" but concluded that it was less probable than ordinary sexual selection with traits transmitted equally to both sexes; 1:277).

Darwin included in this chapter as well a long complicated section on inheritance, which he took to involve two distinct processes: the transmission of characters from parent to offspring, and the development of characters at various periods of life (1:279). He laid out four main principles. The first principle is that characters that develop at a particular stage of life, and are retained for a particular period, appear at and are retained for similar times in offspring (1:280-81). The second principle is that some characters appear periodically at certain seasons of the year, and these characters reappear in offspring at the same seasons (1:282). Examples here include the appearance of white coloration of arctic creatures in winter and the horns of stags and bright colored feathers in birds during mating seasons. The third principle is that while most characters are inherited equally by both sexes, some characters are inherited by just one of the sexes (1:282-85). A fourth principle is that characters that appear earlier tend to appear in both sexes, and those that appear later tend to appear in only one of the sexes – usually the male (1:285). In the dozen pages that followed Darwin gave evidence for each of these principles, referring to reindeer, antelope, pheasants, ducks, elephants, insects, and more.

Darwin conceded that he did not know the mechanisms underlying these principles, but claimed that these principles of inheritance nonetheless reveal *how* sexual selection – in conjunction with natural selection – worked to produce sexually dimorphic characters. Characters useful *only* in the struggle for a mate would be unlikely to appear before they became useful at sexual maturity, because sexual selection would not preserve them and natural selection would eliminate them. These traits would likely be inherited by offspring at sexual maturity as well – in accordance with the first principle. Similarly, and in accordance with the second principle, any characters that were advantageous in the struggle for a mate only during particular seasons would tend to appear only in the seasons in which they were useful, being eliminated by natural selection from the seasons in which they were of no use. In accordance with the third principle, characters that would be advantageous to just one of the sexes in the struggle for a mate would be *more* likely to appear in offspring only in the sex that found the character advantageous. Natural selection would tend to eliminate it in the sex for which it was of no use. This principle was limited though, because it would be countered by the tendency for characters to be equally transmitted to each of the sexes.

After laying out these general principles of sexual selection, Darwin applied them to humans, explaining how human evolution could have been influenced by sexual selection in the production of both racial and sexual differences. Here, as in other organisms, sexual selection involved both combat among males and choice by females, but Darwin placed a somewhat greater emphasis in humans on female display and male choice (1871a, 2:371-72). He argued that racial differences, particularly those pertaining to skin color and body shape and hair, were the products of both male and female choice – the different tastes and senses of beauty of men and women of the various races (2:368, 381-82).

To explain human sexual differences, Darwin appealed to male combat, as well as female and male choice. Past intrasexual male combat explained the greater size and strength of males, as well as the fact that they were more energetic, pugnacious, competitive, and selfish than females, who were more tender, selfless, and perceptive (2:325-26). Some of Darwin's analysis here is controversial. He argued that, given the differences in the "eminence" of men and women in various activities – poetry, painting, music, history, science and philosophy, the mental powers of men must exceed those of women. This claim is moderated somewhat however, by his consideration of the possibility that the appropriate training of some young women could raise them to the same standard as men (which he thought could then be passed on to their female offspring) (2:327-29).

Lurking through much of Darwin's discussion here are his views on the relations between the different kinds of selection. The analogy he drew with artificial selection and domestic breeding is striking, with respect to the two kinds of selection, based on both victory in battle and aesthetic choice. Breeders sometimes subject their animals to combat (in cockfights for instance) and then "select" the superior individuals to breed on the basis of success in combat. And breeders, when they preserve and breed particular individuals on the basis of favored traits, show how tastes can modify organisms. Just as breeders can make their poultry beautiful through artificial selection, so can the females make male birds beautiful through sexual selection (2:370). It may also be that Darwin was led to his views about sexual selection through this analogy (Ghiselin 1969, 219).

The relation between sexual and natural selection was complicated. First, natural selection was part of the explanation for the laws of inheritance. Because many dimorphic and ornamental characters have costs (the peacock's tail makes it vulnerable to predation), they tend to appear only in those individuals in which they are of use and only at the times they are of use. But, second, natural selection also limited the extremes to which sexually dimorphic ornaments could develop. Extravagant tails and horns, for instance, eventually could become so unwieldy and costly that the advantage conferred in the struggle for a mate is outweighed by the disadvantage conferred in the struggle for existence (Darwin 1871a, 1:278-79).

Natural and sexual selection could also be congruent. Those individuals who are most vigorous would have an advantage in the struggle for survival, but they would also have an advantage in the struggle for a mate. This is true for both males and females. The most vigorous will mate earlier and more often, passing on their vigor and health to more offspring (1:261-63). Sexual selection in this case is reinforcing natural selection.

WALLACE

Darwin's contemporaries did not follow him fully in his appeals to sexual selection. Typically they accepted *intra*sexual selection (the struggle among males in combat for possession of females) but had doubts about *inter*sexual selection (the struggle among males to charm females). Alfred R. Wallace was typical in his critical review of *The Descent of Man*. Here he denied that females had the capacity to make the required distinctions and choices (Wallace 1871, 181). Wallace was particularly skeptical that there was a sufficient constancy of preference to produce distinct characters. He contrasted sexual selection with the more constant effect of natural selection:

> To the agency of natural selection there is no such bar. Each variation is unerringly selected or rejected according as it is useful or the reverse; and as conditions change but slowly, modifications will necessarily be carried on and accumulated till they reach their highest point of efficiency. But how can the individual tastes of hundreds of successive generations of female birds produce any such definite or constant effect? (182)

The similar capriciousness of female tastes in humans implied that sexual selection could *not* explain the differences among the human races (180).

Wallace explained sexual dimorphism and ornamentation in terms of development and natural selection. Bright colors were naturally produced in development by some unknown laws (perhaps in proportion to overall vigor). In those cases where bright colors presented a disadvantage, natural selection would mute those colors, and sometimes result in differential transmission in males and females. Female birds with open nests, for instance, would become "dull-coloured," whereas those with covered or hidden nests would retain their bright colors (181). Other ornamental appendages, "beautifully fleshy tubercles or tentacles, hard spines, beautiful coloured hairs arranged in tufts, brushes, starry clusters, or long pencils," were to be explained similarly – in terms of the unknown laws of development and suppression in the struggle for existence (182).

Wallace seemed to have other objections as well. In his *Darwinism*, he expressed relief at finding an alternative to sexual selection.

> The explanation of almost all the ornaments and colours of birds and insects as having been produced by the perceptions and choice of the females, has ... staggered many evolutionists, but it has been provisionally accepted because it was the only theory that even attempted to explain the facts. It may perhaps be a relief to some of them, as it has to myself, to find that the phenomena can be conceived as dependent on the general laws of development, and on the action of "natural selection." (Wallace 1901, 392)

What these other objections were is not exactly clear, but we can see why intrasexual selection *might have been* regarded with suspicion. First, sexual selection was based on what would have seemed licentious and frivolous – the mere sexual desires and preferences of females, in contrast to the obviously utilitarian and virtuous traits that aided survival. Second, familiar human marriage practices that emphasized male choice might have seemed to disprove the significance of female choice in humans. Third, stereotypes of female capriciousness *might have* reinforced Wallace's worries about the necessary constancy of mating preferences. In spite of these worries, Wallace later came to allow for the possibility of choice in human females. In his 1900 essay "Human Progress: Past and Future," he argued that unrestricted female reproductive choice was a force for bringing about progress. On one hand, "the vicious man, the man of degraded taste or of feeble intellect, will have little chance of finding a wife, and his bad qualities will die out with him." "The most perfect and beautiful in body and mind," on the other hand, "will ... be most sought and therefore most likely to marry early" (Wallace 1900, 507).

AFTER DARWIN

According to a now standard history, for the century after Darwin's *Descent of Man*, sexual selection – female choice in particular – was rejected, ignored, downplayed, or subsumed under a *total* natural selection (Cronin 1991, 243-44; Milam 2010, 147-59). There is some truth to this account in that few of those working in the biological sciences followed Darwin in his use of female choice to explain sexually dimorphic, ornamental traits. In part, this was due to a widespread skepticism that nonhuman females had the cognitive resources to make genuine choices (Milam 2010, 27-36). But female choice was not entirely neglected. First, eugenicists and feminists followed Wallace's lead in his 1900 essay on human progress and looked to female mate choice to improve the human race (Milam 2010, 24-27). Second, once female choice was reconceived mechanistically, as a purely physiological, non-cognitive response, it could be incorporated into a variety of other projects by zoologists, experimental biologists at the American Museum of Natural History, theorists associated with the modern synthesis, and ethologists after World War II (Milam 2010). Instead of thinking about female choice as part of a theory of sexual selection, they were more interested in how female reproductive behavior functioned relative to the maintenance of species identity and genetic diversity within a species, the reproductive isolation associated with speciation, and "epigamically" in the simple stimulation by a male of a female to mate. On this view, the peacock's tail is just the way that peahens identify appropriate mates – members of their own species – and get stimulated to copulate, but *without necessarily preferring one tail over another.*

Intersexual selection based on female choice was also subordinated to a *total* natural selection. On this approach, natural selection is the differential survival *and* reproduction of organisms and is to be measured simply by the changing gene frequencies in a population. A gene might be favored either because of an advantage it confers in survival or because of an advantage it confers in reproduction. Many thinkers of the modern synthesis, including Theodosius Dobzhansky, adopted this approach (B. G. Campbell 2006, 76).

There were exceptions though. Edmund Selous, Julian Huxley (early in his career), and then R. A. Fisher all adopted a Darwinian approach to sexual selection (Milam 2010, 36–48). Fisher first addressed sexual selection in a 1915 paper, and then returned to it in his 1930 book *The Genetical Theory of Natural Selection*. In the paper and book, Fisher did three things. First, he treated female taste as a trait itself, and one that can be given an adaptive explanation (Fisher 1930, 136). Second, he introduced the idea of a *fitness indicator*, whereby some traits, such as brightly colored feathers, are "a fairly good index of natural superiority." A preference for such a trait would be useful because offspring with such traits would be superior (Fisher 1915, 187). Third, Fisher argued that sexual selection could produce a positive feedback and generate a runaway process, whereby a trait and the preference for the trait evolve together in mutual reinforcement (Fisher 1930, 136–37).

After the reprinting of Fisher's 1930 book in 1958, sexual selection and female choice gained new respect. John Maynard-Smith and others began to study the actual behavior of organisms with sexual selection in mind. George C. Williams turned to the theoretical side, speculating about the reasons sex evolved, and how that might be relevant to sexual selection. By this time, female biologists had begun to study female choice in the field, generating the empirical data required to evaluate theoretical claims. There was also increased attention to human mating preferences, most significantly by David Buss. Amotz Zahavi worked out his *handicap principle* based on the idea that sexual ornaments could be good indicators of fitness as long as they had high costs. Finally, Robert Trivers's work on parental investment led to questions about the divergent roles and interests of males and females. Females generally invest more in reproduction than males and therefore have different mating interests (Cronin 1991; G. Miller 2000, 33–67; Milam 2010).

These ideas have become widely adopted by theorists and researchers in the biological and human sciences and have become part of recent comprehensive accounts directed toward the educated public, most notably by evolutionary psychologist Geoffrey Miller in his *The Mating Mind*. Miller (2000, 18) extends sexual selection to explain phenomena far beyond the subset of sexually dimorphic traits that Darwin focused on, into many of the distinctively human cognitive and behavioral traits – things that humans are distinctively good at, such as humor, story telling, gossip, art, music self-consciousness, ornate language, imaginative ideologies, religion, and morality.

SEXUAL SELECTION AND THE RETURN OF DESIGN

There is one theme that Darwin did not explore and that has some philosophical significance. Sexual selection represents a return to design. This is not the design of an omniscient, omnipotent creator, though. It is the design of innumerable individual organisms in the preferences they show for mates. If Darwin and his followers are right, we have "designed" ourselves, on the basis of our preferences and choices. If so, evolution is not *just* the random, unguided processes that operate in the struggle for survival. It is also the product of the senses of beauty and taste that operate in the struggle for a mate.

Essay 11

Darwin and Species

James Mallet

ONE WOULD HAVE thought that, by now, 150 years after the *Origin*, biologists could agree on a single definition of species. Many biologists had indeed begun to settle on the "biological species concept" in the late modern synthesis (1940–70), when new findings in genetics became integrated into evolutionary biology. However, the consensus was short-lived. From the 1980s until the present, it seems not unfair to say that there arose more disagreement than ever before about what species are. How did we get into this situation? And what does it have to do with Darwin? Here, I argue that a series of historical misunderstandings of Darwin's statements in the *Origin* contributed at least in part to the saga of conflict among biologists about species that has yet to be resolved. Today, Darwinian ideas about species are becoming better understood. At long last, the outlines of a new and more robust Darwinian synthesis are becoming evident. This "resynthesis" (as it perhaps should be called) mixes Darwin's original evolutionary ideas about species with evidence from modern molecular and population genetics.

WHAT DID DARWIN MEAN BY SPECIES?

Darwin realized he had convincing proof that species were not created but evolved. But this understanding caused a terminological problem that he had to address in his book. Species were defined in the minds of many of his Creation-educated readers as members of real groups: all members of a species were related by descent, whereas no individual was descended from members of another species. A second idea, which had been promoted especially by the French naturalist Buffon, was that the intersterility of species was a protective mechanism with which species had been endowed by the Creator to maintain their purity (Fig. 11.1). Thus, the famous anatomist Richard Owen, a powerful creationist opponent of Darwin, had given this succinct definition in his 1858 treatise on chimpanzees and orangutans: "an originally distinct creation, maintaining its primitive distinction by obstructive generative peculiarities" (as cited by Huxley 1860, 544).

In order that he could make the argument that species evolved under his theory of "descent with modification," Darwin required a new definition of species. In

FIGURE 11.1. Georges-Louis Leclerc, comte de Buffon (1707–88), one of the leading naturalists of the eighteenth century and suspected of having transformist yearnings. Permission: Wellcome

particular, descent must now be allowed to extend not only within species but also across the species boundary, and ultimately to encompass all living things. Common descent could no longer be used simply as a definition of species. If species evolved, we would also expect hybrid sterility to show evidence of continuous evolution across the species boundary. This terminological problem about species did not, apparently, trouble Darwin greatly (except for the matter of hybrid sterility), and he spent only a little space discussing what he meant by species. Perhaps, as a naturalist, he thought that the existence and nature of species would be self-evident to his readers. Even in later editions of the *Origin*, to which he added a glossary, there is no formal definition of species.

Nonetheless, Darwin did, in my view, clearly indicate what he meant by species, and the conception of species in the *Origin* is now generally recognized by philosophers and historians to have been a useful one for his purpose – that is, to demonstrate evidence for their transmutation (A. O. Lovejoy 1959; Ghiselin 1969; Kottler 1978; Beatty 1985; Ruse 1987; McOuat 1996; Stamos 2006; Kohn 2009; Sloan 2009; Ereshefsky 2010a). Darwin's definition was the simplest that allowed for multiple species to originate from a single ancestral species. One of his clearest short statements on species is in the summary at the end of the *Origin*: "Hereafter we shall be compelled to acknowledge that the only distinction between species and well-marked varieties is, that the latter are known, or believed, to be connected at the present day by intermediate gradations, whereas species were formerly thus connected" (Darwin 1859, 485).

In the *Origin*, Darwin devoted a large portion of chapter 2 ("Variation under Nature") to discussing what species and varieties were, and how difficult they can be to distinguish: "Practically, when a naturalist can unite two forms together by others having intermediate characters, he treats the one as a variety of the other, ranking the most common, but sometimes the one first described, as the species, and the other as the variety" (1859, 47). Of course, it is really a statement about varieties, not species: forms lacking morphological gaps between them are varieties; but a species definition is implicit: forms that have gaps between them are separate species.

But then Darwin immediately qualified this statement and, in doing so, unwittingly confused many of his subsequent readers: "But cases of great difficulty, which I will not here enumerate, sometimes occur.... Hence, in determining whether a form should be ranked as a species or a variety, the opinion of naturalists having sound judgement and wide experience seems the only guide to follow" (47). Many subsequent authors have cited the latter sentence as evidence of Darwin's nihilism about species, while ignoring the foregoing statements. In fact, if the unwary reader fails to concentrate, Darwin seems to tack back and forth, with statements such as: "To sum up, I believe that species come to be tolerably well-defined objects, and do not at any one period present an inextricable chaos of varying and intermediate links" (177), which sounds almost like the opposite of what he has said in chapter 2.

Later, in *The Descent of Man* (Darwin 1871a), there is perhaps a rather clearer statement: "Independently of blending from intercrossing, the complete absence, in a well-investigated region, of varieties linking together any two closely-allied forms, is probably the most important of all the criterions of their specific distinctness." Darwin used this definition to argue that all of the races of Man belong to the same species (1:214–15).[1]

It seems quite clear to me, even from the few excerpts cited here, that Darwin never claimed that species did not exist or were "unreal," however many biologists, philosophers of science, and historians of science would have us believe the converse (a more detailed textual analysis is given in Mallet 2010b). Darwin was not arguing that all species are arbitrary. The statement "the opinion of naturalists having sound judgement and wide experience seems the only guide to follow" did not imply that "naturalists of sound mind" were required to use educated guesswork. Darwin was certainly arguing that species were similar to "varieties" but only up to a point. Species differed from varieties in that they lacked morphological intermediates: there were gaps between them. In his view, Darwin had indicated adequately what he meant by species and then moved on. A more important task, and a major one in the *Origin*, was to show that there were many fuzzy borderline cases – these provided evidence for continuous evolution between species.

---

[1] Darwin's bitter opponent Richard Owen (1859), while deprecating the idea that humans evolved from apelike ancestors, nonetheless categorized species the same way as Darwin did: that all human races belonged to the same species, whereas the orang, chimpanzee, and gorilla were separate species. He did this for very much the same anatomical and morphological reasons as Darwin did – the presence or absence of intermediates. The last thing that Darwin would have wanted was to invent a definition of species that played havoc with existing taxonomy. In the *Origin* he needed only to explain how generally accepted taxonomic species, those recognized by "naturalists having sound judgement," could have evolved. He intentionally adopted the practical methods that most naturalists were using in 1859, while separating his definition of species from the creationist baggage it had carried hitherto.

Good evidence for this interpretation is that the pages containing Darwin's most disputed passages about species in chapter 2 all have the header "DOUBTFUL SPECIES" in the first edition (Mallet 2010b). Darwin was merely showing here that, in *doubtful* cases, it is difficult to tell species from varieties, as a necessary prelude to arguments about how species might evolve. He never intended the message, now widely believed to be Darwin's goal by latter-day readers, that *all* species blended together "in an inextricable chaos of varying and intermediate links."

THE MYTH OF "DARWIN'S FAILURE"

It is an extraordinary paradox that what to Darwin was the most important theme of the *Origin* also became the most doubtful in the minds of his readers, even today. Almost everybody, at least by the mid-twentieth century, agreed that Darwin had written a great book, that he had proved that species had evolved from varieties, and that natural selection was an important process in nature. What they found increasingly hard to accept, however, was that Darwin had understood what species were and had made any effort to explain the origin species from varieties or that natural selection was involved (Mallet 2008). By the time of the "modern synthesis," this view hardened into a dogma that Darwin had completely failed:

> Darwin succeeded in convincing the world of the occurrence of evolution and ... he found (in natural selection) the mechanism that is responsible for evolutionary change and adaptation. It is not nearly so widely recognized that Darwin failed to solve the problem indicated by the title of his work. Although he demonstrated the modification of species in the time dimension, he never seriously attempted a rigorous analysis of the problem of the multiplication of species, the splitting of one species into two. (Mayr 1963, 12)
>
> In retrospect, it is apparent that Darwin's failure ... resulted to a large extent from a misunderstanding of the true nature of species. (Mayr 1963, 14).

Ernst Mayr's critique came from the modern synthesis standpoint of his own "biological species concept," in which species were defined as populations reproductively isolated from one another by "reproductive isolating mechanisms" (Plate XVIII). Darwin, argued Mayr, had not understood the fundamental importance of reproductive isolation in speciation implied by the biological species concept. The undoubted primary reason why Mayr found Darwin's pronouncements on species illogical was that Darwin strenuously argued in his chapter "Hybridism" against the importance of hybrid sterility in providing either a useful definition of species or an explanation of speciation: "It can thus be shown that neither sterility nor fertility affords any certain distinction between species and varieties" (Darwin 1859, 248). To Mayr, in contrast, hybrid sterility and other "isolating mechanisms" were the key differences between species and varieties, and the elucidation of their origin constituted an understanding of speciation. Mayr's isolating mechanism of hybrid sterility was to Darwin an incidental by-product of other evolutionary changes between species, that would not have warranted the term "mechanism" at all, because it could not be explained directly by natural selection.[2] Darwin certainly appreciated how species intersterility and reluctance to mate allowed the coexistence of species and discussed that these traits were strongly associated with what taxonomists recognized as separate species (Mallet 2010b). Yet to Darwin it was the failure of direct natural selection to explain the evolution of hybrid sterility, the fertility of many hybrids between "good species," and the existence of some kinds of infertility within species that forced him to abandon an idea that species could be defined via reproductive isolation.

However, let us not just blame Mayr and the modern synthesis for this misunderstanding. The problems for understanding Darwin's ideas about species go back much further than the middle of the twentieth century. The seeds of the difficulty can be seen even in one of the most positive reviews ever published of the *Origin*, by the very man nicknamed "Darwin's bulldog," Thomas Henry Huxley. While generally complimentary about natural selection and the claim that species arose by evolution, he also wrote, "There is no positive evidence, at present, that any group of animals has, by variation and selective breeding, given rise to another group which was even in the least degree, infertile with the first. Mr. Darwin is perfectly aware of this weak point, and brings forward a multitude of ingenious and important arguments to diminish the force of the objection.... but still, as the case stands, this 'little rift in the lute' is not to be disguised or overlooked" (Huxley 1860, 309). This statement forms a conclusion to a long discussion of Darwin's evidence on the nature of species, with which Huxley largely agrees.

But the "rift in the lute" turned out (and was perhaps intended) to be a very British understatement. Wallace (1889, 152) wrote that it was "one of the greatest, or perhaps we may say the greatest, of all the difficulties in the way of accepting the theory of natural selection as a complete explanation of the origin of species." Much later, "the remarkable difference between varieties and species with respect to fertility when crossed" was seen by a major twentieth-century historian of evolutionary ideas as one of the six major difficulties for the acceptance of Darwinian evolution (A. O. Lovejoy 1959).[3]

---

[2] Mayr, unlike Dobzhansky, agreed with Darwin that there was no evidence that sterility and inviability had evolved via natural selection. Nonetheless, he clearly agreed with Dobzhansky that isolating mechanisms were in some sense adaptive, that they were useful to species as a means of keeping them apart from other species (Mallet 2010a).

[3] This paper was originally published for the first Darwin centenary in 1909 and revised for the centenary of the *Origin* in 1959. To my mind, it remains one of the best pieces of scholarship documenting not only precisely what it was that Darwin and Wallace discovered but also the great mystery of why other biologists such as Thomas Henry Huxley did not discover it, even though many of Darwin's conclusions in retrospect immediately seemed quite obvious. As Huxley (1887, 2:197) himself remarked: "My reflection, when I first made myself master of the central idea of the 'Origin' was, 'How extremely stupid not to have thought of that!'"

The problem arises with the second part of Owen's definition "maintaining its primitive distinction by obstructive generative peculiarities." Darwin had argued vehemently against reproductive isolation as a definition of species because creationists, from Buffon onward, had proposed hybrid sterility to be evidence of the Creator's wisdom. Darwin probably felt he had to show that sterility was not, in fact, a valid definition in order to disabuse his readership of the idea. But to those, like Owen and Huxley, for whom it was key to explain hybrid sterility in a theory of speciation, Darwin's belittling of its importance seemed to duck the issue, while his partial explanation seemed weak. Darwin was very clear that his greatest theory, natural selection, failed to explain hybrid sterility. What then caused it? "The foregoing rules and facts ... appear to me clearly to indicate that the sterility, both of the first crosses and of hybrids is simply incidental or dependent on unknown differences, chiefly in the reproductive systems, of the species which are crossed. The differences being of so peculiar and limited a nature, that in reciprocal crosses between two species the male sexual element in one will often freely act on the female sexual element of the other, but not in reversed direction" (1859, 260–61); "sterility of first crosses and of hybrids ... is not a special endowment, but is incidental on slowly acquired modifications, more especially in the reproductive systems of the forms which are crossed" (272).

In other words, Darwin did not know what caused hybrid sterility, although some causes could be ruled out. However, hybrid sterility was far from universal among species and was so scattered and "incidental" that it seemed most unlikely that it was either a naturally selected adaptation or an attribute provided by God to preserve the purity of species. It must instead be "incidental on slowly acquired modifications" – a by-product of evolutionary divergence in general – or a "pleiotropy," to use today's genetic term. Evidence in correspondence from Darwin to Wallace in 1868 indicates that Darwin himself was dissatisfied with his partial explanation, although it was clearly more of a problem for Huxley and others. Today, whatever their view of Darwin's ideas about speciation, evolutionary biologists accept Darwin's opinion that hybrid sterility is not an adaptation. With hindsight, I believe we can forgive Darwin for not explaining sterility: it is only now that its precise causes are becoming understood. Sterility represents a failure in hybrids of normal beneficial interactions among genes that have diverged in different populations for a sufficiently long time. Although such genes are often popularly referred to as "speciation genes," it is now generally recognized that many differences, and probably most of them, that cause negative interactions in hybrids evolved long after speciation is complete and rarely, if ever, cause species to divide (Coyne and Orr 2004; H. A. Orr 2009).

PHYSIOLOGICAL SPECIES VERSUS MORPHOLOGICAL SPECIES

Huxley argued that Darwin's use of the term "species" was indeed useful but that it was based only on morphology. But to Huxley, another very important difference between species was what he called "physiological." "Physiological species" are those that are unable to interbreed successfully (Huxley 1860, 296). It is not entirely clear whether Huxley invented the term "physiological species," which does not appear in Darwin's writings, or whether he co-opted it from other sources that were generally read then. Regardless of the source of the idea, "physiological species" became a touchstone for an argument that dogged evolutionary biology for the next 150 years. A preference for physiological species over Darwinian morphological species was also the major reason for the later rejection of the Darwinian notion of species, as well as of their origin.

Henry Walter Bates, writing in 1863 about *Heliconius* butterflies, alluded, one assumes, to Huxley's critique of the *Origin* in the following terms: "In the controversy which is being waged among Naturalists, since the publication of the Darwinian theory of the origin of species, it has been rightly said that no proof at present existed of the production of a physiological species, – that is, a form which will not interbreed with the one from which it was derived, although given ample opportunities of doing so, and does not exhibit signs of reverting to its parent form when placed under the same conditions with it." Bates argued that his study of *Heliconius* butterflies in Brazil did, however, "tend to show that a physiological species can be and is produced in nature out of the varieties of a pre-existing closely allied one." Bates purported to show that although *Heliconius melpomene* and *H. thelxiope* hybridize in some places, they also "come into contact in several places where these intermediate examples are unknown, and I never observed them to pair with each other" (Bates 1863, 1:256–62). While today's taxonomy does not, I believe, support Bates's argument in the case of *Heliconius*, this passage clearly shows that Huxley's critique and the need to explain "physiological species" were a topic of discussion at the time.

In his own apparent response to Huxley, Darwin's conception of species (1871a, 214–15) added a physiological dimension: "In determining whether two or more allied forms ought to be ranked as species or varieties, naturalists are practically guided by ... the amount of difference between them, and whether such differences relate to few or many points of structure, and whether they are of physiological importance. ... Even a slight degree of sterility between any two forms when first crossed, or in their offspring, is generally considered as a decisive test of their specific distinctness." This added to but did not preclude Darwin's morphological gap argument, still voiced in the same pages (see above).

In March and April 1868, Alfred Russel Wallace and Charles Darwin corresponded extensively on the subject of hybrid sterility. Wallace asked Darwin whether he could imagine that hybrid sterility arose through natural selection and suggested several possible schemes. Darwin, perhaps exhausted by Wallace's youthful enthusiasm, enlisted his more mathematical son George (then at Cambridge) to help; together they rebutted Wallace's arguments. Darwin wrote

I now think there is about an even chance that Nat. Select. may or not be able to accumulate sterility." Prophetically, he ended the discussion with a prediction that, even so, it would be a source of controversy: "However I will say no more but leave the problem as insoluble, only fearing that it will become a formidable weapon in the hands of the enemies of Nat. Selection" (Darwin 1985–, 16:389, letter from Wallace to Darwin, 8 April 1868). As it turned out, this problem led to opposition also from within the ranks of those who called themselves Darwinists.

In 1886 George Romanes, a correspondent and self-avowed "close student" of Darwin's, published a long and discursive paper to suggest a supposedly new mechanism of how Huxley's physiological species separated by hybrid sterility could come into being, a process he called "physiological selection" (Fig. 11.2). Romanes (1886, 370–71) argued that natural selection was incompetent to cause species to diverge: "The theory of natural selection is not, properly speaking, a theory of the origin of species: it is a theory of the development of adaptive structures.... What we require in a theory of the origin of species is a theory to explain [the origin of] the primary and most constant distinction between species ...[:] comparative sterility towards allied forms, with continued fertility within the varietal form."

FIGURE 11.2. George J. Romanes (1848–94) was a very enthusiastic disciple of Darwin. From Mrs. Romanes, *Life and Letters of George John Romanes* (London: Longmans, Green, 1896)

Romanes agreed with another Darwin critic (Wagner 1868, 1873) that if populations were geographically isolated, divergent variations would not be swamped by intercrossing and so could diverge to form separate species. However, Romanes did not believe that all speciation could be due to geographical isolation; physiological selection, in his view could have the same effect of preventing gene flow. According to Romanes (1886, 370–71), if a variation (or mutation) occurs but has no effect within an emerging variety, "such that the reproductive system, while showing some degree of sterility with the parent form, continues to be fertile within the limits of the varietal form, in this case the variation would neither be swamped by intercrossing, nor would it die out on account of sterility. On the contrary, the variation would be perpetuated with more certainty than could a variation of any other kind."

back to a second enquiry: "Let me first say that no man could have more earnestly wished for the success of N. selection in regard to sterility, than I did; & when I considered a general statement, (as in your last note) I always felt sure it could be worked out, but always failed in detail. The cause being as I believe, that natural selection cannot effect what is not good for the individual" (Darwin 1985–, 16:374, letter to Wallace, 6 April 1868). Wallace did, however, touch upon one likely argument that Darwin could not refute, that "disinclination to cross" could be effected by natural selection. Darwin again: "I know of no ghost of a fact supporting belief that disinclination to cross accompanies sterility. It cannot hold with plants, or the lower fixed aquatic animals. I saw clearly what an immense aid this would be, but gave it up. Disinclination to cross seems to have been independently acquired probably by nat. selection; & I do not see why it would not have sufficed to have prevented incipient species from blending to have simply increased sexual disinclination to cross."

Wallace wrote back: "I am sorry you should have given yourself the trouble to answer my ideas on Sterility – If you are not convinced, I have little doubt but that I am wrong; and in fact I was only half-convinced by my own arguments, – and

Wallace (1886), recognizing the similarity of "physiological selection" to earlier ideas he had himself communicated to Darwin for the evolution of intersterility by means of natural selection, wrote a number of articles rebutting Romanes' suggestions. He returned to the theme in his major work intended to update and promote Darwinism thirty years after the *Origin* (Wallace 1889). Perhaps his most cogent criticism was that Romanes had merely asserted the importance of physiological selection (clearly evident also from the Romanes quotation reproduced in the preceding paragraph); he had failed to propose a convincing mechanism whereby it would occur or to provide any empirical evidence for its operation. Wallace (1889, 181–83) introduced a mathematical argument to show that Romanes' assertion did not work, showing that eventually a new and scarcer variety that produced infertile hybrids with the commoner "wild-type" would die out. The argument assumes complete hybrid sterility but works as well with partial sterility.

Nonetheless, Wallace himself reiterated his 1868 ideas in a lengthy and rather diffuse section of nearly 1,800 words earlier in the same chapter. He argued that hybrid sterility could be explained by means of natural selection. This passage is today difficult to interpret, and, as if anticipating the befuddlement of his readers, Wallace used a still rather lengthy footnote (about 850 words) to elaborate a "briefer exposition ..., in a series of propositions." These propositions were almost identical to those in his 1868 letter to Darwin.

Wallace's (1889, 175–78) first idea was that hybrid sterility might arise "in correlation with the different modes of life and the slight external or internal peculiarities that exist between them." If so, sterility would be a by-product of the divergent environments or inherited adaptive change of two emerging varieties and could be stable to swamping. This can be interpreted in today's terms as a pleiotropy argument: a selective adaptation to conditions of life can evolve that outweighs the indirect or pleiotropic disadvantage of the negative side effects of the same genes on hybrid sterility. Wallace essentially reiterated Darwin's (1859) hypothesis for the evolution of hybrid sterility, and this is the one most strongly supported today.

Wallace's second, and major argument for the evolution of sterility should probably be interpreted as a kind of selection on groups rather than Darwinian natural selection on individuals. If in one part of the range of a species, diverging into two varieties under natural selection, hybrids happened to be more sterile, while in another part hybrids among the same two emerging varieties were somewhat less sterile, Wallace (1889, 175) claimed that forms showing greater hybrid sterility would increase more rapidly as a result of their greater genetic purity owing to better adaptation to conditions causing the emergence of the divergent varieties in the first place. This is a tricky argument to make, as it is directly contradicted within each region by the very same Darwinian argument that he used against Romanes, outlined later in the same chapter. It relies on the idea that populations with higher sterility leave more offspring overall (because of the greater purity and better adaptation to local conditions) than populations with lower sterility (and therefore lower purity). Biologists today accept that situations under which interpopulation selection or group selection of this kind outweighs a countervailing force of natural selection within populations will be rare. If we view sterility for what it is, a problem for the individual, we can imagine that sometimes a beneficial adaptation that also causes sterility will evolve *in spite of* sterility, because the benefits of the adaptation outweigh the loss of offspring. This could lead to greater hybrid sterility as a by-product (Wallace's first hypothesis). But by arguing for hybrid sterility as a direct potential advantage for populations, I think that it is correct to say (Kottler 1985; N. A. Johnson 2008) that Wallace was falling into the trap of naïve group selectionism (D. S. Wilson and E. O. Wilson 2007).

A third suggestion by Wallace, again following on from the earlier correspondence with Darwin, was that new varieties would show a correlated "disinclination to pair." Wallace (1889, 172–73, 175–76) argued here that adaptation to different modes of life would also bring about a reduction in tendency to pair between divergent varieties, perhaps simply because organisms specializing in different resources met less often. Darwin, as we have seen, argued that there was no evidence for this. In modern terms, this is arguing for what has been lightheartedly termed a "magic trait" – that is, a pleiotropic effect that automatically aids speciation (Gavrilets 2004). Pleiotropic effects of ecological adaptation on mate choice are today thought to provide an important route to ecological speciation (Drès and Mallet 2002; Hendry, Nosil, and Rieseberg 2007).

There is a fourth and final suggestion, which Wallace could have made in 1868 or 1889 but apparently did not. As Darwin had briefly mentioned in his letter of reply to Wallace, "disinclination to pair" with individuals of a different type would seem likely to be enhanced by natural selection because it would reduce the number of useless offspring that might become sterile. This argument was revived again by Theodosius Dobzhansky in 1940 and became variously known as "reinforcement" (Blair 1955; Levin 1970; Butlin 1985), or the "Wallace effect" (Grant 1966; Murray 1972). Today reinforcement is generally accepted as a possibly common means whereby reproductive isolation is acquired via natural selection (Coyne and Orr 2004; N.A. Johnson 2008).

POST-MENDELIAN IDEAS OF PHYSIOLOGICAL SPECIES

By around the turn of the century, many people were again beginning to argue, in contrast to Darwin and Wallace, that species should be defined physiologically – that is, by means of their reproductive isolation (Cockerell 1897; Petersen 1903; Poulton 1904; K. Jordan 1905). With the rediscovery of Mendelian heredity, William Bateson and the Mendelians approached the understanding of species from a new, experimental genetics viewpoint; sterility could now be investigated in the laboratory. Bateson reiterated the argument that Darwin's definition of species ignored their most important feature, their physiological tendency to produce sterile hybrids (W. Bateson 1913, 1922). Darwin's was an incomplete theory of speciation because it could not explain this important "specificity" of species in nature, as Bateson called it. By 1926, the Russian geneticist Sergei Chetverikov had argued that "the real source of speciation, the real cause of the origin of species is not selection, but [reproductive] isolation" (quoted in Krementsov 1994, 41).

Russian entomologists and geneticists such as Wilhelm Petersen, Sergei Chetverikov, and A. P. Semenov-Tian'-Shanskii, as well as workers in the United States and Europe, who all supported these new ideas on species, were undoubtedly strong influences on the young entomologist and later geneticist Theodosius Dobzhansky (Krementsov 1994). After emigrating to the United States, Dobzhansky wrote the most widely read treatise of this period on the origin of species. This work blended genetic and Darwinian ideas about speciation for the first time and supported the idea of species being definable via "physiological isolating mechanisms": "When such mechanisms have developed [between two diverging races], and the prevention of interbreeding is more or less complete, we are dealing with separate species" (Dobzhansky 1937, 63).

In an important section, "The Origin of Isolation," Dobzhansky argued that hybrid sterility and sexual or psychological isolation could reinforce one another, and that further isolation could in some circumstances be adaptive. As applied to hybrid sterility and inviability, this again appears to be an example of naïve group selectionism (see especially 257–58), even though in the same chapter he also accepted Darwin's argument that hybrid sterility was often a by-product of divergent evolution rather than a directly selected influence on speciation. Dobzhansky was promoting a Darwinian approach to the understanding of speciation, and he seems to have been careful to avoid a direct critique of Darwin's own view of species, which of course differed from his own.

Ernst Mayr (1942, 1963) adopted Dobzhansky's reproductive isolation definition of species, and renamed it "the biological species concept." As we have seen, he did not shy away from arguing that this "new" idea of species was very different from Darwin's and that it demanded an entirely new view of the origin of species. In the opinion of Dobzhansky and Mayr, this new view of species and speciation represented the modern synthesis of Darwinism and Mendelian genetics.

SPECIES CONCEPTS TODAY

We have seen how Darwin failed to convince Huxley, his chief supporter, that it was best not to define species via reproductive isolation. Huxley's invention of the term "physiological species" led first to a resurgence and finally, by the 1960s, an almost complete acceptance of the idea that the fundamental nature of species was reproductive isolation – the very idea that Darwin had tried to disprove. Given that opinions about the importance of reproductive isolation differed, this treatment of species as if they were fundamentally and physiologically distinct from varieties led to a search for alternative fundamental concepts to define species. According to one concept, a phylogenetic species is a distinct form that retains stable morphological or genetic differences, whether or not it is reproductively isolated (Cracraft 1989). A recent version of this idea employs Bayesian statistical analyses of genealogical coalescence to determine the presence of separate, phylogenetic species in a set of individual genomic sequence data. Under this idea of species, one must infer from the genetic data at least a minimal time of separation between a pair of populations to classify them as separate species (Yang and Rannala 2010).

THE VALIDITY OF A DARWINIAN NOTION OF SPECIES IN 2013

Another view, however, is that a Darwinian delimitation of species still today has validity: species are separate "genotypic clusters" when considered in a molecular genetic sense (Mallet 1995). Arguing for two species on the basis of genetic data is equivalent to arguing that there are two sets of individuals each coming from a population with gene frequencies that may differ. In other words, one needs only to disprove the null hypothesis that there is a single population in the array of individual genetic or genomic data in order to prove that the presence of two populations is a better hypothesis; and the method can be extended to multiple populations. If we plot the distribution of individuals along axes representing multilocus gene frequencies, the distribution will be bimodal if there are two species, or single peaked if there is only one. Data can be treated statistically by means of a Bayesian Markov-Chain Monte-Carlo approach (Pritchard, Stephens, and Donnelly 2000; Huelsenbeck and Andolfatto 2007). This procedure is called an "assignment test" because it determines the appropriate number of distinct populations into which to assign each of the genotyped individuals in a sample.

Gene frequencies may of course differ if populations are spatially isolated without necessarily implying speciation, but if distinguishable populations occur together in the same region and yet retain differences at multiple loci, the two populations will generally be accepted to be different species. Intermediates (or hybrids) may occur, but provided they are rare in areas of overlap, these populations can be considered separately delimited species.

Assignment tests are useful in delimiting cryptic species in many groups, such as flowering plants (Larson et al. 2010; Zeng et al. 2010), corals (Pinzón and LaJeunesse 2011), butterflies (Dasmahapatra et al. 2010), or primates such as mouse lemurs (Weisrock et al. 2010). These methods are also useful for identifying genetically distinguishable ecological taxa normally considered below the species level in taxa such as aphids (Peccoud et al. 2009) or social-group forms of mammals such as the orca (killer whale) (Hoelzel et al. 2007). In Darwinian terms, such ecological races represent exactly the "doubtful cases" that Darwin used to suggest that species evolved from varieties.

Today, it seems, we have come full circle from a general disregard for Darwin's view of species to using statistical methods employing a recognizable Darwinian notion of species, although today's methods tend to use genetic rather than purely morphological data. Physiological and biological concepts of species can be seen as explanations for the scarcity of intermediates between species, and so genotypic bimodality makes as much practical sense to those who support phylogenetic or biological concepts of species as it does to those who feel that Darwin was correct about species all along. Perhaps now "we shall at least be freed from the vain search for the undiscovered and undiscoverable essence of the term species" (Darwin 1859, 485). We shall see.

ESSAY 12

Darwin and Heredity

Robert Olby

Charles Darwin's position on the subject of heredity is not the easiest of tasks to establish. Not only was he working on the subject in the shadow of Lamarck's well-known version of the inheritance of acquired characters, but his own views were crucially shaped by what to him were the more important elements in the mechanism he was formulating for the transmutation of species. He never wrote a book specifically on heredity. In his *Origin of Species* there is not even a chapter so entitled. How unlike his cousin Francis Galton, who was to write several books on the subject, and the philosopher Herbert Spencer, who introduced to British biologists the term "heredity" in chapter 8 of his *Principles of Biology* (1864, vol. 1). Compare this with Darwin's treatment of variation. This topic is the subject of three chapters of the *Origin* – the first, second, and fifth. Six years later Darwin published his magnum opus, the two-volume *Variation of Animals and Plants under Domestication*. Here variation is the theme, but this time three chapters are included on heredity: chapter 12 on inheritance, 13 on reversion, and 14 on fixedness of character. Related topics are in chapter 17 on effects of crossing and 19 on hybridism. Of the remaining twenty-five chapters, one is given to his hypothesis of pangenesis, this being Darwin's attempt at a hypothesis that brings together heredity, variation, and other aspects of the broad field of "generation."

As for manuscripts, we find in his Transmutation Notebooks beginning in 1837 frequent notes of sources on inheritance, and his discussion of these sources can be found in the chapters on variation in the *Origin*. The nature of heredity was evidently of considerable concern to him from the early notebooks right on to pangenesis in 1868. How, then, are we to understand Darwin's study and theorizing on heredity in relation to his work as an evolutionist? Was it a "subfield" that he explored "more with an eye to formulating evolutionary explanations than to solving the internal problems of the field" (Glick and Kohn 1996, 47)? If so, we may be able to identify ways in which he exploited and interpreted selected data in his efforts to find a viable hypothesis for the transmutation of species. For if species are to evolve, variations not only must occur but must be heritable, and the strength of that heritability must be lasting. At the same time, he was seeking heritable variations that would show adaptation. Sudden and large variations, verging at times on monstrosities (macromutations), were unlikely to fulfill this requirement. The kind of heritability that he needed was that associated with slight differences in physical characters, physiological constitution, and habits of life.

THE PROBLEM WITH HEREDITY

When Darwin opened his first Notebook on the Transmutation of Species in 1837, inheritance in its scientific sense was hardly considered a subject worthy of treatment in its own right. The first major work to give it that treatment appeared in 1847. It was volume 1 of Prosper Lucas's *Traite philosophique et physiologique de l'hérédité naturelle*. Twelve years later Darwin (1859, 13) disarmed the skeptical reader with his candor by acknowledging, "The laws of inheritance are quite unknown"; and a further nine years later, after giving a long list of human characteristics that are inherited, he confessed the difficulty he experienced "in attempting to reduce these various facts to any rule or law" (Darwin 1868b, 2:16). It was not the case, however, that he avoided drawing any broad conclusions about the nature of inheritance or resisted formulating a hypothesis aimed at accounting for the data. Why, then, was heredity a problem for the great transmutationist?

Darwin was not alone. It was a problem for anyone interested in the subject in mid-nineteenth-century Britain and elsewhere. Consider, for instance, Darwin's contemporary, the philosopher and former physiologist George Henry Lewes. In 1859 he addressed this question with the following proverbial sayings: "That boy is the very image of his mother!" Or : "That boy is remarkably unlike his parents!" And again: "He has his father's talent, or his mother's sharpness." "The sons of remarkable men are generally dunces," and "Men of genius have remarkable mothers." How, he asked, should we understand such contradictory statements? It is as if inheritance were "very much a matter of chance, and that what we usually suppose to be evidence of hereditary transmission, is really nothing more than coincidence." Refusing this conclusion, Lewes found an explanation in the varied relative influence of the two parents in bisexual reproduction. If the "paternal influence is not counteracted," he explained, " we see it transmitted. Hence the common remark, 'talent runs in families.'" And he concluded that *both* parents are *always* represented in the offspring; and although the male influence is sometimes seen to preponderate in one direction, and the female in another, yet this direction is by no means constant, is often reversed, and admits of no absolute reduction to a known formula (Lewes 1859, in Olby 1985, 173) (Fig. 12.1).

But Darwin was aware just how much more complex was the matter than this. No one, he claimed, could say "why the child often reverts in certain characters to its grandfather or grandmother or other much more remote ancestor; why a peculiarity is often transmitted from one sex to both sexes, or to one sex alone " (Darwin 1859, 13). Darwin's list of mysteries runs on to the next page. Contrast that with the Frenchman, Prosper Lucas, who, taking a high-level view of the subject, considered the data of inheritance as the results of the force of

FIGURE 12.1. G. H. Lewes (1817–78) was an English philosopher and critic (and common-law husband of the novelist George Eliot). Nineteenth-century lithograph

imitation, governed by the law of *hérédité* opposing the force of *invention*, this being governed by the law of *inéité*. Invention leads to variability, but heredity restricts such effects within the limits of the species. This hardly solved the problem of heredity, and, as Lucas presented it, Nature, while permitting variation, imposes strict limits to the departure of variations from the species norm. No recipe for a transmutationist!

THE STRENGTH OF HEREDITY

Already on page 5 of the *Origin of Species,* Darwin introduces the reader to "the strong principle of inheritance," and he went on to defend this principle against the skepticism of "theoretical writers." Foremost on his mind was the well-known historian Henry T. Buckle, who had complained about the manner in which claims about inheritance of mental attributes were made, "the usual course being for writers to collect instances of some mental peculiarity to be found in a parent and in his child, and then to infer that the peculiarity was bequeathed. By this mode of reasoning," argued Buckle (1857–61, 1:161), "we might demonstrate any proposition." Darwin (1859, 13) responded forcefully:

> When a deviation appears not unfrequently, and we see it in the father and the child, we cannot tell whether it may not be due to some original cause acting on both; but when, amongst individuals, exposed to apparently the same conditions, any very rare deviation, due to some extraordinary combination of circumstances, appears in the parent—one among several million individuals—and

it reappears in the child, the mere doctrine of chances almost compels us to attribute its reappearance to inheritance. Every one must have heard of cases of albinism, prickly skin [ichtheosis], hairy bodies, & etc., appearing in several members of the same family. If strange and rare deviations of structure are truly inherited, less commoner [sic] deviations may be freely admitted to be inheritable. Perhaps the correct way of viewing the whole subject, would be, to look at the inheritance of every character whatever as the rule, and non-inheritance as the anomaly.

Nine years later Darwin returned to scold writers who, not having "attended to natural history, have attempted to show that the force of inheritance has been much exaggerated. The breeders of animals would smile at such simplicity; and if they condescended to make any answer, might ask what would be the chance of winning a prize if two inferior animals were paired together? Why," he asked, "have pedigrees been scrupulously kept and published of the Short-horned cattle, and more recently of the Hereford breed? Is it an illusion that these recently improved animals safely transmit their excellent qualities even when crossed with other breeds?" Then he wiped the floor with the skeptics, responding: "Hard cash paid down, over and over again, is an excellent test of inherited superiority (Darwin 1868b, 2:3)

BLENDING HEREDITY

Although Darwin emphasized the strength of heredity, he also admitted "how feeble, capricious or deficient the power of inheritance sometimes is" (Darwin 1868b, 2:17). To explore this subject, he turned to the results of experiments in cross-breeding and hybridization. Would the characters of the dissimilar parents be "blended" in the hybrid progeny, yielding "intermediate" offspring, or would the character of one parent be so strong as to become "prepotent" over those of the other, that is, be "nonblending"? (Prepotent is roughly equivalent to the Mendelian term "dominant," but unlike that term, it was often applied to the species or the individual acting as a whole.) This issue of the strength of heredity was of crucial importance, for a novel and advantageous character that blended would, he thought, soon be diluted and ultimately lost by reproduction with the general population. But a prepotent character, by its strength might well survive successive mating with normal individuals and become established. That is why Darwin read widely in the literature on the crossing of varieties and hybridizing of species. It is the reason for his experimental program in these areas, and his appeal to the plant and animal breeders, seeking their know-how and attending their shows.

Darwin placed most confidence in professional sources. The three he trusted most were the botanical hybridists Joseph Kölreuter, Carl von Gärtner, and William Herbert. They showed that in hybridization prepotent characters were seen in all members of the immediate hybrids (or, as we say, the F_1 generation) but only in a proportion of the hybrid offspring (F_2). Characters not prepotent, by contrast, were blended, but in the following generation the original characters might reappear. In the professional medical literature, Darwin noted how strongly prepotent characters like polydactyly or prickly skin converged on the category of monstrosities and were not therefore adaptive. A number of human diseases, he noted, were familial (recurring in successive generations of the same family). Then there were diatheses, or tendencies to the development of specific diseases, that emerged in later life, such as gout, and were also familial, emerging about the same age in successive members of the family. Agricultural sources yielded one generalization, "Yarrell's law," named after the eminent ornithologist William Yarrell, according to which the character longest in breed is strongest.

Of special interest to Darwin was the phenomenon of "reversion," or the return of offspring to a character last seen in a grandparent or, more surprisingly, a character not seen for generations past or even at no time before. In the latter case, Darwin admitted he could find no "proximate cause." Instead, he suggested it belonged to a presumed ancestor in the evolution of the species – even hundreds of thousands of generations ago. Often referred to as "atavism," Darwin called it "distant reversion" to distinguish it from reversion as generally known (the return to a grandparental character). While this explanation implied the extraordinary persistence of heredity in such cases, it hardly followed from the widely used fractional theory of heredity.

HEREDITY IN FRACTIONS

Although admitting the diversity in kinds of hereditary transmission, Darwin frequently applied a test of plausibility to such data on the basis of the commonly held rule of the blending of hereditary "blood" in fractions. Originating from a Spanish American ruling in the eighteenth century to deal with the legal status of half-castes, the rule was that the hereditary contribution of a "black" and a "white" parent to their offspring is half each. Hence the hereditary constitution of that offspring's "blood" will be half black and half white, or "mulatto." If that offspring then has a child by a white parent (i.e., a backcross), the fractional constitution of the child will be one-quarter black three-quarters white. or "terceron," and so on. Darwin applied this formula to the botanical hybridists' claim that to bring a hybrid form back to one of the parental species by "backcrossing" to that species may require as many as twelve generations. At that point, Darwin (1859, 106) noted, "the proportion of blood, to use a common expression, from one ancestor, is only 1 in 2,048, and yet it is generally believed that a tendency to reversion is retained by this remnant of foreign blood." Or take the hybrid plant of the "five-o-clock" *Mirabilis vulgaris* x *M. longiflora*. Even after eight generations of crossing with *M. longiflora*, he reported, the return to *M. longiflora* was incomplete, "although these plants contained only 1/256th part of *M. vulgaris*" (Darwin 1868b, 2:88). These calculations and others he cited did not cause him to reject the

FIGURE 12.2. Gregor Mendel (1822–84), the father of modern genetics. See also Essay 32, "Early Botany." Mendel read the *Origin* soon after a German translation was published, but he never thought of his own work as contributing a major piece to the puzzle of evolutionary causes. From W. Bateson, *Principles of Mendelian Heredity* (Cambridge: Cambridge University Press, 1909)

FIGURE 12.3. Francis Galton, the half cousin of Charles Darwin (they were both grandchildren of Erasmus Darwin), pioneer in studies of heredity and of statistics. He was an enthusiastic "eugenicist," believing that we must apply selection to humankind to preserve and improve the fitness of the species. Permission: American Philosophical Society

fractional theory. The obvious absurdity of a fractional explanation for distant reversions led him instead to offer a novel explanation: in each successive generation there has been a "tendency" to reproduce the character in question, "which at last, under unknown favourable conditions, gains an ascendancy" (Darwin 1859, 161). A tendency is thus not a fraction of the "blood" or a particle. But what, one asks, is a tendency?

Darwin's use of the fractional theory as a basis for judging plausibility indicates his underlying attachment to it. Prosper Lucas, the author he cited on other matters with confidence, had attacked the theory (1847–50, 2:206–15), a point Darwin did not mention. Joseph Gärtner, the plant hybridist Darwin also greatly respected, had classified the numerous hybrids he had formed into three classes: intermediate (*vermittelte*) and blended (*gemischte*); mixed or commingled (*gemengte*); and decided or biased (*decidirte*), that is, hybrids in which the characters of one parent are prepotent, hiding all those from the other. Blending was associated with the first class, but only 12 of Gärtner's 150 different crosses were in his judgment belonging to this class. However, Darwin and others continued to accept the blending theory and its fractional representation. Among these authorities was the highly respected botanist and hybridist Carl von Nägeli, who in 1867 used the fractional theory as the basis for his denial of Gregor Mendel's interpretation of his experiments (Nägeli 1867, in Iltis 1932,193) (Fig. 12.2). The hybridist Max Wichura, like Nageli, analyzed his researches on the hybridization of willow species in terms of the fractional theory. Also, Francis Galton's "Ancestral Law of Inheritance" was structured on the very same theory (Fig. 12.3). Darwin, it appears, was not so out of line as one might have thought, adhering to the fractional or blending rule.

SEXUAL AND ASEXUAL GENERATION

Apart from the absence of a consensus on the issue of blending heredity, had Darwin, perhaps, his own reasons for underestimating the extent of nonblending heredity? There was a relevant issue. It concerns the distinction between sexual reproduction based on the union of male and female germinal material by fertilization and asexual reproduction by budding. When Darwin read his grandfather's book *Zoonomia*, he had been very struck by this contrast. Dr. Erasmus Darwin (1794, 1:487) wrote that "buds and bulbs ... exactly resemble their parents, as is observable in grafting fruit-trees ... whereas the seminal offspring of plants ... is liable to perpetual variation." This was in July 1837, at a time when in a series of notebooks Darwin began seriously to search for a possible mechanism for the transmutation of species. Erasmus Darwin found the contrast "very curious." So did his grandson. "Why," he asked, "is life short. Why such high object generation." The existence of two forms of generation and the contrast between them, he suggested, is due to the fact that the (sexual) germ, unlike a bud, marks a return to the undifferentiated state. In doing so, reasoned Darwin, it leaves behind the accumulated injuries and acquired diseases of the parents. But as germinal material, he reasoned, it is "plastic" – that is, it is highly susceptible to the effects of the changing conditions of life, and these cause hereditary variations in the resulting offspring. Added to this, the germ is required to be fertilized by the male element, likewise present in the undifferentiated state.

Fertilization also brings together the individual differences in constitution and inherited characters of two individuals. Breeding "in-and-in," Darwin knew, can have deleterious effects, but breeding out can restore the future constitution of a family. Here was a possible answer to his grandfather's question of why two forms of generation. Namely, the role of sexual reproduction is to provide fresh combinations of hereditary characters and hereditary variations, thereby restoring vitality to the constitution.

Powerful support for this suggestion came from Darwin's study in 1838–39 of the flowers of the primrose. These plants are all hermaphrodite (bearing flowers with both reproductive organs), but the flowers are of two kinds. Some plants bear flowers called "thrum," the style and stigma (for receipt of the pollen) being short. Other plants have flowers called "pin" because of the long style with its stigma projecting at the mouth of the corolla. These arrangements together with other differences, Darwin realized, serve to prevent self-pollination and ensure cross-pollination. This must result in outbreeding. This knowledge formed the starting point of an experimental program he undertook, leading to papers and three books in later years.

As an aspiring transmutationist Darwin had also to account for the apparent uniformity of the members of a species. If the germ is caused to vary by changes in the conditions of life, and the population is outbreeding, would not the variations cause a breakup of the species into a multitude of different forms? Therefore, he speculated that sexual reproduction among the variant forms serves to blend them, thus producing uniformity. Also, once he had arrived at his conception of natural selection, he felt he needed *slight* changes that can be accumulated over time in an adaptive manner. He was looking for variations that will show adaptation not to brief periods of time but to long ones. Naturally, we see here the expectations of the transmutationist, one who expected evolution to proceed, like geological processes, gradually and yield long-lasting adaptations. The role of sexual reproduction would then be primarily to blend. Changes in the conditions of life would yield the variations.

DARWIN'S SOURCES

Darwin's search for information on inheritance came from a wide variety of sources. On the inheritance of diseases he relied chiefly on the physician Dr. Henry Holland and Mr. W. Sedgwick but also on his own father, Robert Waring Darwin. On plants he mined the works of the German hybridists Joseph Kölreuter and Carl Gärtner, the French hybridist Charles Naudin, and the English horticultural experts Thomas Andrew Knight and the Honorable William Herbert, dean of Manchester. On animals he turned to the surgeon veterinarian William Youatt, author of a well-known series of books on animal husbandry, and to politician and agriculturist Sir John Sebright. These were all professional men with established reputations. But Darwin did not stop here. Yarrell had encouraged him to approach the "fanciers" who breed pigeons to show and sell. Thus he met with men from the working classes and was instructed in their art. James Secord (1981) has explained how Darwin sought their experience, how he listened to their opinions, although their views on the origins of their breeds he rejected. Whereas they claimed their breeds represented just as many original species that once existed, Darwin believed that the wild rock dove was the original source from which all the pigeon breeders' forms have descended. On the other hand, the breeders offered Darwin plenty of support for the strength of heredity and the extraordinary variety of the many breeds, sufficient to constitute a whole genus of species (Fig. 12.4).

REVERSIONS AND ACQUIRED CHARACTERS

Yarrell had earlier drawn Darwin's attention to the results of crossing distinct breeds of pigeons. Breeds with none of the characteristics of the wild rock dove, when crossed, produced offspring with those very features – the slate blue plumage, the two black wing bars, white rump and terminal dark tail bar. Subsequently Darwin read in Boitard and Corbié's great book on pigeons (1824) that crosses between certain breeds of pigeons yielded "only bisets or dove-cot pigeons." This report convinced him to breed pigeons himself, a research project that became a major occupation at the family home, Down House, in the 1850s. Darwin's crosses yielded him offspring with the reported characteristics. Could it be, he suggested, that they represent a return to these long lost wild-type colors? If so, this reversion differed from most reversions, in that it represented a *very* distant ancestor. The fraction of the blood from that ancestor would therefore be infinitesimal, but he described it as a "tendency, for all that

FIGURE 12.4. The ancestral pigeon (formerly known as "rock dove"). From C. Darwin, *Variation under Domestication* (London: John Murray, 1868)

we can see to the contrary, [that] may be transmitted undiminished for an indefinite number of generations" (Darwin 1868b, 1:202). The fractional theory did not therefore apply to it. Now, one might expect that he would have pursued this promising subject from the point of the phenomena of heredity by seeking to correlate it with other results of breeding investigations. But no. It was for him instead key evidence for the descent of all breeds of pigeons from the rock dove (*Columba livia*.).

Heredity was, however, Darwin's chief concern when he compared the size and weight of pigeon skeletons obtained from domesticated and wild birds. He found a clear difference, which he attributed to the different degrees of activity in the wild and domesticated situations. For Darwin, this was evidence of the inheritance of the effects of the different lifestyles – or, as Lamarckians put it, "use and disuse." This form of inheritance Darwin always accepted, although he knew respected authors like Prosper Lucas opposed it. In the *Origin*, Darwin had remarked that acquired characters were "supposed not to be inherited." He was clearly not convinced by the skeptics, but he did draw the line at accepting the inheritance of injuries. Later he was impressed by Brown-Séquard's report that the guinea pigs whose sciatic nerve he had divided gnawed off their gangrenous toes and then bore progeny, thirteen of which were born with deficient toes. Such was the surgeon's reputation that Darwin cautiously accepted this report (Darwin 1868b, 2:24) Clearly, Darwin had always reserved his position on most forms of the inheritance of acquired characters. Only the inheritance of injuries had he rejected until the report from Brown-Séquard made him reconsider the issue.

AGENCIES THAT AFFECT INHERITANCE

By inheritance Darwin sometimes meant the transmission of parental characters to the immediate offspring. Reversion, by substituting grandparental or more distant ancestors, obscures that direct transmission. Another agency that he considered affected hereditary transmission was "changed conditions of life" such as occur in the acclimatization of organisms brought from other parts of the world and established in Europe. Variability, and in some cases even sterility, followed. Like effects often followed also from cross-breeding. In both cases, he urged, slight changes were beneficial, while drastic changes were harmful. This "double parallel" led him to conclude that, when in hybridization and cross-breeding the fertility of the hybrids is reduced and the variability of their offspring increased, the cause is the difference between the previous conditions of life of the two originating species of the cross. Hence, the act of crossing or hybridizing is not in itself the cause of the ensuing variability; rather, it is prior changes in the conditions of life. Far from continuing to accept his grandfather's special status for sexual reproduction, Darwin was now seeking to deny that status. The discoveries of parthenogenesis, graft hybrids, and bud variation all persuaded him that the alleged special status of sexual reproduction should be questioned.

THE PROVISIONAL HYPOTHESIS OF PANGENESIS

In 1865 Darwin prepared a manuscript entitled "Hypothesis of Pangenesis." After major revision, it appeared in the second volume of his book *The Variation of Animals and Plants under Domestication* (Darwin 1868b) (Fig. 12.5). How and why he came to formulate this hypothesis is best explored by turning to the 1865 text, published almost a century later (Olby 1963). There are three parts to this work. In the first part, he states and defends his apostasy over the growing acceptance of a fundamental distinction between sexual and asexual generation – one that he had earlier accepted. In the second, he sets out his hypothesis. In the third, he seeks to demonstrate its explanatory value. Our concern here is only with the role of the hypothesis in accounting for the phenomena of inheritance.

Darwin's growing conviction that there is no fundamental distinction between sexual and asexual generation underlies pangenesis. The referral of hybrid variation to the effects of changing conditions of life was one source for this about-face on his earlier position. Additional support for denying the sexual-asexual distinction came also from the discoveries of ova that develop without sex, as in the parthenogenesis of aphids, and of variations found in graft hybrids and among buds, neither of which involve sexual fertilization.

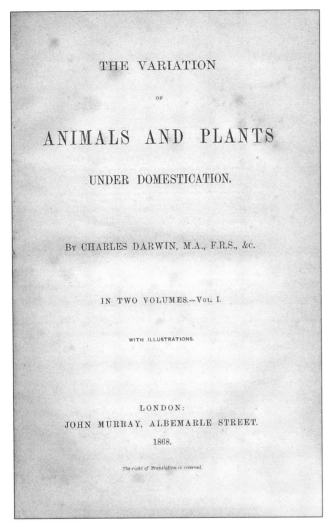

FIGURE 12.5. The title page of *The Variation of Plants and Animals under Domestication* (1868). It was here that Darwin introduced his hypothesis of pangenesis.

Darwin had begun to formulate the idea of pangenesis in 1841 while reading Johannes Müller's great text on physiology, where Schwann's theory of "free cell formation" was described (Hodge 1985). Darwin seems to have been unaware of the rejection of that theory in favor of what by the 1850s was becoming the accepted theory of cell formation - that of Robert Remak, Franz Unger, and Rudolph Virchow, which explained cell formation by division of existing cells. Two features of Müller's account particularly influenced Darwin: the manner in which, according to Schwann, cells could form around granules that float freely in a formative fluid (Schwann's "cytoblastemma"); and the suggestion that cells show what Müller called "elective affinities" for particular tissues and organs, hence becoming appropriately situated in the developing embryo. Darwin wrote that

> protoplasm or formative matter which is diffused throughout the whole organization, is generated by each different tissue and cell or aggregate of similar cells; — that as each tissue or cell becomes developed, a superabundant atom or gemmule as it may be called of the formative matter is thrown off; — that these almost infinitely numerous and infinitely minute gemmules unite together in due proportion to form the true germ; — that they have the power of self-increase or propagation; and that they here run through the same course of development, as that which the true germ, of which they are to constitute elements, has to run through, before they can be developed into their parent tissue or cells. This may be called the hypothesis of Pangenesis. (Darwin, in Olby 1963, 258-59.)

Darwin's gemmules are the specific particles or constituents of the protoplasm. They are thrown off by their respective tissues and are "diffused throughout the whole organization," giving rise to fresh protoplasm, that congregates in buds, and collects in the reproductive organs. "On this view," he explained, "we must believe that the reproductive organs do not by any means exclusively form the generative protoplasm, if indeed they form any of it, but only select and accumulate in the proper quantity, and make it ready for separate existence" (Darwin in Olby 1963, 258).

How, then, does this hypothesis support Darwin's views on inheritance? It confirmed his conclusion from other considerations: "It is not inheritance, but non-inheritance, which is the anomaly." That some characters are not inherited he attributed to reversion or to "the conditions of life incessantly inducing fresh variability." Reversion can be understood in terms of those gemmules that remain latent for any number of generations before becoming developed. Or their expression can be initiated by changes in the conditions of life or by crossing. As for the inheritance of acquired characters, Pangenesis accommodates it, for altered organs will send their kind of gemmules to the reproductive system. On the inheritance of injuries he was cautious. He knew of a case in which the same organ had been removed over several generations but still reappeared. However, "if mutilations are ever inherited," he opined, "... we could in some degree understand the cause." He described the inheritance of the effects of use and disuse as "most perplexing" but supposed that the tissues thus affected "could throw off gemmules endowed with all the qualities which they have acquired" (Darwin in Olby 1963, 259).

He called pangenesis a "provisional hypothesis" not like natural selection, which he referred to as "my theory." But the extent of the phenomena that he could explain with his introduction of the gemmules made him confident. When in 1865 Huxley pointed out that the Virchowian cell theory required that all cells are derived from preexisting cells by division, Darwin simply made his gemmules an additional source of units likewise capable of self-division. This made the hypothesis look as if refashioned only to avoid rejection. However, Darwin was introducing a particulate hypothesis of heredity, and the particles were of many kinds and were associated with particular tissues and characters. The species no longer acted as a whole in heredity.

FRANCIS GALTON'S CRITIQUE

In 1869 Francis Galton decided to test pangenesis, and when he published his results, he remarked that "its postulates are hypothetical and large, so that few naturalists seem willing to grant them. To myself, as a student of Heredity, it seemed of pressing importance that these postulates should be tested" (1871a, 394). Why not test the requirement that the gemmules enjoy free circulation around the body? Accordingly, his plan was to introduce blood from rabbits with colored coats to silver-gray rabbits and report any change in coat color of the silver-gray progeny. The experiments were of three kinds: moderate partially defibrinized transfusions, large transfusions wholly defibrinized, and cross-circulation via the carotid arteries. The latter class he judged the most convincing. It yielded eighty-eight offspring in thirteen litters, but not one rabbit showed any alteration of the coat color. Nor were there any results from the other experiments supportive of pangenesis. In the spring of 1871 these results were published in the *Proceedings of the Royal Society* (Galton 1871a). Darwin (1871b) promptly wrote to *Nature* complaining that he had not specified circulation in the blood stream. Galton (1971b) replied pointing out that Darwin's (1868b, 2:374) words "circulate," "freely," and "diffused" imply a return to the starting point. Where else than the blood system would that be achieved? But he tactfully suggested that he had just been misled by Darwin's language in pangenesis, and the issue was allowed to die (Galton 1871b, 6; Bulmer 2003, 118).

CONCLUSION

One might be tempted to say that heredity was Darwin's "Achilles' heel." More justly it is clear that he elevated the visibility of the subject of heredity and emphasized the distinction between hereditary transmission and expression. In pangenesis, he freed the inheritance of individual characters from the hold of the species acting as a whole, and he stimulated others to theorize on heredity (e.g., De Vries 1889).

How should we interpret Darwin's apostasy over the sexual-asexual distinction and his apparent oblivion over "free" cell formation? Clearly he had not been following developments in cytology, for these had implications for both issues. True, Darwin was both a naturalist and an experimentalist. But as a theoretician, he had overriding concerns. Indeed, one biologist described Darwin's mind as "directed to the conclusions he hoped to reach or confirm" (Darlington 1953, 97).

❦ Essay 13 ❦

Darwin and Time

Keith Bennett

Darwin's theory of evolution by natural selection required time for its operation. Darwin (1859, 287) knew that "it is highly important for us to gain some notion, however imperfect, of the lapse of years." He needed some idea of the total amount of time available and the rate at which evolution took place, but he lacked data on both. Perhaps he was minded of the situation he faced when cataloging the world's coral reefs and developing a theory for their origin, when he had to resort to unquantified phrases such as "slowly sinking" and "prolonged subsidence" (Darwin 1842c). For evolution, he had some relative data on roughly in which order certain taxa had evolved through geological time, but he also lacked detail here, especially with regard to the most recent parts of the geological record, and so he kept a close eye on the rates of appearance of domesticated varieties in relation to the archaeological record (Darwin 1868b). He became entangled with involved discussions on matters for which we now have far more complete data, but where his instincts were broadly correct. On the other hand, he and his contemporaries lacked information on the complexity and rapidity of geological changes (e.g., during the Quaternary period) which might well have made a substantial difference to how he formulated and presented his theory of evolution. In this chapter, I briefly discuss these aspects of how knowledge, or lack of it, influenced Darwin's ideas.

THE AGE OF THE EARTH

The first edition of *On the Origin of Species* (Darwin 1859) predates any significant attempt at a figure for the Earth's age. Darwin's ideas matured in the early years of scientific discussion of topics for which contemporary answers had been provided by the Bible, and interpretations of it (including the suggestion of Buckland [1836] that "millions and millions of years" might have passed between the Creation and the Mosaic narrative). Scientific rationale for understanding the age of Earth was, however, in its infancy (Dalrymple 1991), and Darwin was concerned that objections would be raised against his theory of evolution by natural selection on the grounds that Earth was not sufficiently old, although many geologists were apparently thinking of increasingly long periods of time since the origin of Earth (A. Geikie 1893).

FIGURE 13.1. A diagram from the 1840s showing the denudation of the Sussex Weald. From A. C. Ramsay, On the denudation of South Wales and the adjacent counties of England, *Memoirs of the Geological Survey of Great Britain, and of the Museum of Economic Geology in London* 1 (1846): 297–335.

Darwin discusses examples from the geological record indicating the passage of substantial periods of time, emphasizing repeatedly the slowness of processes involved (rate), and hence the vast amounts of absolute time involved. Then Darwin presents an extraordinary back-of-envelope calculation in which, for the first time, he puts a number to rates and the amount of time involved for one particular episode.

This single calculation concerned the denudation of the Weald, where Upper Cretaceous rocks (chalk) on the top of an uplifted dome in southeast England have been eroded away, exposing underlying Lower Cretaceous rocks (Fig. 13.1). Darwin argues that the sedimentary rocks to a thickness of 1,100 feet have been eroded back 22 miles. He suggests that a cliff 500 feet in height might erode at 1 inch per century, and it is implicit from this (but not stated by Darwin) that a cliff of 1,100 feet would erode proportionately slower, namely 1 inch per 1,100/500 = 220 years. One mile = 63,360 inches, so, assembling these estimates, he argues that the denudation of the Weald would have taken 22 x 63,360 x 220 = 306,662,400 years, "or say three hundred million years" (Darwin 1859, 287), and continues by suggesting briefly that in all probability the real answer is longer.

So far as I am aware, this estimate of Darwin's was the first scientific attempt at the passage of geological time, and the first to place the Earth's age into the realm of at least hundreds of millions of years (presumably much more). As an estimate of the age of the Weald denudation, it is in the right order of magnitude of the total time elapsed since original deposition of the rocks concerned, but falls short of modern estimates of the Earth's age by one order of magnitude. Darwin may have been grasping for any hard evidence of the passage of vast periods of time which his theory needed, but even he seems not to have fully grasped just how much time was potentially available, and he may have felt that suggesting ages as old as hundreds of millions of years was as far as he could go, given a background of popular understanding of the order of thousands of years. He may also have realized the weakness of his estimates ("appallingly naïve," according to Burchfield 1974). In the second edition, the same calculation is presented but followed by a more cautious "perhaps it would be safer to allow two or three inches per century, and this would reduce the number of years to one hundred and fifty or one hundred million years" (Darwin 1860a, 287). An article in the *Saturday Review* (Anonymous 1859) ridiculed Darwin's calculations from a geological viewpoint, concluding that "Mr. Darwin has enormously over-rated the amount

FIGURE 13.2. William Thomson, Lord Kelvin, the physicist who insisted that there is not enough time for a leisurely process of change fueled by natural selection. Permission: Wellcome

of time which can legitimately be demanded to account for the geological phenomena." The writer's criticisms are directed to showing that Darwin's estimates are too generous, although an argument could equally well have been written to claim that the estimates are not generous enough: the data are simply too crude. Darwin was mortified by this and other attacks, so, despite support from some of his friends (Burchfield 1974), he backpedaled, and the offending calculation disappeared from the third edition of the book (1861), although not necessarily from his way of thinking. The fourth edition of *On the Origin of Species* (1866) is noticeably silent on the question of time scales, beyond a passing and lyrical reference to "Let this process go on for millions on millions of years" (210).

Darwin returns to the fray in the fifth edition, with reference to the then recent calculations of Thomson (Lord Kelvin) (Fig. 13.2). Thomson's first publication on the subject seems to have been an abstract read at the 1861 Manchester meeting of the British Association for the Advancement of Science. He makes some estimates of likely cooling rates for the sun (based on data and notions that are hardly better than those in Darwin's Weald calculations), and concludes that it is "most probable that the Sun has not illuminated the Earth for

100,000,000 years, and, almost certain that he has not done so for 500,000,000 years" (W. Thomson 1862). Thomson's views were considered highly authoritative. They dominated thinking on the age of the sun and Earth for four decades (Dalrymple 1991), and his calculations on the age of Earth were endorsed by Croll (1864). Thomson launched a direct attack on those geologists who wanted a long time scale, declaring that " all geological history showing continuity of life, must be limited within some such period of past time as one hundred million years" (W. Thomson 1868, 25), and he was counter-attacked on behalf of geologists by Thomas Huxley (1869), their self-declared "attorney-general."

Darwin (1869, 379) acknowledged these calculations in *On the Origin of Species*, without going into details of the debates:

> Here we encounter a formidable objection; for it seems doubtful whether the earth in a fit state for the habitation of living creatures has lasted long enough. Sir W. Thompson [*sic*] concludes that the consolidation of the crust can hardly have occurred less than 20 or more than 400 million years ago, but probably not less than 98 or more than 200 million years. These very wide limits show how doubtful the data are; and other elements may have to be introduced into the problem.

It was more than twenty years later before A. Geikie (1893) was able to assert that the geological record had to be taken seriously, and that there must be some flaw in the physicists' calculations, though he did not know what it might be. Darwin thus faced attack on his longer time scale from, first, geological and, later, physical considerations. He seems to have had problems reconciling initial justified criticism of his geological calculations and the apparent rigor of the physicists' calculations, on one hand, with his own feeling that much longer periods of time were needed to explain the evolution of life, on the other. The debates of 1868–69 (W. Thomson 1868; T. H. Huxley 1869) appear to indicate that geologists were broadly on Darwin's side by then, as far as the age of Earth was concerned, but could not handle the physicists' arguments. Thomson clearly had no doubts about who was right, declaring that a "hypothesis that life originated on this Earth through moss-grown fragments from the ruins of another" was at least "not unscientific" (W. Thomson 1872). This does have the merit of disposing of the time scale problem, but hardly of the question of how life evolved, or at what rate. The issue of the length of time available for evolution remained, for Darwin, one of the most significant objections he faced for his theory right through to his last edition of *On the Origin of Species* (1872a).

THE ICE AGES

That parts of Earth had once been more extensively glaciated than at present was first brought to scientific attention by Louis Agassiz (1840). He traveled in Britain and presented papers at the annual meeting of the British Association for the Advancement of Science (BAAS) (Agassiz 1841) and the Geological Society of London (GSL) (Agassiz 1842), the latter followed by supporting papers read by Buckland (1842) and Lyell (1842). Collectively, these papers and contemporary discussions established the glacial theory in Britain, and Lyell, among others, adjusted some of his geological interpretations. However, recognizing the phenomenon was one thing; recognizing the complexity, time scale, and significance took longer. In the ninth edition of his textbook (Lyell 1853), the version current for the writing of *On the Origin of Species*, the "glacial epoch" is mentioned only once, and then in the context of not letting it obscure evidence for an earlier period with climate warmer than present:

> It will naturally be asked, whether some recent geological discoveries bringing evidence to light of a colder, or as it has been termed "glacial epoch," towards the close of the tertiary periods throughout the northern hemisphere, does not conflict with the theory above alluded to, of a warmer temperature having prevailed in the eras of the Eocene, Miocene, and Pliocene formations. In answer to this inquiry, it may certainly be affirmed, that an oscillation of climate has occurred in times immediately antecedent to the peopling of the earth by man; but proof of the intercalation of a less genial climate at an era when nearly all the marine and terrestrial testacea had already become specifically the same as those now living, by no means rebuts the conclusion previously drawn, in favor of a warmer condition of the globe, during the ages which elapsed while the tertiary strata were deposited. (Lyell 1853, 75)

Lyell's opinions on the significance of this "glacial epoch," at least as far as the evolution of life is concerned, may have strengthened somewhat in the next few years, for in 1856, at the end of a long letter about uplift of continents, he was writing to Darwin:

> And why do the shells which are the same as European or African species remain quite unaltered like the Crag species which returned unchanged to the British seas after being expelled from them by Glacial cold, when 2 millions? of years had elapsed, and after such migration to milder seas. Be so good as to explain all this in your next letter. (Darwin 1985–, 6:146, letter, 17 June 1856)

Darwin replied promptly, but he commented only on the uplift aspect of the letter (in a postscript on 18 June 1856 of a letter to Joseph Hooker, and in reply to Lyell on 25 June 1856 [Darwin 1985–, 6:147; 6:153–55]), and did not reply directly to Lyell's closing question or comment on it anywhere else at the time. Darwin was well aware of glaciers, not least because he had seen them in Tierra del Fuego (Darwin 1839a). Following the BAAS and GSL meetings, he traveled to Snowdonia to make his own observations of glaciated landscapes, and clearly had no doubt of the existence of former glaciers in Britain, down to sea level (Darwin 1842a), so he was well aware of the phenomenon of a "glacial epoch." He discusses it at several

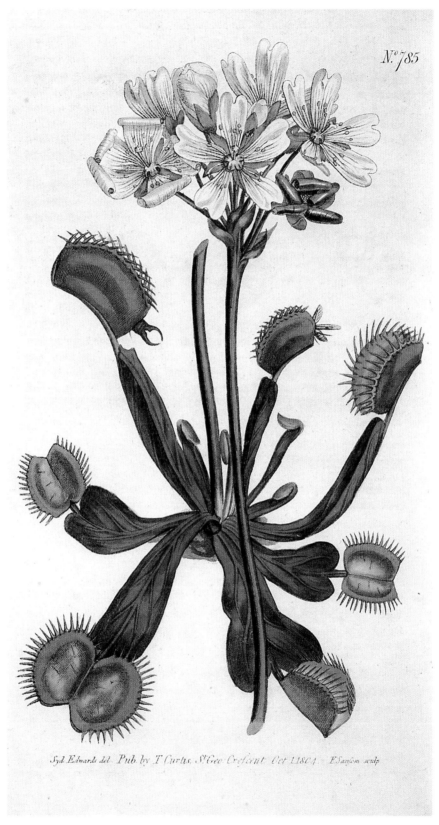

PLATE XIX. In his old age, Darwin had a wonderful time trying to tease out interesting facts about the nature and growth of plants. Insectivorous plants especially caught his attention. From *Curtis's Botanical Magazine* (1804)

1.—Leptalis Theonoë.   *2.—Leptalis Theonoë, var. Melanoë.*   *3.—Leptalis Theonoë, var. Lysinoë.*

1a.—Ithomia Flora.   *2a.—Ithomia Onega.*   *3a.—Stalachtis Phædusa.*

Plate XX. The butterflies in the top row are mimics of the butterflies, the models, on the bottom row. The models are foul tasting and thus repel birds, the major predators. The mimics are not foul tasting but survive by deceiving the predators. Because predators easily learn if they are being deceived, selection keeps the ratio of mimic to model very low (in the order of 1 in a 1,000). From H. W. Bates, *The Naturalist on the River Amazons* (London: John Murray, 1892)

PLATE XXI. *Dazzle-ships in Drydock at Liverpool*, 1919, by Edward Alexander Wadsworth. The artist was a leading figure in the "Vortist" movement, a British pre–First World War school, related to Cubism, that stressed bold colors, sharp lines, and prominently featured industrial subjects. Wadsworth himself was much involved in ship camouflage during the war. Permission: © 2011 Artists Rights Society (ARS), New York / DACS, London

PLATE XXII. Archdeacon William Paley (1743–1805) had a great influence on Darwin, who accepted Paley's claim that the living world is design-like but who wanted to offer an entirely natural cause. Nineteenth-century lithograph

PLATE XXIII. The *Origin of Species* went through six editions between 1859 and 1872. Permission: Wellcome

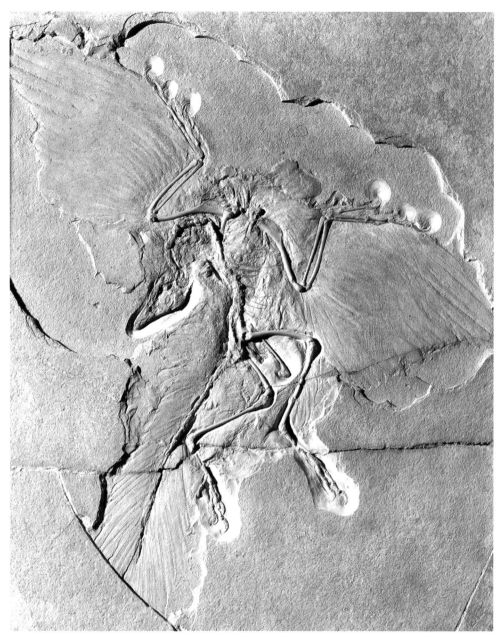

PLATE XXIV. The *Archaeopteryx*, discovered in Germany in 1861, was the archetypal missing link (between the reptiles and the birds). It found its way into later editions of the *Origin*. Permission: Museum für Naturkunde, Berlin

PLATE XXV. The geneticists' animal workhorse, the little fruit fly, *Drosophila*. Permission: Jarmo Holopainen

PLATE XXVI. The peculiarities of the reproduction of the evening primrose, *Oenothera lamarckiana*, misled many early geneticists into thinking that major mutations are the key to evolutionary change. Permission: Martin Young photo

PLATE XXVII. The frontispiece to E. B. Babcock's monograph on the genus *Crepis*. We see the different forms from the most primitive on the left to the most advanced on the right. Chromosomes of each are depicted on the top. From E. B. Babcock, *The Genus Crepis* (Berkeley: University of California Press, 1947)

PLATE XXVIII. John Maynard Smith (1920–2004) was a leading British evolutionary biologist who applied insights from game theory to evolutionary problems, particularly those involving behavior. Much of his work was popularized in Richard Dawkins's *The Selfish Gene*. Like Dobzhansky and Ford before him, and Thomas Henry Huxley even earlier, Maynard Smith's importance lay equally in his great networking abilities and warm encouragement of students and other younger scholars. Permission: University of Sussex

PLATE XXIX. Sometimes, as with dazzle, the deception is not so much a matter of pretending to be another precise organism but of pretending to be doing something other than that which is really happening. The false eye at the tail end of this fish quite confuses the predator, unsure about the direction in which the prey is swimming. Permission: Martin Young photo

Mainland

Beach

PLATE XXX. The strong effects of natural selection – predation by birds (owls, herons, and hawks) and carnivores (foxes and coyotes) – are shown by the camouflaging color patterns of mice from different habitats. Oldfield mice (*Peromyscus polionotus*) can be found in two distinct habitats in Florida – oldfields, which are vegetated and have dark loamy soil; and coastal sand dunes, which have little vegetation and brilliant white sand. Mice that occupy these different habitats have distinct coat colors: mainland mice (*left*) have a typical dark brown coat, whereas beach mice (*right*) largely lack pigmentation on their face, flanks, and tail. Permission: Photo Hopi Hoekstra

PLATE XXXI. William D. Hamilton (1936–2000) is generally considered the greatest evolutionist of his generation. Passionately committed to the ideal of individual selection, he is best known for his work on "kin selection," where adaptations for altruism, giving to others, are seen as benefiting individuals inasmuch as close relatives reproduce. Permission: Science Photos

PLATE XXXII. Edward O. Wilson, longtime Harvard professor and the greatest living authority on ants, has also striven to bring together our understanding of social behavior into one integrated field, "sociobiology." Like his hero Herbert Spencer, he has a taste for wide-ranging, metaphysical speculations. Permission: Edward O. Wilson

PLATE XXXIII. Neil Shubin, a University of Chicago paleontologist, with his discovery, *Tiktaalik*. Permission: University of Chicago

PLATE XXXIV. Darwin's finches. This is one of the illustrations, based on Darwin's collection, painted by John Gould for the official published account of the *Beagle* voyage. This pair is from the species *Geospiza magnirostris*, the large ground finch. From John Gould, *Zoology of the Voyage of H.M.S. Beagle, III, Birds* (London: Smith Elder, 1841)

points in *On the Origin of Species*, and noted that the "glacial epoch" lasted "for an enormous time, as measured by years," and argued that the cold period was simultaneous throughout the world (Darwin 1859, 374). There is also one passage that might be as near as he came to an answer to the question in Lyell's letter. In discussion of changes in geographical distribution during the "glacial period," he writes:

> The arctic forms, during their long southern migration and re-migration northward, will have been exposed to nearly the same climate, and, as is especially to be noticed, they will have kept in a body together; consequently their mutual relations will not have been much disturbed, and, in accordance with the principles inculcated in this volume, they will not have been liable to much modification. (368)

On the other hand:

> Alpine species ... must have existed on the mountains before the commencement of the Glacial epoch, and ... during its coldest period will have been temporarily driven down to the plains; they will, also, have been exposed to somewhat different climatal influences. Their mutual relations will thus have been in some degree disturbed; consequently they will have been liable to modification. (369)

These are extraordinary statements. The whole book is a long argument for a theory about how species evolve through adaptation to changing environments. Then, when faced with evidence for a major environmental change (continental glaciation), Darwin finds a way to argue that there should not have been much change, for the broad mass of organisms, possibly because Lyell had convinced him that that is what the fossil record showed ("Consider the prodigious vicissitudes of climate during the Pleistocene period, which includes the whole glacial period, and note how little the specific forms of the inhabitants of the sea have been affected" [336]). His argument answers Lyell's question about species survival through the "glacial period" but hardly advances the cause of Darwin's own main argument. If there is not much change during environmental changes as dramatic as continental glaciation, what scale of environmental change is needed to bring about "modification"? And why do the altitudinal shifts of Alpine species make them liable to modification but latitudinal shifts of other species do not? One possible explanation of the way Darwin was thinking in 1859 was that he, along with Lyell, had not yet appreciated the scale or intensity of the Quaternary ice ages, and this period was regarded as a detail with little overall significance in the grand scheme of the evolution of life (a view that still exists today). He also had little idea of the time scale of the "glacial period," although the figure of "2 millions? of years" mentioned by Lyell may give some inkling of the kind of time scale that was being thought about in the 1850s, and it was a pretty good estimate for the whole period of Northern Hemisphere glaciation, as it turned out. Coincidentally, in 1859 archaeologists were pushing the antiquity of humans

FIGURE 13.3. James Croll (1821–90), a scientist who developed a theory of climate change based on changes in the Earth's orbit. Nineteenth-century lithograph

back into the ice ages (Gamble and Moutsiou 2011), lengthening their own time scales.

Croll (1864) suggested that geologists of the time were generally reluctant to consider glacial cold seriously because of their wider understanding that Earth had been cooling throughout geological time and thus that any colder episodes were not part of the grand scheme (Fig. 13.3). His 1864 paper marks the beginning of a series of articles arguing that changes in Earth's orbit, through its eccentricity and precession of the equinoxes, were responsible for driving climate change, including periods of glaciation. Further, he argued that understanding the relationship between geological periods and orbitally forced climate change held the prospect of being able to assign absolute ages to geological periods, and he began by suggesting a figure of 100,000 years since the last "glacial epoch." At this stage, detailed calculations of how eccentricity varied through time had not been made, but Croll began the task and published tables of calculated eccentricity data (Croll 1866, 1867a). In these papers, he argued that during periods of high eccentricity, precession would have had the greatest impact, giving periods of maximum cold, and thus glaciation. He noted that the glacial period spanned the interval from 240,000 to 80,000 years ago, preceded by an interval of three major glaciations between 1,000,000 and 700,000 years ago (Fig. 13.4). Croll (1868) suggests that these earlier periods were either the "boulder clay" periods of the most recent glaciation or colder stages within the Miocene, but he inclined toward the latter view because it gave a shorter overall geological time scale. Croll (1867b) realized that Earth's angle of tilt (obliquity) also varies and published on how this would affect climates of higher latitudes.

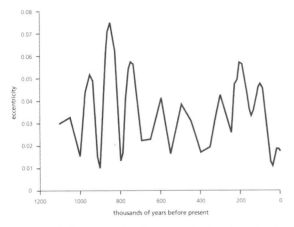

FIGURE 13.4. A diagram based on James Croll's calculations, showing the high points of eccentricity in the past million years

Geologists of the late 1860s would thus have been thinking, first, that there had been a single glacial period and, second, that it should be dated at hundreds of thousands of years ago, probably around 100,000 to 200,000 years ago. Lyell (1867, 1868) incorporated this thinking in his book from 1867. Darwin made no mention of Croll in the 1866 fourth edition of *On the Origin of Species* but uses his work extensively in the fifth (1869). Croll's arguments blend the time scales of eccentricity variations (ca. 100 kyr), which are actually negligible in terms of solar insolation, with the effects of precession of the equinoxes (periodicity ca. 20 kyr) at times of extreme eccentricity, which have substantial consequences for the distribution of solar insolation with latitude and season, and in an opposite sense between the Northern and Southern Hemispheres. Darwin liked this argument, checked it with Croll by letter in 1868 in which he asked explicitly for confirmation of the argument that the Northern Hemisphere would be warm while the Southern Hemisphere is cold, and vice versa (Campbell Irons 1896), and then used it as explanation of the likely behavior of organisms during periods of glacial cold:

> In the regular course of events the southern hemisphere would be subjected to a severe Glacial period, with the northern hemisphere rendered warmer; and then the southern temperate forms would in their turn invade the equatorial lowlands. The northern forms which had before been left on the mountains would now descend and mingle with the southern forms. (Darwin 1869, 456–57)

And this mixing of species by distributional shifts between hemispheres provides a mechanism for "modification":

> But the species left during a long time on these mountains or in opposite hemispheres, would have to compete with many new forms and would be exposed to somewhat different physical conditions; hence they would be eminently liable to modification, and would generally now exist as varieties or as representative species; and this is the case. (Darwin 1869, 457)

Darwin thus now had a mechanism, thanks to Croll, for generating "modification" from oscillating glacial climates covering a wider range of species than just the Alpine species of the first edition but dependent on the climatic oscillations of the two hemispheres being out of phase (shifting his position from the first edition of *On the Origin of Species*), which might be taken as a measure of Croll's influence.

And, that, essentially, is where he leaves it. Wallace (1870b), however, appreciated the more general potential significance of more rapid oscillations of climate on the precession time scales and argued that these would have resulted in general distribution changes of species, in both extinction and rapid modification. He proposed that the precession oscillations at times of high eccentricity would have driven speciation and that this in turn might be used to estimate the length of geological time.

After 1870, Croll consolidated his arguments and calculations in a book (1875), including illustrations of Earth's orbital variations and diagrams of his calculations, but Darwin does not develop his text or theory any further. In 1871, James Geikie started publishing a series of papers in *Geological Magazine*, culminating in a synthesis (1872) with a table indicating how the "glacial epoch" could be subdivided into intervals of warmer and colder climate, and using terms such as "Recent Period," "Post-Glacial Period," "Last Glacial Period," "Last Interglacial Period," and "Great Cycle of Glacial and Interglacial Periods" and he identifies the Norwich Crag with a "Pre-Glacial Period" (Fig. 13.5). For the first time, we begin to see a notion of the geological complexity of the ice ages. Darwin must have been well aware of this, but by this time he

FIGURE 13.5. James Geikie (1839–1915), younger brother of Archibald and his successor to the Murchison Professorship of Geology and Mineralogy at the University of Edinburgh. Nineteenth-century lithograph.

had ceased to make substantive changes to *On the Origin of Species* in particular, or his theory of evolution in general, and turned his attention to worms, orchids, emotions, and other aspects of biology and ecology that can be readily observed. These works include numerous detailed observations and experiments that involve time (e.g., Darwin 1875b; 1880, 1881), but not even in *Descent of Man* (1871a), which he developed during the 1870s, is there reference to the current thinking of Croll, James Geikie, or others on the complexity of the recent past in which humans evolved, even though this book might have been a chance to update the geological thinking of the last editions of *On the Origin of Species*.

SCALING OF THE TREE OF LIFE

On the Origin of Species contains just one figure (see Fig. 6.3), but it is a powerful one, resembling strikingly the torrent of molecular phylogenetic trees that now fill the scientific literature. Darwin (1859) discusses it for no less than eleven pages, indicating the importance that he attributed to this style of presentation of evolution and descent with modification, and it does have many aspects of interest. In terms of the way that Darwin thought about time, a noteworthy aspect is that it is self-scaling (although Darwin does not use that terminology):

> If we suppose the amount of change between each horizontal line in our diagram to be excessively small, these three forms may still be only well-marked varieties; or they may have arrived at the doubtful category of sub-species; but we have only to suppose the steps in the process of modification to be more numerous or greater in amount, to convert these three forms into well-defined species. (120)
>
> In the diagram, each horizontal line has hitherto been supposed to represent a thousand generations, but each may represent a million or hundred million generations. (124)

In other words, although the main discussion treats this as the representation of the evolution of a genus over fourteen thousand generations, it can equally well be taken to represent the evolution of lower taxonomic categories over shorter periods of time, or higher categories over longer periods of time. Darwin must be thinking here, first, that taxonomic categories are somewhat arbitrary, with amounts of difference related to the passage of time, and, second, that available time is more or less continuously available, presumably unbroken by climatic or other environmental changes that would introduce discontinuities into the way that the diagram scales with time and taxonomic level. His use of the phrase "hundred million generations" slips long time-scale thinking into the book. The diagram and its discussion remain essentially the same through to the sixth edition of *On the Origin of Species* (1872a), although the text is rearranged.

CONCLUSION

Darwin's relationship with time was complex, and he had problems with both longer and shorter time scales. His first geological book discussed the formation of coral reefs over patently long periods of time without ever quantifying that time (Darwin 1842c). With *On the Origin of Species*, he begins with a great idea about how life might have evolved, realizes that this would take amounts of time that were almost inconceivably long, does some rough calculations to support this very long time scale, and is immediately shot down by geologists for the crudity of the calculation (which must have been very frustrating). Along comes Thomson with an argument for a short time scale that seems equally weak today, but has all the force of a powerful physicist behind it, and Darwin is unable to stand up to it, beyond repeating to the last that 200 million years "can hardly be considered as sufficient for the development of the varied forms of life" (Darwin 1872a, 286), and hinting at an improbable olive branch proffered by Thomson "that the world at a very early period was subjected to more rapid and violent changes in its physical conditions than those now occurring; and such changes would have tended to induce changes at a corresponding rate in the organisms which then existed" (286). Darwin was certain of the long time scale, and may have done more than most to stand up to Thomson, but he was never able to nail the argument. He did, however, repeatedly emphasize the length of time he thought was required for modifications to take place (e.g., Darwin 1868b).

The situation is very different at the shorter end of the temporal range. Darwin was, mostly, writing in an era where it was assumed that climate changes of the past (which geological evidence showed had occurred) were of long periodicity and slow in rate. Lyell writes, for example, about a "great year" during which the world passes successively through warmer ("summer") and colder ("winter") periods and even, just to be clear about the time scale, writes lyrically about a time in the future when "the huge iguanodon might reappear in the woods" (1830–33, 1:123). Geologists were also generally aware that Earth might be still cooling from an original molten state, which would suggest a generally steady cooling climate throughout geological time, however long that might have been. There was no notion that changes might have taken place more frequently or rapidly, and thus Darwin was placing his theory of evolution of life against a steady-state background (of which his tree figure [Fig. 6.3] is an example), and not even the advent of Agassiz and the glacial theory did much to disturb that. Only during the 1860s did Croll manage to bring home to his contemporaries that matters might not be that simple, and it is noteworthy that his arguments come from astronomy and physics, rather than from the direct evidence of the rocks themselves. By the 1870s, the rock evidence, as presented by James Geikie, was catching up, and revealing glimpses of a far more complex recent geological record than even Croll had imagined.

Much of this passed Darwin by, so far as incorporation in his main thinking and publishing was concerned. For all essential purposes, his theory of evolution of natural selection was developed without knowing the time scale of orbitally forced climatic change, the relationship between that time scale and the longevity of species, and how organisms and species respond to rapid climate change. We now understand that Earth's climate varies at time scales of 20 to 100 kyr, with precession (20 kyr) important at all latitudes, obliquity (40 kyr) dominant at higher latitudes, and eccentricity (100 kyr) only weakly evident, but it may combine with other factors to produce a quasi-100-kyr oscillation within the most recent 700 kyr. We understand the existence of plate tectonics on much longer time scales and appreciate that this interacts with the permanent orbital variations to produce shifts in global climate on these longer time scales, and that these are much more significant than any trend resulting from Earth cooling from its molten origin. Lyell, Darwin, Croll, and James Geikie were all, in their own ways, struggling to make initial sense of what we would now see as an almost impossibly dynamic world.

What would Darwin make of modern understanding of these time scales? He would obviously be delighted to know that Earth is some 4.5 billion years old, with hundreds of millions of years of time for the evolution of multicellular life since the Precambrian. This is exactly what he expected all along, and brilliantly (if prematurely) gave a sense of with his Weald calculations. On the other hand, seeing the complexity and rapidity of climate changes of the past 2 million years or more might well have brought him up short. Major climate change and glaciation with repeated subcontinental scale shifts in distribution might well have excited the biogeographer in him, but surely this should have generated "modification," as Wallace (1870b) argued? Except it did not – both fossil-based paleoecology and molecular phylogenetics agree in placing lineage splits of modern species predominantly on time scales of millions of years ago, not the tens or hundreds of thousands of years ago that would indicate forcing by these climate changes. We will never know how he would have reacted to this knowledge, but with the benefit of hindsight it is one that he should have worried about much more than Thomson's limited age of Earth.

Essay 14

Darwin's Evolutionary Botany

Richard Bellon

BUMBLEBEES, INSISTED A writer signing himself Ruricola (1841) in the *Gardeners' Chronicle*, wrought terrible damage on bean crops by rapaciously drilling holes in the bean flowers in search of nectar. Ruricola advised gardeners to protect their crops from these costly acts of vandalism by eradicating bees' nests as soon as bean flowers bloomed. Charles Darwin (1841) responded four weeks later with a vigorous, if qualified, defense of the bees, "these industrious, happy-looking creatures." The boring did little material damage to the flower, he insisted. The bees' activity perhaps did the plants an injury nonetheless, but in a more indirect, perfidious way than Ruricola imagined. The plants offered nectar to the bees in exchange for transferring pollen from flower to flower. The bees, by lapping up their reward without earning it by brushing over the reproductive parts of the flower, were in effect "picking pockets."

This short communication, written in the summer of 1841, was Darwin's first public remarks on a defining passion of his life. Over the next forty years he published numerous articles and books on the complex relationship between the reproductive organization of flowering plants and their environment. After the publication of the *Origin of Species* in 1859 he promoted evolution as the unifying principle behind his botanical breakthroughs. The German botanist Hermann Müller (1879, 2), one of many naturalists who built a career advancing Darwin's approach, declared that this marriage of evolution and botany provided "the key to the solution of the riddle of the flower." This solution was not, of course, on offer in his response to Ruricola. But in this modest communication, so seemingly inconsequential when laid next to the panoramic generalizations of the *Origin*, we discover the epitome of Darwin's scientific character.

His deep love of outdoor science pours from every line. A sunny day devoted to systematically recording bees flying from flower to flower at London's Zoological Garden was time perfectly spent. The article is dense in original observational detail. He reported exhaustive investigations of insect-flower interactions made over the course of two summers over several locations – no armchair theorizing here. But neither did he find pleasure in one-dimensional empiricism. His impatience with *mere* observation radiates through the entire piece. Discrete observations accumulated into larger conclusions, and larger conclusions guided further observation.

He reported that country bees, unlike their London cousins, had not (yet?) adopted the practice of flower boring. He speculated that this city cunning was acquired knowledge: if true, this had deep implications for the scientific understanding of insect behavior. He noted that bees aggressively pursued their own advantages in direct violation of a mutual duty to flowers that "nature intended" of them. But this intention, he slyly intimated, was simply an artifact of human misassumptions about "the, so imagined, final cause of their existence." The bees' true "final cause" was their own selfish advantage, and their partnership with flowers was merely a precarious bargain subject to constant cheating and renegotiation. Static natural harmony was an illusion; that was the provocative subtext. In retrospect the post-*Origin* reader can disinter an idea buried even deeper. Darwin recognized that small permutations of life, like a few crafty bees breaking their compact with flowers, drove profound coevolutionary change when accumulated and multiplied over geological time. But, of course, he did not even whisper this in 1841.

The place of publication was also significant. The weekly *Gardeners' Chronicle*, then in its inaugural year, opened a spacious commons shared by overlapping communities of farmers, horticulturalists, and naturalists. Darwin was a founding member of the level-headed, practically minded community that congregated around this periodical; years later he listed his occupation for a local directory as "farmer" (Browne 2002, 6). His theorizing, for all its audacity, carried the earthy aroma of the field, garden, and hedgerow.

The *Origin*, if taken in isolation, provides a misleading picture of his character and genius. By distilling his theory to its most fundamental form (an abstract, he called the book), the *Origin* offers an unbalanced view of the relationship between theory and research in his life and science. He did not observe merely to support this theorizing; he also theorized to guide his observations, to aid and advance the homely work that he (and so many other faithful readers of *Gardeners' Chronicle*) loved. Darwin was never happier than when his knees were muddied in the observation of some small but new fact about the natural world.

The story of Darwin and his flowers is not peripheral to his great revolution. It embodies it.

AN ORDER SUBMITTED TO GENERAL LAWS

In order to understand the role of plants in Darwin's life, we have to follow him to the University of Cambridge. In 1829, his second year of enrollment, friends coaxed him to lectures by John Stevens Henslow, the dynamic new professor of botany (Fig. 14.1). Nothing, Darwin (1958, 64) reflected in his autobiography, influenced his career more. He almost instantly fell under the spell of the young, active, disciplined, and reform-minded botanist, whose wide-ranging scientific expertise suited the intellectually omnivorous Darwin perfectly.

FIGURE 14.1. John Stevens Henslow (1796–1861), professor of botany at the University of Cambridge, Darwin's mentor and friend. Darwin attended Henslow's lectures during the years that he was an undergraduate at Cambridge and learned from him the importance of homological thinking in working out relationships. Permission: Wellcome

Henslow's course on botany, the first offered at the university in decades, immersed students in cutting-edge international scientific developments. He assigned the most rigorous overviews of plant science, even if the books happened to be in French (Walters and Stow 2002, 65–66). He published his own textbook in 1835, which codified the lessons he taught to Darwin and others. *The Principles of Descriptive and Physiological Botany* captured his conviction that the study of plants needed to be as broad as possible, in contradiction to a conventional definition of botany as a descriptive exercise that excluded the study of plant function (Henslow 1835, 1–4). By the time Henslow wrote his textbook, botanists had described and classified in the neighborhood of sixty thousand species. At the most basic level, a systematic classification prevented botany from collapsing into chaos under the weight of its diverse materials. But, as Henslow explained, systematists pursued a higher object than cataloging. Systematic botanists searched for the laws underlying the patterns of plant structure, which in turn would reveal the plan "upon which we must feel satisfied that the Author of nature has proceeded in creating all natural objects" (135–36). Thus, the description and arrangement of species was an essential component to a comprehensive understanding of plant life but never as an end in itself. Botany's loftiest goal was to integrate the study of form and function into a comprehensive view of plant life.

The young English professor found particular inspiration in the work of the eminent Swiss botanist Augustin-

Pyramus de Candolle (Sloan 1986; Walters and Stow 2002; Ayers 2008). (Candolle also exerted significant influence on other men of science close to Darwin, including Charles Lyell and Joseph Hooker.) Candolle (who died a few weeks after Darwin's paper on bees and bean flowers appeared) was perhaps the dominant figure in botanical systematics during the nineteenth century (P. F. Stevens 1994). His arrangement of the natural orders of plants provided the foundation for contemporary plant classification. Candolle (1839–40, 2:302–8) defined natural classification in both theory and practice around the search for symmetry, or the general regularity of organization that defined a natural group of plant species. At a superficial glance, he noted, the vegetable world seemed to be nothing but irregularity, with each species existing in isolation, their anomalies obscuring all deeper similarities. But over the generations, careful and extensive investigation of the natural world revealed "an order submitted to general laws" (2:304). Once a botanist determined a group's abstract regular form, he could use the modifications to this basic symmetry – the fusions, degenerations, multiplications, and abortions of organs that gave rise to the characteristics of individual species – to guide his classification (Fig. 14.2). More than this, however, these investigations could, in the fullness of time and research, uncover the regular causes that governed both organic symmetry and the law-bound deviations from it.

Candolle's appeal to Henslow did not rest simply on his sophisticated articulation of the principles of scientific botany. At an even more basic level, Candolle advanced a vision of the philosophical structure of science that accorded perfectly with the ethos of Henslow and his network of scientific reformers at Cambridge. Candolle (1839–40) maintained emphatically that one could never discover the particular symmetry that defined a given taxonomic group except through exhaustive study of particulars. He scorned speculators who disdained the study of facts and subordinated nature to their imprecise metaphysical ideas. Yet he also disparaged simple describers who accumulated isolated facts with no attempt to search for unifying theories. Naïve attempts to shield description from theory were not only barren but the fount of much error. "The simple description of vegetable facts and forms has been singularly improved since the knowledge of some general laws has caused describers to reflect on what they see," he insisted (2:307). Botany, by gradually reducing irregularities to general laws of increasing breadth, thus followed the same basic rules of inductive practice that had governed the progress of more advanced sciences like astronomy and mineralogy (1:iii–viii; 2:302–3, 307–8). For Henslow, eager to reestablish botany as a vigorous intellectual pursuit after its long disappearance from Cambridge, Candolle's attempt to embed botany in the philosophical tradition of more prestigious physical sciences carried obvious appeal.

Henslow's application of Candollian principles led him to correct John Lindley's erroneous interpretation of the structure of the genus *Reseda*, which Lindley graciously accepted. William Whewell (1837, 3:442) used this as an example in his *History of the Inductive Sciences* to highlight the universal

FIGURE 14.2. In his archetypal thinking about flowers, Henslow supposed that the "normal" or ideal type of flower consists of concentric whorls, carrying respectively five sepals, petals, stamens, and carpels. Real flowers can be obtained by reducing the parts of a whorl or by introducing distortions. From J. S. Henslow, *The Principles of Descriptive and Physiological Botany* (London: Longman, Rees, Orme, Brown, Green, and Longman, 1835)

reality of the "principle of developed and metamorphosed symmetry." Yet these principles – as Henslow and Candolle would have been the first to admit – remained constrained and partial. The irregularities of botany remained far from tamed by general theory. Candolle himself in practice often violated his own injunction to integrate anatomical and physiological studies in his classificatory work (P. F. Stevens 1994, 85–86, 89).

In the mid-1820s, Henslow made a series of characteristically careful investigations of cowslips and oxlips, species belonging to the genus *Primula*. These plants are hermaphrodites, but Henslow nonetheless observed two distinct sexual forms. Half had tall male organs and short female ones; the other half, the reverse. Henslow never published these observations (Kohn et al., 2005). Other botanists had noticed the pattern as well, but none granted it any particular significance (Darwin 1877a, 14). His published work on *Primula* concentrated instead on the principles for drawing of species boundaries, an important and contentious controversy in plant classification (Henslow 1830, 1836). The fact that he spoke prominently on the classificatory questions surrounding *Primula* but left his observations of its peculiar sexual organization sitting in the drawer demonstrates that integrating form and function into coherent generalizations was easy to advocate but difficult to accomplish.

Henslow emphasized that the only way out of the difficulty was the precise and long-continued study of particulars, a core principle shared by allies and friends like Whewell, Adam Sedgwick, and John Herschel. Darwin absorbed this lesson deeply. For the rest of his life he drew upon the specific knowledge and skills he acquired under the tutelage of Henslow and his other Cambridge mentors, but above all else he left the university and boarded HMS *Beagle* with the core conviction that the highest calling of science required the synthesis of meticulous investigation with the broad vision of theoretical insight.

A MORE SIMPLE AND INTELLIGIBLE VIEW

After returning from the *Beagle* voyage, Darwin's study of plant fertilization systems fed into his voracious accumulation of facts related to his theory of evolution. But it would be a mistake to interpret this interest as driven solely by theoretical concerns. The study of plants became an enjoyable summer rite, one that eventually swept in the children of his extended family. As his son Francis (1899, x) observed, he simply loved doing it. But the fact that this amusement contributed materially to his theoretical interests was far from incidental. His broader work on the origin of species gave direction and wider meaning to his spring and summer floral investigations; the more theoretical use his studies acquired, the more entertaining they became. Scientific duty, philosophical investigation, and family fun proved a seductive combination.

Most flowing plants are hermaphroditic. Naturalists generally believed that because most individual flowering plants could self-fertilize, they did; insects and other agents actuated plant cross-fertilization only incidentally. Darwin's observations in the late 1830s led him to distrust this assumption. He fell into the close orbit of one of the few botanists who emphasized the importance of insect pollination. Robert Brown's (1833) work on orchids and asclepiads in the early 1830s demonstrated that, at least in these families, cross-fertilization played a much more prominent role than commonly accepted. Darwin became convinced not only that nature abhorred perpetual self-fertilization but that the particular biological mechanisms that promoted crossing in a given species resulted from gradual modifications of inherited structure driven by natural selection. He included in the *Origin* a "short digression" on the intercrossing of individuals, in both the plant and animal kingdoms, although he had space for only a small portion of the "special facts" he had collected (Darwin 1859, 96–101).

Darwin returned to the study of plant reproduction in the spring of 1860. The publication of the *Origin* the previous November had radically changed the context of his investigations. He felt ground down by the thirteen months of writing the *Origin* and disheartened by the viciousness of the controversy it ignited. Progress on the manuscript that would eventually emerge in 1868 as *Variation of Animals and Plants under Domestication* proved painful. His heart was not in it. But working with his eyes and fingers invigorated him, and as summer arrived, he quickly expanded his botany research program in multiple experimental directions. He felt a rush of "boyish delight" with the arrival of each new specimen. At first he felt guilty that his variation manuscript collected dust in his study while he ministered to his experimental charges in the garden. But he soon recognized an opportunity to put plant reproductive physiology on a stable theoretical foundation – and in so doing, he would provide a decisive demonstration

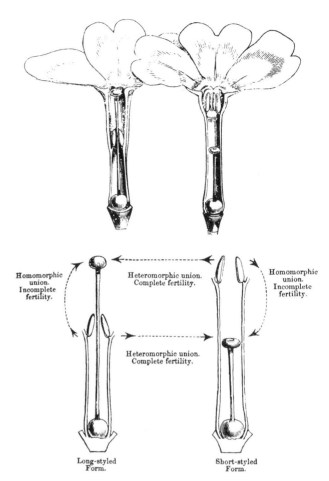

FIGURE 14.3. The possible unions of *Primula* forms and their respective fertilities. From C. Darwin, *Primulae*, *Journal of the Linnaean Society* (1861)

of evolution's value as a tool of inductive biological research (Bellon 2009, 2011).

The need to demonstrate the *Origin*'s inductive legitimacy had become acute in the face of unexpectedly ferocious attacks. Darwin had expected criticism from his scientific colleagues but not personal censure. Many friendly skeptics in fact applauded the book as a bracing and legitimate, if far from proven, attempt to solve some of biology's most pressing problems. Henslow took this line. But others challenged not only Darwin's conclusions but his competence and motives. Sedgwick (1860) excoriated his former pupil both publicly and privately for violating the most fundamental rules of inductive philosophy (Darwin 1985–, 7:396). Samuel Wilberforce (1860) declared Darwin's theory "utterly dishonourable" for unhitching the imagination from the discipline of facts and observation. Most painful of all, Richard Owen (1860) maliciously accused Darwin of repudiating the principle that science should be constrained by "close and long-continued research, sustained by the determination to get accurate results." The *Origin* contained virtually no new facts or observations, Owen charged; instead Darwin fueled an unrestrained fancy (previously disciplined by original work) with the largely misappropriated labors of more cautious men. The *Origin*, in other words, was not simply wrong but was a case of egregious scientific misconduct. Meanwhile, the Linnean Society, the venue where he first announced his theory publicly, prohibited any subsequent formal discussion of it. George Bentham, the new president, justified the ban by insisting that debate over a theory unaccompanied by new facts did not advance the legitimate purposes of the society (Bellon 2009, 380).

The accusations stung. "I can perfectly understand Sedgwick or any one saying that nat. selection does not explain large classes of facts; but that is very different from saying that I depart from right principles of scientific investigation," Darwin protested to Henslow (Darwin 1985, 8:195). His counterattack began, appropriately, at the Linnean Society. He evaded the ban on discussion of his theory by wrapping its (muted but unmistakable) presence in the communication of unambiguously original results. His first paper, read on 21 November 1861, addressed the two distinct hermaphroditic sexual forms of *Primula*, the same finding Henslow had made but left unpublished decades earlier. Darwin (1862a) did more than simply point out the existence of this peculiar dimorphic condition, however. His painstaking experiments provided an explanation. He discovered that full fertility depended on pollen from the other form, or "heteromorphic union." Not only did "homomorphic union," or same-form fertilization (which by definition included self-fertilization), produce significantly fewer seeds than heteromorphic, but the degree of infertility was greater than what resulted in many crosses of distinct species (Fig. 14.3). "The meaning or use of the existence in Primula of the two forms in about equal numbers, with their pollen adapted for reciprocal union, is tolerably plain," Darwin argued; "namely, to favour the intercrossing of distinct individuals" (91–92).

In a quietly matter-of-fact way, he made plain that his experiments and observations rested on the application of his theory of evolution by natural selection. Proponents of species fixity claimed that the Creator endowed sterility to maintain species boundaries, but, Darwin pointed out, this interpretation fit poorly with the evidence from *Primula*. He suggested instead that *Primula* species were in the process of an evolutionary transition "by slow degrees" from hermaphroditism to two distinct sexual forms. This paper, in its sober technicalities, offered a devastatingly effective rejoinder to the attacks that Darwin's evolutionary theory violated the principles of inductive investigation. The point was heightened in the same Linnean meeting by the reading of Henry Walter Bates's (1862) nimble evolutionary explanation of insect mimicry and diversification of Amazon butterflies.

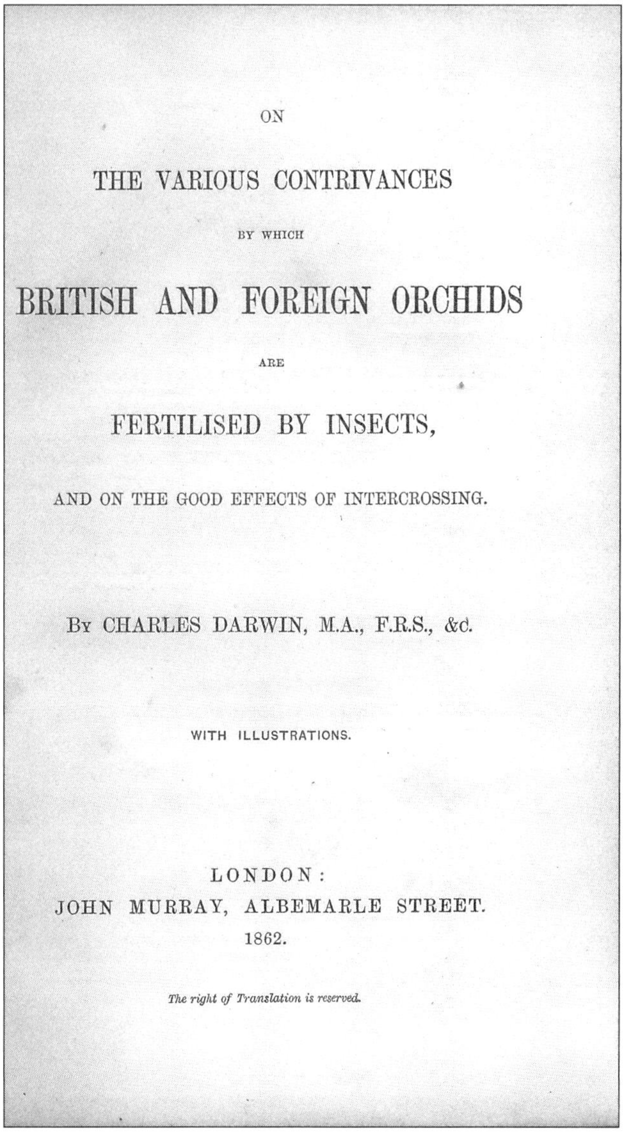
ON

THE VARIOUS CONTRIVANCES

BY WHICH

BRITISH AND FOREIGN ORCHIDS

ARE

FERTILISED BY INSECTS,

AND ON THE GOOD EFFECTS OF INTERCROSSING.

By CHARLES DARWIN, M.A., F.R.S., &c.

WITH ILLUSTRATIONS.

LONDON:
JOHN MURRAY, ALBEMARLE STREET.
1862.

The right of Translation is reserved.

Figure 14.4. The title page to *On the Various Contrivances by Which British and Foreign Orchids Are Fertilized by Insects, and On the Good Effects of Intercrossing* (1862). Darwin intended this little book to be an example of how biological problems could be solved using the theory of evolution through natural selection.

Darwin's (1862b) next paper to the Linnean Society, read the following April, solved the bizarre mystery of an individual orchid plant that bore the flowers of three apparently separate species on the same stock. Lindley, a leading orchidologist, lamented that this strange case shook "to the foundation all our ideas of the stability of genera and species" (Lindley 1846, 178; quoted in Darwin 1862b, 151). Darwin provided a reassuring explanation: the three flowers were the male, female, and hermaphroditic forms of a single species, *Catasetum tridentatum*. Botanists had misclassified the "wonderfully different" sexual forms as separate species because each was adapted to divergent requirements for insect pollination. Darwin's evolutionary analysis demonstrated that "the appearance of these three forms on the same plant now ceases to be an anomaly, and can no longer be viewed as an unparalleled instance of variability" (157).

These two papers were the overture to Darwin's (1862c) masterpiece of evolutionary botany, *On the Various Contrivances by Which British and Foreign Orchids Are Fertilised by Insects, and on the Good Effects of Intercrossing*, published in May 1862 (Fig. 14.4). By the 1860s, taxonomists

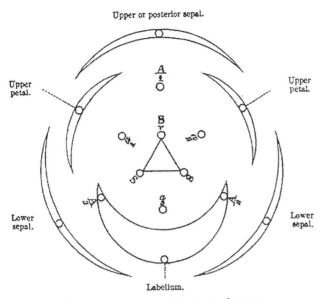

FIGURE 14.5. Darwin's own exercise in botanical archetypal thinking, showing the structural elements shared by all of the approximately 6,000 species of orchids. Unlike Henslow, he denied that this arrangement represented an "ideal type" created by an omnipotent creator, arguing instead that orchids owe their similarity to common descent from a monocotyledonous ancestor. From C. Darwin, *Orchids* (London: John Murray, 1862)

had described 433 genera and approximately 6,000 species in the family Orchidaceae (357). But underlying this remarkable diversity was a plan shared by every species: fifteen organs arranged in five simple and two compound parts (Fig. 14.5). In one species a particular organ acquired exaggerated size, in another it withered into a useless rudiment, and in still others its function changed or became physically confluent with another organ. The breathtaking multiplicity of orchids reduced to a uniform plan illustrated dramatically that, as Henslow (1835, 123) had put it, "an infinite variety of [floral] forms may be supposed to result from a few normal types." Missing – and what Darwin provided – was a general theory to envelope both this symmetry and its modification into a coherent explanatory framework.

Orchids bombarded the reader with example after example after example of minute analysis to illustrate how the form and function of a given species promoted crossing between distinct individuals through insect agency. Darwin demonstrated that the bewildering modifications of the basic orchid organization resulted from the need of each species to coadapt to the pollinators in its particular range: thousands of pollinators living in thousands of different environments – orchids occupied all but the coldest or driest parts of the globe – resulted in thousands of uniquely modified flowers. Elegance of form, opulence of color, and sumptuousness of fragrance were but localized adaptations serving utilitarian reproductive requirements.

Special creationists had of course long used the existence of intricate adaptation as proof of divine design. Darwin attacked this belief aggressively.

At a period not far distant, naturalists will hear with surprise, perhaps with derision, that grave and learned men formerly maintained that such useless organs were not remnants retained by the principle of inheritance at corresponding periods of early growth, but were specially created and arranged in their proper places like dishes on a table (this is the comparison of a distinguished naturalist) by an Omnipotent hand "to complete the scheme of nature." (1862c, 244)

He continued:

Is it not a more simple and intelligible view that all Orchids owe what they have in common to descent from some monocotyledonous plant; ... and that the now wonderfully changed structure of the flower is due to a long course of slow modification, – each modification having been preserved which was useful to each plant, during the incessant changes to which the organic and the inorganic world has been exposed? (306–7)

Darwin's approach in *Orchids* adhered reverently to the method of science he learned in Cambridge. His solution to the puzzle of the *C. tridentatum* provided a particularly beautiful validation of Candolle's (1839–40, 302) pronouncement more than thirty years earlier that a true law, once discovered, would transform apparent irregularities into elegant confirmations. Yet, in doing so, Darwin blew apart some of his mentors' basic theoretical assumptions. The "distinguished naturalist" responsible for the risible dishes-on-a-table analogy was none other than Candolle (1819, 185–86).

Darwin intended his botanical research to reorient the debate over the *Origin* – to serve as a flank movement on his enemies (Darwin 1985–, 10:292, letter from Gray, 2–3 July 1862; 330, letter to William Darwin, 4 July 1862). He reassured his publisher that *Orchids* "will do good to the Origin, as it will show that I have worked hard at details, & it will, *perhaps*, serve [to] illustrate how natural History may be worked under the belief of the modification of Species" (Darwin 1985–, 9:279, letter to John Murray, 24 September 1861). His allies pushed this line publicly. His close friend Joseph Hooker (1862a, 371) insisted in the *Natural History Review* that *Orchids* "cannot fail to secure to its author a more attentive hearing for his ulterior views than these have hitherto gained. Nay further, had Mr. Darwin not investigated this point he would have had no secure foundation for his great hypothesis." He repeated the point in the *Gardeners' Chronicle* (Hooker 1862b). Asa Gray (1862a, 1862b) followed the same strategy in the United States.

This interpretation was not limited to Darwinian partisans. The elderly Oxford professor of botany Charles Daubeny, though skeptical of evolution, pointedly advised an audience at the 1862 meeting of the British Association for the Advancement of Science to read *Orchids* "as it would dispel many notions which had been wrongly entertained with regard to the tendency of [Darwin's] writings" (Ellegård 1990, 72). A reviewer for the *Saturday Review* correctly predicted that "the laborious patience with which Mr. Darwin lays his foundations" in *Orchids* would allow the book to "escape the active, and often angry, polemics which [the *Origin*] aroused" (Anonymous 1862). Bentham's annual presidential address to the Linnean Society in 1862 extolled Darwin's "patient study of the habits of life" in *Orchids* as the ideal of natural history research. His private resolution to ignore the origin of species dissolved, and in his presidential address the following year he made a strong, if qualified, endorsement of Darwin's evolutionary theory (Bellon 2003, 290–91). In Switzerland, Alphonse de Candolle (1862), Augustin-Pyramus's talented son and intellectual heir, responded ecstatically to *Orchids* (Darwin had sent him a presentation copy). His review gushed that he did not know what we should admire most: Darwin's patient, attentive, and never-prejudged observations, or the grandeur of his theoretical views, which elevated their meaning.

Darwin's interest in the reproductive biology of flowering plants lost none of its intensity after the publication of *Orchids*. He was flooded by new cases of sexual dimorphism in hermaphroditic plants, far more than he could personally study in detail (Darwin 1985–, 9:374, letter to Hooker, 18 December 1861; 10:40, letter to Gray, 22 January 1862). In late 1861 he had stumbled across cases of trimorphism in loosestrife (the genus *Lythrum*), which provoked voracious experimentation the following summer. He reported his findings to the Linnean Society in 1864. In Darwin's telling, the plants' remarkable sexual relations were no isolated curiosity but a case study in a grand evolutionary pageant of sexual diversification. He drove the point home by comparing loosestrife with his pre-*Origin* discovery of the complex and diverse sexual forms of barnacles (1864, 194–95). He had carefully embargoed the evolutionary lesson in his depiction of barnacles' reproductive behavior; his research program in plant sexuality allowed him to reveal it retrospectively.

Darwin's botanical research expanded beyond floral reproduction. The completion of his new greenhouse in 1863 provided experimental space to indulge a fascination with climbing plants (Ayers 2008, 81–85). His typically meticulous investigation culminated in a 118-page paper published by the Linnean Society in 1865. This research piled up yet more concrete details to reduce the Candollean laws of plant organization to the higher generalization of evolution by natural selection. In the case of climbing plants, "the object … is to reach the light and free air with as little expenditure of organic matter as possible." The diversity of strategies to achieve this object reflected the modifications of primordial forms and habits in response to the contingencies of environmental requirements (Darwin 1865, 108–9).

As much as he reveled in this original research, duty drove him back to his forlorn manuscript on variation in domesticated species. But even in the theoretical sweep of this project he found innumerable ways, in Janet Browne's words (2002, 202), to transform "his daily activities into scientific knowledge." He drew extensively on the sex lives of flowering plants to explain the importance of crossing, the causes of sterility, and the nature of hybridization (Darwin 1868b, 85–191). If evolution by natural selection had served as the essential analytical tool in his post-*Origin* botanical discoveries, in *Variation* he folded these discoveries back into a synthetic defense of common descent.

Darwin followed *Variation* with two additional grand-scale studies, *The Descent of Man* (1871a) and *The Expression of Emotions in Man and Animals* (1872b). This work done, he returned to botany with a series of books that built upon his research of the previous decade: *Insectivorous Plants* (1875a), a revision of his climbing-plant paper (1875b), *The Effects of Cross and Self Fertilisation in the Vegetable Kingdom* (1876), *The Different Forms of Flowers on Plants of the Same Species* (1877a), a second edition of *Orchids* (1877b), and *The Power of Movement in Plants* (1880) (see Plate XIX). Where *Descent* and *Expression* permeated the barrier between humans and the rest of the animal kingdom, Darwin's botany radically narrowed the gap between plants and animals – a task no less central to his intellectual project, if obviously much less culturally provocative.

He wrote his book on plant movement in collaboration with his son Francis, an accomplished botanist. They advanced the remarkable claim that plant roots were capable of stimulus-response reactions directed by their tip, which "acts like the brain of one of the lower animals; the brain being seated within the anterior end of the body, receiving impressions from the sense-organs, and directing the several movements" (1880, 573). This claim embroiled the Darwins in a bitter personal controversy with the eminent German plant physiologist Julius von Sachs. Sachs, who interpreted root movement in more mechanistic terms, scoffed high-handedly at the suggestion of a rudimentary vegetable brain. Yet the driving force behind the quarrel was not conceptual disagreement, however genuine, but divergent standards of scientific practice. Sachs sought to consecrate laboratories like his own as the exclusive domain for trustworthy experiments in plant physiology; Darwin's relatively simple country-house studies represented a past that desperately deserved burying (De Chadarevian 1996; Ayers 2008, 97–114).

While Sachs's vitriolic response to *Movement of Plants* represents an important moment in the consolidation of authority by laboratory biologists, it was also an anomaly in the overall scientific reception of Darwin's botanical research. Numerous naturalists paid Darwin's botany the ultimate compliment: researchers worldwide turned their attention to fertilization mechanisms and sexual dimorphism and trimorphism in flowering plants. Remarkably, this included older men like John Hutton Balfour and P. H. Gosse who carried out floral investigations explicitly under Darwin's light while clinging to special creationism. Younger investigators (as Darwin noted gleefully to one such recruit) routinely became as "depraved" as he was on the origin of species (Bellon 2009, 384–89). Even Sachs treated Darwin's work on floral reproductive systems with respect. He acknowledged that it stimulated an enormous literature, much of it by German botanists whose standards of practice hewed much closer to his ideals (H. Müller 1873, iv; Sachs 1887, 787–99; Bellon 2009, 393–94).

The outpouring of research was so great that only slightly more than a decade after *Orchids* the German botanist

Hermann Müller (1873) systematized the voluminous but scattered research on plant reproduction systems into the monumental *Befruchtung der Blumen durch Insekten*. The bibliography of its English translation contained hundreds of entries on the fertilization of flowers published in Darwin's wake (H. Müller 1883, 599–630). By the turn of the twentieth century, hundreds of works had become thousands (Knuth 1906–9, 1:212–380).

CONCLUSION

Francis Darwin divided his father's botanical work into two distinct epochs. During the evolutionary period, Darwin used plants to build and test his theory. The physiological period began with *Orchids*, "and then the tables were turned, and the theory served him as a powerful engine to break still further into the secrets of plants" (1899, x). The fact that Darwin was joined by hundreds of other researchers, including several major botanists who devoted their career to floral biology, helped fix the *Origin*'s philosophical legitimacy. "I am quite convinced," he wrote to Bates, "that a philosophic view of nature can solely be driven into naturalists by treating special subjects" (Darwin 1985–, 9:363, letter, 3 December 1861). Without this communal success in using evolution to drive original research into an important special subject – it was never just plant reproduction, of course, but no other field was so closely associated with Darwin's own labor – there could have been no Darwinian revolution (Bellon 2011).

The last thing Darwin wrote for publication before his death was the preface to the English translation of Müller's treatise on the fertilization of flowers. The concluding paragraph urged the "young and ardent observer" to "observe for himself, giving full play to his imagination, but rigidly checking it by testing each notion experimentally. If he will act in this manner he will, if I may judge by my own experience, receive … much pleasure from his work" (1883, x). In these final public words, Darwin encapsulated his approach to the natural world, one that united a body of work that ran from the audacious theoretical sweep of the *Origin* to a homely defense of bumblebees and their role in fertilizing bean plants.

Essay 15

Mimicry and Camouflage

William Kimler and Michael Ruse

EVEN BEFORE CHARLES DARWIN put pen to paper to write up his theory of evolution through natural selection, entomologists knew full well that the colors of insects "deceive, dazzle, alarm or annoy" their enemies (Kirby and Spence 1815–28, 2:219). It was taken as overwhelming proof of the power and beneficence of the good god, a firm plank in the edifice of natural theology (Kimler 1983). In the standard work, *An Introduction to Entomology, or Elements in the Natural History of Insects*, coauthored by the parson-scientist William Kirby, a prize example was the "mimicry" of the Brazilian walking stick insect (Phasma) that so closely resembled the twigs on which it spent its living days. Although no full-blown theory was offered, it was clearly noted that it had a function, because the author "has often been unable to distinguish it [the insect] from them [the twigs], and the birds probably often make the same mistake and pass it by" (Kirby and Spence 1815–28, 2:220).

The explanation of unusual coloration was muddled, however, by natural theology's reliance on design. Resemblance to an object or background as deceptive disguise (camouflage) made sense as a providential aid to the species, but what to make of resemblance merely to another animal? The usual answer for such mimicry (or copying), if not left to a creative god's mysteries, lay in vague, ideal parallelisms or analogies. Darwin's great breakthrough in the *Origin of Species* was to show through his mechanism of natural selection how it is that all such cases of exquisite design can be explained naturalistically, as the result of natural selection brought on by the struggle for existence. However, in the first edition of the *Origin*, Darwin did not take up these problems.

Darwin stayed away from mimicry and camouflage because he very much realized that this is a complex topic and that it demanded more empirical understanding than he (or others) possessed. "I rejoiced that I passed over the whole subject in the Origin, for I should have made a precious mess of it" (Darwin 1985–, 10:540, letter to H. W. Bates, 20 November 1862). But very quickly the young naturalists around and supportive of Darwin picked up the topic, and it soon became one of the great triumphs of Darwinian science, a position that it justifiably keeps down to this day. The truly significant advance was made by Henry Walter Bates, a young collector who had in 1848 traveled to South America with Alfred Russel Wallace (Fig. 15.1).

FIGURE 15.1. Henry Walter Bates (1825–92), the traveling companion of A. R. Wallace and the discoverer of an important ("Batesian") form of mimicry. Permission: Wellcome

They intended to make a name for themselves in natural history but also to investigate a possible theory of evolutionary change. Less impressed by hasty generalization, Bates was less convinced than Wallace by Robert Chambers's *Vestiges of the Natural History of Creation*. Both, however, intended to discover new laws and naturalistic explanations.

Bates spent nearly fifteen years on and around the Amazon, making huge collections of insects. This prepared him for the message of the *Origin* and, returning home at the end of the decade (the 1850s), Bates soon entered into correspondence with Darwin. At once his great knowledge of insect diversity made it possible for him to tread where Darwin had hesitated, and he was quick to extend the mechanism of selection into the area of mimicry. Simply, Bates argued that some insects (his example was the butterfly *Leptalis*, now called *Dismorphia*, in the family Pieridae) closely mimic other insects (his example was the butterfly genus *Ithomia*, in the family Heliconidae, now spelled Heliconiidae) thanks purely and simply to the action of natural selection (see Plate XX). The *Ithomia* are foul tasting to birds, because of the plants on which they feast. The *Leptalis* are to the contrary quite palatable. Normally the *Leptalis*'s mode of protection lies in abundant numbers – it hopes not to be the one taken before it reproduces. However, because there is going to be this very heavy selection, and because there will be ongoing spontaneous new variations, some specimens will evolve to be closer and closer in appearance to the *Ithomia*, simply because that way they will be less likely to be taken. They can never be that abundant, because otherwise the predating birds will learn that they are being deceived and in turn take steps to prevent this. But being comparatively rare, they will survive and reproduce, thanks to their mimicking their poisonous models.

Darwin was delighted with the younger naturalist's findings and thinking. "I thought of your explanation; but I drove it from my mind, for I felt that I had no knowledge to judge one way or the other" (Darwin 1985–, 9:280, letter from Darwin to Bates, 25 September 1861). With this encouragement, "Batesian mimicry" soon made its way into the world (Bates 1862). As Darwin had encouraged, so Bates returned the favor by acknowledging, "The explanation of this seems to be quite clear on the theory of natural selection, as recently expounded by Mr. Darwin in the 'Origin of Species'" (511). It is therefore no surprise that Darwin did his best to spread the word, even to the extent of writing an anonymous review of Bates's work and remarking that it must have pleased Mr. Darwin. More concretely, the well-connected, upper-middle-class Darwin put the rather needy, lower-middle-class Bates in touch with his publisher John Murray, who took up Bates's book on his Amazon travels and then found Bates a full-time job as secretary to the Royal Geographical Society (Clodd 1892). Bates's old companion Wallace (1865) was no less enthusiastic and pushed the Batesian explanation of mimicry, drawing on his own findings in the Malay Peninsula to show just how complex and astounding mimicry can be. Sometimes different members of the same brood mimic different models, because a female can give birth to different types. Of course, the underlying genetics of this case was quite unknown and, as we shall see, led to much acrimony before it was fully untangled.

Despite Darwin's enthusiasm for the new exemplar of adaptation, and how natural selection explained previously unexplainable resemblance, mimicry did not enter the *Origin* in the chapter on natural selection. Darwin did add new reports there to bolster the realism of his hypothetical scenarios of selection in action, but mimicry made its appearance, in the fourth edition of 1866, in the chapter entitled "Mutual Affinities of Organic Beings." If Darwin had known mimicry's meaning in 1859, it would have fit well as an illustration of selection in action. In any case, he was satisfied with his logical argument for selection as a true cause. Bates's new example provided something even more valuable: an empirical study of speciation in action. Resemblances confused taxonomists, tempting them to invoke parallelisms and ideal classifications or the direct action of environment. Darwin saw how mimicry disrupted such notions by showing the path from individual variation to local selection pressure to newly evolved form. They revealed how to separate ecological convergence (analogy) from genealogical similarity (homology), as mimetic species adaptively diverged from their relatives. Mimicry clearly demonstrated the power of natural selection

FIGURE 15.2. Fritz Müller (1821–97), German-born and an emigrant to Brazil, was a keen Darwinian. He discovered the form of mimicry named after him, where unpalatable species resemble each other, thereby warning predators of their offensive nature. From Alfred Müller, *Fritz Müller. Werke, Briefe und Leben* (Jena: Gustav Fischer, 1920)

to produce Darwin's tree of descent and reinvent the meaning of classification.

Toward the end of Darwin's life came another major breakthrough in our understanding of mimicry. One of Darwin's greatest German supporters was the radical scientist Fritz Müller, who had penned a defense of the thinking of the English naturalist that was so effective that Darwin had it translated into English at his own expense (J. F. T. Müller 1869) (Fig. 15.2). Moving to Brazil, Müller discovered that sometimes a butterfly with warning coloration (known technically as "aposematic" coloration) is mimicked by another, even though the latter (unlike the Batesian case) has distasteful adaptations of its own. The reason for "Mullerian mimicry" is simply that the birds, the predators, have to learn to avoid the insects, and by joining a larger group, the mimics are therefore more rapidly avoided. "Now if two distasteful species are sufficiently alike to be mistaken for one another, the experience acquired at the expense of one of them will likewise benefit the other; both species together will only have to contribute the same number of victims which each of them would have to furnish if they were different" (J. F. T. Müller 1879, xxvii).

It was Wallace (1870a, 1889, 1891) who was first to start putting much of this together in a broader context of camouflage generally. (Unlike the older, pre-Darwinian word "mimicry," "camouflage" is a term of the twentieth century, coming over from French in the First World War. But the idea of cryptic disguise was there before.) Seeing both as "protective resemblances" allowed Wallace to stress the primacy of the struggle for existence (in a review essay from 1867, later widely reprinted in Wallace 1870a and 1891). Darwin's own later, most explicit treatment of coloration, in *The Descent of Man*, concerned the use that is made of it by animals competing for mates. Sexual selection in choice of partners focused right in on colors and their patterns, as (primarily) females offered themselves to the males with the most striking or gaudy feathers and wings and the like. Although Wallace accepted Darwinian sexual selection for male combat and even female choice in the case of humans, generally Wallace thought the second version of the mechanism altogether too anthropomorphic, ascribing to animals aesthetic tastes quite alien to them. Rather than seeing males as gaudy, he saw females as drab. This came from the need to conceal themselves against their backgrounds because, when they are raising a brood, they are particularly vulnerable to predators. Males have less need for protection, and so "the most brilliant hues may be developed without any prejudicial effect on the species" (Wallace 1870a, 74). In general, brightly colored animals have other means of protection or survival.

This controversy and related work opened up a whole field focusing on the Darwinian study of animal coloration, and Wallace elaborated by 1889 a far more complete system of animal coloration. Intending to update Darwin's *Origin*, Wallace's own "pure Darwinism" treated sexual selection as antagonistic to natural selection and rejected it as explanation of male coloration. By this time, Wallace had added new ideas about mimicry, camouflage, and brilliant color. Disguise also can be aggressive, as in the tiger's stripes. Bold coloration will be useful as a signal for "easy recognition," either in selecting mates or in warning potential predators. As for gaudy males, that is "probably due to his greater vigour and excitability" (Wallace 1889, 298).

Wallace also had new experimental studies to cite, including the innovative work of the young Oxford biologist Edward Poulton. One of the most surprising features of Darwinian biology before 1890 was how little experimental work was done to support the theory of natural selection. Despite Darwin's early work quantifying the survival of seedlings in planted plots, his later experiments did not include attempts to quantify selection's intensity or demonstrate its direct effect. Perhaps, consciously or otherwise, Darwin could not break from his long-held conviction that the real evidence for evolution and natural selection must always be indirect. Perhaps, too, the logic of selection was so obvious, as in mimicry and camouflage, that no one saw much need for experiments. Few took up the challenge. In the general turn to a more experimental biology, however, a new generation did begin to study the empirical details of predation, survival, and the several categories of coloration. Poulton, taking up the influential chair in entomology at Oxford, made early measurements of differential survival of variously camouflaged insects. He also became a leading promoter of Darwinian adaptation. He lamented that the wonderfully supportive indirect evidence lacked enough direct experimental proof. This he set out to address with an extensive catalog of cases, combining a careful separation of phenomena with a new terminology derived from the Greek. The comprehensive theoretical treatment of *The Colours of Animals* (1890) would guide later work on mimicry, warning colors, and camouflage.

Today, perhaps the most famous experimental support and the iconic textbook case of protective coloration is that of the melanic Peppered Moth (Majerus 1998). These are insects that conceal themselves from predating birds by mimicking the colors of the trees on which they rest, and the Darwinian story is that, as the trees in Britain got darker and darker from the deposits of soot caused by the factories in the Industrial Revolution, the moth colors in tandem got darker and darker. By the 1860s, the variety of coloration of the "pepper and salt moth" was well known among Lepidoptera enthusiasts in Britain, always alert to oddities in their collections.

Modern-day, somewhat anti-Darwinian, American evolutionists have been known to remark, contemptuously, that the British fascination with butterflies and moths is a reflection of the middle-class, inherently amateurish nature of their whole enterprise, not to mention reflecting the Anglican obsession with natural theology or (in the Darwinian case) its secular successor – seeing utility or adaptation everywhere (Lewontin 1974). However true or untrue this may be, it is certainly the case that the kind of work being discussed here was often supplemented and enriched by nonprofessionals, by people who collected and studied insects as a hobby – or who made a living out of those that did.

Such naturalists often had the most detailed evidence. This seems to have been the remarkable case where, it turns out, Charles Darwin's own attention was drawn to industrial melanism. An avid butterfly (and bird's egg) collector, Albert Brydges Farn – otherwise notable for having shot thirty birds in thirty shots on the estate of Lord Walsingham, thus establishing a record "which has probably never been equaled" (Salmon 2000, 176) – wrote the following letter to Darwin on 18 November 1878 (DCP 11747):

My dear Sir,

> The belief that I am about to relate something which may be of interest to you, must be my excuse for troubling you with a letter.
>
> Perhaps among the whole of the British Lepidoptera, no species varies more, according to the locality in which it is found, than does that Geometer, Gnophos obscurata. They are almost black on the New Forest peat; grey on limestone; almost white on the chalk near Lewes; and brown on clay, and on the red soil of Herefordshire.
>
> Do these variations point to the "survival of the fittest"? I think so. It was, therefore, with some surprise that I took specimens as dark as any of those in the New Forest on a chalk slope; and I have pondered for a solution. Can this be it?
>
> It is a curious fact, in connexion with these dark specimens, that for the last quarter of a century the chalk slope, on which they occur, has been swept by volumes of black smoke from some lime-kilns situated at the bottom: the herbage, although growing luxuriantly, is blackened by it.
>
> I am told, too, that the very light specimens are now much less common at Lewes than formerly, and that, for some few years, lime-kilns have been in use there.
>
> These are the facts I desire to bring to your notice.
> I am, Dear Sir,
>
> Yours very faithfully,
> A. B. Farn

Darwin seems not to have replied. By this time he had moved on to new experimental studies and was no longer revising *Origin*. Despite his penchant for querying amateurs for his own purposes, he also was receiving many, many letters from strangers. Even though it was a new, and empirical, study of melanism, it is also true that Farn did not explicitly tie his discovery to predation. It is rather amusing that Darwin, prescient on so many topics of modern evolutionary biology, made nothing of this demonstration of selection in action. Darwin's nonreaction may have been part and parcel of a general feeling that natural selection is never going to be sufficiently effective that we can see its actions in our own life-spans. Gradualism needs masses of time. But we ought to remember as well that the current iconic status of industrial melanism is due to the careful field study and experimentation by H. B. D. Kettlewell in the 1950s. That work was driven by the need for explicit measurements of fitness, to support the new population genetics theory behind the modern synthesis of genetics and natural selection.

Even so, before genetical theory, the entomologist J. W. Tutt in 1890 was making the connection to selection pressure explicitly. "I believe … that Lancashire and Yorkshire melanism is the result of the combined action of the 'smoke,' etc., plus humidity [thus making bark darker], and that the intensity of Yorkshire and Lancashire melanism produced by humidity and smoke, is intensified by 'natural selection' and 'hereditary tendency'" (Tutt 1890, 56). Mimicry and camouflage were the still strong supports for a Darwinian theory with selection more important than direct environmental effects or rules of development.

Moving into the twentieth century, mimicry and camouflage also spread out, somewhat unexpectedly, into new fields outside biology. In both world wars, camouflage became something of intense interest and importance, and although there was the typical resistance of the military mind to outside, innovative suggestions, in both wars the biologists had significant contributions to make. One of the most noteworthy innovations was "dazzle" (Forbes 2009) (see Fig. 15.3 and Plate XXI). Building on Wallace's early perception of the

FIGURE 15.3. The USS *Yorktown* in dazzle camouflage, 1944. U.S. government picture (pre-1955)

value of startle and disruptive color (the tiger's stripes), the idea developed that sometimes, especially when passive inaction is impossible, the best form of deception lies not in imitation but in bold coloration, both the British and the American navies were persuaded to try painting ships, not in the obvious dull ways trying to blend in with sea and sky but in bold, almost-Cubist fashion, with brightly contrasting colors. That way, observers might indeed see the ships but find it very difficult to tell their speeds, directions, and distances. It had to be pointed out gently to one admiral, annoyed at being thus deceived, that that was the whole point of the exercise.

It is nice to be able to note that there was an important feedback from the military to the biological. Today, one of the most significant if sometimes controversial notions in Darwinian biology is that of the arms race, where supposedly lines of organisms compete against each other, thereby improving their respective adaptations (Dawkins 1986). The prey gets faster, and in response the predator gets faster. Darwin knew about this, and indeed the term "race" occurs often, used either by him or by his correspondents. "Of course we believers in *real design*, make the most of your frank and natural terms, 'contrivance, purpose, &c' – and pooh-pooh your endeavors to resolve such contrivances into necessary results of certain physical processes, and make fun of the race between long noses and long nectaries!" (Darwin 1985–, 11:253, letter from Asa Gray to Darwin, 22–30 March 1863). One of Darwin's "imaginary illustrations" of selection had been how fleet deer would select for faster wolves (Darwin 1859, 90). The engineer-naturalist Thomas Belt (1874, 383) updated this image and noted how "the fleetness of both dogs and hares would be gradually but surely perfected by natural selection." Belt's concern, writing on mimicry in the tropics, was to counter anti-Darwinian arguments (including from Wallace himself) that some adaptations appear to be *too* perfect for selection to produce. He saw how perfect mimicry came about through an arms race of disguise and perception.

The idea of an arms race was discussed at length by Julian Huxley in his first book, *The Individual in the Animal Kingdom* (1912), where he talked about the ways in which the British-German naval competition was leading to ever more powerful methods of attack and correspondingly more effective methods of defense. Huxley's interest in color, however, was drawn more to courtship and sexual selection. It was Huxley's protégé Hugh Cott, a biologist who in the Second World War was one of the chief enthusiasts for applying biological principles to military camouflage, who (in his massive *Adaptive Coloration in Animals*) brought language and ideas together.

> The fact is that in the primeval struggle of the jungle, as in the refinements of civilized warfare, we see in progress a great evolutionary armament race – whose results, for defence, are manifested in such devices as speed, alertness, armour, spinescence, burrowing habits, nocturnal habits, poisonous secretions, nauseous taste, and procryptic, aposematic, and mimetic coloration; and for offence, in such counter-attributes as speed, surprise, ambush, allurement, visual acuity, claws, teeth, stings, poison fangs, and anticryptic and alluring coloration. (Cott 1940, 158–59)

By the time that Cott was writing, at the beginning of the Second World War, evolutionary thinking had undergone its biggest change since the time of Darwin. With the incorporation of modern genetics, something that dates from about ten years before, around 1930, truly one had a firm basis for the operation of natural selection (Provine 1971). As is well known, this fusing of selection with new ideas about the processes of heredity was not an easy union. At the end of the nineteenth century (in the absence of a proper understanding of heredity), there was such suspicion of natural selection that the period has been labeled the era of the "Non-Darwinian Revolution" (Bowler 1988). And, when at the beginning of the new century the true principles of heredity were uncovered, or more precisely when the thinking of Darwin's contemporary, the Moravian monk Gregor Mendel, was rediscovered and its value appreciated, far from this being celebrated as a way forward for selection-based thinking, there was somewhat of an inclination to think that this new science was the final nail in the Darwinian coffin. Naturally enough, the new "Mendelians" tended to focus on easily recognized variations, and the assumption was made that these and these alone are the significant factors in evolutionary change. There was renewed enthusiasm for the belief that transformation goes in fits and starts, in jumps – or, as they were called, "saltations" – and along with this came a downplaying of adaptation. As had been recognized by an earlier saltationist, Thomas Henry Huxley – someone who, although a great supporter of Darwin and evolution, had always had doubts about natural selection – adaptation generally demands smooth incremental change (else organisms get out of adaptive focus) and thus, as selection goes, thus goes adaptation. (No great loss in Huxley's opinion. He was on record as saying that the markings and colors of butterflies were without biological significance [Huxley 1893].)

The mimicry-camouflage issue got caught up in this debate – one that flamed heatedly in the first two decades of the twentieth century, as the Darwinian adaptationists fought back strongly against the geneticists (as they now started to be called). One might have predicted this, for the one place where enthusiasm for Darwinian selection never wavered was among precisely those biologists for whom such selection was a real fact of life, namely those biologists working on fast-breeding organisms, often highly vulnerable to the threats of predators and other, hostile, outside forces. In short, those biologists working on such organisms as butterflies and moths. One might think that mimicry of all things would be something even the Mendelians would yield to the selectionists. Not a bit of it! To defend their thinking, they needed all of the resources at their disposal – namely (by the beginning of the century), an incredibly detailed knowledge of mimicry and camouflage in nature backed by an increasing number of breeding studies.

A preliminary shot across the bow had already been fired by William Bateson, who was to become the first and loudest proponent of the new Mendelian genetics. Back in 1894, he was hammering away at the idea of universal adaptation. "We, animals, live not only by virtue of, but also in spite of what we are. It is obvious from inspection that any instinct or organ *may* be of use; the real question we have to consider is *how much* use it is" (W. Bateson 1894, 12). Then, at the hands of Reginald Punnett (1915) – Bateson's protégé and successor at Cambridge as the first Arthur Balfour Professor of Genetics – mimicry was touted as the paradigmatic refutation of the adequacy of a Darwinian approach. Punnett's objections included the worry (often expressed by other critics of Darwinism) that he could not see how a slight variation in a butterfly wing color toward the pattern of another species could possibly be of selective significance; he could see no evidence of significant bird predation; he could not see why often in a species males do not mimic, whereas females do; and, most importantly, he could not see how you could reconcile with Darwinian selection the ever-increasingly empirically verified fact that the different forms or morphs of mimicking species are under tight Mendelian control. How could selection bring about a species where there are two or three different mimicking forms, sometimes produced by the same mother in the same batch? These forms must have been one-shot productions. Punnett did not want to deny that selection might have a cleansing effect when genetics has done its work, but it could not be involved creatively in producing the forms. Rather cruelly breaking into verse about Batesian mimicry, referring to the original poisonous forms as A and the mimics as B, Punnett (1913, 147) quipped:

> *See how the Fates their gifts allot,*
> *For A is happy, B is not,*
> *Yet B is worthy, I dare say,*
> *Of more prosperity than A.*

Poulton (1913, 1914) struck back. There were some things that were just not going to bother an experienced Darwinian. The evidence of predation was clearly there. The differences between males and females is, for someone brought up on sexual selection, almost to be expected, especially because the evidence is that females tend to show greater variability and that they are under greater danger from predators. But although he was prepared to accept the basic principles of Mendelism, Poulton was handicapped by a general lack of understanding or sympathy for genetics. His own emphasis was on the ecological scenario.

It was obvious that to speak to the major charge about the distinctive and significant Mendelian-controlled differences one finds in one family, one needed a supplemental hypothesis about how the effects of Mendelian genes can be modified. One needed some way of showing that a major effect can be built up slowly by stages, and that thus selection can operate all of the way. The crucial work was actually being done at Harvard by the American geneticist W. E. Castle, who was showing that genes can work in just this gradual way desired (Ruse 1996). Unfortunately though, Castle (early in the second decade of the century, at the time Poulton became aware of his work) was arguing that the genes themselves can be modified by selection. But the consensus was that hereditary factors are stable, and Castle's conclusion was controversial. It was not until the end of the decade that, under the influence of his student Sewall Wright, Castle (1917) became fully committed to Mendelian factors and the need for mutation. Gradual alteration of character, he realized, was because other "modifier" genes can affect the expression of the crucial color producing genes. Fortunately for the Darwinians, this was taken up by someone who could work the genetics, a young enthusiast, trained as a physicist, unbelievably gifted in mathematics, and drawn to genetics because of its perceived utility for eugenics. Despite his close acquaintance (and shared efforts in eugenics) with Punnett at Cambridge, Ronald A. Fisher was a committed Darwinian, looking for ways to build a theory of gradual genetic modification. In a paper of 1927, shortly thereafter incorporated into his seminal work *The Genetical Theory of Natural Selection* (1930), Fisher took up Poulton's challenge to Punnett, showing just how, thanks to modifying genes, selection could produce different forms within the same family – different forms being transmitted completely in accordance with Mendelian principles. One final piece of work was needed to put the worries to rest. E. B. Ford was a student at Oxford in the 1920s, where anyone interested in coloration and Darwinism found a network of connections and support from Poulton and Julian Huxley. On the advice of Huxley, Fisher, greatly handicapped for empirical work because he had appalling eyesight, sought out Ford (Huxley's student), and this led to a many-year, very fruitful collaboration, with Ford doing the empirical studies (as well as writing shorter texts that popularized the often mathematically ferocious writing of Fisher). Driven by population genetics, work on mimicry and camouflage also became more quantitative and experimental. Ford's studies of butterflies convinced him from the start that selection pressures in nature are often much greater than Darwinians had hitherto supposed. Hence Ford was able to argue and to back up the belief that even very minor variations – just a slight move in coloration from one form to another – could well be picked up and cherished by natural selection (Carpenter and Ford 1933). Punnett's final objection was answered.

In respects therefore, in the hundred years covered in this discussion, we have an almost perfect cameo of the Darwinian story. On the topic of mimicry and camouflage, there were interesting discoveries and discussions before Darwin published – discoveries and discussions that were, as often as not, framed in a nonscientific, theological context. Darwin provided the evidence for evolution and the mechanism of natural selection. However, although he certainly identified such pertinent characteristics as coloration as things that could be controlled by selection, he did not contribute directly to the mimicry-camouflage debate. This work fell to others, who over a half century and more made very significant contributions to both empirical understanding and (within a Darwinian context) theory. The debate got caught up in the controversies sparked by the arrival of the new Mendelian genetics. Ultimately, neither a naïvely adaptationist view nor a purely laboratory study of genes could explain mimicry. Darwin's great insight to combine ecological pressure and inherited variation was validated. As (with some difficulty but ultimately with great success) genetics was melded with Darwinian selection, the work on mimicry and camouflage was seen to be even more significant and supportive of the Darwinian picture than anyone could have imagined.

Essay 16

Chance and Design

John Beatty

IN THE *ORIGIN*, Darwin explained that he used the term "chance" variation only to signify his (and others') ignorance of the process by which new traits arise. In an enthusiastic review of the book, Asa Gray suggested a friendly amendment: that as long as the cause of variation was unknown, it should be attributed to God. Gray's idea was that God had arranged for particular traits to arise in particular lineages at particular times, to be subsequently accumulated by natural selection. Before Gray's suggestion, Darwin had hoped that evolution by natural selection might be viewed simply as God's way of making new species. But in reflecting on Gray's suggestion, Darwin realized that reconciling evolution by natural selection with any sort of conventional theology was going to be much more difficult than he had imagined (Figs. 16.1 and 16.2).

On the one hand, unless God arranged for just the right traits to arise in just the right lineages at just the right times (as Gray recommended), then evolution by natural selection would not guarantee the existence of any particular evolutionary outcomes, humans included. And surely humans were the end of Creation, no matter what the means. On the other hand, this way of making species required an awful lot of trouble on God's part. Evolution by natural selection initially seemed like such a simple way for God to proceed: he only had to establish a set of laws (to govern population growth, inheritance, etc.) and then sit back and wait for species to make themselves, rather than creating each one separately. But the degree of divine guidance that Gray suggested – presumably for each and every species, not just humans – was as much or more trouble than special creation. Indeed, it was just a complicated form of special creation.

This degree of micromanagement of the direction of evolution by natural selection was incomprehensible to Darwin on other grounds as well. It suggested God's preference for *particular*, well-adapted forms of life rather than just whatever well-adapted forms of life would have resulted from evolution by natural selection without any further guidance on his part. It is one thing to imagine God providing just the right variations at just the right times, to be accumulated by natural selection, thus leading to the evolution of humans. But it is another thing altogether to imagine him going to the same lengths for every other species, providing just the

FIGURE 16.1. Charles Darwin in midlife. Even before he published on evolution, he was a very respected scientist, and undoubtedly this reputation helped to spread and win acceptance for his belief in evolution. From F. Darwin, *Life and Letters of Charles Darwin* (London: John Murray, 1887)

right variations at just the right times to get precisely *these* forms of life and not others. Why *these*? Sheer caprice? And given some of the hair-raising forms of life that have evolved (more on these shortly), it would seem to be pretty dark caprice at that. Better, Darwin responded to Gray, to assume not only that God relied on evolution by natural selection as his means of creating new species, but also that he relied on natural selection of *whatever variations arise – unguided*.

Partly in response to Gray's suggestion about God's role in the production of variation and partly on the basis of his own studies of variation, Darwin came to the conclusion that variation was even more a matter of chance than he had initially acknowledged. It was not just a matter of ignorance concerning the causes of variation, for he was not entirely ignorant. First, he was sure that particular variations are a matter of chance in the sense that they are *not a matter of design*. Whatever the correct explanation of the occurrence of a particular variation, it was *not planned*. Second, Darwin was sure that the interacting causes of variation are so complex that the *prediction of any particular variation was "hopeless"* – as hopeless as predicting the outcome of a *lottery*.

The precise course of variation and the outcomes of evolution by natural selection may even be unforeseen by God, which would be just as well. By relieving God of the control

FIGURE 16.2. Asa Gray (1810–88), professor of natural history at Harvard University and Darwin's great North American champion. But his proposal for reconciling Christianity and evolutionary theory was rebuffed by Darwin in correspondence and in print. From J. L. Gray, *Life and Letters of Asa Gray* (Boston: Houghton Mifflin, 1894)

of variation, and perhaps even the anticipation of it, Darwin absolved God of the determined pursuit of capricious choices, many of them odious.

"DESIGNED LAWS" AND THE "DETAILS LEFT TO CHANCE"

Again, in the *Origin* Darwin (1859, 131) attributed the appearance of new traits to "chance," by which he meant to acknowledge his and others' ignorance as to when, where, and why they arise. The fate of variation was determined, he was sure, by natural selection. But the conditions governing the initial appearance of a trait were obscure. In a review of the *Origin*, Gray (1860, 413–14) sought to plug this hole in Darwin's theory with God:

> [A]t least while the physical cause of variation is utterly unknown and mysterious, we should advise Mr. Darwin to assume, in the philosophy of his hypothesis, that variation has been led [by God] along certain beneficial lines.

Darwin replied in correspondence:

> With respect to the theological view of the question. This is always painful to me. I am bewildered. I had no intention to write atheistically. But I [admit] that I cannot see as plainly as others do, and as I should wish to do, evidence of design and beneficence on all sides of us…. [For instance] I cannot persuade myself that a beneficent and omnipotent God would have designedly created the Ichneumonidæ with the express intention of their feeding within the living bodies of Caterpillars. (Darwin 1985–, 8:224, letter to Gray, 22 May 1860)

Ichneumonids are parasitic wasps that deposit their eggs in insects of other species, where they develop and consume their host from the inside out, eventually emerging (a la *Alien*, the movie) to repeat the hoary cycle. Was Gray suggesting that God had provided just the right variations at just the right times in just the right lineages to ensure the existence of the repugnant ichneumonids? What could possibly have been going through God's mind to have settled on this particular form of life, and then to have jigged nature so carefully to ensure its existence? (On the Darwin-Gray correspondence, see Lennox 2010 and Hunter 2009. On the "dark" side of evolution, see also Müller-Wille 2009.)

Darwin's favorite criticism of special creationists was to show how much caprice they had to impute to God in order to explain many phenomena that did not evince his trademark benevolence and wisdom. All special creationists can say in such cases, Darwin complained, is that "it has pleased the Creator to cause a being of one type to take the place of one of another type" (e.g., 1859, 185–86; 1875c, 1:9). Better not to implicate God than to imply his capriciousness. In the case at hand, it was better to suppose that he had not so closely attended to the order in which variations arise, and hence the exact outcomes of evolution by natural selection, than to suggest that he was hell-bent on the existence of parasitic wasps, of all things.

Perhaps, as well, God did not guide the course of variation in such a way as to ensure the evolution by natural selection of humans. Darwin chose a roundabout way of putting this point to Gray (in the same letter as above, 22 May 1860): "Not believing this [that God guides the course of variation to ensure the existence of parasitic wasps], I see no necessity in the belief that the eye was expressly designed." However unsettling the conclusion that *we* were not designed, it was nonetheless more satisfying to Darwin than the alternative of a disturbingly capricious God. Better to credit God with designing *general* laws of nature that ensure the evolution of well-adapted forms of life, but not with designing any *particular* forms. As Darwin continued in his private response to Gray, "I am inclined to look at everything as resulting from designed laws, with the details, whether good or bad, left to the working out of what we may call chance (Darwin 1885–, 8:224, letter to Gray, 22 May 1860)."

In referring to the "details left to chance," Darwin was employing a concept of chance (and, by implication, chance variation) beyond that of ignorance. This was the concept of *chance as opposed to design*; to attribute something to chance, in this sense, was to deny that it had been the object of intention, human or divine. In the context of natural history, one contrasted chance with divine design. This was, for instance, the way William Paley treated "chance" in his influential text *Natural Theology* (1802), which Darwin read carefully in his student days:

> What does chance ever do for us? In the human body, for instance, chance, i.e. the operation of causes without design, may produce a wen, a wart, a mole, a pimple, but never an eye. (1809, 62–63)

Things left to chance in this sense, being unintended, are not the sorts of things for which one assigns blame. For example, in the passage just quoted, Paley implicitly praises God for eyes and does not hold him responsible for wens, warts, moles, and pimples. Darwin employed the same distinction – between what God intended and deserves praise for, and what he left to chance and cannot be blamed for – but Darwin applied the distinction differently. God deserves praise for the general laws of nature, which he carefully designed to order to ensure well-adapted forms of life. But he deserves neither praise nor blame for the particular variations that arise and are accumulated by natural selection leading to the evolution of particular traits (eyes) and particular forms of life (parasitic wasps and humans). Again, this may be unsettling with respect to *our* importance in God's mind, but it absolves him of responsibility for many more repugnant forms of life:

> [I]t may not be a logical deduction, but to my imagination it is far more satisfactory to look at such instincts as the young cuckoo ejecting its foster-brothers, – ants making slaves, – the larvae of ichneumonidae feeding within the live bodies of caterpillars, – not as specially endowed or created instincts, but as small consequences of one general law leading to the advancement of all organic beings, – namely, multiply, vary, let the strongest live and the weakest die. (Darwin 1872a, 234)

Cuckoos lay their eggs in the nests of other species; their chicks hatch first and eject the eggs of their foster siblings. Slave-making ants raid colonies of other species of ants, steal the young, and press them into the service of their own colonies. Why in the world would God have settled on such macabre forms of life, which he then pursued intently by providing just the right variations at just the right times and places for natural selection to accumulate so that evolution would result in these outcomes? (With regard to the slave-making ants, keep in mind that Darwin was a fervent abolitionist. See Desmond and Moore 2009.)

But there was a problem with the idea of God leaving the order of variation – or anything else for that matter – to "chance" in this sense. And Darwin was aware of the problem. God is after all omniscient as well as omnipotent. So having designed the laws of nature, and having decided upon an initial arrangement of matter, he would surely have foreseen the results, and he would have had a chance to "revise and resubmit" (so to speak). If at that point God did not revise, then he effectively ordained the outcomes. As Darwin sometimes put it, "to foresee is the same as to preordain" (Darwin 1985–, 8:106, letter to Gray, 24 February 1860). In that case, though, what could God ever truly leave to chance and shed responsibility for?

Darwin employed two examples to illustrate how God might leave certain things, including variation, to chance and thus be acquitted of responsibility for them. Continuing his response to Gray:

> The lightning kills a man, whether a good one or bad one, owing to the excessively complex action of natural laws. A child (who may turn out an idiot) is born by the action of even more complex laws. (Darwin 1985–, 8:224, letter to Gray, 22 May 1860)

What is the point here? Darwin did not elaborate. But he seems to be stressing the complexity of the causal processes in question: the *"excessively complex* action of the natural laws" that govern lightning, and the *"even more complex"* combination of laws that govern inheritance and variation. He seems to be suggesting that the complexity of these interactions distances God from the consequences. As if the consequences – a good man, a *particular* good man being struck by lightning, and an innocent child, a *particular* innocent child being born with a serious disadvantage – were too unpredictable to attribute to design, and hence too unforeseeable to be blameworthy.

In this regard, it is worth saying a bit more about the degree of complexity of the causal processes influencing variation, from Darwin's point of view. At the time of his correspondence with Gray, Darwin was already well into a massive study of the origin of variation. He focused on variation in domesticated species, in order to take advantage of the watchful eyes and testimony of a multitude of breeders; the results were finally published in his two-volume work, *The Variation of Animals and Plants under Domestication* (1868b, 1875c). His working hypothesis was that variations arise in offspring when environmental changes affect the reproductive organs of their parents, and that particular sorts of environmental changes lead to the appearance of particular new traits. This hypothesis was only partly borne out. Ultimately, he distinguished between two classes of variation: those that are predictable given particular changes in the environment (he called these "definite" variations), and those that are unpredictable (what he called "indefinite" or "fluctuating" variations). He believed that most variations were of the latter sort (e.g., Darwin 1872a, 5–8; 1875c, 2:260–82). In the case of indefinite variation, he came to believe that not only the environment but also the particular constitution of the parents was relevant to the outcome. This would explain the appearance of "nearly similar modifications under different conditions, and of different modifications under apparently nearly the same conditions" (1875c, 2:281). Given that no two members of a species are constitutionally identical, and that no two members of a species face identical environments, it would be extremely difficult to foresee what variations would arise from generation to generation.

> No doubt each slight variation must have its efficient cause; but it is as hopeless an attempt to discover the cause of each, as to say why a chill or a poison affects one man differently from another. (282)

As one reviewer commented:

> The one strong impression that affects the reader ... is that of the endless complication of the phenomena in question, and the (perhaps hopeless) subtlety and occultness of the immediate causes. At the first glance, the only "law" under which the greater mass of the facts ... can be

grouped seems to be that of Caprice, – caprice in inheriting, caprice in transmitting, caprice everywhere, in turn. (Anonymous 1868, 362)

Indeed, for Darwin variation was like a lottery. The only way to increase one's chances of winning a fair lottery is to purchase more tickets. And the only way to increase the probability of a particular variation occurring, Darwin argued, was to increase the size of the population. This is why commercial breeders are so much more successful than individual breeders, and why rich breeders are so much more successful than their poorer counterparts.

> [A]s variations manifestly useful or pleasing to man appear only occasionally, the chance of their appearance will be much increased by a large number of individuals being kept. Hence, number is of the highest importance for success. Of this principle Marshall formerly remarked, with respect to the sheep of parts of Yorkshire, "as they generally belong to poor people, and are mostly in small lots, they never can be improved." (Darwin 1859, 41; 1875c, 2:221, 230, 234–35)

And similarly, "Lord Rivers, when asked how he succeeded in always having first-rate greyhounds, answered, 'I breed many, and hang many'" (1875c, 2:221).

In the closing pages of the study, Darwin revisited his earlier conversation with Gray, but this time publicly. It was much better to assume – in the case of artificial selection, as well as natural selection – that God left the order of variation to chance, than to assume that he arranged for particular variations to occur in particular lineages at particular times, to be subsequently accumulated. To make the point, Darwin employed an analogy involving an "architect" who constructs a building using only stones fallen from a nearby precipice. The architect selects stones with appropriate sizes and shapes for the uses to which they are put: large, rectangular stones for the foundation, light flat stones for the roof, etc. That there are stones of these sizes and shapes at the base of the cliff is a consequence of a complex interaction of various laws governing the composition and formation of the cliff face, its erosion, gravity, and more. We might say that God designed these laws. But would we also say that he foresaw and preordained that stones of precisely those sizes and shapes would be formed in falling, in order that the architect could construct precisely the building that he (God) had in mind? Why *that* building rather than some suitably functional building that the architect could construct without all that guidance? And given the complexity of the causal processes involved in producing the stones, and the complexity of prearranging things so that just the right shapes and sizes were produced, God must have been awfully determined to get precisely the outcome in question. But is it really conceivable that he attends to such details with such care? Darwin urged chance or "accident," not design, as the more reasonable alternative in explaining the shapes and sizes of the stones.

If, on the other hand, one insists that God attends to every detail, including the architect's materials, then does he also provide the breeder's materials? Did he arrange for just the right variations to occur at just the right times in order for dog breeders to create (via artificial selection) bull-fighting terriers that clamp down their huge jaws over the bull's snout, not letting go until the bull nearly suffocates?

> Did He [God] cause the frame and mental qualities of the dog to vary in order that a breed might be formed of indomitable ferocity, with jaws fitted to pin down the bull for man's brutal sport? (Darwin, 1875c, 2:430–31)

Again, better to blame this outcome on the capriciousness of variation, together with the combined caprice and malice of dog breeders, than on God. Although again, God's omniscience makes problematic the idea that he had not foreseen what would happen. And by allowing it, had he not preordained it?

"GOD DOES NOT PLAY DICE" OR?

Darwin's attempted solution to this problem was to stress that the processes governing variation were so complex, and the outcomes so unpredictable as to render it incomprehensible why God would have chosen such means to guarantee such *particular* ends. Surely the choice of such means was more consistent with God being not so concerned with the *particular* outcomes of evolution.

How unconventional an idea was it, that God might have left the world so largely to chance? Interestingly, the idea was favorably entertained by – of all people – Paley. This might be surprising given the way Paley (1809, 62–63) expressed his aim in *Natural Theology*: "I desire no greater certainty in reasoning, than that by which chance is excluded from the present disposition of the natural world." But he immediately followed by allowing that God left such things as wens, warts, moles, and pimples to chance. Paley did not see these outcomes only as blemishes on an otherwise thoughtfully designed world. Their unsightliness was not his only reason for excluding them from God's design. These are also irregularly or unpredictably occurring phenomena. They are the unforeseen – certainly unintended – consequences of the interaction of multiple laws of physiology and development.

Paley went even further, arguing not only that God *left* some things to chance but that he also *intentionally designed* the *appearance* of chance, and also *chance itself* into the world. In the last chapter of his book, Paley posed the question: "Why, under the regency of a supreme and benevolent Will, should there be, in the world, so much, as there is, of the appearance of *chance*?" (513). He gave a number of different answers.

One reason involves God's intention to create a world that is difficult for us to predict. Uncertainty can be useful to us. For example, "It seems to be expedient, that the period of human life should be *uncertain*" (517). If mortality was predictable, then it would lead to a growing horror on the part of each of us as we approached our calculated time, "similar to that which a condemned prisoner feels on the night before his execution" (517). That's no way to live productively. As another example, uncertainty concerning seasonal changes and the weather has a beneficial effect by rewarding attention to climatic details, thus promoting the progress of agricultural science. Indeed, Paley contended, agriculture languishes most in those regions where conditions are most predictable (518). Of course, in this case, Paley acknowledges, there needs to be some "mixture of regularity and chance" in order to bring about the desired effect. Similarly, one presumes, mortality would have to involve some mixture of regularity and chance in order to promote the progress of medicine. In these cases, God has built the appearance of chance into the world, even if he has not really left the world to chance.

A second reason for the appearance of so much chance in the world involves the possibility that God actually designed *chance, and not just the appearance of chance*, into the system. Paley raised this possibility by way of analogy with cases in which humans employ lotteries for purposes of fairness. Among equals, he noted, there is often no better way to assign privileges and duties than by lottery. In these instances, we intentionally rely on chance:

> Work and labour may be allotted. Tasks and burthens may be allotted.... Military service and station may be allotted. The distribution of provision may be made by lot, as it is in a sailor's mess; in some cases also, the distribution of favours may be made by lot. In all these cases, it seems to be acknowledged, that there are advantages in permitting events to chance, superior to those, which would or could arise from regulation. In all these cases, also, though events rise up in the way of chance, it is by appointment that they do so. (516–17)

But there are some assignments in society that it would not be so reasonable to decide by lot. Every complex economy depends on a variety of occupations, and these should be filled by people with suitable abilities. Some people are cut out to be laborers; others are cut out to be land or factory owners. It is not unfair that a person cut out to be a leader enjoys a leadership position. But because such differences in abilities are associated with differences in privilege and prestige, it would be unfair to have them distributed in any other way than by lot. So God designed the world with this sort of chance variation in ability built in.

> It appears to be ... true, that the exigencies of social life call ... for a mixture of different faculties, tastes, and tempers.... Now, since these characters require for their foundation different original talents, different dispositions, perhaps also different bodily constitutions; and since, likewise, it is apparently expedient, that they be promiscuously scattered amongst the different classes of society: can the distribution of talents, dispositions, and the constitutions upon which they depend, be better made than by chance? (521–22)

One envisions the goddess Fortuna, veiled like Justice, carrying her roulette wheel of life. Only now it is the world itself that is a roulette, serving the supposedly just ends of God. Our fate is in many respects, according to Paley, "the drawing of a ticket in a lottery" and is thus "left, to chance, without any just cause for questioning the regency of a supreme Disposer of events" (520).

CONCLUSION

Darwin's God created a world in which some things were undesigned, and some of those things, like the production of variation, were as unpredictable as a lottery. Darwin did not go as far as Paley, who suggested that the production of some variations (variations in abilities, among humans) were *designed to be governed by chance*. But at least with Paley, there was a respectable precedent for the idea that God *left* some things – including the production of variation – to chance.

The idea that the production of variation was left to chance was in keeping with Darwin's extensive studies of variation. And it was satisfying to Darwin on theological grounds as well, insofar as it absolved God of capriciousness that bordered on the obscene.

In closing, it is worth noting one more reason why, according to Paley (1809, 523–24), there appears to be so much chance in the world: by making the world effectively, if not in principle unpredictable, God leaves room for a certain amount of unnoticeable, supernatural intervention in securing very particular ends. This is what Gray had in mind in his recommendation to Darwin. That is, the lack of discernible regularities concerning variation does not represent a mere lack of design but rather the height of design. Against this background of irregularity, God can unsuspectingly intervene in the fate of the world – supernaturally injecting just the right variations at just the right times to be subsequently selected, thus ensuring just the right forms of life.

For Darwin, this would have involved God masking his own capriciousness with the pretense of capriciousness in nature – hiding his own preferences for parasitic wasps, cuckoos, slave-making ants, and pit bulls, together with humans, behind the elaborately contrived unpredictability of evolution.

Which was unacceptable.

Darwin's God was not capricious. Nature on the other hand?

Essay 17

Darwin and Teleology

James G. Lennox

A TELEOLOGICAL EXPLANATION IS one in which some property, process or entity is said to exist or be taking place *for the sake of* a certain result or consequence. For example, after returning from a run, someone might ask, "Why did you go for a run?" If you answer, "In order to keep fit," you are explaining your run by pointing to a consequence of running, keeping fit. Or, someone might ask, "Why do hawks and owls have sharp, hooked beaks and talons?" If one answers, "Those sharp talons and hooked beaks are for the sake of capturing and eating their prey," these traits are explained by reference to the (valuable) consequences for the organism of having those traits. It is not just that they have these traits and these traits serve a valuable function for these birds – they have these traits *because* they serve this valuable function.

Teleological explanations have played a central role throughout the history of the life sciences. Biological textbooks invariably suggest that teleological explanations were expunged from the physical sciences in the seventeenth century and finally, thanks to Charles Darwin, from the biological sciences in the nineteenth. And yet the same textbooks often explain adaptations by reference to natural selection in language that sounds suspiciously teleological. "That color pattern is present in the males of that population of fish because it increases their attractiveness to female mates without increasing their visibility to predators."

Moreover, explanations that at least appear to be teleological are not restricted to the observable, phenotypic adaptations of vertebrate behavior. Notice the explanatory structure implicit in the following quotation from Albert Lehninger's *Bioenergetics: The Molecular Basis of Biological Energy Transformations* (1971, 110; emphasis added).

> Thus photo-induced cyclic electron flow has *a real and important purpose*, namely, *to* transform the light energy absorbed by chlorophyll molecules in the chloroplast into phosphate bond energy.

A common response to passages such as this is to say that the use of the term "purpose" is merely a kind of shorthand for a more complicated mechanical explanation, not evidence of a commitment to teleology. Yet this passage is embedded in a detailed description of the mechanisms of photosynthesis, and historically the discovery of the process described led to a quest for its purpose. Biochemists did not feel that they understood cyclic electron flow until they figured out its biological function.

Such dismissive responses are likely due to two primary concerns raised by such explanations of natural, biological phenomena. First, there is a concern that a teleological explanation implies that something that *will* result only *after* a process has occurred is the cause of that process – and thus that such explanations imply temporally backward causation. Second, there is a concern that such explanations imply some sort of conscious, or anyway cognitive, agency – either in the form of an external, perhaps divine, agent, or in the form of an inherent drive or vital power. Much philosophical effort has been devoted in the past fifty years or so to making sense of natural teleology as a distinctive mode of explanation *without* accepting either of those implications.

Are explanations of adaptations by appeal to natural selection *teleological* explanations? Many philosophers of biology would answer affirmatively, whereas most practicing evolutionary biologists would answer in the negative. Part of the reason for this discrepancy is simply a matter of terminology: many of those answering in the negative would, I suspect, answer affirmatively to a question like the following: "Is it appropriate to explain the presence of a trait in a population by appealing to its value in enhancing fitness?" But to explain the presence of a trait by appealing to its value consequence for its possessors is to offer a teleological explanation of that trait, at least as such explanations are typically understood.

Lying behind this confusion are, I believe, deeper historical and philosophical issues that are complicated and significant for understanding contemporary misunderstandings about evolutionary biology. I spend a good part of this essay on teleology clearing up some historical misunderstandings and then make use of these historical results to clarify the sense in which an entirely naturalistic understanding of natural selection may nevertheless be robustly teleological.

At the most abstract level, three distinct positions have been defended regarding the legitimacy of teleological explanation in natural science, and all have their roots in ancient Greece. Plato's *Timaeus* and *Laws* argue that much about the natural world can be accounted for only by supposing the operations of an intelligent and beneficent God (in fact one with a penchant for mathematics), a view I refer to as "unnatural teleology." Unnatural teleology takes the application of teleological explanation to natural phenomena to depend on the natural world being an artifact of a divine, extranatural, intelligent agent. Aristotle's *Physics* and biological writings defend what I term "natural teleology," according to which the natures of living things act for the sake of their own development and preservation. On this view, there may be a place for the use of analogies drawn from human crafts in thinking about teleology, but teleology is an entirely natural phenomenon. Finally, the Greek atomists argue against the legitimacy of teleological explanation in nature, the anti-teleology position. The unnatural "intelligent design" model found in Plato was melded to Christianity in the medieval period, and various medieval commentators on Aristotle attempted to downplay the differences between Plato and Aristotle on this score in the interests of integrating Aristotelian philosophy and Christian theology.

The story gets more complicated in the early modern period, primarily because of three distinct, nonatomistic voices arguing against the legitimacy or value of teleological reasoning in natural science. René Descartes injects a new, skeptical argument against the use of teleology in natural science – it is presumptive to think we can discern God's plans by studying his creation. Francis Bacon argued that final causes are of value only in the study of human affairs; in the study of nature, they are "barren virgins." Baruch Spinoza argued against teleology on grounds of a thoroughgoing determinism – natural events did not happen for the sake of some end but were inevitable manifestations of God's nature.

But teleology was not without powerful allies in the seventeenth century. Robert Boyle in *Disquisition about the Final Causes of Natural Things* and John Ray in his *The Wisdom of God as Manifest in the Works of Creation* develop a Christian version of Platonic unnatural teleology into the form that comes to be known as natural theology, the form of teleological reasoning that Darwin studied carefully in the writings of William Paley in his years at Cambridge (Fig. 17.1). In his unpublished autobiography, Darwin reports that as an undergraduate he did not question Paley's premises and was thoroughly convinced by his logic (Plate XXII). By contrast, in the eighteenth and early nineteenth centuries there were also defenders of teleology who aimed to distance their defense of final causes from an unnatural source, in particular in Germany (relying on Kant's *Critique of Teleological Judgment*) and France (Georges Cuvier's principle of "conditions of existence" is often associated with final causes in the literature of this period). Because the context for this entry is evolution, in the following section I focus on the interaction between Charles Darwin and those more or less explicitly influenced by natural theology.

CHARLES DARWIN AND TELEOLOGY

As already noted, Darwin was well aware, through his study of Paley's *Natural Theology*, of the importance of biological adaptation to the argument from design that has a modern lineage going back through Robert Boyle and John Ray to ancient roots in Plato. Among Darwin's private Species Notebooks is an abstract he made, with critical comments, of *Proofs and Illustrations of the Attributes of God from the Facts and Laws of the Physical Universe, being the Foundation of Natural and Reveal Religion* of John Macculloch (Barrett et al. 1987). In this work, likely written in late 1838 shortly after his encounter with Malthus gave him the key to natural selection, Darwin is clearly testing natural selection's ability to explain the adaptations adduced by Macculloch as evidence of the Creator. At one point he comments: "The Final cause of innumerable eggs is explained by Malthus, – [is it anomaly in me to talk of

FIGURE 17.1. Although the great chemist Robert Boyle (1627–91) was a leading spokesman for the Scientific Revolution, he always insisted that organisms call for final-cause understanding. Permission: Wellcome

Final causes; consider this! –] consider these barren Virgins" (Barrett et al. 1987, Macculloch, 58r). Darwin had likely seen Francis Bacon's disparaging reference to final causes as barren virgins while reading William Whewell's contribution to the Bridgewater Treatises. This comment points to Darwin's awareness, from early on, of the following question: Once we have a reasonable naturalistic explanation of biological adaptation, are we able to dispense with teleology? The evidence I am about to discuss, from Darwin's later work, is that Darwin answered in the negative. (For more on this, see Lennox 1993, 2010, and Ruse 2003, ch. 6)

In 1862 Charles Darwin presented the results of his research on sexual dimorphism in the genus *Primula* to the botanical section of the Linnean Society. In the published version, he wrote:

> The meaning or use of the existence in *Primula* of the two forms in about equal numbers, with their pollen adapted for reciprocal union, is tolerably plain; namely, to favour the intercrossing of distinct individuals. With plants *there are innumerable contrivances for this end; and no one will understand the final cause of the structure of many flowers without attending to this point*. (Barrett 1977, 2:59; emphasis added)

It will be noted that Darwin describes himself as engaged in a *teleological* enquiry, a search for the *final cause* of a particular feature of these two varieties of plants. And he refers to the different mechanisms to encourage intercrossing in different plants as contrivances that are present *for the sake of that end*. Darwin (1862c) had earlier written a well-received study of the "contrivances" found in orchids to promote fertilization by insects. That work was much admired by Asa Gray, a self-taught American botanist who in 1842 had been designated the Fisher Professor of Natural History at Harvard College. Darwin and Gray began corresponding on botanical topics in 1855, and in 1857 Darwin revealed to Gray, a reform Presbyterian, that he was working on a book that will present a new theory of species transformation – and was pleasantly surprised by Gray's cautiously positive reaction. Emboldened, later that year Darwin sent Gray a brief sketch of his theory. This sketch was then incorporated into the material presented, along with Alfred Russel Wallace's "On the Tendency of Varieties to Depart Indefinitely from the Original Type," to the Linnean Society in 1858 – Darwin's first public presentation of his theory of natural selection. A year later he was to publish the work that was to introduce evolution by natural selection to the biological sciences, *On the Origin of Species*.

The *Origin* characterizes natural selection as a goal-directed, teleological force. In introducing the concept in chapter 4, for example, he speaks of it "daily and hourly scrutinizing throughout the world, every variation, however slight; rejecting that which is bad, preserving and adding up all that is good" (1859, 84); and he goes on to tell us that "natural selection can act only through and for the good of each being." And, in a later appreciation of Darwin, Gray (1874, 80) makes direct reference to the overtly teleological character of the botanical work published after the *Origin*, urging Darwin's readers to "recognize Darwin's great service to natural science in bringing back to it Teleology; so that, instead of Morphology vs. Teleology, we have Morphology wedded to Teleology." We find similar language already in his 1862 review of Darwin's monograph on orchid fertilization, in which Gray (1862b, 428–29) applauds Darwin for having "brought back teleological considerations into botany." In response, Darwin (DCP, 9483, letter from Darwin to Gray, 5 June 1874) underscores his agreement with Gray's characterization of his theory as teleological: "What you say about Teleology pleases me especially and I do not think anyone else has ever noticed the point." And though "Darwin's Bulldog," Thomas Henry Huxley (1893, 86), is ambivalent about Darwin's obsession with adaptation, he makes much the same point as Gray: "The apparently diverging teaching of the Teleologist and of the Morphologist are reconciled by the Darwinian Hypothesis" (on Gray and Huxley, see Ruse 2003, ch. 7).

Nevertheless, in corresponding with others, Gray shows himself aware that he and Darwin ground their teleology in very different ways. In a letter to Alphonse de Candolle in 1863, for example, Gray admits he recognizes that Darwin does not accept the inference from the presence of ends in nature to an unnatural designer:

> Under my hearty congratulations of Darwin for his striking contributions to teleology, there is vein of petite malice, from my knowing well that he rejects the idea of design, while all the while he is bringing out the neatest illustrations of it! (Jane Loring Gray 1893, 498)

And Darwin is equally aware that there is a deeper disagreement behind their common endorsement of teleological reasoning. In an 1861 letter to Gray, Darwin reports on Sir John Herschel's first public response to the *Origin*, in a new edition of Herschel's *Physical Geography*.

> [He] agrees to certain limited extent; but puts in a caution on design, so much like yours that I suspect it is borrowed. – I have been led to think more on this subject of late, & grieve to say that I come to differ more from you. It is not that designed variation makes, as it seems to me, my Deity "Natural Selection" superfluous; but rather from studying lately domestic variations & seeing what an enormous field of undesigned variability there is ready for natural selection to appropriate for any purpose useful to each creature. – (Darwin 1985–, 9:162, letter, 5 June 1861)

This exchange reveals deep disagreement over whether a significant element of chance, in the form of that "enormous field of undesigned variability," is compatible with a teleological account of adaptive modification. As with a number of Darwin's closest scientific allies, Gray saw natural laws as laws of "intermediate causes," instituted, and perhaps maintained, by God. Insofar as he could interpret the operation of natural selection as an agent for adaptive design, he was prepared to endorse Darwin's theory – biological adaptation was achieved by divinely instituted laws of nature. Darwin suggests, in his autobiographical remarks about his changing religious views, that when he wrote the *Origin* he shared this view, and its frontispiece quotations from Bacon and Whewell, both stressing the idea of natural laws having a divine source, are further evidence for this (Darwin 1859, ii). But that requires that the production of variation, an important part of Darwin's mechanism, also be due to divinely instituted natural law. And by that Gray does not simply mean there must be deterministic laws of variation; he means that there must be *design* in the production of variation. The primary meaning of "chance" for Gray is "undesigned" or "useless." He cannot rest content with Darwin's field of undesigned variability.

Darwin's choice of the word "contrivance" to characterize his orchid adaptations is likely related to his views about the origins of variation: variations arise for reasons unrelated to an organism's adaptive needs – they "chance to occur," as he often puts it. Adaptation thus results from the differential survival and preservation of those variations that *happen* to be advantageous. Near the close of his book about adaptations for fertilization in orchids, Darwin gives the following example of a case where a very simple "optimal design" solution to an adaptive problem appears to be passed over in favor of a more complicated, suboptimal, solution, in a species of the genus *Malaxis*. He supposes that at one point in the past its ovarium was oriented so that the labellum hung downward, but at a certain point in its history it became advantageous to have the labellum in the more typical, upward position.

> [T]his change, it is obvious, might be simply effected by the continued selection of varieties which had their ovarium a little less twisted; but if the plant only afforded varieties with the ovarium more twisted, *the same end* could be attained by their selection until the flower had turned completely round on its axis: this seems to have occurred with the Malaxis, for the labellum has acquired its present upward position, and the ovarium is twisted to excess. (Darwin 1862c, 349–50)

Notice again that Darwin sees himself engaged in a teleological inquiry, a search for the end to be served by this adaptation. But selection must attain that end by making use of whatever chance variations are actually present, which leads to this oddly "suboptimal" *contrivance*. Asa Gray good-naturedly responded that "we believers in real design make the most of your 'frank' and natural terms, 'contrivance, purpose,' etc., and pooh-pooh your endeavors to resolve such contrivances into necessary results of certain physical processes" (Darwin 1985–, 11:253, letter, 22–30 March 1863). Note Gray's self-description: he is among the believers in *real* design, Darwin a mere pretender.

Darwin hits on a craft analogy to make the role of chance in his "two step" mechanism clear. He compares biological variations to rocks of all shapes and sizes that have broken away from a cliff and lie about on the ground, and the builder who selects those which are appropriate for various roles in the building of a house to natural selection. He first expresses it in a letter to Gray in August 1863, while he is working on *Variations in Animals and Plants under Domestication*; and in the last two pages of that book, he mines this analogy for all it is worth and reveals that the lengthy correspondence with Gray has helped him to differentiate two notions of chance that are not clearly distinguished in the *Origin*. Darwin first expands the analogy, imagining rock fragments of various shapes and sizes accumulating, as a consequence of erosion, at the base of a precipice. An architect then selects those with shapes and sizes best suited to play various roles in a building he is erecting. These rock fragments were not *designed* for these roles – they are *selected* for them. He points out that "the fragments of stone ... bear to the edifice built by him the same relation which fluctuating variations ... bear to the varied and admirable structures ultimately acquired by their modified descendants" (Darwin 1868b, 2:430). He then argues that ignorance of the cause of each variation does not detract from the explanatory power of selection. "[I]t would be unreasonable" to claim "that nothing had been made clear ... because the precise cause of the shape of each fragment could not be told" (2:430–31). He goes on to distinguish two senses of what is "accidental." In one sense, the shapes of the fragments are not accidental because "the shape of each depends on a long sequence of events, all obeying natural laws." "But," he goes on, "in regard to the use to which the fragments may be put, their shape may be strictly said to be accidental" (2:431).

Darwin (1868, 2:432) is explicitly accounting for adaptations as a consequence of chance variation and the "paramount power of selection."

> If we assume that each particular variation was from the beginning of all time preordained, the plasticity of organisation, which leads to many injurious deviations of structure, as well as that redundant power of reproduction which inevitably leads to a struggle for existence, and, as a consequence, to the natural selection or survival of the fittest, must appear to us superfluous laws of nature.

Darwin, then, thinks of himself as providing teleological explanations, but without any backing from theology, and in conjunction with a view of the sources of variation being "accidental" with respect to the organism's well-being. The critical issue for answering whether selection explanations are in some significant sense teleological depends on how we interpret Darwin's discussion. I present here a reading of Darwin's explanatory method based on an earlier paper (1993) that argues that in fact it is teleological.

As we have seen, Darwin became very interested in "contrivances" in plants that encouraged cross-fertilization and discouraged self-fertilization, recognizing in this a fruitful mechanism for the production of new variation in populations. This was a major focus of his book on insect pollination in orchids, but as we have seen, it was the sole topic of his paper on dimorphism in *Primula* (in Barrett 1977, 2:45–63), research later expanded into a monograph on dimorphism in plants generally. This paper is entirely focused on presenting the results of Darwin's search for the final cause of this dimorphism in *Primula* (primroses and cowslips). He explicitly characterized that work as the search for its final cause. I earlier highlighted the robust teleological language in this paper. The "end" in question is that of promoting crosses between distinct plants. But the teleological investigation Darwin is involved in is more fundamental than that: he is interested in knowing what value is achieved by mechanisms that promote intercrossing and discouraging self-fertilization, mechanisms that are quite widespread in the plant kingdom. On the basis of a careful analysis of the argument in that paper, I was able to abstract the following *argument schema*.

1. Dimorphism is present in *Primula veris*. (Variation of interest [V] is present in Organism of interest [O].)
2. Dimorphism has the effect of increasing heteromorphic crosses and decreasing homomorphic fertilization. (V has a certain Effect [E].)
3. Heteromorphic crosses are more fertile and produce more vigorous offspring than homomorphic fertilizations. (E is advantageous to O.)
4. Natural selection would thus favor increased dimorphism in *Primula veris*. (Therefore V in O would be selectively favored.)

Conclusion: Dimorphism is present in *Primula veris* *because* it promotes intercrossing. (Therefore V is present in O *because* of E.)

Darwin, without a blink, refers to the promotion of intercrossing as the "Final Cause" of the dimorphic condition of *Primula*. Is this merely a careless mode of expression, or does this reasoning reveal a legitimate sense in which the reproductive consequences of sexual dimorphism are the *cause* of its presence in *Primula*? It is unlikely that Darwin would have used such a loaded expression unreflectively; as we have seen, this has been a topic of reflection for him since the 1830s. And, indeed, there is a clear sense in which Darwin *has* identified the "Final Cause" of the trait in question. The various environmental "checks" to population expansion, which Darwin thinks of as the principal mechanisms promoting adaptive evolutionary change, bias reproductive frequencies on the basis of whether the consequences of particular variations are advantageous or disadvantageous to their possessor's living to sexual maturity and reproducing.

If a variation functions, in a particular environment, to increase its relative frequency in subsequent generations, that variation is selectively favored *for* that function. Darwin's explanation thus has the form of what has come to be termed, following Larry Wright's (1976) analysis of teleology, "consequence etiology." The effect of the character trait under investigation provides an advantage to those individuals that possess it, and that advantageous effect increases its possessor's chances of surviving and reproducing. It is thus also clear that there is a significant *value* component to Darwin's understanding of teleology. Those traits which provide a relative *advantage* (to adopt Darwin's language) to the organisms that have them are selectively favored.

TELEOLOGY AND EVOLUTION AFTER DARWIN

During the period after the integration of Mendelian genetics with evolutionary theory that took place during the 1920s and 1930s, there was little discussion of this topic, but in the 1950s the question of whether the explanation of adaptations by reference to natural selection was teleological came to the fore again in two distinct forums. In the context of debates about the unity of the natural sciences, questions were raised regarding the possibility of reduction of biological explanations to chemical explanations, which would require that apparently teleological explanations be reducible to explanations by reference to chemical and physical laws. The issue then became central to a number of books devoted directly to the philosophical foundations of biology, by Morton Beckner (1959), Thomas Goudge (1961), David Hull (1974), and Michael Ruse (1973), which in turn provoked monographs devoted entirely to the topic of teleological explanation.

In parallel with this philosophical discussion, and in part provoked by it, a suggestion was made, and then widely adopted, that the word "teleology" should be replaced by "teleonomy" in evolutionary contexts. The grounds for this move appears to have been that the word "teleology" was too closely associated with either vitalism or natural theology; and yet there was a need to acknowledge the teleological character of many explanations in the biological sciences. Thus, in work by Ernst Mayr, George Gaylord Simpson, Theodosius Dobzhansky, George G. Williams, and Francisco Ayala, to name a few, various defenses of the importance of teleological explanation to biology were offered, often under the banner of "teleonomy." In particular, in the concluding chapter of his highly influential 1966 *Adaptation and Natural Selection: A Critique of Some Current Evolutionary Thought*, George C. Williams called for a science of teleonomy. As he put it at the time, "Pittendrigh (1958) suggested that the explicit recognition of the functional organization of living systems be called teleonomy. This term would connote a formal relationship to Aristotelian teleology, with the important difference that teleonomy implies the material principle of natural selection in place of the Aristotelian final cause. I suggest that Pittendrigh's term be used to designate the study of adaptation" (Williams 1966, 258).

In his recent book, *Not by Design: Retiring Darwin's Watchmaker*, John Reiss argues that this form of teleology still permeates evolutionary biology today, which he regrets. That regret reflects his sensitivity to the fact that evolutionary biologists, or at least their popular advocates, all too easily jump from apparent functionality to the conclusion that selection-based teleology was at work, a leap he correctly sees as unwarranted. He is also arguing against gene-centric neo-Darwinians (such as Williams and Richard Dawkins) from the standpoint of evolutionary developmental biology. This antiadaptationist stance, popularized by Richard Lewontin and Stephen Jay Gould, has its historical roots as well. Darwin was educated during the debate between Georges Cuvier and Etienne Geoffroy Saint-Hilaire, and one issue dividing these two great French naturalists was the relative importance of form (morphology) versus functional adaptation (teleology) in accounting for the characteristics of living things. Richard Owen and Darwin's Bulldog T. H. Huxley disagreed about many things, but they were in agreement about the relative importance of "morphology," or the study of form, independently of adaptation, for evolutionary systematics. And Huxley, despite the apparent agreement noted earlier, read the *Origin* quite differently than did Asa Gray: on his first reading of the *Origin*, he claimed that what struck him most forcibly was that "Teleology, as commonly understood, had received its deathblow at Mr. Darwin's hands" (Huxley 1896, 82). The crucial words here are, perhaps, "as commonly understood." For he also said that "there is a wider Teleology, that is not touched by the doctrine of Evolution, but is actually based upon the fundamental proposition of Evolution" (86). This wider, selection-based teleology is not, to this day, what people commonly understand by teleology. We can reconcile Huxley's comments, then, if his point is that the teleology associated with natural theology had received its deathblow but that a selection-based teleology survives. Nevertheless, it is clear that he saw it playing a far more limited role in the evolutionary process than did Charles Darwin or Asa Gray.

The debate over the legitimacy of teleology has waxed and waned in the historical development of evolutionary biology, but it has never gone away. And with the hoped-for integration of developmental biology and evolutionary biology may come a reconsideration of the goal-directed character of development as well.

❧ Essay 18 ❦

The Evolution of the *Origin* (1859-1872)

Thierry Hoquet

Darwin had been elaborating his theory of evolution since 1837 and was consciously working on his "Big species book" at least since 1854, when he had to write *On the Origin of Species* within little more than one year. Consequently, he bluntly presents it as an abstract, with all its contingencies. We usually see "abstract" as a positive quality, for it led Darwin to keep a clear line of argumentation; but he certainly perceived it as a fault. Having amassed hundreds of pages of material, Darwin at first decided to publish his book, only to avoid being forestalled by A. R. Wallace, under the title "the abstract of an essay on the origin of species and varieties through natural selection." In the first edition of the *Origin*, Darwin refers constantly to a "longer work" that he was planning to complete. But this was eventually pushed aside by other projects and Darwin's involvement in the debates launched by the *Origin*. Instead, Darwin dedicated a lot of time to a careful reworking of his 1859 text, which makes the *Origin* a book with different versions.

During Darwin's own life, no less than six successive editions were published by John Murray. In this essay, I follow Morse Peckham's system of reference: [a] for the first edition (November 1859); [b] for the second (January 1860); [c] for the third (April 1861); [d] for the fourth (December 1866); [e] for the fifth (August 1869); [f] for the sixth (February 1872). This evolution of the *Origin* was the textual process through which 75 percent of the book underwent modifications, while its global length increased by one-third. All those changes are documented in Peckham's *Variorum* text – a book that changed our view of the *Origin*, although it is an unreadable maze of additions and corrections (Peckham 1959, noted hereafter as *Var*). Almost everything changed during the long life of the *Origin*, including its birthdate (from "October 1st, 1859" in [a] to "November 24th" in [d]) and its title (the initial "*On*" is dropped in [f]). Some critics even suggest that dramatic modifications changed the general meaning of the *Origin* and the role that Darwin attributed to natural selection in evolutionary processes. Unluckily, those modifications still belong to what Darwin may have called "a grand and almost untrodden field of inquiry" (*a*486). We know what kind of misprints distinguish the first printing from its followers ("speceies" on page 20, line 11), but little has been written about the textual evolution of the *Origin* (with the exception of Vorzimmer 1972 and H. P. Liepman 1981). It is still hard to get a grasp on what is *really* (i.e., theoretically, intellectually, conceptually, or even socially) going on in those various editions. This essay deals with a question that should haunt the spirit of any reader of Darwin's masterpiece: Are there any major changes of doctrine in the *Origin*, especially regarding the role of natural selection in the transformation of species? What edition should one read? (See Plate XXIII.)

SIX DIFFERENT EDITIONS

Arguments in favor of [a] are the following. Darwin described his book as "one long argument" (*a*459). It seems that, throughout the process of revisions, this "one" argument became less readable, that Darwin made his case for natural selection less concise and pithy, more equivocal and diffuse. Darwin was led on the wrong path by critics, whose objections were committed to barren hypotheses or archaic frames of mind. And even in the cases where the book was actually "bettered" in the sense of clarified or sharpened, it seems that we hardly need those latter versions which lack the freshness of the first edition.

Only a couple of weeks separate [a] from [b]. John Murray, Darwin's publisher, was more than 250 copies short for the orders received at his autumn sale, and he asked Darwin, then on a water cure in Ilkley, Yorkshire, to bring revisions to his text (Dixon and Radick 2009).

The "Ilkley edition" contains only 7 percent of the total variants of the five subsequent editions (*Var*773). Darwin suggested that it was "merely a reprint of the first with a few verbal corrections & some omissions" (Darwin 1985–, 7:411, letter to John Murray, 2 December 1859), or "only Reprint; yet I have made a *few* important corrections" (Darwin 1985–, 7:440, letter to Asa Gray, 21 December 1859).

As to [c], major changes include a "Historical Sketch," where Darwin acknowledged the achievements of his predecessors, and a postscript on Asa Gray's favorable review, which suggests that Darwin wanted to reconcile his theory with natural theology – but it eventually disappears in [d]. Within the body of the text, the changes (14 percent of the total changes) appear mostly in chapter 9 (more than half), and in chapter 4. They include Darwin's reaction to some major reviews (by Owen, Bronn, and Harvey).

The fourth edition [d] represents 21 percent of the total variants. It gives titles to many previously unnamed sections and pays attention to new discoveries and to new objections (Falconer).

The fifth edition [e] of 1869, with nearly 30 percent of the changes and an increase of more than 20 percent in size, is among the much-revised ones. It is noteworthy on at least two different accounts: for the introduction of Spencer's phrase "survival of the fittest" and for the answer to some important objections (Fleeming Jenkin).

The sixth edition [f] is usually regarded as the last, just after the publication of the *Descent of Man* (1871) raised great interest in Darwin's works. Designed for a wider audience, it is smaller and cheaper than its predecessors and includes a glossary. A new chapter, on "Miscellaneous Objections to the Theory of Natural Selection" (ch. 7), consists of parts taken from chapter 4 together with additional material, aiming chiefly at a rebuttal of St. George Mivart's attacks (whose *Genesis of Species* was published in 1871). The big issue with [f] is with Darwin's "Lamarckianism": Did Darwin diminish the role of natural selection, and put more stress on the "other means of modification," such as habit, use and disuse of parts, and direct effect of external conditions? With the 1872 edition, Darwin stopped modifying the *Origin*. A last printing corrected by Darwin was issued in 1876, with minor changes. Darwin's incentives for changing his text are numerous. First, he wanted to introduce recently discovered data in his book: the bird-reptile *Archaeopteryx* fossils (*d*367) (Plate XXIV); and the concept of mimicry, after Bates's 1862 paper (*d*503). Ernst Haeckel is accountable for the introduction of the phrase "phylogeny, or the lines of descent of all organic beings" (*e*515, *Var*676). Other publications like Spencer's *Principles of Biology* (1864) had an influence on Darwin's maturing views on the question of variations (compare *a*131–32 with *e*165–66, *Var*276). Darwin was considerably affected by reviews and critics, and they were powerful incentives in the constant process of revising the book. He also used subsequent editions as a means to correct some blunders. For instance, his figures on the denudation of the Weald (*a*287, *Var*484) disappear from [c]).

More than a historical fact, the variation of the *Origin* throughout its successive editions has become a theoretical and historical problem. Yet, though the *Origin* considerably mutated and evolved from 1859 to 1872, do those changes matter? Contemporary readers of the *Origin* want to know what to think of Darwin's ideas on a few key questions like religion, variations, inheritance, and progress.

COMMITMENT TO RELIGIOUS VIEWS

Is the *Origin* more and more critical of theological language or, on the contrary, more and more infused with occurrences of the Creator? Edition [a] refers to "the laws impressed on matter by the Creator" (*a*488, *Var*758) and twice includes the Pentateuchal verb "to breathe," which suggests a paramount Power (*a*484, *a*490). Those instances were never removed, and [b] adds after them the mention "by the Creator." But, to be true, Darwin was altogether very uncomfortable with the verb "to breathe." The latter addition of the Creator, in the final paragraph, remains unchanged throughout the subsequent editions. But Darwin seems to be vacillating with the other occurrence, and [c] deletes the Creator again, adding in its place a long development on the principle of natural selection (*c*519, *Var*753).

Other additions are ambiguous. Darwin had stated that "two individuals must always unite for each birth" (*a*96, *Var*185) – to which he adds a parenthesis: "with the exception of the curious and not well-understood cases of parthenogenesis" (*b*96). What motivates this biological addition? Gillian Beer (1996 xxiv) cunningly suggests, it is a disguised theological addition, parthenogenesis standing for Virgin birth.

The truth is that Darwin was editing for a theological audience as is suggested by the transformation of the phrase "natural selection will account for the infinite diversity in structure and function of the mouths of insects" (*a*436) into "natural selection acting on some originally created form will ..." (*b*435, *Var*678). He puts forward any hint of support from theologians (like Charles Kingsley, *b*481, *Var*748). He adds on the verso of the half title a third quotation, which is placed between those from Whewell and Bacon (*Var*40): it comes from Joseph Butler's *Analogy of Revealed Religion*, a text that was a constant resource for theologians at the time. With its reference to an "intelligent agent," it might easily be read as part of Darwin's move to mollify religious readers. It may as well stress that Darwin's own method had recourse to analogy – namely, that of artificial and natural selection (Jon Hodge, personal communication).

THE POWER OF NATURAL SELECTION

In [a], Darwin refers to a black bear seen swimming, "with widely opened mouth, thus catching, like a whale, insects in the water." He states that he could "see no difficulty in a race of bears being rendered, by natural selection, more and more aquatic in their structure and habits, with larger and larger mouths, till a creature was produced as monstrous as a whale" (*a*184, *Var*333). In [b], Darwin adds in the first sentence "*almost* like a whale," and he simply deletes the second one, probably after having been ridiculed by Charles Lyell. But Darwin was not to be left in peace with this bear-whale story. In 1860 Owen compared the two versions and violently attacked Darwin's attempt to amend his text: to him, the wording of [b] is not better than the one of [a]; if [a] was vague, [b] clearly evinces Darwin's cowardness (Owen 1860, 519). Owen is indignant: when Darwin modifies his text, he changes only details, instead of revising and clarifying the whole argument. Owen attacks Darwin for adding "*almost*" when the problem with the bear-whale case is much worse: What can be effected by natural selection?

Very early in the life of the *Origin*, Darwin had to explain that he does not understand natural selection "as an active power or Deity" (*c*85, *Var*165). The great fault of the term "natural selection" is that it tends to personify nature: one should never forget that it is a metaphorical expression. Darwin tries to be more accurate, because "several writers have misapprehended or objected to the term natural selection" (*c*84, *Var*164). Strikingly enough, Darwin bluntly avows that a great part of the criticisms made to his beloved term are founded in this misunderstanding: "In the literal sense of the word, no doubt, natural selection is a misnomer" (*c*85, *Var*165); later, Darwin goes as far as calling it "a false term" (*e*93). Nature does not *literally* select but does so only in a metaphorical sense. During the same time, Darwin also tried to explain to Heinrich G. Bronn, the German translator of the *Origin*, the rationale behind the term "natural selection," noting that "its meaning is *not* obvious & each man could not put on it his own interpretation"; besides, the phrase "at once connects variation under domestication & nature" (Darwin 1985–, 8:83, letter to Bronn, 14 February 1860).

Personifications also pervade the term "nature." Darwin suggests in an addition how it is "difficult to avoid personifying the word *Nature*" and he strongly reminds us that Nature means only "the aggregate action and product of many general laws, and by laws the sequence of events as ascertained by us" (*c*85, *Var*165).

In [d], Darwin deals with the issue of natural selection producing gradually "utter and absolute sterility" between two species (*d*311, *Var*444). This issue is crucial for the question of "the origin of species," or what Darwin terms in 1869 "species in process of formation" (*e*318). Darwin's sentence is quoted by Francis Darwin (1886, 407) in a polemic against George Romanes. In 1886 Romanes claims that natural selection is not enough to explain the formation of interspecific sterility and that another process is required, what he calls "physiological selection." This suggestion entails a huge debate on the "Darwinian" character of Romanes' hypothesis. In this context, does Darwin's sentence support Romanes' physiological selection? Or, on the contrary, does it destroy Romanes' claim to innovation? Through this example, we see that not only [a] has played a role in the reception of Darwin's ideas. Darwinian scholars, including Darwin's own son, were always very apt to dig in the various layers of the *Origin* to find the unexpected gold nugget that supports their own claims – or ruins the others' claims to be Darwin's true heirs or to originality. The diversity of the various editions provided readers with plenty of material to play one Darwin against the other (Hoquet 2009).

Besides, it seems that Darwin became more committed than before to isolation as a necessary condition for speciation – a point he had previously denied (*a*105). Here the German naturalist Moritz Wagner has undoubtedly played a role. By 1868, Wagner was convinced that isolation was the all-important factor accounting for the origin of species, and he even termed it "the law of migration" (*Migrationsgesetz*). Darwin partly acknowledges the importance of this factor, while clearly stating that he "can by no means agree with this naturalist, that migration and isolation are necessary for the formation of new species." On the contrary, if "an isolated area be very small ..., the total number of the inhabitants will be small; and this will retard the production of new species through natural selection, by decreasing the chances of the appearance of favourable individual differences" (*e*120, *Var*196–97). In this case, we see Darwin trying to please a critic, while staying firm on his major argument (the importance of natural selection).

Edition [e] really matters for the fate of natural selection, because Darwin introduced for the first time Herbert Spencer's phrase "the survival of the fittest"; it appears as an equivalent of natural selection, first in chapter 3, where it is said "more accurate and ... sometimes equally convenient" (*e*72, *Var*145), and then in the heading of chapter 4. The introduction of Spencer's phrase is due to Wallace's influence. Wallace had crossed out "natural selection" in his copy of the *Origin* and had substituted for it "survival of the fittest." In a letter dated 2 July 1866, Wallace very vividly argues that "natural selection," although crystal clear to some readers, is nonetheless a stumbling block for many others (Darwin 1985-, 14:227). But *survival of the fittest*, far from bringing more clarity, suggests that natural selection is only a tautology, the fittest being precisely defined as *those which survive*.

THE NATURE OF VARIATIONS

Are minute variations necessary to the Darwinian process, or does Darwin progressively though reluctantly accept to take "sports" into account? Does he move from a theory where variations are minute and continuous to a more saltationist account on variation?

Thomas Henry Huxley has been very critical of the principle *Natura non facit saltum*, both in private letters and in published reviews (Darwin 1985-, 7:391, 23 Nov. 1859). Consequently, Darwin makes two changes in [b]: on page *a*194 (*Var*361), he changes the phrase "that old canon in natural history" into "that old but somewhat exaggerated canon"; and on page *a*210 (P383), he simply deletes the sentence that referred to the *Natura non facit saltum*.

Edition [c] devotes a special development to "various good objections" raised by Bronn (*c*139, *Var*230). Bronn thinks that the Darwinian theory requires "that all the species of a region" should be "changing at the same time." Darwin always thought that this is an unnecessary supposition: Darwin replies that "it is sufficient for us if some few forms at any one time are variable." Bronn also remarks that "distinct species do not differ from each other in single characters alone, but in many"; and he asks how it comes that "natural selection should always have simultaneously affected many parts of the organisation?" To this, Darwin replies that "probably the whole amount of difference has not been simultaneously effected; and the unknown laws of correlation will certainly account for, but not strictly explain, much simultaneous modification" (*c*140, *Var*231). Correlation of growth and new emphasis on the laws of variation is therefore Darwin's common answer to many objections.

William Harvey also raised influential objections on the problem of saltations and monstrosities. For Harvey, the origin of species means nothing until the origin of variation is better understood. As he writes to Darwin (Darwin 1985-, 8:322, 24 August 1860), "until something more is known of the inciting causes of the Variation & Correlation of Organs, ... I can only regard Natural Selection as one Agent out of several; – a handmaid or wetnurse – so to say – but neither the housekeeper, nor the mistress of the house." On the same kind of issues, "infinitesimally small inherited modifications" (*a*95) simply becomes "of small inherited" (*c*100, *Var*185).

The nature of variation impacts on the status of natural selection. Edition [e], for instance, contains an important (and much discussed) attempt to reply to Fleeming Jenkin's "able and valuable article in the 'North British Review' (1867)" (*e*104, *Var*178). Jenkin objected that there are absolute limits to variation, that a new form of a living entity would be swamped, and that the earth is much younger than Darwin assumed (Fig. 18.1).

Jenkin's review increases Darwin's awareness on the problem of variations: "I did not appreciate how rarely single variations, whether slight or strongly-marked could be perpetuated" (*e*104, *Var*178). Jenkin had taken the case of "a highly-favoured white," shipwrecked on an island, who fails in "blanch[ing] a nation of negroes." Darwin interprets this as a convincing case *against* single variations and rethinks the respective roles of individual differences (occurring in several organisms) and of single variations (rare and discontinuous forms of change). Jenkin's review apparently led Darwin to put less stress on natural selection, for instance when he writes: "The conditions might indeed act in so energetic and definite a manner as to lead to the same modification in all the individuals of the species without the aid of selection" (*e*105, *Var*179). Darwin deemphasized sports and placed more emphasis on the normal range of variability. Indeed, Darwin's insistence on the individual level can easily be perceived in many additions to or modifications of [e]. At the beginning of chapter 4, "an endless number of strange peculiarities" (*a*80, *Var*163) becomes "peculiar variations" (*e*91) and then "slight variations and individual differences" (*f*62).

Whereas (*a*102, *Var*192) reads "A large amount of inheritable and diversified variability is favourable, but I believe mere individual differences suffice for the work," (*e*117) now reads: "A great amount of variability, under which term individual differences are always included, will evidently be favourable." Other examples of Darwin's focus on individual variation can be found in *e*94 (*Var*166) or *e*104 (*Var*178). The limits and scope of variation are obviously of great concern to him, and he deals with the question whether "many changes would have to be effected simultaneously" (*e*225, *Var*342). Darwin confesses that "this could not be done through natural selection"; but, safely relying on his 1868 work on *Variation*, he considers that this is a superfluous condition, noting that "it is not necessary to suppose that all the modifications were simultaneous, if they were extremely slight and gradual" (*e*225, *Var*342).

On the issue of continuous variation, in spite of all the hesitations often attributed to him, Darwin clearly dispenses with the objection that new species can appear by saltations, and he reaffirms in [f] his commitment to continuous variation, saying that a "conclusion, which implies great breaks or discontinuity in the series, appears to [him] improbable in the highest degree" (*f*201, *Var*264).

PROGRESS

Did Darwin accept the idea of tendencies in evolution, especially toward a degree of superior "highness" in organization? This is clearly a matter of concern to him. As early as

FIGURE 18.1. Fleeming Jenkin (1833–85), Scottish engineer and effective critic of Darwin's theory. Nineteenth-century lithograph.

[b], Darwin complements, in the summary of chapter 4, the sentence "This principle of preservation, I have called, for the sake of brevity, Natural Selection" with the remark that natural selection "leads to the improvement of each creature in relation to its organic and inorganic conditions of life" (*b*127, *Var*271) – a sentence that he develops again in [c] into "and consequently, in most cases, to what must be regarded as an advance in organisation" (*c*144). Darwin also adds, in the discussion on geological succession, that the best definition of highness is greater division of physiological labor and, consequently, that natural selection "will constantly tend" to make later forms "higher" than their progenitors (*b*336, *Var*547).

Darwin adds a whole new section in [c]: "On the degree to which Organisation tends to advance" (*c*133–37, *Var*220–26). There he clearly distinguishes "highness" from "progress." On the one hand, he asserts: "If we look at the differentiation and specialisation of the several organs of each being when adult (and this will include the advancement of the brain for intellectual purposes) as the best standard of highness of organisation, natural selection clearly leads towards highness" (*c*134, *Var*222). On the other hand, Darwin clearly denies progressive development, stating that "natural selection includes no necessary and universal law of advancement or development – it only takes advantage of such variations as arise and are beneficial to each creature under its complex relations of life" (*c*135, *Var*223).

A notable feature of [f] is that the word "evolution" is finally introduced in the *Origin*. Previously, the *Origin* contained only the word "evolved," at the closing of the book. Now "evolution" occurs eight times, and Darwin refers to "the theory of evolution through natural selection." Usually, the absence of the term is attributed to two different sets of reasons: to avoid confusion with the use of the word by Herbert Spencer, and to avoid confusion with its embryological meaning of "development." Consequently, it might seem that the introduction of the term means that Darwin was ready to accept more confusion on those accounts. But it might as well be contended that, by the 1870s, the term "evolution" was much more commonly in use and that Darwin was ultimately making things more clear rather than more confused.

(LAMARCKIAN) INHERITANCE?

The formidable – though unclear – question of "Darwin's Lamarckianism" is certainly the main reason for avoiding later editions of the *Origin* (Darlington 1950). But it is an anachronistic question, mainly due to Weismann's refutation of "Lamarckian inheritance," and I think it should be avoided by any means. "Lamarckian" mechanisms generally include what Darwin called "use and disuse" or "direct effect of external conditions." Are those factors or forces more generally active or efficient in the last edition than they were in the first?

In the *Origin*, Darwin clearly states that natural selection can help discard Lamarckian explanations (*a*242). Nonetheless, the obvious signs of Darwin's leanings toward "Lamarckian" factors are numerous, even in [a]. Direct effects of environment on organisms are admitted in several passages, such as "we must not forget that climate, food, &c., probably produce some slight and direct effect" (*a*85), or, about some "habitual action" that became inherited: "I think it can be shown that this does sometimes happen" (*a*209). Those various instances of Lamarckian themes in the *Origin* seem to coalesce in a key sentence at the close of the introduction: "Furthermore, I am convinced that Natural Selection has been the main but not exclusive means of modification" (*a*6).

Those other "means of modification" are chiefly the action of a changing environment. H. P. Liepman (1981) has documented a clear shift of emphasis in the role devoted to natural selection in the editions [e] and [f]: "Up to the 5th edition, the alterations are mostly supportive to the theory of accumulation of modifications by natural selection, but in the last two editions non-selective forces come into play."

However, the evidence brought by the supporters of the "Darwin's Lamarckianization" thesis is rather frail. Can we claim that Darwin gave "extra stress to the direct action of the conditions of life" just because, where the first edition reads: "We should remember that climate, food, &c., probably have some little direct influence on the organisation" (*a*196, *Var*363), Darwin changes *little* into "some, *perhaps a considerable*, direct influence"? It is impossible to conclude from this that Darwin significantly changed his views. There are some obvious and probably not insignificant changes, such as, in chapter 1: "Habit also has a decided influence" (*a*11, *Var*83), which ends up being "Changed habits produce an inherited effect" (*f*8); or the next sentence: "In animals it has a more marked effect," which finally reads: "With animals the increased use or disuse of parts has had a more marked influence." But evidence of the contrary could also be called for, such as this sentence of the new chapter 7 where Darwin refers to "the inherited effects of the increased use of parts, and perhaps of their disuse," being "strengthened by natural selection": "How much to attribute in each particular case to the effects of use, and how much to natural selection, it seems impossible to decide" (*f*188, *Var*253). The emphasis on a "tendency to vary in the same manner" is strong, in this passage and others such as:

> There can also be little doubt that the tendency to vary in the same manner has often been so strong that all the individuals of the same species have been similarly modified *without the aid of any form of selection*. (*f*72, *Var*179).

This tendency to vary certainly leads Darwin from selection. But again, against Mivart's belief that species change requires "an internal force or tendency," Darwin is very clear:

> [T]here is no need to invoke any internal force beyond the tendency to ordinary variability, which through the aid of selection by man has given rise to many well-adapted domestic races, and which through the aid of natural selection would equally well give rise by graduated steps to natural races or species (*f*201, *Var*264).

In fact, in [f] more than ever before, Darwin is facing the accusation that he made natural selection an all-powerful operator. As we have seen, this was already the case in [c]. But this constant accusation provoked clear changes in various passages. For instance, [a] stated that "species have changed and are still slowly changing by the preservation and accumulation of successive slight favourable variations" (*a*480). From [b] to [d], the passage reads "that species have been modified, during a long course of descent, by the preservation or the natural selection of many successive slight favourable variations" (*Var*747). In [e], the end becomes "a long course of descent, chiefly through the natural selection of numerous successive, slight, favourable variations" (*Var*747). But [f] adds to this, that selection has been "aided in an important manner by the inherited effects of the use and disuse of parts; and in an unimportant manner ... by the direct action of external conditions, and by variations which seem to us in our ignorance to arise spontaneously."

Does such an addition express some important change in Darwin's perception of his theory? Strikingly enough, Darwin takes a special pain to refer these later additions to the sentence that closes the introduction in [a]: that "natural selection has been, the main but not the exclusive means of modification" (*Var*747). Why see ruptures, when Darwin himself indicates continuities? From Darwin's own perspective, nothing has changed: he is just trying to make clearer a point that he has always made but that has been constantly overlooked. It is only from a "Darwinian" vantage point (which equates Darwin to natural selection and only to that) that Darwin can be accused of having changed his theory. But Darwin-the-man seems to be quite at ease with those changes of inflection in the perception of his theory and the reading of his book.

FROM NATURAL SELECTION TO THE LAWS OF VARIATION?

Far from being more and more Lamarckian, Darwin simply stresses the power of variations, something acting simultaneously with the power of natural selection. The laws of variation put constraints on natural selection, and they also entail a refutation of the pan-utilitarian reading of the organism. It is probably this lifelong interest in the laws of variation that leads Darwin to consider in depth the question of some so-called Lamarckian factors, such as the effects of changed conditions of life.

Finally, while looking for the changes that did happen, we should not overlook some changes that never happened. Notably enough, the "provisional hypothesis of pangenesis," developed in the 1868 *Variation*, never made its way in the *Origin*. On a general level, no full-scale revision of the structure of the argument was ever attempted. The creation of an additional chapter in [f] is only a way to unify scattered objections in one single body, while considerably lightening chapter 4.

But, as is often the case in Darwinian processes, minute and sometimes insensible modifications might have dramatically altered the meaning of the whole. Darwin's interest in variation progressively led him to put more stress on other factors "aiding" natural selection: the *Origin*'s momentum progressively shifts from chapter 4 to chapter 5. During this process, publications such as the 1868 *Variation of Animals and Plants* show his attempts to secure a considerable amount of raw material, on which he can confidently rely. It seems that the main incentive for changes was Darwin's desire to address objections and critiques. However, this change of focus from natural selection to the other means and to the laws of variation does not equate to a "Lamarckianization" of his thinking. The idea that the *Origin* became increasingly Lamarckian might have been forged by supporters of the modern synthesis in the 1930s and 1940s. As to Darwin, he hoped that the study of the causes of variation would resolve most of the difficulties arising from the *Origin*. But it turned out that the difficulties were rather amplified by this shift of emphasis.

Essay 19

Alfred Russel Wallace

John van Wyhe

Alfred Russel Wallace (1823–1913) was an English naturalist who famously conceived of the principle of evolution by natural selection independently of Charles Darwin in 1858 (Fig. 19.1). Wallace is often incorrectly referred to as working class. In fact, he was the son of a solicitor with inherited property sufficient to generate an income of £500 per annum (Wallace 1905,1:7). Thus, according to the conventions of the day, Wallace's father was a gentleman. The family's financial circumstances, however, declined so the Wallace family moved from London to a village near Usk, on the Welsh borders, where Wallace was born in Kensington Cottage on 8 January 1823. As far as Wallace could later remember, the family kept one servant. Wallace is also sometimes described as Welsh. This is also incorrect. His parents were English. As a small boy in Usk, Wallace could remember, because of his blonde hair, that "I was generally spoken of among the Welsh-speaking country people as the little Saxon" (1:29). Wallace also referred to himself as "English" or an "English naturalist" many times in his publications (C. S. Smith 1998).

When Wallace was six years old, the family moved to Hertford, north of London, where he lived until he was fourteen. Here Wallace attended Hertford Grammar School, where he followed a classical education, not unlike Darwin's at Shrewsbury School, including Latin grammar, classical geography, and "some Euclid and algebra" (Wallace 1905). During his last year in Hertford, the family's finances further declined so that Wallace was obliged to tutor other students to pay his fees. Wallace was deeply conscious of this fall in status before his peers. He later described the shame of this and other cost-saving measures imposed by his parents as a "cruel disgrace," "exceedingly distasteful," and perhaps "the severest punishment I ever endured" (1:58). Wallace left school in March 1837 aged fourteen, just as Darwin was becoming a transmutationist.

WORKING LIFE

Wallace left home to join his elder brother John, an apprentice builder in London. Here Wallace observed working-class men or artisans for the first time. He clearly saw them as a different type of person, as is clear from his careful recollections of

FIGURE 19.1. Alfred Russel Wallace in old age. From A. R. Wallace, *My Life* (London: Chapman and Hall, 1905)

their language, dress, and behavior in his autobiography (Wallace 1905). His long association with working-class people adds to the modern misconception that Wallace was working class. However the designation of Wallace as working class by some modern commentators is in ignorance of the meanings and definitions of social class in Victorian Britain. There is a vast scholarly literature on the subject that shows that class was by no means simply a product of financial wealth (see, e.g., Cannadine 1999).

Like other Victorians of his generation, Wallace described a society composed variously of "the higher classes," "the middle classes," "tradesmen and labourers," "peasantry," and the "lowest class of manufacturing operatives" (Wallace 1905). Caught between the usual groupings, Wallace seems to have gone through life with the impression of watching all "classes" from the outside, though he clearly felt the greatest affinity with middle-class peers. This, in addition to his formative experiences in a radical working-class context, left him with a sense that the social arrangement of his country was deeply flawed.

Wallace spent his London evenings in a "hall of science" or mechanics' institute. In this context he encountered the socialist ideas of Robert Owen (Fig. 19.2). Wallace was deeply impressed by Owen's utopian social ideals – with his stress on environment determining character and behavior. Hence, if the social environment were improved, so would the morals and well-being of the workers. The hall of science also introduced Wallace to the latest views of religious skeptics and secularists. Although Wallace's parents were perfectly orthodox members of the Church of England, Wallace became a skeptic. From 1837 he joined his brother William as an apprentice land surveyor, first in Bedfordshire. It was a very good time to be a surveyor. The year before the Tithe Commutation Act was passed. It replaced the ancient system of the payment of tithes in kind with monetary payments based on the average value of tithable produce and productivity of the land. The valuation process required accurate maps. Wallace liked the instruments of surveying and the mathematics involved. He began to read about mechanics and optics, his first introduction to science. His days in the open air of the countryside led him to an interest in natural history. From 1841 he took up an amateur pursuit of botany, although he had no one to guide or encourage his nascent scientific interests.

In 1843 his father died and with a decline in the demand for surveyors, his brother no longer had sufficient work to employ Wallace. After a brief period of unemployment in early 1844, Wallace, although barely qualified, worked for over a year as a teacher at the Collegiate School at Leicester.

In these years, Wallace read some very influential works for his future life. Alexander von Humboldt's *Personal Narrative* (1814–29) and Darwin's *Journal of Researches* (Darwin 1845; van Wyhe 2002–) introduced Wallace to the exciting prospect of scientific travel. Another major influence on Wallace's nascent scientific views was Charles Lyell's *Principles of Geology* (1830–33). Thomas Malthus's *Essay on the Principle of Population* (1826) would later contribute to Wallace's independent discovery of natural selection. Wallace also read the anonymous *Vestiges of the Natural History of Creation* in 1845 (Secord 2000). The argument in *Vestiges* for the progressive physical "development" of nature and species, Darwin's numerous remarks suggesting that species change (Darwin 1845), and Lyell's lengthy dismissal of Jean-Baptiste Lamarck's transmutation, despite a masterful exegesis of the paleontological evidence for "the gradual birth and death of species," all contributed to Wallace accepting, from about 1845, that species were not fixed but could change. However, it should be stressed that the there is and was no homogeneous idea of evolution. Instead there were very many different conceptions of biological change. The genealogical descent and branching pattern that Darwin had developed since 1837 does not appear in Wallace's private documents until the mid-1850s (Barrett et al. 1987).

Most of the naturalistic framework of *Vestiges* was in fact derived from a work Wallace had already read, the phrenologist George Combe's *Constitution of Man* (1828) (van Wyhe 2004). Both works portrayed the world as governed by universal and beneficent natural laws tending toward progress. Combe's phrenological laws of mind were described as the most recently discovered laws of nature. Combe elaborated a system of hierarchically arranged natural laws: physical, organic, and moral. These three classes mapped onto man's constitution as described by phrenology. By combining these with a "law of hereditary descent," Combe argued that the

FIGURE 19.2. Robert Owen (1771–1858) was an early socialist and a great influence on Wallace. Nineteenth-century lithograph

human race would ascend the scale of improvement in organic and mental spheres (van Wyhe 2003). Hence the progress of nature was just as applicable to the human mental faculties as organic ones. Therefore, a "doctrine of natural laws," rather than religion, would lead to future scientific and social progress. These themes appeared again and again in Wallace's later writings as these formative experiences led him to adopt much of the rationalist, skeptical, and naturalistic outlook of his Owenite working-class environment with an optimistic faith in physical and social progress through the unimpeded operation of beneficent natural laws (Durant 1979).

Another lifelong influence Wallace encountered in Leicester was mesmerism (Winter 1998). He experimented by mesmerizing some of his students, to cause rigidity of the limbs, a trance state, suggestion, as well as phrenomesmerism. In phrenomesmerism it was believed possible to excite the behavior of a particular phrenological organ by touching the specific spots on a mesmerized person's head. As Wallace (1905, 1:236) wrote in his autobiography,

> The importance of these experiments to me was that they convinced me, once for all, that the antecedently incredible may nevertheless be true; and, further, that the accusations of imposture by scientific men should have no weight whatever against the detailed observations and statements of other men, presumably as sane and sensible as their opponents, who had witnessed and tested the phenomena.

This was perhaps the earliest instance of Wallace's lifelong characteristic of convincing himself by a few coincidences that an explanation was true and then never again doubting it or losing his belief. The fact that his mesmerized subjects were familiar with the phrenological map of the head, for example, never entered his written consideration to explain the actions of his subjects.

It is hardly surprising that, as a young man interested in natural science reading works on the most intriguing scientific questions of the day at the Leicester town library, Wallace there met another budding young naturalist, an enthusiastic entomologist named Henry Walter Bates. Bates introduced Wallace to his next scientific pursuit: the collecting of insects, particularly beetles.

Wallace's brother William died in March 1845, causing Wallace to leave the school to attend to William's surveying firm in Neath, together with his brother John. The business did not succeed. Wallace next worked as a surveyor for a proposed rail line for a few months. Then he and John attempted to establish an architectural firm, which produced a few successful projects, such as the building for the Mechanics' Institute of Neath. The director of the Mechanics' Institute invited Wallace to give lectures there on science and engineering. In late 1846 Wallace and his brother John bought a cottage near Neath where they lived with their mother and sister Fanny.

AMAZON, 1848–1852

In April 1848 Wallace and Bates sailed for Brazil to earn a living as natural history specimen collectors. They initially stayed in Para (now Belém). After collecting Amazonian specimens together for nine months, Wallace and Bates continued separately. Wallace focused particularly on collecting in and exploring the Upper Rio Negro. The principal scientific result of his time on the Amazon was an appreciation of the biogeographical boundaries, particularly broad rivers, that separated different species. Thus, Wallace employed a similar mode of regional demarcation to his earlier surveying work (J. R. Moore 1997).

In 1852 Wallace was returning home when disaster struck. His ship caught fire and sank destroying almost the entirety of his notes and personal collection. Fortunately the collection had been insured by Wallace's agent Samuel Stevens for £200. If Wallace collected any notes or material for his interest in the origin of species, none has survived, and he never referred to any in his later writings.

Wallace's subsequent publications therefore suffered from the dearth of data he was able to bring home. His first book *Palm Trees of the Amazon and Their Uses* (1853) described the distribution and uses of the palms he had observed and was illustrated from his own sketches. The book was criticized by some contemporaries because of its scanty detail, inaccuracies in some of the drawings, and sometimes amateurish descriptions, all resulting from his lack of training as a botanist. His other book fared better. *A Narrative of Travels on the Amazon and Rio Negro* (1853), although also criticized for its dearth of particular data, was better received and sold better. Wallace also read papers before scientific societies and made important connections in the London scientific community.

SOUTHEAST ASIA, 1854–1862

After only eighteen months in England, Wallace again set off for the tropics to work as a specimen collector. As Bates remained in the Amazon basin, Wallace headed instead for Southeast Asia. He had been advised that British cabinets were particularly lacking in specimens from those regions and hence it would be a profitable collecting ground. Wallace was also keen to observe one of the world's few species of great apes, the orangutan, and the different human races in the region. The scientific connections made during his time in London allowed him to appeal for financial assistance to the Royal Geographical Society which in turn secured government funding to pay for a first-class passage to Singapore and a second-class ticket for a young assistant named Charles Allen. Wallace arrived in Singapore on 18 April 1854.

Over the next eight years Wallace made dozens of expeditions procuring 125,000 specimens, including insects, birds, shells, and mammals. In 1855, while living in Sarawak on the island of Borneo, Wallace wrote his first theoretical paper on species: "On the Law Which Has Regulated the Introduction of New Species" (Wallace 1855). In this essay Wallace argued, "Every species has come into existence coincident both in time and space with a pre-existing closely allied species." Although a clear and lucid exegesis of the paleontological and biogeographical data of the time, the paper did not explicitly state that species transmuted one into another. It instead made the case of geological succession. Wallace used intentionally vague language that new species were somehow created according to the model of preceding species. He was testing the waters. Therefore, it was possible for some readers, such as Darwin, to conclude that Wallace referred to a series of supernatural creations in particular times and places. Hence, only much later in *Origin of Species*, Darwin (1859, 355) wrote, "I now know from correspondence, that this coincidence [Wallace] attributes to generation with modification." Others, less accustomed to accepting the evidence for transmutation, such as Lyell, found the implications of the Sarawak paper more novel and suggestive. Lyell opened his own species notebooks (L. G. Wilson 1970). Lyell also urged Darwin to publish his views in outline first rather than continuing to complete his studies and publish on a large scale (van Wyhe 2007). Hence, Darwin began on 14 May 1856 "by Lyells advice" a more condensed version of his original plan (van Wyhe 2006). This condensed version is still known as the "big book" and would have extended to three volumes (R. C. Stauffer 1975, 11). By the spring of 1858, Darwin had completed more than ten chapters, covering two-thirds of the topics later discussed in *Origin of Species*.

In 1858 Wallace was living on the island of Ternate in the Moluccas, the fabled spice islands, west of New Guinea, and then part of the Dutch East Indies. It was here that Wallace conceived of an explanation for the origin of new species that was strikingly similar to Darwin's. According to his own much later recollections, he was suffering from a recurrent bout of fever when the idea came to him. Years before, he had read Malthus's observations that the inevitable geometrical population human growth was prevented only by severe checks. Hence, remembering the argument of Malthus, Wallace conceived of "a general principle in nature" that permitted only a "superior" minority to survive "a struggle for existence" (Darwin and Wallace 1858).

Wallace elaborated this theory in his so-called Ternate essay "On the Tendency of Varieties to Depart Indefinitely from the Original Type." As he wrote in the essay itself,

> The numbers that die annually must be immense; and as the individual existence of each animal depends upon itself, those that die must be the weakest – the very young, the aged, and the diseased, – while those that prolong their existence can only be the most perfect in health and vigour – those who are best able to obtain food regularly, and avoid their numerous enemies. It is, as we commenced by remarking, "a struggle for existence," in which the weakest and least perfectly organized must always succumb. (Darwin and Wallace 1858, 56–57)

Many species have one or more daughter varieties. How these were formed is not stated in the essay. However, as the environment slowly changed as Lyell had argued, a species might become unsuited to its environment and die out. One of its daughter species might, however, be well suited to the new environment and prosper. It could never return to the original parent form as this was now inferior in that environment. This process, reiterated over vast geological time, would account for the origin of new species and the fact that some species had common ancestors.

What happened next has been surrounded by confusion and conspiracy theories for decades. However, there is no evidence for any of the accusations against Darwin. Wallace sent his essay to Darwin, whom he knew to be preparing a large work on evolution, in case it might interest him, with the request that it be forwarded on to Lyell if sufficiently interesting. The essay was largely written against Lyell, but using his own style of reasoning. Wallace hoped to convince Lyell that evolution was the inevitable outcome of the gradual laws of nature.

The single greatest mystery in this story is the date that Wallace sent the essay to Darwin. The Ternate essay is dated February 1858. The original manuscript and its covering letter do not survive. If the essay was sent to Darwin on the next monthly mail steamer after February, as Wallace recollected over a decade later, this would have been 9 March 1858. A letter to Frederick Bates sent on this steamer still survives and bears postmarks showing that it arrived in London on 3 June 1858 (see McKinney 1972). Davies (2008) has shown that all the intermediate mail steamer connections fit for these dates. Darwin's letter to Lyell, which claimed receipt of Wallace's letter and essay on the same day, has been dated to 18 June 1858 (Darwin 1985–, 7:107).

Hence, several writers have asked, if both the Bates and Darwin letters left Ternate on the same ship, how could Darwin receive his on 18 June (as he claimed) and not 3 June? This apparent discrepancy has been the source of great confusion. The reason these two weeks are of consequence is that some commentators believe that Darwin delayed forwarding Wallace's essay to Lyell in order to appropriate, unacknowledged, ideas from Wallace's manuscript into his own (Brackman 1980; J. L. Brooks 1984; Davies 2008).

However, the conspiracy theorists have failed to realize that Wallace wrote his lost letter in reply to a letter from Darwin received on that very same 9 March steamer. There is no evidence from his surviving correspondence that Wallace could reply by the same steamer while in the Moluccas. Furthermore, the date of receipt of Wallace's letter and essay by Darwin on 18 June 1858 is exactly the right day for the mail steamer that left Ternate in early April and, through an unbroken series of mail steamer connections, arrived in London on 17 June (van Wyhe and Rookmaaker 2012).

The recurring accusations that Darwin did or could have borrowed ideas, such as the principle of divergence, from Wallace's writings were conclusively refuted in an important essay by David Kohn (1981). Kohn showed that what many writers mistakenly call an idea of "divergence" between Darwin and Wallace is two different things, which Kohn called "taxonomic divergence" and "a principle of divergence." Taxonomic divergence is the observation that "taxa can be arranged in a branched-hence diverging-scheme" (Kohn 1981, 1105). Darwin made this observation as early as 1837, and this is reflected in his famous Notebook B family tree sketch, which depicts daughter species diverging off a central ancestral trunk. Taxonomic divergence was also mentioned in one line of Wallace's Sarawak paper (1855), but no explanatory principle was given.

A "principle of divergence," according to Kohn, explains "how divergence occurs." Darwin developed this by the mid-1850s and clearly described it in a letter to Asa Gray in September 1857. The same treatment of divergence appeared in Darwin's draft chapters for *Natural Selection* (R. C. Stauffer 1975). After these documents were written, Darwin received Wallace's Ternate essay. The essay contained only one statement on how divergence occurs: "But this new, improved, and populous race might itself, in course of time, give rise to new varieties, exhibiting several diverging modifications.... Here, then, we have progression and continued divergence." As Kohn demonstrated, there were fundamental differences between Wallace's 1858 continued divergence and Darwin's much longer 1857 principle of divergence. Wallace "offered an explanation that is ecologically static, where a new species forms only by the extinction of its parent. There is none of the creation of new evolutionary opportunities by the subdivision of the environment that characterized Darwin's principle of divergence" (Kohn 1981, 1106).

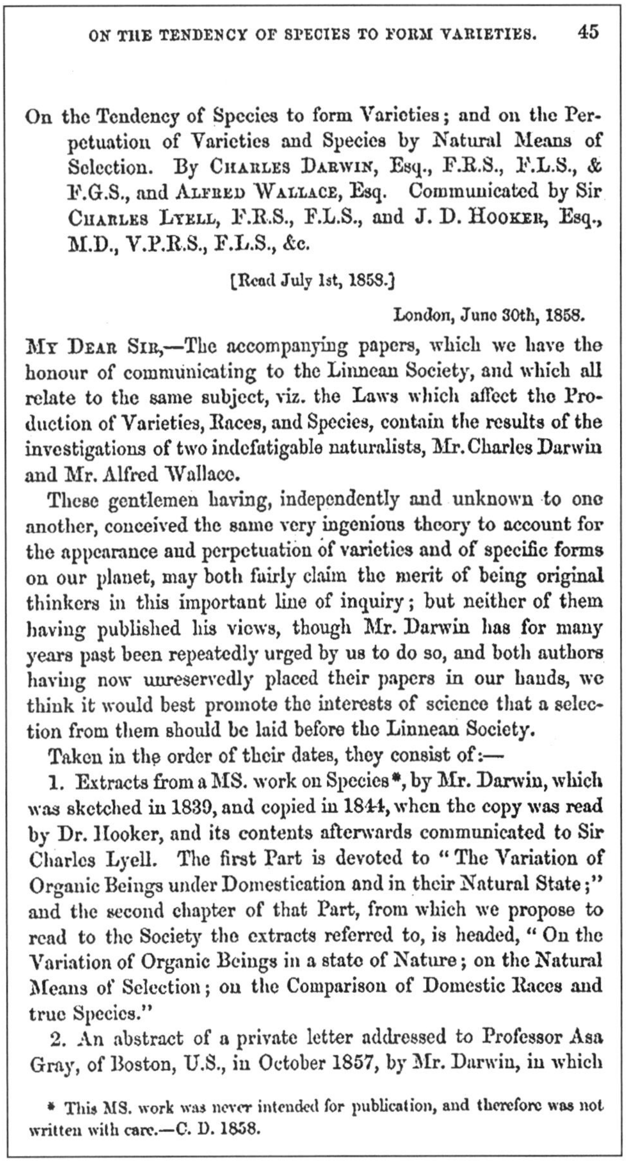
ON THE TENDENCY OF SPECIES TO FORM VARIETIES. 45

On the Tendency of Species to form Varieties; and on the Perpetuation of Varieties and Species by Natural Means of Selection. By CHARLES DARWIN, Esq., F.R.S., F.L.S., & F.G.S., and ALFRED WALLACE, Esq. Communicated by Sir CHARLES LYELL, F.R.S., F.L.S., and J. D. HOOKER, Esq., M.D., V.P.R.S., F.L.S., &c.

[Read July 1st, 1858.]

London, June 30th, 1858.

MY DEAR SIR,—The accompanying papers, which we have the honour of communicating to the Linnean Society, and which all relate to the same subject, viz. the Laws which affect the Production of Varieties, Races, and Species, contain the results of the investigations of two indefatigable naturalists, Mr. Charles Darwin and Mr. Alfred Wallace.

These gentlemen having, independently and unknown to one another, conceived the same very ingenious theory to account for the appearance and perpetuation of varieties and of specific forms on our planet, may both fairly claim the merit of being original thinkers in this important line of inquiry; but neither of them having published his views, though Mr. Darwin has for many years past been repeatedly urged by us to do so, and both authors having now unreservedly placed their papers in our hands, we think it would best promote the interests of science that a selection from them should be laid before the Linnean Society.

Taken in the order of their dates, they consist of:—

1. Extracts from a MS. work on Species*, by Mr. Darwin, which was sketched in 1839, and copied in 1844, when the copy was read by Dr. Hooker, and its contents afterwards communicated to Sir Charles Lyell. The first Part is devoted to "The Variation of Organic Beings under Domestication and in their Natural State;" and the second chapter of that Part, from which we propose to read to the Society the extracts referred to, is headed, "On the Variation of Organic Beings in a state of Nature; on the Natural Means of Selection; on the Comparison of Domestic Races and true Species."

2. An abstract of a private letter addressed to Professor Asa Gray, of Boston, U.S., in October 1857, by Mr. Darwin, in which

* This MS. work was never intended for publication, and therefore was not written with care.—C. D. 1858.

FIGURE 19.3. The opening page of the Darwin-Wallace announcement of evolution through natural selection. Permission: Wellcome

Darwin was greatly surprised to receive Wallace's essay with its stress on a struggle for selection that sounded so similar to his own explanations. He forwarded Wallace's essay the same day to Lyell and asked for advice. Concerned that their friend would lose his priority in the idea of natural selection of twenty years, Lyell and J. D. Hooker had extracts from Darwin's manuscripts from 1844 and 1857 and Wallace's draft essay read before the Linnean Society of London on 1 July 1858. These documents were published together in the society's proceedings in August 1858 (Fig. 19.3). Both events, despite their retrospective importance, were largely overlooked by contemporaries and were certainly too brief to engender any scientific revolution (Moody 1971; England 1997). Even Lyell and Hooker themselves were not yet fully convinced of Darwin's views, and hence neither could have had the slightest idea that he was unveiling the greatest theory in biology, as modern commentators now see the event.

Had Darwin not forwarded Wallace's essay for publication, Wallace would probably never have been credited as co-discoverer of natural selection at all because Wallace did not plan to publish on the subject until his return to England,

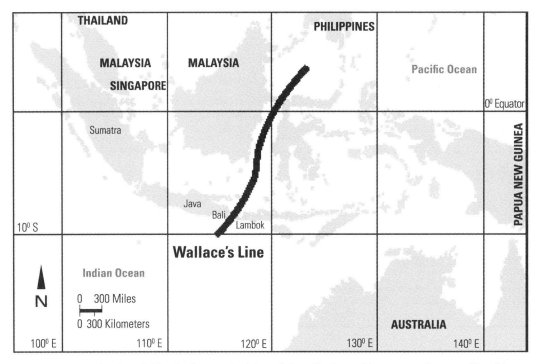

FIGURE 19.4. The sharp dividing line between the Asian and Australian fauna of the eastern and western sides of the Malay Archipelago proposed by Wallace

as he wrote to the ornithologist Alfred Newton in 1887: "I *had* the idea of working it out, so far as I was able, when I returned home" (F. Darwin 1892, 190). In 1857 letters to Darwin and H. W. Bates, Wallace also indicated his intention to prepare a work on species after returning, when he would have access to essential English libraries and collections (Darwin 1985–, 6:457). Wallace returned home only in 1862, an estimated two years after Darwin would have completed and published his big book on species (van Wyhe 2007).

After Wallace's return to Britain in 1862, he was, for the first time in his life, financially secure. Stevens had invested his money well. However, over the next several years, Wallace lost his savings through the demands of a needy family and a series of bad investments. (Raby 2001) He tried unsuccessfully to secure full-time employment. Instead, he earned money by writing, giving occasional lectures, and correcting exam papers, the only regular paid job of his later life.

In late 1864, Wallace was devastated when his fiancée suddenly broke off their engagement. "I have never in my life experienced such intensely painful emotion" (Wallace 1905, 1:410). A few months later, in 1865, he began attending spiritualist séances. Like mesmerism and phrenology before, Wallace claimed he approached the subject with initial skepticism but soon became entirely convinced that the "phenomena" produced by mediums such as table rappings, spirit writings, and apparitions in dark rooms must be genuine and never again doubted the correctness of his conclusion, despite numerous cases of mediums publicly exposed as frauds. The following year he published "The Scientific Aspect of the Supernatural" (1866) and suggested that spiritualism merited scientific investigation. Spiritualism opened a new avenue for Wallace's belief in the possibility of human progress. It also gave him an explanation for what he believed were human abilities not needed for survival in a savage state and therefore not capable of explanation by natural selection.

In 1866 Wallace married Annie Mitten, the daughter of his botanist friend William Mitten. They had three children, two of whom survived to adulthood. In 1869 Wallace published his most famous book *The Malay Archipelago* recounting his travels in Southeast Asia. It was his most successful work both financially and critically. It is still in print and continues to enthrall readers with its tales of adventure and a deep appreciation for tropical natural history. In it he popularized his famous generalization of a sharp line between the fauna of Australia and Asia, now known as the Wallace Line, as he described it: "We have here a clue to the most radical contrast in the Archipelago, and by following it out in detail I have arrived at the conclusion that we can draw a line among the islands, which shall so divide them that one-half shall truly belong to Asia, while the other shall no less certainly be allied to Australia" (Wallace 1869a,1:13). It is important to remember that it was already common knowledge that Asian fauna inhabited the western side of the archipelago and Australian forms the eastern. Wallace attributed his line to two great sunken continents, one Asian, the other greater Australian. The islands of the archipelago were the scattered fragments that remained. But these preserved evidence of two former ancestral homes for the two faunas.

The book was also heavily anthropological, focusing on the races, languages, and other cultural details he observed (Fig. 19.4). He divided the peoples also into to two main types, the Malayan and Papuan races. These too were roughly segregated east and west.

Also in 1869–70 Wallace published new proposals about the origins of human beings, which marked one of his greatest differences with Darwin (Wallace 1869b; 1870a: 332–71). His account was in fact based on the argument from ignorance. He could not see how natural selection could bring about several attributes of human beings, such as a moral sense and high intelligence, as he assumed these were not needed in a savage state of existence in early human prehistory as he did not believe they were needed by the "savage" peoples he had visited in Brazil and Southeast Asia. Therefore, he reasoned, natural selection could not have done so. Building on this assumption, Wallace asserted that this was evidence that a "Higher Intelligence" had intervened in the course of human evolution. These views were not well received by the new Darwinian community.

In the 1870s Wallace returned to his earlier surveyor's perspective with further publications on biogeography. In 1876 he published one of his most important books: *The Geographical Distribution of Animals*. Following Sclater (1857), Wallace divided the world into six main regions. Wallace discussed all of the known factors that determined the dispersal of living and extinct terrestrial animals including elevation, vegetation, land bridges, ocean depth, and glaciation.

Tropical Nature, and Other Essays (1878) was mostly reprinted material. It included Wallace's response to Darwin's theory of sexual selection to explain the origin of some animal coloration. Wallace argued that endless reiterations of female choice could not bring about male colorations and other features such as Darwin had argued for the feathers of the Argus pheasant. Instead, Wallace (1878, 365) imagined the "greater vigour and activity and the higher vitality of the male" led to more vivid coloration.

In 1870 Wallace took up the published wager of a flat-earth advocate. Although Wallace demonstrated, using his old surveying equipment, that a six-mile stretch of the old Bedford canal was indeed slightly convex, his opponent refused to accept the results and spent the rest of his life libeling and persecuting Wallace. It was, Wallace (1905, 2:364) recalled, "the most regrettable incident in my life" and "cost me fifteen years of continued worry, litigation, and persecution, with the final loss of several hundred pounds."

Island Life (1880) was one of Wallace's most successful books. It surveyed the problems of the dispersal and speciation of plants and animals on islands that he categorized, following Darwin, as oceanic or continental. The latter type Wallace subdivided into "continental islands of recent origin," like Great Britain, and ancient continental islands, such as Madagascar. Unlike Darwin's theories of erratic spread to account for the discontinuous distribution of types, Wallace favored theories of continuous spread followed by selective extinctions, thus creating the appearance of gaps.

After 1880 Wallace's attention was increasingly spread across ever wider interests including a land nationalization campaign, anti-vaccination campaign, urban poverty, socialism, private insane asylums, militarism, and life on other planets. At the end of the 1880s, Wallace dropped his adherence to the individualism of Herbert Spencer and returned to the Owenite socialist fold (G. Jones 2002). This huge spread of interests in social and other matters depleted his scientific output.

From 1886 to 1887, Wallace traveled on a lecture tour across the United States. His lectures outlined the theory of evolution by natural selection and the evidence that supported it. These lectures formed the basis of one of his most important books, *Darwinism* (1889). The book was perhaps the clearest and most convincing overview of the evidence for evolution produced in the nineteenth century, second only to *Origin of Species*, and remains an outstanding overview even today. Wallace was more strictly selectionist than Darwin, who had allowed a role for other causes of change. However, the supernatural speculations regarding mankind's origins in the final chapter were either ignored or lambasted by contemporary reviewers. Some of the harshest words ever published about Wallace, in fact, were in reference to these views. The Darwinian acolyte G. J. Romanes (1890) wrote: "It is in the concluding chapter of his book, much more than in any of the others, that we encounter the Wallace of spiritualism and astrology, the Wallace of vaccination and the land question, the Wallace of incapacity and absurdity." The accusation of belief in astrology was incorrect.

The Wonderful Century (1898) discussed the achievements of the nineteenth century and, at even greater length, its problems. *Land Nationalisation* (1882) was a handbook on land reform aimed at telling the "landless classes" how to recognize their rights regarding landownership: "to teach them what are their rights and how to gain these rights" (Wallace 1882, vii).

Man's Place in the Universe (1903) argued against the existence of human beings on any other planet in the solar system (particularly given recent speculation about Mars) or indeed anywhere else in the universe but Earth. In 1905 he published his autobiography *My Life*; it remains the principal biographical source on Wallace. *The World of Life* (1910) was his final word on spiritualism and his view that humanity was placed on Earth for a reason. His last two books were on social issues and the land question. *Social Environment and Moral Progress* and *The Revolt of Democracy* appeared in 1913. The two base causes of poverty and starvation in a land of superfluous wealth were "land monopoly and the competitive system of industry" (Wallace 1913b, 1). Here again was Wallace's belief in removing social obstacles so that natural progress could ensue.

CONCLUSION

Wallace will no doubt remain an endearing, colorful, confusing, and controversial figure in the history of science. He is now often described, especially by commentators outside professional history of science, as overlooked, forgotten, and overshadowed by Darwin. Some recent Wallace admirers even describe him as among the most famous Victorian scientists during his lifetime or at his death. This is certainly incorrect if we refer to the views of contemporary Victorians. While Wallace achieved considerable fame and reputation for his independent discovery of natural selection and his scientific works, especially *The Malay Archipelago*, he never approached anything like the level of fame or respect attributed to Lyell, Richard Owen, William Whewell, Louis Agassiz, T. H. Huxley, Hooker, or Darwin. The oft-repeated view that Wallace was somehow the victim of a Victorian class-based glass ceiling is equally false. Several of his contemporary men of science such as Huxley, born over a butcher's shop, were from humbler origins than Wallace (Desmond 1997).

Wallace's many heresies, as they were seen by more orthodox men of science at the time, clearly contributed to his mixed reputation. The unusually broad range of his literary output remains hard to appreciate. Michael Shermer (2002, 16–17) categorized the topics addressed by Wallace's publications as follows:

Book Topics	*Article Topics*
Evolution, 27%	Biogeography and natural history, 29%
Social commentary, 27%	Evolution and origins of life, 27%
Biogeography, 14%	
Botany, 9%	Social commentary, 25%
Natural history, 9%	Anthropology, 12%
Origins of life, 9%	Spiritualism and phrenology, 7%
Spiritualism, 5%	

There need not necessarily be some hidden consistency underlying his many interests. But if there is, it is likely to be Wallace's deeply held belief that the overall leitmotif of nature is progressive change. Where this is inhibited, such as in the social and political arrangements of his time, artificial impediments should be removed so that natural progress could follow. Wallace did enjoy a rise in fame in the last years of his life, but this was by outliving his contemporaries and becoming the only remaining prominent man of science from the Victorian age.

❧ Essay 20 ❧

Darwin and Humans

Gregory Radick

D ARWIN WENT PUBLIC with his views on human evolution in *The Descent of Man, and Selection in Relation to Sex* (1871) and *The Expression of the Emotions in Man and Animals* (1872). By that time, he had been researching the subject on and off for decades, sometimes in unexpected directions. While on the *Beagle*, for example, he had met a surgeon who reported that the lice infesting Sandwich Islanders on his whaling ship were very distinctive and, furthermore, that when these lice crawled onto white men, the lice soon died. Darwin made a note about the story, adding: "If these facts were verified their interest would be great. – Man springing from one stock according his *varieties* having different parasites" (CUL DAR 31.315). That was in 1834, before Darwin believed that species evolve. He was nevertheless wondering how to connect the fact (as it seemed) that the human races, originating from a single stock, formed mere varieties within a single species, with the fact (as it seemed) that those races were so different physiologically as to sustain different species of lice. In 1844, and again in 1865, he quizzed England's leading louse expert, Henry Denny, about it all – in the interim attempting to get Denny some lice from American blacks. In the *Descent*, Darwin cited Denny in a paragraph-long discussion of the matter. On the whole, Darwin judged, the facts about lice – and the surgeon's observations had since been confirmed more generally – seemed to support the ranking of the different human races as distinct species (Darwin 1871a, 1:219–20; Radick and Steadman forthcoming).

BEFORE THE *DESCENT* AND THE *EXPRESSION*

What explains such a prolonged concern with the human races, their status, and their origins? The short answer is that Darwin was born into an era when questions about race had become entangled with questions about slavery – and into an activist family that regarded slavery as an abomination. Both of his grandfathers, Erasmus Darwin and Josiah Wedgwood, backed the campaign to end slavery, with Wedgwood's potters producing the campaign's great emblem: an image of a kneeling, enchained black slave who asks "Am I not a man and a brother?" By Charles's time those words had acquired a natural-historical resonance. At issue was whether the

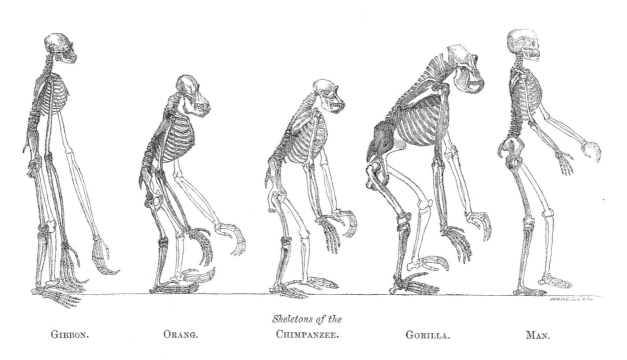

FIGURE 20.1. The frontispiece of Thomas Henry Huxley's *Man's Place in Nature* (London: Williams and Norgate, 1863), drawn deliberately to show the close relationship of humans with the apes

different human races originated from a single stock – the brotherhood-of-man view – or whether each race had a separate origin. The latter view came to be identified with the slavers and their interests; for if blacks belonged to a different and lower species than whites, the moral case against black slavery became less straightforward. Conversely, the common-ancestry answer became a taken-for-granted part of antislavery argumentation. Darwin seems to have absorbed wholesale the argumentation and its associations. When, in 1850, he learned that the U.S.-based naturalist Louis Agassiz had spoken on the separate origins of the human races, Darwin wrote to a friend about Agassiz's upholding "the doctrine of several species, – much, I daresay, to the comfort of the slaveholding Southerns" (Darwin 1985–, 4:353, letter to W. D. Fox, 4 September 1850). Darwin's books on humankind would update a common-origins case deriving from abolitionism's heyday (Desmond and Moore 2009; E. Richards forthcoming).

Race is, of course, just one of the topics addressed in those books. They also set out to show that the human species is the modified descendant of a previous, lower, extinct species and to explain how human bodily and mental characteristics had evolved. Before starting work on writing the books in the late 1860s, Darwin's most intensive theorizing on human evolution had taken place as part of his more general theorizing about "transmutation" (as he then called it) in the late 1830s, in a series of small private notebooks. The ones mainly concerned with humankind were Notebook M, begun in July 1838, and a follow-up notebook, N, begun in October of that year. "M" stood for "metaphysics," which then named

a broad inquiry into the nature of mind. By mid-1838, Darwin was already committed to the view that new species emerge gradually, with humans no exception. Reflecting on discussions with his friends and relatives (especially his physician father Robert) and on his readings in medicine, natural history, and philosophy, Darwin ranged widely over the continuities between humans and nonhuman animals – in their capacities for reason, moral action, communication, emotion and its physical expression, and so on – which, for him, showed the animal origins of the human mind. In another notebook, C, he challenged anyone to compare the humanlike qualities of a domesticated orangutan, with "its expressive whine," "its intelligence when spoken [to]," and "its affection," with the brutishness of the "savage," "roasting his parent, naked, artless, not improving yet improvable," and still to "dare to boast" of the "proud preeminence" of humankind (Barrett et al. 1987, C, 79). (The rest of the entry makes plain that the savage Darwin had in mind was Fuegian; his shock at the extraordinary looks, sounds, and ways of the tribal peoples he encountered when the *Beagle* reached Tierra del Fuego in 1832 never deserted him.) The explorations of this period pushed very far indeed, extending, in Notebook C (Barrett et al. 1987, C, 166), to the material basis of mind, and the possibility that even religious faith was nothing but an effect of the brain's organization (J. Hodge 2009, 59–63).

Apart from Darwin's notebook theorizing of the 1830s, the main corpus on which he drew in the *Descent* and the *Expression* was the large and mostly public one that accumulated in the 1860s in the wake of the *Origin of Species* (1859). Although not explicit in the *Origin*, the easily inferred conclusion for humankind – that humans are the evolved descendants of apelike progenitors – struck commentators not merely as unlikely but, in undermining Christian teachings and the moral striving they inspired, unwelcome. The debates were many and complex (R. J. Richards 1987, ch. 4); but if we consider their significance for Darwin's own theorizing, they fall into three clusters. First, there were battles over human-animal continuity and evolutionary kinship, with the Darwinian case put most elegantly in the London naturalist Thomas Henry Huxley's *Evidences as to Man's Place in Nature* (1863) and most comprehensively in his German counterpart Ernst Haeckel's *Natürliche Schöpfungsgeschichte* (1868) (Fig. 20.1). Second, there were attempts, by the Scottish mill-owner William Greg and the London geographer and mathematician Francis Galton (Darwin's cousin) among others, to work out the conditions of continued moral and intellectual progress in civilization, given what they saw as shortsighted tendencies to protect the weak and to mate without regard to inheritable quality. Third, there were Darwin's disagreements with Alfred Russel Wallace, "co-discoverer" of natural selection. To Darwin's dismay, Wallace had begun publicly doubting that natural selection had brought about human mental faculties and privately doubting that sexual selection – which Darwin would call upon to account for human racial divergence – had quite the explanatory reach that Darwin thought (Cronin 1991, chs. 5–8).

THE DESCENT OF MAN, AND SELECTION IN RELATION TO SEX (1871)

To read the *Descent* and the *Expression* is to keep company with an author who, for all the demanding intellectual territory his books explore, takes care to provide clear maps at the outset. In the introduction to the *Descent*, Darwin announces the three questions that will occupy him throughout (Darwin 1871a, 1:2–3). The first is "whether man, like every other species, is descended from some pre-existing form"; the second concerns "the manner of his development," that is, the process whereby "man" evolved from apelike progenitors (for ease as well as accuracy, Darwin's gendered language will mostly be used from here); and the third takes up "the value of the differences between the so-called races of man," that is, whether the different races should count as varieties of one species or as different species, and how such differentiation came about (Fig. 20.2) In answering the first two questions Darwin observes the same division of labor: body first, then mind. He starts with bodily signs of man's evolutionary past (ch. 1), from the many close similarities with ape bodies, to the appearance in a human embryo – in early stages, scarcely distinguishable from a dog embryo – of gill slits and other features absent from the adult human but present in the adult forms of lower species, to the many uselessly rudimentary structures and capacities that characterize at least some humans. The next two chapters make a complementary case for man's mental powers, considered in general (ch. 2) and with special attention to the moral sense (ch. 3), the whole

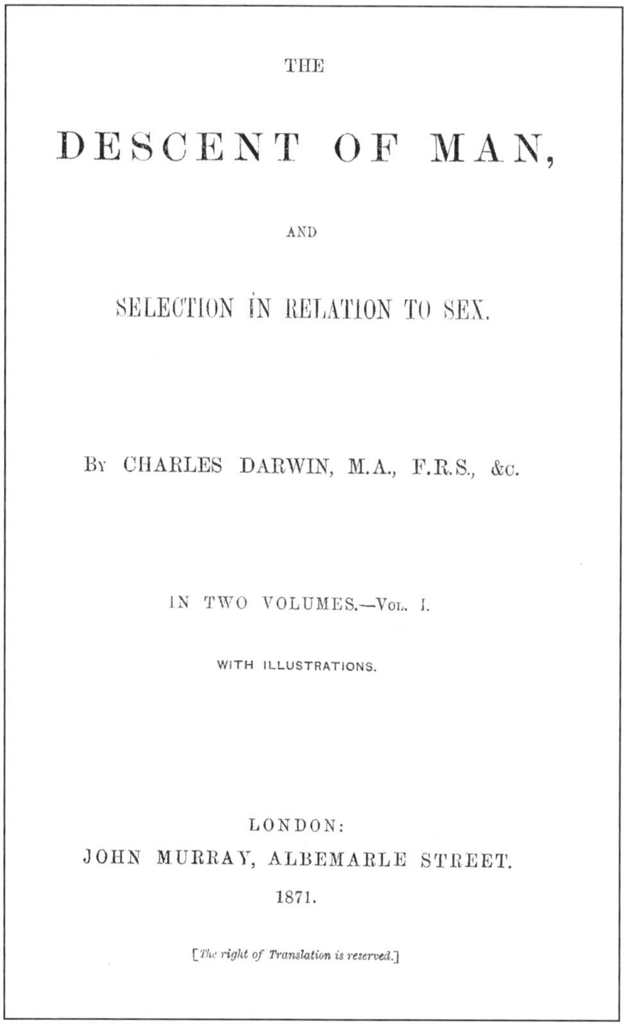
THE

DESCENT OF MAN,

AND

SELECTION IN RELATION TO SEX.

BY CHARLES DARWIN, M.A., F.R.S., &c.

IN TWO VOLUMES.—VOL. I.

WITH ILLUSTRATIONS.

LONDON:
JOHN MURRAY, ALBEMARLE STREET.
1871.

[*The right of Translation is reserved.*]

FIGURE 20.2. The title page of the *Descent of Man* (London: John Murray, 1871), where Darwin sheds light "on the origin of man and his history." Note that the book also covered sexual selection in very great detail, something much bound up with Darwin's wanting to find naturalistic alternatives to Wallace's claim that human evolution was driven by spirit forces.

showing that "the difference in mind between man and the higher animals, great as it is, is certainly one of degree and not of kind" (1:105) – a pattern well explained by man's evolutionary origin but otherwise mysterious.

With the argument for man as the product of an evolutionary process concluded, Darwin turns to his second question, about the nature of the process. Again dealing with man's body first (ch. 4), Darwin surveys the evidence that man is no different from other species in showing inheritable variation and in experiencing, at least from time to time, a struggle for existence brought on by Malthusian overpopulation. Given these facts, Darwin reasons, it follows that man is subject to natural selection, and so natural selection – acting as the main but not exclusive modifying agency – may have generated man's characteristic anatomy. Darwin's reconstruction of how that happened pivots on the survival advantages that, after man's progenitors had dropped from the trees, probably accrued to those individuals who showed greatest specialization of the

FIGURE 20.3. Although fun was made of Darwin, for the English it was always rather gentle. They were and are immensely proud of Charles Darwin. From the *Hornet* (1871)

feet for locomotion and the hands for prehension. In the next chapter (ch. 5), Darwin considers the parallel and – when it came to man's enlarging brain and skull – interacting development of his mental powers under (mainly) natural selection, emphasizing the advantages to individuals of high intelligence and to tribes of the moral habits and codes that make for success in struggles with other tribes. Here Darwin also examines the Greg-Galton points about whether civilized mercies and freedoms thwart continued progress under natural selection, concluding that, although they sometimes do, the factors promoting progress tend to counterbalance. A further chapter (ch. 6) provides a deep genealogy, proceeding backward in time from an apelike African ancestor all the way to the ascidian-like progenitor of the vertebrates (Fig. 20.3).

Aside from the book's conclusion, the remaining fifteen chapters bear on Darwin's third question, about the races of man. On a quick glance, this purpose is not obvious; for the bulk concern sexual selection, considered as a set of evolutionary principles (ch. 8), and then as the main agency behind differences between the sexes in a range of animals, starting with mollusks, annelids, crustaceans, and spiders (ch. 9) and proceeding up the animal scale of complexity, through insects, fish, amphibians, reptiles, birds, and mammals, reaching man only at the end (chs. 10–19). But as Darwin explained in his introduction, he had reckoned that, because his main answer to his question about racial differences was going to be sexual selection, and because he had nowhere previously set out a detailed, general argument for it as an evolutionary process, he would take the opportunity here. And, indeed, the sex-differences chapters are book-ended by chapters on race in man. In the first (ch. 7), Darwin undertakes a balanced discussion of the classificatory, varieties-or-species debate, finding that some considerations (such as the observations concerning lice) favor a ranking of the human races as distinct species, whereas others (such as their grading into each other) favor a ranking as mere varieties. Following Huxley, Darwin declares that, whatever ranking one decides upon, anyone persuaded about the principle of evolution will admit that, given the many similarities among the races, even in the most unimportant characters, the races must have descended from "a single primitive stock" (1:229).

But what brought about racial divergence? Darwin rapidly proposes and rejects a number of possibilities, including natural selection, on the grounds that "not one of the external differences between the races of man are of any direct special service to him" (1:248–49). Having eliminated the alternative explanations, Darwin introduces sexual selection and embarks upon the massive theoretical and empirical detour that culminates in a final pair of chapters on man. The first (ch. 19) catalogs what Darwin takes to be the main differences between men and women, notably the greater strength of men in body and mind, and seeks to show how the processes of sexual selection – above all, men battling for the most attractive women – might have produced those differences. But Darwin's discussion here is complex and takes in, for example, a conjecture about how primeval courtship ultimately led to the high musicality of the human voice, in males and females, and to the connections we still experience between our emotional lives and musical voices (we hear the latter and we are moved; we are moved and our voices go up and down in pitch). Finally, in the book's penultimate chapter (ch. 20), Darwin extends this account, by way of some interesting twists, to the formation of the different races of man (Millstein 2012). Beauty is the key. As local standards of beauty came to prevail in different human groups, men sought women – and, to a lesser extent, women sought men – who most closely conformed to the local standard, in facial features, skin color, and so on. In his closing paragraphs, he acknowledges that, for some, an evolutionary origin for man will be "highly distasteful." But, he goes on, no one can doubt that before man was civilized, he was uncivilized; furthermore, no one who has seen what Darwin had seen of man's uncivilized state in Tierra del Fuego at that first contact – the men "absolutely naked and bedaubed with paint, their long hair ... tangled, their mouths frothed with excitement" – has much to defend when it comes to the supposedly threatened dignity of the species. Returning to the theme of that long-ago notebook entry, and recalling some of the humanlike animal feats described earlier in the book, he drives the point home:

> For my own part I would as soon be descended from that heroic little monkey, who braved his dreaded enemy in order to save the life of his keeper; or from that old baboon, who, descending from the mountains, carried away in triumph his young comrade from a crowd of astonished dogs – as from a savage who delights to torture his enemies, offers up bloody sacrifices, practises infanticide without remorse, treats his wives like slaves, knows no decency, and is haunted by the grossest superstition. (Darwin 1871a, 2:404–5)

THE EXPRESSION OF THE EMOTIONS IN MAN AND ANIMALS (1872)

What became Darwin's second book on man grew from an essay initially intended for the *Descent*. There he described his interest in emotional expression as twofold (1871a, 1:5). First, he saw a challenge to his case for man's evolutionary origin in the view that "man is endowed with certain muscles solely for the sake of expressing his emotions." Second, he "wished to ascertain how far the emotions are expressed in the same manner by the different races of man," because, as he went on to explain in the first chapter on race, nothing showed common ancestry more clearly than close similarity in lots of unimportant details, and the different human races were nearly identical in the ways they expressed their emotions. But Darwin reserved his evidence on this matter for *The Expression of the Emotions in Man and Animals* (1872). Undoubtedly carrying forward the argument of the *Descent*, the *Expression* nevertheless has its own ambitions and character, not least because Darwin regarded the book as exemplifying a new and more rigorous approach to the collection of data on emotional expression (Browne 1985a; see also Dixon 2003, 175–76). He stressed the importance of six kinds of evidence: observations on infants; observations on the insane (like infants, prone to strong emotional expression); answers to questions about what emotion is being expressed in a photograph; the study of great art (though in practice this features little in the book); observations on men and women of different races (Darwin sent out a questionnaire to missionaries and others); and observations on animals.

Even a casual reader will notice two further and more pronounced contrasts with the *Descent*. Most obviously, there are the many photographs, of sometimes dramatically emoting infants, boys, girls, men, and women. Photographs in books were still unusual in this period, and Darwin went to considerable trouble and expense to acquire and reproduce the ones in the *Expression*, many of them specially commissioned (Prodger 2009). The other contrast lies with the explanations on offer. For all its centrality in the *Descent*, natural selection in the *Expression* is marginal. Instead, Darwin introduced three new principles, expounded in the first three chapters. There is the "principle of serviceable associated habits," according

FIGURE 20.4. This pouting chimpanzee, from Darwin's *Expression of the Emotions in Man and in Animals* (London: John Murrray, 1872), is intended to show the similarities between man and the higher apes.

to which movements that somehow gratify or relieve a state of mind become habitual under that state of mind and then stay habitual, in the individual and in the lineage, even after the movements have ceased to gratify or relieve. There is the "principle of antithesis," which holds that such habits tend to bring into being their opposites; so – to use an example from later in the book – because indignation in humans came to be expressed by squared shoulders, clenched fists, and other elements of a fighting posture (for, in the past, fighting brought relief), the opposite feeling of helplessness, or impotence, came to be expressed by the opposite movements of shrugged shoulders, open palms, and so on. Finally there is the "principle of the direct action of the nervous system," in which strong emotions generate excess nervous energies, which, in dissipating, cause various movements.

Darwin goes on to put these principles to work, in explaining emotional expression in the lower animals (chs. 4 and 5) and in man (chs. 6–13), with each of the human emotional expressions, from weeping to blushing, provided with a close anatomical and psychological description, an explanation in terms of some combination of the three principles, and a summary of the evidence for cross-racial universality. Emotional expression comes to be subsumed within Darwin's general case for humankind's evolutionary origin in part by his identifying expressive continuities with animals (for instance, chimps and humans pout when sulky), and in part by his relying on the same three physiologically grounded principles to do all the explaining (Fig. 20.4). In the conclusion (ch. 14), he even speculates on the evolutionary history of expression, noting, to return to an earlier example, that the indignation posture in humans – and so the antithetical posture of helplessness – could not have entered our expressive repertoire until after our progenitors had started walking upright. The reconstruction ends with some remarks on the lessons to draw from this history – remarks very similar to ones made at a comparable reconstructive moment in the *Descent* (1:213). For Darwin, there was nothing inevitable about human evolution taking exactly the form it did. Even small differences early on might have led to big differences later. In the *Expression*, he put the point vividly: if man had breathed water instead of air, his face would now be no more expressive than his hands or limbs are.

Scientists and historians have long wondered about Darwin's curiously "non-Darwinian" handling of emotional expression. There is, most conspicuously, his heavy reliance on the inheritance of acquired habit, or so-called Lamarckian inheritance (which features in the earlier books, though nowhere near as much; see Radick 2002, 10–13). But there is also his near-total indifference to the possibility that, like so many of the traits discussed in the *Descent*, emotional expressions might have been useful either in the struggle for life or in the struggle for mates. Three observations about Darwin's notebook theorizing of the 1830s may offer clues to an explanation. One is that his theory of natural selection emerged only *after* his expression theorizing was already well advanced. The second is that this early expression theorizing – including the germs of the three principles – drew on the work of Darwin's evolutionist grandfather Erasmus, for whom habit and its (often useless) persistence were of central importance. The third is that, for all the breadth of Darwin's notebook theorizing on expression, there was no engagement in those years with a topic that would matter hugely in the *Expression*: race. Through the 1860s, as Darwin collected data on human emotional expression from around the world, his old theorizing on expression as nonadaptive seems to have acquired a new significance, spelled out in the *Expression*'s conclusion. The remarkable sameness of emotional expressions across the human races suggested, he wrote, "a new argument in favour of the several races being descended from a single parent-stock" (Darwin 1872b, 361), itself already mostly human in character before the races diverged. For it was most improbable, he went on, that natural selection could have generated such similarity, verging on uniformity, in evolutionarily separate lineages. With emotional expression, therefore, we must be dealing with something nonadaptive, beyond natural selection's scope (Radick 2010a).

NINETEENTH-CENTURY RESPONSES

Both books sold well, and the *Descent* especially so. In 1874 Darwin brought out an expanded and lightly rearranged second edition. (A second edition of the *Expression* was published posthumously.) In a new preface, Darwin wrote of the "fiery ordeal" through which the *Descent* had passed (Darwin 1874, v). Some of the reactions were certainly overheated. The book's appearance had come just before the election of the Paris Commune, and the reviewer for the *Times* of London (8 April 1871, 5) saw in Darwin's unsettling vision of human change a dangerous encouragement to the revolutionaries. Others were scandalized by Darwin's frankness about human sexuality and declared the book obscene – a judgment Darwin

FIGURE 20.5. This cartoon from the humorous weekly *Punch*, in 1861, shows just how quickly Victorians picked up on the implications of evolution for our own species, a focus that exists to this day. Had Darwin's been a theory applicable only to warthogs, one doubts that this *Encyclopedia* would have been compiled.

and his publisher had been worried enough about beforehand that they had toned down some of the sexier discussions (Dawson 2007, ch. 2). On the whole, however, the reception of the books on man was a more muted affair than the reception of the *Origin* (Ellegård 1990, ch. 14), in part no doubt because the main issues had been so well aired, in high and not-so-high culture, throughout the 1860s. (A famous 1861 cartoon showed an ape wearing a sign: "Am I a man and a brother?" [Fig. 20.5].) Nevertheless, the responses, public and private, were voluminous, and Darwin took them seriously, incorporating a number of them in the pages of the 1874 *Descent*, on everything from whether man's suffering from some of the same diseases afflicting the lower animals favors the evolutionary theory to the correct lessons for the origins of human reason from an experimental study of learning in a pike.

Indeed, though evolution swept through all of the human sciences in the late nineteenth century, the one most profoundly reshaped in the wake of Darwin's books on man was psychology (Boakes 1984; R. Smith 1997, ch. 13). Sigmund Freud and William James are perhaps the best remembered of a generation of psychological inquirers and psychiatrically engaged medical men for whom the idea of mental evolution became foundational for understanding the human mind

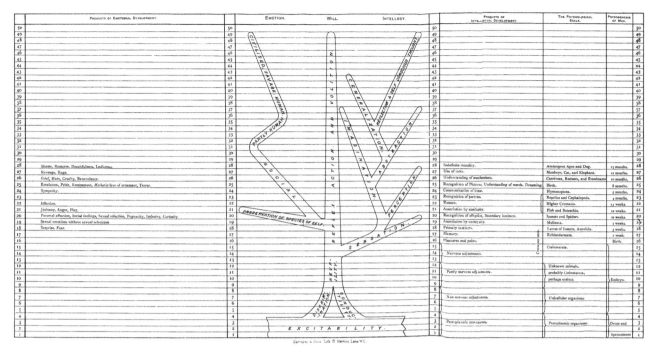

FIGURE 20.6. A tree of mental development drawn by G. J. Romanes to show how humans develop through stages similar to those of adult lower animals. From G. J. Romanes, *Mental Evolution* (London: Kegan Paul, Trench, 1883)

(Sulloway 1979, ch. 7; R. J. Richards 1987, ch. 9; Adriaens and De Block 2010). With the growth of interest in mental evolution came new kinds of psychological inquiry, notably into the minds of children (who were widely thought to "recapitulate" the evolutionary emergence of the human mind) and animals. Darwin's only other important contribution on man after 1874 was his article "A Biographical Sketch of an Infant," published in *Mind* in 1877 and drawing on a diary he had kept thirty-seven years before on the development of emotional expressions, reason, the moral sense, and so on in one of his own children (Darwin 1877c). It was not Darwin but his younger ally George John Romanes, a comparative physiologist by training, who, with a series of books in the 1880s, became the first great champion of the Darwinian study of animal minds. This moment in intellectual history is well summed up in an image from Romanes's *Mental Evolution in Animals* (1883): in the center is a tree, with nervous excitability at the base and self-consciousness at the top; along the side are scales correlating psychological faculties, kinds of animals, and ages of the human child (Fig. 20.6). (Romanes' next work, *Mental Evolution in Man* [1888], is said to have been the most heavily annotated book in Freud's library.)

What of the evolutionary future? At the end of the *Descent*, in words virtually unchanged across the two editions, Darwin indicated qualified support for a couple of proposals that, over the succeeding decades, would come to be known as "eugenics" (progress through selective breeding) and "social Darwinism" (progress through competitive struggle in human society). On behalf of selective breeding, he advised that men and women "ought to refrain from marriage if in any marked degree inferior in body or mind"; but he immediately added that "such hopes [of progress] are Utopian and will never be even partially realised until the laws of inheritance are thoroughly known." On behalf of competitive struggle, he advocated "open competition for all men" and the abolition of any laws or customs that prevented "the most able ... from succeeding best and rearing the largest number of offspring"; but he straightaway disowned an extreme interpretation, insisting that, important as natural selection had been, other agencies had been more important in ensuring human moral progress, including reason and religion (Darwin 1871a, 2:403). Neither proposal was original to Darwin; both would go on to inform and inspire some of the most appalling policies of the twentieth century, most egregiously in Hitler's Germany (Paul 2009). In the nineteenth century, however, in Darwin's Britain as elsewhere, there was notable enthusiasm for his tying of moral progress to reason, and relatedly, for his notion that, because cooperation in the past had given ancestral humans the competitive edge, their descendants were naturally disposed to the altruistic giving of "mutual aid," in the Darwinian anarchist Peter Kropotkin's famous phrase (Dixon 2008).

Enmeshed with eugenics and social Darwinism were Darwinian views on sex and race. In the books on man, and the *Descent* especially, Darwin assigned different kinds of people to different positions in an evolutionary hierarchy: men higher than women; white civilized races higher than the other races – and the higher the race, Darwin suggested, the greater the gap between men and women (E. Richards 1983, 74–75). At certain points – a notable example involving race is the discussion in the *Descent* of the origin of language – the *explaining* of these widely accepted hierarchies comes across as another of the advantages that Darwin saw in his evolutionary theory over the theory of special creation, on which these patterns simply had to be accepted as part of the Creator's plan (Radick 2008). Even so, Darwin regarded the state of women and "savage" peoples such as the Fuegians as improvable; and some of his nineteenth-century readers extracted legitimation from his writings for campaigns for sexual and racial equality (Erskine 1995; Radick 2010b). These readings did not, however, have anything like the public prominence of those emphasizing the permanence of evolved differences and the comparative lowness of the nonmale and the nonwhite, interpreted as occupying lower stages in a progressive evolutionary scheme.

By the end of the nineteenth century, empire, Darwinism (a much more diffuse thing than Darwin's own views), and anthropological race ranking often marched together (Brantlinger 2003). Especially in the equanimity with which he contemplated the ongoing and future exterminations of lower races by higher ones, the Darwin of the *Descent* can, for the present-day reader, be uncomfortably of his imperial age.

TWENTIETH- AND TWENTY-FIRST-CENTURY RESPONSES

As the evolutionary science of humankind has evolved, so have responses to Darwin's contribution. At the 1909 Darwin centennial, the German-born, New York-based anthropologist Franz Boas – easily the best-informed and, eventually, the most influential anthropologist of his generation – delivered a mixed verdict. Boas praised Darwin for clarifying major problems and making undeniable the case for the evolutionary emergence of man from a lower form. And on a range of subsidiary topics, Boas reckoned, Darwin's views had been vindicated, from the existence of intermediate fossils linking humans and the apes to the notion that customs and beliefs can get established in human groups without anyone's consciously deciding to establish them. But, Boas went on, there was another side of the balance sheet. For one thing, where Darwin had thought that some human races were anatomically closer to the animals than other races – "the essence of savagery," he wrote in the *Expression*, apropos the exaggerated protruding of lips observed in the sulky children of savage races, "seems to consist in the retention of a primordial condition" (Darwin 1872b, 235) – up-to-date anthropologists, Boas reported, recognized that no race was more animal-like than any other. They had likewise become doubtful about mental evolution as a uniform, progressive process taking place at different rates in different groups, insisting instead that it was history, not biology, that had put some groups further up the scale of civilization than others. And the evidence for human evolution generally as the steady, gradual affair that Darwin had envisaged was distinctly lacking (Boas 1909).

By the time of the next big Darwin celebrations, in 1959, a newly "synthetic" – and avowedly antiracist – Darwinian theory enjoyed much broader support among biologists as well as anthropologists. Yet Darwin on man continued to elicit less-than-worshipful views. The historian Gertrude

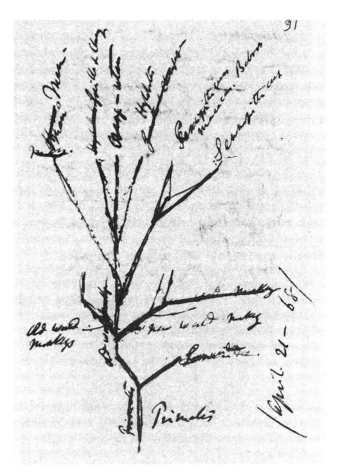

FIGURE 20.7. A sketch made by Darwin (but not published) of the human family tree. Note that he puts humans off on their own, whereas today we would put humans very close to the chimpanzees, even more than chimpanzees are to gorillas. Permission: Cambridge University Library, http://darwin-online.org.uk/content/frameset?viewtype=image&itemID=CUL-DAR80.B91&pageseq=1, CUL-DAR 84.91

Himmelfarb, in her *Darwin and the Darwinian Revolution* (1959), gave low marks to Darwin's accounts in the *Descent* of the emergence of morality, religion, and other distinctly human attainments – and low marks to their author too. "Darwin's failures of logic and crudities of imagination," she wrote, "emphasized the inherent faults of the theory; a finer, more subtle mind would only have obscured or minimized them. The theory itself was defective, and no amount of tampering with it could have helped" (308). It was only over the next fifty years that the more positive evaluations now so familiar took hold, thanks in no small part to the rise of a new era of Darwinian-anthropological enthusiasm (Degler 1991). Starting with human ethology in the 1960s, and continuing with sociobiology in the 1970s and evolutionary psychology in the 1980s and 1990s, the scientists involved throughout stressed their links with Darwin and his works on man. The ethologist Konrad Lorenz (1965) contributed an introduction to a reprint of the *Expression*. Soon-classic work in sociobiology by Robert Trivers and others appeared in the early 1970s in volumes commemorating the centenaries of the *Descent* and the *Expression* (B. Campbell 1972; Ekman 1973).

The editor of the volume on emotional expression, Paul Ekman, later brought out a third, "definitive" edition of the *Expression*, complete with a postscript in which he recounted his battles with the Boasian anthropologist Margaret Mead over the Darwinian versus cultural basis of human emotional life (Darwin 1872b).

More strictly historical studies too have played their part. Three books in particular merit close study from anyone wishing to pursue the subject. Although dated in several ways, *Darwin on Man: A Psychological Study of Scientific Creativity* (1974), by the psychologist Howard E. Gruber (and with a foreword from Gruber's mentor, Jean Piaget), remains an insightful and remarkably thorough exploration, especially strong on Darwin's notebooks and other manuscripts to do with humans, much of it reproduced – including a sketch Darwin made in 1868, but never published, of the primate family tree (H. E. Gruber 1974, 197; CUL DAR 84.91) (Fig. 20.7). The other two books are by historians of science whose approaches neatly exemplify what used be called "internalist" and "externalist" approaches to science. Setting out to show how wrong Himmelfarb was, Robert J. Richards's *Darwin and the Emergence of Evolutionary Theories of Mind and Behavior* (1987) remains the best guide to Darwin's ideas on the evolution of mind and morals: what they were, how he came to develop them, how they resonated with debates in his day and afterward, and why they continue to bear attention in our own day (see also R. J. Richards 2009a). An otherwise very different kind of book, Adrian Desmond and James Moore's *Darwin's Sacred Cause: Race, Slavery and the Quest for Human Origins* (2009) – the most important publication of the recent Darwin anniversary year – is at one with Richards in taking issue with the notion that, as Himmelfarb once put it, Darwinism "de-moralized man" (quoted in R. J. Richards 1987, 6). For Desmond and Moore, the whole Darwinian project, culminating in Darwin's argument in the *Descent* for the common ancestry of the human races, is inseparable from the intensely moral antislavery politics to which his family was devoted, and within which the "unity of man" was an article of faith.

Opinion on Desmond and Moore's claims is far from settled. (One of the most severe critiques is from Richards [2009b].) But their provocation has helped to make "Darwin and humans" one of most exciting areas in Darwin scholarship right now. Adding to the ferment is the publication in progress of the Darwin correspondence volumes covering the years when he wrote, published, and responded to controversies over the *Descent* and the *Expression*. The creative tracking of routes through the letters, and through the copious and little-examined manuscript material now available online from the Darwin archive at Cambridge University, is bound to teach us much about the making and meaning of these books and their legacies. The possibilities can be glimpsed in recent studies on developments within the evolutionary sciences (Burkhardt 2005; Radick 2007; Borrello 2010; Milam 2010) and well beyond them (Bender 1996, 2004; Dawson 2007). At the same time, renewed discussion of how far responsibility for the black spots on biology's political record can be laid at Darwin's door has stimulated new sophistication about larger issues of historical influence and its assessment (Weikart 2004; Bowler 2008, 565–66; R. J. Richards 2008, appendix 2). The challenge now is to absorb all of these innovations, and the best of the older scholarship, in ways that at once enhance our readings of Darwin's texts and open up new ways of connecting them to their multiple contexts, and to our own.

❧ Essay 21 ❦

Darwin and Language

Stephen G. Alter

Charles Darwin's views on language were inseparable from his views on the evolution of humanity's brain capabilities as well as on the origins of racial distinctions – topics that will form a significant share of this discussion. Our starting point, however, is Darwin's fundamental theory of how language originated. It was universally acknowledged in Darwin's day, as in our own, that language use was a key aspect of what it means to be uniquely human, and so Darwin was obliged to explain how language could have emerged via gradual and naturalistic means, as an essential part of human evolution. On this and related topics we find Darwin working with a few simple ideas that he held throughout his career, even though he elaborated those ideas in increasingly complex ways.

DARWIN ON THE ORIGIN OF LANGUAGE

Exposition in *Descent*

Darwin's main views on the origin of language appear in a ten-page section on "Language" found in chapter 2, on the "Mental Powers of Man," of his book *The Descent of Man* (1871a, 1:53–62). The section begins with preliminary observations on how communication among higher animals often approximates language. There are also remarks about language's hybrid nature: it is part instinct and part invention. Darwin was careful, however, to qualify his use of the latter term: "No philologist [i.e., linguist] now supposes that any language has been deliberately invented; each has been slowly and unconsciously developed by many steps" (1:55). As to the specific means of origination, Darwin said: "I cannot doubt that language owes its origin to the imitation and modification, aided by signs and gestures, of various natural sounds, the voices of other animals, and man's own instinctive cries" (1:56). Here Darwin built on two standard eighteenth-century conjectures about the way early humans could have formulated their first words. Both theories involved vocal mimicry: the difference lay in what was said to be imitated. One emphasized sounds in nature such as animal cries; the other emphasized humans' own spontaneous grunts, groans, and mating calls (Stam 1978). Darwin in essence combined these two perspectives.

Darwin (1871a, 1:56) cited support from recent works that updated these theories: the Anglican churchman F. W. Farrar's *Chapters on Language* (1865) and the gentleman-scholar Hensleigh Wedgwood's *On the Origin of Language* (1866). (Wedgwood was Darwin's cousin as well as the brother of Darwin's wife.) These writers, especially Wedgwood, taught that vocal mimicry presented a scientific *vera causa*: it was a mode of coining new words (as in the use of onomatopoeia) that could be observed operating independently of the particular outcomes it was called on to explain. A related assumption was that the types of forces causing change hold constant over time, this being an essential feature of the "uniformitarian" reasoning emergent in natural history, especially in geology. Darwin's particular contribution was to set these notions about speech origins into an evolutionary context, involving man's prehuman ancestors. "Instinctive cries" were mainly those used by higher animals to gain mates, employed either to express desire for targeted females or to warn off rival males. The reflexive imitation of those cries eventually would have produced words expressive of the relevant emotions: desire, jealously, or anger (1:56). It was likewise, according to Darwin (1:57), with other sounds from nature:

> As monkeys certainly understand much that is said to them by man, and as in a state of nature they utter signal-cries of danger to their fellows, it does not appear altogether incredible, that some unusually wise ape-like animal should have thought of imitating the growl of a beast of prey, so as to indicate to his fellow monkeys the nature of the expected danger. And this would have been a first step in the formation of a language.

We may surmise that Darwin regarded such a scenario to be influenced by group selection: those communities having members who used vocal mimicry to send warning signals to their fellows would be more likely to survive en masse, preserving with them the incipient talkers and their similarly endowed progeny (1:159–61).

Development of Vocal Capability

Darwin's account of the development of the physical aspects of speech was a more complicated affair, involving a mix of sexual selection, use inheritance (Lamarckianism), and the deployment of old capabilities for new ends. His views on this subject have to be pieced together from various passages in his writings, including his 1872 book *Expression of the Emotions in Man and Animals*. At the broadest level, Darwin (1872b, 355–56) distinguished between an early, involuntary use of the voice for mating purposes and a later, more intentional production of articulate sounds for the purpose of general communication. The most fluent use of the vocal apparatus would have been for singing during courtship. The greater reproductive success of the best singers would have spread their skill over successive generations to a continually increasing share of the population: this was the working of sexual section (1871a, 2:330–37).

Yet Darwin also suggested how other mechanisms would tend to produce this same end. Vocal tones used in mating would become associated with strong (sexual) emotion and would be reinforced through rewarding outcomes. Vocal behavior of this kind would therefore become habitual even during the individual user's lifetime. The resulting vocal strengthening would then increase transgenerationally on the principle of the inherited effects of use. Reciprocally, when moved by any strong emotion, higher animals would tend to use the voice in a musical way, with pitch varied according to the kind of emotion: this pattern would further develop vocal strength and dexterity (1871a, 1:56, 57; 1872a, 84–88). Indeed, musical tones manifestly had become part of actual speech, as when a rising pitch signals an interrogatory. The instinctive character of this feature was shown in its use by infants, suggesting an inheritance from the prehuman origins of speech (1871a, 2:336–37; 1877c, 293). Finally, once sufficient vocal capability had been gained in this musical sense, the vocal organs could then be turned to the purpose of speech, on the oft-seen principle by which organs or instincts originally adapted to one purpose were turned to a wholly distinct use (1871a, 1:139; 2:335).

It is unclear whether Darwin had in mind a particular sequence in which inherited habit and sexual selection operated in this process, although there is evidence (Radick 2002) that he regarded habit as needing to come first, this because he was pessimistic about the spontaneous appearance of adaptive variations capable of being selected.

Darwin's Notebooks on the Origin of Language

Darwin arrived at his basic views on speech origins (although not everything on vocal development) in the years immediately following his voyage on HMS *Beagle*, that is, at the same time (1837–39) that he formulated both his general theory of evolution by natural selection and his main ideas about human descent. The largest number of comments on language in his theoretical notebooks of this period concerned present-day animal communication: monkeys that uttered signal cries, dogs that understood verbal commands, the intelligent look of an orangutan when spoken to – all suggested the kind of untrained potential that could have evolved into articulate speech (Barrett et al. 1987, C79, C104, M31–32, M58, M97, M153, and N94). A second category of notes speculated on the mechanics of speech origins, especially the imitative reproduction of involuntary cries. (On the use of imitation: Barrett et al. 1987, N18, N20, N65, N107, OUN ["Old and Useless Notes"] 5; on mimetic poetry: N31, N39, N127; on gesture language: N102.) Darwin's reading program in the post-*Beagle* years included a substantial amount of material on these subjects, nearly all by either English or Scottish authors – most famously Adam Smith, James Burnett (Lord Monboddo), John Horne Tooke, and Dugald Stewart – even if some of the original theorists had been Frenchmen (Condillac and Rousseau). Darwin concluded that one should not "overrate" language as a mark of distinction between man and animal, for linguistic understanding involved nothing more than the association of sound and meaning. He also considered it possible that animals could learn to connect a verbal sign with an entire conceptual category (Barrett et al. 1987, respectively M96–97, N20, N62.)

THE COEVOLUTION OF MIND AND LANGUAGE

The Essential Concept

Intertwined with Darwin's theory of linguistic origins was the idea that articulate language and humanity's unique cognitive abilities had coevolved, each reinforcing the other in an ascending spiral of development. The striking feature of this thesis was the notion that incipient speech had helped to stimulate the evolution of the human brain, what Robert J. Richards (2002c) has aptly termed "the linguistic creation of man." Darwin said in chapter 2 of *Descent*: "We may confidently believe that the continued use and advancement of this power [of speech] would have reacted on the mind by enabling and encouraging it to carry on long trains of thought" (1871a, 1:57; also 2:390–91).

As with all evolutionary scenarios, the reciprocal emergence of language and the human brain could not be demonstrated: one could look only for aftereffects in the present-day interdependence of language and thought. Following his usual practice, Darwin turned this inherent limitation into a scientific virtue: proceeding according to the *vera causa* principle, he took much of his evidence in support of coevolution from observable experience. He noted that "a long and complex train of thought can no more be carried on without the aid of words, whether spoken or silent, than a long calculation without the use of figures and algebra." Indeed, he said, it appeared that "even ordinary trains of thought almost require some form of language." Darwin also pointed to recent studies of aphasia (partial speech loss) as well as to the anatomist Carl Vogt's discussion of how the several mental functions were localized in specific areas of the human brain, research Darwin saw as highlighting the "intimate connection" between the brain and speech (1871a, 1:57–58; Radick 2000) (Fig. 21.1).

Coevolution in the Notebooks

Darwin conceived of the essential coevolution thesis early on. He declared in one of his post-*Beagle* notebooks that Benjamin H. Smart's treatise *Beginnings of a New School of Metaphysics* (1839) "give[s] my doctrines about origin of language – & effect of reason. Reason could not have existed without it" (Barrett et al. 1987, 599). This is to say that Smart (1839, 3–5, 21–22) confirmed views that Darwin had already adopted: the theory that speech arose from natural cries as well as the idea that "reason could not have existed" without the parallel development of language.

The notebooks also suggest that Darwin viewed mind-language coevolution within an intellectual framework

FIGURE 21.1. Carl Vogt (1817–95) was praised by Darwin for finding that certain mental functions are correlated with certain specific parts of the brain. Permission: Wellcome

provided ultimately by John Locke's *Essay concerning Human Understanding* (1690). Locke said that words served as "signs of internal conceptions," yet he also suggested that a verbal sign for an idea aided cognitive reflection on that idea (Locke 1975, book III, 1.2 [p. 402]). Darwin noted remarks in this vein appearing in the astronomer John Herschel's *Preliminary Discourse on the Study of Natural Philosophy* (1831) (Barrett et al. 1987, N60), yet he connected those remarks with the subject of speech origins – a first step toward the coevolution thesis (1987, N60).

Descent's Revised Thesis: Monogenetic Coevolution

Darwin added a crucial stipulation to his coevolution theory when he came to write the *Descent of Man*, this in response to a competing version of the concept that appeared in the 1860s, chiefly in the writings of the zealously pro-Darwinian zoologist Ernst Haeckel. The issue was whether coevolution had occurred before or after the protohuman tribe split into distinct racial groups. In his *Entstehung des Menschengeschlechts*, Haeckel (1868, 65–66) drew on the recent work of the linguist August Schleicher to reach the following conclusion: "It can be proved with certainty, from many facts, that humanity's protolanguages developed after the various races had already separated. Prehistoric humans, whom we regard as the forerunners of the five to ten human races, did not possess human languages" (author's translation). Haeckel reiterated this point later that year in his *Natürliche Schöpfungsgeschichte*, adding that "the origin of human language must, more than anything else, have had an ennobling and transforming influence upon the mental life of man, and consequently upon his brain" – essentially the coevolution thesis (Haeckel 1883, 302, 300; R. J. Richards 2002c). In sum, mind and language had coevolved not once but multiple times, in each case within a distinct racial community. (Alfred Russel Wallace [1864, clxiv-clxvi] presented essentially this same thesis, yet Wallace's essay impressed Darwin mainly with its explanation of how natural selection could account for primeval man's mental development. Darwin [1871a, 1:158] therefore had nothing but praise for this work.)

By contrast, Darwin in *Descent* taught that coevolution had occurred only once, before the rise of distinct racial groups (Alter 2007a, b). (For an alternate interpretation, see R. J. Richards 2002c and 2009a, 110.) In chapter 7, "The Races of Man," Darwin (1871a, 1:229) famously declared that all races were "descended from a single primitive stock." It is imperative to see that this viewpoint included a combined mental and linguistic aspect. In the same chapter, Darwin weighed the idea that language had developed only after humans became "widely diffused" geographically – that is, after the distinct racial groups had begun to form. Darwin responded: "But without the use of some language, however imperfect, it appears doubtful whether man's intellect could have risen to the standard implied by his dominant position at an early period" (1:234-35).

Darwin thus believed that a polygenetic version of coevolution, such as Haeckel's, precluded any convincing explanation of primeval man's initial triumph over rival apelike populations. Only through early speech-and-brain development could humanity's ancestors have spread at the expense of other simian tribes. Triumphing over neighboring tribes would, in turn, have been prerequisite to geographical dispersion and, finally, racial diversification. Darwin did not say that language itself was necessarily monogenetic: he left open the possibility that multiple languages had been spoken in the predispersal period. But he did insist that there had been a common "*use* of some language" (1:234) at that time. It was essential, then, to assume a sufficient degree of mind-language coevolution *at* the predispersal stage in order to account for primeval man's emergence *from* that stage – this being a necessary step toward full human status for humanity as a whole.

In advancing this thesis, Darwin cited works by mid-Victorian Britain's leading anthropologists, Edward B. Tylor and John Lubbock. Tylor (1865) and Lubbock (1867) sought to identify those mental traits and cultural attainments that were common to all races and hence must have had their origins in the predispersal stage. The result was a short list: certain tools and weapons, and probably the use of fire. Darwin (1871a, 1:136-37, 231-34) added the basic use of language. As he also noted, linguistic researchers had concluded

that the most primitive units of speech were grammarless *roots* that named simple objects and "obvious" relations: these rudimentary words would thus have been used "by the men of most races during the earliest ages" (1:61).

This belief that mind-language coevolution had taken place before racial dispersion did not mean that Darwin was any less of a believer in mental inequality based on race. But it did mean that, in Darwin's view, the evolutionary origins of language had played little part in the inception of the mental hierarchy among the races. Darwin believed in that hierarchy, with mental differences reflected in (among other things) the scale of grammatical elaboration found in the various families of languages (Stocking 1968, 113-14; Radick, 2002, 2008). Yet he regarded those grammatical differences as a later effect, not an initial cause, of mental evolution. In the same way, the anthropologists on whom Darwin drew saw the various races progressing – however unequally – from an original state of mental unity (Darwin 1871a, 1:180-84).

The Problem of Romantic Language Theory

It was of course essential to Darwin's theory of speech origins to suggest that the most basic steps toward language could have been made by originally speechless beings. This argumentative necessity produced a counterthesis to much of the evidence supporting coevolution. For, even as he stressed in *Descent* how present-day thought was dependent on words, Darwin (1871a, 1:58) also argued nearly the opposite: "Nevertheless a long succession of vivid and connected ideas, may pass through the mind without the aid of any form of language." Retriever dogs, for example, apparently were able to reason to some extent.

Darwin encountered a problem in this connection because of a superficial resemblance between his essentially Lockean coevolution idea and the theme in German-romantic philosophy that language use was indispensable to the development of human mental powers (Harris and Taylor 1997, 71-84). The outstanding exponent of this tradition in the English-speaking world was the Oxford Sanskritist Friedrich Max Müller, who was also a strong opponent of the idea of human evolution (Fig. 21.2). Müller taught that language was essential to the ability to conceive of general categories, which he said was a uniquely human trait; hence the existence of language told against the idea that human mental capabilities could have evolved from the brain of an apelike ancestor. Muller set forth these themes in his famous London lectures on "The Science of Language" (1861) and again in lectures on "Mr. Darwin's Philosophy of Language" (1873).

Darwin responded in the revised (1874) edition of *Descent* by redoubling his emphasis on the notion that rational thought was *not* dependent on speech. Human infants of ten to eleven months old, as well as deaf-mutes, quickly learned to connect certain sounds with certain general categories. Surely the most intelligent animal species must likewise have the power of abstract thought, "at least in a rude and incipient degree," although manifestly without the aid

FIGURE 21.2. The philologist Max Müller was a strong opponent of human evolution, but in the *Descent* Darwin used some of his ideas to suggest that language evolves in much the same way as do organisms. This photograph, taken in 1857, was by Charles Lutwidge Dodgson, better known as Lewis Carroll. Permission: © Mark Gerson / National Portrait Gallery, London

Africa, and Australia, as well as by isolated groups such as the Laplanders and Basques. These, he said, were proof of humanity's decline from an original high state of mental culture, supposedly a sign of supernatural creation.

Darwin countered by inverting Wake's standard of linguistic perfection. The study of organic nature revealed that complexity was not necessarily a sign of high development: rather, a naturalist "justly considers the differentiation and specialization of organs as the test of perfection. So with languages, the most elaborate and symmetrical ought not to be ranked above irregular, abbreviated, and bastardized languages" (Darwin 1871a, 1:61–62). Darwin then offered evidence from recent linguistic science that ratified the naturalist's standard of perfection. As Darwin surely was aware, the research cited by C. S. Wake dated from near the turn of the nineteenth century. Darwin countered (implicitly) with the more up-to-date work of Franz Bopp and his school, which showed that the grammatical affixes found in the familiar Indo-European languages had been produced over time by the joining of previously independent words, a process of simplification. (Darwin would have seen Bopp's findings summarized in Schleicher [1863, 1869], cited in Darwin [1871a, 1:56].) As Darwin remarked, "Philologists *now admit* that conjugations, declensions, &c., originally existed as distinct words, since joined together" (1871a, 61; emphasis added; Pedersen

of language (1874, 1:104–5). When he came to defend himself against Müller, therefore, although he avoided outright contradiction, Darwin undercut some of his own argument supporting the coevolution idea (Alter 2008). This ironic outcome was perhaps unavoidable given the inherently complex relationship between language and thought, especially viewed in light of the nineteenth century's meager research base. (Darwin's protégé G. J. Romanes [1888, 83, 290, 369] continued to fight Max Müller on the language-thought issue after Darwin's death.)

LANGUAGE EVOLUTION AND RACIAL HIERARCHY

At the end of the language section in chapter 2 of *Descent*, Darwin addressed the traditional theory that modern-day primitive societies had degenerated from the civilized condition in which all of mankind supposedly had been created. This notion had recently been revived by certain writers anxious to reassert a creationist view of human origins (Stocking 1987, 149–50, 179–80). Darwin (1871a, 61) focused on the linguistic component of this thesis, particularly as set forth by Charles Staniland Wake in his book *Chapters on Man* (1868). Drawing from older linguists such as Friedrich Schlegel, Stephen de Ponceau, and Wilhelm von Humboldt, Wake (1868, 97–102) stressed the highly regular and detailed grammars of the languages spoken by the natives of America,

FIGURE 21.3. John Lubbock (1834–1913) was a banker, politician, and archaeologist. The Lubbocks were neighbors of the Darwins and there was some little tension when the Darwins wanted to buy a piece of land next to their property – the "sandwalk," where Darwin would take a prelunch, daily constitutional – and they felt the Lubbocks struck too hard a bargain. Permission: Wellcome

1959). To this Darwin (1871a, 1:62) added an appeal to John Lubbock's discussion (1870, 278) of how the simplification of language was actually a mark of progress. (Darwin would have seen the same message in Farrar [1865, 52–55].) Hence technologically primitive peoples who spoke complex languages should still be ranked among the least developed mentally. This meant that they represented the earliest stages of human evolution and that their existence helped to confirm the idea of humanity's evolutionary origins (R. J. Richards 1987, 204; Radick 2002, 2008) (Fig. 21.3).

DARWIN'S USE OF LANGUAGE-SPECIES PARALLELS

Heuristic Analogies

A significant number of Darwin's comments about language consisted of comparisons emphasizing similar change processes in the linguistic and bioevolutionary spheres (Beer 1989; Taub 1993; Alter 1999). Most of these comparisons pertained not to the origin of speech but to the evolution of species in general. Obviously they offered no real evidence of the evolutionary process: they were intended only to enhance the plausibility of Darwin's theory by habituating readers to thinking in terms of gradual transformation and branching descent. The early nineteenth century had seen a revolution in linguistic scholarship that highlighted the branching descent of widely divergent tongues from common ancestors: the Indo-European family of languages was the case most studied. This research suggested apt analogies by which to represent aspects of Darwin's theory: hence not only Darwin but a number of his scientific friends, including Charles Lyell, T. H. Huxley, and Asa Gray, made creative use of language-based illustrations in their writings.

Analogies of this kind appeared in Darwin's notebooks (Barrett et al. 1987, respectively N 64–65, OUN 6), his unpublished "species book" manuscript (1975, 384), and the *Origin* (1859, 40, 310–11, 422–23, 455). Perhaps the most famous such image in the latter work supported the idea of classifying all organic forms, living and extinct, by genealogical relationship. Darwin (1859, 422–23) compared the benefits of this arrangement with the advantages to be gained if all the languages ever spoken could be classified according to their ethnological history – that is, on the basis of a hypothetical "perfect pedigree" of human racial groups. Darwin later drew a similar analogy (with languages again representing species) in *Descent*'s chapter, "The Affinities and Genealogy of Man" (1:188–89).

"Darwinian" Linguistic Development

A largely different role was served by an entire series of comparisons between language and species appearing in *Descent*, chapter 2. Both kinds of phenomena could be shown to have developed "through a gradual process," in that both manifested descent from common ancestors, genealogical classification, morphological homology and analogy, correlated growth, the struggle for existence, extinction, and natural selection (Darwin 1871a, 1:59–60). The argumentative function of these parallels is not immediately obvious: they can be read either as additional analogies representing bioevolutionary processes in general or (reversing direction) as serving to advance Darwin's account of how the faculty of language in particular evolved (Beer 1983, 54–55; Alter 1999, 100). As Gregory Radick (2002, 2008) has shown, Darwin likely intended these parallels mainly as the latter, as indicated by their position immediately prior to the discussion of primitive peoples who spoke apparently sophisticated languages. The point was to prepare readers to adopt a naturalistic view of language formation, this again according to the *vera causa* strategy employed in ways great and small throughout Darwin's writings. Readers attuned to that evidential standard would perceive that the parallels showed how bioevolutionary concepts could handily "explain" the natural development of languages from simple beginnings. Why not then apply a naturalist's view as well when ranking languages on a scale of perfection? Darwin thus made new use of old material, grafting this message about language development onto the analogical kind of usage seen elsewhere. He thus embodied in his own argumentation the principle of novel functions for features originally adapted for a "quite distinct purpose" (Darwin 1871a, 2:335).

At least one of the comparisons in *Descent*, chapter 2, however, likely did double duty, functioning in part along the lines seen earlier. Darwin (1874, 106, no. 67) praised the "very interesting parallelism" between species and languages that made up chapter 23 of Charles Lyell's *Geological Evidences of the Antiquity of Man* (1863). Yet because Darwin knew that Lyell used this analogy to suggest that evolution had resulted from divine superintendence, he therefore noted: "A language, like a species, when once extinct, never, as Sir C. Lyell remarks, reappears"; likewise, "single words, like whole languages, gradually become extinct" (Darwin 1871a, 1:60). In this way, Darwin hinted at the wholesale waste of organic forms attendant on the selection process, a phenomenon presumably incompatible with supernatural design (Alter 1999, 56–68). Through this subtle polemic, Darwin continued a pattern in which various writers tried to turn the language-species analogy each to his own advantage.

CURRENT STUDIES

Updates on the subjects discussed in this essay appear in recent works. Radick (2007) recounts the history of post-Darwinian research on animal language; Radick (2008) does the same for the relationship between language and race. Pennock (1999) reconsiders the language-species parallel, and Pinker (2010) continues the study of language-brain coevolution. An overview of the work of cognitive science on language and evolution appears in Sterelny (2009), and recent research on language evolution in a variety of fields is considered in Fitch (2009, 2010).

Essay 22

Darwin and Ethics

Eric Charmetant

It is often felt that Darwin's views on ethics betray his great contributions to science. When he makes comments about women and the Irish and others, he reveals all of the prejudices of his Victorian class. Even worse, he is committed to a form of Spencerian evolutionary ethics, somewhat misnamed "social Darwinism." However, properly understood, Darwin on morality is much richer and more rewarding than in this general perception. To see this, we must go back well beyond the *Descent of Man*, published in 1871, and usually the only source to which people refer. While recognizing that Darwin's genius is more than just the sum of its parts, for full understanding we must look at his family background and education as well as other sources.

THE EARLY YEARS (1809–1836)

On his father's side, Charles Darwin came from a medical family. His grandfather Erasmus Darwin (1731–1802) was a well-known physician. He was also a poet and inventor, a member of the Lunar Society of Birmingham, a group of businessmen and industrialists that included Joseph Priestley (the chemist) and James Watt (the inventor). On his mother's side, Darwin came from a family of industrialists, for his grandfather was Josiah Wedgwood (1730–95), another member of the Lunar Society, who founded the great pottery factory famous for its ceramics and porcelain. Both sides of Darwin's heritage were socially concerned and liberal, and members of the family (the Wedgwoods particularly) were in close contact with British intellectual life. Darwin's mother, Susannah, knew the poet Samuel Taylor Coleridge, for instance. Particularly distinctive of both the Darwins' and the Wedgwoods' social concerns was a strong commitment to the abolition of slavery. Indeed, Erasmus Darwin wrote poetry condemning slavery, and Josiah Wedgewood financed the Sierra Leone Company, which was established to create a homeland in West Africa for liberated slaves.

After his early schooling, Charles Darwin, following in the footsteps of his older brother Erasmus, enrolled in 1825 in the University of Edinburgh in order to study medicine. He soon lost interest in his formal studies, but he benefited greatly from lectures in natural history by Robert Jameson and also more informally from meetings of the Plinian Society, a gathering of people interested in the world of animals and plants. Giving up medicine in 1827, Darwin was next pushed by his father into a course of studies designed to qualify him for a life as an Anglican priest. To this end, Darwin went to the University of Cambridge from which institution he graduated with a bachelor's degree in 1831. Although Darwin did reasonably well in his studies, he was never very committed to the priesthood. He was much more interested in natural history, and his ample free time made this pursuit easily possible.

Next Darwin spent some five years as the naturalist on board HMS *Beagle*. Some of his experiences on this trip clearly colored his lifelong beliefs about human beings and about the ways in which people treat each other. For instance, in Brazil Darwin encountered the brutality of slavery. In Olinda (near Recife), Darwin heard the screams of a tortured slave; this haunted him for the rest of his life. In Argentina, Darwin saw the brutality of Gauchos like General Juan Manuel de Rosas, who claimed that "Indians were pests to be eradicated, like rats" (Desmond and Moore 2009, 90). One consequence of this was that, for all of his future Victorian prejudices, Darwin was always very suspicious of claims that white people are superior to peoples of other colors. The extermination of the copper-colored Indians by the white Gauchos certainly did not represent progress: Gauchos were "a little superior in civilisation" but "inferior in every moral virtue" (Darwin 1988, 181).

Somewhat countering this, however, was Darwin's experience with the inhabitants of Tierra Del Fuego, the land at the bottom of South America. The differences between these "savages," who lived by hunting and gathering, and the average, civilized Englishman, a citizen of the leading world power of its time, were profound. Darwin (1890, 216) wrote in his *Journal of Researches*: "I could not have believed how wide was the difference between savage and civilised man: it is greater than between a wild and domesticated animal, inasmuch as in man there is a greater power of improvement." It is true that the Fuegians had shown themselves capable of adaptation to another society. This was proved by the effects of a British education on three of them, brought back to London after a previous voyage by the captain of the *Beagle*, Robert Fitzroy. However, once brought back home, the Fuegians quickly reverted to their old manners and habits, forgetting all that they had learned in England. This showed the ship's naturalist just how close we humans are to the lower forms of animals.

THE DEVELOPMENT OF DARWIN'S VIEWS ON ETHICS (1837-1840)

On his return to England, Charles Darwin embarked on a very heavy course of reading on a very broad range of subjects. Although not a great number of the many books mentioned as having been read (in Notebook C) were on or about philosophy, Darwin's diet did include accounts of the lives and writings of David Hume, Adam Smith, and Dugald Stewart. This was all secondary literature with the exception of Hume's *Enquiry concerning Human Understanding* (1748, revised 1758). At the same time, Darwin was participating fully in the intellectual life of London, particularly through his membership in the Athenaeum Club, which he joined in 1838. This institution was devoted to discussion and the development of the sciences, literature, and arts. In other words, although with good reason we usually think of the years 1837-39 as the time when Darwin was thinking hard about the problems of the transmutation of species, this was also a time when Darwin was broadening his intellectual horizons, making of himself the cultivated English gentleman he was to become. Particularly pertinent to our inquiry here were two authors whose writings helped introduce Darwin to issues in ethics and who stimulated him to think for himself on these issues: William Paley and James Mackintosh.

Archdeacon William Paley (1743-1805) was justly famous as the author of textbooks. We know that Paley's *Natural Theology* was an important influence in Darwin's later thinking about organisms and design. When Darwin studied for his undergraduate degree in Cambridge (1828-31), two other books by Paley were required reading. In the first year and also at the end of the course of studies, students had to answer a series of questions on Paley's *The Evidences of Christianity* (1794). Then there was the other work, *The Principles of Moral and Political Philosophy* (1785), which was to have a great influence on Darwin's thinking about moral philosophy (Fig. 22.1).

Paley argued strongly that our moral sense is entirely a function of imitation and childhood training. In no way is it biologically innate. Moreover, Paley argued that a moral maxim is not necessarily binding in every circumstance. For instance, the duty of truth telling could be suspended when faced with an enemy, a robber, or an insane man. Underlying Paley's attack on moral innateness was the burning moral and psychological worry that, by arguing that some moral dictate is given or natural, one is thereby trapped into endorsing the naturalness of practices that, by the end of the eighteenth century, civilized and decent people would regard as abominable. Paley was particularly concerned with Aristotle's characterization of slavery as natural, obviously unacceptable to someone like Paley himself.

With respect to the utilitarianism, that is to say the belief that morality is a direct function of the promotion of happiness, Paley ([1785] 2002, 25) united theology and philosophy: "*Virtue* is 'the doing good to mankind, in obedience to the will of God, and for the sake of everlasting happiness.'" How does one discover the will of God? Paley believed there are two complementary ways: first, through the explicit declarations of the scriptures (e.g., the Ten Commandments) and, second, in the revelation of God's intentions through his works, that is to say in nature. So, with respect to the second point, Paley argued that the disposition or capacity that human beings have for pleasure shows clearly that God wants human beings to be happy. In other words, inquiring into what increases or decreases happiness is at the same time a way of discovering

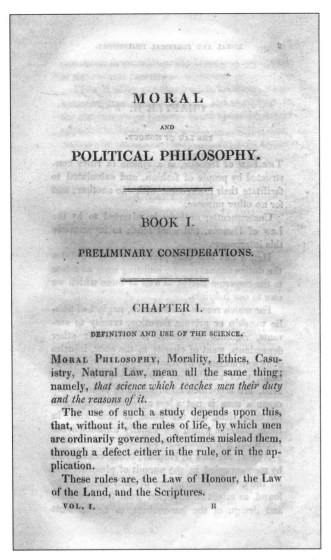

FIGURE 22.1. First page of William Paley's *Moral and Political Philosophy* (1785). Like all of Paley's writings, it was an influence on Darwin, not the least because of its strong attack on slavery.

the will of God. Up to this point, therefore, Paley remains true to a theological form of what is known as "eudemonism" (a capacity for causing happiness).

On top of this, however, Paley added the criterion of utility in order to justify the morality of an act: "'Whatever is expedient is right.' But then it must be expedient on the whole, in the long run, in all its effects collateral and remote, as well as in those which are immediate and direct; as it is obvious, that, in computing consequences, it makes no difference in what was or what distance they ensue" (47). So, in order to avoid having to justify the legitimacy of a dictatorship by the criterion of utility, Paley invites us to consider not just immediate actions and consequences but also long-term effects. If a tyrant kills someone, whatever the immediate consequences, he is thereby legitimizing murder, and this is not a good thing. It should be noted that for Paley happiness is always relative, and a person is "happy" if he or she is happier than the majority of other people. It should also be noted that happiness is never a function simply of the pleasures of the senses (linked to the body or to the spirit) or of the avoidance of sorrow or pain. For Paley happiness in the positive sense is related to social being and affection, to the exercise of one's body and intellectual abilities, and to health. Happiness also depends on the building of habits. In short, Paley's utility is a long way from the hedonistic utilitarianism of Jeremy Bentham.

Darwin would certainly have discussed this book extensively while a student because, although the book was part of the required reading, by the end of the 1820s some Cambridge professors (with whom Darwin had become intimate) were challenging strongly Paley's theological utilitarianism and also his criticism of an innate moral sense. In particular, Adam Sedgwick, professor of geology, and William Whewell, at that time professor of mineralogy (but later to be professor of moral philosophy), were highly critical of Paley's moral philosophical views. Nevertheless, this thinking of Paley had a great influence on Darwin. In a page taken from Notebook M, dated 8 September 1838, Darwin referred explicitly to Paley's rule in order to underline his own understanding of what is good: "I am tempted to say that those actions which had been found necessary for long generation, (as friendship developed animals in the social animals) are those which are good & consequently give pleasure, & not as Paley's rule is those that on a long run *will* do good period. – alter *will* in all cases to *have* & *origin* as well as *rule* will be given" (Barrett et al. 1987, M, 132e).

Note, however, that the right or good action for Darwin does not come, as in Paley, from looking at the immediate or future consequences of an act but from a retrospective glance at the past. Thus, for Darwin only an action found necessary over several generations is good. One consequence of Darwin's formulation of the principle of utility therefore is that, unless an action has already occurred many times in the past, it is not possible to evaluate its moral worth. Moreover, pleasure alone is never the criterion of the morality of an act. It is only one consequence of the goodness of an act.

In addition to Paley, whose writings were of such interest to Darwin during his years of study in Cambridge, a (probably *the*) major influence on Darwin's development of his moral thinking was James Mackintosh (1765–1832) (Fig. 22.2). A medical doctor, a disciple of the Scottish philosopher Dugald Stewart (1753–1828), lawyer, and judge in India, Mackintosh was a man of exceptional eloquence and culture. Darwin met Mackintosh through the Wedgwood family before he left for the voyage of the *Beagle*. He was much struck by the force and interest of the older man's conversation. In turn, Mackintosh took notice of and spoke favorably of this interesting young man.

In 1829 James Mackintosh agreed to finish the article on the history of metaphysics, moral philosophy, and political philosophy being written by the philosopher Dugald Stewart (who died unfortunately before finishing the job) for the introductory volume of the seventh edition of the *Encyclopedia*

FIGURE 22.2. Sir James Mackintosh (1765–1832), Scottish jurist and philosopher. From his writings, Darwin learned about the history of recent moral philosophy. Permission: © Mark Gerson / National Portrait Gallery, London

Britannica. He was asked especially to add to the ethical and political discussion for the eighteenth century. This Mackintosh did around 1829–30. Mackintosh's contribution was so detailed and lengthy that it was reprinted as a separate piece, with a foreword by William Whewell (dated 1 July 1835), and published in 1836 under the title *Dissertation on the Progress of Ethical Philosophy, Chiefly during the Seventeenth and Eighteenth Centuries*. Starting in August 1838, Charles Darwin read the second edition of this book (printed in 1837 and identical to the first edition). It was to become his major source of information about moral philosophy, as is made clear from his abundant notes dating from May 1839 to March 1840. (The comments are to be found in a folder labeled "Old and Useless Notes." This is a title Darwin scribbled on them many years later and should not be taken literally.)

Mackintosh's moral thinking is set out along five main lines: that we have a practical reason, which is close to our feelings; that there exist benevolent affections; that the moral conscience or moral sense is what really counts in morality; that we can have a theory of moral development; and that there is a convergence between our moral sense and any objective criteria of morality. According to Mackintosh ([1836] 1991, 199), moral feelings themselves are never the actual means of action, which is always something done by the will itself; however, they always remain in contact with the will, and no external circumstances can ever break this relationship. This is because feelings and will both come from within the human spirit. This "peculiar relation of the conscience to the will" makes it possible to give the moral sense or moral conscience its power over other social affections and faculties. It also leads to the immutability and independence of the moral sense.

One obvious objection to Mackintosh's thinking is that he does not offer a criterion of morality, as do such philosophers as Paley, Jeremy Bentham, and Immanuel Kant. Fully aware of this objection, Mackintosh examined the objects of the moral sense and concluded that they all contribute to happiness: "Every principle of which consciences compose has some portion of happiness for its object. To that point they all converge. General happiness is not indeed one of the natural objects of conscience, because our voluntary acts are not felt and are perceived to affect it" (382–83). That is, social affections or feelings promote the happiness of others, and conversely anger can be used to prevent what is harmful to others. At the same time, private passions play a role in integrating the personal dimension in general happiness. In contrast to Paley, who underlines the difference between an innate moral sense and his own theological utilitarianism, or to Jeremy Bentham's morality, which separates the criterion of utility based on the maximization of happiness from a criterion based on sympathy, Mackintosh wants to show that the moral sense and the criterion of utility come together. Nothing, however, guarantees that the moral sense ultimately will achieve the optimum of happiness or maximize general happiness.

Darwin seized on this idea of a convergence between a moral sense and happiness. What he wanted to do distinctively was to anchor this convergence in zoology: "Two classes of moralists: one says rule of life is what *will* produce the greatest happiness. – The other says we have a moral sense. – But my view unites both & shows them to be almost identical. What has produced the greatest good or rather what was necessary for good at all is the instinctive moral senses" (Barrett et al. 1987, OUN, 30). For Darwin, following Mackintosh, utility was not primary in his ethical thinking. Where he made an original contribution was in connecting our moral sense to an instinct that had appeared useful or beneficial in the past and which tends toward the greatest happiness. For Darwin, there can never be any guarantee about the future, because any convergence results from the experiences of history in a particular range of environments. One could never be sure that this instinct could evolve and become adapted in a completely new environment.

THE MIDDLE YEARS: 1840–1871

It should be noted also that the analogy between human behavior and animal behavior was (by 1840) central to Darwin's views on morality (Barrett et al. 1987, M, 75; N, 3). For him, this underpinned our thinking about moral evolution and its transmission: "The change of our moral sense is strictly analogous to change of instinct amongst animals" (Barrett et al., OUN, 30v). But Darwin remained unclear how one should explore and extend this analogy, and for many years there seems to have been little progress on this issue. Instead, Darwin was more concerned to develop and lay out

FIGURE 22.3. Frances Power Cobbe (1822–1904) was a British feminist and ardent antivivisectionist. From Cobbe, *Life of F. P. Cobbe as Told by Herself* (Boston: Houghton, Mifflin, 1894)

his general theory of evolution. Even in the *Origin of Species*, there was little on ethics because (as is well known) Darwin stayed away from humans in that work.

From 1860 on, however, Darwin was starting to think hard about the evolution of humankind and the possibility of a book on the subject. He was very busy with other tasks, for instance, with the little book on orchids, with responding to reactions to the *Origin* and revising that work through several editions, and with the drafting and writing of his book on the *Variation of Animals and Plants under Domestication*. But by February 1868, Darwin had started to write on the subject of humankind, and by the end of spring 1869 two chapters had already been written. Nothing at that time showed that Darwin intended to deal in any detail with human intelligence and morality; the explanation of sexual dimorphism and the origin of human races seem then to be his only objectives.

Obviously, though, if Darwin did write on humans, the issue of our moral nature was going to arise and need discussion. This point was brought home forcibly in the summer of 1869. Frances Power Cobbe (1822–1904) was a leading intellectual and militant for the cause of women in Victorian society (Fig. 22.3). An enthusiast for Kant's thinking, she gave Darwin a partial translation of the German philosopher's works on moral philosophy. Darwin thanked her, writing: "It has interested me much to see how differently two men may look at the same points. Though I fully feel how presumptuous it sounds to put myself even for a moment in the same bracket with Kant – the one man a great philosopher looking exclusively into his own mind, the other a degraded wretch looking from the outside through apes and savages at the moral sense of mankind" (Cobbe 1904, 487–88).

In the end, Darwin's empirical projects, starting from behaviors in order to reach an understanding of our moral sense, proved little influenced by Kant's ethical starting point: the presence of moral law in every human being. It clearly stimulated his thinking – in the finally appearing work on humankind, the *Descent of Man* (1871), there is reference to Kant's ideas. Darwin (1871a, 1:70–71) even quoted one of Kant's more purple passages in the *Critique of Practical Reason* (Ak. V, 86) : "Duty! Wondrous thought, that workest neither by fond insinuation, flattery, nor by any threat, but merely by holding up thy naked law in the soul, and so extorting for thyself always reverence, if not always obedience; before whom all appetites are dumb, however secretly they rebel; whence thy original?" In truth, however, Darwin was never really on that track. The real influences on Darwin lay back in the 1830s and earlier.

THE DESCENT OF MAN, AND SELECTION IN RELATION TO SEX (1871)

The *Descent of Man* begins by showing how close human beings are, biologically speaking, to animals. This is done by showing how the organs of humans are very similar to corresponding structures in the lower animals. Then, moving on to human intellectual faculties, Darwin underlined our resemblances to animals in these respects also. Emotions, curiosity, imitation, attention, memory, imagination, reason, progressive improvement, tool use, self-awareness, language, the feeling in beauty, and belief in God – all of these are shown to have their animal correspondences.

Next Darwin looked at the intellectual faculties of humans and of the lower animals in order to explain and arrive at the moral sense that humans have. He recognized that having a moral sense is what distinguishes humans from other animals; but, nevertheless, he wanted to root this sense in a sociability that is to be found not only in humans but also in the lower animals. To this end, Darwin argued that there is no absolute reason why the lower animals, if certain conditions be met, should not themselves develop a moral sense. "Any animal whatever, endowed with well-marked social instincts, would inevitably acquire a moral sense or conscience, as soon as its intellectual powers had become as well-developed, or nearly as well developed, as in man" (1871a, 1:71–72).

For this development or evolution to occur, first the social instincts of an animal had to lead that animal to take pleasure in the company of its fellows and also to feel a sympathy for them and a desire to help them when in need. Second, there had to be the development of a sufficiently powerful memory, so that the individual involved could remember past actions. This development of memory was essential, because the feelings of satisfaction or dissatisfaction related to performing social actions had to be sufficiently strong and well remembered to avoid being overcome by short-term instincts or desires like hunger, thirst, and migration. Third, in order to develop a morality, there had to be a development of language, so that an individual could not only convey his desires and opinions to others and thus influence their behaviors but also know the opinions of others in regulating his own behaviors. This allows for a development and regulation of behaviors in the group.

Darwin stressed that the development of a moral sense by other species would not necessarily mean that throughout the animal kingdom there would be one uniform moral sense. Somewhat amusingly, he suggested that were bees to acquire a moral sense, then they would feel a necessity or duty of killing their brothers, and for queens to kill their fertile daughters. This was the thin end of the wedge that truly divided Darwin from Kant. For Kant, morality had a kind of necessity, stemming from the conditions required for rational beings living together. Darwin, to the contrary, seeing continuities in the animal kingdom between lower animals and humans, was bluntly to assert that if things had gone otherwise, we might think that killing each other is the highest moral duty.

Darwin was an empiricist, and his thinking about ethics was that of an empiricist. It is this that links him right back to the philosophers of his younger years. Although they were not thinking in evolutionary terms, they too were thinking in terms of the actual facts of (the psychology of) human nature. The heart of Darwin's ethical vision was a sense of sympathy (remember Mackintosh's "benevolent affections"): it is this which moves the will, and it is this which controls actions in the group, seeking the approval of others and avoiding rejection. Darwin argued that the intensity of sympathy that one individual has for another is directly proportional to the familiarity between the two people. Darwin (1874, 129) argued that, among animals, one observes a similar phenomenon (of sympathy), although somewhat less strong than in humans: here, "sympathy is directed solely towards the members of the same community, and therefore towards known, or more or less beloved members, but not to all the individuals of the same species." Darwin noted that this sense of sympathy is not restricted purely to social species like the insects. We see it also among tigers, although it is indeed true that they show sympathy only toward their own cubs.

The rise of civilization was another important element in Darwin's ethical vision: evolution for humans occurs less in the body than in the mind. In their evolution, humans had to show considerable ingenuity, and very often this led to competition between tribes for the exploitation of resources. Thus, the most powerful tribes gained an advantage over others. Darwin (1871a, 1:160) wrote: "At the present day civilised nations are everywhere supplanting barbarous nations, except where the climate poses a deadly barrier; and they succeed mainly, though not exclusively, through the arts, which are the products of the intellect. It is, therefore, highly probable that with mankind the intellectual faculties have been gradually perfected through natural selection."

It should be stressed that Darwin did not see the development of civilization as something opposed to his mechanism of natural selection. Rather, it is something that focuses on the powers of the intellect and that occurs at the level of intergroup competition. Darwin worried that the most courageous individuals would have fewer descendants. Altruistic people may well die sacrificing themselves for others and hence not leave any offspring. To get around this problem, Darwin invoked the development of reason. The person who helped others within the tribe received assistance in return: "From this low motive he might acquire the habits of aiding his fellows; and the habit of performing benevolent actions certainly strengthens the feeling of sympathy, which gives the first impulse to benevolent actions. Habits, moreover, followed during many generations probably tend to be inherited" (1871a, 1:163–64). Hence, approval or disapproval of actions within the group contributed much to the development of moral virtues. Shame and remorse about egoistic or bad actions encouraged the development of concern for others within the group.

What of Darwin's philosophical understanding of the criterion of morality? As with his just-discussed understanding of the nature of morality, in 1871 Darwin's thinking was still firmly based on his evolutionary interpretation of James Mackintosh's views on ethics, formulated some thirty or more years previously. Social instincts and, above all, moral behaviors are related to the general good of the group. "The term, general good, may be defined as the means by which the greatest possible number of individuals can be reared in full vigour and health, with all their faculties perfect, under the conditions to which they are exposed" (1871a, 1:98). Happiness therefore comes out as a secondary dimension of good actions. Nevertheless, in the case of human beings, we usually find that the general good and happiness converge: "No doubt the welfare and the happiness of the individual usually coincide; and a contented, happy tribe will flourish better than one that is discontented and unhappy" (1:98).

LIMITS TO DARWIN'S APPROACH TO MORALITY

Let us end this discussion with some brief remarks. The first point is that, in the *Descent of Man*, Darwin seems to reconcile, or least to be happy to live with, many different and partly incompatible approaches to ethics: virtues, moral sense, utilitarianism, extension of the moral community to all humans, and even the stoic control of thoughts. How should we understand this apparent neutrality and ecumenical attitude towards morality? At least part of the answer, something that emphasizes Darwin's originality, lies in the fact that he is concerned first with understanding and describing the evolutionary roots of the moral sense. Only secondarily, if at all, does he care about contributing to debates between moral philosophers. Although, having said this, Darwin's normative neutrality is truly only apparent. It is clear that Darwin defends a genuine "altruistic" origin and status for morality. He is therefore opposed to views of moral selfishness as expressed in the thought of such earlier philosophers as Thomas Hobbes and Bernard Mandeville.

The second point is that Darwin surely underestimates the importance of deliberation in morality. The examples that he gives in the animal world are of impulsive actions of help and assistance toward other members of the group. A critic might rightly object that without more emphasis on deliberation we do not have actions that are truly ethical. Could a moral sense truly emerge in other animals without deliberation? Darwin (1871a, 1:87) is aware of the objection, "that some persons maintain that actions performed impulsively, as in the above cases, do not come under the dominion of the moral sense, and cannot be called moral." But even having made this admission, he insists only on the difficulty of separating impulsive and moral actions. There is, for instance, no reference in Darwin's writing to the moral philosophy of Aristotle, in which he might have found a way to link impulsive actions with the moral habits of a virtuous man.

Third, given that the environment of the group plays such a crucial role in the formation of good moral action, could one not say that Darwinian morality arises only from social pressure? What about broader extensions of human morality, for instance, to the lower animals? No social pressures seem involved here. These are questions not really explored by Darwin even though, in the end, he does say that "sympathy beyond the confines of man, that is humanity to the lower animals, seems to be one of the latest moral acquisitions" (1871a, 1:101). This may indeed be true, but Darwin's thinking about morality gives us no definitive answers why this should have occurred; rather, it is left as an exercise for the reader.

Criticisms notwithstanding, there is much of great interest in Darwin's fusion of British moral philosophy with his theory of evolution through natural selection. It is not surprising that there is much interest in returning to his thinking and trying to use it in our philosophical discussions today.

❦ Essay 23 ❦

Social Darwinism

Naomi Beck

"In the distant future I see open fields for far more important researches.... Light will be thrown on the origin of man and his history" (Darwin 1859, 488). This statement, which appears in the concluding chapter to the *Origin of Species*, was Darwin's only mention of human evolution in the entire book. He was well aware of the difficulties his biological propositions would encounter from believers in special creation and therefore thought it wise to leave the delicate question of human evolution aside for the time being. Darwin was nonetheless fully conscious that his theory *would* lead to important insights in this domain and would probably revolutionize the way we think about ourselves and our cultures. Enter social Darwinism.

The term social Darwinism, which came into fashion after 1940 (Hodgson 2004), has been used mainly to decry doctrines that justify some form of individual, social, or racial superiority through evolutionary principles with which Darwin's theory is identified, such as the struggle for existence and natural selection. It has also been employed in reference to teleological explanations of the causes of human progress that often carry with them value judgments concerning the degree of civilization attained by various peoples. Yet many of the positions typically attached to social Darwinism do not correspond to this stereotypical description. Even among the main proponents of evolutionary theory in the nineteenth century – Darwin, Wallace, Huxley, and Spencer – there were important disagreements concerning the process of evolution in humans and its results. This article offers an examination of their claims, as well as some related and antagonistic viewpoints, in an effort to tease out the various and complex meanings of social Darwinism. By tuning the microscope to grasp the finer details, a surprisingly different picture from the one usually conveyed by this blanket term will emerge.

The context of our story is composed of two related elements: on the one hand, the debate over wealth distribution and landownership, and on the other, the question of the relationship between evolution and ethics. I intentionally leave aside other subjects associated with social Darwinism, for example, racism and imperialism, for the sake of a more focused analysis. Another reason for concentrating on the relationship between evolution, economics, and ethics resides in the predominance of this

FIGURE 23.1. An older Herbert Spencer, whose influence flowed out from Britain increasingly into North America and on to the rest of the world. From David Duncan, *Herbert Spencer* (London: Williams and Norgate, 1911)

issue in the public debate from the nineteenth century up to the present day.

SURVIVAL OF THE FITTEST, PROGRESS, AND CAPITALIST COMPETITION

The most appropriate thinker with which to begin our examination is not Darwin but rather his contemporary Herbert Spencer. Nowadays, Spencer is an almost forgotten figure, yet his reputation during the second half of the nineteenth century rivaled that of Darwin. More importantly, while Darwin was first and foremost a naturalist, Spencer was a philosopher, and his main interest lay from the outset in politics and social progress, or more generally human evolution. Spencer wrote on these subjects close to a decade before the publication of the *Origin of Species*. His first book, *Social Statics* (1851), was an attempt to develop a science-based morality and uncover the conditions essential to human happiness. According to Spencer, the most important of these conditions was liberty, because, without the liberty to exercise the faculties, any living organism would suffer or, in the extreme case, die. This "physiological truth" led Spencer (2009, 39) to declare the law of equal freedom as the principal moral rule: "Every man may claim the fullest liberty to exercise his faculties provided always he does not trench upon the similar liberty of any other." From this first law, Spencer derived all other forms of individual liberty, such as the right of free speech and the right of property, and specified their political applications.

Spencer (2009, 151) argued that when a government tries to alleviate social suffering, for instance with poor laws destined to help the underprivileged, the result would be greater misery: "Blind to the fact, that under the natural order of things society is constantly excreting its unhealthy, imbecile, slow, vacillating, faithless members … unthinking, though well-meaning, men advocate an interference which not only stops the purifying process, but even increases the vitiation – absolutely encourages the multiplication of the reckless and incompetent by offering them an unfailing provision, and discourages the multiplication of the competent and provident by heightening the prospective difficulty of maintaining a family." This is the principle known as "survival of the fittest," an expression coined by Spencer in an article published a year after *Social Statics*. He used it to describe the mechanism employed by nature to assure the survival of the only part of the population able to adapt to conditions of existence (Fig. 23.1).

Darwin adopted Spencer's expression in later editions of the *Origin of Species*, in conjunction with "natural selection," as a way to clarify his original metaphor. The two thinkers however had very distinct views on evolution, especially its relationship to progress. For Spencer, evolution and progress were synonymous. In 1857 he published an article with the telling title "Progress, Its Law and Cause," in which he claimed that a universal law of evolution is accountable for all change in nature, human beings, and society. Shortly afterward, Spencer announced his intention to publish a full-fledged *System of Synthetic Philosophy*, which promised to demonstrate in multiple volumes the workings of the universal law of evolution in biology, psychology, sociology, and ethics. The enterprise won him worldwide reputation as the thinker who provided a link between biological and social development. Spencer established this biosocial connection through an organic analogy between living organisms and social "super-organisms." He maintained that the same principles govern the progress of both types of organisms: growth leads to increasing division of labor, which in turn engenders greater complexity of structure. He added, however, an important caveat to this description in order to accommodate his political position in favor of individualism and restricted government intervention.

According to Spencer, in biological organisms, the emergence of a nervous system and the development of a brain, which functions as a central regulating organ of the body, are the signs of a highly evolved animal. In the social organism, the presence of a central coercive authority is instead the sign of a low phase of evolution, a transitory state that Spencer termed the *militant* type. As societies grow in dimension, and the division of labor becomes more important, the *industrial* type emerges, in which economic competition replaces the

FIGURE 23.2. Scottish-born Andrew Carnegie (1835–1919) founded what was to become US Steel. One of America's richest men and an ardent philanthropist, he was much influenced in his thinking by Herbert Spencer. Nineteenth-century photograph

violent struggle for existence as the motor of further progress. In Spencer's openly teleological account of social development, evolution has a goal, defined as a society governed by the law of equal freedom and regulated through its economic systems of production and distribution, without any need for government intervention other than for the maintenance of justice and protection against outside aggressions. In the name of this view of social development, Spencer condemned most social reforms as measures that either hinder natural progress or vainly attempt to accelerate it.

Proponents of free-market competition, such as American magnate and philanthropist Andrew Carnegie, held Spencer's views in great esteem. In *The Gospel of Wealth*, Carnegie ([1889] 2009, 186) attempted to justify the great social inequalities of modern industrial society as necessary for the progress of humanity, claiming that, "while the law [of competition] may be sometimes hard for the individual, it is best for the race, because it insures the survival of the fittest in every department. We accept and welcome, therefore, as conditions to which we must accommodate ourselves, great inequality of environment; the concentration of business, industrial and commercial, in the hands of a few; and the law of competition between these, as being not only beneficial, but essential for future progress of the race" (Fig. 23.2). Carnegie proposed to alleviate the severity of crude capitalism through increased inheritance taxes and large-scale philanthropy (Bannister 2006). Spencer himself, however, did not feel entirely comfortable with this use of evolutionary theory to endorse cutthroat competition.

In a little known speech, delivered on the occasion of a visit to the United States in 1882, Spencer beseeched the audience to promote the "gospel of relaxation" instead of the "gospel of work." Lamenting the harsh consequences of a merciless struggle for wealth, Spencer warned his listeners of the ill effects an intense race for riches would have on their physical and mental constitutions: "Nature quietly suppresses those who treat thus disrespectfully one of her highest products, and leaves the world to be peopled by the descendents of those who are not so foolish." Observing his fellow Americans, prematurely aged, and often suffering from depression, Spencer was distressed by the toll that material development took on American civilization. In his eyes, "Americans have diverged too widely from savages," and their "high-pressure life" has reached an extreme that risked leading to degeneration instead of further progress (Youmans 2008, 29–31, 35). Despite this criticism, Spencer remained faithful to free-market competition throughout his life. Many of his admirers, however, changed their position radically when faced with the great inequalities in wealth distribution. This was the case of Spencer's good friend and the co-discoverer of the theory of evolution by natural selection, Alfred Russel Wallace.

THE PERFECT SOCIAL STATE AND HUMAN SELECTION IN INDUSTRIALIZED SOCIETIES

In a report on research conducted in the Malay Archipelago, Wallace (1869a, 456–57, emphasis in original) wrote: "We most of us believe that we, the higher races, have progressed and are progressing ... [but] if we continue to devote our chief energies to the utilizing of our knowledge of the laws of nature with the view of still further extending our commerce and our wealth, the evils which necessarily accompany these when too eagerly pursued, may increase to such gigantic dimensions as to be beyond our power to alleviate. We should now clearly recognise the fact, that the wealth and knowledge and culture of *the few*, do not constitute civilisation, and do not of themselves advance us towards the 'perfect social state.'" The notion of a "perfect social state" came from Spencer's philosophy, which had considerable influence on Wallace. The latter concluded his famous discourse on "The Origin of Human Races and the Antiquity of Man" (1864) with a reflection on the future of humanity that predicted a society governed by the law of equal freedom and composed of "a single homogeneous race, no individual of which will be inferior to the noblest specimens of existing humanity." However, in the same lecture, Wallace (1864, clxviii–clxix) also maintained that natural selection stopped modifying humans' bodily structures at some point in the past while continuing to act on their intellectual and mental faculties. This meant that, when applied to humans, the survival of the fittest was in fact the survival of those who were more fit for the "social state." In other words, natural selection in humans leads to the displacement of the less morally advanced individuals by those with superior "sympathetic feelings," who readily help the sick and less fortunate members of society.

Faithful to this view, Wallace was greatly impressed by American socialist Henry George's treatise on *Progress and Poverty*. George ([1879] 2005, 4:265–66) argued that the great advances in material development did not deliver their awaited benefits. In fact, he claimed: "Progress simply widens the gulf between rich and poor." George denounced the "prevailing belief" that society moves forward through a struggle for existence that spurs people to new efforts and inventions, in which the more capable and industrious prosper and propagate their kind. This misconception, he contended, puts a scientific cachet on opinions popular among capitalists and leads to a sort of "hopeful fatalism: progress is the result of slow, steady, remorseless forces. War, slavery, tyranny, superstition, famine, and poverty are the impelling causes that drive humans on. They work by eliminating poor types and extending the higher." As counterevidence to this view, George called on the voice of history, with its many examples of civilizations that have advanced and then regressed. Progress, he concluded, was not an inevitable necessity. Moreover, the obstacles that bring it to a halt are caused by the course of progress itself. George ended his essay with a warning: unless the evils arising from unequal and unjust distribution of wealth were removed, they would expand until they swept us back to barbarism. His practical suggestion was to make land a common property by appropriating rent revenues through taxation. He predicted that a single tax on land would make all other taxes unnecessary, thereby reducing the gap between workers who earn wages and landowners who would no longer be able to charge rent and would have to find alternative ways to make a living.

George's ideas struck a chord with Wallace, who was already a member of the land reform movement, which led a campaign to transfer landownership to the state. In 1881 Wallace became president of the Land Nationalisation Society, and a year later published an essay that endorsed George's position. This brought Wallace into direct conflict with Spencer, who at this point turned his back on claims made in *Social Statics*. In that early treatise, highly praised by George and Wallace, Spencer argued that the right of all to use the earth, a right limited only by the equal rights of fellow individuals, forbids private property in land. However in the polemical collection of essays *Man versus the State*, Spencer ([1884] 1981, 39) criticized the land nationalization movement for disregarding the just claims of existing landowners, who have the right to enjoy the fruits of their, and their ancestors' past efforts. The proposed reform, Spencer retained, "goes more than half-way to State-socialism," and in so doing enslaves individuals to society.

Spencer attempted to justify his volte-face in part by stressing that the views advanced in *Social Statics* pertained to the "perfect social state," in which humans' intellectual and moral advancement would achieve its highest point as a natural result of prolonged existence in a free-market society. In Spencer's opinion, we were still far removed from this ideal, for otherwise it would have realized itself naturally, without any need for the external inducement of a reform movement. This argument left many of Spencer's erstwhile followers, including Wallace, unconvinced. Wallace did not only go half-way to state socialism but went the whole way. In 1890, he published an article on "Human Selection," which, as he tells in his autobiography (1905, 2:209), he regarded as his most important contribution to the science of sociology and to the study of the causes of human progress. His aim was to show that by following a rational social organization, which recognizes the equal rights of all members of society to land and to an equal share of the wealth produced, human evolution would naturally progress in accordance with our most cherished ideals.

Wallace began his essay by quoting Francis F. Galton's studies on eugenics and August Weismann's research as conclusive evidence against the principle of heredity of acquired characteristics. It was clear, he asserted, that the beneficial influences of education, hygiene, and social refinement, which an individual may enjoy during his or her lifetime, did not have a cumulative effect, and therefore only selection could improve the stock of humanity. Wallace (1890, 328–31, emphasis in original) was critical of Galton's proposed scheme for "human betterment" through selective breeding not on ideological grounds but because he believed that it was an indirect and inefficient method to achieve the desired result: "What we want is not a higher standard of perfection in the few but a higher average, and this can be best produced by the elimination of the lowest of all and a free intermingling of the rest." Prima facie, this view seems to resonate with a hardhearted interpretation of the survival-of-the-fittest principle, but Wallace stipulated that for selection to take a beneficial course, direct intervention of a specific kind was necessary: "It is my firm conviction ... that when we have cleansed the Augean table of our existing social organisation, and have made such arrangements that *all* shall contribute their share of their physical or mental labour, and that all workers shall reap the *full* reward of their work ... we shall find that a system of selection will come spontaneously into action, which will steadily tend to eliminate the lower and more degraded types of man, and thus continuously raise the average standard of the race."

Women held a special place in Wallace's system of selection. Thanks to better education, extended to both sexes until the age of twenty-one, and followed by three years in the "industrial army" before entering into the public service, the marriage age would be pushed back. This would put a check on the rapid increase of population and thereby reduce the severity of the struggle for existence. Furthermore, under the new social conditions, which would render every woman independent and provide her with proper intellectual preparation, female choice of partners would be more exacting. Young women would reject the idle, selfish, diseased, and "all men who in any way fail in their duty to society," leaving the unfit unable to reproduce. Wallace (1890, 332–37) insisted that this "weeding-out system," the social equivalent to natural selection, was in tune with the noblest attributes of humankind, such as the propensity to save the lives of the suffering and those who are maimed in body or mind.

Wallace explained that in "hitherto imperfect civilisation," the development of our moral character has been to some extent antagonistic to the process of extinction of the unfit. In the society of the future, this defect would be remedied through conditions that would encourage reproduction among the more capable men and women. Rather than a diminution in our humanity, the number of the less fortunate would diminish from generation to generation. If we leave aside the question of the validity of Wallace's rather optimistic analysis, one thing remains clear. In his eyes, natural selection was a "wholesome process," responsible not only for the elimination of the unfit but also for the development of the moral characteristics of our species and the pronounced expression of emotions such as compassion and sympathy. This was his position already in 1864, when he claimed that natural selection favored in humans the sense of justice, cooperative behavior for the sake of protection and assistance, and other traits that benefit the community. "For it is evident," Wallace then declared, "that such qualities would be for the well-being of man; ... Tribes in which such mental and moral qualities were predominant, would therefore have an advantage in the struggle for existence over other tribes in which they were less developed, would live and maintain their numbers, while the others would decrease and finally succumb" (Berry 2002, 182). This idea would become the essence of Darwin's theory of community selection in humans. We now return to the thorny question of the origins and evolution of human morality, which posed a potential threat to Darwin's theory of natural selection.

COOPERATION, STRUGGLE, AND MORAL BEHAVIOR

In the *Descent of* Man, Darwin was faced with the following dilemma. On the one hand, it seemed that the principle of survival of the fittest could not favor the rise of pro-social behavior. Imagine, as Darwin did, a society made of selfish people. The individual willing to sacrifice herself or himself would die and not leave any offspring behind. Thus, on average, altruistic individuals would perish more often than the others, and there would be a natural selection against altruism. On the other hand, Darwin ([1871] 1981, 162) thought, as Wallace did, that "when two tribes of primeval man, living in the same country, came into competition, if the one tribe included (other circumstances being equal) a greater number of courageous, sympathetic and faithful members, who were always ready to warn each other of danger, to aid and defend each other, this tribe would without doubt succeed best and conquer the other." Thus, what appeared to be a winning strategy in the struggle for existence on the individual level, namely selfish behavior, was a losing strategy on the group level.

In order to solve this conundrum, Darwin advanced the hypothesis that as humans' reasoning powers evolved, combined with accumulated experience, individuals learned that helping others increases the chances of getting help in return. From this "low motive" (163), humans acquired the habit to help, which in turn strengthened preexisting feelings such as sympathy. Throw into the mix the development of communication skills – especially the language of praise and blame – and the set-up was right, thought Darwin, for selection to favor pro-social behavior within the group. Because groups possessing social and moral qualities in the highest degree would spread and be victorious over other groups in ongoing tribal wars, these qualities would tend to become more pronounced and diffused. As Darwin put it, "At all times throughout the world tribes have supplanted other tribes, and as morality is one element in their success, the standard of morality and the number of well-endowed men will thus everywhere tend to rise and increase" (166). By preserving the groups that exhibit cooperative behavior, natural selection could act indirectly on the individual and promote altruistic traits. Darwin's comments on the extermination and replacement of the "savage races" by the "civilised races" (201), often used to point a blaming finger at his improper views, should be understood in the context of his theory of community selection and its central role in his account of the evolution of morality.

Darwin prophesized that as civilization developed and small tribes, which predominantly consisted of related members, were united into larger communities, the social instincts and sympathies of humans would extend, as reason and learning advanced, to include a widening circle of humanity and perhaps other sentient beings. In light of this belief, it may seem that Darwin, like Spencer, perceived evolution to be synonymous with progress. Yet Darwin's cautionary attitude made him hesitate to assign a specific direction to the evolutionary process. "We must remember," he admonished, "that progress is no invariable rule. It is most difficult to say why one civilised nation rises, becomes more powerful, and spreads more and more widely, than another, or why the same nation progresses more at one time than at another." Darwin pointed to history's examples, namely the ancient Greeks, which given their high intellectual powers and great empire should have, according to the principle of natural selection, increased in number and stocked the whole of Europe. "Here we have a tacit assumption," Darwin remarked, "so often made with respect to corporeal structures, that there is some innate tendency towards continued development in mind and body. But development of all kinds depends on many concurrent favourable circumstances. Natural selection acts only in a tentative manner. Individuals and races may have acquired certain indisputable advantages and yet have perished from failing in other characters" (177–78). Darwin conceded, nevertheless, that it is "a truer and more cheerful view" to regard progress as general and "that man has arisen, though by slow and interrupted steps, from a lowly condition to the highest standard as yet attained by him in knowledge, morals and religion" (184).

Darwin furthermore suggested that the "obscure ... problem of the advance of civilisation" depended on an increase in population and the portion within it of benevolent members with high intellectual and moral faculties. Notice that population pressure, a condition that follows from increase in numbers and which leads to the struggle for existence, was for Darwin an indispensable factor in moral progress. He maintained that had humans not been subjected to the struggle for existence, and to the natural selection that follows from it, they would never have attained to "the rank of manhood" (180). This opinion was not shared by his faithful "bulldog" Thomas Henry Huxley. Famously, Huxley exclaimed in an essay on "The Struggle for Existence in Human Society": "From the point of view of the moralist the animal world is on about the same level as a gladiator's show." Similarly to Darwin, Huxley (1894, 199–200) did not think that evolution signified a constant tendency to increased perfection or progress and declared that "retrogressive is as practicable as progressive metamorphosis." However, Huxley also argued that while society is undoubtedly part of nature, it is desirable and even necessary to consider it apart "since society differs from nature in having a definite moral object; whence it comes about that the course shaped by the ethical man – the member of society or citizen – necessarily runs counter to that which the non-ethical man – the primitive savage, or man as a mere member of the animal kingdom – tends to adopt. The latter fights out the struggle for existence to the bitter end, like any other animal; the former devotes his best energies to the object of setting limits to the struggle" (203). Huxley believed that the origin of the problem lay in unlimited multiplication, which by exacerbating the struggle for existence tends to destroy society from within. The only solution to this predicament was to control the continual free fight by deliberately opposing nature.

Huxley reiterated this conclusion with greater conviction in a famous lecture on evolution and ethics: "Let us understand, once for all, that the ethical progress of society depends, not on imitating the cosmic process, still less in running away from it, but in combating it." According to Huxley, in the course of our development, the idea of justice underwent a gradual sublimation from punishment and reward according to acts, to punishment and reward according to dessert. As a result, the conscience of humans began to revolt against the moral indifference of nature. Huxley denounced "fanatical individualism" for misunderstanding the nonmoral character of natural evolution and deplored the fallacy that arose from the "unfortunate ambiguity" of the phrase "the survival of the fittest"; whereby "fittest" received the connotation of "best" or "good" in a moral sense. He then continued to claim that laws and moral precepts should be directed to the end of curbing nature and to reminding the individual of his or her duty to the community in making peaceful and protected existence possible. Social organization should aim "not so much to the

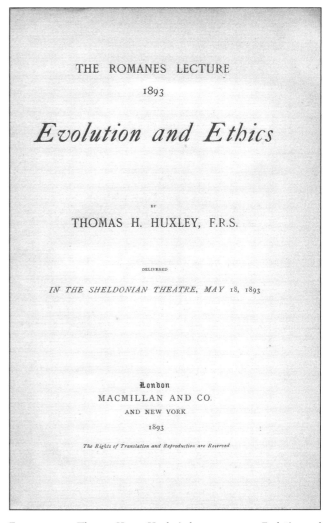

FIGURE 23.3. Thomas Henry Huxley's last great essay, *Evolution and Ethics*, delivered as the Romanes lecture in 1893. Controversial to the end, many of its readers were astounded that Huxley could seemingly turn on the worldview that he had so long championed and claim that evolution is no guide for moral behavior. Huxley argued that far from betraying Darwinism he was reading it truly, a stance that is endorsed by many today. From T. H. Huxley, *Evolution and Ethics* (London: Macmillan, 1893)

survival of the fittest, as to the fitting of as many as possible to survive" (Ruse 2009a, 80–83) (Fig. 23.3).

Huxley was undoubtedly one of the most critical voices against attempts to draw a connection between biological and cultural evolution. He saw nature as a formidable power, red in tooth and claw, yet believed, somewhat contradictorily, that humans' intelligence would provide enough stamina to counter this cosmic force. In the meantime, an alternative understanding of the nature of the evolutionary process emerged, carrying with it a very different message. It is most commonly associated with the view of Russian zoologist and anarchist Peter Kropotkin, though he was by no means the only one, or even the first, to enounce it (Fig. 23.4). Kropotkin argued that mutual aid and support were as much a law of nature as the struggle for existence. He distinguished between two different aspects of the struggle for existence: the exterior war of the species against the harsh environment and other

FIGURE 23.4. Prince Peter Kropotkin (1842–1921) argued that an innate disposition for mutual aid was the main factor in evolution and supported political views congenial to his own anarchism. From calling card, about 1890

species, and the intraspecies war for means of subsistence. The latter, Kropotkin claimed, was often greatly exaggerated. He brought forth as evidence his own observations made in Siberia of many adaptations for struggling in common against the adverse circumstances of the climate or against various enemies. Kropotkin ([1902] 2008, 5, 12, 137) concluded that the animals that acquired habits of mutual aid were "undoubtedly the fittest" and the most highly developed. These findings applied also to human beings, whose history Kropotkin reviewed, asserting "the ethical progress of our race, viewed in its broad lines, appears as a gradual extension of the mutual aid principles from the tribe to always larger and larger agglomerations." Huxley's gladiatorial view was simply a "very incorrect representation of the facts of Nature."

Faced with the grim reality of the 1914 hostilities, Kropotkin wrote a preface to a reprint of *Mutual Aid: A Factor in Evolution* in which he condemned the use of the struggle for existence as an explanation for the war horrors. Though clearly the evidence for his theory was at this point far from convincing, Kropotkin did not lose faith. This might prompt us to ask whether some of the other viewpoints surveyed previously relied on stronger foundations and to ponder the essence of social Darwinism. Our survey shows that under the auspices of the theory of evolution the most disparate conceptions of progress and diametrically opposed political positions were heralded. Today there is still great disagreement as to how evolutionary principles apply to the human domain and what practical conclusions we can gain from understanding them. We know more about biology, and we have better tools to study the particularities of our species. Yet Darwin's prediction seems to hold: "light will be thrown," and we still have much to learn from further research into the history, psychology, and social behavior of our species. Evolution is too complex a theory to yield quick or simple answers, and this complexity is at the core of many partial interpretations and abuses of it. It is also what makes the theory of evolution so fascinating: we know that it must provide invaluable insights if only for the reason that we are part of the living world.

❧ Essay 24 ❦

Darwin and the Levels of Selection

Daniel Deen, Brian Hollis, and Chris Zarpentine

In the mid-1820s, Charles Darwin was in medical school at the University of Edinburgh. There he met the evolutionist Robert Grant. Grant was interested in zoophytes, organisms that were considered plantlike animals. He and others hoped these organisms might help bridge the gap between the two kingdoms. Darwin accompanied Grant on collecting trips to the Firth of Forth, and it was through this work that he had his first brush with scientific scholarship. Darwin delivered a short report to the Plinian Society, a natural history club, on his observations of the "ova" of *Flustra*, a seaweed-like aquatic invertebrate.

A few years later, while aboard the *Beagle*, Darwin's interest in zoophytes continued. In his account of the voyage, he offered the following reflective description of one of these species, *Virgularia patagonica*:

> Each polypus, though closely united to its brethren, has a distinct mouth, body, and tentacula. Of these polypi, in a large specimen, there must be many thousands; yet we see that they act by one movement; that they have one central axis connected with a system of obscure circulation; and that the ova are produced in an organ distinct from the separate individuals. Well may one be allowed to ask, what is an individual? (1839b, 117)

This is not the only time Darwin writes of individuality in this work. Later, in a brief discussion of what Darwin calls "compound animals," he writes:

> With regard to associated life, animals of other classes besides the mollusca and radiata present obscure instances of it. The bee could not live by itself. And in the neuter, we see an individual produced which is not fitted for the reproduction of its kind – that highest point at which the organization of all animals, especially the lower ones, tends – therefore such neuters are born as much for the good of the community, as the leaf-bud is for the tree. (1839b, 262)

Early on, Darwin had stumbled upon the issue of biological individuality. While these passages do not appear in either Darwin's zoological notes or his diary from the *Beagle* voyage, the passage discussing "associated life" seems to derive from remarks in the Red Notebook, which Sulloway (1982b) has dated to around 15 March

1837: "Considering all individuals of all species as [each] one individual [divided] by different methods, associated life adds one other method where the division is not perfect" (this is now transcribed and edited in Barrett et al. 1987, RN 132). This passage appears adjacent to remarks commonly thought to indicate Darwin's first endorsement of transmutation. However, a manuscript (CUL DAR 5:98–99) transcribed in Sloan (1985) shows Darwin making similar remarks between February and April 1836. "I think there is much analogy between Zoophites & Plants/ the polypi being buds; the gemmules the inflorescence/ which forms a bud & young plant" (Sloan 1985, 107). These passages show Darwin considering the organization of different kinds of individuals prior to or simultaneous with his early evolutionary speculations and certainly before his discovery of natural selection. We may, then, interpret Darwin's reference to the "good of the community" in natural theological terms. In the *Origin of Species*, we see Darwin expressing similar ideas but from the vantage point provided by the theory of natural selection.

Scholars have pointed out how Darwin's thinking about individual reproduction influenced his thinking about the origin of species (e.g., M. J. S. Hodge 1985; Sloan 1985, 2009). However, what has been less studied is how Darwin's lifelong thinking about individuality influenced his perspective on whether natural selection could sometimes act for the good of the community. In contemporary terms, this issue falls under the levels of selection debate, which has generated discussion among philosophers and evolutionary biologists. Unfortunately, during the twentieth century Darwin's actual views have been obscured by contemporary theorists of various persuasions claiming him as their own.

DARWIN AND THE CONTEMPORARY DEBATE

Traditional Darwinian thought evokes a struggle for existence among individuals motivated by self-interest. This characterization immediately raises a problem, which we call the "classical problem of altruism": how can we reconcile seemingly altruistic behavior, where there is no obvious individual advantage, with this notion of natural selection maximizing individual fitness? Suppose we have a group of altruists in a certain area. While these altruists will do well among themselves, they will always do worse than "freeloaders" who accept the benefits bestowed by altruists without incurring any cost. Thus, altruists will always be subject to invasion by selfish types and should decrease to the point of extinction.

One possible solution to this problem is that selection might operate on differences in fitness between groups of individuals. This would provide an advantage to self-sacrificial behavior for the group over other groups consisting of selfish individuals. This idea is known as group selection and has been the subject of considerable debate. Much of the criticism of group selection can be traced back to the work of V. C. Wynne-Edwards (1962). He argued that the often limited size of natural populations can be explained as adaptive population regulation, which is accomplished via group selection favoring populations that exercise restraint.

In response to this kind of group-level thinking, G. C. Williams (1966) argued that group selection, though theoretically possible, would be slow acting and rare and thus should be invoked only as an explanation of last resort. The mathematical development of inclusive fitness theory by W. D. Hamilton (1963, 1964a, 1964b) offered a "solution" to the problem of altruism by accounting for the benefits of self-sacrificial behavior received by related individuals. This began a period of increasingly gene-centric views of the scope of natural selection (e.g., Dawkins 1976).

Though these criticisms of group selection persuaded many subsequent thinkers, the idea did not disappear. By focusing on what he called "trait-groups," David Sloan Wilson (1975) described a model by which altruistic traits could evolve by group selection. Empirical work demonstrated an evolutionary response to artificial selection at the group level (Wade 1977). Despite the finding that group- and individual-centered views of selection are formally equivalent (Grafen 1984), in the past ten years the debate has been revived, in part because of interest in the study of "major transitions in evolution" (Maynard Smith and Szathmáry 1995). These involve shifts in the biological hierarchy (e.g., shifts from single-celled to multicellular life or individual ants to social superorganisms). Disagreement persists. Some theorists have defended multi-level selection theory, arguing that natural selection acts on entities at various levels of the biological hierarchy (Damuth and Heisler 1988; Okasha 2006). Others, more focused on individual-level adaptation, have argued that looking up and down the biological hierarchy for explanations is unnecessary and unhelpful (West et al. 2007).

Of course, Darwin did not discuss these issues in contemporary terms. This has not stopped contemporary theorists on both sides of the debate from claiming Darwin as their own. Philosopher and historian Michael Ruse (1980) offers an individual selectionist interpretation, claiming that "apart from some slight equivocation over man, Darwin opted firmly for hypotheses supposing selection always to work at the level of the individual rather than the group." Similarly, philosopher Elliott Sober (1984) dismissed Darwin's use in the *Origin* of the phrase "profitable to the community" as merely a "verbal slip." Stephen J. Gould (2002) has said that "Darwin labored mightily to encompass the entire domain of evolutionary causation within a single level – natural selection working on organisms."

Others have offered alternative interpretations of Darwin's view. Historian Robert J. Richards (2002b; see also Richards 1987) claims that, in the *Origin*, Darwin accepted that "natural selection operated not on the individual workers to provide their unusual traits but on the whole hive or community, which would contain relatives of the workers. And in the fifth edition of the *Origin*, he extended the idea of group selection to any assemblage of social animals, including human beings." Mark Borrello (2005) agrees, writing: "It is apparent that he conceived of the mechanism of natural selection as acting at the level of the community." After coauthoring a book-length defense of group selection with D.S. Wilson (1998), Elliott Sober (2010, 84) has now come to a similar conclusion: "To explain the traits of individuals that promote sociality, he was happy to endorse group selection."

So what was Darwin's view concerning the levels of selection? Careful reading of Darwin indicates that he was aware of the special difficulty posed by self-sacrificing behaviors. Here we discuss three cases that Darwin treated at great length: nonreproductives in the social insects, sterility of hybrids between species, and human intelligence and morality. While there is evidence that Darwin was aware of the problem of altruism at an abstract level, he approached questions about the adaptive benefits of traits in a way sensitive to the natural history of the organism in question, though he was not particularly sensitive to whether these benefits accrue to individuals or groups. Darwin never gives a full answer to the question posed in the *Voyage*: What is an individual? But over the course of his life, he increasingly came to see social communities as a kind of individual. Embracing the idea of higher-level individuals renders much of the debate over individual and group selection moot and focuses attention on how the struggle for existence plays out at various levels of the biological hierarchy.

NATURAL SELECTION: ITS PRIORITY AND VARIETY

Early in the *Origin*, Darwin (1859, 6) wrote:

> I can entertain no doubt ... that the view which most naturalists entertain, and which I formerly entertained – namely, that each species has been independently created – is erroneous. I am fully convinced that species are not immutable ... [and] that Natural Selection has been the main but not exclusive means of modification.

Darwin's primary purpose in the *Origin* was to show that species have changed over time and that natural selection has been the driving force of this change. Thus, the important distinctions for him were between natural selection and special creation and between natural selection and other evolutionary mechanisms (e.g., Lamarckism, which Darwin distinguished from natural selection and endorsed in some cases but not others). One reason for the diversity of opinions about Darwin's view is that he was less concerned than contemporary theorists with distinguishing between selection acting at the level of the individual organism and selection acting on higher-level entities, because both are examples of natural selection. Nonetheless, he did endorse a higher-level selective explanation in some cases, while in others he did not.

SOCIAL INSECTS AND STERILE CASTES

The social insects provided a particularly important test case of Darwin's hypothesis of natural selection (Fig. 24.1). Ants and bees had long been recognized by natural theologians as prime examples of the Creator's handiwork. William Kirby

FIGURE 24.1. A hive of bees, the epitome of animal sociality and in dire need of Darwinian explanation

and William Spence (1818–28, 121), authors of an entomology text that Darwin had praised and annotated, introduced their natural history of hive bees with this statement:

> The glory of an all-wise and omnipotent Creator, you will acknowledge, is wonderfully manifested by the varied proceedings of those social tribes [insects] of which I have lately treated: but it shines forth with a brightness still more intense in the instincts that actuate the Hive-bee, and which I am next to lay before you.

Indeed, Darwin (1859, 207) began his chapter on instinct in the *Origin* by stating, "so wonderful an instinct as that of the hive-bee making its cells will probably have occurred to many readers, as a difficulty sufficient to overthrow my whole theory." He struggled with the social insects right up to the publication of the *Origin* (see R. J. Richards 1987, 142–52). Darwin wanted to see whether natural selection could provide a satisfactory alternative to special creation. In his discussion, we can distinguish two kinds of explanation: adaptive explanations (*why* does a certain trait exist?) and mechanistic explanations (*how* can a certain trait be passed from one generation to the next?).

Darwin discussed the difficulties raised by sterile castes, as well as the stinger and hexagonal cell-making instinct of the bee. Darwin (1859, 202) noted, "Natural selection will never produce in a being anything injurious to itself, for natural selection acts solely by and for the good of each." How then could natural selection produce a stinger that when used by the bee "inevitably causes the death of the insect by tearing out its viscera?" Darwin explained that if it had originally evolved as "a boring and serrated instrument" and was then modified into a stinger, it becomes easy to understand that, "if on the whole the power of stinging be useful to the community, it will fulfill all the requirements of natural selection, though it may cause the death of some few members." This problem is analogous to the classical problem of altruism. Darwin's adaptive explanation for the stinger involved natural selection acting at the level of the community.

Darwin gave a similar explanation for the bee's cell-making instinct, though it may not be immediately apparent why this case should be analogous. As Darwin knew, hexagonal cells for the storage of honey allow the cells to "hold the greatest possible amount of honey, with the least possible consumption of precious wax in their construction" (224). They also function as a public good, requiring that "a multitude of bees all work together" (231), and, as Darwin found out from one of his informants, "a prodigious quantity of fluid nectar must be collected and consumed by the bees in a hive for the secretion of the wax necessary for the construction of their combs" (233–34). Because the storage of honey is crucial for the survival of bees through the winter, Darwin saw that natural selection would favor the most efficient communities: "that individual swarm which wasted least honey in the secretion of wax, having succeeded best, and having transmitted by inheritance its newly acquired economical instinct to new swarms, which in their turn will have had the best chance of succeeding in the struggle for existence" (235).

Thus, Darwin was able to explain one of the presumed triumphs of natural theology by invoking natural selection acting at the level of insect communities. In addition to offering this adaptive explanation, Darwin offered a mechanistic explanation. He drew on his extensive knowledge of natural history – provided by his research, his correspondence with informants, and his own experiments – to show how natural selection might have "taken advantage of numerous, successive, slight modifications of simpler instincts" (235). He traced a gradual path of increasingly efficient cell construction from the humble-bees, through the Mexican bee *Melipona domestica*, all the way to "the extreme perfection of the cells of the hive-bee" (225).

In the final section of the chapter on instinct, Darwin focused on what he described as "one special difficulty, which at first appeared to me insuperable, and actually fatal to my whole theory" (236). The difficulty was, of course, sterile castes in insect communities. In addressing this issue, however, Darwin actually identified three distinct problems. The first problem concerned the adaptive explanation for sterile castes. This is essentially the classical problem of altruism. Yet, it was the one Darwin thought he could most easily resolve, considering its difficulty "not much greater than that of any other striking modification of structure" (236). As in the stinger and the cell-making instinct, Darwin explained, "if such insects had been social, and it had been profitable to the community that a number should have been annually born capable of work, but incapable of procreation, I can see no great difficulty in this being effected by natural selection" (236).

The other two problems, which Darwin considered to be more difficult, required mechanistic explanations rather than adaptive ones. The first was that neuters "often differ widely in instinct and in structure from both the males and fertile females, and yet, from being sterile, they cannot propagate their kind" (236). Closely related to this, the third problem, which Darwin referred to as "the climax of the difficulty," was that "the neuters of several ants differ, not only from the fertile females and males, but from each other, sometimes to an almost incredible degree" (238). Of course, Darwin knew that various traits can become correlated, for "we have innumerable instances, both in our domestic productions and in those in a state of nature, of all sorts of differences of structure which have become correlated to certain ages, and to either sex" (237). He wrote: "I can see no real difficulty in any character having become correlated with the sterile condition of certain members of insect-communities: the difficulty lies in understanding how such correlated modifications of structure could have been slowly accumulated by natural selection" (237). Darwin quickly saw how the benefits of a sterile caste could accrue to the community (an adaptive explanation) but not how the traits of these sterile individuals could be passed on to subsequent generations. These problems were exacerbated by his ignorance of the mechanisms of inheritance.

Darwin's solution to these problems involved the notion of *family selection*. He explained: "This difficulty though appearing insuperable, is lessened, or, as I believe, disappears, when it is remembered that selection may be applied to the family, as well as to the individual, and may thus gain the desired end" (237). As he so often does in the *Origin*, Darwin drew on an analogy with artificial selection. A good steak, Darwin noted, requires the marbling of fat and flesh. Obviously cattle breeders cannot breed from an individual that has been slaughtered, "but the breeder goes with confidence to the same family" (237–38). He continued:

> I have such faith in the powers of selection, that I do not doubt that a breed of cattle, always yielding oxen with extraordinarily long horns, could be slowly formed by carefully watching which individual bulls and cows, when matched, produced oxen with the longest horns; and yet no one ox could ever have propagated its kind. Thus I believe it has been with social insects: a slight modification of structure, or instinct, correlated with the sterile condition of certain members of the community, has been advantageous to the community: consequently the fertile males and females of the same community flourished, and transmitted to their fertile offspring a tendency to produce sterile members having the same modification. (238)

Darwin's brilliant insight was that selection could act on parents and favor the tendency to produce a certain kind of sterile offspring.

He applied this to the third difficulty, pointing out that "natural selection, by acting on the fertile parents, could form a species which should regularly produce ... one set of workers of one size and structure, and simultaneously another set of workers of a different size and structure" (241). By beginning with sterile offspring that vary along a "graduated series," if the "extreme forms" had been "the most useful to the community" they would slowly become more numerous until there were no individuals born in the middle of the series, and there were simply two (or more) distinct sterile castes (241).

Darwin again used an analogy to emphasize how sterile castes could provide benefits to their community:

> We can see how useful their production may have been to a social community of insects, on the same principle that the division of labour is useful to civilized man. As ants work by inherited instincts and by inherited tools or weapons, and not by acquired knowledge and manufactured instruments, a perfect division of labour could be effected with them only by the workers being sterile; … And nature has … effected this admirable division of labour … by the means of natural selection. (241–42)

Darwin recognized that without culturally transmitted knowledge and technology there is great difficulty in producing a division of labor without caste sterility. If members of the castes were able to reproduce, emerging differences between types would be lost through interbreeding. As Darwin pointed out, not only does natural selection offer an alternative to special creation in this case, but Lamarckism cannot explain sterile castes, for "no amount of exercise, or habit, or volition, in the utterly sterile members of the community could possibly have affected the structure of instincts of the fertile members" (242).

In these difficult cases involving social insects, Darwin offered adaptive explanations in terms of selection acting at the level of the community. The more difficult problems for him involved mechanistic explanations: how could the traits of sterile castes be transmitted to the next generation? The idea of family selection provided the solution to these problems. Indeed, Darwin himself expresses surprise at the power of natural selection: "I am bound to confess, that, with all my faith in this principle, I should never have anticipated that natural selection could have been efficient in so high a degree, had not the case of these neuter insects convinced me of the fact" (242).

HYBRID STERILITY AND SPECIES SELECTION

Darwin saw the cases of insect sterility and hybrid sterility – the tendency of crosses between two species or varieties to be sterile – as structurally analogous. This tendency can benefit incipient species by preserving differences that would otherwise be lost through interbreeding. As we saw, Darwin accepted the higher-level explanation in the case of insect sterility. Interestingly, he rejected it in the case of hybrid sterility.

In both the *Origin* and *The Variation of Animals and Plants under Domestication*, Darwin held the view that "sterility which almost invariably follows the union of distinct species depends exclusively on differences in their sexual constitution" (1868b, 2:184–85; cf. 1859, 272) and are not "endowed through an act of creation" (1868b, 2:188; cf. 1859, 272). He compared it to the capacity for grafting between different species or varieties, which "is so entirely unimportant for its welfare in a state of nature, I presume that no one will suppose that this capacity is a specially endowed quality" but rather that it is "incidental on differences in the laws of growth of the two plants" (1859, 261; cf. 1868b, 2:188).

In *Variation*, Darwin mentioned his own brief flirtation with the higher-level alternative, that natural selection had favored mutual sterility because it benefited insipient species by preventing their blending. Ultimately Darwin rejected this possibility and offered several reasons (see Ruse 1980, 623–24). Most importantly, he thought that "it could have been of no direct advantage to an individual animal to breed badly with another individual of a different variety, and thus leave few offspring" (1868b, 2:186). He wrote:

> With sterile neuter insects we have reason to believe that modifications in their structure have been slowly accumulated by natural selection, from an advantage having been thus indirectly given to the community to which they belonged over other communities of the same species; but an individual animal, if rendered slightly sterile when crossed with some other variety, would not thus in itself gain any advantage, or indirectly give any advantage to its nearest relatives or to other individuals of the same variety, leading to their preservation. (1866, 2:186–87)

Alfred Russel Wallace disagreed and in February 1868 wrote to Darwin that, "given a differentiation of a species into two forms each of which was adapted to a special sphere of existence, – every slight degree of sterility would be a positive *advantage*, not to the *individuals* who were sterile, but to *each form*" (Darwin 1985–, 16:171, 24 February 1868)

Darwin wrote back saying he disagreed. He had gone through the reasoning "over & over again on paper with diagrams" and two of his grown children (whom he described as "acute reasoners") tried to convince him of the soundness of it on more than one occasion but always ended up coming back to Darwin's view (Darwin 1985–, 16:196, 27 February 1868). Wallace sent Darwin a manuscript outlining his view that selection explains hybrid sterility by acting on "forms" or "sterile varieties" (Darwin 1985–, 16:219–22). Darwin did not immediately respond in detail but reported that considering the issue "made my stomach feel as if it had been placed in a vice.… Your paper has driven 3 of my children half-mad – One sat up to 12 oclock over it" (Darwin 1985–, 16:278–79, 17 March 1868) (Fig. 24.2).

Wallace's reasoning was as follows. Suppose a species has two forms (or varieties). Hybridization between these forms would not permit speciation. However, if partial sterility arises in a particular area, then the hybrids will not increase, and pure individuals of each form will supplant the hybrids. Darwin's son George, who had just made second wrangler in mathematics at Cambridge, wrote up a response that Darwin forwarded to Wallace (enclosed in Darwin 1985–, 16:291–92, 21 March 1868) George explained that, in fact, this area where partial sterility arises is subject to outside invasion by hybrids and that the pure forms will actually decrease.

FIGURE 24.2. George Darwin (1845–1912), Charles Darwin's brilliant mathematical son, who often helped his father with calculations. Close to Kelvin, George Darwin was the one who conveyed to his father the bad news that modern physics denies the time needed for the slow process of natural selection. Permission: Wellcome

George's line of reasoning here anticipates the classical problem of altruism. Just as altruists are subject to invasion by selfish types (unless there is correlated interaction of types), pure forms are subject to invasion from hybrids and, like altruists, will eventually decrease to the point of extinction. Wallace tried to respond to George's notes but did not seem to understand the objection. He reiterated the higher-level selectionist idea in the form of what he suggested was a "strong general argument" (Darwin 1985, 16:302, 24 March 1868). Because it is known that the degree of sterility varies in nature, why shouldn't selection tend to increase it to the point of complete sterility? Wallace added, "If Nat. Select. can not do this how do species ever arise, except when a variety is isolated" (Darwin 1985–, 16:303)

Finally, in early April, Darwin responded to Wallace's manuscript. He came down in agreement with his son and against the higher-level explanation: "The cause being as I believe, that natural selection cannot effect what is not good for the individual, including in this term a social community" (Darwin 1985–, 16:374, 6 April 1868) Darwin went on to make some more detailed remarks on Wallace's manuscript but noted that "it w^d take a volume to discuss all the points," concluding: "Life is too short for so long a discussion – We shall, I *greatly* fear, never agree." Wallace pressed the issue in one more letter but admitted that he was probably wrong if Darwin remained unconvinced (Darwin 1985–, 16:389, 8 April 1868).

Darwin's endorsement of George's line of thought indicates that his appreciation of the problem was quite contemporary. In the passage from *Variation* quoted earlier, Darwin saw the structural similarity between neuter insects and hybrid sterility but rejected the selective explanation in the latter case. The benefits that accrue to the community in the case of sterile castes of insects are simply not present in the case of hybrid sterility. In his exchange with Wallace, Darwin explicitly endorses the view that social communities are another kind of individual upon which natural selection acts.

What about Wallace's challenge: if natural selection does not explain sterility, then how does speciation occur except by geographical isolation? While he did not respond to this issue in his reply to Wallace, it must have seemed a familiar complaint to Darwin. In the fourth edition of the *Origin* (1866), Darwin had significantly expanded the chapter on hybridism in order to respond to just such a problem put to him by T. H. Huxley.

In his review of the *Origin*, Huxley (the "Objector-General on This Head" [Darwin 1985–, 14:437, 22 December 1866]) wrote that "there is no positive evidence, at present, that any group of animals has, by variation and selective breeding, given rise to another group which was, even in the least degree, infertile with the first" (1860, 567). Until then, Huxley claimed, there was no proof that natural selection explained the origin of species.

In the fourth edition, Darwin rejected the view that mutual sterility is necessary for speciation: "the physiological test of lessened fertility, both in first crosses and in hybrids, is no safe criterion of specific distinction" (1866, 323; cf. 1868b, 2:183–84). In a letter to Huxley the next year, Darwin was even more explicit, adding after his signature, "Nature never made

species mutually sterile by selection; nor will man" (Darwin 1985–, 15:15, 7 January 1867). Darwin responded to Huxley's worry not by showing how selection can produce mutual sterility but by denying that mutual sterility is necessary for speciation.

Darwin's (1859, 112) own explanation involved what he called the principle of divergence: "[T]he more diversified the descendants from any one species become in structure, constitution, and habits, by so much will they be better enabled to seize on many and widely diversified places in the polity of nature." Curiously, he drew on the same economic metaphor he had when talking of the social insects: "The advantage of diversification in the inhabitants of the same region is, in fact, the same as that of the physiological division of labour in the organs of the same individual body" (115–16). In the case of the diversification of inhabitants according to the principle of divergence, Darwin invoked a division of labor but not higher-level selection.

Recently, Stephen J. Gould (2002, 64) has claimed that Darwin "reluctantly admitted a need for species selection to resolve the problem of divergence." We can now see that there is something seriously wrong with this interpretation. Darwin did not think selection led to mutual sterility of varieties, despite considering higher-level selection. He thought sterility was a by-product of incidental modifications of the reproductive system. A fortiori, he denied that selection needs to bring about mutual sterility for speciation to occur. While mutual sterility might be an important indicator of species, it is not strictly necessary. The account Darwin proposed, the principle of divergence, was an explanation of the pattern that selection will generate given the existence of multiple niches; it was not a mechanism of "species selection."

MORALITY AND INTELLIGENCE

In *The Descent of Man*, Darwin turned his attention to humans, placing them on the tree of common descent and explaining the path by which they came to acquire their unique traits. Darwin (1871a, 1:70) recognized that "of all the differences between man and the lower animals, the moral sense or conscience is by far the most important," but he maintained that these differences were a matter of degree, not kind. While he accepted a variety of selective processes at work, he recognized tribe-level selection acting in the evolution of human intelligence and moral sense.

Darwin offered the following account, which he considered to be "in a high degree probable" and which formed the basis of his discussion of human morality: "[A]ny animal whatever, endowed with well-marked social instincts, [the parental and filial affections being here included,] would inevitably acquire a moral sense or conscience as soon as its intellectual powers had become as well developed, or nearly as well developed as in man" (1871a, 1:71–72; bracketed text added in the second edition, 1874, 98).

Because Darwin was interested in both the continuity of human traits with those of the "lower animals" and how humans came to have their current traits, it is important to keep distinct explanations of the *origin* of traits from explanations of the *elaboration* of those traits. For example, in the *Descent*, Darwin reiterated the view expressed in the *Origin* (1859, 207), "In what manner the mental powers were first developed in the lowest organisms, is as hopeless an enquiry as how life first originated" (1871a, 1:36). Yet he offered an account of how natural selection could have elaborated upon these to produce the mental powers present in humans. Similarly, as we have seen in the case of the bee stinger, Darwin understood that traits adapted for one purpose can be co-opted for another purpose.

The first step in Darwin's explanation of the moral sense was the development of the social instincts. Darwin (1871a, 1:80) took them to be an outgrowth of the "parental and filial affections" through a process of individual selection, with the result that "the individuals which took the greatest pleasure in society would best escape various dangers; whilst those that cared least for their comrades and lived solitary would perish in greater numbers." As for the parental and filial affections, he offered no explanation for their origin, but suggested that they had arisen through natural selection (80–81).

Once the social instincts had emerged, they could be elaborated upon. Darwin wrote, "With strictly social animals, natural selection sometimes acts indirectly on the individual, through the preservation of variations which are beneficial only to the community" (155). He then mentioned his own explanations for "pollen-collecting apparatus, or the sting of the worker-bee, or the great jaws of soldier-ants" as having evolved in this way (155). While he did not give any indication as to whether he considered humans to be "strictly social," he did invoke selection at the level of human tribes to explain human intelligence and moral sense.

Darwin explained how selection acting on both individuals and tribes could have contributed to the elaboration of these mental powers in humans. He wrote that, "in the rudest state of society, the individuals who were the most sagacious, who invented and used the best weapons or traps ... would rear the greatest number of offspring" (159). In addition, he suggested that tribes including more "sagacious" individuals would "increase in number and would supplant other tribes" and such tribes would also be "further increased by the absorption of other tribes" (159).

Darwin cited evidence from human prehistory indicating that "from the remotest times successful tribes have supplanted other tribes" (160). Beneficial variation in intelligence could spread by individual selection, but Darwin also explained how such traits could benefit tribes through a kind of cultural inheritance mediated by imitation: for example, "if some one man in a tribe, more sagacious than the others, invented a new snare or weapon ... the plainest self-interest, without the assistance of much reasoning power, would prompt the other members to imitate him; and all would thus profit" (161). Darwin also invoked his idea of family selection, for even if such individuals left no children, "the tribe would still include their blood-relations; and it has been ascertained by agriculturists that by preserving and breeding from the family of an

FIGURE 24.3. Is a tribe just a collection of individuals, or are a tribe's members related in a manner akin to an insect colony? From A. R. Wallace, *The Malay Archipelago* (London: Macmillan, 1869)

animal, which when slaughtered was found to be valuable, the desired character has been obtained" (161) (Fig. 24.3).

Darwin thought a moral sense would inevitably arise from the social instincts and the development of intelligence. However, he thought that the elaboration of the human moral sense would depend largely on selection between tribes: "When two tribes of primeval man, living in the same country, came into competition, if the one tribe included (other circumstances being equal) a greater number of courageous, sympathetic, and faithful members, who were always ready to warn each other of danger, to aid and defend each other, this tribe would without doubt succeed best and conquer the other" (162).

However, it is here that we must be careful to distinguish the elaboration of the moral sense from its origin. After offering the preceding account of the elaboration of the moral sense in humans by selection acting on tribes, Darwin raised the problem of its origination:

> But it may be asked, how within the limits of the same tribe did a large number of members first become

endowed with these social and moral qualities, and how was the standard of excellence raised? It is extremely doubtful whether the offspring of the more sympathetic and benevolent parents, or of those which were the most faithful to their comrades, would be reared in greater number than the children of selfish and treacherous parents of the same tribe. He who was ready to sacrifice his life, as many a savage has been, rather than betray his comrades, would often leave no offspring to inherit his noble nature. (163)

While Darwin clearly saw the problem, W. D. Hamilton (1972, 193) has suggested that he did not offer a solution to this problem. But Darwin did offer two possible explanations, both depending on human mental powers.

The first explanation, sounding very much like reciprocal altruism, proposes that, "as the reasoning powers and foresight of the members became improved, each man would soon learn from experience that if he aided his fellow-men, he would commonly receive aid in return" (Darwin 1871a, 1:163). If this reasoning was followed habitually, the disposition for "beneficial actions" could eventually be inherited (163–64). According to Darwin's second explanation, if "even dogs appreciate encouragement, praise, and blame," then it seems plausible that "primeval man, at a very remote period, would have been influenced by the praise and blame of his fellows" and that "members of the same tribe would approve of conduct which appeared to them to be for the general good, and would reprobate that which appeared evil" (164–65). This would have led them to "the foundation-stone of morality": "To do good unto others – to do unto others as ye would they should do unto you" (165).

While Darwin was less certain about his account of the origination of the moral sense, he did appreciate the problem that it posed and tried to address it. In reading the following oft-quoted passage in which Darwin invokes tribal selection, it is important to remember that he is speaking here not of the origin of the moral sense but rather of the "advancement of the standard of morality":

> It must not be forgotten that although a high standard of morality gives but a slight or no advantage to each individual man and his children over the other men of the same tribe, yet that an advancement in the standard of morality and an increase in the number of well-endowed men will certainly give an immense advantage to one tribe over another. There can be no doubt that a tribe including many members who, from possessing in a high degree the spirit of patriotism, fidelity, obedience, courage, and sympathy, were always ready to give aid to each other and to sacrifice themselves for the common good, would be victorious over most other tribes; and this would be natural selection. (166)

Thus, Darwin argued that human intelligence aided by habitual reciprocity and an existing love of praise and blame could explain the origination of a moral sense.

This distinction between the origin and development of the moral sense arises in a letter to George, discussing a recent article by Henry Sidgwick. Darwin praises its clarity but takes issue with Sidgwick's claim that "moral men" arise in a tribe by accident, failing to mention Darwin's own account of the origin of such individuals: "I have endeavoured to show that such men are created by love of glory, approbation &c&c. – – However they appear the tribe as a tribe will be successful in the battle of life, like a hive of bees or nest of ants" (CUL DAR 210.1:52). Once the basics of moral sense were established, tribal selection would raise the standard of morality and spread it across the world. Darwin (1871a, 1:166) continues: "At all times throughout the world tribes have supplanted other tribes; and as morality is one element in their success, the standard of morality and the number of well-endowed men will thus everywhere tend to rise and increase."

CONCLUSION

Charles Darwin's interest in individuality began before he boarded the *Beagle* and continued throughout his life. Much of his thought on the matter became integrated into his theory of pangenesis, but it also heavily influenced his thinking about the ways natural selection might act on groups. Indeed, Darwin is sometimes strikingly explicit about this: he sees social communities as another kind of individual, even comparing human tribes to bee hives and ant nests. In large part because of this, he accepted a role for higher-level selection in the evolution of sterile castes in social insects, and human intelligence and moral sense. As we have indicated, Darwin was exposed through his son George to a line of reasoning strikingly similar to the classical problem of altruism. He accepted that selection might act on insect communities, which he saw as a kind of individual, but he resisted explanations of hybrid sterility along these same lines. In the case of human moral sense, he saw the problem of origination and attempted to explain it. However, Darwin was most troubled not by *whether* natural selection could explain traits that benefit the group, but *how exactly* selection could contribute to the origination and elaboration of these traits. His theorizing on this was always informed by his rich knowledge of the natural history of the organisms at hand. After several decades of theorists claiming Darwin as their own, one wonders whether we may still have something to learn from the man himself.

Essay 25

Darwin and Religion

Mark Pallen and Alison Pearn

Darwin's attitude to religion can be summarized as thoughtful but detached. There is no evidence that he ever had any strong religious feelings or a sudden crisis of faith. Although he gradually lost any belief in Christianity "as a divine revelation," he described himself variously as a theist or agnostic but never as an atheist, drawing a careful distinction between the neutral "unbelief," or lack of belief, of agnosticism and the positive "disbelief" of atheism. Toward the end of his life, Darwin wrote that disbelief in "Christianity as a divine revelation" had crept over him at a very slow rate "but was at last complete." The rate was so slow that he felt "no distress," and he had "never since doubted even for a single second" that his conclusion was "correct" (*Recollections*).[1] Emma Darwin and Francis Darwin both referred to the importance to Darwin of the distinction between disbelief and unbelief in letters discussing publication of the *Recollections* written after his death; Emma Darwin considered that the use of the word "correct" was misleading and that Darwin intended to convey that he himself never "altered his opinion" rather than that he thought the position untenable (CUL DAR 210.8: 42 and 219.1: 179).

Darwin accepted that others found it possible to believe both in evolution and in a deity, and he respected that position. He had a profound respect for the views of others and was generally reticent about his own, both from a natural aversion to causing unnecessary distress and, more pragmatically, because he regarded conflict as counterproductive. Typical is this response to an inquiry as to whether natural selection was compatible with belief in a personal God: "My opinion is not worth more than that of any other man who has thought on such subjects ... I thank you

---

[1] In his final decade, Darwin wrote an autobiography that he called "Recollections of the Development of My Mind and Character." He did not publish this but after his death his son Frank, preparing the *Life and Letters*, included a somewhat bowdlerized version – omitting passages that his family thought should remain private – calling it Darwin's "Autobiography." In 1958, Darwin's granddaughter Nora Barlow published an unexpurgated version of what she too called the "Autobiography." The original "Recollections" was finally published in 2010 by James Secord, as part of a volume of Darwin's writings. The catalog number for the manuscript is CUL DAR 26. Page numbers in this essay refer to the published version, Darwin 2010.

for your Judgement & honour you for it, that theology & science should each run its own course & that in the present case I am not responsible if their meeting point should still be far off" (Darwin 1985–, 14:423, letter to M. E. Boole, 14 December 1866). When his son George proposed writing an essay on religion and the moral sense, Darwin urged caution: "The evils are giving pain to others, & injuring your own power & usefulness." Had John Stuart Mill made his own religious views public, Darwin argued, he would "never have influenced the present age in the manner in which he has done" (DCP, 9105, letter to G. H. Darwin, 21 October [1873]).

There is little direct documentary evidence for Darwin's personal beliefs. The primary evidence is contained in his private notebooks on transmutation and metaphysics kept in the late 1830s (Barrett et al., 1987), in his correspondence, and in the autobiographical *Recollections of the Development of My Mind and Character*. The notebooks were for personal use and are thus likely to reveal his inner convictions during his early adulthood. Darwin's correspondence has to be used with care – Darwin was more open with some correspondents than with others. Nevertheless, it provides contemporary evidence for his state of mind at various stages in his life. Darwin's *Recollections* were intended for his family, not for publication, and are most reliable as evidence for his thinking in the period in which they were written. A combination of reminiscence and philosophical musing begun in 1876 when Darwin was sixty-seven years old and with later additions up to the time of his death, the *Recollections* include a substantial section entitled "Religious Belief," which is the most complete statement by Darwin of his own view of the course of his personal relationship with religion.

FAITH AND FAMILY

Charles Darwin grew up exposed both to liberal dissenting Christianity and to nonbelief. Darwin's mother, Susannah Wedgwood, followed the family tradition of Unitarianism. Unitarians deny the Anglican doctrine of the "Holy Trinity" (the three-person nature of God) and therefore the divine nature of Jesus; they remained a proscribed sect in England until 1813. The influence of Darwin's Unitarian heritage, with its emphasis on religious tolerance and the value of inner conviction, rather than dogma, as a guide to conduct is clearly traceable in his later thinking.

Darwin's father, Robert Waring Darwin, like his grandfather Erasmus Darwin, although notionally a Unitarian, was skeptical of received religion, his beliefs verging on atheism. However, the family's skepticism and Unitarianism were coupled with a pragmatic acceptance of the established church as an integral part of a stable society and of the need for outward conformity in order to secure social and professional advancement – up until 1834, for example, subscription to the thirty-nine articles of the Church of England was required in order to hold a government post. The thirty-nine articles defined the dogma of the Anglican Church, asserting the three-person nature of God, the authority of the Old and New Testaments, the authority of the church, and the doctrines of original sin and justification by faith.

Darwin's sisters were baptized as Unitarians, but Charles and his elder brother, Erasmus Alvey Darwin, were baptized in the Anglican parish church of St. Chad's, Shrewsbury. Although Charles spent a few months at a Unitarian day school, at the age of nine he was sent to Shrewsbury School, an Anglican boarding school. In many ways, this blend of skepticism, religious dissent, and pragmatic conformity characterized Darwin's relationship with the church for the rest of his life. Charles's sisters remained practicing Christians and encouraged him to read the Bible; his brother, like their father, appears to have been an atheist and became the center of a radical social circle that included leading Unitarians such as James Martineau, John James Taylor, and Frances Power Cobbe, and freethinkers such as James's sister Harriet Martineau. (See *Recollections*, 392 for Darwin's statement that his father and brother were both unbelievers, along with "almost all" his best friends. See Fig. 25.1.)

EDINBURGH: SOCIALIZING WITH SKEPTICS

As a younger son, Charles needed a career. Options were limited for someone of his social class, the three principal professions being medicine, the law, and the church. Charles's first experience of university was at Edinburgh, where he studied medicine. Darwin's letters confirm his later recollection of his young self as uncritically Christian but lacking any great religious drive: while at Edinburgh he wrote to his sister Caroline with naïve enthusiasm: "I have tried to follow your advice about the Bible, what part of the Bible do you like best? I like the Gospels," and from Cambridge wrote to a bereaved friend of the "pure & holy comfort" of the Bible (Darwin 1985–, 1:39, letter to C. S. Darwin, 8 April [1826]; 1:83, letter to W. D. Fox, [23 April 1829]). However, the intellectual climate in Edinburgh was liberal. Darwin was befriended by the evolutionist and radical dissenter Robert Grant and later recalled Grant's passionate promotion of Lamarckian evolution. Although Grant's more overtly materialist views became public only later in his career, Lamarckism with its championing of transmutation and spontaneous generation and denial of design in nature, was considered dangerously close to atheism. Darwin also joined the undergraduate Plinian Society, known for its radicalism and discussion of materialism, and his brother, with whom he shared lodgings, was reading works by skeptical philosophers, including David Hume and Voltaire (Secord 1991a; Browne 1995).

CAMBRIDGE: THE CHURCH AS A POSSIBLE CAREER

After two years at Edinburgh, Charles abandoned medicine. According to his later recollection, it was his father who suggested the church as an acceptable alternative. There was a family precedent: Darwin's uncle Josiah Wedgwood, himself

FIGURE 25.1. The gates of hell, as pictured by the French illustrator Gustave Doré, for an edition of Dante's *Divine Comedy*. Darwin, like many of his fellow Victorians, worried about the theological implications of Christianity. He was not alone in finding many of the doctrines as conducive to nonbelief at least as much, if not more, than the challenge of science. From *Dante's Divine Comedy* (Boston: Ticknor and Fields, 1867)

a Unitarian, was patron of the Anglican parish church of Maer in Staffordshire and in 1825 had secured the appointment of another nephew, Allen Wedgwood, as vicar (Litchfield 1904).

The usual route to ordination in the Anglican Church was first to obtain a degree from either Oxford or Cambridge University, both entwined with the political establishment and the established church, and both requiring subscription to the thirty-nine articles in order to graduate. In the first half of the nineteenth century, 50 percent of Cambridge and Oxford graduates went on to be ordained, but theology as a separate undergraduate degree subject did not exist at Cambridge before 1871. Until then, most Cambridge students, including Darwin, studied for a general or "ordinary" degree, which entailed examination on a small number of set texts in mathematics (or "Natural Philosophy"), theology, "Moral Philosophy" and "Belles Lettres." Darwin was examined on Euclid, Newton, and Locke, on classical texts and on logic, on the New Testament, and on William Paley's *View of the Evidences of Christianity* (1794) and *Principles of Moral and Political Philosophy* (1785) (*Cambridge University Calendar*).

YOUNG ADULTHOOD: FROM CONFORMITY TO DOUBT

As a student, Darwin voiced some scruples about subscribing to the thirty-nine articles but nevertheless did so. Until the last year of the *Beagle* voyage, he was apparently resigned to the idea of becoming a clergyman, a potentially congenial way of life that would allow him to continue his pursuit of natural history. Darwin recalled that at the outset of the *Beagle* voyage, he regarded himself as "quite orthodox" and was teased by the crew for citing biblical authority to defend his position in an argument; during the course of the voyage, however, his geological observations forced a growing realization that the Old Testament account of the creation of the world was "manifestly unreliable" (*Recollections*). That Darwin was aware of arguments against the literal truth of the Old Testament is evident from his discussion, in a letter to his sister Caroline a few months after his return, of the astronomer Herschel's view that the biblical chronology of six thousand years since the creation of man was too short a time to account for the divergence between Chinese and Caucasian languages (Darwin 1985–, 2:8, letter to C. S. Darwin, 27 February 1837).

It was in London in the late 1830s and early 1840s, in the period around his marriage, that Darwin seems to have thought most intensely about religion; in his later *Recollections*, he identified the years 1837 and 1838 as particularly important, and this is borne out by his *Journal* entries. For example, the entry for 1838, "All September read a good deal on many subjects: thought much upon religion. Beginning of October ditto." This was intellectually an extremely productive period during which he read widely and intently. He kept a series of notebooks to record his most private lines of research, research that led ultimately to his species theory and to his later publication on its consequences for the place of human beings in nature. It was in these notebooks that he first sketched out a theory of the origin of religious belief: "[P]eople say I know it, because I was always told so in childhood, hence the belief in the many strange religions." In considering the possible origins of morality and conscience in humans, Darwin speculated that impulses to anger and revenge, far from being the result of the biblical fall of man, could be explained as the vestiges of animal behavior, concluding with a crack of triumphant humor, "Our descent, then, is the origin of our evil passions!! – The Devil under form of Baboon is our grandfather!" (Barrett et al. 1987, M123). Speculating on the physical basis of mind, he ended with the exclamation, "Oh you Materialist!" (C166). "Materialism" in the nineteenth century denoted a belief that all phenomena have a physical basis, including those generally thought of as metaphysical, that is the mind and the soul. Although often associated with atheism, this was not, however, a position that precluded the possibility of belief in a Creator.

From 1838 to 1851 in a brown leather-covered notebook, Darwin maintained lists of books-to-be-read and books that he had read (Darwin 1985–, 4: appendix 4). These lists reveal an interest in books about religious belief, both those that argued for a Creator and those that argued against. In an entry for 29 September 1839, Darwin recorded reading *Dialogues concerning Natural Religion* and *Natural History of Religion* by the empiricist philosopher and skeptic David Hume. Darwin also mentions Hume in his Notebook M, and Hume's influence on Darwin has been perceived in the arguments on the origins of morality in *Descent* (Lewens 2007, 162) and in the similarities between the topics covered by Hume in his *Dialogues* and Darwin's own discussion of religion in the *Recollections* (Keynes 2001, 278–79; see also Huntley 1972).

MARRIAGE

By the time of his marriage in 1839 Darwin's religious doubts were strong enough that his father advised him to keep them from his wife, advice Darwin evidently did not follow (*Recollections*, 397). Charles's wife, Emma Wedgwood, had also grown up in a Unitarian household but, like Charles, was confirmed in the local Anglican church. According to the later reminiscences of her children, Emma remained a Unitarian: she read the Bible with them, led family prayers, and taught them a Unitarian creed, although all of the children were baptized into the Church of England and attended services at the local parish church (Keynes 2001, 113–21). Emma is sometimes credited with a hard-line Christianity at odds with her husband's views, but she had an avowed distaste for religious fervor and was pragmatic about Sunday observance. Furthermore, her own beliefs do not appear to have been static, her faith becoming less "vivid" in old age (Litchfield 1904, 2:190).

The little remaining evidence suggests that it was in the spirit of Unitarianism that Emma approached both Charles's religious doubts and the consequences of the theory of natural selection for religious belief as a whole. From a series of letters written around the time of their marriage, it is clear that Charles had made his doubts known to her: she suspected that his brother's doubts had informed his own views and differed from Charles in her interpretation of the Christian view of suffering, believing that it could be seen as morally uplifting. His "honest & conscientious doubts," however, she thought could not be a sin, and she expressed the hope that although they did not agree "upon all points of religion," they might nevertheless sympathize a good deal in their "feelings" on the subject (Darwin 1985–, 2:169, letter from Emma Wedgwood, [23 January 1839]). She concluded a letter written after their marriage by expressing her unhappiness at the thought that they might not "belong to each other forever," but her annotations to the section on "Religious Belief" in Darwin's *Recollections* make it clear that she did not accept the doctrine of eternal damnation as a necessary consequence of Christian dogma. After Darwin's death she wrote: "Nothing can be said too severe upon the doctrine of everlasting punishment for disbelief – – but few would call that 'Christianity'" (*Recollections*, notes to 392).

Darwin, with many skeptics among his own family and friends, came to view this doctrine as "damnable." "I can hardly indeed see," he wrote, "how anyone ought to wish Christianity to be true: for if so, the plain language of the text seems to show that the men who do not believe … will be everlastingly punished" (*Recollections*, 392). It has been suggested on the basis of this passage and other circumstantial evidence that the death of Darwin's daughter Annie in 1851 was the final impetus for Darwin's loss of belief in Christianity (J. Moore 1989). However, there is no direct textual support for this from Darwin's own writings or from anything written by his contemporaries. While there is no doubt that Annie's death represented the lowest of the low points in Darwin's life, it seems likely that Darwin's rejection of Christianity was already complete by the time of Annie's death (van Wyhe and Pallen, 2012).

DARWIN AND RELIGION IN LATER LIFE

Although Darwin had apparently ceased to have any belief in the personal God of Christian doctrine by the early 1840s, he remained interested in the arguments for and against religious belief. He recognized critical distinctions between theism, Christianity, and the teachings of the Anglican Church: a lack of belief in the doctrine of the Anglican Church did not entail disbelief in Christianity; a loss of belief in Christianity as divine revelation did not necessarily entail denial of the possible existence of a designer deity or First Cause.

Among Darwin's letters, the single body of correspondence with the most sustained discussion of religion is that with the Harvard professor of botany and devout Presbyterian Asa Gray. Gray was instrumental in the publication of a U.S. edition of *Origin* and wrote a series of reviews that were vital to the promotion of Darwin's ideas in North America. Darwin seized on Gray's reviews as a defense against charges of irreligion and paid to have them republished as pamphlets in Britain. Their correspondence around the time of the publication of *Origin* reveals the subtlety of Darwin's thinking. Acute awareness of the cruelty of nature made it impossible for him to agree with Gray that, in its detail, the world provided evidence of design by a beneficent and omnipotent God, yet he denied that this equated to atheism. A sense of wonder at the complexity of the universe "& especially the nature of man," inclined him to look at everything as "resulting from designed laws, with the details, whether good or bad, left to the working out of what we may call chance." The laws themselves, he conceded, might have been "expressly designed by an omniscient Creator, who foresaw every future event & consequence." "But," he concluded, "the more I think the more bewildered I become" (Darwin 1985–, 8:224, letter to Asa Gray, 22 May [1860]).

In 1871 he wrote to Francis Abbot: "My views are far from clear.... I can never make up my mind how far an inward conviction that there must be some Creator or First Cause is really trustworthy evidence" (Darwin 1985–, 19:551, letter to F. E. Abbot, 6 September 1871). He corresponded sympathetically in the mid-1870s with Charles Voysey about Voysey's establishment of a magazine to promote theism and subscribed to the Toledo, Ohio, *Index*, a freethinking periodical established by Abbot. Perhaps the best summary of his views is that from a letter to John Fordyce, an author of works on skepticism, written in 1879. "It seems to me absurd to doubt that a man may be an ardent Theist & an evolutionist.... What my own views may be is a question of no consequence to any one except myself. But as you ask, I may state that my judgment often fluctuates. Moreover whether a man deserves to be called a theist depends on the definition of the term." "In my most extreme fluctuations," he concluded, "I have never been an atheist in the sense of denying the existence of a God" and, he continued, "I think that generally (& more and more so as I grow older) but not always, that an agnostic would be the most correct description of my state of mind" (DCP, 12041, letter to John Fordyce, 7 May 1879. Darwin was an enthusiastic early adopter of the term "agnostic," which had been coined by his close friend and supporter Thomas Henry Huxley in the early 1870s and sought to draw a distinction between atheists who positively denied the existence of any deity and those who held, as Darwin clearly did, that anything beyond the material is simply unknowable.

Much of Darwin's discussion of religion in his *Recollections*, dating from the late 1870s onward, revisits the arguments for and against belief in a benevolent designer deity that had characterized his discussions with Asa Gray. He states that the argument from apparent design in nature fails thanks to the discovery of natural selection, which also provides an explanation for his perception that on balance there is more that is good than bad in the world. Pain and suffering are better explained by the workings of natural selection than by the argument that suffering is morally strengthening: although humans might conceivably benefit in this way, Darwin protested, the suffering experienced on such a large scale by animals could have no such purpose. He dismisses the argument for the existence of God from inner convictions because "all men of all races" have not had "the same inward conviction of the existence of one God." The widespread and almost instinctual belief in an immortal soul he explains as a comforting defense mechanism when faced with the inevitable extinction not only of the self but eventually of the species and the world. The grandeur of that world, however, sometimes compels him "to look to a First Cause having an intelligent mind in some degree analogous to that of man," for which he says he deserves to be called a "Theist." Darwin does not define "theism" in this context; his use of the term "First Cause" suggests what might more usually be defined as "deism," that is belief in a nonpersonal supreme being with no continuing active presence in the universe. However, a few lines later, he argues that finite, evolved minds fail when contemplating the infinite and therefore concludes, "The mystery of the beginning of all things is insoluble by us; and I for one must be content to remain an Agnostic."

Darwin's eventual unequivocal rejection of Christianity is increasingly clear in statements made toward the end of his life. In 1880 he responded to one inquiry: "I am sorry to have to inform you that I do not believe in the Bible as a divine revelation, & therefore not in Jesus Christ as the son of God" (DCP, 12851, letter to F. A. McDermott, 24 February 1880).

SCIENCE, NATURAL THEOLOGY, AND RELIGION

The academic consensus in early nineteenth-century Britain was that the study of nature and the study of religion were in harmony. According to the tenets of "natural theology," the study of one illuminated the study of the other, and the wonder and perfection of the natural world were evidence of divine design. One of the most influential texts promoting this view was William Paley's *Natural Theology, or Evidences of the Existence and Attributes of the Deity* (1802). As a student, Darwin was convinced by Paley's arguments and in later life he continued to admire Paley's style. In fact, Paley's influence can be traced in both the language and structure of *On the Origin of Species*, in the style of evidence (detailed observations of the natural world) and argument (rigorous relentless logic), but both put to very different ends by Darwin. "The old argument from design in nature, as given by Paley," Darwin wrote, "which formerly seemed to me so conclusive, fails, now that the law of natural selection has been discovered" (*Recollections*, 50).

The closing words of the first edition (1859) of *Origin* – "There is grandeur in this view of life, with its several powers, having been originally breathed into a few forms or into one" – echo not only Paley's language but the "divine breath of life" in Genesis, chapter 2, and Darwin reinforced the biblical connotation by altering the sentence in all subsequent editions to read: "originally breathed by the Creator." There is another occurrence of the phrase in the second edition in a sentence that was removed from later editions: "Therefore I should infer from analogy that probably all the organic beings which have ever lived on this earth have descended from some one primordial form, into which life was first breathed by the Creator" (1860a, 2:484). Privately Darwin explained the use of the terms "Creator" and "creation" to mean "'appeared' by some wholly unknown process" and regretted that he had "truckled to public opinion" in his use of a "Pentateuchal term" (Darwin 1985–, 11:278, letter to J. D. Hooker, [29 March 1863]). Neither, however, did Darwin intend to suggest that natural selection ruled out a deity ultimately responsible for natural laws (Lewens 2007, 97).

Although there are allusions to Darwin's views on the argument from design in both *Origin* and *Variation*, Darwin did not directly address the implications of his theories for religious belief in print until 1871, when he published *The Descent of Man and Selection in Relation to Sex*. *Descent* and the companion volume *Expression of the Emotions in Man and Animals*, published the following year, were designed to demonstrate that humans are part of a continuum with all living things and address the grounds on which separate or special creation of human beings had been claimed, including a capacity for religious belief.

Darwin denied any universal tendency in humans to religious belief but, in doing so, was careful to emphasize its positive contribution: although "belief in God is not universal in Man," Darwin (1871a, 1, 255) nevertheless regarded it as "ennobling." He drew a firm distinction between belief and morality. Morality he argued developed naturally from the social instincts that humans share with other animals; social instincts lead to cooperation and, by gradual steps, to sympathy first for immediate family and finally embracing all living creatures: "[T]he social instincts ... naturally lead to the golden rule," that is, to the New Testament injunction to behave toward others as you would wish them to behave toward you (255). All creatures could be capable of developing notions of right and wrong, but, as there are different kinds of social organization, there can be no absolute standard of morality.

FIGURE 25.2. St. George Jackson Mivart (1827–1900), student of Huxley and Roman Catholic convert, managed to alienate both the Darwinians and his fellow Catholics. Permission: Wellcome

Religious belief, Darwin argued, had developed from the superstitious postulation of the existence of spirits in order to explain the unknown. He illustrated this with a story about his dog, which, when a parasol was blown around the garden by the wind, reacted as if there were a human intruder; the dog must, he thought, have "reasoned to himself in a rapid and unconscious manner, that movement without any apparent cause indicated the presence of some strange living agent." He analyzed religious devotion as a complex feeling consisting of "love, complete submission to an exalted and mysterious superior, a strong sense of dependence, fear, reverence, gratitude, hope for the future," reminiscent of the love of a dog for its master (246).

Miracles, used by Paley and others as evidence for divine intervention, Darwin by the 1870s characterized as phenomena not correctly observed, or not yet explained by science; in a lengthy addition to the sixth edition of *Origin* refuting St. George Jackson Mivart's arguments in favor of special creation of species, Darwin commented that to admit such arguments is to "enter into the realms of miracle, and to leave those of science" (1872a, 204; see also *Recollections*, 392) (Fig. 25.2).

Darwin was careful to point out that establishing that there was not a universal belief in God is "wholly distinct" from the question "whether there exists a Creator or Ruler of the universe."

This, he said, without commenting on his own position, "has been answered in the affirmative by the highest intellects that have ever lived." However, in the second edition of *Descent*, Darwin moderated this to "some" of the highest intellects.

DARWIN, THE CHURCH, AND CHURCHMEN

Clergymen were a significant group in both Darwin's social and scientific circles. As a Cambridge undergraduate, Darwin was required to attend Anglican services in his college chapel regularly. His two most influential university teachers, John Stevens Henslow, professor of mineralogy and botany, and Adam Sedgwick, professor of geology, were both Anglican priests, as Cambridge University fellows at the time were expected to be. Although Sedgwick later campaigned for Cambridge to admit non-Anglicans, both he and Henslow were devout in their faith. Despite being shaken by Sedgwick's vehement opposition to *Origin*, Darwin remained friends with both for the rest of their lives. Within Darwin's immediate family, his cousin, contemporary at Cambridge, and lifelong friend and correspondent, William Darwin Fox, took holy orders while Darwin was on the *Beagle* and settled down as rector of Delamere, Cheshire. Perhaps the best known of Darwin's clerical supporters after publication of *Origin* was Charles Kingsley, the author of *The Water Babies* (Fig. 25.3).

Clergymen formed one of the largest professional groups among Darwin's correspondents – around two hundred of a network of nearly two thousand. The study of natural history was a common practice among the clergy, and most who were writing to Darwin were doing so to give him information. A typical example was Octavius Pickard Cambridge, vicar of Bloxworth and the leading expert on spiders, who in 1874 responded to an inquiry from Darwin to give him information on the proportion of the sexes in spider populations (DCP, 9299, letter from O. P. Cambridge, 17 February 1874).

Darwin lived for the last forty years of his life in the village of Down (now Downe) in Kent. (On the Darwins and Downe parish, see Browne 1995, 2:452–56, and P. White 2010.) Although there is no evidence that he ever attended services in either the parish church or any of the local dissenting chapels, he took an active role in parish affairs and regarded the local Anglican Church as an essential part of the social fabric, on one occasion contributing twenty-five pounds to the church restoration fund. After an uncertain beginning, the Darwins became close friends with the local high-church Anglican clergyman, John Brodie Innes. Darwin gave Innes a presentation copy of the first edition of *Origin*, and although Innes never accepted natural selection, he helped Darwin with research and defended Darwin's integrity to his clerical colleagues. In return, Darwin helped maintain the integrity of the church: when two curates scandalized the village with sexual and financial improprieties, Darwin informed Innes of the problems and took over management of the accounts, urging Innes to find a better replacement, fearing that "the Church will be lowered in the estimation of the whole neighbourhood" (Darwin 1985–, 16:871, letter to J. B. Innes, 1 December 1868).

FIGURE 25.3. The Reverend Charles Kingsley (1819–75), author of *The Water Babies*, was an enthusiastic Darwinian. From Mrs. Kingsley, *Life and Letters of Charles Kingsley* (London: Kegan Paul, 1877)

Not all relations with clergymen were cordial. George Ffinden, appointed in 1871 and the last vicar of Down during Darwin's lifetime, was openly disapproving of Darwin and alienated not only Charles but Emma, who abandoned Down church for the neighboring parish of Keston. Privately, Darwin expressed dislike of the higher clergy, and among his most vocal antagonists were leading churchmen such as Samuel Wilberforce, bishop of Oxford, who famously spoke out against the implications of Darwin's theories for human ancestry at the 1860 meeting of the British Association for the Advancement of Science. Also privately, Darwin supported more liberal elements within the established church: he contributed anonymously to a defense fund for John William Colenso, bishop of Natal, who was threatened with removal from office for his published doubts about the historical authenticity of the Pentateuch.

Darwin was never, during his lifetime or afterward, subject to any formal denunciation by the Anglican Church. On the contrary: the church moved swiftly to honor him on his death, overriding his own wishes to be buried in Down. Instead, he was buried with full honors in Westminster Abbey, where nine pallbearers, including the Reverend Frederic Farrar, canon of Westminster, carried Darwin to his final resting place in one of the most prestigious churches in England a few yards from the grave of Isaac Newton.

ESSAY 26

Darwinism in Britain

Peter Bowler

CHARTING THE COURSE of evolutionism in Britain can be seen as an exercise in trying to understand the emergence of and response to what became known as "Darwinism." Thomas Henry Huxley coined the term and tried to control how it was used – he was, of course, known as "Darwin's bulldog" because of his aggressive support for the theory. But Darwinism certainly didn't entail complete acceptance of the program outlined in the *Origin of Species*, because even Huxley would not have been a Darwinian on those terms (Fig. 26.1). To understand what was going on in the context of the time, we must be aware that the meaning of the term "Darwinism" has also changed over time. In the modern world it usually refers to the theory of evolution by natural selection. But in the late nineteenth century many evolutionists who did not believe that natural selection was the main mechanism of evolution called themselves "Darwinians." The true value of the selection theory was recognized only in the twentieth century, so the contemporary reception of Darwin's theory has to be understood in terms of a much broader debate over what evolutionism entailed.

To many ordinary people, Darwin simply became a symbol or figurehead for a generalized evolutionary philosophy, probably entailing notions of progress and the struggle for existence. In his later life, his face became familiar to all thanks to the publication of portraits and caricatures – often emphasizing certain apelike aspects of his features – published in popular magazines (Browne 2002). Even at this level there were ambiguities, though. Our vision of the initial debate over the *Origin of Species* has been shaped by the negative reaction of conservative religious thinkers and by Huxley's strident anticlericalism, both fueling the claim that evolution and Christianity are incompatible. But because evolution was popularly supposed to entail progress, it was accepted by many liberal clergymen, whose views were thus not so far removed from those of the less aggressive secularists. By the 1870s opponents of materialism such as Samuel Butler began to sideline the liberal Christian approach, accelerating the process by which the more naturalistic implications of Darwinism came to the fore and belatedly endorsing the definition that Huxley had hoped to establish (Fig. 26.2).

Reactions within the scientific community were similarly complex and can be understood only by recognizing that there were alternatives to natural selection

FIGURE 26.1. Thomas Henry Huxley was a brilliant teacher and – a great virtue in pre–PowerPoint days – a gifted blackboard artist. Permission: Wellcome

FIGURE 26.2. Samuel Butler (1835–1902) the novelist (*Erewhon*, *The Way of All Flesh*) became a bitter critic of Darwinism and has inspired later critics from George Bernard Shaw to Karl Popper. Nineteenth-century photograph

available to late nineteenth-century evolutionists, alternatives that seem outdated today but which were perfectly plausible at the time (Bowler 1983, 1988). Darwin himself accepted a limited role for one of these mechanisms, the Lamarckian theory of the inheritance of acquired characteristics. Later in the century the rival theories were promoted by outright opponents of the selection theory during what became known as the "eclipse of Darwinism" – although none of the opponents doubted the basic idea of evolution. The rediscovery of Gregor Mendel's laws of heredity in 1900 led to the creation of genetics and ultimately to a synthesis with the selection theory, although several of the early geneticists were themselves opponents of Darwinism.

The assumption that natural selection was the main focus of debate among evolutionists in the late nineteenth century is one of many myths that have grown up around the history of the subject, fueled by our modern recognition of the true power of the selection mechanism. This emphasis on the debate over natural selection has marginalized not only rival theories but also evolutionary debates that did not center on the actual mechanism of change. Recent history has challenged this and other distortions of our perception of the period, which were created to serve the interests of those involved – either at the time or later. The anatomist Richard Owen is frequently dismissed as an opponent of evolutionism because he wrote a critical review of the *Origin*, although he was in fact a supporter of non-Darwinian evolutionism (Rupke 1994). Similarly, it is widely believed that Huxley trounced Bishop Samuel Wilberforce in the famous 1860 debate over Darwin's book at the British Association in Oxford. But later studies have shown that this impression of the debate was created by Huxley's followers to support the claim that naturalistic science was triumphing over religion (e.g., Jensen, 1988). We must be ever alert to probe the evidence for misconceptions accepted as fact.

LEVELS OF DEBATE

There were several interacting levels of debate over evolutionism. The most obvious division is that between science and religion, although these are not black and white alternatives. Many scientists were still deeply religious (if not conventional Christians), and many religious believers were liberal enough to look for some compromise with the latest developments in science (J. R. Moore 1979; Bowler 2007). What is now called Young Earth Creationism reemerged in the early twentieth century, so Darwin did not have to face significant opposition from people who thought the world is only a few thousand years old (Lord Kelvin argued for a much shorter time scale than the Darwinians needed, but even he conceded 100 million years). The Catholic Church was probably a more powerful source of opposition, although it had only limited influence in England. Too close a focus on the current "conflict" between evolutionism and religion obscures the relative ease with which liberal religious thinkers in the late nineteenth century came to terms with the general idea of evolution – although they certainly found it much harder to accept the theory of natural selection.

On the whole, religious believers tended to be conservative in their social opinions, but there was a whole range of liberal and radical positions pushing for a change to the status quo. Even the Church of England, traditionally a bastion of the aristocratic social order, had a liberal wing open to the latest scientific advances. Conservative religious thinkers favored a static social order within a static, divinely ordered universe. This position had already been challenged in Robert Chambers's *Vestiges of the Natural History of Creation* in 1844 (Secord 2000). Chambers showed how progressive evolution in the natural world could be seen as a model for social progress. His message appealed to the rising middle classes that spearheaded the Industrial Revolution. By the time Darwin published, Herbert Spencer had begun to argue that the motor of social progress was the cumulative effect of individuals seeking to improve themselves in a competitive environment (Francis 2007). Darwin's theory did not map directly onto this progressionist model, but he concluded the *Origin* with passages that made the link seem plausible. Most ordinary people assumed that evolution entailed progress, at least in the long run. The fact that Spencer coined the term "survival of the fittest" encouraged the association between Darwinism and the ideology of progress, although Spencer also invoked the Lamarckian mechanism.

A parallel transformation was underway in the relationship between the scientific community and the rest of society. Darwin was a gentleman-amateur – he had no formal scientific training, and he did not earn his living from his scientific work. Some of his supporters came from the same background, but the scientific community was increasingly dominated by men (and they were all men) who were professionals in the modern sense of the term. Both Huxley and Owen were dependent on their positions in museums and colleges, and most of their younger followers were in the same position. Owen had forged a connection with the religious and social establishment and was thus anxious to avoid confrontation. Huxley was younger and more radical – he wanted scientists to replace clergymen as sources of authority in a modernized economy. He thus relished confrontation with the church and welcomed Darwinism as a weapon in the fight against the subordination of science to religion. This motive explains why he welcomed the *Origin*, even though he did not think it provided an adequate explanation of how evolution worked. It was just too good a rhetorical tool in his debate with the church (Fichman 1985).

Despite his enthusiastic support for Darwin, Huxley made no use of the theory of evolution in his scientific work until the late 1860s (Bartholomew 1975; Di Gregorio 1984). Here we must note another difference between his background and Darwin's. Huxley was a morphologist, interested in the form – that is, the internal structure – of animals. His specialization was comparative anatomy, linked to embryology and later to the reconstruction of fossil specimens. Many of the scientists who took up the cause of evolutionism were also morphologists, which is why they saw the main goal of their work as being the reconstruction of the history of life on earth from anatomic and embryological clues as well as from the fossil record (Bowler 1996). Darwin had only a limited engagement with this project, because to a large extent it proceeded without the need for detailed discussion of the evolutionary mechanism. Biogeographers such as Alfred Russel Wallace and the botanist J. D. Hooker also wanted to reconstruct the history of life by tracing the migrations of animals and plants around the globe. Here again there was little need to worry about the actual mechanism of evolution, although biogeographers were more likely to follow Darwin in believing that adaptation to new environments was a key factor. The technical biological literature of the period abounds with analyses of the key steps in the development of life through geological time. But discussions of the evolutionary mechanism, including the many challenges to the plausibility of natural selection, were conducted in the pages of general periodicals and books (Ellegård 1958). There was no recognizable discipline of evolutionary biology in the modern sense of the term, and in that sense Darwin failed to convert the scientific community to the program he sketched out in the *Origin* (Fig. 26.3).

REACTIONS TO THE *ORIGIN*

It is often assumed that Darwin's book hit the public like a bolt from the blue. But, in fact, Chambers's *Vestiges* had accustomed his many readers to the general idea of progressive development up to and including the human race, which is why Darwin's attempt to avoid raising the issue of human origins failed. Religious conservatives were certainly aroused, as they had been by Chambers's book, and there is a story that Darwin was pointed out as the most dangerous man in England by a clergyman. When Bishop Samuel Wilberforce rose at the 1860 meeting of the British Association to challenge Darwin's arguments, he did so as the representative

FIGURE 26.3. Joseph Dalton Hooker (1817–1911), Darwin's closest scientific friend. From L. Huxley, *Life and Letters of Joseph Dalton Huxley* (London: John Murray, 1918)

FIGURE 26.4. As always, Benjamin Disraeli (1804–81), prime minister, leader of the Conservative Party, and great favorite of Queen Victoria, sensed the opportunity to affirm his allegiance to the established powers and at the same time to make a good joke, as is shown in this famous cartoon from *Punch* by John Tenniel (best known for his illustrations of the *Alice* books by Lewis Carroll). Permission: Wellcome

of conservative Anglicans who saw that evolutionism threatened the unique status of humanity and that natural selection threatened the belief that the world has been designed by a wise and caring God (Fig. 26.4).

Huxley's supporters subsequently created the impression that his response demolished the bishop's arguments, although modern studies of accounts written by those who were there do not support this version of the events. Wilberforce had been coached by the anatomist Richard Owen, who was not averse to evolutionism but who recognized the many scientific arguments that could be raised against natural selection. Owen's position also reminds us that not all scientists shared the radical social opinions of Huxley's faction. Many retained some form of religious belief and were thus as anxious as the clergy to find reasons for doubting the validity of natural selection – although this does not mean that their technical objections were invalid by the standards of the time. There was no clash between science and religion but rather a series of skirmishes between conservative, liberal, and radical camps in both the scientific and religious communities.

The interpenetration of scientific and broader objections points us toward deep divisions in worldview, which could play out at many levels. Those who objected to natural selection focused on what they saw as the theory's tendency to undermine any sense of the world as a harmoniously ordered system. This was obviously a position with religious implications, but many scientists too hoped to find an underlying structure in the world and feared that Darwin's reliance on "random" – that is, undirected – variation as the raw material of selection meant that the whole evolution of life would be open-ended. It was hence not shaped by any coherent plan and had no meaningful goal. As the astronomer and philosopher of science Sir John Herschel said, natural selection was the "law of higgledy-piggledy." Many of Darwin's efforts to win over members of the scientific community were devoted to showing that selection by the environment could direct evolution along adaptive channels – but his efforts left many believing that selection itself was some kind of purposeful force with a long-range goal. In the end he was able to convince many that evolution did occur but not that natural selection was its motive force (Ellegård 1958; Hull 1973).

Darwin was largely successful in convincing naturalists that evolution was best depicted as a branching tree rather than a ladder ascending to a single goal. It is easy for us today to forget just how new the idea of divergent evolution was at the time. Darwin was one of the first to appreciate that this model allowed a much better understanding of how species are related, and the *Origin* marshaled a wealth of arguments from taxonomy, comparative anatomy and embryology, and paleontology to illustrate this point. Wallace, Hooker, and others contributed more evidence from biogeography to support the claim that evolution made sense of the relationship and distribution of species if one assumed that populations derived from a common ancestor became separated and then diverged away from each other. Owen had already demonstrated divergence within classes from the fossil record. As this evidence was assessed in the course of the 1860s, most naturalists came to accept the concept of branching evolution, although not all were convinced that adaptation to different environments (let alone natural selection) would explain the whole structure of the tree of life.

To the extent that adaptation was accepted as an evolutionary process, there was an alternative to natural selection readily available. Even before Darwin published, Herbert Spencer had begun to argue that the so-called Lamarckian process of the inheritance of acquired characteristics could explain adaptive evolution. In this process, named after the French biologist J. B. Lamarck, animals exposed to a new environment could modify their bodily structure by adopting new habits, and if the resulting modifications were inherited (which many at the time thought possible), they would accumulate to give an adaptive evolutionary trend. Lamarckism in effect resolved the difficulty pinpointed in Herschel's description of natural selection as "higgledy-piggledy" – it provided a way of directing variation along useful channels. Instead of depending on random or undirected variations, most of which had to be eliminated, the whole species would vary in the same, purposeful direction. Spencer's vision appealed to many precisely because it avoided the implication that evolution was a purposeless process of trial and error. And because he assumed that the struggle for existence encouraged individuals to adopt more effective behavioral strategies, his Lamarckism was easily absorbed into Darwinism – especially as Darwin himself allowed a minor role for the Lamarckian effect.

Clergymen too were encouraged to adopt this version of Darwinism, for all that Spencer himself was a secularist. Adaptation was identified with progress, and the assumption that nature is inherently progressive seemed compatible with the liberal Christian assumption that God's purpose is being worked out in nature. Charles Kingsley's *The Water Babies*, originally published in 1862, is frequently held up as evidence of how a liberal Anglican clergyman could lend support to Darwinism. Kingsley (1889, 273) certainly saw the Creator acting through law rather than miracle: Tom, the water baby, is told that Mother Nature does not need to act directly to create new species – she "makes them make themselves." But the message the book conveys is that effort and initiative are required to succeed in life and assure progress to a higher state. The book certainly promoted an ideology of struggle, but it is struggle as the spur to self-improvement, not as the agent of selection. Kingsley's vision, taken up by a host of popular science writers later in the century, was really a synthesis of Spencer's Lamarckism and muscular Christianity.

Many naturalists, Huxley included, were not convinced that adaptation was the sole key to evolution. This attitude fueled some opposition to Darwinism, but for Huxley the key point was the link with a naturalistic worldview: whether variation was directed or undirected, its causes were purely natural. The opponents wanted to preserve a role for God's designing hand in nature by arguing that evolution was driven by inbuilt trends. Owen proposed his theory of "derivation" in 1868 based on the idea that evolution was the unfolding of a divine plan through law-bound processes rather than a sequence of miracles. His disciple, the Catholic anatomist St. George Jackson Mivart, turned against the Darwinian program and developed a whole series of arguments against the model of divergent, adaptive evolution, presented in his *Genesis of Species* of 1871. Curiously, Huxley had been on good terms with Mivart at the start of the latter's career, but when it became clear that Mivart's antiselectionist arguments were being used to suggest a divine purpose built into evolution, Huxley had him ostracized from the Darwinian camp (J. W. Gruber 1960). This episode illustrates that Darwinism (even in the loose form linked to Spencer's philosophy) was becoming identified with scientific naturalism. It was not so much adherence to the selection theory that defined a Darwinian in Huxley's eyes as adherence to naturalism. To an increasingly vocal group of critics including Mivart and eventually the novelist Samuel Butler, Darwinism implied materialism. This paved the way for the emergence of anti-Darwinian ideas explicitly linked to efforts intended to retain elements of the old teleological worldview. The fact that liberal clergy such as Kingsley were able to identify with a form of Spencerianism was deliberately ignored by these critics.

When Huxley coined the term "Darwinism," he no doubt hoped to control its meaning. But, in fact, the response to Darwin's theory had to be negotiated at many levels within both the scientific community and the general public. By the 1870s, Darwin had triumphed in the sense that almost everyone accepted the general idea of evolution, and for many ordinary people Darwinism meant little more than evolutionism with Darwin as its figurehead. Perhaps it would be understood in terms of the synthesis with Spencer's ideology of progress through struggle. In the scientific community, however, what counted as Darwinism reflected different positions on how much of the program sketched out in the *Origin* one accepted, along with one's professional interests and loyalties. Hardly anyone thought that natural selection was the sole agent of evolution, certainly neither Darwin nor Huxley. Most accepted the outline of Darwin's model of divergent adaptive evolution, although they suspected that local adaptation could not explain all of the major steps in the emergence of wholly new types. Huxley and the exponents of scientific naturalism were relatively successful in identifying Darwinism with that program, although the efforts of liberal religious thinkers such as Kingsley counted against this move. Those who openly challenged that program increasingly began to see themselves as opponents of Darwinism.

In one respect, though, the scientific naturalists succeeded in making it less acceptable for scientists to appeal openly to divine agency as an explanatory tool. In the 1860s and early 1870s, it was still possible for conservative thinkers such as Owen and Mivart to imply that the pattern unfolding in the history of life was somehow implanted into the laws of evolution by the Creator. In the later decades of the century, even the anti-Darwinians avoided this implication. The Creator's powers were transferred into nature itself, with the life force being portrayed as a creative force in its own right.

HUMAN ORIGINS

If the question of design troubled religious thinkers, a second area of concern was the origin and status of humanity. Darwin tried to head off discussion of this topic by virtually ignoring it in the *Origin*, but thanks to earlier debates, everyone already knew that one of the most contentious implications of evolutionism was the animal ancestry of humankind. Christians were used to thinking of humans as distinct from the "brutes that perish" thanks to their possession of an immortal soul and moral awareness. To accept that we had evolved from animals, this rigid distinction had to be abandoned. The higher mental and moral faculties would have to be derived from the lower mentality of animals. Far from believing that we had fallen from an original state of grace, we would have to see ourselves as the as yet imperfect end products of a progressive development rising through the whole animal kingdom. Even some of Darwin's supporters, including Charles Lyell and Alfred Russel Wallace, found this hard to accept (Greene 1959; Turner 1974).

The popular assumption was that we had evolved from the great apes, a particularly frightening prospect given that new discoveries in Africa were sensationalized to present the gorilla as a ferocious beast. This impression served to reinforce fears that if evolution worked through a struggle for existence, there would be little room for any but the most ruthless instincts in humans or animals. The link to the apes featured almost immediately in a highly publicized spat between Huxley and Owen over the closeness of the anatomical relationship between humans and apes. Owen argued for the traditional view that there were significant differences, while Huxley's *Man's Place in Nature* of 1863 emphasized the close relationship between all the primates, humans included. Huxley is popularly supposed to have won the debate, although the true picture is rather more complex (Rupke 1994). Later accounts, especially translations of Ernst Haeckel's works, stressed how an apelike ancestor (but not one of the living apes) could have evolved the various distinctive human characters. The question of anatomical relationship was in one sense secondary – what concerned people was whether the higher mental powers of humans could have evolved naturally, and what this would entail for traditional values.

There was, in fact, a framework already falling into place for tackling these issues. Spencer's *Principles of Psychology* of 1855 had invoked the Lamarckian effect to explain how our mental faculties and instincts could have been shaped by generations of our ancestors learning to cope with a more complex social environment. Progress toward higher levels of intelligence and sociability was an inevitable consequence of evolution. This message was driven home by a revolution in the understanding of human prehistory, which took place simultaneously but independently of the Darwinian revolution in biology. Geologists had maintained that, whatever the age of the earth, humans had appeared only in the last few thousand years, just as Genesis implied. But around 1860 they suddenly began to take seriously the evidence of stone tools being found in ancient deposits alongside the remains of extinct animals (Van Riper 1993). This evidence was summed up in Charles Lyell's *Antiquity of Man* in 1863. Darwin's neighbor Sir John Lubbock coined the terms "paleolithic" and "neolithic" to denote the old and new stone ages, implying a progressive increase in technological sophistication and hence, by implication, in culture. Even without fossil evidence, it was easy to imagine our distant Stone Age ancestors as being more apelike than their modern descendants. Physical anthropologists routinely depicted the "lower" human races as apelike, allowing them to be seen as relics of the intermediate phase in human evolution. The new worldview was based firmly on the idea of progress, and it was this which made it acceptable to Darwin's contemporaries.

Darwin himself accepted much of this model, although his *Descent of Man* rocked the boat by arguing that natural and sexual selection played major roles in the emergence of human characteristics. Darwin recognized that it was not enough to suggest that superior intelligence was a survival factor, because this did not account for the difference between humans and apes. Perhaps the adoption of an upright posture had preceded the increase in human intelligence – a perceptive insight that was ignored by most of his contemporaries because it did not fit the model in which mental progress was the driving force of human evolution. Darwin did, however, follow Spencer in appealing to Lamarckism to explain how we evolved the social instincts that are the foundation of our moral sense, although he also invoked a form of group selection based on the assumption that the most cooperative tribes would eliminate their rivals. For most ordinary readers, it was the model of self-improvement highlighted by Spencer and Kingsley that made the prospect of an animal ancestry bearable. Struggle promoted thrift, initiative, and industry – all aspects of the Protestant work ethic – and built them into the human constitution. Much of what was later criticized as "social Darwinism" derived from this application of the ideology of self-improvement to the question of human origins.

THE DEBATE OVER NATURAL SELECTION

Historians have focused a great deal of attention on the debates over natural selection because from a modern perspective it seems important to understand how the objections were eventually overcome to provide the basis for the modern synthesis of Darwinism and genetics (Hull 1973). It is worth remembering, though, that some of the issues described here also generated suspicions about the wider Darwinian program. To those morphologists who thought that new structures might be generated by forces internal to the organism, the true source of evolution was the directed variations that produced the new characters, not selection. From this perspective it might be possible for two species independently to evolve similar structures (parallel evolution), and in this case the relationship would not indicate common descent. Furthermore, the assumption that variation trends could produce structures without selection implied rejection of Darwin's claim that the struggle for existence was a relentless force that would ensure the survival of only those structures which were adaptive. In what Julian Huxley later called the "eclipse of Darwinism" around 1900, theories of evolution by orthogenesis (directed variation) and saltations (sudden, abrupt variations) brought the anti-Darwinian perspective pioneered by Mivart to center stage (Bowler 1983).

There were also objections raised against the selection theory even by those who accepted a major role for adaptation in evolution. These were often inspired by the religious motivations noted previously, but in the absence of any clear understanding of variation and heredity, they had genuine plausibility at the time. Many found it hard to grasp the idea that selection of "random" variations could have a positive effect, although most accepted that harmful variants would be weeded out. Darwin often complained about how difficult it was to get people to understand his theory, suggesting that it was far from being a natural extension of current thought patterns. He used the analogy between artificial and natural selection as an explanatory tool, but this had several pitfalls. Some readers could not shake off the assumption that nature must be a purposeful agent, just like the human breeder seeking to achieve a goal. The analogy also left the theory open to the charge that breeders had never produced a new species, leaving many (Huxley included) to suggest that major new variants (saltations) would be needed to pass beyond the limit imposed by normal variability. Even this assumption was vulnerable without a clear understanding of heredity, which at the time included assumptions rejected by modern genetics. The critique published by the engineer Fleeming Jenkin in 1867 has been widely noted by historians because it disturbed Darwin himself and seemed to imply the need for a new model of heredity to make the selection theory work. The frequently repeated claim that natural selection cannot work if heredity is a process in which parental characters blend together is an oversimplification, but Jenkin's attack is still noteworthy as an example of the confusion surrounding variation and heredity at the time.

Significantly, Jenkin was a friend of the physicist William Thomson, later Lord Kelvin, whose arguments to limit the age of the earth to around 100 million years were also intended to undermine the plausibility of the selection theory (Burchfield 1975). Many of the non-Darwinian theories suggested in the

FIGURE 26.5. August Weismann (1834–1914) was the most formidable opponent of Lamarckian inheritance. From G. J. Romanes, *An Examination of Weismannism* (Chicago: Open Court, 1899)

late nineteenth century were advanced in part because it was thought that they would explain how evolution could work more rapidly than natural selection alone would allow.

The selection theory did have its defenders, of course, most notably Alfred Russel Wallace, who rejected the Lamarckian theory that even Darwin accepted as a supplement to natural selection. By the 1880s, August Weismann's new ideas on heredity were also widely known for their implication that Lamarckism was untenable (Gayon 1998) (Fig. 26.5). Wallace and Weismann became known as "neo-Darwinians," even more Darwinian (i.e., selectionist) than Darwin himself. Darwin's cousin Francis Galton endorsed Weismann's model of "hard" heredity (in which the characters transmitted cannot be influenced by environmental effects). His "law of ancestral heredity" retained the idea that characters blend together over many generations. He applied the principle of hard heredity to human affairs by calling for a eugenics policy that would deny the "unfit" the right to reproduce.

Galton and the neo-Darwinians were opposed by the "neo-Lamarckians," who wanted a much greater role for the inheritance of acquired characteristics, including both Spencer and critics of scientific naturalism such as Samuel Butler. In the

1890s the neo-Darwinian camp acquired new members who tried to put the selection theory on a firm basis by providing hard evidence of both the range of natural variation in populations and the effects of selection by the local environment. W. F. R. Weldon joined with the statistician Karl Pearson, a disciple of Galton, to study these phenomena in local populations of snails and crabs. They demonstrated the effectiveness of natural selection, although on a scale so small that the anti-Darwinians could dismiss it as trivial. Far from indicating that the selection theory was implausible on the basis of the old notion of blending heredity, Pearson's detailed analysis of the selection effect relied on a modified version of Galton's law.

Weldon and Pearson were thus not involved in the events that generated the new model of heredity that became known as genetics; indeed, Weldon was trying to construct a rival theory when he died in 1906. It was supporters of the saltationist alternative to natural selection who provided the impetus toward the new theory, assuming that if characters were produced as coherent units they must also be transmitted as such. William Bateson, who translated Mendel into English, was a saltationist and a prominent opponent of the idea that evolution was determined by adaptation. Not surprisingly, neo-Darwinians such as Weldon and Pearson were suspicious. Only in the 1920s and 1930s did the new understanding of heredity become synthesized with the selection theory to give the foundations of modern Darwinism – a very different version of the theory from that debated in Darwin's own time.

Darwinism in the United States, 1859–1930

Mark A. Largent

As explosive as was Darwin's theory of evolution by natural selection in Britain, it initially received a cool response from American naturalists. This was partly because it did not engage the middle class in the United States the way it did in Britain, and partly because in the first half of the nineteenth century the United States was an intellectual and scientific backwater. Nonetheless, Darwin's work served an important formative role in the establishment of the scientific enterprise in the United States.

Before and throughout the Darwinian revolution, science in the United States was a profoundly practical endeavor pursued primarily for its economic potential. In its emergence in the eighteenth, development in the nineteenth, and maturation in the twentieth century, American science was intricately bound to the development of new technologies and was justified almost entirely on its ability to generate practical economic, moral, or military benefits, especially when it was funded with public money. Tocqueville, the French political theorist who toured the United States about the same time that Darwin voyaged around South America, posited that Americans took up science "as a matter of business, and the only branch of it which is attended to is such as admits of an immediate practical application" (Tocqueville 2003, 65). American science thus contrasted sharply with the European scientific tradition in which science was generally pursued by wealthy gentlemen and usually for its own sake.

The intensely practical nature of American science and the relative immaturity of the American scientific community also shaped the uses that American naturalists made of Darwin's theory and of evolutionary science generally. From the initial American responses to *Origin*, through the training of the first generation of American naturalists after Darwin's revolutionary work, and well into the twentieth century, Darwinism in the United States was put to explicitly practical uses. Ultimately, when it became a target for critics in the 1920s, Darwinism's practical benefits to science were judged against the challenges it posed to cherished political, cultural, and social values in the United States.

THE EARLY RECEPTION OF DARWIN IN THE UNITED STATES, 1859–1873

Two particularly dominant figures in nineteenth-century natural history, Asa Gray (1810–88) and Louis Agassiz (1807–73), framed the initial reception of Darwin in the United States (Fig. 27.1). After Darwin explained his theory to him, Gray accepted the reality of evolution and believed that Darwin's work explained a great deal of how evolution operated. After the *Origin* was published, Gray served as Darwin's American lieutenant by arranging for favorable reviews and widespread distribution of the book in the United States and by securing for Darwin the book's American royalties. Agassiz, on the other hand, was an ardent opponent of evolutionary theories generally and believed that Darwin's theory was unsupported by an empirical study of the natural world. Both were professors at Harvard University, both stood as competing figures on the validity of Darwin's theory in the United States, and the two educated a great number of nineteenth-century American naturalists.

Gray had trained as a physician, but eventually left medicine for botany and was appointed professor of natural history at Harvard in 1842. At Harvard, Gray amassed an immense book and plant collection as he tried to catalog the nation's botanical resources. He and Darwin had met briefly in 1838 when Gray visited England, but it was not until 1855 that they began what would become a long-running correspondence. For the next quarter century, Darwin and Gray exchanged about three hundred letters, many of them dealing explicitly with the subjects of evolution, design in nature, and religion. While Darwin largely abandoned his earlier religious training and appreciation for the argument from design, Gray steadfastly maintained a theistic viewpoint and attempted to reconcile Darwin's theory of natural selection with Christian theology by portraying evolution as a method by which God altered the natural world.

Gray originally had little interest in transmutationist ideas. However, when he began his comparative botanical studies in the 1850s, he was at a loss to explain the striking similarities that many plants shared. While Darwin developed an appreciation for the possibility of evolutionary explanations by way of biogeography, Gray became an evolutionist through his taxonomic studies. Gray saw in Darwinism a useful application to the complicated task of classifying plants as well as a way to integrate his scientific studies with his religious beliefs. He was, therefore, a Darwinian in only a limited respect. While he believed that evolution was indeed a process found in nature and he accepted that natural selection was an important part of evolutionary change, his ardent belief in a theistic Christian worldview compelled him to insist that design was still a critical component of the natural world. As such, Gray believed that the new variations on which natural selection acted were provided by an omnipotent, omniscient God. "Variation," he wrote in an article in *Atlantic Monthly* in 1860, "has been led along certain beneficial lines" (Gray 1860, 148). Unlike those who cast evolution as a purely materialist or atheistic process, Gray asserted, "Agreeing that plants and animals were produced by Omnipotent fiat does not exclude the idea of natural order and what we call secondary causes" (131). For Gray, Darwin's theory of evolution by natural selection was just such a secondary cause, and in letters to Darwin, Gray explained that he intended to baptize the *Origin* by ridding it of its apparent materialism and portraying evolution as a tool that God employed to craft the world as he wished.

FIGURE 27.1. Swiss-born Louis Agassiz (1807–73) was the leading ichthyologist of his day. Although he was close to Cuvier, the transcendental morphology (*Naturphilosophie*) of his German education was always the significant factor molding his science. In midlife he moved to Cambridge, Massachusetts, becoming a professor at Harvard, where he led the opposition to evolution. From E. C. Agassiz, *Louis Agassiz* (Boston: Houghton Mifflin, 1885)

Gray's advocacy of Darwin's work in the United States was met with staunch opposition by Agassiz, who had immigrated to America in 1847 and joined Gray at Harvard. Born in Switzerland and educated at several universities in Europe, Agassiz had extensive knowledge of botany. He moved to Paris in the early 1830s to work with Alexander von Humboldt (1769–1859) and Georges Cuvier (1769–1832), where he developed expertise in geology and zoology. It was from Cuvier's earlier skirmishes with Lamarck over the question of the permanence of type that Agassiz first developed his sullen resistance to evolution. His most significant works merged zoology with geology in the study of fossilized fish, which led to his assertion that large parts of the earth had once been shrouded in ice during what came to be called an Ice Age. In the fall of 1846, he toured the United States with the intention of investigating its natural history and geography and giving a course of twelve lectures on "The Plan of Creation as Shown in the Animal Kingdom" in Boston. A year later, Harvard appointed him professor of zoology and geology.

Agassiz's rejection of Darwin's theory of evolution by natural selection was based on three specific scientific arguments. First, Agassiz insisted that Darwinism encouraged its adherents to selectively interpret facts from nature, rather than inducing conclusions from them. Already in his earlier *Essay on Classification*, with evolution in mind, Agassiz (1857, 62, n. 8) had written: "I must protest now and forever against the bigotry spreading in some quarters, which would press upon science doctrines not immediately flowing from scientific premises and check its free progress." Second, he believed that, while organisms within a species vary from one to another, the variation exists only within a narrow range. Finally, Agassiz asserted that the fossil record, with which he was so authoritative, did not demonstrate the progressive evolutionary change expected if evolution indeed occurred.

Agassiz's ardent rejection of Darwin's theory, combined with his considerable social and scientific influence, ultimately delayed the widespread acceptance of Darwinism for at least a decade. After Agassiz's death in 1873 and Gray's retirement from Harvard that same year, the emerging generation of American biologists slowly turned away from Agassiz's staunch rejection of transmutationism and began a gradual acceptance of both evolution and of Darwin's theory of evolution by natural selection.

THE FIRST GENERATION OF AMERICAN EVOLUTIONISTS

Among the few American naturalists whose careers bridged the time before *Origin* and through Americans' initial reaction to Darwin's theory was James Dwight Dana (1813–95). Trained at Yale in the 1830s, Dana was responsible for developing much of our early knowledge about volcanism on the Hawaiian Islands. His scientific reputation was rivaled only by his resolute Christian faith, which earned him undying favor from American clergymen. For most of his career, Dana accepted, like Agassiz, that the physical world underwent some form of development but rejected transmutationist views when applied to living things. Throughout the 1860s, Dana continued to lecture his classes at Yale about the errors of transmutation. There was, he wrote in as late as 1870 in his *Manual of Geology*, "no lineal series through creation corresponding to such methods of development" (Dana 1870, 602). Dana's transformation into an evolutionist occurred more than a decade after the publication of the *Origin*, and by 1874 he had adopted an appreciation for evolution that looked very much like Gray's. For both men, their deep Christian faith had powerfully encouraged their adoption of an evolutionary worldview as they came to see evolution as a tool employed by God to enact his will on earth. In his final edition of *Manual of Geology*, Dana (1870, 603–4) wrote that "the evolution of the system of life went forward through the derivation of species from species according to natural methods not yet clearly understood, and few occasions for supernatural intervention." However, he still refused to include humans in the evolutionary change because he believed that

FIGURE 27.2. Alpheus Hyatt (1838–1902), a student of Agassiz, accepted a form of Lamarckian evolution. Nineteenth-century lithograph after photograph

reason and will were direct gifts from God, and in a footnote he added, "There is here no discordance with the Biblical account of creation."

By the middle of 1870s, after Agassiz's death, most American naturalists came to accept evolution as true and Darwin's work as an important, if not the principal, explanation for how it occurs. Ironically, many of the most influential biologists of the first generation of the American Darwinists were themselves trained by Agassiz, including his own son, Alexander Agassiz (1835–1910). Even while they turned away from him, most of his former students credited Agassiz's commitment to a careful study of the natural world as the source of their ultimate conversation to evolution. Among the most significant of these were Alpheus Hyatt (1838–1902), Burt Wilder (1841–1925), David Starr Jordan (1851–1931), Joseph Le Conte (1823–1901), Nathaniel Southgate Shaler (1841–1906), and Alpheus Packard (1839–1905). Their adoption of evolution and rejection of Agassiz's worldview, however, often had not come easily. In his autobiography, Jordan (1922, 1:114) remarked that he "went over to the evolutionists with the grace of a cat the boy 'leads' by its tail across the carpet!" Likewise, Wilder, a Cornell anatomist and zoologist, explained that he adopted an evolutionary worldview only "when forced to decide for himself what should be said to earnest and thoughtful students" (Loewenberg 1933, 692) (Fig. 27.2).

DARWINISM AT THE TURN OF THE TWENTIETH CENTURY

Throughout the last decades of the nineteenth and the first decades of the twentieth century, Darwin's theory remained the most prominent explanation for the widely accepted notion of evolution. Many convinced Darwinists were among the growing numbers of American biologists interested in studying evolutionary theory in the early twentieth century. Most worked in agricultural research settings, including land-grant universities and state agricultural extension stations, where they had firsthand experience in the efficacy of selection, both artificial and natural. It is sometimes said that Darwinism was "in eclipse" in the decades from about 1880 to 1940, a term employed to suggest that Darwin's work was underappreciated or even dismissed by most professional biologists (Largent 2009, 17). This was clearly not the case, as a careful examination of the work of people like Liberty Hyde Bailey (1858–1954), Leon Cole (1877–1948), Maurice Bigelow (1872–1955), and Frank Lillie (1870–1955) all demonstrate that Darwinism was critical, if not central, to their notions about evolution. American Darwinists viewed themselves as much more moderate than the European neo-Darwinists, most notably August Weismann (1834–1914), who believed that a completely random process of mutation produced the variation on which natural selection operated. At the other end of the spectrum, were a handful of neo-Lamarckians, such as the Harvard psychologist William McDougall (1871–1938), who conducted a number of studies in an attempt to demonstrate the inheritance of acquired characteristics.

But, an increasing number of competitors to Darwinism did emerge around the turn of the century, and there was some discussion among American naturalists and intellectuals about whether Darwin's theory of evolution by natural selection alone was adequate to explain the process of evolution or if it needed to be discarded, revised, or complemented by additional theories. Stanford entomologist Vernon Kellogg (1867–1937) surveyed the subject in his 1907 book *Darwinism To-Day* and described the early twentieth century as a time in which there was "a most careful re-examination or scrutiny of the theories connected with organic evolution, resulting in much destructive criticism of certain long-cherished and widely held beliefs, and at the same time there are being developed and almost feverishly driven forward certain fascinating and fundamentally new lines, employing new methods, of biological investigation" (Kellogg 1907, 1–2) (Fig. 27.3). Kellogg's *Darwinism To-Day* was the most comprehensive description of the state of Darwinism at the start of the twentieth century. He began it by explaining how, while there was much fruitful debate among naturalists about the details of how evolution operated, there was little or no contestation about whether evolution itself was a natural phenomenon. That is, most if not all American biologists accepted evolution to be true, but there was considerable discussion about the details of the mechanisms by which evolution occurred.

FIGURE 27.3. Vernon Kellogg (1867–1937) was a leading evolutionist and humanitarian, whose pacifism was destroyed by his encounter with social Darwinian German officers in the early years of the Great War. Permission: Belgian American Educational Foundation

At the start of the twentieth century, even the most convinced American Darwinists, including Kellogg himself, believed that there remained some lingering questions about Darwinism that additional research would hopefully address. Chief among them was the source of the new variations on which selection operated. While Darwin's theory could explain how nature, unguided by a sentient planner, could select for certain traits and against others, it could not explain how the accumulation of selection would push members of species to acquire characteristics that were fundamentally different from those possessed by at least some of their ancestors. Kellogg (1907, 375) concluded *Darwinism To-Day* by explaining, "The selection theories do not satisfy present-day biologists as efficient causal explanations of species-transformation. The fluctuating variations are not sufficient handles for natural selection; the hosts of trivial, indifferent species differences are not the result of an adaptively selecting agent." Advocates of the two principal competing theories, mutationism and orthogenetic evolution, accepted certain aspects of Darwin's theory of evolution by natural selection but discarded other portions of his theory and replaced them with alternative mechanisms.

Evolutionary theories centering on mutationism had emerged in the work of the American naturalist William Keith Brooks (1848–1908). Brooks, a student under both Louis and Alexander Agassiz, spent his career at Johns Hopkins University and trained the first significant cohort of American biologists. His students included Edward Grant Conklin

(1863–1952), T. H. Morgan (1866–1945), and Edmund B. Wilson (1856–1939). An early American Darwinist, by the mid-1870s Brooks was investigating and writing about potential hereditary mechanisms that would work cooperatively with Darwin's theory of evolution by natural selection. His early work investigated Darwin's theory of pangenesis and the notion of saltations, which were unusual and often radically different variations that may occasionally appear within a species. In *The Law of Heredity*, Brooks (1883, 328) suggested, "There are many reasons for believing that variations under nature may not be so minute as Darwin supposes, but that evolution may take place by jumps or saltations.... A slight change in one generation may thus become in following generations a very considerable modification, and there is no reason why natural selection should not be occasionally presented with great and important saltations." Brooks's ideas on heredity and saltations complemented those offered by both Francis Galton (1822–1911) and Thomas Henry Huxley (1825–95).

In the 1890s, researchers began to more seriously consider the possibility that saltations might be the source for the new variations on which natural selection could operate. Researchers like the Dutch biologist Hugo de Vries (1848–1935), who introduced the term "mutation" to describe the introduction of fundamentally new variations, and the British geneticist William Bateson (1861–1926) both believed that new, discontinuous variations could serve as the source for new variations. Their views rejected Darwin's claim that natural selection acted "only by the preservation and accumulation of infinitesimally small inherited modifications" (Darwin 1859). In its place they asserted that evolution occurred by the chance occurrence of a mutation, which was then eliminated or preserved through natural selection. In contrast to the neo-Darwinists, mutationists denied natural selection any creative powers.

Nineteenth-century mutationists could offer a solution to the question of the source of new variations on which natural selection could operate, but they were still subject to the problem of return to average, another of the shortcomings that plagued Darwinism at the turn of the twentieth century. In his 1867 review of *Origin*, the University of Edinburgh engineer Fleeming Jenkin (1833–85) had persuasively argued that any mutation that might arise in a population, no matter how adaptively advantageous it might be, would be washed out in subsequent generations as an organism that possessed it reproduced with organisms that did not.

At the turn of the century, the mutationists' case was given a significant boost by the rediscovery of Gregor Mendel's (1822–84) work. In 1866, in the wake of the publication of *Origin*, Mendel had published a paper reporting the results of his research on heredity in pea plants, which was ignored by almost every biological researcher until his work was rediscovered in 1900 by three European scientists, Carl Correns (1864–1933), Erich von Tschermak (1871–1962), and de Vries. Mendelism provided the mutationists with a hereditary pathway that explained how the new variations that arise through

FIGURE 27.4. Edward Drinker Cope (1840–97) was one of the great American fossil hunters in the second half of the nineteenth century. He was also prominent in the American neo-Lamarckian group. From H. F. Osborn *Impressions of Great Naturalists* (New York: Scribner's, 1928), 166

mutation are preserved by natural selection and passed intact from one generation to another. By the end of the first decade of the twentieth century, Mendelism had been taken up by the mutationists, who viewed their theory as a competitor to mainstream Darwinian explanations of evolutionary change. It was in this context that many early twentieth-century evolutionists viewed Darwinism and Mendelism as competitors, a situation that was not resolved until the modern evolutionary synthesis in the mid-twentieth century.

The second evolutionary theory that attempted to address perceived shortcomings with Darwinism in the early twentieth century was orthogenetic evolution. Introduced by the German zoologist Wilhelm Haacke (1855–1912), orthogenesis hypothesized that variations appear one after another to move evolution in a particular direction toward an ideal or nearly ideal evolved state. Orthogenesis addressed the two principle shortcomings identified with Darwinism, the unknown source of new variations and the problem of the return to average, through the recurring introduction of similar, new variations.

In the United States, orthogenesis found considerable support among paleontologists, who saw in the fossil record evidence of unilinear, progressive evolutionary change.

The first among these was the American paleontologist Edward Drinker Cope (1840–97), who had little formal scientific training and used inherited wealth to pursue a scientific career through extensive fieldwork in the Midwest and West (Fig. 27.4). Best known for his intense personal feud and fossil-finding competition with Othniel Charles Marsh (1831–99), Cope merged orthogenesis and neo-Lamarckianism by arguing that organisms drove their own evolution by developing new behaviors and changing how they chose to use their physical or mental traits (Fig. 27.5). He believed he had found evidence for linear progress in the fossil record that he was quickly assembling throughout the last quarter of the nineteenth century. Cope's ultimate influence on evolutionary thought in America was relatively limited, but he was responsible for mentoring Henry Fairfield Osborn (1857–1935), who adopted some of Cope's ideas and had a great deal more influence on American biology (Fig. 27.6).

Osborn was a geologist and paleontologist, who, after studying at Princeton, was jointly hired by Columbia University and the American Museum of Natural History in New York. He accumulated one of the world's finest fossil collections for the museum and did much to promote paleontology to the average American through his museum exhibitions. Osborn accepted Cope's belief that later generations diverted from their ancestral forms as they intentionally adapted themselves to new environments. Subsequent generations, he believed, diverged to follow certain evolutionary traits along a unilinear evolutionary pathway. However, he did not take up any of Cope's Lamarckian explanations; instead, he believed that organisms possessed genetic traits that would appear under certain environmental conditions. Unlike Cope who believed that organisms would acquire and pass along new

FIGURE 27.6. Henry Fairfield Osborn (1857–1935), aristocrat and paleontologist, sometime student of T. H. Huxley, directed the American Museum of Natural History and argued for non-Darwinian evolution. He was a strong opponent of the creationists in the 1920s. From H. F. Osborn, *Great Naturalists* (New York: Scribner's, 1928)

traits, Osborn asserted that the capacity for a new trait was already present in an organism's ancestors and would appear when the conditions were appropriate.

Despite competition from a myriad of alternative and complementary theories, in the first decades of the twentieth century Darwinism remained the most popular explanation for evolution among American professionals and the public alike. It was a central component to the theories that emerged. Even while they sought new theories to explain shortcomings in Darwinism, most American evolutionists still accepted the efficacy of natural selection to perpetuate better-adapted traits and to extinguish less-adapted traits. This was largely due to the fact that, despite its recognized inadequacies, Darwinism was useful as a research agenda, as a tool for organizing a complex natural world, and as an explanation for any number of biological questions.

DARWINISM BEYOND BIOLOGY

Beyond biology, evolutionary thought proved useful in the social sciences in the United States as well as in addressing economic and political issues. It is perhaps ironic that Darwinism had its greatest influence on social and political thought at a time when biologists were so actively offering alternative and complementary ideas about evolution. Nonetheless, both

FIGURE 27.5. Cope's great fossil hunting rival O. C. Marsh convinced Huxley that the first horse was Eohippus. To celebrate, Huxley drew this cartoon, complete with rider. From L. Huxley, *Life and Letters of Thomas Henry Huxley* (London: Macmillan, 1900)

Darwin's method and his theory of evolution by natural selection significantly influenced American intellectuals throughout the late nineteenth and early twentieth century.

The emergence of what has come to be called social Darwinism is perhaps the most frequently discussed influence of Darwinism in American social and economic thought. Closely associated with the writings of the British philosopher and sociologist Herbert Spencer (1820–1903), social Darwinism encompassed ideologies and political and economic theories that placed special emphasis on the notion of the survival of the fittest. In the United States, social Darwinism generally served as a justification for laissez-faire economic and political policies and was most closely associated with the work of the Yale sociologist William Graham Sumner (1840–1910). Even though it bears Darwin's name, social Darwinism actually owes little to Darwin; it actually originated among opponents to the social, political, and economic policies that were later labeled social Darwinian (Bannister 1979, xxv).

Darwinism also powerfully influenced pragmatism, the American philosophical movement that emphasized the practical consequences of any particular ideology or belief. Pragmatists like Charles S. Peirce (1839–1914) and William James (1842–1910) saw in Darwin's theory a biological exhibition of pragmatism. Natural selection, they believed, effectively rid species of traits that were not adapted to the environments and conditions in which they lived. "Darwin, while unable to say what the operation of variation and natural selection in any individual case will be, demonstrates that in the long run they will adapt animals to their circumstances" (Peirce 1992, 11). Similarly, John Dewey (1859–1952) drew heavily on Darwin's work to assert that higher mental functions were products of evolution and on Darwinian natural selection to posit a program for the empirical study of human's place in nature.

Progressivism, the American political movement that was so powerfully influential in the twentieth century, also drew inspiration and justification from Darwinism. Whereas their laissez-faire adversaries, the social Darwinists, emphasized the efficacy of nature to ultimately select those traits that were most advantageous to species' survival, the progressives borrowed from the first half of Darwin's analogy between artificial and natural selection. They saw in Darwin's explanation of the efficacy of selection in the hands of plant and animal breeders justification for sweeping social, political, economic, and educational reforms. Just as the educated eye of the breeder could identify and select for the most desired traits to alter species as they wished, well-trained leaders could guide the nation's social and political evolution toward an identified ideal state. For this, science and education were necessary both to identify the best reform strategies and to educate the nation's citizens in hopes of improving their moral and economic conditions.

Finally, Darwinism's influence on American social thought can be clearly seen in literature from the early twentieth century. The two most notable authors influenced by Darwin's work were Jack London (1876–1916) and Theodore Dreiser (1871–1945). London had learned about Darwinism in an extension class on evolution offered by David Starr Jordan, and he incorporated themes of struggle, survival, and the brutality of nature in many of his writings, especially in the short story "To Build a Fire" (1902) and in the novel *White Fang* (1906). In sharp contrast to the harsh view of nature offered by London, Dreiser portrayed it as relatively harmonic and often compassionate. Nonetheless, Darwinism played every bit as much an influence in shaping Dreiser's depictions of nature as it did London's.

Be it in economics, politics, or literature, Americans turned to Darwinism to either guide or justify particular worldviews or practices. We frequently see figures on both sides of a debate drawing from Darwinism for either inspiration or support for their positions. For example, the Progressives and the social Darwinists, whose ideologies and aims could not be more at odds, both appealed to Darwinism and drew from its authority in both public and professional venues. However it was employed, the fact that Darwin's name and his theory of evolution by natural selection were so frequently invoked is a demonstration of his influence in the United States.

THE RISE OF AMERICAN ANTI-DARWINISM

For Americans in the early twentieth century, as had been the case throughout the later half of the nineteenth, Darwinism's appeal rested on its usefulness. Darwinism provided naturalists with a fundamentally new research program as well as new methods for conducting their research. It gave social scientists a naturalistic foundation on which they could construct the new sciences of psychology, sociology, and anthropology, and it revitalized earlier work done by economists and political theorists by providing them with analogies and justifications drawn directly from nature. Darwinism inspired a new generation of American authors, and it captured the public's attention by making them consider and reconsider humans' relationship to nature and to God.

From the start, Americans linked both evolution generally and Darwinism specifically with their ideas about the inherent goodness and inevitability of progress. A powerful theme underlying the Western adoption of an evolutionary worldview has been the notion that progress is inherent in nature and that society could similarly advance (Ruse 1996, 284). As Michael Ruse (2009c, 2) has aptly put it, "Evolution is the child of the idea of Progress, the belief that through our own efforts humans can make a better life here on earth." Progress, in this respect, meant social and political advancement in the form of political reformism as well as biological evolution, as there exist higher- and lower-order organisms, and, over time, individuals and species alike improve themselves, evolving ever higher. The sacredness of ideas about progress was precisely what made Darwinism so attractive to Americans. In the twentieth century, faith in progress and its alliance with

Darwinism proved highly effective in the hands of progressive reformers, who believed that they were speeding along progress by the application of scientifically valid principles and methods.

However, Darwinism was vulnerable to criticism when it undermined long-cherished ideals in American culture. In the wake of the social and political upheaval of World War I, the 1920s and 1930s witnessed a series of challenges to evolutionary theory generally and to Darwinism specifically. The war challenged Enlightenment assumptions about the ascendance of reason and the ability for science and technology to improve the human condition. Those aspects of human life that were most emblematic of the Enlightenment – science, democracy, the rise of nation-states, and the development of sophisticated technology – were what made the First World War so terribly devastating. For many Americans, Darwinism represented the height of the corrosive materialism that they believed had ultimately led to the Great War.

The emerging cohort of American antievolutionists had good reason to be concerned about a potential link between Darwinian evolutionary theory and the host of social and political ills that they believed were caused by or exacerbated by it. For the first three years of the war, Americans had steadfastly maintained their neutrality. When the United States finally joined the war effort, President Woodrow Wilson had to convince Americans of the necessity of ending their neutrality. A small army of authors was enlisted to write about the atrocities that Germany was committing in the war. Among the most influential of these authors was Vernon Kellogg, the Stanford entomologist who had written *Darwinism To-Day*. During America's years of neutrality, Kellogg had left Stanford to join his former student Herbert Hoover in distributing food and clothing to civilians trapped in German-occupied Belgium and northern France. Kellogg had lived with members of the German military, and his 1918 book *Headquarters Nights* detailed evening conversations he had with his German colleagues. He reported that German intellectuals had adopted a perverted form of Darwinism to justify their militaristic aggression and imperialism. "The creed of the *Allmacht* (total sufficiency) of natural selection based on violent and fatal competitive struggle," Kellogg (1918, 28) wrote, "is the gospel of the German intellectuals; all else is illusion and anathema." Kellogg's work found wide readership among those Americans who were inclined to go to war with Germany as well as with those who saw in Darwinism threats to American values.

The emerging generation of antievolutionists found their leader in William Jennings Bryan (1860–1925), Wilson's former secretary of state. Bryan had resigned in 1915 in protest over differences with Wilson over American responses to the war. Whereas Wilson believed that, despite efforts to maintain neutrality, the best course of action for the United States was to prepare a large standing army, Bryan was a convinced pacifist. After the war ended, persuaded by authors like Kellogg that

FIGURE 27.7. Breeding better families (and avoiding the bad ones) became somewhat of an obsession in the early years of the twentieth century. Permission: American Philosophical Society

Darwinism was somehow responsible for it, Bryan joined an emerging campaign to prevent public schools from teaching that human beings descended by way of evolution from lower order animals (Larson 1997, 40–42). His efforts ultimately led to his role in the 1925 Scopes Trial.

Darwinism also became socially and politically problematic when it was linked to the American eugenics movement (Fig. 27.7). There is little in Darwinism itself that lends support to eugenics; instead, the linkage resulted from the fact that most early twentieth-century American evolutionary biologists and geneticists supported the eugenics movement to some degree. The movement grew increasingly popular in the United States through the 1920s and 1930s, and biologists' advocacy for eugenics was an important part of the professionalization of biology in the United States (Largent, 2008, 39). Criticism of the American eugenics movement first emerged in the 1920s from many of the same figures – including Bryan – who feared the acidic effects of Darwinism on American society. Today, allegations of an overt link between Darwinism and eugenics play a significant role in creationists' attacks on evolution.

On the eve of the modern evolutionary synthesis, which ended decades of disputes among biologists and produced an effective explanation about how Darwinian selection and modern genetics operated cooperatively, Darwinism was deeply engrained in American scientific, social, political, and intellectual life. It had, over the course of the previous seven decades, found its place in American science and literature, influencing both profoundly. The truest mark of its impact in the United States has been the ferocity with which both its advocates and its opponents have debated the accuracy and significance of the *Origin*.

❦ Essay 28 ❦

The German Reception of Darwin's Theory, 1860–1945

Robert J. Richards

When Charles Darwin wrote in the *Origin of Species* (1859, 482) that he looked to the "young and rising naturalists" to heed the message of his book, he likely had in mind individuals like Ernst Haeckel (1834–1919), who responded warmly to the invitation (Haeckel 1862, 1:231–32n) (Fig. 28.1). Haeckel became part of the vanguard of young scientists who plowed through the yielding turf to plant the seed of Darwinism deep into the intellectual soil of Germany. As Haeckel would later observe, the seed flourished in extremely favorable ground. The German mind, he would write (1868), was predisposed to adopt the new theory. The great philosopher Immanuel Kant (1724–1804), for instance, was on the verge of accepting a transmutational view in his Third Critique ([1790] 1957, 538–39), though he stepped gingerly back from the temptation. Johann Wolfgang von Goethe (1749–1832), about the same time, dallied with transmutational ideas, or at least Haeckel would convince Darwin that the Englishman had an illustrious predecessor. Jean-Baptiste de Lamarck's (1744–1829) conceptions had taken hold among several major German thinkers in the first few decades of the nineteenth century in a way they had not in England and France. Among those ready to declare themselves for the new dispensation was Rudolf Virchow (1821–1902), Haeckel's teacher at Würzburg – though this very political scientist would prove Haeckel's nemesis later in the century. So Haeckel's estimate of the ripeness of German thought was not off the mark. Darwinism took hold in the newly unified land, though not without some struggle; at last, it became the dominant view in the biological sciences. But with its success, did it also foster the malign racist ideology that transfixed Adolf Hitler (1889–1945)?

EVOLUTIONISM BEFORE DARWIN

In his *Critique of the Faculty of Judgment* (*Kritik der Urteilskraft*, [1790] 1957), Kant argued that the naturalist could provide a mechanistic understanding of organisms only up to a point. The researcher could deploy physical laws to explain, for example, the refraction of light rays by the various media of the vertebrate eye; yet the composition and special layout of cornea, lens, and humors so as to focus an image on the retina bespoke a purposeful arrangement. The investigator could ultimately

FIGURE 28.1. Ernst Haeckel (1834–1919), the author of the "biogenetic law," was Darwin's greatest German supporter, but it is debated how much his thinking was genuinely Darwinian and how much it owed to older traditions that stressed morphology. Permission: Wellcome

construe the operations of the eye only by postulating the *idea* of the whole as the cause of its design; such postulation would imply, at least heuristically, an *intellectus architypus* – an intellect whose ideas were creative. At first Kant rejected any notion of a gradual development of organisms over time; at least he did so when his former student Johann Gottfried Herder (1744–1803) had suggested this possibility in his *Ideas for a Philosophy of the History of Humanity* (*Ideen zur Philosophie des Geschichte der Menschheit*, 1781–84). However, the work of Johann Friedrich Blumenbach (1752–1840) finally convinced Kant that it was conceivable, at the limits of understanding, to unite mechanism with teleology in the explanation of organisms (R. J. Richards 2002b, 229–37). It could be, for instance, that the mechanical deformation of the vertebrate skeleton might produce all the various vertebrate forms. The several osteological patterns did evince purposiveness, but they might have arisen naturally from a resourceful mother earth, as it were, and developed through time under physical forces. Kant cautioned that this possibility yet required the naturalist further to assume that the original seeds themselves had a purposive core. This transformational hypothesis, Kant thought, would be "a daring adventure of reason." He concluded, however, that there was little empirical evidence to support the view and that one would wait in vain for a Newton of the grass blade.

Though Kant had initially rejected the speculations of Herder in harsh and dismissive tones, the Weimar community

FIGURE 28.2. The great German poet Johann Wolfgang von Goethe (1749–1832) was always interested in science, and by the end of his long life was embracing some form of morphology-based transformism. Permission: Wellcome

was more hospitable to such ideas. Charlotte von Stein (1742–1827), Goethe's great love, recalled her friend had imagined that human beings were once fish. Later in the 1820s, in his series *Zur Morphologie* (1817–24), Goethe proposed a scenario in which the giant megatherium, whose fossil remains were unearthed in South America, had been transformed into the modern sloth (R. J. Richards 2002b, 476–86). He argued that the common pattern of bones that underlay the various vertebrate skeletons could have been transformed through interactions with the environment. He consequently supposed that Kant's daring adventure of reason was more than groundless fantasy (Goethe 1989, 98–99) (Fig. 28.2).

Goethe's speculations may have been fueled by the rapid translation into German of Erasmus Darwin's (1731–1802) *Zoonomia, or The Laws of Organic Life* (1794–96), which had a long section proposing the natural transformation of simple creatures, originally created by God, into the variety of living species populating the globe. The number of German biologists succumbing to the transformational hypothesis increased during the first decades of the nineteenth century as Lamarck's ideas gained traction in Germany, even while they faltered in Britain and France. For instance, in the first volume of his *Zoologie* (1708–14), Friedrich Tiedemann (1781–1861) argued that the paleontological evidence indicated a parallel between human embryological development and the history of no-longer-living organisms.

> From the oldest strata of the earth to the most recent, there appears a graduated series of fossil remains, from the most simply organized animals, the polyps, to the most complex, the mammals. It is evident too that the entire animal kingdom has its evolutionary periods [*Entwickelungsperioden*], similar to the periods which are expressed in individual organisms. (Tiedemann 1808–14, 1:64–65)

The great embryologist Karl Ernst von Baer (1792–1876), however, objected strongly to this hypothesis. He declared in his celebrated *Developmental History of Animals* (*Entwickelungsgeschichte der Thiere*, 1828–37), citing his earlier Latin disputation (1823), that "the law proclaimed by naturalists is foreign to nature, namely 'that the evolution [*Evolutionem*] which each animal undergoes in its earliest period corresponds to the evolution which they believe to be observed in the animal series.'" (von Baer 1828–37, 1:202–3). Von Baer thus rejected the parallel between the embryological "evolution" of an animal and the supposed historical evolution of species. Even before Cuvier's famous lampoon of his colleague Lamarck in the eulogy at his death, von Baer had struck the comic note:

> One gradually learned to think of the different animal forms as evolving [*entwickelt sich*] out of one another – and then shortly to forget that this metamorphosis was only a mode of conception. Fortified by the fact that in the oldest layers of the earth no remains from vertebrates were to be found, naturalists believed they could prove that such unfolding of the different animal forms was historically grounded. They then related with complete seriousness and in detail how such forms arose from one another. Nothing was easier. A fish that swam upon the land wished to go for a walk, but could not use it fins. The fins shrunk in breadth from want of exercise and grew in length. This went on through generations for a couple of centuries. So it is no wonder that out of fins feet have finally emerged. (von Baer 1828–37, 1:200)

Despite the objections of zoologists like von Baer, two strains of evolutionary thought arose in Germany during the first half of the nineteenth century. One followed the direction given by individuals like Tiedemann and developed a naturalistic account of species change. So, for example, Rudolf Virchow (1862, 31) maintained, in a lecture of 1858, that it was scientifically necessary, on the basis of paleontological evidence, to assume the "transmutability of species" (*die Uebergangsfähigkeit von Art in Art*). He was proud to have made that judgment prior to Darwin's publication of the *Origin of Species*. Later he would become less convinced of the scientific probity of evolutionary ideas, especially as applied to human beings. More religiously minded scientists traveled a second path. One such was Heinrich Georg Bronn (1800–62), for whom the paleontological evidence suggested that extinct species had been progressively replaced. The replacement of one species with an improved one, he argued, followed general laws relating the local environment to particular kinds of adaptation. Bronn yet maintained that replacement was not transformation in the Lamarckian sense; he looked to a Divine source for the progressive changes in species over vast periods of time (Rupke 2005). The history of biological thought before the publication of the *Origin of Species* does indicate that the community of German researchers was more predisposed to be receptive to the new theory than naturalists of other nations. Yet the introduction of Darwin's conception also produced hesitation, modification, and objection.

THE *ORIGIN OF SPECIES* AND ITS EARLY ADVOCATES

The *Origin of Species* was published by John Murray in November 1859, with a second, lightly corrected edition in December. Darwin had been contacted by H. G. Bronn with a request to supervise a translation into German. Bronn himself translated the second edition of the book in lightning-fast order. *Über die Entstehung der Arten* appeared in June 1860. The translation was quite adequate, with only a few infelicities (Gliboff 2008, ch. 4). Bronn, however, appended an essay of critical analysis to his translation that set the tone for the German reader. He was quite admiring of Darwin's accomplishment, recognizing in him a naturalist of considerably ability, especially as his ideas moved in the direction of Bronn's own. But he also pointed out the difficulties, especially the notion of lawless variation and the assumption of a spontaneous generation at the beginning of life on earth. The criticism that evoked the most positive response, however, was Bronn's (1860, 503) observation that Darwin showed only that transmutation of species was possible; he had not provided the evidence that it was actual. Bronn, whose main empirical concern had been paleontology, did not fully appreciate the *Origin*'s several conceptual strands that, when woven together, yielded "one long argument" (Darwin 1859, 459). Bronn's request for evidence inspired two ardent disciples: Ernst Haeckel and August Schleicher (1821–68).

Haeckel, who trained as a medical doctor at Würzburg and studied under Virchow, pursued research in marine biology. While he was preparing his prize-winning work on radiolaria, he read Bronn's translation of the *Origin*, and thought his own study of these microscopic marine organisms provided the kind of empirical evidence Darwin's theory required: the relationships of species within families of these creatures bespoke genealogy, and the transitional species between families confirmed it (Haeckel 1862, 1:231–33). Later his three-volume study of sponges provided greater and more abundant evidence, he believed, of the natural origin of these transformed invertebrates (Haeckel 1872).

Perhaps the most convincing evidence for Darwin's theory came in Haeckel's early study of siphonophores, complex colonial organisms. The research on these creatures occurred while Haeckel and several assistants spent about four months in the Canary Islands during the winter of 1866–67. Just before the trip, Haeckel stopped in England, where he visited an array of naturalists, including Thomas Henry Huxley in London and Darwin at his village of Downe (R. J. Richards 2008, 173–75). The research led to a prize-winning tract *On the Developmental History of Siphonophores* (*Zur Entwickelungsgeschichte der Siphonophoren*, 1869).

Haeckel's experiments and dissections of siphonophores preceded by twenty years the similar experiments by his two students, Wilhelm Roux (1850–1924) and Hans Driesch (1867–1941). In one set of experiments, he followed the development of siphonophore eggs from species of ten different genera. He altered the ambient light, water salinity, temperature, and movement to determine if these disturbances caused alteration in development. The environmental changes did have significant effects on development, causing some embryos apparently to revert to the morphology of ancestor species or to cross over to related species forms. In another set of experiments, he carefully divided the cells of very young embryos into two, three, or four groups to see if the separated cells would continue to develop. Like his students Roux and Driesch, he got independently developing embryos, some continuing their growth for almost a month. The embryonic clones, as we would call them, were complete but usually smaller than normal embryos. These latter experiments, like those of Driesch, showed early embryonic cells to be totipotent. The former set of experiments seemed to reveal the evolutionary history of siphonophores (R. J. Richards 2008, 185–96).

Schleicher, an eminent linguist and Haeckel's colleague at Jena, also took up Bronn's challenge to find evidence for Darwin's theory. He explored the history of language, where linguistic fossils could be found that indicated descent with modification. In 1863 Schleicher published his investigations in a little tract entitled *Darwinian Theory and the Science of Language* (*Die Darwinsche Theorie und die Sprachwissenschaft*), which Darwin himself arranged to have published in English. Schleicher argued that language and mind were two sides of the monistic coin; he maintained that human mental evolution could be gauged by the complexity of language spoken. Haeckel's description of the hierarchy of the various races of mankind was deeply in Schleicher's debt, as was Darwin's own argument for the evolution of human mind in the *Descent of Man* (1871).

Jena was the first significant redoubt for *Darwinismus*. In addition to Haeckel and Schleicher, Carl Gegenbaur (1826–1903) had cast his lot with the new theory, though initially with some hesitation. In the first several volumes of his monograph series *Investigations in the Comparative Anatomy of Vertebrates* (*Untersuchungen zur vergleichenden Anatomie*

FIGURE 28.3. Carl Gegenbaur (1826–1903), an ardent Darwinian and close collaborator with Ernst Haeckel. Permission: American Philosophical Society

der Wirbelthiere, 1864, 1865, 1872), Gegenbaur demonstrated the homologous relationships of the vertebrate skeleton but did not mention Darwin's conception (Fig. 28.3). Only in the second edition (1870) of his *Foundations of Comparative Anatomy* (*Grundzüge der vergleichenden Anatomie*) did he proclaim:

> From the standpoint of descent theory, the "relationship" of organisms has lost its metaphorical meaning. When we meet a demonstrable agreement of organization through precise comparison, this indicates an inherited trait stemming from a common origin. The task becomes to trace, step-by-step, the various paths the organ has followed by reason of acquired adaptation; it no longer suffices to derive each relationship from some remote similarity. (Gegenbaur 1870, 19)

Because of Gegenbaur and Haeckel, the small university at Jena drew some of the next generation's most significant biologists: the "golden" brothers Oscar (1849–1922) and Richard Hertwig (1850–1937), Anton Dohrn (1840–1909), Hermann Fol (1845–92), Eduard Strasburger (1844–1912), Vladimir Kovalevsky (1842–83), and Nikolai Miklucho-Maclay (1846–88). After Gegenbaur's departure from Jena, they still came to study with Haeckel: Arnold Lang (1855–1914), Richard Semon (1859–1918), Wilhelm Roux (1850–1924), and Hans Driesch (1867–1941). When Gegenbaur moved to Heidelberg in 1873, there quickly formed around him another group of students who would extend his research in evolutionary morphology (Nyhart 1995). Among this number were: Max Fürbringer (1846–1920), Georg Ruge (1852–1919), Friedrich Maurer (1859–1936), Hermann Klaatsch (1863–1916), and Ernst Göppert (1866–1945).

The work of the Jena evolutionists seems to have encouraged their own mentors and teachers to move in Darwin's direction. Rudolf Leukart (1822–98), the great invertebrate morphologist at Giessen, affirmed in a review of Haeckel's *General Morphology of Organisms* (*Generelle Morphologie der Organismen*, 1866) that, while he did not completely agree in all particulars with Haeckel, he was with him on the "main question of descent" (Nyhart 1995, 175n). Haeckel's own teacher at Würzburg, Albert Kölliker (1817–1905), was more restrained in his support. Like many others in Germany and Britain, he became convinced of evolution in the wake of the *Origin of Species* but rejected natural selection as the means by which this occurred. In his 1864 article "On the Darwinian Theory of Creation" ("Ueber die Darwin'sche Schöpfungstheorie"), Kölliker complained of Darwin's "teleological" mode of arguing, contending that the Englishman assumed that all traits of an organism were "the best" and that the general harmony of the organic world derived from natural selection. Kölliker (1864, 184) rather thought a general law, presumably of divine origin, was necessary to explain the "great developmental plan that drove the simplest forms to ever more variable unfolding." Other professional biologists, like the embryologists Wilhelm His (1831–1924), Ludwig Rüttimeyer (1825–95), and Alexander Goette (1840–1922), were ready to accept the notion of the transformation of species but balked at the specific proposals of the Darwinians, especially Haeckel's biogenetic law that ontogeny recapitulated phylogeny.

Though Darwin had assumed that patterns of phylogenetic transformation would be preserved in the sheltered maternal environment of the embryo – which, as he claimed in the *Origin of Species* (1859, 450), would be left as "a picture, more or less obscured, of the common parental form of each great class of animals" – that conception never became the central principle for him that it did for Haeckel. In his many publications and lectures, Haeckel would illustrate the biogenetic law with a comparative analysis of the embryological development of phylogenetically related organisms, showing that at very early stages embryos were quite similar, expressive of the morphology of the common ancestor, and only in latter stages did they diverge from shared patterns, just as their ancestors had. Both Haeckel's colleague Gegenbaur and his friend August Weismann (1834–1914) endorsed the biogenetic law (Fig. 28.4).

Weismann became what Darwin's British disciple George Romanes (1848–94) called an "ultra-Darwinian." He parted from Haeckel – and Darwin himself – over the inheritance of

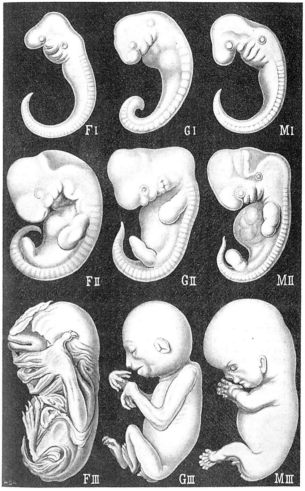

FIGURE 28.4. Haeckel's famous diagram showing the growth of the individual (ontogeny) mimics the history of the group (phylogeny). From Haeckel's 1905 Berlin Lecture series, *Der Kampf um den Entwickelungs-Gedanken*

acquired characteristics. Both Darwin and Haeckel believed that traits acquired by parents could alter the hereditary substance and be passed to offspring. Natural selection could operate on such traits as well as on those that spontaneously arose as small variations; such variations would result from the impact of the environment on the sexual organs of the parents. Weismann (1889, 419–48), by contrast, demonstrated that five generations of mice whose tails were cut off and then bred together nonetheless gave birth to offspring with tails intact. He argued that the germ-plasm, carried in the genital organs, had only a one-way connection with the somato-plasm, which gave rise to manifest bodily features: the germ-plasm guided development of the organism but remained unaffected by changes in the body of the creature. There was a certain sense in which the germ-plasm, in Weismann's view, was immortal, carried along through the hereditary line.

OBJECTIONS TO DARWINIAN THEORY BY GERMAN BIOLOGISTS

Haeckel's biogenetic law became the point of attack by other biologists who more or less accepted the idea of transmutation. The three aforementioned embryologists – Rüttimeyer, His, and Goette – became Haeckel's most vitriolic critics. They especially objected to the recapitulation hypothesis, mostly from a desire to protect the newly emerging field of professional embryology from the ingressions of evolutionary theory; this kind of territoriality continued to fuel studies of embryology in the twentieth century (see De Beer 1940 and Oppenheimer 1967). Indeed, these embryologists issued an indictment of fraud against Haeckel, a charge that would haunt him through his later years and provide grounds for suspicions about evolutionary theory more generally. Rüttimeyer (1868), in an early review of Haeckel's *Natural History of Creation*, noticed that illustrations of embryos at the very earliest stages were strikingly similar: Haeckel had used the same woodcut three times to depict what he called the sandal stage of development in the embryos of a dog, a chicken, and a turtle. Haeckel argued that at the earliest stages it was impossible to discriminate the embryos. In later editions of his book, he employed just one illustration of an embryo at this very early stage and claimed it might as well be the depiction of a dog, chicken, or turtle because they cannot be distinguished. Though Darwin and Huxley supplied moral support to Haeckel, the damage was done, and the charge of fraud was frequently repeated by enemies of evolutionary theory in Germany.

Wilhelm His directed the most probing and relentless attack against Haeckel. In *Our Corporeal Form and the Physiological Problem of Its Origin* (*Unsere Körperform und das physiologische Problem ihrer Entstehung*, 1874), His argued for the primacy of proximate mechanical causes – as opposed to remote evolutionary causes – for the understanding of embryological development. He took the opportunity, as well, to remind his readers of the fraud perpetrated by one of evolution's leading exponents and of that individual's continuing malfeasance. His claimed that Haeckel exaggerated the length of the tail of the human embryo to make it more apelike.

His's insistence on appealing only to proximate, potentially observable causes conformed to the epistemological dicta of Haeckel's former teacher and later opponent, Rudolf Virchow. In a famous confrontation at a meeting in Munich in 1877, Virchow utterly rejected Haeckel's proposal that evolutionary theory be taught in the German lower schools. Virchow claimed that authentic science should avoid speculation and rely only on observable and experimentally justifiable causes. Evolutionary theory supposed the spontaneous generation of life in the early seas and the transition from apelike creatures to man, neither of which could be demonstrated. But the real danger of evolutionary theory, Virchow (1877) urged, was its connection with socialism, the fuel that ignited the Paris Commune a few years earlier. Both Darwin and Huxley thought this political indictment in Bismarck's Germany was vicious and unfair; but it obviously carried weight. Haeckel (1878) himself would argue that evolutionary theory had no political implications; one could draw such implications only when the theory was wedded to antecedent philosophical and political doctrines.

THE RELIGIOUS OBJECTIONS TO EVOLUTIONARY THEORY

The most vocal opposition to evolutionary theory came from religious dogmatists. In particular, members of the Keplerbund (an organization of Protestant naturalists) objected to the antireligious and anti-Christian conclusions that Haeckel and others had drawn on the basis of evolutionary theory. Eberhard Dennert (1861–1942), a lower-school teacher and founder of the Keplerbund, unleashed a torrent of pamphlets and books in opposition to Darwinian ideas. Typical was his *On the Deathbed of Darwinism* (*Vom Sterbelager des Darwinismus*, 1905), which pitted Darwin's version of evolution against that of others, with the implication that the whole enterprise was uncertain. Dennert (1905, 6) concluded that we had "no clear and exact demonstration of evolutionary doctrine." Several of the books and articles of the Keplerbund were translated into English and became the basis for tracts in the collection called *The Fundamentals* (1910–15), from which the religious movement in the United States received its name.

The response to evolutionary ideas in the Catholic community took an unexpected turn. In the wake of the German liberals' reaction to Pope Pius IX's brief against the modern world – the *Syllabus errorum* (1864) – Bismarck took the opportunity to curb the growing power of the Catholic Center Party. He promoted what Virchow called a *Kulturkampf* against the Roman Church, which ultimately led to the expulsion of the Jesuits from Germany in 1872. By the end of the century, however, the hostilities had quieted to the extent that rumor even had the Emperor ready to convert to Catholicism. Haeckel was called from retirement by friends to combat the resurgent ultramontanist threat. In a series of lectures he gave in Berlin in 1905, he disclosed what he thought a Jesuit plot. Father Erich Wasmann, S.J., had published a book entitled *Modern Biology and the Theory of Evolution* (*Die moderne Biologie und die Entwicklungstheorie*, 1904) (Fig. 28.5). Wasmann, a research biologist who specialized in ants and beetles, had surprisingly argued that evolutionary theory was supported by empirical facts. His work on an order of beetles that lived in ant nests, the myrmecophile, had convinced him of a view that he had previously rejected. He investigated several species of these beetles and discovered that some had taken on the color of the various ant species with which they lived and that others, even more remarkably, seemed to have evolved to resemble ants and were treated accordingly by their hosts. But a Jesuit who endorsed evolutionary theory! Haeckel thought there had to be sinister motivation involved, something Jesuitical. Wasmann did reserve to divine power the existence of man's soul and rational faculties, even if his body arose from apelike ancestors. His effort at reconciliation

FIGURE 28.5. For all that Father Eric Wasmann, S.J., endorsed evolutionary thinking, he earned the skepticism and hostility of Haeckel, who thought he was up to something devious and sinister. From *Berliner Tageblatt*, 7 February 1907

ultimately became the way the Vatican decided to avoid a repetition of the Galileo affair.

EVOLUTIONARY THEORY AND NAZI BIOLOGY

Several recent critics have alleged that Darwinian theory was foundational to Hitler's racism and Nazi biology more generally. Daniel Gasman (1971, 40) claimed that "Haeckel ... was largely responsible for forging the bonds between academic science and racism in Germany in the later decades of the nineteenth century." According to Gasman (1998, 26), Haeckel had virtually begun the work of the Nazis: "For Haeckel, the Jews were the original source of the decadence and morbidity of the modern world and he sought their immediate exclusion from contemporary life and society." Richard Weikart, in his book *From Darwin to Hitler* (2004, 6), argues that "no matter how crooked the road was from Darwin to Hitler, clearly Darwinism and eugenics smoothed the path for Nazi ideology, especially from the Nazi stress on expansion, war, racial struggle, and racial extermination." In the 2008 film *Expelled*, promoting the doctrines of intelligent design, the purported connection between Darwinism and Hitler is made part of the religiously conservative effort to undermine evolutionary theory. In the film, Weikart and the philosopher David Berlinski discuss the issue; and the latter asserts that "if you open *Mein Kampf* and read it, especially if you can read it in German, the correspondence between Darwinian ideas and Nazi ideas just leaps from the page."

Before indicating the factual misrepresentations of these indictments of Darwinian theory, a few conceptual considerations are in order. First, even if Hitler was a dedicated reader of the *Origin of Species* and drew inspiration from the book, that has no bearing on the truth of the basic premises of Darwinian theory or the moral character of Darwin and his followers. Hitler and the Nazis endorsed modern chemistry and its uses in the extermination camps, which of course hardly precludes the truth of that science or morally taints all chemists. It can only be rampant ideological confusion to suggest that somehow Darwin and Haeckel, both dead long before Hitler came to power, are responsible for the crimes of the Nazis or that the alleged connection with Nazi biology invalidates evolutionary theory. Second, the theory fundamental to the Nazi social hygienists, as well eugenicists in Britain and the United States, was Mendelian genetics, which in the early part of the century was seen as a replacement for Darwinian theory. Yet, none of those railing against Darwinism suggests that somehow genetics has been falsified or morally corrupted by the Nazi employment of that science. Finally, the charges made by Gasman, Weikart, Berlinski, and other members of the Discovery Institute, the Seattle organization that defends Intelligent Design, reduce the complex motivations of Hitler and the Nazis to monistic simplicity; Gasman, Weikart, and the rest ignore the economic, political, and social causes operative in the Germany of the 1930s, as well as the deeply rooted anti-Semitism that ran back to Luther and medieval Christianity.

There is little doubt that Charles Darwin and Ernst Haeckel, as well as most evolutionary thinkers of the nineteenth century believed in a hierarchy of races, with the criteria being intelligence and moral character. In this assumption, however, they did not differ from most other thinkers of the period (R. J. Richards 2002a). The preevolutionary scientists Carolus Linnaeus (1707–78), Johann Friedrich Blumenbach (1752–1840), and Georges Cuvier (1769–1832) – all of whose works subsequently directed thought about the distinction of human races – ranked those races in a hierarchy, with Europeans in the top position. James Hunt (1833–69), founder of the Anthropological Society of London and no friend of the Darwinians, declared in his presidential address to the society that Africans constituted a distinct species, much closer to the apes than to Europeans (Hunt 1864). There was, thus, nothing unique about evolutionists' recognizing such hierarchies; the assumption of a progressive racial gradation pervaded European cultural life and certainly did not derive from evolutionary theory. In respect to anti-Semitism, however, the facts speak well of Darwin and Haeckel. In Darwin's case, rather, they do not speak at all: he mentioned Jews only once or twice in letters that betray no taint of anti-Semitism. Haeckel, when queried about anti-Semitism by the journalist Hermann Bahr (1894), declared that he did not share that prejudice, though some of his students did. He recognized that Germany and some other countries barred Jewish immigrants from the East, particularly Russia, because they refused to be assimilated; and he thought such restrictions justified, not because they were Jews, but because they would not conform to conventional norms. He concluded his discussion with an encomium to the educated (*gebildeten*) Jews who had always been vital to German social and intellectual life: "I hold these refined and noble Jews to be important elements in German culture. One should not forget that they have always stood bravely for enlightenment and freedom against the forces of reaction.... We cannot do without their tried-and-true courage" (Bahr 1894, 69). Some Nazi apologists did make an effort to recruit Haeckel, as well as other German cultural giants – Beethoven, Humboldt, Goethe – posthumously to the Nazi side. Yet because Haeckel was at times regarded as a friend of Jews and because of his materialistic monism, Nazi Party officials claimed his work in no way formed a foundation for *volkische Biologie* and demanded that any such suggestion cease. His books were banned by Nazi officials in Saxony, along with those by Jewish authors (R. J. Richards 2008, 269–76).

Did Hitler have any knowledge of Darwinian evolutionary theory (R. J. Richards 2013)? Darwin's name does not appear in any of Hitler's writings. But perhaps the racial views expressed in *Mein Kampf* ([1925–27] 1943) yet indicate the influence of Darwin, as Berlinski urges. But only the dogmatically robotized would find in those tedious pages anything resembling Darwinian theory. Hitler (1943, 312) makes no claim that the human species arose from lower animals; his notions of racial homogeneity and a "general drive to racial purity" (*allgemein gültigen Triebes zur Rassenreinheit*) are foreign to a theory that requires variation and transmutation; his assertions that religion is not in conflict with "exact science" (294) and that it forms the foundation for morality (293) deny the efforts of Darwin and Haeckel to replace religious dogma with exact science and to demonstrate the origin of morality in the natural selection of community groups. Hitler's notions of struggle or battle (*Kampf*) among the races seems antithetic to Darwin's conception that struggle occurs primarily and most strongly *within* a variety or race and only distantly among distinct varieties or species. Indeed, because Hitler characterizes the Jews as alien (*fremde*) and having racial features completely distinct from the Teutons, any struggle, by Darwinian lights, ought to be mitigated or eliminated. Hitler's ideas about the degenerate quality of Jews and the dangers of racial mixing come more likely from the anti-Darwinian Huston Stewart Chamberlain (1855–1927), the Germanophilic Englishman: he married Richard Wagner's daughter Eva, became a friend and correspondent of Hitler, and was greatly admired by Alfred Rosenberg (1893–1946), the individual responsible for elaborating Nazi racial theory. Chamberlain's masterwork, *Foundations of the Nineteenth Century* (*Die Grundlagen des Neunzehnten Jahrhunderts*, 1899), quoted by Hitler in *Mein Kampf* (1943, 296), devotes considerable space (1899, 1:323–459) to explaining the alien (*fremde*) and inferior status of the Jews and why racial mixing would cause the degeneration (*Entartung*) of the superior, pure German race (1:325). Chamberlain thought the existence of Jews "a crime against the holy laws of life" (1:374). This mystically besotted historian called for "a struggle of life and death" (*ein Kampf auf Leben und Tod*, 1:531) against the non-German races; but like that of his disciple Hitler, his notion of struggle was a common trope and owed nothing to Darwinian natural-selection theory, which he compared to the theory of phlogiston (2:805).

By the beginning of the 1930s, Darwinism had reached a nadir. The geneticist and formidable historian of biology Erik Nordenskiöld (1936, 476–77) had declared it dead; its romantic speculations had been replaced by real science, laboratory genetics. Not death, of course, but a slumber. Awakened by the unexpected congress with genetics, the "modern synthesis" of the mid-1930s and early 1940s laid the grounds for the flourishing of the biological sciences today.

❦ Essay 29 ❦

Darwin and Darwinism in France before 1900

Jean Gayon

Among the nations with a major scientific tradition in the nineteenth century, France certainly resisted the penetration of Darwin's evolutionary ideas the most. Darwin himself observed this, in a letter he sent to the French anthropologist Armand de Quatrefages, ten years after the publication of the *Origin of Species*:

> It is curious how nationality influences opinion; a week hardly passes without my hearing of some naturalist in Germany who supports my views, and often puts an exaggerated value on my works; whilst in France I have not heard of a single zoologist, except M. Gaudry (and he only partially), who supports my views. But I must have a good many readers as my books are translated. (Darwin 1985–, 18:141, letter, 28 May 1870)

Some years later, Ernst Haeckel was more radical. In his popular book, *The History of Creation*, he insisted on the crucial role of Lamarck in the origins of evolutionary ideas, but observed that, despite this precedent, the French naturalists had simply ignored Darwin:

> In no civilized country of Europe has Darwin's doctrine had so little effect and been so little understood as in France, so that in the further course of our examination [i.e., Haeckel's book] we need not take French naturalists into consideration. (Haeckel 1883, 118)

There is some truth in Darwin's and Haeckel's statements. To be crude, the French had Claude Bernard and Louis Pasteur; they did not have Darwin. Bernard's and Pasteur's methods and theories put France at the highest possible level in experimental biology at the very time that Darwin published his *Origin of Species*, while French biologists showed high reluctance to adopt, imitate, or even simply understand the kind of biology that Darwin initiated. I could even go a little further: the French ignored Darwin *because* they had Bernard and Pasteur. Again, there is a great deal of truth in such a diagnosis (Figs. 29.1 and 29.2).

However, things were a bit more complicated. In reality, the reception of Darwin and the fate of Darwinism in France were a succession of paradoxical events that I

FIGURE 29.1. Claude Bernard (1813–78), the great French physiologist, whose work as an experimentalist set standards that made his countrymen feel that the efforts of English naturalists could be ignored. Permission: American Philosophical Society

will describe at a large historical scale. Here, I will first examine the reception of Darwinism in France before 1900, where two parallel stories have to be considered, inside and outside of the natural sciences.

RECEPTION OF DARWIN, 1859–1900

One can hardly say that Darwin was ignored in France. *The Origin of Species* was translated into French no fewer than three separate times between 1859 and 1900 (Darwin 1862d, 1873, 1876b), in a total of nine editions (compare eight editions in German and two in Italian). True, the first German edition appeared in 1860, the same year as a Dutch translation, and two years before Royer's first translation. However, this was not a great delay. Of Darwin's other works, fifteen of them were translated into French in the same period, in a total of thirty-one editions, compared with thirty-three editions of the German translations of the same books, and twelve Italian editions. On the whole, Darwin was almost as much translated into French as into German, and almost as quickly, with one major exception: *The Voyage of the* Beagle was translated into German in 1844, and only thirty years later (1874) into French. This means that Darwin was known in Germany before the publication of the *Origin*, whereas he was almost totally unknown in France. (See the appendix to this essay for the dates of French translations of Darwin's works.)

Therefore, the issue of the reputation of Darwin in France is not a question of diffusion of his works. Obviously, Darwin was widely translated and commented upon. The right question is: By whom was he recognized? A more careful survey of the translations shows that "none of the French translations or prefaces to the major Darwinian works was by a noted French man of science" (R. E. Stebbins 1988, 129). This is especially true of the *Origin*: Clémence Royer was Swiss and definitely not a naturalist (Fig. 29.3); Jean-Jacques Moulinié was a young scientist, a pupil of Carl Vogt in Geneva whose slight fame is entirely for his translation of the *Origin*; Barbier was a professional translator in Paris. Here, the comparison with Germany is striking: two major German biologists, Bronn and Carus, translated and prefaced the first editions of the *Origin*. Similar observations apply to the first Dutch translation (by Winkler, 1860) and the first Russian translation (by Rachinsky, 1864). In these three cases (and in others, in other countries, later on), renowned academic naturalists devoted considerable time to translating and introducing *The Origin of Species* to their professional colleagues. Nothing of the kind occurred in France for any of Darwin's major books bearing on evolution (*Origin, Variation, Descent*). This fact alone indeed testifies to some kind of resistance of the French scientific establishment to Darwin.

The resistance of French biologists is well illustrated by the story of Darwin's election as corresponding member of the Paris Academy of Science. It took eight years and nine votes (for nine positions) to get him elected. His case was examined at least seven times in the Section of Anatomy and Zoology between 1870 and 1873. Famous zoologists, such as Henri Milne-Edwards and Armand de Quatrefages, defended his case, although they acknowledged that they themselves did not accept Darwin's theories; others, opposed to Darwin's candidacy, argued that Darwin had not added much to science in terms of demonstrable facts. It was only in 1878 that the Botanical Section finally elected Darwin, immediately after the American botanist Asa Gray. He was elected *not* because of his theories but in light of his specific and, indeed, significant contributions to botany. When he died in 1882, Armand de Quatrefages, probably his principal defender among French naturalists, made this comment before the academy (fig. 29.4): "There were two men in Charles Darwin: a naturalist, observer, experimenter as the case may be, and a theoretical thinker. The naturalist is exact, sagacious, patient; the thinker is original and penetrating, and often just, often also too daring" (from R. E. Stebbins 1988, 150).

Thus it is simply not true that Darwin was unknown in France. His work was extensively translated and diffused, and he was honored by the scientific community, though later than in Germany, for instance. What, then, did occur?

FIGURE 29.2. Louis Pasteur (1822–95), chemist and microbiologist, famous for his work refuting spontaneous generation, and another great French scientist whose labors contributed to the ignoring and belittling of work across the English Channel. Permission: American Philosophical Society

Let us look again at Haeckel's judgment: "In no civilized country of Europe has Darwin's doctrine had so little effect and been so little understood as in France." Haeckel did not say that Darwin was neglected or unknown but that he had "little effect" and was "little understood" among naturalists.

Similarly, Darwin himself did not complain that *no* French author had supported his views but only that he had not heard a single "zoologist" supporting his views. This is the key to the real story. There was indeed nothing like a stereotypical French reaction as a whole to Darwin. Rather, different

London in 1864. The members of the Société d'anthropologie were not all Darwinians, but they believed that evolution was a necessary dimension of anthropological research. In 1868 a member of the society invented the word "transformisme" (transformism), which became widely used by the French as an alternative to "évolution." In 1871 the society elected Darwin as foreign member without any difficulty, in contrast with the long and difficult process of Darwin's election to the Paris Academy of Sciences. If there was an academic milieu in which Darwin was greeted unproblematically, it was definitely anthropology, not the biological sciences.

But, again, we must be cautious here. We should make a distinction between those anthropologists who were professional naturalists and those who were not, who were, for example, lawyers, economists, and sociologists. Among the former, we find prominent scientists, such as Paul Broca (1824–80) and Armand de Quatrefages (1810–92). They actively defended Darwin because they conceded that species evolve, but they were critical of natural selection as a major factor in evolution. Quatrefages (1870), who was highly respected by Darwin, and who fairly diffused Darwin's ideas on evolution in general and on the evolution of man, was nevertheless opposed to the application of the principle of natural selection to the question of the origins of man.

In fact, the real supporters of Darwin in France were another kind of "anthropologists," those involved in "social Darwinism." The term "social Darwinism" seems to have been used for the first time in 1877 by the British historian Joseph

FIGURE 29.3. Clémence Royer (1830–1902) translated the *Origin* into French, upsetting Darwin with her fiery introduction that put him firmly in the ranks of the freethinkers, and going so far as to add an emendation to the title that the work was about *des lois du progrès* (the laws of progress). This addition was removed in later editions, but Darwin was moving on to find other translators. Nineteenth-century cartoon

groups of people, belonging to different disciplines and scientific communities, held different attitudes.

Who were those who really supported Darwin, and who made his name familiar to the general public? The answer is quite simple. All of them had something to do with anthropology. This story begins with the first translation of the *Origin of Species* in 1862 by Clémence Royer (1830–1902). Royer was not at all a naturalist. Teaching at the University of Geneva, she was a prolific essayist who wrote on economics, sociology, and moral and political philosophy. Clémence Royer was the first female member of the Société d'anthropologie de Paris (Paris Society of Anthropology), founded in 1859. She translated the *Origin* because she wanted to provide a biological basis to her own political philosophy and to feminism (Blanckaert 1991; Harvey 1997). After Royer's translation, Darwin began to be quoted by members of the society, especially after the publication of Wallace's famous paper "The Origin of Human Races and the Antiquity of Man Deduced from the 'Theory of Natural Selection,'" read at the Anthropological Society of

FIGURE 29.4. Armand de Quatrefages (1810–92) was Darwin's French champion, for all that he had doubts about Darwin's theorizing. Permission: Wellcome

Fisher (1877, 250). But Fisher's use was purely marginal, with no theoretical intention. In fact, the first author who deliberately used the term to designate a particular way of thinking seems to have been the French political thinker Emile Gautier, a young anarchist who began using the expression in a series of talks in Paris in 1879 and gathered his ideas in a book published in 1880 under the title *Le Darwinisme social*. He defined "social Darwinism" as a doctrine that applied natural selection to human societies and used it as a means of justifying social inequality. Gautier refused this supposed doctrine and proposed replacing the "struggle for existence" by a principle of universal solidarity. After the publication of Gautier's book, the expression spread to other countries, used mainly in a pejorative way by sociologists who opposed the extrapolation of Darwin's natural selection to human societies.

However, if one applies retrospectively the concept of social Darwinism *avant la lettre*, then Clémence Royer was probably the first systematic "social Darwinist." Darwin was scandalized by Royer's preface, which went far beyond the book itself. Royer claimed that Darwin's theory embodied not only a "philosophy of nature" but also a "philosophy of humanity": "Never anything as ambitious has been designed in natural history: it is a universal synthesis of economic laws, the natural social science *par excellence*" (Royer, in Darwin 1862d, LXII). Royer asserted that Darwin's book demonstrated the falsehood and the utopian character of the idea of equality, at the level of both individuals and races: "Nothing more obvious than the inequality of the various human races; nothing more patent than the inequalities between various individuals of the same race" (LXI). "The data of the theory of natural selection [*la théorie de l'élection naturelle*] forbid us to doubt that the superior races are intended to supplant the inferior races" (LXI). Darwin's theory, Royer says, "favors a political regime of unlimited individual freedom, that is to say a regime of free competition of forces and faculties" (LXII). She also claims that Darwin's theory shows the fundamental errors of Christianity: "exaggeration of charity and fraternity" and "sacrifice of the strong in favor of the weak" (LVI). Molina (1992), who offers one of the most lucid comments on Royer's preface, insists that these ideas were not the product of Royer's imagination alone. In fact, the publisher of the first translation of *The Origin of Species* (Guillaumin) specialized in the publication of books and journals on economics. The presentation of the book in the publisher's catalog was perfectly explicit about the kind of public that was targeted: "[T]he book is not only destined for botanists, zoologists, and physiologists; it is also destined for philosophers and economists. Mr. Darwin's theory is no more no less the law of progress, mathematically formulated and extended to nature in general; it is an extension of Malthus's law to all living species, an extension that results in the most unexpected moral and political consequences" (quoted in Molina 1992, 377). An extensive survey of economic literature in France would indeed show that, at the very time when the French naturalists raised doubts about the scientific value of Darwin's theory, this theory was simultaneously celebrated and appropriated by economists who found there a justification for their liberal view of economics. Of course, not all French economists believed this. Others were convinced that the biological sciences were inappropriate for economic thinking, either because they thought that economic theory needed to be established upon foundations of its own (the marginalist school) or because they adhered to a general vision of society based upon "solidarity." The important point, however, is that economics and the social and political sciences offered a favorable context for the reception of Darwin in France. In that context, Darwin was a real stakeholder, even if we may think today that this was not the "real Darwin." Politics and economics, rather than natural science and religion, were the decisive context of the reception of and controversy about Darwin in France.

Let us now return to the reception (or rather nonreception) of Darwin among the French naturalists. All the historians who have examined this subject (Conry 1974; Farley 1974; R. E. Stebbins 1988) have come to the same conclusion: although Darwin was widely read and discussed, no significant French biologist before 1900 incorporated Darwin's major hypotheses into an active research program. There is no single explanation for this. Among a wide array of explanations that have been discussed, I would like to emphasize one.

First and foremost, French science in the second half of the nineteenth century was dominated by a positivist view of science. This was particularly true of biology, where it was reinforced by a close connection between biological research and medicine. For nearly a century, the best of French biology was published in the *Comptes rendus des séances hebdomadaires de la Société de biologie*. The Société de biologie (Society of Biology) was founded in 1844 by a group of brilliant biologists, who all adhered to one or another version of positivism. Some were declared disciples of Auguste Comte (the father of positivism); others adhered to the idea that science should not consider the origins or remote causes of phenomena but just describe them or explain them in terms of "actual" or "proximate" causes, with the help of the experimental method. Claude Bernard and Louis Pasteur were two founding members of this Society of Biology. Both of them were explicitly antagonistic to any biological research that aimed at explaining the phenomena of life in terms of origins:

> There are many great questions being discussed these days: unity or multiplicity of human races; creation of Man thousands years of centuries ago; fixity or slow transformation of species from one to another; matter reputed eternal rather than created; the idea of God being useless, etc. These are all questions that cannot be solved. I take a much more humble role in tackling a problem that can be solved experimentally. (Pasteur 1864, quoted in R. E. Stebbins 1988, 134–35)

> In place of making unrealizable hypotheses on the origin of things, on which one can discuss or experiment only in a sterile and blind manner, the experimenter proceeds otherwise. (Claude Bernard, quoted in R. E. Stebbins 1988, 136).

This attitude toward evolution was widely shared among the French biologists of the second half of the nineteenth century. Like many of their counterparts in British learned societies, the majority of French biologists thought that Darwin's theory belonged to the realm of speculation, went beyond the facts, generated idle controversy, and was thus "non-scientific" (Burkhardt 1988). There was nothing exclusively French in this attitude. The difference from other countries, especially England, is that Bernard and Pasteur became exemplars for the majority of French biologists around 1860, at the very time Darwin published his *Origin of Species*. While "the French had Bernard and Pasteur, they did not have Darwin," one could equally have written, "The British had Darwin; they did not have Bernard and Pasteur." Historians can never insist enough upon the extreme admiration that French biologists have had – and still have today – for Bernard and Pasteur. Together with positivism, this was certainly a major factor in their reluctance to work along Darwinian lines.

Another factor that played a significant role was neo-Lamarckism. The French did not invent this term, which was first coined and adopted by American naturalists. Laurent Loison, a young and gifted historian of science, has wonderfully reconstructed the history of French neo-Lamarckism (Loison 2010). The French neo-Lamarckians (Perrier, Giard, Bonnier, Le Dantec) often claimed to be "Darwinians" in the sense that they admitted "the fact of evolution." Most often, they also admitted the existence of natural and sexual selection. But they denied that natural selection was the main evolutionary process. Their basic doctrine was significantly different from that of the American neo-Lamarckists. It was crudely materialistic and mechanistic. They believed that organisms were directly modified by the action of the external milieu (essentially the physical milieu) and transmitted these modifications to their progeny. This doctrine never generated a fruitful experimental program and was thus essentially sterile, as Claude Bernard himself might have observed. But it fitted well with the primacy of experimental biology and the idea that only "actual" or "proximate" causes should be taken into consideration.

A third factor in resistance to Darwinism should be mentioned. Not all French biologists were materialists like the neo-Lamarckians. In fact, most were not. They were positivists, not materialists. And, like a majority of British biologists, they very often adhered to a general notion of nature as a finalized or purposive process. In her masterly book, *The Introduction of Darwinism in France* (1974), Yvette Conry gave hundreds of examples of the attachment of French biologists to the idea of the "harmony of nature," often combined with an acceptance of the "fact" of evolution. This is why they often claimed that they were "Darwinians" in the sense that they admitted that species genealogically derive from one another but denied that the "struggle for existence" and "natural selection" were the main drivers of evolution. This tendency was particularly true of a number of paleontologists. Indeed, the first paleontologist in the world who ever made real genealogical trees, upon the basis of real fossils, was Albert Gaudry (1827–1908) in 1866 (Gaudry 1862–67) (Fig. 29.5). In contrast with Haeckel's phylogenetic trees, Gaudry's trees were intended to represent real trees based on paleontological data, not speculative genealogies based on morphological data. Gaudry initiated a rather brilliant school of paleontology emphasizing the necessity of not only describing the succession of fossils in stratigraphic data but also of making hypotheses about their genealogy (what Gaudry called "philosophical paleontology"). But Gaudry was opposed to Darwin's explanatory theory of evolution: "It is the proper formation of paleontologists to supply proofs of the doctrine of evolution; it does not fall to them to explain the processes by which the author of the world has produced this modification" (Gaudry 1877, quoted in R. E. Stebbins 1988).

Not all French paleontologists before 1900 believed that reconstituting phylogenies was part of their work. This was in fact the subject of a major controversy (see Tassy 1991). In the case of "descent with modification," however, Darwinism was incorporated into real scientific practice. Nothing similar can be observed for the mechanism of natural selection.

Although Darwin was widely known, translated, and discussed in France between 1859 and 1900, Darwinism was never really "introduced," in the sense that no French biologists took Darwin's theory of natural selection as a basis for real scientific work. All historians who have worked on this question come to this conclusion. (The one possible exception, the young Lucien Cuénot, who used Darwin's ideas to explain the origin of phagocytosis and immune cells, renounced Darwinism around 1900 when he became a Mendelian [Limoges 1976].) The positivist and experimentalist mode of French science was probably the most important obstacle to such an introduction, in a context of biological research dominated by physiologists (Claude Bernard) and microbiologists (Louis Pasteur). The wide diffusion of Darwinism resulted primarily from the interest in Darwin generated among anthropologists, social scientists, and economists, for ideological reasons ("social Darwinism"). However this early popularity finally turned to be also a major obstacle to a real acceptance of Darwin into French science, because the main stream of social, political, and economic science in France that emerged at the end of the nineteenth century, indeed one of its most significant claims to glory (the sociology of Durkheim and Mauss, the economics of Walras), was based upon the claim to the methodological autonomy of these human sciences, and the rejection of any sort of biological foundation. Beside this, Darwin generated strong adhesion among certain sorts of naturalists, especially in botany (for reasons unrelated to Darwin's evolutionary theory), and paleontology (where the theory of descent of modification, but not the explanatory mechanism of natural selection, enjoyed considerable success).

APPENDIX

Here is the list (dates refer to the first English edition and to the first French translation; chronological order;

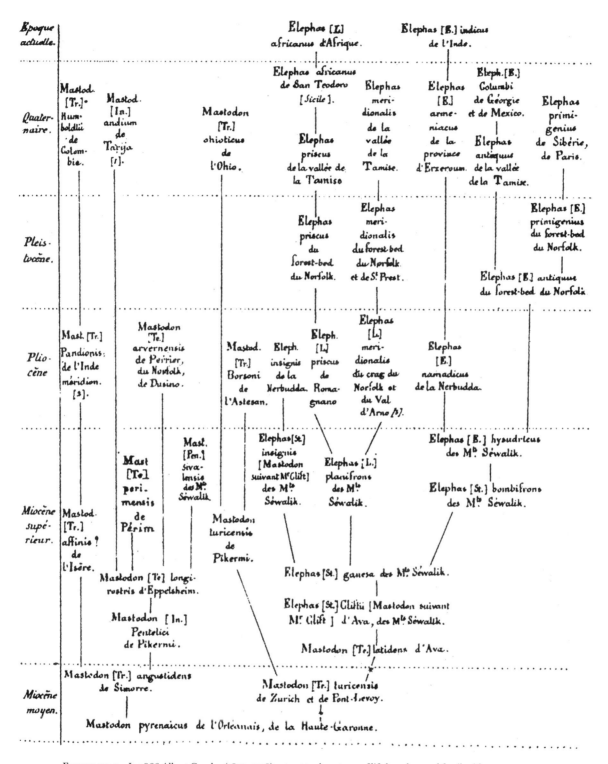

FIGURE 29.5. In 1866 Albert Gaudry (1827–1908) set out to draw trees of life based on real fossil evidence.

titles abbreviated): *Origin of Species* (1859–62); *Variation* (1868–68); *Fertilization of Orchids* (1862–70); *Descent of Man* (1871–72); *Expression of Emotions* (1872–74); *Voyage of the Beagle* (1839–74); *Climbing Plants* (1865–76); *Insectivorous Plants* (1875–77); *Cross and Self-Fertilization* (1876–77); *Different Forms of Flowers* (1877–77); *Coral Reefs* (1842–78); *The Power of Movement in Plants* (1880–82); *Vegetable Mould and Worms* (1881–82); *Essay on Instinct* (1883–84); *Life and Letters* (1887–88). (Source: Conry 1974, 438.)

❧ Essay 30 ❦

Encountering Darwin and Creating Darwinism in China

Yang Haiyan

Just as a strong white light fragments into colorful beams through a prism, so Charles Darwin has various images throughout the world. He not only appears as a scientific sage, the founder of modern evolutionary biology; he also has wider cultural images. He can be a liberal or a conservative; an abolitionist or a racist; a moralist or a devil's chaplain (Kjærgaard 2010, 105–22). His banner can be waved by socialists for mutual aid, as well as by capitalists for jungle rule. Finally, and inevitably, he was a Victorian gentleman. These different, sometimes contradictory images show diverse appropriations of Darwin in the various contexts in which he has been encountered. My aim in this essay is to investigate the Chinese encounter with Darwin and the appropriation of his theories in changing political and social contexts.

The image of Darwin in the People's Republic of China, founded in 1949, was clearly embodied in the "Meeting in Commemoration of Great Figures of World Culture"[1] held on 27 May 1959 at Beijing to celebrate the 150th anniversary of the birth of Darwin and the 200th anniversary of the birth of – amazingly – Robert Burns, the plowman poet from Scotland. What a combination, a gentlemanly capitalist and a spokesman of the proletariat! Bing Zhi, the president of the Zoological Society of China, lectured on "A Century of Charles Darwin's *Origin of Species*," and Zheng Zuoxin, the secretary-general of the Zoological Society of China, delivered an address "In Commemoration of the Great Naturalist: Charles Darwin." According to their address, Darwin was first of all a "great materialistic scientist" who was buried in Westminster Abbey with Isaac Newton, another "great materialistic scientist" (for if Newton can be counted a materialist, why not Darwin?); Darwin was also a scientist who "broke through the shackle of religion," and his *Origin* "sets forth a new outlook on the universe, which overthrows the superstitious allegation that God is the creator." One reason we should pay homage to Charles Darwin is "his resolute

---

[1] Quotations related to this meeting are from materials held at the East Asian History of Science Library, Needham Research Institute, Cambridge. Thanks to Mr. John P. C. Moffett, the librarian, for introducing me to those materials. Some of them have been published in China. For example, Zheng Zuoxin's speech was published first in *Guangming ribao* (Guangming Daily, no. 3584 [27 May 1959]: 3), then in *Shengwu xue tongbao* (Bulletin of Biology, no.11 [1959]: 495–97).

endeavor in overthrowing the conservative and reactionary forces." According to the chairman, "The establishment of Darwin's theory drove away the ignorance and superstition of mankind"; "Darwin's theory overthrew the metaphysical allegations and shook the dominating rule of the idealist world outlook in people's minds. He established historical and materialistic views in the field of biology." Pairing a capitalist and a proletarian was, in Beijing, in 1959, very strange indeed. Because Karl Marx and Friedrich Engels praised Darwin, and his science was seen as progressive, his capitalist background was overlooked. He was criticized for using in his theory Thomas Malthus – who saw poverty as "natural" rather than the product of an unjust capitalist order – but the criticism was gentle: Darwin had only "failed to see through the reactionary essence" of Malthus's principle of population. Scientist-materialist-atheist represented the official image of Darwin in new China, and it still prevails in the language of political discourse.

However, when Darwin and evolution were first introduced into and appropriated in China at the turn of the century, the political situation was very different. The official image mentioned in 1959 was barely recognized; Darwin's theories were wide open to interpretation.

Let's first go back to the late Qing dynasty era, from 1840 to 1911. In an urgent memorial presented to the imperial throne in 1872 to justify building steamships for coastal defense, Li Hongzhang (1823–1901), a Chinese scholar-official, warned that the Qing dynasty was confronting an upheaval unlike any for the past three thousand years. For many centuries, China saw itself as the center of civilization and viewed outsiders as "barbarians." Although neighboring warlike "barbarians" at first triumphed in battle, they were quickly assimilated into the Chinese culture. But then came the loss of successive wars with the Western imperialist powers, beginning with the First Opium War (1839–42). Chinese elite intellectuals gradually realized that they faced a different and advanced civilization, strong not only in guns and ships but also in political organization and economic power. The encounter with the West was a danger because they felt the survival of the Middle Kingdom and its people were imperiled; the encounter was also an opportunity because, by learning from the West, the empire might progress into a wealthy and powerful modern state. Darwin and evolution were imported and communicated in China in this critical political context.

FIGURE 30.1. The news story in *Shen bao* (Shanghai Journal, no. 404, 21 August 1873) that reported the publication in 1872 of Darwin's book, *The Expression of the Emotions in Man and Animals*

THE EARLIEST INFORMATION

As far as we know, Darwin's name was first published in China in 1871 in the Chinese translation of the sixth edition of Charles Lyell's *Elements of Geology* (1865), which briefly referred to Darwin's theory. Two years later, a short news story in *Shen bao* (Shanghai Journal, no. 404 [21 August 1873]: 2) reported the publication in 1872 of Darwin's book, *The Expression of the Emotions in Man and Animals* (Fig. 30.1). *Gezhi huibian* (Chinese Scientific and Industrial Magazine), founded by John Fryer (1839–1928), mentioned the possibility of the ape ancestry of man in the autumn issue of 1877 (pp. 6–7) in an article entitled "Hundun shuo" (The Theory of Chaos). In the "Bowu xinwen" (Scientific News Items) in the spring issue of 1891 (pp. 32), the *Chinese Scientific and Industrial Magazine* quoted the English physicist John Tyndall on Darwin's theory ("thousands upon thousands present kinds of animals and plants in fact derived from a few kinds") as a sign of scientific progress.

These few references did not stir up discussion on evolution among Chinese people for several reasons. First, at that time, dynasty officials promoted education in Western science and technology with a Chinese essence at its core. Under this policy, the Manchu government took an overriding interest in practical knowledge and techniques, especially military ones. So evolution as a new theoretical discovery in the world of plants and animals attracted limited attention. Second, Darwin's provocative concept of "struggle for existence" or "survival of the fittest" was almost absent in those texts. Indeed, translators of English-language science books – typically Protestant missionaries – deliberately avoided the notion for religious reasons. Joseph Edkins, for example, promoted natural theology while omitting the concept of the "fittest survive" in his translation of a *Botany* textbook by Joseph Hooker (a friend of Darwin) in 1886 (Elman 2005, 327–30). Another reason for the Chinese silence about Darwin is the absence of a religious tradition opposing animal origins of humans and the vast time periods that this required. Chinese Christian converts did show a hostile attitude toward evolution (B. Zhang and Wang 1982, 43–50), but their attack began from the start of the twentieth century, and its influence was quite limited. Last but not least, Chinese people did not discuss Darwin simply because they did not read him or much about him – the press and journalism still had a very limited impact on the country by the end of the nineteenth century, and the literacy rate was very low. Consequently, at this stage Darwin was something of a nonevent.

Yet there were exceptions. In answer to the extra-theme question for the 1889 Spring Civil Service Examination, Zhong Tianwei (1840–1900), a middle-aged alternate county governor in Guangdong Province, in an account of Darwin's theory of evolution, clearly described the principle that "the strong survive and the weak perish," or "the unfit gradually die out and the fit exist forever." He commended it as the "natural principle of the heavenly way" (Zhong [1889] 2009, 342; Elman 2005, 345–51). Zhong Tianwei probably picked up his ideas on a trip to Europe in 1880–82. His essay won the fourth place, though exhibiting a better knowledge of Darwin than the other three. Such an understanding of evolution would become commonplace in the coming decades.

THE REAL SENSATION

After the shock and humiliation of defeat in the Sino-Japanese War, Yan Fu (1854–1921) wrote a newspaper article, "Yuan qiang" (Whence Strength), which appeared in March 1895 in *Zhi bao* (*Chih pao*, or Tianjin Newspaper) (Fig. 30.2). Among the first Chinese to be educated in England, he had studied naval science there from 1877 to 1879, mainly in the Royal Naval College, Greenwich, and had eagerly absorbed knowledge of Western philosophy and social science. In the article, Yan Fu introduced the concepts of "struggle for existence" and "natural selection" from Darwin's *Origin*. His own terms in a later revised version of the article were respectively *wu jing* (things compete) and *tian ze* (nature chooses [lit. heaven chooses]). Drawing on the yellow-white-brown-black race categories, he argued that the previous neighboring warlike "barbarians" who became eventually assimilated into the Chinese culture were not an alien race at all; instead, people

FIGURE 30.2. Yan Fu (1854–1921) was instrumental in bringing Darwin's ideas to China at the end of the nineteenth century.

from the West who threatened the survival of Chinese people were really a different and superior race. He identified the origin of Western wealth and power; lamented China's weakness, poverty, and backwardness; and clearly stated his ultimate solution. This was to transform the Chinese people mentally, physically, and morally: to open their minds, strengthen their bodies, and harmonize their virtues. In this way, individuals' energy could be liberated in order to make communal bonds and thus a stronger *qun* (collective, or Spencer's social organism), which would better survive in the struggling world (Yan [1895] 1986, 10–11, 14).

This article and others published in the same period "articulate all the basic assumptions which are to underlie his translation efforts of the next few years" (Schwartz 1964, 43). After partly publishing first in his own periodical *Guowen huibian* (Collection of National News or The Light Seeker, one of the first journals founded by the Chinese people themselves) from December 1897 to February 1898, Yan Fu's paraphrased translation of Huxley's Romanes lecture "Evolution and Ethics" and a later "Prolegomena" (with an extensive commentary occupying one third of the book) appeared in 1898 with the Chinese title *Tianyan lun* (The Theory of Evolution [lit. The Theory of Heavenly Evolution]).[2] With

---

[2] *Tian*, the Chinese concept with complicated meanings, appears in the title of the Chinese edition.

the help of the eloquent and emotional journal essays of Liang Qichao (1873-1929), one of the Reform Movement (1895-98) leaders and at the same time the founder of Chinese journalism, the main ideas of *Tianyan lun* were disseminated effectively and caused an immediate sensation among literati and young students. Darwin's name now entered the vocabulary of every intellectual household. The Chinese equivalents of "struggle for existence," "natural selection," and "survival of the fittest" became popular slogans; evolutionary cosmology grew so familiar that key words from Yan Fu's translation were even adopted for people's names. In the 1970s, Cao Juren (2003, 371), a scholar of modern Chinese intellectual history, reported after reading more than five hundred autobiographical memoirs from this period that almost all the authors were influenced by Yan's *Tianyan lun*, including leading intellectuals such as Lu Xun and Hu Shi. In the next several decades, Darwin's name was quoted by almost all sides: reformers or revolutionists, nationalists or communists.

Huxley's (1894, 16-17) horticultural metaphor of colonization struck Yan Fu sharply, warning him of the doom of Chinese people in the international struggle for existence. At the same time, Huxley's emphasis on human endeavor to create favorable conditions of existence, restrain ruthless self-assertion, and strengthen the social bond (43, 35-36, 81-82) appealed to him greatly, though the opposition between the cosmic process and the ethical process, which lay at the core of Huxley's original argument, was disagreeable. For Yan Fu, and for Wu Rulun who wrote the preface, *tianyan* (evolution) is an ubiquitous principle leading to an all-encompassing process, taking in the biological, intellectual, moral, social, and political realms (Yan [1898] 1998, 57, 60, Wu's preface 1). As long as one followed the universal convention – "Everyone has freedom, but it shall be bound by the freedom of others" (or "enlightened self-assertion") – self-assertion and self-restraint, hence cosmic nature and ethical nature, could be reconciled (187, 433). In his commentary, incorporating Spencer's progressive social organism model and the key causation between improved brain and diminished reproductive potential, which demolishes the Malthusian natural inequality, a kind of optimism replaced Huxley's pessimism: evil stops proceeding, and good is arriving day by day; a well-organized, united, and strong society is achievable (422, 196-97).

Tianyan lun is thus like an assemblage of those ideas of Huxley and Spencer under Yan Fu's deliberate selection. What he paraphrased in the main text is not a "foil" to what he presented in his comments; neither did he intend to do so, as Schwartz (1964, 111) thought was the case. On the contrary, Huxley's advocacy of human action to combat the cosmic process struck a chord with Yan Fu and his contemporaries against the background of imperial expansion. According to one of Yan Fu's commentaries in a later translation of Spencer's *The Study of Sociology*, *tian* in *tianyan* means neither God nor sky but that things can develop according to causality, though causality is itself unchangeable (Yan [1903] 1981, 298, n. 5). *Tianyan*, coined by Yan Fu and the core concept of his worldview, is a universal way for continuous change in the cosmos with ethical significance. This advocacy, like a wedge to keep the door wide open, facilitates Yan Fu's aim to accommodate Huxley's dualism to Chinese traditional Confucian-Taoist conflict over the relationship between *ren zhi* and *tian xing*, which are roughly parallel to the ethical process and the cosmic process (Yan [1898]1998, 433), and eventually assimilated Confucianism into evolution in order to urge the government and people to act immediately to eliminate artificial checks to progress. According to James Pusey (1983, 173), Yan Fu made the best choice he could make in translating Huxley's text. This emphasis on human determinationism and self-strengthening goes well with Spencer's upbeat message, which is viewed as a spur to realize people's potential (Ruse 1998, 73-74). *Tianyan lun* is filled with a heart-stirring paean to the progress of society and humanity on the condition that right action is taken. The hope is that a reformed China will not just survive international competition but also in fact prosper as a result of it.

Not all joined with Yan Fu and his followers in their optimistic views on evolution. There were Chinese observers who chose to see the dark side of evolution. For example, Zhang Taiyan, in an article in *Min bao* (People's Journal, no. 7 [1906]: 1–14), acknowledged with Huxley that good and evil evolved together. Zhou Zuoren, a younger brother of Lu Xun, found "the theory of evolution is great, but too cruel." As he put it, "If we use strong and weak as the standards, and regard competition as the vital link, how can the world be in peace?" (Z. Zhou [1906] 1961, preface 497). It is noteworthy that in the devastating aftermath of World War I, the early advocates of evolution had their doubts too. Both Yan Fu and Liang Qichao attributed the bloodiest wars to the misuse of "the struggle for existence and the survival of the fittest" (Pusey 1983, 439-40).

Evolution meant human evolution and social evolution at this stage. Yan Fu and other late Qing commentators showed little interest in biological knowledge. Political, social, and moral concerns were paramount for them; Darwin's painstaking methods and the details of his arguments were all but irrelevant. Yet the poverty of their understanding of evolution itself during their active assimilation, appropriation, and creation indicates just how cultural heritage is transmitted and constituted, for good or for ill. It happens not only when knowledge travels across different cultures but also when it is communicated within one culture.

What Yan Fu constituted through *Tianyan lun* is a unique cosmology with a complicated structure responding to his own intellectual needs (Z. Wang 2002, ch. 3; H. Wang 2004, ch. 8). For him and his contemporaries, materialism and anticreationism, which according to the chairman of the 1959 Beijing commemoration were the main reasons why the Chinese people should praise and commemorate Darwin, did not appear at the top of their agendas. Even Darwin's image as a scientist was alien. At the beginning of the twentieth century, "professional" scientists in China were few and far between. The field of scientific knowledge had not been broken up into well-defined, distinct disciplines in institutional spaces. Chinese intellectuals still felt obliged to advise those

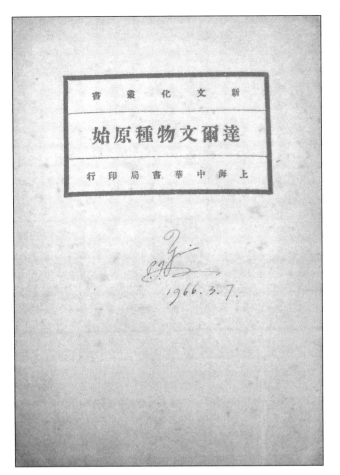

FIGURE 30.3. Title page of the Chinese translation of *The Origin of Species* by Ma Junwu (1920)

FIGURE 30.4. Ma Junwu (1881–1940) was the earliest Chinese translator of Darwin's *The Origin of Species*.

in power directly, and the scientific knowledge and methodology were held to be relevant for their moral bearings primarily. The abolishment of the Civil Service Examination in 1905 started channeling their energy in different directions, and the establishment of a new education system in the very beginning of the Republic of China shaped the expectations of the next generation. The intrascientific interest in evolution began to grow, especially with the return of those young pioneers trained overseas in biology.

THE FOCUS ON DARWIN

The complete translation by Ma Junwu (1881–1940) of Darwin's *Origin* was published relatively late, in 1920, with the Chinese title *Wuzhong yuanshi* (Figs. 30.3 and 30.4). Parts of it, however, had been translated by the same person at the very beginning of the twentieth century. The influence of *Tianyan lun* can be revealed clearly in a poem composed by Ma in 1900, which starts as: "A vast expanse of past to present, viewing *tianyan* in action; the fierce struggle for existence prevails on earth" (Ma 1985, 13). In 1902 a follower of reformists and an overseas student in Japan, Ma Junwu, first translated Darwin's "Historical Sketch of the Progress of Opinion on the Origin of Species" (added to *Origin*'s third edition in 1861). This article was published in Liang Qichao's propagation vehicle – *Xinmin congbao* (Journal of a New People, no. 8 [22 May 1902]: 9–18) (Fig. 30.5). Then his translation of chapters 3 and 4, "Struggle for Existence" and "Natural Selection," were published in 1902. Soon after, the Chinese version of the first five chapters and the "Sketch" came out together as the first volume of the whole book in the spring of 1904.

The gap between 1904 and 1919 (the year Ma finished the whole translation) is worth noting. He studied first metallurgy (1907–11) and then agricultural chemistry (1913–16) in Germany. During the interval of 1911–13 and from 1916 onward, he was occupied with politics as a congressman of the Republic of China. The political turmoil made him leave his position several times, and when he began to work as the general engineer in a gunpowder factory in Guangzhou in 1918, he finally had some leisure time. Being busy all the time was one reason why he was slow to finish the whole translation (Ma 1920, "Translator's Preface"). On another occasion, Ma told a translator that doubts about the facts in the second half of Darwin's *Origin* caused his delay – those facts "are still waiting for confirmation by my own experiments" (J. Zhou [1937] 2009, 24). It is more likely an excuse than an explanation, though. The doubts were real: the authenticity of Darwin's natural selection was doubted especially when Mendel's laws were rediscovered and elaborated, a development that influenced Chinese intellectuals. However, it is hard to imagine that Ma would do biological experiments to prove Darwin's theories. Maybe the right question is not why delay but why continue. Ma himself actually answered it: "Civilized countries in the whole world have all translated this book. Since our country can not afford not to be a civilized country now, we must translate it even just for the country's dignity" (Ma 1920, "Translator's Preface"). Correcting the mistakes in his previous translation was another reason, though some major errors were still there. For example, Darwin's sense of consolation in the last sentence of chapter 3 disappeared in the Chinese translation – "no fear is felt" was translated as "don't fear (war of nature)" in both versions, and "death is generally prompt" as "death is inevitable for every species" and "death is so fast that no one can escape indeed" respectively (Darwin 1872a, 61; Ma [1902] 2009, 153; 1920, 104).

There were serious attempts, however, to introduce Darwin's theory *and* understand the scientific details of evolution, especially by those new biologists with an overseas education. Still at Cornell University pursuing his doctoral degree in entomology, Bing Zhi, along with other Chinese students, cofounded the Science Society of China in 1914, with the aim of promoting modern science among the Chinese people through their own journal – *Kexue* (Science). The translation by Bing Zhi of the part about dogs in the first chapter of *The Variation of Animals and Plants under Domestication* was published in the second, third, and sixth issues of the first volume (1915). Other translations of evolutionary texts and many articles and news items about Darwin and evolutionary theories flourished in *Science*, and other journals as well. In 1921 the fourth issue of *Bowu zazhi* (The Magazine of Natural History), founded by the Society of Natural History in Peking Normal College, gathered four articles to commemorate the 112th anniversary of Darwin's birthday (Fig. 30.6). The next year, the third volume of *Min duo* (People's Bell) devoted its fourth and fifth issues to evolution, with sixteen articles in all, including a detailed chronicle of Darwin and a reading list of evolutionary works (almost two hundred items). Because of increased literacy and the vernacular language revolution, ordinary Chinese readers (still few in number, mainly in cities) had their first access to detailed accounts of Darwin's theories. However, it is noteworthy that this encounter was a critical one. Standing beside Darwin, were Gregor Mendel, William Bateson, Hugo de Vries, and others. The introduction of the new genetics complicated the picture. Thomas Hunt Morgan was thought as "the first person to successfully oppose Darwin's theory" (Qian 1919, 1211). Mechanisms other than natural selection were quite popular. According to one author, the fact of evolution and the theory of evolution should be carefully distinguished. One should know "the *Origin of Species* is insufficient to represent all evolutionary theories" (Chen 1922, 2).

Although working biologists focused on their own specialties to unearth the potential of evolution, they did

FIGURE 30.5. Title page of *Xinmin congbao* (Journal of a New People, 1902), where the *Origin* in translation started to appear

almost no original research on the mechanisms of evolution. Ideologically, Darwinian slogans like "struggle for existence" backed up their aspiration of saving China through science; practically, the general concepts and narratives of evolution were localized through their own investigating, collecting, and excavating. Rich resources of Chinese flora, fauna, and fossils were researched, especially with the opening of the Biological Laboratory of the Science Society of China (1922), the Fan Memorial Institute of Biology (1928), and the Paleontology Research Laboratory (1928) and the Cenozoic Research Laboratory (1929) of the Geology Survey of China (Pei 1930, 1127–33; Hu 2005, 35–83; J. Zhang 2005, 197–233; Schmalzer 2008, 17–54). Their work was viewed as an effort to nationalize science and build up their identity as Chinese scientists. Darwin's name was still shining, yet his most original contribution was somehow out of focus, except later on in a few important researches made by Tan Jiazhen (C. C. Tan, 1909–2008), who obtained his PhD in 1936 from Morgan's lab at Caltech, under the direction of Theodosius Dobzhansky. Tan's work on the chromosome structure of *Drosophila*

FIGURE 30.6. A group gathered in 1921 to celebrate the 112th anniversary of Darwin's birthday. From *Bowu zazhi* (The Magazine of Natural History), no. 4 (1921).

was shaped heavily by Dobzhansky's interest in combining genetics and evolution; in turn, his work was incorporated in Dobzhansky's *Genetics and the Origin of Species*, a widely read classic for modern *evolutionary synthesis*. Tan's later work on the geographical and seasonal variations of ladybird beetles in China offered experimental evidence for the newfound population genetics (Tan 1987, 202, 243; Schneider 2003, 75–78). As for theoretical population genetics, Li Jingjun (C. C. Li, 1912–2003), a 1940 Cornell PhD in plant breeding and genetics and then a postdoc in mathematical statistics, had his textbook *An Introduction to Population Genetics* published in English by National Peking University Press in 1948. However, it was never on sale publicly because of the turmoil at that time (Majumder 2004, 103). Consequently, it almost had no influence in China, except for his immediate students.

The understanding of Darwin and evolution became modernized promptly but quietly. However, the institutionalization of evolutionary research involving modern Darwinism had no chance to happen properly in China because of the Japanese invasion and the civil war afterward. When peace finally came, unfortunately Soviet Lysenkoism or creative Soviet Darwinism was introduced and promoted in China, which halted the research and education in Mendelian genetics and modern Darwinism. Li Jingjun had to leave under pressure in 1950 and went to the University of Pittsburgh. His textbook, published by the University of Chicago Press in 1955, was welcomed as the first of its kind by working biologists.

Viewed as a scientific base for the official Marxist ideology, Darwinism had a privileged position in the science and education policies of the Communist Party newly in power, only its meaning was far from Darwin's main ideas and much further from synthetic evolutionism. Textbooks for the final-year students in middle school and for college students, *Da Er Wen zhuyi jichu* (Foundations of Darwinism) and *Da Er Wen zhuyi jiben yuanli* (Basic Principles of Darwinism) respectively, came out in 1952 and 1953, with substantial content devoted to Michurinism and political discourse. Learning this kind of Darwinism was a serious nationwide political task until the late 1950s. In the 1959 commemoration mentioned earlier, the intertwining of Darwinism and dialectical materialism was still clearly present.

For the great majority of Chinese people, Darwin was, and perhaps remains, a familiar stranger: "familiar" because of the communication of *Tianyan lun* during the first decades of the twentieth century and also because of the later popularizing of "Darwinism" as an important part of the school curriculum and of several nationwide training programs during 1950s; "stranger" because, in the same period and beyond, Darwin's theories were little understood and investigated, until very recently among scientists. According to the international Darwin survey conducted by the British Council in 2009, 67 percent of adults in China believe that "life on Earth, including human life, evolved through natural selection, in which no God played a part." This extraordinary world-leading result should not be taken to imply that the Chinese understand or care much about Darwin's science. Like their Western counterparts, they believe, or disbelieve, only what they are taught.

Darwinism in Latin America

Thomas F. Glick

DARWINISM WAS RECEIVED in Latin America always in relationship, whether explicit or not, with positivism, a term first used by the social philosopher Saint-Simon to refer to scientific method and its extension to philosophy: scientific knowledge was viewed as "positive." As reformulated by Auguste Comte, positivism came to be a system of thought in which science was the *only* source of authority. It was *not* a philosophy of science, had no universal notion of truth, and did not promote specific methods or laws. In Europe, it was envisioned as a kind of capstone to the scientific revolution. In Latin America, however, positivism (in its Comtean form) *preceded* the instauration of science; therefore, it was programmatic, and one of the programs was science (Fig. 31.1).

SCHOOLS OF THOUGHT

Positivism came in two varieties, Comtean ("social positivism") and Spencerian ("evolutionary positivism"). Social positivism promoted a more just society through the application of science. Evolutionary positivism was associated with Herbert Spencer and, of course, with Darwin. In Spencer's writings there was a stress on universal progress as a continuous, unilinear evolution from a primitive nebula to human civilization. He used the term "evolution" as a synonym of progress even before the publication of the *Origin of Species* (1859), and Darwin's theory simply gave substance to his view of a general evolutionary process characterized by the passage from the homogeneous to the heterogeneous, from the simple to the complex.

In virtually all Latin American countries, Spencer prepared the way for Darwin, and the assimilation of Darwin's theory via the Spencerian corpus was normative particularly for lawyers and medical doctors who constituted a hefty proportion of the elite. An older generation of historians of Latin American philosophy had assumed that Comtean positivism acted to block acceptance of the Spencerian version, and this was true in Comtean strongholds like Brazil and Mexico, early on. But in both cases Comte was replaced by Spencer, and as a result, Darwinism was debated, then assimilated. Although positivism was influential everywhere, the specific nature of

FIGURE 31.1. Auguste Comte (1798–1857) was the father of positivism, a philosophy that sees society as progressing through three stages: the religious, the metaphysical, and finally the positive or scientific. From J. F. E. Robinet, *Notice sur l'oeuvre et sur la vie d'Auguste Comte* (Paris: Dunod, 1860)

the Darwinian discourse that emerged, however, was sharply conditioned by the ethnic and racial composition of local populations.

"Darwinism" in nineteenth-century Latin America, usually meant social Darwinism because, as a general rule, the debate was about human society and the primary receivers of Darwinism, lawyers. In only Brazil and Cuba can one point to a "Darwinian" program in biology, while in Uruguay there was a Darwin-tinged debate over selection in stock breeding. The only country in the region where there was no debate over Darwin's work in the nineteenth century was Paraguay, where, in the twentieth century, this seeming failure was infused with political significance as public figures seized upon the nonreception of Darwin as evidence of a defective cultural and educational structure.

BRAZIL

Brazil was famed in the nineteenth century as a paradise for naturalists. The roster of those who did significant work there includes Louis Agassiz, Henry Bates, Alfred Russel Wallace, Charles Darwin, and Fritz Müller. Persons interested in physical anthropology and race were also drawn to Brazil, because it was the largest laboratory in the world for the study of miscegenation: fully half the population was mestizo (i.e., mulatto). It was widely believed that Brazilian Indians – in particular, a group known as the Botocudos – were among the most primitive, if not the *most* primitive – on earth. De

Gobineau – to name a racial theorist – visited Brazil, and even those who did *not* visit Brazil, used Brazilian data (Virchow is a case in point).

What makes Brazil different from other Latin American countries where Darwin was received is a national obsession with race. There were quite a few polygenists, but monogenists, whether because of Darwin or the Bible, adapted polygenist language to stress racial differentiation. Out of this ideological mélange came a double law of miscegenation: (1) it is adaptive: Europeans must miscegenate (with blacks) to ensure survival in the tropics; (2) it is degenerative.

According to Silvio Romero, the most important intellectual historian in late nineteenth-century Brazil, the fin-de-siècle intellectual movement began in the mid-1870s with the confluence of republican politics and the replacement of Comtean positivism by Darwinism, Spencer, and German monism. More exactly, the French defeat in the Franco-Prussian War stimulated the abandonment of French intellectual models, Comtean positivism in the first place, at the same time as the war with Paraguay focused attention on the perceived need to define Brazilian nationality.

In any case, Darwinism leads ineluctably into the thicket of Brazilian high culture of the early republican era because the two main shapers of that culture, Tobias Barreto at the school of law in Recife and his student Romero, were outspoken Darwinians. What "Darwinian" means in this context is that Barreto and Romero (until the late 1890s), following Haeckel, thought that the struggle for existence covered *all* processes of culture and social life and that literary production, for example, had no meaning or value outside of the environment that produced it. Environment here means climate plus culture, culture being a readout of race.

Romero took literally Haeckel's insistence on the total coincidence of human beings and their environment. In order to survive in the tropics, the Europeans *had* to miscegenate, because the hybrid *mestizo* was better adapted to the environment. But then the resulting mestizo society is in a bind, because it is culturally degenerate. Romero argued that writers developed ontogenetically in the context of the phylogeny of their ancestors, as they adapted to their particular environments. Romero had picked up cultural recapitulationism from evolutionist social theorists like Gabriel Tarde and Sumner Maine, but abandoned this line in the 1890s, just as Barreto was abandoning Haeckel for a more philosophically nuanced neo-Kantianism.

Fritz Müller (1822–97), arguably the most significant Darwinian biologist of the nineteenth century after Darwin himself, emigrated to Brazil in 1852 and settled in Santa Caterina state in a German colony called Blumenau. From 1867 to 1874 he did research mainly on economic botany for the provincial governor's office, and from 1876 to 1891 he worked as "traveling naturalist," mainly in his home region for the National Museum in Rio de Janeiro. He was hired by its director, Ladislao Netto, an evolutionist interested in botany.

Müller read the *Origin of Species* in Bronn's German translation of 1860, was convinced by it, and set out to support

Darwin with findings from his own research, on comparative morphology and embryology. The result was his famous book, *Für Darwin*, published in 1863, a study of crustaceans that made the case for common descent and also illustrated the adaptive nature of dimorphism. Because he saw that the genius of Darwin's concept of natural selection as the mechanism of evolution lay in the interaction between organisms and their environment, he then abandoned morphology for natural history, that is, he became a field biologist. With a new research program, over the next twenty years or so he performed a series of observations and experiments on the interrelationships of insects and flowers, in constant epistolary communication with Darwin. At the same time, he was studying insect coloration, describing what is now known as "Müllerian mimicry" – an important piece of research in a tricornered relation with the research of Wallace and Bates. Mimicry, of course, is a phenomenon that highlights selective pressures operating on discrete populations. He also performed a series of observations and experiments (some suggested to him by Darwin) that confirmed Darwin's conclusions on the nature of adaptation in plant fertilization, dimorphism, heterostyly (plants that have styles of more than one form [the style is the elongated portion of the pistil that plays a key role in pollination]), climbing plants, heliotropism – the whole series of questions that Darwin studied in his botanical works. As early as 1866, Darwin wrote Müller, "It is quite curious how, *by coincidence*, you have been observing the same subjects that have lately interested me" (Darwin 1985–, 14, 323, letter to Müller, 25 September 1866; emphasis added).

Müller had no direct influence on the reception of Darwinism in Brazil. He was, however, a friend of the German publisher Carl von Kosseritz, who had emigrated to Brazil in 1850 and played a proactive role in the dissemination of Darwinism there through his German-language and Portuguese newspapers. He became involved in a Kulturkampf with German Protestants around 1870, when he announced, "I am a frank supporter of the Jena School [i.e., Haeckel], a Darwinian and scientific materialist, and I had the courage to make known my opinions in a country which, insofar as public education is concerned, is basically Catholic and metaphysical." A student of Haeckel's, Wilhelm Breitenbach, was for a while a science writer for von Kosseritz's newspapers; in a letter, he informed Darwin that he was the first person to popularize Darwinism in Brazil.

Between 1875 and 1900, there were three great centers of Darwinian irradiation: the law school in Recife, the medical school in Bahia, and National Museum in Rio de Janeiro. All were self-consciously evolutionist, Darwinian in their emphasis on selection in human populations, Haeckelian in their insistence on a hierarchy of races (they also pick up from Haeckel the Lamarckian notion of the direct impact of environment on phylogeny), and Spencerian in their opposition to metaphysics. They also shared the obsession for defining the nation that characterized all Brazilian positivists. Moreover, there were no universities in Brazil until 1920. So these institutions substituted for universities. The National Museum had

FIGURE 31.2. Raimundo Nina Rodrigues (1862–1906), a physician from Brazil, argued that different races are at different levels of evolutionary development and hence should be subject to different penal laws. Nineteenth-century photo

the feel of a research university, with natural and social science departments and, until the 1890s, the nation's first and only experimental biology laboratory.

Darwinian biology had been introduced at the Bahia Medical School in the 1870s by a German professor, Otto Wücherer, and Haeckelian insights were applied to public health issues by his student Raimundo Nina Rodrigues (Fig. 31.2). If, for Romero, miscegenation was generally positive, for Nina Rodrigues it was reprehensible, the source of physical, mental, and cultural degeneracy. Believing races to be fundamentally different, Nina pushed for craniological identification of racial groups pursuant to a reform of the legal code according to an evolutionary logic: you cannot, he wrote in 1894, punish races at different levels of evolution by a single standard. "It is said that we have proven among us the mestizos are incapable of perfectibility," the *Bahia Medical Gazette* editorialized in 1886, "imprisoned as they are in an advanced state of decline." One solution was to let the weaker mestizos (and all Indians, culturally inferior by definition) select out, naturally.

The third center was the National Museum in Rio de Janeiro. Under the directorships of Ladislao Netto (1874–93), a Darwinian botanist, and his successor João Batista Lacerda (1895–1915), the National Museum enjoyed a golden age, with a staff of talented foreign naturalists. Emilio Goeldi was subdirector of zoology. And among the group of "traveling naturalists" were the two German biologists, Müller and Hermann von Ihering, the latter an evolutionist but not a believer in the efficacy of natural selection. Lacerda and von Ihering were both hard-line racists. Lacerda was a polygenist who, following Nina Rodrigues, wanted to "whiten" blacks and mestizos through European immigration. Von Ihering claimed that all hybrids were decadent and somewhat later caused a scandal by advocating that an Indian group inhabiting an area targeted for a new road should be exterminated. Darwin is cited as the source for the claim that hybrids (= mestizos) are *both* more resistant *and* unstable.

The emperor, Don Pedro, played a double game, on one occasion saying he recommended that his young courtiers immerse themselves in the *Origin of Species*, while on another he denounced evolution as a fantasy. Nevertheless, while visiting England he begged Joseph Hooker to arrange a meeting with Darwin, which the former avoided. The republicans who overthrew his empire were modernizers and Darwinians (Domingues et al., 2003, 2009).

CUBA

In 1859 Cuba was an outlier of the Spanish empire; therefore, it is useful to recap the fate of Darwinism in Spain. Darwinian ideas had been successfully excluded from the country by obscurantist regimes between 1859 and 1868, when liberals took power though a revolution. Nowhere else in Europe was ideological polarization over Darwin as great. Catholics were uniformly opposed with the exception of a handful of harmonizers who followed the lead of St. George Mivart; virtually everyone to the left of center was favorable. Positivists were overwhelmingly Spencerian and anticlerical. One cannot point to any Darwinian research programs at all among nineteenth-century Spanish Darwinians, although the neuroanatomist Santiago Ramón y Cajal published influential texts with organisms discussed in phylogenetic series. In a number of well-publicized cases, local bishops condemned Darwinian writings, but such strictures had no effect except to bring ridicule upon the church. The standard Spanish translation of the *Origin of Species* was published in Madrid in 1877, *after* the publication of the *Descent of Man*.

The Darwin debate in Cuba whirled around positivist groups although, as Pruna and García González (1989) observe, Cuban positivism was a somewhat incoherent mixture of contradictory doctrines. There was also a circle of neo-Hegelians, mainly lawyers generally favorable to evolution.

Cuba was perhaps the only other Latin American country where one can perceive the lineaments of a Darwinian research program in biology. That is not so surprising because in the last decades of the nineteenth century it was scientifically and technologically precocious with respect to the metropolis. The University of Havana was a Darwinian stronghold. The anatomist Carlos de la Torre was openly Darwinian, writing on the role of natural selection in his 1883 dissertation, "The Geographical Distribution of the Terrestrial Malacological Fauna of the Island of Cuba." One of Torres's students, Arístides Mestre, wrote his dissertation on coloration in animals, using Cuban fauna to illustrate mimetism and sexual selection. By the end of the century, all of the professors teaching natural history at the University were evolutionists.

The Anthropological Society of Cuba, founded in 1877, was another center of Darwinian debate, some members arguing the innate inferiority of blacks (slavery was not abolished until the 1880s). One of the members, Francisco Calcagno, wrote a novel about the search for the Haeckelian missing link (*En busca del eslabón*, 1888), presumed to be halfway between simians and Africans. The common "scientific" view was that inferior races exercise a pernicious influence on higher ones and that, therefore, blacks should be both isolated and educated to a higher level of culture.

URUGUAY

Uruguay presents an example revelatory of the rich complexity of Darwinism in Latin America (see Glick 2001). It is the only country in the region whose brand of positivism can be described as Darwinian (because of the literal application of Darwinian precepts) rather than Spencerian, although Spencer was massively influential as well. There was no indigenous population, so the kind of social Darwinism that emerged was a literal application to economies and institutions.

Moreover, although the standard debate between religionists and secularists took place here, as everywhere, the first phase of debate took place among cattlemen starting in 1872. The Rural Association was distinctive in several ways. First it had English and French members who were secular in outlook, Second, the statutes forbade religious or ideological debate on its precincts. Therefore, the debate between supporters of selection and crossing, respectively, as to the best method for improving the indigenous "creole" herd, was based on objective considerations, the selectionists freely quoting Darwin in their favor. Even though it was recognized that the creole herd had been adapted to local pastures by natural selection, the wealthy proponents of crossing through importation of expensive breeding stock won the debate; crossing, however, eventually proved to have been just an expensive fad.

Much of the early promotion of Darwinism by Uruguayan positivists took place at the Ateneo del Uruguay in Montevideo where, according to one of the Darwinian leaders, José de Arechavaleta (professor of chemistry at the University of the Republic and of zoology at the Ateneo), most of the lecturers (by 1881) were evolutionists. In his own course at the Ateneo, he used a text written by Alfred Giard, who mixed Lamarckian and Darwinian mechanisms freely, endorsing the struggle for existence and natural selection, along with the inheritance of acquired characteristics. Arechavaleta was such a loyal devotee of Ernst Haeckel that he claimed to have discovered the bathybius (an organism bridging life and nonlife) whose existence Haeckel had hypothesized.

Lorenzo Latorre, a positivist dictator who ruled between 1876 and 1880, installed Darwinians in the education ministry and the rectorship of the university, converting the latter, as one Catholic deputy complained, into a "Darwinian

FIGURE 31.3. The Uruguayan dictator Lorenzo Latorre (1844–1916) pushed Darwinism as part of his positivist philosophy. "Rojo y Blanco," no. 10, Montevideo, 3 March 1901.

dictatorship" (Fig. 31.3). As law and medical professorships turned over, Darwinians were appointed to succeed believers. In the Faculty of Law, Martin C. Martínez taught natural law according to Darwinian and Spencerian perspectives. He called himself an "explanatory naturalist," believing that Darwinian mechanisms explained social phenomena in a literal, and not merely figurative, way. Both the history of law (particularly that of property) and the development of social institutions could be understood through the action of natural selection.

VENEZUELA

References to Darwin appear in the acts of the Sociedad de Ciencias Físicas y Naturales, in Caracas, between 1867 and 1878, generally introduced by Adolfo Ernst, a German-born naturalist. It was Ernst who gave the first lectures on Darwin in his department of natural history at the University of Caracas. In 1893 Pablo Acosta Ortiz, professor of anatomy, introduced modern anatomy following Darwinian norms. Luis Razetti, who succeeded Acosta in the same chair in 1896, was an outspoken evolutionist. In 1904 Razetti maneuvered the Academy of Medicine into a statement of public support for evolution by introducing a motion in such a way that the members had a choice of voting in favor of the motion or declaring themselves unscientific ideologues. As had been the case earlier in Uruguay, Venezuela was ruled by a positivist dictator, Cipriano Castro, who backed the publication of Razetti's materialistic book, *Qué es la vida?* (What Is Life?) in 1906 (Glick 1984).

MEXICO

In Mexico, the leader of the positivist revolution was the Comtean Gabino Barreda, who had studied in France with Comte (1847–51) and later founded the key positivist institution in Mexico, the Escuela Nacional Preparatoria. He opposed Darwin on evidentiary grounds, which seemed to him too scant to support such a broad theory, and favored Lamarckian evolution on philosophical grounds. Nevertheless, the Comteans' reticence with respect to Darwin and the Spencerians' campaign in his favor, were simultaneous and took place in the Preparatory School itself, where Justo Sierra defended Darwin and Wallace in 1875. Two years later, a full-blown debate took place under the auspices of the Methodophile Association, a Comtean society named after Barreda and where Barreda, while not damning Darwin, questioned his evidence and was answered by Porfirio Parra, who asserted that Barreda had failed to understand Darwin's metaphorical language. The church got involved in the polemic in 1878 when objections were raised to the supposedly Darwinian content of Sierra's recently published *Compendium of Ancient History*. Prehistory was a key battleground in many countries between the church and Spencerian historians, who liked to begin large-scale surveys with hominization and prehistory (Argueta 2009). Roberto Moreno demonstrated that Darwinism found its way into official education by the end of the decade and that resistance to the new doctrine was stronger among Comteans than among Roman Catholics. After the turn of century, naturalists and philosophers argued that Mexican Amerindians were more highly evolved than Europeans, a line that reached it peak in *La raza cósmica* (The Cosmic Race, 1948), by José Vasconcelos.

ARGENTINA

The Darwinian "narrative" in Argentina was heavily racialist, in spite of the prior extinction of most local indigenes at the time it was developed. It was argued that extinction is a natural event, and so the slaughter of Argentina's indigenous population could be rationalized by recasting it in Darwinian terms. The extinction of the Amerindians was now a mark of progress, as if a less favored race had lost out by virtue of a "natural" law (the Darwinian struggle for life). The "culture of extinction" was pervasive, from government policy to the way in which museums were organized, history written, and nationhood conceptualized (Novoa and Levine 2010).

Darwin was elected a corresponding member of the Argentine Scientific Society in 1877 after almost a decade of

FIGURE 31.4. Florentino Ameghino (1854–1911) made massive collections of fossils in Argentina and promoted the New World origins of humankind. Nineteenth-century photograph

Henri Gervais) and in 1880–81 published an influential and controversial volume titled *La antigüedad del hombre en el río de la Plata* (The Antiquity of Man in the Plate River Basin) in which he argued that *Homo sapiens* had emerged in Argentina (Fig. 31.4). As a result, he was backed by the government in spite of Ameghino's open evolutionism.

POSITIVIST PRESIDENTS

Because of the vogue for Spencerian perspectives, there were quite a few positivist presidents in the various republics: Lorenzo Latorre in Uruguay and Cipriano Castro in Venezuela were authoritarians who put the weight of the state on Darwin's side. The Argentine Sarmiento was an emblematic Darwinian. In Bolivia, José Manuel Pando, a geographer and president between 1899 and 1904, was an arch social Darwinist believing that Amerindians were inferior beings and their elimination, via natural selection, was a necessary concomitant of civilization. Another Bolivian geographer-president (1921–25) was Bautista Saavedra, who believed that indigenous Aymará communities, based on extended families, represented an anachronistic form of social organization and ought not to survive.

COMMON CURRENTS

In spite of the wide variation in response, common traits across countries can be detected. First is the salience of race wherever there were indigenous and/or black populations. Moreover, debates over emerging national identities were imbued with social Darwinian explanation. Second is the universality of positivism among Latin American intellectuals: if Spencerian, its effect on the reception of Darwinism was reinforcing, whereas Comteans were almost always in opposition. Third, the opposition of the Catholic Church, while present everywhere, tended to be along the lines of the Italian reception, where strong secular scientific institutions blunted the church's influence with respect to Darwin. Fourth, faculties of medicine and law were Darwinian strongholds. Fifth, the books of Darwin, Haeckel, and Huxley were read in French, even though some libraries owned English editions. Sixth, foreign Darwinians were influential in some countries – the German ex-Jesuit Theodor Wulf in Ecuador, the German botanist Adolfo Ernst in Venezuela, the Spanish radicals José Arechaveleta and Francisco Suñer in Uruguay, and in Brazil the Italian Swiss Emilio Goeldi.

public debate. Domingo Faustino Sarmiento, president from 1868 to 1872, was an outspoken Darwinian. Days after Darwin's death he was the principle speaker at a massive public homage to Darwin at the Teatro Nacional, attended by three thousand people. The second speaker was Eduardo Holmberg, a physician and Darwinian paladin who, in 1875, had published a polemical novel titled *Dos partidos en lucha* (Two Parties in Battle [i.e., pro- and anti-Darwinists]). The fictitious polemic was resolved by Darwin himself, who (in the novel) arrives in Buenos Aives in August 1874, greets President Sarmiento, and then addresses a scientific meeting and suggests an experiment that might resolve the debate.

The Argentine paleontologist Florentino Ameghino studied fossil mammalogy in Paris in the 1870s (with Paul and

❦ ESSAY 32 ❦

Botany: 1880s–1920s

Dawn Mooney Digrius

According to Howard S. Reed (1942, 3), "among the events of the nineteenth century which indicated the impetus given to biological studies by the publication of Darwin's *Origin of Species* (1859) none was more significant than the rise of international congresses." Indeed, international congresses were a "symbol of the new freedom that science found after the emergence of the great ideas presented by Darwin." For the botanical sciences, the introduction of Darwin's theory of descent with modification created new challenges and fostered key developments that forever altered its practices. The world was expectant of new discoveries, the integration of ideas, the unification and simplification of terminologies, improvements in record keeping and documentation, and frequent international gatherings to present ongoing or novel programs of research to the larger community of botanists (see Reed 1942). This essay examines key movements and figures in the botanical sciences from the years between the 1880s and the 1920s, highlighting the impact of Darwin on the botanical sciences.

DARWIN'S BOTANY BY 1880

Botany provided key evidence to support Charles Darwin's argument for descent with modification in the *Origin of Species* (1859), stemming from botanical experiments and observations by Darwin himself, as well as research from plant breeders such as T. A. Knight, plant distribution information from the Candolles, and advice from Joseph Dalton Hooker (Morton 1981, 415). Once the *Origin of Species* was completed in 1859, Darwin's attentions were focused increasingly on botanical subjects, including publications on the fertilization of orchids (1862), climbing plants (1865), insectivorous plants (1875), fertilization (1876), flowers on plants of the same species (1877), and in 1880 the power of movement in plants.

As Soraya de Chadarevian (1996, 17) has noted, Darwin's results in his work on movement "contradicted the observations and explanations of the same phenomena offered by the German plant physiologist Julius von Sachs" in 1868. The argument between Sachs and Darwin centered on the manner by which a botanist should conduct experiments. Their disagreement not only reflects a monumental shift in

FIGURE 32.1. Julius von Sachs (1832–97), the leading German plant physiologist in the second half of the nineteenth century, was very critical of Darwin's experimental skills, yet never wanted even to speculate on the causes of evolution. Permission: American Philosophical Society

the nature of botanical practice but outlines the evolution of the botanical sciences toward a more professional pursuit. Indeed, what can be deduced from this controversy is that by 1880 botany was changing (Fig. 32.1).

Sachs, who had initially been under the tutelage of Jan Evangelista Purkyne in Prague, obtained his *Hablitation* or teaching qualification at the University of Prague in plant physiology, a subject not previously offered before Sachs's qualification. After obtaining positions at the Agricultural Division of the School of Forestry in Tharandt and the School of Agriculture in Bonn, Sachs held the position of professor of botany at Würzburg from 1868 until his death in 1897 (de Chadarevian 1996, 29, 27).

J. Reynolds Green, in his *History of Botany, 1860–1900* (1909, 18), considered Sachs to be the "father of modern botany" because of his influence on the science in the classroom and the laboratory. Many future botanists studied under Sachs at Würzburg, including Josip Baranetsky, F. O. Bower, Julius Brefeld, Francis Darwin, Karl Goebel, Emil Godlewski, Emil Heinricher, Georg Klebs, Pierre-Marie-Alexis Millardet, J. W. Moll, Hermann Müller-Thurgau, Atsusuke Nagamatsz, Fritz Noll, Wilhelm Pfeffer, Karl Prantl, Johannes Reinke, D. H. Scott, Christian Stahl, Sydney Howard Vines, Hugo de Vries, Marshall Ward, Mikhail Woronin, and Julius Wortman (Noll 1898, 7).

Darwin, in contrast, had obtained his botanical training through his close relationship with John Stevens Henslow at Cambridge. Long walks and field trips with his botanical mentor provided a good foundation in natural history. However, Darwin had not been through the rigorous scholarship in botany that Sachs had experienced; he had "not been academically trained" in botany and "did not possess the detailed knowledge of plant systematics," which, at the time was the hallmark of a "true botanist" (Morton 1981, 415). Darwin himself acknowledged his ignorance in botany in an early letter to J. D. Hooker, beseeching that Hooker write "as well for Botanical ignoramuses as for great Botanists" (Darwin 1985–, 2:420, letter to Hooker, 12 December 1843).

Their differences in training had a profound effect on Sachs's opinion of Darwin's work. For example, skill was a factor that highlighted not only the debate between these two men but also the significant move toward professionalization. Sachs stressed experimental skill not only in microscopy but also in one's competence in plant physiology. Thus, in this debate over the movement in plants, Sachs intended to criticize Darwin's scientific practice as well as to establish exact standards in said practice. Sachs believed that not only had proper methods of scientific investigations of plants declined because of Darwin's influence but botany had also suffered as a result because Darwin had "merely gathered facts from literature" and could neither "conduct experiments, nor use the microscope" (de Chadarevian 1996, 31).

These were not the only disagreements that Sachs had with Darwin. While accepting the doctrine of descent, as outlined by Darwin, Sachs had "abandoned faith in all attempts, including especially the theory of natural selection, to explain the evolutionary process" (C. E. Allen 1933, 344). Why would Sachs not fully support Darwin's views? Part of the reason is that he believed that the "natural system is explicable only by descent; how descent is to be explained, nobody knows" (ibid.). The question, however, of relationship led him, like Darwin, into discussions of organic evolution, yet Sachs was not in complete agreement on how organic evolution took place. Botanists, it seems, were not as keen to follow Darwin, largely owing to questions regarding speciation. Sachs represents a key figure here, for he neither saw eye to eye with Darwin regarding natural selection nor agreed on how one does botany in order to understand the evolutionary process. Experimentation, training in the proper use of technological tools such as the microscope and the auxanometer, and the "knowing eye" of the practitioner were all critical in providing useful results. Instrumentation and skill, both centered in the laboratory and not the country house, were key to a new perspective in botanical science. Thus, Sachs and botanists like him, not Darwin, ushered in the "new" botany.

THE "NEW" BOTANY

In his discussion of the impact of the "new" botany on American agriculture, Richard A. Overfield (1975, 164, 165) noted that many of the practitioners influenced by the move toward a more professional botany saw themselves as scientists, and specifically botanists, more so than agriculturalists. Over and over again, these men stated that "research must replace mere observation and collection," with botany no longer a "pleasurable pastime of identifying plants" but rather a science based on the "experimental method of inquiry." The aforementioned debate between Sachs and Darwin clearly illustrates this move. Advocates of professionalization in botanical science such as Sachs wished to break free from the narrow constraints of taxonomy and instead make botany an "evolutionary study of the origin and relationship of species." Thus, new areas of exploration in the plant sciences were promoted through the "new" botany: paleophytology, physiology, pathology, ecology, cytology, and chemical studies of carbon and nitrogen fixation in plants. An examination of the tables of contents from several histories of botany that highlight the period after the publication of the *Origin of Species* reveals that these innovative avenues of research were now possible largely owing to the distinct influence of its practitioners (see Sachs 1890; J. R. Green 1909; Reed 1942; Morton 1981).

With this shift toward professionalization, there came the establishment of credentials and academic programs to train specialists. In correlation, botany saw the institution of more professional societies, journals, and funding avenues for experimentation and research. For example, the Botanical Society of America was formally recognized in 1893, beginning as an offshoot of the Botanical Section of the American Association for the Advancement of Science. The British Association for the Advancement of Science had been established in 1831, adding a Botanical Section in the latter part of the 1800s. In addition, universities were now becoming the geographical locales of training. Chairs of Botany had been established at many European universities before the nineteenth century, however, reflective of a move toward professionalization, and such programs eventually added laboratories. Sachs and Williamson installed botanical laboratories at Würzburg and Manchester, and the Jodrell Laboratory at Kew Gardens in London was installed in 1876.

What is seminal about this shift toward the "new" botany was its emphasis on skill, training, experimentation, and professionalization. The new crop of botanists that emerged from the laboratories of the "new" botany secured the "necessary foundation on which professionalization advanced" (Overfield 1975, 169). Sachs, in his stress on the development of experimental skill, saw the laboratory as "precisely the place where one could acquire experimental skill by practical training ... where standards were established" (de Chadarevian 1996, 37). This innovative attitude toward botanical practice influenced not just botany alone but also specific pursuits within the botanical sciences. Paleobotany, like botany, now required a new outlook and a new locale for practice. William Crawford Williamson would be a major contributor, along with Sachs, in this move.

PALEOBOTANY EMERGES AS AN AREA OF STUDY

Just as the "new" botany was taking hold during the second half of the nineteenth century, paleobotany, or the study

FIGURE 32.2. William Crawford Williamson (1816–95) was the leading paleobotanist of his day, strongly supporting a Darwinian approach to the study of fossil plants. From F. W. Oliver, *Makers of British Botany* (Cambridge: Cambridge University Press, 1913), pl. 21

of fossil plants was coming into its own. Led by the British naturalist William Crawford Williamson (1816–95), paleobotany developed out of geology and botany and became an established area of study beginning in the 1850s. Williamson, a medical man based in Manchester, had a long historical connection to the study of fossil plants. His father, John Williamson, had been director of the Scarborough Museum and frequently worked alongside noted geologists William Smith (1769–1839), his nephew John Phillips (1800–74), and Sir Roderick Impey Murchison (1792–1871). As a young boy, Williamson the younger accompanied this group of men on geological excursions along the Yorkshire coast, spending his evenings identifying and drawing specimens collected during these adventures (see Williamson 1896; Watson and Thomas 1986). At sixteen years of age, William Crawford Williamson was invited to illustrate specimens for John Lindley and William Hutton's *Fossil Flora of Great Britain* (1831–37), a task he undertook at his kitchen table (Fig. 32.2).

Where botany evolved from a gentleman's pursuit into a professional area of expertise in the botanical sciences, paleobotany came as a result of Williamson's, like Sachs's, insistence that the practice become more standardized and specialized.

Beginning in the 1830s, some paleobotanists, such as Henry Witham (1779–1844), had called for the closer observation of fossil plants through use of the microscope. However, this practice was slow to take hold until the 1860s. It was through Williamson's persistence that the only true means by which one could identify and classify fossil plants was by observations of their internal structure, and not on the basis of the observations of their external character, that paleobotany became, according to Williamson, scientific (see Digrius 2007). The microscope was tapped as the primary tool for observation, which guaranteed accurate results (within the realm of the aberrations that may still exist with some instruments) in seeing the internal characters of fossil plants.

Why was Williamson so keen on establishing technological methods and professional training for paleobotany? Part of the reason was his relationship with Charles Darwin. As early as the 1840s, Williamson and Darwin traded specimens, and Williamson was responsible for analyzing sediment samples Darwin took in South America (Darwin 1985–, 3: 324, letter to Williamson, 23 June 1846). Williamson became a strong supporter of Darwin's theory and its application to the study of fossil plants, in contrast with the noted French paleobotanist Adolphe Théodore Brongniart (1801–76). Brongniart was trained under his mentor Georges Cuvier (1769–1832), who was a proponent not of transmutation but of catastrophism. The British school of paleobotany distinguished itself from the leading French school of paleobotany during the middle part of the nineteenth century by its heavy reliance on microscopy and its theoretical leanings against catastrophism (see Digrius 2007). In all of his published works from the 1860s onward, Williamson adopted a Darwinian approach in his researches, establishing the British school of paleobotany under this framework in Manchester at the Owens College (now the University of Manchester), and trained the future evolutionary paleobotanists Albert C. Seward (1863–1941), Dukinfield Henry Scott (1854–1934), and Robert Kidston (1852–1924).

How do we know that the evolutionary paleobotanical framework was heartily espoused by Williamson? A. C. Seward was responsible for editing the 1909 centennial publication recognizing Darwin, *Darwin and Modern Science* (1909), after becoming professor of botany at the University of Cambridge in 1906. In the preface, Seward (1909, v) noted that it was hoped that the publication of the essays written in honor of Darwin's centennial would "serve the double purpose of illustrating the far-reaching influence of Darwin's work on the progress of knowledge." Above all what Seward also intended in this volume was to reflect upon the "present attitude of original investigators and thinkers towards the views embodied in Darwin's works," scientific and societal. Williamson's influence on paleobotany was not felt in Britain alone. Gaston de Saporta (1823–95) was a frequent correspondent of Williamson's, as well as a strong supporter of Darwinian paleobotany in France. Hermann Solms-Laubach (1842–1915), a noted German paleobotanist, counted Williamson as a friend and agreed with him that paleobotany should be professional, technological, and evolutionary.

DARWIN'S INFLUENCE ON BOTANY OUTSIDE OF CONTINENTAL EUROPE

In Russia, as in Britain and in Germany, the botanical sciences evolved into professional areas of study beginning with the establishment of the Imperial Botanic Garden in St. Petersburg about 1714 (Shetler 1967, 23). However, by 1823 the reorganized Botanic Garden saw its first botanist as director, F. E. I. Fischer (1782–1854). The nineteenth century saw rapid expansion of the botanical garden in St. Petersburg, making it the "major botanical institution of Russia" but also a major international botanical center in Europe (Shetler 1967, 30). During the 1860s, Darwin began correspondence with Russian scientists who wished to incorporate Darwinism into their "particular branches of science" (J. A. Rogers 1960, 379). For botany, the most important figure in bringing Darwin to Russia (and to Russian botanical science) was Kliment Arkeedevich Timiriazev (1843–1920). As Rogers (1973, 499) noted, Timiriazev first encountered Darwin as a student at St. Petersburg University through his professor, "who told his class that the theory was new, but sound." After traveling abroad to complete his education in 1868, Timiriazev took a position as teacher of botany at the Petrovsky Academy of Agriculture and Forestry, vowing to be a strong supporter of Darwin's work. Just at the time that Darwin was working on his book *The Power of Movement in Plants*, Timiriazev met him at Down House, where, after viewing Darwin's experiments, he spoke with the man for roughly two hours (501).

Until the 1880s, Timiriazev was the main propagandist for natural selection in Russia, a place where opposition to Darwin was slow to develop. Not only a strong supporter of Darwin, the botanist also felt that the task of the plant physiologist is not to describe but to explain and that his role is not as an observer but as an active experimenter (see his 1943 obituary in *Nature* 151, 611). Thus, just as Sachs and Williamson had done, Timiriazev in Russia contributed greatly to the professionalization and evolutionary focus in the botanical sciences.

THE RECOVERY OF MENDEL AND THE RISE OF GENETICS IN BOTANY

In February 1865, the same month that Charles Darwin was reading W. C. Spooner's work on blended characters, Gregor Mendel read his paper "Versuche über Pflanzen-Hybriden" before the Natural History Society of Brünn. This paper examined the nature of inherited characters of the plant *Pisum* (edible pea), with Mendel conducting artificial pollination in order to tease out "what developmental laws govern the propagation of hybrids" (Gliboff 1999, 225). Because Mendel was keen to put forward his theory in terms of mathematics, more like a physicist than a natural historian, Mendel's mathematical expressions of the combinations of traits in *Pisum* helped establish the methods of planning genetic experiments and predicting the appearance of new combinations of hereditary

FIGURE 32.3. Hugo de Vries (1848–1935), Dutch botanist and student of von Sachs, was one of the rediscoverers of Mendel's laws of heredity. Permission: American Philosophical Society

characters, becoming known as Mendel's Law of Independent Assortment (Digrius 2008, 95). Interestingly, Charles Darwin, it seemed, was not aware of Mendel's work. Darwin, while cognizant of the role of heredity in variation, was not keen on the laws governing inheritance and admitted that they were not fully known or understood. Mendel was, in essence, investigating the relationship of heredity in the origins of variety, at odds with Darwin's conception of a steady flow of variations, to which small changes were constantly occurring.

This, according to Darwin, gave rise to new species slowly over time. Unfortunately for Darwin, Mendel's contributions to understanding the nature of heredity were kept largely isolated within a small group of naturalists in Central Europe and did not figure into his theoretical perspectives. The significant contributions to the history of the botanical sciences with the rediscovery of Mendel would have to wait. A happy accident, independently occurring among three European botanists, Hugo de Vries, Carl Correns, and Erik von Tschermak in 1900, would, however, usher in a new phase of research in botany (99).

As the story goes, Hugo de Vries (1848–1935) obtained a copy of Mendel's paper in the early part of 1900. Before 1900, de Vries had been conducting hybridization experiments on *Zea mays* (corn) and other plant species such as *Silene alba* (white campion) and *Papaver somniferum* (poppy). What de Vries found in his experiments was that in each case he saw a 3:1 ratio (same as expressed in Mendel's experiments), meaning that, in every four plants in the third generation, three would manifest the dominant character and one would manifest the recessive (Digrius 2008, 94). Results from these hybridization experiments were presented before the Royal Horticultural Society in 1899 and published in the *Comptes Rendus de l'Academie des Science* in Paris under the title "Sur la loi des disjunction des hybrids" (de Vries 1900). However, no mention of Mendel appeared in the work. It was this publication that sparked the Dutch botanist Martinus Beijerinck to send de Vries a copy of Mendel's paper, with the notation that "I know that you are studying hybrids, so perhaps the enclosed reprint of the year 1865 by a certain Mendel which I happen to possess is still of some interest to you" (Olby 1966, 127). The rest, as we say, is history (Fig. 32.3).

Independently of de Vries, Carl Correns (1864–1933) had been conducting experiments on *Pisum* and *Zea mays* in Tübingen, after gaining access to its botanical garden and serving as *Privatdozent* beginning in 1892. Like Mendel, Correns found that there were some instances where hybrids were intermediate between the parent types, arguing that de Vries's law of dominance was not universally acceptable. Correns's attitude may have stemmed from his belief that Mendel should receive credit for propagating such laws of heredity, and not de Vries (Fig. 32.4).

Erich von Tschermak (1871–1962), the third member of the rediscovery triad, undertook breeding experiments at various locations on vegetables and ornamental plants (Digrius 2008, 104). While at Gent, Tschermak read Darwin's work on the cross- and self-pollination of plants, spurring on further studies. In January 1900, Tschermak completed his dissertation on his crossbreeding experiments and was now in the sphere of influence of de Vries, whom he had visited previously. De Vries sent Tschermak a copy of his 1900 paper published in the *Comptes Rendus,* of which Tschermak responded by publishing his own studies on crossbreeding in the *Zeitschrift für das landwirtschaftliche Versuchswesen in Österreich* in order to promote the idea that Mendel's fundamental principles of inheritance must be applied in order to achieve stable and uniform combinations of different characters of parental genotypes by crossing experiments (Fig. 32.5).

William Bateson (1861–1926) is another integral figure in the rediscovery of Mendel. Bateson had read the republished Mendel paper in 1900 and soon after published *Mendel's Principles of Heredity* (1902) (Fig. 32.6). The goal, according to Bateson, was to stave off critiques of Mendel's work and to promote Mendelism in the biological sciences. Bateson made it a central part of his researches to support

FIGURE 32.4. Carl Correns (1864–1933), German botanist, was one of the rediscoverers of Mendel's laws of heredity. Permission: American Philosophical Society

FIGURE 32.5. Eric von Tschermak (1871–1962), Austrian agronomist, was one of the rediscoverers of Mendel's laws of heredity. Permission: American Philosophical Society

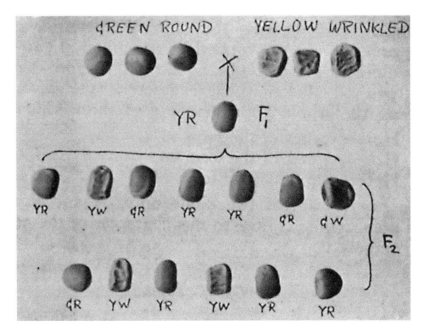

FIGURE 32.6. Mendel's laws as illustrated by William Bateson in his translation of Mendel's *Principles of Heredity*. From W. Bateson, *Mendel's Principles* (Cambridge: Cambridge University Press, 1909)

Nikolai Ivanovich Vavilov's works on the geographical origins and evolutionary history of economic plants (1915-40) stemmed from his travels in Europe and collaborations with William Bateson (1861-1926) (Fig. 32.7). Vavilov's belief that related genera and species were defined by similar morphological features and variations in morphological and physiological characters, all of which are an expression of phytogenetic relationships influenced evolutionary theory and plant breeding (Morton 1981, 453). Charles E. Bessey's 1915 paper on the phylogenetic taxonomy of flowering plants, connected *Ranunculus* and its nearest relatives as closely representative of primitive angiosperms (flowering plants), leading to increased work on the nature of the origin of angiosperms, something that Darwin tagged "the abominable mystery."

Questions of speciation had arisen soon after the publication of *Origin of Species*, and there was no general agreement among plant researchers as to what is or is not a species. Sachs did not accept natural selection as the mechanism for species change. He knew that change happened, but the manner in which it happened was not known. Garland Allen (1969, 63) notes that de Vries felt that "the term species itself was often used in two different senses: one was the systematists' species, the other the 'real' species." The systematists' had defined species as arbitrary units useful only to make sense of the seemingly orderless nature of the natural world. According to de Vries, distinctions systematists made between species were wholly dependent upon "trivial or non-adaptive characters" and "had no reality in nature." In contrast, real species existed, and were identified by marked character differences. Mutations, as de Vries found with his work on *Oenothera*, solved a key issue in presenting an example of how a process of heredity could also explain evolution (G. E. Allen 1969, 63, 65). De Vries, having studied under Sachs, may have also been influenced by Sachs determinate view regarding natural selection. Furthermore, de Vries, like Correns, was a botanist and "worked in the botanical evolutionary tradition which favored speciation by hybridization . . . whatever the details turned out to be, Mendelism would provide a more plausible mechanism for speciation than Darwinian gradual selection, clearing the way for the ultimate triumph of the much maligned botany-based tradition" (Depew and Weber 1995, 223, 224).

The reception of de Vries's work at the time (1901) was highly favorable to many biologists, mainly because it had highlighted an innovative approach to understanding a problem that Darwin's natural selection had left unanswered. However, plant breeders were not supportive of de Vries's mutation theory, nor was there full agreement among botanical practitioners as a whole. While experimentalists such as the Danish botanist Wilhelm Johannsen (1857-1927) heartily

and bolster Mendelian hereditary ideas from attack and to work them into the general understanding of how evolutionary change occurred. He noted that Mendel's experiments were first undertaken in the hope that they would make the problem of speciation easier to explain. Bateson (1930, 17) argued that the consequence of the application of Mendel's principles shaped a vast medley of seemingly capricious facts recorded on heredity and variation and brought them into an orderly and consistent whole. For the botanical sciences, the introduction of Mendelian genetics allowed practitioners to "penetrate the molecular organization of living systems" (Morton 1981, 450).

RESEARCHES IN BOTANY 1900s-1920s

Because of the rediscovery of Mendel, botany, "in common with other divisions of biological science, was profoundly affected by closer integration with the physical sciences" by the early decades of the twentieth century (Morton 1981, 450). This was facilitated largely by the innovation of techniques such as chromatography, ultracentrifugation, isotope labeling, and electron microscopy, as well as by the movement of hard scientists into the biological sciences. With these new methods and microscopically trained practitioners, studies of plants at the cellular and molecular level were now possible. As we have seen, in the years between 1900 and 1929, cytological studies of heredity were firmly established in the botanical sciences. Morton (1981, 453) noted that "the combination of cytological studies with systematics brought new concepts in the study of plant speciation when it was shown that polyploidy (more than two paired sets of chromosomes) and hybridization have played a frequent part in plant evolution."

FIGURE 32.7. Nicolai Ivanovich Vavilov (1887-1943), Russian botanist, who contributed much to the understanding of the geography of plant evolution. He died of malnutrition in prison, arrested for opposing the non-Mendelian speculations of Stalin's favorite, Trofim Lysenko. Permission: American Philosophical Society

supported an alternative to Darwin, others did not. As Paolo Palladino (1993) argued in his work on Mendelian genetics and plant breeding, "the desire to increase agricultural productivity was the principle reason for the promotion of the Mendelian theory."

Thus, what can be said regarding the period between 1900 and 1920 is that no definitive agreement was attended to regarding either the question of species or of the adherence to or rejection of natural selection by plant researchers.

Ronald A. Fisher (1890-1962) contributed greatly to the botanical sciences and the promotion of evolutionary theory. After taking a post at the Rothamsted Agricultural Field Station in 1919, Fisher applied statistical analyses to experiments conducted at this, the oldest agricultural research institute in the United Kingdom (1837), that revolutionized agricultural research. Fisher described the statistical methods for evaluating the results of small sample experiments in order to minimize the disturbances due to the heterogeneity of soils and the unavoidable irregularity of biological material. Part of the reason for Fisher's interest in these matters was that he was intrigued by the "controversy on the hereditary determination of continuously variable characters which had raged between the Mendelian geneticists and the Darwinians" (Clarke 1990, 1447). Fisher believed that Mendelism and Darwinism must fit together, thus using statistical analyses to show "that the properties of continuous variation were compatible" with the small differences revealed through Mendelian inheritance, provided the effects were additive (1447). Fisher's *The Genetical Theory of Natural Selection* (1930) represents the unification of Darwin's ideas and a quantitative way to work with data to support it. For botany, the ability to utilize mathematical tools in the discussion of evolution ushered in a vast range of research topics that now had quantitative methods to support them. Fisher believed that "maximization leads to constantly renewed equilibrium," and therefore "change is gradual over an array of continuous variation"; what this means with respect to debates over natural selection, especially for botanists, is that his "commitment to the primacy of selection, to gradualism, and to equilibrium ensures its intellectual continuity" (Depew and Weber 1995, 251).

Statistical analyses were not the only new approach to be applied to botanical researches. The genetical turn after rediscovery of Mendel led to new avenues for the study of plants. George Ledyard Stebbins (1906-2000) is considered one of the foremost botanists of the twentieth century (Smocovitis 1997, 1625). While at Harvard University between the years 1924 and 1931, Stebbins was "like any ambitious young researcher ... interested in new ideas, and in his case new approaches to taxonomy" (Smocovitis 2005, 400). Because cytological studies were proving useful in taxonomic works, Stebbins chose to study the cytology of the genus *Antennaria* in the Asteraceae family for his PhD dissertation. Experimental taxonomists like Stebbins added a new dimension that field and herbarium studies, however rigorous, could not duplicate; these botanists were concerned primarily with ecological and genetic problems rather than with classification (Hagen 1984, 257).

Also reflecting this turn toward genetics and botany, Ernest Brown Babcock (1877-1954), in his "Genetics and Plant Taxonomy" (1924) suggested the significance of applying experimental genetics to taxonomy. Stemming from this call to incorporate genetical studies into the botanical sciences, an entire session at the 1926 International Congress of Plant Sciences was organized to bring together botanists, cytologists, taxonomists, and geneticists to encourage the cooperation of these practitioners in their research pursuits. Illustrative of this call, Babcock and Harvey Monroe Hall (1874-1932), in their study of hayfield tarweeds (1924), combined methods drawn from genetics, cytology, ecology, and comparative morphology (Hagen 1984, 259; see also Babcock and Hall 1924). Similarly, Babcock's work on *Crepis* remained, according to Stebbins, the foremost attempt to explain the evolution of a genus of plants primarily on a genetic basis (see Stebbins 1968).

Working with R. A. Fisher, C. D. Darlington, and J. B. S. Haldane at the John Innes Horticultural Institute in Britain, Edgar Anderson (1897-1969) attempted through a National Research Fellowship to answer the question of what a species is. His work on irises, according to Kleinman, "strengthened the case that hybridization has been an important factor in the evolution of species." Stemming from his curiosity about the role of hybridization in species formation, Anderson's works on *Nicotiana*, *Iris*, and *Zea mays* reveal that by the 1920s the intersection of genetics and taxonomy "anticipated and contributed to the developing synthesis in evolutionary theory" (Kleinman 1999, 304-5, 300).

William Bertram Turrill (1890–1961), another experimental taxonomist, had established transplant studies at the British Ecological Society, including genetical and general botanical researches on *Ranunculus*, *Saxifraga*, *Centaurea*, and *Silene*. Like Babcock and Anderson, Turrill heavily supported the unification of cytological and taxonomic methods, and "far more intensive and wider-ranging studies of plants, covering all aspects of botanical investigation, were required before any lasting improvement could be effected in their recognition as species, grouping into genera, and in their classification generally" (C. E. Hubbard 1971, 692). What this shift toward a more inclusionary and genetically centered botany allowed was the introduction of the modern evolutionary synthesis to the biological sciences. And, owing largely to the genetic turn in botany that took place in the years between 1900 and the 1920s, botanists like Stebbins, Babcock, Anderson, and others contributed to the foundational aspects of research that brought the botanical sciences later into the synthesis (Smocovitis 1997, 1635).

CONCLUSION

The contributions of the botanical sciences noted here illustrate how it all "hangs together and bears on the one great problem in biology – the evolution of life" (Stopes 1912, 9). The botanical sciences were greatly influenced by the work of Charles Darwin, with his idea of descent with modification. During the years between the 1880s and the 1920s, significant changes were made in the practice of botany that reflected its evolutionary path. Practitioners of botanical science became trained professionals mastering microscopic and experimental methods necessary to better understand the nature of species change in the vegetation of the earth. Situated within the "new" botany, laboratories became the geographical locales of experimentation, and stress was placed on skill and observation. As the debate between Darwin and Julius von Sachs revealed, as botany evolved, so too did the nature of its practice. No longer could a botanist be considered a botanist without having been connected to a research station, laboratory, or university-centered program.

In addition to the shift toward a more professional level of practice, the role of technological and methodological tools was critical to moving botany forward into the twentieth century. The microscope, and its proper application, was a required and necessary vehicle for observations of plants, both living and fossilized. As William Crawford Williamson stressed, no classifications could be valid without the application of microscopy to the study of plants. The emphasis on the microscope assisted in ushering in an exciting phase of research questions for practicing botanists.

The rediscovery of Mendel in 1900 represented a monumental shift in the history of biology. Without this, it is unlikely that evolutionists could have made as many strides in understanding the workings of living organisms as they did. Mendel's seemingly innocuous fiddling with pea plants, completely overlooked in his time, was recognized as one of the greatest breakthroughs of evolutionary biology. For botany in particular, Mendel's rediscovery enabled investigations of inheritance and origins of plants not previously undertaken. And the genetic turn in botany would not have occurred if it were not for the contributions of de Vries, Correns, and Tschermak.

Finally, botany earned a solid footing in the modern evolutionary synthesis largely because of the stress placed on the unification of genetics and taxonomy. With the genetic turn, botanical researches focused on investigations of the relationship between hybridization and species formation, and as a result a comprehensive synthesis of plant evolution incorporating the implementation of genetics was established.

Thus, the years that encompass the period ranging from the publication of Darwin's *The Power of Movement in Plants* (1880) to the experimental taxonomic works of Stebbins, Babcock, and Turrill represent a monumental shift in the history of botany that reflects the influence of Darwinian evolutionary theory and its subsequent iterations based on the rediscovery of Mendel in 1900 and a necessary adherence to natural selection. This period created the necessary foundation for the coming modern evolutionary synthesis and presented new avenues for botanical researches that far superseded the questions available to practitioners in the years before professionalization in botany occurred. Exciting challenges faced the botanists of the twentieth century, and practitioners emerging out of the research programs of the early decades of that century forged innovative paths in evolutionary botany that still resonate to this day.

❧ Essay 33 ❦

Population Genetics

Michael Ruse

O<small>N</small> 8 M<small>AY</small> 1900, the English biologist William Bateson was on the way to the Royal Horticultural Society to read a paper on the topic "Problems of Heredity as a Subject for Horticultural Investigation" (Fig. 33.1). Several years later, his wife tells the story: "He had already prepared this paper, but in the train on his way to deliver it, he read Mendel's actual paper on peas for the first time. As a lecturer he was always cautious, suggesting rather than affirming his own convictions. So ready was he however for the simple Mendelian ratios that he at once incorporated it into his lecture" (B. Bateson 1928, 73). Mendelism had arrived on the scene!

MENDELIANS AND BIOMETRICIANS

Bateson was one of a number of biological researchers who, in the 1880s, had started their careers as morphologists or embryologists, much interested in evolutionary questions (Ruse 1996). However, it was not long before he and his fellows realized that truly they were getting nowhere. Although, out in the American West, fabulous fossil finds were being made almost daily, overall the fossil record was not strong enough to support detailed investigations of life's past. Closer to home, Haeckel's biogenetic law was simply not sufficiently accurate or powerful to allow for reliable tracings of phylogeny. Morphology and embryology did not suffice. In Bateson's own words: "Morphology was studied because it was the material believed to be most favourable for the elucidation of the problems of evolution, and we all thought that in embryology the quintessence of morphological truth was most palpably presented. Therefore every aspiring zoologist was an embryologist, and the one topic of professional conversation was evolution." To no avail. "Discussion of evolution came to an end primarily because it was obvious that no progress was being made. Morphology having been explored in its minutest corners, we turned elsewhere" (B. Bateson 1928, 390). Bateson himself turned to the potential building blocks of change, producing a massive tome on the subject in the early 1890s (*Materials for the Study of Variation*). There is little surprise, therefore, that when he first encountered Mendel's work, he embraced it with great enthusiasm (Bateson 1902). And so it was for the rest of his

FIGURE 33.1. William Bateson (1861–1926), British champion of Mendel and bitter opponent of the Darwinian selectionists (the "biometricians"). Permission: Wellcome

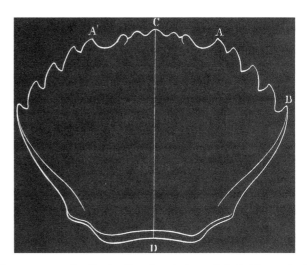

FIGURE 33.2A. A stylized outline of a crab. A morphological feature that is possibly biologically significant and easy to quantify is the "frontal breadth," the distance from A' to A. From W. F. R. Weldon, Attempt to measure the deathrate due to the selective destruction of *Carcinus moenas* with respect to a particular dimension, *Proceedings of the Royal Society* 57 (1895): 360–79.

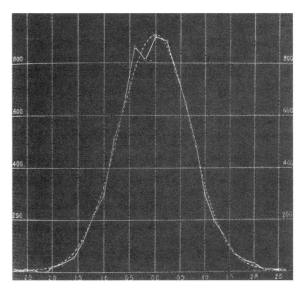

FIGURE 33.2B. Distribution of frontal breadths in 8,069 female crabs from Plymouth Sound, old and young. The curve is normal, and hence there is no reason to look for the working of selection splitting the group. From W. F. R. Weldon, On certain correlated variations in *Carcinus moenas*, *Proceedings of the Royal Society* 54 (1893): 318–29.

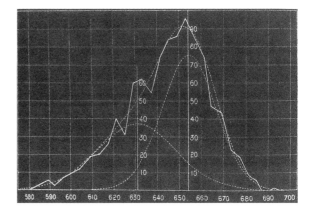

FIGURE 33.2C. Distribution of the frontal breadth of 1,000 Naples crabs. Through trial and error, Weldon showed that this breaks down into two normal curves (summing the ordinates), suggesting two different groups of crabs. From ibid.

life, as Bateson pushed hard to develop the new science of what became known as "genetics."

Bateson and his school were ardent evolutionists (Provine 1971). Even when it was not the central focus, it was always in the back of people's minds. With the coming of Mendelism, it was now thought that here was the time for a fresh approach to the topic. But it was not a Darwinian approach. At most, natural selection was believed to have a cleansing action after the real activity had occurred. Perhaps directly as a function of the fact that the variations studied by the early geneticists tended to be fairly significant and easy to spot, the belief was that evolutionary change goes in jumps – that is to say through "saltations." There was of course a long and honorable history of such thinking. Thomas Henry Huxley for one inclined that way.

However, the Mendelians did not have it all their own way. Others who turned against morphology in the 1880s went on a somewhat different route. The butterfly specialist Edward B. Poulton (1890) is a good example. He moved directly into Darwinian selection studies, and – very much in the tradition of Henry Walter Bates and Alfred Russel Wallace before him – his interests were in such problems as adaptive coloration and camouflage. Another who went along a similar path was a sometime teacher of Bateson, Raphael Weldon (1898), who turned with great enthusiasm to the study of selection and its effects in nature. Significantly, given the all-important role that mathematics plays in professional science, Weldon (1893, 329) was convinced absolutely that we must take such an approach to the problems of evolution: "It cannot be too strongly urged that the problem of animal evolution is essentially a statistical problem: that before we can properly estimate the changes at present going on in a race or species we must know accurately *(a)* the percentage of animals which exhibit a given amount of abnormality with regard to a particular character; *(b)* the degree of abnormality of other organs which accompanies a given abnormality of one; *(c)* the difference between the death rate percent in animals of different degrees of abnormality with respect to any organ; *(d)* the abnormality of offspring in terms of the abnormality of parents, and vice versa."

Putting his philosophy into practice, Weldon (1893) did a detailed study of crabs off the coast of Naples in Italy (Fig. 33.2). He found that the frontal breaths of female crabs, when plotted on a graph, give a somewhat strange curve. Using brute trial and error, Weldon broke this down into two normal curves – the distributions one might expect from random variation in two regular groups. He therefore argued that what we have are two different types under different selection pressures. Turning for help to his incredibly mathematically gifted colleague Karl Pearson (1894), who devised a technique for separating the curves mathematically, he was able to show that this surmise seems to be correct. Not content with this, however, Weldon turned then to experimentation. Off the coast of Devon, he ran experiments showing how the frontal breaths of crabs are tied adaptively to the muddiness of the water in which the crabs live. In a highly sophisticated way, Weldon performed some of the first selection experiments demonstrating that Darwin's mechanism is indeed a real force in nature. Moreover, they revealed that Darwin was wrong in thinking that selection is too slow to be observable in action.

> I hope I have convinced you that the law of chance enables one to express easily and simply the frequency of variations among animals; and I hope I have convinced you that the action of natural selection upon such fortuitous variations can be experimentally measured, at least in the only case in which anyone has attempted to measure it. I hope I have convinced you that the process of evolution is sometimes so rapid that it can be observed in the space of a very few years. (Weldon 1898, 902)

THE HARDY-WEINBERG LAW

The "biometricians," as Weldon and Pearson and their supporters were called, became bitter enemies of the Mendelians. The latter thought that genetics was everything. The former were less interested in the underlying causes of variation and more in the adaptive effects. It was not long, however, before saner forces started to prevail, and we soon find that there were those who suspected strongly that Mendelian genetics and Darwinian selection both hold pieces, albeit only partial pieces, of the whole picture. What was needed for such a fusion, given that selection is inherently a populational rather than an individual process, was an extension of Mendelian thinking to groups. Such an extension soon came, thanks to the mathematician G. H. Hardy in England and the physician Wilhelm Weinberg in Germany. In 1908, independently, they came up with the formula that now bears their names. In a large population, with random breeding, if there are no external, intervening factors, the organisms will go to and stay at the distribution: "$p^2AA + 2pqAa + q^2aa = 1$." (A and a are two forms, "alleles," of the same gene. p is the proportion of A alleles and q of a alleles. They are the only alleles of this kind in the population, so $p + q = 1$. AA is an individual with two A alleles, aa is an individual with two a alleles, and these are known as "homozygotes." Aa has one of each kind of allele and is a "heterozygote." If A masks the effects of a, then A is said to be "dominant" and a is "recessive.")

The importance of the Hardy-Weinberg principle or law is that it states an equilibrium position (Ruse 1973). It functions in the genetics of populations much as Newton's first law functions in mechanics: it says that if nothing happens, then nothing happens. This gives a background of stability against which one can now, in both sciences, introduce new, disruptive forces and follow their effects. And we do find that, even at the start of the second decade of the twentieth century, evolutionists were starting to put together Mendelism and selection by making the Hardy-Weinberg law the principal premise in evolutionary theorizing. With the law in place, one could then follow the effects of natural selection on populations – not to mention the effects of other disruptive factors like the arrival of new variations, or what came to be known as "mutations."

Here, for instance, is Oxford's comparative morphologist Edwin S. Goodrich explaining in a little book, published in 1912, how with the background of Mendel we can avoid the worry that so bedeviled Charles Darwin, namely about how a rare-but-useful new variation or mutation can be preserved in a population. It does not get swamped out, in a generation or two, by breeding. "The relative scarcity of the mutation at the start does not prevent that a number of individuals interbreeding at random, some with and others without a certain factor, will give rise to a population of impure heterozygotes and pure homozygotes in which the proportion of the three classes will be in equilibrium so soon as the square of the number of heterozygotes equals the number of pure 'dominants' multiplied by the number of pure 'recessives.' If this proportion is not already present at the beginning it will soon become established, and will continue, provided there is no selection to disturb the equilibrium" (Goodrich 1912, 69).

What should be remarked on is the near-paradoxical fact that Goodrich's book, *The Evolution of Living Organisms*, was published in a series for the general reader. Although he spent his whole long career on studies of homologies taken as evidences of evolution, he never once thought to incorporate

FIGURE 33.3. Thomas Hunt Morgan (1866–1945) integrated our knowledge of the units of heredity with our knowledge of the physical aspects of the cell, into the "classical theory of the gene." Permission: American Philosophical Society

selection-based thinking into his work or his lectures. If anything demonstrates the rather low grade, or popular status, of evolutionary thinking even fifty years after the *Origin of Species*, it is this.

ADVANCES

With respect to moving toward a more mature or professionally based evolutionary theory, the second decade of the twentieth century was important for a number of factors. Two are particularly important. Both are American, showing the growing scientific status and significance of that country.

First, there was the development of the "classical theory of the gene" by Thomas Hunt Morgan and his students at Columbia University in New York City (T. H. Morgan et al. 1915). Working with fruit flies, and relying on the now-accepted fact that genes have a material basis and live on the threadlike entities, the chromosomes, in the nuclei of cells, they mapped the order of the genes in great detail. Importantly, they showed that new variations can be of various magnitudes, including some with very little magnitude at all (Fig. 33.3). Although, as it happens, Morgan himself never became a full-blown Darwinian evolutionist, he and his co-workers prepared the way for those who wanted to argue that natural selection can be truly effective, because there really are those tiny variations on which it must work. Populations do have just the variations that Darwin asked for in the *Origin*.

Second, almost by default, there were important studies showing the lasting effects of selection. Working at Harvard, the geneticist William E. Castle showed how selection can affect the coat color of rats (Castle and Phillips 1914). This was no great surprise, but what was a surprise was when he showed that once a line of rats has been selected in one particular direction, for instance toward a dark coat, these effects can be lasting (Fig. 33.4). Relaxing selection does not mean that, in a generation or two, everything reverts to the original state. One should note that this discovery was somewhat ironical, because Castle set out in his experiments to show that geneticists like Morgan were wrong in thinking that, mutation apart, genes persist unchanged from generation to generation. Castle wanted to show how the units of heredity can themselves be changed by selection. In the end, he realized that he had not done this but that, almost by default, he had proved something at least as important, if not more so (Castle 1917).

RONALD A. FISHER AND SEWALL WRIGHT

The time was now getting ripe for a full-scale synthesis of Darwinian selection and Mendelian genetics. This occurred in the 1920s, culminating around about 1930. Particularly important work was done by Ronald A. Fisher (Fig. 33.5) and J. B. S. Haldane (Fig. 33.6) in Britain and by Sewall Wright (Fig. 33.7) in America. They took and expanded the program sketched by Goodrich (Haldane was Goodrich's student), showing in full mathematical detail how one can have a theory of evolution where selection plays a large role, a role backed and guaranteed by the workings of Mendelian genes in populations. However, although the "population geneticists" produced theories that were formally mathematically equivalent, there were significant and lasting differences between the different visions of the evolutionary process. This applies particularly to the dominant figures of Fisher in Britain and Wright in America.

Ronald Fisher, who trained at Cambridge as a mathematician, had always a keen interest in evolution (Box 1978). He became very friendly with Darwin's youngest son, Leonard Darwin, and it is clear that Darwinism was an emotional crusade as much as an exercise in mathematical modeling. Through the 1920s, Fisher worked at an agricultural station in England, and it was then that he developed powerful statistical techniques, an achievement for which, to this day, he

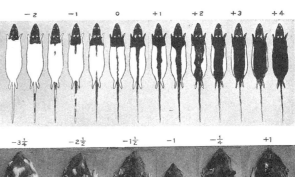
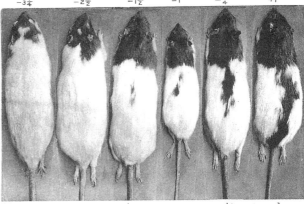
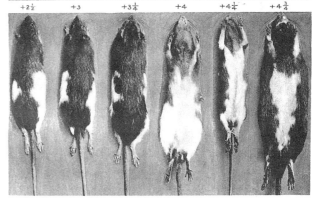

FIGURE 33.4. The effects of selection on rats. Arbitrary grades used for classification (*top row*). Selection for light color (*middle row*). Selection for dark color (*bottom row*). Viewed from their top side, the dark ones were almost entirely black. From W. E. Castle and J. C. Phillips, *Piebald Rats and Selection: An Experimental Test of the Effectiveness of Selection and of the Theory of Gamete Purity in Mendelian Crosses* (Washington D.C.: Carnegie Institution of Washington, 1914)

is justifiably famous. But at the same time, he kept up a keen interest in evolution. He intervened on the side of Poulton when he was criticized by the Mendelians for an inadequate understanding of the heredity behind mimicry and camouflage (Kimler 1983). This work culminated in 1930 with the publication of what many regard as the most important work in evolutionary theory since the *Origin*: *The Genetical Theory of Natural Selection*.

As an undergraduate, Fisher had taken a course in gas theory with the physicist James Jeans, and the idea of huge numbers of molecules buzzing about in a container fed right into Fisher's model of evolutionary change. For Fisher, evolution was the selection of small variations in large populations. We have a large pool of genes, and natural selection

FIGURE 33.5. Ronald A. Fisher (1890–1962) is often with reason characterized as the greatest Darwinian since Charles Darwin himself. Cursed with very bad eyesight, he trained to do mathematics entirely without visual aids and could produce proofs of such dense brilliance that successors spent years trying to show their validity. Permission: American Philosophical Society

grinds away, eliminating the less fit and promoting the fitter. There is therefore a kind of constant, upward progress, although this is ever in danger of being eaten away by changing conditions, themselves a function of the environment or other organisms. Fisher expressed this vision through his "fundamental theorem of natural selection": "The rate of increase in fitness of any organism at any time is equal to its genetic variance in fitness at that time." In other words, there is going to be change, and it will depend on the amount of variation that is available for the action of natural selection. Explicitly, Fisher likened this theorem to the second law of thermodynamics, which likewise suggests that there is a direction to events that will continue, unless external forces impinge to disrupt progress. Note, however, that Fisher's law pointed upward, whereas the second law predicts decay and decline.

Sewall Wright had a very different take on the nature and process of evolution (Provine 1986). He was educated at Harvard, actually working with Castle, and then went for some ten years to work at the United States Department of Agriculture. In the middle of the 1920s, he took a position at Chicago where he remained until he retired. Undoubtedly influenced by his time at the USDA – where breeding involves working first with just a few animals rather than a whole variety – Wright saw the key evolutionary events taking place, not in large populations but only after such populations fragment into small groups. In such tiny sets, there is scope for significant change. Then, when the groups reassemble to make one large population, the new, worthwhile features can spread through this whole population (Wright 1931, 1932).

To illustrate his theory, Wright introduced what was to prove a very powerful metaphor, that of the "adaptive landscape" (Fig. 33.8). The peaks represent points of adaptive fitness, and the valleys the lack of such fitness. Wright saw evolution being a matter of small groups moving from one peak to another. It is tempting to think of the adaptive landscapes as cast in stone, as it were. But, like Fisher, Wright saw things in constant motion, and so evolution is not just an internal matter of moving between peaks but also an external matter of landscapes being dramatically changed by outside factors – the environment or, indeed, other organisms. But how, one might ask, can a group at the top of one peak possibly move to another peak, even though the new home would be far adaptively superior? It would mean going down into the valley, which is nonadaptive. Here Wright introduced his most distinctive idea, pointing out that, in small populations, random forces might well overcome the effects of natural selection. In other words, key change might come about through nonadaptive forces, in particular what Wright

FIGURE 33.6. J. B. S. (Jack) Haldane (1892–1964) was the most brilliant man anyone had every met, capable of doing ground-breaking mathematical genetics in the morning and turning out polished essays for the general reader in the afternoon. It was said that he had ruined many a conference by posing questions no one else could answer and then giving solutions of such penetration that all further conversation was deemed pointless. Permission: American Philosophical Society

FIGURE 33.7. Sewall Wright (1889–1988) moved from agricultural breeding patterns to formulate his distinctive "shifting balance theory of evolution." His training with Castle meant that he always thought of his work on evolution as a sideline to his rather humdrum work on breeding patterns in guinea pigs. Permission: American Philosophical Society

called "genetic drift." Evolution, therefore, is a matter of processes in a temporal line: fragmentation, drift, combining, and subsequent selection. Wright called this theory his "shifting balance theory" of evolution.

UNDERLYING PHILOSOPHIES

Fisher and Wright had very different visions of the evolutionary process. One was Darwinian through and through, and the other rather less so, particularly since it is clear that Wright thought that drift is responsible for the really creative moves in evolution. Given that these two mathematical biologists had immense influences on the evolutionary studies of their respective countries – in the case of Fisher through his younger associate E. B. Ford, and in the case of Wright through his younger associate Theodosius Dobzhansky – expectedly we find that, for all there was much sharing of ideas and concepts, British and American evolutionary biology had significant differences, differences that persist perhaps even to this day.[1] For a start, there was always much more of an individualistic approach to selection in Fisher (who, for instance, took a great interest in sexual selection), whereas Wright was more holistic, more thinking in terms of groups, with significant change not involving one against all. Although there was much subsequent ironing out of differences – for instance, when Dobzhansky in the 1940s found that supposedly paradigmatic instances of drift were in fact tightly controlled by selection (Ruse 1999b) – there is still a flavor of national divide on this issue.

It is therefore worth asking if there were underlying feelings or motives – what we might call different philosophies – leading Fisher and Wright to their different conceptions of evolutionary biology. It seems that indeed there were. In the case of Fisher, the fundamental starting point is that he was as committed a member of the Church of England as he was a Darwinian (Hodge 1992). Fisher truly thought that God is working his purpose out and that this applies completely and absolutely to the evolutionary process. Darwinian evolution is God's evolution.

> To the traditionally religious man, the essential novelty introduced by the theory of the evolution of organic life, is that creation was not all finished a long while ago, but is still in progress, in the midst of its incredible duration. In the language of Genesis we are living in the sixth day, probably rather early in the morning, and the divine artist has not yet stood back from his work, and declared it to be "very good." Perhaps that can be only when God's very imperfect image has become more competent to manage the affairs of the planet of which he is in control. (Fisher 1947, 1001)

This leads us straight to what was, after Darwinism and Christianity, the third leg of Fisher's world picture. Eugenics. From his youth, Fisher was ever an ardent believer in the possibility and need of selective breeding among humans. He worried nonstop that the working classes were having far too many children and the middle classes far too few. He advocated a kind of reverse child allowance, to be paid by the state, to encourage the middle classes to breed. As part of his commitment to his eugenic vision, Fisher himself married a very young girl chosen expressly because she would be of good breeding material and proceeded to have a very large family.

It all fits together. For all the troubles with achieving progress, Fisher thought that ultimately the fundamental theorem would prevail. There has been an upward progress in evolution from the blob to the human. This was powered by God and was his form of creation. Now, however, because we are not breeding in the right way, the human species is threatened with decline and decay. At some level, civilization has made us soft, particularly the middle classes, who now selfishly refuse

---

[1] A not entirely unfair reason why Haldane always takes third place in the story is that he did not have people to carry on his vision of the evolutionary process. Much later, the very influential John Maynard Smith became his student and colleague.

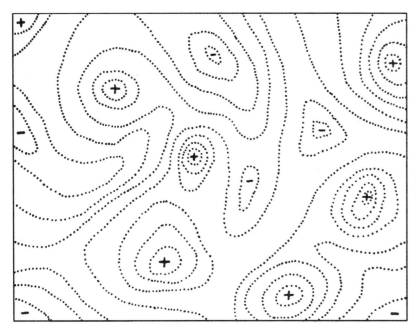

FIGURE 33.8. After "natural selection," and before the "selfish gene," the adaptive landscape is the most famous metaphor in the history of evolutionary biology. From S. Wright, The roles of mutation, inbreeding, crossbreeding and selection in evolution, *Proceedings of the Sixth International Congress of Genetics* (1932), 1:356–66

help its status. Hence, although the philosophy was there, some care was taken to conceal or downplay it. Yet, every now and then, bursts of Spencerian sunshine come through the clouds: "Evolution as a process of cumulative change depends on a proper balance of the conditions, which, at each level of organisation – gene, chromosome, cell, individual, local race – make for genetic homogeneity or genetic heterogeneity of the species" (Wright 1931, 158). And what is the result of all of this? "Changing conditions such as more severe selection, merely shifts all gene frequencies and for the most part reversibly, to new equilibrium points in which the population remained static as long as the new conditions persist."

DARWINIAN EVOLUTIONARY BIOLOGY?

Whatever the philosophies of the population geneticists – or, perhaps more properly, because of the philosophies of the population geneticists – evolutionary theorizing would never be the same again. It was now firmly on the path to full professionalization. No longer would it be a museum subject, used only for teaching and entertainment, spinning unverifiable hypotheses in the sky – or, more appropriately, in the rocks. It was now being mathematized, the sure sign that the subject is on the way to professionalization. It was now being readied for the empiricists to move in and to test and experiment.

Was it still recognizably Darwin's theory? In certain respects it obviously was. No one wanted to deny selection's significant role. No longer could one dismiss selection simply as something that cleaned up after the real work had been done. But, in respects, it was not uniformly or wholeheartedly Darwinian. In Britain, although one doubts Darwin would have sympathized strongly with Fisher's Christian commitments – and although, while Darwin, like almost everyone else, worried that the poor were breeding too much, there is no reason to think he would have been as enthusiastic about eugenics as was Fisher – Darwin would have seen that, thanks to Fisher, his theory of the *Origin* was now on the way to being a fully functioning paradigm. In America, somewhat less so. Natural selection did have a significant role in Sewall Wright's evolutionary thinking. It cannot be said to have had the only role or even the truly crucial role. For Wright, the really creative moments in evolution come about through chance rather than through selection.

Admittedly, whether Wright personally thought that drift ultimately was only chance is another matter. He was what is known as a "panpsychic monist," meaning that he thought the whole of the material world has intelligence.

to do their bit. Fisher therefore saw eugenics as our God-given duty, our part of making evolution fully triumphant, as the deity intended.

By his own admission, Sewall Wright was no less influenced by his world philosophy. Moreover, this was a philosophy that reflected his national origins as much as did the philosophy of Fisher reflect his national origins. Herbert Spencer was the great influence in American intellectual circles from the 1880s on, and the young Wright felt the full blast. His father was an ardent Spencerian, as were his teachers before and when he was at Harvard. In particular, Wright fell under the sway of the biochemist L. J. Henderson, a very great supporter of the Englishman's philosophy. Wright absorbed the ideas fully and, brilliantly fusing them with the lore on animal breeding picked up at the USDA, produced the shifting balance theory of evolution (Ruse 1996).

For Spencer (1862), it is all a matter of disruption, then upward progress, and finally the reachievement of equilibrium. In the process, we go from simple, or what Spencer called "homogeneous," to the complex, or what Spencer called the "heterogeneous." Wright's theory is this theory translated into the language of genetics. A group of organisms are in balance on an adaptive landscape. Then they get fragmented, and through drift there is an increase in heterogeneity. The small groups come together into one whole population, and there has been an upward movement to a higher form of peak. In this way, therefore, Wright was as committed to progress as was Fisher. As with Fisher, though, by now there was an increasing awareness that explicit talk of progress within one's science did nothing to

So probably Wright thought that ultimately everything is guided rather than chance. He was probably closer to Asa Gray on this matter than he was to Charles Darwin. However, unlike Gray and like Darwin, Wright realized that there is no place for this kind of talk in science. As far as the shifting balance theory is concerned, it is drift and drift alone that is important.

Darwinian or not, things could never be the same again. The population geneticists left their mark. The question now is what use their successors would make of their legacy.

Essay 34

Synthesis Period in Evolutionary Studies

Joe Cain

The "evolutionary synthesis" is a phrase widely used for a period in evolutionary studies between 1920 and 1950 when important theoretical developments took place. The period also saw new types of interdisciplinary collaborations develop. These new associations reset the priorities of evolutionary studies for more than fifty years. Contributors came from every country with a significant scientific community and from nearly every discipline in the life sciences. The phrase "evolutionary synthesis" also refers to a period of discipline formation. This involved new community infrastructure, such as new professional societies and journals, dedicated to evolutionary studies. Those at the heart of these organizations who built this infrastructure quickly rose to prominence in the community and found themselves in a strong position to shape outside impressions of community activity.

When these promoters said they had invented the modern science of evolutionary biology, everyone who knew their work understood what they meant. Later, when the same people wrote their history, they were absolutely certain they walked in Darwin's footsteps. Some said this was because, by and large, they agreed with his theory of natural selection and because they accepted his other major conclusions about evolution. However, the connections to Darwin went far deeper than any agreement about natural selection.

This essay describes key layers of the evolutionary synthesis. It explores the rich connections between champions of the evolutionary synthesis and Darwin's program for science. We begin by discussing one of the most distinct layers in the period: the rise of mathematical population genetics. Next we look at the mathematical modelers who coexisted with many other types of studies of evolutionary causes. Then we examine ways researchers took up the problem of how the diverse range of evolutionary factors balanced in nature.

Because describing what happened is different from explaining why, we next place the champions of synthesis into key historical contexts. The rules for science were changing rapidly between 1920 and 1950. Success in the synthesis period must be seen as an adaptive response by some researchers to those changing rules. After examining changing rules for what counts as a good method, we consider the changing sense of focus in the life sciences. These discussions offer compelling connections to Darwin and his program for science. Finally, we show how the champions of synthesis consolidated their developments and solidified the foundations of evolutionary biology for the next fifty years.

MATHEMATICAL POPULATION GENETICS

The core decade of the evolutionary synthesis was the 1930s. Several layers of activity were developing during this period toward a robust and complex understanding of evolutionary processes. One of the most distinctive layers in this decade involved rapid developments in mathematical models for population genetics. This section describes some benchmarks on that layer. Arguments here that the power of natural selection makes this a twentieth-century extension of Darwin are easy to believe but superficial. The continuity runs much deeper.

Basic mathematical models for genetic change in populations appeared in the 1900s and 1910s. These were used by geneticists to predict the outcome of breeding experiments. The models were built on the assumptions of simple Mendelian genetics. As genetics became more sophisticated during the 1920s and 1930s, so did the mathematical models. During the 1930s, internal discussion about models was complemented by the exploration of applications. How did these models match data from natural populations? What processes did these models predict as important for evolution, and were these really found in nature? Key contributors identified with the 1930s are R. A. Fisher and J. B. S. Haldane in Britain and Sewall Wright in the United States (Provine 1971, 1978).

Fisher's mathematical work, summarized in his 1930 book, *The Genetical Theory of Natural Selection*, demonstrated the impact natural selection could have in shifting the relative frequency of advantageous traits (or alleles) compared to disadvantageous ones. To the surprise of many, Fisher's models suggested even tiny selective advantages – advantages far smaller than most people thought nature would notice – could produce significant changes in a few generations. For example, an advantage of only 1 percent would be enough in Fisher's model for a trait to become universal in a population in only 350 generations. A trait providing a 10 percent advantage would become universal in approximately 50 generations. That's a blink of an eye on evolutionary time scales. Fisher's work had a major influence on evolutionary studies, especially in genetics and ecology in the United Kingdom. It offered an enormous boost for those who thought natural selection was the dominant agent of evolutionary change.

Fisher's work was not without criticism. The models he used made many simplifying assumptions. The genetic system he used was little more than what Mendel proposed in the 1860s. Processes known to occur in every population, such as mutation and migration, were minimized. Most important, Fisher built his models on the assumption that the populations involved were infinitely large, and mating within the population was perfectly random. He had good reasons to make these assumptions; nevertheless, biologists who did not understand the mathematics also failed to appreciate the limitations these imposed on Fisher's conclusions.

In many ways, Sewall Wright was Fisher's rival. He also worked on mathematical models of genetic changes in evolution, but he used different simplifying assumptions and different mathematical tools. Fisher lived in England; Wright lived in America. They probed each other's work in a long correspondence. They also maintained a healthy and critical dialogue in publications.

In terms important to the evolutionary synthesis, the key difference between Fisher and Wright related to their assumptions about the size and structure of populations. Whereas Fisher assumed populations were infinitely large and breeding was random, Wright assumed the effective size of a population (the number of potential mates an individual might have access to) could change significantly, and sometimes change quite quickly. The effect on his modeling was profound. When very large populations were used in models, results were the same from one trial of the model to the next. However, as smaller populations were introduced into the models, results became increasingly unpredictable one test to the next. Sewall Wright argued this element of randomness (sometimes called "genetic drift") was a matter of fundamental importance to evolutionary processes. When populations were small, random events – luck – would smother any gains made by selection. Survival and evolutionary success might have nothing to do with being better adapted. Wright developed a theory of "shifting balance" in which selection was presented as the main driver of evolution in large populations, and randomness was presented as the main driver of evolution in small populations. In midsized populations, the two processes engaged in a tug of war.

Wright argued his mathematical models were more faithful to the ecological realities of a species' life history. His work proved immensely popular in the United States. The emphasis he gave to small effective breeding populations and to randomness appealed to field naturalists, who thought much of the variation found in nature had no adaptive value – that is, it did not evolve by and was not refined by natural selection. The theoretical possibility of drifting also was used by Wright and others to explain paradoxical cases of evolutionary history in which a group experienced periods of relatively poor adaptedness after periods of peak fitness, or when they shifted from one reasonably good adaptive solution to another. For Wright, sometimes selection drove evolution; however, sometimes characters drifted randomly, and there was nothing selection could do about it.

In the 1930s, the mathematical theorists collaborated with laboratory and field biologists to expand, test, and apply their models to natural populations. For instance, Fisher collaborated with laboratory and field researchers under the direction of E. B. Ford (Oxford). This produced a research school on ecological genetics that combined the mathematical models, laboratory genetics, and considerable information about natural populations, usually butterflies and moths. Sewall Wright had several important collaborations, too, though none was more important than with the *Drosophila* geneticist and field naturalist Theodosius Dobzhansky. As with Ford and Fisher, the Wright-Dobzhansky collaboration put theory and data into a dialogue in which theory influenced experiment design and field work, then the data produced led to further developments of the mathematical models and working assumptions.

In addition to Fisher-Ford and Wright-Dobzhansky, other similar collaborations between model builders and experimenters or naturalists took place in the Soviet Union, France, Germany, and Italy during the 1920s and 1930s and around the world during the 1940s and 1950s.

In sum, one important layer of the evolutionary synthesis involved mathematical models. It's not that these models trumpeted the exclusive power of natural selection; they didn't (at least, not all of them did). Their importance rather came from the underlying appeal to some of the most sophisticated scientific methods of the day and from their attention to analyzing the piece-by-piece causal mechanics of evolutionary processes. Mathematical modeling carried a sense of power and precision that felt hard to resist. Confidence in the modeling process boosted not only conclusions drawn by the models but also the interest in studying evolutionary causes. Mathematics helped turn speculation into reasoned judgment. Though the computational steps were taken for granted by nonmathematicians, the working assumptions of these models and their range of application were carefully explored, and much debated, over the 1930s and 1940s. People considered the conclusions carefully because they trusted in mathematics.

FIELD NATURAL HISTORY MEETS LABORATORY METHODS

Not even the mathematical modelers thought mathematics could do everything. If nothing else, into the 1930s the study of models and scenarios boosted confidence in the study of causes, the variety of causes, and the results of their interplay. A second layer of the evolutionary synthesis involved the study of evolution in natural populations. This combined research methods developed in laboratories (largely from genetics and cytology), together with those developed for field studies (largely from population monitoring, natural history, and systematics). Most of these methods worked in isolation from the mathematical modeling. It was the growing recognition that parallel pursuits were underway in many disciplines that drove widespread support for synthesis as the 1930s moved on. Researchers wanted to know what others were doing – in part to prevent being marginalized by rivals (J. Cain 1993).

In the 1910s and 1920s, experimentalists and naturalists also mostly held each other in suspicion. Each community developed its own working knowledge of evolutionary studies, including separate methods, assumptions, explanations, and data sets (Mayr 1980b, 1982).

In the 1930s this isolation changed. Some researchers in each community took an interest in the other group's work. They set about building bridges and cultivating interdisciplinary connections. This involved reconciliation and translation. It also involved a certain amount of reeducation, breaking of outdated stereotypes and restoring communication across various barriers. By the end of the 1940s, interactions were routine and firmly embedded. A newly synthesized community had emerged. Key contributors in America were Theodosius Dobzhansky, Ernst Mayr, Edgar Anderson, Julian Huxley, George Simpson, and G. Ledyard Stebbins (Fig. 34.1).

Dobzhansky had an energy and charisma that was hard to match (Adams 1994). A Russian émigré to the United States, he worked in several of the world's most important genetics laboratories. During the 1930s, Dobzhansky developed a research program focusing on genetic diversity in *Drosophila*. Combining techniques invented by many people, this program was developed first for laboratory studies. Dobzhansky loved field work, and he put these techniques to work studying genetic diversity in natural populations. This started in the mountains of California but eventually extended across most of the Americas. He encouraged others to collect most everywhere else, too. Once Dobzhansky obtained a good sense of the frequencies of various genetic features, he was in an ideal position to monitor the effects of natural selection, drift, migration, mutation, isolating mechanisms, and other processes. For instance, he monitored seasonal changes in the frequency of certain gene combinations, surmising this was caused by shifting demands of natural selection. Dobzhansky quickly concluded this model system allowed him to observe evolution within and between natural populations. In 1937 he published a summary of this work, *Genetics and the Origin of Species*. The title nicely captures the importance Dobzhansky attributed to his work (Plate XXV).

In this second layer of the evolutionary synthesis, Dobzhansky is said to have created a synthesis between mathematical, laboratory, and field studies in genetic research on microevolution (including evolution within populations, the formation of subspecies, and the origin of new species) (Lewontin 1981). He translated Wright's mathematical theory into terms nonmathematicians could understand, and he combined it with the most up-to-date understanding of genetics and field techniques. Everyone in the community read *Genetics and the Origin of Species*. As the story goes, it not only taught people new facts about evolution and genetics but also showed them the value of mathematical theory, and it gave evolutionary studies a much needed boost of confidence.

Ernst Mayr was an ornithologist. Trained to be a museum curator and systematist in Germany, he emigrated to New York City in 1931 for a job at the American Museum of Natural History (Haffer 1997). At the museum, he had responsibility to organize part of the world's largest collection of birds. Not only a museum curator, Mayr also was skilled in field natural history. He was trained to think about species as clusters of varying local populations, sometimes divided into subspecies or local races. In part, he was taught, this variation was governed by natural selection to local circumstances. It also was

FIGURE 34.1. Theodosius Dobzhansky (1900–75) surrounded by his students. (Richard Lewontin is second from the left, Bruce Wallace stands just behind, Francisco Ayala in priest's gear is on the far right, and the hirsute individual at the back is Leigh Van Valen. Dobzhansky's only female student, also shown, was Lee Ehrman.) The photograph underlines the fact that not only was Dobzhansky important for his ideas and discoveries but also for the way in which he attracted students and passed on his thinking to the next generation. Permission: American Philosophical Society

shaped sometimes by random processes or by various types of environmental influences. Crucially, Mayr was trained to think of subspecies as populations on the road to becoming a new species. What was needed to push transformation far enough along to create a new species was isolation and some force driving local differentiation.

In the late 1930s, Mayr developed a theory of speciation based on these two processes. He used geographical isolation as the key. No matter how it occurred, if a local population became physically isolated from the rest of a species, it became a prime candidate for evolution. Selection might drive adaptation to its particular microenvironment. Sampling might randomly mean some characters were universal or entirely absent. Perhaps other processes took place, too. Whatever its cause, the isolated offshoot might come to diverge from the rest of the species. Physical isolation was stage one. Divergence was stage two. Stage three involved biological isolation. When the physical barrier disappeared and the various populations of a species came back into contact, interbreeding was a distinct possibility. If that occurred, then distinctive features locally would blend back into the species as a whole. Far more interesting to Mayr were the cases in which the isolated offshoot happened also to develop differences that prevented interbreeding, such as new behaviors or glitches in genetic or developmental processes. These "isolating mechanisms" prevented assimilation when the two populations came back into contact. This meant the offshoot now would be on an evolutionarily distinct path. In Mayr's thinking, a new species had come into being.

Mayr already was knowledgeable about genetics, development, and ecology before he met Dobzhansky, and once they became close friends after 1935, they influenced each other greatly. In many ways, the friendship between this geneticist and this naturalist is the heart of the interdisciplinary bridge building associated with the evolutionary synthesis.

Mayr wrote a great deal about the historical significance of synthesis in evolutionary studies. It always centered on his interactions with Dobzhansky, then expanded outward to include other bridge builders. For instance, George Simpson's (1944) *Tempo and Mode in Evolution* was used as a bridge from the Dobzhansky-Mayr work to paleontology (Fig. 34.2). G. Ledyard Stebbins's (1950) *Variation and Evolution in Plants* became a bridge from Dobzhansky and Mayr to botany.

example, though historians disagree on which of the three alternatives best applies to this example (Ruse, 1996).

NEW RULES FOR WHAT MAKES A GOOD METHOD

Describing what happened in history is different from explaining why it happened when it did. The next two sections shift to explanation. Champions of synthesis in the 1930s and 1940s were not working in isolation. Outside forces were pressing into evolutionary studies. These changed the rules and the external measures of success everyone had to meet. The innovations these champions pursued should be seen as adaptive responses to those outside forces. Just as Darwin transformed the study of evolution in a world of changing standards about science, so too did the champions of synthesis.

One part of the context explaining why the evolutionary synthesis took on the character it did relates to changing expectations about method (J. Cain 2009). It is not important that new methods were invented. It is important that standards were changing for judging good versus poor methods. This can be seen in the preference for mathematical over other types of models for scenario building. It can be seen in calls for a test rather than for more data collecting. It can be seen throughout the contributions promoted as exemplars within the evolutionary synthesis.

In short, experimental methods came to dominate the life sciences in the first half of the twentieth century. This emphasized testing, standardization, intervention, and control. It encouraged researchers to prefer prediction testing and to isolate variables. It encouraged comparison of results with theoretical modeling. Proof came to rely on the ability to replicate phenomena more or less at will. Success came when a researcher could announce the discovery of a new general explanatory concept or heuristic.

These experimental methods stood in stark contrast to older methods in zoology and botany (Rainger, Benson, and Maienschein 1988; Kingsland 1991, 1997). Those traditional methods placed emphasis on comparison and massive accumulation of data. Conclusions were expected to come through induction and generalization, expressed as descriptive empirical "laws" (e.g., Bergmann's "rule" that body size is inversely proportional to habitat temperature). A hypothesis was judged in terms of its consistency with accumulated data and its capacity for consilience.

This transition in expectations about methods took hold during the mid-nineteenth century in disciplines such as physiology. By the start of the twentieth century it had driven transformations in many areas: heredity became genetics, natural history became ecology and ethology, development became embryology. With its descriptive, narrative phylogenies and its vague appeals to causes, evolutionary studies during the 1900s and 1910s look increasingly weak and old-fashioned.

The pressure to change methods was core to the evolutionary synthesis because one thing being celebrated in the claims to innovation during this period was the shifting in

FIGURE 34.2. George Gaylord Simpson (1902–84), the greatest paleontologist of the twentieth century. An expert in fossil mammals of North and South America, Simpson spent his career trying to shift the attention of paleontologists away from single fossils and toward Darwinian evolutionary theory and big themes in the history of life. Permission: American Philosophical Society

A long list of these bridges has been proposed, covering fields as diverse as protozoology and human behavior. They also link many countries. As any historian might have guessed, participants in the synthesis period disagreed on who initiated a project, which projects came sooner than others or were the most important, and what was particularly special about each one.

The many bridge-building projects have several features in common. Most importantly, they build from a foundation of the chromosome theory of genetics as it was developed in the 1930s. This was combined with "population thinking," which required an appreciation for diversity and relative frequencies within groups. They also drew from a strong desire to understand evolution as it occurred in nature rather than in artificial laboratories or abstract theoretical arguments. Interestingly, some bridges either failed, had no builders, or had their construction actively discouraged. Developmental biology is an methods within evolutionary studies from old to new. When champions of synthesis praised new studies, they often spoke of how evolutionary studies were now coming to stand on more secure, modern, and scientific footings (Hagen 1981, 1984, 2009; Ilerbaig 2009).

Consider the growing use of mathematics and numbers in evolutionary studies. Between 1920 and 1950, research becomes increasingly quantitative in both data and analysis. Mathematical models receive wider audiences. Statistical tools play increasingly prominent roles in comparison, hypothesis testing, and expressions of confidence. Part of the value of quantification comes from the sense of objectivity and precision. The value of statistical tools rested on the way they gave both perspective to comparison and explicitness when expressing the strength of conclusions: a hypothesis with a 10 percent measure of confidence was different from one with a 95 percent measure.

Throughout the 1930s, the community of researchers in evolutionary studies favored methods more like those elsewhere in experimental biology. A simple example demonstrates how subtle and profound this shift was. The biological species concept was promoted in the synthesis period as a central conceptual tool. At its heart is the idea that the ultimate test of species status for any group is interbreeding. If two populations successfully interbreed, then they are parts of one species; if they do not, then they are different species. A great deal of attention has been given to precisely how this idea might be expressed and applied, but this misses the forest for the trees. The key importance of the biological species concept comes from its use as an objective standard of evidence. It is a clear, simple, explicit, and decisive test. Even though it might be hard, even impossible, to use in many circumstances, this idea of a test for species status helped promoters of systematics argue that their work was more science than art, more objective than subjective.

This shift in standards produced a movement known in Britain as "new systematics" and in the United States as "experimental taxonomy" (Winsor 1995; Kleinman 1999) (Fig. 34.3). The central idea bringing this group together was the search for standardized and objective criteria when naming new taxonomic groups. One tool involved counting chromosome numbers and sets. Another used blood chemistry. Hybridization tests were popular tools for "new systematists." So were statistical measures of central tendency, such as comparisons of means and standard deviations. Some experimental taxonomists focused their attention on variability and sought methods for testing just how far a single species might vary in different environmental extremes. They set up a series of growing experiments with conditions varying in controlled fashion. They knew looks could be deceiving. They wanted to know just how much variation could arise in the phenotype of organisms given a single genotype grown in different environments.

In this growing preference for experimental methods, the mathematical population genetics promoted by researchers like Fisher and Wright found a receptive audience. A few

FIGURE 34.3. Julian Huxley (1887–1975), the oldest grandchild of Thomas Henry Huxley, was an ardent evolutionist, much interested in the new systematics, and through his *Evolution: The New Synthesis* (1942) he became an important figure in spreading the updated version of Darwinism melded with Mendelian genetics. Paradoxically, he always had a yen for the neo-Aristotelian vitalism of the French philosopher Henri Bergson and, with him, reached back to the progressionism of Herbert Spencer. Permission: © UNESCO

people pushed this research program forward as an end in itself. Many more used the models for scenario testing, consuming the predictive tools of modeling to guide research and experimentation. This created a dialogue between modeler and experimenter. No one took the mathematical work alone as gospel. At the same time, the modeling proved an important source for idea generation, and validation often was claimed when data collected matched predictions from the theory.

Laboratory biologists developed a series of important surveillance tools, too. One of the most sophisticated drew attention to the overall shape and pattern of single chromosomes. Developed in the early 1930s by John Patterson at University of Texas for use in fruit flies, this technique revealed a pattern of alternating light and dark banding in chromosomes, with the same chromosome showing different structures in different individuals. (Compare the banding patterns for bar codes on different packages for a similar effect.) Researchers measured the frequency of each pattern in different populations of a species as a way not only to measure the overall variation but also to track changing to the relative frequencies of different banding patterns.

Excitement for synthesis in evolutionary studies during the 1930s and 1940s is closely tied to shifting expectations about method. On offer, it seemed, were increased objectivity, improved confidence, and a clearer sense of the ground's solidity. This seemed to put the science back into evolutionary studies. The same sense of improvement was common among Darwin's promoters half a century before. Supporters like Thomas Henry Huxley, Joseph Hooker, and John Lubbock used Darwin as a role model for science precisely because he showed a disciplined and thorough focus on methods and because he stood for objective weighing of evidence.

SHIFTING FOCUS: OBJECT TO PROCESS

Shifting expectations about methods is one key to explaining the shape and timing of the evolutionary synthesis. Another relates to a shift in the focus of study: What knowledge should scientists produce as a result of their studies? The goal of science could be to know *what* nature is and *where* things can be found. Alternatively, it could be to know *how* nature works and *why* some outcomes occur when others do not. The evolutionary synthesis consolidated the shift in evolutionary studies from what/where questions to how/why questions. This offers another deep connection to Darwin.

Darwin was unlike most life scientists of his day. Though he became a specialist in several groups of organisms, including corals, barnacles, orchids, and carnivorous plants, when he described himself as a scientist, Darwin gave priority to studying processes over things. No matter what particular thing he studied, Darwin did so to relate that information back to his investigations into the laws, causes, and mechanisms of nature. His work on coral reefs, for example, illustrated processes in geology. His work on barnacles revealed the legacies of common ancestry. Orchids were applications of natural selection. Darwin undertook his study of objects with thoroughness. At the same time, he did this work to support his study of process. He thought that was what the best scientists should do. Another shift in helping to explain the significance of synthesis in evolution during the 1930s related to this shift from object to process.

Between approximately 1900 and 1960, zoological and botanical disciplines previously the realm of natural history were being converted into new disciplines in academic biology (J. Cain 2010). Disappearing were disciplines such as mammalogy, ornithology, and herpetology. Appearing were disciplines such as ecology, biogeography, ethology, and evolution. Biologists tend to study organisms only to create instances or illustrations of general phenomena. The peppered moth serves as an illustration of natural selection; the peacock, courtship display and sexual selection. Compare the academic life sciences in 1850 with 1950, and one of the most significant changes to occur is this shift in emphasis from object to process. (This shift has its modern roots in Paris during the Enlightenment, but that is outside the scope of this essay.)

The 1930s saw a sharp rise of interest in the study of general processes involved in the formation of new species (J. Cain 2000). This consolidated into "speciation studies" by the end of the decade. Themes included the causes of variation, the causes of divergence, mechanisms for isolation, and types of selection. Each involved a myriad of subdivisions. Isolation, for instance, could be characterized in terms of geography, ecology, season, or physiology. These could be subdivided further. For instance, geneticists focused on the causes of variation. Mutation was one part of this, but hardly the only part. A great deal of attention was focused on processes of chromosome change: deletions, duplications, recombinations, and exchanges. Also, a focus was given to changes in the number of chromosomes within a cell, normally through the multiplication of whole chromosome sets. Doubling and tripling of chromosome sets could be identified from field samples. The same processes could be created in the laboratory using chemicals, such as colchicine. Newly discovered processes were related back into the accumulating directory of potentially relevant explanatory tools.

Some of the traditionally object-based taxonomists, but certainly not the majority, turned with delight from object to process. This helped especially with classification of groups less than a whole species (such as varieties or races) and of genera that seemed to have many species not much different from one another. Thinking about these groups as caught at one instant in an ongoing evolutionary process brought clarity to the taxonomist's job. They used the process approach to project forward and backward in time and hence proposed classifications organized by evolutionary history. Admittedly, these were speculative. But new tools were being developed that added weight to these proposals. Critics were dubious, but the clarity brought by evolutionary taxonomy seemed, to its defenders, worth the risk.

The case of a "ring species" illustrates the point (Fig. 34.4). Imagine that a species of birds has a range that loops around the Arctic Circle. Local populations show some geographical distinctiveness (perhaps enough to count as subspecies), but the neighboring populations overlap such that they form a continuous series around the Arctic. Interbreeding occurs between every pair of adjacent populations except in one location. In that contact zone, the continuous ring of gene flow is broken. This case posed a contradiction for systematists because it presents a good biological species (gene flow can take place throughout the species because interbreeding is continuous in one direction); however, barriers to gene flow are starting to develop because at least one contact zone shows isolation. That suggests the presence of two species. But where should the line be drawn along this otherwise continuous loop? Evolutionary taxonomists argued ring species like this showed evolution in progress. They introduced concepts like "species complexes" and "superspecies" to preserve a formal ambiguity within an otherwise clear understanding of process.

Systematics and speciation studies in the 1930s were rich with similar examples. Just as *Drosophila* became a model organism for transmission genetics, a long list of other genera served as tools for researchers who studied the evolutionary processes of divergence and isolation. These included plants (e.g., *Crepis*, *Iris*, and *Tradescantia*), insects (*Cynips*), reptiles

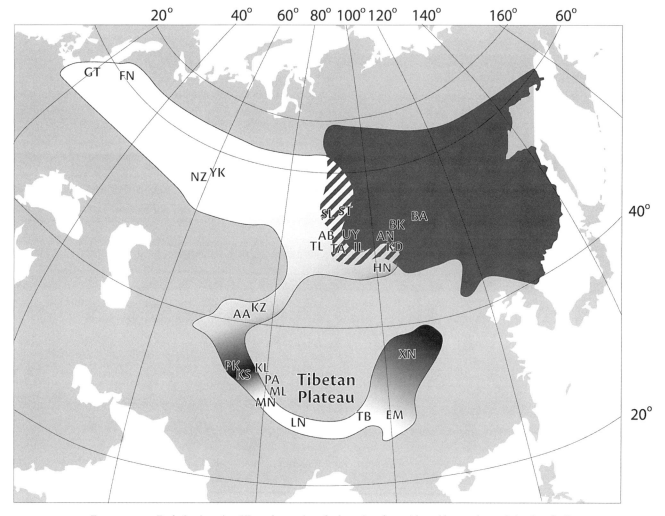

FIGURE 34.4. Evolution in action. We see here a ring of subspecies of greenish warblers, each population interfertile with those adjacent, but with the end groups unable to cross-fertilize. The gap in China is human-caused (in the past millennia) by destruction of the forests in which the warblers live. Redrawn from D. E. Irwin, S. Bensch, and T. D. Price (2001), Speciation in a ring, *Nature*, 409, 333–337.

(*Plethodon*), birds (*Junco*), and mammals (*Peromyscus*, and *Equus*). The messier the case, the better.

No one made the study of these general processes their full-time work. However, over the arc of the synthesis period, growing numbers of researchers were encouraged, pressured, and rewarded for contributing. Some disciplines were full of researchers confident in the fundamental value of their conclusions. Other disciplines had champions who struggled for the smallest amounts of recognition. They also struggled to convince colleagues in their own fields to see the contribution as worthwhile. These different scenarios gave different meanings to "synthesis."

Dobzhansky frequently spoke about the need to shift from a "static" understanding of evolution to a "dynamic" one. Mayr campaigned for systematists to shift from descriptive taxonomy to examine underlying causes. Simpson encouraged paleontologists to shift from asking "what and when" to "how and why." Julian Huxley encouraged zoologists to forget about things almost entirely and study processes. (This cost him his job as director of the London Zoo because he regularly neglected the animals under his care.) Each of these is an example of the broader sweep of change in the life sciences from objects to process. The evolutionary synthesis represents both a resurgence of interest in the subject of evolution itself and a shift in evolutionary studies away from narratives about the transformation of things to an analytical study of the mechanisms by which that transformation might occur.

INFRASTRUCTURE EMBEDS SYNTHESIS

Science has a social infrastructure that facilitates and shapes interactions among researchers. Professional societies and journals are examples of this infrastructure. The choices scientists make about the shape and inner workings of this infrastructure reveal a lot about what they want to accomplish, or prevent, and how they want to go about it. As developments in the synthesis period gained momentum, campaigners sought to consolidate these gains through new infrastructure. The idea was to make the work of their consensus easier to continue: studying common problems and cooperating in the creation of solutions. All these efforts promoted programs at the heart of the synthesis: study evolutionary processes rather than narratives, study its causes and mechanisms, use rigorous methods, and build only on the most up-to-date biological knowledge. The people at the heart of these organizational efforts also used these opportunities to embed some of their own priorities, too.

Infrastructure building started in the 1930s. In Britain, the Association for the Study of Systematics in Relation to General Biology came together around 1935. In the United States, local interdisciplinary groups gathered in the San Francisco Bay area, at the University of Chicago, at the American Museum of Natural History, and at the Smithsonian Institution. In 1939, enough momentum built up to lead Dobzhansky, Huxley, Mayr, and Alfred Emerson to launch the Society for the Study of Speciation. Owing to the start of the war, however, this proved a false start. In 1942 Mayr took over another effort to bring together the many interdisciplinary groups in the United States, via the National Research Council's Committee on Common Problems of Genetics, Paleontology, and Systematics. After the war, this committee sponsored a conference on the current state of evolutionary studies, and the resulting publication of papers from that conference functioned as a benchmark for advocates for the new synthetic approach. The same momentum inspired Mayr, Dobzhansky, Simpson, and others to formalize the community network into a new professional society, the Society for the Study of Evolution, in 1946. This group organized annual meetings and came to speak on the subject at larger gatherings of biologists (Fig. 34.5). In 1947 it also launched a new research journal, *Evolution*, with Mayr as its first editor. Though the postwar society and journal were distinctly American in flavor, organizers worked hard to build international connections across war-torn Europe and Japan as well as South America. Owing to the Cold War and to political control by Lysenkoists, interchange with Soviet biologists proved extremely difficult.

These various organizations and their activities had two major results. One involved sharply increased levels of certain activities. In 1920 most studies of evolution focused on its course: phylogenies, narratives, relations, and overall patterns. The study of process was kept at a loose and generic level. Most leaders in evolutionary studies strongly discouraged students and colleagues from speculating on mechanisms or causes. The whole subject seemed simply too hard to prove.

By 1950, evolutionary studies seemed completely transformed. Research focused mainly on the analysis of causal processes and mechanisms, what Dobzhansky called the "physiology of evolution." The results of this research filled the pages of journals like *Evolution*. It defined university courses on evolution, museum displays, and popular writing on evolution. A collective view developed that "evolutionary biology" now existed as a distinct and new discipline. Darwin and the *Origin of Species* became heroes to the champions of this new discipline. This was less because they had returned to the substance of his theories and more because he had championed the study of causal processes and because his book built on a foundation of cautious deliberation and solid methods.

The other key point relates to interdisciplinarity. Specialists always argue over which subject is most fundamental, or first among equals. In 1920 such arguments were intense in academic biology, and the typical pattern of behavior was self-sufficiency. Interdisciplinary exchanges were the exception rather than the rule. Reasons differed in different cases, but the 1930s saw increasing numbers of interdisciplinary collaborations develop. Some involved pairs of researchers; others involved whole local communities organized by a central leader. People did not join these interchanges out of some abstract commitment to interdisciplinarity. They did so because they saw distinct, practical benefits for their own work. Botanical taxonomists worked with experts in cytology, for example, because counting chromosome numbers seemed a useful tool for delimiting new species. Paleontologists learned statistics, for another example, because it helped them assess the confidence they could have in their decisions.

Across a wide range of biology during the 1930s, new laboratory and field techniques, new analytical tools, and new ideas were becoming available, and opportunities beckoned. Broadly speaking, pressure increased on researchers interested in evolutionary problems to tie old and new together and to reassess the potential for progress. Opportunities abounded for researchers to co-opt tools developed in their disciplines for quite different purposes and apply them to fundamental questions like the causes of evolution. The result was unexpectedly productive. No one predicted in 1930 that by 1940 there would be an explosion of interest in topics like speciation, population genetics, or a new systematics. In these areas, being a "synthetic" worker meant either knowing about developments in related fields or working with those who did.

One consequence of these many interactions was not only the removal of idiosyncrasies within disciplines but also constant pressure to get up-to-date on developments elsewhere in biology. It simply was no longer acceptable to develop views that contradicted well-established phenomena in another discipline. Likewise, it was clear to everyone that all disciplines were developing quickly. Whatever they had learned a decade before quickly was becoming obsolete. Books like Dobzhansky's 1937 *Genetics and the Origin of Species* and Mayr's (1942) *Systematics and the Origin of Species* functioned as introductions to a discipline's new developments as much as anything else.

CONCLUSION: TRUE TO DARWIN'S LEGACY?

The evolutionary synthesis is best understood not as a single event or as a small set of landmark books. It is much more diffuse, and it involved a convergence of developments along several layers of activity. Make no mistake. The synthesis period represents a major transition in the history of evolutionary studies.

A comparison of 1920 with 1950 shows several key transitions. First, a shift from an object focus to a process focus places renewed emphasis onto evolutionary studies. Second,

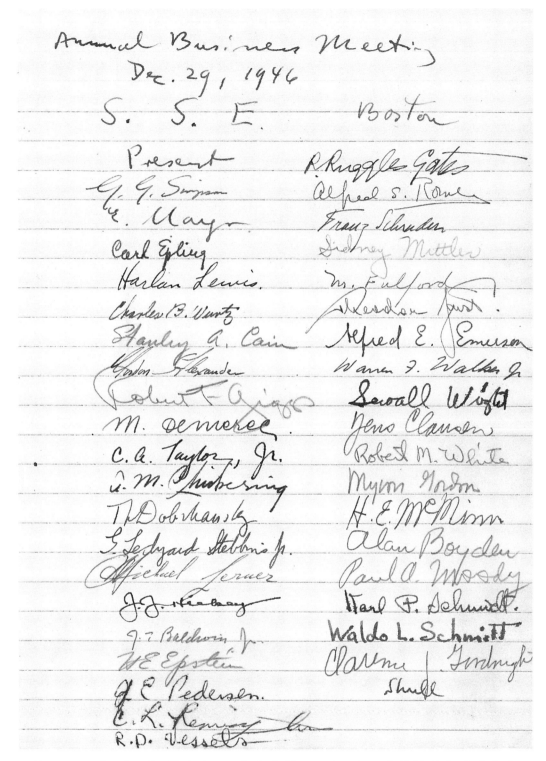

FIGURE 34.5. Attendance list of the business session for the "first annual regular meeting" of the Society for the Study of Evolution in Boston, Massachusetts, December 1946. Permission: American Philosophical Society

within evolutionary studies, the focus shifts from the study of descriptive patterns and narratives to causes, heuristics, and mechanisms. Third, standards for judging good methods were changing away from observational and inductive methods toward analytical and experimental ones. By the 1930s, all three transitions were well underway, and researchers in many specialties were engaged in pushing forward many boundaries. The cumulative effect was a collective and confident belief that a new period in evolutionary studies was underway. When researchers presented the evolutionary synthesis as the rebirth of evolutionary biology as a science, they tapped directly into Darwin's legacy. This legacy went far beyond any of Darwin's particular answers to life's questions and emphasized instead his role as a model scientist.

Campaigners for synthesis constructed a narrative of history in which Darwinism in its original form went into "eclipse" near the end of the nineteenth century and was nearly lost owing to the strength of rivals, notably a self-sufficient Mendelian genetics and a general disdain in biology for evolutionary studies as speculative and largely fanciful (Smocovitis 1999; Largent 2009). Recovery came in several waves over the next decades. The first wave was a return of confidence in methods and a resurgence in evolutionary studies as a legitimate topic during the 1930s. This was coupled with a renewed emphasis on evolutionary causes and mechanisms, such as natural selection. The climax of Darwin's recovery and vindication occurred in the 1950s. In the end, there was more to the synthesis period than a vindication of Darwin's theories and his approach to science. However, promoters of the evolutionary synthesis as a distinct moment in history tied themselves firmly to all the key innovations associated with Darwin himself.

When studying periods of major change in history, most people reach for the language of revolution. Radical shifts and sharp changes have a drama and clarity that are hard to resist. Stories about revolution help to draw sharp lines and help keep history simple. Keeping history simple is not always a bad thing; however, in this case, simplicity diminishes a far more interesting and diverse story.

Essay 35

Ecological Genetics

David W. Rudge

ECOLOGICAL GENETICS IS a field of biology at the intersection of genetics, ecology, and evolution that emerged during the early 1930s in Britain and is particularly associated with the research of E. B. Ford, H. B. D. Kettlewell, A. J. Cain, P. M. Sheppard, C. A. Clarke, and their numerous intellectual descendants (Fig. 35.1). Ford (1964, xi), the self-identified inventor of the field, defined it as "the experimental study of evolution and adaptation, carried out by means of combined field-work and laboratory genetics." It is devoted to the study of the genetics of adaptations, that is, traits affecting survival and reproduction, and the ecological processes that affect the distribution and evolution of these genes in natural populations.

CONNECTION TO DARWIN

There is an obvious sense in which Charles Darwin is the progenitor of the modern biological field known as ecological genetics. Darwin's *Origin of Species* drew attention to the ubiquitous presence of heritable variation in nature, the power of natural selection as an explanatory agent in accounting for the origin and maintenance of biological adaptations, and the importance of studying living organisms by means of biogeographical patterns of distribution. Darwin was also an early and important pioneer and advocate of the systematic use of experimental methods in the study of natural history. Ecological geneticists certainly see themselves as intellectual descendants of Darwin in that they often position their work as providing experimental confirmation of his theory of evolution by natural selection.

The actual connection between Darwin and the origins of ecological genetics is more complicated. Darwin's *Origin of Species* emphasized the slow accumulation of slight variations over geological time periods, a consideration that led him and his contemporaries to despair of the prospect of studying natural selection in the field. Darwin believed the process of natural selection was in principle simply too slow to detect in the span of a human lifetime. The *Origin of Species* also embraced a blending theory of inheritance (i.e., the overall appearance of offspring is intermediate to that of its parents) and identified natural selection as only one of several possible evolutionary mechanisms. In the first edition of the *Origin of Species* (and increasingly

FIGURE 35.1. Edmund Brisco ("Henry") Ford (1901–88) was a highly eccentric Oxford professor, but he proved as adept in Britain at building a group of working evolutionists, as did Theodosius Dobzhansky in America. By the kind permission of the Warden and Fellows of All Souls College

in subsequent editions), Darwin relied on such factors as "use and disuse" and what he referred to as "correlation of parts" to account for some of the variation one finds in nature. Indeed, in contrast to Darwin, ecological geneticists are often referred to as "neo-Darwinists" because their writings stress the primacy of natural selection to the exclusion of other possible mechanisms of evolutionary change.

OXFORD SCHOOL OF ECOLOGICAL GENETICS

Ecological genetics developed first in Britain as a direct consequence of multiple developments in the wake of Darwin's theory. Chief among these were (1) the rediscovery of Mendel's work on genetics in 1900, which led to widespread recognition of the advantages of a particulate theory of inheritance; (2) the rise of population genetics, which drew attention to the power of statistical analyses of variation in nature; (3) R. A. Fisher's (1930) mathematical work, which demonstrated the first two developments were compatible with Darwin's theory of evolution by natural selection; (4) the discovery by naturalists of multiple examples of natural selection taking place before their eyes, the most famous of these being the phenomenon of industrial melanism; and (5) J. B. S. Haldane's (1924) mathematical analysis of this particular example.

These developments set the stage for a gifted group of individuals loosely associated with E. B. Ford's subunit of genetics at the University of Oxford, which initially included among others Bernard Kettlewell, Arthur J. Cain, and Philip Sheppard. (Cain and Sheppard later became associated with the University of Liverpool, where they were joined by Cyril Clarke.) It is sometimes referred to as the "Oxford School of Ecological Genetics," an invisible college of individuals sharing common interests and approaches to the study of variation in nature. Students who worked with them were expected to choose some naturally occurring variation to study, work out the genetics of the variation by laboratory breeding, document the distribution of alleles for that gene in natural populations, search for possible selective agents that might account for these distributions, and manipulate conditions to demonstrate the effects of selection. It should be noted that much of the research conducted by Ford and others associated with the Oxford School of Ecological Genetics was funded by the Nuffield Foundation, a British charitable trust established in 1943 to promote social well-being through research. Their success at obtaining continued funding depended crucially upon their skill at drawing out potential medical implications of their research.

In a later memoir, E. B. "Henry" Ford (1901–88) traced his interest in the study of variation in field populations to some studies of variation in the Marsh Fritillary, *Melitaea aurinia*, begun in the summer of 1917 with his father. This early work and his initial exposure to Darwin's *Origin of Species* led Ford to become a devout evolutionist. Ford met R. A. Fisher while an undergraduate at Oxford University, during which time he became convinced by Fisher's mathematical proofs that small selective advantages could be important in evolution. Ford recognized, as had other naturalists before him, the enormous potential of natural populations of lepidoptera for the study of natural variation, highlighting their relatively short life spans and the fact that wing color, banding, and spotting patterns were easily observed markers of underlying genetic variation. He also drew attention to how collections assembled by past lepidopterists and amateur collectors could be used for the study of variation in nature over time.

The general principles of ecological genetics are set out in Ford's magnum opus, *Ecological Genetics* (1964), anticipations of which may be seen in his earlier elementary textbook entitled *Mendelism and Evolution* (1931). Much of Ford's research, and that of his associates at Oxford, can be seen as a series of attempts to test and confirm Fisher's theories and, in particular, the statistical methods that Fisher was developing expressly for the purpose of field research. Fisher and Ford (1947) devised a technique for marking, releasing, and recapturing insects, which they and others since have used for the estimation of population size, demonstration of the power and magnitude of natural selection in wild populations, and also differential average survival between populations of known sizes.

OVERVIEW OF ECOLOGICAL GENETICS RESEARCH

Research by ecological geneticists has focused on three distinct but related problem areas left in the wake of Darwin's *Origin of Species*: documenting the presence and magnitude of natural selection, accounting for the persistence of multiple morphs (forms) of the same trait in wild populations (aka *polymorphism*), and testing whether "genetic drift" (a theoretical alternative mechanism for evolutionary change) actually occurs in nature.

Natural Selection

Darwin's theory of natural selection drew attention to the fact that if certain conditions exist in nature, "favored" forms will increase at the expense of less adaptive forms. Thus,

(1) if members of a population vary in ways that affect their ability to survive and/or reproduce in a given environment; and
(2) if these variations are to some extent heritable; and
(3) if there is a competition in nature for resources, owing to the fact that members of the population reproduce in excess of those that can possibly survive; then
(4) it follows that favored forms will increase in frequency in the population inhabiting that environment over time.

In the *Origin of Species*, Darwin summarized the overwhelming direct evidence available in his time that each of the conditions (1–3) exist in nature. He was unable to provide direct evidence for his conclusion (4) but did note that it followed as a logical probabilistic consequence whenever these conditions all obtain. He also drew an analogy between the results of domestic breeders (artificial selection), reasoning that if, for instance, pigeon fanciers can produce entirely new varieties in the relatively short span of time the pigeon has been domesticated by man, this is evidence of what can and indeed has occurred in nature. Darwin's formulation of natural selection involving numerous slight variations among members of a population accumulating gradually over geological time periods led him to reluctantly conclude that a more direct demonstration of the power of natural selection was simply not possible.

Members of the Oxford School of Ecological Genetics rejected Darwin's views on inheritance in favor of a particulate model in terms of genes. While they retained Darwin's conception of natural selection as the primary mechanism of evolutionary change, acting continuously on slight selective advantages, their entire research program was based on the conviction that it is possible to document the action of natural selection in the field. Ford and his colleagues W. H. Dowdeswell, E. R. Creed, and K. G. McWhirter conducted a long-term study of the Meadow Brown butterfly, *Maniola jurtina*, and in a series of papers demonstrated that variation in the number of spots on the hind wings can be accounted for by natural selection.

FIGURE 35.2. *Biston betularia*: one typical and one *carbonaria* resting on blackened and lichen-free bark in an industrial area (the Birmingham district). Plates 14 and 15 in E. B. Ford, *Ecological Genetics* (New York: Wiley, 1975), reproduced with kind permission from Springer Science+Business Media B.V.

Among the best-known but by no means the most important studies that became part of the corpus of research done by Ford and his associates aimed at providing direct evidence of the presence and magnitude of selection in nature are a set of investigations on the phenomenon of industrial melanism.

Toward the end of the nineteenth and early twentieth centuries, it was clear that large-scale air pollution associated with the Industrial Revolution was having a dramatic effect on the countryside of large manufacturing districts. Trees in once pristine forests became darker owing to the dying off of lichen (which previously gave their trunks a pale appearance) and the gradual accumulation of soot. Coincident with these changes, naturalists noted that heretofore rare dark forms in many moth species were becoming more common in these industrial districts. The phenomenon of industrial melanism refers to the rapid rise in the frequency of dark moths in the affected districts that appeared to be a direct consequence of large-scale air pollution (Fig. 35.2).

Discovery of the phenomenon in the wake of the publication of Darwin's *Origin of Species* led many to speculate that it might be a consequence of natural selection. J. W. Tutt, building off the work of others, popularized the notion that the reason why dark moths were becoming more common in polluted districts was because visual predators, such as birds, had more difficulty spotting them than their pale counterparts. E. B. Ford thought the rise in frequency of the dark form might have to do with an alleged physiological advantage associated with the gene for dark coloration, but agreed with Tutt that the inability of birds to spot moths when they rest on soot-darkened surfaces would explain why a similar spread had not occurred in unpolluted districts.

Bernard Kettlewell (1907–79), a gifted naturalist who left medical practice to pursue his lifelong hobby as a researcher in Ford's newly formed subunit of genetics at Oxford, is widely hailed as establishing that birds preferentially remove the more inconspicuous form of the moth in polluted and unpolluted settings. Using Ford and Fisher's technique, Kettlewell marked known quantities of dark and pale peppered moths with a dab of paint, released them in a heavily polluted area near Birmingham and, then, over the course of several nights attempted to recapture as many as possible using a combination of assembling and mercury vapor light traps. Kettlewell reasoned that, all things being equal, the recapture rates should be the same. If, on the other hand, one form was better able to survive than the other (e.g., the dark form was better able to hide from birds than the pale form), it would have a higher recapture rate. This is indeed what Kettlewell found in both of the polluted settings. As expected, Kettlewell found the reverse when he conducted a complementary experiment in an unpolluted wood: here the recapture rate for the pale form was higher. The results of Kettlewell's first investigation were greeted by some skepticism, particularly by naturalists who doubted birds were significant predators on moths. Kettlewell is widely regarded as having clinched the argument by having an associate, the renowned ethologist Niko Tinbergen, film the order of bird predation. In this way, Kettlewell documented that birds representing multiple species with very different search behaviors had the same difficulty spotting moths when they rest on their correct (matching) background as humans do. Over the years these investigations, conducted in the early 1950s, have been severely criticized. Some of these concerns have had to do with what at the time were reasonable assumptions on Kettlewell's part, such as his conviction moths spend the day motionless on tree trunks in plain sight. (Judith Hooper [2002], a popular science writer, has actually gone so far as to suggest Kettlewell committed fraud, a completely baseless accusation [Rudge 2005].) It should be recognized, nevertheless, that the basic outline of the explanation we associate with Kettlewell has been confirmed by at least eight field studies since. There is no doubt among researchers who work on the phenomenon that it has occurred primarily as a result of differential bird predation. Contemporary research has established that the phenomenon is more complicated than textbooks imply and has drawn attention to the role of other factors, such as sulfur dioxide concentrations and differential migration (Majerus 1998).

C. A. Clarke and P. M. Sheppard continued work on the peppered moth in the vicinity of Liverpool. A later study by B. S. Grant, D. F. Owen, and C. A. Clarke has documented a similar rise and predictable fall in the frequency of dark peppered moths in Britain and the United States, following the advent of clean air legislation (B. S. Grant, Owen, and Clarke 1996). Industrial melanism continues to be an active area of research, with recent work extending to numerous other species where this change has been observed to occur.

Polymorphism

Darwin's theory of natural selection draws attention to how interactions between the members of a population with one another and their environment over generational time will probabilistically lead adaptive forms of traits to increase in frequency at the expense of less adaptive forms. Natural selection, so construed, is a process that continually removes less adaptive forms. (Darwin recognized that there must be some process that continually introduces variation in nature [what biologists now refer to as random mutations in the genes that code for traits] but was unable to do more than speculate as to the causes.) Darwin's theory seems to imply in general that when one examines a particular trait in a population, a single most adaptive form should typically be the most common. Thus, only under very rare circumstances, such as when a change in the environment leads one form to replace another (e.g., the phenomenon of industrial melanism mentioned previously), will two or more forms of the same trait coexist in a given population.

Critics of Darwin's theory of evolution by natural selection increasingly drew attention to numerous instances of field observations of natural populations in which two or more forms of a trait appeared to stably coexist. While the concept of polymorphism was well known before Ford's research (by E. B. Poulton, among others), Ford (1964, 84) is often identified as the first to rigorously define genetic polymorphism as "the occurrence together in the same locality of two or more discontinuous forms of a species in such proportions that the rarest of them cannot be maintained merely by recurrent mutation." Ford carefully distinguished between neutral polymorphisms (i.e., the persistence in a population of adaptively neutral forms, discussed further in the next section) and selected polymorphisms (i.e., the persistence of two or more forms, each of which is maintained by selection). Ford accounted for polymorphisms in part by drawing attention to increasing evidence provided by genetics studies that genes may have more than one effect on the constitution of an organism (i.e., pleiotropism). (For example, with regard to the phenomenon of industrial melanism discussed previously, the gene responsible for dark coloration was also thought to confer a physiological advantage that made the moth hardier.) Work by geneticists also drew attention to how individual genes might be affected by other genes possessed by the organism associated with other traits, the "genetic environment" of the gene. These considerations ultimately led Ford to conceive of heritable variation in nature in terms of adaptive gene complexes and ultimately the evolution of supergenes.

The evidence for polymorphism among insects and other invertebrates led Ford to believe ABO blood groups in humans represent an example of a balanced polymorphism. He predicted these different blood groups would therefore be characterized by different susceptibilities to disease, a prediction that subsequent work has shown to be true.

Ford also argued that the most common reason why two forms for the same trait persist in a population is because of a phenomenon known as heterozygote advantage. In sexually reproducing organisms, each individual has two genes for each trait (alleles), one from each of its parents. Heterozygote advantage refers to the possibility that when an individual inherits two different alleles the expression of these two alleles will give it an advantage in terms of survival or reproduction, over individuals who have two copies of the same allele. Perhaps the best-known example of this is the blood disease sickle-cell anemia, which involves a mutation to the gene that codes for the blood protein hemoglobin (Allison 1954a, 1954b). Sickle-cell anemia is an inherited disease among (descendants of) peoples who live in tropical environments throughout the world where malaria is common, such as Africa and India (Fig. 35.3). Individuals who inherit two normal alleles for hemoglobin are healthy but susceptible to malaria, a deadly disease transmitted by mosquitos. Individuals who inherit two mutated alleles are anemic, suffer episodes of pain, and are more vulnerable to infection. Left untreated, they normally have very short life-spans. Individuals who inherit one normal and one mutated allele for hemoglobin often display none of the symptoms of sickle-cell anemia, but are less susceptible to malaria. Thus one can appreciate why, in areas where malaria is present, both the normal and sickle-cell alleles for the gene coding for hemoglobin production would persist.

Genetic Drift

A central problem Darwin took on in the *Origin of Species* was that of accounting for the persistence of several well-known traits in nature, despite the fact that they appear, on the surface, to be nonadaptive – for example, the gaudy display of a male peacock. To account for this specific example, Darwin made reference to a special type of natural selection he termed "sexual selection." He conjectured an exaggerated male tail display had evolved in peacocks because they were preferentially chosen by females, who use the gaudy display as a surrogate for assessing the overall "fitness" (i.e., if a potential mate with such an unwieldy tail display can survive despite the fact that his tail makes him more vulnerable to predators, this suggests his other heritable characteristics must be above average). Darwin accounted for other examples of "nonadaptive" traits with reference to such factors as "the correlation of parts," which refer to a trait that has arisen for reasons other than its current role, however selectively advantageous it might be.

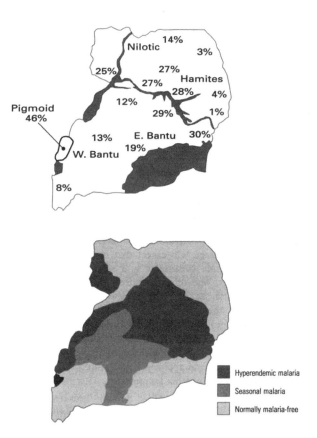

FIGURE 35.3. Distributions of the sickle-cell gene in proportion to the normal genes in the indigenous peoples of Uganda (*top*) and of types of malaria in Uganda (*bottom*) (around 1949). Based on A. C. Allison, Protection by the sickle-cell trait against subtertian malarial infection, *British Medical Journal* 1 (1954): 290.

Darwin (1859, 197) pointed out, for example, that the sutures in the skulls of young mammals may facilitate birth, but they probably first arose in reptiles simply as a consequence of the laws of growth. This being said, the general tenor of Darwin's work strongly suggested that whereas there is no necessity that variations that regularly arise in nature are adaptive, he clearly implied that nonadaptive traits will be quickly weeded out under the scrutiny of natural selection acting over long periods of time. This was widely interpreted as suggesting that truly nonadaptive traits should be very rare in nature.

In the years following the publication of Darwin's *Origin of Species*, naturalists discovered examples of polymorphisms that persist in populations without any obvious adaptive value. One example that figured prominently among critics of natural selection was the highly polymorphic land snails, *Cepaea nemoralis*, which exhibit differences in both the color and the number of bands on their shells. Whereas Darwin's theory suggested that ultimately the most advantageous type should become the most common, fossil records indicated the snail had been polymorphic with regard to its shell banding pattern since at least the Pleistocene (Figs. 35.4 and 35.5).

Sewall Wright (1889–1988), in a series of papers written in reaction to Fisher's genetic theory of natural selection, also openly questioned a central assumption of Fisher's

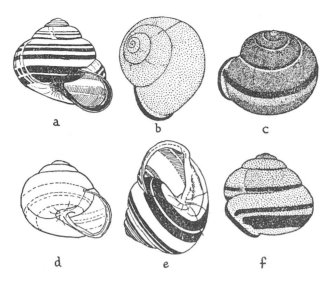

FIGURE 35.4. The many forms of the land snail *Cepaea nemoralis*, including (a) yellow shell with five bands; (b) pink, no bands, dark lip; (c) brown, one central band; (d) yellow, translucent bands; (e) yellow, bands, light lip; and (f) pink, missing some bands. From A. J. Cain, and P. M. Sheppard, Natural selection in *Cepaea, Genetics* 398 (1954): 89–116. Permission: Genetics Society of America

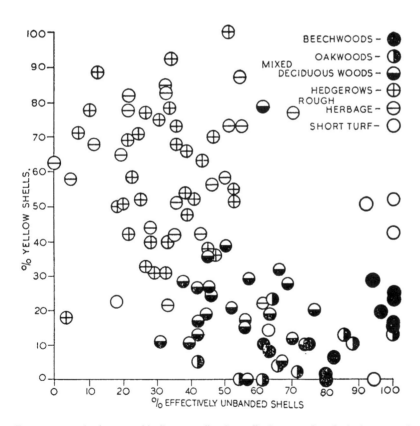

FIGURE 35.5. As shown on this diagram, yellow is an effective camouflage for hedgerows and meadows, as is banding, whereas pick and brown and uniformity are effective in beech woods where the undergrowth tends to be uniform and darkish. From A. J. Cain, and P. M. Sheppard, Natural selection in *Cepaea, Genetics* 398 (1954): 89–116. Permission: Genetics Society of America

models – namely, that populations in nature were generally large and, in the case of sexually reproducing organisms, provided equal access to all available mates. Wright developed an alternative mathematical model, which he proposed was more realistic because it incorporated the idea that species in nature were composed of numerous small populations. Wright's model accordingly emphasized the potential role of genetic drift, or chance fluctuations in the frequency of genes due to sampling error, and it was widely taken to imply that much of the variation we find in nature might be nonadaptive.

These issues came to a head with the publication of the French ecologist Maxime Lamotte's (1951) paper, which reported the results of a large field study in France of banding patterns in the land snail, *Cepaea nemoralis*. Lamotte's analysis drew attention to the fact that in addition to its apparently inexplicable variation, the land snail represented an ideal test case for Wright's theory. It was known to live in numerous small isolated populations with little migration between groups – precisely the conditions that Wright's model suggested would lead to genetic drift becoming an important factor. Lamotte's analysis attempted to identify correlations between the diverse habitats in which the snail is found and banding types by comparing broken shells to proportions of the types among living snails. His analysis acknowledged that selection due to predation was present but exceedingly small.

He concluded that the residual variation left unaccounted for must be due to sampling error – that is, genetic drift.

This challenge posed by Lamotte's analysis was taken up by Arthur J. Cain (1921–99) and Philip M. Sheppard (1921–76). Cain and Sheppard tested Lamotte's hypothesis by studying the various extreme habitats in which the land snail occurred in Oxford. They found patterns of yellow and banded shell morph frequencies in grasslands, with significantly higher proportions of brown, pink, and unbanded shells in woodlands, and attributed these differences to predation by wood thrushes, which smash snails on stones. In later experiments using snails with marked shells, they were able to demonstrate that predation by thrushes has resulted in differential selection among morphs (A. J. Cain and Sheppard 1954; A. J. Cain 1954). Cain and Sheppard extended this work in a lengthy set of papers starting in 1968 that worked out the genetic basis for this polymorphism. Cain and Sheppard's research has been followed up by numerous additional studies, which suggest that their work needs to be revised in light of other selection-controlled factors (Jones et al. 1977). While it has not conclusively demonstrated that random genetic drift is a negligible factor in the evolution of land snail shell banding patterns, it is often cited as an example of assuming a character is selectively neutral without carefully studying the effects of the variation on the organism and how it interacts with its environment.

Ford, Sheppard, L. M. Cook, and D. A. Jones conducted a study of variation aimed at detecting the presence of genetic drift in the scarlet tiger moth, *Panaxia dominula*. In a long-term study conducted from 1939 to 1946, they demonstrated that a decline in the frequency of a dark form (*medionigra*) must be due to natural selection because the populations involved were too large for the change to be the result of random drift.

It should be noted that, although these studies (and others like them since) draw attention to the challenge of documenting genetic drift in the field, they do not establish that genetic drift never occurs in nature. Clearly drift must play some role, at least to the extent that the actual sizes of wild populations depart from the assumptions of Fisher's models. Like other debates in biology, the question posed by the theoretical possibility of genetic drift is one of relative importance.

The heyday of ecological genetics from the 1950s onward occurred in relative isolation from a revolution taking part in the life sciences in the wake of the newly formed discipline of molecular biology. Ford and other members of the Oxford School of Ecological Genetics initially fought against an increasing trend that shifted funding, academic positions, and awards from traditional areas of biology into the emerging molecular disciplines (Ruse 1996). In subsequent years, practitioners of ecological genetics have nevertheless come to embrace these developments. This shift is reflected in successive editions of Philip Sheppard's (1958) *Natural Selection and Heredity*. Sheppard's influential book provided an accessible introduction to the synthetic theory of evolution, drawing attention to how natural selection can be understood in terms of Mendelian genetics. The first edition of this work, published five years after James Watson and Francis Crick's ground-breaking discovery, makes no mention of DNA. The fourth edition (1975) in contrast not only discusses DNA but includes a chapter on protein evolution, drawing attention to significant advances by molecular biologists (Ruse 2009b).

❧ Essay 36 ☙

Darwin and Darwinism in France after 1900

Jean Gayon

The incorporation of Darwinism – its theory of modification of species through natural selection – occurred in research programs in France in the 1930s with the development of a remarkable and unique school of genetics of experimental populations. Around the same time, however, France witnessed another remarkable episode, perhaps the most impressive example of a durable and late opposition of French science to evolutionary theory: the general aversion of French paleontologists to phylogenies in the years 1900–50. (For a detailed account of these episodes, see Gayon and Veuille 2001; Gayon 2006, 2009.) As will be seen, images play an important role in this story.

THE FRENCH PALEONTOLOGISTS' AVERSION TO REPRESENTING PHYLOGENIES (1900-1950)

This is a rather strange story, well known to paleontologists, but that has escaped the attention of historians of science. A quantitative enquiry into the three French periodicals that published almost the entire production of French paleontology in the years from 1900 to 1950 gave the following results.

Case 1: *Annales de paléontologie* (1906–1950)

Let us first consider the *Annales de paléontologie* (Annals of Paleontology). Founded in 1906, this was the very first periodical devoted entirely to paleontology in France. In the first issue of the journal, the editor Marcellin Boule (known for his work in human paleontology) stated that "philosophical paleontology" should be a priority for the authors. "Philosophical paleontology" was a term Albert Gaudry used as a synonym for "evolutionary paleontology." "Philosophical" meant that paleontologists should not only describe the presence of fossils in stratigraphic layers but should dare to make phylogenetic inferences. In 1866 (the year when Haeckel coined the expression "phylogenetic tree" in Germany), Gaudry was the very first scientist to publish phylogenetic diagrams representing the hypothetical genealogy of real fossil groups clearly identified from the fossil record (see Fig. 29.5). Boule's 1906 preface was in

FIGURE 36.1. In his prime, Pierre Teilhard de Chardin (1881–1955) was the best French paleontologist of the day. He was a close friend of George Gaylord Simpson, who much appreciated his abilities and achievements. Courtesy of the French Jesuit archives, Vanvesg

fact preceded by a letter from Gaudry, still active, who urged paleontologists to follow the path that he and his pupils had laid down for more than forty years. However, although Boulle headed the journal until 1942, this wish was not fulfilled. Of the approximately two hundred articles published in the period from 1906 to 1950, only four offered phylogenetic conjectures, each being accompanied by a graphic representation. Pierre Teilhard de Chardin wrote the first two, in 1915 and 1921 (Fig. 36.1). Teilhard, who was certainly the most brilliant French paleontologist of the first part of the twentieth century, drew a distinction between diagrams representing morphological affinities (Fig. 36.2) and diagrams representing genealogical relationships (Fig. 36.3). The latter unequivocally represent events of modification, splitting, and common descent. The graphic conventions used in Teilhard's phylogenetic diagrams resemble those used by Gaudry in 1866.

In 1933 Jacques Mercier published another branching diagram in an article devoted to the crocodilians. Much more cautious than Teilhard's representation, the diagram looks like a hierarchy of candelabras (Fig. 36.4). It is as much a table marking the presence of fossils in given strata as an explicit genealogical conjecture. It might also suggest the instantaneous formation of new species. Significantly, there is no caption for the diagram in the article, a rather strange feature in a carefully edited scientific journal.

In 1936 Colette Dechaseaux presented an even more prudent diagram, in an article on fossil Pectinidae scallops) (Fig. 36.5). The diagram has no real nodes, except in four special cases where dotted lines indicate a possible phylogeny. Although Dechaseaux admitted in the text that her diagram could be taken in some cases as representing a hypothetical phylogeny, she did not write "phylogeny" in the figure's caption, but "repartition."

Four phylogenies over two hundred articles and forty-six years is not many, especially given the explicitly expressed view of the editor of the journal. We may therefore conclude that the authors publishing in the *Annales de paléontologie* deliberately avoided phylogenetic conjecture. In the rare case when they offered such conjectures, they seem to have been increasingly cautious. Only Teilhard de Chardin offered diagrams that were openly consonant with Darwin's "descent with modification."

Case 2: *Bulletin de la Société géologique de France* (1901–1950)

This journal (Bulletin of the French Geological Society) was one of the major scientific French periodicals from the 1850s. I have examined the fourth and fifth series of this journal, from 1901 to 1950, with the same criteria as in the previous case: Did the authors make phylogenetic conjectures, and, if so, did

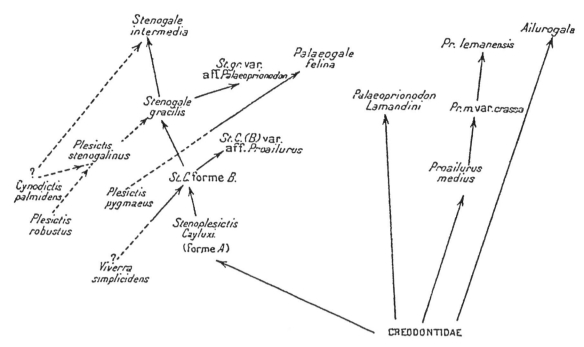

FIGURE 36.2. Morphological relationships as portrayed by Teilhard. From P. Teilhard de Chardin, Les Carnassiers des phosphorites du Quercy, *Annales de paleontology* 9 (1915): 103-92

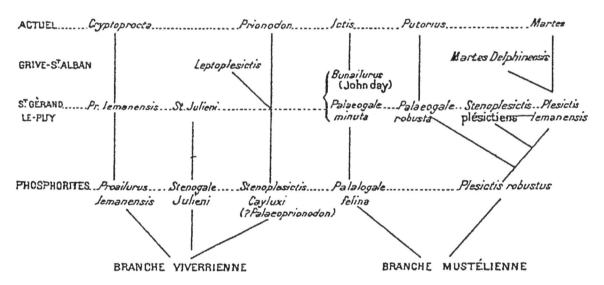

FIGURE 36.3. Genealogical relationships as portrayed by Teilhard. From P. Teilhard de Chardin, Les Carnassiers des phosphorites du Quercy, *Annales de paleontology* 9 (1915): 103-92

they offer a graphic representation? This journal had many more articles than the *Annales de paleontology*, even in paleontology. Most of the professional articles in paleontology in the first half of the twentieth century were published there. I have found nine articles presenting phylogenetic conjectures, but only four of them offered a diagram.

Between 1901 and 1926, three articles made phylogenetic conjectures, in brief terms, and with no diagram. In 1927 an article on a group of bivalves by Guillaume contained a rather strange figure (Fig. 36.6). Some species seem to transform into other species; there are two cases of splitting, but the whole diagram expresses extreme cautiousness (dotted lines, question marks). Obviously, the intention is to maintain the phylogenetic conjectures within the limits of the available stratigraphic data.

René Abrard's 1929 article on Nummulites (unicellular protozoa) is unique in the entire literature that I have examined between 1900 and 1950 in all French periodicals. From the first to the last page, the author is overtly an evolutionist. The title of the article itself ("Contribution à l'étude de l'évolution des Nummulites" [Contribution to the Study of the Evolution of Nummulites]) includes the word "evolution" – a unique case.

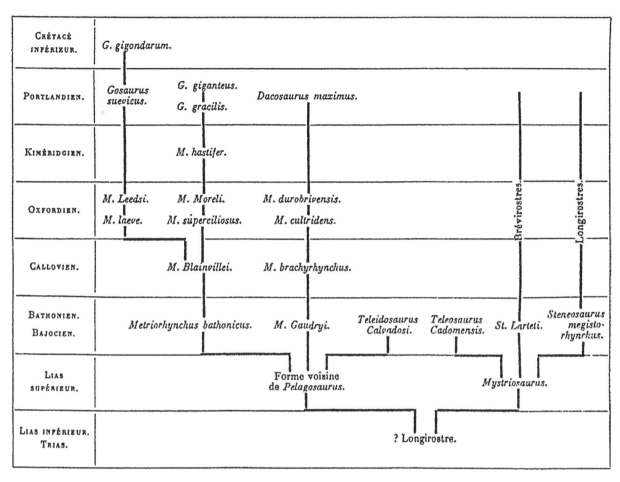

FIGURE 36.4. The relationships between crocodilians. From J. Mercier, Contribution à l'étude des Métrionhynchnidés (Crocodiliens), *Annales de paléontologie* 22 (1933): 91–120

The several diagrams look like the one represented in Figure 36.7. All of them represent explicit phylogenetic conjectures, with precise stratigraphic indications. Each of the diagrams in each figure represents a unique tree, a unique pattern of common descent. The third table of Figure 36.7 is the most interesting as the three diagrams also illustrate a pattern of convergent evolution. But, again, this article is exceptional.

In 1934 Colette Dechaseaux devoted an article to Gryphaea (extinct oysters). She gave several stratigraphic tables, one of which resembled a phylogenetic conjecture (Fig. 36.8). This diagram can hardly be called a tree. Dotted lines represent possible "mutations" from one to another. Again, extreme caution seems to have been the rule.

In 1939, in another article on bivalves, Dechaseaux devoted a paragraph explicitly to "phyletic relationships." The diagram represented in Figure 36.9 is intended to connect Dechaseaux's conjectures with stratigraphic data. No real branching is represented, no species change, but, just as in the previous diagram, such processes *might* be imagined. In fact, the figure represents no more than a table of the presence of fossils in stratigraphic layers.

The last phylogenetic diagram came in Jean Roger's 1944 article, "Phylogeny of Octopod Cephalopods." In this paper,

all diagrams except for one are tables of the presence of fossils in stratigraphic layers. But the article ends with a figure representing a phylogenetic conjecture (Fig. 36.10). This diagram is absolutely unique in the entire French paleontological literature that I have looked at in the first half of the twentieth century. No stratigraphic data are given there. A real tree is given, which resembles the genealogical trees that zoologists had been proposing since Haeckel. This tree represents a gradual process and successive branching. It makes one think of the genealogical diagrams that were so frequent (and, to be frank, so speculative) in the international literature of the time, especially in the American paleontological literature (e.g., Osborn).

Case 3. *Travaux du laboratoire de géologie de la Faculté des sciences de Lyon* (1921-1943)

This publication was not, properly speaking, a periodical but a temporary series handled by a brilliant local paleontologist, Frédéric Roman. Within twenty-two years, thirty-nine rather extensive studies were published in this series. Two of them proposed phylogenetic conjectures. The first of them was a gigantic table of "filiation and repartition" of Jurassic

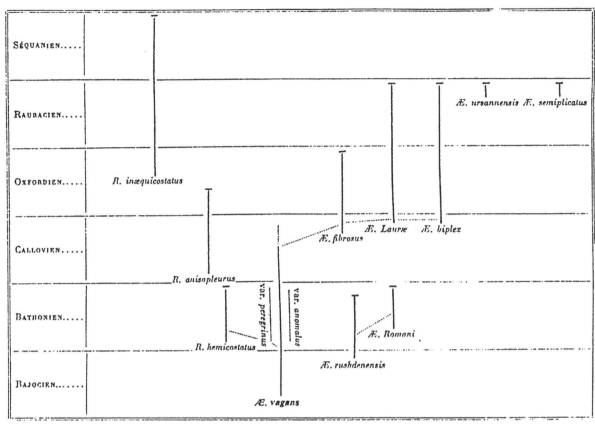

FIGURE 36.5. Representation of relationships between members of a group of Pectinidae (scallops) by Dechaseaux 1936. The original caption does not say "phylogeny" but "repartition." From C. Dechaseaux, Pectinidés jurassiques de l'Est du Bassin de Paris. Révision et biogeography, *Annales de paleontologie* 25 (1936): 1–146

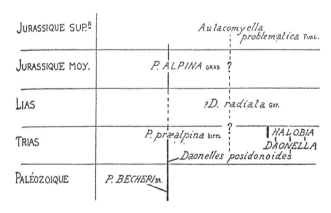

FIGURE 36.6. Table summarizing the "history of Posidonomyids" (a group of bivalve mollusks). From L. Guillaume, Révision des Posidonomyes jurassiques, *Bulletin de la Société zoologique de France* 27, no. 4 (1927): 217–34

belemnites. The table is so big that it is not possible to reproduce it. Quite strangely, the text does not offer the slightest commentary on it. The second was Jean Viret's 1939 study of fossil rodents of the region of Montpellier. The figure resembles many of the phylogenetic diagrams found in the international literature of the time (Fig. 36.11). It is overtly gradualist and conjectural (as noted by the dotted lines). Stratigraphic information is given.

Let us sum up this inquiry. First, we observe that, of approximately six hundred French professional articles in paleontology within a span of fifty years, only fifteen offered phylogenetic conjectures. Among these, just ten offered a diagrammatic representation. We also see that, with one exception, all diagrams plotted genealogical conjectures against stratigraphic information. This comes as no surprise in geological journals. But the systematic character of this behavior seems to have been peculiar to French paleontologists. A general feature of almost all diagrams is the use of various conventions that make as explicit as possible the methodological uncertainties of phylogenetic conjectures. This should be related with the fact that evolutionary theory is almost totally absent (even allusively) in all papers in the three periodicals. Finally, around 1940, diagrams resembling the zoologists' trees seem to have begun to become acceptable. How can one account for such a pattern in the French paleontological literature in the years 1900–50, at a time when it became almost impossible to find in advanced scientific countries a biologist or a paleontologist who did not believe in the "general fact of evolution"? Here are some comments and possible explanations.

(1) It would simply be untrue to say that French paleontologists did not believe in evolution in the most general sense

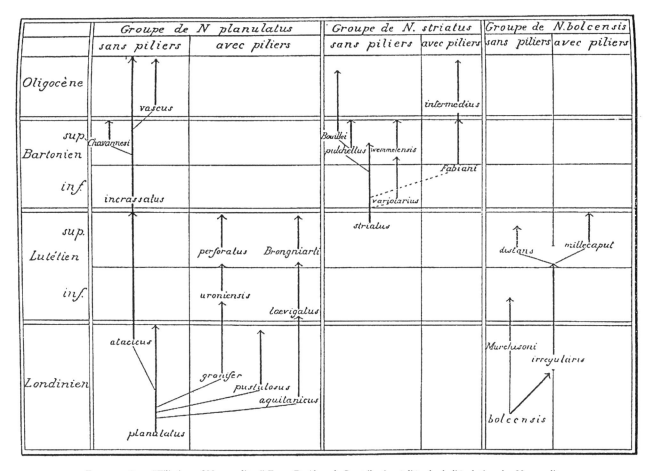

FIGURE 36.7. "Filiation of Nummulites." From R. Abrard, Contribution à l'étude de l'évolution des Nummulites, *Bulletin de la Société zoologique de France* 28, no. 4 (1928): 161–82

of "descent with modification." In fact, most if not all of them did.

(2) Figure 36.12 offers a comparative table of the quantity of phylogenetic conjectures made in journals published in France and in other countries between 1900 and 1930. This table has been compiled from the reviews offered in *L'Année biologique*. In the French case, the table gives the number of phylogenetic conjectures made by, respectively, biologists and paleontologists in all journals considered. (Note that, for thirty-five years, *L'Année biologique* certainly published one of the best and most exhaustive records in natural history and biology; I stopped my count at 1930, when this periodical began to decline.) This table shows that French scientists in general do not seem to have made fewer phylogenetic conjectures than other scientists did. But the relative rarity of phylogenies made by French paleontologists should be underscored.

(3) The period from 1900 to 1950 corresponds to a general assertion of the autonomy of paleontology relative to "natural history" throughout the world. In the late nineteenth century, French paleontologists were much bolder than they were later: some made phylogenies, others not, in a context of open controversy about the fact of evolution in general. After 1900, French paleontologists, who were becoming more professional, decoupled from biological research. The main issue for stratigraphers was the measurement of geological time as precisely as possible. Evolution was no longer a priority question. In this context, Teilhard de Chardin was a major exception. As strange as it may appear, he was the key figure who progressively restored the interest of French paleontologists in evolution.

(4) Positivism played a major role in the French paleontologists' reluctance to make phylogenetic conjectures. Félix Bernard (1863–98) published the first French textbook in paleontology in 1895. As noted by Stephen Jay Gould, this treatise written by a young and soon-to-be prematurely deceased professor in natural sciences remains one of the most impressive of all times and all languages. This treatise was read again and again by successive generations of French paleontologists. In this book, the author explained that finding genealogical relationships between groups is essential to paleontology as a science but requires extreme caution, because phylogenies are always conjectural. This book of more than a thousand pages contains hundreds of diagrams, concerning all possible groups. Félix

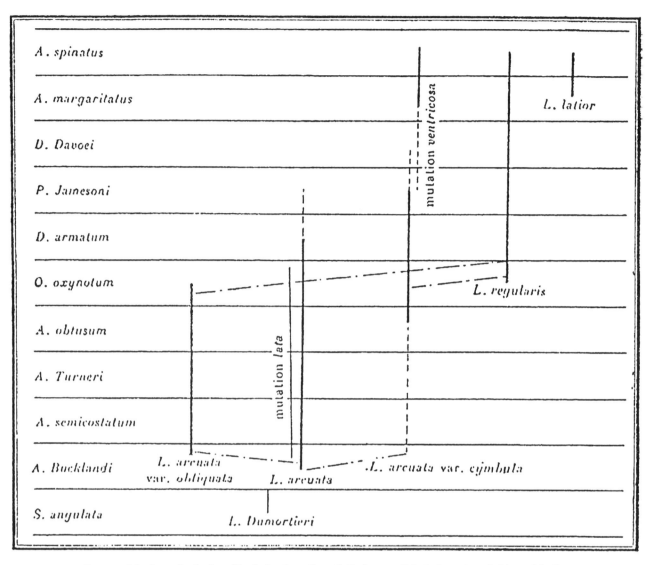

FIGURE 36.8. Stage distribution of liassic Gryphaea. From C. Dechaseaux, Principales espèces de Liogryphées liasiques. Valeur stratigraphique et remarques sur quelques formes mutantes, *Bulletin de la Société Géologique de France* 4, no. 5 (1934): 201–12

Bernard carefully avoided presenting genealogical trees when formulating phylogenetic conjectures. According to Bernard, in paleontology, diagrams should represent only the presence of fossils in given stratigraphic layers and be empirically undeniable. Figure 36.13 gives a typical example of the dozens of "tables of stratigraphic repartition" that Bernard gave in his book for each group. Imagine a ninety-degree rotation of this diagram, and you will have the implicit methodological model that explains the strange aspect of most of the phylogenetic diagrams presented earlier in this paper. Here, I believe, is the main reason why French paleontologists stopped representing phylogenies in the years 1900–50. Positivism, which was so dominant at that time in both the scientific world and the political scene, was the major obstacle that prevented a true assimilation of Darwin's thinking among paleontologists (and biologists): it was too conjectural a theory.

NATURAL SELECTION IN THE LAB (L'HÉRITIER AND TEISSIER, 1932-1937)

France has the reputation of being "the only major scientific nation that did not contribute significantly to the evolutionary synthesis" (Mayr and Provine 1980, 320). This is not entirely true. In the 1930s, an influential school of mathematically trained population geneticists developed in Paris. It flourished in a peculiar and privileged part of the French university system, the École Normale Supérieure, and was led by two young biologists, Philippe L'Héritier (1907–94) and Georges Teissier (1900–72). Later on, L'Héritier became the leading figure who introduced and institutionalized genetics in France. As for Teissier, he was already known in the 1930s as one of the most competent biometricians in the world. The contribution of L'Héritier's and Teissier's work in population genetics was evident in Sewall Wright's work. The third

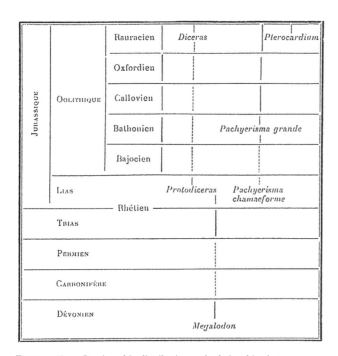

FIGURE 36.9. Stratigraphic distribution and relationships between some bivalves. From C. Dechaseaux, Megalodon, Pachyerisma, Protodiceras, Diceras, Pterocardium et l'origine des Diceras, *Bulletin de la Société Géologique de France* 5, no. 5 (193): 207–18

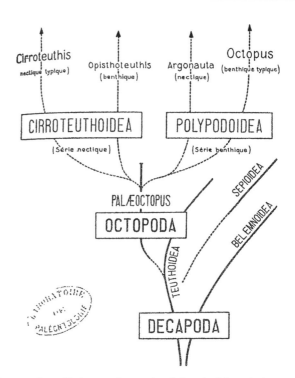

FIGURE 36.10. Phylogeny of octopod cephalopods. Of note is the occurrence of the word "Phylogeny." From J. Roger, Phylogénie des Céphalopodes Octopodes: Palaeoctopus newboldi (Sowerby, 1846) Woodward, *Bulletin de la Société Géologique de France* 5, no. 5 (1944): 83–98

volume of his book *Evolution and the Genetics of Natural Populations* (1977) devoted the major part of the chapter on "natural selection in the laboratory" (one-tenth of the volume) to L'Héritier's and Teissier's results from the 1930s. Sewall Wright had in fact visited the Parisian team just before World War II, and he was responsible for informing Theodosius Dobzhansky about the exceptional tool developed by Teissier and L'Héritier for experimentally studying the evolution of population of *Drosophila*, the "population cage." This episode is intrinsically interesting for the history of population genetics in general. In France, specifically, it played a major role in the history of the acceptance of Darwinism. After Teissier's and L'Héritier's seminal work, it became simply impossible to say that Darwin's theory of natural selection was not amenable to mathematization and experimentation. At the very time of their discoveries, Teissier's and L'Héritier's work was widely recognized in France and abroad. After World War II, the two biologists became key figures in the reconstruction of French science. Teissier (a major figure of the Communist Party) was director general of the CNRS for three years immediately after the war and professor at the University of Paris. L'Héritier (a Catholic) became one of the most renowned geneticists of his time, the author of major textbooks, and an institutional organizer of the development of genetics in research and education. Teissier and L'Héritier were the two key figures who "naturalized" Darwin in France, both scientifically and institutionally.

Philippe L'Héritier designed population cages in 1932, just two years after the publication of Fisher's *Genetical Theory of Natural Selection* (1930), and one year after Wright's "Evolution in Mendelian Populations" (1931). In 1932 both Wright and Fisher spoke at the Sixth International Congress of Genetics, in Ithaca, New York. L'Héritier, then twenty-five, attended the meeting. Mathematically trained, he was excited by what he heard. L'Héritier was apparently the first French biologist to benefit from a Rockefeller grant, long before Boris Ephrussi and Jacques Monod. He left France in 1931 with the aim of learning genetics and of finding a research project that he could continue in France after returning from the United States. L'Héritier (1981, 335–36) recalled the episode:

> During this stay in the US, I discovered the existence of population genetics, which I had never heard of before. I read Fisher's book, which had just appeared. I also discovered Sewall Wright's work. I met and heard these men and heard them speak at one of the first international Congresses of Genetics, held at Cornell in July 1932. I also met founding fathers of modern genetics and evolutionary biology, Theodosius Dobzhansky and H. J. Muller.... I discovered the famous fly, Drosophila. I had never seen it before.

The idea of population cages soon occurred to him. Before returning to Europe, he paid a visit to Woods Hole:

> One day, while walking on an American beach, I realized it would be possible to breed the fly, not in small bottles where only one generation could be observed, but in boxes in which food would be periodically renewed. This was the origin of the famous population cages, or

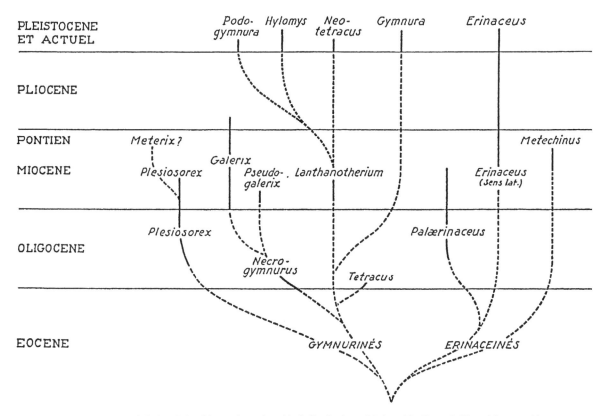

FIGURE 36.11. Phyletic relationships and stratigraphical distribution of Erinaceids. From J. Viret, Monographie paléontologique de la faune des vertébrés des sables de Montpellier. III. Carnivora Fissipedia, *Travaux du Laboratoire de Géologie de la Faculté des Sciences de Lyon* 37 (1939): 1–26

"demometers" [*démomètres*]. This word, "the measure of populations," corresponded to my initial idea of comparing the ability of flies from different origins or strains to establish themselves, demographically speaking, in a given milieu.

In October 1932, L'Héritier returned to France with this idea in mind, and some wild types of *Drosophila* in his luggage. He intended to use them for a PhD thesis on quantitative variation in *Drosophila melanogaster* that he effectively wrote – the first doctoral dissertation in population genetics ever defended in a French University (and even in genetics), in 1937. The invention of "population cages" was first made in this context. Meanwhile, L'Héritier (1934) wrote a mathematically subtle book, entitled *Genetics and Evolution: Analysis of Some Mathematical Studies on Natural Selection*, which was a discussion of Fisher's and Wright's models.

L'Héritier himself, a skilled woodworker, built the population cages. The boxes were maintained at a constant temperature and contained twenty food vials (yeast). Every day, a vial of fresh food was introduced to replace the oldest one. The twenty vials corresponded to the development of two generations of *Drosophila* at twenty-six degrees Celsius. From time to time, adults were anesthetized using carbon dioxide and sprinkled in the dark onto a light sensitive paper. A photograph was made, which made it possible to count the entire imaginal population. Between 1933 and 1937, L'Héritier counted 500,000 flies for his PhD research. At the same time, L'Héritier carried out population genetics experiments with Teissier.

Georges Teissier was a little older than L'Héritier. In the 1930s, he was already known for his biometrical studies on the growth of organisms and the growth of populations. Teissier was in fact, with Julian Huxley, the co-discoverer of the power law of allometric growth (Gayon 2000); Teissier and Huxley had extensive exchanges on the subject. This means that both Teissier and L'Héritier were part of the original international network of geneticists that initiated the modern synthesis. As the editor of a collection of scientific books published in Paris by Hermann, Teissier was also the director of a series of monographs entitled "Biometry and Biological Statistics," in which major figures involved in the mathematical theory of evolution published exceptionally important papers, among them Lotka, Volterra, D'Ancona, Gause, and Sewall Wright himself. In his book on Wright, Provine says that Wright's monograph in this series was the clearest and most rigorous account that he ever gave of his shifting balance theory of evolution.

These biographical facts about L'Héritier and Teissier help us understand how Darwinism in its purest form – that is, the study of competition and natural selection – was finally unequivocally developed in France. L'Héritier and Teissier both had strong mathematical backgrounds. Together, they

Volume	Total number of publications	In French	Among which:	
			Biologists	Palaeontologists
1895	4	1	1	0
1896	21	2	1	1
1897	10	1	1	0
1898	18	2	1	1
1899-1900	14	1	1	0
1901	7	1	1	0
1902	11	2	0	2
1903	14	3	1	2
1904	11	3	3	0
1905	4	2	2	0
1906	14	2	2	0
1907	12	4	4	0
1908	17	6	4	2
1909	23	11	9	2
1910	10	2	1	1
1911	13	5	5	0
1912	24	4	1	1
1913	18	2	0	2
1914	18	2	2	0
1915	9	3	3	0
1916	13	3	1	2
1917	20	4	3	1
1918	17	3	3	0
1919	18	4	3	1
1920-21	2	0	0	0
1921-22	13	1	1	0
1922-23	17	9	9	0
1923-24	11	6	1	5
1925-26	26	7	5	2
1926-27	14	3	3	0
1927-28	8	1	1	0
1929	10	1	1	0
1930	15	5	5	0
Total	456	106	79	25

FIGURE 36.12. A comparative table of the quantity of phylogenetic conjectures made in journals published in France and in other countries between 1895 and 1930. Permission: Jean Gayon drawing

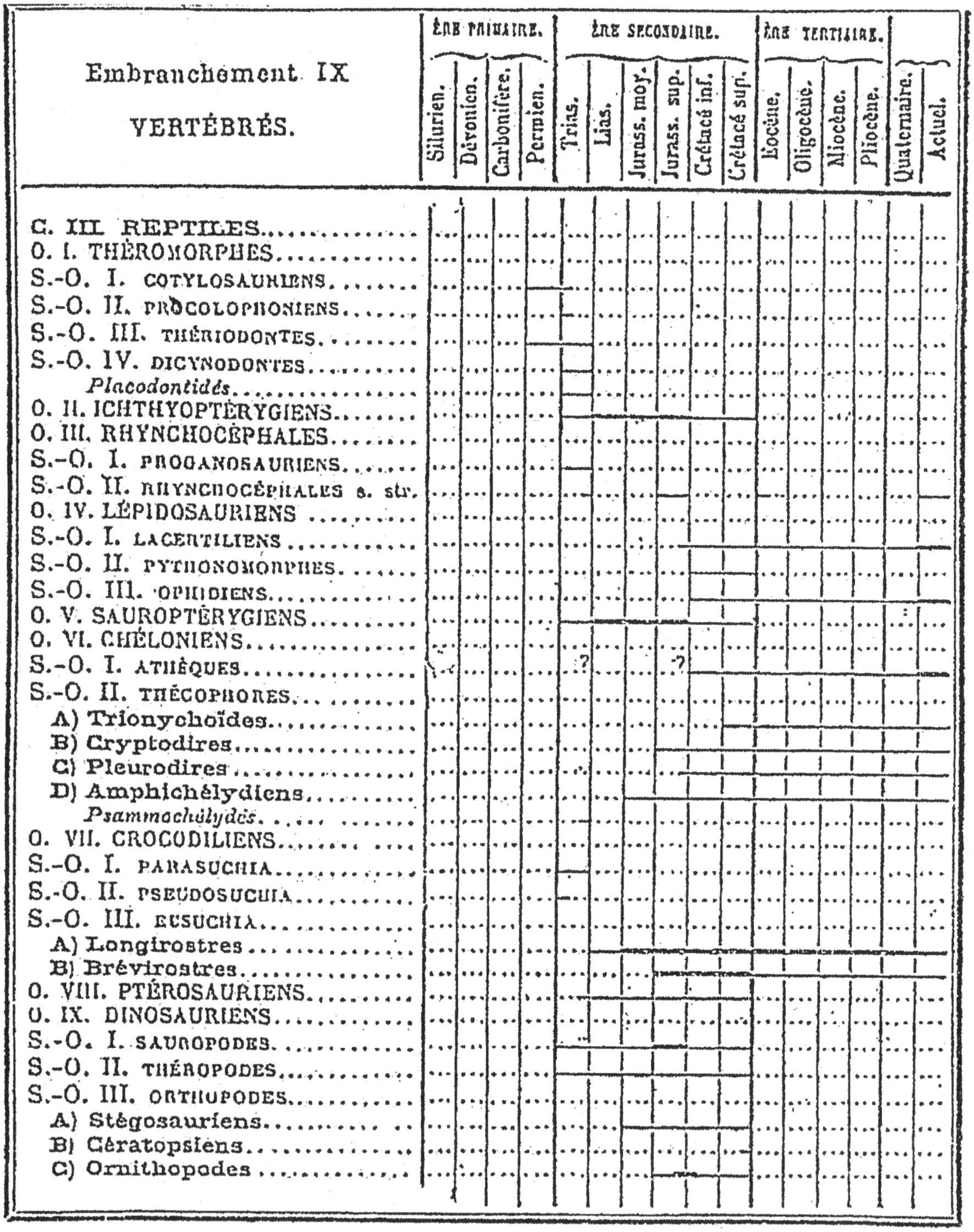

FIGURE 36.13. One of the numerous "tables of stratigraphic repartition" given by Félix Bernard, *Éléments de paleontologie* (Paris: J.-B. Baillière, 1895)

decided to put mathematical models of natural selection to the test in the lab. Mathematics and the experimental method were the two vectors of the adoption of Darwinism in France.

It was the exceptional work of L'Héritier and Teissier – realized in less than five years (1932–37) and published in a series of ten joint papers – that led to a major school of population genetics and to the conversion of French biologists to Darwinism after World War II. (For a detailed exposition, see Gayon and Veuille 2001.)

The first key paper was published in 1934 and had the term "natural selection" in its title (L'Héritier and Teissier 1934). This paper reported an experiment on the confrontation of two alleles (*bar* and *wild*) in the same cage. The dynamic of the polymorphic population was followed over five months, until the experiment was accidentally ended. The experiment showed a gradual increase of the wild type. It also showed that the selection coefficient changed as a function of the frequency of the two types. This was the first demonstration ever given that natural selection at the genetic level can be studied in the lab.

The same experiment was then repeated, but over a longer period of 600 days rather than 150, and in two different boxes. In each case, the result was the gradual elimination of the mutant. However, the mutant *did not* become entirely extinct. An equilibrium was asymptotically reached. The authors' conclusion was that the superiority of the normal gene tends to disappear when the mutant becomes rare in the population. Reading the notebooks of the experiment, the modern reader is struck by the unexpected outcome that the two young scientists observed: the nonelimination of an unfavorable gene, the establishment of a polymorphic equilibrium, and the gradual change of selection coefficients. (This showed, by the way, that Fisher's and Wright's original models were insufficient because they posited constant selective coefficients [L'Héritier and Teissier 1937a].)

In 1937 a similar experiment confronted a wild type strain and an *ebony* strain (L'Héritier and Teissier 1937b). Over two years, the frequency of the mutant decreased, but much more slowly than in the experiment with the *bar* mutant. L'Héritier and Teissier stopped the experiment after calculating that it would have taken seven years for the frequency of *ebony* to reach 1 percent. Furthermore, they hypothesized that the mechanism leading to an equilibrium was not the advantage of rarity (as in the previous experiment) but a heterozygote advantage, a mechanism that had been first suggested by Fisher from purely mathematical considerations in 1922.

In parallel with these experiments on natural selection in a given species, L'Héritier and Teissier also used their population cages to test Volterra's and Gause's models of competition between different species. In his "Lessons on the Mathematical Theory of the Struggle for Life," delivered in French in Paris in 1928 and published in 1931, Volterra had claimed, on the basis of purely mathematical considerations, that the competition of two species for the same resource would always end up with the elimination of one of the two species. L'Héritier and Teissier's experiment refuted that prediction.

These experiments and others showed that evolution could be studied by comparing mathematical models with experimental data obtained in carefully controlled conditions.

In 1937 Philippe L'Héritier, Yvette Neefs (Teissier's companion), and George Teissier carried out another remarkable experiment. They took a population cage to Roscoff (Brittany), and introduced a mixture of wild-type *Drosophila* and *vestigial* mutants. These mutants are characterized by the absence of wings. They left the box *open* for forty days and compared the results with the same type of flies in a closed population cage. The experiment showed that, in the closed box, larval competition quickly eliminated the wingless flies, while in the open air, larval competition was counteracted by the loss of wild-type flies (dispersed and killed by the wind). The result was quite complex: a fluctuation of the proportions was observed and, again, with no total elimination of one type or the other. Larval competition was detrimental to the mutant, but the absence of wings protected them at the adult stage (Fig. 36.14).

The laboratory notebook of the experiment shows that the experiment had been carefully prepared. The box was covered every night with a sheet, and opened again every morning. The physical conditions (rain, average wind speed) were recorded. The experimental purpose was clearly to introduce well-controlled physical factors in addition to the biotic factors that were acting in all experimental populations introduced into the population cages.

The resulting paper begins with a long citation from Darwin's *Origin of Species* on the evolutionary cause of the disappearance of wings in island insects. In short, Darwin predicted that wingless insects should have the best chance of surviving because they are not blown out to sea. In their concluding paragraph, L'Héritier and Teissier insisted that no mutation is disadvantageous in itself. Advantage or disadvantage depends on the environmental context. The paper on "insect apterism and natural selection" was ridiculed by a number of anti-Darwinians. But it played a decisive role in the diffusion of key Darwinian ideas in France.

Thus, it took nearly eighty years for Darwin's theory of natural selection to be wholly incorporated into the practice of French biologists. For the first time since 1859, with L'Héritier and Teissier, two French biologists could be labeled as standard "Darwinians" who took Darwin's theory of natural selection not as an object of rhetorical discussion but as a working paradigm. Immediately after his return to France, Teissier began recruiting students and built a brilliant school of population genetics, which durably established Darwinism as a segment of normal science in France and put an end to an extreme case of scientific resistance to Darwinism in a single country.

I have here deliberately ignored other aspects of the history of evolutionary thinking in France in the twentieth century. Let me mention briefly some of them. Till the 1950s, neo-Lamarckism was the dominant mood of most of the biologists who occupied key positions in French universities. This was a major cause of the durable aversion to Darwinism (for a masterly description, see Loison 2010). Correlatively, the rise of Darwinism occurred in two principal contexts.

One was applied mathematics: Darwinism began to be interesting to the French when it became obvious that population biology (theoretical ecology and population genetics) required a high level of mathematic competence. L'Héritier's and Teissier's work was not isolated: they were supported by famous mathematicians (Borel, Darmois, Malécot) who had realized in the 1930s that evolutionary theory could benefit from their skills in probability theory.

The other context that favored the rise of Darwinism was microbiology. Since the 1890s, Darwin had generated

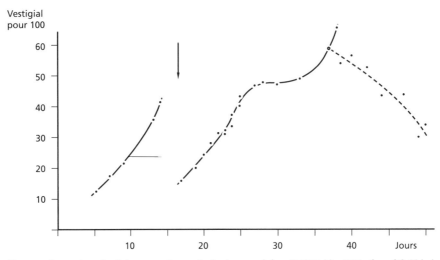

FIGURE 36.14. Apterism in insects and natural selection, graph from P. L'Héritier, Y. Neefs, and G. Teissier, "Aptérisme des insectes et sélection naturelle," *Comptes Rendus des Séances et Mémoires de la Société de Biologie* 204: 907–9. The ordinate represents percentage of vestigial (or wingless) phenotype; the abscissa, number of days. At the beginning of the experiments, two populations of homozygous Drosophila (wild and wingless) are placed in a population cage open to the air in Roscoff (Brittany, a windy place on the Atlantic coast). The vestigial gene is recessive. Equal numbers of wild and wingless flies (both homozygous) are present in the population cage. After the first hatching, wingless flies are rare, because they are severely disadvantaged in terms of largval competition. But the proportion increases up to more than 40 percent after twelve days because the wild adult parents have been progressively eliminated. Then comes the second generation of flies; because of crosses, the population now includes heterozygote flies (+vg). They have wings, but they contribute to an increase of the frequency of the vg allele and therefore to an increase of the proportions of wingless flies (vgvg). However, a plateau is attained, which corresponds to a period with no wind. When the wind comes back, the number of wingless flies increases again. The third curve (*dotted line*) is a counterexperiment: the open population box is displaced to a room within a building; windows are open, but there is no more significant wind. The proportion of vestigial phenotype drops down. The article concludes: "From the facts given here, it is legitimate to conclude that Darwin's hypothesis is entirely justified by experimental data." This conclusion mirrors a quotation from Darwin's second edition of *Origin of Species* given at the beginning of the article in Clémence Royer's translation. Darwin's original sentence about the wingless beetles of the island of Madeira described by Wollaston notes that "these several considerations have made me believe that the wingless condition of so many Madeira beetles is mainly due to the action of natural selection, but combined probably with disuse. For during thousands of successive generations each individual beetle which flew least, either from its wings having been ever so little less perfectly developed or from indolent habit, will have had the best chance of surviving from not being blown out to sea; and, on the other hand, those beetles which most readily took to flight would oftenest have been blown to sea and thus have been destroyed" (Darwin 1860a, 135). This graph was reproduced in Sewall Wright's *Evolution and the Genetics of Natural Populations* (1977).

continuous interest among the Pasteurians. In the 1940s and 1950s, this tradition led to the rise of the French school of molecular biology. All members of this school, especially their leaders, the future Nobel Prize winners André Lwoff, Jacques Monod, and François Jacob, were overt Darwinians. Darwin rather than Lamarck fit better with their discoveries on the regulation of genetic expression. These three biologists also took very strong positions at the time of the Lysenko affair. And each of them pled vigorously in favor of Darwinism in their popular writing.

Mathematics, experimentation, microbiology, and physiology – these were the matrices that finally fostered a deep scientific adhesion to Darwinism in France.

ACKNOWLEDGMENTS

I thank Abigail Lustig as warmly as I can for her linguistic revisions on my two essays and for her fruitful suggestions. I am indebted to Michel Veuille for the pages devoted to the French school of population genetics (see Gayon and Veuille 2011).

Essay 37

Botany and the Evolutionary Synthesis, 1920–1950

Vassiliki Betty Smocovitis

Though it would be hard to consider him a botanist in the strict sense of the term, Charles Darwin used plants in at least three interrelated ways: in his thinking about evolution, in his own researches, and in his professional life as a whole. By the end of his long and productive career, he had completed no fewer than six books, published between 1862 and 1880, exclusively devoted to botanical subjects, in addition to botanical articles published in the weekly *Gardner's Chronicle* and journals like the *Agricultural Gazette* (Ornduff 1984; Browne 2003; Ayres 2008; Kohn 2008). Even his magnum opus, *On the Origin of Species*, which drew on examples from as many types of living organisms as Darwin could find, relied heavily on plant examples to ground his famous argument (Smocovitis 2009). In their habits, mating systems, morphological structures, adaptations, distribution patterns, and even behavior, plants provided some of the best evidence in support of his theory of descent with modification by means of natural selection. After 1860, in fact, Darwin turned increasingly to botanical subjects of research.

DARWIN'S BOTANICAL WORK

Darwin's botanical works were voluminous and impressive, to be sure, but his contributions remained underappreciated or incompletely understood, until the second half of the twentieth century. This was due to several reasons. For one thing, Darwin was taxonomically promiscuous, flitting from organism to organism as his curiosity dictated or in search of appropriate examples in support of a generalizable theory of evolution. He lacked the kind of single-minded devotion to plants (or, indeed, to any one organismic system, let alone to a taxonomic group) that characterized contemporaries like Asa Gray and Joseph Hooker, both of whom were renowned in their day as systematic botanists. Darwin's methodology, furthermore, lacked the kind of experimental rigor that was increasingly associated with late nineteenth-century botanical sciences generally and the German export of the "new" botany in particular, which stressed laboratory practice and relied heavily on microscopy and other instrumentation. Although he was a skilled experimentalist, devising remarkably novel and clever techniques, as well as a keen observer, Darwin came across as an old-fashioned gentleman-naturalist, puttering around in his own backyard garden or private greenhouses. He was no match for the likes of German plant physiologists like Julius von Sachs, who used the latest techniques requiring precision instrumentation in far more sophisticated laboratory settings within institutions devoted exclusively to scientific research. Without surprise, Sachs criticized, if not belittled, some of Darwin's efforts in botanical research (Heslop-Harrison 1979; de Chadarevian 1996). Caught between the full-time systematists whose botanical knowledge was incomparably greater than Darwin's and the proponents of this "new" botany whose experimental methods appeared to be more rigorous, Darwin's contributions to botany therefore resided in a peculiar place that did not really gain status or legitimacy until the middle decades of the twentieth century with the emergence of a new field known as "plant evolutionary biology" (Smocovitis 2009). Even Darwin's contributions to the area of "plant ecology," or "plant population biology," or "plant evolutionary ecology," areas that were clearly articulated in the *Origin*, had to wait until those fields came to fruition in the latter half of the twentieth century (Harper 1967). His equally keen insights into the biology of invasive species, of which plants compose a significant number, have been recognized only in recent years (Hayden and White 2003).

It did not help that plants themselves seemed to resist any attempt to formulate a coherent theory of plant evolution. Indeed, some of the very same qualities that made them such useful study organisms for Darwin and others also made them problematic. They were frequent hybridizers, which meant they lent themselves to crossing experiments, opening the doors to understanding mechanisms of speciation, or even to understanding what constituted a proper species, but that also meant that until hybridization itself was properly understood in genetic terms it would remain the source of unresolved problems for understanding plant evolution. Plants reproduced using a staggering variety of means, involving highly specialized forms and structures, and their mating systems included both self- and cross-fertilization as well as varying asexual means (Briggs and Walters 1997). Until such reproductive mechanisms were understood alongside patterns of hybridization and in genetic terms, they were oftentimes just as confusing as they were useful to the student of evolution. Plants were also highly plastic, possessing open or indeterminate systems of growth and could adapt themselves readily to shifting environments; this meant that plants showed elaborate variation patterns but that distinguishing genotype from phenotype could also be difficult. The abundance of reproductive strategies, distinct mating systems, and the complex interplay between all these made understanding plant evolution difficult. Having few easily fossilized "hard parts," moreover, the fossil history of plants was difficult to reconstruct and subject to the vagaries of preservation, which often meant that seeds, roots, leaves, and other structures were frequently disassociated from the whole plant if they were fossilized at all. As a result of all of these properties, phylogenetic histories of plants were often difficult to reconstruct, counterexamples prevailed, and any attempt to formulate a general theory of plant evolution seemed impossible. Thus, despite Darwin's fine efforts, a majority of botanists remained advocates of neo-Lamarckism, in toto or in part, and often supported theories distinct in the plant world like evolution by means of hybridization (associated later with J. P. Lotsy and others) or mutation theory (associated later with Hugo de Vries).

PLANTS, POST-DARWINIAN DEVELOPMENTS AND THE RISE OF MENDELIAN GENETICS

Plant evolutionary biology did not properly come of age until the 1930s and 1940s during the period designated by historians as the "evolutionary synthesis" (Mayr and Provine 1980; Smocovitis 1996). It was during this time that Darwinian understanding of evolution was integrated with Mendelian genetics in a manner that could account for the origins of biological diversity. The synthesis also eliminated alternative mechanisms of evolution in the plant world that had been popular at the time. Darwin of course, knew nothing about genes and incompletely understood heredity. In his 1868 *Variation of Animals and Plants under Domestication*, he famously came up with his "provisional hypothesis of pangenesis" to offer an explanation for the mechanism of heredity, which was crucial to grounding his theory because it stressed the inheritance and preservation of favorable variation. It was rapidly challenged and gained no real adherents. Unbeknown to him was the presence of an Augustinian monk named Gregor Mendel, working from Brno, then in Moravia, part of the Austro-Hungarian Empire (now the Czech Republic), whose crucial experiments on the garden pea, or *Pisum sativum*, would have provided precisely the particulate (rather than blending) theory of heredity that Darwin needed. Mendel's crucial insights providing the first supportable mechanistic and materialistic theory of heredity, unfortunately lay in obscurity, having been published in the little known *Proceedings of the Brno Society for Natural History*. Mendel himself published little more beyond these experiments on peas, having taken on comparable experiments in the hawkweed *Hieracium* at the suggestion of the German botanist Carl Naegeli (Stern and Sherwood 1996). Failing to replicate the discrete ratios he obtained with *Pisum*, Mendel eventually gave up his experimental studies into understanding heredity and turned to other interests, as well as administering the monastery (Orel 1996). Mendel did not know that *Hieracium*, a member of the Compositae family, had a far more complex genetic system than his original choice of *Pisum*; difficulties understanding the pattern and process of evolution at the genetic level for plants like *Hieracium*, in fact, would not be resolved until the decades of the 1930s and 1940s, during the period of the evolutionary synthesis.

Not until 1900 was Mendel's work "rediscovered" by three researchers working independently: Carl Correns in Berlin, Eric Von Tschermak in Vienna, and Hugo de Vries in Holland. All three were keenly interested in plants, and all three instantly recognized the importance of what came to be known as "Mendel's laws" of segregation and of independent assortment. Promoted by advocates like the English biologist William Bateson, who famously called the new science of heredity "genetics" at the 1906 meeting of the Royal Horticultural Society (the meeting was originally entitled "Conference on Hybridization and Plant Breeding," but renamed the "Third International Conference on Genetics" after Bateson's famous address), genetics was presented as the new experimentally rigorous science that would unlock the mysteries of inheritance. Bateson also formally recognized Gregor Mendel as the founding father of the new discipline and began his campaign to draw in an increasing number of plant breeders, both agricultural and horticultural, along with a large number of younger workers to the new science that promised control and improvement of plant stock (the science related to the improvement of humans known as "eugenics" had been earlier introduced by Darwin's first cousin, Francis Galton).

The turn of the century as a whole witnessed astonishing developments in the botanical sciences, as biologists turned to plants for study, or as experimental organisms in a number of new areas of research. No longer interested only in the study of plants in and of themselves, in sorting or classifying them, or in understanding their basic morphology, physiology, or phylogenetic history, these workers sought to derive generalizable theories of heredity that also had direct applications in agriculture or horticulture. The rise and development of genetics were thus inextricably linked both conceptually and institutionally to horticulture, plant breeding, plant hybridization, and plant genetics. That so many genetic concepts came from plants is an indicator of their utility as study organisms to geneticists. Easy to grow, fecund, subject to experimental conditions in the laboratory, the garden, the greenhouse, and also in the wild, and oftentimes having few chromosomes that were large in size, plants had the added bonus of being free of ethical or moral concerns raised by experimentation. Taking advantage of increasing emphases on agriculture and horticulture in institutes and university settings, the field of plant genetics boomed in the first few decades of the twentieth century, as workers entered this promising area of research (Smocovitis 1988, 2009).

Initially, however, the generation of abundant but conflicting data that drew attention to them and that opened doors to understanding the genetic basis for evolutionary change proved to be confusing, if not even antithetical to Darwin's views of evolution in terms of slow, gradual change operating at the level of individual differences. In the early years of the twentieth century, for example, some plant workers were followers of "mutation theory," set forth by Hugo de Vries, one of Mendel's rediscoverers. Noticing sudden morphological changes in the evening primrose, *Oenothera lamarckiana*, de Vries thought that sudden, drastic mutations were responsible for generating new species (see Plate XXVI). The origin of variation, according to his theory developed between 1901 and 1903, lay in mutation. While natural selection still selected the most favorable of the new species, it had only an eliminative role; mutation had the creative role. The peculiar behavior of *Oenothera* drew an industry of workers and followers in the first two decades of the twentieth century who studied the chromosome behavior of the plant and performed crossing experiments in studying the pattern and process of hybridization. Not until the interval between 1917 and 1922, however, did plant geneticists like Otto Renner and Ralph Erskine Cleland demonstrate that the plant was a permanent translocation heterozygote that threw up two different chromosome complexes depending on mode of fertilization; these were not, in fact, mutations leading to new species but by-products of *Oenothera*'s unique chromosomal configurations. In its day, however, "mutation theory" was wildly popular as an alternative to standard Darwinian evolution. Unlike Darwinism, which reflected an older natural history tradition and appeared descriptive, "mutation theory" was an experimental, laboratory science that was more rigorous. It seemed more promising in terms of methodology but also more legitimate in terms of it appearing to be a "hard" science.

The mystery of *Oenothera* genetics took years to resolve and led a number of workers to challenge the central tenets of Darwinian selection theory. For this reason, some scientists and historians like Ernst Mayr (1980b), C. D. Darlington (1980), and G. Ledyard Stebbins (1980) suggested that it was a hindrance to the development of proper understanding of evolution in plants and that it served as a "roadblock" for understanding plant evolution; this view, however, is erroneous. By drawing attention to the complex interplay of mechanisms at the chromosomal level that were expressed morphologically and determined by patterns of mating and hybridization, *Oenothera* drew attention to the power and utility of plants as tools for understanding patterns and modes of speciation. That contribution was in fact critical in the next two decades of research for evolutionary workers who combined such studies of chromosome behavior, hybridization, morphology, and knowledge of taxonomy with patterns of geographical variation in wild or natural populations of plants that were simultaneously studied in the experimental garden and in the laboratory (Smocovitis 1988).

PLANTS AND THE SYNTHESIS OF GENETICS, SYSTEMATICS, AND PALEONTOLOGY (1920–1950)

Beginning with late nineteenth-century advances in cytology, attention was drawn more and more to the chromosomes, their number, their pattern, and their behavior during critical processes of plant division. Plants were easily subjected to cytological study. With their comparatively large chromosomes that were fewer in number than in animals, they became the focus of attention for chromosome studies. The karyotype (the stable complement of chromosomes that are found in each species), for example, was first invented and determined for a number of economically important plants that were members of the grass family, or the Gramineae in

Russia (Avdulov 1931). The phenomenon of polyploidy, or the spontaneous doubling or multiplication of chromosome sets, was discovered in a series of fits and starts in plants like *Primula kewensis* and the *Oenothera lamarckiana* (Lutz 1907; R. R. Gates 1909; Farmer and Digby 1912; Newton and Pellew 1929). Not until 1917, however, with Ovjnid Winge's pathbreaking study "The Chromosomes: Their Numbers and General Importance" did phenomena like polyploidy and the behavior of chromosomes in the reproductive process draw attention to the relationship between polyploidy, hybridization, and speciation. Winge's suggestion that chromosomal doubling might enable the formation of interspecific hybrids was subsequently demonstrated in a number of species. Artificial allopolyploidy (meaning the doubling or multiplication of chromosome sets that resulted from interspecific hybridization) was demonstrated in *Nicotiana glutinosa* and *N. tabacum* hybrids (R. E. Clausen and Goodspeed 1925) and in an artificial plant – arguably a new species called *Raphanobrassica* – resulting from a cross between a radish and cabbage that was made shortly thereafter (Karpechenko 1927). Along with the realization that polyploids were already recognized in some species, the way opened up for studies that would enable a phylogenetic reconstruction of taxonomic groups. Thus, the rediscovery of Mendel, the establishment of the chromosome theory of heredity (which made chromosomes the material carriers of heredity), the rising interest in *Oenothera* genetics, and the development of new microscopic, staining, and sectioning techniques in an increasing number of institutional settings ranging from agricultural universities to horticultural institutes all led to a surge in the use of plants in cytogenetic studies (or studies that combined cytology with genetics). Most important for evolutionists was the recognition that chromosome numbers and characteristics could be used to determine relationships in closely related groups. The integration of such studies was the first step in deriving a general theory of evolution in plants.

One of the first studies to explore speciation and evolution in plants using such novel methods and insights was German geneticist's Erwin Baur's (1932) studies to understand the genetics of the common snapdragon *Antirrhinum majus* and species relationships in its close relatives. Baur studied patterns of geographical variation and isolation as well as hybridization between different populations he thought might constitute taxonomic species, and then performed crosses between these groups and tracked their generations. Baur calculated the number of gene differences needed to separate two plant species in the genus and recognized that hybrids were fertile but also exhibited intermediate degrees of reproductive isolation that would now be considered semispecies. His work on *Antirrhinum* was published in 1932, but because Baur died shortly thereafter, his work was never completed. It was, nonetheless, the first such synthetic study and would pave the way for projects in the late 1920s and 1930s that led to understanding of plant evolution (Stebbins 1980; Smocovitis 1988; Harwood 1993).

Plant taxonomy, or more correctly systematics, only slowly began to incorporate Darwinian evolution in the 1920s. Not all taxonomists accepted or applied the more dynamic population-oriented Darwinian emphasis in their taxonomic work (Smocovitis 1988). Plant taxonomists preferred to focus on nonadaptive characters in their classification schemes. Because they were not as variable, they were thought to be reliable and reflective of proper differences in types. To such taxonomists, patterns of variation frequently got in the way of proper taxonomic study. The influence of Linnaeus, the arch-taxonomist, reigned supreme, as taxonomists worked mostly with herbarium-type specimens for their studies, rather than relying on variation patterns in natural populations of plants of any character that varied too much. In the 1920s a number of systematists began to call for reform of such static approaches to taxonomic methods and emphasized instead the study of variation and the study of plant populations in natural conditions that took into account local adaptations to the environment. The emphasis also began to shift to the different responses to the environment, drawing attention to the characters that did not respond (or the genotype-phenotype distinction). Göte Turesson, in Sweden, for example, drew on earlier European transplantation (often reciprocal transplantation) studies to discern genotypic versus phenotypic responses to the environment along varied altitudinal gradients of the Swedish landscape. Among other notable contributions, Turesson first articulated the concept of the ecotype, particular ecologically adapted forms that showed distinctive morphological and physiological characteristics, which were preserved when plants were transplanted from varied environments. Such forms could be seen as stages in the evolution of plant species. The science of genecology, which Turesson perfected and which focused on plant variation, contributed greatly to the understanding of such variation patterns in natural plant populations. So too did the efforts of British workers like J. B. Turrill, who explored how plants adapted to various edaphic (or predominantly soil) conditions (Smocovitis 1988).

Turesson's ecological work combined with Baur's genetical insights, along with a host of other researchers' efforts in the 1920s, came to fruition in the mid-1930s in the San Francisco Bay area as a cluster of specialists in the botanical sciences began to bring interdisciplinary perspectives to understanding evolution (Hagen 1984; Smocovitis 1988, 1997, 2006, 2009). In the 1920s University of California at Berkeley systematist Harvey Monroe Hall along with ecologist Frederic Clements called for the reform of static herbarium-based taxonomic methods and began to stress instead environmental variation and schemes that reflected the true phylogenetic history of plants. They met with some resistance from traditional plant taxonomists but nonetheless continued in their efforts, publishing what became a taxonomic manifesto in 1923, *The Phylogenetic Method in Taxonomy*. Both Hall and Clements had attempted the kinds of reciprocal transplant experiments done by Turesson and other Scandinavian genecologists. In the late 1920s Hall went further by formally instituting a long-term California-based study that focused on transplant studies. The goal would be to understand plant evolution in terms of all the interdisciplinary approaches available at that time; in addition to workers versed in classical taxonomy, he also called for those with knowledge of ecology, genetics, and physiological ecology to shed light on the mechanisms responsible for adaptation to varying environments. He was instrumental in bringing Danish geneticist and genecologist Jens Clausen to California as part of a Stanford University-based Carnegie Institution of Washington's interdisciplinary efforts to understand plant speciation and evolution. Clausen joined with taxonomist David Keck and physiologist William Hiesey to form what became "a mythic collaboration" in the history of twentieth-century botanical science (Smocovitis 1988, 1992; Craig 2005). Throughout the 1930s and 1940s, the "Carnegie team," of Clausen, Keck, and Hiesey performed a series of elegant transplant studies along altitudinal gradients located at Stanford (30 feet above sea level), Mather (4,600 feet), and Timberline (10,000 feet). The most famous of these studies involved transplant experiments on *Achillea* species (Fig. 37.1). In addition to shedding light on adaptive responses to the environment, providing a more precise understanding of mechanisms of plant speciation, and clarifying what counted as a proper species in the plant world, they provided the best evidence that quashed belief in Lamarckian inheritance in the plant world and once again restored belief in Darwinian selection. This research also clearly demonstrated that plant ecotypes – and other such locally adapted forms – were so dynamic that they should be seen as "stages in the evolution of plant species" (Clausen 1951) (Fig. 37.2). Clausen, Keck, and Hiesey's efforts drew the attention of leading evolutionary workers at the time, including Theodosius Dobzhansky, who closely tracked the research of the Carnegie team and eventually began to share field sites at Mather in his own efforts to understand the genetic basis of evolutionary change in natural populations of *Drosophila pseudoobscura*.

At least as important as the Carnegie team, were the efforts associated with Berkeley geneticist Ernest Brown Babcock, who was also a collaborator and friend of Harvey Monroe Hall (Fig. 37.3). A talented administrator and visionary, Babcock created the first department with the name genetics in the United States; it would be dedicated to the new science of heredity and run out of the College of Agriculture. In his own research efforts, Babcock tried to emulate the success of Thomas Hunt Morgan's pioneering efforts in transmission or classical genetics on the basis of his work on *Drosophila melanogaster* (G. E. Allen 1975). Seeking a plant counterpart in 1915, Babcock began an enormous international study to understand the genetic properties of the genus *Crepis*, a common roadside weed that showed astonishing ability to colonize varied environments. The genetic project on *Crepis* began to shift in the early 1920s as Babcock began to encounter the same kinds of problems that had plagued Darwin, Mendel, and other workers: the plant was simply too complex in its patterns of hybridization, speciation, and reproduction. In particular, the interaction of hybridization, polyploidy, and apomixis (a form of asexual reproduction) seemed mystifying. The basic phylogeny of the genus – and its closest relatives – was largely unknown as a result. Ever the visionary, Babcock shifted toward a phylogenetic – and therefore evolutionary study – of the genus *Crepis*, employing the same kinds of interdisciplinary techniques drawing on cytology, genetics, ecology, and morphology to study the systematics of the entire group. Most importantly, Babcock was one of the first to apply insights from the fossil history of the plant to discussions of geographical distribution, centers of origin, and dispersal patterns in order to reconstruct the phylogenetic history of the group. In 1947 Babcock published *The Genus Crepis*, the first comprehensive phylogenetic history of any genus that integrated techniques and insights from genetics, ecology, morphology, and paleontology with systematics (Babcock 1947; Plate XXVII). It became the classical example of the new, more dynamic emphasis on understanding plant evolution that became known as "biosystematics," which focused on mechanisms of speciation and saw species in nonessentialistic terms (Camp and Gilly 1943).

Babcock's crowning achievement was unquestionably the 1947 monograph, but the foundations were laid earlier in 1938 with the publication of a monograph on the American species (Babcock and Stebbins 1938). Working with the Harvard-trained geneticist George Ledyard Stebbins Jr., Babcock resolved the crucial problem of the interaction of hybridization, polyploidy, and apomixis. They recognized that certain plant genera consisted of a complex of reproductive forms that centered on sexual diploids and that had given rise to polyploids; sometimes, as in *Crepis*, these were apomictic polyploids. Polyploids that combined the genetic patrimony of two species usually had a wider distribution pattern. The articulation of the polyploidy complex was considered pathbreaking work at the time. Not only did it demonstrate in detail the complex interplay of hybridization, polyploidy, and apomixis in a geographical context but also offered insights into species formation, polymorphy in apomictic forms, and knowledge of how these complex processes could inform an accurate phylogentic history of the genus. Stebbins extended these efforts further in subsequent review articles and in 1947 published a classic review article that synthesized knowledge bearing on polyploidy in plants (Stebbins 1947) to complement his earlier review of the apomixis literature (Stebbins 1941).

The research efforts on *Crepis* and Babcock's energetic management of resources drew increasing attention to the fusion of plant genetics, systematics, and paleontology in the 1930s. He was also instrumental in bringing a staggering assortment of students, collaborators, and international visitors to the Bay area. By the 1930s he – and the Carnegie team – had helped to make the Bay area the hub of evolutionary activity and one of the premier centers of plant evolution in the world. They helped organize an informal group of systematists located in the Bay area who worked on a range of organisms but who wished to share perspectives and the new methodologies coming from genetics. It was initially called

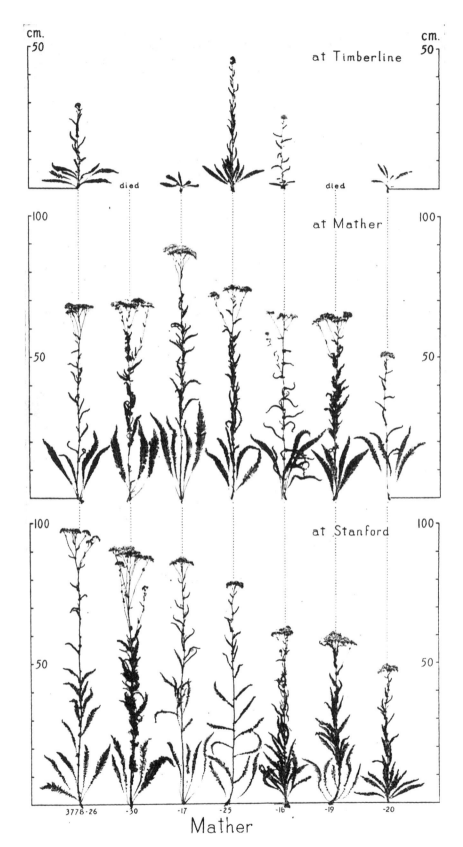

FIGURE 37.1. The responses of *Achillea* at different localities (based on altitude) showing the effects of environment on plants of the same genetic type (the plants considered vertically are clones). From J. Clausen, D. D. Keck, W. M. Hiesey, *Experimental Studies on the Nature of Species* (Washington, D.C.: Carnegie Institute, 1940)

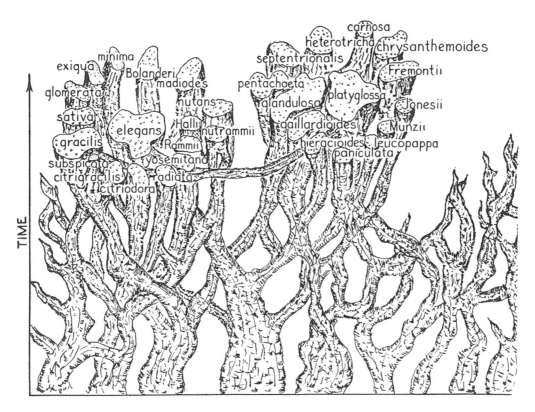

FIGURE 37.2. Hypothetical reticulate evolutionary relationships between the species of the genera *Madia* (*left*) and *Layia* (*right*) in their present status and projected back in time. Permission: Reprinted from *Stages in the Evolution of Plant Species*, by Jens Clausen © 1951 by Cornell University. Used by permission of the publisher, Cornell University Press

FIGURE 37.3. Ernest Brown Babcock (1877–1954), leading American plant geneticist and evolutionist in the first half of the twentieth century. With G. L. Stebbins he resolved long-standing problems in understanding plant genetics that had baffled Mendel and wrote the first modern phylogenetic history of a plant group. Permission: V. B. Smocovitis, photograph courtesy Harlan Lewis

"the Linnean club," but later took on the name of the "biosystematists" in the Bay area (Fig. 37.4). The group also attracted frequent visitors to the Bay area keen on understanding plant evolution, including Edgar Anderson from the Missouri Botanical Garden, whose own efforts to understand variation and evolution in plants like *Iris*, led him to devise novel methods to measure variation but also to appreciate mechanisms of introgressive hybridization (a way of introducing genetic material through species crosses that enabled recombination) (Kleinman 1999). Another visitor was Carl Epling, from the University of California at Los Angeles, who was also keen on understanding variation in plants and who, with the mathematical population geneticist Sewall Wright, was concentrating his efforts on fine-focused studies of population genetics in *Linanthus parrye* (Smocovitis 2006).

By far the most frequent and ultimately influential visitor to the Bay area in the 1930s was Theodosius Dobzhansky, the Russian émigré geneticist, then located at the California Institute of Technology and beginning his celebrated studies on natural populations of *Drosophila pseudoobscura* (Lewontin et al. 1981). In 1937 Dobzhansky published what became the foundational work in the synthesis between Darwinian selection theory and Mendelian genetics. *Genetics and the Origin of Species* announced the arrival of the field of evolutionary genetics, which ushered in what Julian Huxley described in 1942 as the "modern synthesis" of evolution. Dobzhansky's book served as the catalyst for other important books

FIGURE 37.4. The biosystematists in the Bay Area. Placerville Forest Genetics Station, 1946. *Standing*: H. E. McMinn (Mills), G. F. Ferris (Stanford), E. G. Linsley (Berkeley), H. Graham (Mills), L. Adams (Stanford), C. Y. Yang (unknown), E. B. Babcock (Berkeley), W. E. Castle (ex-Harvard), R. H. Weidman (station), R. Goldschmidt (Berkeley), G. S. Meyers (Stanford). *Sitting*: R. C. Miller (California Academy of Sciences), G. L. Stebbins (Berkeley), C. O. Sauer (Berkeley), H. L. Mason (Berkeley), I. L. Wiggins (Stanford), L. Constance (Berkeley), N. Mirov (station), P. Stockwell (station), W. Cummings (station), and H. Kirby (Berkeley). Members of Carnegie Institution of Washington group, J. Clausen, D. D. Keck, and W. Hiesey were not present at the meeting. Permission: V. B. Smocovitis, photograph courtesy Lincoln Constance

published with Columbia University Press that attempted to integrate newer approaches; they were all associated with the New York–based lecture series, named after Morris K. Jesup. In 1941 plant evolutionist Edgar Anderson was invited to give the lectures with the Harvard zoologist Ernst Mayr. While Mayr published his set of the lectures under the title *Systematics and the Origin of Species from the Viewpoint of the Zoologist* in 1942, Anderson never completed the publication of his set of lectures. The viewpoint of the botanist was therefore still needed in the emerging new synthesis.

Thus, in 1945, through the recommendation of Dobzhansky, who was well acquainted with Ledyard Stebbins, L. C. Dunn at Columbia invited Stebbins, the young "spark plug" of the Department of Genetics at Berkeley, to deliver the next series of lectures devoted expressly to plant evolution (Smocovitis 1988, 1997, 2006). Stebbins took the lecture notes from his Berkeley course and, revising them, delivered the Jesup Lectures of 1946. They were wildly successful at integrating the newer systematics not just with genetics but also with paleobotany and ecology, as well as in resolving long-standing issues in plant evolution. He upheld the importance of most of the tenets emerging from Dobzhansky's synthesis that was critical to establishing a consensus between a number of fields. Drawing on his own research, but especially on all the efforts of plant workers before him, he carefully explained the interplay of hybridization, polyploidy, and apomixis in particular plant groups, but then drew on the recent work of C. D. Darlington's (1939) notion of genetic systems to argue that they themselves could be understood as genetic systems subject to selection. He stressed the centrality of natural selection but left plenty of room for stochastic processes like genetic drift and nonadaptive evolution. He also upheld Dobzhansky's and Mayr's emerging notion of the biological species concept (or the BSC), which defined species in terms of potentially breeding populations. This was vitally important because mechanisms of plant speciation were enormously more complex than in animals. He also brought the microevolutionary picture, or the view of plant evolution at the genetic level, in line with the macroevolutionary picture by integrating insights from paleontology and paleobotany, especially the fossil-rich history of the flowering plants, or the Angiosperms, as it was emerging from the work of his Berkeley colleagues like Ralph Works Chaney. Altogether, the new picture of plant evolution as it emerged in Stebbins's novel synthesis effectively killed Lamarckism in the plant world and other alternatives like mutation theory that had been popular at the turn of the century. As a substitute, he offered the neo-Darwinism that was reflected in the works of Dobzhansky. His book *Variation and Evolution in Plants*, published in 1950, offered such a comprehensive picture of plant evolution that it opened up an entirely new field of research, known as plant evolutionary biology (Fig. 37.5). Peter Raven (1974) described it as "the single most influential book in plant systematics this century." At 643 pages in length, it was the longest – and the last – of the great works that constituted the historical event designated as the evolutionary synthesis. The book, which remains a heavily cited text, also formed the conceptual framework for a new field known as plant evolutionary biology, drawing on a new generation of students who began to appreciate Darwin's original botanical studies.

According to Ernst Mayr and William B. Provine, the publication of *Variation and Evolution in Plants* signaled that botany had been brought into the wider evolutionary synthesis and that George Ledyard Stebbins (after 1950 he dropped the use of the "junior" after his name) served as one of it architects, alongside Dobzhansky, Mayr, George Gaylord Simpson, and Julian Huxley. But it is also clear that not all botanists, or even geneticists, ecologists, systematists, or paleontologists accepted the new methods, approaches, or insights as they

FIGURE 37.5. G. Ledyard Stebbins Jr. (1906–2000), the leading plant evolutionary biologist of the twentieth century, and one of the great synthesizers, along with Theodosius Dobzhansky, Ernst Mayr, and G. G. Simpson. Permission: V. B. Smocovitis, photograph courtesy Harlan Lewis

were emerging from the integration of genetics and Darwinian selection theory in the 1930s and the 1940s. Nonetheless, a new area of research that integrated all these approaches did emerge by 1950, and plant evolutionary biology finally recognized the centrality of Darwin's botanical efforts.

Essay 38

The Emergence of Life on Earth and the Darwinian Revolution

Iris Fry

How did life emerge on the ancient earth? Since the middle of the twentieth century, scientists have applied a plethora of experimental and theoretical approaches in an attempt to answer this question. Yet it is still considered one of the most challenging questions facing science today. Holding different theories and favoring different scenarios, all researchers, however, are united in a conviction that the organization of the first living systems out of chemical building blocks was a natural process. Moreover, scientists have no doubt that it was an evolutionary process.

The evolutionary view presents a radical departure from the previous conception of the origin of life held by both laypersons and naturalists for most of human history. On the basis of everyday experience, people were obviously aware that various organisms were being sexually generated from their parents. Since the rise of monotheistic religions, the general belief was that God originally created the "founding fathers" of the major types of living beings that kept perpetuating their fixed kind generation after generation. In parallel, people were also convinced that plants and many animals repeatedly arise under the influence of moisture and heat not from parents but rather from mud, tree bark, excrement, and decaying plants and animal matter. This belief in "spontaneous generation" accompanied humanity from antiquity till the modern age, having been sustained in different epochs by both religious and materialistic lines of reasoning (Farley 1977; Fry 2000).

Complex interaction of empirical, philosophical, and cultural changes engendered gradual decline during the past centuries of the beliefs in the separate creation of fixed species and in spontaneous generation. Not surprisingly, these changes also contributed to the rise of evolutionary ideas that culminated in Darwin's evolutionary theory. Here, I shall limit the historical discussion to the nineteenth century, with special attention to ideas on the origin of life held by Darwin and some of his contemporaries. This will be the springboard for discussion of the crucial contribution of evolutionary ideas to the beginning of origin-of-life research in the twentieth century and to the current state of this research.

SPONTANEOUS GENERATION, EVOLUTION, AND THE ORIGIN OF LIFE

The belief in the spontaneous generation of insects and many other organisms was abandoned in the seventeenth and eighteenth centuries. However, the question of the spontaneous generation of microorganisms was debated well into the nineteenth century. The French chemist Louis Pasteur conducted in the 1860s decisive experiments showing that no microbes were generated in various organic solutions under conditions of strict sterilization. The results of these experiments won the support of the majority of the scientific community, especially in France. At the same time, several respected scientists, notably Felix Pouchet in France and Henry Charlton Bastian in England, presented contradicting experimental evidence supporting spontaneous generation of microorganisms (Fig. 38.1).

Not only was the very possibility of spontaneous generation an open question; it was not clear whether simple forms of life could be generated only from organic materials or also from inorganic ones. The term "heterogenesis" was used to describe the generation of organisms from organic materials. The common belief was that organic matter could be uniquely produced only within organisms and heterogenesis was seen in most cases as the spontaneous generation of life from decaying dead organisms. It thus was not conceived to be as philosophically radical as the claim for "archebiosis" or "abiogenesis," the generation of living beings from "not-living" – that is, inorganic – starting materials, especially on the ancient earth (Strick 2000).

The connection between the issues of spontaneous generation and evolution goes back to the beginning of the nineteenth century and the ideas of the French biologist Jean-Baptiste Lamarck. Fifty years before Darwin, in his book *Zoological Philosophy* (1809), Lamarck put forward a theory of the transformation of species during the very long history of earth. As part of his theory, Lamarck assumed repeated events of spontaneous generation of the simplest life forms through a materialistic process. Unlike Darwin's later hypothesis of a single origin of the evolutionary tree, Lamarck believed that these repeatedly formed simple systems evolved along several evolutionary scales into all the complex forms of life known to us. This, according to Lamarck ([1809] 1984), guaranteed the diversity of life observed in nature.

Debates over the possibility of repeated events of heterogenesis and abiogenesis or of such occurrences on the ancient earth were intertwined with controversies over evolution. Not surprisingly, Darwin's suggestion that all life might have descended from a common root, focused attention on the ancient earth.

THE PUBLIC AND THE PRIVATE DARWIN ON THE ORIGIN OF LIFE

In the *Origin of Species*, Darwin (1859, 484) raised the hypothesis that "all the organic beings which have ever lived on this earth may be descended from some one primordial

FIGURE 38.1. H. Charlton Bastian (1837–1915) was an enthusiast for abiogenesis (life from inorganic matter) to such an extent that his advocacy combined with his lack of solid evidence led to his being ostracized by leaders of the scientific community, including T. H. Huxley. From *Popular Science Monthly* (1875)

form." Yet, in the *Origin*, Darwin hardly discussed the question of the origin of this primordial form. Furthermore, departing from the secular tenor of his book, Darwin ended the first edition of the *Origin* by saying that life has been breathed into a few primordial forms. In the second edition, he went further, adding that life's powers were "originally breathed by the Creator into a few forms or into one" (Darwin 1860a, 484).

Darwin's most ardent supporters were highly critical of this move. T. H. Huxley, "Darwin's bulldog," protested against the mixing of science and genesis (1900, 1:244, letter to C. Kingsley, 22 May 1863). According to the German biologist Ernst Haeckel (1862, 1:231–32, n. 1), "The chief defect of the Darwinian theory is that it throws no light on the origin of the primitive organism.... When Darwin assumes a special creative act for this first species, he is not consistent, and, I think, not quite sincere."

In 1863, in a letter to his close friend the botanist Joseph Hooker, Darwin expressed his long-felt regret for giving in to public opinion and using the biblical terms of the creation of life. In fact, he added, his intention was to refer to the origin of life as "some wholly unknown process" (Darwin 1863). Toward the end of his life, Darwin openly admitted that because the question was beyond the scope of science, he "intentionally left the question of the origin of life uncanvassed" (DCP 13747, letter from Darwin to George Charles Wallich, 28 March 1882).

Darwin's (1860a, 481) later-to-be regretted use of biblical language was a concession he made to Reverend Charles Kingsley, the first clergyman to suggest that evolution could be reconciled with religious belief. Darwin was probably affected as well by the fact that Lamarck's ideas on the transmutation of species and spontaneous generation, considered in France as a materialistic threat to religion, were also vehemently attacked in England during the first half of the nineteenth century (Desmond and Moore 1991). The context of Darwin's use of biblical language, as well as Haeckel's and Huxley's reactions, indicate how attitudes to the question of the origin of life, no less than to the question of evolution, were largely shaped by philosophical, religious, and political factors.

In one of his most famous comments on the origin of life and spontaneous generation, included in a private letter to Hooker in 1871, Darwin contemplated the hypothesis of an abiogenetic process that might have taken place on the ancient earth "in a warm little pond." Darwin was uniquely aware of the difference between conditions on the earth then and now. He pointed out that for life to have had emerged, organic materials had first to be synthesized from inorganic substances under the influence of various sources of energy. This process could have occurred, he realized, only on a sterile earth, that is, on an earth that was devoid of life. In the present, "such [organic] matter would be instantly devoured, or absorbed" by living organisms (Darwin 1985–, 19:53–54, letter to Hooker, 1 February 1871).

Notwithstanding the fact that Darwin did not commit himself publicly to a natural origin of life on the ancient earth, he clearly contemplated the idea, and the question preoccupied his thoughts. In a response to an attack on his theory by the renowned biologist Richard Owen, who rejected evolution through natural selection and supported a Lamarckian process of repeated heterogenesis of microscopic life during each geological period, Darwin published in 1863 a letter in the journal *Athenaeum*. Not only did Darwin present a detailed summary of the evidence for his theory; he also made some explicit and important comments on the origin of life: "But let us face the problem boldly. He who believes that organic beings have been produced during each geological period from dead matter must believe that the first being thus arose. There must have been a time when inorganic elements alone existed on our planet." Darwin then added that because "at present" we are profoundly ignorant as to how such a process could have happened, Owen's proposals are of no merit. Moreover, Darwin (1863b) pointed out that these simple forms of life, like the marine microorganisms foraminifera "with their beautiful shells," seen fit by Owen and others to arise spontaneously from inanimate matter, were in fact organized enough to have required a process of organization, that is, of evolution, in order to reach their known form.

Darwin's realization of the need for evolution even of the simplest forms of life foreshadowed future scientific developments. Indeed, only when an unwavering evolutionary conception of the gradual organization of life was adopted in the twentieth century could the doctrine of spontaneous generation be completely abandoned.

THE OPARIN-HALDANE HYPOTHESIS: AN EVOLUTIONARY BREAKTHROUGH

Toward the end of the nineteenth century, Darwin's supporters were convinced that a natural emergence of life was a necessary requirement of a general evolutionary worldview. Unlike Darwin's awareness of the complexity of the problem and its intractability to science at the time, many Darwinians suggested a simple passage between non-life to life, emphasizing the similarities between physical and biological systems (Haeckel [1899] 1902). This view was supported by the dominant theory of the cell in the 1860s and 1870s that regarded protoplasm – "the physical basis of life" – as a rather basic homogeneous stuff made of protein (Geison 1969). Because primitive organisms were perceived as naked lumps of protoplasm, leading Darwinians claimed a nonproblematic abiogenesis on the ancient earth.

This position was losing ground as the nineteenth century turned into the twentieth: new cytological studies revealed the crucial function of the cell nucleus and its role in cell division; the rise of the discipline of biochemistry and the isolation of specific cellular enzymes made the complex nature of the cell highly apparent. The realization of the extreme complexity of even the simplest organisms led many biochemists to despair of solving the problem of the origin of life and "to prefer to let the riddle rest" (Henderson [1913] 1970). Several scientists, mainly physicists, raised the hypothesis of the eternal existence of life in the universe, side by side with matter, and its delivery to earth on comets, meteorites, or cosmic dust particles. By assuming that life and matter were separate entities, these so-called panspermia (from Latin, for "seeds of life everywhere") theories, tried to explain away the question of the origin of life (Kamminga 1982).

The pioneering ideas of the Russian biochemist Alexander Oparin, first published as a booklet in 1924 in the Soviet Union, challenged this impasse and confusion. Oparin offered a detailed scenario for the emergence of life and suggested a set of conditions on the ancient earth that could have enabled such a scenario. A 1936 book of his, a broader and updated version of the original booklet, became known in the West mainly after the Second World War. Elements common to the Oparin's scenario and to an independent theory published by the British biochemist and geneticist J. B. S. Haldane in 1929, became later known as the Oparin-Haldane Hypothesis. This hypothesis, and particularly Oparin's ideas, triggered in the 1950s the establishment of an empirical field of research on the emergence of life on earth (Fry 2006) (Figs. 38.2 and 38.3).

The philosophical significance of the hypothesis was its insistence on the evolutionary nature of the origin-of-life process. Oparin and Haldane emphasized both continuity and novelty, referring to the continuous, gradual development from chemical building blocks to primitive organized systems

FIGURE 38.2. Alexander Ivanovich Oparin (1894–1980) was a pioneer in formulating a detailed evolutionary theory of the origin of life from inorganic materials to organic molecules up to primitive metabolic systems. His ideas were also inspired by the dominant Marxist ideology in the Soviet Union. Oparin believed that dialectical materialism made it possible to emphasize both the material basis of life and life's emergent unique qualities. Permission: Photo Researchers

FIGURE 38.3. J. B. S. Haldane (1892–1964) is important in the history of evolutionary theory both for his work on population genetics and his speculations about the origin of life. Haldane and several other origin-of-life pioneers were also Marxists at certain stages of their life. Permission: © Mark Gerson / National Portrait Gallery, London

as well as to the new, unique features of these systems. This philosophical message was embodied in their empirical scenarios. On the basis of developments in geochemistry and astronomy, both Oparin and Haldane claimed that a necessary precursor to the origin of life was the utilization of energy sources like heat, lightning, and ultraviolet radiation for an extensive synthesis of organic compounds out of the inorganic constituents of a reducing (devoid of oxygen) primordial atmosphere. Upon dissolving in the primordial ocean, these organic compounds formed a "hot dilute soup" and underwent chemical evolution to produce various organic polymers, similar to proteins and polysaccharides (Oparin [1924] 1967, [1936] 1953; Haldane [1929] 1967).

Oparin ([1936] 1953) the biochemist saw a metabolism based on the cellular organization of proteinlike enzymes as the defining characteristic of life. Relying on colloidal chemistry, a major biochemical approach during the early decades of the twentieth century, he envisaged the emergence of colloid droplets made of organic polymers in the primordial soup. According to this view, these droplets, or "coacervates," selectively absorbed organic building blocks from the environment and demonstrated primitive metabolism catalyzed by primitive enzymes. Oparin further proposed that upon growing in size the coacervates divided and gave rise to a new generation of "protocells." He believed that such primitive division and the transmission of the "parent's" internal organization to the variant offspring formed the basis for natural selection and for the evolution of more efficient enzymes and complex metabolism.

Haldane's origin-of-life scenario was inspired by the contemporaneous development of genetics and the discovery of viruses. Whereas Oparin hypothesized that the metabolic coacervates were the intermediate link between inanimate matter and life, Haldane argued that self-reproducing organic polymers were the "first living or half-living things." Following the then-suggested comparison between a virus and a gene, Haldane compared the virus's reproduction within its host cell to the primitive reproducing molecules in the "vast chemical laboratory" of the "soup." It is historically noteworthy that more than a decade before the introduction of Haldane's gene-inspired scenario, the Harvard psychophysiologist Leonard Troland (1914) formulated a theoretical model for the emergence of the first autocatalytic molecule. Troland claimed that such a molecule could by its very autocatalytic nature and by the vast time allowed for its emergence overcome the high improbability involved in its chance-like appearance in the primordial ocean.

While Oparin's prebiotic infrastructure for the evolutionary emergence of life was a multimolecular metabolic system, Troland's and Haldane's was a single genetic polymer. This distinction, later to divide the origin-of-life community between metabolism-first and gene-first supporters, was still implicit. Meanwhile, the Oparin-Haldane hypothesis pertaining to the synthesis and chemical evolution of organic compounds in the reducing primordial atmosphere and in the primordial soup provided the framework for the beginning of the empirical study of the origin of life. In the early 1950s, on the basis of the Oparin-Haldane hypothesis, the young American doctoral student Stanley Miller, guided by his mentor Harold Urey at the University of Chicago, simulated in a glass apparatus the primordial atmosphere and ocean. The Miller-Urey landmark experiments, in which out of basic inorganic chemicals the synthesis of amino acids, the building blocks of proteins, and other organic compounds was achieved, inspired numerous laboratories to explore prebiotic chemistry and the origin of life (Fry 2000, 79–83) (Fig. 38.4).

EVOLUTION PRIOR TO LIFE: IS IT POSSIBLE?

The challenge facing researchers since the 1950s was not only to account for the prebiotic synthesis of the relevant building blocks but also to understand their subsequent organization. It was Darwin's (1859, 186–89) realization that the mechanism of natural selection, based on the processes of reproduction and mutation, was the major natural means to overcome the huge improbabilities involved in producing an organized complex biological system. Still unaware of the molecular basis of life, Darwin offered this explanation of the evolution of adaptively organized living forms as an alternative to Divine Design upheld by natural theology.

As we saw, early twentieth-century pioneers contended that the first living systems on earth were the product of a gradual evolutionary process. Not only did they suggest hypothetical intermediary links in this evolutionary chain; they also

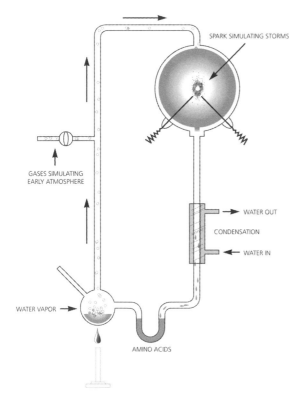

FIGURE 38.4. A stylized version of the famous experiment (of 1953) in which Stanley Miller and Harold Urey simulated the earth's early atmosphere and oceans and produced amino acids naturally

involve the replication and mutation of a nucleic-acid template, these experiments were designed and carried out within the gene-first framework. These experiments revealed that populations of isolated viral RNA molecules capable of replicating, mutating, and then reproducing these mutations responded to applied selective pressures by evolving and adapting to the new "environmental conditions" (Spiegelman 1967). The American biochemist Sol Spiegelman has thus managed to demonstrate the evolution in the test tube of a shorter RNA "species" that replicated more rapidly, was stable at higher temperatures, and was more resistant to degrading enzymes relative to the original viral RNA.

It was clear, however, that such molecular evolution depended on the participation of a highly complex replicating enzyme, utterly irrelevant to a prebiotic environment. Further experiments by the English chemist Leslie Orgel (1994) at the Salk Institute failed to achieve self-replication of RNA-like molecules without such an enzyme. Thus, a major "chicken-and-egg" problem presented itself: How could the interdependence between information-carrying nucleic acids and catalytic proteins, as manifested in each living cell, been circumvented on the primordial earth? Which came first, genes or enzymes? And how could the one arise and function without the other? The gene-first versus metabolism-first division, only hinted at in the early twentieth century, became a clear fault line that has separated the origin-of-life community until this very day (Fry 2000).

GENE-FIRST AND METABOLISM-FIRST INFRASTRUCTURES

The crucial question under consideration is whether a gene-like molecule or a primitive metabolic system could have arisen on the ancient earth by regular physical and chemical means and could have later evolved via natural selection? An unexpected discovery in the early 1980s signified a major breakthrough favoring the gene-first approach. Challenging the common knowledge that only proteins can function as enzymes, RNA molecules manifesting catalytic activity were discovered in extant cells (Joyce and Orgel 2006). These RNA enzymes that catalyze the cutting and joining of segments of RNA were aptly named by their discoverers "ribozymes."

Promptly, it was proposed that these ribozymes are a relic of an ancient RNA world in which RNA molecules could have functioned as both chicken and egg. This RNA-world theory is being continuously strengthened by the discovery of additional natural ribozymes engaged in diverse crucial activities in present cells. Notable among them is the ribosomal RNA catalyzing the formation of peptide bonds in protein synthesis. Additional ribozymes are being generated and isolated in evolution-in-the-test-tube experiments (Joyce and Orgel 2006).

Some supporters of the gene-first approach believe that RNA was the first template molecule to arise prebiotically. Interestingly, several of the fiercest critics of this RNA-first conception are gene-first proponents who regard the RNA

assumed that some sort of natural selection had to be active prebiotically. However, when the nature of the genetic material as well as the molecular basis of evolution became known, in the 1960s and 1970s, and the intricate relationship between nucleic acids and proteins in replication and metabolism was revealed, origin-of-life researchers faced considerable new challenges: How could natural selection have operated prior to the existence of living systems capable of reproducing, mutating, and competing for resources? Wasn't it the case that replication and metabolism depended on an organization that must have been the *product* of evolution and could not have emerged by prebiotic chemistry?

Following theoretical and empirical work, an awareness grew beginning in the late 1960s that any group of entities, not only living systems, could evolve through natural selection, provided it conformed to a set of specific conditions: reproduction, variation, inheritance of these variations, relative advantages conferred by some of these variations, and competition (see Cairns-Smith 1986).[1]

The first experiments demonstrating Darwinian evolution in the laboratory were conducted in the 1960s and 1970s. Based on the known mechanisms active in extant cells, which

---

[1] The Scottish chemist Graham Cairns-Smith (1986) suggested since the 1960s that clay minerals could have functioned on the early earth as "mineral genes," demonstrating replication, variations, and natural selection, eventually evolving into organic genes.

world as a crucial but definitely not the earliest step on the way to life. Leslie Orgel enumerated the many obstacles preventing the "invention" of RNA or a similar template by prebiotic chemistry. Being nevertheless convinced that only a genetic infrastructure and not a metabolic one could have formed the basis for the evolution of life itself, Orgel explored much simpler genetic systems, even polymers made of amino acids. Other RNA-later options are being explored by others (Fry 2010).

Metabolists, on the other hand, argue that the synthesis and replication of RNA or RNA-like polymers were extremely improbable under prebiotic conditions. They contend that metabolic cycles made of small organic molecules, such as amino acids or lipid-like building blocks, could have arisen more easily and could have enabled an evolutionary process (Segré et al. 2001). Instead of looking for a simpler template molecule, they suggest an alternative mode of reproduction and variation altogether as a basis for natural selection. They follow in principle Oparin's idea that during the origin of life the "memory" or "inheritance" guaranteeing the preservation of adaptive advantages from "parent" to "offspring" depended on the reproduction not of a single molecular template but of a metabolically organized whole. Rare changes in such systems constituted the required variation. The German organic chemist Günter Wächtershäuser (1992) argues for the highly determined emergence of autocatalytic cycles, out of which rare "catalytic branch products" resulted in variations of the original cycle. Any "branch product" feeding back into the cycle and into its own production guaranteed "a memory effect," that is, the basis for heredity of a variant cycle.

Several current metabolic theories also reject the emergence of life in the primordial soup, suggesting instead a scenario taking place in the high-temperature, high-pressure environment of submarine hydrothermal vents. According to this new paradigm, the first systems on the way to life developed autotrophically on the surface of sulfur-metal minerals that provided both catalytic aid and chemical energy to form organic molecules and metabolic cycles (Cody et al. 2004; Russell and Martin 2004).

Recent experiments by RNA-first proponents that managed to overcome long-standing hurdles seem to encourage this line of work. Metabolic theories, however, face a much more difficult experimental challenge. Though the synthesis of key biomolecules and the establishment of several metabolic reactions under relevant submarine conditions were achieved (Cody et al. 2004; Huber and Wächtershäuser 2006), demonstrating experimentally the closing of an autocatalytic cycle or its evolvability is enormously difficult. Thus, at present, metabolic theories are based solely on computer models and simulations.

THE ORIGIN OF LIFE AND THE EVOLUTIONARY WORLDVIEW

Growth of knowledge in areas such as geology, astronomy, prebiotic chemistry, molecular biology, and evolutionary biology allows researchers in the origin-of-life field today to perform sophisticated experiments or computer simulations and to get answers to various pointed questions. However, despite advancements in several directions, difficulties still loom large, and no decisive, full-scale answer has yet been attained. It is not surprising that creationists of all stripes, including the intelligent design variety, present this state of affairs as evidence that the only solution lies in a purposeful design of the first cell by a supernatural agent (Behe 1996).

Evidence against the creationist claim that living systems are "irreducibly complex" and could not, in principle, have evolved gradually through natural selection was already provided by Darwin (1859, 186–94) in his discussion of "organs of extreme perfection and complication" and of "modes of transitions" and more recently, on the basis of molecular data, by evolutionary biologists (see, among many, Musgrave 2004). Yet, because so far no complete scientific scenario of the origin of life was experimentally demonstrated, the origin-of-life question is often regarded even by adherents to science as the "soft underbelly of evolutionary biology" (Scott 1996). Considering the lack of full empirical validation of origin-of-life theories, some scientists and philosophers of biology prefer at the moment not to make the natural emergence of life "an issue" in debating religious believers (Ruse 2001; De Duve 2002).

Indeed, the presence or absence of empirical validation is a crucial issue. It is the aim of scientists to corroborate hypotheses and, where possible, to gather supporting empirical evidence. Unlike religious claims about the role of a supernatural agent in nature, which in principle are not open to empirical assessment, scientific hypotheses are testable and, at least in the long run, can be refuted or confirmed. But this empirical distinction of science does not fully describe its unique nature. Science today is characterized by the interaction between specific empirical claims and a broader philosophical naturalistic worldview that eschews purposeful explanations. This interaction, originating and developing first in the physical sciences in the seventeenth century, is now also the hallmark of biology. The notion that the phenomena of life can be explained naturally on the basis of evolution became a possibility after Darwin's *Origin* in 1859. It was subsequently established in the first half of the twentieth century as the framework that gives a unified sense to everything in biology (Dobzhansky 1973).

Philosophical conceptions, unlike empirical claims, cannot be refuted or confirmed. However, based to a large extent on the empirical achievements of the natural sciences during the past few hundred years, the evolutionary naturalistic worldview is by now strongly substantiated (Fry 2009, 2012). Though origin-of-life scientists are busy devising theoretical and empirical means in order to check their various theories, the question whether life emerged naturally from chemical compounds is by now long considered settled. This is why the postulate of the natural origin of life is not a "soft underbelly" of evolutionary biology. Darwin's impact on the study of the origin of life was both philosophical and empirical. Establishing the natural evolution of life on earth suggested also a natural origin of life. The mechanism of natural selection – Darwin's major answer to the evolution of biological complex organization – proved essential also in dealing with the origin of such organization. Thus, it is the philosophical and empirical strength of the Darwinian revolution that underlies the confidence shared by researchers that sooner or later the origin of life on earth will be accounted for.

Essay 39

The Evolution of the Testing of Evolution

Steven Hecht Orzack

WE OWE TO CHARLES DARWIN and Alfred Russel Wallace not the discovery of evolution but the creation of a powerful causal explanation for the small and large facts about nature that fascinate us as children and as adults. The principle of natural selection provides at least part of an explanation for a myriad of traits ranging from the shape of an orchid blossom to the visual acuity of a hawk's eye. Darwin's 1859 presentation of the principle in *On the Origin of Species* is a synthesis of observation and explanation that should amaze any curious reader. Wallace's (1870a) discovery of natural selection is no less amazing, especially given the more difficult circumstances that surrounded his work (see Berry 2002). Their discovery is a human achievement of the first order, which should transcend political and social divisions. But it is best to understand the world as an "is" and not as a "should." Most people believe some sort of divine explanation for the kind of facts about nature that I just mentioned; if they are aware of Darwin's and Wallace's discovery, they dismiss it as false. Few, if any, of these people will read *this* essay. You, the reader, are likely aware of the truth and importance of Darwin's and Wallace's discovery or are open to their discovery being correct. Even so, in order to understand how the testing of evolution has evolved, it is useful to examine the nature of Darwin's and Wallace's achievement, especially because it is often misunderstood even by those who recognize that it is correct.

In this context, it is worth comparing the "before" and the "after" of their discovery. To understand the "before," it is most meaningful to consider claims about nature that at least appear to be based on substantial assemblages of facts (as opposed to claims as to the "fact" of, say, divine creation that were simply appeals to faith and referenced few, if any, facts). With this restriction, all claims we consider are arguably science, inasmuch as there is marshaling of evidence (as opposed to just an appeal to belief). Given this restriction, one finds a wide variety of claims that range from the claim that the facts are explained by divine creation (Paley 1802) to those that invoke natural causes. The latter include Cuvier (1830), who argued that species do not change and that numerous "revolutions" were responsible for their appearance and disappearance in the fossil record, and Lyell (1830–33), who argued that species do change and that gradual natural processes governed appearances and disappearances.

As noted, these "before" claims are scientific inasmuch as they were presented as claims based on evidence. We could say that some of the before claims were also scientific because they included a hypothesis (divine creation) that is readily falsifiable (see Sober 2008). But this is a retrospective judgment, as Paley, Cuvier, and Lyell never acknowledged what evidence would falsify their *own* claim; instead, each marshaled evidence only in support of a favored claim and evidence against other claims.

What about the "after" of Darwin and Wallace's discovery? Of course, we have much the same assemblages of facts about nature. We also have a hypothesis as to how the natural processes can help explain this assemblage, in the form of the hypothesis of natural selection.

So far, the "before" and "after" look similar, at least if we consider claims like those put forth by Cuvier and Lyell, in that we have facts and a proposed explanation based on natural causes. What was different was the appearance of a changed consciousness about testing. In a chapter entitled "Difficulties on Theory," Darwin (1859, 189) addresses a variety of potential challenges to his own claim, the hypothesis of natural selection:

> If it could be demonstrated that any complex organ existed, which could not possibly have been formed by numerous, successive, slight modifications, my theory would absolutely break down.

To my knowledge, there is no analogous pre-Darwinian statement in biology about what information would necessitate rejection of a claim being advanced. Of course, you may wonder what kind of alternatives to, say, divine creation might even have crossed the mind of, say, Paley? Indeed, there were alternatives, at least inasmuch as one considers alternative creation scenarios (e.g., Christian vs. Hindu). But the strangeness of the question underscores what Darwin's contribution really was. In 1859, we see him marshaling substantial amounts of data, synthesizing these data, describing how they support his hypothesis, *and* providing explicit evidentiary grounds for rejection of his hypothesis. Taken together, these four elements constitute the first landmark in the evolution of the testing of evolution. In addition, while some elements of this quartet had appeared previously, one can view Darwin's ensemble contribution of these elements as a substantial advance of the process of doing science, even if one were to reject his biological conclusion. This is something an advocate of divine creation should applaud. (The ultimate genesis of Darwin's inclusion of a criterion for "self"-falsification may well have been his reading of Whewell's text [1837], which effectively amounts in part to be Whewell's inspirational guide to scientific discovery; for example, he writes "One of the most important talents requisite for a discoverer, is the ingenuity and skill which devises means for rapidly testing false suppositions as they offer themselves" [1837, 413]. Darwin is known to have studied this book during the time he discovered natural selection, and it is easy to imagine him being inspired; it is ironic that Whewell did not accept the fact of evolution, much less the fact of natural selection after 1859 [see Ruse 1975a].)

It is this distinction between the content of the science and the doing of the science that is often obscured when we consider Darwin and Wallace. The content and the doing are distinct, and a creationist's rejection of the former should at least be accompanied by acceptance of the latter as an amazing manifestation of human endeavor and creativity. Rejecting Darwin's and Wallace's contribution to the doing of science because its content is unacceptable makes no more sense than not admitting the beauty of, say, Handel's *Messiah* because it was inspired by the false notion that Jesus Christ is the Messiah.

It is of note that some modern proponents of divine creation appear to surpass, say, Paley with respect to acknowledgment of falsifiability. For example, in Yahya's (2007, 426) presentation of an Islamic version of divine creation we find:

> Pictured is a 50-million-year-old wasp preserved in Baltic amber. Like all other living things, wasps, which have remained the same for 50 million years, show that evolution never happened, and that God created them.

If we put aside the logic of the conclusion (as well as the falsehood that wasps have not changed), the structure of the inference defines an outcome that would falsify the hypothesis of divine creation (wasps *not* remaining the same over the past 50 million years).

Darwin and Wallace were careful to make the case for the influence of natural selection on trait evolution – nothing more and nothing less. We are now beyond testing of the claim that natural selection has *some* influence on the evolution of nearly all traits. The confirmed status of this hypothesis is a consequence of the general evolutionary research program sparked by Darwin and Wallace and of the work of many less-famed scientists who have documented the influence of natural selection on this or that trait. The net result is a compelling enough ensemble of results (encompassing traits small and large in a wide variety of species) such that we can appropriately conclude that natural selection is ubiquitous.

How was this conclusion about natural selection reached? Almost immediately after 1859, numerous scientists started to examine Darwin's and Wallace's claim about natural selection and/or interpret various biological phenomena in light of the hypothesis of natural selection. Some of the notable contributions of this kind in the decades leading up to 1900 include H. W. Bates (1862), Spencer (1862), Huxley (1863a, 1863b), F. Müller (1864), Haeckel (1868a), F. Müller (1879) (the first mathematical model of natural selection), Düsing (1884), Galton (1889), Geddes and Thomson (1889), and K. Pearson (1892). See Glick (1988) for a description of the wide variety of receptions and treatments of Darwin's and Wallace's claims about evolution and natural selection (see also Nyhart 1995).

However, none of these works presented evidence for the action of natural selection in natural populations in the form of observations of the differential fates of individuals; instead, these works continued to be typological in their treatment of species and populations. There continued to be pluralism with respect to acceptance of the claim that natural selection can have a marked influence on populations of, say, vertebrates (see Ruse 2003), in contrast to a broad acceptance among scientists of the fact of evolution (which, of course, was not a fact discovered by Darwin and Wallace). Kellogg (1907) described then-current questions about and objections to the claim that natural selection is a ubiquitous influence on natural populations. Many of these skeptical questions appeared justified at the time, given the lack of observational and experimental analyses of individuals and of the selective consequences of their differences. This lack likely arose mainly from the research tradition then dominant in biology, which emphasized broader-scale comparison of ensembles of species, as opposed to detailed analyses of individuals within species. As research practices changed, more and more studies of individuals appeared that could underwrite assessments of the realized influence of natural selection on populations. This shift was likely motivated by the "challenge" provided by Darwin and Wallace, by an increasing number of individual-focused experimental analyses in genetics and physiology and other closely related disciplines (cf. G. E. Allen 1978; Pauly 1987), and by an increasing emphasis on quantitative analysis after 1900 or so (influenced particularly by Francis Galton and by Karl Pearson, who published nineteen papers between 1894 and 1916 in a series entitled "Mathematical Contributions to the Theory of Evolution").

Among these studies was one by Bumpus (1899), who claimed that the morphological differences observed after a winter storm between the surviving and the dead House Sparrows indicated that natural selection had acted to eliminate those individuals most deviant from an "ideal" type. (Controversy over this conclusion concerns the exact nature of the process of natural selection involved, though some effect of natural selection is accepted; see Buttemer 1992.) Detailed studies of individual variation and natural selection to appear in the next two decades include Weldon (1895, 1902), K. Pearson (1903) (one of the nineteen papers noted previously), Poulton (1908), Harris (1911), Elderton and Pearson (1915), F. E. Lutz (1915), and Punnett (1915). Few, if any, such studies appeared in the next fifteen years (perhaps as a consequence of World War I and its aftermath). Thereafter, more studies appeared, including East (1932), Dubinin et al. (1934, 1936), Olenov et al. (1937), Gordon (1939), Gershenson (1945), Dubinin and Tiniakov (1946), Dobzhansky (1947) (see Lewontin et al. 1981 for a presentation of Dobzhansky's numerous publications describing the influence of natural selection on populations of *Drosophila*), A. J. Cain and Sheppard (1954), and Kettlewell (1955). Until the end of World War II, most studies concerned the influence of natural selection on mutations in laboratory populations or its influence on mutations released into natural populations. After the war, many more studies of variants found in natural populations appeared (publication of some studies from the 1930s was delayed by the war). These analyses, along with influential appeals for integrated treatments of individual differences and natural selection (Fisher 1930; Haldane 1932; Mather 1943) helped convince many biologists by the 1950s that natural selection does occur in nature, that it can (but need not) be very potent, and that large-scale evolutionary patterns could be governed mainly by natural selection. Once-correct assessments like that of Gordon (1939, 278) that "there is justification for the statement by Robson and Richard (1936) that 'the direct evidence for the occurrence of natural selection is very meagre and carries little conviction'" were no longer valid. The ascent of the stance that natural selection is ubiquitous is another landmark in the evolution of the testing of evolution, inasmuch as biologists were thereby licensed to assess not whether natural selection could help explain a trait of interest but how it helped to explain a trait of interest.

It is essential to note that acknowledging this transition is distinct from concluding that it happened mainly as a result of epistemic considerations or that the transition fostered (as well as embodied) a "modern synthesis" of scientific beliefs about evolution; these controversial claims are discussed in Mayr and Provine (1980) and Smocovitis (1996) and references therein.

Studies of the influence of natural selection on naturally occurring variation continue today. The need to do such studies reflects the need to more precisely quantify the qualitative and quantitative features of natural selection, as opposed to fulfilling any further need to document the fact of selection itself. An overview of many studies done until the mid-1980s can be found in Endler (1986). Some of the many more recent notable studies of the influence of natural selection on populations are Herre (1987), Grant and Grant (1989), Gillespie (1991), Reznick et al. (1996), Kreitman (2000), Schluter (2000), Reznick and Ghalambor (2005), Sabeti et al. (2007), Siepielski, DiBattista, and Carlson (2009), and Kingsolver and Diamond (2011). New studies on the influence of natural selection on traits in natural populations appear routinely in the pages of the journals *Evolution* and the *Journal of Evolutionary Biology*.

Acceptance of the ubiquity of natural selection has meant one thing to some scientists and another thing to other scientists. It is for this reason that further landmarks in the evolution of the testing of evolution are much less widely agreed upon. In fact, what are regarded by some as landmarks are not even acknowledged by others.

The reason why the landscape of understanding has become so balkanized is that "ubiquitous" covers a multitude of sins; the fact of natural selection licenses a variety of hypotheses as to the influence of natural selection on the evolution of a trait. For some biologists, this fact means that natural selection is powerful (even if we do not always readily recognize its power). To this extent, all or nearly all traits of organisms are viewed as having had only one important influence during their evolutionary history: natural selection. An important enabling attitude for this approach is that the influence of natural selection may be subtle and so apparent failure to find the influence of natural selection reflects only the inadequacy of the analysis (cf. A. J. Cain 1989). For other biologists, "ubiquitous" means natural selection influences almost all traits but nothing more; natural selection may not be the most important influence for a given trait. Instead, other evolutionary forces may be more important. Biologists believing in this pluralism invoke genetic constraints (such that the best possible trait cannot breed true), the lack of genetic variation, the influence of traits on each other, random change of genetic variation (genetic drift), and the legacy of past evolution (often termed phylogenetic inertia) as influences on a trait's current evolution that sometimes separately and sometimes in combination match the influence of natural selection or even exceed it. Because these biologists believe that natural selection is just one of several possibly potent influences on a trait's evolution, an apparent failure to find that natural selection has an important influence on the trait may be something real and not be a reflection of inadequacy of the analysis.

This division of opinion underlies the debate over adaptationism. Those scientists who endorse the attitude that one can focus solely on natural selection are termed "adaptationists" and those who endorse the attitude that one must focus on several evolutionary forces are termed "pluralists." This debate is long-standing (e.g., Fisher 1930; Wright 1931; Maynard Smith 1978; Mayr 1983; Gould and Lewontin 1979; Parker and Maynard Smith 1990; Orzack and Sober 2001). Orzack and Forber (2010) contains a description of many other important contributions to this debate. A description of how one might interpret the same biology from the two different perspectives is provided by Millstein (2007) (Plate XXVIII).

The lack of uniform interpretation of the fact that natural selection is ubiquitous is not a consequence of a lack of intelligence or of quality of thought. One can readily find excellent scientists who routinely ignore all evolutionary influences except for natural selection (e.g., Parker and Maynard Smith 1990; Grafen 1998); one can also readily find excellent scientists who invoke natural selection only along with other evolutionary influences (e.g., Kimura 1983).

To this extent, modern evolutionary biology consists of largely separate research programs. Beyond formal acknowledgment, most biologists and other scientists studying behavioral, morphological, and life history traits typically invoke only natural selection in their work; in contrast, most biologists studying population genetics and molecular evolution typically invoke some combination of natural selection, genetic drift, and genetic constraints. Despite this discrepancy, each of these research programs is "internally" evolving toward improved hypothesis testing (even as compared to just twenty years ago). This is not a trivial achievement.

At the same time, the divisions among research programs are problematic. This is not a matter of a cognitive "division of labor" that might arguably lead to improved understanding because the truth is "triangulated" from several different research directions (cf. Kitcher 1993). Instead, beyond an acknowledgment of the scientific standing of the "other," many practitioners of any one research program do nothing in order to understand how the other research programs might connect with what one is doing. It is essential to note that differences in the kinds of traits studied (those readily observable by the naked eye as opposed to molecular traits, such as DNA sequences or amino acid sequences) do not necessarily underwrite a different understanding of what the ubiquity of natural selection implies about the evolutionary forces to be considered in an analysis. The traditional focus on natural selection in analyses of observable traits has been expanded. For example, some important recent analyses demonstrate the degree to which nonselective processes such as genetic drift can strongly influence observable traits (Lande 1976; Lynch 1990; Bell, Travis, and Blouw 2006). In addition, the importance of accounting for shared ancestry of related species when assessing adaptive hypotheses (i.e., accounting for phylogenetic inertia so that the number of independent evolutionary events in support of a particular hypothesis is not overestimated) has become widely accepted (Felsenstein 1985, 2004; Harvey and Pagel 1991; Hansen and Orzack 2005). Similarly, the traditional focus on genetic drift in studies of molecular evolution has been expanded. For example, genetic drift as the cause of most changes in amino acid sequences of proteins has been challenged by Gillespie (1991), who argues that natural selection is the major cause.

But how are different understandings of "natural selection is ubiquitous" any different than, say, the different understandings that arise in physics and biology? The fact that these disciplines rarely intersect with one another in areas of study where they could jointly provide better insight than each does separately might just be an unavoidable inefficiency of the human doing of science. From this standpoint, different research programs are perhaps nothing to be concerned about. In fact, this attitude fails here. Why? Biologists, no matter what their specialty, in one way or another view their work as both stemming from and reinforcing Darwin's and Wallace's claim about the ubiquity of natural selection. For, say, a behavioral ecologist, Darwin and Wallace are taken to have demonstrated nothing less than the "fact" that traits are the best they can be (so that one can consider only natural selection when analyzing this or that trait). In contrast, for, say, a population geneticist, Darwin and Wallace are taken to have demonstrated nothing more than the "fact" that natural selection influences traits (so that one must consider it along with other evolutionary influences when trying to understand why this or that trait is the way it is).

These are dramatically different invocations of "the" fact of natural selection established by Darwin and Wallace. Orzack and Sober (1994a) untangled some of the mixture of claims arising from the varied ways in which biologists interpret Darwin's and Wallace's discovery and distinguish among three hypotheses: natural selection is ubiquitous (with no necessary implication as to its power to influence trait evolution), natural selection is an important influence on a trait's evolution (with possibly other forces also being important), and natural selection is the only important influence on a trait's evolution. These hypotheses differ substantially in what they imply about the nature of the data needed to confirm them. But these evidentiary differences and the normative implications they underwrite for the *practice* of testing evolutionary hypotheses are generally not acknowledged because the multiple interpretations of Darwin's and Wallace's claim are not acknowledged.

The practical consequences of confusion in regard to which hypothesis about natural selection can be tested are illustrated in the elegant study by Milinski (1979) of the foraging behavior of a small fish, the stickleback (see also Orzack and Sober 1994b). His investigation concerned whether its foraging behavior matches that predicted by optimal foraging theory, which predicts the best strategy for a predator to exploit spatial heterogeneity of prey (e.g., abundant prey in one location as compared to few prey in another location). His conclusion that the observed distribution of the fish matches the predicted distribution was based on "snapshot" behavioral assessments of ensembles of individuals, as opposed to assessments of whether all or almost all individuals each manifests the correct set of behaviors over time. The former kind of data in which the trait expressed by each individual is unknown cannot serve as a basis for a claim for optimality. After all, a claim that a trait is optimal and that it has evolved as a result of the process of natural selection described by Darwin and Wallace is based upon the assumption that individuals (not groups) compete with one another. It is a claim that a particular trait is superior to any other (plausible) trait that an individual could possess; as a result of the process of natural selection, there will be a single trait (possibly consisting of a set of behaviors) present in the population (except for rare mutant behaviors).

In the case of the stickleback, the match between the observed distribution and predicted distribution is consistent with at least two different scenarios with respect to the expression of foraging behavior by an individual. In the first, an individual always occupies the same position relative to the available resource, but the ensemble of individuals expresses an overall distribution that matches the prediction. In the latter, each fish expresses the predicted distribution over time (given multiple opportunities). These alternatives underwrite very different conclusions about the power of natural selection to influence foraging behavior (note that both of these conclusions are consistent with natural selection being ubiquitous). When only the ensemble manifests the predicted distribution, we can reasonably conclude that natural selection has had an important influence on the evolution of this trait (we cannot say more because we have not resolved the evolutionary cause of the fixed differences among individuals; it could be that they are due to, say, genetic constraints, and, if so, this would imply that an evolutionary force other than natural selection had an important influence on the trait). In contrast, when the *individual* manifests the correct distribution over its lifetime, we can reasonably conclude that natural selection is the only important influence on the trait and that it is optimal relative to plausible alternatives (there are no individual differences in need of explanation by other evolutionary forces). These are very distinct conclusions about natural selection, but their substantial distinctions and what each implies about the kind of data needed to support each conclusion are rarely acknowledged. To his credit, Milinski acknowledged that his results

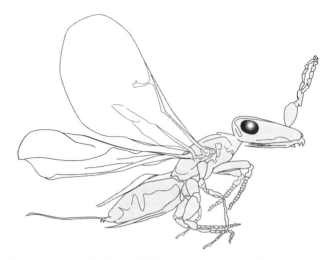

FIGURE 39.1. Optimality model. One common way of studying adaptation is to predict the best trait for an organism to possess in a given situation and then to compare this "optimal" trait and the observed trait. William Hamilton reasoned that when matings occur within small local groups and siblings of one sex compete more for mates than do the other sex, the optimal sex ratio evolved in response to this "local mate competition" should be biased against the sex in which mate competition is greater. Some species of wasp lay eggs in figs, from which the offspring emerge to mate locally, with males competing more for mates than do females. The sex ratios of some of these wasp species have been shown to be (anti-) male-biased, which provides evidence of the influence of natural selection arising from local competition for mates. Permission: Martin Young

were consistent with the possibility that individuals express fixed behaviors. At the same time, he did not acknowledge the conflict between this empirical outcome and the process of natural selection that he assumes influences this trait (1979, 36).

This kind of ambiguity has very much hindered the evolution of the testing of evolution. For example, Orzack and Sober (1994a) examined published tests of the optimality of hundreds of traits. They could identify only a handful of studies that contained the proper type of data and analyses that would allow the investigator to conclude that the trait is or is not optimal (see Brockmann, Grafen, and Dawkins 1979 for an analysis supporting a claim of optimality and Orzack, Parker, and Gladstone 1991 for an analysis supporting a claim of non-optimality). At best, the data and/or analyses in all of the other studies were structured in a way that would allow one to conclude only that natural selection was an important influence on the trait's evolution. This is not a trivial accomplishment, but it is something very distinct from the claim for optimality that was present in most of these studies (Fig. 39.1).

The problem is that the division of received opinions about the power of natural selection *coincides* with the division between research programs; the division of opinions is not expressed in such a way that any given program considers as plausible all of the different stances about the power of natural selection described here. Instead, different programs in evolutionary biology agree (correctly) that Darwin and Wallace provided a unitary insight about natural selection but disagree substantially as the substance of the unitary insight.

Of course, each program claims to embody *the* true insight of Darwin and Wallace.

Given this dissonance, is it true that the post-Darwin-and-Wallace evolution of the testing of evolution is characterized by stasis? No. Neither is it true that there has been no progress *within* evolutionary research programs; as noted, it is manifestly true that they have yielded many substantive insights. Nonetheless, it is indisputable that less progress has been made *within* research programs, especially behavioral ecology, because of confusion about the exact nature of the hypothesis about natural selection under test; too often the exact evidentiary basis for a claim that a trait is optimal has been left ambiguous (see examples in Orzack 1993). In addition, it is apparent that the evolution of the testing of evolution has been hindered because the divisions among research approaches cover over substantive evolutionary questions that, as a result, remain poorly resolved at best. All too often, the divisions are explained by claims that "the biology is obvious and it tells us to do things differently over here." To most behavioral ecologists it is "obvious" that natural selection is the only important influence on a trait like foraging behavior, whereas to most molecular evolutionists it is "obvious" that natural selection is only one of several important influences on the evolution of DNA sequences.

If we are to live up to the monumentality of Darwin's and Wallace's achievement, if we are to understand the "grandeur" in their view of life, if we are to understand how "endless forms most beautiful and most wonderful have been, and are being, evolved," we need to analyze and understand evolutionary phenomena large and small with a consistent and less-balkanized conceptual framework, one that honors the multiplicity of hypotheses that arise from the fact of natural selection and does not sustain confusion about what Darwin and Wallace really discovered. We owe Darwin and Wallace nothing less.

Essay 40

Mimicry and Camouflage: Part Two

Joseph Travis

WITH THE ACCLAIM FOR Darwin's postulate of evolution through natural selection came the excitement and the challenge of explaining ever-more complicated natural phenomena in Darwinian terms. And as biological explorers continued to describe new observations, particularly from tropical habitats, the number of challenges grew rapidly. An avalanche of letters among these naturalists exchanged ideas and hypotheses, with Darwin's correspondence itself revealing extensive musing on an array of patterns and their possible emergence from natural selection on individuals.

Few observations were as intriguing as those on mimicry and camouflage. While some observations seemed easy to explain – caterpillars that blended with their leafy backgrounds would be less likely to be preyed upon than caterpillars that contrasted with their backgrounds – others were more difficult. Mimicry was one of those more difficult challenges. The early history of Darwinian evolution, as a science, is tightly entwined with the arguments about whether mimicry could be readily explained in Darwinian terms (see Essay 15, "Mimicry and Camouflage").

It is important to distinguish two distinct phenomena of organismal coloration and pattern (Ruxton, Speed, and Kelly 2004). Crypsis occurs when it is difficult to distinguish an organism from its background. This can happen when an organism's color or pattern causes it to blend visually into its background, when its shape and color make it resemble an object in its background, or when its pattern and color break the outline of its shape against its natural background and make it difficult to recognize. Our usual understanding of "camouflage" embraces one or the other of these descriptions (Figs. 40.1 and 40.2). *Mimicry* occurs when the features of one species resemble those of another and, through that resemblance, confer some survival advantage on the mimic.

Crypsis is the more easily explained phenomenon. Everyday experience suggests that cryptic coloration or patterning, which is rampant in nature, can enhance the probability of survival. Darwin made this point in chapters 4 and 6 in the first edition of *The Origin of Species* in 1859 and there was little controversy over it. Indeed, a paper in 1859 on cryptic coloration in desert larks was the first paper to support the Darwin-Wallace theory of natural selection. The Cambridge ornithologist Alfred

Newton was a quick convert to the new theory and convinced the Reverend Henry Tristram to interpret coloration in birds as a case of natural selection (I. B. Cohen 1985).

For the earliest Darwinian scientists studying crypsis, or any aspect of animal coloration for that matter, the proof was in a sensible ecological scenario, with the assumption of sufficient inherited variability. The Darwinian view was that natural selection was so effective because it had a great deal of raw genetic material with which to work. Darwin's demonstration of the lability of domestic stocks in *The Variation in Plants and Animals under Domestication* (1868) provided the foundation for this view, even though neither Darwin nor his earliest defenders truly understood inheritance. A contrasting point of view was that natural selection, as opposed to artificial selection, would not be capable of driving sustained evolution because most natural variants were inherited as large effects and natural selection would rapidly exhaust the supply of available variation. In modern terms, the debate was whether evolution via natural selection was more likely to be limited by the strength and consistency of selection or the availability of mutational variation.

Only with the discovery of Mendelian genetics in 1900 did the problem become how color and pattern variations would be inherited. With the discovery of particulate inheritance, the challenge for the later generation of Darwinians was to diagnose whether heritable variation included a few large distinctions in pattern or color or many small variations, which, in their cumulative effect, would transform the appearance of an organism. This problem was important because the answers could inform a fierce debate over which factors would limit the rate of evolution and, consequently, how effective natural selection could be as a transformational mechanism. In early field studies of ecological genetics in the 1940s, crypsis quickly became one of the first phenomena to be studied intensively in the light of Darwin's description of natural selection (Cain and Provine 1992).

FIGURE 40.1. Simply being the same color as the background is not enough, as is shown by this conspicuous white cock standing against a light sky. From H. C. Cott, *Adaptive Colouration in Animals* (London: Methuen, 1940)

While studies of mimicry would also contribute substantially to the debate on variation and selection, mimicry posed its own challenges for Darwinian explanation. First and foremost, why would one organism resemble another? Second, even if one could discern the advantage of being mimetic, the means through which such intricate visual patterns could be inherited required explanation. And if one could be certain of the "why" and the "how" of mimetic patterns, a deeper consideration of the paths through which such mimicry could arise, spread, and be refined raised a substantial number of difficult questions.

The question of why was answered relatively quickly compared to the others. As described in Essay 15, Henry Walter Bates deduced the likely reason for the form of mimicry that bears his name. In Batesian mimicry, a harmless or palatable species mimics one that is dangerous or unpalatable, thereby gaining protection from predators or other enemies who otherwise avoid the dangerous or unpalatable model. Bates studied tropical butterflies carefully and noted that some species in the family Heliconiidae flew

FIGURE 40.2. The "obliterative shading" on this bush buck breaks up the appearance of solidity, thus making the animal far more difficult to perceive. From H. C. Cott, *Adaptive Colouration in Animals* (London: Methuen, 1940)

languidly and without any visible attempt at eluding predators. He noted that there were other species that looked like the languid fliers in the colors and patterns of their wings but that were in the family Pieridae and did not exhibit the same pattern of flight, instead exhibiting furtive, elusive patterns. Bates noted that these butterflies did not share other aspects of their morphology with the languid fliers – as one might expect from species in a completely different taxonomic family – and in fact, wing patterns aside, closely resembled other pierid butterflies that flew furtively. Bates's patient observations convinced him that the languid fliers were protected from bird predation, presumably by noxiousness, and that the other butterflies were protected only through their mimetic patterns. In effect, the mimics bore the wing colors and patterns of their models but the morphology and behavior of other butterflies.

Starting in the 1950s, quantitative and experimental work proved Bates's Darwinian explanation of these striking observations to be correct: mimetic species do gain a statistical refuge from predation by imitating noxious ones. But soon after Bates's initial ideas became public, the impression made by those ideas, along with the increasing number and diversity of examples of apparent mimicry that were being discovered, led to a rush of interest in the subject. As it turns out, mimicry is as rampant in nature as crypsis (Ruxton, Speed, and Kelly. 2004). There are innumerable cases drawn from almost every animal group in which a harmless species mimics one that is dangerous in some way (Joron 2003). Besides the many examples in butterflies, the roster includes the mimicry of stinging bees by harmless flies, the mimicking of venomous snakes by nonvenomous ones, and the imitation of distasteful salamanders by palatable ones. There are many cases of aggressive mimicry, in which the mimic is the dangerous species, such as when predaceous fireflies mimic the mating flashes of nonpredaceous firefly species so as to lure them as prey (Lewis and Cratsley 2008). Some cases of mimicry involve noxious or dangerous species resembling one another (called Müllerian mimicry; Sherratt 2008). There are many cases of parasitic mimicry in which the eggs of avian nest parasites like cowbirds resemble those of the host species in whose nest the parasitic eggs are placed (Klippenstine and Sealey 2008). And while visual mimicry has received the most study, mimicry also occurs in other sensory modes, as when parasitic cuckoo chicks mimic the begging calls of their host bird species (Ranjard et al. 2010). Inasmuch as Batesian mimicry provided the major support for Darwin, I focus on this variety for this essay.

There is considerable diversity in patterns of Batesian mimicry. Within the swallowtail butterflies (family Papilionidae), for example, there are mimetic species in which both genders mimic a single unpalatable model, species in which males imitate one noxious model and females another, species in which mimicry of a single model is limited to females, species in which each gender is mimetic but there are two or more models being imitated by each gender, and species in which only females are mimetic but there are multiple mimetic forms each imitating a different model (Kunte 2009a). Some of these patterns occur in the same species; for example, populations of the African swallowtail *Papilio dardanus* in Madagascar and nonmimetic populations on the African mainland exhibit striking female-limited mimicry.

It is easy to see why mimicry, with its variety and complexity, attracted so much attention from both early proponents and early opponents of the Darwinian paradigm. If Darwin's notion of natural selection could explain mimicry, there might be nothing it could not explain. But if it could not, then the odds of its being a general explanation for the intricacies of nature would be low; if it failed with mimicry, how likely was it to be the correct explanation for complicated, highly integrated structures like the vertebrate eye? Indeed, mimicry has been a favorite topic for illustrating almost every aspect of Darwinian evolution, from the early days of the modern synthesis (J. S. Huxley 1943) to the present (Charlesworth 1994; Mallet and Joron 1999).

Darwin devoted little of his own investigative skill toward explaining mimicry, declaring, in correspondence, his lack of confidence in his own knowledge of the subject and deferring to the expertise of Bates and others. This is an odd position to be taken by anyone who had spent considerable time in the tropics, where diverse examples of mimicry offer themselves for inspection on a daily basis. It is especially odd for Darwin, whose keen eye and inquiring mind rarely overlooked a striking natural phenomenon. Darwin recognized how completely Bates had covered the subject, and his own interest was drawn more to sexual selection of coloration. Darwin was engaged in an ongoing debate with Alfred Russel Wallace about the limits of selection and the explanation of sexual dimorphism (see Essay 15, "Mimicry and Camouflage"). What did capture his attention were the numerous cases of sex-limited mimicry, in which females mimicked one or more other species, while males retained the colors and patterns of their close relatives. In *The Descent of Man, and Selection in Relation to Sex* (1871, ch. 11), Darwin recounted the observations of Bates, Wallace, Trimen, and Riley on mimicry in butterflies, reminding his reader that mimetic butterflies gain their advantage by imitating brightly colored species that are noxious to the taste and that predators normally avoid. He then proceeded to describe the curious, striking phenomenon of sex-limited mimicry and offered his wholly original insight into its evolutionary cause: females are mimetic because it protects them from the attacks of predators, but males are not because sexual selection through female choice favors the original colors and patterns.

That Darwin would seize upon the contrast between natural and sexual selection to explain sex-limited mimicry ought not to surprise his careful readers; Darwin saw sexual selection as an extremely powerful force. But a little thought about this hypothesis for sex-limited mimicry will indicate how dramatic a hypothesis it is. The idea is *not* that only females benefit from mimicry but that sexual selection through female choice for the traditional (or what we would now call the ancestral) pattern is so strong that the mating advantage of the nonmimetic pattern in males overwhelms the survival advantage that would be conferred by the mimetic pattern. This hypothesis places an enormous amount of confidence in the power of sexual selection;

more specifically, it places an enormous amount of confidence in the power of female choice among males.

Considered in this light, Darwin's hypothesis may seem far-fetched. This might have been particularly so in the decades after his death when more and more effort was being devoted to documenting the survival advantage of mimicry and the inheritance of mimetic patterns in butterflies. It is important to remember that while the practitioners of the emerging discipline of ecological genetics were compiling an impressive roster of studies of natural selection on visible polymorphisms, there was very little effort devoted to quantitative studies of sexual selection. To be sure, sexual selection was receiving attention (e.g., Bateman 1948; L. Levine 1958; O'Donald 1974; O'Donald, Wedd, and Davis 1974), but not until the 1980s was there a substantial body of data on sexual selection in diverse systems comparable to the accumulated data on natural selection.

Yet Darwin was correct. Over a period of about thirty years, a series of investigations on mimetic butterflies by many research groups confirmed his hypothesis (reviewed in Kunte 2009b). Females benefit more from mimicry than males; female butterflies are more vulnerable to predators because the load of eggs they carry constrains the aerodynamics of their flight. Males gain a survival benefit from mimicry, but the advantage is not as great because, not carrying egg loads, the aerodynamics of their flight is much less constrained and their flight can be more elusive. Females in species with sex-limitation demonstrate very strong discrimination against males that deviate from the modal non-mimetic pattern. Of course, Darwin's hypothesis for sex-limited mimicry would not seem far-fetched today because we have thousands of studies that demonstrate that sexual selection is at least as strong and often much stronger than selection via survival advantages. But in the light of history, the verification of Darwin's hypothesis for sex-limited mimicry could be considered among the most compelling proofs of his genius.

Neither crypsis nor mimicry is passé as a subject of research in modern biology. New insights from vision physiology have revealed how an animal can be conspicuous to individuals of its own species but cryptic to predators (e.g., Cummings et al. 2008). New experimental work has shown that cases of apparent Müllerian mimicry may in fact have features more akin to Batesian mimicry; when two noxious or dangerous species are not equally so, the one not as well defended may be parasitic Batesian mimics of the better-defended one (Rowland et al. 2010). New theoretical insights have clarified the ecological conditions that favor the different varieties of mimetic dimorphism and polymorphic female mimicry (Kunte 2009b). And a deeper understanding of the genetic control of variation in color and pattern has reopened the old debate over the size of the allelic effects that are the fuel for adaptive evolution (for a sample of views in the modern debate, see H. A. Orr and Coyne 1992; Charlesworth 1994; Hoekstra and Coyne 2007; Nadeau and Jiggins 2010) (Plate XXIX).

Two recent case studies are especially interesting in light of Darwin's interest in sex-limited mimicry and crypsis. They illustrate that there remains much to learn about these subjects. But perhaps more importantly, they illustrate that mimicry and crypsis remain fertile areas of inquiry from which all of evolutionary biology has much yet to learn.

As noted, the swallowtail butterflies, family Papilionidae, exhibit a dizzying array of mimetic patterns that includes almost all conceivable possibilities from monomorphic, non-mimetic species through species with female-limited mimetic polymorphisms. Vane-Wright (1971) postulated two evolutionary trajectories to account for this diversity, only one of which would lead to female-limited mimicry and mimetic polymorphisms in females. He also postulated that sexually monomorphic mimicry and simple female-limited mimicry, once each had evolved, would be readily and repeatedly interconverted. That is, it would be easy for an ancestor with one pattern to produce descendants with the other. Kunte's (2009a) careful mapping of mimetic patterns onto a robust phylogeny of the Papilionidae suggests that nothing could be further from the truth. She showed that monomorphic mimicry and female-limited mimicry have evolved repeatedly and independently in different parts of the swallowtail butterfly family tree. Moreover, there were only a very few transitions between them; ancestors with one of these patterns rarely gave rise to descendants with the other. This result has forced us to reconsider our understanding of the evolutionary pathways through which complex adaptations can evolve.

One of the great examples of putative cryptic adaptation is the Florida beach mouse. A little more than eighty years ago, Sumner (1929a, 1929b) documented the striking distinctions between the pale, almost white color of mice in beach populations and the darker color of mice from populations of the same species in nearby forested habitats. While Sumner entertained a variety of hypotheses for this well-documented distinction, opinion settled intuitively on the selective advantage of crypsis as the best explanation: light animals blended better with the background of the beach environment, while darker animals blended better with that of the forest. Oddly enough, there were no quantitative tests of this hypothesis until quite recently when Hopi Hoekstra's research group published its extensive examination of these mice (Mullen et al. 2009; Vignieri, Larson, and Hoekstra 2010, and papers cited therein) (Plate XXX). Hoekstra and her colleagues have shown that predators in each habitat act as agents of natural selection against mice that stand out against their background. In each habitat, mice either lighter or darker than their background are more likely to suffer predation. Moreover, the level of relatedness among populations did not appear to constrain their ability to evolve to match their backgrounds; closely related populations in different habitats can be quite different, and distantly related populations found on separate beaches have converged on the same bright color. Finally, they found an association between the pelage hues of the mice and allelic variation at a single pigmentation gene. Not only have Hoekstra and colleagues resolved the explanation for a long-known pattern; they have implicated a single specific gene as critical for a wide-ranging pattern of adaptation and helped reopen one of evolutionary biology's longest-running debates.

Essay 41

The Tree of Life

Joel D. Velasco

COMMON ANCESTRY IS one of the pillars of Darwin's theory of evolution. Today, the tree of life, which represents how all life is genealogically related, is often thought of as an essential component in the foundations of biological systematics and so therefore of evolutionary theory – and perhaps all of biology itself. It is an iconic representation in biology and even penetrates into popular culture.

Massive amounts of time, effort, and money are being put into understanding and reconstructing the tree. Yet there are serious debates as to the usefulness and even the very existence of the tree. Here I will attempt to critically evaluate the merits of some of these worries. In doing so, we will see that questions about the tree and the foundations of systematics can be answered in the light of a wide range not only of empirical considerations but of philosophical considerations as well. A historically informed picture of how and why we got to where we are today is important for understanding these debates; however, here I can give only the briefest of introductions to the history of the tree as it has been used in systematics before turning to contemporary and future considerations.

A POTTED HISTORY

Many authors before Darwin had considered the possibility or even promoted the idea that some species were directly genealogically related to each other (Fig. 41.1). Some, including Jean Baptiste Lamarck, had even proposed treelike structures to capture these relationships (Ragan 2009). But it was Darwin who revolutionized our understanding of the diversity of life with his *On the Origin of Species* (1859). It is in the *Origin* that we first see the importance of genealogy on a grand scale where Darwin convincingly argues that common ancestry explains both the striking similarities between different species and the apparent naturalness of a groups-within-groups hierarchical classification. In the *Origin*, Darwin (1859, 129–30) introduces the metaphor of the tree of life, which connects all life through common descent:

> The affinities of all the beings of the same class have sometimes been represented by a great tree. I believe this simile largely speaks the truth.... The green and budding twigs may represent existing species; and those produced during

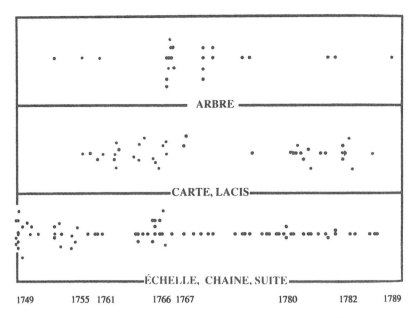

FIGURE 41.1. Before the idea of evolution took firm root, it was by no means obvious that a tree was the best way of portraying life's history. As can be seen from this chart mapping his different metaphors, Georges Buffon toyed with trees, maps, and chains (the dots record the usages of the respective images; the y-axis shows multiple usages in the same year), and indeed trees became less prominent in his thinking in later years. Drawing, inspired by G. Barsanti, Buffon et l'image de la nature, in *Buffon 8*, ed. J. Gayon (Paris: Vrin, 1992)

former years may represent the long succession of extinct species.... the great *Tree of Life* ... covers the earth with ever-branching and beautiful ramifications. (emphasis added)

To help us understand descent with modification, which is essential for his theory of natural selection, Darwin gives us a figure – the only figure in the entire *Origin* – to which he then repeatedly refers (116) (Fig. 6.3). This tree represents real genealogical history and is not simply a classification scheme representing subordination of groups within groups, such as the diagrams previously given by Linnaeus, among others.

This idea of a tree that connects all life has been part of the biological literature since Darwin, but it would require twin revolutions in methodology and in the types of data available before serious attempts could be made at building truly universal phylogenies. By the 1950s, despite great advances in the knowledge of the phylogeny of eukaryotes, bacteriologists had generally given up on the idea of that it was possible to build a comprehensive phylogeny for most groups of bacteria. Morphological and physiological data just seemed too sparse and often conflicted (Sapp 2009). But in the early 1960s, Emile Zuckerkandl and Linus Pauling, among others, suggested that molecules such as genes, amino acids, or proteins could be used to track phylogenetic history. Zuckerkandl and Pauling (1965a, 1965b) proposed that some changes might occur at a constant rate forming a "molecular clock," which would aid in phylogenetic reconstruction as well as in determining the timing of evolutionary events.

At the same time, Carl Woese was working on the evolution of the genetic system itself. To examine the early evolution of life, one needs to know the broad-scale phylogenetic history of all life. In 1977, after painstakingly cataloging numerous rRNA sequences (and then searching for further kinds of data to validate their findings), Woese and George Fox announced that they had discovered a third kind of life: what they called the Archaebacteria. Despite being prokaryotic, the Archaebacteria lacked the typical signature found in all bacterial rRNA and, in addition, also shared many deep similarities with eukaryotes, such as the way that they performed transcription and translation. Over the next thirteen years, Woese and colleagues produced the first universal phylogenies (Fox et al. 1980; Pace, Olson, and Woese 1986; Woese 1987) and eventually proposed the three-domain model in which the Archaebacteria were renamed the Archaea, as opposed to the Bacteria and the Eucarya (Woese, Kandler, and Wheelis 1990). Today, the most common representations of the tree are akin to the phylogenetic tree depicted by Woese et al. in Figure 49.5. While some of the details of the tree are no longer accepted, this division of life into three great domains – the Bacteria, the Archaea, and the Eucarya – has been generally (though not universally) accepted as can be seen in Figure 41.2 taken from the back cover of an evolution textbook (Barton et al. 2007).

Woese was not concerned primarily with classification but was trying to answer a particular question: What is the correct evolutionary branching sequence for "major groups" of taxa? More recent reconstructions of the tree may add more taxa or use more or different kinds of data and may come to conclusions different from Woese's, but fundamentally they are working on the same project. While this certainly seems like a perfectly objective task, it depends on the idea that there is a unique, objectively correct tree of life. If there is not, then what purpose is there for us to infer *the* tree?

WHAT IS THE TREE?

A standard way to describe the tree is to propose that it is a universal phylogenetic tree depicting the genealogical relationships of all species through time. Thus, the tree of life is meant to be universal, to be a phylogeny, and to be a tree. Critics have directly or indirectly attacked each of these three apparently essential features.

What does it mean to say that the tree of life is a tree? Modern depictions of the tree of life do not look at all like *biological* trees, such as in the familiar drawings of Ernst Haeckel, but rather are *phylogenetic* trees, which are trees in the mathematical sense of a special kind of object in graph theory. It is conceptually helpful to think of a tree as a set of directed branches connecting nodes where there is a root node with

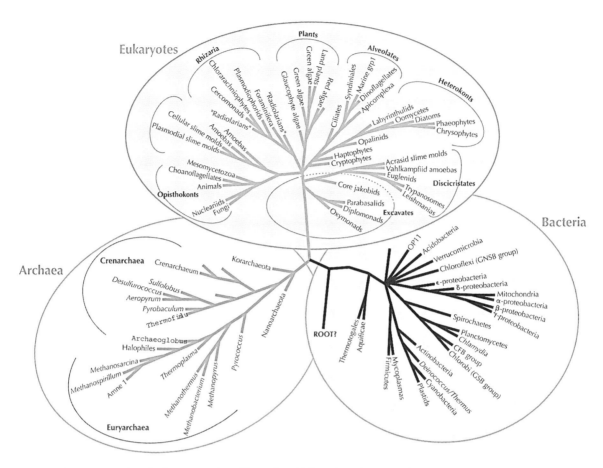

FIGURE 41.2. A textbook representation of the modern thinking about the tree of life. As can be seen, there are three great Domains – the Bacteria, the Archaea, and the Eucarya. This is based on the work of Carl Woese in the 1970s. Permission: Sandie Baldauf

no parental nodes, interior nodes that have exactly one parent and two (or sometimes more) offspring nodes, and leaf tips that are nodes that have one parent but no offspring nodes. Important features of trees for systematics is that between any two points there is a unique path on the tree and that each node (other than the root) has exactly one parent. To say that the tree of life is a tree is to say that it is a phylogenetic tree in this sense.

To say that the tree is universal implies that the tree should depict the relationships between all living things. Are viruses alive? Canonical representations of the tree typically do not mention viruses, but it is worth mentioning viruses in this context, because they place limitations on those wishing to defend the tree. One must either embrace the idea that the tree is not universal or deny that viruses are alive or have the kind of evolutionary history that the tree is supposed to be tracking. The typical assumption is that the tree must connect all species but that viruses (along with mobile genetic elements like transposons and plasmids) do not form species, at least in the way relevant for inclusion on the tree. But do all *organisms* form species in the relevant way? Many bacteriologists, as well as systematists and philosophers of systematics, deny that prokaryotes form species (Gevers et al. 2005; Ereshefsky 2010b; Lawrence and Retchless 2010). Worse, many of those who do accept that there is a good species concept that applies to prokaryotes will deny that these groups are phylogenetic groups and have branching histories.

It is sometimes thought that a simple change in how we describe the tree can solve this problem. The tree of life shows how *organisms* (or perhaps genomes instead) are genealogically related. But many organisms are not related to each other in a treelike hierarchy of descent. Rather, they form a reticulated network. This is even clearer for genomes where recombination is present. The defender of the tree needs to say something about how, at the appropriate level of description (perhaps when talking about populations or lineages or clades of organisms directly), these entities can form a tree. It is not clear how this can be done, and the burden of proof is surely on the defender of the tree here.

HYBRIDS

If we do manage to muddle through the species problem and say that the tree can connect all species, we then have the empirical question of just how treelike this evolutionary history is. There is massive reticulation in the form of gene flow through hybridization and introgression between species. While some have attempted to minimize the problem, we now know that even in the best-behaved groups (plants and animals) hybrids regularly form. Mallet (2005) surveys a variety of studies on hybrids and concludes that at least 25 percent of plant and 10 percent of animal species form hybrids with other species in nature. This usually leads to introgression and therefore gene flow between species. This problem is far worse with populations at the tips (now any migration is reticulation) or any kind of lineages.

Of course, like Darwin, we can allow that some hybridization is consistent with the tree. But how much reticulation is it reasonable to allow? This is a difficult question and can be realistically answered only in a context where we know what the purpose of the tree is. If the tree is supposed be allow us to make inferences about genetic history, similarity, biogeography, and other factors, then it is okay if it sometimes leads to errors – any possible model will do that – but it must have a good balance of simplicity, explanatory power, predictive power, and perhaps other less easily describable virtues. If systematists were aided in their research by using the tree, that would count in its favor. If they were positively misled, that would count against its use. Exactly how these have to be balanced against each other is a perennial question in the philosophy of science and one that is unlikely to have a general answer; rather, it needs to be examined carefully in the particular case at hand.

LATERAL GENE TRANSFER

The problem of reticulation might plausibly be thought to be manageable in eukaryotes, but when we generalize to all forms of reticulation, we face what is arguably the most serious problem for the tree: the phenomenon of lateral gene transfer. Lateral gene transfer (LGT), also called horizontal gene transfer, is the name for any instance of a variety of processes where genetic material moves from one organism to another by some process other than reproduction. This includes transformation, transduction, and conjugation.

It is now widely agreed that LGT has been, and still is, a major force in evolutionary history (Gogarten, Doolittle, and Lawrence. 2002; Dagan, Artzy-Randrup, and Martin 2008). The epistemological question of what can be inferred about genetic history is a serious one, given that genes do not in general track the same history and that, as we go deeper in time, any trace of signal may be lost. But the metaphysical question is serious as well – what could the tree be tracking, since clearly the history of *all* genes is not a single tree. It is not clear exactly what this means for the tree because different proposals about what the tree is will be affected differently. (For arguments that widespread LGT undermines the tree concept and possibly traditional phylogenetics as a whole, see Bapteste et al. 2004; Bapteste et al. 2005; and Bapteste and Boucher 2008.) Before looking at different responses to lateral transfer, we first consider more potential problems for the tree.

ENDOSYMBIOSIS

Another source of problems for the tree is endosymbiosis. In endosymbiosis, one organism comes to live inside another, and eventually its descendants become obligate symbiotes. Over evolutionary time, they reach the point where they are so tightly interconnected, often because of extensive LGT between host and symbiote, that it is appropriate to think of the host plus symbiote as one integrated organism. For example, most eukaryotic cells contain many mitochondria in the cytoplasm surrounding the nucleus of the cell. Mitochrondria are clearly functional parts of the cells today and are not organisms in their own right. But mitochondria have their own genomes, and it is now clear that historically they are closely related to various groups of the alpha proteobacteria. Likewise, the chloroplasts that give plants and other organisms such as some algae the ability to photosynthesize were once free-living cyanobacteria. A natural way to depict these genealogical relationships is with a fusion of lineages of very distant branches on the tree as in Figure 41.2. Endosymbiotic events have occurred a number of times in the history of life (Lane and Archibald 2008), but even with a strict understanding of "new lineage," if in some respects these events might be rare, they could hardly be more important. If there is any sense to be made of "key" events in evolutionary history, the origins of mitochondria and of chloroplasts surely count. Any purported universal phylogeny that fails to represent these events is lacking in a very important respect.

THE ROOT OF THE TREE

A major feature of the tree is its root. The root it typically thought to represent LUCA: the last universal common ancestor. Understanding the root is essential for studying the evolution of various ancient biological features, such as the genetic code, protein synthesis, cellular membranes, and, indeed, the cell itself.

As with the tree, different authors have a different conception of what would count as a LUCA, and different conceptions lead to different conclusions about its existence. In phylogenetics with trees, it is assumed that each descendant node gets its traits through common descent with modification. Thus allowing for mutational or other changes, the genes present in organisms today would have to have their ancestors in LUCA. But if LUCA is a single organism with a single genome, this leads to the absurd conclusion that LUCA contained genes for nearly all types of biochemical reactions known in bacteria and archaea today and had a genome larger than any known prokaryotic genome today. This is what Doolittle et al. (2003) termed "the genome of Eden." Such an entity surely never existed.

Different genes have genealogical histories that coalesce in the past at vastly different times. In describing his view of early life, Woese (1998, 6858) says, "The universal ancestor is not an entity, not a thing. It is a process characteristic of a particular evolutionary stage," from which he believes multiple

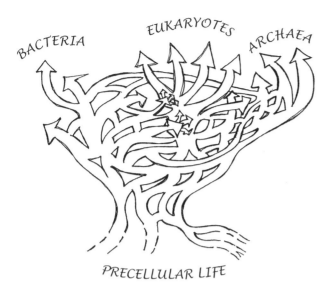

FIGURE 41.3. Does the existence of lateral gene transfer, where genetic information can hop from one branch directly to another, make the traditional tree of life otiose? Was Buffon on the right track after all? W. Ford Doolittle sees more of a net in life's history than a conventional tree. Permission: W. F. Doolittle

communities and independent lineages emerged. Theobald (2010, 220) argues for universal common ancestry and the existence of LUCA but clarifies what he means, saying, "Rather, the last universal common ancestor may have comprised a population of organisms with different genotypes that lived in different places at different times." While Doolittle and others do not consider this "population" dispersed in space and time worthy of being called an "ancestor" of anything, even granting that we should call it LUCA, it is clear that it would be inappropriate to depict this as a single node (the root) on the tree of life. If we attempted to use this tree as we would any other phylogenetic tree (say the tree of primates), we would be led to make mistaken inferences about evolutionary history. As the critics would say, a tree without a root is no tree at all.

If we attempt to represent lateral gene transfer, endosymbiosis, and the base of the tree all on the same diagram, even a very conservative picture will look something like Figure 41.3, which has been drawn by W. Ford Doolittle (see also Doolittle 2000). This is certainly not a phylogenetic tree, but whether that is an essential feature of anything appropriately called the tree of life is not clear.

SAVING THE TREE?

Given the multitude of problems, clearly defenders of the tree must deny that the tree represents the history of species, is universal, and is fully branching with no reticulations at all. But given that is not the case, it is no longer exactly clear what the tree is. A first pass might be as simple as saying that the tree is an idealization that is not perfect but still gets it mostly right and is extremely useful. This view makes the existence of the tree dependent on things like the extent of actual reticulation.

I would guess that this is the most common view of practicing biologists – especially those who work on eukaryotes. But the problems are serious, and the extent of reticulation, especially in prokaryotes, far too great to simply idealize away.

But what of those who work explicitly on reconstructing the tree to include prokaryotes? What exactly are they reconstructing? Not a phylogenetic tree that just idealizes out a few reticulations. Here, different, incompatible views of the tree have been proposed.

One idea might be that, as a practical matter, we simply need some kind of reference tree on which to base our classifications and to locate clades so that we can make sense of such things as lateral gene transfer in the first place. The 16S SSU rRNA gene is often used this way. We know that many gene histories disagree. Why not just pick one for a reference tree? For a great many taxa, we have sequenced the 16S gene, and so it has become the default classification tool for prokaryotes. To say that something is an alpha-proteobacteria or a haloarchaen is just to say that its RNA sequence fits in at a certain place in the universal 16S tree. While this may be practical for classification, it provides no defense of anything like the Darwinian hypothesis that there is a unique tree. Deciding which tree is *the* tree cannot depend on conventional choices by us. A plausible tree could be one of two things – a tree that is as reliable for phylogenetic inferences as possible or a tree that represents the actual genealogical history of some kind or other but not the full genetic history.

Galtier and Daubin (2008) explicitly stick to the idea that the tree is a tree of species. On their view, LGT is not obviously a problem metaphysically, though it would be if it meant there were no species (Lawrence 2002). Similarly, if we are building a tree of organisms, it might seem that LGT is only an epistemological issue. Organisms come from other organisms. This history is often referred to as "the tree of cells" because it tracks the cellular history and not necessarily the history of the genes inside the cells. Cicarrelli et al. (2006) claim that the tree is a tree of organisms. Given this, it is easy to see why they simply remove from their data set genes that they have reason to believe have been transferred. If they do not represent the organism's genealogy, why include these misleading data?

But even if there is some genuine tree of species or a tree of cells, this leaves open just how useful it is to reconstruct it and raises a serious question as to what extent the tree could play in the foundational role it is sometimes claimed to play – which may be relevant to whether this object is properly called the tree of life. For example, as Galtier and Daubin (2008) themselves point out, their species tree may not be consistent with *any* single gene's history. Cicarrelli et al. (2006) are criticized by Dagan and Martin (2006) for producing a "tree of one percent" because their tree is based on only thirty-one genes, which are consistent with at most 1 percent of the typical prokaryotic genome of more than three thousand genes.

In order to preserve the idea that the tree is supposed to represent something like a dominant pattern, Koonin, Wolf, and Puigbo (2009) examine whether there is a "statistically significant trend" in the forest of life that represents all genetic history. They argue that there is and that this could plausibly be called a tree of life. Wu et al. (2009) construct a genome tree based on a concatenation of all the gene data they have and compare this to the known rRNA tree. Although they do not explicitly present it this way, one could reasonably say that the tree of life is this genome tree, which represents something like an average signal that may not be the actual signal of any particular gene.

If we wanted the tree to play the role of representing the history of all genes or genomes, then it is now clear that no such tree can play that role. Some have proposed alternate names and conceptualizations to play the "represent everything" role such as Koonin et al.'s "forest of life." Other suggestions include the "ring of life" (Rivera and Lake 2004), the "net of life" (Kunin et al. 2005), or a "web," "coral," or "potato of life" (Olendzenski and Gogarten 2009). If they are right, then it seems that the Darwin's hypothesis was wrong.

THE FUTURE OF THE TREE

So should we continue to talk of the tree of life and attempt to reconstruct it? There are two kinds of factors relevant to this question. One set of factors relies on the empirical facts. Just how common is the tree of life in LGT, and what kinds of patterns does it produce? What was the early evolution of life like? Was there some single universal common ancestor? What is the extent of hybridization and introgression between sexual species? The answers to each of these questions can tell us something about the utility of talk of the tree and the importance of reconstructing it.

But another equally important set of factors concerns questions about what the tree is supposed to represent, about how we do and ought to use the tree in biological inferences. Does defending the tree entail defending a particular history of life and perhaps even a particular view about what counts as life? Or does it mean defending a particular set of practices? Or perhaps defending the explanatory power and heuristic uses of a particular model? Depending on how these questions are answered, empirical facts like the extent and pattern of LGT may or may not dictate abandoning the tree.

Critics of the tree are certainly right that the tree has been used for many different and sometimes inconsistent purposes. Some practices, such as assuming that one gene will have the same broad-scale genealogy of another, are bad practices. A weak reading of the pluralism defended by pattern pluralists such as Doolittle and Bapteste (2007) is surely right: just as there are a multitude of evolutionary processes besides natural selection, there are a multitude of genealogical patterns besides the single tree pattern. But this is consistent with the tree being one of those patterns – and perhaps a very important one at that. But the critics of the tree want to claim something stronger – that the history and usage of the phrase the "tree of life" dictate that it is the unique pattern or at least a special kind of universal pattern. Further, any specific way of understanding the tree, such as the tree of cells, simply fails to have the power to play the role that the tree was supposed to play.

Is this stronger view correct? Phylogenetic trees really are of central importance in a variety of contexts; trees will continue to be built, and the phrase "tree of life" has a special kind of importance (appearing in both the Bible and Darwin). But it is now clear that different understandings of what the tree is supposed to be and how it can be used come apart, and so the phrase must be used more carefully and only in restricted contexts. What about the prospects for "universal" tree building? Research programs investigating questions about the origin of life, the genetic code, the cell, the eukaryotic cell, and the connections between apparently very disparate forms of life will continue. It is clear that this research will involve phylogenetic trees as well as patterns other than trees and processes other than vertical descent. What this research will uncover is unknown, but we can be certain that it will be a fascinating story of the deep evolutionary connections between all humans, the *Escherichia coli* in our guts, the archaea living in hydrothermal vents deep in the ocean, the roses in our gardens, and the penguins in the Antarctic. At least in this respect, we can surely claim that Darwin was right.

Essay 42

Sociobiology

Mark E. Borrello

THAT DARWIN'S CONTRIBUTIONS toward understanding the evolution of social behavior were significant is undeniable. In the *Origin of Species* there was detailed treatment of "social evolution," and this thinking led to much discussion, something given fresh impetus and further fuel in the *Descent of Man*. Darwin's chapters in *Descent* on "The Comparison of the Mental Powers of Man and the Lower Animals" and "On the Development of the Intellectual and Moral Faculties during Primeval and Civilised Times" led to a great deal of speculation and comment with regard to the possibility of social and perhaps moral instincts in lower animals. Yet, as we shall see, although Darwin explicitly engaged this issue, its significance in evolutionary thinking, especially regarding the evolution of behavior, waxed and waned.

SOCIAL INSECTS AND SOCIAL INSTINCTS

Passages from both the *Origin* and *Descent* illuminate Darwin's position with regard to selection acting on traits involved in social behavior. The following oft-quoted passage, from chapter 3, "The Struggle for Existence," illustrates the breadth of action that Darwin (1859, 62) assigns to the struggle leading straight into the mechanism of natural selection: "I should premise that I use the term struggle for existence in a large and metaphorical sense, including dependence of one being on another and including (which is more important) not only the life of the individual, but success in leaving progeny." The problem is that (seemingly) success is at the individual level and the group gets overlooked and lost. This is no recipe for social behavior. Darwin's solution, however, involved the idea that selection could act at a level above the individual: a family, a colony, a social group, or a community. Later in the *Origin*, where Darwin is dealing mostly with the social insects, we see how this insight comes into play. Darwin recognizes the difficulty that the neuter insects with their distinct morphology and habits present to his theory and thus, in typical Darwinian style, he does his best to explain and diffuse this potentially devastating case.

> How the workers have been rendered sterile is a difficulty; but not much greater than that of any other striking modification of structure; for it can be shown

that some insects and other articulate animals in a state of nature occasionally become sterile; and if such insects had been social and it had been *profitable to the community* that a number should have been annually born capable of work, but incapable of procreation, I can see no very great difficulty in this being affected by natural selection. (236)

Notice incidentally, how (despite the lack of a clear hereditary theory) this passage illustrates Darwin's commitment to the mechanism of selection. Although Darwin was always committed to the idea of the inheritance of acquired characteristics (Lamarckism so called), he saw the sterility of the social insects as something demanding a selective explanation. One simply cannot pass on sterility directly. Again and again in the *Origin* we find passages about the evolution of various castes among the social insects, emphasizing the importance of selection.

> I believe that natural selection, by acting on the fertile parents, could form a species which should regularly produce neuters, either all of large size with one form of jaw, or all of small size having jaws of widely different structure; or lastly, and this is our climax of difficulty, one set of workers of one size and structure, and simultaneously another set of workers of a different size and structure; – a graduated series having been first formed, as in the case of the driver ant, and then the extreme forms, from being the most useful to the community, having been produced in greater and greater numbers through the natural selection of the parents which generated them; until none with an intermediate structure were produced.
>
> Thus as I believe, the wonderful fact of two distinctly defined castes of sterile workers existing in the same nest, both widely different from each other and from their parents, has originated. We *can see how useful their production may have been to a social community of insects,* on the same principle that the division of labour is useful to civilised man. (241–42, emphasis added).

Interestingly, the preceding passages come from the chapter on instinct in the *Origin* but make no explicit reference to the inheritance of instinct. In the case of the social insects, instinct is clearly recognized as an important factor in the evolution of the social systems. This idea is more carefully developed in *Descent of Man*. However, as the following quote shows, one can get some indication of Darwin's position with regard to instinct in the social insects in the *Origin*.

> Thus, I believe it has been with social insects: a slight modification of structure, or instinct correlated with the sterile conditions of certain members of the community has been *advantageous to the community*: consequently the fertile males and females of the same community flourished, and transmitted to their fertile offspring a tendency to produce sterile members having the same modification. (238, emphasis added)

The final part of the *Origin* pertinent to our inquiry comes in chapter 6, "Difficulties on Theory." In this chapter, Darwin discusses various phenomena that he recognizes as potentially contradictory to his theory. However, through his extended application of the mechanism of natural selection – that is to say, the extension of the mechanism to communities or social groups over and beyond just the individual – Darwin's theory can encompass even the most (apparently) self-destructive of adaptations: "[W]e can perhaps understand how it is that the use of the sting should so often cause the insect's own death: for if on the whole the power of stinging be *useful to the community*, it will fulfil all the requirements of natural selection, though it may cause the death of some few members" (202, emphasis added).

Moving on to *The Descent of Man*, we see Darwin shifting his emphasis from the social insects to the social instincts. Generally, he continues to use social insects for the model of the evolution of social instincts; however, he also includes the social behavior of primates and other higher animals. This shift in emphasis represents the increasing interest in the wider implications of Darwinian theory, especially as applied to our own species. Here, Darwin draws explicitly a connection between the moral faculties of man and the social instincts of the lower animals. This, we shall see, has direct relevance to later discussions of the evolution of social behavior.

Darwin's most straightforward presentation of the evolution of the social instincts comes in chapter 3, "Comparison of the Mental Powers of Man and the Lower Animals." In this passage, he argues that the inheritance of the social instincts is of the utmost importance to the later development of human society and furthermore that the development of these instincts is for the good of the community over and above the advantage of the individual. "Finally, the social instincts which no doubt were acquired by man, as by the lower animals, *for the good of the community*, will from the first have given him some wish to aid his fellows, and some feeling of sympathy" (Darwin 1871a, 1:103, emphasis added). In the following chapter, "On the Manner of Development of Man from Some Lower Form," reaffirming this kind of thinking, Darwin points out that in the case of the social animals, selection can act indirectly on the individual through higher-level selection. "With strictly social animals, natural selection sometimes acts indirectly on the individual, *through the preservation of variations which are beneficial only to the community*. A community including a large number of well-endowed individuals increases in number and is victorious over other and less well-endowed communities; although each separate member may gain no advantage over the other members of the same community" (1:155, emphasis added). Darwin goes on to illustrate the point with the example of the social insects, describing pollen-collecting behavior and the sting of worker bees, as well as the jaws of the soldier ants. These apparatuses and behaviors are of no direct advantage to the individual: rather, they serve the community and are maintained by natural selection acting on the level of the community.

Again and again in the *Descent*, one sees unambiguous evidence of the importance that Darwin assigned to the social instincts. "All this implies some degree of sympathy, fidelity and courage. Such social qualities, *the paramount importance of which to the lower animals is disputed by no one*, were no doubt acquired by the progenitors of man in a similar manner, namely, through natural selection, aided by inherited habit" (Darwin 1871a, 1:162, emphasis added). All of this leads to one clear conclusion. If these instincts are as important to the evolution of social groups as Darwin insists, and if the selection of these instincts often occurs at a level above that of the individual, then higher-level selection is an important factor in evolutionary theory. It plays a significant and indispensable role in the evolution of social behavior.

SOCIAL EVOLUTION

The last words of the passage just quoted – "aided by inherited habit" – show that even though Darwin was quite sure in his belief that natural selection gives the key to social behavior and its evolution, even he was not entirely convinced that other factors – Lamarckism in particular – had no role at all to play in evolution, including the evolution of social instincts and behavior. In *The Non-Darwinian Revolution: Reinterpreting a Historical Myth* (1988), Peter Bowler shows in great detail how complex a story is that of causal thinking about evolution in the late nineteenth and early twentieth centuries. Bowler's point – one that is well taken – is that most evolutionists in the half century or so after Darwin were, with respect to causes, not genuine Darwinians, despite their frequent claims to the contrary. In detail, Bowler describes the continuing influence of Lamarck's ideas about the importance of use and disuse; Haeckel's idea of recapitulation, which was closely linked to the idealist and transcendentalist origins of the developmental view of nature; speculations about jumps (or "saltations"); and a plethora of other scientific and philosophical concepts about change, all of which illustrate the intellectual heterogeneity that reigned during this period. Also important during this period was thinking about cultural or social evolution, what has come to be known as "social Darwinism," something that in truth was often closer to "social Lamarckism." Bowler makes the important point that if social thinkers wanted people to strive to get ahead, Darwinian theory gave them no grounds for making any effort. On the Darwinian account, either they had the advantageous traits or they didn't. On the other hand, the Lamarckian notion of inheritance of acquired characteristics made quite a bit more sense. This point applies especially to the philosopher and biologist Herbert Spencer. Traditionally he is presented (for instance, in Richard Hofstadter's classic *Social Darwinism in American Thought*) as an ultra social Darwinian. This is simply not true. As Bowler points out, and as is reaffirmed by other leading historians (notably Robert J. Richards in his definitive *Darwin and the Emergence of Evolutionary Theories of Mind and Behavior*), Spencer's ideas about evolution improving society were far more Lamarckian than Darwinian.

Evolutionary ideas were still very much in a formative stage at the turn of the century. And with regard to the evolution of social behavior, to say that the ideas were in their formative stages (and continued to be for many years after the turn of the century) is something of an overstatement of the facts. All of this is well illustrated, especially (what, for want of a better term, one might call) the "fluidity" of the thinking regarding social behavior, by J. Arthur Thomson's popular and influential book *Concerning Evolution*, published in 1925. Thomson, a professor of natural history at Aberdeen University (and the translator of August Weismann's work), derived this book from a series of lectures presented at Yale University the year previously. His object, according to the preface, was to show that the evolutionary view of nature and of man provided an enriching and encouraging account of the world and of human beings, contrary to popular understanding. Significantly, and in line with many others at the time, Thomson's thinking about evolution was far from exclusively Darwinian. In a section headed "Self-Regarding and Other-Regarding," Thomson (1925, 120) quoted Spencer on the importance of mutual aid. "As Herbert Spencer said: 'From the dawn of life altruism has been no less essential than egoism. Self-sacrifice is no less primordial than self-preservation.'"

Not that Thomson was unwilling to mix up Spencer with elements of Darwinian thinking. Throughout his book, Thomson emphasized the importance of what he called Darwin's "subtlety" with regard to the idea of the struggle for existence. He introduced the notion of sieves acting on different aspects of an organism and at different levels (i.e., a sieve of the quest for food, a sieve of the physical environment, a sieve of the animate environment, a sieve of courtship). Although these ideas are not developed into a systematic, theoretical structure – one certainly gets little sense that Thomson was sensitive to the issues that so concerned and absorbed Darwin – they indicate nevertheless Thomson's sympathy for the idea of selection acting at multiple levels in the evolution of social behavior. Thomson (1925, 141) made specific reference to selection acting at the level of society in his chapter on organic evolution: "Moreover, under the shelter of society there is a possibility of new departures which would be speedily eliminated by the sieves which apply to ordinary, more or less individualistic, life. At different levels of animal society there will be a different pattern of sieve." Clearly, altruistic behavior, which would be difficult to explain by a selective sieve operating at the individual level, could be effectively explained given another selective sieve operating at the level of the societal group. The altruistic group, having the higher fitness due to cooperative effort, would outlive the group of nonaltruist selfish individuals. (For a contemporary discussion of nested selective sieves, see Sober 1984, esp. 97–102.)

SUPERORGANISMS

In a lecture delivered at the Marine Biological Laboratory at Woods Hole in 1910, Harvard entomologist William Morton Wheeler made a compelling argument for the consideration of the ant colony as an organism. It is important to point out that Wheeler was not merely analogizing. He was arguing along lines hypothesized by the late nineteenth-century German evolutionist August Weismann, who had distinguished (in an anti-Lamarckian way) between the germ cells (the cells of heredity) and the somatic cells (the cells of the body). Wheeler presented the queen as the germ-plasm and the workers as the soma. He went on to stipulate, in the course of his address, that the division of labor among the two classes of the nutritive-worker division and the protective-soldier division clearly resembled the differentiation of the personal soma into entodermal (interior) and ectodermal (exterior) tissues.

Wheeler (1911, 325) concluded the paper with the assertion that we must pay closer attention to the innumerable cases of symbiosis (organisms working together for mutual benefit), parasitism (one group of organisms exploiting other organisms), and coenobiosis (or xenobiosis, organisms working together for their own ends) to explain the evolution of social behavior: "Since in all of these phenomena our attention is arrested not so much by the struggle for existence, which used to be painted in such lurid colors, as by the ability of the organism to temporize and compromise with other organisms, to inhibit certain activities of the aequipotential unit in the interests of the unit itself and of other organisms; in a word, to secure survival through a kind of egoistic altruism." (For a nice philosophical analysis of this kind of thinking, see Mitchell 1995; for an updated version of Wheeler's position, see Hölldobler and Wilson 2008)

This interest in the colonial organism was not Wheeler's alone. Wheeler mentioned a two-volume work by Driesch on the *Philosophy of the Organism* (1908), whose vitalism he found particularly unscientific. Julian Huxley had written a short (Spencer-influenced) work on the subject *The Individual in the Animal Kingdom* (1912). Further, many of the general texts of the time explored the issue of individuality and sociality. David Starr Jordan and Vernon Kellogg's chapter on "Mutual Aid and Communal Life among Animals" in *Evolution and Animal Life* (1908), included discussions of colonial organisms such as the Portuguese man-of-war and asexually reproducing animals such as the hydra. In the *Principles of Biology*, Spencer (1864, 250) admitted that individuality is problematic for the biologist; however, he advised, contra Wheeler, that we must "accord the title of individual to each separate aphis, each polype of a polypedom, each bud or shoot of a flowering plant, whether it detaches itself as a bulbi or remains attached as a branch."

As should be clear from the presentation of the panoply of opinion here, there was a wide range of ideas regarding the evolution of social behavior in the early twentieth century and a number of different ways that Darwin's theory was invoked (or, to be candid, ignored) in support of those ideas. Nevertheless, despite having been of interest to Darwin himself and many of his successors, for various still-not-fully-understood reasons (although surely not unconnected to the fact that so much of the thinking was so non-Darwinian), social behavior was not a focus of the mathematical population geneticists whose work formed the foundation of evolutionary thinking and research for the second half of the twentieth century. Though the resulting "modern synthesis" was meant to unify biology under the theoretical umbrella of neo-Darwinism, the evolution of behavior, for our purposes especially social behavior, was left to the side.

Indeed, as historian Richard Burkhardt (1992, 145) noted, "Julian Huxley, in 1925, believed that the time had come to gather data from 'field observation, animal psychology & behavior, genetics, and comparative psychology … [and consider] the problem [of behavior] from a truly broad & unitary biological standpoint.' He failed, however to carry through on the project. What is more, in the broad, synthetic book that he eventually did write, *Evolution, the Modern Synthesis*, he neither made behavior a part of the synthesis nor offered guidelines to suggest how that might be accomplished."

THE ETHOLOGISTS ON SOCIAL BEHAVIOR

Ethology is the mid-twentieth-century European field of inquiry built on ideas of the biological explanation of animal behavior. One might expect such an enterprise to have much to say about the evolution of social behavior. This was not so. The classical ethologists, led by Konrad Lorenz and Niko Tinbergen, were more focused on the evolution of behavioral instinct and confined their studies essentially to the behavior of individual organisms (Fig. 42.1). For Lorenz in particular, social behaviors were merely manifestations of the aggregate of myriad, individual-behavioral instincts. Showing his Continental training by focusing on morphological notions such as analogy and homology, he argued that they are as applicable to characters of behavior as they are in those of morphology. Lorenz (1974a, 233) also defended the deduction of function from behavioral analogies, arguing: "Since we know that the behavior patterns of geese and men cannot possibly be homologous – the last common ancestors of birds and mammals were lowest reptiles with minute brains and certainly incapable of any complicated social behavior – and since we know that the improbability of coincidental similarity can only be expressed in astronomical numbers, we know for certain that it was a more or less identical survival value which caused jealousy behavior to evolve in birds as well as in man."

It is clear from this just-quoted passage that Lorenz thought ethologists could provide an evolutionary account of social behavior. But what exactly was the relationship between ethology and evolutionary theory, especially Darwinian evolutionary theory? Critics complained that Lorenz provided only a proximate (or physiological) explanation for social behavior but left the ultimate (evolutionary) explanation aside. He would have disagreed, being satisfied that his account of the evolution of individual instincts could account for the social behaviors; but as historian R. W. Burkhardt (1983, 436–37) points out: "The relations between ethology and evolutionary theory are not as straightforward as Lorenz might like … Lorenz in particular emphasized the study of behavior and the way in which special structures and behavior patterns had evolved in the service of intraspecific communication."

FIGURE 42.1. Konrad Lorenz (1903–89) was one of the twentieth-century leaders in the study of animal behavior (ethology). He was particularly well known for his discovery of the importance of imprinting, where early life experiences remain rooted in the adult. Permission: Photo Researchers

Lorenz interpreted behavior as a particulate trait useful for phylogentic analysis in the same way that morphological traits had been used. He emphasized the invariability of instinctive behavior and its mechanistic stimulation. In 1963, Lorenz's fellow ethologist Niko Tinbergen, in a justly celebrated paper, insisted that biological explanations must answer four questions – about function or adaptation, about history or phylogeny, about (proximate) causation, and about development or ontogeny. In these terms, Lorenz was interested in phylogeny and ignoring or downplaying function or adaptation. Moreover, Lorenz's preoccupation with ritualization, which, while accepting the virtues of aggression, looks for means to contain its vices, approached the challenge largely from the viewpoint of the individual. Even when he was dealing with groups, he was not interested in the traditional evolutionary issues about whether behavior is for the individual or the group. Aggression is a key element of each individual in Lorenz's community, and every individual is an isolated actor. In 1935 Lorenz (1970, 218) wrote that "such co-operation of individuals in a colony is based entirely upon instinctive behavior patterns, just as in the case of the insects, and is nowhere based upon the traditionally acquired behavior patterns or upon the insight that co-operation in furthering the colony is advantageous to the individual."

One scientist who did take social behavior seriously, famously (or perhaps notoriously) opting strongly for a view of natural selection working for the group even against the interests of the individual – stressing that selection leads to population homeostasis and the avoidance of true aggression – was the Oxford-trained ornithologist Vero Copner Wynne-Edwards (Fig. 42.2). But this was certainly no universal belief of the ethologists (i.e., those ethologists who grasped the issues at stake). Tinbergen in particular was skeptical of Wynne-Edward's argument for the evolution of social behavior by group selection. A look at Tinbergen's 1965 paper "Behavior and natural selection" shows his unease with Wynne-Edwards's interpretation of social behavior. Wynne-Edwards's book *Animal Dispersion in Relation to Social Behaviour* (1962), Tinbergen (1965, 536) writes, "contains two main theses: first, many animals have developed means, usually behavioral, of preventing overcrowding; second, many of these means are 'altruistic' – that is, beneficial to the population as a whole but not individuals – and as such can only be explained as consequences of group selection. While I believe Wynne-Edwards' first thesis to be sound – even though he seems to apply it to many phenomena that may well have other functions – his second thesis has the weakness of being based on negative evidence, on lack of analytical data." Tinbergen

FIGURE 42.2. Vero Copner Wynne-Edwards (1906–97) gave the classic statement on group selection, inspiring a generation of critical debate. Permission: © Mark Gerson / National Portrait Gallery, London

FIGURE 42.3. Richard Dawkins, the great popularizer of the gene's-eye view of Darwinian evolution through such works as *The Selfish Gene* (1976) and *The Blind Watchmaker* (1986). Permission: Photo by Lisa Lloyd

also argued that Wynne-Edwards's definition of "altruism" was overly broad. He suggested that if it could be shown by concrete analysis that such forms of social interaction could arise as the result of conventional natural selection, such a theory, though not of course disproved in principle, would lose the only type of support that Wynne-Edwards marshaled in its favor. And this is exactly what Tinbergen and others – notably Oxford ornithologist David Lack and American ichthyologist George C. Williams – proceeded to do. Indeed, Cambridge-based Robert Hinde (1985, 194), another ethologist, later recalled his experience at Oxford: "David Lack was no longer especially interested in behavior, but he taught me much about science, gave me a background interest in behavioral ecology, and made it difficult for me to ever think in other than individual selectionist terms." He continued, referring to later criticisms that the "view that all ethologists were then group selectionists is nonsense."

SOCIOBIOLOGY

The 1960s saw a strong surge in the study of the evolution of social behavior, a study that was firmly Darwinian in the sense of making adaptation the crucial object of inquiry and selection the key mechanism of change. Indeed, so strong was the growth that the field took on the new name of "sociobiology." It was firmly centered on an individualistic perspective, at least in part in reaction to what were taken to be the inadequacies of group thinking most particularly as manifested by Wynne-Edwards's *Animal Dispersion in Relation to Social Behavior*. One of the most significant developments contributing to the development of sociobiology was a paper by William D. Hamilton (1964a, 1964b) in which he pointed out that success in the struggle in genetical terms means simply doing better than others at increasing the proportion of one's genes in future generations (Plate XXXI). Almost paradoxically, one might do this by proxy as it were, through the success of one's relatives given the fact that they share (a proportion of) the same genes as oneself. If behavior can lead to a sibling having more than twice the number of offspring that a solitary individual could have, then selection will favor that behavior even if the individual does not reproduce at all. Although more recent reflection doubts that it is as effective as it seemed then, extremely influential was Hamilton's application of this insight to the sociality of the hymenoptera (ants, bees, and wasps) where, through atypical breeding practices, sisters are more closely related than mothers and daughters, thus apparently explaining how it can be in the interests of sterile workers to raise fertile sisters. (Because the interactions are between relatives, Hamilton's mechanism became known as "kin selection.")

Others working in the same vein included Williams, who penned the highly influential *Adaptation and Natural Selection* (1966); the English evolutionist (and student of J. B. S. Haldane) John Maynard Smith (1982), who applied game theory to problems of social behavior (and whose ideas were popularized by Richard Dawkins in the *Selfish Gene*, 1976); and in America at Harvard Robert Trivers (1971), who offered other mechanisms, including "reciprocal altruism," essentially you scratch my back and I'll scratch yours (Fig. 42.3). It was stressed that taking an individual perspective on selection had nothing to do with (human) social interests – it was not a replay of traditional social Darwinism – but followed from such things as the problem of cheating: if an organism helps others (in the language of sociobiology, shows "altruistic" behavior) without return and others do not (they "cheat" by taking and not giving), very quickly the altruist will be eliminated because the cheater will benefit from its own efforts and those of others. Moreover, empirical evidence backs this individualistic perspective. Generally, given that males do little work in raising offspring, populations can do with far fewer males than females, but numbers almost always tend to the equality of the sexes, simply because, if males become rare, then it is in the biological interests of individual parents to have male offspring, no matter what the group really needs.

All of this set the stage for Harvard-based Edward O. Wilson, the world's leading authority on ants, to a sizable volume titled *Sociobiology: The New Synthesis* (1975) (Plate XXXII). This work represented Wilson's attempt to understand all of animal behavior (ultimately including human behavior) in terms of evolutionary adaptiveness. *Sociobiology* created a storm of contention almost immediately. The controversy was perhaps the most publicly debated episode in the field of biology since the Scopes trial in Dayton, Tennessee, in 1925 (Segerstrale 2000). Although Wilson had many supporters, he also suffered a great deal of criticism from within the scientific community and without. Biologists criticized Wilson's methodology and adaptationist reasoning, philosophers of science challenged his attempt to "biologicize ethics," and sociologists, educators, feminists, and others deplored his apparent ignorance of the social and political implications of his work (see, for instance, Lewontin, Rose, and Kamin 1984; Kitcher 1985). In his autobiography Wilson wrote that the reviews of *Sociobiology* "whipsawed it with alternating praise and condemnation" (E. O. Wilson 1994, 330).

Along with recollections of the wide-ranging response to his theory, Wilson (1994, 330) also recalled something about his methodology: "In order to use models of population genetics as a more effective mode of elementary analysis, I conjectured that there might be single, still unidentified genes affecting aggression, altruism, and other behaviors." Wilson's sympathy for the gene's-eye view is quite apparent in this passage. His commitment to genic (i.e., gene-based individual) selection was such that he rejected most of Wynne-Edwards's thinking. Wilson (1975, 110) argued that "one after another of Wynne-Edwards' propositions about specific 'conventions' and epideictic displays were knocked down on evidential grounds or *at least matched with competing hypotheses of equal plausibility drawn from models of individual selection*" (emphasis added).

Wynne-Edwards's reactions to *Sociobiology* were predictable. He had found a neo-Darwinian willing to address selection theory in broader applications. But thanks to Wilson's commitment to explanations of social behavior in terms of kin-selection theory (and like mechanisms), group selection continued to be belittled. This Wynne-Edwards (1982, 1096) considered a fundamental defect of sociobiology as then conceived: "[Wilson] attaches no significance to conventional competition or conserving resources, and not much to population regulation. His key properties of social existence, including cohesiveness, altruism, and cooperativeness, are sufficiently imprecise for him to suggest that some of the primitive colonial invertebrates, such as corals and siphonophores, come the closest of all animals to producing perfect societies." Regretfully, "lacking a valid definition of society, Wilson's synthesis sometimes sheds more confusion than light."

At the time, most people thought that Wynne-Edwards was completely off track; and, to be frank, today most people continue to think just that. In the past thirty years, sociobiology, by that name or some alternative – behavioral ecology is popular for the animal world and evolutionary psychology for the human – has proved to be one of the most exciting and fruitful areas of evolutionary biology. It is also firmly Darwinian, based through and through on selection arguments, generally stressing the virtues of social behavior for the individual. But, most fascinatingly, E. O. Wilson of all people has recently revised his position with respect to group selection. In a recent paper with his longtime collaborator Bert Hölldobler, Wilson now argues that "group selection is the strong binding force in eusocial evolution" (E. O. Wilson and Hölldobler 2005). Indeed, even more recently Wilson has coauthored a paper with David Sloan Wilson, another enthusiast for group selection, going even farther down this path. In "Rethinking the Theoretical Foundation of Sociobiology," the authors say flatly: "Current sociobiology is in disarray.... Part of the problem," they continue, "is a reluctance to revisit the pivotal events that took place during the 1960s, including the rejection of group selection and the development of alternative theoretical frameworks to explain the evolution of cooperative and altruistic behaviors" (D. S. Wilson and Wilson 2007).

In a way, one might have expected that there would be this resurgence of more inclusive forms of selection. After all, kin selection itself puts an emphasis on community beyond the single individual, so perhaps it was just a matter of time before forms of group thinking would reemerge – and Darwin is cited as an authority for so doing! In Wilson's own case, perhaps he is simply showing the power of phylogeny, for he was the student of Frank Carpenter, who was in turn the student of William M. Wheeler. Be this as it may, what cannot be disputed is the living relevance of Darwin's own work in today's inquiries. (For more detail on the issues discussed in this essay, see Borrello 2010.)

ESSAY 43

Evolutionary Paleontology

David Sepkoski

MUCH OF THE PHYSICAL evidence for evolution comes from paleontology. Before the arrival of molecular genetics, fossils were just about the only evidence available that evolution had actually taken place, and some individual specimens have come to have iconic status for their role in confirming predictions of evolutionary theory (*Archaeopteryx*, Lucy, *Tiktaalik*, etc.) (Fig. 43.1 and Plate XXXIII). Darwin, of course, was very much interested in the fossil record, and indeed his geological and paleontological observations both during his *Beagle* voyage and afterward played a formative role in shaping his ideas about evolution (see Brinkman, Essay 4 in this volume). However, Darwin also worried a great deal about how the fossil evidence supported his theory of evolution; in *Origin*, he set aside an entire chapter to discuss the "imperfections" of the geological record, and in general it is fair to say he regarded the fossil record as a disappointment at best and a serious liability at worst. Darwin's assessment of the fossil record, then, cast a long shadow over the subsequent development of the professional discipline of paleontology.

Up until the time of the modern evolutionary synthesis in the 1940s, paleontology was generally regarded by biologists as a discipline suited mostly to the collection and description of empirical evidence – fossils – but not one that could make unique contributions to our understanding of the patterns and processes of evolution. In the mid-twentieth century, however, some paleontologists began to resist this "descriptive" label for their discipline and to promote an approach to the history of life and the fossil record that was explicitly theoretical and evolutionary. By the 1970s, this approach came to be known as "paleobiology," and today it is one of the central viewpoints in the discipline. Evolutionary paleontology or paleobiology attracted prominent adherents, such as the late Stephen Jay Gould, and became associated with signature theoretical innovations – for example, punctuated equilibria, "species selection," and a "hierarchical" view of macroevolution. One sign of the increasing acceptance of paleontology's contributions to evolutionary theory was the renowned geneticist John Maynard Smith's (1984, 402) comment, in 1984, that "the palaeontologists have too long been missing from the high table [of evolutionary biology]. Welcome back."

But the paleobiological approach to evolution has also been a source of conflict. One frequent criticism of paleontological theories like punctuated equilibria is

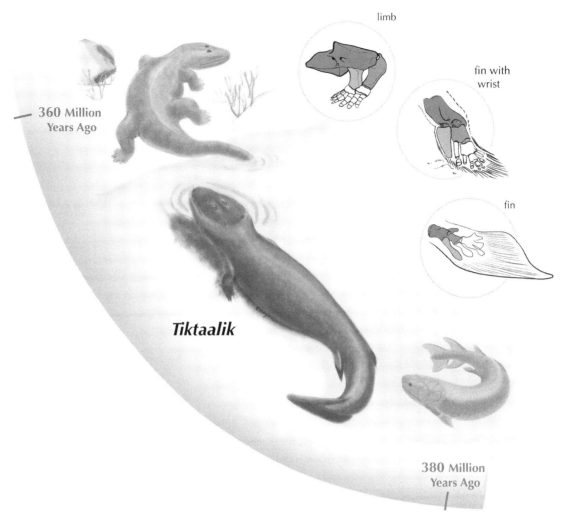

FIGURE 43.1. *Tiktaalik*, a "missing link" to rival *Archaeopteryx*. Its remains were discovered early in the twenty-first century; it is about 375 million years old and is the link between the water-dwelling fish and the land-based tetrapods (four-legged animals). Permission: University of Chicago, Shubin laboratory

that they contradict tenets of Darwinian evolutionary theory. The widespread – if often incorrect – impression among biologists was that paleontologists were attempting to dismantle Darwinism. During his career, Gould was frequently a lightning rod for this kind of criticism, and he often fanned the flames of controversy by apparently deliberately antagonizing evolutionary biologists. In one infamous example, Gould (1980, 120) wrote that the modern synthesis was "effectively dead, despite its persistence as textbook orthodoxy." In another case, Gould and coauthor Richard Lewontin (1979) charged evolutionary biologists with an overreliance on adaptation as a source of evolutionary novelty, which appeared to imply that Darwin's central mechanism of selection was in jeopardy.

This essay explores two related questions. In the first instance, it considers what influence, if any, Darwin had on the development of paleontological approaches to evolutionary theory. We briefly examine what Darwin had to say about the fossil record, and the effect this had on the professional development of paleontology in the late nineteenth and early twentieth centuries. We find, on the whole, that Darwin's influence was considerable, though somewhat negative. Second, we examine paleontologists' attempts to get out from under Darwin's shadow and to position paleontology as a source of theoretical insights into evolution. This process began around the time of the modern synthesis and has transformed the discipline of paleontology in important ways. The question we ask, however, is whether more recent paleontological (or paleobiological) theories of evolution genuinely break with the logic or substance of Darwinian evolutionary theory. In other words, is paleontology – or, at any rate, the theoretical paleontology practiced by scientists like Gould – still Darwinian?

THE PROBLEM: DARWIN'S DILEMMA

In the *Origin of Species*, Darwin (1859, 84) tells us that natural selection is a process that is "daily and hourly scrutinising, throughout the world, every variation, even the slightest;

rejecting that which is bad, preserving and adding up all that is good; silently and insensibly working, whenever and wherever opportunity offers, at the improvement of each organic being in relation to its organic and inorganic conditions of life." So slow is the process, indeed, that it is all but unobservable. According to Darwin, "we see nothing of these slow changes in progress, until the hand of time has marked the long lapse of ages, and then so imperfect is our view into long past geological ages, that we only see that the forms of life are now different from what they formerly were." It is only the geological record, in other words, that provides the resolution in which the grand sweep of evolution can be recognized; natural selection explains evolution as a process, while paleontology gives us a sense of evolution as a pattern.

Yet herein lies a dilemma: by Darwin's theory, the fossil record ought to show the slow, continuous intergradation of one species into another, over millions and millions of years. Because of this, transitional fossils – "missing links" – should abound, from fish to reptiles to birds to mammals. However, as Darwin was only too aware, the known fossil record in the mid-nineteenth century contained precious few of these transitional forms and offered very little support to the claim that species transformed very slowly and gradually. Rather, the fossil evidence recorded sharp discontinuities between lineages, and transitional forms were very rare. One of Darwin's greatest anxieties was that the incompleteness of the fossil record might be used as evidence against his theory and that the discontinuities in that record might even bolster arguments for the spontaneous, special creation of species promoted by natural theologians. As with all potentially negative evidence, Darwin addressed this concern in the *Origin*, where he spent a great deal of energy apologizing for the sorry state of the fossil record.

Because he was committed to the transformational view of evolution, Darwin's only recourse was to argue that the fossil record made sense only if we assume it is woefully incomplete. In a perfect world, we would see in the fossils a complete documentation of slight transition from one form to another; however, owing both to the erratic nature of geological preservation and to the limited investigation of geologists, many of those forms are missing. Chapter 9 of the *Origin*, "On the Imperfection of the Geological Record," deals squarely with this problem. According to Darwin (1859, 301), "we have no right to expect to find in our geological formations, an infinite number of those fine transitional forms, which on my theory assuredly have connected all the past and present species of the same group into one long and branching chain of life." Rather, "we ought only to look for a few links, some more closely, some more distantly related to each other; and these links, let them be ever so close, if found in different stages of the same formation, would, by most palæontologists, be ranked as distinct species" (301–2). In this manner, Darwin attempted to turn a liability into a virtue; nonetheless, he conceded, "I do not pretend that I should ever have suspected how poor a record of the mutations of life, the best preserved geological section presented, had not the difficulty of our not discovering innumerable transitional links between the species which appeared at the commencement and close of each formation, pressed so hardly on my theory" (302).

Famously, the metaphor Darwin selected to characterize the fossil record was that of a great series of books from which individual pages had been lost and were likely unrecoverable:

> For my part, following out Lyell's metaphor, I look at the natural geological record, as a history of the world imperfectly kept, and written in a changing dialect; of this history we possess the last volume alone, relating only to two or three countries. Of this volume, only here and there a short chapter has been preserved; and of each page, only here and there a few lines. Each word of the slowly-changing language, in which the history is supposed to be written, being more or less different in the interrupted succession of chapters, may represent the apparently abruptly changed forms of life, entombed in our consecutive, but widely separated formations. On this view, the difficulties above discussed are greatly diminished, or even disappear. (310–11)

An important point to make for the further development of evolutionary theory within paleontology is that Darwin's view of the incompleteness was not simply exculpatory. In other words, Darwin *needed* the fossil record to be incomplete in order to justify his view that evolution is a process of very gradual transformation. It is thus only after evolution came into the picture that the incompleteness of the fossil record became a significant issue. Darwin's theory revolutionized paleontology, because the fossil record became the only evidence to show that evolution had occurred and for demonstrating the pattern of life's history. Darwin's dilemma, however, was that he was embarrassed by paleontology as much as he needed it. Even while potentially elevating the status of paleontological evidence, he simultaneously undermined the discipline of paleontology, because he essentially predicted that the fossil record could never be sufficiently complete for paleontology to offer theoretical contributions to evolutionary theory on its own.

PALEONTOLOGY AFTER THE ORIGIN

In the several decades following publication of the *Origin*, Darwin's dismal assessment of the fossil evidence cast a long shadow over paleontology. Essentially, Darwin had left paleontologists with only three options with respect to evolutionary theory: they could put aside any theoretical ambitions and focus on purely descriptive studies of fossil morphology and stratigraphic placement; they could accept the Darwinian interpretation and try to simply locate some of those "few links" Darwin had predicted would help validate his theory; or they could reject Darwinian evolution entirely and instead focus on alternative theories of evolution that were more welcoming to a discontinuous fossil record. Most professional paleontologists, from the 1880s to the 1930s or 1940s, tended toward some version of option 1, which meant treating their empirical work as essentially agnostic toward evolutionary theory. This attitude became even more entrenched in the early twentieth century, when more and more paleontologists were employed by a petroleum industry that valued paleontology simply as a means to locate oil (Rainger 2001). The least popular choice was the second option, although some of Darwin's closest supporters – including T. H. Huxley – actively incorporated fossil evidence into defending Darwinism.

However, for paleontologists with active theoretical ambitions, the third option proved quite attractive. Indeed, non-Darwinian theories of evolution such as Lamarckism and orthogenesis were championed by many prominent paleontologists of the early twentieth century, including O. C. Marsh, Edward Drinker Cope, Henry Fairfield Osborn, William Diller Matthew, George Mivart, Othenio Abel, Louis Dollo, K. A. von Zittel, and Otto Schindewolf. These non-Darwinian theories were especially popular among paleontologists in the United States and Germany, although for different reasons. In any case, theories such as Lamarckism and orthogenesis promoted the view that evolution was driven not by selection but rather by some internal directing force. This idea appealed to paleontologists because many well-documented fossil lineages – particularly among large vertebrate animals such as mammals or dinosaurs – appear to exhibit fairly linear trends that (it was assumed) could not be accounted for purely by natural selection.

While it is not accurate to conclude that all paleontologists rejected Darwinian evolution or that paleontology had become a completely theoretically sterile discipline, it is fair to say that by the early 1900s the status of paleontology within evolutionary biology was marginal. This would prove especially costly for paleontologists as, during the first decades of the twentieth century, Darwinism began to gain traction among biologists thanks to the emergence of population genetics. In addition to having little empirical data to contribute to this enterprise, paleontologists were also often isolated in geology and museum collections departments, where they had little regular interaction with experimental biologists. Additionally, because some paleontologists had flirted with explicitly non-Darwinian evolutionary mechanisms like Lamarckism and orthogenesis, paleontological theories of evolution were viewed with some suspicion by Darwinian-minded biologists. This suspicion undoubtedly contributed to a lasting impression that paleontologists were all too comfortable entertaining heterodox views about evolution. In the ensuing modern evolutionary synthesis, paleontology was in serious danger of being left out in the cold.

THE LEGACY OF G. G. SIMPSON

As the modern evolutionary synthesis emerged in the 1930s and 1940s under the guidance of biologists like Ernst Mayr and Theodosius Dobzhansky, a more welcoming environment for paleontology prevailed. For example, the great American vertebrate paleontologist George Gaylord Simpson become one of the major framers of the synthetic view and took a leading role in organizing the institutional framework for the modern synthesis (such as the Society for the Study of Evolution). Even so, it was still fairly clear that paleontologists were expected to follow the party line, which elevated certain practices (experimental genetics, population biology) and mechanisms (natural selection, genetic drift) and excluded others (orthogenesis, saltationism).

Despite being a major framer of the synthesis, however, Simpson was also one of the most aggressive early advocates for paleontology's autonomous theoretical role within evolutionary biology. In 1944 Simpson published a slim volume titled *Tempo and Mode in Evolution*, which was essentially a manifesto for a new kind of theoretical, quantitative approach to interpreting the fossil record. Simpson's approach was innovative, but it was also inspired by contemporary innovations in biology and genetics that had contributed to the synthesis. In particular, Dobzhansky's *Genetics and the Origin of Species* had a deep influence on Simpson's thought: "The book profoundly changed my whole outlook and started me thinking more definitively along the lines of an explanatory (causal) synthesis and less exclusively along lines more nearly traditional in paleontology." It not only "opened a whole new vista to me of really explaining the things that one could see going on in the fossil record and also by study of recent animals" but also allowed Simpson to relate his own paleontological work to the exciting new developments in genetics (Simpson, quoted in Mayr 1980a, 456). Simpson's reading of Dobzhansky especially encouraged him to think about the history of life in terms of the genetics of populations of once-living organisms, and the major argument of *Tempo and Mode* is that what happens on the Darwinian population level both explains and is explained by transformations in the fossil record. Paleontology, in other words, could be used for exploring the mechanisms that drive evolution, and not just for documenting the physical historical record itself. Simpson's great insight was that paleontology's major claim for importance and autonomy within evolutionary biology was the added dimension of time: he described *Tempo and Mode* as a work in "four dimensional" biology and emphasized that the temporal (or historical) dimension of paleontology offered a unique and critical perspective to evolutionary theory.

One of Simpson's most novel and exciting proposals was that the fossil record has something unique to say about macroevolution, or the large scale patterns in the history of life. According to the view propounded by the modern synthesis, major evolutionary patterns are simply the extrapolated effects of microevolution, or the dynamics of natural selection in individual populations. Simpson broke with this view by suggesting that evolution operated at three causally related but distinct tiers. Microevolution was the process that explained the evolution of populations via natural selection and genetic drift. This tended to be the *only* level of evolution recognized by synthetic biologists, but Simpson added two more: the second, macroevolution, showed how microevolution accumulated to produce broader patterns of evolutionary change, much as the synthesis proposed. However, as a paleontologist,

Simpson was impressed by the fact that major faunal transitions (such as the appearance of entirely new species) often appear very abruptly in the fossil record. Rather than dismissing this as an artifact of the incompleteness of that record, he accepted sudden change as a valid evolutionary phenomenon and created a third tier, which he termed "mega evolution," wherein major taxonomic transformations take place. To explain these abrupt transitions, Simpson invented a process he called "quantum evolution," which described how in small, isolated populations evolution could take place so quickly that transitions did not show up on the fossil record. While he did not suggest that anything other than the standard Darwinian mechanisms of mutation and natural selection were required to produce quantum evolution, he nonetheless emphasized that this sudden change might constitute an independent evolutionary process.

In addition to providing a rationale and a rallying cry for more theoretical paleontological studies of evolution, the other great contribution *Tempo and Mode* made was to introduce greater quantitative rigor into paleontology. Fairly or unfairly, the dismissive attitude shown by biologists towards paleontology had much to do with paleontology's very descriptive orientation; "real" science, according to an old prejudice, is *quantitative*, and in that department the paleontology of the late nineteenth and early twentieth centuries simply was not sufficient. In just 217 pages of text, *Tempo and Mode* helped to change all that by introducing mathematical models and techniques for understanding the population genetics of fossils that mirrored the sophistication of those used in the study of living populations. Simpson's effect on future generations of paleontologists cannot be overstated. Although he toned down some of the more radical proposals (such as quantum evolution) in later years, Simpson's work remained an inspiration for decades to paleontologists interested in mining the fossil record for unique insights into the patterns and processes of evolution. Simpson also, through his contributions to the formation of the Society for the Study of Evolution and its journal *Evolution*, helped give paleontologists a provisional seat at what Maynard Smith later termed the "high table" of evolutionary theory. Much work would be left to his intellectual descendants, but in *Tempo and Mode* and other publications, Simpson showed what an independent, theoretically autonomous paleontology might look like.

PUNCTUATED EQUILIBRIA AND THE GROWTH OF PALEOBIOLOGY

By the 1960s, the Simpsonian approach to studying the fossil record had begun to catch on among younger paleontologists. Increasingly during this period, this approach was referred to by its proponents as "paleobiology," to distinguish it from the more traditional descriptive paleontological study of individual taxa and stratigraphy. Two such younger paleontologists were Niles Eldredge and Stephen Jay Gould, who had met in the mid-1960s as fellow students at Columbia University under the guidance of Norman Newell, an invertebrate paleontologist who was another important early advocate for paleontology (Fig. 43.2). As Gould (1989a, 118) later described it,

> Niles and I went to study with Newell because we were primarily interested in evolution – a direction that was, at the time, still a rarity in palaeontology.... Now imagine the frustration of two hyperenthusiastic, idealistic, non-cynical, ambitious young men captivated with evolution, committed to its study in the detailed fossil record of lineages, and faced with the following situation: the traditional wisdom of the profession held (quite correctly) that the fossil record of most species showed stability (often for millions of years) following a geologically unresolvable origin. "Evolution," however, had long been restrictively defined as "insensibly graded sequences" – and such hardly existed. Niles and I had one advantage in combating this frustration. We had been well trained in the details of modern evolutionary theory.... We had long discussions about whether insights from evolutionary theory might break the impasse that traditional explanations for the fossil record had placed before our practical hopes – for why would one enter a field where intrinsic limitations upon evidence had wiped out nearly all traces of the phenomenon one wished to study.... Eventually we (primarily Niles) recognized that the standard theory of speciation – Mayr's allopatric or peripatric scheme (1954, 1963) – would not, in fact, yield insensibly graded fossil sequences when extrapolated into geologic time, but would produce just what we see: geologically unresolvable appearance followed by stasis.

What Gould was describing is the origin of one of the major theories of modern paleobiology – and one that would propel Gould and the movement to prominence: punctuated equilibria. This theory, which was based on ideas Eldredge had formulated in his dissertation and an earlier paper, essentially argued that the appearance of discontinuity in the fossil record was not due to incompleteness of data, as theorists from Darwin down to the modern synthesis had argued, but rather reflected a genuine evolutionary pattern. Here Eldredge and Gould resuscitated an idea similar to Simpson's "quantum evolution" to suggest that, in most cases, the appearance of new species was a process that took place very rapidly and was thus unlikely to leave much trace in the fossil record. These rapid episodes of speciation, they argued, "punctuated" long periods of "stasis" during which little evolutionary change accumulated. Eldredge and Gould (1972, 96–97) explicitly contrasted their view to the traditional assumption of "phyletic gradualism," which is essentially the slow, transformational view of evolution that was central to Darwin's Dilemma for the fossil record, and they presented their theory as a direct response to Darwin's claim that the fossil record was unreliable (Fig. 43.3):

> Many breaks in the fossil record are real; they express the way in which evolution occurs, not the fragments of an imperfect record. The sharp break in a local column

FIGURE 43.2. Stephen Jay Gould and Niles Eldredge celebrating the ninetieth birthday of their teacher Norman Newell. Permission: Niles Eldredge, photo by Gillian Newell

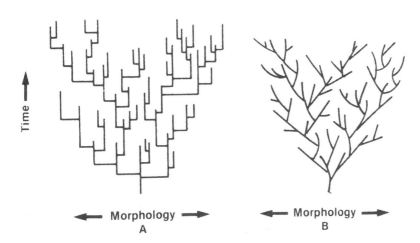

FIGURE 43.3. The different visions of life history. Punctuated equilibrium on the left; phyletic gradualism on the right. No evolutionary biologist wants to deny that there is something to both pictures, particularly when it is realized that a jump in the fossil record might take thousands of years. The question is, How predominant are the rival patterns?

accurately records what happened in that area through time. Acceptance of this point would release us from a self-imposed status of inferiority among the evolutionary sciences. The paleontologist's gut-reaction is to view almost any anomaly as an artifact imposed by our institutional millstone – an imperfect fossil record.... We suspect that this record is much better (or at least much richer in optimal cases) than tradition dictates.

The theory of punctuated equilibria would acquire iconic status in paleontology and, indeed, in the wider community of evolutionary biology. Almost from the very beginning, the theory was controversial. On an empirical level, many scientists – paleontologists and biologists both – questioned whether the pattern of stasis followed by rapid evolution that Eldredge and Gould described could be widely documented in actual lineages. This empirical debate has gone on since the publication of the first paper in 1972, and it is still lively. But the more controversial aspect of punctuated equilibria had to do with whether it presented an explicit challenge to Darwinism. In the first paper, Eldredge and Gould made it very clear that they were describing a *pattern* of evolution and not proposing a new evolutionary mechanism. Stasis, they argued, could be explained by the neo-Darwinian mechanism of "stabilizing selection," while periods of rapid evolution could be accounted for using Ernst Mayr's theory of "allopatric speciation," in which small, peripherally isolated populations undergo rapid genetic "revolutions." But in later presentations of the theory, Gould and Eldredge (and particularly Gould) often insinuated that punctuated equilibria was a more radical challenge to Darwinian evolutionary theory. In particular, many biologists suspected that Gould favored Richard Goldschmidt's discredited idea that evolutionary change was sometimes produced by major genetic "saltations" that produced "hopeful monsters." Despite his insistence that the pattern of punctuated equilibria could be produced solely by traditional Darwinian mechanisms, this suspicion haunted Gould throughout his career.

IS MODERN PALEONTOLOGY DARWINIAN?

As the example of punctuated equilibria demonstrates, there is a case to be made that some paleontologists – Gould perhaps most prominently – have been willing to explore interpretations of the fossil record that sit uneasily with the defenders of strict neo-Darwinism. This was especially the case during the 1970s and 1980s, when paleobiology experienced a phase of rapid expansion and heightened visibility as the result of aggressive self-promotion by a group of younger theoretical workers including Gould, David Raup, Thomas J. M. Schopf, and Steven Stanley. One result of this phase was the establishment of new institutional footholds for theoretical, evolutionary paleontology, including a new journal (*Paleobiology*, established in 1975), more university and museum appointments for paleobiologists, and greater attention for paleobiology within the wider community of evolutionary biology (Sepkoski 2009; Sepkoski and Ruse 2009). It may have been advantageous for paleobiologists to be aggressive in their theoretical interpretations of evolution during this period in order to emphasize the disciplinary autonomy of paleontology and to attract maximum attention for their cause (Sepkoski 2012). Examples of this approach include simulations of evolution as a random process (e.g., Raup et al. 1973), critiques of adaptationism in evolutionary explanations (Gould and Lewontin 1979), and proposals for a hierarchical theory of macroevolution (Gould 1985). This does not necessarily mean, however, that paleontology – or even paleobiology – was genuinely non-Darwinian (Fig. 43.4).

Some of the most committed supporters of paleobiology did, at times, certainly give the impression that they opposed the received view of Darwin as enshrined in the modern synthesis. Perhaps the most infamous example was Gould's 1980 essay "Is a New and General Theory of Evolution Emerging?" in which he is reputed to have pronounced the death of neo-Darwinism. Here is what Gould (1980, 120) actually wrote:

> I well remember how the synthetic theory beguiled me with its unifying power when I was a graduate student in the mid-1960's. Since then I have been watching it slowly unravel as a universal description of evolution. The molecular assault came first, followed quickly by renewed attention to unorthodox theories of speciation and by challenges at the level of macroevolution itself. I have been reluctant to admit it – since beguiling is often forever – but if Mayr's characterization of the synthetic theory is accurate, then that theory, as a general proposition, is effectively dead, despite its persistence as textbook orthodoxy.

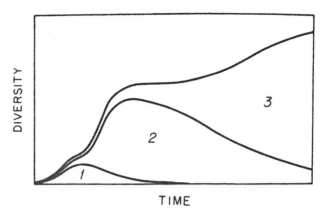

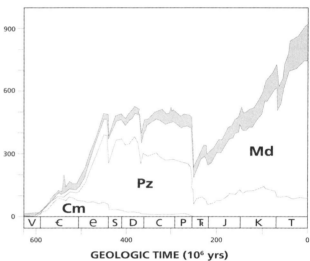

FIGURE 43.4. John J. Sepkoski Jr. (1948–99), one of a new breed of paleontologists as comfortable with a computer as with a pickax, made massive inventories of frequencies of taxa as revealed by the record, finding interesting patterns as particular types swelled up and then reached a plateau until a new type appeared. In one sense, this seems to invite a Spencerian interpretation in terms of dynamic equilibrium, but obviously the patterns can be given a Darwinian underpinning in terms of the conquering of new (empty) niches. Permission: Paleontological Society

Note, however, that Gould's target here is a particular *interpretation* of Darwinism, specifically Ernst Mayr's definition that "the proponents of the synthetic theory maintain that all evolution is due to the accumulation of small genetic changes, guided by natural selection, and that transspecific evolution is nothing but an extrapolation and magnification of the events that take place within populations and species" (Mayr 1963, 586, quoted in Gould 1980, 120). The argument Gould made here and throughout the rest of his career is that the traditional synthetic interpretation of evolution is insufficiently pluralistic to encapsulate the extraordinary causal complexity of evolution. Gould envisioned a "new Synthesis" that would preserve the central logic of Darwin's evolutionary theory but which would leave room for an expanding understanding of processes and interpretations that fall outside of the traditional mold of Darwinism. As Gould (2002, 1339) put it in his final, expansive treatment of evolutionary theory in 2002, his aim was "to expand and alter the premises of Darwinism, in order to build an enlarged and distinctive evolutionary theory that, while remaining in the tradition, and under the logic, of Darwinian argument, can also explain a wide range of macroevolutionary phenomena lying outside the explanatory power of extrapolated modes and mechanisms of microevolution."

It should also be stressed that despite the radical language used by Gould and others, much of the most important work of recent paleobiology has confirmed Darwin's central insights and expanded our understanding of evolution along fairly traditional Darwinian lines. For example, one of Darwin's greatest anxieties concerned the absence of a fossil record earlier than the Cambrian (some 550 million years ago), when complex life seemed to simply appear, as if from out of nothing. Paleobiology has definitively put that fear to rest by exposing a remarkable fossil record of evolution back to the earliest microbial stages of life, extending the fossil record some seven times longer – or almost two billion years – than was previously known (Schopf 2009). Paleobiology has also helped to settle questions about how the earliest complex life diversified and "exploded" in the Cambrian oceans to produce the lineages that are antecedent to the modern phyla (Gould 1989b), and time and again it has confirmed – in perfect accordance with Darwin's expectations – the fossil evidence of the major transitions in the evolutionary history of life (Shubin 2008).

The larger theoretical landscape of paleobiology allows for interpretive viewpoints that expand well beyond some of the basic premises Darwin laid down in *Origin of Species*, but the same may be said for evolutionary biology more generally. Paleobiology was no more a repudiation of Darwinism than is molecular genetics, or evo-devo, or any of the other countless developments in evolutionary biology that have come about since 1859. As other essays in this volume attest, modern evolutionary theory is remarkable both for the continuing validity of Darwin's original insights and for its adaptability to 150 years of continued investigation of evolutionary phenomena. Like all scientific theories, evolutionary theory adapts and evolves, and paleobiology has contributed to this ongoing process – and to a robust and pluralistic definition of what it means to be "Darwinian."

Essay 44

Darwin and Geography

David N. Livingstone

A GOOD DEAL OF CHARLES DARWIN'S endeavors were bound up with geography. First and foremost he followed in the steps of scientific geographer-travelers like Alexander von Humboldt when he embarked on his five-year round-the-world voyage on the *Beagle* in 1831. Later in life he recalled that during his final year at Cambridge he had "read with care and profound interest Humboldt's *Personal Narrative*," and he kept it constantly by his side throughout the *Beagle* mission, not just for its landscape evocations but for its scientific insights on subjects as diverse as polished syenitic rocks in the Orinoco, atmospheric conditions in the tropics, crocodile hibernation, connections between earthquakes and weather conditions, and miasmas in the torrid zone (Darwin 1958b, 67; Egerton 1970) (Fig. 44.1). Not surprisingly, shortly after his return home, he was elected to Fellowship of the Royal Geographical Society in 1838, having been nominated by the British diplomat, traveler, and geologist, Woodbine Parish. J. Stuart Wortley, agriculturalist and politician, and the geologist Charles Lyell, added their signatures in support (Fig. 44.2). Later in 1840 he served for a year on the society's council.

Beyond the external fabric of Darwin's life and institutional affiliations, of course, Darwin, Darwinism, and geography have intersected in numerous other significant ways. Here I propose to take three cuts at the subject: I turn first to the role of geography in Darwin's thinking and experience; then I trace something of the influence Darwin exerted on the geographical tradition; and finally, as an experiment with Darwinian resonance, I suggest that Darwinism itself might well be considered an intellectual species that endured different fortunes in different cultural environments and thus displayed its own geographical distribution.

DARWIN'S GEOGRAPHY: PUZZLES AND PLEASURES

Matters of geographical distribution were critical to Darwin's entire project, and he devoted two chapters to the subject in the *Origin of Species*. Here Darwin worked especially hard to undermine the theory of multiple creations of flora and fauna in specific regions of the globe. And while he did not name any particular proponent of such inflationary creationism, there can be little doubt that he had in mind the

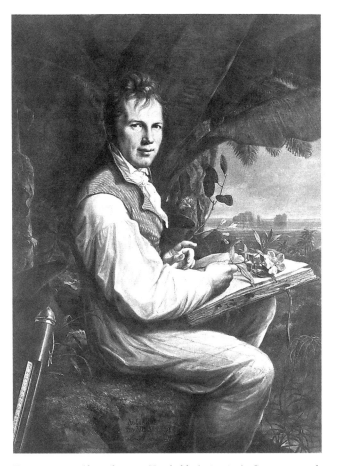

Figure 44.1. Alexander von Humboldt (1769–1859), German naturalist, whose writings about his travels in South America (1799–1804) greatly inspired the young Darwin, who took von Humboldt's work as a model for his own travel writing. Permission: Wellcome

Died 1882
Resigned
Removed

COPY OF CERTIFICATE OF CANDIDATE FOR ELECTION
from Registers and Council Minute Books.

Charles Darwin Esq. M.A. Trinity Coll. Cambridge
is desirous of becoming a Member of the Royal Geographical Society of London; and we the undersigned do recommend him, as likely to become a useful and valuable Member.

Woodbine Parish — *Signature of Member having personal knowledge of the Candidate.*

Cha Lyell — *Signature of one or more other Members.*

Date of Election 1838

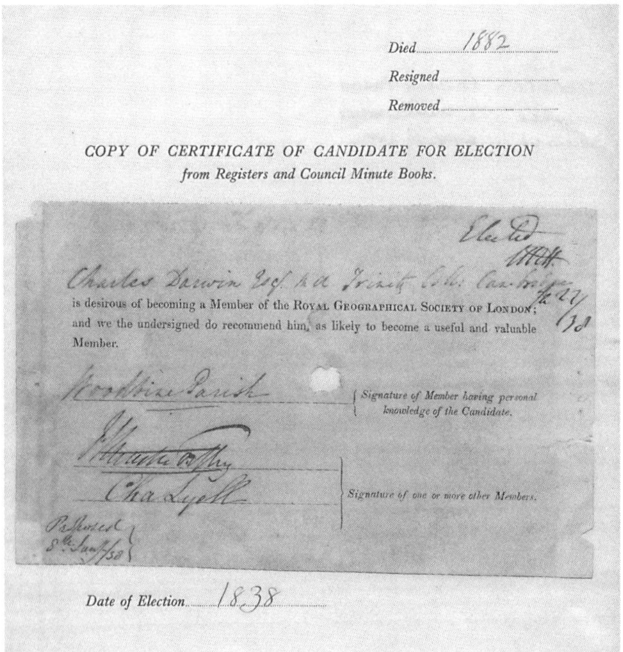

Figure 44.2. Darwin's Fellowship Certification for the Royal Geographical Society. Permission: Royal Geographical Society

interventions of Louis Agassiz. In 1850 Agassiz (1850a, 1850b) had taken up the question of animal and human geographical distribution arguing that there were distinct zoological provinces – an arctic, a European temperate, an African, a tropical Asiatic, and so on – in which the Creator had placed discrete species. Not only had animals and plants originated in centers of creation across the globe, but according to Agassiz (1850a, 193) they remained "within fixed bounds in their geographical distribution," and any blurring of their transcendental individuality was biologically – and, at least as far as human races were concerned, socially – repugnant.

Darwin's solution to the complexities of geographical distribution could not have been more different. Creationist expansionism held no appeal. And, indeed, multiplying acts of creation seemed to him to rub the facts of geography the wrong way. If "the same species can be produced at two different points," he pondered, "why do we not find a single mammal common to Europe and Australia or South America?" (Darwin 1859, 352). And if species were specially created to fit them to particular geographical provinces why was it that when "certain parts of South America" were compared "with the southern continents of the Old World, we see countries closely corresponding in all their physical conditions, but with their inhabitants utterly dissimilar"? (372). The processes of migration, descent with modification, adaptation to environment, and natural selection were more than sufficient to make sense of the seemingly intractable patterns of global biogeography.

Darwin began by noting that despite the "parallelism in the conditions of the Old and New World, how widely different are their living productions!" (347). Alongside this "first great fact" (346) of the geography of life was another equally striking geographical reality: "barriers of any kind" were "related in a close and important manner to the differences between the productions of various regions" (347). A third foundational feature of life's spatiality was "the affinity of the productions of the same continent or sea, though the species themselves are distinct at different points and stations" (349). In combination, these arrangements persuaded Darwin that the "deep organic bond" (350) linking living things together was inheritance. Inheritance, coupled with "modification through natural selection" (350), delivered a far more convincing account of the geography of life than the polygenist – multiple origins –hypercreationism that elaborated, calling in "the agency of a miracle" (352), numerous centers of independent creation across the face of the earth.

Darwin's theory was of the widest applicability. The puzzling peculiarities of remote island biogeographies, for example, with their scarcity of kinds, conspicuous absences of certain classes, affinities with their nearest mainland occupants, and the like, were only "explicable on the view of colonisation from the nearest and readiest source, together with the subsequent modification and better adaptation of the colonists to their new homes" (408). Of course, Darwin's theory required a persuasive account of the means of dispersal through which organic distribution could be effected, and he devoted much effort to establishing just what these might be – climate change, elevation and subsidence of landmasses, seaborne seeds, accidental transmission by birds, drift wood, icebergs, and so on. It all confirmed Darwin's conviction that the enigmas of plant and animal geography could be solved not by gratuitously resorting to separate acts of creation in different zoogeographical regions but "on the theory of migration ... together with subsequent modification and the multiplication of new forms" (408).

Of course, the puzzles of global geography had preoccupied Darwin long before *On the Origin of Species* saw the light of day. In an 1845 letter to Joseph Dalton Hooker, for instance, he had remarked that he looked forward to the day when Hooker would be acclaimed as the greatest European authority on "that grand subject, that almost key-stone of the laws of creation: Geographical Distribution" (Darwin 1985–, 3:140, letter to Hooker, 10 February 1845). And earlier again, in the late 1830s, once his theory of natural selection had begun to fall into shape, he "immediately began to use case studies from biogeography as part of the argument he would construct to defend the theory" (Bowler 2009, 155). To be sure, Darwin's thinking on particular dimensions of the subject, not least on dispersal and speciation, changed over the years (Browne 1983), but the fundamental importance of geographical distribution never waned: he allocated it sixty-five pages in the first edition of the *Origin*, which, as Bowler (2009) points out, constituted 13 percent of the entire work, and in later editions, save for minor adjustments, the chapters remained essentially unchanged.

Geography, of course, was not just a creative cause of puzzlement to Darwin; it was also a source of pleasure. For Darwin was a lover of the natural world. And, indeed, it was precisely because of Darwin's repeated resort to the language of beauty and wonder that Michael Ruse (2003, 335) was prompted to speak "of the genuine love and joy" that evolutionists sometimes sense in their encounters with the organic world. Darwin's narrative of the *Beagle* voyage is thus replete with glorious landscape evocations showing the emotional depth of his love of nature. Fashioned in conversation with Alexander von Humboldt's writings, Darwin (1845, 503) confessed that his own landscape sensibilities were derived from "the vivid descriptions in the *Personal Narrative* of Humboldt, which far exceed in merit anything else which I have read."

The entry in his *Journal of Researches* for 29 February 1832, at Salvador, Brazil, just the day after he set foot on the continent for the first time, already discloses his profound emotional involvement with plant geography:

> Delight ... is a weak term to express the feelings of a naturalist who, for the first time, has wandered by himself in a Brazilian forest. The elegance of the grasses, the novelty of the parasitical plants, the beauty of the flowers, the glossy green of the foliage, but above all the general luxuriance of the vegetation, filled me with admiration.... To a person fond of natural history, such a day as this brings with it a deeper pleasure than he can ever hope to experience again. (Darwin 1842c, 11)

By April he was so entranced by the forests around Rio de Janeiro that at a height of five or six hundred feet when "the landscape attains its most brilliant tint," he confessed that to the naturalist "every form, every shade, so completely surpasses in magnificence all that the European has ever beheld in his own country, that he knows not how to express his feelings" (32). And by the time his voyage was nearing its end, his intoxicated delight at the glories of tropical scenery in Bahia, Brazil, verged on the apophatic:

> When quietly walking along the shady pathways, and admiring each successive view, I wished to find language to express my ideas. Epithet after epithet was found too weak to convey to those who have not visited the intertropical regions the sensation of delight which the mind experiences.... In my last walk I stopped again and again to gaze on these beauties, and endeavoured to fix in my mind for ever an impression which at the time I knew sooner or later must fail. The form of the orange-tree, the cocoa-nut, the palm, the mango, the tree-fern, the banana, will remain clear and separate; but the thousand beauties which unite these into one perfect scene must fade away; yet they will leave, like a tale heard in childhood, a picture full of indistinct, but most beautiful figures. (496)

Such outpourings of emotional reaction to a sequence of ecstatic encounters with tropical geography, of course, did nothing to negate Darwin's pursuit of what might be called the analytics of landscape phenomenology. Because, for Darwin, a landscape "could become monotonous without scientific understanding of its different features" (K. Smith 2006, 80), he devoted some thought to the ways in which the discrete elements of landscapes merged into holistic visions. As he put it at the end of his journal: "I am strongly induced to believe that, as in music, the person who understands every note will ... more thoroughly enjoy the whole, so he who examines each part of a fine view, may also thoroughly comprehend the full and combined effect. Hence, a traveller should be a botanist, for in all views plants form the chief embellishment" (Darwin 1845, 502–3). Indeed there are hints here that Darwin was already working toward a scientific explanation of landscape appreciation. Contemplating the capacity of natural scenery to evoke strong emotion, he mused that just as "the love of the chase is an inherent delight in man – a relic of an instinctive passion," so "the pleasure of living in the open air, with the sky for a roof and the ground for a table, is part of the same feeling; it is the savage returning to his wild and native habits" (505). In years to come, such inklings would be used by others to provide an evolutionary account of landscape sensibilities. Jay Appleton (1975, vii), to take just one example, was sure that "pleasurable sensations in the experience of landscape" were directly related "to environmental conditions favourable to biological survival." If "a landscape 'component' appears beautiful, its beauty," he speculated "... derives from the contribution which it seems, *actually or symbolically*, to be capable of making to our chances of biological survival in the environment of which both we and it form a part" (Appleton 1975, 243; Dutton 2009).

For Darwin, these puzzles and pleasures were intimately interconnected. Elucidating biogeographical puzzles was a pleasure, and explaining landscape pleasure was a puzzle. Evolution made sense of both. Natural selection was the secret hand behind the mosaic of global distributions. And it was an intellectual delight to ascertain, as he put it in *The Origin of Species* (1859, 61), that "Natural Selection ... is as immeasurably superior to man's feeble efforts, as the works of Nature are to those of Art." At the same time the forces of selection explained the earliest stirrings of human appreciation of natural beauty. All in all, his theory deciphered problems and delivered delights: "There is grandeur in this view of life, with its several powers, having been originally breathed into a few forms or into one; and that, whilst this planet has gone cycling on according to the fixed law of gravity, from so simple a beginning endless forms most beautiful and most wonderful have been, and are being, evolved" (490).

DARWINIAN GEOGRAPHY: IMPACT AND INSPIRATION

According to David Stoddart (1966, 683), "Much of the geographical work of the past hundred years ... has either explicitly or implicitly taken its inspiration from biology, and in particular from Darwin." After all, as he explains, "Many of the original Darwinians, such as Hooker, Wallace, Huxley, Bates and Darwin himself, had been actively concerned with geographical exploration." In Darwin's own case, original work on physical geography included his influential account of the formation of coral reef systems (Darwin 1842; see also Stoddart 1976). Recognizing three types of reef – the fringing reef, the barrier reef, and the atoll – Darwin proposed that they constituted a developmental sequence on a submerging landmass, often a volcanic island, from fringing reefs via detached barrier reefs to coral lagoon islands (see Fig. 3.5). In this way Darwin inferred a temporal sequence from spatial arrangements – he deduced time from space (see Fig. 44.3). Fringing reefs were initially formed on coastlines in clean tropical seas; on stable coastlines, growth continues outward from the shore forming a barrier reef, while on submergent coastlines a circular atoll forms as the landmass falls beneath the waves. This evolutionary sequence was picked up by the American geomorphologist William Morris Davis (1850–1935) whose ideas about landscape evolution from youth, through maturity, to old age – the cycle of erosion, as he described it – resonated with Darwin's atoll hypothesis, which he strongly favored (Davis 1884; Chorley, Beckinsale, and Dunn 1973). The fact that he also referred to "inorganic natural selection" to explain aspects of landscape morphology further attests to his Darwinian sympathies (Davis 1895). Such considerations, rotating around denudation chronology and landform evolution, have often been taken as the locus classicus of Darwinian physical geography.

But Darwin's influence on geographical research cannot be restricted to geomorphological specifics like his thoughts on marine action, reef formation, glacial erosion, continental elevation, the role of earthworms in denudation, and the like (Kennedy 2006). It was also mediated through the evolutionary atmospherics that he did so much to invigorate. In point of fact, the impression that Darwin left on geography relied less on the impact of his own geographical findings than on the inspiration of his grand evolutionary narrative. At the time, the label "Darwinism" was widely taken to denote "evolution," and many enthusiasts for evolution unselfconsciously merged Darwinian and Lamarckian mechanisms. This was certainly the case with several of the founding figures of modern geography as a university discipline.

W. M. Davis, just mentioned, in some ways the father of professional American geography, is noteworthy in this regard. For while he directly registered the influence of Darwin's theory of atoll formation and was much enamored of Darwinian vocabulary, he was influenced in a more general way by the neo-Lamarckian ethos of Louis Agassiz's Harvard, where he studied with Nathaniel Southgate Shaler (Livingstone 1987). Ideas about the direct, heritable impress of the physical environment on organisms – particularly humans – gripped him, not least because it seemed to provide warrant for the fledging discipline of geography in the United States. His determinist-sounding reference to "the geographical relations of physiographic controls and ontographic responses," which marked out a territory that geography as an enterprise could inhabit, had clear Lamarckian resonance (D. W. Johnson 1909, 15). So too did his reference to organic "structures, processes, and habits" as "responses to physiographic causes" in a 1903 exposition of the "modern principle of evolution" (D. W. Johnson 1909, 52). As a student of Shaler, all of this is understandable, for Shaler (1893, 146) had categorically insisted that organisms "adapt themselves in an immediate manner to the peculiarities of their environment" and that those "conditions which surround them make an impression on their bodies which is transmitted to their progeny."

In Germany, the professionalization of academic geography owed much to Friedrich Ratzel (1844–1904), who likewise registered the force of Darwinian impulses (Livingstone 1992b). Having come under the influence of Moritz Wagner's migration theory, with its Lamarckian overtones, Ratzel developed an organismic concept of the state and used it to elaborate his theory of *Lebensraum* (living space). He had been exposed, too, to Ernst Haeckel's Darwinismus at Jena during the late 1860s, and his first publication, *The Nature and Development of the Organic World* (*Sein und Werden der organischen Welt*, 1869), was modeled very largely on Haeckel's *General Morphology*. Later, in 1881, when he laid out the foundations of human geography as a new discipline in the first volume of *Anthropogeographie*, it was the influence of Wagner that dominated. Wagner's insistence on the

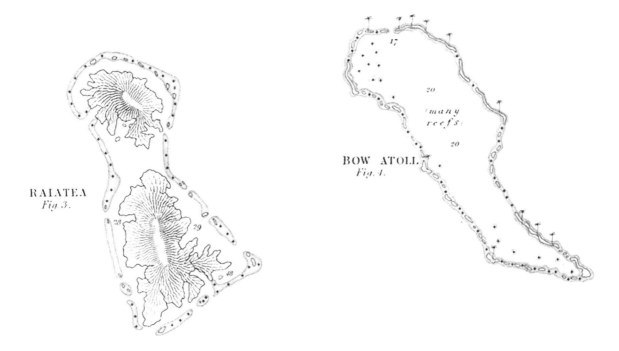

FIGURE 44.3. Coral reefs. From Darwin's *The Structure and Distribution of Coral Reefs* (London: Smith, Elder, 1842): "Shewing the Resemblance in Form between Barrier Coral-Reefs Surrounding Mountainous Islands, and Atolls or Lagoon-Islands."

importance of migration and geographical isolation in the processes of speciation and his emphasis on the formative influence of environmental circumstances provided Ratzel with a rationale for *Anthropogeographie*. In his hands, the Wagnerian principles of diffusion, migration, and *Raum* were woven together to provide a network of natural laws within which the spatial arrangements, cultural characteristics, and social functionings of human society could be understood. The naturalistic bias in this vision is certainly clear, for Ratzel squarely positioned the human species within the orbit of secular evolution, and this inclined him toward an environmental determinist account of human geography. It was out of this naturalistic ethos that his concept of *Lebensraum*, which he most fully articulated in the *Politische Geographie* of 1897, was born. Now he expanded on the biological analogy of the state as an organism that inevitably underwent demographic growth to the point where resource exhaustion or territorial expansion was inevitable. In expounding these Malthus-like principles, Ratzel believed he had disclosed the natural laws of the territorial growth of states, and he happily identified the contemporary colonial thrust of the European powers in Africa as the manifestation of their quest for "living space."

In many ways the British counterpart of Davis in the United States and Ratzel in Germany was Halford John Mackinder (1861–1947) – geography professor, scientific traveler, member of Parliament, British high commissioner, and university principal (Blouet 1987; Kearns 2009). It was his methodological *pronunciamento*, "On the Scope and Methods of Geography," at the Royal Geographical Society in 1887 that delivered a good deal of the intellectual rationale for the campaign to have the subject recognized as a discrete university discipline in Britain and earned him a readership in geography at the University of Oxford. For Mackinder, the subject's intellectual core lay in its capacity to keep nature and culture connected within a single explanatory system. Evolution provided the foundation. Mackinder's initial training had been in the biological sciences, and when he turned to human society, he acknowledged the inspiration of Walter Bagehot's *Physics and Politics: Thoughts on the Application of the Principles of Natural Selection and Inheritance to Political Society* (1872). The deterministic construal of nature's effects on human society that he found there provided Mackinder (1904, 421) not only with a causal, scientific, rationale for a discipline whose raison d'être was to trace "geographical causation in universal history" but also with a tool that could readily be mobilized for global geopolitics. Understanding the geographical forces shaping world history, he was convinced, was the first step toward managing an overseas empire and critical for the development of a patriotic geostrategic policy in the aftermath of the First World War.

Beyond these modern founding fathers and their canonical texts, the geographical tradition has registered the inspiration of Darwinian motifs in many different ways (Livingstone 1992b; N. Roberts, 2011). The Russian geographer, climatologist, and ichthyologist Lev Semyonovich Berg (1876–1950) developed his theory of "nomogenesis," which interpreted evolutionary change as due to inherent orderly processes fundamental to organic nature and which, by emphasizing mutations, allowed for the possibility of evolutionary "jumps." The *Animal Geography* (1913) of Marion Newbigin (1869–1934),

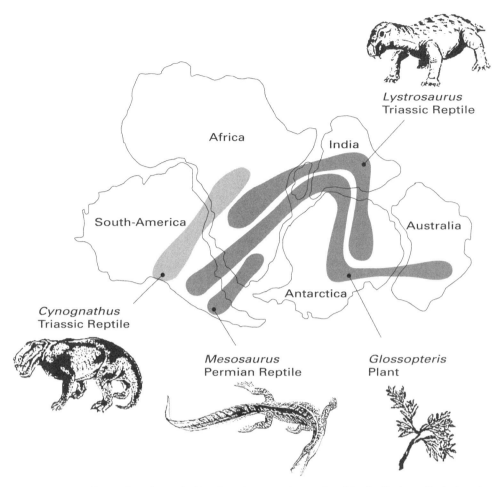

FIGURE 44.4. Highly significant for our thinking about the nature of the earth and the distributions of its denizens has been the relatively new theory of plate tectonics, seeing the earth as a machine, with the heat within causing great plates to emerge from the depths, move across the surface of the planet, before disappearing below.

geographer and biologist, displayed numerous moments of Darwinian inspiration. So too did the geographical anthropology of Herbert John Fleure (1877–1969), who was elected a Fellow of the Royal Society for his contributions to anthropometric cartography. For him it was the interplay of racial type, evolutionary mechanisms, anthropometric localization, and psychosocial factors that were of central importance to his conception of a humanized geography. Patrick Geddes (1854–1932), the Scottish botanist, town planner, and polymath, who had studied with Haeckel and influenced a generation of geographers, synthesized Darwinian and Lamarckian mechanisms in his writings on biological and urban evolution alike, and he used Spencerian ideas about social evolution to underwrite his reformist approach to city planning, museum culture, and environmental awareness. Across the Atlantic, the environmental determinist writings of Ellsworth Huntington (1876–1947) at Yale and Ellen Churchill Semple (1863–1932), who taught at the University of Chicago and Clark University, were also suffused with evolutionary motifs derived indiscriminately from Darwinian and neo-Lamarckian sources. Similarly the plant geographer Frederic Clements (1874–1945), who worked hard to highlight any hint of Lamarck in Darwin's own writings (Clements 1907), marshaled evolutionism in support of his scheme of vegetational succession and ecological climax even though he held to a belief in plant polygeny. In the case of the Australian geographer Griffith Taylor (1880–1963), it was the evolutionary significance of climate change that provided him with the scientific scaffolding around which he could construct his racialized account of paleoanthropology.

In more recent times, judgments about the lasting legacy of Darwin on the geographical tradition have been rather controverted. The remarkable absence of celebration within the discipline during the 2009 bicentenary of Darwin's birth has recently prompted some to wonder just how much of Darwin's influence continues to linger in the subject (Castree 2009). The silence, it has been suggested, might have to do with the racial and imperial tinge of the writings produced by many of the pioneer geographers referred to above, as well as to the outlawing, during the mid-twentieth century, of the environmental determinism they often espoused. Alongside this, a turn to other sources of geographical inspiration, from Marxist social theory to continental postmodernism, might also provide some explanation. At the same time, the value of Darwinian thought-forms to a range of recent research themes in geography – from the connections between Quaternary climate change and hominid evolution to the human impact of global warming, from the creation of cultural landscapes during the Holocene era to the challenges of biodiversity loss and the geopolitics of biotechnology, to name but a few (see Figs. 44.4 and 44.5) – seem as relevant now as they were in the period of modern geography's professionalization. Besides, as Finnegan (2010) tellingly notes, some of the fashionable thinkers to whom geographers have recently resorted – Derrida, Deleuze, Guattari, Foucault, Bhabha – owed much in different ways to their dialogue with Darwin's naturalistic biophilosophy.

THE GEOGRAPHY OF DARWINISM: DIFFUSION AND DIVERGENCE

As an exploratory coda to conclude this chapter, a self-referential application of geographical analysis to Darwinism itself – as a conceptual species – is surely in keeping with the Darwinian spirit. For the fact of the matter is that Darwin's theory, like any species, has fared very differently in different environments (Livingstone 2005). As it diffused, it diverged.

Among the Charleston naturalists in South Carolina, for example, Darwin's ideas about human origins and species transmutation were profoundly troubling to naturalists like John McCrady, who were dedicated to the idea of racial superiority. Like Louis Agassiz, he insisted that the different races constituted different species. Each race had a separate point of origin, and any blurring of its transcendental individuality was repulsive to both nature and culture. McCrady thus repeatedly insisted that it was simply impossible to conceive that the white and black races could have descended from the same origin. To him, and to other members of his scientific circle, Darwin's monogenetic theory of the origin of species was nothing less than a subversive threat to southern culture (Stephens 2000).

Half a world away in New Zealand, it was different. For here Darwin's theory was read as underwriting the runaway triumphs of colonialism. Just as the European rat, honey-bee, goat, and other invader species had displaced their New Zealand counterparts, Wellington audiences were told during the late 1860s, so the supposedly vigorous races of Europe were wiping out the Maori. It was a law of nature: inferior peoples disappeared when confronted with a superior race. Numerous spokesmen pushed this line. And it was precisely sentiments of this stripe that led John Stenhouse (1999, 81) to observe that "New Zealanders embraced Darwinism for racist purposes."

The very principle – struggle – that made Darwinian theory attractive to Wellington audiences was precisely what most perturbed the St. Petersburg naturalists in nineteenth-century Russia. Here, Karl Kessler instigated a tradition of research dedicated to identifying what he called mutual aid – cooperation – in evolutionary history. He was profoundly critical of Darwin's ideas about the struggle for existence and sought ways of reasserting the survival value of cooperation, not least in harsh environments. Later that idea was championed by Peter Kropotkin, who elaborated in detail on the role of mutual aid in biological and social history alike. This viewpoint, of course, fitted Russian collectivist ideology at the

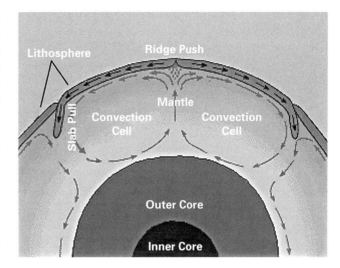

FIGURE 44.5. The idea of continental drift, that the continents move around the face of the earth, is an old one, but it was not until the creation of the causal theory of plate tectonics around 1960 that the idea could be generally accepted. At once, many of the anomalous distributions of fossil and living organisms are explained without need of imaginary land bridges or implausible methods of transport, becoming now great support for Darwinian thinking. From U.S. Geological Survey.

time – an ideology deeply critical of Thomas Malthus, whose atomistic conception of society had already been castigated as a cold, soulless, mechanistic product of English political economy (Todes 1989). In Russia, the St. Petersburg naturalists worked hard to extract Darwinism's Malthusian teeth.

The race question, which, in one form or another, shaped the way Darwin was read among the cognoscenti in both Wellington and Charleston, is conspicuous more by its absence in the debate over Darwinism in the South African *Cape Monthly Magazine* in the 1870s. Established to advance the virtues of intellectual enlightenment, social progress, and the spread of civilization in the Cape, it aspired to involve itself in the global scientific conversation (Dubow 2006). Perhaps for that reason, the early assessments of Darwin in its pages by figures like Langham Dale (archaeologist and superintendent-general of education), William Bisset Berry (Queenstown surgeon and politician), and William Porter (former attorney-general and first chancellor of the Cape University) were all marked by liberal sentiments. Support for the theory was judicious, criticism cautious. And the temperate tone that interlocutors adopted was entirely in keeping with the progressive, Enlightenment aspirations of the Cape's literati, whose eyes were firmly fixed on science's metropolitan horizon.

This thumbnail sketch, I hope, is sufficient to demonstrate that the fate of Darwin's theory was different in different locations. And in each place, Darwin and Darwinism were made to mean different things: from an assault on collectivism to a justification for colonial supremacy, from a subversive attack on racial segregation to a symbol of progressive enlightenment. How well the theory survived was crucially dependent on how well it fitted the prevailing environment. Surely Darwin could not have been surprised.

Essay 45

Darwin and the Finches

Frederick Rowe Davis

Like many other naturalists, Charles Darwin did not find the finches to be very interesting. During his five-week visit to the Galapagos Islands, Darwin saw many finches and collected some of them, but they were so different in outward appearance that he failed to recognize that they all came from the same family. Instead, he initially called one a finch, another a blackbird, and another a grosbeak. After his return to England, the ornithologist John Gould (1839), who analyzed and described Darwin's ornithological collection, convinced Darwin that the finches merited more interest. In the first edition of the *Voyage of the* Beagle (1839), Darwin noted the similarities among the finches (see Plate XXXIV).

The biological importance of the finches had made an impression on Darwin in the years since his brief encounter with them: "These birds are the most singular of any in the archipelago," but in most respects of form and function, they remained uninteresting. Nevertheless, the beaks of the various species did capture Darwin's (1839c) attention: "It is very remarkable that a nearly perfect gradation of structure in this one group can be traced in the form of the beak, from one exceeding in dimensions that of the largest gros-beak, to another differing but little from that of a warbler."

Darwin returned to the possible implications for natural history of this "nearly perfect gradation of structure" later in his account of the wildlife of the Galapagos, and he tentatively suggested that the different finches may have been confined to different islands; unfortunately, he failed to label his specimens to island (Sulloway 1982a, 1984). Darwin visited five islands of the Galapagos, but the captain and crew of HMS *Beagle* conducted hydrogeographic surveys of most of the islands, and Darwin drew upon their natural history collections also. From these collections, Darwin and the ornithologist Gould speculated that the finches differed from island to island.

DARWIN ON THE FINCHES

Several scholars have argued that other taxa played a much greater role in the development of Darwin's thinking on evolution, most notably the mockingbirds, which he labeled according to island, and the tortoises, which also varied from island

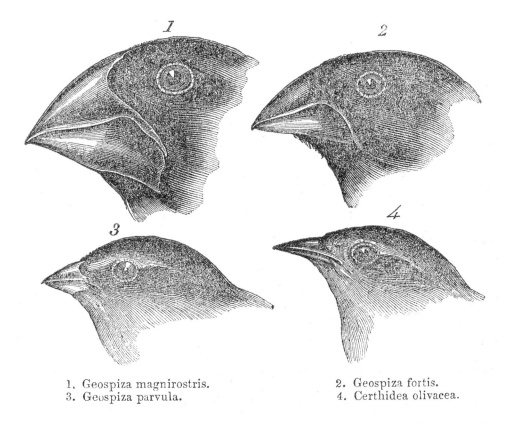

1. Geospiza magnirostris.
2. Geospiza fortis.
3. Geospiza parvula.
4. Certhidea olivacea.

FIGURE 45.1. Darwin's illustration of four of the finches of the Galapagos, showing the very different beaks, adaptations for different food stuffs. From C. Darwin, *Journal of Researches into the Natural History and Geology of the Countries Visited during the Voyage Round the World of H.M.S. "Beagle" under the Command of Captain Fitz Roy, R.A. Second Edition, Corrected, with Additions* (London: John Murray, 1845)

to island, according to respectable residents (Sulloway 1982a, 1984).

By the time he was preparing the second edition of the *Voyage of the* Beagle (1845), Darwin's thoughts regarding the finches had undergone considerable development. As in the first edition, he noted the similarities within the group, but he placed much stronger emphasis on the differences reflected in the beaks:

> The most curious fact is the perfect gradation in the size of the beaks in the different species of *Geospiza*, from one as large as that of a hawfinch to that of a chaffinch, and (if Mr. Gould is right in including his sub-group, *Certhidea*, in the main group), even to that of a warbler. The largest beak in the genus *Geospiza* is shown in Fig. 1, and the smallest in Fig. 3; but instead of there being only one intermediate species, with a beak of the size shown in Fig. 2, there are no less than six species with insensibly graduated beaks. (Darwin 1845, 379–80)

Moreover, Darwin included an illustration of four of the finches that showed the range of bill size from the largest to the smallest as well as two of the species with intermediate bills (Fig. 45.1).

Darwin (1845, 380) even speculated on the potential significance of the finches for biology: "Seeing this gradation and diversity of structure in one small, intimately related group of birds, one might really fancy that from an original paucity of birds in this archipelago, one species had been taken and modified for different ends." Such a statement strongly suggests that Darwin's view of evolution or transmutation had developed considerably during the six years between the first and second edition of *The Voyage of the* Beagle.

Frank Sulloway (1982b) developed a fine-grained analysis of Darwin's thoughts regarding evolution in the years following his return to England. Sulloway argued that Darwin certainly did not recognize the variability of finches while he was visiting the Galapagos. The different forms of mockingbirds (genus: *Nesomimus*) and tortoises, however, did strike Darwin as significant. More than a year after his voyage ended, when John Gould described the species in his collection, Darwin realized the potential importance of the different species of finches to his ideas regarding transmutation.

Sulloway identified a key passage in Darwin's *Ornithological Notes*, in which he drew upon evidence from the mockingbirds and tortoises to suggest the possibility of species variability. Sulloway found little evidence to support the myth that the finches played a significant role in Darwin's emerging ideas regarding the mutability of species, at least not until after Gould had described the species in his collection.

Sulloway's (1982a) exhaustive analysis of Darwin's writings and specimens also revealed that Darwin failed to record the islands of origination for the finches. Only one bird specimen, a vagrant bobolink (*Dolichonyx oryzivorus*), still has the original label written in Darwin's hand, a fact that Sulloway determined by examining all of Darwin's specimens held in the British Museum. Standard ornithological practice dictated the replacement of original labels with museum labels. No specific reference to island appears on the bobolink's label. Moreover, Darwin pointedly recommended that collectors number each specimen and that they note the specimen's collection locality in a master catalog, which would not be necessary if the tag included specific locality information.

When Gould shocked Darwin with his taxonomy of the finch species, Darwin endeavored to reconstruct the localities where he had collected his finch specimens. Collections made by Captain Fitzroy and Darwin's servant facilitated this effort. Nevertheless, Sulloway (1982a) determined Darwin's reconstruction to be imprecise at best, with his designations often amounting to little more than educated guesses based on other *Beagle* collections. Ironically, when Darwin published his suspected localities in the *Zoology of the* Beagle, they became inscribed in the ornithological record, as subsequent ornithologists at the British Museum relabeled some of the finches to conform to Darwin's conjectural localities.

Other than the rather provocative statements in the second edition of the *Voyage of the* Beagle, Darwin did not include the finches in his subsequent writings on evolution, or anywhere else for that matter. He mentioned the Galapagos Islands six times in the *Origin of Species* and drew specific attention to the evolutionary significance of islands and animal and plant colonizers, but not the finches (Sulloway 1982a, Wyhe 2012). Despite Darwin's reticence on the subject, other naturalists strove to fill the remaining gaps in the natural history of finches beginning with the taxonomy of the closely related group (Donohue 2011).

DARWIN'S FINCHES IN THE AGE OF SURVEYS

In the aftermath of Darwin's introduction of the finches to the community of scientists, a number of expeditions targeted Galapagos avifauna and the finches in particular. Darwin collected sparingly, returning to England with just thirty-one finches and sixty-four birds in total from Galapagos. Darwin's fairly limited objective was to obtain representative (or "type") specimens of each animal (and plant) he encountered. Subsequent expeditions sought to represent variations within and between finches.

In 1868, Habel collected 460 specimens, which Salvin described in 1876. The British Museum of Natural History became the repository of the collections of the *Beagle* and Habel. In 1897 the ornithologist Robert Ridgway described the collections of the *Albatross* (1888) and George Baur (1891). The *Albatross* collections remained at Ridgway's home institution, the United States National Museum (Smithsonian), while Baur's collection and that of the Webster-Harris expedition (3,075 specimens) became part of the Rothschild collection, which was eventually acquired by the American Museum of Natural History in New York. The Hopkins-Stanford expedition amassed another large collection that Robert E. Snodgrass and Edmund Heller described in 1904. But an expedition by the California Academy of Sciences obtained the largest collection (8,691 specimens) in 1905 and 1906. Harry S. Swarth published a monograph on this collection in 1931.

After enumerating the remarkable extent of finch collecting efforts, David Lack (1947) noted: "As a result of all these visits, Darwin's finches are more adequately represented by museum specimens than almost any other group of birds." Robert Kohler (2006) has reframed this period in the history of biology as the "Age of Survey."

Even before he visited the Galapagos, George Baur (1891) argued that the islands fell into one of two categories: continental and oceanic. He went on to argue that the fauna of continental islands was "harmonic" with the neighboring continents, while that of oceanic islands was "disharmonic." Through the lens of environmental action, Baur believed he could explain the origin of species and argued that the view of the "neo-Darwinians" had not received any support.

Drawing in part on Baur's collections from the Galapagos, the ornithologist Robert Ridgway (1897) described the many forms of birds from the Galapagos. Unlike Baur, who sought to confine the number of forms per island, Ridgway represented the other taxonomic extreme in designating numerous forms as full species. He expressed a degree of self-consciousness regarding his tendency toward splitting.

Ridgway acknowledged that analysis of additional specimens might cause his new names to be "degraded" but that such decision must result from experience rather than individual opinion. His conclusion left no doubt as to his taxonomic inclination: "I am sure that all who have had equal experience in the laborious and time consuming task of dissecting and reconstructing synonymies will bear me witness that the real promoter of chaos and enemy of order is the 'lumper,' and not his much maligned co-worker, the 'hair-splitter'" (1897, 467–68).

Nevertheless, Ridgway recognized that many of his designations of local or insular forms as species would not survive the analysis of additional specimens. He believed that previous taxonomists had been too conservative in designating species. In all, Ridgway added twenty-five full species to the list of Darwin's finches, including six in the genus *Certhidea*, twelve in *Geospiza*, and seven in *Camarhynchus*.

Unlike Baur, Ridgway generally hesitated to make broader claims regarding the evolutionary significance of the finches. Still, his approach to the finches annoyed Baur, who challenged him. In a departure from his more typical status as splitter, Ridgway reduced the genus *Cactornis* to a synonym of *Geospiza*. Noting that both *Cactornis* and *Geospiza* had different representatives on different islands, Baur (1897) argued that lumping the two genera was not natural. More problematic, in Baur's view, Ridgway arranged the different species of finches in a single line to show the gradual connection between the different forms. Baur organized the finches into several parallel series, arguing that species remained true on different islands and never intergraded on the same island, which was consistent with his essential point that natural selection played no role in the development of the finches.

Other naturalists did make general claims regarding the evolution of the finches during the Age of Surveys. In one of the first ecological studies of the finches, Robert E. Snodgrass and Edmund Heller preserved the stomachs of 209 specimens of *Geospiza* between December 1898 and June 1899 with a plan to determine whether bill size correlated with food type. Snodgrass hypothesized if the various sizes and shapes of bills within the *Geospizae* were adaptations, then it would be possible to differentiate diets (seed size) by species. The two naturalists separated the seeds by size (they did not attempt to determine the plants whence the seeds came), but found no patterns of consumption. Thus, Snodgrass (1902) concluded: "The evidence, then, seems to be in favor of the general conclusion that *there is no correlation between the food and the size and shape of the bill*. If this is true, then we must look elsewhere for an explanation of the variation of the *Geospiza* bill."

Despite ever-growing collections of finches, the taxonomy of the birds remained somewhat elusive. Ridgway elected to designate many of the insular forms as full species; moreover he divided the finches across four genera: *Geospiza*, *Platyspiza*, *Camarhynchus*, and *Certhidea*. Other taxonomists rejected Ridgway's designation and preferred to lump all of the finches into *Geospiza*. Noting the remarkable diversity of the finches, Harry S. Swarth (1929) of the California Academy of Sciences argued that the problem of classifying the finches lay at the level of the family. Rather than trying to force the finches into the New World finches (Fringillidae) or the New World warblers (Mniotiltidae), he included the finches in a family of their own, Geospizidae, which Swarth believed would accommodate Ridgway's several genera as well as intermediate forms.

Percy Lowe (1936) coined the term "Darwin's Finches" in an address to the British Ornithologist's Union in 1935 to mark the centenary of Darwin's visit to the Galapagos. On the basis of his anatomical studies of finches, Lowe challenged Swarth's contention that the finches should be placed in an independent family, arguing that Fringillidae was the appropriate designation for the finches. To explain the diversity of finches, Lowe proposed that the group represented a hybrid swarm, but he called for actual breeding experiments because Lord Rothschild had already suggested that if the extensive collections in existence could not provide the information, none would.

DAVID LACK ESTABLISHES A DARWINIAN ICON

By the 1930s, ornithologists had reached general consensus regarding the systematics of Darwin's finches. From December 1938 to April 1939, the Oxford ornithologist David Lack led an expedition to the Galapagos to study three aspects of the biology of the finches: breeding behavior, ecology, and hybridization. Because scientists had barely scratched the surface of the first two subjects, Lack felt the expedition was a success despite the failure of the captive breeding experiments that might have shed light on hybridization. After the expedition, which was supported by several grants from the Royal Society, the Zoological Society of London, and the Elmgrant Trustees, Lack spent the remainder of 1939 at the California Academy of Sciences, where he wrote his first detailed study of the finches.

Even though Lack divided the elements of his study of the finches into the three topics, his findings had bearing on the other elements of the biology of finches. For example, Lack rejected Swarth's justification for a new family for the finches, but he accepted Swarth's taxonomy of the finches in most respects. However, his careful study of the breeding behavior revealed that the finches were very similar to each other. Similarities in breeding behavior corroborated anatomical findings that suggested a close relationship between the species. Ecological studies revealed three aspects of finch natural history: the near absence of food competitors, almost complete absence of predators, and the existence of several islands that provide partial but not complete isolation for island forms. Each of these ecological factors influenced speciation. But Lack (1945, 135) concluded: "Differences between closely related species are nonadaptive except that bill characters serve in species recognition. The main genera show adaptive radiation." Lack (1940) also published a brief article on his findings in the journal *Nature*, and he again denied evolutionary agency. Sewall Wright's random drift provided a better explanatory framework.

With the title *Darwin's Finches*, David Lack established the name in the popular scientific imagination for the group of birds that scientists had come to call the subfamily Geospizinae. Yet Lack opened the book with a bleak introduction to the landscapes of the Galapagos. Nonetheless, Lack (1947) acknowledged the intellectual interest of the archipelago and then proceeded to enumerate the many unpleasant nuisances found on the islands including rats, fleas, ants, mosquitoes, and scorpions, not to mention dysentery and disagreeable human settlers and Indians.

If Lack (1947, 11) found the Galapagos dismal, his initial impressions of the finches were even less inspirational: "Darwin's finches are dull to look at, not only in their orderly ranks in museum trays, but also when they hop about the ground or perch in the trees of the Galapagos, making dull unmusical noises." And yet, in the remarkable variety of the finches' beaks, Lack found something interesting to report.

However, Lack's views on the ecology and evolution of the finches had changed since his earlier writings. He had accepted Geogii F. Gause's (1934) contention that no two species can occupy the same ecological niche in the same place as a consequence of natural selection. If two species did in fact live in the same habitat and ate the same types of food, competition would inevitably result, and one species would eliminate

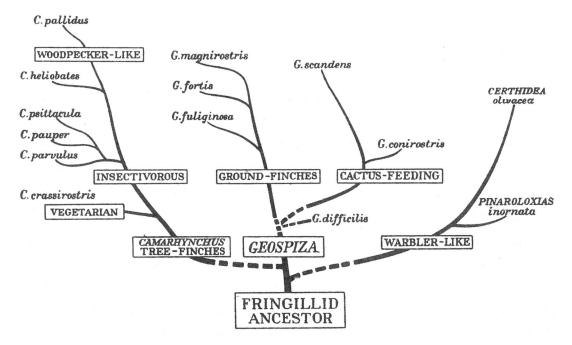

FIGURE 45.2. David Lack's (1910–73) tree showing his hypothesis about the evolutionary history of the finches. From D. Lack, *Darwin's Finches: An Essay on the General Biological Theory of Evolution* (Cambridge: Cambridge University Press, 1947). Permission: Lack estate

the other. Given that there were three species of ground finches (*Geospiza*) and at least two tree finches (*Camarhynchus*) living together in the same habitat on the Galapagos, Lack theorized that another factor prevented these species from competing.

Lack explained that geographical isolation appeared to be the critical factor in speciation. To bolster his case, Lack surveyed bird species and found that in every species that maintained continental and island populations, the island races differed from each other more strikingly than the continental races. Moreover, the more isolated an island was, the greater the degree of differentiation within the land birds. Thus, Lack (1947, 119) concluded: "The primary cause of geographical variation in birds would seem to be not adaptation, but isolation." Darwin's finches provided Lack's case in point. The islands that were the most isolated had the highest proportion of endemic forms, while more central islands had fewer distinct forms: "Hence in Darwin's finches there is a marked correlation between the degree of isolation and the tendency to produce peculiar forms."

Lack also wondered about forms that were originally geographical races of the same species, which met later in the same region, remained distinct, and formed new species. How might these forms compete? Lack offered four possibilities: one much better adapted form swamps the other and exterminates it; one form has an advantage in the region where it meets another form, but the other has an advantage in an adjacent region; one form may have an advantage in a section of the original habitat and another in the rest; and one form proves better adapted to taking one food, and the other to obtaining other foods. Lack (1947) noted that the differences in food habits were often associated with marked differences in size, including the size of beak. Darwin's finches provided at least two cases of each type of ecological isolation. For example, *Geospiza conirostris* replaced *G. scandens* on three outer (geographically remote) islands. *Geospiza scandens* bred in the same habitat with other ground finches but fed on *Opuntia*. *G. magnirostris*, *G. fortis*, and *G. fuliginosa* occupied the same habitat but their foods were different, at least in part. Lack suspected a similar state of affairs for three species of tree finches, though their dietary preferences remained unknown (Fig. 45.2).

Lack's earlier writings on the finches, though transcending the taxonomic studies of the Age of Surveys to make claims regarding evolutionary significance, seem to have been missing a theoretical framework with which he could integrate the data from his ecological studies of the breeding and feeding behavior of the finches with systematic data from the large collection of finches at the California Academy of Sciences. We have seen the influence of Gause's ecological niche concept for Lack's understanding of evolutionary patterns. Lack also cited Julian Huxley's writing on the modern synthesis (1942) and Ernst Mayr's *Systematics and the Origin of Species* (1942). In this way, the unification of biology provided a new theoretical rigor that enriched Lack's *Darwin's Finches* (and countless biology textbooks).

In 1956 William Brown and E. O. Wilson identified Darwin's finches (à la Lack) as a striking case of character displacement. Citing Lack's data regarding *Geospiza fortis* and *G. fuliginosa*, they noted that, on most islands where the two species occur together, the two species could be separated by the measurement of beak depth, and a random sample of this character resulted in two completely distinct distribution

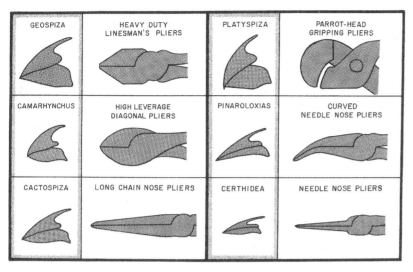

FIGURE 45.3. Comparison of shapes of bills with pliers. It is hard to imagine a more striking example of the extent to which the design metaphor pervades Darwinian evolutionary biology. From R. I. Bowman, Evolutionary patterns in Darwin's finches, *Occasional Papers of the California Academy of Sciences* 44 (1963): 107–40. With permission of the California Academy of Science

After two expeditions to the Galapagos, Robert Bowman (1963) argued that the major patterns of differentiation in Darwin's finches were related to adaptations for food getting. Bowman correlated the finches' beaks to various kinds of pliers (Fig. 45.3). Although Darwin had confused the finches with similar continental families, Bowman deliberately compared the individual finches with continental families on the basis of the niches field by the different species. This group of songbirds represented no less than seven continental families, which Bowman illustrated with a chart that depicted "ancestors of the Geospizinae" (Fig. 45.4).

The key to the differentiation within the finches was the kind of food they consumed. In this and other respects, Bowman challenged the interpretations of David Lack, who had identified isolation and interspecific competition as the primary engine of adaptation among the finches. On the basis of the field observations and anatomical analyses of the muscular structures supporting the finch beaks, Bowman argued that the evolutionary pattern of the Galapagos finches hinged on differing vegetation between islands and related structural modifications of seeds and behavioral reactions of insects to escape widespread aridity with profound effects on feeding adaptations in the finches. In addition, predation (by Galapagos Hawks and Galapagos snakes) had been selective forces, which led to adaptive differences in plumage between genera, species, and even populations. Finally, Bowman (1961) pointed to the "genetic constitution of ancestral colonists" as a constraint in the ability of finches to evolve to exploit all ecological niches rather than the presence of "ecological equivalents."

THE GRANTS WITNESS EVOLUTION IN REAL TIME

In 1971 Peter R. Grant embarked on a new research project. First, he sought to clarify whether population variation in the size of traits such as beaks was adaptive, as suggested without convincing evidence by Leigh Van Valen. From Lack he knew that Darwin's finches could shed light on this question. Second, following Brown and Wilson, Grant found evidence for character displacement lacking (i.e., the tendency for differences between ecologically similar species to be enhanced where they occur together as a result of natural selection minimizing competition between them). His decision to examine the related process of character release (changes in morphology and ecology of a species resulting from the absence of restraints from a competitor species) led him to the classic case of two species of Darwin's finches as described by Brown and Wilson (1956). Moreover, a graduate student had proposed a study of Darwin's finches on the basis of the

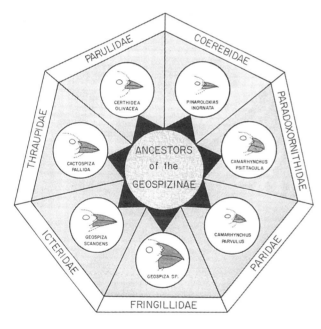

FIGURE 45.4. The pattern of adaptive radiation in Darwin's finches. From R. I. Bowman, Evolutionary patterns in Darwin's finches, *Occasional Papers of the California Academy of Sciences* 44 (1963): 107–40. With permission of the California Academy of Science

curves. However, a comparable sampling on the smaller islands of Daphne and Crossman revealed a single unimodal curve that fell between the curves of *fortis* and *fuliginosa* on the larger islands. Lack's careful analysis of beak to wing proportions revealed that *fortis* occurred on Daphne while the Crossman population was *fuliginosa*. Brown and Wilson (1956) accepted Lack's conclusion that each species had converged toward the other species, thereby filling the ecological vacuum created by its absence.

conflicting conclusions of Lack and Bowman concerning the same material. Initially, Grant (1985) focused on the Medium Ground Finch (*G. fortis*) population on Daphne Major, which was small enough so that he could uniquely color band every individual, and the birds were variable in both body size and bill shape (Plate XXXV).

On the basis of the Daphne Major population and through detailed analysis, Grant determined that, though ground finches (*Geospiza*) occurred together in nonrandom combinations, members of all pairs of coexisting species differed by at least 15 percent in at least one bill dimension with minimal overlap in frequency distributions. By evaluating the nature of the food supply (seed characteristics), Grant revealed a polymodal frequency distribution during the dry season, which in turn determined adaptive peaks. Grant's (1985) long-term study of the Medium Ground Finch (*G. fuliginosa*) and the Small Ground Finch (*G. fortis*) provided evidence of both character displacement (an evolutionary process) and differential colonization (an ecological process).

More generally, Grant asked why there were only thirteen species of finches on the Galapagos, rather than twice that many or more. And why did the species range from 8 grams to 40 grams, rather than from 5 to 100? Drawing on the evolutionary history of the finches, Grant suggested that isolation of the islands, ecological differences, and the passage of time all contributed to finch speciation from one to many, but limited ecological opportunity coupled with relatively limited time and the presence of other bird species all placed constraints on the continued diversification of finches.

A few years later, Grant, along with his wife and scientific collaborator, Rosemary, and graduate students, initiated a study of *G. conirostris* on Genovesa, one of the most isolated islands of the entire Galapagos Archipelago. Over the course of eleven years, the Grants and their students examined the ecology, behavior, and genetics of the population predominantly through direct observation, because experimentation was limited by National Park regulations and the natural state of the population. The Grants found that rainfall was the key environmental factor that influenced the population and its variation. During the eleven years, there were two major droughts and two El Niño years. In 1983 the plants, arthropods, and finches all reproduced throughout the prolonged period of rainfall brought about by El Niño. In 1985, during a drought, all reproduction ceased. Both events placed selective pressures on the finches and, more specifically, their beaks. Their observations (B. R. Grant and Grant 1989) challenged long-held views that evolution occurs over the course of prolonged periods of time and the corollary that changes are generally imperceptible. In fact, using data from DNA, the Grants have calculated that it could take as little as two hundred years for one finch species to evolve into a new species (K. T. Grant and Estes 2009). Widely regarded as one of the most important studies in ecology and evolutionary biology (see Travis 1990), the Grants' long-term study of Darwin's finches has demonstrated that evolution can be studied in real time. Given the right circumstances, scientists can study evolutionary change as it happens.

In addition to revealing the ecology and evolution of Darwin's finches, the conservation of these geographically constrained birds has been a recurring theme in the Grants' work. Others have also recognized the need for conservation efforts. The Mangrove Finch (*Cactospiza heliobates*) was the last of finches to be described, and the entire population was restricted to some of the mangrove forests bordering Isabela and Fernandina, but the Grants determined that the latter population may no longer exist and the Isabela birds are under threat from an introduced wasp and habitat destruction (P. R. Grant and Grant 1997). Critically endangered and limited to highland regions of Floreana, the Medium Tree Finch (*Camarhynchus pauper*) also suffers from habitat destruction as well as introduced predators and a parasite (O'Connor et al. 2010). Local extinctions of wider ranging species such as the Warbler Finch (*Certhidea fusca*) on Floreana have also occurred (P. R. Grant et al. 2005). Each of these cases offers a cautionary tale for the continued survival of Darwin's finches.

CONCLUSION

More than 150 years passed between Darwin's first suggestion that the Galapagos finches would serve as an excellent model for evolution in nature and the full realization of that prediction. After Darwin's initial confusion, John Gould described most of the species, but the complexity of speciation vexed biologists during the Age of Survey. Like Darwin, David Lack initially misinterpreted the significance of the finches, but he eventually recharacterized the finches as a model of adaptation by natural selection. The longitudinal studies conducted by Peter and Rosemary Grant (as well as numerous students and collaborators) revealed that, in the right circumstances, it is possible to measure evolution as it occurs. Studies of the finches' conservation status suggest that rare and widespread species alike are threatened by anthropogenic change in the Galapagos. The development of Darwin's finches as an evolutionary icon began in a fog of confusion, but a long series of empirical and theoretical studies revealed and continues to provide rich insights into evolution and ecology.

ACKNOWLEDGMENTS

Many individuals gave me timely advice and support as I wrote this study. Special recognition goes to Michael Ruse, Melissa Wiedenfeld, K. Thalia Grant, Gregory Estes, Mark Barrow, Elizabeth Dobson, Andando Tours, the captain and crew of S/Y *Sagitta*, Alejandro Villa, Spenser, Dan, and Judith Davis. The research for this essay was generously supported by a Developing Scholar's Award at Florida State University.

❧ Essay 46 ❧

Developmental Evolution

Manfred D. Laubichler and Jane Maienschein

As part of the 2009 Darwin celebrations, we have seen the emergence and widespread acceptance of a standard narrative of the history of evolutionary biology that construes a more or less direct line from Darwin to present-day evolutionary developmental biology, or evo-devo (Mayr 1982; Larson 2004; Carroll 2005; Zimmer 2006, 2009; Ruse and Travis 2009). It is a story of completions and syntheses that not only celebrates Darwin's genius but also implies an implicit progression of ideas, with inclusion of new empirical facts and methodological approaches within the general framework of Darwinism leading to an increasingly more complete understanding of the evolutionary process. This narrative involves both scientific and public discourses. It can be found in textbooks of evolutionary biology and in popular accounts of evolution; it is also the basis the many efforts to construct a more inclusive evolutionary worldview.

But the standard narrative "From Darwin to Evo Devo" is also woefully incomplete as it leaves out several important traditions within the history of evolutionary biology (Laubichler and Maienschein 2007). These neglected traditions are not fringe ideas with no relevance to current understanding of evolutionary processes that can therefore be relegated to the dustbin of history. Quite the contrary. The ideas and approaches that are part of a complementary tradition – namely, to explain the evolution of organisms in reference to the developmental mechanisms that first generate phenotypes and phenotypic variation – have informed some of the most important current evolutionary biology research programs, those in developmental evolution (devo-evo) and synthetic experimental evolution (SEE) (Wagner, Chiu, et al. 2000; Davidson 2006; Davidson and Erwin 2006; Laubichler 2007; Erwin and Davidson 2009).

In this essay we briefly discuss the standard narrative and contrast it with one of the complementary traditions focused on the role of developmental mechanisms in explaining phenotypic evolution. We then argue that a more inclusive understanding of the history of evolutionary biology can better inform present discussions and contribute to a broader synthesis of twenty-first-century evolutionary biology, one that also reflects more fully the richness of Darwin's original vision. We also argue that understanding these multiple trajectories within the history of evolutionary biology sheds light on currently emerging transformations of evolutionary biology into a causal mechanistic science.

THE STANDARD NARRATIVE OF THE HISTORY OF EVOLUTIONARY BIOLOGY: FROM DARWIN TO EVO-DEVO

In its most basic form, the narrative begins with Charles Darwin, although some versions include pre-Darwinian conceptions of phenotypic transformations such as those of Lamarck, Goethe, or Geoffroy de Saint-Hilaire, all of whom emphasized the importance of internal, organismal, or developmental factors. Darwin himself is generally situated within the nineteenth-century British context. The intellectual environment includes such debates as the age of the earth, the new conception of Lyellian geology with its emphasis on actualism and uniformitarianism, expertise of animal and plant breeders, natural historians exploring the far reaches of the emerging empire, the whole package of continental science and *Naturphilosophie* (courtesy of Robert Grant, Darwin's mentor during his short-lived stint as a medical student in Edinburgh), Whewell's philosophy of science, Adam Smith's theories of economics, and Malthus's insights into the dynamics of populations.

Furthermore, as Darwin's biographers have shown in great detail, these intellectual concerns existed in a symbiotic relationship with the social and economic transformations of nineteenth-century Britain. There is now a widespread consensus that Darwin truly was a child of his time, and this is exactly why he could, in turn, affect it as much as he did (Desmond and Moore 1991; Browne 1995).

All these concerns shaped the intellectual challenge Darwin tried to answer: How can we explain the patterns of organismal diversity, their distribution in space and time, and the incredible adaptations of organisms to the challenges presented by their environment? His answer, first formulated shortly after returning from his *Beagle* voyage, was a genuine and novel synthesis of various ideas and observations. It culminated, after two decades of refinement, in two canonical insights: descent with modification and natural selection.

Conceptually, Darwin's theory represents a breakthrough by combining two types of observations into a common explanatory framework. Organisms vary, at least part of this variation is passed on through generations, and organisms compete for limited resources as a consequence of Malthusian dynamics. Many chapters in this volume deal with these issues in more detail, so here we can summarize this first stage in the standard narrative as follows: Darwin's explanation of evolution includes the origin of variation, inheritance of variation, and the fate of specific variants competing for resources within populations. In his writings, he made suggestions about the first problem, developed an idiosyncratic (and wrong) theory of inheritance, and brilliantly applied the logic of natural selection to analyze the consequences of this mechanism for a whole range of phenomena (from the evolution of reproductive division of labor and the existence of ornaments and displays to the patterns of the fossil record).

The second part of the standard narrative focuses on the problem of inheritance, the issue that Darwin did not solve. Without a clear understanding of the material basis of heredity – in the form of distinct factors or genes – a variety of theoretical possibilities had been discussed, including several versions of neo-Lamarckism (to which Darwin himself was at least partially sympathetic, as he accepted the possibility of the inheritance of acquired characteristics).

The establishment of the rules of inheritance based on the Mendelian concept of discrete factors of inheritance that are passed on intact across generations, along with the subsequent discovery that these factors are localized on chromosomes, reinvigorated evolutionary thought. Building on mathematical and statistical methods that had been developed to describe and analyze variation and inheritance within populations, mathematical population genetics developed as a new foundation for evolutionary theory. It allowed exploration of the consequences of natural selection within populations, through the analysis of formal models, and it explicitly connected evolutionary and genetic analyses. It also solved the question of inheritance, insofar as it relates to patterns of transmission and the dynamics of genes/alleles within population (Provine 1971).

But population genetics introduced some substantial changes to the structure of evolutionary theory. The gene, as the unit of stable transmission between generations, now occupied a privileged position in accounts of the evolutionary process. And, while the combination of natural selection with gene-based views of inheritance provided a better understanding of the short-term consequences of evolutionary change, this early twentieth-century vision did not address the problem of the origin of variation. It simply assumed that variants of existing genes emerge as mutations and that the traces and consequences of these mutations can be studied and observed in populations. There were debates about the quality and size of mutational events (as in the debate between Mendelians and biometricians), but these were largely about the consequences and less about the causes of these different types of mutations.

This part of the standard narrative actually ignores several important lines of research. Here we discuss one such alternative that focused on the role of genes in developments, though there are others, such as those in cytology and cytological genetics (Laubichler 2003; Laubichler and Maienschein 2004, 2007; Laubichler, Aird, et al. 2007; Laubichler and Davidson 2008). In the progressive narrative of the history of evolutionary biology, this period mainly stands for two developments: the solution of the inheritance problem and the emergence of a mathematical approach that further emphasized what Ernst Mayr referred to as "populational thinking" (Mayr and Provine 1980; Mayr 1982).

For our purpose of contrasting the standard narrative with a complementary history of developmental evolution, we need to emphasize one important assumption that characterized

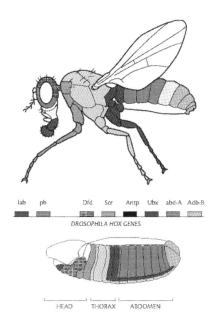

FIGURE 46.1. Hox genes. The genes controlling development in the fruit fly (*top*) and a comparison with homologous genes controlling development in other organisms including humans (*bottom*). This shows that the causes of development are not something newly created for each animal but part of a shared kit, where the genes function rather like Lego pieces – build one way and you get a fruit fly, build another way and you get a human. Permission: Sean B. Carroll et al., *From DNA to Diversity* (Oxford: Blackwell Scientific, 2001), 23, fig. 2.5, and 27, fig. 2.8

the population genetic (and also quantitative genetic) models developed within the modern synthesis framework. These models continued to emphasize the importance of the gene as the fundamental evolutionary unit. They also reinforced an additional formal assumption of population genetics, namely that the structure of the genotype-phenotype map (a technical term connecting genotype and phenotype or fitness values) is simple and that we can therefore describe the dynamics of evolutionary change solely on the level of genotypes.

Given the technical constraints of the time, such as the limited computational powers, this assumption of linearity was essential. And while it seemed to hold in some cases, the structural limitations of these models also contributed to an inherent discontent with the modern synthesis. Another central feature of mid-twentieth-century evolutionary biology connected with the formal structure of population genetic models has been its focus on adaptation and an implicit commitment to more or less gradual patterns of evolutionary change. Both of these assumptions were challenged in the 1970s and 1980s, mostly by paleontologists and developmental biologists, who contributed to the emerging field of evolutionary developmental biology representing the latest episode in the standard narrative of the history of evolutionary theory (Gould 1977; Laubichler and Maienschein 2007).

While early concerns of evo-devo focused on such issues as developmental constraints, punctuated equilibria in the fossil record, and heterochrony or life history evolution, today's version also includes results from comparative genomics and developmental genetics that have revealed the high degree of conservation of what has come to be known as the "genetic toolkit for development" (Carroll 2005; Carroll, Grenier, et al. 2005) (Fig. 46.1).

Within the standard narrative, evo-devo is mostly seen as a completion of the modern synthesis, the details of which are still unfolding (Carroll 2005; Zimmer 2009; Pigliucci and Müller 2010). This view allows commentators to contextualize some of the arguments within the evo-devo and evolutionary biology communities, such as those about the genetics underlying phenotypic change as necessary debates en route to a consensus. In this view, the main theoretical innovation of evo-devo lies in its treatment of the genotype-phenotype map, which is now seen as more complex and representative of the known facts of developmental genetics. This focus on the genotype-phenotype map also continues the well-established collaboration between theoretical and empirical work in evolutionary biology. While developmental geneticists uncover empirical details of how genes affect the development of phenotypes, theorists explore the formal consequences of epistasis and complex genotype-phenotype maps within the framework of population genetic models. The emerging twenty-first-century evolutionary theory is thus very much like its twentieth-century predecessor, only better in the sense that it more adequately represents the known facts of development. But this also means that the primacy of evolutionary dynamics (in the form of population genetics) for all explanations of evolutionary transformations remains intact.

What can be seen, even from this very brief sketch, is that the standard narrative emphasizes a clear progression in the development of evolutionary theory, one grounded in Darwin's original conception and continuously incorporating new perspectives within a framework of population-based adaptive dynamics that has remained more or less unchanged for at least a century.

A BRIEF HISTORY OF DEVELOPMENTAL EVOLUTION: FROM DARWIN TO SYNTHETIC EXPERIMENTAL EVOLUTION

The standard history of evolutionary theory sketched in the preceding section is but one narrative organizing a whole range of complex historical developments connected with the idea of evolution. Alternative positions have been discussed, but for the most part these accounts have focused on critiques and challenges to the Darwinian mainstream that did not add up to a similarly substantive and successful research program (Bowler 1983, 1988). While it is, for instance, interesting to understand how neo-Lamarckian ideas persisted as a challenge to a Darwinian consensus, these ideas did not contribute much to our current understanding of evolutionary biology, which in turn might explain the widespread appeal of the standard narrative.

The alternative history we explore here is different in that it has, for the most part, not been considered in the context of evolutionary biology, even though, as we will argue, these developments have made substantial and central contributions to our understanding of the evolutionary process.

As with the standard narrative, we begin our story with Darwin. The same caveat, that many of the concerns discussed here have an important history before Darwin and that these antecedents influenced Darwin's thinking in important ways, is true here as well (Desmond and Moore 1991; R. J. Richards 1992; Browne 1995). We have already seen that Darwin recognized the origin of variation as an important problem. He collected numerous data on the specific patterns of variation found in populations, both natural and artificially selected, and offered several possible explanations for the existence of these variants. Among those, mechanisms of development were especially important. To quote just one of many passages throughout his oeuvre: "Our ignorance of the laws of variation is profound.... Changes of structure at an early age will generally affect parts subsequently developed; and there are very many other correlations of growth, the nature of which we are utterly unable to understand" (Darwin 1859, 167–68).

FIGURE 46.2. Race horses and draft horses. In the *Origin*, Darwin was particularly interested in the way in which natural selection tears apart the adults of organisms with very similar embryos. His hypothesis that natural selection works on variations that appear only later in individual development was given strong support by his studies of the practices of animal breeders. They are indifferent to juvenile features but select for desired adult features. A prime example is that of horse breeders, some of whom want strong workers (*top*) and others of whom want fast racers (*bottom*). Nineteenth-century etching from S. Sidney, J. Sinclair, and W. C. Arlington Blew, *The Book of the Horse* (London: Cassell, Peter and Galpin, 1893)

Darwin clearly recognized that developmental processes, such as the timing of embryological events or the correlations of growth, are essential components of any explanation of the origin of phenotypic variation. This close connection between embryology and evolutionary ideas was characteristic for much of the nineteenth century, both before and after the publication of the *Origin*, as has been pointed out repeatedly. We can therefore summarize that for Darwin as well as many of his contemporaries, development was an integral part of any explanation of phenotypic transformation and evolution and that the problem of the origin of variation was considered a major challenge for which developmental mechanisms offered possible solutions (Fig. 46.2).

In subsequent decades the question of "generation" or *Entwicklung* became one of the prime research areas of the emerging experimental biology. We use two nineteenth-century terms to highlight the fact that during this time the problem of the origin of organismal forms was seen as a more inclusive process that involves questions of inheritance, development, and evolution (Laubichler and Maienschein 2004). Historical scholarship has mostly focused on the highly influential theoretical ideas of August Weismann and debates triggered by his rejection of the inheritance of acquired characteristics, the concept of the separation of germline and soma, and his broadening of the action of selection, for which George Romanes coined the term "neo-Darwinism." Weismann's proposals paved the way for the subsequent separation of the different dimensions of "generation." Once inheritance, through continuity of the germline, was conceptually separated from development (or the processes of differentiation and morphogenesis), it became possible to connect evolutionary transformations with patterns of hereditary transmission (Laubichler and Rheinberger 2006). This conceptual insight led to the establishment of population genetics.

But evolutionary concerns also influenced the work of numerous cell and developmental biologists during this period even though they often did not discuss their work explicitly in those terms. Here, as an example, we briefly discuss the work of Theodor Boveri, arguably one of the most influential experimental biologists of his time (Laubichler and Davidson 2008) (Fig. 46.3). Boveri's major research questions were all related to "generation" in its inclusive sense. He studied among other things the behavior and functional role of chromosomes, fertilization, heredity – culminating in the chromosomal theory of inheritance, the structure of the egg, and the role of cytoplasm and nucleus in development and differentiation. His accomplishments in these areas are well documented. Less known are Boveri's conceptual and institutional contributions to the tradition of developmental evolution.

Boveri summarized his views in his *Rektorratsrede* of 1906. In this inaugural speech, entitled "Organisms as Historical Beings," Boveri (1906) argues that any explanation of the evolution of organisms must begin with an understanding of developmental processes because these represent the constructive mechanism that generates organisms; these processes are controlled by a highly structured system of hereditary materials located within the nucleus; and these phenomena have to be studied experimentally, with one experiment standing out: "to transform organisms in front of our eyes." Boveri thus mapped out the research program of experimental developmental evolution: analyzing the ways the system of hereditary factors controls development and studying how changes to this system transform organisms before our eyes. This approach to evolution clearly focused on the primacy of understanding the origins of phenotypic variation experimentally before any attempt to study the consequences of natural selection. Boveri also realized that for such an ambitious research program to work, a new type of research institution would be needed. When he was asked to develop the plans for the Kaiser-Wilhelm-Institut für Biologie, he organized it to

FIGURE 46.3. Theodor Boveri (1862–1915), along with the American biologist Walter Sutton, was the discoverer of the fact that the units of inheritance are carried by the chromosomes. He made other significant discoveries, including the fact that cancer starts with the disruption of the chromosomes in a single cell, causing uncontrolled division. Permission: American Philosophical Society

support such a program of experimental developmental evolution (although he did not use this term).

Illness, however, prevented Boveri from finally accepting the post of founding director. Even though his successor Carl Correns made only a few changes, a good deal of Boveri's vision was realized, partly as a result of the appointment of Richard Goldschmidt, arguably one of the most colorful characters in the history of biology. Here we are mostly concerned with his long-term study of physiological gene action in development and some of the conceptual conclusions he drew for understanding phenotypic evolution. Basically, Goldschmidt (1940) realized that mutations can have different phenotypic or morphological effects depending on which part of the developmental machinery they affect. And while some of his wilder speculations about large-scale rearrangements of the chromosomes and the dissolution of the gene as a unit did not work out, his recognition that regulatory mutations can have large-scale effects proved insightful. His approach also championed experimental analysis of how genes control development, a research program that his successor at the Kaiser Wilhelm Institute for biology, Alfred Kühn, also pursued. (Kühn succeeded Goldschmidt in 1936 after he was forced out of Germany.)

Kühn, working with the flour moth *Ephestia*, attempted to fully characterize the causal chain of biochemical events that connects a gene (for eye color) with its phenotypic effect. He and co-workers discovered that gene action is based on two interacting pathways with multiple parts (one pathway of gene products, the other of substances) that finally result in the phenotype. He later generalized this conception and argued that (1) each phenotype is the consequence of a complex network of interactions between genetic elements representing a "developmental physiological equilibrium"; (2) each phenotypic variant is the product of a different equilibrium state; and (3) evolution is the product of a series of transformations of developmental physiological equilibria. The details of Kühn's conception were highly speculative. He did not have much empirical evidence for his somewhat vague notion of "equilibrium," but it was also a clear expression of the logic of developmental evolution, and as such proved to be inspirational for some younger German and Swiss biologists (Kühn 1955; Laubichler and Rheinberger 2004).

The most immediate source of the present-day conception of developmental evolution dates back to the late 1960s and early 1970s. A seminal paper by Eric Davidson and Roy Britten, "Gene Regulation for Higher Cells: A Theory," published in *Science* in 1969, refocused many of the conceptual ideas related to differentiation and gene expression from earlier periods and connected them with the rapidly advancing field of molecular biology (Britten and Davidson 1969) (Plate XXXVI). The resulting theory provided a clear and logical formulation of how developmental processes are controlled by gene activity, how regulation of gene activity is the underlying mechanistic cause for differentiation, and how regulatory changes in gene expression are the direct cause for phenotypic variation.

From the very beginning, the evolutionary implications of the Britten-Davidson model, as it became known, were obvious. Britten and Davidson (1971) already discussed those in the original article as well as in a follow-up paper published two years later. And while traditional developmental biologists took some time to fully accept the regulatory- and genome-based reorientation of their field, some evolutionary theorists immediately recognized the implications of this proposal and its emphasis on regulatory networks. They developed a range of theoretical models that suggested that the structure of the genome and patterns of interactions between genes would show evidence of their evolutionary history. For example, these models proposed that those genes involved in fundamental developmental processes that arose early in evolution would be more conserved or have a higher "burden" (Riedl 1975; G. P. Wagner and Laubichler 2004)

As well as the theoretical implications of the Britten-Davidson model, empirical evidence for some of its conclusions began to emerge as well. One of the implications is that major phenotypic changes would more likely be the consequence of mutations in the regulatory sequences than those that code for structural proteins. This conclusion was supported by the observation that many functional proteins have highly pleiotropic effects, which would place their genes under strong stabilizing selection, whereas it was thought that the regulatory regions controlling their expression during development could be more variable. One of the first empirical studies supporting this idea was the classic paper by Mary-Claire King and Alan Wilson (1975) comparing human and chimpanzee macromolecules. After surveying the available molecular evidence, they concluded that the observed phenotypic differences between these two species must be the result of regulatory mutations, in line with predictions of the Britten-Davidson model.

After these early theoretical and empirical discoveries, research into the molecular mechanisms of developmental evolution continued, despite some substantial technical difficulties. By the late 1970s and early 1980s, the question of the relationship between development and evolution had become a major concern within evolutionary biology.

The 1981 Dahlem conference, organized by John Bonner (1982), represents a major landmark. It also highlights the substantial differences between the two approaches that would soon be known as evolutionary developmental biology and developmental evolution. The former takes the phenomenology of evolutionary and developmental patterns as its starting point and asks how developmental processes can add to the explanation of these observations. The best examples of this trend are the notions of developmental constraints, where developmental processes are thought to explain the fact that the observed phenotype space is spotty and mostly empty, and heterochrony, where changes in the timing of developmental processes are employed to explain large-scale and correlated changes in phenotypes. Both examples exemplify the structure of evolutionary developmental biology whereby developmental processes are incorporated into the standard framework of evolutionary theory, either as limits to variation (constraints) or as highly pleiotropic effects of genes affecting developmental timing.

Developmental evolution, on the other hand, emphasized the underlying genomic and regulatory mechanisms as the foundation of all evolutionary change. The main conceptual differences between these two approaches were that developmental evolution focused on (1) the genome as an integrated regulatory system rather than the single (or multiple) gene locus paradigm of standard evolutionary theory; (2) the mechanisms generating phenotypes and phenotypic variation as the primary step in all explanations of evolutionary change; and (3) a causal-mechanistic and experimental approach to the problem of evolution. Beginning in the 1990s, as the experimental repertoire of molecular biology expanded and available sequences brought about a more genome-based biology, empirical research in developmental evolution began to catch up with its conceptual and theoretical insights (E. H. Davidson 1990).

The culmination of decades worth of detailed empirical and conceptual work in this area has been the transformation of the early ideas of regulatory networks proposed in the Britten-Davidson model into the fully characterized gene regulatory networks (GRNs) of today (E. H. Davidson 2001, 2006). The majority of the early and foundational work in this area has been done, by Eric Davidson and his collaborators, with the purple sea urchin as a model system. Today, molecular, genomics, and bioinformatics tools make it possible to study developmental GRNs in an increasing number of organisms, thus allowing for a comparative analysis of GRNs. Such analysis is essential for a GRN-based explanation of evolutionary transformations. In this context several important findings have refined earlier conceptual ideas about the role of genomic regulatory systems in evolution. It is now clear that the modular and hierarchical structure of GRNs has important implications for understanding the origin and patterns of phenotypic variation. Different elements in the GRN have different variational properties; for example, those elements responsible for highly conserved body plan features (kernels) are generally more conserved than more downstream elements of the network (E. H. Davidson 2006; E. H. Davidson and Erwin 2006, 2009; Peter and Davidson 2009). This seems to confirm earlier ideas that the evolutionary history is also inscribed into the genomic developmental systems that control the development of organisms.

Eric Davidson (2006) has proposed a classification of genomic regulatory elements that includes the most conserved control elements or kernels; a set of multipurpose modules that are used in a variety of contexts (switches, plug-ins, and input-output devices); and the differentiation gene batteries, those sets of genes that characterize the specific cell state. Functional as well as comparative analysis of these different GRN elements reveals that changes in different parts of the network correspond to qualitatively different phenotypic and evolutionary transformations. These insights have reinvigorated a developmentally based approach to phylogenetic history. As kernel differences tend to map onto phylum- or superphylum-level morphological features, the evidence suggests that those which are part of the regulatory elements evolved before the separation into distinct lineages and body plans. Subsequent evolution of different parts of the network led to the elaboration of these body plans, while adaptive evolution and speciation tend to be caused by changes in downstream differentiation gene batteries or the ways these are deployed (E. H. Davidson, Peterson, et al. 1995; Erwin and Davidson 2002, 2009; E. H. Davidson and Erwin 2006, 2009). Many details of these evolutionary scenarios are still unknown, but the conceptual framework of developmental evolution has already transformed the way we research and interpret the origin and evolution of GRNs and of phenotypic evolution more generally. And as the evidence for conserved genes and a relatively low number of open reading frames continues to accumulate, the importance of regulatory evolution at multiple layers of control systems only increases.

The developmental evolution perspective also provides an interpretative framework for numerous molecular details that are being revealed in the context of genomic-based approaches. The majority of these findings, such as the discovery of multiple families of regulatory RNA molecules, points to an increasing importance of regulatory processes in both development and evolution.

The latest episode, currently unfolding, in the history of developmental evolution is, in many ways, the most radical transformation of evolutionary biology since the formulation of population genetics early in the twentieth century. Previous experimental approaches to the study of evolution involved either long-term selection experiments, direct manipulation of phenotypic characters in order to measure their contributions to fitness, or the exposure to mutagens to increase the mutation rate. Today, on the basis of insights into the structure of GRNs a new kind of experimental approach to evolution is emerging, *synthetic experimental evolution* (Erwin and Davidson 2009). This approach studies the phenotypic consequences of targeted changes to GRNs, which will allow us to reengineer major phenotypic transformations in evolutionary history. The preconditions for these types of experiments are all within reach: (1) comparative analysis of GRNs of a species that has acquired a novel phenotypic character as well as of related species that represent the ancestral condition; (2) identification and experimental verification of those changes to the structure of the GRN that are causally sufficient to generate the novel phenotype; (3) targeted insertion of these GRN elements into the genome of the species representing the ancestral condition; and (4) testing the prediction that a rewired GRN will generate a phenotype similar to the one that has been acquired during the evolution of the derived lineage. Taken together, these experimental approaches enable us to study the developmental basis of evolutionary transformations and investigate how different kinds of phenotypic variation are generated.

As a consequence of these developments, developmental evolution is now becoming a causal mechanistic and experimental science that is closely aligned with two transformative paradigms of twenty-first-century biology: systems biology and synthetic biology. An emphasis on genomic regulatory systems (such as GRNs) is only a first step in the direction of a more inclusive systems focus within developmental evolution. The consequences of a number of additional regulatory systems – from microRNAs to epigenetic systems – for developmental evolution are also being investigated. However, all these additional layers of regulatory control are being anchored by the regulatory genome, which therefore occupies a privileged position within both developmental and evolutionary processes. Additional connections to systems and synthetic biology are methodological and include a close connection between targeted experimental interventions designed to reengineer functional control circuits and mathematical and bioinformatical approaches.

With regard to our original question how these developments change the narrative of the history of evolutionary biology, we can conclude that developmental evolution represents a theoretical and conceptual departure from standard evolutionary explanations and their focus on the adaptive dynamics of populations. Developmental evolution is a return to the more inclusive focus of Darwinism, with its emphasis on both the origin of variation as well as the fate of variants within populations. While the latter has been well studied during the past century, the former involves significant conceptual and methodological changes that represent a departure from mainstream twentieth-century evolutionary biology.

CONCLUSION

Our brief exploration of multiple pathways in the history of evolutionary biology offers a much richer and more diverse picture than the standard narrative "From Darwin to Evo-Devo" suggests. Furthermore, many of these previously neglected episodes shed new light on current developments in evolutionary theory. The possibilities enabled by the emerging causal-mechanistic understanding of phenotypic evolution arguably represent the most dramatic transformation of evolutionary theory in decades. Yet the roots of this conceptual reorientation of evolutionary biology can only partially be found within the mainstream history of the field as it has been portrayed so far. Rather the antecedents of much of this work fall within the alternative tradition sketched here. As a result, the history of evolutionary biology increasingly resembles our current understanding of evolutionary history; we no longer see a linear or simple branching pattern, with one "progressive" trunk and major and minor branches diverging from it. Quite the opposite, we see the same reticulate pattern and horizontal as well as vertical transmission of ideas (genes/memes) that characterizes the majority of evolutionary events (especially in the microbial domain).

Having a more complete understanding of the history of evolutionary theory since Darwin has important practical implications. Early twenty-first-century evolutionary biology emphasizes the need for synthesis, either as a "completion of the modern synthesis" or in the form of a new synthesis. Both of these calls require the integration of radically different conceptual frameworks, experimental traditions, fundamental assumptions, and epistemologies in order to achieve a more inclusive understanding of the evolutionary process. Without successful integration of the different domains of developmental and evolutionary biology, the developmental evolution or evo-devo project will fail. After an initial period of enthusiasm, the field has now entered a phase of routine data generation. What is still largely left undone is the hard work of conceptual integration that is required for a true theoretical synthesis.

In this context critical historical perspectives are essential. They will enable us to evaluate the often-hidden assumptions of certain models and concepts. The examples discussed here can serve as an illustration. Do we build our new synthesis of twenty-first-century evolutionary biology within the conceptual framework of population-based adaptive dynamics – which implies that developmental mechanisms feature as an explanation of the genotype-phenotype map – or do we represent the evolutionary process within a causal-mechanistic framework that can be captured by the following logical structure: (1) all phenotypes are the product of developmental mechanisms; (2) all phenotypic variation is therefore a consequence of a corresponding variation in the developmental process; (3) understanding these developmental processes provides a causal-mechanistic explanation for the origin of phenotypic variation (Darwin's first question); (4) the subsequent fate of phenotypic variation can be analyzed within the population-based framework of adaptive dynamics.

To fully appreciate the differences between these two proposed versions of a twenty-first-century synthesis of evolutionary theory we need to understand how each of those viewpoints emerged historically and what epistemological assumptions guide the integration of developmental and evolutionary perspectives. As we have seen, the possibilities of synthetic experimental evolution represent a significant addition to the standard experimental repertoire of evolutionary biology, which is no longer confined to comparative and functional analysis or selection experiments. The ability to reconstruct major phenotypic transitions in evolutionary history through the manipulation of the underlying developmental mechanisms turns evolutionary biology into a mechanistic science. One consequence of this emerging transformation of evolutionary biology is that the standard distinction between proximate and ultimate causes no longer serves as the most obvious way to separate explanatory paradigms within (evolutionary) biology.

Evolutionary biology thus continues to evolve. And insofar as embryological considerations were already central to the earliest formulations of evolutionary theory, the current resurgence of developmental approaches reveals some of the deep conceptual structures at the core of evolutionary thought. Darwin, for once, would not be surprised and, we suspect, would be intrigued by the new experimental approaches and the possibility of causal-mechanistic explanations of phenotypic evolution.

PLATE XXXV. The husband and wife team of Peter and Rosemary Grant has made a forty-year-long study of Darwin's finches on the little Galapagos islet of Daphne Major. Permission: Rosemary Grant

PLATE XXXVI. Eric Davidson (*left*) and Roy Britten (*center*), authors of a seminal paper on developmental evolution, with Gary Felsenfeld (*right*) from the National Institutes of Health. Permission: Manfred Laubichler

PLATE XXXVII. *The Hungry Lion Throws Itself on the Antelope*, 1905, by "Le Douanier," Henri Rousseau. Self-taught, the artist never visited tropical climes and got his inspiration from zoos and botanical gardens. Permission: Fondation Beyeler, Riehen/Basel

PLATE XXXVIII. Thomas Lovejoy is an ecologist who has devoted his life to the interface between science and public policy. He has long been concerned with the fate of the Brazilian rainforests and works through the United Nations and other organizations to halt the destruction of the earth's habitats. Permission: Heinz Center

PLATE XXXIX. England-residing, Canadian-born Brian Goodwin (1931–2009) was one of the most forceful critics of orthodox Darwinian thinking, calling for a more holistic evolutionary vision, one giving scope for D'Arcy Thompson–like physico-chemical processes to self-organize, thus providing the complex adaptations supposedly produced by natural selection, without the harsh competition demanded by that mechanism of change. Permission: Christel Ankersmit

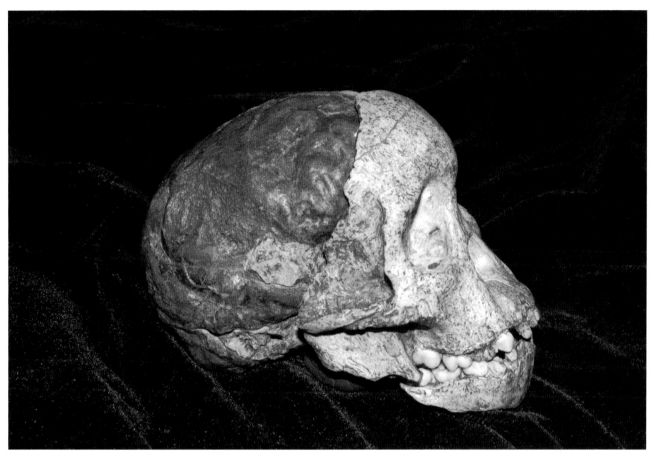

PLATE XL. The Taung baby, the first discovered example of an Australopithecine. Permission: Bernhard Zipfel

PLATE XLI. Lucy. *Australopithecus afarensis*, now believed to be about 3.2 million years old, with Donald Johanson, the man who discovered her in 1974. Permission: New York Times

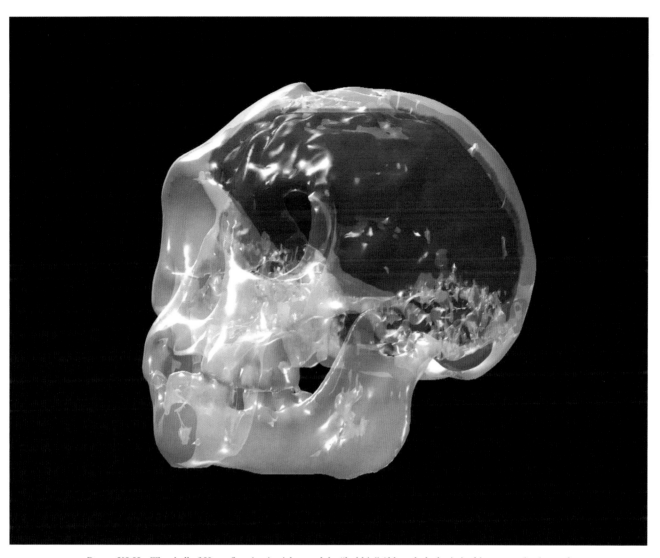

PLATE XLII. The skull of *Homo floresiensis*, nicknamed the "hobbit." Although the brain is chimpanzee size (around 400 cubic centimeters as opposed to our 1,200 cubic centimeters), it shows features (notably a prefrontal cortex as large as ours) that suggest significant intelligence. Courtesy of Kirk Smith, Mallinckrodt Institute of Radiology, Washington University School of Medicine

PLATE XLIII. Kanzi, born in 1980, is a male bonobo (*Pan paniscus*) who shows significant linguistic abilities. Permission: Great Ape Trust

PLATE XLIV. David Hull (1935–2010) and Michael Ruse (b. 1940) have both argued that Darwinian theory throws important light on the major questions of philosophy. That this was a viewpoint often regarded with scorn by their fellow professionals was worn as a badge of pride and one of many factors that bound them in a deep, lifelong friendship. Permission: Lizzie Ruse

PLATE XLV. Frans de Waal, a Dutch-born primatologist, who has done much to persuade people that the higher primates show genuine cooperative behavior. Permission: de Waal, photo by Catherine Marin

CHAPTER 27. MAN: FROM SOCIOBIOLOGY TO SOCIOLOGY

Let us now consider man in the free spirit of natural history, as though we were zoologists from another planet completing a macroscopic catalog of social species on Earth. In this view the humanities and social sciences shrink to specialized branches of biology; history, biography, and fiction are the research protocols of human ethology; and anthropology and sociology together comprise the sociobiology of a single primate species. Homo sapiens is ecologically a very peculiar species. It occupies the widest geographical range and maintains the highest local densities of any of the primates. An astute ecologist from another planet would not be surprised to find that only one species of Homo exists. Modern man has pre-empted all the conceivable hominid niches. Two or more species of hominids did coexist in the past when the Australopithecus apes and possibly an early Homo lived in Africa. But only one evolving line survived into late Pleistocene times.

Modern man is anatomically unique. His erect posture and bipedal locomotion are not approached in other primates that occasionally walk on their hind legs, including the gorilla and chimpanzee. The skeleton has been profoundly modified to accommodate the change: the spine is curved to distribute the weight of the trunk more evenly down its length; the chest is flattened to move the center of gravity back toward the spine; the pelvis is broadened to serve as an attachment for the powerful striding muscles of the upper legs, and reshaped into a basin to hold the viscera;

PLATE XLVI. The opening manuscript page of the notorious final chapter on humans of Edward O. Wilson's *Sociobiology: The New Synthesis*, a work starting with the cry to take philosophy from the philosophers and "biologicize" it. After initial resentment, some philosophers started to take this plea seriously. Permission: Edward O. Wilson

PLATE XLVII. *The Genesis Flood*, the "young-earth creationists'" answer to the *Origin*. It swung evangelicals from acceptance of an aged earth to a time span of ten thousand years or less. Permission: Presbyterian Publishing Company

PLATE XLVIII. Phillip Johnson (b. 1940), the éminence grise of the intelligent-design movement. A longtime law professor at Berkeley, his anti-Darwinism came as a result of a midlife conversion to evangelical Christianity. Permission: Kathy Johnson

PLATE XLIX. Gilbert Keith Chesterton (1874–1936), another convert from the Anglicans to the Catholics, opposed modernism (especially as espoused by his friend George Bernard Shaw) and, with this, developed a dislike of evolution. From *Vanity Fair*, 21 February 1912

PLATE L. The "zoo rabbi," Natan Slifkin (b. 1975), and friend, a black-and-white ruffled lemur. Although Slifkin's training was in the ultra-Orthodox tradition, his belief in an old earth and evolution through natural selection has brought on his head the wrath of the ultra-Orthodox community. Permission: Natan Slifkin

PLATE LI. Thanks especially to the devastating critique of unwonted group selection thinking and like practices, *Adaptation and Natural Selection* by George C. Williams (1926–2010) was one of the most influential works on evolutionary thought in the past half century. Pictured here in Iceland, one of his favorite countries, Williams's approach to medicine was always grounded in a deep appreciation of the extent to which evolutionary processes have left their marks on every aspect of organic existence and flourishing. Permission: Doris Williams

PLATE LII. Randolph Nesse (b. 1948), a practicing physician (psychiatrist), brought his understanding of medicine to fuse with the evolutionary understanding of George Williams. Permission: R. Nesse

❧ Essay 47 ❧

Darwin's Evolutionary Ecology

James Justus

"Nothing in biology makes sense except in the light of evolution." Dobzhansky's (1964, 449) sweeping generalization is provocative but also partial. Ecology casts the same indispensable light in biology, particularly on evolution. Nowhere is this clearer than in the origin of evolutionary theory. Although the term "ecology" was not coined until 1866 (Haeckel 1866), ecological insight is at the core of Darwin's theory. It is reflected both in the theory's concepts – for example, adaptation and natural selection – and in its compelling accounts of biological phenomena, such as the transmutation of species and the fit between organisms and environments. That evolutionary biology's chief architect is Darwin is well known. The foundational role his work had in ecology and that an ecological perspective underpins the theory of natural selection are less appreciated.

RECONCEPTUALIZING THE ENVIRONMENT

Perhaps the most theoretically fertile issue at the intersection of ecology and evolution is the adaptive fit between organisms and their environment. Seeing that relationship as the key to evolutionary dynamics required a reconceptualization of how the environment impacts organisms and the environment itself.

At the turn of the nineteenth century, philosophical and scientific conceptions of the environment reflected a romantic zeitgeist. Thoreau's *Walden*, for example, exemplifies the view (Fig. 47.1). In it, organisms and their environments are coupled components of an encompassing, harmonious system, each complementing the other in a providential symbiosis. The same underlying theological commitment to a beneficent and coherent order in the living world arguably compelled the impressive systematicity (and occasional biological misstep) of Linnaeus's classification system. But by the mid-nineteenth century a less idyllic, more brutal view of the environment was challenging the prevailing romanticism (Worster 1994). Tennyson's grim characterization of nature as "red in tooth and claw" captured the new sentiment, and would find scientific expression and vindication in Darwin's theory.

As it did for most aspects of the theory, the *Beagle* voyage helped catalyze the relevant insights. Before visiting the Galapagos Islands, Darwin (1845) observed the

FIGURE 47.1. Henry David Thoreau (1817–62), American transcendentalist and hero of back-to-the-land enthusiasts. He is best known for his book *Walden*, recording his time spent in a cabin on the edge of Walden Pond, in Massachusetts. Lithograph, 1854

striking ecological effects European settlement – particularly introduced domesticated animals – had had on Argentinian grasslands. During an excursion through the Blue Mountains west of Sydney, Darwin noted the same destructive impact introduced English greyhounds had had on native Australian marsupials (Nicholas and Nicholas 2002, 48). While digging for fossils in Patagonia Darwin also uncovered several extinct mammal species that differed significantly from existing fauna (Darwin 1845). The dramatic impact that introduced species could have and the evidence that radical alterations in species composition had occurred in the past led Darwin to question the presupposed stable "harmony" between organisms and their environments proposed by scientific predecessors such as Linnaeus and Humboldt.

The peculiar properties of species in the Galapagos reinforced this belief and seeded doubt about the divine basis of biological patterns. What could possibly explain the perplexing and apparently capricious anomalies in the archipelago if a logical, beneficent, and preordained purpose exists behind biological associations and behaviors? The more egregious examples included tortoises and other lizards (rather than ungulates) functioning as the dominant herbivores and the small archipelago's seemingly inordinate number of finches (thirteen species) exhibiting a similarly incongruous diversity (see Grant 1986). Although unobserved by Darwin, the behavior of one finch in particular epitomized the pernicious struggle between species that other naturalists had largely missed: *Geospiza difficilis* makes an ecological livelihood by pecking seabirds to drink their blood and by cracking their eggs to consume the developing material within (Schluter and Grant 1984).

Other parts of the *Beagle* voyage generated different puzzles. On the same excursion west of Sydney, Darwin was struck by the remarkable similarity between the conical pitfall of an Australian lion-ant (more commonly now known as an "ant-lion") and the pitfalls of different species in England, and the behavioral and physiological similarities between different Australian and English birds (Nicholas and Nicholas 2002, 53–54). These phenomena are straightforwardly identifiable as convergent evolution today, but for Darwin they were perplexing: Given the similar properties involved, why wouldn't an intelligent creator deploy the same species? Rather than constitute difficult but ultimately surmountable challenges to the existing paradigm, these phenomena seemed to demand a new understanding of the mechanisms responsible for the living world.

THE MALTHUSIAN CALCULUS AND STRUGGLE FOR EXISTENCE

Being free of rosy preconceptions is one thing. Understanding how the austere conditions environments impose on organisms yield a mechanism of evolutionary change is another. Although there are seeds of understanding about the struggle for existence during Darwin's visit to Australia and before he read Reverend Malthus's influential work (Nicholas and Nicholas 2002, 30), in his autobiography Darwin stated that Malthus provided the crucial insight (Darwin 1958b, 120). It was Darwin's brilliance to recognize it as such.

According to Malthus, humanity's biological destiny is a tragic predicament. Human populations grow geometrically. Food supplies can at best increase arithmetically. Our reproductive tendency is therefore incompatible with what is ecologically sustainable. For Malthus, the point was edification. The need to attempt to elude this potential tragedy with hard work confirmed the virtue of a Protestant ethic. For Darwin, the point was scientific. Malthus's biological predicament pinpointed the inescapable struggle for limited resources confronting all organisms. In such a struggle, variations between individuals in a population – which Darwin had documented in extensive field studies and experimental work (Darwin 1859, ch. 1–2) – could favor some organisms over others. For example, one finch's slightly larger beak could convey a selective advantage, perhaps by improving its ability to crack thicker, more nutritious seeds. The extra energy consumed might then yield higher fecundity. If this advantageous variation is heritable, Darwin realized it could spread in a population across generations. Seeded by Malthus's mathematical presentation of the conflict over resources and nurtured by a perspective that emphasized rather than dismissed the biological salience of variation (Sober 1980; Ruse 1999a, ch. 7), this principle of natural selection emerged as the principal, organizing idea in Darwin's "one long argument" on the origin of species (Darwin 1958b, 140). As Haeckel (1869a) recognized, ecology provided the relevant window into the biological struggle underlying selection – "in a word, ecology is the study of all those complex interrelations referred to by Darwin as the conditions of the struggle for existence" (as quoted in Stauffer 1957).

An accurate account of the biological mechanisms responsible for individual variation and heritability would have to wait until the work of Gregor Mendel and ultimately the much later discovery of DNA. But the deficiencies of Darwin's pangenesis theory of heredity are a minor footnote to his monumental achievement. And they have no bearing on the novel ecological insight at its core and the other significant ecological innovations Darwin introduced. Darwin's conceptualization of natural selection as a form of niche dynamics is an example of the latter. Rather than focus directly on species-environment relations, Darwin frequently emphasized that species realize different function roles in ecological systems, which he often labeled "places" and later ecologists termed "niches" (Worster 1994). Species' efficiency in utilizing and expanding their niche and the nature of relationships between inhabitants of different niches would then explain why some species succeed and others fail. This more abstract representation of natural selection significantly expanded its explanatory scope.

SELECTION AND SPECIATION IN THE ECONOMY OF NATURE

A shift in Darwin's views about the mechanisms of speciation illustrates the additional theoretical resources an ecological perspective afforded the theory of natural selection (see Vorzimmer 1965). In his first efforts to formulate the theory (e.g., Darwin 1844), Darwin thought speciation required geographic isolation, at least in the countless cases where multiple species appeared to have emerged within a single region rather than migrated to it. In particular, for two species to evolve from a population composed of just one, a barrier initially partitioning the population into separate groups seemed necessary. Only with this separation could selection then act to cause divergence among the subpopulations.

This type of speciation mechanism was suggested by the inheritance mechanism Darwin initially favored: blending inheritance. Blending inheritance is a form of evolutionary regression to the mean. On this view, phenotypic traits of individuals in subsequent generations are a "blend" of those in ancestral generations, that is, descendant phenotypes are in some way an average of parental phenotypes. For example, baboon offspring of parents with fang lengths of 5 and 10 centimeters would possess fangs of intermediate length, usually close to the mean (7.5 cm). As such, parental phenotypes determine the range of phenotypic variation offspring can exhibit and the range must decrease across generations. Like the decaying fluctuations of a damped oscillator, even the selectively advantageous atypical phenotypes that surfaced in one generation would be gradually bred out of existence in subsequent generations.

This posed serious problems for attempts to explain how species can emerge *within* a population by appeal to natural selection (called sympatric speciation today [Coyne and Orr 2004]). For Darwin, speciation was the culmination of the gradual accumulation of favorable adaptations in organisms. But with blending inheritance, if a favorable adaptation emerged in a population – for example, an atypically long neck in an otherwise healthy giraffe – the initial advantage it conferred would gradually be swamped out in subsequent generations through blending. To retain the adaptation, a geographical barrier was therefore required to isolate parts of a population, to partition the effects of selection, and thereby to prevent reversion to the phenotypic mean – hence, Darwin's early view that speciation required geographical isolation. But Darwin's extensive travels and field studies had established precisely the opposite. It appeared many (perhaps most) species had emerged in the absence of geographical barriers. The Galapagos Island finches were a striking example. The sea between islands in the archipelago presented little obstacle to migration for these aeronautical organisms, yet many morphologically distinct species seemed to have emerged from one or a small number of wayward founders of a single species. If inheritance is a blending process and physical barriers did not exist to partition selection's effects, by what mechanism did these species emerge?

Darwin's niche concept provided resources for answering this question. The key insight was generalizing what counts as a species' environment to recognize the significant selective impact of intra- and interspecific interactions between organisms, interactions that may change in form and intensity over time. As realizers of particular functional roles in an ecosystem (i.e., niches), organisms face more than just a static environment composed of a suite of abiotic factors such as precipitation, temperature, and nutrient availability. Their niche is also the product of intraspecific interactions and is shaped by relationships with other species occupying different niches in the overall dynamics of the ecological system, which are also often evolving. These interactions and interspecific relationships could, in turn, produce reproductive barriers that would prevent blending across generations.

In the *Origin*, Darwin (1859, 103) highlighted precisely this kind of possibility, noting that "within the same area, varieties of the same animal can long remain distinct, from haunting different stations, from breeding at slightly different seasons, or from varieties of the same kind preferring to pair together." This pithy statement is pregnant with insights about the ecological processes that can catalyze sympatric speciation. The first two cases Darwin mentions are different forms of *niche partitioning*. "Haunting different stations" involves *spatial* partitioning. Rather than being imposed by a physical barrier,

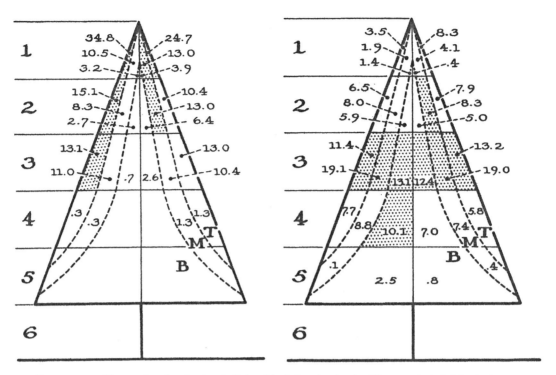

FIGURE 47.2. Diagrams based on drawings by Robert MacArthur showing the different ecological niches in the same tree occupied and used by two different bird species. Permission: Ecological Society of America

organism behavior reinforced by natural selection can spatially partition a niche and produce the impediments to reproduction that fuel speciation.

The ecologist Robert MacArthur (1958) was one of the first to rigorously document this kind of phenomenon in his rightfully famous dissertation research on New England warblers (Fig. 47.2). The objective of the study was to determine how so many behaviorally and physiologically similar bird species could coexist in boreal forests, which seemed to contradict prevailing ecological theory. With such similar properties, it seemed that interspecific competition would be especially strong between the birds and eventually lead to the extirpation of all but one species. A quarter century before MacArthur's work, the Russian ecologist Georgyi Gause (1934) observed similar dynamics between *Paramecium* and yeast species, respectively. In constant ecological conditions (e.g., nutrient levels, water temperature, turbidity) and in the absence of refugia for *Paramecium* that would mitigate the effects of interspecific competition, one species inevitably outcompeted the other to extinction. On this basis, Gause generalized the *competitive exclusion principle*: species with identical niches, that is, two species that would compete for exactly the same resources, cannot coexist. Drawing upon Darwin's work a few decades before, the early ecologist Joseph Grinnell (1917) arrived at the same kind of exclusionary principle (see Hardin 1960). The intuitive appeal of the idea and its apparently exceptionless status across many different biological systems has prompted its honorific designation as a law of ecology, Gause's Law. Apart from the extreme case of exclusion, the degrees and types of niche overlap that permit coexistence has become an important focus in contemporary attempts to explain species distribution patterns and dynamics in biological communities (Abrams 1983). An apparently recalcitrant counterexample to Gause's principle was originally identified by MacArthur's adviser, G. E. Hutchinson (1961): the seemingly inordinate number of plankton species, given their simple, homogeneous niche space. Known as the "paradox of the plankton," this issue also remains an active area of contemporary ecological research and is yet to be conclusively resolved (see Tilman, Kilham, and Kilham 1982).

For the warblers, MacArthur uncovered the mechanism that eluded the exclusionary outcome: different species bred and fed in distinct parts of coniferous trees and, furthermore, exhibited strong territoriality toward those parts. These behaviors effectively divided the homogeneous arboreal habitat into disparate sections, thereby (spatially) partitioning the niche space. Currently, this process curtails competition and allows the extant set of warblers to coexist. But it also likely produced the intraspecific variation originally required for speciation to occur.

"Seasonal breeding" involves *temporal* niche partitioning, and MacArthur (1958) was again one of the first to demonstrate its ecological import. He found that nesting times differed across the warbler species. This affects the same minimization of competition as spatial partitioning. Ecologically, niche partitioning is a quite general phenomenon. Besides time and space, similar dynamics emerge for any factor that influences behavior in a way that hinders reproduction. Of course, niche partitioning can also have a nonbehavioristic basis. As thoroughly documented by the ecologist and evolutionary biologist David Lack (1947), Darwin's finches provide a striking example. The different seed sizes and shapes that account for marked differences across Galapagos Island finches beaks are the primary drivers of niche differences between them, minimizing competition and thereby permitting coexistence. And behaviorally induced niche partitions can evolve into nonbehavioral partitions through the influence of natural selection. What begins solely as behavioral partitioning – preferences for different habitat types, for example – may evolve into a physiological barrier to reproduction as selection reinforces differences between individuals in the different habitats. This ossification of behavioral differences into physiological differences can obviously occur only for organisms with sufficiently rich behavior – behavior that can catalyze niche partitioning – but for such organisms it constitutes a powerful vehicle by which natural selection can induce evolutionary change.

The final clause of Darwin's prescient passage recognizes another way in which organism behavior can drive evolutionary change. In contemporary terminology, the preferential pairing in question is called "assortative mating." Occurring when individuals in a population form nonrandom reproductive pairings, assortative mating comes in positive and negative forms. With positive assortative mating, similar individuals (or similar with respect to a specific trait) are more likely to pair; with negative assortative mating, dissimilar individuals (or dissimilar with respect to a specific trait) are more likely to pair. For example, humans positively assortatively mate by race: individuals of one race are more likely to mate with one another than with individuals of a different race. On the other hand, the alleged attraction between opposites among humans would constitute negative assortative mating.

Positive assortative mating with respect to heritable traits tends to increase their variance, thereby providing a greater range on which selection can act. This alone, like niche partitioning, can initiate and sustain the reproductive barriers necessary for speciation (see Kondrashov and Shpak 1998). It can also provide the basis for divergent selection that leads to genetic and phenotypic differences within a species, and ultimately speciation. As a specific, female-driven form of assortative mating, sexual selection is usually responsible for sexual dimorphism in many vertebrate species. Among *Homo sapiens*, Darwin (1871a, ch. 2) correctly observed that the larger average size of males than females has been one evolutionary outcome.

For all these processes – spatial and temporal niche partitioning and assortative mating – reproductive barriers between subpopulations are created not through physical isolation but rather through isolation in niche space. Focusing on evolutionary dynamics at this more abstract level of ecological representation recognized the significant role species play shaping the selective forces impinging on themselves and other species. By doing so, Darwin avoided the daunting problem that blending inheritance seems to preclude the emergence of new species. More important than this historical point, the niche perspective revealed the underappreciated power of natural selection as a mechanism of evolutionary change, long before

Joseph Grinnell (1917) is usually credited with having rigorously introduced the niche concept into biology. Besides this insight about *evolutionary* dynamics, the next section shows that Darwin's conceptualization of the natural world in terms of niches also had implications for how the ecological dynamics of biological communities, which occur on much shorter time scales, should be understood.

COMPETITION, COMMUNITY STRUCTURE, AND THE "BALANCE OF NATURE"

Darwin (1859, 73), with most of his scientific contemporaries, was committed to the idea of a "balance of nature":

> Battle within battle must ever be recurring with varying success; and yet in the long-run the forces are so nicely balanced, that the face of nature remains uniform for long periods of time, though assuredly the merest trifle would often give the victory to one organic being over another.

The commitment to some type of balance was a staple of the schools of natural philosophy from which biology emerged, long before the term "ecology" was even coined (Egerton 1973). Darwin and other early ecologists continued this tradition by attempting to derive the existence of a "natural balance" in biological populations from organismic metaphors and anthologies with physical systems, although the analogical and metaphorical content often differed (see Kingsland 1995). For example, the ecologist Frederic Clements (1916) is best known for claiming to find functional integration within biological communities that resembled the physiological integration within individual organisms, and which justified conceptualizing communities as a kind of superorganism with analogous homeostatic properties (Fig. 47.3). But Darwin (1859, 115–16) employed the same metaphor several decades before, with a much less problematic aim:

> The advantage of diversification in the inhabitants of the same region is, in fact, the same as that of the physiological division of labor in the organs of the same individual body.... No physiologist doubts that a stomach by being adapted to digest vegetable matter alone, or flesh alone, draws more nutriment from these substances. So in the general economy of any land, the more widely and perfectly the animals and plants diversified for different habits of life, so will a greater number of individuals be capable of there supporting themselves. A set of animals, with their organization but little diversified, could hardly compete with a set more perfectly diversified in structure.

Although this conclusion likely holds for communities in relatively constant environments and thereby provides a plausible explanation of the greater species diversity found in the tropics than in more environmentally turbulent temperate regions (see Rosenzweig 1992), later ecologists would show that specialization often constitutes a handicap in fluctuating environments that favor adaptable generalists (e.g., Pianka 1970).

FIGURE 47.3. Much influenced by Herbert Spencer, Frederick Clements (1874–1945) regarded ecological systems as being group-type organisms, growing, thriving, and reproducing like individual organisms. The mature form is called a "climax formation," and shown here is a climax forest, on Mount Rainier, in the state of Washington. From F. Clements, *Plant Succession* (Washington, D. C.: Carnegie Institute, 1916, 106)

There were, however, important differences between Darwin and most of his predecessors' views on the character of this balance. To appreciate these differences, two threads in Darwin's view should be distinguished. Perhaps the most important was what he thought the causal forces responsible for the putative balance were. Most scientists before Darwin, Charles Lyell being the clear exception (Pearce 2010), did not fully appreciate the extent inter- *and intra*specific competition shaped communities (Bowler 1976). For them, a balance of nature was the result of a predetermined harmony that competition would only undermine. Similarly, Lamarck thought the use and disuse mechanism of evolutionary change would generally minimize potential competition between organisms. As illustrated in the proverbial example, the advantaged giraffe's longer neck would not facilitate it outcompeting lesser-endowed giraffes. Rather, it would expand the giraffe's resource pool (and thus its niche), which would decrease its utilization of the resources accessible to other giraffes, thereby decreasing competition.

Darwin's balance of nature was undergirded by a much more harsh but realistic dynamics. Interspecific competition, for example, constrained the populations composing biological communities by limiting organisms' access to the resources they need to metabolize and ultimately reproduce. This curtailed populations' geometric tendency to increase through the same mechanism grounding the competitive exclusion principle. Other forms of interspecific interaction have similar consequences. Predators and parasites, for instance, inhibit prey and host populations. Intraspecific competition produces the same inhibitory effect within a species, and it can inhibit other species through interspecific relationships. For example, intraspecific competition among prey limits predator populations.

But, as Darwin was well aware, these inhibitory relationships do not alone account for the kind of dynamic balance ostensibly exhibited in the natural world. The problem was the differential power and scope of intra- and interspecific competition. Intraspecific competition is fully general: it arguably occurs in all biological populations (although see G. Cooper 2001). But its power to restrain population growth is governed by the availability of resources. When resources are plentiful, little check on growth occurs. On the other hand, interspecific competition (predation, parasitism, etc.) can suppress population growth more effectively than intraspecific dynamics, but it is not universal: not all species seem to be connected in inhibitory interspecific relations. Thus, although intraspecific competition would limit *all* populations when resources were scarce and interspecific interactions would *sometimes* suppress growth further, if these were the only checks on populations, it seemed many species would exhibit unrealistic rates of growth for unrealistic periods of time.

For Darwin, the potential problem stemmed from underappreciating a second important thread in his concept of a balance of nature: the extent of inhibitory interspecific relationships throughout the biological world. Darwin (1859, 80) saw nature as a vastly complicated and intricately complementary set of ecological interdependencies between species: "[H]ow infinitely complex and close-fitting are the mutual relations of all organic beings to each other and to their physical conditions of life." Although most species do not interact directly, Darwin believed they do indirectly through chains of intermediaries; the result is a "web of complex relations" (73) in which all species are ecologically connected. A specific species' position in the web indicates what other species curb or enhance its growth. Darwin described examples of several such food webs, perhaps the best known (and engaging) being the ecologically serpentine relation between a clover species (*Trifolium pratense*) and the common cat:

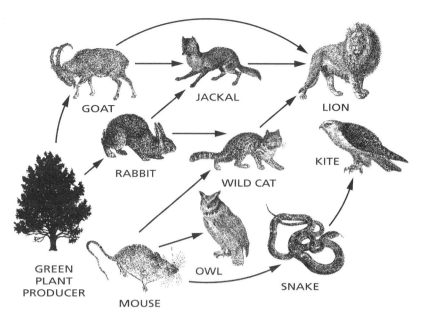

FIGURE 47.4. A "food web" diagram, showing the connections between and dependences of different parts of a community, in this case a forest.

> From experiments I have tried, I have found that the visits of bees, if not indispensable, are at least highly beneficial to the fertilization of our clovers; but humble-bees [bumble-bees] alone visit the common red clover (*Trifolium pratense*), as other bees cannot reach the nectar. Hence I have very little doubt, that if the whole genus of humble-bees became extinct or very rare in England, the heartsease and red clover would become very rare, or wholly disappear. The number of humble-bees in any district depends in a great degree on the number of field-mice, which destroy their combs and nests.... Now the number of mice is largely dependent, as every one knows, on the number of cats; and Mr. Newman says, "Near villages and small towns I have found the nests of humble-bees more numerous than elsewhere, which I attribute to the number of cats that destroy the mice." Hence it is quite credible that the presence of a feline animal in large numbers in a district might determine, through the intervention first of mice and then of bees, the frequency of certain flowers in the district! (73–74)

Notice that the types of relationships differ in this food web, and some are beneficial. The humble-bee–clover relationship is mutualistic and is obligate for the clover according to Darwin. This relationship benefits both species, but it also ties the clover's fate to the influence of field mice and in turn to the cats. Not all parts of the complex set of ecological relationships in nature exemplify an antagonistic struggle for survival. But the population-suppressing effects of those struggles are propagated throughout the web *via* those relationships.

Investigating the structure of different food webs and how their properties might generate stability at the population and community level remains an active area of ecological research today (see J. Cohen, Briand, and Newman 1990) (Fig. 47.4). Darwin's allusion to the indispensability of the bees is particularly prescient. It foreshadows the influential concept of a *keystone species*—a species in a community that plays a far more significant role in community dynamics relative to its abundance than other species—first developed in Thomas Paine's (1966) experiment work establishing the starfish *Pisaster ochraceus* as a keystone species of coastal marine communities in the northwest United States. Some of this work has also challenged Darwin's claim that there are high degrees of connectedness in biological communities (McCann et al. 1998), or that a high degree of connectedness enhances stability (May 1974a).

Unlike previous accounts that assumed a static, predetermined pattern or structure, Darwin's web-based balance of nature concept was rooted in the struggle between individual organisms to survive and reproduce. Species were precisely balanced at their current population levels through a complex array of checks and balances finely honed by natural selection. As the final clause in Darwin's (1859, 73) claim about balance indicates, the exact character of the balance could change as species evolved, so in this sense the niche structure of a community was not fixed. But even this kind of balance requires an equilibrium assumption: population levels at a given time reflect the homeostatic processes of a biological community at a point equilibrium. This assumption was implicit in almost all ecological theorizing until the mid 1970s, and it made the first mathematical models of ecological systems analytically tractable. But this predominant focus has also been supplanted with a recognition that nonequilibrium models with complex dynamics such as chaos, limit cycles, and so-called strange attractor sets may best represent many types of ecological systems (May 1974b; DeAngelis and Waterhouse 1987). These and other shortcomings, however, do little to muddy Darwin's record of ecological insights. As briefly and incompletely recounted in this essay, his contributions helped set much of the agenda of contemporary ecological science.

ESSAY 48

Darwin and the Environment

David Steffes

CHARLES DARWIN IS WELL known for his portrayal of the endless struggle in nature, the view that Tennyson immortalized as "nature red in tooth and claw" (Fig. 48.1). In addition to the fangs and claws of predators, Darwin's "red view" also incorporated more subtle mortality factors, such as competition for food or mates, crowding, disease, parasitism, and climate flux (Plate XXXVII). Darwin detailed the central features of his red view in chapter 3 of *Origin of Species* (derived from chapter 5 of his unpublished "big species book"), entitled "The Struggle for Existence," wherein he argued that the rate of population increase was so great that it regularly outstripped nature's resources, compelling a constant struggle that had to be alleviated through a compensatory rate of death (Darwin 1859, 60–89; Darwin 1975, 172–212). Darwin proposed that the primary agent responsible for subduing this growth was "natural selection," which drew upon the environment's factors of mortality to eliminate the less favorable individuals and populations, thereby shaping the kinds and numbers of organisms found in nature. For Darwin, environment not only imposed the limits to growth but also applied the pressures to mold the population's fitness. He explained that it was because of environmental "checks" that natural selection was "daily and hourly scrutinizing [for the] improvement of each organic being in relation to its organic and inorganic conditions of life" (Darwin 1859, 84). Environment, according to the red view, was an engine of "warfare" among and within species, or what his friend T. H. Huxley characterizes as a gladiatorial blood sport, in which "the creatures are fairly well treated and set to fight, where the strongest, the swiftest, and the cunningest live to fight another day" (T. H. Huxley 1894, 200; La Vergata 1990).

THE PERCEIVED INCOMPATIBILITY OF DARWINISM AND AN ENVIRONMENTAL PERSPECTIVE

Darwin's red view has become so iconic in the past century that we now scarcely bother to consider whether Darwin devised a complimentary "green view" of nature: where the natural environment is not merely a source of struggle and death but also a source of tremendous biological diversity and ecological sustainability. Perhaps we

FIGURE 48.1. Alfred Lord Tennyson (1809–92), favorite Victorian poet and coiner of "nature red in tooth and claw." The phrase occurs in the poem *In Memoriam*, published nearly a decade before the *Origin*, and refers to the hopeless vision Tennyson abstracted from Lyell's *Principles of Geology*, where there are endless cycles of purposeless life forms. It was learning of the progressivist vision of Robert Chambers's evolutionary tract *Vestiges of the Natural History of Creation* that gave Tennyson his spirit to finish the poem. From W. E. Smyser, *Tennyson* (New York: Eaton and Mains, 1906)

have been seduced by enthusiastic "red Darwinists" such as Richard Dawkins and G. C. Williams to believe that nature is primarily an "arms race" among selfish survival machines, the violence and malevolence of which demand an "extreme condemnation of nature," as Williams put it (Dawkins 1976; Williams 1988, 383–85). The popular perception of modern evolutionary biology is that it operates based solely on this red view, such that Darwinism is "useful only for explaining (and justifying) individualistic selfish greed" and therefore cannot be counted on to aid ecologists, conservationists, and other perpetrators of "green" agendas (Penn 2003, 277). Furthermore, since the 1970s, the perception of Darwinism as an enemy to green agendas has also been reinforced by certain intellectual factions within the environmental movement. Some groups, particularly within ecofeminism, ecosocialism, ecotheism, and postmodernists, have expressed opposition toward Darwinism, blaming Darwin for unwarranted applications of his theory during the late nineteenth and twentieth centuries, including the use of Darwinism by industrialists and social Darwinists to justify the stripping of nature as an exercise of their right as "the fittest"; and the use of Darwinism by modern scientists and philosophers to reduce the living world to mechanist-materialist processes and parts, thereby excluding intrinsic and aesthetic values from nature, and the view that nature is "sacred."

However, belying this perception of Darwinism as incompatible with environmental thinking is the fact that Darwinian biology has been in cahoots with ecologists and conservation biologists since the 1960s, aiding them in their investigations of biological diversity, environmental change, and the human-environment interaction. In the 1960s and 1970s, just as Western culture was entering its "Age of Ecology," the actual science of ecology was in the midst of a merger with evolutionary biology, forming new fields such as "evolutionary ecology," "island biogeography," and "invasion ecology" (Futuyma 1986; E. O. Wilson 1994, 238–59; Haila 2002; D. M. Richardson and Pyšek 2008). These fledgling sciences sought to integrate different biological approaches in order to investigate the changing environment and its impact on the evolution of biological diversity. Monographs such as Richard Levins's *Evolution in Changing Environments* (1968) and Robert MacArthur and E. O. Wilson's *Theory of Island Biogeography* (1967) demonstrated this shift in focus toward environment-based research: Levins examined the effects of environmental heterogeneity on the fitness of a species as it varied over time, while MacArthur and Wilson considered the ways in which environment's structure (patchy islands, fragments) contributed to the patterns of fluctuation in multispecies assemblages, or what are typically called "communities" (Fig. 48.2). Jared Diamond and Martin Cody's 1975 volume, *Ecology and Evolution of Communities*, confirmed that a new wave of "Darwinian ecologists" had set out to understand the processes by which the environment traded off ecological stability, and how these processes impacted diversity and future evolutionary opportunities. They were motivated by a growing concern for earth's fragile biosphere, which had come under siege from habitat destruction, nuclear testing, chemical pollution, and human population expansion. Pioneering field experiments like Thomas Lovejoy's Amazon

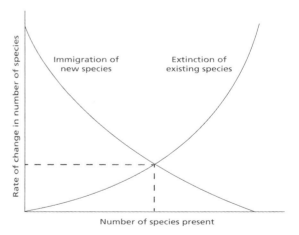

FIGURE 48.2. The MacArthur-Wilson theory of island biogeography argued that the number of species on an island reaches equilibrium as the number of invaders equals the number going extinct. Actual numbers are a function of the distance of the island from the mainland and the size of the island.

rainforest project furnished the first conclusive evidence that human activities were causing unprecedented diversity loss with their reconfiguration of the environment. In response to these findings, yet another new scientific field – "conservation biology" – was formed to steer existing research programs in Darwinian ecology toward the problem of diversity loss and to employ this research to aid policy decisions on environmental issues (Soulé 1986; Simberloff 1988).

Given the integral role that Darwinian biology has had in these contemporary efforts to study environmental issues, it seems only reasonable to wonder whether Darwin himself may have offered a "green view" of the natural word. There is certainly a case to be made. However, one has to be careful not to transpose contemporary themes from environmentalism onto Darwin's thinking in *Origin*. For instance, even though Darwin introduced numerous protoecological ideas in *Origin* and inspired later ecologists to refer to him as "the first evolutionary ecologist" and "greatest of all ecologists," it should be recognized that *Origin* scarcely dealt with the dynamics of communities or ecosystems – the core units of environmental research in the latter half of the twentieth century (quotes from Harper 1967, 247; Rosenzweig 1987, 3; see also Justus, Essay 47 in this volume). Historians have shown that both "community ecology" and "ecosystem ecology" took shape in the early to mid-twentieth century under the auspices of ecologists who were not enthusiastic about Darwinism and that these fields only later merged with Darwinism in the 1960s after the rise of evolutionary ecology (Futuyma 1986; Hagen 1992; Kingsland 2005). Likewise, it should be recognized that, although Darwin's *Origin* and *Descent of Man* (1871) firmly situated mankind within the realm of nature (portraying humans as higher animals), Darwin did *not* suggest what sort of relationship humans should maintain with their environment; he offered neither a conservation ideal, in the sense of the fostering a "harmony between men and land" (as Aldo Leopold proposed in 1949), nor an environmental ethic, in the sense of seeking to establish value in different aspects of the natural environment (as done by Arne Naess, Peter Singer, Garrett Hardin, Charles Birch, and others environmental philosophers).

To make the case for a "green Darwinism," one must focus on the theme of biological diversity. Darwin wrote prolifically in his journals and notebooks on the splendor of the diversity exhibited by the natural world and later made diversity a central theme in *Origin*, where it tied together his interests in natural history, taxonomy, and biogeography. Darwin's fascination with the German *Naturphilosophie* seems to have inspired a romantic vision of nature that emphasized the vital powers of environment as responsible for the aesthetic beauty of the living world (R. J. Richards 2002b). The icon of this romantic nature was the tropical forest environment, which was first introduced to Darwin by the writings of the Prussian naturalist Alexander von Humboldt and then revealed in full glory to Darwin during his own voyage to South America and the Pacific in the 1830s. Over a century later, amid the environmental crisis of the Western world, the theme of biological diversity again rose to prominence, forming the basis of both conservation biology (beginning in the 1970s) and the "biodiversity" movement (beginning in the 1980s). Once again, the tropical environment became the center of attention, this time as part of scientific and public scrutiny of the destruction of rainforests. Like Darwin, the pioneers of this modern biodiversity movement (E. O. Wilson, Thomas Lovejoy, Peter Raven, and others) have been at pains to introduce the general public to the complexity of evolutionary and ecological processes that have shaped the diversity of life in the tropics (Takacs 1996). Unlike Darwin, however, their intention has been to convey the price that the earth will have to pay for the loss of diversity as a result of deforestation and other human activities. Nevertheless, it is the theme of biological diversity that builds a bridge between the environmental perspective of Charles Darwin himself and the perspective of twentieth- and twenty-first-century advocates of "biodiversity" and conservation biology.

FINDING SHADES OF GREEN IN DARWIN'S *ORIGIN*

In composing *Origin*, Darwin sought to depict not just a "tooth and claw" wilderness but also the sublime of an enchanted natural world that had been crafted by powerful environmental forces (R. J. Richards 2002b, 514–54). He acquired this deeper, aesthetic appreciation for nature from the romantic philosophy of the *Naturphilosophen*, and from the writings of the romantic naturalist-geographer Alexander von Humboldt. It was Humboldt's reverent account of the tropical wilderness of South America in his *Personal Narrative of Travels* that inspired Darwin to embark on his own voyage on the *Beagle* during the 1830s. Humboldt foretold of the naturalist's experience in the tropics that, "on no other part of the globe is [the naturalist] called upon more powerfully by nature, to raise himself to general ideas on the cause of phenomena, and their natural connection" (R. J. Richards 2002b, 523). The lush wilderness revealed "that luxuriance of vegetation, that eternal spring of organic life, those climates varying by stages as we climb the flanks of the Cordilleras, and those majestic rivers which a celebrated writer [Chateaubriand] has described with so much precision" (ibid.). Darwin eventually experienced tropical paradise for himself during his 1830s voyage and later drew upon this important experience to summarize nature's extraordinary diversity and complexity in *Origin* (1859, 489–90): "[I]t is interesting to contemplate an entangled bank, clothed with many plants of many kinds, with birds singing on the bushes, with various insects flitting about, and with worms crawling through the damp earth, and to reflect that these elaborately constructed forms, so different from each other, and so dependent on each other in so complex a manner, have all been produced by laws acting around us" (Fig. 48.3). Darwin's experience in the tropics of South America and the Pacific convinced him that even with the "the war of nature" and its requisite "famine and death," the natural environment was still worthy of being commended as view, more than his red view, accounted for the causes of taxonomic differences and laid the groundwork for the study of the origin of species (Ospovat 1981; Winsor 2009). The *Beagle* voyage provided Darwin with crucial evidence that species had undergone ecological and geographical divergence, and consequently, that the natural environment played a key role in emergence of new types. As Darwin reclined in his study at Down House during the 1850s, contemplating the process by which divergence unfolded, he reflected back on his observations of the Galapagos finches and tortoises and slowly recognized the subtle ways in which "organic conditions of life" had modified the characteristics of species over time (Darwin 1859, 111–26; Ospovat 1981, 194–98). He saw that a "complex web of relations" between and among species had caused a differentiation in organic functions and habitats and that this differentiation had opened the door for the gradual emergence of evolutionary novelty. Marveling at the transformative power of the environment over the course of time, Darwin (1859, 490) remarked in *Origin* that "there is a grandeur to this view of life," a splendor that had been missing from traditional accounts of a static nature.

Thus, for Darwin, nature's plenum of diversity had been the result of dynamic environment, not the stable "balance of nature" so often credited by traditional sources in natural history (Egerton 1973). Darwin discussed only a limited stability for nature. He disliked the word "equilibrium," which suggested that there was "far too much quiescence" (Darwin 1975, 187–88). It did not fit with his belief that the economy of nature was always being pressed by "ten thousand sharp wedges, many of the same shape and many of different shapes representing different species, all packed closely together and all driven in by incessant blows" (Darwin 1859, 67). Darwin accounted for the human preoccupation with balance by pointing out that people were generally ignorant of the innumerable "wedges" driven into nature. He indicated that it was "causes quite inappreciable by us" that had determined "whether a given species shall be abundant or scanty in numbers" (Darwin 1975, 188). Furthermore, he explained that, because of the power of selection to adapt living forms to each other, certain groups of species had become delicately synchronized in communities over time, presenting *the appearance* of harmonious balance: "[I]n the long-run the forces are so nicely balanced that the face of nature remains uniform for long periods of time" (Darwin 1859, 73). However, Darwin stressed that, in the shorter time scales, only very slight environmental alterations were needed to inflict serious losses, even in the stoutest of communities. He noted that "the merest trifle would often give the victory to one organic being over another," rattling the previously stable web of life and thus pushing some species to rarity or extinction (73). Briefly sympathizing with the common outlook on nature, Darwin admitted that, "when one views the contented face of a bright landscape or a tropical forest glowing with life, one may well doubt [the struggle for existence]." Nevertheless, he hoped that the logic of his arguments would sway rational individuals to put aside their tendency to see a "balance" everywhere

FIGURE 48.3. The Brazilian rainforest as experienced by the young Charles Darwin, a vision filtered through his enthusiastic reading of the romantic writings of Alexander von Humboldt. From C. Darwin, *Journal of Researches into the Natural History and Geology of the Countries Visited during the Voyage Round the World of H.M.S. "Beagle" under the Command of Captain Fitz Roy, R.A. Second Edition, Corrected, with Additions* (London: John Murray, 1845)

the source of vitality and beauty, and to be revered in romantic poetry for its magnificent splendor. The complexity of the environment, after all, had given rise to a profusion of living forms, and ultimately "the production of the higher animals" (490).

Inspired by his Humboldtian vision of a lush and fertile landscape, Darwin fashioned a "green" view of nature, portraying the environment as a positive force (or set of forces) in the creation, preservation, and advancement of the diversity of life. The green vein of Darwinism was not exclusive from the red vein; green Darwinism was also predicated on natural selection (and hence population forces) but was centrally concerned with selection's impact in ecological and geographical contexts, where it was responsible for producing local adaptive specializations and, over the long course of time, the divergence of taxa (the phyletic branching of species, genera, families, and so forth). Darwin's green and to accept that "the doctrine that all nature is at war is most true" (Darwin 1975, 175–76).

This is not to say that Darwin was unaware that, in some cases, the natural environment exhibited great stability in its "complex web of relations" and maintained a relatively constant proportion in its local diversity. In the *Origin* (1859, 74–75), he described a forest system that exhibited this sort of stability: "[E]veryone has heard that when an American forest is cut down, a very different vegetation springs up; but it has been observed that the trees now growing on the ancient Indian mounds, in the Southern United States, display the same beautiful diversity and proportion of kinds as in the surrounding virgin forests." Darwin suggested that a very delicate balance had been sustained throughout the "long century" in which this fragmented forest had thrived and was probably the result of a stubborn persistence in the proportions of the forest's trees, insects, snails, birds, and other organisms affecting the local environment, as well as the physical conditions in which these creatures lived. Yet he still believed that the forest sat on the edge of a knife, always subject to transformation under the appropriate conditions. Darwin never wavered from his conviction that "whatever the number of a species in any country may be, the average being determined by a complex struggle, that number will steadily decrease, if we add without any compensation the least additional cause of destruction, until the species becomes extinct" (Darwin 1975, 175). An ecologist from the twentieth century would have found the example of the American forest an interesting case study, but for Darwin, such ecological systems were not the focus of his "green view." Rather, Darwin emphasized the historical development of biological diversity, and the environment's role in shaping this diversity through evolutionary processes. His protoecological ideas were all associated with this emphasis upon the development of diversity (Ospovat 1981, 194–98; Pearce 2009).

"GREEN DARWINISM" IN THE TWENTIETH CENTURY: ECOLOGY, BIOGEOGRAPHY, AND THE EMERGENCE OF BIOLOGICAL DIVERSITY RESEARCH

During the 1960s and 1970s, community ecology aligned with the nascent fields of evolutionary ecology and island biogeography, fostering a new line of research in which ecologists could investigate biological diversity patterns and their environmental determinants. One of the major objectives was to study the ways in which the shifting environment impacted ecological stability; how the loss of stability – in some cases, a total ecological collapse – affected both the present status of diversity and the future availability of evolutionary opportunities. Yale ecologist Thomas Lovejoy set out to investigate this complex relationship between diversity and environment beginning in the mid-1960s, conducting extensive field research on the ecology the Amazonian rainforest. Lovejoy coordinated with U.S. and Brazilian governments to study "fragments" of rainforest left over from deforestation and agricultural encroachment.

These fragments were patches of suitable habitat surrounded by nonsuitable habitat, ecologically similar to oceanic islands, which had a limited carrying capacity for species numbers owing to their limited quantity of niches in which species could specialize (Pimm 1998; Haila 2002). Lovejoy knew that his contemporaries E. O. Wilson and Robert MacArthur were in the midst of building theoretical models on island biogeography that would describe the behavior of biotic communities in small fragmented environments. He wanted to test the applicability of the Wilson-MacArthur island models in the case of forest fragments, seeing if "forest island dynamics" existed. He also sought to determine whether their models could be implemented to evaluate the extent of fluctuations in forest diversity, predicting the degree to which immigration, emigration, and extinction conspired to account for the number of species in these fragments. Lovejoy especially wanted to know whether fragments could sustain a "minimum viable population" for their species, creating a delicate balance in the ecosystem that would preserve diversity (Pimm 1998, 23–24) (Plate XXXVIII).

By the mid-1970s, Lovejoy was able to conclude that fragments failed to sustain the native species of palms, vines, bees, butterflies, birds, primates, and so forth, and that the consequence of further deforestation (further fragmentation) would be the rapid extinction of the rainforest's flora and fauna. Lovejoy had proved what many environmentalists and conservationists had already feared: that the destruction of rainforests would pose a serious problem for the earth as a whole. Rainforests were "mega-diversity habitats" that sustained more than half the world's biota. To cut or burn even a small section of these habitats would result in countless extinctions, and it was already well known by the 1970s that vast tracts of forests had been wiped out by farming and industry. Because the diversity in rainforests represented such a large portion of the word's biota, Lovejoy (1980) was able to use his Amazon diversity measures to devise some of the earliest estimates for the *global* rate of extinction. He was joined in the 1980s by fellow diversity researchers E. O. Wilson and Peter Raven, who helped to promote his extinction estimates in order to draw greater attention to the need for new rainforest conservation measures. At the 1986 U.S. National Forum on Biodiversity, Wilson (1988, 8) warned that "forests are being destroyed so rapidly that they will mostly disappear within the next century, taking with them hundreds of thousands of species into extinction." He estimated that the deforestation of the world's rainforests was on course to extinguish a quarter of the earth's species by the midpoint of the twenty-first century – extinction at an astonishing rate of fifty thousand species per year (Stevens 1995). Raven (1988, 121) pointed out that if rainforest extinctions continued at this rate, there would soon be a mass extinction event equal to those marked in geological history, most notably the one that led to the demise of the dinosaurs 65 million years ago. Even as it stood, the damage done to global diversity would require a long term fix. Wilson (1984, 121) noted that it would "take millions of years to correct … the loss of genetic and species diversity by the destruction of natural habitats."

Lovejoy, Wilson, and Raven promoted habitat destruction as the biggest threat to diversity in the twenty-first century (Lovejoy 1996, 7–14). They emphasized not only the disappearance of rainforests but also the ruin of earth's barrier reefs, freshwater lakes, coastal wetlands, and other hubs of organic relations. Yet habitat destruction was not the only factor threatening global diversity. Ecologists in the 1980s also took notice of the alarming rise in numbers of exotic species being introduced into new habitats. These alien invaders rapidly transformed their new environment and homogenized its populations, destroying the preexisting diversity. The problem of introduced species, or what Charles Elton termed "invasive species" in his classic 1958 book, led to the offshoot of an entirely new branch of evolutionary ecology in the 1980s called "invasion ecology" – a close ally to conservation biology (Elton 1958; Richardson and Pyšek 2008). The description of exotic species as "invasive" has always been somewhat misleading because in many cases, these species were introduced quite accidentally through human migration or commerce. Darwin himself considered the problem of invasions in chapter 5 of the long manuscript on "Natural Selection" (1856–58), where he offered numerous examples of invasions and their disastrous consequences for small, undiversified genera in native lands (Darwin 1975, 172–212; see also Justus, Essay 47 in this volume). Darwin emphasized "the enormous increase of birds, fish, frogs, snails & insects, when turned out in new countries," and promised that "of the rapid increase of plants run wild, numerous instances could be given" (Darwin 1975, 178). Contemporary ecologists showed that species introduced through Western imperialism – particularly Britain's "Second Empire," from 1800 to 1945 – were responsible for some of the highest extinction rates in modern history. The Australian continent was especially hard hit, accounting for more than half of the world's mammal extinctions between 1800 and 2000 (T. Griffiths and Robin 1997; Dunlap 1999; C. Johnson 2007). Smaller "capitals of extinction" also sprang up in Hawaii and Florida, where human settlers facilitated the impact of introduced species by constructing distinctly "human habitats" swept clean of native predators and other natural defenses (Simberloff, Schmitz, and Brown 1997).

During the 1980s and 1990s, biological diversity experts became the torch bearers for Darwin's green view of nature, or what may be called "green Darwinism." They boosted scientific and public awareness of the value of nature's "entangled bank," which had fallen under threat from human agency. E. O. Wilson imbedded within his 1992 narrative *The Diversity of Life* a very similar reverence for the tropical wilderness to what Darwin exhibited in his journals and notebooks. As one reviewer described, "Wilson takes us by the hand and leads us through the wilderness of diversity – explaining along the way how species evolve, adapt, specialize, colonize, hybridize, recreate new versions of themselves, radiate out to new locations, become new things in often symbiotic combination with other new things, then transmogrify themselves into something else and move on again to fill other niches" (Watkins 1992). Popular narratives like Wilson's were important tools in the biodiversity movement. Diversity advocates believed that people would be profoundly influenced by nature's multitude when confronted with it; thus, if society would not go to see the rainforest, then biodiversity experts had to bring the rainforest to society through compelling stories and descriptions (Takacs 1996).

It should be noted that biodiversity experts also differed from Darwin in some aspects. For instance, they veered from Darwin's unshakable faith in the capacity of the environment to generate diversity time and time again. In *Origin*, Darwin expressed great confidence that the natural environment would continue to provide copious evolutionary opportunities and, hence, continue the flow of diversity in the living world. He cited the case of the old Indian ruin to illustrate that, in some situations, the natural environment would keep the same proportions of kinds in diversity even after a significant environmental disturbance, presumably because of the strong ecological web of relations (Darwin 1859, 74–75; the ecologist C. S. Hollings would later call this phenomenon "ecological resilience"). More often than not, however, environmental fluctuation would cause one sort of fragile balance to pass away in favor of another sort of fragile balance, providing a new window of evolutionary opportunities from which selection could bring forth diversity. Darwin believed that through the "wedging" of environmental forces, diversity would rebound again and again, although its proportions of kinds would ultimately vary. Lovejoy, Wilson, and Raven discovered in their research that Darwin's faith in this perpetual diversification of life had been misplaced; the effectiveness of evolutionary processes in generating diversity could not be ensured in a world dominated by humans. Wherever significant human activities were involved (such as in deforestation, or introduced species), the loss of ecological stability was often so pronounced, and so rapid, that it not only diminished the current diversity of species but also undercut the resilience of the ecological system, thereby degrading the potential for future evolutionary novelty and diversity. Quite simply, human activities rendered the natural environment infertile for the evolution of diversity, or at least to the degree that diversity had flourished since the last Ice Age.

❧ Essay 49 ❦

Molecular Biology: Darwin's Precious Gift

Francisco J. Ayala

Darwin and other nineteenth-century biologists found compelling evidence for biological evolution in the comparative study of living organisms, in their geographical distribution, and in the fossil remains of extinct organisms. In the *Origin of Species*, Darwin dedicates five chapters to the evidence for evolution: two chapters to the geological record, or, as we are more likely to say nowadays, to paleontology; two chapters to biogeography; and one chapter to comparative anatomy and embryology. Since Darwin's time, the evidence from these sources has become stronger and more comprehensive, while biological disciplines that have emerged recently – genetics, biochemistry, ecology, animal behavior (ethology), neurobiology, and especially molecular biology – have supplied powerful additional evidence and detailed confirmation.

Darwin surely would have been pleased by the enormous accumulation of paleontological evidence, including the discovery of fossils of organisms intermediate between major groups, such as *Archaeopteryx*, intermediate between reptiles (dinosaurs) and birds, and *Tiktaalik*, intermediate between fish and tetrapods (Ahlberg and Clack 2006) and the numerous fossils and diverse species of hominins, intermediate between apes and *Homo sapiens* (e.g., Dalton 2006; T. D. White et al. 2006; Cela-Conde and Ayala 2007). But there are good reasons to believe that Darwin would have been most pleased and most impressed with the overwhelming evidence for evolution and precise information about evolutionary history provided by molecular biology, a source of evidence and document of history that Darwin could not have even imagined.

Molecular biology, a discipline that emerged in the second half of the twentieth century, nearly one hundred years after the publication of the *Origin of Species*, undoubtedly provides the strongest evidence yet of the evolution of organisms. Molecular biology proves evolution in two ways: first, by showing the unity of life in the nature of DNA and the workings of organisms at the level of enzymes and other protein molecules; second, and most important, by making it possible to reconstruct evolutionary relationships that were previously unknown, and to confirm, refine, and time all evolutionary relationships from the universal common ancestor up to all living organisms. The precision with which these events can be reconstructed is one reason why the evidence from molecular biology is so useful to evolutionists and so compelling.

THE UNITY OF LIFE

The molecular components of organisms are remarkably uniform – in the nature of the components as well as in the ways in which they are assembled and used. In all bacteria, archaea, plants, animals, and humans, the instructions that guide the development and functioning of organisms are encased in the same hereditary material, DNA, which provides the instructions for the synthesis of proteins. The thousands of enormously diverse proteins that exist in organisms are synthesized from different combinations, in sequences of variable length, of twenty amino acids, the same in all proteins and in all organisms. Yet several hundred other amino acids exist. Moreover, the genetic code, by which the information contained in the DNA of the cell nucleus is passed on to proteins, is virtually everywhere the same. Similar metabolic pathways – sequences of biochemical reactions – are used by the most diverse organisms to produce energy and to make up the cell components. Many other pathways are theoretically possible, but only a limited number are used in organisms, and the pathways are the same in organisms with extremely different ways of life.

The unity of life reveals the genetic continuity and common ancestry of all organisms. There is no other rational way to account for their molecular uniformity, given that numerous alternative structures and fundamental processes are in principle equally likely. The genetic code may serve as an example. Each particular sequence of three nucleotides (called a "triplet" or "codon") in the nuclear DNA acts as a code for exactly the same particular amino acid in all organisms. For example, in any given gene of any organism, the codon GCC determines that the amino acid alanine will be incorporated in the protein specified by the gene, the codon GAC determines the incorporation of the amino acid asparagine, and so on. The universal correspondence between the DNA language (codons) and the protein language (amino acids) is no more necessary than it is for any two spoken languages to use the same combination of letters for representing the same particular concept or object. If we find that certain sequences of letters – planet, tree, woman – are used with identical meanings in different books, we can be sure that the languages used in the books are identical and that they must have had a common origin (Fig. 49.1).

INFORMATIONAL MACROMOLECULES

DNA and proteins have been called "informational macromolecules" because they are long linear molecules made up of sequences of units – nucleotides or amino acids – that embody evolutionary information. Comparing the sequence of the components in two macromolecules establishes how many units are different. Because evolution usually occurs by changing one unit at a time, the number of differences is

	Second Position				
First Position	U	C	A	G	Third Position
U	UUU ⎱ Phe UUC ⎰ UUA ⎱ Leu UUG ⎰	UCU ⎱ UCC ⎰ Ser UCA ⎱ UCG ⎰	UAU ⎱ Tyr UAC ⎰ UAA Stop UAG Stop	UGU ⎱ Cys UGC ⎰ UGA Stop UGG Trp	U C A G
C	CUU ⎱ CUC ⎰ Leu CUA ⎱ CUG ⎰	CCU ⎱ CCC ⎰ Pro CCA ⎱ CCG ⎰	CAU ⎱ His CAC ⎰ CAA ⎱ Gln CAG ⎰	CGU ⎱ CGC ⎰ Arg CGA ⎱ CGG ⎰	U C A G
A	AUU ⎱ AUC ⎰ Ile AUA ⎰ AUG Met	ACU ⎱ ACC ⎰ Thr ACA ⎱ ACG ⎰	AAU ⎱ Asn AAC ⎰ AAA ⎱ Lys AAG ⎰	AGU ⎱ Ser AGC ⎰ AGA ⎱ Arg AGG ⎰	U C A G
G	GUU ⎱ GUC ⎰ Val GUA ⎱ GUG ⎰	GCU ⎱ GCC ⎰ Ala GCA ⎱ GCG ⎰	GAU ⎱ Asp GAC ⎰ GAA ⎱ Glu GAG ⎰	GGU ⎱ GGC ⎰ Gly GGA ⎱ GGG ⎰	U C A G

FIGURE 49.1. The genetic code: correspondence between the sixty-four possible codons in messenger RNA and the encoded amino acids. Notice that the DNA of a gene is first transcribed into messenger RNA, which is in turn translated into the amino acids that make up proteins. Three nucleotides (A, C, G) are the same in DNA and RNA, but the fourth nucleotide is different; RNA has uracil (U) rather than thymine (T). Permission: Ayala drawing

an indication of the recency of common ancestry. Thus, the inferences from paleontology, comparative anatomy, and other disciplines that study evolutionary history can be tested in molecular studies of DNA and proteins by examining the sequences of nucleotides and amino acids. The authority of this kind of test is overwhelming: each of the thousands of genes and thousands of proteins contained in an organism provides an independent test of that organism's evolutionary history.

Molecular evolutionary studies have three notable advantages over comparative anatomy, paleontology, and the other classical disciplines: precision, universality, and multiplicity. First, *precision*: the information is readily quantifiable. The number of units that are different is easily established when the sequence of units is known for a given macromolecule in different organisms. It is simply a matter of aligning the units (nucleotides or amino acids) between two or more species and counting the differences. The second advantage, *universality*, is that comparisons can be made between very different sorts of organisms. There is very little that comparative anatomy or paleontology can say when, for example, organisms as diverse as yeasts, pine trees, and human beings are compared, but there are numerous DNA and protein sequences that can be compared in all three. The third advantage is *multiplicity*. Each organism possesses thousands of genes and proteins, every one of which reflects the same evolutionary history. If the investigation of one particular gene or protein does not satisfactorily resolve the evolutionary relationship of a set of species, additional genes and proteins can be investigated until the matter has been settled.

The resourcefulness of molecular biology in studying evolution can be noted in other ways as well. The widely different rates of evolution of different sets of genes opens up the opportunity for investigating different genes in order to achieve different degrees of resolution in the tree of evolution. Evolutionists rely on slowly evolving genes for reconstructing remote evolutionary events, but increasingly faster evolving genes for reconstructing the evolutionary history of more recently diverged organisms.

Genes that encode *ribosomal* RNA molecules are among the slowest evolving genes. They have been used to reconstruct the evolutionary relationships among groups of organisms that diverged very long ago: for example, among bacteria, archaea, and eukaryotes (the three major divisions of the living world), which diverged more than 2 billion years ago, or among the protozoa compared with plants and with animals, groups of organisms that diverged about 1 billion years ago. Cytochrome *c* evolves slowly, but not as slowly as the ribosomal RNA genes. Thus, it is used to decipher the relationships within large groups of organisms, such as among animals, in comparisons, for example, between humans and fishes, or between humans or fishes and insects. Fast-evolving molecules, such as the fibrinopeptides involved in blood clotting, are appropriate for investigating the evolution of closely related animals – with the primates, for example, the evolutionary relationships among macaques, chimps, and humans.

It is now possible to make an assertion that would have delighted Darwin and would surely shock creationists and other antievolutionists, and perhaps startle many scientists and most of the general public: gaps of knowledge in the evolutionary history of living organisms no longer need to exist. Molecular biology has made it possible to reconstruct the "universal tree of life," the continuity of succession from the original forms of life, ancestral to all living organisms, to every species now living on earth. The main branches of the tree of life have been reconstructed on the whole and in great detail. More details about more and more branches of the universal tree of life are published in scores of scientific articles every month. The virtually unlimited evolutionary information encoded in the DNA sequence of living organisms allows evolutionists to reconstruct all evolutionary relationships leading to present-day organisms, with as much detail as wanted. Invest the necessary resources (time and laboratory expenses) and one can have the answer to any query, with as much precision as one may want (Fig. 49.2).

MOLECULAR EVOLUTIONARY TREES

DNA and proteins provide information not only about the branching succession of lineages from common ancestors (*cladogenesis*) but also about the amount of genetic change that has occurred in any given lineage (*anagenesis*). Molecular evolutionary trees are models or hypotheses that seek to

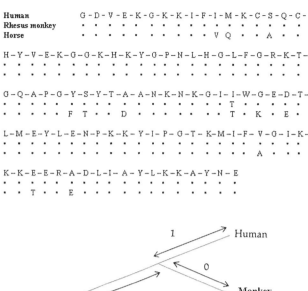

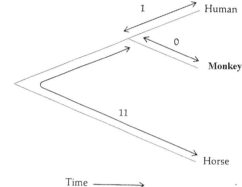

FIGURE 49.2. Cytochrome *c*, a protein involved in cell respiration. *Top*. The 104 amino acids in the cytochrome *c* of humans are shown on top (using conventional one-letter representations for each amino acid). Dots indicate amino acids identical to those in human cytochrome *c*. At one position rhesus monkeys have threonine, while humans have isoleucine. Human and horse differ by 12 amino acids; monkey and horse differ by 11 amino acids. Ayala drawing after W. M. Fitch and E. Margoliash, Construction of phylogenetic trees, *Science* 155 (1967): 279–84. *Bottom*. Evolutionary tree of human, rhesus monkey, and horse, based on their cytochrome *c*. The one difference between human and monkey (see top figure) is due to a change in the human lineage. This conclusion is reached because monkey and horse (as well as other animals) have the same amino acid, threonine, at this position, while humans have a different one (isoleucine). Even if nothing else was known about the evolutionary relationships between the species, we would conclude that human and monkey have a more recent common ancestor than human and monkey compared to horse, because horse cytochrome *c* differs much more from human and monkey cytochrome *c* than these two differ from each other. Permission: Ayala drawing

reconstruct the evolutionary history of taxa – that is, species, genera, families, orders, and other groups of organisms. The trees embrace information about both dimensions of evolutionary change, cladogenesis, and anagenesis. It might seem at first that quantifying anagenesis for proteins and nucleic acids would be impossible, because it seems to require comparison of molecules from organisms that are now extinct with molecules from living organisms or from other extinct organisms. Organisms of the past are sometimes preserved as fossils, but their DNA and proteins have largely disintegrated. Nevertheless, comparisons between living species provide information about anagenesis.

Consider, for example, the protein cytochrome *c* involved in cell respiration. The sequence of amino acids in this protein

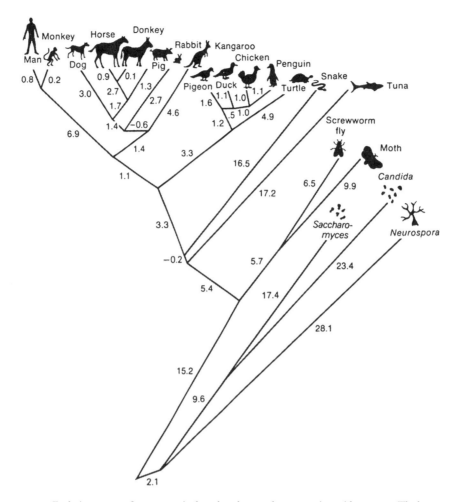

FIGURE 49.3. Evolutionary tree of twenty species based on the cytochrome *c* amino acid sequence. The last common ancestor of yeast and humans (the point at the bottom from which the branches diverge) lived more than 1 billion years ago. The numbers along the branches estimate the nucleotide substitutions occurring in the span of evolution represented by the branch. Although fractional (or negative) numbers of nucleotide substitutions cannot occur, the numbers along the branches are those that best fit the data. More detailed studies make it possible to determine the exact number of changes along each branch. This figure illustrates the attribute of universality: the species compared are extremely divergent morphologically and otherwise. Ayala drawing after W. M. Fitch and E. Margoliash, Construction of phylogenetic trees, *Science* 155 (1967): 279–84

is known for many organisms, from bacteria and yeasts to insects and humans; in animals, cytochrome *c* consists of 104 amino acids. When the amino acid sequences of humans and rhesus monkeys are compared, they are found to be different at position 58, but identical at the other 103 positions. When humans are compared with horses, twelve amino acid differences are found, and when horses are compared with rhesus monkeys, there are eleven amino acid differences. Even without knowing anything else about the evolutionary history of mammals, we would conclude that the lineages of humans and rhesus monkeys diverged from each other much more recently than they diverged from the horse lineage.

Moreover, it can be concluded that the amino acid difference between humans and rhesus monkeys must have occurred in the human lineage after its separation from the rhesus monkey lineage. This conclusion is drawn from the observation that, at position 58, monkeys and horses (as well as other animals) have the same amino acid (threonine), while humans have a different one (isoleucine), which therefore must have changed in the human lineage after it separated from the monkey lineage. The amino acid sequences in the cytochrome *c* of twenty very diverse organisms were ascertained in 1967. Counting the amino acid differences between the twenty species resulted in the evolutionary tree shown in Figure 49.3.

THE MOLECULAR CLOCK OF EVOLUTION

Molecular evolution has the important attribute of *precision*: differences between DNA molecules or between proteins can be precisely quantified and expressed as, for example, the number of nucleotides or amino acids that have changed. Rates of evolutionary change can therefore be fairly precisely established with respect to DNA and proteins. Studies of molecular evolution rates have led to the proposition that macromolecules may serve as evolutionary clocks.

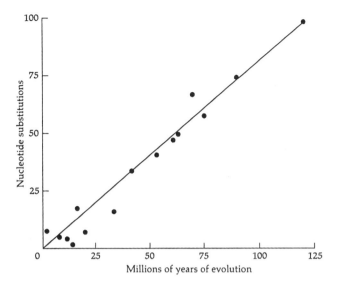

FIGURE 49.4. The molecular clock of evolution: numbers of nucleotide substitutions for seven proteins in seventeen species of mammals. The numbers of substitutions have been estimated for each comparison between pairs of species whose ancestors diverged at the time indicated in the abscissa, as previously known from other information. Each dot represents the number of substitutions for the seven proteins added up. The line has been drawn from the origin to the outermost point and corresponds to a rate of 0.41 nucleotide substitutions for every million years for all seven proteins combined. The proteins are cytochrome *c*, fibrinopeptides A and B, hemoglobins α and β, myoglobin, and insulin-c peptide. Most points fall near the line, except for some representing comparisons between primates (points below the line at lower left), in which protein evolution has occurred at a lower than average rate. Ayala drawing after W. M. Fitch, Molecular evolutionary clocks, in *Molecular Evolution*, ed. F. J. Ayala (Sunderland, Mass.: Sinauer Associates, 1976), 160–78

It was first observed in the 1960s that the number of amino acid differences between homologous proteins of any two given species seemed to be nearly proportional to the time of their divergence from a common ancestor. If the rate of evolution of a protein or gene were approximately the same in the evolutionary lineages leading to different species, proteins and DNA sequences would provide a molecular clock of evolution. The sequences could then be used to reconstruct not only the sequence of branching events of a phylogeny but also the time when the various branching events occurred (Fig. 49.4).

Consider, for example the tree of twenty species shown in Figure 49.3. If the substitution of nucleotides in the gene coding for cytochrome *c* occurred at a constant rate through time, we could determine the time elapsed along any branch of the phylogeny simply by examining the number of nucleotide substitutions along that branch. We would need to calibrate the clock by reference to an outside source, such as the fossil record, that would provide the actual geologic time elapsed in at least one specific lineage or since one branching point. For example, if the time of divergence between insects and vertebrates is determined to have occurred 700 million years ago, other times of divergence can be determined by proportion of the number of nucleotides or amino acid changes.

The molecular evolutionary clock is not expected to be a metronomic clock, like a watch or other timepieces that measure time exactly, but a stochastic (probabilistic) clock, like radioactive decay, where the *probability* of a certain amount of change is constant, although some variation occurs in the actual amount of change. Over fairly long periods of time, a stochastic clock is quite accurate. The enormous potential of the molecular evolutionary clock lies in the fact that each gene or protein is a separate clock. Each clock ticks at a different rate – the rate of evolution characteristic of a particular gene or protein – but each of the thousands and thousands of genes or proteins provides an independent measure of the same evolutionary events.

Evolutionists have found that the amount of variation observed in the evolution of DNA and proteins is greater than is expected from a stochastic clock – in other words, the clock is overdispersed, or somewhat erratic. The discrepancies in evolutionary rates along different lineages are not excessively large, however. So it is possible, in principle, to time phylogenetic events with considerable accuracy, but more genes or proteins must be examined than would be required if the clock were stochastically constant in order to achieve a desired degree of accuracy. The average rates obtained for several proteins, taken together, become a fairly precise clock, particularly when many species are studied.

This conclusion is illustrated in Figure 49.4, which plots the cumulative number of nucleotide changes in seven proteins against the dates of divergence of seventeen species of mammals (sixteen pairings) as determined from the fossil record. The overall rate of nucleotide substitution is fairly uniform. Some primate species (represented by the points below the line at the lower left of the figure) appear to have evolved at a slower rate than the average for the rest of the species. This anomaly occurs because the more recent the divergence of any two species, the more likely it is that the changes observed will depart from the average evolutionary rate. As the length of time increases, periods of rapid and slow evolution in any lineage will tend to cancel one another out.

THEORETICAL BASIS FOR THE CLOCK: NEUTRALITY VERSUS SELECTION

The theoretical foundation of the molecular clock of evolution is the neutrality theory of molecular evolution. The neutrality theory asserts that most amino acid substitutions in a protein, as well as most nucleotide substitutions in a gene, are neutral, that is, functionally equivalent and thus not subject to the vagaries of natural selection. If molecular evolution is neutral with respect to adaptation, the rate of evolution is expected to occur with a constant probability, because the rate of amino acid or nucleotide replacement along evolving lineages would be determined by mutation rate and time elapsed, rather than by natural selection. Natural selection is rather fickle, subject to the vagaries of environmental change and organism interactions, whereas mutation rate for a given gene is likely to remain constant through time and across lineages. The number of amino acid replacements, as well as the number of nucleotide substitutions, between species would, then, reflect the time elapsed since their last common ancestor. The time of remote events, as well as the degree of relationship among contemporary lineages, could be thus determined on the basis of amino acid, or nucleotide, differences (Zuckerkandl and Pauling 1965a).

Early investigations showed that the evolution of the globins in vertebrates conformed fairly well to the clock hypothesis, which allowed reconstructing, for example, the history of globin gene duplications (Zuckerkandl and Pauling 1965b). Fitch and Margoliash (1967) would soon provide a "genetic distance" method that was effectively used for reconstruction of the history of twenty organisms, from yeast to moth to human, based on the amino acid sequence of a small protein, cytochrome c (see Fig. 49.3).

A mathematico-theoretical foundation for the clock was provided by Kimura (1968), who developed a "neutral theory of molecular evolution," which was formulated with great mathematical simplicity. Notably, the theory states that the rate of substitution of adaptively equivalent (neutral) alleles, k, is precisely the rate of mutation, u, of neutral alleles, $k = u$. The neutrality theory predicts that molecular evolution behaves like a stochastic clock, such as radioactive decay, as stated earlier, with the properties of a Poisson distribution, in which the mean, M, and variance, V, are expected to be identical, so that $V/M = 1$. The index of dispersion, measuring the deviation of this ratio from the expected value of 1, is a way to test whether observations fit the theory.

As pointed out earlier, experimental data have shown that often the rate of molecular evolution is "overdispersed," that is, that the index of dispersion is often significantly greater than 1, which is the value expected. Numerous experiments have shown that deviations from rate constancy occur between lineages, say between rodents and mammals, as well as at different times along a given lineage, both factors having significant effects (see, e.g., Rodriguez-Trelles et al. 2001, 2006). Consequently, several modifications of the neutral theory have been proposed, seeking to account for the excess variance of the molecular clock.

Four subsidiary hypotheses that have been proposed to fix the clock are: (1) most protein evolution involves *slightly deleterious* replacements rather than strictly neutral ones; (2) certain *biological properties*, such as the effectiveness of the nucleotide error-correcting polymerases, vary among organisms; (3) the *population size* hypothesis proposes that organisms with larger effective population size have a slower rate of evolution than organisms with smaller population size, because the time required to fix new mutations increases with population size (thus, the rate of divergence between, say, humans, dogs, and elephants would be lower than between, say, ants, butterflies, and crickets); (4) the *generation-time* hypothesis. Protein evolution has been extensively investigated in primates and rodents, with the common observation that the number of amino acid replacements is greater in rodents. In plants, the overall rate at the *rbc*L locus is more than five times greater in annual grasses than in palms, which have much longer generations (Gaut et al. 1992). These rate differences could be accounted for, according to the generation-time hypothesis, by assuming that the time rate of evolution depends on the number of germ-line replications per year, which is several times greater for the short-generation rodents and grasses than for the long-generation primates and palms. The rationale of the assumption is that the larger the number of replication cycles, the greater the number of mutational errors that will occur.

From a theoretical, as well as operational, perspective, these and other supplementary hypotheses have the discomforting consequence that they invoke additional empirical parameters, often not easy to estimate. It is of great epistemological significance that the original proposal of the neutral theory was highly predictive ($k = u$, and $V/M = 1$) and, therefore, eminently testable. The supplementary hypotheses lead, nevertheless, to certain predictions that can be tested. The generation-time, population size, and biological properties hypotheses uniformly predict that rate variations observed between lineages or at different times will equally affect (in direction and magnitude) all genes of any particular organism, because these attributes are common to all genes of the same species. The "slightly deleterious" hypothesis predicts that the rate of evolution will be inversely related to population size, and thus it reduces to the population size hypothesis.

Extensive investigations undertaken as tests of these four supplementary hypotheses, as well as of the more general or null hypothesis underlying the molecular clock hypothesis have concluded that inferences about the timing of past events (and about phylogenetic relationships among species) based on molecular evolution are subject to much greater variation than expected from the neutral theory, even when the four subsidiary hypotheses mentioned are taken into account (Rodriguez-Trelles et al. 2001, 2006).

It is fair to conclude that there is no molecular clock in the general sense that we can assume that any given gene evolves at a nearly constant rate over time and across lineages. But this conclusion does not imply that the timing of evolutionary events cannot be determined using molecular data. It rather means that caution should be used in assuming that molecular evolution is proceeding in any particular case in a clock-like manner. As I pointed out earlier, molecular investigations have three obvious advantages, in degree if not completely in kind, over phenotypic traits and paleontological data: namely, precision, universality, and multiplicity. Every one of the thousands of genes in the makeup of each organism provides information about the evolutionary history of any taxon, and differences can be more precisely quantified, measured as they are in terms of distinct units, such as amino acids or nucleotides.

There are many evolutionary issues concerning both timing and phylogenetic relationships between species for which molecular sequence data provide the best, if not the only, dependable evidence. The large-scale reconstruction of the universal tree of life is a case in point: the phylogenetic relationships among archaean and bacterial prokaryotes and between them and the eukaryotes were first determined with

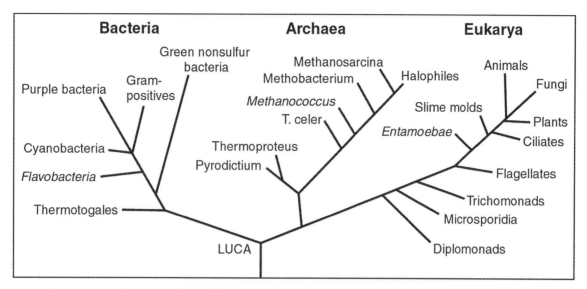

FIGURE 49.5. The universal tree of life, reconstructed with rRNA (ribosomal ribonucleic acid) genes. The Last Universal Common Ancestor (LUCA) is at the bottom. Branches represent different kinds of organisms. There are three major groups of organisms: bacteria, archaea, and eukaryotes. Bacteria, archaea, and most eukaryotes are microscopic. Plants, animals, and fungi are multicellular (macroscopic) branches of eukaryotes. Ayala drawing after C. R. Woese 2000 Interpreting the universal phylogenetic tree. *Proceedings of the National Academy of Sciences, U.S.A* 97 (2000): 8392-96

DNA sequences encoding ribosomal RNA genes. The multiplicity of genes opens up the possibility of combining data for numerous genes in assessing the time of particular evolutionary events, or the phylogeny of species. Because of the time dependence of the evolutionary process, the multiplicity of independent results is expected to tend to converge (by the so-called law of large numbers) on average values reflecting with reasonable accuracy the time elapsed since the divergence of taxa (Fig. 49.5).

THE HUMAN GENOME PROJECT

A contemporary development that would have greatly delighted Darwin is the determination of the DNA sequence of the human genome, an investigation that was started under the label the Human Genome Project, which opens up the possibility of comparing the human DNA sequence with that of other organisms, observing their similarities and differences, seeking to ascertain the changes in the DNA that account for distinctively human features.

The Human Genome Project was initiated in 1989, funded through two U.S. agencies, the National Institutes of Health (NIH) and the Department of Energy (DOE), with eventual participation of scientists outside the United States. The goal set was to obtain the complete sequence of one human genome in fifteen years at an approximate cost of $3 billion, coincidentally about $1 per DNA letter. A private enterprise, Celera Genomics, started in the United States somewhat later but joined the government-sponsored project in achieving, largely independently, similar results at about the same time. A draft of the genome sequence was completed ahead of schedule in 2001. The government-sponsored sequence was published by International Human Genome Sequencing Consortium (2001) in the journal *Nature* and the Celera sequence was published by Venter et al. (2001) in the journal *Science*. In 2003 the Human Genome Project was finished, but the analysis of the DNA sequences chromosome by chromosome continued over the following years. Results of these detailed analyses were published on 1 June 2006, by the Nature Publishing Group, in a special supplement entitled *Nature Collections: Human Genome*.

The draft DNA sequence of the chimpanzee genome was published on 1 September 2005, by the Chimpanzee Sequencing and Analysis Consortium in *Nature*, embedded within a series of articles and commentaries (Anon, 2005). The last paper in the collection presents the first fossil chimpanzee ever discovered (McBrearty and Jablonski 2005).

In the genome regions shared by humans and chimpanzees, the two species are 99 percent identical. The differences may seem very small or quite large, depending on how one chooses to look at them: 1 percent is only a small fraction of the total, but it amounts to a difference of 30 million DNA nucleotides out of the 3 billion in each genome.

Twenty-nine percent of the enzymes and other proteins encoded by the genes are identical in these species. Out of the one hundred to several hundred amino acids that make up each protein, the 71 percent of nonidentical proteins differ between humans and chimps by only two amino acids, on average. If one takes into account DNA stretches found in one species but not the other, the two genomes are about 96 percent identical, rather than nearly 99 percent identical as in the case of DNA sequences shared by both species. That is, a large amount of genetic material, about 3 percent or some 90 million DNA nucleotides, have been inserted or deleted since humans and chimps initiated their separate evolutionary ways, about 8–6 million years ago. Most of this DNA does not contain genes coding for proteins, although it may include tool-kit genes and switch genes that impact developmental processes, as the rest of the noncoding DNA surely does.

Comparison of the two genomes provides insights into the rate of evolution of particular genes in the two species. One significant finding is that genes active in the brain have changed more in the human lineage than in the chimp lineage (Khaitovich et al. 2005). Also significant is that the fastest-evolving human genes are those coding for *transcription factors*. These are switch proteins that control the expression of other genes; that is, they determine when other genes are turned on and off. On the whole, 585 genes have been identified as evolving faster in humans than in chimps, including genes involved in resistance to malaria and tuberculosis. (It might be mentioned that malaria is a severe disease for humans but not for chimps.)

Genes located on the Y chromosome, found only in the male, have been much better protected by natural selection in the human than in the chimpanzee lineage, in which several genes have incorporated disabling mutations that make the genes nonfunctional. Also, there are several regions of the human genome that contain beneficial genes that have rapidly evolved within the past 250,000 years. One region contains the *FOXP2* gene, involved in the evolution of speech.

Other regions that show higher rates of evolution in humans than in chimpanzees and other animals include forty-nine segments, dubbed human accelerated regions or HARs. The greatest observed difference occurs in *HAR1F*, an RNA gene that "is expressed specifically in Cajal-Retzius neurons in the developing human neocortex from 7 to 19 gestational weeks, a crucial period for cortical neuron specification and migration" (Pollard et al. 2006; see also K. Smith 2006).

Extended comparisons of the human and chimpanzee genomes and experimental exploration of the functions associated with significant genes will surely further advance our understanding, over the next decade or two, of what it is that makes us distinctively human – what it is that differentiates *H. sapiens* from our closest living species, chimpanzees and bonobos – and provide some light on how and when these differences may have come about during hominid evolution of the human species.

❦ Essay 50 ❦

Challenging Darwinism: Expanding, Extending, Replacing

David J. Depew and Bruce H. Weber

WHAT COUNTS AS an alternative to the Darwinian view of biological origins depends on what one takes Darwinism to be. To advocates of "special creation," Darwinism refers to its author's claim that all organisms evolved from a single common ancestor. Among evolutionists themselves, however, Darwinism refers to natural selection, Darwin's distinctive mechanism for explaining species transformation and common descent. Creationist challenges are not our topic. Challenges to natural selection as causally explaining it are.

As it happened, natural selection inserted itself into an already lively debate about evolutionary mechanisms (Desmond 1989; Secord 2000). Darwin's was a powerful idea, but not powerful enough to knock out all of its rivals, which have reasserted themselves whenever a *version* of Darwinism that has organized evolutionary research for a time – and there have been several – runs into trouble. Trouble can come from new empirical discoveries, from the discrediting of false claims or assumptions on which Darwinians had relied, or from research methods that prove flawed.

Darwinism's currently established but aging version, the modern evolutionary synthesis, has been facing challenges of all three sorts since the 1980s. The modern synthesis was first articulated just before and during World War II on the basis of technical work done earlier (Dobzhansky 1937; J. S. Huxley 1942; Mayr 1942; Simpson 1944). It claimed to unify all biological fields by integrating natural selection with genetics, about which neither Darwin nor nineteenth-century Darwinians knew anything. Since then, challengers have called either for "expanding" the modern synthesis by allowing natural selection to operate at various levels of the biological hierarchy; for "extending" it to incorporate fields previously shunted aside, notably developmental biology; or for replacing the synthesis altogether. It is an open question whether in the last case the contemplated replacement would be a new, post-synthesis form of Darwinism or would be nonselectionist enough to count as truly post-Darwinian. Because post-Darwinism is likely to mean that one or more of Darwinism's old rivals would have staged a comeback in a new form, we should, before reporting on recent proposals for expanding, extending, or replacing its current version, first situate the Darwinian research tradition as a whole among its most persistent alternatives.

DARWINIAN AND NON-DARWINIAN APPROACHES TO EVOLUTION

Darwin recognized Lamarckism as a rival evolutionary theory. Lamarckians think that, impelled by life's internal drive toward complexity, organisms actively adapt by rising to challenges posed by environmental change. Active adapting has evolutionary consequences because, on the Lamarckian view, characteristics acquired in single lifetimes can be passed on to offspring.

Although Darwin conceded Lamarckian inheritance – a concession withdrawn by Darwinians at the end of the nineteenth century – he thought Lamarckism could not be generally correct because it made adaptation too easy. Because *ex hypothesi* lineages could transform themselves whenever the environment posed difficulties, Lamarck could not explain the vast amount of extinction in the evolutionary record. According to Darwin's alternative theory of natural selection, active adapting is more a result of the process of adaptation than its cause. Adaptations for Darwinism are made out of heritable variations that arise independently of their subsequent utility in improving reproductive rates. Over multiple generations, variations that at first merely happen to have a positive effect on reproduction rates under the stress of competition for scarce resources are preserved, amplified, and gradually shaped into adaptations. Hence, Darwin's theory is variational, not transformational; stresses external over internal causes; and interposes an element of chance between the problems organisms encounter and their solution.

Important and often ignored consequences follow: (1) Individual organisms do not adapt. That happens only to populations over transgenerational time. (2) Individual organisms do not evolve. They merely develop. (3) Whatever adaptation and diversification occurs is hostage to whether the right kind of variation happens to crop up. The extent of extinction is proof positive that, *contra* Lamarck, the requisite variation is not always available. For this reason, today's self-identified Lamarckians – and there are some – rest their case on putative phenomena such as "directed" or "influenced" variation: variation whose occurrence is biased toward filling a need; "epigenetic" – that is, nongenetic – forms of inheritance that are more open to environmental influences than genes; and evidence that populations adaptively respond to environmental challenges without waiting around for genetic mutation (Jablonka and Lamb 2005).

A second stream of non-Darwinian evolutionary theorizing treats speciation and the origin of higher *taxa*, not adaptation, as the test of an evolutionary theory's worth. It regards the origin of organic forms not as the result of niche

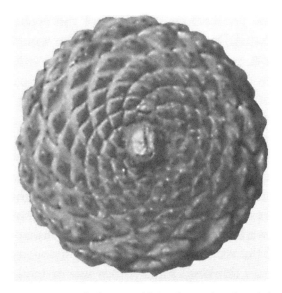

FIGURE 50.1. The flowers and fruits of many plants have their parts in complex pattern known as "phyllotaxis." This structure is governed by successive numbers in the Fibonacci formula (made famous by the *Da Vinci Code*), where the value of a member of a sequence is given by the sum of the last two members of the sequence ($m_i = m_{i-1} + m_{i-2}$). Thus 0, 1, 1, 2, 3, 5, 8, 13, 21, ... In the pinecone case, we have an 5, 8, pattern. For Brian Goodwin and like thinkers, this is proof that the laws of physics and mathematics unaided produce organic complexity. Darwinians respond that no one denies the importance of the structural constraints of nature, but this does not preclude the possibility and necessity of natural selection. Based on Asa Gray, *Structural Botany* (New York: Ivison, Blakeman, Taylor, 1878)

diversification by adaptation, Lamarckian or Darwinian, but as internally caused by sudden fortuitous, often nonadaptive "leaps" or "saltations" (from Latin *saltus*, "jump"). "Process structuralists," as they are called, are contemporary saltationists (Goodwin 1994) (Fig. 50.1 and Plate XXXIX). But strains of saltationism can be found even in Darwin's first and greatest, or at least loudest, defender, Thomas Henry Huxley. Moreover, saltationism went into the making of early twentieth-century mutation theory, from which genetics developed. Early geneticists, as well as their Darwinian opponents, thought of genetics and natural selection as mutually contradictory because both parties assumed that mutations can have lasting effects only if they are large and sudden. The modern evolutionary synthesis was a synthesis in part because it showed with persuasive mathematical prowess that natural selection can reward small genetic changes whose effects, though minuscule at first, can spread in ways that render one population better adapted than a closely related one (Provine 1971). This process can be brought into view, however, only by shifting with help from statistics and probability theory to a perspective that sees organisms as members of interbreeding populations. Twentieth-century "genetic Darwinism" has dominated evolutionary science because the makers of the modern synthesis and their successors showed by "population thinking" that traits marking off species and other *taxa* are in most cases adaptive, not saltational (Mayr 1942; Simpson 1944; Lack 1947; but see Gould 1983 and Singh, Xu, and Kulanthinal 2012). Until the 1980s, this research program tended to put saltationism into eclipse.

So far we have seen that Lamarckian explanations rely on an intuitive sense of the verb "adapt" as something organisms do, that Darwinian natural selection explains the adaptedness of populations by highlighting two other facts about organisms – that they vary and compete for resources – and that saltationist mutationism makes more use of variation than of competition. The currently most powerful alternative to Darwinism depends, however, on another fact about organisms, from which a fourth evolutionary tradition springs. Organisms develop. German embryologists have long tended to think of evolution as due to changes in the self-organizing, self-differentiating, self-formative properties of organisms seen as developmental systems. These changes are both spontaneous *and* adaptive. Nineteenth-century versions of evolutionary developmentalism – some of them conflated with Darwinism through the influence of Herbert Spencer, Ernst Haeckel, and others – were based on treating the course of evolution as itself a large-scale developmental process, thereby underestimating the factors of chance variation, competition, and adaptation stressed by Darwin. By means of statistical descriptions of phenomena and probabilistic forms of explanation, the modern synthesis integrated these three factors in ways that made the fact that organisms develop irrelevant to the evolutionary process. What evolve are at best the "genetic programs" that determine how individuals develop, not individuals as such.

No one today proposes to revert to ontogeny-phylogeny parallelism. Yet in our own time discoveries have suggested that developmental modification, not genetic mutation, is the proximate source of the variation over which natural selection ranges (Gilbert and Epel 2009). The consequences of this seemingly innocent perception are in dispute. It may serve merely to correct the modern synthesis's acquired tendency, after Watson and Crick had discovered the structure of DNA, to look for the sources of variation only in point mutations in structural genes. Changes in regulatory sectors, and hence in developmental ones, are more likely suspects, especially when it comes to evolution above the species level. But more radical thoughts can easily surface. Developmental innovations may be part of a cyclical process in which organisms, considered as complex adaptive systems, evolve by spontaneously reorganizing themselves. This view is not necessarily non-Darwinian. Selection may be an essential ingredient in this process. But it can quickly become so if the line Darwinism of all stripes draws between development and evolution is erased.

EXPANDING THE MODERN SYNTHESIS

Darwinians themselves, and not just their enemies, have long made a habit of thinking that the fate of Darwinism depends on factual claims and conceptual assumptions that, when removed, turn out not to have undermined Darwinism at all, as anticipated, but instead to have made room for novel, more powerful articulations of its basic idea, natural selection (Depew and Weber 1995). We have already glanced at a good example. The pseudoconflict between genetic mutationists and defenders of natural selection proved that turn-of-the-twentieth-century rumors suggesting that mutationism had consigned Darwinism to its "deathbed" were greatly exaggerated. If you shift to population thinking, you can see how effective small mutations can be (Provine 1971).

The late 1960s and early 1970s witnessed an equally instructive case of a difficulty that turned out to be temporary. Population genetic Darwinism got a big boost when the small genetic variations on which the modern synthesis was predicated were shown to arise from expectable but inherently random and nondirectional copying errors in sequences of the four bases of DNA, A, C, T, and G, that translate into amino acids and, when assembled, proteins. There were, however, some surprises. It was, for example, a surprise – intensified by the reductionist expectations of the logical empiricist philosophy of science popular at the time – that most "point" mutations in DNA are selectively neutral. They are so neutral, in fact, that the rate at which bases mutate yields a rough-and-ready "molecular clock" that can be used to establish the age of an evolutionary branch (Fig. 50.2). The discoverers of this fact called it "non-Darwinian evolution" (King and Jukes 1969; on molecular clocks, Kimura 1983). It was non-Darwinian, however, only if one failed to recognize that, living things being structured into a hierarchy, natural selection can in principle operate at more than one level. If selection is blind to mutations that have no effect on a protein's structure or function, it can certainly "see" those that do – as well as potentially fitness-enhancing variation that arises at the level of organisms, colonies, cooperative groups, perhaps even species. If you get rid of the reductionist assumption that natural selection must work at the lowest, molecular level, you get rid of the problem.

Actually, two rival ways of incorporating new findings in molecular genetics arose in the 1970s to meet the challenge of neutral evolution: the idea of a hierarchically expanded synthesis and "selfish-gene theory." The two are still at each other's throats.

Selfish-gene theory accommodates neutral mutations to Darwinism by taking a gene's-eye view of the target of natural selection (Dawkins, 1976, 1986). DNA makes as many copies of itself as it can, Richard Dawkins argues, because that is just what self-replicating molecules do. Naturally, too, DNA sequences vary as a result of copying errors at a rate as statistically regular as we find in medieval manuscripts copied for centuries by tired or bored monks. Indeed, the rate of mutation is highest in sectors of DNA that do not code for protein or perform other functions at all. But only coding sectors of DNA are genes. The rest is "so-called junk DNA (but see the ENCODE Project Consortium 2012)." Genes, by being translated into proteins that fold up to make cell types and tissue, make phenotypes (observable traits) and the organisms that bear them. Some phenotypes enable the organisms that carry them to interact with environments in ways that reproductively outperform others. This has the effect of increasing the representation of the genes that code for more effective phenotypes. Driven by the inherent more-making tendency of DNA, the

FIGURE 50.2. Motoo Kimura (1924–94), seen here with his family, was a Japanese theoretical population geneticist who devised the "neutral theory of molecular evolutionary change," arguing that at the level of the molecule selection is inefficient and it is genetic drift that causes most change. This is the insight behind the "molecular clock." Kimura never denied the importance of natural selection at the higher, physical, and behavioral levels. Permission: American Philosophical Society

course of evolution is thus an ongoing arms race that maximizes adaptedness. This is a decidedly adaptation*ist* theory.

Selfish-gene theory commended itself not only because it made neutral mutation seem natural but because it also suggested a novel way of explaining the cooperative behavior of social animals. This inconvenient fact had led a puzzled Darwin to depart from his otherwise resolute commitment to competition among individuals by hypothesizing selection between groups. There are group-level traits, he argued, and some groups, especially internally cooperative ones, are better than others (Darwin 1871a). But from a gene's-eye perspective, selfish-gene theorists pointed out, it matters not a whit what or how many genetically related bodies are useful in maximizing self-replication. If the cooperative phenotypes of genetically related individuals yield higher rates of replication, then selection will favor them. People as well as ants might be altruistic – but only if their genes are selfish.

Stephen Jay Gould's and Niles Eldredge's proposal for an "expanded synthesis" contrasts with Dawkins's genocentrism in two ways. (1) It allows selection to operate at every level of the biological hierarchy, not just that of DNA or, as in the orthodox modern synthesis, of organisms in populations (Gould 1980; Eldredge 1985). (2) Unlike selfish-gene theory, it makes as much of genetic drift and gene flow as of mutation and natural selection. These various evolutionary forces, Gould claims, had been allowed by the original synthesis to work in different combinations at different levels of the biological hierarchy, some of which might be more open to chance than to adaptive natural selection. But by the 1960s, Gould (1983) argued, this original pluralism about evolutionary forces and levels had been "hardened" into a selection-centered adaptationism, setting the stage for selfish-gene theory.

Gould and Eldredge wanted to return to pluralism about levels and forces. Within this framework, they argued that group-level properties make for cooperative behaviors even if some individuals remain more "hawkish" than "dovish" (Sober and Wilson 1998). More distinctive of their contributions as paleontologists to the "expanded synthesis," however, is how they proposed to explain an observed statistical pattern in the fossil record that Simpson, before the synthesis hardened, had called "quantum evolution" and Gould and Eldredge now dubbed "punctuated equilibrium" (Eldredge and Gould 1972; Gould and Eldredge 1977). Speciation, when it occurs, is rapid and not the effect of slow adaptive divergence. To explain this, they imagined that species, considered as spatiotemporal units that exist between two points of evolutionary branching rather than as classes marked off by defining characteristics – an essentialist view left over from preevolutionary thought – are selected from larger branches that exhibit higher speciation rates and life expectancies than others (Gould 1980; Eldredge 1985). In countenancing "species selection," Gould and Eldredge were defying the presumption of the founders of the modern synthesis that macroevolution – evolution at and above the species level – is merely extrapolation over longer time frames from microevolutionary processes (Dobzhansky 1937). But Gould (1989b) also left more room for the role of pure chance in the macroevolutionary process than any of the formulators of the modern synthesis would have tolerated, thereby adding a saltationist, and thus far non-Darwinian, element to macroevolution.

Advocates of genic selectionism and of the expanded synthesis have spent a lot of time calling each others' Darwinian credentials into question. Genic selectionism's opponents complain that its conception of organisms as aggregates of separately adapted traits betrays Darwin's own more integrated sense of the organism, which the modern synthesis at its best retains. For their part, genic selectionists contend that Gould was not a Darwinian at all, because he explicitly repudiated two of its central planks: adaptationism and gradualism (D. Dennett 1995). Admittedly, Gould had a saltationist streak. Nonetheless, at least from the perspective of evolutionary developmentalism, of whose history he was a pioneering student (Gould 1977), it is precisely Gould's Darwinism that stands out. He argued that what deflects natural selection away from one hierarchical level to another are constraints that have accumulated in the course of evolution (Gould 2002). In saying this, Gould was paying a handsome compliment to adaptive natural selection by portraying it as a "force" that tries as hard as it can to find its way around inherited constraints, as farmers try to plow around tree stumps too deeply rooted to dig up. Moreover, he argued that the stumps are themselves largely the result of previously successful evolutionary novelties that selection has entrenched in the developmental programs of organisms. His point was that natural selection usually cannot evolve the best possible traits for a given environment. So adaptation*ism* is wrong as a basic presumption. Yet, in appealing to what natural selection *would do if it could*, Gould buried the basic assumptions of the modern synthesis, and even its adaptationist interpretation, in very shallow graves indeed.

Our judgment is that Gould's expanded and Dawkins's genocentrically contracted frameworks are both Darwinian and, if not fully consistent with the modern synthesis, at least fall well within the long shadow it casts.

EXTENDING (OR REPLACING) THE MODERN SYNTHESIS

The modern synthesis was up and running before Crick and Watson found the mechanism for generating genetic variation. When molecular genetics arrived, the founders of the synthesis were nervously enthusiastic about it (Mayr 1959; Dobzhansky 1964). They were enthusiastic because it showed that point mutations in DNA are the ultimate, if remote, source of the variation assumed by the population genetic theory of natural selection. They were nervous, however, because Crick's and Watson's "greedy reductionist" philosophy of science threatened to put "naturalists" like themselves out of business by reducing population genetics to or replacing it with molecular genetics. They were also nervous because they realized that the population genetic core of the modern synthesis depended on an expectation that molecular genetics would turn up no facts that contradict its basic principles.

This expectation was never more than a hope, and potentially a vain one. The definition of evolution that has joined population and molecular geneticists in common cause – evolution is change in gene frequencies caused by natural selection operating on random mutations in DNA sequences – requires dividing mutations into those occurring in structural genes, which are its usual textbook examples, and those affecting regulatory sectors of the genome, which tell structural genes when to start and stop manufacturing enzymes and other structural gene products. Mutations in "enhancer" and "promoter" sectors of the genome, which determine the timing, placement, and rate at which gene products are produced during the developmental process, have raised hopes of finding better explanations of what Darwin called the "endless forms most beautiful" that distinguish protists from plants; invertebrates from vertebrates; and, within these classes of animals, mollusks from arthropods or mammals from placentals. A perception is now widespread even among self-described Darwinians that, contrary to what the makers of the modern synthesis anticipated, the architecture on display in our classification systems will never be explained simply by extrapolating from selection operating on point mutations in amino acid codons or by treating mutation in regulatory sectors as acting in precisely the same way as mutation in structural genes (Carroll 2005; Pritchard and Gilad 2012; but see Hokstra and Coyne 2007; Lynch 2007).

We use the phrase "extended synthesis" to identify a diverse array of hypotheses that make the developmental process the proximate source of the variation on which selection works (West-Eberhard 2003 [using "expanded" for Pigliucci's "extended"]; G. Müller 2007; Pigliucci 2007; Carroll 2008 [using "expanded"]; Pigliucci and Müller 2010). Extending the modern synthesis to include developmental biology, which its population genetic founders pushed aside because, to their mind, it failed to make a sufficiently clean cut between development and evolution, might just as easily have been called expanding the synthesis if Gould had not already used that term to refer to his theory of multilevel selection. Some advocates of a developmentally extended synthesis welcome multilevel selection (Pigliucci and Kaplan 2006). But they do not share Gould's presumption that development is a constraint on selection's naturally optimizing path. On the contrary, they think that the developmental process is what makes evolution possible (and presumptively adaptive) because only organisms, considered as developmental cycles, can generate the sort of variation on which natural selection works as it evolves phylogenetic order (Alberch 1991; Pigliucci 2007; Gilbert and

Epel 2009; Pigliucci and Müller 2010; and, on developmental cycles, Oyama, Griffiths, and Gray 2001). On this view, adaptive evolution by natural selection is not a substrate-neutral process, working just as well on computers as on organisms (*contra* Dennett, 1995). And natural selection is itself best considered a phenomenon that emerges from, but is still tied to, more basic chemical and physical selective processes (Weber and Depew 2001).

Empirical discoveries about how genes and phenotypes interact in the developmental process have motivated calls for an extended synthesis. Among these are: (1) Changes in developmental genes are more important than changes in structural genes (Barroso 2012). (2) Developmental genes are highly conserved across surprisingly divergent clades. Genes such as Homeobox, for example, which undergirds segmentation and bilateral symmetry, form part of a relatively compact "tool kit" from which many different kinds of organisms can be made (Gilbert and Epel 2009). (3) Developmental genes express themselves very differently in different contexts. The cellular environment, which is itself open to influences from the wider environment, can affect the timing, placement, and rate at which enhancers and promoters go to work making enzymes and other structural gene products. This makes for much wider "phenotypic plasticity" than the modern synthesis typically posits (Pigliucci 2001; West-Eberhard 2003; Gilbert and Epel 2009). (4) Mutations, mostly in duplicate copies of promoter and enhancer gene segments, are only one source of developmental variation. Some phenomena that occur in the epigenetic (developmental) process, including chemical marking of DNA by methyl groups, are heritable even in the strictest sense (Jablonka and Lamb 1995, 2005). Micro RNA segments also direct some of the traffic. (5) Sometimes phenotypically plastic gene expressions are sufficiently recurrent across generations, and hence sufficiently heritable in a broad sense, to sustain adaptive behaviors that are only later stabilized by genes. This phenomenon can be seen in the so-called Baldwin effect, which allows learned behaviors to be heritably reconstructed across generations in ways that eventually come to be supported genetically (West-Eberhard 2003; see Weber and Depew 2003). The phenomenon can also be seen in "niche constructionism," which, inspired by Darwin's beautiful little treatise showing how earthworms create their own (and incidentally our) species-specific environment, generalizes the Baldwin effect well beyond animals whose social way of life exhibits learned behavior and hence social inheritance (Darwin 1881; Odling-Smee, Laland, and Feldman 2003).

Whether scientists who make much of these five calls can be loyal Darwinians, as they often claim to be, depends once again on how we draw boundaries around Darwinism. If Crick's central dogma of molecular biology, which declares that information moves from DNA to RNA to proteins and not the other way around (Crick 1970), means that random genetic mutations must precede and directionally guide anything that will count as evolutionary change, then reversals of this temporal order will count as non-Darwinian. It is true that the makers of the synthesis, even before they compromised with molecular geneticists, tended to stipulate that in all but the most exceptional cases the evolutionary process does indeed start with genetic mutation. They take the Baldwin effect, for example, where it is not a Lamarckian fantasy, merely to be uncovering genetic variation that is already there. Still, there is nothing in population genetics or the modern synthesis that absolutely demands a "genes first" stipulation. Indeed, the persuasive uses to which this stipulation was put – in the 1940s, to immunize the infant synthesis from reverting to anything relating to developmentalism in order to facilitate population thinking; in the 1950s, to block genetic Darwinians, some of whom were left liberals, from sympathizing with the Soviet Union's sponsorship of Lysenko's brand of Lamarckism; in the 1960s and 1970s, to cement a working alliance between population and molecular geneticists by blocking any way of getting around Crick's central dogma – are yesterday's news. Exceptions to the central dogma, unless the later is reconceived very restrictively, are so numerous by now that they can no longer be ignored. If one abandons these old, nearly dead controversies and allows epigenetic and phenotypic change to take the lead in evolution conceived as a cyclical process of gene-environment interaction in and through the developmental process, wide phenotypic plasticity will appear as natural selection's finest and most adaptive product and its platform for further "evolvability" (West Eberhard 2003; Pigliucci 2001, 2007).

Even if this is Darwinian, is it the Darwinism of the modern synthesis? Asking natural selection to evolve adaptations out of small genetic changes in regulatory rather than structural sectors of the genome surely counts as continuing, and so extending, the Darwinism of the modern synthesis (Carroll 2005). We may also see as extensions of the synthesis proposals in which genomes evolve wide norms of reaction, developmental plasticity, and a capacity for future evolvability (Pigliucci 2001, 2007; West-Eberhard 2003; Pigliucci and Kaplan 2006; Pigliucci and Müller 2010). In fact, this was Dobzhansky's own view, now being helped along by the maturation of developmental genetics. If, however, genomes have evolved norms of reaction so plastic that they can produce phenotypes able to adjust themselves to nearly every environment and, by stabilizing those environments, to facilitate the recurrence of these adjustments in the next generation, talk about extending the synthesis begins to reach its conceptual boundaries. The process of adaptation starts to acquire a Lamarckian feel (Jablonka and Lamb 1995, 2005; Jablonka 2006; Walsh 2006). To be sure, natural selection need not lose its importance. But as in the pre-Weismann period, when it coexisted with the heritability of acquired characteristics, it will no longer appear as the creative factor in evolution. Instead, it will reappear in the eliminative, pruning, purifying, stabilizing, or normalizing roles that it played in Darwinian research programs before the modern synthesis (Gilbert and Epel 2009). It will preserve the *presumptively* adaptive phenotypes that the developmental process serves up. Those advocating this approach can call themselves Darwinian with good conscience and solid historical precedent. But those among them who call for a new rather than an expanded or extended evolutionary synthesis are right (Gilbert and Epel 2009, 398). The modern synthesis is bounded by its view of selection as a creative force in evolving relative adaptedness. In crossing that boundary, some, though not all, versions of the new developmentalism gesture toward replacing the synthesis.

WILL A NEW SYNTHESIS BE NON-DARWINIAN?

We have noted that new developmentalists, even if they recast natural selection as executioner of the unfit rather than creator of the fit, can still call themselves Darwinians. But they can do so only if they retain a gap between what produces variation in the developmental cycle and what is responsible for differentially retaining it. That, as we saw at the outset, is what marks off the boundaries of the otherwise quite diverse Darwinian tradition. By this standard, one can certainly remain a Darwinian, even if one sees that genes are only one "developmental resource" among others in the massive process of feedback-driven parallel processing by which organisms come to be (P. Griffiths and Gray 1995; Oyama, Griffiths, and Gray 2001; Moss, 2004). We agree that the mutual interactions among various developmental resources are so extensive, indeed so self-organizing, that the genome should no longer be seen as a set of instructions like a card-punched Jacquard loom or the early computer programs from which the questionable notion of "genetic program" got its inspiration. Nonetheless, if one's estimate of the actively adaptive and self-organizing character of development blurs or erases the line between the production in each developmental cycle of variation and its retention, one will be opting for a "new and general theory of evolution" that is post-, non-, or (in polemical contexts) anti-Darwinian.

Just as population genetic Darwinians used representational devices afforded by statistical mechanics and thermodynamics to talk about population dynamics, so today's self-identified non-Darwinians (as well as some of their Darwinian peers) are making use of new mathematical forms of representation and analysis to help revive old, often pre-Darwinian traditions in evolutionary theory. Like their predecessors, these proposals tend to minimize the role of natural selection. The "process structuralists" who think of the appearance of new organic types in the tree of life as resulting from sudden, spontaneous, saltationist reorganizations of the genome look for help in mathematical "catastrophe theory" (Goodwin 1994). New developmentalists who downplay selection have been looking to Alan Turing's reaction-diffusion model to represent and explain morphological formation (G. Müller and Newman 2003). Those who stress the self-organizational (or for some "autopoietic") character of embryonic development can look for help to the dynamics of cellular automata and genetic algorithms (Maturana and Varela 1991; Kauffman 1993). Similar mathematical tools have promising application to the ecological coevolution of organisms and environments. Developing organisms, because they are actually highly structured and bounded ecological systems, are driven by the same end-directed energetic processes we find in ecological systems, while ecological systems themselves exhibit nascent developmental trajectories (Ulanowicz 1997, 2009). We need not go into the details of such overtly non-Darwinian projects. But any Darwinism of the future, we are sure, will have to make itself empirically adequate and epistemologically secure by interacting with these rivals. It will have to explore, for example, how selection and self-organization are related in complex developmental systems (Depew and Weber 1995). The answer to this question is open, making it antecedently as likely as not that the evolutionary theory of the future will be Darwinian, although in new and unanticipated ways.

❧ Essay 51 ❧

Human Evolution after Darwin

Jesse Richmond

Darwin's primary goal in *The Descent of Man* was to convince his readers that the general principles of the evolutionary theory he had laid out in *The Origin of Species* were equally applicable to humankind as they were to "lower" forms of life – namely, that our species, like all others, was descended from another, preexisting species, and that this process had been accomplished in large part (though not solely) through the agency of natural selection. Conspicuously absent from *Descent* was any account of the *actual* forms through which humankind's line of evolutionary descent had passed. The reason for this was simply that, on the matter of human evolution, the fossil record remained silent, and Darwin was too cautious a scientist to venture into lines of argument for which he saw little supporting evidence. He also knew that the sparseness of the fossil record was not in itself sufficient reason to reject evolution. So, in the absence of any fossilized remains of ancestors, Darwin restricted himself to such genealogical inferences as could be made by comparing humans to other living forms in the light of his evolutionary principles.

As convincing as Darwin may have been using the evidence he had at hand, absence breeds curiosity. The scientific study of human evolution after Darwin has been animated in large part by the desire for the direct, material evidence of our species' evolutionary ancestry that was still lacking at the time of Darwin's death in 1882. Pervasive talk of "missing links" throughout the twentieth century testified to the hold that the absent ancestors had on both professional and public minds: Darwin had shown in principle that humans had an evolutionary history, but now the task was to populate that history. This is not to say that the work done by students of human evolution since Darwin has been only to slot newly discovered fossil ancestors into a theoretical framework set in stone by the Great Man. Quite the contrary, Darwin's model of human evolution was challenged, defended, and modified on a number of fronts simultaneously to the influx of previously unknown fossil evidence. The most significant point of contention, as with Darwin's theory of evolution more generally, was the ability of natural selection to accomplish all that he claimed it could. Developments in theory and new fossil discoveries had, it might be said, a dialogical relationship in the more than a century since Darwin's death: new fossils were interpreted in the light of existing theory, and theory modified in the light of new fossil discoveries. The relative influence of each in the case of individual scientific opinion depended to a large degree on the training and interests of the individual in question.

Even while Darwin was still alive, not all evolutionists were so cautious as he when it came to reconstructing human phylogeny and speculating about the bodily structure of ancestral forms. Chief among enthusiasts of this stripe was Ernst Haeckel (1834–1919), professor of comparative anatomy at the University of Jena. Haeckel believed that it was possible to reconstruct the history of life even in the absence of fossils, using the nascent field of embryology for evidence about the nature of ancestral forms. Also unlike Darwin, Haeckel held a view of evolution as a progressive process, moving through a hierarchy of stages on the path from microorganism to human being. In fact, he believed that humans experienced twenty-two distinct stages over the course of their evolutionary development. During the twentieth stage, humankind's ancestors had attained a level of organization equal to that of the living species of apes. So, he reasoned in *Natürliche Schöpfungsgeschichte* (published in 1868, translated into English as *The History of Creation* in 1876), before humans attained their current stage (the twenty-second), their ancestors would have had to pass through a stage that split the difference between ape and human characteristics. The two most important characteristics separating humans from apes, according to Haeckel, were an upright stance and the ability to use language. He believed that the upright stance would have evolved before language, and so he posited the former existence of a bipedal, silent ape-man that he called *Pithecanthropus alalus*. Because there were no known fossil remains of such a creature and no living species to approximate this stage of evolution, *Pithecanthropus* literally represented a "missing link" in Haeckel's reconstruction of human phylogeny.

Intrigued by Haeckel's postulated "missing link," an enthusiastic young Dutch anatomist named Eugene Dubois (1858–1940) decided to leave a fledgling academic career in the Netherlands and go in search of the hypothetical ancestor (Fig. 51.1). In 1887 Dubois traveled to the island of Sumatra to take up his search in the Dutch East Indies. The choice was made partly on a pragmatic basis – it was simply easier for a Dutchman to travel to and work in a Dutch colony than elsewhere in the tropics – but it was also a reasoned choice. Darwin believed that humankind's closest living relatives were the African apes, and so he reasoned in *The Descent of Man* that it was more likely that humankind's extinct ancestors would be found in Africa rather than elsewhere. Dubois, in contrast, believed that the African apes were farther off the human line of descent than were the Asian apes, and so by Darwin's own reasoning Dubois found it more likely that the missing link was an Asian species.

Remarkably, Dubois succeeded. In 1891 and 1892, while excavating on the island of Java near the village of Trinil, Dubois's workmen uncovered a molar, a skullcap, and a femur

FIGURE 51.1. Eugene Dubois (1858–1940), the Dutch anatomist who (inspired by Ernst Haeckel) discovered the first "missing link" in 1891. Nineteenth-century photo

that, in Dubois eyes, fulfilled the criteria for Haeckel's missing link. The skullcap seemed to indicate that this creature had possessed a brain somewhat larger than those of the living apes, though still far smaller than those possessed by modern humans. More crucially, Dubois believed that the femur clearly showed that the creature had walked upright. After initially naming the creature *Anthropopithecus*, Dubois adopted Haeckel's generic nomenclature but substituted his own specific name to emphasize the fact of the creature's upright gait: *Pithecanthropus erectus* (Fig. 51.2).

Dubois returned to Europe in 1895 with the *Pithecanthropus* remains, where he argued that the creature he had discovered represented the direct ancestor of modern human beings. He received some support, but scientists were in general reluctant to give their full assent to Dubois's strong phylogenic claim. Some were outright dismissive, such as Germany's preeminent anti-Darwinian biologist Rudolf Virchow, who argued that Dubois had discovered not a human ancestor but an extinct form of giant gibbon. Dubois, perhaps believing that he would be celebrated by all of scientific Europe when he returned with his missing link, resented the muted response that greeted *Pithecanthropus*. In retaliation, he locked away the specimens, refusing to allow other scientists to examine them until he was forced to do so in 1923 (see Theunissen 1989).

Meanwhile, another fossil discovery had redirected anthropological controversy back to prehistoric Europe. In 1912 the amateur archaeologist Charles Dawson discovered some fragments of human skull in a gravel pit near the village of Piltdown in Sussex, England (Fig. 51.3). A subsequent search,

FIGURE 51.2. Dubois's discovery, nicknamed "Java Man," now assigned to the taxon *Homo erectus*. Permission: Wellcome

FIGURE 51.3. Piltdown man, the most notorious fraud in the history of science, now known to be a combination of human and orangutan bones. Suspicion is strong that Charles Dawson was primarily responsible for the fraud, although a younger worker Martin Hinton – whose effects discovered after his death contained suggestive stained bones – may also have been involved. Angus Wilson wrote an entertaining novel, *Anglo-Saxon Attitudes* (1956), inspired by Piltdown. Permission: Wellcome

in which Dawson was joined by A. S. Woodward of the British Museum, resulted in the discovery of more skull fragments, a jawbone with several teeth, some animal bones and primitive stone tools. Woodward's reconstruction of the skull showed a being with a cranial capacity intermediate between humans and apes and, significantly, without the prominent brow ridges of Dubois' *Pithecanthropus* or the European Neanderthals. In contrast, the jaw was similar to that of a modern ape, though the teeth seemed to show a pattern of wear more consistent with that of humans than of apes. The creature was given the name *Eoanthropus dawsonii* by the neuroanatomist Grafton Elliot Smith of University College London, who found in "Piltdown Man," as *Eoanthropus* came to be called, striking confirmation of his belief that brain growth and modernization had been driving forces behind human evolution, coming before the development of a characteristically human jaw or gait. The picture of human evolution prompted by the Piltdown discovery was thus in opposition to that prompted by Java Man, whose relatively apelike skull and humanlike femur seemed to indicate that bipedalism had arisen before the evolution of a very large brain. For some anthropologists, the apparent incongruity of the Piltdown skull and jaw was too much for them to believe that the two had come from the same creature. In the 1940s, further suspicion was aroused when Kenneth Oakley of the British Museum demonstrated, using a new method based on measuring the comparative fluorine content of fossils, that the *Eoanthropus* remains were much more recent than the animal bones that had been found with them. Subsequent investigation revealed that while the skull was human, the jaw was that of an ape whose teeth had been filed to give them the appearance of human-like wear. All of the *Eoanthropus* remains had been stained to give them the appearance of age to match that of the animal bones. Piltdown had been a deliberate fraud (see F. Spencer 1990).

Fraud or not, the Piltdown episode revealed a tendency among many students of human evolution at the beginning of the twentieth century to focus narrowly on the brain as the trait that most essentially characterized the evolutionary differentiation of humans from their nearest relations in the animal world. Darwin had warned in *The Descent* against focusing on even extreme differences in just one or a few traits when determining relationships among groups. He was especially concerned that those who would deny the close genealogical relationship between humans and apes were placing undo importance on the large difference in brain size and paying not enough attention to the multitudinous points of similarity in less conspicuous structures. And yet, after his death, even those who were prepared to accept that humans and apes shared an ancestry were not prepared to take Darwin's advice and consider the brain as just one among many traits that could be used to more precisely determine just when and how humans had become differentiated from their nearest kin.

The first decades of the twentieth century also saw a move away from Darwin in theoretical matters. This movement was not limited to theories of human evolution: a significant number of scientists were questioning whether natural selection was indeed the primary driving force behind evolutionary change in general, as Darwin had claimed (Bowler 1983). In its place, some evolutionary scientists substituted alternative mechanisms. For instance, the theory of "orthogenesis" posited the existence of an innate biological drive toward evolutionary change in a predetermined direction, regardless of any environmental pressures affecting the species. Alternatively, some preferred a version of the Lamarckian mechanism of the inheritance of acquired characters that recognized a role for the environment in evolutionary change but not as the agent of selection, as Darwin had characterized it. Both mechanisms allowed for similarity of structure between species to be interpreted in a very different way from how it would have been under a strictly Darwinian theory. To Darwin, the many points of structural similarity between humans and the apes indicated that they had lately (in evolutionary terms) shared a common ancestry. In contrast, to an orthogenecist, those same similarities could be the result of both humans and apes having independently followed parallel courses of evolution from different points of origin; while to a Lamarckian, the points of structural similarity could have arisen through adaptation to similar environmental circumstances in a process of evolutionary convergence. In a particularly striking instance of the application of non-Darwinian evolutionary theory to the descent of human beings, the British anatomist Frederic Wood Jones proposed in 1919 that humans had not descended from any ancestor shared with chimpanzees or gorillas, nor had they ever passed through any "apelike" stage, but had instead evolved independently from a very distant ancestor that most resembled, among living Primates, the diminutive tarsier. Such traits as were shared by humans and apes, but not tarsiers, could be attributed, according to Jones, to convergent evolution (Bowler 1986).

Such non-Darwinian visions of human evolution maintained their appeal for many scientists well into the 1930s, by which time developments, both evidential and theoretical, had begun to militate against such approaches. One such development had its origin in the mid-1920s, though it did not begin to seriously alter opinions among many scientists until the next decade. In 1924, Raymond Dart (1893–1988), the newly appointed chair of anatomy at the University of the Witwatersrand in Johannesburg, South Africa, came into possession of a small skull belonging to an immature apelike creature that had been blasted out of the walls of a limestone quarry near the village of Taung (Plate XL). In the report that followed, Dart – who had no problem with the idea of apish ancestry – described the skull as possessing a remarkable combination of humanlike and apelike characters, and he made the provocative claim that *Australopithecus africanus*, as he named the species, represented the first evolutionary step that human ancestors had taken after diverging from the common ancestor with apes. Among the morphological features that Dart took as an indication of a step in the direction of the human and away from the ape were a somewhat enlarged brain, smaller teeth and a more humanlike dental arch, and a bipedal gait. This last Dart inferred from the position of the foramen magnum – the hole in the base of the skull through which the spinal column attached to the brain – because the skull had not been accompanied by any postcranial remains. Dart was aware that, if he was correct about the evolutionary significance of *Australopithecus*, it would represent a vindication of sorts for Darwin's argument that the earliest human ancestors to become differentiated from the common ancestor with the apes would be found in Africa, and he took the opportunity to construct a Darwinian tale of an Africa-based human evolution: the first of our apelike ancestors to come out of the dense African jungles onto the open savannahs, such as existed on the South African Highveld, would have experienced new environmental pressures under which bipedalism conferred a distinct advantage. Once on two feet with hands free, those individuals in possession of manual dexterity and the intelligence to use it would have a further advantage. Dart (1925, 1926) was confident that in *Australopithecus* he had discovered the true missing link.

Back in Europe, however, Dart's claim for the ancestral status of *Australopithecus* did not fare well. For one thing, the specimen came from an immature individual, and it was known that immature apes more closely resemble humans in much of their cranial morphology than they do when fully grown. Thus, it could be argued, as it was by Sir Arthur Keith (1866–1955), conservator of the Museum at the Royal College of Surgeons, that Dart had discovered a very interesting extinct ape closely allied to the chimpanzee, but one that bore little significance to human evolution. Further, Keith was among those who believed that brain growth had led the way in human evolution, and from that perspective Dart's relatively small-brained creature failed to fit the ancestral mold as well as larger-brained claimants like Java Man and Piltdown Man.

Despite Darwin's hypothesis of an African ancestry, geography also worked against Dart's claim. Many students of human evolution had by this time adopted the view that human evolution had played out largely in Central Asia. One prominent advocate of this view was Henry Fairfield Osborn (1857–1935), director of the American Museum of Natural History. Osborn believed that Central Asia had played host to the evolution of much of the modern mammalian fauna and that humans would be no exception to the rule. To that end, the museum sponsored a series of expeditions to Central Asia with the express intention of finding the missing link. While those expeditions succeeded in significantly advancing the paleontology of dinosaurs, they turned up no potential human ancestors. That distinction went to Davidson Black (1884–1934), a Canadian biologist who traveled to China separately, but with the same belief that Asia held the key to human evolution. He was rewarded in 1927 with a hominid tooth found at Chou Kou Tien, near Beijing, and further cranial and postcranial remains discovered there over the next two years. Black named the creature *Sinanthropus pekinensis*. The specimen shared many characteristics with the Javanese *Pithecanthropus*, though Black preferred to claim for his discovery a more direct evolutionary connection to modern humans. *Sinanthropus* caused a sensation, and consequently Dart's *Australopithecus* was pushed further to the margins of scientific attention.

One of the few to adopt Dart's position as his own was the paleontologist Robert Broom (1866–1951), who was also living and working in South Africa. In the early 1930s, when most other students of human evolution were fawning over *Sinanthropus* and Dart had largely given up trying to convert others to his view, Broom continued to publish on the merits of an Australopithecine ancestry and to actively search for more specimens. His searches began to bear fruit in 1936, when he discovered a second specimen in a cave near Johannesburg. From then until the end of his life fifteen years later, Broom and his associates produced a near constant stream of new specimens from several caves in the same general vicinity. The cranial remains showed that the adult form retained some of the humanlike features that Dart had recognized in the Taung Child, while the postcranial remains, especially parts of the pelvis, provided evidence that these creatures had indeed been bipedal. Additionally, the variation within the sample led Broom to believe that there must have existed several distinct forms of these South African ape-men, and he created two new genera in addition to Dart's *Australopithecus* to house them. The culmination of his efforts was a monograph (Broom and Schepers 1946) that for the first time summarized the evidence that had been collected up to that point and made it available to a worldwide audience. The evidence convinced a now aged Arthur Keith (1946) that he had erred two decades previously in dismissing Dart's claim that *Australopithecus* was a human ancestor.

Curiously, the African location and distinctly apelike characteristics of the Australopithecines were not for Broom evidence for a Darwinian interpretation of human evolution. Broom was skeptical in general of the ability of natural selection acting on randomly generated variation to produce the kind of large-scale evolutionary change that Darwin had claimed for it. Instead, he developed an idiosyncratic evolutionary theory in which "spiritual agencies" inhabiting each and every creature were responsible for producing individual variations, while an overarching "intelligence" was responsible for directing an overall trend in such variations toward a designed goal. That goal, according to Broom (1931), was the evolution of human beings, and having achieved its goal, evolution had in all significant respects ceased. Broom used his conception of a goal-directed evolutionary process to argue in the 1946 monograph that the human line of descent, including the Australopithecines, had diverged from the primate stem at a very early stage, and thus did not share a common ancestry with the apes. The many points of similarity between the Australopithecines and the apes, according to Broom, were the result of evolutionary parallelism.

The influence of Broom's theoretical ruminations about the goals and mechanisms of evolution was minimal, but his monograph forever altered the way students of human evolution conceived the phylogenic relationship between Australopithecines and modern human beings. In Britain, the newly synthesized evidence from South Africa had an especially strong impact on the Oxford professor of anatomy W. E. Le Gros Clark (1895–1971). Clark had long been interested in how humans fit into the broader context of primate morphology and evolution, and the fossils from South Africa provided him with a crucial piece of the puzzle. Indeed, Clark was so impressed with the potential significance of the Australopithecine material that he took the trouble of traveling to South Africa in 1946–47 in order to examine the specimens for himself – something that few scientists from Europe had ever bothered to do.

After examining the Australopithecine remains in South Africa, Clark, along with both Dart and Broom, traveled to Nairobi to attend the first Pan-African Congress on Prehistory, an event organized by the Kenyan-British prehistorian Louis B. Leakey. In Clark's report to the conference on his recent study of the South African fossils, he publicly declared for the first time his belief that the Australopithecines were part of the human line of descent, rather than that of the living apes as Dart's critics had long contended. He also made the novel move of describing them as part of the superfamily *Hominoidea*, a term recently coined by the eminent paleontologist George Gaylord Simpson to contain humans, apes, and their extinct ancestors. In addition to the Australopithecines, Clark cited another, more primitive branch of extinct African Hominoidea that was known from the environs of Lake Victoria courtesy of Leakey, among others. These Miocene forms, which included the genera *Proconsul* and *Limnopithecus*, represented for Clark evidence for the early radiation of the Hominoidea, allowing for a remarkably rich, multistage picture of that group's evolution in Africa. In Clark's (1952) analysis, Darwin's cautious speculation about humankind's African roots, made without the benefit of any fossil record, had been transformed into a more confident belief in the African evolution of the higher Primates, including humans, as a group.

The emergence of Africa as a favored locus of human evolution was not the only way in which Darwin's views were regaining influence in the study of human evolution during the postwar years. Scientists had become interested in reconceiving human evolution as a process driven primarily by natural selection, as opposed to the non-Darwinian mechanisms that had been favored by evolutionary theorists for much of the early twentieth century. This came as part of the wider neo-Darwinian synthesis that had been upending biology in recent years by using the insights of population genetics to demonstrate that natural selection had the power to drive the sort of evolutionary change that Darwin had claimed it could.

The consequences of the neo-Darwinian synthesis for human evolutionary studies were made evident at a 1950 conference on the "Origin and Evolution of Man" held at the Biological Laboratory at Cold Spring Harbor on Long Island. The conference was organized by the geneticist Theodosius Dobzhansky, one of the main advocates for the synthesis, and the anthropologist Sherwood Washburn, who wanted to bring the theoretical insights of the synthesis to bear on the problems of his discipline. Other synthesis notables present at Cold Spring Harbor were George Gaylord Simpson of the American Museum of Natural History, chief paleontologist to the neo-Darwinians, and Ernst Mayr, whose well-known work attempted to bring biological systematics into line with the Darwinian principles of the synthesis. In his presentation to the conference, Mayr lamented what he saw as the unprincipled way in which fossil hominids had been categorized. In Mayr's view, taxonomic categories ought to reflect the biological realities that structured relationships between groups in an evolutionary process driven by natural selection. Instead, scientists like Broom had categorized fossil hominids to reflect small morphological differences without reference to the underlying biological significance of those differences, resulting in the creation of a huge number of separate taxa. For Mayr (1951), the fact that all known fossil hominids, for all their differences, apparently shared with modern humans the trait of bipedal locomotion – the ecological importance of which was difficult to underestimate – argued for bringing all of these forms together as separate species within a single genus: *Homo*.

Mayr's proposal to radically reform hominid taxonomy was just one aspect of the efforts by proponents of the new synthesis to bring all aspect of biology into line with their emphasis on natural selection. However, the synthesis was not the only force looking to reshape the study of evolution around midcentury. Following on the discovery of the double-helical structure of DNA in 1953, many researchers became interested in how evolutionary change occurred at the molecular level and its relation to changes at the organismal level. With respect to Primates, one of the first to investigate evolutionary problems with molecular techniques was Morris Goodman. In the early 1960s, Goodman, theorizing that stronger immune reactions would indicate more distant evolutionary relationships, performed experiments to determine the immunoreactivity of blood proteins between primate species. His results indicated a closer evolutionary relationship among humans and the African apes than any of those had with orangutans. Another pioneer of molecular techniques in the study of primate evolution was Emile Zuckerkandl, who introduced the name "molecular anthropology" to denote the new field. Working with the famous biochemist Linus Pauling, Zuckerkandl introduced the concept of the "molecular clock," which stated that the number of molecular differences between the hemoglobin protein of two species could be used to determine the amount of time that had passed since they last shared a common ancestor. Later in the decade, the idea of the molecular clock became the basis for Motoo Kimura's neutral theory of molecular evolution, which claimed that the vast majority of evolutionary change at the molecular level was not driven by natural selection but by a selectively neutral process of genetic drift (see G. Morgan 1998).

Because taxonomies and phylogenies created using the techniques and theories of molecular biology deemphasized the role of natural selection, they were generally opposed by the proponents of the synthesis, and molecular evolution was sometimes derided as anti-Darwinian. Neither Goodman nor Zuckerkandl, however, viewed his contribution as a denial of the role of natural selection in evolution, but rather as a technique to improve scientists' ability to make phylogenic determinations (Sommer 2008). Theoretical disagreements notwithstanding, the importance of molecular biology to the study of human and primate evolution increased greatly over the next several decades, as techniques for the sequencing of proteins and eventually DNA itself made possible direct comparisons of the molecular makeup of different species. Among the most significant findings that resulted was that humans were more closely related to chimpanzees than any of the other apes and that the two had shared a common ancestor more recently than anyone had expected. More recently, there has been progress in extracting genetic material from fossil hominids, especially Neanderthals, and using molecular analysis to determine their evolutionary relationship to modern humans.

Molecular biology has without a doubt drastically changed the study of human evolution, but it has not, as some have worried, made older, fossil-based approaches obsolete. Indeed, the usefulness of the molecular clock is dependent on paleontological data for "calibration." Further, no amount of molecular analysis can predict what the fossil record might turn up next. Even as Goodman and Zuckerkandl were introducing molecular techniques into the study of human evolution in the early 1960s, important new fossils were emerging from Louis Leakey's excavations at Olduvai Gorge in northern Tanzania. Some of these resulted in the creation of a new taxon, *Homo habilis*, representing the oldest and most primitive member of our species' genus and further evidence of the centrality of Africa to human evolution (Leakey, Tobias, and Napier 1964). The 1970s saw even more spectacular discoveries in Africa. Beginning in 1973, researchers working in the Afar region of northern Ethiopia uncovered a large number of hominid fossils that were dated earlier than any previous hominid remains. Among these was a partial skeleton that represented the most complete remains of an individual early hominid yet known. The skeleton, affectionately nicknamed "Lucy," showed that bipedalism had already evolved in this early hominid, named *Australopithecus afarensis* (Plate XLI). If there were any lingering doubts about the morphological evidence for the early evolution of bipedalism, an almost

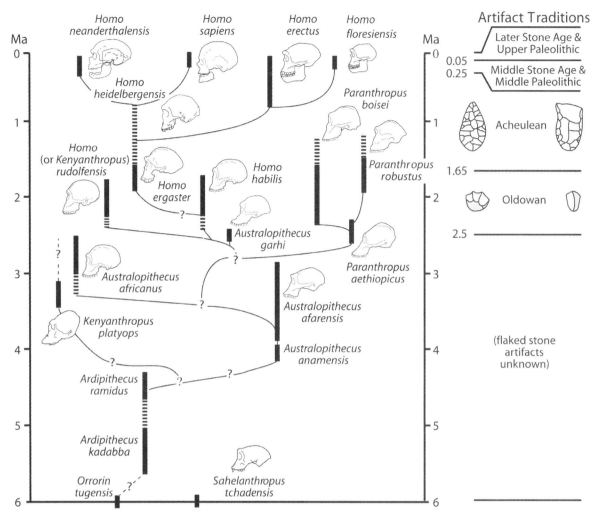

FIGURE 51.4. Human phylogeny. As can be seen, we now have an ever-more-detailed picture of human ancestry and of our close relatives.

simultaneous discovery put them to rest. At Laetoli, near Olduvai in northern Tanzanian, a team led by Mary Leakey uncovered a sample of hominid remains of similar age and morphology to those from Ethiopia and, preserved in a layer of three and a half million year old volcanic ash, a set of hominid footprints clearly demonstrating a bipedal gait. Many scientists came to consider *Australopithecus afarensis* as the common ancestor of all later species of the genera *Australopithecus* and *Homo* (see Johanson and Edey 1981). In the 1990s, the Afar region yielded even older remains belonging to the genus *Ardipithecus*, which some scientists have argued was already bipedal at a point not to far removed in time from the last common ancestor with chimpanzees (Gibbons 2009) (Fig. 51.4).

The upright posture, then, has become the trait that most scientists see as the definitive marker of early human evolution. However, the question of *why* bipedalism evolved has remained vexed. Part of the difficulty has been that the question of cause actually comprises several distinct questions, something that scientists have not always appreciated in their efforts to provide simple, all-encompassing reasons for why our ancestors adopted an upright gait. One question concerns the nature of the ancestors from which the earliest bipedal hominids evolved. Given the overall anatomical and genetic similarity of humans to the African apes, one possibility is that the direct antecedents of the earliest bipedal hominids were, like present-day chimps and gorillas, knuckle-walkers. This possibility is favored by many modern paleoanthropologists, not least because the African apes are known to occasionally adopt an upright stance and to use bipedal locomotion over short distances, demonstrating that the raw material is present in knuckle-walkers for natural selection to potentially effect a transition to bipedalism. However, not all students of human evolution in the post-Darwinian period believed that the forerunners or the first bipeds were necessarily chimplike. For instance, Arthur Keith (1923) believed that bipedalism in our ancestors had emerged from brachiating ancestors – that is, from apes that moved through the trees by swinging with their arms in the manner of gibbons. Among the supporting evidence for this hypothesis is that effective brachiation requires that the spine be aligned vertically with respect to the pelvis, thus fulfilling one biomechanical precondition for bipedal locomotion. Gibbons were also known to shamble on two legs for short distances.

Whatever the circumstances in which the first steps were taken, however, there remains the question of why creatures engaging in occasional, unsteady bouts of bipedalism evolved into creatures relying almost exclusively on a highly developed upright gait. Here, scientists have put a lot of stock into hypotheses of changing climate and ecology. Bipedalism, so goes this explanation, was selected for in response to the pressures of life on the open plains; our ancestors were forced into open country by receding forests during dry periods and needed a way to adapt the existing primate body plan, forged in the trees, for life in a treeless environment. This scenario was favored by those in the early twentieth century who sought human ancestors on the steppes of Central Asia, and also by Raymond Dart, who found in it good reason for how the parched savannahs of highland South Africa could have fostered the evolution of *Australopithecus*. A big part of adapting to new ecological circumstances is, of course, securing a new food supply. Modern apes are primarily frugiverous, and their fruit is provided by their lush, forested habitat. On the savannah, there is little fruit but much protein is available in the form of the various ruminants that inhabit it. Humans eat much more meat than do our closest relatives, and thus many scientists around midcentury argued that the gradual improvement in our ancestors' bipedal gait was associated with the benefits it conferred on their ability to hunt. The "Man the Hunter" scenario was popularized by Raymond Dart in conjunction with the writer Robert Ardrey (1961), who saw in *Australopithecus* the origin of a creature who found in his newly freed hands the ability to kill (not only prey, but enemies, as portrayed in the opening sequence of the film *2001: A Space Odyssey*).

These causal hypotheses for the evolution of bipedalism have all come in for significant criticism. More recent fossil evidence has shown that hominids had already adopted an upright stance while still living in a comparatively wooded environment, so the barren savannah could not have played so decisive a role as it did in Dart's imagination. Also, analyses of the fossil remains of prey animals have suggested that early hominids were more likely scavenging the carcasses left by quadrupedal carnivores rather than chasing after and spearing the animals themselves. Other hypotheses have entered the fray. The paleoanthropologist Owen Lovejoy (1981) famously suggested that the major selective advantage conferred by bipedalism was the freeing up of the forelimbs for carrying things – especially males carrying food back to their mates and offspring. Lovejoy's hypothesis has proved no less controversial, for, among other things, appearing to assume that the traditional gender roles of the modern human family were operative at this early date.

While the debate continues over the cause of bipedalism in humans, some students of human evolution have come to adopt the view that seeking a singular cause is wrongheaded and bound to fail in the face of expanding evidence. Rather, they suggest, we should expect that our mode of locomotion evolved through a gradual series of steps, which did not necessarily all occur for the same reasons. Natural selection may have effected change in the posture and mode of locomotion of our ancestors in different ways and for different reasons at each step, and the many of the causal factors that have been suggested over the decades may have played roles at one time or another (Stanford 2003). In a sense, this approach is most faithful to Darwinian theory, as it takes seriously Darwin's insistence on a gradualist conception of evolution in which the relation of function to structure often changes over time.

As the foregoing examples of fossil discoveries and theoretical debates amply demonstrate, scientists' understanding of the early stages of human evolution before the appearance of our own species have been in almost constant flux right up to the present day. In contrast, in the latter decades of the twentieth century, scientists' vision of the most recent period in human evolution remained comparatively constant; since the disappearance of the Neanderthals in Europe around thirty thousand years ago, it was thought, *Homo sapiens* have been the only hominid species in existence. However, that all changed with the 2004 discovery on the island of Flores in the Indonesian Archipelago of the remains of *Homo floresiensis* (Plate XLII). These creatures, of which the partial remains of nine individuals were found, grew to heights only just exceeding one meter and had brains more comparable in size to chimpanzees than to human beings. The most remarkable fact is that the most recent of these remains have been dated to just thirteen thousand years ago, long after all non-*sapiens* species of *Homo* were thought to have gone extinct. The discoverers of the remains have maintained that *H. floresiensis* evolved from Asiatic populations of *Homo erectus* long before the arrival of *H. sapiens* in the region, but also that *H. sapiens* and *H. floresiensis* likely coexisted on the island for tens of thousands of years after the arrival of the former more than forty thousand years ago. However, critics have disputed the separate species designation, arguing that the remains are those of severely pathological *H. sapiens* who suffered from microcephaly or extreme thyroid deficiency (see Morwood and Van Oosterzee 2007).

The debate over the identity of *H. floresiensis* is not settled, but it has already highlighted the enduring nature of at least one important Darwinian theme: the ability of islands to produce the most striking examples of evolution at work. When he explained island endemism as a predictable effect of natural selection at work on isolated populations, Darwin was thinking more about floras and finches than about human ancestors. Yet the evidence that the very same evolutionary process had been shaping diversity near our own branch on the tree of life is testament to Darwin's paramount argument in *The Descent of Man*: the story of our species is not a story apart, but part of the story of all life on earth.

Essay 52

Language Evolution since Darwin

Barbara J. King

In England, the month of March 1838 brought with it an early-spring chill. At this time at the London Zoo, Jenny the orangutan was quartered inside the Giraffe House, a location heated to a degree appropriate for a tropical ape. Into Jenny's cage one day walked a man, age twenty-nine and two years back home in England after participating in a naturalist's dream, a five-year, seagoing scientific expedition during which he observed, studied, and collected specimens of the world's fauna.

The man, of course, was Charles Darwin. Darwin was, on that spring day, twenty-one years away from publishing a book on evolution, and thirty-three years away from committing to paper his thoughts on *human* evolution (Zimmer 2007). He was already keenly curious, though, about behavior of nonhuman primates. Darwin "watched Jenny gaze at herself in a mirror. She used bits of straw like tools.... Others might believe they were vastly different from an orangutan, but Darwin didn't." In a letter to his sister, Darwin even concluded that Jenny "certainly understood every word" of the zookeeper's language directed at her (Zimmer 2007, 5).

Darwin, years later, would stuff his book *The Expression of the Emotions in Man and Animals* with examples of communicative behavior by apes and a wide variety of other species besides, as he struggled to understand how language in his own species might have evolved (Fig. 52.1). As always for Darwin, the behaviors of "lower animals" formed a foundation for understanding how our human behaviors might have developed. In *Expression* and also in *The Descent of Man*, he posed questions about, and created scenarios for, the relationship of animal communication to human language, and the gradual sequence of language development during human prehistory.

These same two issues preoccupy scholars today in anthropology, biology, linguistics, neuroscience, philosophy, and psychology. Approaches to the study of language and language origins have changed massively, however, since Darwin's day, via advances in theory and technique. Scientists now study nonhuman primates in the wild and captivity with techniques (such as rigorous experimental playback of vocalizations or close analysis of videotaped gesture behaviors) more ingenious than Darwin could have anticipated. Further, they study aspects of the anatomical, behavioral, and cultural evolution of the hominins – human ancestors who lived after the split between the ape and human lineages and creatures of whom Darwin could have known next to nothing – for clues to the processes of language origins and changes over time.

Despite these inevitable scientific advances in methods, it's true also that we still may learn about language evolution from the close observation of individual apes, as Darwin once did from Jenny. Kanzi is a bonobo now living at the Great Ape Trust in Iowa in the United States, who produces and comprehends utterances made with human symbols called lexigrams (Plate XLIII). Lexigrams are abstract images located on a computer keyboard. An example would be a series of lines that look something like a Japanese letter, which is the lexigram symbol for "potato" (other examples may be found at http://www.greatapetrust.org/science/history-of-ape-language/interactive-lexigram/).

Kanzi, unlike those apes, including Washoe the chimpanzee, who had their hands molded by humans as they learned aspects of American Sign Language for the deaf, was not explicitly tutored in the lexigram system. He was present in the same room – though scampering around as infants will tend to do – when his mother took "language lessons" from researchers. It soon became apparent to psychologist Sue Savage-Rumbaugh that infant Kanzi responded to requests and engaged in limited two-way conversations using the lexigrams. His production and comprehension apparently extended to abstract concepts such as "good" and "bad" rather than being limited merely to symbols for concrete items like foods or objects.

From early days forward, Kanzi was treated by his human caretakers not only as a thinking and feeling ape but also as an ape capable of participating in emotion-based cultural routines. He hiked in the woods, played games, and engaged in emotional interactions, as Par Segerdahl, William Fields, and Sue Savage-Rumbaugh (2005, 20) explain:

> [Kanzi] acquired language in the context of climbing trees, tracking forest paths, searching, finding, preparing and eating food, chasing others and being chased, tickling and being tickled, frightening others and being frightened, pretending to bite others and pretending to be bitten, comforting others and being comforted, giving food to others and receiving it, being aggressive towards others and making friends again.

It is arguable – and, indeed, fiercely argued – whether Kanzi has indeed "acquired language" as these researchers claim. While he has mastered a broad vocabulary and uses aspects of language in creative ways that clearly go well beyond any framework of mere stimulus-response, neither he nor any other ape utters or comprehends sentences with complex syntax or creates narrative stories steeped in actions remembered from the past or imagined of the future.

The work with Kanzi does suggest that key capabilities of language are neither unique to the human brain nor innate in humans in the sense of a species-specific instinct; it points to language-related capabilities that underpin language, are plastic, and maybe be significantly altered by a primate's nurturing context.

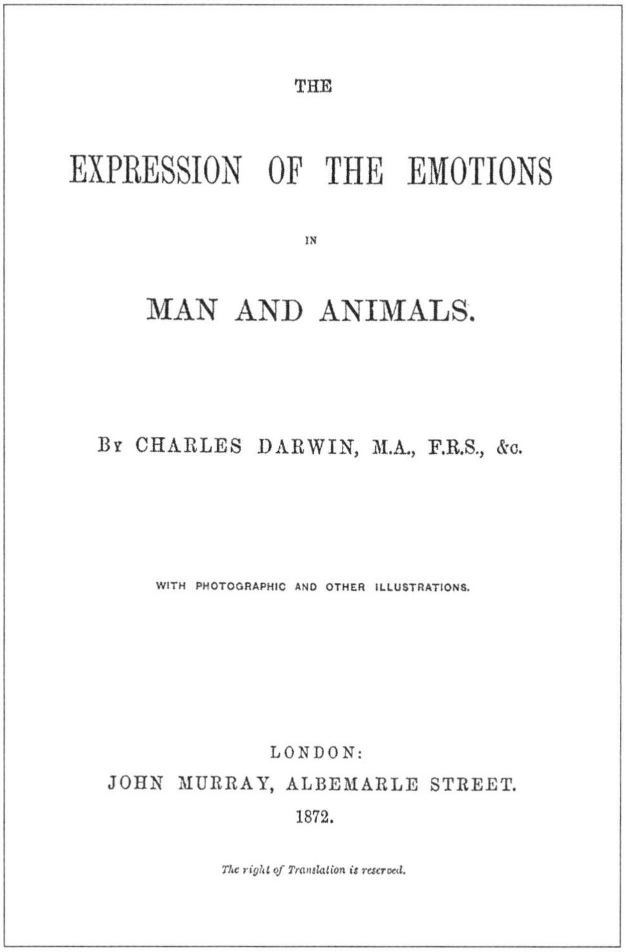
THE

EXPRESSION OF THE EMOTIONS

IN

MAN AND ANIMALS.

BY CHARLES DARWIN, M.A., F.R.S., &c.

WITH PHOTOGRAPHIC AND OTHER ILLUSTRATIONS.

LONDON:
JOHN MURRAY, ALBEMARLE STREET.
1872.

The right of Translation is reserved.

FIGURE 52.1. The title page of Charles Darwin's *The Expression of the Emotions in Man and Animals* (London: John Murray, 1872)

That the deep roots of human language abilities exist in other species is a conclusion reinforced by research with a variety of other primates in the wild and captivity – and, given what we know from his writings, is one with which Darwin would almost certainly have felt comfortable. A supernova explosion of scholarship in this arena and others, across multiple disciplines focused on the definition, origins, and evolution of language, renders this chapter's task – a review of approaches to language origins undertaken in a Darwinian spirit in the past fifty or so years – an exercise in exquisitely constrained selectivity. No comprehensive review can be managed.

Instead, a dual focus may serve to track Darwin's own twin foci. The study of monkeys' and great apes' vocal and gestural abilities may illuminate the relationship of animal communication and human language. Conclusions about the meaning of certain types of material culture in our hominin ancestors may shed light on stepwise changes in evolving language in prehistory.

Readers hungry for more thorough reviews – both in the designated areas themselves and beyond them into, for

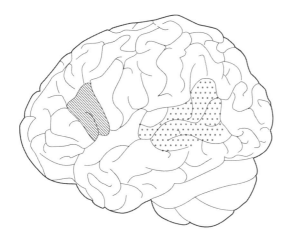

FIGURE 52.2. The human brain showing some of the areas associated with language ability

example, brain, vocal tract, and other anatomical studies of primate speech and language; agent-based modeling of language; and children's language acquisition applied to evolutionary questions – may consult Arbib (2005); Armstrong, Stokoe, and Wilcox (1994); Armstrong and Wilcox (2007); Bickerton (2000, 2009); Burling (2005); Chomsky (1975a, 1975b); Deacon (1997); Dunbar (1993, 1998); Falk (2004); Fitch (2010); Hurford, Studdert-Kennedy, and Knight (1998); B. J. King (1999); Kirby, Cornish, and Smith (2008); Knight, Hurford, and Studdert-Kennedy (2000); P. Lieberman (1984, 2006); Locke and Bogin (2006); Noble and Davidson (1996); Pinker (1994); Pinker and Bloom (1990); Savage-Rumbaugh, Shanker, and Taylor (1998); Shanker and King (2002); and Tomasello (1995, 1999) (see Fig. 52.2).

THE DARWINIAN SPIRIT

Research undertaken in a Darwinian spirit involves an evolutionary framework. Significantly, the twentieth century's leading figure in language theory insisted that there was no use in seeking evolutionary roots of language in other species. For decades, Noam Chomsky famously and explicitly rejected an evolutionary framework for language (Fig. 52.3). With his focus on universal grammar (UG), a set of rules unique to humans enabling language to flower, Chomsky (1965) set in place a highly influential program of research that depended critically on a complete break between human language and animal communication.

The concept of universal grammar, at the heart of Chomsky's early theoretical work, was "a set of innate linguistic elements, operations, and constraints on operation that determine the form(s) that a natural language can take" (Bates, Thal, and Marchman 1991, 30). For Chomsky, UG is, as its name implies, universal in all humans; is clearly demonstrable, in that children speak at a young age with far more sophistication than could ever be learned by listening to or imitating their elders (the poverty of the stimulus argument); and is domain specific, in the sense that acquiring language was completely unrelated to general perceptual or sensory systems in the brain and thus not a process derivable from those other systems (Bates et al. 1991).

Simply put, Chomsky's formulation of UG set the agenda in linguistics and related domains for decades to come. UG adherents engaged tensely with those who wished linguistics to shift focus to real-world conversational interactions and who felt that the poverty of the stimulus argument gave short shrift to the complexities of intergenerational learning dynamics. From this latter camp, Michael Tomasello offered robust evidence that indicates children do learn language rather than acquire it via maturation of the UG. Noting that "children have at their disposal much more powerful learning mechanisms" than granted by the UG camp, Tomasello (2003, 5) described socially mediated processes in which children participate, including pattern finding, joint attention, and imitation.

The key point for our purposes is these two towering figures, Darwin and Chomsky, each from a different century and each of enormous historical impact, seem at first blush to be polar opposites in the realm of language theorizing. As Alter (Essay in 21 in this volume) makes clear, Darwin concluded that language as a Rubicon between animals and humans should not be overstated; Chomsky, on the other hand, thought the significance of the language barrier separating humans and all other animals could not possibly be exaggerated.

In one sense, though, the two are not so far apart. Darwin was fascinated by questions of language learning versus language innateness, the very same questions that motivated Chomsky. Darwin though looked to dog snarls, monkey cries, ape gestures, and early humans' grunts to understand the evolutionary roots of human language, whereas Chomsky looked at abstract rules in the head that allow language to emerge in only our species and no other.

In leapfrogging centuries to compare Darwin and Chomsky, then, a key question emerges. What *is* language? All animals, from frogs to buffalo and beetles to apes, communicate, using chemicals, body postures, vocalizations, gestures, and/or facial expressions. Language is different – the key unresolved issue being what precisely is the difference that makes a difference in allowing us to distinguish language from communication (G. Bateson 1972). Even now, in the twenty-first century, no agreement is reached about what language is, or the role of symbols and syntax in defining it. What is clear is that when language researchers assert their preferred "difference that makes a difference," the framework of choice is overwhelming an evolutionary one.

Indeed, by 2002 and in a seismic shift from his earlier work, Chomsky too embraced an evolutionary framework. Writing with coauthors Marc Hauser and Tecumseh Fitch, Chomsky warned of an all-too-typical failure to distinguish between two sets of questions: those about language as a communicative system and others about the computations underlying

FIGURE 52.3. Noam Chomsky (b.1928), the American linguist, who devised the hypothesis that all human languages exhibit the same "deep structure." Although this clearly lends itself to an evolutionary interpretation, Chomsky (unlike his students) was loath to give this a Darwinian interpretation. His negativity has softened with time. Permission: Noam Chomsky.

this system. The trio separated the broad faculty of language from the narrow faculty of language. The faculty of language in the broad sense (FLB), they said, amounts to some biological capacity of humans that allows us to master any human language without explicit instruction. Hauser, Chomsky, and Fitch (2002, 1570) wrote:

> The empirical challenge is to determine what was inherited unchanged from [the common ancestor with some apes], what has been subjected to minor modifications, and what (if anything) is qualitatively new. The additional evolutionary challenge is to determine what selectional pressures led to adaptive changes over time and to understand the various constraints that channeled this evolutionary process.

Hauser et al. focused their attention on recursion as the uniquely human defining aspect of language. Recursion refers to self-embedded structures within sentences, such as when one noun phrase contains another within it ("the chimpanzee who was ranked alpha ate the most fruit"). Their recursion theory reignited the debate about the nature and evolution of language (for one example, see Pinker and Jackendoff 2005). The comparative focus of Chomsky's work with Hauser and Fitch sits squarely in line with much other language research, yet no merging of minds has occurred regarding the central importance (or unique status) of recursion.

MONKEYS AND APES

A central question in recent decades has concerned the balance between emotional and referential communication in monkeys and apes, and how that balance might shed light on scenarios for human language evolution. If entirely emotional in nature, the calls of monkeys and apes – alarm calls, food calls – would reflect arousal rather than specific content about (i.e., specific reference to) the environment. Because language itself is highly referential, this question has taken on evolutionary importance.

A turning point in primatologists' understanding of this issue came when vervet monkeys at Amboseli National Park, Kenya, were shown to have predator-specific alarm calls (Seyfarth, Cheney, and Marler 1980; see also Cheney and Seyfarth 1990). Previously, it had been widely assumed that monkeys' alarm calls expressed only fear, a kind of universal terror at the winged or loping predators that can drive up mortality on the African savannah. The calls' specificity was demonstrated via the playback technique (which itself has a long and fascinating history; see Radick 2007).

Taped alarm calls made by vervet monkeys during predator sightings or attacks were played back to the monkeys in the absence of predators, and the listeners' responses recorded. Vervet monkeys hearing an alarm call originally uttered in the presence of an eagle responded in ways quite distinct from those who heard an alarm call uttered originally in the presence of a leopard, for instance. The calls, the researchers reasoned, must therefore have contained specific information and were not expressions of only fear.

Bolstered by this outcome, researchers fanned out to primate research sites in order to carry out playbacks of vocalizations with various other monkey species. Diana monkeys too, for instance, exhibit referential (also called representational) vocalizations. In West Africa, these monkeys produce distinct alarm calls for classes of predators, again distinguishing – as had vervet monkeys – between leopards and eagles (Zuberbühler, Noe, and Seyfarth 1997; Zuberbühler 2000). Not only alarm calls, though, are referential in nature. On the island of Cayo Santiago off Puerto Rico, rhesus macaque monkeys emit screams with discrete meanings such that the screams are best described as representational signals (Gouzoules, Gouzoules, and Marler 1984). In the absence of any kind of aggressive event ongoing around them, the mother monkeys who heard played-back screams of juveniles acted in ways that could only have emerged from the information contained in the screams. These monkeys reacted less strongly, for instance, when screams emitted came from altercations without, rather than with, physical contact. "The scream vocalizations appear to have, as referents," wrote the researchers, "the type of opponent and the severity of aggression in agonistic encounters: these are external referents" (1984, 190).

A reasonable expectation, given results across a diversity of monkey species, would be that apes, with their closer evolutionary relationship with humans and arguably greater cognitive abilities, would demonstrate equal or greater evidence of referential calling. For many years, however, comparable results on apes were not forthcoming. Indeed, Slocombe and Zuberbühler (2005, 1779) noted explicitly the largely "negative results" in the search for referential communication in apes, remarking that they had been "taken to suggest that ape vocalizations are not the product of their otherwise sophisticated mentality and that ape gestural communication is more informative for theories of language evolution."

These researchers showed, however, that a captive male chimpanzee changed his foraging behavior on the basis of information he derived from other chimpanzees' acoustically distinct "rough grunts" played back to him. The rough grunts, produced in feeding contexts, were in this case given in the presence of one food highly prized (bread) and one food not much favored (apples). The subject male searched more locations and searched longer after hearing the calls given for the preferred food. Slocombe and Zuberbühler (2005, 1779) note that studies of single animals in the past "have crucially contributed to our understanding of a species' cognitive capacities," and studies of context-specific calls by wild apes point suggestively toward referential capacities (Crockford and Boesch 2003; Slocombe and Zuberbühler 2006; also see Slocombe, Townsend, and Zuberbühler 2009).

Limited data also hint at the possibility that monkeys and apes may use a sort of basic syntax – perhaps an evolutionary forerunner to recursion – in their vocal communication.

Studying the vocal combinatorial behavior of Campbell's monkeys, Ouattara, Lemasson, and Zuberbühler (2009) discovered that a suite of six vocalizations expands, via combinatorial behavior, to nine distinct call sequences. Among the principles that govern the monkeys' choices of concatenating were these: combine two meaningful sequences into a more complex one with a different meaning; add meaningless calls to a meaningful sequence and change the meaning; and add meaningful calls to a meaningful sequence and refine the meaning (Ouattara et al. 2009, 22029).

What about apes? Clarke, Reichard, and Zuberbühler (2006, 1), following research with white-handed gibbons in Thailand, reported "the first evidence of referential signalling in a free-ranging ape species, based on a communication system that utilises combinatorial rules." The gibbons sing elaborate duet songs in the mornings. Clarke et al. focused on structural differences between those songs and calls made in response to predators (in this case, artificial predators of both terrestrial and aerial varieties). The two types of calling differed in precise ways, not in terms of the repertoire of call notes but rather in terms of how they were assembled. Clarke et al. (2006, 9) conclude, "Not unlike humans, gibbons assemble a finite number of call units into more complex structures to convey different messages, and our data show that distant individuals are able to distinguish between different song types and infer meaning." The role of syntax is this process was identified as key.

Clearly, an evolutionary platform existed in the primate lineage for the origins of speech. Referential and syntactic capacity in the study of monkey and ape vocal capacities yield results Darwin might have marveled at but not in the sense of great surprise. Darwin never erected a barrier between the vocalizations of other primates and those of evolving humans. Note the prescience with which Darwin thought, in this key passage from *The Descent of Man* (also cited by Alter), that alarm calls might play a central role in our understanding the origins of language, though in a framework of vocal imitation:

> Since monkeys certainly could understand much that is said to them by man, and when wild, utter signal-cries of danger to their fellows; and since fowls give distinct warnings for danger on the ground, or in the sky from hawks (both, as well as a third cry, intelligible to dogs), may not some unusually wise ape-like animal have imitated the growl of a beast of prey, and thus told his fellow-monkeys the nature of unexpected danger? This would have been a first step in the formation of language. (1871, 1:57)

Breakthroughs in understanding the importance of nonhuman vocalizations for the origin of language should not, however, eclipse key clues emerging from gesture studies. As Slocombe and Zuberbühler (2005) point out, it is not clear that vocal behavior in monkeys and apes is the result of a conscious choice to inform others. In this regard, it's significant that at least among great apes, gestures are intentional, and via nonverbal deixis, the ability to direct someone's attention to a specific location, may also be referential in nature as with pointing (see review in Leavens, Racine, and Hopkins 2009, but see also Slocombe, Waller, and Liebal 2011).

Captive orangutans modify their gestures according to their human audience's response (Cartmill and Byrne 2007). The orangutans were offered highly prized and not-so-highly prized foods in an experimental situation. They had clear opinions about what food they wished most to eat and gestured to zoo staff to make those preferences clear. When the humans responded as if they somewhat understood, by giving half the food desired to the orangutans, the apes repeated the gestures that they had been using all along. However, when the humans conveyed no inkling that they grasped what the orangutans wanted and offered only undesirable food, the apes switched tactics and began to use different gestures altogether. This process of active gestural negotiation between partners, dynamic and contingent, is more readily visible to the observer than is the case with vocalizations (Tanner and Byrne 1999; B. J. King 2004; Tanner 2004; Leavens et al. 2009). The study of vocalizations may lend itself to a rather linear approach, with a sender and receiver. By contrast, with the growing tendency to videotape one's subjects and watch the results in slow motion, gestural studies can point up the incredible complexity that goes beyond a linear analysis.

Primatologists offer much data to further ground the quest for language origins in the natural behavior of humankind's closest relatives (but see Pinker 1994; Bickerton 2009). Two conclusions may be offered, one methodological and one substantive, vis-à-vis the nexus of contemporary primate studies and Darwin's own research approach. First – and this conclusion seems to require continual rediscovery over the centuries – no single research method is naturally superior. Rigorous experimentation and keen natural observation as well as quantitative report and qualitative description are all worthy methods and may work effectively in concert to inform us about the roots of language in other primates. Even the anecdotal evidence has its place in illuminating rare behaviors.

Second, while Darwin was no perfect prophet in predicting the relationship of monkey and ape behavior to origins of language, he was prescient in some key ways. Tecumseh Fitch (2010, 398) concludes:

> Despite a few statements, that, today, can be recognized as errors (e.g., the idea that some monkeys can imitate vocalizations), a reader today cannot fail to be impressed by the broad sweep of data Darwin considers (ethological, neural, physiological, and comparative/evolutionary) and his mastery of the logical and theoretical issues involved in language evolution. Summing his theory up in modern terms, Darwin recognizes the distinction between the evolution of the language *faculty* and of a particular language, seeing the former as crucial.

Although there is no consensus, much research suggests that deep primate roots exist in the human ability to use speech and gesture in meaning making. How these roots were transformed into the fully semantic and syntactic capabilities of all human societies today is as contentious now as it was in Darwin's day. Whether the process involves symbols, syntax, and recursion is an ongoing debate (see Deacon 1997). Attempting to organize this diversity, Leavens et al. (2009) divide existing language theories into representational and epigenetic categories.

In representational theories, mental states are paramount, with a focus on the increasing ability in human ancestors to manipulate intentions, desires, wants, and beliefs. In the epigenetic view, the focus is instead on epigenetically heritable ontogenetic contexts characterized by lengthy periods of infant and juvenile dependency with the emphasis not on fixed and abstract mental representations changing over time but instead on the degree of joint attention or other dynamic processes changing. I. Davidson (2003,140), summarizing his work on language evolution with Noble, makes a similar strong argument for focusing on the ontogenetic context: "In our view, language and mindedness are learned at our mothers' breasts through interactions which involve joint attention between mother and infant.... The circumstances of joint attention arose from the evolutionary emergence of bipedalism and prolonged infant dependency, leading to changed circumstances for learning and transmission of knowledge."

Whether one accepts Leavens's divide or categorizes the diverse theorizing with different labels, the ongoing conversation is heated and healthy. It is also linked with Darwin's own writings. As we have seen, Darwin seized the opportunity to observe the behavior of Jenny the orangutan at the London Zoo and to observe and read about other apes and monkeys (as well as other species), as he contemplated the evolution of language.

During Darwin's lifetime, the study of human evolution via fossils and material culture was in its bare infancy. As a nod toward the inventive ways scholars are now exploring the period of human evolution itself for clues to language, the next section explores some work that proceeds in a Darwinian spirit but, in its specifics, was necessarily unanticipated in Darwin's writings.

HOMININ MATERIAL CULTURE

Vocal tracts and brain areas for language do not fossilize. Because of this biological fact, the treasure trove of thousands of human-ancestor fossils accumulated by paleoanthropologists since the late 1800s helps us only in limited ways to understand directly the language capacities of hominins. Archaeologist Davidson (2003, 144), following a review of the subject, concludes that anatomical evidence generally serves "as a poor guide to speech abilities, and has contributed little to the understanding of the emergence of language" (but see references listed in Davidson's introduction for some counterarguments to this claim).

A cache of nonanatomical information comes from material-culture remains, ranging from the tools (of stone and bone), art images and objects, and early markings made by

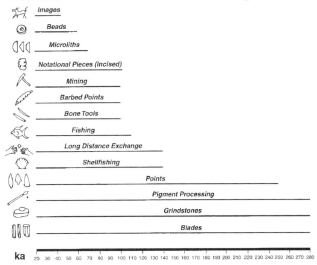

FIGURE 52.4. Human symbolism as revealed through ornaments. From S. McBrearty and A. S. Brooks, The revolution that wasn't, *Journal of Human Evolution* 39 (2000): 453–63. Permission: Sally McBrearty

hominins. To the extent that these objects give insight into the cognitive abilities of hominins broadly speaking– or perhaps even their linguistic abilities more directly – language theorists may engage with an entirely different line of inquiry than the comparative primatology one just described. Even here, although in the nineteenth century this line of thinking was nascent, links do exist with Darwin's writing, for, as Alter (Essay 21 in this volume) notes, a key idea for Darwin was that language and cognition coevolved in our ancestors, "each reinforcing the other in an ascending spiral of development."

Analysis of stone-tool technology, which began around 2.6 million years ago in East Africa, or cave paintings and portable art, starting within the past 40,000 years in Europe, Australia, and Africa, may help identify the origins and evolution of language. One key question has been whether standardization of form in tools has been imposed via hominin agency – that is, was independent of the mechanics of production or use (Ingold and Gibson 1994; see I. Davidson 2003). Our focus here, however, shall be a more recent inquiry into the nature of personal ornamentation, jewelry, and inscribed marks in an earlier time period in Africa: around 75,000 years ago. The finding of artifacts that indicate advanced material culture at such a date – so much before the earliest cave images and centered so clearly in Africa – has been central in understanding the evolution of human symbolism (McBrearty and Brooks 2000) (Fig. 52.4). Taking the perspective that study of the evolution of cognition must include direct evidence from the archaeological record (Davidson 2010), scholars have been assessing the artifacts to see if language on the part of their makers can be inferred from them.

A case-study site is Blombos in South Africa, an early *Homo sapiens* occupation area. Here, found by archaeologist Christopher Henshilwood and his associates, exists a cache of artifacts that for some scholars points toward modern cognitive and linguistic abilities (Henshilwood and Dubreuil 2009).

One set of artifacts included sixty-five beads, perforated and worn by our ancestors around the neck or wrist as jewelry. Found in groups of 2–17, the beads show wear patterns that indicate that they were suspended from thread and worn in prolonged contact with human skin. Most importantly, each discrete bead group was similar in size, shade, use wear, and nature of the perforation. In addition, two pieces of ochre were engraved, using similar preparation techniques, engraving techniques, and final designs.

It is the combination of the symbolic use of materials and the broad standardization in their creation – seen not only in the beads and the incised ochre but also in bifacial points and bone tools at Blombos – that form the crux of an argument for language. As d'Errico and Vanhaeren (2009, 37) put it, referring specifically to the items of personal decoration, "We argue that symbolic items with no utilitarian purpose, created for visual display on the body, and the meaning of which is permanently shared by the members of a community, represent a quintessential archeological proxy for the use of language or, at least, of an equally complex communication system."

The idea of community-identity being asserted through material culture, which then can be used as a proxy variable to argue for language, has, as noted, a venerable history. Some who argue for language abilities evident in the Blombos material remains now go so far as to posit that recursion is made visible through the manufacturing process (Henshilwood and Dubreuil 2009), a claim that highlights both the power of the Hauser et al. (2002) framework and the willingness of archaeologists to give strong readings to material culture.

Others, however, argue equally strongly against this archaeological-proxy approach. Botha (2009), for instance, finds the Blombos accounts impoverished because they do not specify step-by-step sequences that might account for any linkage between the properties of material culture and properties attributed to language evolution.

CONCLUSION

What can be said of the study of language origins since the time of Darwin? Darwin's dual focus – on the empirical evidence regarding the communication abilities of species other than humans, and the speculative scenarios of the time period of human evolution itself – continues but is now grounded in a wealth of hard-won information about nonhuman primates and hominin material culture. Yet, despite this grounding in the visible processes of behavior and the tangible products of culture, language origin remains one of the single most contested areas in all of evolutionary theory. With every piece of new research, the questions multiply rather than decrease.

Moving forward, we should expect a certain wildness, a barely controlled chaos in the approach to how we engage in language-evolution scholarship. Such scholarship seeks to understand all the components of language reviewed here and more, pulling in data on screaming monkeys and lexigram-arranging bonobos, on gesturing chimpanzees and jewelry-making hominins (and much more). What Darwin did, as much as give us a platform for how to think about shared language processes with other primates and about human evolution scenarios, is to ground us in a panoply of questions that attack a fascinating evolutionary issue from all sides.

ACKNOWLEDGMENTS

Several passages of this article were closely adapted or quoted from my article in *AnthroNotes* 29, no. 1 (Spring 2008): 1–7.

❧ Essay 53 ❦

Cultural Evolution

Kenneth Reisman

WHAT KINDS OF THINGS FALL within the scope of evolutionary theory? One view holds that evolutionary theory concerns how populations of biological entities (such as genes, organisms, and species) change over successive generations. Another view maintains that evolutionary theory deals with how populations of *any* kind change over successive generations. This latter view is implicit in the vast literature on cultural evolution, a literature that addresses the evolution of such diverse things as religious beliefs, scientific theories, social norms, vocabulary, technologies, agricultural techniques, and corporate practices. Most generally, "cultural evolution" refers to the various ways that an evolutionary perspective may be applied to the study of culture and society.

As the term "evolution" has acquired various meanings, the term "cultural evolution" is also employed in several ways. For nineteenth-century thinkers such as Herbert Spencer and Lewis Henry Morgan (Fig. 53.1), "evolution" was synonymous with "progress" or "development." These thinkers posited natural laws of cultural evolution whereby human societies all pass through the same sequence of developmental stages, from primitive to civilized. Their theories of cultural evolution bear little connection to evolution as Darwin understood it, and they have long been discredited.

The recent literature on cultural evolution tends to use the term "evolution" in the contemporary biological sense and to draw upon aspects of neo-Darwinian theory. As the present volume is devoted to Darwin, this essay will focus on these neo-Darwinian approaches, the most prominent of which have been *memetics* and *dual-inheritance theory*. Memetic theories are based on the concept of a selection process; these theories assume that genes have cultural analogues called "memes," and that memes evolve by a process of natural selection much as genes do. By invoking natural selection in this way, memeticists aim to provide hidden-hand explanations for the origin of religious beliefs, survival skills, and other aspects of culture. Dual-inheritance theory has emerged from the practice of model building in mathematical biology. Theorists working in this tradition build and test models of how human genetic evolution has influenced human culture, and vice-versa.

FIGURE 53.1. Lewis Henry Morgan (1818–81), pioneering American cultural anthropologist, known for his work on kinship and social structure. Nineteenth-century lithograph

Are evolutionary concepts such as natural selection, inheritance, and drift useful for understanding how cultures change? This is perhaps the central question surrounding the literature on cultural evolution. Advocates of cultural evolutionary theories argue that there are substantial similarities between biological evolution and cultural change. They suggest that studying culture from an evolutionary perspective will bring new rigor, unity, and insight to the social sciences (Mesoudi, Whiten, and Laland 2006). Critics tend to dismiss these supposed similarities as superficial (Fracchia and Lewontin 1999).

This essay surveys some of the motivations for adopting an evolutionary perspective on culture. It addresses explanations for adaptive design in culture, ways in which natural selection may operate on culture, proposed explanations for human irrationality, the interaction of genes and culture, and how applying evolutionary methods to culture can shed light on human history.

ADAPTIVE DESIGN

Our species has a unique capacity to flourish under widely varying ecological conditions. Traditional Inuit peoples possessed highly effective technologies for surviving in the frigid Arctic, including the use of kayaks and harpoons for hunting by sea. The Nuer people of East Africa have an entirely different repertoire of skills and technologies for coping with their arid desert climate, relying on a multitude of products derived almost exclusively from cattle. What has enabled humans to adapt to such a wide range of environments? In short, culture and learning.

Culture enables an individual to benefit from the knowledge and experience of others. Made of sealskin and driftwood, an Inuit kayak is an intricate artifact requiring hundreds of independent decisions for its proper construction (Fig. 53.2). The myriad discoveries and incremental refinements that have contributed to the design of the kayak could not plausibly have been discovered entirely from scratch, by one person in a lifetime. Rather, they must have accumulated piecemeal over many Inuit generations.

Many animals have some capacity for culture, in the sense that they exhibit shared patterns of behavior that individuals learn from their conspecifics. Examples of such "behavioral traditions" include the acquisition of food preferences in Norway rats, the transmission of birdsong in starlings and cowbirds, the manufacture of Pandanus leaf tools among New Caledonian crows, and the use of twig tools in termite foraging among chimpanzees (Avital and Jablonka 2000). Yet, human culture is in a category of its own. We have more routes for the social transmission of information than other species, including imitation, explicit instruction, and speech. Owing to symbolic communication, we are capable of transmitting an infinitely larger range of different informational states. With a few notable exceptions (such as the transmission of birdsong), we transmit culture with higher fidelity than other species. Most importantly, humans have a far greater capacity than other species for cumulative cultural transmission (Tomasello 1999). The behavioral traditions of other species involve, at most, an accumulation of a few distinct elements of behavior. Humans, by contrast, have a potentially limitless ability to combine new cultural innovations with prior ones. This has enabled us to develop and perpetuate technologies, systems of belief, and social organizations with a tremendous degree of complexity.

The capacity for cumulative cultural transmission explains how a human group can develop a complex repertoire of skills and artifacts over many generations, but it does not, in itself, explain why this repertoire may be adaptive. There are many intricate ways to construct kayaks that will leak and sink. What explains why the Inuit kayaks are so functional – so buoyant, watertight, sturdy, fast, and maneuverable? To explain this, we must appeal to *learning*; there must have been some mechanism that tested many boat designs in the arctic environment and made it likely that the better performing, more functional designs would be retained. The most obvious form of learning that contributed the adaptive design of the kayak would have come from the ingenuity of Inuit themselves; Inuit boat builders would have thought of and implemented occasional improvements to their kayaks.

Yet there is another factor that may have played a role. At this point, it is useful to draw a parallel with Darwinian theory. Darwin's aim in the *Origin* (1859, 3) was to explain "how the innumerable species inhabiting this world have been modified

FIGURE 53.2. A Native American (Inuit) boat, a kayak, for traveling in the sub-Arctic, the manufacture and use of which represents many generations of experience and knowledge. From Marcelle Sexe, *Two Centuries of Fur Trading* (Paris: Revillon Freres, 1923)

so as to acquire that perfection of structure and co-adaptation which most justly excites our admiration." In particular, Darwin wanted to explain the adaptive fit between species and their environments in a way that would not require the intervention of an intelligent designer. His solution was natural selection.

Just as natural selection provides a hidden-hand explanation for the adaptive fit between organisms and their environments, in principle it can also provide a hidden-hand explanation for the adaptive fit between culture and the demands of different human environments. Kayak designs conducive to sinking were likely to lower the reproductive success of their riders. Kayak designs that promoted buoyancy, speed, and agility would have increased the reproductive success of their riders, and so it is plausible that these designs were more likely to be preserved in the population. While little is known about the evolutionary history of kayaks, a study of Polynesian canoe evolution by Rogers and Ehrlich (2008) supports the hypothesis that there was selection for functional canoe designs.

In summary, the adaptive modification of culture can result from two types of learning processes. There is learning that takes place in *individual minds*, as agents cope intelligently with the demands of their environments. There is also learning that can take place at the level of a *population*, as the more successful individuals survive or reproduce at higher rates and thereby exert a greater cultural influence on future generations. The importance of natural selection on culture remains a matter of debate. Steven Pinker (1997, 210) rightly remarks that "a group of minds does not have to recapitulate the process of natural selection to come up with a good idea." Our minds, having evolved by natural selection, are extraordinary machines for learning and problem solving. That said, society-level selection processes on culture resulting in the appearance of "design without a designer" do occur.

FORMS OF SELECTION ON CULTURE

The conditions for evolution by natural selection are highly general (Lewontin 1970). Given a population of individuals capable of reproduction, the population will evolve by natural selection if the following three conditions are met: there must be variation in the physiological or behavioral phenotypes of individuals (i.e., the principle of variation); different phenotypes must have different propensities for survival and reproduction (i.e., the principle of differential fitness); and offspring phenotypes must tend to resemble parent phenotypes more than those of unrelated individuals (i.e., the principle of inheritance).

These conditions leave room for selection and culture to interact in many ways. What produces the requisite phenotypic resemblance between parents and offspring? Genetic transmission can do so, but cultural transmission can as well. What is responsible for differences in fitness among phenotypic variants? The natural environment is one source of fitness differences, but the cultural environment can be too. Consider amylase, an enzyme in our saliva that breaks down starch. Individuals from populations with traditionally high-starch diets tend to have more copies of the amylase gene than those from populations with traditionally low-starch diets, suggesting that the preferred diets of some cultural groups has led to genetic selection for greater expression of amylase (Perry et al. 2007). What makes the conditions for natural selection especially general is that they do not specify what kinds of things may compose a population. In biological evolution, a selection process may operate on populations at different scales including genes, cells, organisms, and demes. In human cultural evolution, selection processes can potentially act on at least three different types of populations: populations of *individual humans*, *cultural groups*, or *memes* (Godfrey-Smith 2009).

In the first form, natural selection acts on a population of individual persons exhibiting phenotypic variation in a cultural trait. Inheritance is realized by cultural transmission from parents to their biological offspring. The kind of reproduction that matters here is biological reproduction, and thus we measure the fitness of a cultural trait in terms of its effect on a person's rate of survival and reproduction. This form of natural selection on culture is really a special case of natural selection acting on any phenotypic trait; it will favor the evolution of biologically adaptive behaviors even if the members of the population are not aware of these adaptive effects. Plausible examples include the evolution of incest taboos, diet, and survival skills (Durham 1991).

In the second form, natural selection acts on a population of human groups exhibiting variation in a cultural trait. As with bees, ants, and other superorganisms, cultural groups are typically more than the aggregate of the persons that compose them. They often manifest a degree of coherence and social complexity that render them as individuals in their own right (Pagel and Mace 2004). There is also growing recognition that groups can be units of selection (Sober and Wilson 1998). A group of a given trait type (e.g., English-speaking, Muslim, pastoralist) can "reproduce" its type in two different ways. A new group can be formed by fission from a parent group and, in so doing, may be of a type that is similar to the parent group. A dramatic example of this has been the creation of the many distinct Polynesian chiefdom societies from a single, ancestral society (Kirch 1984). A group may also transmit its cultural type to another, existing group. Either way, the fitness of a cultural type is measured in terms of how well it promotes group survival and how frequently it reproduces. Many scholars consider group-level cultural selection of social norms to be an important mechanism in the evolution of large-scale cooperation (Henrich 2004).

In the third and most debated form, it has been proposed that natural selection may act on a population of cultural units.

In his influential book *The Selfish Gene*, Richard Dawkins coined the term "meme" as an analogue of "gene" to describe a unit of culture that evolves by natural selection (other authors, such as the psychologist Donald Campbell, had written about similar ideas prior to Dawkins). Memes can be ideas, artifacts, or behaviors – any aspect of culture that may be represented as a population of individually reproducible units. Dawkins (1976, 192) writes that "examples of memes are tunes, ideas, catch-phrases, clothes fashions, ways of making pots or of building arches." These disparate elements of culture might be said to "reproduce" by means of social learning processes such as imitation (particularly if the meme in question is an idea or a behavior) or template copying (if the meme in question is an artifact).

Are there really such things as memes? One objection to the meme hypothesis is that the culture of a society is not made up of isolable units, akin to genes (Bloch 2000). Another objection is that ideas and behaviors do not reproduce (or replicate) themselves at all; rather, they are constructed anew by each learner on the basis of complex inferences (Sperber 1996). These objections rightly caution us about drawing facile comparisons between culture and genes. All the same, we must also be wary of blanket statements regarding culture. The concept of culture encompasses a hodgepodge of disparate phenomena, each with its distinctive mechanisms of transmission. For example, the *word* "justice" is transmissible as a discrete unit, and its proper pronunciation in a given dialect can be learned with high fidelity upon hearing it just once. The *meaning* of "justice" has very different properties; it is not transmitted in a discrete fashion, difficult to infer on the basis of a few utterances, and indeed continues to be disputed by philosophers.

Many aspects of culture are not meme-like – they are not readily individuated as individuals in a population, and the relevant mechanisms of transmission are too complex to be considered forms of reproduction or replication – but others are more so. The meme perspective is perhaps most plausible in the case of language grammar evolution. Empirical studies suggest that linguistic rules evolve to be more learnable. For example, Lieberman and his collaborators (2007) have studied the evolution of the English past tense over the past twelve hundred years and have shown that irregular verbs forms have gradually and systematically been replaced by regularized versions.

An objection, however, can be raised to all of these supposed forms of natural selection. It is sometimes said that cultural evolution is Lamarckian, not Darwinian (S. J. Gould 1991). The meaning of "Lamarckian" is not generally agreed upon, but it often implies that acquired characteristics are passed directly to offspring or that the mechanism of variation is biased toward the production of adaptive variants. Arguably, cultural evolution is Lamarckian because behavioral characteristics may be passed directly from one person to another via imitation learning, and because the generation of cultural variation is heavily influenced by higher-order reasoning and adaptive biases in learning. Of these two points, biases in the mechanism of variation are especially difficult to reconcile with Darwinian theory as they reduce the explanatory power of natural selection; if adaptive variants are more likely to be generated, then selection is not strictly necessary to explain adaptive evolution.

So is cultural evolution Lamarckian or is it Darwinian? The answer is that under different circumstances it can be one or the other, or some mix of both. When the direction of evolution is attributable to biases in the generation of cultural variation, then it will be Lamarckian; and when it is attributable to differential selection, it will be Darwinian. Many cases fall in between. Moreover, because learning can be viewed alternately as a mechanism of variation or as a mechanism of selection, many cases may be difficult to classify. This result conflicts with strictly Darwinian interpretations of cultural evolution (e.g., Mesoudi, Whiten, and Laland 2004) but is consistent with a pluralistic view that regards natural selection as one mechanism, among others, that can explain cultural change. Some philosophers and biologists have advocated for a similar pluralism with respect to biological evolution (e.g., Jablonka and Lamb 2005).

For evolution by natural selection to occur there must be a population of entities capable of reproducing. In theory, cultural evolution by natural selection can occur in several types of populations of reproducing entities: populations of individual humans, populations of cultural groups, and populations of memes. However, this point comes with certain qualifications. Not all cultural phenomena are readily understood in terms of entities that reproduce, and thus not all are subject to natural selection. Even for those aspects of culture that can be understood this way, much cultural change is attributable to factors other than selection, such as individual learning, migration, or drift.

IRRATIONAL BEHAVIOR

Selection processes can reinforce each other, but they can also come into conflict. When natural selection acts on a population of persons exhibiting phenotypic variation, the traits with the highest fitness will be those that make *humans* most likely to survive and reproduce. When natural selection acts on a population of memes, the memes with the highest fitness will be those that are most effective at perpetuating *themselves*. In theory, the relationships between memes and their human hosts may be mutually beneficial, neutral, or parasitic.

Are some memes parasitic? This unsettling prospect has been raised by Richard Dawkins, Daniel Dennett, and Susan Blackmore, among others. In a recent article, Blackmore (2009, 267) writes:

> Memetics provides a completely different way of thinking about human evolution from other theories. The fundamental difference is that culture is seen not as an adaptation of benefit to early hominids and their genes, but as a parasitic second-level replicator that appeared when our ancestors became capable of imitation.

Parasitism is a type of symbiotic interaction where one organism, the parasite, benefits at the expense of another. In standard biological theory, the currency in which benefits and expenses are measured is biological fitness. Yet Blackmore is suggesting that memes are parasites in a different sense. As title of her book *The Meme Machine* implies, Blackmore suggests that memes somehow *control* or *program* us, subverting our normal preferences and decision-making capacities and inducing us to act in the service of meme propagation.

The best-known purported examples of parasitic memes are religious concepts, including God, hell, and faith. Of the idea of a God, Dawkins (1976, 193) writes,

> The survival value of the god meme in the meme pool results from its great psychological appeal. It provides a superficially plausible answer to deep and troubling questions about existence. It suggests that injustices in this world may be rectified in the next.... These are some of the reasons why the idea of God is copied so readily by successive generations of individual brains. God exists, if only in the form of a meme with high survival value, or infective power, in the environment provided by human culture.

Hence, the suggestion is that religious memes are parasites because they have evolved to exploit frailties in our mental architecture which cause us to perpetuate false but superficially appealing beliefs.

While provocative, this claim is difficult to reconcile with the many alternative explanations for the origin and evolution of religious thought. For example, Pascal Boyer (2001) and Scott Atran (2002) contend that the key features of religious thought are not due to cultural evolution at all but instead are by-products of psychological mechanisms that evolved for other reasons. Alternatively, David Sloan Wilson (2002) contends that religions are adaptive at the group level and that they have evolved as a result of group selection. At best, meme evolution plays a small role in a larger story to be told about the origins of religion.

In all, there is little evidence for the existence of parasitic memes. The most promising candidates as memes – linguistic structures – tend to be either beneficial for their human bearers or relatively innocuous. The grain of truth in the idea of parasitic memes is that we, as humans, are imperfect decision makers. Psychologists have documented numerous inconsistencies and biases in way that we make decisions (Kahneman, Slovic, and Tversky 1982). Political scientists have long observed that under some circumstances one social class can exert ideological dominance and control over another (Gramsci 1971). We can and do fall prey to specious ideas, but this occurs because we are imperfect products of evolution, not because we are plagued by parasitic memes.

GENE-CULTURE INTERACTION

Why are most people right-handed? If right-handedness confers an advantage, then why aren't we all right-handed?

FIGURE 53.3. Male-female ratios across the globe

Conversely, if left-handedness confers some advantage, then why are left-handers so rare? These questions are difficult for traditional evolutionary theory to address because the transmission of handedness from one generation to the next cannot be explained by genetic factors alone. Consider blood type. Blood type is transmitted by a single gene with three alleles (A, B, and O). A child receives one copy of the gene from each parent. Her blood type (either A, AB, B, or O) will depend only upon which two alleles she has received. By contrast, the mechanism of heredity for handedness is far more complex, likely involving a mix of genetic and cultural factors. One important motivation for studying culture from an evolutionary perspective is to extend traditional evolutionary reasoning so that we can understand the evolution of behavioral traits such as handedness that are influenced by both genes and culture.

Dual-inheritance theory (DIT), also called gene-culture coevolution, is a branch of evolutionary theory that has been developed for this purpose (Boyd and Richerson 1985; Durham 1991; Laland, Odling-Smee, and Feldman 2000; Richerson and Boyd, 2005). DIT is not, in fact, a specific theory but rather a family of models and modeling techniques for understanding gene-culture interactions. It is best understood as an extension of the mathematical models used in population genetics. In textbook population genetic models, a population is represented in terms of gene frequencies, and evolution is defined as any change in those frequencies over time. DIT builds on these models by assuming that culture constitutes an additional channel of inheritance so that a population is now defined both by the frequencies of various genes and by the frequencies of different cultural variants. In DIT models, cultural evolution can come about from standard evolutionary forces, such as natural selection and drift, as well as nonstandard forces, such as biases in social learning. This framework makes it possible to explore many different forms of gene-culture interaction.

DIT has been applied in several different ways (Laland 2008). First, it has been used to investigate why social learning improves human adaptability and what sorts of learning biases are adaptive. Humans appear to be primed with various social learning biases, including a bias to imitate successful individuals, and a bias to conform with the majority. Although such biases can give rise to maladaptive fads, DIT models have shown that they are adaptive in a broad range of changing environments. Second, it has been deployed to investigate the inheritance of various behavioral traits, such as IQ or left-handedness. The findings have suggested lower genetic heritabilities and higher degrees of cultural transmission than previously thought. Third, it has been used to investigate specific instances of human evolution that involve both genes and culture, such as the evolution of lactose absorption, or of the skewed male to female sex ratio in China. The message to emerge from many of these studies is that culture and genes do interact, and that culture can have long-ranging effects on genetic evolution (Fig. 53.3).

RECONSTRUCTING THE PAST

Darwin argued in the *Origin* that all organisms are united by descent from a common ancestor, a hypothesis he conveyed through the metaphor of the "tree of life." Phylogenetics is the branch of biology concerned with inferring the evolutionary relationships among organisms and populations, commonly represented in the form of evolutionary trees. Some scholars have been drawing upon the methods of phylogenetics to make similar historical inferences about cultures.

Do elements of culture form lineages, and can these lineages be reconstructed from data? In the case of language evolution, the answers appear to be yes. Just as DNA contains discrete units of inheritance, languages contain discretely inherited units of vocabulary and phonology. Just as different species exhibit homologous characters due to common ancestry, different languages contain homologous words, or "cognates," which can be used to infer phylogenies. For example, terms that mean "water" exist in English (*water*), German (*wasser*), Swedish (*vatten*), and Gothic (*wato*), reflecting descent from the proto-Germanic term *water* (Atkinson and Gray, 2005).

Application of phylogenetic methods to data on cognates have now produced highly resolved phylogenetic trees of the major language families (Pagel 2009). While interesting in their own right, these trees can also be used to make inferences about historical population movements. Gray and Jordan (2000) used their tree of Austronesian languages as evidence that the human settlement of the Pacific some six thousand years ago began in Taiwan and subsequently spread south into Indonesia and east into Polynesia and Hawaii.

Material artifacts can also form lineages. The historian of technology George Basalla (1988, 209) claims that "every novel artifact has an antecedent." With modern artifacts, lineages may be difficult to trace because a novel invention will often have many different antecedents. Recent patent filings in the United States cite about nine other similar patents on average (Hall, Jaffe, and Trajtenberg 2001). With archaeological artifacts, such as stone tools, a longer time scale of analysis can bring phylogenetic patterns to light. Phylogenies have been constructed for Acheulean hand axes produced many thousands of years ago, though these do not have the detailed resolution of linguistic phylogenies (Lycett 2009) (Fig. 53.4).

To what extent are cultural phylogenies treelike? In other words, is most cultural change the result of branching processes, in which one cultural lineage branches into several new ones, or blending processes, in which several cultural lineages merge into one? Arguably, cultural lineages are not treelike because blending processes are common in culture. Yet this argument is not decisive. When we consider whole human groups, a largely treelike pattern of cultural descent can be maintained over long periods. For one thing, groups can branch, and subsequent interaction between them can be limited by geographic barriers (as occurred in the colonization of the Pacific islands). For another, groups appear to

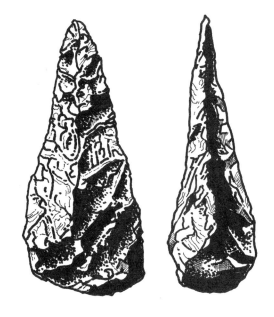

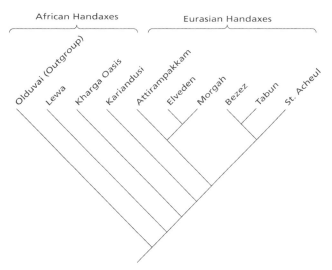

FIGURE 53.4. A phylogeny showing the cultural evolution of Stone Age ax heads. Permission: Stephen J. Lycett, Understanding ancient hominin dispersals using artefactual data: A phylogeographic analysis of Achulean handaxes, *PloS One* 4 (2009): e7404

have mechansims that can preserve a largely treelike pattern of descent even when some cross-cultural interaction occurs (Shennan 2002).

CONCLUSION

Evolutionary models are a valuable addition to the toolkit of the social sciences. As this essay has demonstrated, an evolutionary perspective has the potential to address several classes of questions about culture and society, including questions about the causes of adaptive design, about the interactions between genes and culture, and about human history. Yet, for all its potential, the evolutionary perspective is useful only under limited circumstances.

Evolutionary models are typically premised on certain basic assumptions. These include, for example, that there is a well-defined system of inheritance or transmission that operates similarly for all members of a population, and that there is a set of forces that acts consistently over successive generations to drive change. How broadly do these assumptions hold true of cultural and social phenomena? Arguably, grammar evolution and kayak design – cases for which such assumptions are plausible – represent special cases rather than the norm.

ACKNOWLEDGMENTS

The author wishes to thank Patrick Forber, Michael Ruse, and Daniel Dennett for helpful comments on an earlier draft of this essay.

Essay 54

Literature

Gowan Dawson

DISEMBARKING FROM HMS *Beagle* on the lengthy inland expeditions that would prove so crucial for his subsequent evolutionary theorizing, Charles Darwin was conscious of the need to travel light. In the cramped quarters onboard, he had access to the ship's "immense stock" of books, "upwards of 400 volumes!" that were ingeniously "stowed away in dry and secure places" in the poop cabin where Darwin worked and slept, and which included his own much-prized personal copies of Charles Lyell's *Principles of Geology* (1830–33) and Alexander von Humboldt's *Personal Narrative of Travels to the Equinoctial Regions of the New Continent* (1814–29). When on dry land among the immense vistas of South America, though, Darwin had to confine himself to just one book, and even then he was contravening Robert Fitzroy's strict directive, "Books are never on any account to be taken out of the Vessel." With the *Beagle*'s "complete library in miniature" comprising almost "all travels, & many natural history books," Darwin's choice of a travel reading was notable (Darwin 1985–, 1:553–54). As he recalled four decades later, "in my excursions during the voyage of the *Beagle*, when I could take only a single small volume, I always chose Milton."

Paradise Lost (1667), which Darwin (2002, 48) observed was "my chief favourite," was one of only three works of imaginative literature known to have been on board the *Beagle*, the others being Samuel Richardson's sentimental epistolary novel *The History of Sir Charles Grandison* (1753) and Harriet Martineau's didactic short stories *Poor Laws and Paupers Illustrated* (1833–34) (Darwin 1985–, 1:562–63, "Books on the *Beagle*"). In comparison, the shelves of the *Beagle*'s poop cabin groaned under the weight of more than one hundred titles on travel, geology, and natural history. This relative paucity of shipboard literary reading notwithstanding, that Milton's verse, with its epic vision of the luxuriant satiety of divine creation, was the sole volume that Darwin could never countenance leaving at sea indicates the significance that such imaginative works had for a young naturalist experiencing the beauty, profusion, and unruliness of the Southern Hemisphere for the very first time. As Gillian Beer (1985, 553–57) and George Levine (2006, 139–41) – although both working on the erroneous assumption that the availability of other books on the *Beagle* was limited – have argued persuasively, Darwin's intensive reading of Milton in the tropical forests of

South America made him especially sensitive to the productive superabundance and fertility of the natural world, and coalesced creatively with his earlier reading of Thomas Malthus's jeremiads on overpopulation. Literature, then, was no less formative in Darwin's early theorizing than the other, better-represented genres in the *Beagle*'s well-stocked library.

Once back in Britain, Darwin's (2002, 48) literary reading was by no means restricted to Milton, and instead he voraciously consumed the works of nineteenth-century authors such as Jane Austen, Samuel Taylor Coleridge, and William Wordsworth, whose hefty *The Excursion* (1814) he claimed to have read "twice through." The recognition that this exposure to a diverse range of imaginative literature, as recorded in the extensive reading lists in Darwin's notebooks, was not merely incidental but in fact had profound implications for the development of both his style of writing and, more significantly, his actual ideas has meant that the study of Darwin and literature has assumed a prominent role in the so-called Darwin industry over recent decades. An earlier generation of critics had shown that in *The Temple of Nature* (1803) Darwin's grandfather Erasmus used rhyming couplets to articulate the Lamarckian evolutionary ideas that he had formulated in his prose work *Zoonomia* (1794), as well as analyzing how nineteenth-century writers such as Alfred Tennyson or Thomas Hardy, among many others, incorporated both earlier versions of evolutionism like that of Robert Chambers's *Vestiges of the Natural History of Creation* (1844) and Darwin's own thinking in their poems and novels (Stevenson 1932; Henkin 1940; Primer 1964). Since the 1980s, however, the emphasis has shifted decisively to the dynamic, reciprocal interchange between Darwin and the literary culture that was such a significant element of his early reading. Of course, hardly any major writers of the later nineteenth century – as well as many of the twentieth and twenty-first centuries – were left untouched by the profound implications of Darwin's version of evolutionary theory, but the intellectual "traffic," as Beer (2009, 5) puts it, was distinctly "two-way."

Even when, in later life, Darwin (2002, 84) lamented that "I cannot endure to read a line of poetry" and now found Shakespeare "so intolerably dull that it nauseated me," his hyperbolic response to this (hardly uncommon) decline in aesthetic sensitivity resonates with such an intense sense of loss and pained revulsion – it is nothing less than the physical "atrophy of ... part of the brain" - that it actually emphasizes just how vital what he called the "higher tastes" had been to him in his formative years. In any case, his continued enjoyment of "moderately good" romantic novels that "do not end unhappily" and featured "a pretty woman," which were "read aloud" to him by his wife Emma, importantly influenced the sentimental manner in which the theory of sexual selection was depicted in *The Descent of Man* (1871), with "young rustics at a fair, courting and quarrelling over a pretty girl" (Darwin 1871, 2:122; 2002, 84; Dawson 2007, 32). This essay, therefore, will explore both sides of the dynamic process of interchange between Darwin and literature, as well as going beyond the nineteenth century to examine more recent Darwinian impacts on literary writing.

THAT UNIVERSAL STRUGGLE

On 21 June 1838 two ambitious young men, sharing the same initials and born only three years apart in 1809 and 1812, were elected to the most prestigious gentleman's club in nineteenth-century London, the Athenæum (Flint 1995, 153). Just over twenty years later they had become perhaps the two most famous men of their age, globally celebrated representatives of, respectively, British science and literature. While Charles Darwin and Charles Dickens (Fig. 54.1) seem never to have actually met or corresponded (Dickens was instead a close friend of Darwin's bitter rival Richard Owen), their lives and, in particular, their writings were intertwined in ways that are highly revealing of the mutual interchange between Darwinism and literary culture. The intensely reciprocal nature of the relationship between these two eminent Victorians in fact affords an especially illuminating example of the broader concerns of this essay, and it will thus be examined in detail. As with many other volumes in the same collection, Dickens probably never read the copy of *On the Origin of Species* (1859) in his personal library, but it is evident that, from the early 1860s, he was familiar with Darwin's contribution to the long-running disputes over organic development. As editor of the weekly magazine *All the Year Round*, Dickens published articles – including a conspicuously evenhanded review of the *Origin* – that both quoted Darwin at length and lent qualified support to the controversial new theories he propounded. An anonymous essay entitled "Species" from June 1860 even employed Darwin's own unattributed words, dealing with the nobility of the conception of a deity creating a few original forms capable of self-development, as if they were those of the article's unnamed author (Levine 1988, 128–29).

From this period, Dickens's fiction, which since his very first novels in the mid-1830s had registered the social and personal impacts of Malthusian political economy, became acutely concerned with the fierce struggle for existence in every sector of Victorian society. In the famous opening scene of *Great Expectations* (1860–61), which was published in *All the Year Round* only five months after its sympathetic review of the *Origin*, Pip, recalling his infant self sitting in the marshy churchyard containing the expressive tombstones of his younger siblings, reflects that they "gave up trying to get a living exceedingly early in that universal struggle" (Dickens 1996, 3). This, of course, directly echoes Darwin's (1859, 62) invocation of the undeniable "truth of the universal struggle for life" in the *Origin*'s third chapter. Unlike his lethargic brothers and sisters who perished because they failed to remove their hands from "their trouser-pockets," Pip is later compelled to raise his fists in combat with Herbert Pocket. In this literal enactment of the struggle for survival, Pip trounces the pale and unhealthy-looking Herbert, whose surname already aligns him with Pip's fatally apathetic siblings. Feeling only a "gloomy satisfaction in my victory," however, Pip and

FIGURE 54.1. Throughout his writing, Britain's great novelist Charles Dickens (1812–70) used Darwinian metaphors and references – barnacles in *Little Dorrit*, dinosaurs in *Bleak House*, the struggle for existence in *Great Expectations*. This illustration is of his last public reading of his work. From George C. Leighton, *Illustrated London News* 56 (1870)

his "brave and innocent" adversary are subsequently reconciled, and Herbert shows himself to possess the true qualities of a gentleman. The adult Pip looking back on this ignominious incident expresses the "hope that I regarded myself while dressing, as a species of savage young wolf, or other wild beast" (Dickens 1996, 3, 92–93). As with his initial reflections on his siblings' premature surrender in life's "universal struggle," the analytical language in which Pip retrospectively describes the events of his childhood casts them in a distinctly Darwinian light, although, paradoxically, the bestial aggression intrinsic to this new scientific vocabulary represents something entirely antithetical to the rationality and ethical self-knowledge finally achieved by the novel's adult narrator.

Great Expectations both registers the indubitable logic of Darwin's vision of a world of brutal and limitless competition, and at the same time laments the losses, especially of humane and compassionate values, entailed in accepting the necessity of aggression and dominance. Another novel published at the very same time, Anthony Trollope's *The Struggles of Brown, Jones, and Robinson* (1861–62), also alluded to the *Origin* in ironically celebrating "competition, that beautiful science of the present day, by which every plodding carthorse is converted into a racer" (Trollope 1981, 43; cf. Darwin 1859, 445). Like Dickens in *Great Expectations*, Trollope in this satire on the nineteenth-century advertising industry readily conflated the Darwinian struggle for existence – echoed in the novel's very title – with more societal and commercial forms of contest, and again not without regret at the passing of a more gentle way of life (Dawson 2004, 135). The representation of society as a site of perpetual competition, fluidity, and change is one of the most consistent tropes in later Victorian literature, contrasting with the static, ordered, and hierarchical social order depicted in the literature of earlier periods, and in Dickens's and Trollope's fiction of the early 1860s this theme was given sharpened focus, along with a suggestive new vocabulary in which to articulate it, by Darwin's recently published treatment of the incessant struggle for life.

Having putatively regarded himself as a "savage young wolf" after trouncing Herbert, Pip is repeatedly cast as a member of the same species by the vengeful Orlick ("speak, wolf!"), although it is in fact the violent rustic himself who most closely resembles a predatory animal, with his "mouth snarling like a tiger's." As Orlick's drunken threats of violence intensify, he no longer represents something merely resembling a tiger in Pip's "excited and exalted state of … brain," but instead he actually becomes a ferocious "tiger crouching to spring!" (Dickens 1996, 424–25, 427). In the often violent struggle for existence depicted in *Great Expectations*, the boundaries between human and animal become increasingly blurred in a way that Darwin, conscious of the need to curb the more radical implications of his evolutionism, had felt unable to make explicit in the *Origin*.

By Dickens's next novel, *Our Mutual Friend* (1863–64), the representation of Victorian society as a primeval "Dismal Swamp" filled with lowly "amphibious human-creatures" all competitively "extracting a subsistence out of tidal water" comes to dominate the dark and complex narrative. Alongside "all the jobbers who job in all the jobberies jobbed … [who] may be regarded as the Alligators of the Dismal Swamp," the "slime and ooze" also plays host to a diverse menagerie of other creatures, including the amputee Silas Wegg, who, in another nod to the ruthless consequences of failing in the struggle for life, is described as "like some extinct bird." So pervasive are these bestial comparisons in *Our Mutual Friend* that Gaffer Hexam, the Thames boatman who scavenges the river for corpses, becomes overdetermined with animal imagery. This sharp-eyed "half savage" begins as a "bird of prey" and is admired as "like the wulturs" by Rouge Riderhood, but his "ruffled crest" is not recognized by his supercilious social superiors, who dismiss him as merely "vermin … [a] water rat!" There is no fixity in any of these animalistic identities, and instead they continually mutate into new forms, although seemingly without any sense of a progressive evolutionary ascent. Even the kindly Nicodemus Boffin, when posing as an avaricious miser and confronting the treacherous Wegg, "eyed him as a dog might eye another dog who wanted his bone; and actually retorted with a low growl" (Dickens 1997, 80, 211, 13, 32, 14, 172, 483). Dickens's last completed novel offers a brooding vision of a Darwinian world in which the bonds of human sympathy have been supplanted by a dog-eat-dog struggle for subsistence among animalistic creatures perpetually transmuting into new, more bestial forms, and, unlike Darwin's stridently optimistic peroration to the *Origin*, with little prospect of moral or physical progress.

While this identifiably Darwinian imagery in Dickens's later fiction has been noted by numerous critics (Fulweiler 1994; Flint 1995), it is important to recognize that the novelist had in fact been investing humans with animalistic characteristics, as well as depicting a volatile society that could suddenly break out into instinctual and primitive violence, since the very beginning of his career. As a keen reader of Dickens's early novels, Darwin himself evidently noticed such vivid depictions of the animalistic brutality that was only partially sublimated by the advance of human civilization. Almost four decades later in *The Expression of the Emotions in Man and Animals* (1872), he used a passage from *Oliver Twist* (1837–39) to show how human expressions were derived from animal behavior, observing that "Dickens, in speaking of an atrocious murderer who had just been caught, and was surrounded by a furious mob, describes 'the people as jumping up one behind another, snarling with their teeth, and making at him like wild beasts'" (Darwin 1872b, 243). Darwin's "Books to Be Read" and "Books Read" notebook shows that in the early 1840s he also read *The Pickwick Papers* (1836–37), *Barnaby Rudge* (1840–41), and *Martin Chuzzlewit* (1843–44), and these novels would likewise have reinforced the interchangeability of human and animal behaviors, as well as revealing the impacts of Malthusian economics on vulnerable individuals and groups, in this crucial period for Darwin's theorizing (Darwin 1985–, 4:464, 467). The increasing density and complexity of Dickens's plotting in these lengthy novels, in which an often bewildering multiplicity of diverse characters is shown to be connected in various unexpected ways, also had important parallels with Darwin's emerging conception of a world teeming with organisms that are linked below the surface in a vast interconnected ecological system. Indeed, Beer (2009, 6) has argued that "*The Origin of Species* seems to owe a good deal to the example of … Dickens, with its apparently unruly superfluity of material gradually and retrospectively revealing itself as order." Intriguingly, then, the same conception of the natural world that would later give a sharpened focus to Dickens's portrayal of societal competition in his last novels was one that his earlier fiction may have actually played some part in shaping.

Dickens's early fiction arguably had a still more significant role in the formulation of Darwin's evolutionary ideas, even if it was one that involved relieving his mind of any onerous scientific considerations. As Darwin recorded in his Notebook M for 12 August 1838, "At the Athenæum Club. was very much struck with an intense headache <<after a good days work>> … my head got better when reading an article by Boz. –… and read so intently as to be unconscious of all around." With a work such as *Sketches by Boz* (1836), Darwin reflected, there is "no strain on the intellectual powers," thereby permitting a level of rapt concentration that rendered the reader unconscious of both his surroundings and even the physical ailments brought on by intensive scientific study (Barrett et al 1987, M, 539). It is unlikely, however, that Darwin's decision to read one of Dickens's early picaresque tales constituted merely a casual means of escaping from his exhausting intellectual labors. Rather, it was while at the Athenæum during the summer of 1838, where the young Boz himself might also have been found, that Darwin adopted a "new plan" of "only working about two hours at a spell" recommended by the geologist Charles Lyell, who advised him that "as your eyes are strong, you can afford to read the light articles and newspaper gossip" and then "after lying two hours fallow the mind is refreshed, and ... in five minutes your fancy will frame speculations which it will take you the two hours to realise on paper." Reading even the most seemingly undemanding literature was an essential element of such a regime for the self-conscious regulation of intellectual energies, and with the "new plan answer[ing] capitally," as Lyell soon heard back, the emergence of Darwin's evolutionary understanding of the natural world during the late 1830s might well have been materially assisted by Dickens's absorbing comic fiction (Darwin 1985–, 2:97, letter to Lyell, 9 August 1838; 101, letter from Lyell, 6 and 8 August 1838).

THE INEXTRICABLE WEB OF AFFINITIES

While Dickens seems not to have actually read the *Origin* and, as with his contemporaries Trollope and William Makepeace Thackeray, instead gained his knowledge of its controversial new theories as the editor of a journal eager for topical copy, the subsequent generation of Victorian novelists paid much closer attention to Darwin's momentous tome (Dawson 2004, 134–40). George Eliot, after reading "Darwin's Book on the 'Origin of Species'" within days of its first publication, proclaimed that it "makes an epoch," although she also expressed disappointment that it was "ill-written and sadly wanting in illustrative facts," which would "prevent the work from becoming popular" (Haight 1954–78, 3:227). Thomas Hardy, who was only nineteen in 1859, claimed to have "been among the earliest acclaimers of *The Origin of Species*" and in later life regularly included Darwin, along with Thomas Huxley and Herbert Spencer, in lists of the thinkers who had influenced him most (Glendenning 2007, 72). The work of both writers, as several critics have shown in detail, is suffused with Darwinian language and imagery, and the influence of the *Origin* seems even to extend to the formal structure of their fiction (Glendenning 2007; Beer 2009).

The central organizing metaphor of Eliot's most famous novel *Middlemarch* (1871–72) is that of the web of human community, as when the omniscient narrator speaks of "unravelling certain human lots, and seeing how they were woven and interwoven" within the "particular web" of an English provincial town. The image of the web, of course, has several sources, including textiles and those woven by spiders, but Eliot (1994, 141, 148) makes clear that her use of it in *Middlemarch* also derives from contemporary science, for in the novel the experimental doctor Tertius Lydgate spends his time studying "certain primary webs or tissues, out of which the various organs [of the human body] are compacted." In fact, Eliot seems to have borrowed the image from the allegedly "ill-written" *Origin*, where Darwin had argued that "plants and animals ... are bound together by a web of complex relations." The numerous characters in *Middlemarch* are certainly connected together in a no less complex web of affinities and relationships (Beer 2009, 156–61). As the narrative progresses, it is revealed that the destinies of the very different Nicholas Bulstrode and Will Ladislaw are unexpectedly linked together in a web of relations that neither man had previously been aware of. Similarly, Lydgate's success as a doctor will be determined not by his own individual skill and talent but rather by the actions of others, like Rosamond Vincy and Bulstrode, in the social medium of Middlemarch in which he has to mix. Showing how our individual destinies are all linked together in what Darwin (1859, 73, 434) termed an "inextricable web of affinities" is also central to the principal moral aim of Eliot's fiction, for by structuring her narrative in accordance with this Darwinian analogy she seeks to extend her reader's sympathies beyond the narrow focus of the individual and onto the mutual interdependence which shapes our lives.

Hardy's novels, especially *The Woodlanders* (1886), *Tess of the D'Urbervilles* (1891), and *Jude the Obscure* (1895), postulate an evolutionary continuum between the human and animal that goes far beyond the blurring of the same boundaries in Dickens's later fiction (Fig. 54.2). Their tragic protagonists are portrayed as subject to exactly the same processes of natural and sexual selection as all other organisms, with Arabella Donn in *Jude* "a complete and substantial female animal – no more, no less." Even when, as with the same novel's eponymous antihero, they aspire to a more elevated, spiritual condition, this only marks them out, ironically, as the victims of a debilitating nonadaptive over-evolution, for "at the framing of the terrestrial conditions there seemed never to have been contemplated such a development of emotional perceptiveness among the creatures subject to those conditions as that reached by thinking and educated humanity" (Hardy 1998, 39, 419, 342). This last passage should alert us to the important point that the orthogenetic forms of evolution that underlie much late nineteenth-century fiction were not necessarily Darwinian, at least from a modern perspective, and that some writers, Samuel Butler and George Bernard Shaw most prominently, used their literary works to advance explicitly anti-Darwinian, neo-Lamarckian understandings of evolutionary change (Bowler 1988) (Fig. 54.3).

Hardy's own writing, though, is at the same time acutely sensitive to the crucial role of chance and fortuity in Darwin's much less directive version of evolution. This is particularly the case in the stark, innovative poetry that Hardy only began publishing after the storm of controversy over *Jude* induced him to give up writing novels. In "Hap" ([1898] 1976, 9) the speaker considers that it would be preferable to blame his suffering on a malevolent deity, before having to acknowledge, in the final stanza:

> *But not so. How arrives it joy lies slain,*
> *And why unblooms the best hope ever sown?*
> *Crass Casualty obstructs the sun and rain,*
> *And dicing Time for gladness casts a moan ...*
> *These purblind Doomsters had as readily strown*
> *Blisses about my pilgrimage as pain.*

FIGURE 54.2. Thomas Hardy (1840–1928) was deeply influenced by evolutionary ideas, giving a much greater role to Darwinian themes of chance and lack of purposeful direction than other evolution-influenced creative writers (like Samuel Butler and George Bernard Shaw). Permission: Photo about 1910 from George Grantham Bain Collection, Library of Congress

FIGURE 54.3. H. G. Wells (1866–1946), the novelist and socialist, was a student of Thomas Henry Huxley. His immensely popular story *The Time Machine* (1895) picks up on Victorian worries about decline and fall, portraying the future human race as divided into two groups, the beautiful but rather stupid Eloi living above ground and the intelligent but vile Morlocks (who farm and eat the Eloi) living in caves underground. From A. Newcomb and K. M. H., Blackford, *Analyzing Character* (New York: Blackford, 1922)

The cynicism of the speaker's tone is the only possible response to the arbitrariness of this godless, Darwinian universe, while the use of awkward, jarring idioms such as "purblind Doomsters" or the abstract proper noun "Casualty" suggests the difficulty of even articulating such a capricious view of existence within the conventional, melodious poeticisms of Victorian verse (Holmes 2009, 80–81). Its unflinching adherence to even the most disconcerting implications of Darwinism was one of the key factors that invested Hardy's poetry with such striking modernity in the early twentieth century, and which ensured that it retained a powerful impact on later poets like Philip Larkin and Ted Hughes.

Darwin's influence on literary writers extended both into the twentieth century and across significant portions of the globe. In the frozen wastes of northern Canada in the winter of 1897 the young American Jack London devoured the *Origin* alongside *Paradise Lost* (recalling Darwin's own reading in South America sixty years earlier) (Fig. 54.4). This inhospitable landscape, to which London had come to join the Klondike gold rush, would later become the setting of his best-selling Northland adventure stories, all of which bore the traces of his reading of Darwin in their portrayal of rugged individuals – canine as well as human – struggling to adapt to a harsh and perennially changing environment. The domesticated dog taken to Canada in *The Call of the Wild* (1903), for instance, must make himself "fit to survive in the hostile Northland environment" by relying on "his adaptability, his capacity to adjust himself to changing conditions, the lack of which would have meant swift and terrible death." This, though, entails "the decay or going to pieces of his moral nature, a vain thing and a handicap in the ruthless struggle for existence," for in "the Northland, under the law of club and fang, whoso took such things into account was a fool" (London 1981, 62–63). Along with the *Origin*, London also greatly admired Spencer's application of the doctrine of the survival of the fittest to human society, and he became an adherent of the ruthless American strain of social Darwinism that emphasized the evolutionary superiority of the Anglo-Saxon and Teutonic races, although also complicated by his simultaneous reading of both Nietzsche and Marx (Bannister 1979, 220–25). In subsequent novels like *The Sea-Wolf* (1904) London protested at those who "read Darwin ... misunderstandingly" and "conclude that the struggle for existence sanctions" solipsistic individualism, wanton violence, and rapacious capitalism, and instead he endorsed a socialistic version of social Darwinism that stressed the necessity of cooperation (London 1992, 62; Berliner 2008). His hugely popular tales of masculine physicality in the primeval North American wilderness nevertheless gave a seductively adventurous dimension to the most brutal elements of Darwin's conception of the natural world.

The increasing significance of evolutionism in contemporary culture, initially brought about by developments in genetics and more recently by a concomitant rise of religious fundamentalism, has ensured that Darwin has remained an

FIGURE 54.4. Social Darwinian themes of struggle and survival are the bedrock of the writings of American author Jack London (1876–1916), seen especially in his still-popular *The Call of the Wild* (Boston: L. C. Page, 1903), the story of a domestic dog thrust into the harsh conditions of the Canadian North.

FIGURE 54.5. Ian McEwan (b. 1948) has explored Darwinian themes in several of his novels, notably *Enduring Love*. Permission: © Mark Gerson / National Portrait Gallery, London

important presence in modern literature. He is an obvious point of reference in the genre of so-called neo-Victorian novels such as John Fowles's *The French Lieutenant's Woman* (1969), A. S. Byatt's *Insects and Angels* (1992), and Roger McDonald's *Mr Darwin's Shooter* (1998), where the nineteenth-century trauma at the godless universe adumbrated in the *Origin* affords an apposite parallel with contemporary existential crises over the condition of humanity (Gutleben 2001, 204–16). The writer who has engaged most extensively with Darwin's legacy in contemporary culture is Ian McEwan, whose novels as well as nonfictional writings regularly tackle the fraught relationship between neo-Darwinian science and other areas of intellectual inquiry (Fig. 54.5). While in *Enduring Love* (1997) the principal characters, Joe Rose, a science journalist, and Clarissa Mellon, a scholar of Romantic poetry, cannot reconcile their respective commitments to evolutionary psychology and literary humanism despite their love for each other, in *Saturday* (2005) the neurosurgeon Henry Perowne accepts evolution as an alternative creation myth that has the advantage of being demonstrably true, but is brought to a recognition of the affective aspects of his biochemical expertise by the poetic resonance of a line from the *Origin*, "There is grandeur in this view of life," that reverberates through his mind after inattentively reading a biography of Darwin (rather than the *Origin* itself) in the bath (McEwan 2005, 55, 255; Greenberg 2007). The resonant, literary quality of Darwin's language, instilled, as was seen earlier, by his own enthusiastic reading of fiction and poetry, gives a new, deeper meaning to the complex circuitry of the brain – itself an intricate web of connections – that Perowne operates upon (Amigoni 2008). *Saturday* implies that, just as Darwin's evolutionary conception of the natural world was forged, at least in part, by his receptiveness to imaginative literature like *Paradise Lost*, so modern incarnations of evolutionary science, no matter how abstract or esoteric, cannot themselves be separated from the affective qualities of literary forms of writing.

Essay 55

Darwin and Gender

Georgina M. Montgomery

We hold that Evolution's plan,
To give as little as she can,
Is sometimes trying.
Fair share of brains, indeed, we win;
But why not throw the swimming in,
Why not the flying?

– May Kendall

The 1870s were characterized by debate of the "woman question" – or, more accurately, *questions*. What were the moral, intellectual, and physical capabilities and limitations of women? What roles should women be afforded in Anglo-American society? From which social arenas should they be excluded? This political climate shaped the contents of Charles Darwin's *Descent of Man, and Selection in Relation to Sex* (1871), while also being significantly impacted by it. Antifeminists and feminists alike saw the opportunity to use the power of scientific authority, and specifically the power of Darwin's name and theory of sexual selection, to promote what were often diverse views of woman's place in nature and society.

Beginning in the late 1970s, historians of science began to correct a past blind spot in scholarship and university courses on the Darwinian Revolution by including analysis of gender issues, specifically in relation to the content of *Descent of Man*. For example, Ruth Hubbard (1979), Evelleen Richards (1983), and Rosemary Jann (1997), among others, investigated the extent to which Darwin should be identified as sexist and highlighted feminist responses to the *Descent of Man* written by nineteenth-century women such as Eliza Burt Gamble (1841–1920), May Kendall (1861–1943), and Charlotte Perkins Gilman (1860–1935).

For feminists in the nineteenth century, the *Descent of Man* represented an opportunity to appropriate the power of Darwin's observations and theory of sexual selection to demonstrate women were not inferior to men, as Darwin clearly argued in the *Descent of Man*, but in fact at least man's equal, if not his superior, especially in matters of altruism and social worth. These women's voices were often muffled by a

FIGURE 55.1. "That troubles our monkey again." Cartoon illustrating the unease many Victorians felt about the implications of Darwin's ideas for our understanding of the "angel in the house," as Coventry Patmore's immensely popular, sentimental poem (published in 1854) characterized his wife. From *Fun* magazine (1872)

mainstream view of women as intellectually inferior to their male counterparts and bound by their biology to the domestic sphere. The works of intellectuals such as Herbert Spencer and Carl Vogt sustained these limited views of womanhood, conceptions that were further preserved by popular culture. For example, images of primates in *Punch* and other popular periodicals, coupled with living animal and human subjects in so-called freak shows, reflected nineteenth-century concerns with gender and race, as well as humans' place in the animal kingdom (Fig. 55.1).

During the twentieth century and indeed still today, the scientific and cultural clout of Darwin's name and works continues to be interpreted and utilized in myriad ways to promote diverse political perspectives. Issues of gender and evolution continue to be discussed in the modern field of evolutionary biology, in the pages of popular scientific books, on the stages of comedy clubs, and on the small and large screen. For instance, representations of the wild man and coy female, grounded as they are in Darwinian concepts of gender and reproduction, are now so much part of the popular psyche that they often go unquestioned (Lancaster 2003). Thus it has been left to feminist voices within evolutionary biology, accompanied primarily by scholars in cultural studies, women's studies, and LGBTQ studies, to demonstrate how Darwinian views continue to influence scientific studies and popular representations of men and women.

DARWIN AND THE DESCENT OF WOMAN

Darwin briefly defined sexual selection in the *Origin of Species* (1859) as a mechanism that could account for traits that were not advantageous for the struggle for existence but instead increased successful reproduction. As such, sexual selection focused on competition between males and female mate choice. The success of the male was determined not only by his "general vigor" but also "on having special weapons confined to the male sex" (Darwin 1859, 88). For females, making the evolutionarily correct choice of mate required discerning taste in regard to which male she found worthy of copulation. In other words, males were active, competitive, and eager to mate whereas females were coy and careful. Nevertheless, within the animal kingdom, females did exercise a degree of agency through their choice of mates, although Darwin's description of female choice is often interpreted as emphasizing female passivity rather than agency (Hubbard 1979; E. Richards 1983).

In the *Descent of Man*, female choice, and any degree of agency that went with it, failed to transfer into Darwin's (1871, 2:371) discussion of human evolution: "Man is more powerful in body and mind than woman, and in the savage state he keeps her in a far more abject state of bondage, than does the male of any other animal; therefore it is not surprising that he should have gained the power of selection." Thus, according to Darwin, in modern, civilized, society the male had become the chooser of his mate. Despite Darwin being commonly understood as a great revolutionary, the human male and female he described fulfilled, rather than challenged, the gender ideals of Victorian culture. Darwin saw man as "the rival of other men; he delights in competition, and this leads to ambition which passes too easily into selfishness" (2:326). Despite men's failure to be highly cooperative and altruistic, or rather because of these unfortunate traits, he had achieved "a higher eminence" in all areas when compared to the achievements of women. In Darwin's view, the results of this apparent superiority were plain to see: "If two lists were made of the most eminent men and women in poetry, painting, sculpture, music ... history, science, and philosophy ... the lists would not bear comparison.... We may also infer ... that if men are capable of a decided pre-eminence over women in many subjects, the average mental power in man must be above that of woman" (2:327). Thus, for Darwin it was nature, not nurture, that determined the qualities of men and women.

For Ruth Hubbard (1979, 16), prejudice lay at the root of Darwin's conception of gender: "For although the ethnocentric bias of Darwinism is widely acknowledged, its blatant sexism – or more correctly, androcentrism (male-centered) – is rarely mentioned, presumably because it has not been noticed by Darwin scholars, who have mostly been men." Certainly, Darwin's words reflect an incredibly limited view of womanhood – and manhood, for that matter – and one that fully conformed to his cultural context. Nevertheless, as Hubbard correctly points out, the fact that Darwin's conception of gender reflects his time does not mean it should escape critique, both because Darwin's words and works played a significant role in nineteenth-century debates of the "woman question" and because Darwinian evolution continues to shape contemporary discussions of gender and sexuality.

For Evelleen Richards (1983, 60), Darwin's conception of men and women was founded on his rigorous application of naturalistic observation rather than sexism per se: "It is not only historically inaccurate to impute an anti-feminist motive to Darwin, but unnecessary.... Darwin's conclusions on the biological and social evolution of women were as much constrained by his commitment to a naturalistic or scientific explanation of human mental and moral characteristics as they were by his socially derived assumptions of the innate inferiority and domesticity of women." Using evidence from his notebooks, Richards demonstrates that Darwin's understanding of the evolution of men and women resulted from his observations of human societies and the anthropomorphism he applied to his animal observations. And although Richards concedes prejudice resulting from cultural context played a role in Darwin's theory of sexual selection, she argues that the theory should not be "primarily" seen as a "political ploy" (98). This conclusion is further supported by evidence that Darwin did not "engage actively in sexual discrimination" and therefore should not be called an antifeminist.

In recent years, Darwin scholars have increasingly turned their attention to Darwin's relationships with women, including family members and amateur scientists with whom Darwin corresponded. Like Richards's analysis, the work of Joy Harvey (2009) and the Darwin and Gender section of the Darwin Correspondence Project has added layers of complexity concerning Darwin's view of women. Darwin had 115 female correspondents. Some of these women were family members or part of his social circle, others were not known to him until they wrote "with something to offer, an observation, an unusual plant, a query, a book, or, sometimes, a religious concern" (Harvey 2009, 201). Surprisingly, given the way in which he described women in the *Descent of Man*, there are several examples of Darwin including observations provided by female observers in his publications, with and

FIGURE 55.2. Lydia Becker (1827–90) was a Victorian popular-science writer and feminist, who corresponded with Darwin and held him in regard, even though she disagreed with his views on the inferiority of women's intellects. From Helen Blackburn, *Women's Suffrage* (London: Williams and Norgate, 1902)

without crediting them by name, and of his active encouragement of women to publish their scientific findings and pursue admission into various scientific societies. For instance, during the early 1860s Lydia Becker (1827–90), British botanist, suffragist, and popularizer of science, exchanged letters with Darwin concerning the plant *Lychnis dioica* (Fig. 55.2). Becker provided Darwin with observations, specimens, and a copy of her book on botany, while Darwin provided Becker with a paper on climbing plants to be read at the first Manchester Ladies' Literary Society meeting and a form of participation in the scientific community, the majority of which was closed to women (Lightman 2007; Harvey

2009). It is certainly interesting to note that although Becker published works challenging the essentialism of women that "anticipated and rejected ... Darwin's position on the intellectual inferiority of women ... she never spoke out publicly against" him – a silence that leads Bernard Lightman (2007, 160) to suggest that, "perhaps it was Becker's previous relationship with Darwin that led her to hold her tongue and pen in the 1870s" when several nineteenth-century feminists publicly critiqued the description of the evolution of woman laid out in the *Descent of Man*.

THE WORDS OF WOMEN: RESPONSES TO AND USES OF THE *DESCENT OF MAN*

Several feminists, speaking with diverse voices, responded to the scientific community's argument that woman was inferior to man and that this inferiority was substantiated only by Darwin's theory of sexual selection. Although scholars such as Penelope Deutscher (2004) have begun to call for an integration of discussions of race and the works of non-Western women into analyses of feminist responses to Darwinian evolution, most publications on this topic have focused on the works of a small number of white, Anglo-American feminists and particularly Charlotte Perkins Gilman, Antoinette Blackwell (1821–1921), and Eliza Burt Gamble (R. Love 1983; Fausto-Sterling 1997; Jann 1997; Kohlstedt and Jorgensen 1999; Vandermassen, Demoor, and Braeckman 2005; Milam 2010). Of these three women, Gilman is probably the best known (Fig. 55.3). Works such as Gilman's *Women and Economics* (1898) and *Herland* (1915) were widely read in their day and, indeed, continue to be read by undergraduates studying the history of science and/or women's history. Using the power of scientific theory and terminology, Gilman spoke out in defense of not simply female equality but superiority, especially in regard to cooperation and altruism. In these books Gilman wove together her own interpretation of Darwinian evolution and socialism to paint a vision of a world in which women were freed from society's shackles in order to fulfill their evolutionary potential. In contrast to Darwin's own view, Gilman argued that evolution did not limit a woman but in fact held the power to enable her to become all she could be for society. As Rosaleen Love (1983) reminds us, Gilman's call for social reform was part of a chorus of voices demanding change during the late nineteenth and early twentieth centuries, a chorus composed of men, such as Utopian novelist Edward Bellamy, as well as women.

Like Gilman, Eliza Burt Gamble was a feminist and socialist (Jann 1997; Deutscher 2004;Vandermassen et al. 2005). A Michigan woman who pursued a career as a teacher, Gamble contributed to the nineteenth-century women's movement in varied ways, including presentations and writings in support of women's suffrage and critiques of Darwin's *Descent of Man* (Peck 2010). Her 1894 book, *The Evolution of Woman: An Inquiry into the Dogma of Her Inferiority to Man*, was a thorough analysis of Darwin's failure to see the female choice that was so evident in the animal kingdom among modern,

FIGURE 55.3. Charlotte Perkins Gilman (1860–1935): "There is no female mind. The brain is not an organ of sex. Might as well speak of a female liver." From *The Living of Charlotte Perkins Gilman: An Autobiography* (New York: D. Appleton-Century, 1935)

civilized, human society. With eloquent and compelling prose, Gamble argued that Darwin's failure to follow through with his analogy between animals and humans, a fundamental part of the evolutionary argument he laid out in the *Origin of Species* and the *Expression of the Emotions in Man and Animals* (1872), was a significant flaw in his analysis of human gender and sexuality. Thus, Gamble "shrewdly identified ... the issue in sexual selection that has attracted feminist criticism ever since" (Jann 1997, 152). After quoting Darwin in regard to sexual selection and female choice, Gamble turned his argument on its head. As Jann (1997, 155) demonstrates, "For Gamble matrilineality and matriarchy represented the logical – and natural – extension of female control of sexual relations in the animal kingdom. She in effect took Darwin's equation of choice with control and applied it to females." Such use of Darwin's own words followed by her own rebuttal was a tactic Gamble used to great effect in *Evolution of Woman* to persuade the reader of the truth of each aspect of her argument that, in regard to evolution and moral character, woman surpassed man (Fig. 55.4).

Staying true to the essentialism of her day, Gamble (1894, 65) deviated from Darwin's interpretation of men and women,

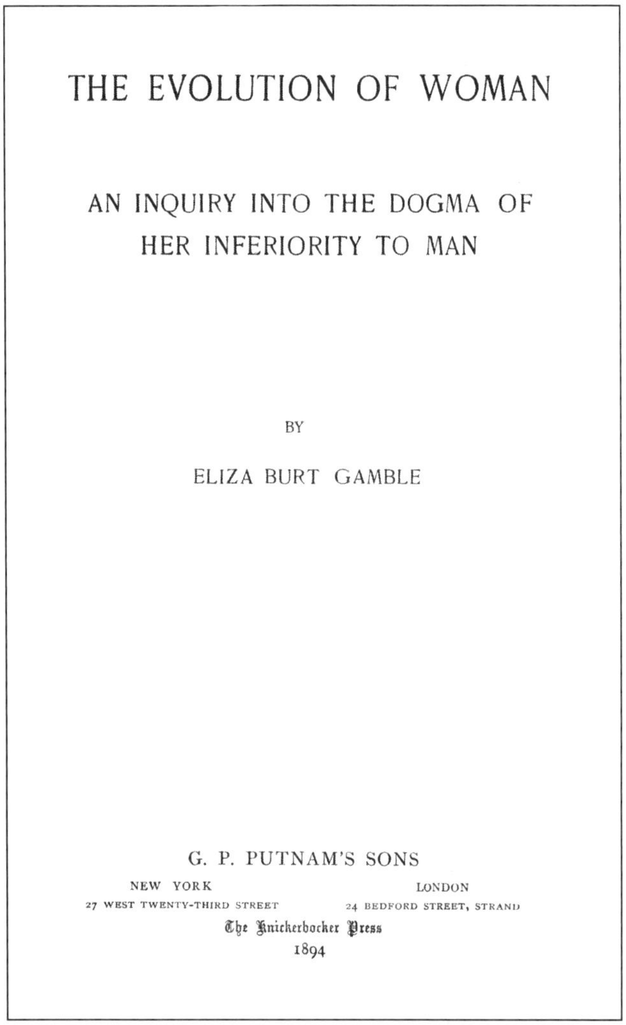
THE EVOLUTION OF WOMAN

AN INQUIRY INTO THE DOGMA OF
HER INFERIORITY TO MAN

BY

ELIZA BURT GAMBLE

G. P. PUTNAM'S SONS
NEW YORK — LONDON
27 WEST TWENTY-THIRD STREET — 24 BEDFORD STREET, STRAND
The Knickerbocker Press
1894

Figure 55.4. Eliza Burt Gamble (1841–1920) shared Alfred Russel Wallace's conviction that female choice is an important factor in human evolution and that, in fact, in major respects females are superior to males. Title page, *The Evolution of Woman* (New York: Putnam's Sons, 1894)

identifying females as holding the potential to push human society further forward: "We have seen that all the facts which have been observed relative to the acquirement of the social instincts and the moral sense prove them to have originated in the female constitution, and as progress is not possible without these characters, it is not difficult to determine within which of the sexes the progressive principle first arose. Even courage, perseverance, and energy, characters which are denominated as thoroughly masculine, because they are the result of sexual selection, have been and still are largely dependent on the will or desire of the female." This potential for progress would be fulfilled only when social barriers to women's education were removed. Thus, for Gamble, women's capabilities were determined not only by naturalistic means, as Darwin had argued, but also by culture. When these hurdles were surmounted, woman's ultimate superiority would become clear: "So soon as women are freed from the unnatural restrictions placed upon them through the temporary predominance of the animal instincts in man, their greater powers of endurance, together with a keener insight and an organization comparatively free from imperfections, will doubtless give them a decided advantage in the struggle for existence" (66). This fundamental fact of human society had, according to Gamble, completely escaped the eye of Darwin, the great observer, "[Darwin] seems to have entirely forgotten that all the avenues for success have for thousands of years been controlled and wholly manipulated by men, while the activities of women have been distorted and repressed in order that the "necessities" of the male nature might be provided for" (68).

GENDER AND THE "GORILLA CRAZE"

Issues of evolution and gender were not only popularized through the books of nineteenth-century feminists but also literally embodied in the form of women and primates displayed for the entertainment of the Anglo-American public in the late nineteenth and early twentieth centuries (Churchill 2010; Hamlin 2011). Even before publication of the *Origin of Species*, nonhuman primates occupied a significant cultural space, representing as they did the apparent bridge between humans and the animal kingdom, especially in regard to gender and sexuality (Schiebinger 1993). After 1859, primates, and particularly gorillas, increasingly attracted attention as characters in literary works such as *The Fall of Man, or The Loves of the Gorillas, A Popular Scientific Lecture upon the Darwinian Theory of Development by Sexual Selection, by a Learned Gorilla* (1871) and motion pictures such as *King Kong* (1932–33) and *Planet of the Apes* (1968) (Bernstein 2001; Browne 2005b; Voss 2009; Jones 2010). These imaginary primates reflected not only society's real anxiety about the human-animal boundary but also beliefs about the naturalness of man's sexual aggression and the worryingly close connection between women and humans' animal origins.

The life of Julia Pastrana, a Mexican woman displayed by her husband as an "ape-woman," "baboon-woman," or "nondescript," demonstrates the intersection of concerns about the human-animal boundary and womanhood. In 1857 Pastrana was the subject of an article in the prominent medical journal, the *Lancet* and exhibited as a touring "freak show." Janet Browne and Sharon Messenger (2003) contextualize the display of Pastrana as the "missing link" between man and beast within the broader "gorilla craze" of the mid- to late twentieth century. With her beard and jaw-line, Pastrana simultaneously embodied questions about humans' evolutionary origins and women's animality, while also exemplifying the racist nature of the Victorian freak show, which sought to display non-Caucasian "bodies that did not fit the norm" (Churchill 2010, 128).

Racial discourse often intersected with issues of gender and evolution in Victorian popular culture. For example, Jeannette Eileen Jones (2010, 203) has analyzed the poem "The Missing Link," published in the *Boston Commercial Bulletin* and *Ward's Natural Science Bulletin* in 1880, to demonstrate how images of gorillas and African women served to simultaneously "emphasize [black women's] hypersexuality" and "proximity to animals." The poem chronicles a wedding between a black woman and a gorilla that produces "Mr. Darwin's missing link" between man and the higher apes. Jones contextualizes this literary image of black women within debates about the "Negro question" occurring in America during the late nineteenth century, stating that "the imagined black woman occupied a critical space in this configuration of race, gender, and sexuality in evolutionary discourses designed to scientifically prove the 'Negro's' biological incapacity for governance." Similarly, exhibits of African men, such as William Henry Johnson, who was displayed by P. T. Barnum under the name "Zip" or "What Is It?" were identified as evidence of the "missing link" between man and monkey. Thus, the exhibit of men such as Johnson served to "not only reinforce stereotypes about the supposed uncivilized nature of African men, but played on deeply entrenched ideologies about the 'simple,' 'childish' nature of people of color as a species" (Churchill 2010, 133).

Identification of the "missing link" in female bodies continued into the early twentieth century. The body of a mountain gorilla, for example, was greeted by the American press as a missing link between humans and other primates while also embodying stereotypical views of femininity. Congo, named after the country where she was captured, was brought to the United States and later studied by psychobiologist Robert Mearns Yerkes. As the first female mountain gorilla to come to America, Congo was greeted by publicity. Congo's womanhood and humanlike nature dominated these popular accounts:

> The first thing one is apt to notice about Congo is her eyes. She has a great furry head with enormous mouth and semi-prognathous jaw: a flat, broad, deeply ribbed nose, and her particular eyes. The face is ape, but the eyes are indubitably human.... When standing, she is more nearly erect than any other animal except man. Her head is larger and more nearly in proportion. Her shoulders, back, chest, hips and buttocks uncannily resemble those of the human. The arms and legs, however, are those of an ape.... Again, the hands and feet ... are so nearly human that they take one's breath. Congo not only has a thumb which easily reaches the base of her little finger ... but a palmist could tell her fortune, for the "heart line," "head line," and "life line" are there just as distinctly as they are on your own hand. (Sparks 1926, 19–20)

Along with such descriptions, Congo was identified as "typically female," with particular attention paid to her modesty and the affection she displayed toward males, including a primate companion and Yerkes (Montgomery 2009).

THE WILD MAN AND COY FEMALE: FROM PRIMATOLOGY TO *PLAYBOY*

The use of nonhuman primates such as Congo to navigate questions of human gender and sexuality can also be found in the more recent past. For example, studies of primate

FIGURE 55.5. Sarah Blaffer Hrdy (b. 1946) is a leading human sociobiologist who has made major contributions to our understanding of family dynamics, broadly conceived, including such topics as infanticide and cooperative mothering. Permission: S. B. Hrdy, photo by S. Bassoul

behavior conducted shortly after World War II were used to naturalize male domination and sexual aggression in human societies while glossing over the role of females in primate social groups. Drawing on evidence from other areas of the animal kingdom, the female of the species continued to be stereotyped as choosy, coy, and reluctant to mate for anything other than reproductive purposes. This depiction of the coy female was only amplified by the emergence of sociobiology in the 1970s. As primatologists Sarah Blaffer Hrdy and George Williams have pointed out, sociobiology was founded in large measure on the theory of sexual selection, applying the theory for the identification of universal, rather than diverse and complex, gender characteristics: "Sexual selection theory is one of the crown jewels of the Darwinian approach basic to sociobiology. Yet so scintillating were some of the revelations offered by the theory that they tended to outshine the rest of the wreath and to impede comprehension of the total design, in this instance, the intertwined, sometimes opposing, strategies and counter strategies of both sexes which together compose the social and reproductive behavior of the species" (Hrdy and Williams 1983, 7).

Hrdy (2002, 173) identifies the concept of the coy female as "the single most commonly mentioned attribute of females in the literature on sociobiology" (Fig. 55.5). The endurance of the myth – for it is a myth, as we will see momentarily – of the coy female is particularly surprising. The 1970s and 1980s, when sociobiology gained a significant foothold in biology departments and in the popular media, was also a period when long-term field studies, particularly those led by female primatologists, clearly demonstrated great diversity in primate sexual behavior (Sperling 1991; Strum and Fedigan 2000). Studies of infanticide and female aggression, for instance, revealed how primates often broke the confines of the image of the coy female. These conclusions were further reinforced by studies of other animal behaviors, such as those that revealed bird species previously believed to be in monogamous "marriages" were in fact consorting with several individuals of the opposite sex. Despite this scientific evidence, the coy female identified by Darwin in the nineteenth century, and further promoted by sociobiologists more than a hundred years later, continues to be a central stereotype applied to females today.

Part of the explanation of the endurance of the myth of the coy female is that any complexity and variety in animal sexual behavior has all too frequently been lost when applied to discussions of human sexuality in the popular media. Take for example the *Playboy* article, "Darwin and the Double Standard" (S. Morris 1978). After referencing Darwin's name in the title, scientific authority, this time in the form of sociobiology, is directly referenced in the subtitle, "it has been said that a man will try to make it with anything that moves – and a woman won't. Now the startling new science of sociobiology tells us why" (109). The article goes on to use Darwin's name and evolutionary theory as the foundation upon which to build a scientific justification for man's apparent genetic compulsion to have sex with as many women as possible. Thus, according to Morris, a Darwinian understanding of gender and sexuality had been given a genetic basis by "a new branch of science – known as sociobiology" (160).

The *Playboy* article may appear as a one-off, titillating example of Darwin's name being called upon to justify stereotyping of both men and women – for the sexual identity of males is also being stereotyped by Morris – but in fact it is just one of a slew of cultural references to Darwinian evolution when attempting to naturalize human sexuality. Indeed, anthropologist and cultural studies professor Roger Lancaster (2003, 80) argues that Darwin's theory of sexual selection "laid the groundwork for much of what would subsequently come to be understood as 'natural' about desire." As a result of Darwin's failure to see variation in sexual behavior, the kind of sexual desire revealed in the *Descent of Man* failed to "even contemplate what every farm boy (and surely every naturalist) knows about the prevalence of nonprocreative and nonheterosexual sex in the animal kingdom" (85). Thus, for Lancaster, Darwin's naturalizing of heterosexual sex for the means of reproduction cascaded into all areas of our intellectual and cultural life to "become part of the warp and woof of heteronormative cultures on both sides of the Atlantic" (90).

Certainly not all applications of Darwinian approaches to gender and sexuality have helped to construct such restrictive understandings of men's and women's sexual behaviors. Kimberly Hamlin (2009), for example, identifies Darwinian evolution, and specifically the theory of sexual selection, as a key inspiration for more progressive, indeed "radical," approaches to human sexuality. For Hamlin, Darwin and his theory of sexual selection was a significant influence on the work of early sexologists, such as Alfred Kinsley, who sought to reveal the diversity that characterized human sexual behavior, thus freeing men and women from the ignorance and guilt that often limited and/or closeted their sexual identities.

In conclusion, just as nineteenth-century feminists saw multiple meanings and applications in Darwin's *Descent of Man*, contemporary scientists and science writers have used Darwin's name and his theory of sexual selection to promote a range of images of gender and sexuality. Unfortunately, much of this diversity has been diluted, if not lost completely, in many popular accounts of human behavior. Thus, caricatures of femininity and masculinity continue to endure, sustained at least in part by selective and dated use of Darwinian evolution. Like Gilman and Gamble, scholars in the scientific community and social sciences continue to speak out against the myth of the coy female and the restricted view of women – and men – that such stereotypes create. One can only wonder how Darwin would respond to these twenty-first-century debates concerning the nature of human gender and desire, and the mass of scientific data upon which they are based.

Essay 56

Evolutionary Epistemology

Tim Lewens

As Darwin's readers have often noted, he was enthusiastic about the explanatory reach of his theory. Everyone knows of Darwin's promissory note in the *Origin* (1859, 448): "In the distant future I see open fields for more important researches. Psychology will be based on a new foundation, that of the necessary acquirement of each mental power and capacity by gradation." Of course, evolutionary approaches to psychology have been prevalent for some time now; however, theorists from philosophy and the sciences have wondered if knowledge itself – understood as a state of mind, to be explained in scientific terms – might also be approached from the perspective of an evolutionarily informed psychology. This essay gives a selective overview of some of the projects we might collect under the heading of "evolutionary epistemology" and of the likely limits to using evolution to shed light on what knowledge is and how much of it we have.[1]

One of the best-known essays on Darwin's broader impact is John Dewey's "The Influence of Darwinism on Philosophy," originally delivered as a lecture in 1909. Darwin taught us that species were malleable, ephemeral; that there was no hard-and-fast distinction between good species and mere varieties; and that we should expect no rigorous answer to be had to the questions of when, precisely, a new species has been created or what sort of a thing a species is (Fig. 56.1). As Dewey (1910, 5) notes near the beginning of his essay,

> In laying hands upon the sacred ark of absolute permanency, in treating the forms that had been regarded as types of fixity and perfection as originating and passing away, the "Origin of Species" introduced a mode of thinking that in the end was bound to transform the logic of knowledge, and hence the treatment of morals, politics, and religion.

Dewey's concern is to undermine the assumptions that knowledge must recognize what is permanent about things and that things themselves must conform to strict

---

[1] Many of the original sources referred to in this article, including those by Nietzsche, Dewey, James, and Lorenz, as well as further relevant material by Plantinga, Popper, and Kuhn are collected in Ruse (2009a). Further discussion of the implications of Darwin's thinking for the growth of empirical and moral knowledge can be found in Lewens (2007).

FIGURE 56.1. John Dewey (1859–1952), the great American pragmatist philosopher, was deeply influenced by Darwinian ideas. Permission: Southern Illinois University at Carbondale

types or species. He goes on to claim that notions of development, design, and purpose must be reevaluated in a Darwinian light. Dewey's comments on transformations to the "logic of knowledge" do not concern epistemology per se but are an effort to encapsulate the entire philosophical impact of a Darwinian worldview.

My focus in this essay is much narrower than Dewey's. Other commentators have started this task of narrowing the field of discussion by distinguishing two rather different projects for evolutionary epistemology (e.g., Bradie 1986). I begin with a rough account of this now standard taxonomy and then make some suggestions about how it might mislead. First, there is what we might call the *traditional* evolutionary project. Knowledge, so the story goes, is a human capacity. But like any organic capacity, it has a history. This means that we can understand knowledge better if we look at it from an evolutionary perspective, just as the evolutionary perspective helps us to a better understanding of other elements of human anatomy and physiology. Second, there is what we might call the *extended* evolutionary project. Traditional evolutionary accounts focus on competitive struggle and selection between organisms, and the evolutionary advantages that knowledge, or the capacity to know, may bring. Extended evolutionary accounts instead lean on the thought that elements of knowledge themselves compete: we can think of alternative scientific theories, for example, as engaged in a struggle for acceptance within the scientific community, with the result that theories are subject to a form of selection that is decoupled from that affecting their organic hosts.

DARWIN'S EVOLUTIONARY EPISTEMOLOGY

Darwin himself had rather little to say in his published works about either form of evolutionary epistemology, but we can discern several hints. Extended evolutionary accounts assume that natural selection is the sort of process that does not occur only in populations of organisms; evolutionary epistemology of this sort tells us that selection occurs also in populations of scientific theories. Extended evolutionary accounts presuppose that selection can be characterized in a fairly abstract way, such that entities of many different types – organisms, computer viruses, tools, moral values – can be said literally to undergo selection.

In both the *Origin* and *The Descent of Man*, Darwin seems to endorse the view that selection occurs whenever there are entities that struggle for existence, regardless of what other properties they might have. Darwin does not talk about struggle between theories or beliefs; instead, he talks about struggle between words in a language. In *The Descent of Man*, he quotes the linguist Max Müller with approval:

> A struggle for life is constantly going on amongst the words and grammatical forms in each language. The better, the shorter, the easier forms are constantly gaining the upper hand, and they owe their success to their own inherent value. (Darwin 1871a, 1:60)

Darwin summarizes: "The survival or preservation of certain favoured words in the struggle for existence is natural selection" (60–61). This process is not similar to natural selection, analogous to natural selection, or reminiscent of natural selection – it *is* natural selection, even though it does not act on organisms.

Darwin says even less about traditional evolutionary accounts of knowledge, but his notebooks give some clues. As he puts it in his Notebook M (where M stands for Metaphysics): "Plato ... says in *Phaedo* that our 'necessary ideas' arise from the preexistence of the soul, are not derived from experience. – read monkeys for preexistence" (Barrett et al. 1987, M128). Darwin is suggesting that inheritance from early ancestors might explain how we come to know some things innately. But we should recognize what Darwin does *not* say here. When Darwin writes "read monkeys for preexistence," he says nothing about natural selection. This highlights a series of problems when we try to pin down the distinction between what I have been calling "traditional" and "extended" evolutionary accounts. Darwin's own theorizing about the origins of knowledge often alludes to a phenomenon known at the time as "use-inheritance." Darwin believed throughout his career that habits, initially produced intentionally, might through constant practice become automatic and might eventually reappear (as a result of modifications to the brain) in offspring. So when Darwin writes of the earlier experience of species eventually appearing as items of innate knowledge, we should not assume that selection acting on blind variation has anything to do with the story.

Darwin's own accounts of "the necessary acquirement of each mental power and capacity by gradation" often depart considerably from what we today might think of as a Darwinian account of knowledge. This is especially apparent in *The Descent of Man*. Part of this book is devoted to an explanation of how what Darwin calls "the moral sense" – our sense that some actions are morally right, others wrong – emerged in humans. In places the story focuses on natural selection acting on groups of humans to promote a sense of sympathy – that is, a capacity to feel injury to others as though it were injury to ourselves, and a consequent disposition to help others when they are in pain or distress. But as Darwin's account moves on, he tells of how the sense of sympathy became enlarged to encompass not only members of our immediate communities but also members of other tribes, other countries, other races, and even other species. Darwin also tells of how our tendencies to act in ways that promote the welfare of others have been refined through patient observation of the effects of our actions on others; moreover, when he then moves on to describe our instinctive tendency to help others, the efficacy of this tendency is explained by reference to use-inheritance.

In a sense, then, Darwin offers an evolutionary epistemology of moral knowledge: he gives a historical explanation for our ability to act in ways that are both unreflective and reasonably effective in promoting the general good. That explanation is also gradualist: the refinement of the moral sense and the accumulation of moral knowledge are achieved by small increments over long periods of time. But Darwin's own evolutionary epistemology is not an instance of a traditional evolutionary account, because it often makes no reference to natural selection, and it makes frequent use of Lamarckian inheritance. Nor can it be understood as an instance of an extended evolutionary account: in these sections of *Descent*, Darwin makes no effort to argue that the accumulation of moral knowledge is a form of natural selection acting at the level of moral principles or rules of thumb.

These observations are of more than scholarly interest. If we note simply that human capacities for knowledge, or specific items of knowledge, have emerged over time, and that they can be explained in a scientific manner, we in fact offer very little to constrain what sort of scientific story we can tell. Moreover, there is frequently a form of slippage when commentators move from observing that the capacity to know, as well as specific items of knowledge, have "evolved" in the sense that they have a history of emergence from some prior state of nonexistence, to the quite different and much more contentious observation that these same capacities and items have "evolved" in the sense that natural selection acting on (for example) genetic variation is the specific process that explains their emergence.

EVOLUTION AND THE A PRIORI

On the one hand, many philosophers have found it hard to imagine how humans could acquire knowledge about the world through any process other than learning from perceptual experience of that world. These are the empiricists. On the other hand, it has seemed to many either that it is a matter of observable fact that we do have elements of innate knowledge or that some form of knowledge must be innate in order to explain how we could possibly go about observing the external world in an intelligible and productive manner. These are the rationalists. Darwin's Notebook M entry, which suggests that perhaps individual humans have inherited knowledge acquired by their simian ancestors, hints at a reconciliation between these two traditions.

Darwin's suggestion was taken up with enthusiasm by a group of philosophically inclined biologists in the German-speaking tradition, most prominently Konrad Lorenz and Rupert Riedl. As Riedl (1983, 46) puts it, "By distinguishing phylogenetic and individual acquisition of knowledge, a rationalist-empiricist synthesis is made possible." We have already seen that Darwin's own rationalist-empiricist synthesis often leans on use-inheritance as a mechanism, not only in the early notebook days but in his more mature works, too. For Riedl and Lorenz, the mechanism that allows species to acquire knowledge of their environments that is a priori from the perspective of the individual but a posteriori from the perspective of the evolutionary lineage is natural selection acting on genetic variation. For both theorists, the evolutionary approach allows us to give a scientific grounding to talk of Kantian categories, understood as a priori cognitive structures that give shape to experience of the natural world.

The Lorenz-Riedl view has two clear advantages over Darwin's use-inheritance mechanism. First, and most obviously, we no longer think that use-inheritance can possibly work in the way Darwin imagines. Darwin's picture of use-inheritance has it that through continued exercise of habits in parental generations, modifications are passed to the brains of developing offspring in such a way that the habit in question develops independently of learning. If such processes occurred, they would indeed account for innate knowledge. But they do not occur. So adaptation by natural selection offers a far more plausible mechanism, whereby the cognitive dispositions of a species might be gradually shaped to match the species' environment. Second, some rationalists have worried that learning from the environment cannot plausibly explain the initial appearance of some cognitive dispositions – most obviously the ability to learn itself. If we are focusing on explaining how these sorts of dispositions appeared, then use-inheritance, which appeals to the inheritance of habits that were learned by earlier generations, will not help. Adaptation by natural selection offers a mechanism that has the potential to explain the emergence of learning itself.

FIGURE 56.2. Friedrich Nietzsche (1844–1900) was no friend of Darwinian ideas and yet in respects showed more insight into the implications of Darwinism for philosophy than any of his contemporaries. From F. W. Nietzsche, *Werke* (Leipzig: C. G. Naumann, 1897)

NIETZSCHE ON THE GENEALOGY OF KNOWLEDGE

The Lorenz-Riedl view has limitations of its own. To say that some proposition is known entails that the proposition in question is true. But why should we assume that propositions that our ancestors found it advantageous to believe are also likely to be true? This theme was investigated in some detail by Nietzsche in Book 3 of *The Gay Science* (Fig. 56.2). Consider, by way of an example of Nietzsche's (1882, §111) own evolutionary epistemology, his reflections on the origin of logic:

> *Origin of the logical.* – How did logic come into existence in man's head? Certainly out of illogic, whose realm originally must have been immense. Innumerable beings who made inferences in a way different from ours perished; for all that, their ways might have been truer. Those, for example, who did not know how to find often enough what is "equal" as regards both nourishment and hostile animals – those, in other words, who subsumed things too slowly and cautiously – were favoured with a lesser probability of survival than those who guessed immediately upon encountering similar instances that they must be equal. The dominant tendency, however, to treat as equal what is merely similar – an illogical tendency, for nothing is really equal – is what first created any basis for logic.

Nietzsche is offering a genealogical account of entrenched belief, but it does not amount to an evolutionary account of knowledge because on Nietzsche's view the struggle for survival favors error and distortion over truth. It is intriguing that Darwin and Darwinism are not mentioned in any of the passages in *The Gay Science* where Nietzsche considers the apparent benefits of belief in falsehoods. Darwinism is explicitly mentioned much later in the book, where Nietzsche (1882, §349) tells us that it is characterized by "its incomprehensibly onesided doctrine of the 'struggle for existence.'" These comments are entirely negative: "The whole of English Darwinism breathes something like the musty air of English overpopulation, like the smell and distress and overcrowding of small people.... The struggle for existence is only an exception, a temporary restriction of the will to life."

Where Nietzsche does offer a genealogy of knowledge, he often begins by discussing the value of falsehoods – what he sometimes calls "basic errors" – in promoting survival of the organism, but he then moves on to note the contribution truth and falsehood make to other changing goals, and here once again his skepticism about the general applicability of the notion of struggle for existence becomes apparent. For example, Nietzsche (1882, §110) draws attention to cases where

> new propositions, though not useful for life, were also evidently not harmful to life: in such cases there was room for the expression of an intellectual play impulse, and honesty and skepticism were innocent and happy like all play. Gradually, the human brain became full of such judgements and convictions, and a ferment, struggle and lust for power developed in this tangle. Not only utility and delight but every kind of impulse took sides in this fight about "truths."

After a time, "not only faith and conviction but also scrutiny, denial, mistrust, and contradiction became a power" and in the end "knowledge became a piece of life itself, and hence a continually growing power – until eventually knowledge collided with those primeval basic errors: two lives, two powers, both in the same human being" (§110). It is hard to assimilate Nietzsche's notion of conflicting powers here to anything recognizably Darwinian, and this, in part, accounts for the very lively debate among scholars about the extent to which Nietzsche deserves to be called a Darwinian at all (J. Richardson 2004; J. E. Johnson 2010).

TRUTH AND ADAPTATION

In lieu of engaging with this scholarly debate in further detail, let us instead look at the question of whether evolution by natural selection is likely to have given us knowledge. Alvin Plantinga (1993, ch. 12) has taken the pessimistic response so far as to argue that a form of philosophical naturalism – namely, a view that combines belief in evolution with denial of the existence of gods or other supernatural beings – is self-undermining. Plantinga claims that if we have evolved without divine assistance, then we should doubt the reliability of our cognitive faculties, and we should therefore doubt philosophical naturalism itself on the grounds that it is one of the outputs of those faculties.

On the other side of this debate, many eminent biologists have felt that natural selection leads us to accurate belief about the world around us. George Gaylord Simpson (1963, 84) puts the argument like this:

> The fact is that man originated by a slow process of evolution guided by natural selection. At every stage in this long progression our ancestors necessarily had adaptive reactions to the world around them. As behaviour and sense organs became more complex, perception of sensations from those organs obviously maintained a realistic relationship to the environment. To put it crudely but graphically, the monkey who did not have a realistic perception of the tree branch he jumped for was soon a dead monkey – and therefore did not become one of our ancestors. Our perceptions do give true, even though not complete, representations of the outer world because that was and is a biological necessity, built into us by natural selection. If it were not so, we would not be here!

Simpson offers no arguments: he simply takes as obvious what Nietzsche felt to be eminently contestable – namely, that perception must be "realistic" if it is to aid our survival.

Gerhard Vollmer, a German philosopher much influenced by Riedl and Lorenz, notes that, in saying that a representation of the world is adaptive, we do not thereby say it is true. In his terms, we do not thereby say that the representation is perfectly "isomorphic" with the world it depicts. He goes on to remark, "This process of mutation and selection, of trial and error elimination, of conjectures and refutations, of hypotheses and tests, leads to a partial isomorphism.... A total isomorphism is neither needed nor possible. But we cannot predict from evolutionary principles alone the extent of this isomorphism. It might be very good or rather poor" (1983, 80). The problem here is that Vollmer's comments are intolerably vague. If I believe a tomato is cubic, is my belief "partially isomorphic," on the grounds that I attribute some shape or another to the tomato, albeit the wrong one? Or what about Nietzsche's concern that nothing is really equal, only similar. Is belief in equality then "partially isomorphic" too? Unless we say in some detail what we mean by "isomorphism" between a proposition and the world it represents, we leave open the possibility that every proposition – including every outright falsehood – is "partially isomorphic" with the world. And we then fail to make progress on the question of whether evolution has given us beliefs that are close to the truth.

Another way to respond to these worries is rather more radical: we can simply deny any distinction between what is adaptive and what is true. This move was suggested by Lorenz (1941, 124–25) himself, who argued:

> Our categories and forms of perception, fixed prior to individual experience, are adapted to the external world for exactly the same reasons as the hoof of the horse is already adapted to the ground of the steppe before the horse is born and the fin of the fish is adapted to the water before the fish hatches.

On Lorenz's view, the horse's hoof is a representation of the steppe, in virtue of being adapted to the steppe. This is a highly misleading way of putting things. We can ask whether representations – claims about how the world is – are true or false, and how accurate they might be. It simply makes no sense to ask whether a horse's hoof is true, or whether it is an accurate representation of the steppe, even though, of course, someone who studies the adaptations of the horse's hoof might be able to make plenty of good inferences about the character of the environment in which it evolved.

The distinction between the truth of a representation and the adaptive value of a representation is important to maintain. If we abandon it, we can no longer meaningfully ask whether falsehood may be adaptive. Such questions are perfectly tractable, so long as we are examining well-specified circumstances. Consider two individuals, one who believes all snakes are deadly, the other who believes some snakes are harmless. The second believes truly, the first believes falsely. The cost of running away from harmless snakes in the mistaken belief that they are deadly can be much lower than the cost of getting bitten by a deadly snake in the mistaken belief that it is harmless. And so, false belief can be favored. This will not always be the case – it will depend, for example, on how many deadly snakes are around – but the general point remains that we can model the adaptive benefits of overgeneralization in interesting ways, and if we deny any distinction between what is true and what is adaptive, we render this sort of project unintelligible.

Some readers may have lingering doubts about even making use of a notion of truth that is so distinct from the notion of adaptation. This skepticism may draw on worries about thinking of truth as "correspondence" between a representation and the world. Indeed, there is a persuasive set of reasons to be concerned about the so-called correspondence theory of truth. For while it seems reasonable to say that truth consists in correspondence with the facts, it is entirely unclear of what the supposed correspondence relation consists. Do we really think, for example, that the belief that snow is white, which resides in people's brains, in some way resembles, or has a structural similarity with, the white stuff that falls from the sky? Moreover, is the belief that snow is not green supposed to bear the same relationship of "correspondence" with some worldly state of affairs of snow's not being green? A plausible response to all this holds that there is really no such thing as the correspondence theory; rather, in saying that "snow is white" corresponds with the facts, we say nothing more than that snow is white. Similarly, in saying that "snow is not green" corresponds with the facts, we say nothing more than that snow is not green.

So-called minimalist theories of truth, which have been exceptionally influential among philosophers over the past

FIGURE 56.3. Karl Popper (1902–94) always had a somewhat ambiguous relationship to Darwinism. He saw its great importance and yet notoriously labeled it a "metaphysical research programme" rather than genuine science (a charge that he later retracted somewhat). Permission: © Mark Gerson / National Portrait Gallery, London

fifty years or more, tend to hold that truth is not a substantial property of representations at all (Horwich 1998). Rather, when we say it is true that snow is white, instead of attributing the property of truth to a proposition, we simply say that snow is white. If we are moved by minimalism, we are likely to reject Vollmer's talk of isomorphism between representations and states of the world, for just the same reason that we will reject talk of correspondence – unless, of course, the claim "the representation that snow is white is isomorphic with the world" is just another way of saying that snow is white. The important point to note here, though, is that while minimalism will make us suspicious of this way of talking, it does not impugn the distinction between a proposition's truth and its adaptive value. That distinction makes perfectly good sense even given minimalism; it still makes perfectly good sense, for example, to ask whether, given that some snakes are harmless, it might nonetheless be more advantageous in fitness terms to believe that all snakes are deadly. That, for the minimalist, is an example of what it means to ask whether truth is adaptive.

THE EVOLUTION OF THEORIES

It is time to turn to what I have been calling extended evolutionary accounts of knowledge. It is common for students beginning courses in the philosophy of science to be asked to contrast the views of Karl Popper and Thomas Kuhn (Figs. 56.3 and 56.4). Where Kuhn and Popper are in agreement, however, is in their use of evolutionary analogies to illustrate their very different conceptions of scientific progress.

For Kuhn, the evolutionary analogy is first used in an attempt to explain his notion of progress without goals. Toward the end of his best-known book, *The Structure of Scientific Revolutions* (1970, 170–71), Kuhn explains:

> We may ... have to relinquish the notion, explicit or implicit, that changes of paradigm carry scientists and those who learn from them closer and closer to the truth ... The developmental process described in this essay has been a process of evolution from primitive

beginnings – a process whose successive stages are characterized by an increasingly detailed and refined understanding of nature. But nothing that has been or will be said makes it a process of evolution *toward* anything.

Readers might wonder how on earth to make sense of this. What could it possibly mean to say that science makes progress, unless it makes progress toward the truth? It is at this point that Kuhn invokes Darwin:

> The *Origin of Species* recognized no goal set either by God or nature. Instead, natural selection, operating in the given environment and with the actual organisms presently at hand, was responsible for the gradual but steady emergence of more elaborate, further articulated, and vastly more specialized organisms. Even such marvelously adapted organs as the eye and hand of man ... were products of a process that moved steadily from primitive beginnings but toward no goal. (172)

Kuhn might be making the mild claim that, for Darwin, progress without a goal is intelligible; hence, we should not assume that Kuhn's own conception of progress is unintelligible. But he may be alluding to something much stronger – namely, that goal-free progress makes sense in science because science, like organic evolution, is a selection process of sorts (Fig. 56.5). My own view about this is that the answer is underdetermined.[2] Indeed, it is a typical feature of analogy that users of analogy need not spell out explicitly how strong they intend the analogy to be. This allows analogies to have a heuristic function in their own right, as their strengths and weaknesses are explored. In line with this, Kuhn seemed to lean more and more heavily on the evolutionary analogy as his career moved on, making allusions to similarities in mechanism between science and selection.

The sort of project that treats evolution by natural selection as an informative analogue for scientific change has a long history. An early and very explicit statement of the view comes from William James (1880, 441): "A remarkable parallel, which to my mind has never been noticed, obtains between the facts of social evolution and the mental growth of the race, on the one hand, and of zoological evolution, as expounded by Mr Darwin, on the other." The attractions of evolutionary epistemology of this sort are not hard to appreciate. Scientific hypotheses are exposed to rigorous experimental test. If their fit with recorded data is good, the hypothesis lives to fight another day; indeed, it is likely to increase its representation in the community of scientists. If the fit with data is bad, the hypothesis is likely to be rejected. It is tempting, then, to think of this process as a selective struggle, where hypotheses live or die depending on how well they are adapted to an environment constituted by empirical data.

Karl Popper felt that an evolutionary idiom was a fine vehicle to articulate his falsificationist methodology of science. And Popper (like Lorenz) also used the analogy between theory testing and selection in both directions: just as theory testing is rather like adaptation to an environment, so we can

FIGURE 56.4. In his *The Structure of Scientific Revolutions*, Thomas Kuhn (1922–96) argued that scientists work within "paradigms" and that revolutions are switches from one paradigm to another. Crucially, these switches also bring about changes to the standards by which theories should be evaluated, and to the meanings of theoretical terms. The result is the "incommensurability," that is, lack of common measure, of theories across paradigm shifts. Courtesy MIT Museum

understand biological adaptation even in very rudimentary organisms as a process whereby (for example) the amoeba makes some conjecture about the nature of its environment, and the hypothesis is either preserved or rejected depending on its fit with that environment. The great advantage of the scientific method over biological adaptation is that we, like the amoeba, gradually become better adapted to our surroundings, but, unlike the amoeba, we have no need of generations of wanton death and destruction in order to achieve adaptation:

> The critical attitude might be described as the result of a conscious attempt to make our theories, our conjectures, suffer in our stead in the struggle for the survival of the fittest. It gives us a chance to survive the elimination of an inadequate hypothesis – when a more dogmatic attitude would eliminate it by eliminating us.... We thus obtain the fittest theory within our reach by elimination of those which are less fit (by "fitness" I do not mean merely "usefulness" but truth). (Popper 1963, 68–69)

[2] My thinking on these matters has been influenced by Vashka dos Remedios, especially by her unpublished Cambridge Part III essay "Kuhn's Evolutionary Epistemology" (University of Cambridge, Department of History and Philosophy of Science, 2010).

FIGURE 56.5. Does the theory of Charles Darwin (shown here in old age, in a celebrated portrait by Julia Margaret Cameron) constitute a paradigm switch, meaning that, although it works and is an advance in some sense, it gets us no closer to some absolute truth about reality? Permission: Wellcome

Where Popper's view falls down, it seems to me, is in the attempt to argue that the selection process leads to theories that are "fit" in the sense of being true. Of course, theories that have a good fit with the data are selected. But Popper says very little to combat traditional concerns about underdetermination of theory by data: on the face of things, a great many incompatible theories, some of which must be false, might all generate the same observational predictions. Under these circumstances, it is hard to see how a good "fit" with the data can be thought indicative of truth. Indeed, it is ironic that some philosophers have used the Darwinian analogy to argue for the opposite of Popper's conclusion. Bas van Fraassen, for example, offers the following rebuttal of the argument that truth of a theory is the only plausible explanation for a theory's success in accounting for data. For van Fraassen (1980, 40), the mere fact of selection is enough to explain success, regardless of truth:

> I claim that the success of current scientific theories is no miracle. It is not even surprising to the scientific (Darwinist) mind. For any scientific theory is born into a life of fierce competition, a jungle red in tooth and claw. Only the successful theories survive – the ones which in fact latched onto the actual regularities in nature.

Evolutionary considerations by themselves fail to adjudicate between realist and antirealist views. This remains the case when we move to more formal efforts to offer "extended" evolutionary models. Popper is quite content to use evolution in a rather loose, analogical manner to lay out his views on theory change. But others – most notably Donald Campbell (1974) and David Hull (1988) – have argued for a stronger evolutionary epistemology. In both cases, they argue that selection can be understood in a quite general, abstract manner, such that organic evolution and theory change are different instantiations of the same broad type of process (Plate XLIV).

FORMAL EVOLUTIONARY EPISTEMOLOGY

Campbell (1974, 421), who was directly inspired by Popper, asserts the following basic claims for his evolutionary epistemology:

1. A blind-variation-and-selective-retention process is fundamental to all inductive achievements, to all genuine increases in knowledge, to all increases in fit of systems to their environment.
2. In such processes there are three essentials: (a) mechanisms for introducing variation; (b) consistent selection processes; and (c) mechanisms for preserving and/or propagating the selected variations....
3. The many processes which shortcut a more full blind-variation-and-selective-retention process are in themselves inductive achievements, containing wisdom about the environment achieved originally by blind variation and selective retention.
4. In addition, such shortcut processes contain in their own operation a blind-variation-and-selective-retention

process at some level, substituting for overt locomotor exploration or the life-and-death winnowing of organic evolution.

Campbell's first claim might seem suspect. Perhaps knowledge is sometimes generated by selection among a set of ideas. But is it always so generated? Consider a mathematical deduction. Unless we belittle the achievements of mathematicians, this looks like a "genuine increase in knowledge"; however, it may follow a linear chain of one idea leading to the next – no variation, no selective retention. Campbell could reply that this is no counterexample – selection, as he specifies, underlies all inductive achievements, and this is not an inductive achievement. Even so, might there not be chains of reasoning that deliver new knowledge about the external world, based only on a balance of probabilities, and which involve no selective processes? Might we not correctly determine, from the evidence available, that the Butler did it? Lord Arbuthnot is on the kitchen floor, in a pool of blood. The Butler is standing over him, professing his guilt, holding a bloody knife in his hands. It is not a matter of deductive proof that the Butler is guilty – it is logically possible, after all, that he has been framed and brainwashed – but we can infer his guilt from our evidence without, apparently, allowing a range of alternative hypotheses to struggle against each other.

Once again, Campbell could accept the terms of the example, but point to his condition 3: if we are able to reason successfully that the Butler did it, then that is because we bring in additional pieces of knowledge – the knowledge that a confession under such circumstances is generally reliable, the knowledge that knives inflict mortal wounds – and these pieces of knowledge, or the rules of thumb that produced them, were produced through selection. At this point, Campbell's position begins to look less radical than one might think at first glance. He is not denying that deduction can be a source of knowledge. He agrees that inductive inference may not always involve selection, but he does claim that the rules of thumb that underlie inductive inference will themselves, at some level, have a selective explanation.

Campbell's position amounts to an endorsement of what we might call the "no free epistemic lunch" principle. This is the very plausible view that, if some piece of knowledge exists, then at some time in the history of that knowledge a risk must have been run in its acquisition. Not all knowledge-acquiring processes must be selective in nature – we might simply observe the world and infer correctly. However, to do this demands that we have, for example, reliable sensory apparatus and reliable inference rules. These must come from interaction with the world, too, and at some point we will arrive at the question of how an ability to navigate the world was acquired where there was none before. At some point, we must run a risk of being wrong – that is the nature of ignorance – hence the transition from ignorance to knowledge must, at some time, involve a selective process. For, unless we are exceptionally lucky, we cannot get from ignorance to knowledge without some of our guesses being false. Generalizing the point to cover noncognitive adaptation, whatever our criterion of rightness and wrongness, we cannot acquire dispositions to behave in generally the right way without initially making some wrong moves.

THE SCIENCE WARS

I want to close with a speculative suggestion. An evolutionary perspective may help us to broker a truce between two images of the sciences that are often needlessly opposed to each other. *Scientific realists* argue that science is in the business of telling us the truth about the world around us. Indeed, they usually add that science does not merely aim at truth but makes reliable progress toward it. *Sociologists of knowledge* argue that sociological factors should be invoked not just when we want to explain why scientists make mistakes, but why they believe in successful theories, too. One might think these two views must be at odds with each other, on the grounds that social factors can only be distorting, while the pursuit of truth must be untainted by social influence. But the evolutionary perspective suggests a variety of ways in which science might be subject to social influence in all kinds of ways, while making reliable progress to the truth all the same.

The evolutionary view encourages us, for example, to note how social background conditions can constrain the likely range of hypotheses that scientists are able to test. The sources of scientific variation depend in part on local scientific traditions. These depend, in turn, on the problems that have seemed particularly pressing in the past. And the urgency of scientific problems is sometimes a function of political pressure. So the range of scientific variation can sometimes be socially constrained and socially explained.

The evolutionary view also makes room for an interesting reconciliation between those who see science as progressing toward the truth and those who see the overt choices of scientists as unconcerned with truth, either because they prefer theories on grounds of aesthetic virtues or because the ideological preferences of scientists sometimes influence theory choice. For those who see science as progressive, what is important is that science is organized in such a way that the theories that are selected tend, as a matter of fact, to get closer to the truth. This does not rely on scientists themselves choosing theories because they are closer to the truth. Just so long as aesthetic, or even ideological, factors are locally correlated with truthfulness, conscious decisions to select according to these criteria will lead to a growth in knowledge. Even if Darwin was attracted to the theory of natural selection because of its portrayal of nature in economic terms, thereby suggesting that the capitalist order that had brought prosperity to his family was in some sense the natural order, his theory can still fit the facts just so long as there are salient resemblances between economic and biological change. Of course, this is not to say that every ideologically motivated choice in science is truth conducive; however, it does show that one should not assume that if a theory choice appears to have a political element about it, it is therefore productive of error.

Finally, the evolutionary view suggests a more positive role for the social organization of science in generating truths. Evolution by natural selection is not merely a process of "generate and test"; to describe it as such, or to equate it as Popper does with trial-and-error learning, suggests that cumulative evolution is explained merely by a process in which bad variants are thrown away and the good ones are kept for further experimentation. This downplays the role of population structure itself in the explanation of adaptation. Darwin (1859, 41) notes in his discussion of artificial selection how, merely by increasing in numerical representation, the chances that a population generates further adaptive variants will increase: "[A]s variations manifestly useful or pleasing to man appear only occasionally, the chance of their appearance will be much increased by a large number of individuals being kept; and hence this comes to be of the highest importance to success." Similarly, if we take a population perspective on scientific change (e.g., Kitcher 1995), we can begin to look at how the dissemination of scientific theory through reputation, the cohesion of research groups, cross-group collaboration, and so forth either fosters or hinders the generation and transfer of knowledge.

ACKNOWLEDGMENTS

I am grateful to Michael Ruse for comments on an earlier draft of this essay and to Beth Hannon for help in preparing the final version. The research leading to these results has received funding from the European Research Council under the European Union's Seventh Framework Programme (FP7/2007–2013)/ERC Grant agreement no 284123.

Essay 57

Ethics after Darwin

Richard Joyce

Through most of the twentieth century, the influence of Darwin on the philosophical field of ethics was negligible. Things changed noticeably in the last couple of decades or so of that century, and now "evolutionary ethics" – which had lain dormant since Darwin's contemporary Herbert Spencer – is a lively and hotly debated topic. There are several Darwinian theses that might have bearing on moral philosophy.

i. Humans are the product of natural selection.
ii. (i) + Humans have been forged by that process to be social organisms.
iii. (ii) + Among the mechanisms that govern that human sociality is an innate moral sense.

The first two are beyond serious question, but the last – moral nativism – can be reasonably doubted. It is a plausible counterclaim that the human tendency to engage in moral assessment (of oneself and others) is not a discrete psychological adaptation but a learned cultural trait that depends on psychological capacities that evolved for other purposes. Darwin himself, however, arguably endorsed all three theses; he possibly advocates (iii) in *The Descent of Man*:

> I fully subscribe to the judgment of those writers who maintain that of all the differences between man and the lower animals, the moral sense or conscience is by far the most important.... [A]ny animal whatever, endowed with well-marked social instincts, the parental and filial affections being here included, would inevitably acquire a moral sense or conscience, as soon as its intellectual powers had become as well, or nearly as well developed, as in man. (Darwin 1874, 1:98)

What bearing might these theses have on ethics? It is important to start out distinguishing two programs passing under the name "evolutionary ethics." The first is the *empirical* enterprise of exploring the evolutionary origins of the human moral sense, drawing evidence from primatology, developmental psychology, evolutionary biology, and so on. But though often called "evolutionary ethics," this is not a field of ethics in the traditional sense (any more than the investigation of the origin of the human musical sense is a kind of musical production). By contrast, *philosophical*

FIGURE 57.1. Henry Sidgwick (1838–1900) opened the philosophical critique of evolutionary ethics. His main target was Herbert Spencer, but he included Darwin in his repudiation, arguing that natural selection could never produce the altruism demanded by morality. Interestingly his skepticism about evolutionary ethics did not extend to psychic phenomena. He is seen here with the well-known medium Eusapia Palladino. Permission: © Mark Gerson / National Portrait Gallery, London

evolutionary ethics proposes that facts about human evolution can help address certain perennial problems in moral philosophy, such as how we ought to act or whether our moral judgments are justified. When we ask what bearing these theses might have on ethics, we are asking what impact they may have on ethics as a philosophical subject.

In 1876 (in the first issue of the new academic journal *Mind*), the great utilitarian Henry Sidgwick declared that "the theory of Evolution ... has little or no bearing upon ethics" (1876, 54) (Fig. 57.1). Around the turn of the twentieth century two influential attacks on philosophical evolutionary ethics seemed to settle the matter in Sidgwick's favor.

FIGURE 57.2. G. E. Moore (1873–1958), Sidgwick's student, continued the onslaught on evolutionary ethics (especially Spencer), labeling it an egregious instance of the "naturalistic fallacy" in full flight and fancy. Permission: © Mark Gerson / National Portrait Gallery, London

Giving the 1893 Romanes Lectures at Oxford, "Darwin's bulldog" Thomas Huxley argued that even if the human moral sense is the product of natural selection, this affords it no particular justification. "Goodness and virtue," he proclaimed, demand self-restraint and the helping of one's fellows, whereas the process of natural selection demands "ruthless self-assertion" ([1893] 2009, 82). Moral considerations require that we *combat* the activity of the "gladiatorial theory of existence" provided by Darwin. Huxley, however, seriously underestimated the extent to which natural selection can produce cooperative traits (which is curious, given that he acknowledges that moral sentiments are themselves the product of evolution).

A decade later the Cambridge philosopher G. E. Moore (1903) drew attention to what he called the "naturalistic fallacy" – an error supposedly committed by any attempt to derive ethical conclusions from scientific data (Fig. 57.2). Moore's presentation of the naturalistic fallacy does not place a restriction on deriving normative claims from empirical data per se; rather, it asserts that the quality of goodness is indefinable and therefore any attempt to define it in some other terms (including evolutionary terms) is doomed to failure. Moore says that same thing about the quality of yellowness. But why should we agree that goodness (or yellowness) is indefinable? Moore seeks to convince us with his "Open Question Argument," which is as follows. Suppose we try to define goodness by reference to some natural property – let's say some evolutionary property E. Thus when we ask of something, x, (1) "Is x good?" we are asking (2) "Does x have E?" Suppose we are inclined to answer the last question affirmatively; we can then sensibly ask a further question: (3) "Is it good that x has E?" – which, accordingly, would be the same as asking (4) "Does the fact that x has E itself have E?" And at this point Moore throws up his hands and pronounces that in asking (3) we clearly do not mean anything "so complicated" as (4).

Moore's view was influential for decades, but it is both widely misunderstood and dubious. A comprehensive assessment of this argument cannot be pursued here, but the standard objection should be mentioned. Goodness might be identical to some naturalistic property – including some property pertaining to human evolutionary origins (we will continue to call it "E") – while this is unobvious to competent speakers. If an ancient Greek, ignorant of molecular chemistry, asks whether x is water, he should not be interpreted as asking whether x is H_2O – yet this observation does not undermine our confidence that water is identical to H_2O. In the same way, the fact that the question "Is x good?" should not be interpreted as "Does x have E?" (and "Is it good that x has E?" should not be interpreted as "Does the fact that x has E itself have E?") does not undermine the possibility that goodness is in fact identical to the property E.

The naturalistic fallacy is frequently confused with the claim that one cannot validly derive an "ought"-claim from a set of premises that are purely descriptive ("You can't get an *ought* from an *is*") – an injunction that is also widely assumed to sink evolutionary ethics. But much the same objection applies. The evolutionary ethicist may claim that goodness (say) is identical to E without supposing that conclusions about what is good can be logically derived from premises that mention only E – any more that one can validly derive the conclusion "x is H_2O" from premises couched entirely in "water" terms.

The impact of Huxley's and Moore's arguments had a lot to do with the virtual abandonment of philosophical evolutionary ethics through the first half of the twentieth century (and, in Moore's case, the abandonment of moral naturalism more generally). There was a glimmer of interest in 1943 when Huxley's grandson, Julian Huxley, gave a much more positive account of evolutionary ethics in *his* Romanes Lectures, but the later Huxley's view was not very influential, due no doubt in part to a certain obscurity surrounding his reasoning and, indeed, his intended positive thesis. (Huxley's view is effectively dissected by C. D. Broad in a critical notice the following year.) The philosophical advances that really helped break the spell cast by Moore came in the form of midcentury progress in conceptual understanding of identity statements and analyticity. By the 1980s robust forms of moral naturalism were being offered by philosophers, the advocates of which felt entirely unhindered by Moore's worries. These changes modified the landscape of moral philosophy in a way that rendered it much friendlier toward the prospects of evolutionary ethics in its philosophical sense.

At the same time, advances in evolutionary biology were rendering theses (ii) and (iii) more plausible – that is, promoting *empirical* evolutionary ethics. Despite the attention that Darwin paid to the natural selection of social traits in general, and to the human moral sense in particular, the profusion of cooperation evident in nature continued to be seen as a challenge for Darwinian thinking. Natural selection, one might be tempted to assume (as did Thomas Huxley), is a process that will always favor self-serving behavior over self-sacrifice. Yet when we look around us, we find a natural world teeming with examples of helpful organisms: from the bee's suicidal sting to vampire bats sharing blood. This challenge has been referred to as the "paradox of altruism."

It was not until William Hamilton's work on kin selection in the 1960s that a comprehensive solution began to crystallize. Kin selection essentially presupposes a gene's-eye view on evolution, appreciating that a gene carried by organism O1 might further its reproductive chances if O1 sacrifices its interests for the advantage of organism O2, provided that O2 also carries a copy of that gene. Hamilton's theory of kin selection was complemented a few years later by Robert Trivers's work on reciprocal altruism. In this case, O1 acts in a helpful and seemingly self-sacrificing manner toward O2 because there is a high probability of O2 repaying the favor at a later date, with net gain for both parties. ("I'll scratch your back if you scratch mine.") Darwin (1871a, 1:161) appreciated both evolutionary forces, though only vaguely. Writing of helpful and inventive individuals in prehistoric tribes, he points out: "Even if they left no children, the tribe would still include their blood-relations." A few paragraphs later he writes that as reasoning powers increase, "each man would soon learn that if he aided his fellow-men, he would commonly receive aid in return" – a tendency that, he makes clear, may be inherited. Reciprocity may also be indirect, where O1 helps O2 and receives a proportionally greater benefit from O3 (and others). Darwin's frequent acknowledgments of the importance of *reputation* (our love of praise and dread of blame) – which he says clearly was "originally acquired ... through natural selection" (156) – is in effect an appreciation of the importance of indirect reciprocity in the evolution of human sociality.

Kin selection and reciprocal altruism are by no means the end of the story of the evolution of cooperation – both theories have been refined and complemented by descriptions of further evolutionary processes leading to cooperation (e.g., mutualism) – but it is fair to say that by the mid-1970s it had become accepted that the abundance of cooperative behavior observed in nature poses no major difficulty when it comes to providing a Darwinian explanation of the mechanisms productive of those behaviors (Fig. 57.3 and Plate XLV).

Against this background, E. O. Wilson's *Sociobiology* had a major impact when it appeared in 1975 (Plate XLVI). Wilson undertakes to explain how natural selection leads to cooperation – a perfectly reputable ambition when applied to ants and zebras, but one that proved incendiary when applied to humans. The leading concern seems to have been that in providing an evolutionary explanation for human traits – including such things as aggression and sexual preferences – one somehow provides a *justification* for these behaviors. The fear was frequently expressed, though never properly

FIGURE 57.3. Recognition of the social interactions between the primates has been a major factor in convincing even philosophers that Darwinism might have something of importance to say about moral behavior. Permission: Photo by Frans de Waal

explained, that sociobiological theories underwrite certain political systems.

While Wilson focused on the idea of behaviors as adaptations, subsequent thinkers came to focus on the psychological mechanisms underlying those behaviors. The shift in emphasis is important, for organisms with the same suite of psychological adaptations may behave very differently if placed in different environments. This change of emphasis heralded a change in name in the 1990s, from "sociobiology" to "evolutionary psychology." Evolutionary psychology was pioneered by psychologist Leda Cosmides and anthropologist John Tooby, whose preferred case study was the hypothesis that the human mind contains a "cheater detection module" for governing social exchanges (Barkow, Cosmides, and Tooby 1992). Because of this decision to focus on such a "moralistic" human trait, the growth of evolutionary psychology encouraged work in empirical evolutionary ethics, which in turn stimulated discussion in philosophical evolutionary ethics.

Philosophical evolutionary ethics can be divided roughly into two antagonistic programs. First, it has been argued that Darwinian thinking applied to humans can serve to *vindicate* morality – either morality in general or some specific set of moral norms. Second, one might draw the opposed conclusion that moral nativism in fact *undermines* morality, providing grounds for some form of moral skepticism. These will be discussed in turn.

Suppose moral nativism is true. This shows that morally assessing aspects of one's environment (and oneself) enhanced the reproductive fitness of our ancestors. And from this one might draw the conclusion that morality is useful, and thus justified (see Campbell 1996). But such an argument is invalid, and the conclusion is, in any case, misleading. It is invalid because of a fallacious tense shift: from the fact that morality *was* useful, it does not follow that it *is* useful. If one wants to show that morality *is* practically justified, then examining the ways in which moral thinking was useful in the Pleistocene may provide some insight, but it is strictly superfluous; rather, contemporary data are needed. More importantly, the conclusion that morality is practically useful is not the kind of justification in which moral philosophers are typically interested. Metaethics is concerned with whether moral judgments are *epistemically* justified, not whether they are *instrumentally* justified. When we seek epistemic justification for a belief, we inquire into the grounds for holding the belief to be true. Ernie's holding a certain belief might bring him reassurance and happiness, and thus might be instrumentally justified – but if Ernie holds this belief irrespective of any supporting evidence, then it is not epistemically justified.

One might object that moral judgments are not beliefs, and thus the question of their *epistemic* justification does not arise – that instrumental justification is the only kind that matters for morality. But, then, the line of reasoning from nativism to vindication would require supplementation by a preliminary argument demonstrating that moral judgments are not beliefs (i.e., an argument for noncognitivism). Metaethics has debated the merits and pitfalls of such arguments for decades. One might, however, think that Darwinian considerations can be pressed into service here, to settle the metaethical debate over whether moral judgments are beliefs. If it were shown, for example, that moral judgment emerged in our ancestral lineage because of a payoff that relied on emotional arousal (e.g., guilt or punitive anger), then one might suppose that noncognitivism is corroborated. But the success of such an argument, while it cannot be excluded, faces serious challenges. After all, cognitivism is not the view that emotions play *no role* in moral judgment. Moral judgments may be prompted by emotion, may produce emotion, may have evolved precisely because of their emotional components, and yet, for all that, moral judgments may be beliefs. (Compare the hypothesis that we have an innate fear of snakes. Fear is an emotion, but it doesn't follow that the associated judgment "This snake is dangerous!" is anything other than a belief.) So evidence that morality evolved because of its *emotional* adaptiveness cannot be taken as evidence supporting noncognitivism.

Putting noncognitivism aside, the kind of justification of morality in which metaethicists are interested is *epistemic*. Can Darwinism applied to humans help supply such justification? Several attempts have been made.

If moral nativism is true, then certain mechanisms pertaining to moral judgment have evolutionary functions. This allows one to speak of these mechanisms fulfilling their functions "well" or "poorly," of what they are "supposed" or "ought" to do. Philip Kitcher (2011), for example, argues that the evolutionary function of morality is to encourage social cohesion when natural altruistic sentiments fail. If this is correct, then moral systems (and moral beliefs) can be assessed according to whether they fulfill or deviate from this function.

The Aristotelian virtue ethicist will make a similar teleological claim, but not necessarily one that pertains to the proper functioning of an innate moral sense, but rather one that pertains to the flourishing of a human understood as a complex organism. Just as biology provides understanding of what it is to be a flourishing frog, as opposed to a diseased or unhealthy frog, so too it can in principle provide the same with respect to humans. Given that humans are social creatures – that is, given thesis (ii) – the virtues, it is claimed, are those character traits that are conducive to, or constitutive of, human flourishing. Thus, the virtue ethicist takes a Darwinian premise about what kind of evolved creatures we are and strives to produce a normative output – one that favors such things as friendliness, benevolence, and so forth – and thus hopes ultimately to provide epistemic justification for such claims as "One ought to be friendly" (Casebeer 2003).

The principal problem with such attempts to vindicate morality using evolutionary data is that whatever normative language legitimately follows appears to be the wrong sort to underwrite the kind of practical guidance we require of *morality*. Consider: the function of a hammer is to bang in nails, but if I find it convenient to use a hammer to prop open the garage window, there is nothing fishy about my action; it is not even an instance of the new function I have assigned the hammer *overcoming* requirements imposed by the hammer's "real" function. The hammer's real function may license assertions like "A good hammer bangs in nails well" and "This hammer is supposed to bang in nails," but it turns out that this normative language is really quite toothless when it comes to making claims upon our practical deliberations independent of our standing interests. In the same way, moral systems may have the evolutionary function of promoting cooperation when altruism falters, and this may thus be what moral systems are "supposed" to do, but if a society chooses to use its moral system for some other end (in support of militaristic imperialism, say), then the "real" function of moral systems carries no weight per se to cast doubt on that decision. (That's not to say that there is nothing wrong with such a decision; we just need to look somewhere other than biological functions in order to locate grounds for criticism.)

The same point applies when we consider someone electing to cultivate personality traits other than those virtues conducive to flourishing. Such a person need not have given up on the aim of flourishing but rather has chosen a vision of flourishing other than that laid down by biology. Perhaps this person has embraced the kind of flourishing that goes along with being a Buddhist monk (which presumably diverges spectacularly from what it took to be a fine human specimen in the Pleistocene) and thus cultivates the kind of character traits necessary for this end. That we could legitimately say of this person that her vision of flourishing is not the one she is "supposed" to be pursuing (qua human organism) might sound impressive, but it is not obviously any more of a genuine criticism of her behavior or character than is the observation that in propping open the window with a hammer one is using the object in a manner for which it is not intended. The problematic consequences of this failure to derive genuine normative criticism from evolutionary function becomes apparent when we note that it applies as much to the person who chooses a life of violent crime as it does to the Buddhist.

Another possible route from nativism to epistemic vindication is via epistemological reliabilism. True beliefs are far more likely to enhance reproductive fitness than false beliefs; therefore, on those occasions that natural selection produces some discrete belief-forming mechanism, it is likely that the resulting beliefs will be true (Carruthers 1992, 111ff.). Thus, beliefs that are fixed or prewired by natural selection can be considered the product of a *reliable* process and hence are, according to the theory of process reliabilism, epistemically justified.

The prospects of any such attempt to vindicate morality are only as good as the prospects of the theory of process reliabilism upon which it depends – and such theories are controversial. One of the problems of reliabilism is that it is difficult to specify precisely *which* process any given belief is the product of, for invariably it is simultaneously the product of numerous processes. If moral beliefs are the output of some kind of "moral sense," then it is natural to assume that when we try to identify "the process" that produced them, we should not look to natural selection in the general sense but rather to the particular evolutionary trajectory of the innate faculty in question. It might be correct that *in general* we are better off with true beliefs than false, but it need not be correct that *when it comes to moral beliefs* we are better off with truth than falsity. A false belief about the value of benevolence may be adaptive in a way that a false belief about the behavior of predators is not. Indeed, this observation segues naturally into discussion of the second program of philosophical evolutionary ethics: that moral nativism *undermines* morality.

Contemporaries of Darwin already felt uneasy about the possible undermining influence his views might have on moral authority. One called his position "dangerous" and expressed concern that moral nativism "aims … a deadly blow at ethics" (Cobbe 1872, 10). Another wrote that if Darwin's views on moral nativism were true, "or should they come to be generally accepted, the consequences would be disastrous indeed! We should be logically compelled to acquiesce in the vociferations of [those] who would banish altogether the senseless words 'duty' and 'merit'" (Mivart 1871, 232). The general worry is that if humans assess the world in moral terms only because doing so helped our ancestors produce more babies than their competitors, then these judgments appear not to carry the binding authority over our actions that we usually think they do. Moreover, if it is true not only that evolutionary origins deprive moral judgments of their authority but also that invoking such authority is the whole point of having a moral system – indeed, that such authority is a necessary feature of our basic moral concepts – then it appears that moral nativism reveals our moral concepts to be bankrupt: they imply an authority that they cannot supply.

Darwin himself gives no hint of having ever been tempted by such skeptical thoughts. He held that moral thinking is both practically necessary for human society and one of the most striking of human adaptations. He considered that any social creature granted sufficient intelligence would evolve a moral sense – though he also conceded that the content of that morality may differ dramatically among species. But one looks in vain for any satisfying metaethical statement from Darwin (understandably enough); rather, one is forced to infer from his seemingly untroubled attitude toward morality that he was unaware of, or had little patience for, the possibility that moral nativism might debunk morality.

In recent years, the debunking argument of evolutionary ethics has been explored by several philosophers. According to Michael Ruse (1986), in order for moral judgments to serve their evolutionary function (roughly, encouraging cooperation), they must be imbued with objectivity. This is a thesis about the *content* of moral judgments; it does not follow that the actions in question (or any other actions) *are* objectively required. In fact, Ruse thinks, moral nativism provides grounds for doubting that any actions are objectively morally required, for supposing that they are so is entirely unnecessary. Nativism may explain why humans make judgments about moral objectivity, but to go further – to suppose that the judgments are *true*; that is, that some actions *are* objectively required – involves populating our conception of the world with properties that play no explanatory role. (It is not merely that they are not needed to explain our moral judgments, but they are not needed to explain *anything* – for what explanatory role could they play independently of anyone making a moral judgment?) Humans have been set up by natural selection to believe in objective moral properties (Ruse thinks) – and have been set up to do so irrespective of whether there are any such properties – so there are no grounds for believing in them at all.

Ruse here wields Ockham's razor to cut objective moral properties from our conception of the world. In fact, we should distinguish two razors (following Sober 2009). Suppose our evidence fails to discriminate between "X exists" and "X does not exist." The Razor of Denial states that we should deny the former and affirm the latter; the Razor of Silence states that we should suspend judgment about both. The debunking argument of evolutionary ethics is more plausible when construed in the latter manner. It is unlikely that moral nativism can show that our moral judgments are all *false*; but that it might show them to be all *unjustified* is an argument with more promise. This argument is pressed by Richard Joyce (2006).

It is important to note that this argument has promise only if moral nativism is understood in a certain way – namely, that the ancestral adaptiveness of moral judgment was secured independently of any truth-tracking relation between these judgments and moral facts. A comparative illustration may help. Suppose that humans are prewired by natural selection to divide their social environment into in-groups and out-groups. The supposition implies that such thinking was reproductively useful. But *why* was it useful? The only plausible answers presuppose that our ancestors' environment did actually contain in-groups and out-groups. It is important to see that the hypothesis of moral nativism may be crucially different in this respect. The most plausible accounts of why it was reproductively useful to our ancestors to categorize aspects of their social world as good, bad, evil, obligatory, and so on (e.g., that such categorization strengthened social cohesion) *nowhere presuppose* that the environment contained such things as goodness, badness, evil, and obligatoriness.

However, even if the nativist hypothesis nowhere explicitly mentions any actual moral properties, it remains possible that such properties are identical to, or supervene upon, those properties that are explicitly mentioned. Ruse's use of Ockham's razor, for example, seems to assume that to allow the existence of objective moral properties would admit an extra ontological layer into the world (and thus should be disallowed if unnecessary). But this is not obviously so. An opponent can counter that objective moral properties were implicitly present all along in the evolutionary worldview accepted by Ruse, just as H_2O was implicitly present in ancient Greek explanations involving water. This debate then moves to the question of whether it is plausible to claim that objective moral properties may be identical to, or supervene upon, those naturalistic properties recognized by science. It is noteworthy that Joyce's (2006, ch. 6) evolutionary debunking argument has to be supplemented with an attempt to undermine the prospects of moral naturalism on purely metaethical grounds – an undertaking to which Darwinian thinking has no obvious contribution to make.

Sharon Street (2006) comes to a similar conclusion to Joyce, though starting out with a different understanding of moral nativism: whereas Joyce is willing to speculate that natural selection left the *content* of the moral faculty pretty much open, Street supposes that the content of morality has been "deeply influenced" by Darwinian forces. She then poses the moral realist with a dilemma focused on the relationship between these evolved evaluative tendencies and objective moral values. Either (A) there is no relation at all, in which case the chances that natural selection has guided us to approximately correct evaluative judgments are vanishingly small; or (B) there *is* a positive relation, and our evolved moral faculty "tracks" real moral properties. The problem with the latter is that it is an empirically doubtful claim; as noted earlier, the most plausible accounts of the evolution of the moral faculty see its adaptiveness in terms of enhancing social bonds, not in tracking truths.

Street's argument targets moral *realism*, understood as the thesis that moral truths hold independently of our attitudes. But she is not targeting moral truths per se; she leaves open the door to moral facts that are in some sense constructed by us. Similarly, Ruse's (1986, 253) conclusion is that we have no reason to believe in *objective* moral properties – seemingly allowing the possibility of *nonobjective* moral properties: "[T]he illusion lies not in morality itself, but in its sense of objectivity." By contrast, Joyce's skeptical attack is leveled at moral facts *tout court* – subjective as much as objective.

Evolutionary ethics – both the empirical and the philosophical programs – barely existed for the best part of the century following Darwin's death. In the past few decades it has mushroomed into a rich interdisciplinary field concerned with both the explanation and justification of a fundamental aspect of the human organism.

Essay 58

Darwin and Protestantism

Diarmid A. Finnegan

In one original formulation, the term Protestant identified those "protesting" or asserting the basics of Christian belief as discerned in the Bible and in the face of perceived distortions. "The Bible alone" (*sola scriptura*) quickly became a rallying cry for different kinds of religious agendas and political causes. In a more radical guise, it also made individual believers the final arbiters of religious truth. One consequence of this unstable mix of biblicism and individualism has been the proliferation of Protestant denominations. While many have retained a firm commitment to the Bible as their final authority others, particularly since the rise of biblical criticism in the nineteenth century, have given increasing priority to reason, experience, or individual conscience. Such diversity makes a discussion of the relations between Protestantism and Darwin at once problematic and absorbing. What we manifestly do not have is a single relationship between a well-defined religious movement and a fixed set of scientific or philosophical ideas (Figs. 58.1 and 58.2).

This complexity is indicated here by looking at a sample set of theologians and thinkers from Anglican, Baptist, Lutheran, Presbyterian and Dutch and Swiss Reformed denominations. This denominational diversity alerts us to the range of views included within Protestantism but also points to Protestant perspectives not explored here (on Quakers, for example, see Cantor 2005; and for wide-ranging surveys, see J. R. Moore 1979; Livingstone 1984; and J. H. Roberts 1988). Of course, denominational affiliations do not map neatly onto other ways of registering diversity within Protestantism. More conservative or evangelical and more liberal or progressive versions of Protestantism cut across denominational lines. While it is true at some general level that most evangelicals have been more hostile to Darwin than liberal Protestants, this claim masks a messier history. In the end, no register of Protestant diversity provides a straightforward basis for predicting responses to Darwin (on this, see Livingstone 1992; J. H. Roberts 1999; and J. R. Moore 2001). As the following examples demonstrate, Protestant intellectuals of various stripes, as well as reacting in fairly general terms to Darwin, mobilized Darwin's writings to defend or to adjust certain theological claims and approaches. How and why they did this cannot be accounted for by referring to denominational identity or to a crude evangelical-liberal divide.

FIGURE 58.1. Charles Hodge (1797–1878), for many years principal of Princeton Theological Seminary, was the leading Presbyterian theologian of his day. Although he respected Darwin as a thinker and as a man, he thought that Darwinian evolutionary theory could end only in atheism. From A. A. Hodge, *The Life of Charles Hodge* (London: Nelson, 1881)

FIGURE 58.2. Benjamin Breckinridge Warfield (1851–1921), a distant relative of Wallis Warfield (later the Duchess of Windsor), was a deeply committed Calvinist, but found this quite compatible with a form of guided evolution. "Calvin doubtless had no theory whatsoever of evolution; but he teaches a doctrine of evolution." From B. B. Warfield, *The Power of God unto Salvation* (Philadelphia: Presbyterian Board of Publication, 1903)

After examining the efforts of several prominent Protestant theologians in the period between 1859 and 1918 to grapple with the ramifications of Darwin's work, particularly for theological anthropology, we take a summary look at the twentieth century, beginning with Karl Barth. Unlike in the earlier period, Darwin was not engaged with in any sustained fashion but nevertheless remained an important figure to think with as well as against.

CLOSE ENCOUNTERS: DARWIN AND PROTESTANT THEOLOGY BEFORE 1918

Contrary to what is often thought, the immediate reaction to *On the Origin of Species* (1859), at least among Protestant elites, was by no means uniformly negative. Responses ranged from warm embrace and cautious acceptance to qualified rejection and outright dismissal. Major public controversies were rare, even after the appearance of *The Descent of Man* in 1871, and Darwin commanded respect as an expert naturalist even among his most vocal religious opponents. It is worth noting, too, that for many readers *On the Origin of Species* did not spark but rather resolved, or at least mitigated, a religious crisis (see Secord 2000, 511–14).

Respect for Darwin and for his evolutionary ideas continued to be expressed by Protestant theologians and scientists after his death. This is particularly evident in reviews by prominent Protestants of the three-volume *Life and Letters of Charles Darwin* published in 1887. In a lengthy assessment, the conservative Princeton theologian Benjamin Breckinridge Warfield (1851–1921) designated Darwin's *Life and Letters* a literary descendant of Augustine, Bunyan, and Rousseau and pored over the spiritual changes that had slowly overtaken Darwin's "noble soul" (Warfield 1888, 570). In his own extensive reflections on the *Life and Letters*, the Anglican theologian Aubrey Moore (1848–90) confessed that he found Darwin's account of his slow movement toward agnosticism "intensely interesting" (Moore 1889, 216). Neither Warfield nor Moore thought Darwin's loss of Christian faith was due to a necessary conflict between Darwin's scientific conclusions and Christian belief. For Warfield (1888, 575), it was the "entire doctrine of evolution" that had eroded Darwin's always rather undeveloped religious convictions. Moore, on the other hand, put the withering of Darwin's faith down to a practical neglect of religious truths. But both reviewers also noted Darwin's apparent failure to grasp the reality of God's constant or immanent action in the world. This theological blind spot had forced Darwin to choose between "special creation" and natural cause as a complete explanation of the living world. Asa Gray (1810–88), the American botanist, orthodox Presbyterian, and longtime defender of Darwin, agreed. In his review of *Life and Letters*, Gray (1887, 402) commented in closing that a "fuller recognition of Divine immanence" might have eased some of Darwin's doubts.

The nature of divine action was only one of a number of theological reflections that clustered around the personage of Darwin in this period. Moore (1889, 200), for example, noted several other areas of theological concern, including the challenge to the argument from design, a sharpening of the problem of evil, and a radical questioning of "man's place in nature." It was the final difficulty that attracted most comment by Protestant theologians grappling with the implications of Darwinism in the late nineteenth and early twentieth centuries. Indeed, even Protestant theologians who remained deeply skeptical of Darwin's theory of evolution argued that because God constantly acted with or through secondary or natural causes – denying this was tantamount to deism – admitting evolutionary change was of little theological consequence. The perceived threat to Christian and Protestant understanding of human origins and human nature was another matter entirely.

To illustrate this, it is instructive to turn to the Dutch dogmatician Herman Bavinck (1854–1921), a theologian who drew the circle more tightly than most around what could be considered fact in Darwin's science. In the section on humans as the bearers of God's image (*imago Dei*) in the second volume of his *Reformed Dogmatics*, Bavinck devoted considerable attention to the pernicious effects of Darwinism on theological understandings of human nature. Here, Bavinck (2004, 559) countered materialist and supernaturalist accounts of the image of God found in modernist and Roman Catholic theology respectively. The image of God was not properly defined either as a capacity for moral development (as in accounts of theologians inclined towards a qualified naturalism) or as a superadded ability to receive God's grace (as in Roman Catholicism). Instead, it referred to the "whole person," soul *and* body, the latter being "a marvelous piece of art from the hand of God Almighty." For Bavinck the understanding of *imago Dei*, proposed by theologians willing to go a certain distance with Darwin on human evolution, was inconsistent. They allowed for the evolution of humans but always with some exception – a moral disposition, for example – that permitted the reintroduction of a direct act of creation that obviated their objection to special creationism and introduced difficulties not present in the more traditional account.

Bavinck was not alone in defending a single divine origin for the physical and spiritual aspects of human nature. James Iverach (1839–1922), professor of apologetics at the Scottish Free Church College in Aberdeen, shared Bavinck's concern. Iverach (1894, 175) objected in particular to the account of human origins proposed by Alfred Russel Wallace and St. George Mivart that made "man a highly organized animal to which somehow a spiritual nature had been superadded." But Iverach's defense of the "unity of man" and its divine cause was very different from Bavinck's. Iverach had no objection to Darwin's theory of natural selection once it was admitted that it relied by necessity on a presiding intelligence. The powers of selection Darwin ascribed to nature were to Iverach full proof of the fact that evolution was God's creative method. It followed that there was no need to appeal to divine interference or "special creation" to account for the emergence of the body or the soul of the first humans. God's continuous creative action was the only thing required. Iverach's view was shared by his fellow Free Churchman Henry Drummond, who made a similar argument for a more popular audience in his book *The Ascent of Man* (1894).

Darwin's account of human evolution had a direct bearing, then, on theological accounts of the *imago Dei* and on the supposed origins of the body and soul of the first human. A more indirect influence can be discerned in related discussions about the origin of the souls of Adam's descendants. Writing in the wake of Darwin a number of theologians returned to the age-old debate between "creationists," who held that the soul was created *in utero*, and the traducianists, who argued that the soul like the body was inherited in some way from a child's parents. The Calvinist theologian Charles Hodge (1797–1878) noted that the traducian theory – not unlike Darwinism – led to the eclipse of God's action in the world, resulting either in deism or in the restriction of divine agency to second causes. Traducianism for Hodge (1872, 74) came too close to the "mechanical theory of the universe" that denied God any role in creation and led ultimately to atheism. This of course echoed Hodge's (1874, 89) conclusion that Darwinism – defined as natural selection "conducted by unintelligent causes" – excluded any role for God. Representing a different generation of theologians, the American Baptist systematican Augustus H. Strong (1836–1921), found Hodge's creationism unconvincing. One of the benefits of the traducian theory was that it underlined the importance of "mediate creation" or God acting through natural laws of propagation for both an individual's body and soul. Crucially, this was consistent with the proposal that humans were a "product of natural evolution," an idea that Strong (1907, 492), with certain qualifications, was happy to countenance.

Discussions about the origin of individual souls frequently occurred alongside another dimension of theological anthropology that for more conservative Protestants was nonnegotiable, namely the doctrine of the fall. Some saw in Darwin's account of human evolution a serious threat to the belief that the first human pair had existed in a state of original righteousness. The catastrophic event of the fall did not seem to fit into an evolutionary scheme that emphasized gradual change and incremental improvement. On this scheme, humans had a "savage" and indistinct rather than an elevated and instantaneous beginning. The Scottish Free churchman James Orr (1844–1913) pointed to the dangers of such a view. In his 1905 treatise *God's Image in Man*, he opposed theologians who, too enamored with evolution, denied a state of original holiness and talked instead of a hypothetical primitive innocence. For Orr, the widespread rejection of a more purely Darwinian theory of evolution that had emphasized undirected and gradual change allowed some scope for an account of human origins that made such accommodationist moves unnecessary. Even if these "non-Darwinian" evolutionary schemes did not go far enough for Orr, they offset some of the damaging effects of Darwin's emphasis on insensible gradations and accidental variations.

Others were less persuaded that Darwin's contributions to the study of human origins endangered the doctrine of the

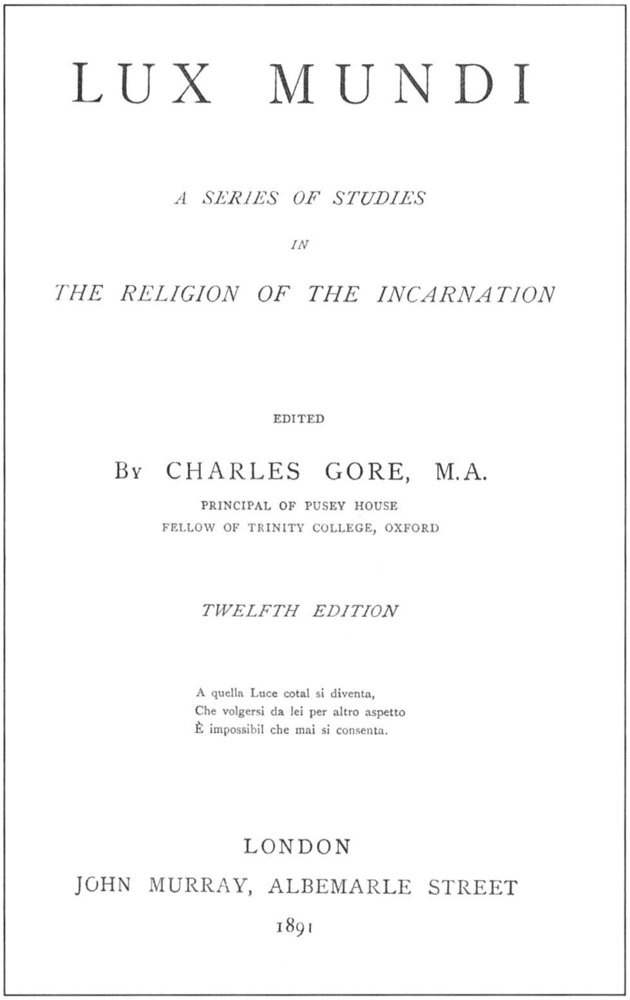

> LUX MUNDI
>
> *A SERIES OF STUDIES*
> *IN*
> *THE RELIGION OF THE INCARNATION*
>
> EDITED
>
> BY CHARLES GORE, M.A.
> PRINCIPAL OF PUSEY HOUSE
> FELLOW OF TRINITY COLLEGE, OXFORD
>
> TWELFTH EDITION
>
> A quella Luce cotal si diventa,
> Che volgersi da lei per altro aspetto
> È impossibil che mai si consenta.
>
> LONDON
> JOHN MURRAY, ALBEMARLE STREET
> 1891

FIGURE 58.3. Title page of *Lux Mundi* (1891). In this collection of essays, Anglo-Catholic theologians showed that they could readily accept the findings of science. Included was Aubrey Moore (1848–90), famous for declaring of Darwin, "Under the disguise of a foe he did the work of a friend."

fall. Aubrey Moore (1890b, 62, 63), for example, noted that all that was required by the doctrine was the belief that the first human parents were in "happy communion with God" (Fig. 58.3). The fall occurred in the "moral region" and in a way that could not be detected by scientific investigation. This rather benign rendition of the doctrine meshed with Moore's opposition to Calvinism with its emphasis on the total corruption of human nature after the fall. James Iverach, on the other hand, found ample evidence in human history to paint a darker picture consistent with a qualified Darwinism and a moderate form of Calvinism. The corruption of humanity evident to any observer suggested to Iverach (1894, 183) that the Darwinian concept of struggle for existence was "derived not from the cosmos but from the more virulent form of human competition." Nature was not as violent and cruel as Darwin had made out (a position that Henry Drummond [1894] also championed). Instead, it was the appearance of rational and self-conscious beings that had significantly increased the level of "selfishness, ruthlessness and ferocity" in the world suggesting that humans had "fallen from the level of higher animals" (Iverach 1894, 183). Something had gone significantly awry, and theology had an explanation to hand. None of this was to deny, however, the evolutionary origins of humanity.

Darwin could also be called upon in elaborations of the doctrine of original sin. Deliberations about the transfer of a corrupt nature and a guilty condition from Adam to the rest of the human race provide one example. In his wide-ranging discussion of this subject, James Orr (1905) engaged with scientific accounts of heredity in order to defend the possibility that Adam's fallen nature was inherited by all of his descendants. He attempted to find a middle way between August Weismann's neo-Darwinism and neo-Lamarckism to maintain that the propagation of a corrupt nature from Adam squared with the science of inheritance. This was among the aspects of Orr's thesis that B. B. Warfield found wanting. Following his theological mentor, Charles Hodge, Warfield (1906, 558) noted that we are guilty of Adam's sin on account of the "principle of representation" – there was no need to bring in a theory of heredity. In Warfield's view, Adam stands in the dock as our representative or "federal head," and just as he is condemned, so are we. This "judicial" or forensic view does not need any theory of the physical propagation of Adam's guilt. Augustus Strong (1907), on the other hand, categorically rejected a federalist line of thought and defended instead a realist account of original sin. Following Augustine, this account allowed that all humans actually participated in Adam's sin and were thus directly and justly culpable for it. In defending this realist version of the doctrine, Strong called upon recent theories of inheritance for support. Here Darwin appears as a stepping-stone to August Weismann's account of inheritance, which, unlike Darwin's, entirely repudiated any notion of the inheritance of acquired characteristics. For Strong, in contrast to Orr, Weismann's "neo-Darwinism" confirmed a real connection with Adam's rebellion. Strong's (1907, 631) case rested on the argument that only the guilt of the human species as a whole was inheritable. Weismann had demonstrated that character is determined by the entire "stream of humanity." In the same way, individuals had not acquired and were thus not culpable for the sins of their parents. It was only the radical apostasy of Adam that, in being a sin of the whole human species, could be propagated to all in a manner apparently consistent with and supported by Weismann's modified Darwinism. Notably, this view required a monogenist account of human origins. Hodge's federalism, on the other hand, did not logically require this even if he held firmly to monogenism on other grounds.

The defense of the doctrine of original sin was also a defense of a theodicy that some felt was no longer tenable in a post-Darwinian age. Herman Bavinck was aware of this and defended a notion of the fall that rendered it the source of the physical evil that infected the natural world. To sustain this argument, Bavinck (2006) suggested that animals were largely vegetarian before the fall, and it was only in a post-fall world that they had descended into carnivory. The flesh-eating larva of the ichneumon wasp, which Bavinck cited and which Darwin

(1985–, 8:223–24, letter to Asa Gray, 22 May 1860) famously found incompatible with a notion of direct divine design, was by implication correlated with the post-fall "curse" that affected the whole of nature. In support, Bavinck (2006, 181) quoted Darwin's suggestion in *Variations of Animals and Plants* that certain animals could substantially adjust their diet in changed circumstances. This was not entirely ironic. Bavinck was keen to praise Darwin's powers of scientific observation while denying the truth of his "speculations."

Asa Gray's position on natural evil differed markedly from Bavinck's. As the recipient of Darwin's letter about the Ichneumonidae, Gray was acutely aware of Darwin's own doubts about God's beneficence. But rather than make predation, disease, and excessive pain a consequence of the fall, as Bavinck did, Gray (1876, 378) made it a necessary part of the economy of an evolving creation. By being necessary, it became at least explicable. Aubrey Moore (1889, 198) followed Gray's lead and found in Darwin's own theory a "hint" toward a rationale for the existence of waste, suffering, and imperfection in the living world. Others, however, maintained a more traditional account of natural evil even while acknowledging that death and pain were realities during the long ages before the fall. Augustus Strong repudiated Darwin's new theodicy and denied that, on balance, "happiness decidedly prevails" (Darwin 1958, 88). Instead, Strong (1907, 403) argued that the excess of "evil in creation ... has some cause and reason in the misconduct of man." To sustain this view, Strong argued that prehistoric pain and imperfection anticipated the sin of the first humans (which God knew would occur) and subsequently functioned as a means of discipline and redemption.

These examples of the ways in which Darwin was called upon to do theological work are indicative rather than exhaustive. They alert us to the fact that the response of Protestant theologians to Darwin cannot be reduced to a sliding scale between antagonism and accommodation. From another angle, it is also clear that the responses cannot be adequately explained simply in terms of the clash or consonance of free-floating ideas. The interplay between text and subtext or ideas and ideologies must also be considered. Two quick examples will suffice here. First, Bavinck's "organic" understanding of the *imago Dei* was relevant to his involvement in Abraham's Kuyper's Anti-Revolutionary Party (see Harinck 2008) and, second, Gray's evolutionary theodicy dovetailed with his justification of the extreme violence of the American Civil War (see Desmond and Moore 2009, 325–26). This is not meant to imply a simple case of political cause and theological effect or, indeed, the reverse. But it does underline the need to attend to precisely where and why certain connections between Darwin and Protestantism were forged and defended. Space precludes offering any conclusions on this front. What we can presume here is that evolutionary ideas associated with Darwin had become part of the bone and sinew of Protestant theological reflection before the First World War. A rather different picture emerges from a consideration of representatives of Protestant responses to Darwin after 1918.

LONGER SHADOWS: DARWIN AND TWENTIETH-CENTURY PROTESTANT THEOLOGY

It has famously been said that Darwin's *On the Origin of Species* (1859) came "into the theological world like a plow into an ant-hill" (A. D. White 1896, 70) and that, decades later, Karl Barth's (1886–1968) *The Epistle to the Romans* (1922) "fell like a bomb on the playground of the theologians" (Adam 1926, 276–77). Barth, of course, entered the fray as a Christian thinker. One of his purposes was to undermine the conviction that God could be readily identified with human or natural history. We might say, then, that if Darwin undercut the scientific relevance of theology, Barth denied the theological relevance of science. The trouble with this claim is that it obscures more than it reveals. As we have seen, Darwin stimulated theological reflection just as much as he overturned theological convictions. Karl Barth, the twentieth century's most famous Protestant theologian, used science as an ally as well as an enemy in constructing his theological project. His attitude toward Darwin's contested legacy provides the relevant case in point.

There is no doubt that Barth opposed the efforts of nineteenth-century Protestant theologians to square at least some of Darwin's evolutionary ideas with Christian belief. This opposition stemmed in part from Barth's denial that knowledge of God was possible apart from revelation. For Barth, in the face of nature and without God's word in Jesus Christ, we are left in a state of profound uncertainty. Barth's famous "Nein" to natural theology came some time after his explosive commentary on Romans, but his radical aversion to reading natural knowledge of God from his works was evident early on. Peter Bowler (2001) correctly presents Barth's theology as an obstacle to the efforts of modernist theologians in early twentieth-century Britain to reconcile evolution and religion. The kind of synthesis between a postmaterialist version of evolution and the Christian faith that liberal Anglican and nonconformist theologians attempted to forge could find no support in Barth.

For all that, it would be mistaken to think that Barth simply dismissed Darwin or was unduly suspicious of scientific developments. Instead, Barth found an unlikely ally in Darwinian skepticism about finding God through nature. This tactical alliance is most sharply apparent in the second edition of Barth's *The Epistle to the Romans*. Here Barth allows a more secularist interpretation of a Darwinian universe to do its worst and dissolve attempts to tie theology to natural knowledge. This was a much more radical expression of Aubrey Moore's (1890a, 99) conviction that "under the disguise of a foe [Darwin] did the work of a friend." It was not, as with Moore, that Darwin helped to highlight God's immanence in nature but rather that Darwin – or at least his more skeptical followers – demonstrated God's absence. This did not mean that Barth thought Darwin had got to the real truth about the universe. In Barth's view, science by its very nature could not disclose this. Instead, Darwinian skepticism confirmed the fact that without divine initiative in revelation no knowledge of God was possible. In a less polemical way Barth made the same point in the section of his *Church Dogmatics* concerned with theological anthropology. Here he pointed to the inevitable failure of previous attempts to defend the "special position" of humans in a post-Darwinian world. Although there were things to commend in these efforts, Barth argued that they already presupposed what they purported to prove. Without any prior theological commitments, there was no reason to believe that the mind or will of humans marked them out as God's image bearers. It was equally possible to consider these features as a "disease" or the "cause of all man's sufferings" (Barth 1960, 89). Barth's argument had at least two consequences for the relationship between Protestant theology and Darwin. The first more negative consequence was that any direct point of contact between Darwinism and theology was refused, making dialogue more difficult. The second more positive consequence was that science in general and biology in particular was freed from artificially propping up theological or metaphysical claims.

Few Protestant theologians interested in the relationship between Darwin and theology followed the more radical aspects of Barth's line of thought. Thomas F. Torrance (1913–2007), often considered the leading British theologian of the twentieth century, is one exception. Nevertheless, his extensive efforts to conduct a dialogue with the natural sciences that was faithful to Barth's theological project were orientated toward the physical sciences and only occasionally touched on topics more directly connected with Darwin (e.g., Torrance 1981, 122–23). Another exception was the Baptist theologian Bernard Ramm (1916–92). Ramm wrote extensively on the relationship between science and religion, and one of his earliest books, *The Christian View of Science and Scripture* (1954), proved instrumental in shaping the direction of the American Scientific Affiliation, an organization for evangelical Christians involved in science. It was only later that Ramm incorporated Barth's theological method into his own efforts to come to terms with the challenges to Christian belief associated with Darwin. According to Ramm, Barth provided the best way forward in the face of Darwin's "across the board challenge to the traditional Christian theology of creation" (Ramm, 1983, 40). In Ramm's opinion, Barth provided a more promising way forward than the approaches adopted by the large number of American evangelicals enamored with various forms of creationism. At the same time Barth also helped to avoid the "concessionism" found among theologians who had departed from the tenets of a more conservative form of Protestantism.

Around the same time as Ramm, the analytical philosopher Alvin Plantinga (b. 1932) found in Barth's rejection of natural theology confirmation of a much wider antipathy toward the enterprise in Reformed theology. Building on this antipathy, Plantinga (1983) argued that Christians are within their epistemic rights to hold that certain theological claims are "properly basic" and do not require the support of external evidence or additional arguments. Applying this "reformed epistemology," Plantinga (1993) later argued that a belief in evolution is incompatible with a belief in naturalism. Starting with "Darwin's doubt" about the cognitive reliability of an evolved mind, Plantinga has argued that the belief that evolution is true is itself in conflict with the belief that the truth-tracking capacity of human cognition evolved by purely natural processes. If, however, we start with the belief that humans were created in God's image, then we have grounds to suppose that our cognitive faculty is reliable. Evolution – at least as orchestrated in some way by God – can then be defended as a firm conclusion of reliable minds. Needless to say, this argument has attracted considerable critical attention (e.g., Beilby 2002).

Other Protestant thinkers concerned with coming to terms with Darwin's legacy resisted Barth's emphasis on the primacy of revelation. The American Baptist theologian Langdon Gilkey (1919–2004) provides one notable example (Fig. 58.4). Perhaps best known for his role in the 1981 Arkansas creation trial, Gilkey addressed at some length the problem of doing theology after Darwin. As early as the 1960s, Gilkey diagnosed a failure to engage positively with the natural sciences in Barth and in the neo-orthodox theology he had learned from Reinhold Niebuhr (1892–1971). This lack of engagement, in Gilkey's (1970, 18) view, was not sustainable: Darwin not only had finally "removed religious truth from the area of matters of fact" but had also called into question direct appeals to divine revelation. Much of Gilkey's later work attempted to respond to this radical challenge by excavating the "traces of ultimacy" (62) latent in the very scientific culture that had apparently erased all forms of religious meaning. Arguing that the mythical was ineradicable even in a post-Darwinian world, Gilkey urged theologians to engage closely with the "actuality of scientific inquiry" (135). By so doing, a place for religious belief, albeit in a much chastened form, could be and should be retained. This argument, as Gilkey himself acknowledged, owed much to Barth's friend and theological protagonist Paul Tillich (1886–1965).

Wolfhart Pannenberg (b. 1928) provides a final example of a Lutheran theologian who, in a way different from Gilkey's, has criticized Barth's approach and engaged more directly with the natural sciences in general and Darwin in particular (Fig. 58.5). In dealing with the scientific developments associated with Darwin, Pannenberg (1994, 122) has turned to the concept of continuous creation characterized as a "more precise definition of creation out of nothing." Creation, in Pannenberg's judgment, should not be regarded as a one-off event occurring at some point in the distant past but as the maintaining in being of the whole universe – past, present, and future. Pannenberg has also argued that the emergence of novelty in the living world points to God's creative activity, a claim couched in a larger argument about the relationship between time and eternity that emphasizes the importance of God's action "from the future" (146). In this way Pannenberg has tried to come to terms with, and co-opt for theological ends, Darwin's emphasis on the role of contingency in the evolution of life.

FIGURE 58.4. Langdon Gilkey (1919–2004), doughty fighter against creationism, argued for a "neo-orthodox" position: science and religion properly understood cannot conflict, because they speak to different issues. Permission: University of Chicago News Office

FIGURE 58.5. Wolfhart Pannenberg (b. 1928) advocates a "theology of nature" rather than a "natural theology," believing that nature illuminates our understanding of God rather than proves his existence. From the German Federal Archive

CONCLUSION

It goes without saying that the four figures outlined here were not the only twentieth-century theologians to find in Darwin resources to restate or reformulate beliefs or approaches considered central to Protestant theology. In the 1980s both Gilkey and Pannenberg joined a larger community of scholars who regarded, and continue to regard, a dialogue between evolution and religion as an urgent challenge. The Lutheran theologians Philip Hefner and Ted Peters, the United Church of Christ pastor Robert J. Russell, and the Presbyterian minister and philosopher Holmes Rolston III have been prominent among the Protestant participants. These theologians, along with Ian Barbour, Philip Clayton, Nancey Murphy (one of the few women participants), Arthur Peacocke, John Polkinghorne, and Jeffrey Schloss, have been among the most important Anglophone Protestant voices in a revivified engagement with the theological implications of living in a post-Darwinian age.

By and large, this loosely constituted group has had a liberal, Anglophone complexion. Other voices involved in ongoing Protestant engagement with Darwin can readily be found. The work of the BioLogos Foundation founded by the geneticist Francis Collins supports theological rapprochements with evolutionary science that have a moderately evangelical hue. The process theology formulated by John B. Cobb and David Ray Griffin, among others, has presented a God who develops with an evolving nature and thus departs more markedly from prevailing definitions of Protestant orthodoxy. It goes without saying that other prominent Protestant theologians have adopted a more reactionary position often connected with a defense of some version of creationism (see Numbers, Essay 59 in this volume).

Evidence of the increasingly global nature of Protestant engagement with Darwin's legacy can also be identified. The increasing reach of creationism can be mentioned in this respect. More productive engagements with Darwin by non-Western Protestant theologians are also noteworthy. The Korean American theologian Paul Chung's (2009) attempt to bring Darwin, Christianity, and Buddhism into conversation provides one recent example. The explosive growth of Pentecostalism is another globalizing movement that future theological engagements with Darwin may well take more seriously (see J. K. A. Smith and Yong 2010). But whatever form such encounters take we can be sure the relations between Protestant theology and Darwin will continue to defy easy summary and will present to the future historian a complex morass of theological reflection, political commitment, and cultural circumstance.

Essay 59

Creationism

Ronald L. Numbers

Charles Darwin's primary goal in writing *On the Origin of Species by Means of Natural Selection* (1859) was to discredit what came to be known as creationism. Twelve years after publishing this book he explained that he had "had two distinct objects in view": "firstly, to show that species had not been separately created, and secondly, that natural selection had been the chief agent of change." Admitting that he may have exaggerated the power of natural selection, he took comfort in having at least "done good service in aiding to overthrow the dogma of separate creations" (Darwin 1871a, 1:146–47). Indeed, his primary scientific accomplishment was convincing his fellow naturalists that evolution was a fact of nature – and doing so within about fifteen years.

Despite believing that attributing the structure of animals to "the *will* of the Deity" was "utterly useless" scientifically (H. E. Gruber 1974, 417–18), Darwin did not himself entirely shun appeals to the Creator. Near the end of the *Origin* he wrote:

> I believe that animals have descended from at most only four or five progenitors, and plants from an equal or lesser number.
>
> Analogy would lead me one step further, namely, to the belief that all animals and plants have descended from some one prototype. But analogy may be a deceitful guide.... Therefore I should infer from analogy that probably all the organic beings which have ever lived on this earth have descended from some one primordial form, into which life was first breathed. (Darwin 1859, 483–84).

In the second edition of the *Origin*, he modified the last phrase to read "was first breathed by the Creator" – and inserted another credit to "the Creator" in the last sentence of the book (Darwin 1860a, 484, 490). He quickly lamented this decision. As he later explained to a friend, he had "long regretted that I truckled to public opinion, and used the Pentateuchal term of creation, by which I really meant 'appeared' by some wholly unknown process. It is mere rubbish, thinking at present of the origin of life; one might as well think of the origin of matter" (F. Darwin 1985–, 11:278, letter to Hooker, 29 March 1863). However, despite his expressed regret, in subsequent editions he never deleted the final reference to the Creator (Darwin 1876, 429).

FIGURE 59.1. Buckland's cave. In 1823 William Buckland (1784–1856), professor of geology at the University of Oxford, published what he believed was definitive evidence of Noah's Flood, namely a cave in Yorkshire with the remains of extinct organisms, supposedly showing that they had been rushed into place and drowned on the spot. No one was convinced, and by the time Darwin went to university in 1828, his mentors like Adam Sedgwick denied the universality of the Flood, thinking it at most a local, Mideast event. From W. Buckland, *Reliquiae Diluvianae* (London: John Murray, 1823)

CREATION AND CREATIONISM

Until well into the twentieth century, critics of evolution tended to identify themselves as antievolutionists rather than creationists. Three factors help to explain this practice. First, the word already possessed a well-known meaning unrelated to the creation-evolution debate. Because early Christianity theologians had attached "creationism" to the doctrine that God had specially created each human soul – as opposed to the traducianist teaching that God had created only Adam's soul and that children inherited their souls from their parents. Second, even the most prominent scientific opponents of organic evolution differed widely in their views of origins. Some adopted the biblical view that all organisms had descended from the "kinds" divinely created in the Garden of Eden and preserved on Noah's ark (Fig. 59.1). Others, such as the British geologist Charles Lyell (1797–1875), advocated the spontaneous but nonsupernatural appearance of species in regional "*centres* or *foci* of creation." Still others followed the leading American antievolutionist, the Harvard zoologist Louis Agassiz (1807–73), in arguing for repeated plenary creations, during which "species did not originate in single pairs, but were created in large numbers" (Numbers 1998, ch. 2; cf. Rupke 2005). Third, even Bible-believing fundamentalists could not agree on the correct interpretation of the first chapter of Genesis. A majority probably adopted the ruin-and-restoration view endorsed by the immensely popular *Scofield Reference Bible* (1909), which identified two creations (the first "in the beginning," the second associated with the Garden of Eden) and slipped the fossil record into the vast "gap" between the two events. Another popular reading of Genesis 1, advocated by William Jennings Bryan (1860–1925), the leading antievolutionist of the time, held that the "days" mentioned in Genesis 1 represented immense ages, each corresponding to a section of the geological column or perhaps to a period in the history of the cosmos. Only a handful of those writing against evolution insisted on what later came to be known as young-earth creationism but was then called

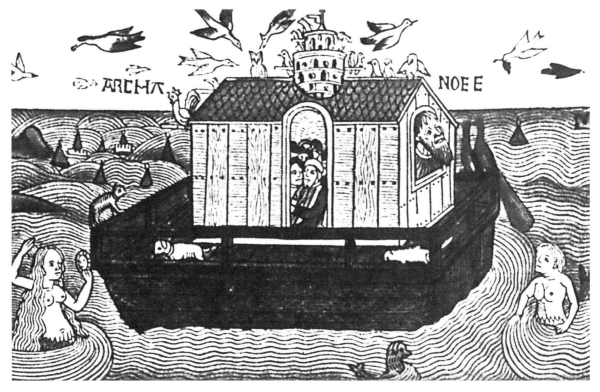

FIGURE 59.2. Why are creationists so obsessed with Noah's ark and the Flood rather than other Genesis events? Because attributing the fossil record to the year of the Flood seemed to be the best alternative to accepting the long history of life suggested by the geological column. A woodcut from the 1483 bible published by Anton Koberger in Nüremberg.

"flood geology": a recent special creation of all "kinds" in six twenty-four-hour periods and a geologically significant flood at the time of Noah that buried most of the fossils (Numbers 1998, 52–53) (Fig. 59.2).

Flood geology was the brainchild of the scientifically self-educated George McCready Price (1870–1963) (Fig. 59.3). A Canadian by birth, Price had converted to Seventh-day Adventism as a youth and had accepted the writings of the Adventist prophetess, Ellen G. White (1827–1915), as divinely inspired. Throughout her life White had experienced religious dreams and trancelike visions, which she and her followers believed to be divine. During one episode she claimed to have been "carried back to the creation and ... shown that the first week, in which God performed the work of creation in six days and rested on the seventh day, was just like every other week." She also endorsed a six-thousand-year-old earth and a worldwide catastrophe at the time of Noah that had buried the fossils and reshaped the earth's surface (Numbers 2006, 90; 2008). There was nothing novel about White's history, except its timing. By the middle of the nineteenth century, when she began writing, almost all evangelical expositors on Genesis and geology had conceded the antiquity of life on earth and the geological insignificance of Noah's flood (Stiling 1991).

As a young man full of religious zeal, Price dedicated himself to providing a scientific defense of White's outline of earth history. Although he could scarcely tell one rock from another, he read the scientific literature voraciously – and critically. Early on it struck him that the argument for evolution "*all turned on its view of geology*," which provided the strongest evidence for both the antiquity of life and its progressive development. The more he read, the more he became convinced that the vaunted geological evidence for evolution was "a most gigantic hoax." Guided by Mrs. White's "revealing word pictures of the Edenic beginning of the world, of the fall and the world apostasy, and of the flood," he concluded that "the actual facts of the rocks and fossils, *stripped of mere theories*, splendidly refute this evolutionary theory of the invariable order of the fossils, *which is the very backbone of the evolution doctrine*" (Numbers 2006, 91–92). In 1906 Price published a booklet titled *Illogical Geology: The Weakest Point in the Evolution Theory*, in which he offered a $1,000 reward "to any who will, in the face of the facts here presented, show me how to prove that one kind of fossil is older than another" (Price 1906, 9–11).

Before his death in 1963, he would author some two dozen books, the most systematic and comprehensive being *The New Geology* (1923). In it, he restated his "great" law of conformable stratigraphic sequence, which he modestly described as "by all odds the most important law ever formulated with reference to the order in which the strata occur." According to this law, "Any kind of fossiliferous beds whatever, 'young' or 'old,' may be found occurring conformably on any other fossiliferous

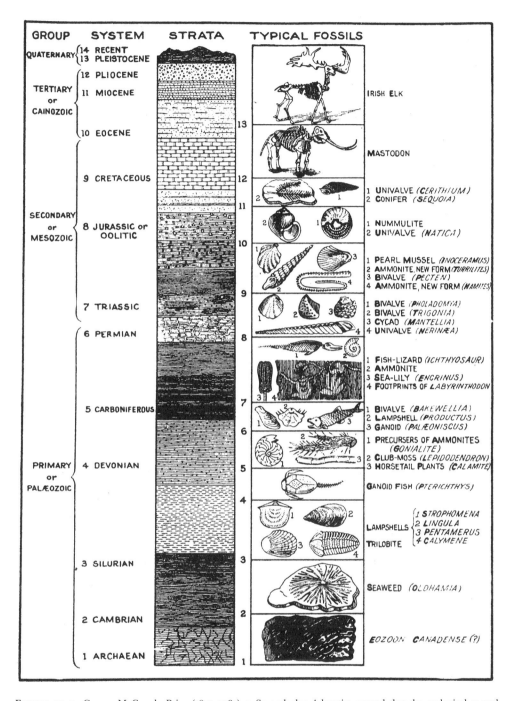

FIGURE 59.3. George McCready Price (1870–1963), a Seventh-day Adventist, argued that the geological record is an artifact of the rising flood waters rather than evidence of evolution over a long time period. Compare this to Richard Owen's paleontological column of 1861 (see Fig. 9.4), something that by then Owen himself would have been giving an evolutionary interpretation. From G. M. Price, *Predicament of Evolution* (Nashville: Southern Publishing Association, 1925)

beds, 'older' or 'younger'" (Price 1923, 637–38). To Price, so-called deceptive conformatives (where strata seem to be missing) and thrust faults (where the strata are apparently in the wrong order) proved that there was no natural order to the fossil-bearing rocks, all of which he attributed to Noah's Flood. Throughout his life, Price saved his sharpest barbs for so-called uniformitarian geology, because, in his opinion, "the modern theory of evolution is about 95% due to the geology of Lyell and only about 5% to the biology of Darwin" (Numbers 1995, x). Despite repeated attacks from the scientific establishment, Price's influence among non-Adventist fundamentalists grew rapidly. By the mid-1920s the editor of *Science* could accurately describe him as "the principal scientific authority of the Fundamentalists" (Anonymous 1926), and Price's byline was appearing with increasing frequency in a broad spectrum of religious periodicals. Nevertheless, few fundamentalist leaders, despite their appreciation for Price's critique of evolution and defense of a biblical flood, gave up their allegiance to the "gap" and "day-age" theories for his flood geology.

ORGANIZED CREATIONISM

As the American antievolution movement petered out in the late 1920s, a few diehards tried to keep the protest alive by organizing a new society. Their efforts, however, immediately ran into two obstacles: a paucity of trained scientists and the continuing disagreement over the meaning of Genesis 1. Price had never finished college or even taken an advanced course in science. Other antievolution activists with some exposure to science were Harry Rimmer (1890–1952), a Presbyterian evangelist and self-described "research scientist" who had briefly attended a homeopathic medical school; Arthur I. Brown (1875–1947), a Canadian surgeon whose handbills described him as "one of the best informed scientists on the American continent"; S. James Bole (1875–1956), a professor of biology at Wheaton College, who had earned a master's degree in education and would in 1934 receive a PhD in horticulture from Iowa State College; and Bole's colleague on the Wheaton faculty, L. Allen Higley (1871–1955), a chemist (Numbers 2006, ch. 4).

In 1935 Price, Rimmer, and Higley joined with a few others to create "a united front against the theory of evolution." The resulting society, the Religion and Science Association, quickly dissolved, however, when the members fell to squabbling about the age of the earth. As one frustrated antievolutionist observed in the 1930s, fundamentalists were "all mixed up between geological ages, Flood geology and ruin, believing all at once, endorsing all at once." How, he wondered, could evangelical Christians possibly turn the world against evolution if they themselves could not even agree on the meaning of Genesis 1 (Numbers 2006, ch. 6)?

A few years after the demise of the Religion and Science Association, Price and a small number of mostly Adventist colleagues in Southern California, where he had retired, organized a Deluge Geology Society, which for several years in the early 1940s published a *Bulletin of Deluge Geology and Related Science*. The group consisted of "a very eminent set of men," bragged Price. "In no other part of this round globe could anything like the number of scientifically educated believers in Creation and opponents of evolution be assembled, as here in Southern California." By far the best-trained scientist in the society was a Missouri Synod Lutheran, Walter E. Lammerts (1904–96), who had earned a PhD degree in genetics at the University of California, Berkeley, and was teaching horticulture at its southern branch in Los Angeles. The society's most exciting moment came in the early 1940s, when it announced the discovery of giant fossil footprints, believed to be human, in geologically ancient rocks. This find, one member predicted, would demolish the theory of evolution "at a single stroke" and "*astound the scientific world*!" But even this group of flood geologists, who all agreed on the recent appearance of life on earth, divided bitterly over the issue of "*pre-Genesis time for the earth*," that is, whether the inorganic matter of the earth antedated the Edenic creation. About 1947, the society died (Numbers 2006, ch. 7).

By this time a more ecumenical society of evangelical scientists had appeared on the scene: the American Scientific Affiliation (ASA). Created in 1941 by associates of the Moody Bible Institute, the association at first took a dim view of evolution. By the end of the decade, however, the presence of well-trained young scientists who embraced theistic evolution (or its intellectual sibling "progressive creationism") was dividing the association. The most influential of the insurgents were J. Laurence Kulp (1921–2006) and Russell L Mixter (1906–2007). Kulp, a Wheaton alumnus who had earned a doctorate in physical chemistry from Princeton University and then completed the course work for a second PhD in geology, had established himself at Columbia University as an early authority on radioisotope dating. As one of the first evangelicals with advanced training in geology, he spoke with unique authority. Worried that Price's Flood geology had "infiltrated the greater portion of fundamental Christianity in American primarily due to the absence of trained Christian geologists," he set about exposing its abundant scientific flaws. In an influential paper first read to ASA members in 1949, he concluded that the "major propositions of the theory are contraindicated by established physical and chemical laws." Mixter, meanwhile, was pushing for greater acceptance of the evidence for limited organic evolution. While teaching biology at Wheaton College, he earned a doctorate in anatomy from the University of Illinois School of Medicine in Chicago in 1939. Before long he was nudging creationists to accept evolution "within the order" and assuring them that they could "believe in the origin of species at different times, separated by millions of years, and in places continents apart" (Numbers 2006, ch. 9).

THE CREATIONIST REVIVAL

In 1954 Bernard Ramm (1916–92), a theologian-philosopher associated with the leadership of the ASA, brought out a book audaciously called *The Christian View of Science and Scripture*. Damning "hyperorthodox" Christians for their "narrow bibliolatry" and "ignoble" attitude toward science, this avatar of neo-evangelicalism urged Christians to quit getting their science from Genesis and adopt the progressive creationism so popular within the ASA. He dedicated his book to one of the founders of the ASA and thanked Kulp for vetting the book for "technical accuracy." Ramm aimed his harshest rhetoric at the Flood geology of Price, whose growing influence among fundamentalists he regarded as "one of the strangest developments of the early part of the twentieth century." Despite Price's manifest ignorance, his brand of creationism had come, at least in Ramm's imagination, to form "the backbone of much of Fundamentalist thought about geology, creation, and the flood" (Ramm 1954; Numbers 2006, 208–11).

Many evangelicals, including Billy Graham (b. 1918), hailed Ramm's book, but fundamentalists tended to respond angrily to what they regarded as an arrogant and heterodox attempt to equate progressive creationism with *the* Christian view. Ramm's attack provoked one young fundamentalist, John C. Whitcomb Jr. (b. 1924), a Princeton-educated Old Testament scholar teaching (and working on a doctorate) at the fundamentalist Grace Theological Seminary, into turning his dissertation into a spirited response to Ramm and a defense of "the position of George M. Price." When Whitcomb approached the Moody Press about publishing his study, the editor recommended that the biblical scholar recruit a trained scientist as coauthor. He eventually found an acceptable, if not perfect, partner: Henry M. Morris (1918–2006), a fundamentalist Baptist who had earned a PhD in hydraulics from the University of Minnesota and had just taken over as head of the large civil-engineering program at Virginia Polytechnic Institute (Numbers 2006, 212–17) (Plate XLVII).

As defenders of Price's Flood geology, Whitcomb and Morris faced the difficult – perhaps impossible – task of not being dismissed as "crackpots" for trying to promulgate his theory. Early on, Morris suggested to Whitcomb that it might be best "simply to point out Price's arguments as a matter of historical record, and then leave your main emphasis on the Scriptural framework and the geological implications thereof." Later, as he and Morris neared the end of their project, Whitcomb shared his own concerns about being identified with the disreputable Price and his strange church (Numbers 2006, 215, 223–24):

> I am becoming more and more persuaded that my chapter on "Flood Geology in the Twentieth Century" will hinder rather than help our book, at least in its present form. Here is what I mean. For many people, our position would be somewhat discredited by the fact that "Price and Seventh-Day Adventism" (the title of one of the sections in that chapter) play such a prominent role in its support. My suggestion would be to supply for the book a fairly complete *annotated bibliography* of twentieth-century works advocating Flood-geology, without so much as a mention of the denominational affiliation of the various authors. After all, what *real* difference does the denominational aspect make?

In the end the authors camouflaged their intellectual debt to Price by deleting all but a few incidental references to him and all mention of his Adventist connections. The authors virtually ignored Darwin. Their primary goal was to wean Bible-believing Christians from the compromising day-age and gap theories, not to convert evolutionists to creationism.

Although one critic accurately described *The Genesis Flood* as "a reissue of G. M. Price's views brought up to date," it created a sensation within the evangelical community. Two years after its appearance a small group of Christian scientists energized by Whitcomb and Morris's stand – and increasingly annoyed by the ASA's drift toward evolution – walked out of the ASA and founded their own hyperorthodox society, the Creation Research Society (CRS). Leading this effort, both administratively and financially, was the Lutheran geneticist Lammerts, who until this time had maintained a low creationist profile. The initial eighteen-man CRS steering committee imprecisely reflected the theological composition of the emerging young-earth creationism movement: six Missouri Synod Lutherans, six Baptists (four Southern, one Regular, and one independent), two Seventh-day Adventists, and one each from the Reformed Presbyterian Church, the Christian Reformed Church, the Methodist Church, and the Church of the Brethren. The committee included five biologists with PhDs earned at major universities, two more biologists with master's degrees, and one biochemist with a doctorate from the University of California. There were no physicians in the group and only one engineer, Morris. Twelve of the eighteen lived in the Midwest, four in the Southwest, one in California, and one in Virginia (Numbers 2006, 239–59).

Despite a common commitment to young-earth creationism, disagreements soon arose. One of the most significant was over the issue of speciation. As biologists discovered more and more species, it became clear to creationists that Noah's ark could not have accommodated representatives of each one. Thus many of them adopted the solution of a former student of Price's, Frank Lewis Marsh (1899–1992), who argued that the Genesis "kinds" should not be equated with species but with families or what he called *baramins*. This solved the problem of space on the ark but created another one: how had the kinds preserved on the ark produced so many genera and species, and in only forty-three hundred years. It seemed likely, for example, that the *Canidae* family – including domestic and wild dogs, wolves, foxes, coyotes, jackals, and dingoes – had descended from a single kind. Morris and most of his colleagues embraced rapid "microevolution." However, as a geneticist, Lammerts knew that that was scientifically impossible, that there must have been a second creation to repopulate the earth after the deluge. Unfortunately for him, the Bible never mentioned such an event; so his supernatural solution never caught on (Numbers 2004).

For a young-earth creationist organization, the CRS grew rapidly. On the occasion of its tenth anniversary it boasted a membership of 1,999, with 412 of them holding advanced degrees in science (Numbers 2006, 259). By this time, society leaders were switching from "flood geology" as the name of choice for their model of earth history and substituting the labels "creation science" and "scientific creationism" (268–70). In truth, there was little difference between the old and the new, except that scientific creationism made no mention of biblical events and persons, such as the Garden of Eden, Adam and Eve, and Noah's flood. However, the focus on the flood remained the same. H. M. Morris made this clear in a book titled *Scientific Creationism* (1974, 252):

> The Genesis Flood is the real crux of the conflict between the evolutionist and creationist cosmologies. If the system of flood geology can be established on a sound scientific basis, and be effectively promoted and publicized, then the entire evolutionary cosmology, at least in its present neo-Darwinian form, will collapse.

> This, in turn, would mean that every anti-Christian system and movement (communism, racism, humanism, libertinism, behaviorism, and all the rest) would be deprived of their pseudo-intellectual foundation.

Driving the switch in labels was a desire to have a product acceptable for use in public schools, especially in California, which was revising its guidelines for teaching science. Tellingly, *Scientific Creationism* appeared in two almost identical versions: one for public schools, stripped of all references to the Bible, and another for church schools, which retained biblical references and added a chapter on "Creation according to Scripture."

In 1968 the U.S. Supreme Court struck down as unconstitutional the last of the laws from the 1920s outlawing the teaching of evolution. This forced creationists to abandon any thought of making the teaching of evolution illegal and turn their attention to writing legislation that would allow the teaching of "creation science" alongside that of "evolution science." The creationists sought scientific status for their views in order to circumvent the constitutional separation of church and state, which had implications for the teaching of religion in schools. The Bill of Rights in the U.S. Constitution forbade Congress from passing any "laws respecting an establishment of religion, or prohibiting the free exercise thereof." Before World War II, the Supreme Court had interpreted this narrowly in its literal sense; in the late 1940s, however, it held that the Constitution had erected "a wall of separation" between church and state. At a time when public-opinion polls were revealing that "half of the adults in the U.S. believe God created Adam and Eve to start the human race," the movement for "balanced treatment" enjoyed a large reservoir of popular support (Larson 2003). In the end only two states, Arkansas and Louisiana, adopted the two-model approach. In 1982 a federal judge in Arkansas, having been tutored by the philosopher Michael Ruse (b. 1940) on the demarcation criteria that allegedly distinguished science from nonscience, declared the Arkansas law to be infringement of the constitutional requirement to keep church and state separate; three years later a court in Louisiana reached a similar decision (Ruse 1988). The U.S. Supreme Court ratified these judgments in 1987, while allowing, in the words of one justice, that "teaching a variety of scientific theories about the origins of humankind to schoolchildren might be validly done with the clear secular intent of enhancing the effectiveness of science instruction" (Larson 2003, 180)

INTELLIGENT DESIGN

The Supreme Court's decision dashed the hopes of creation scientists who had expected their stripped-down version of creationism to pass constitutional muster, but it did little to dampen the widespread antipathy toward evolution in America. Few found the decision more disappointing than two creationist authors, Dean H. Kenyon and Percival Davis, who had drafted a manuscript tentatively titled *Biology and Creation* in anticipation of the demand for a high-school textbook when the court ruled for creationism. Their optimistic publisher calculated a financial bonanza of "over 6.5 million in five years." When the court virtually wiped out the market for creationist texts, Kenyon and Davis quickly sanitized their manuscript by substituting *Of Pandas and People* for the original title and replacing the words "creation" and "creationists" with the euphemisms "intelligent design" and "design proponents." As they defined it, intelligent design (ID) provided a frame of reference that "locates the origin of new organisms in an immaterial cause: in a blue-print, a plan, a pattern, devised by an intelligent agent" (Numbers 2006, 375–76).

Of Pandas and People may have begun as a conventional creationist work, but it put into play a new slogan in the ongoing campaign against evolution: intelligent design. The intelligent-design *movement* began in the early 1990s with the publication of an antievolution tract, *Darwin on Trial* (1991) by a Berkeley law professor, Phillip E. Johnson (b. 1940) (Plate XLVIII). Unlike Whitcomb and Morris, who had pretty much left Darwin alone, Johnson zeroed in on the British naturalist and – especially – his followers. Upset by the anti-Christian stridency of Darwinists such as Richard Dawkins, the Presbyterian layman set out to expose what he saw as the logical weaknesses of the case for evolution, particularly the assumption made by its advocates that naturalism is the only legitimate way of doing science. Ever since investigators of nature in the early nineteenth century had shifted from natural philosophy (which allowed for appeals to the supernatural) to science (which did not), practitioners, regardless of religious persuasion, had refrained from invoking divine or diabolical forces when explaining the workings of nature. In short order, explaining nature naturally became the defining characteristic of science, for Christians as well as for atheists. In contrast to metaphysical naturalism, which denied the existence of a transcendent God, this methodological naturalism supposedly implied nothing about God's existence (Numbers 2003). Johnson vehemently disagreed. Professing to see little difference between methodological naturalism and scientific materialism, he set out to resacralize science or, as one admirer put it, "to reclaim science in the name of God" (Vardiman 1997). If the evidence warranted a supernatural explanation, Johnson argued, then invoking intelligent design should count as a legitimate scientific response (Numbers 2006, 380).

Johnson aspired to pitch a tent big enough to accommodate all antievolutionists who were willing to set Genesis aside (at least temporarily) and focus on the purported scientific evidence against evolution. Although a few young-earth creationists sought shelter in the tent, Morris and other Bible-based creationists resented the effort of the intelligent designers to marginalize their views and to avoid "having to confront the Genesis record of a young earth and global flood" (Numbers 2006, 377–78). In the mid-1990s the founder of the right-of-center Discovery Institute in Seattle invited ID theorists to establish an institutional home within the institute called the Center for the Renewal of Science and Culture. With a year or so, they had raised "nearly a million dollars in grants."

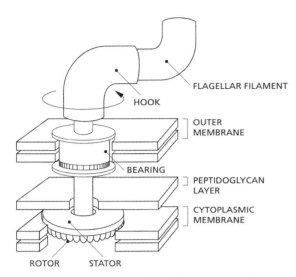

FIGURE 59.4. The motor of the bacterial flagellum, a tail-like appendage from certain single-celled organisms, used for locomotion, and – as argued by Michael Behe in his *Darwin's Black Box* (New York: Free Press, 1996) – supposedly an example of irreducible complexity and hence inexplicable by natural Darwinian processes

The most generous donor was Howard Fieldstead Ahmanson Jr. (b.1950), heir to a fortune made in the savings-and-loan business. A devotee of Rousas J. Rushdoony, the leader of the theocratic Christian Reconstructionists, Ahmanson, like his mentor, sought "the total integration of biblical law into our lives" (Numbers 2006, 382; cf. Worthen 2008).

By this time several younger men had joined Johnson as the public face of the movement, among them Michael J. Behe (b. 1952), a Catholic biochemist at Lehigh University. In 1996 the Free Press of New York released Behe's *Darwin's Black Box: The Biochemical Challenge to Evolution*, the first antievolution book in seven decades published by a mainstream publisher (O'Toole 1925). In his book, Behe argued that biochemistry had "pushed Darwin's theory to the limit ... by opening the ultimate black box, the cell, thereby making possible our understanding of how life works." The astonishing complexity of subcellular organic structure – its "irreducible complexity" – led him to conclude that intelligent design had been at work. "The result is so unambiguous and so significant that it must be ranked as one of the greatest achievements in the history of science," he concluded grandiosely. "The discovery of [intelligent design] rivals those of Newton and Einstein, Lavoisier and Schroedinger, Pasteur and Darwin" (Behe 1996, 15, 193, 232–33). The tip of the hat to Darwin was no slip. In contrast to most of his colleagues in the movement, Behe did not rule out the possibility of divinely guided evolution (Fig. 59.4).

More typical of attitudes toward theistic evolution within the ID camp was that of another rising star, the mathematician-philosopher William A. Dembski (b. 1960). "*Design theorists are no friends of theistic evolution*," he declared.

> As far as design theorists are concerned, theistic evolution is American evangelicalism's ill-conceived accommodation to Darwinism. What theistic evolution does is take the Darwinian picture of the biological world and baptize it, identifying this picture with the way God created life. When boiled down to its scientific content, theistic evolution is no different from atheistic evolution. (Dembski 1995, 3, 5)

On the origin of organic forms, his position did not vary much from that of the scientific creationists. While acknowledging that organisms had "undergone some change in the course of natural history," he believed that such changes had "occurred within strict limits and that human beings were specially created" (5). As an expert in probability theory, Dembski focused on the unlikelihood of organisms arising by accident, and especially on a method for detecting intelligence, his much-maligned "explanatory filter." Like Johnson, Dembski attacked evolution as part of a much larger strategy to revolutionize the way science was practiced. "*The ground rules of science have to be changed*," he declared quixotically. "We need to realize that methodological naturalism is the functional equivalent of a full blown metaphysical naturalism" (7–8).

Intelligent design emerged as front-page news in 2005, after a group of parents in the Dover, Pennsylvania, filed suit against the school board for promoting ID in ninth-grade biology classes. The religiously conservative board had instructed teachers to tell their students about the weaknesses in Darwin's theory and direct them to *Of Pandas and People*. The case, like the creation-science trials of the 1980s, hinged on whether the recommendation of ID theory constituted the teaching of religion and therefore violated the U.S. Constitution. Behe appeared as the star witness for the defense but scarcely helped his side when he lamely, but honestly, conceded that ID "does not propose a mechanism in the sense of a step by step description of how these structures arose." In the end, the judge condemned the school board for its actions – memorably declaring it a "breathtaking inanity" – and ruled that ID was "not science" because it invoked "supernatural causation" and failed "to meet the essential ground rules that limit science to testable, natural explanations." A conservative Christian himself, the judge rejected as "utterly false" the assumption "that evolutionary theory is antithetical to a belief in the existence of a supreme being and to religion in general" (Numbers 2006, 391–94).

INTO ALL THE WORLD

As late as 2000, the American paleontologist Stephen Jay Gould (1941–2002) confidently assured non-Americans that they had nothing to fear from American-style creationism. "As insidious as it may seem, at least it's not a worldwide movement," he said. "I hope everyone realizes the extent to which this is a local, indigenous, American bizarrity" (Numbers 2006, 399). Gould, a great scientist, proved to be a false prophet. Even as he spoke, creationism was becoming a truly global phenomenon, successfully overcoming its "Made in America" label and flourishing not only among conservative Protestants but also among pockets of Catholics, Eastern Orthodox believers, Muslims, and Jews. Conservative Protestants, however, continued to lead. And no one in the twenty-first century occupied a bigger leadership role than the charismatic former high-school biology teacher from Australia Kenneth A. Ham (b. 1951), an associate of Morris at the Institute for Creation Research. Seven years after moving to the United States in 1987, he launched his own creationist ministry, Answers in Genesis (AiG), headquartered in northern Kentucky, just south of Cincinnati. Within a decade, AiG had emerged as the most dynamic creationist organization worldwide, with Ham alone speaking to more than 100,000 people a year. In 2007, to great fanfare, AiG opened an impressive $27 million Creation Museum, which attracts hundreds of thousands of visitors annually (Numbers 2006, ch. 18).

Over a century after the scientific community had embraced organic evolution, many laypersons continued to scorn Darwin's notion of common descent. In the United States, where polls since the early 1980s have shown a steady 44–47 percent of Americans subscribing to the statement that "God created human being pretty much in their present form at one time within the last 10,000 years or so," nearly two-thirds (65.5 percent), including 63 percent of college graduates, according to a 2005 Gallup poll, regarded "creationism" as definitely or probably true (J. M. Jones 2005; D. W. Moore 2005).

ACKNOWLEDGMENTS

This essay is adapted from Ronald L. Numbers, "Scientific Creationism and Intelligent Design," in *The Cambridge Companion to Science and Religion*, ed. Peter Harrison (Cambridge: Cambridge University Press, 2010), 127–47.

Essay 60

Darwin and Catholicism

John F. Haught

From the last third of the nineteenth century until the middle of the twentieth, official Catholicism, like other sectors of Christianity, had expressed considerable hostility to "Darwinism." Early Catholic resistance to evolution usually followed from the impression that "Darwinism" is inseparable from "naturalism," "materialism," "rationalism, "socialism," and other creeds taken to be atheistic. From the time of Pope Pius IX (1792–1878) and his promulgation of "The Syllabus of Errors" (1864) until the mid-twentieth century, conservative church officials usually suspected that evolution is especially allied with materialism, the belief that mindless matter is ultimately "all there is" and that therefore God does not exist (Fig. 60.1).

Such a suspicion was not entirely without foundation. During Pius IX's papacy (1846–78) both Karl Marx and Ernst Haeckel had interpreted Darwin as supporting their own distinct versions of materialism. And in his diaries even Darwin revealed at times his own temptations to materialism. Furthermore, philosophical materialists well into the twenty-first century enthusiastically embraced evolution not only for scientific but also for philosophical reasons. By that time, "postmodern" criticism had called attention to the ideological bias that often accompanies putatively objective discourse, but several highly celebrated biologists and scientific thinkers (e.g., E. O. Wilson, Richard Dawkins, and Daniel Dennett) continued to stitch their evolutionary ideas tightly into a materialist belief system, thus making evolution, at least as they interpreted it, religiously indigestible on any terms. Hence, it is not surprising that scientifically unsophisticated popes and theologians in the late nineteenth and early twentieth century, unable to distinguish clearly between science and materialist beliefs, were often appalled by Darwin's evolutionary theory.

Nevertheless, from the very start a few Catholic scientists and theologians accepted Darwinian evolution as doctrinally harmless (Fig. 60.2). The most famous of these was the biologist St. George Jackson Mivart (1827–1900), a convert to Catholicism and a close acquaintance of Darwin. Mivart initially expressed enthusiasm for Darwin's theory, but he became convinced later on that natural selection could not be a sufficient explanation of life's diversity, thus bringing an end to his friendship with Darwin. Before 1950 several other Catholics, notably John Zahm

FIGURE 60.1. Although he started as a liberal when elected pope in 1846, the troubles of the church caused by the reunification of Italy moved Pius IX (1792–1878), the longest-reigning pope in the history of the papacy, to move decisively away from "modernism." This intensified church opposition to science in general and evolution in particular. Nineteenth-century lithograph

(1851–1921), an American priest and professor at the University of Notre Dame, and later Pierre Teilhard de Chardin (1881–1955), a French Jesuit priest and geologist, defended nonmaterialist versions of evolutionary science, but officials in the church held even these to be doctrinally dangerous. Indeed, until the last half of the twentieth century, there was little enthusiasm among Catholics for Darwinian ideas. Highly visible Catholic opponents of evolution included G. K. Chesterton and Hilaire Belloc (Plate XLIX).

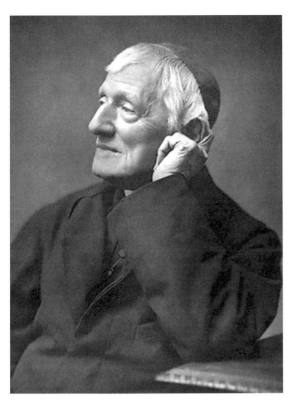

FIGURE 60.2. Showing why not all Roman Catholics were enthused by his conversion to their religion, John Henry Newman (1801–90), the greatest English theologian since the Reformation, was sympathetic to science and endorsed a proposal for Oxford University to give an honorary degree to Charles Darwin (which the now-aged scientist declined on grounds of health). Herbert Rose Barraud, carbon print on card mount, 1887; published 1888

Opposition to Darwin on the part of the Vatican and Catholic bishops was nourished especially by Irish Catholicism, which, because of its widespread influence on Catholics in North America, Australia, and elsewhere, had set the minds of all but a few Catholics against evolution (O'Leary 2009, 13–19). The opposition to Darwin by many scientists during the late nineteenth and early twentieth centuries only reinforced Catholic suspicions.

During the same period, the Vatican's suppression of modernism was inseparable from a general distrust of evolution. Modernism was a broad and ambiguous label for what the Vatican took to be a secularist, relativist, and rationalist modern culture's debunking of sacred traditions of faith and morals. To church officials, as well as to many theologians in the early twentieth century, Darwinian evolution had become central to the modern intellectual hostility to Christianity. Because evolution seemed inextricably part of the modernist agenda, the church vehemently chastised, and at times excommunicated, several renowned Catholic intellectuals who had enthusiastically accepted a dynamic worldview deeply influenced by evolutionary science. One of the Catholic thinkers who incurred the Vatican's censorship, though not excommunication, during this period was Teilhard de Chardin.

WARMING UP TO EVOLUTION

In spite of its initial resistance to Darwinism, however, the Catholic Church never officially condemned evolution. In fact, Catholic catechisms, papal statements, and theological reflection, especially since 1950, have been comparatively hospitable to the theory. In 1950, a watershed moment in the relationship between Catholicism and Darwin's science occurred when Pope Pius XII's encyclical *Humani Generis* finally acknowledged officially that evolution, including that of the human body, is a theory that Catholics may legitimately explore. Pius and subsequent popes, however, have uniformly insisted on the direct creation of the human soul by God in order to protect the church's constant belief, in accordance with the book of Genesis, that human beings are created in the "image and likeness" of God.

Such a qualification notwithstanding, since the mid-twentieth century most Catholic thinkers and educators have become fully reconciled to Darwin's science (although not to Darwinian materialism). Indeed, by the late twentieth century several Catholic thinkers had made evolution central to their theologies of nature (e.g., Edwards, Delio, and Haught). In the mid-1960s, moreover, *Gaudium et spes,* one of the most important documents promulgated by the Second Vatican Council (1962–65), provided unmistakable evidence that official Catholic teaching by that time had begun to assimilate evolutionary themes. Then, in 1996, John Paul II issued an official statement maintaining that the evidence for evolution is strong and that it is "more than a hypothesis."

Several events in modern Catholic history had already eased the way toward the church's explicit settlement with evolutionary biology. The first of these was an instruction by Pope Leo XIII, in his encyclical *Providentissimus Deus* (1893), that Catholics should not look for scientific information in the scriptures. The faithful should not read the Bible with the objective of gaining insights that scientific research can in principle make available on its own. Much earlier, Augustine of Hippo (354–430) had similarly cautioned readers of his *De Genesi ad Litteram* not to take literally the cosmology of Genesis because drawing scientifically precise pictures of nature is not the Bible's concern. If Christian educators were to insist that prospective converts take the cosmology of biblical creation accounts literally, Augustine wrote, this would only lead them to ignore the Bible when it speaks of more important matters. So, by acknowledging officially that the text of Genesis does not provide a strictly scientific account of origins, Pope Leo XIII confirmed explicitly what countless educated Catholics, including Galileo (1546–1642), had already held since antiquity.

Second, and more fundamentally, the deliverance of scripture from having to function as a source of literal scientific truth had already been prepared for in nineteenth-century biblical scholarship (mostly Protestant). The new exegesis, unlike the previously allegorical and plain readings of sacred texts, took into account the diverse historical situations and distinct literary genres that had shaped the various books of the Bible, including Genesis. Generally speaking, however, popes, bishops, and Catholic theologians remained suspicious of modern critical methods of inquiry into the Bible until around 1943. It was then that Pope Pius XII's encyclical *Divino Afflante Spiritu* officially permitted Catholic scholars to take advantage of literary criticism and other scientific approaches to the understanding of ancient biblical documents.

Divino Afflante Spiritu is significant not only for liberating modern Catholic interpretation of scripture but also for facilitating the eventual Catholic reception of evolution. Indeed, the whole issue of the relationship of Christian faith to evolutionary biology is inseparable from modern developments concerning the interpretation of scripture. Literary and historical criticism of the early chapters of Genesis and other creation accounts in the Bible implied that the biblical authors could not possibly have intended to give a historical or scientific account of origins in the modern sense of the terms history and science. Thus, the 1943 encyclical, along with more recent instructions from the Pontifical Biblical Commission, have rescued Catholic theology from the impossible burden of trying to reconcile modern science and historical inquiry with a literalist interpretation of the Bible.

By contrast, even into the early twenty-first century more than half of non-Catholic American Christians, along with some of the more prominent materialist exponents of evolution, still interpreted Genesis literally, thus setting up a logically meaningless contest between faith and science based on the false assumption that, if the Bible is supposed to be "inerrant," then it should at least be an infallible source of scientific truth. This literalist set of assumptions became the intellectual foundation not only of creationism but also of the "new atheism" that sought to strip the world completely of all vestiges of religious faith. Meanwhile, since the middle of the twentieth century Catholic scientists and theologians had become increasingly at peace with the new nonliteralist biblical criticism and, as an indirect result, had embraced evolutionary biology, geology, and paleontology.

A third and even more fundamental factor facilitating the eventual reception of evolution by Roman Catholics goes right to the heart of what, at least originally, distinguished Catholicism from Protestantism. This distinctiveness consists of the Catholic emphasis on the importance of tradition and sacraments, as compared to the Protestant preoccupation with the "Word" of God in scripture. Catholicism, both East and West, characteristically seeks contact with God not only by way of the written Word of Scripture but also through a more broadly construed "deposit of faith" mediated by a long teaching tradition, a hierarchical Church, and a highly accessible system of sacraments. In most varieties of Protestantism, the biblically based Word of God, not ecclesiastical structures, church officials, or tradition, has provided the main access to divine revelation. Understandably, then, the enshrinement of the written text of the Bible, especially by Protestant fundamentalists, often entails a reluctance to tamper in any way with the "plain" sense of scripture. As a result, the "word" of Darwin has often seemed especially irreconcilable with the inerrant Word of God in scripture.

After 1950, Catholic thought, by remaining relatively successful in avoiding biblical literalism, became increasingly comfortable with the view that God creates through or by way of evolutionary processes, although on this matter there is still much room for theological clarification. Catholic officials and theologians also repudiated what has come to be called "scientific creationism." The latter, which comes in several varieties, tries to reconcile contemporary biology, geology, paleontology, and astrophysics with a literal interpretation of biblical accounts of origins. Needless to say, scientific creationism significantly twists and distorts contemporary science in order to achieve this impossible synthesis. Furthermore, scientific creationism, by expecting biblical literature to be scientifically accurate, not only robs science of its own autonomy but also trivializes biblical teachings by placing them in the context of mundane scientific discourse.

In summary, then, Catholic teaching and theology now rejects any interpretations of the Bible that try to turn it into a resource that might potentially provide information that science is quite able to discover on its own. This is a point that Galileo had already made in the early seventeenth century in his "Letter to the Grand Duchess Christina." Catholic teaching has consistently maintained that there can be no contradiction between authentic science and divine revelation, and this principle demands that the propositions of theology never be placed in a competitive relationship with evolutionary claims.

AN UNFINISHED TASK

The acceptance of evolution by Catholics, however, must not be mistaken for a mature appropriation of what Darwin and his scientific followers have discovered about the natural world. Most Catholic thought in the early twenty-first century shares with other Christian denominations the trait of still not being deeply transformed by evolutionary findings, especially those regarding human ancestry, morality, intelligence, and religion. Reacting to the threats of modern rationalism and scientific materialism, Catholic theology until recently has devoted an extraordinary amount of energy simply defending belief in God, creation, redemption, human freedom, prophetic teachings, and the sacredness of human life against the materialist interpretations of biology by much of the modern and contemporary intellectual world. As a result of this excessively apologetic concern, truly constructive Catholic theologies of evolution have been relatively few.

This is ironic because Catholicism claims to possess a strong tradition of "natural theology," which presumably looks for an *intelligible* relationship between divine action on the one hand and specific features of the universe and life on the other. However, partly owing to the influence of Cartesian dualism and mechanistic philosophy, modern Catholic theology, like Christian theology in general, has portrayed God as acting most characteristically in the arena of human history or in the privacy of personal existence rather than in nature and evolution. By relating God almost exclusively to human personality and human history, pre–Vatican II theology unfortunately contributed to a sense of distance between human beings and their natural habitats. This severance is reinforced by traditional Catholicism's excessively otherworldly conceptions of human destiny. Significant strains of post–Vatican II Catholic theology, on the other hand, have slowly and unevenly adopted a fresh evolutionary accent. Some Catholic theologians, many of them women, have begun to highlight the ecological, genetic, and cosmological connections that tie the human species intricately into the whole story of life (e.g., Delio 2008; E. Johnson 2008; Deane-Drummond 2009).

Much earlier in the twentieth century, the philosopher Henri Bergson (1859–1941), who was sympathetic toward Catholicism, had been influential in Catholic circles as he struggled to situate evolution in a nonmaterialist intellectual setting. However, Bergson's ideas on evolution suffered from being too unscientific and dualistic for most Catholic scientists and philosophers, including Jacques Maritain and Teilhard de Chardin (Fig. 60.3). Even by the early twenty-first century, evolution had still not become a universally shared concern of Catholic systematic theologians and ethicists.

Occasionally, scattered Catholic officials still question the evidence for, and implications of, evolution. For example, on 7 July 2005 Cardinal Christoph Schönborn of Vienna, who had directed the development of the *New Catholic Catechism*, contributed an essay to the *New York Times* claiming that evolution might be true in the sense of descent from common ancestors but that "neo-Darwinian" accounts portraying evolution as a blind and unguided process are irreconcilable with Christian faith. His essay, which also disparaged John Paul II's important 1996 endorsement of evolution, immediately drew considerable criticism from Catholic scientists and theologians who had already comfortably appropriated contemporary evolutionary accounts of life. Schönborn's statement seemed to be a step backward in the modern Catholic Church's efforts to come to terms with Darwin. The cardinal, however, later qualified his views, claiming that what he really opposed was not evolutionary biology but materialist interpretations of it (O'Leary 2006, 213–18).

On the other hand, Pope John Paul II's favorable remarks on evolutionary biology, delivered at a meeting of the Pontifical Academy of Science in 1996, go considerably beyond all present and past lukewarm ecclesiastical concessions. Indeed, perhaps nothing provides more dramatic evidence of the church's emerging hospitality to evolution than Pope John Paul II's bold endorsement of evolutionary research after a century and a half dominated mostly by ecclesiastical suspicion.

BERNARD LONERGAN AND TEILHARD DE CHARDIN

To date, the most sophisticated example of Catholic thought's concern for a metaphysical framework that can make proper sense of Darwinian evolution is that of the Jesuit philosopher and theologian Bernard Lonergan (1904–84). Reflecting on modern science in the light of both medieval and modern thought, Lonergan's magisterial book *Insight* (first published in 1957) presents a sophisticated worldview that can logically ground evolutionary biology without bringing along the obscurantism of materialist metaphysics. Combining classical and statistical scientific methods into a "genetic method" that outlines an empirically rich worldview that he called "emergent probability," Lonergan demonstrates that the seemingly unintelligible "accidents" that occur at any level of world process can become intelligible when understood from successively higher or "emergent" points of view. Genetic method is able to make an intelligible place for contingent occurrences (such as those that happen in biological evolution) without having to embrace either Bergsonian dualism or materialist monism. Darwinian evolution, Lonergan argued, fits comfortably into the intelligible world of emergent probability. By contrast, the typically materialist philosophical interpretations of biology only lead the mind back to the unintelligible world of mindless matter. So far, however, very few biologists and philosophers of science have taken the effort to become familiar with Lonergan's difficult but original ideas (Fig. 60.4).

After *Humani Generis*, several other Catholic theologians, especially Karl Rahner, sought an understanding of God and Christ consistent with evolutionary science. Both Lonergan's and Rahner's efforts, however, were preceded by the synthesis of Christianity and evolution undertaken long before 1950 in previously unpublished works of the Jesuit geologist and paleontologist Pierre Teilhard de Chardin (1881–1955).

In dealing with the topic of "Darwin and Catholicism," it is Teilhard's name that comes most readily to mind (see Grumett 2005). Some critics do not think of Teilhard as a true Darwinian, but such suspicion arises mostly from those who assume that philosophical materialism cannot be disassociated from evolutionary biology. As far as his own understanding of the *scientific* evidence for evolution is concerned, Teilhard was as Darwinian as any other claimants to that label. He fully accepted the scientific data present in the fossil record, geographical distribution, comparative anatomy, and other empirically available tributaries leading to evolutionary theory. Nor did he object to the notion of natural selection as such. Rather, what he rejected was the nonscientific assumption that only a materialist metaphysics can adequately render evolutionary research intelligible. Materialism, Teilhard unceasingly declared, only makes the evolution of life ultimately unintelligible, in contrast to a Christian theological understanding of Darwin's discoveries. While spurning materialist interpretations of evolution, as any Catholic would have to do, Teilhard accepted Darwinian ideas of variation and selection, the statistical play of large numbers, and the necessity of deep time, all of which he took to be explanatory factors in evolutionary change.

No other Catholic in the post-Darwinian period has done more to integrate evolutionary biology, geology, paleontology, and cosmology into a theological vision than has Teilhard, even though professionally he was not technically a theologian. At the same time, it is doubtful that any Catholic scientist since Galileo has suffered more from ecclesiastical censorship in his own lifetime. Ironically, however, even though Teilhard

FIGURE 60.3. Pierre Teilhard de Chardin claimed that his theological interpretation of evolution was consistent with both science and Catholic teaching. Nevertheless, his superiors forbade him to publish *The Phenomenon of Man*, and even though the atheist Julian Huxley and the Russian Orthodox Theodosius Dobzhansky both regarded this work highly, it earned the undying scorn of the general scientific community. Courtesy of the French Jesuit archives, Vanves

was forbidden by the Vatican to publish his major works while he was alive, his influence can be seen, only ten years after his death, in the document *Gaudium et spes*, the Second Vatican Council's *Pastoral Constitution on the Church in the Modern World* 1965. Among the innovative aspects of this official document is the following statement: "The human race has passed from a rather static concept of reality to a more dynamic, evolutionary one. In consequence there has arisen a new series of problems ... calling for efforts of analysis and synthesis" (Paul VI 1965, 5).

The council's call for fresh analysis and synthesis clearly exhibits the influence of Teilhard with whose writings many theologians at the council had recently become familiar. After the First World War, Teilhard had already set forth a novel and creative theological interpretation of evolution, features of which are clearly registered in *Gaudium et spes*. In numerous unpublished writings, including his best-known books, *The Phenomenon of Man* (1955) and *The Divine Milieu* (1960), Teilhard had already set forth a religiously inspiring synthesis of evolution and Christianity. Unfortunately, during his lifetime the Vatican had prevented him from publishing most of his revolutionary writings, and even as late as 1962 the Vatican warned seminary faculties to protect students and prospective priests from exposure to "the dangers" of Teilhard's thought. This warning (*monitum*) served only to arouse curiosity on the part of Catholic thinkers, and by the time the council convened, some of the experts who advised the participating Catholic bishops had already steeped themselves in Teilhard's newly released publications. Through Teilhard, as it turns out, Darwin's science has had a significant influence on the contemporary shape of Catholic thought.

Teilhard died in 1955 in New York City, relatively unknown except in scientific circles, where he was considered to be one of the most accomplished experts on the geology of the Asian continent. He had been free to publish scientific papers as a geologist and paleontologist, but the church had not given him permission to publish *The Phenomenon* and most of his other writings on science and faith. After his death, however, friends and supporters submitted his manuscripts to eager publishers, so that by the end of Vatican II in 1965 his ideas had become widely disseminated in theological circles and the intellectual world in general. By 1965 *The Phenomenon* (retranslated into English for Sussex Academic Press in 1999 with the title *The Human Phenomenon*) had even become a best seller for Harper & Row Publishers in the area of religious thought. Renowned Catholic theologians of the twentieth century such as Karl Rahner and Henri de Lubac had been significantly influenced by Teilhard's contributions. By the early twenty-first century, few Catholic thinkers considered Teilhard theologically controversial, although his ideas had begun to lose some of their earlier popularity, partly because of conservative antipathy to Vatican II. Nevertheless, both Pope John Paul II and Benedict XVI referred to Teilhard with warmth and admiration.

Why, though, had Teilhard's writings been considered "dangerous" in the first place? Mostly, it seems, because his

FIGURE 60.4. Bernard Lonergan (1904–84), a Canadian Jesuit priest, argued that the seemingly meaningless nature of Darwinian evolution can be seen to be part of an overall plan when viewed from a higher "emergent" level of understanding. Permission: Lonergan archives, Marquette University

views on "original sin" had alarmed some members of the church's hierarchy as well as Teilhard's Jesuit superiors. Going beyond the story of Adam and Eve, which could no longer be taken literally after Darwin and modern biblical scholarship, Teilhard had attempted, even in some of his earliest essays on Christianity and evolution, to provide a deeper understanding of what the "Fall" of humankind could possibly mean in a post-Darwinian world. He attempted to connect the universal human experience of, and complicity in, evil, both physical and moral, to the unfinished state of a still emerging universe. Because, as the very notion of evolution implies, the world is still in process, Teilhard concluded that the cosmos cannot be perfect at any present moment. And because an imperfect universe would inevitably have a dark side, logically speaking, tragedy and evil can in principle gain a foothold in it. All the more urgent, then, is the theological necessity, after Darwin, of magnifying the healing significance of the Redeemer to cosmic proportions (Teilhard de Chardin 1969).

In Teilhard's Christian interpretation of the still emerging universe, original sin is not literally traceable to a primordial sin of Adam and Eve, biblical figures that modern exegesis had already demythologized. Rather the human predicament traditionally referred to as "original sin" is really the broken state of the world (social, political, economic, biological, and cosmological) into which each person is born and which stains every human being, calling out for final redemption. As a Christian, Teilhard believed that only God, as incarnate in the "Cosmic Christ," could bring ultimate healing and meaning to life's evolution, to the individual's misery, to human history, and ultimately to the entire unfinished universe. Instead of doing away with the need for redemption, as Teilhard's early religious critics had feared, his new interpretation of original sin only magnifies the scope and significance of the redemption that Christians believe to have been wrought by Christ. Teilhard's theological reinterpretation of nature, sin, and salvation had proved too extreme for the still biblically literalist Vatican censors in the early twentieth century. However, by 1965, only a decade after Teilhard' s death, his own church had become comfortable enough with evolution to even adopt some of Teilhard's new ideas on faith and evolution as officially its own.

According to Teilhard, an "unfinished universe" means that the world is still coming into being and that the cosmos and the life process remain open to a future of ongoing creation "up ahead." The world is still undergoing a dramatic transformation. Late nineteenth- and early twentieth-century Catholicism had still assumed that the universe is essentially static and unchanging, but the Darwinian revolution, along with contemporary cosmology, has led to a reconfiguration of the traditional Christian theological understanding of divine creation and redemption in the writings of many Catholic thinkers.

Evolution, along with recent cosmology, entails for Teilhard a new and wider understanding of the Christian doctrine of creation as well. First, in the light of evolution, the universe could not literally have come into being in a state of finished perfection. Second, the figure of Christ and the meaning of redemption have to do with the healing and fulfillment of the earth and the whole universe rather than with the harvesting of souls from the earth. And, third, after Darwin Christian hope gets a whole new horizon, not one of expiating an ancestral sin and nostalgically returning to an imagined paradisal past, but one of supporting the adventure of life, of expanding the domain of consciousness, of building the earth, of participating in the ongoing creation of the universe in whatever small ways are available to each person.

This understanding also requires, Teilhard thought, a rethinking of the meaning of "God" and Christian virtue. God must henceforth be thought of as residing "up ahead" (as the goal of evolution) rather than exclusively "up above." And real virtue means responding to the divine invitation to participate in the ongoing creation of the universe with renewed zest and an ever deepening faith, hope, and love.

Catholic faith in the age of evolution, Teilhard insisted, should not be childish credulity that feeds on scientific ignorance, but a faith that nourishes itself on a steady flow of new scientific information. In this respect, it is not without theological significance for Catholicism that science has discovered biological evolution and, deeper even than that, a still emerging universe. The new picture of a Darwinian life story enmeshed in a cosmic process of transformation alters the whole context of human life religiously and intellectually. It means, at the very least, that the question of the cosmic future will continue to draw our attention and energy. Teilhard's sense of the future stems from the fact that he was one of the first scientists in the twentieth century to notice that after Darwin and Einstein the entire cosmos, and not just the biological and human periods, is a still-unfolding story. Therefore, he claimed that Catholic faith must now awaken to the fact that the earth is not just a stage for the human drama but a small part of a vast universe that turns out to be a great drama itself (Teilhard de Chardin 1969).

In keeping with this new picture of evolution and the emerging universe, the Second Vatican Council's exhortations, made explicit in *Gaudium et Spes,* take on a significance that would not have been anticipated in a pre-Darwinian or pre-Einsteinian era. "A hope related to the end of time," the council declares, "does not diminish the importance of intervening duties but rather *undergirds* the acquittal of them with fresh incentives" (Paul VI 1965, 21). What does this mean? A Teilhardian interpretation would recognize here a belated response by the church to the Darwinian revolution as well as to the Marxist complaint that Christian hope has traditionally been too escapist and dismissive of the importance of action in the world.

Secular critics had long accused Catholics and other Christians of focusing so intently on the individual's life after death that Christianity was unable to motivate the faithful to participate fully in what Teilhard called the "building of the earth." After Darwin and contemporary cosmology, however, informed people realize that the universe is still a work in progress. Consequently, a cosmically and biologically reenergized Christian hope will turn human lives toward participation in the ongoing work of creation rather than simply waiting to be rescued from "this veil of tears." The Christian hope for final fulfillment, according to *Gaudium et Spes,* is not a reason for passivity in our lives here and now. On the contrary, as Teilhard proposed, an evolution-informed Catholic faith can in the council's terms, provide a " fresh incentive" to contribute to the great work of bringing the whole story of life and the universe to fulfillment. Thus, both directly and indirectly, the contributions of Charles Darwin have had a major impact on the recent shaping of Catholic thought.

❧ Essay 61 ☙

Judaism, Jews, and Evolution

Marc Swetlitz

THE PUBLICATION OF DARWIN'S *On the Origin of Species* (1859) sparked little interest among Jews during the 1860s. When it was discussed, most rabbis opposed Darwin, citing human intellectual and moral uniqueness, the absence of observed speciation, and the variety of technical issues debated by scientists. Some rabbis maintained that science alone would decide the truth of Darwin's theory; science and religion addressed different human concerns and different aspects of reality. This position had roots in traditional Judaism and philosophical trends that shaped nineteenth-century German Jewish thought (Faur 1997; Swetlitz 1999; Cantor 2005; Efron 2007).

Attention increased significantly in the 1870s. The publication of Darwin's *The Descent of Man* (1871) and publicity given to scientists proclaiming the materialist implications of evolution were important factors. Rabbis discussed the implications of evolution for belief in God and conceptions of humanity and produced a diversity of views, ranging from outright rejection to enthusiastic embrace and all flavors in between. Moreover, diversity existed within Jewish communities, as well as between the emerging Reform, Conservative, and Orthodox movements. The resulting tensions and debates about the nature of Judaism helped to shape discussions about evolution.

Most Reform rabbis adopted versions of theistic evolution, giving God a role in evolution and interpreting human uniqueness as the possession of more of God's spirit. This aligned with their existing view of Judaism as progressively changing, a view forged in early nineteenth-century Germany. Reform rabbis embraced critical study of history and Jewish texts and reinterpreted Jewish views of God and revelation. In the 1870s and 1880s, the American rabbis Kaufmann Kohler and Emil Hirsch preached in support of evolution and pointed to the science of evolution as evidence for the superiority of their brand of progressive Judaism. Such views regarding evolution and Judaism became common among British Reform Jews as that movement expanded in the late nineteenth and early twentieth centuries.

At the same time, Rabbi Isaac Mayer Wise, president of the Reform rabbinical seminary and leader of the American Reform movement, vigorously opposed all evolutionary theories (Fig. 61.1). He had a special animus for Darwin's theory of natural selection, arguing that it was materialist, denied human freedom, and led to an

FIGURE 61.1. Isaac Mayer Wise (1819–1900), a leader of the American Reform Jewish movement, opposed Darwinian evolutionary theory, thinking it brutalized humankind. From I. M. Wise, *Reminiscences* (Cincinnati: Leo Wise, 1901)

ethic of "homo-brutalism." His opposition was deeply connected to his views about Judaism. Wise rejected biblical criticism and believed morality did not progress; Moses received the perfect moral law from God at Sinai. By the early twentieth century, opposition to evolution disappeared among Reform Jews – some changed their minds, others passed away. Theistic evolution, with natural selection as a mechanism for progress, became the common position and a model for Conservative and Reconstructionist Jews as these movements coalesced in the early to mid-twentieth century (Swetlitz 1999; Cantor 2005; Cantor and Swetlitz 2006).

Throughout the twentieth century, most Reform, Conservative, and Reconstructionist rabbis adhered to forms of theistic evolution and took comfort in the new physics, which they viewed as supporting nonmaterial agency in the world. The evolutionary synthesis of the 1940s and 1950s, with its emphasis on natural selection, caused concern for rabbis because scientists like George Gaylord Simpson used evolutionary theory to argue that God had no role in evolution. Not all scientists agreed, and rabbis turned for support to others, for example, Theodosius Dobzhansky, who argued that faith and Darwinism could be integrated. In the 1960s, a new generation of Jewish theologians emerged, influenced by European religious thought that viewed science as irrelevant to faith. In addition, a focus on the Holocaust and Israel as theological topics and a general disenchantment with science led rabbis to give little attention to evolution. Only recently have a few Reform, Conservative, and Reconstructionist rabbis and philosophers returned to this topic, motivated by both ethical concerns about the environment and a recognition that Jewish writers had not yet addressed the tough issues posed by modern science to teleology, ethics, and human freedom (Green 2003; Swetlitz 2006; Samuelson 2009).

More traditional, Orthodox Jews have also expressed a variety of views about Darwin and evolution, a diversity that continues even today. Orthodox Jews continue to be challenged by broader concerns – how much to accommodate religious life, practice, and belief to the norms of modern, secular culture – that form the backdrop for discussions about evolution. And the diverse positions on the relationship between science and Judaism in the Talmud, medieval Jewish philosophy, and Jewish mysticism continue to be resources for addressing scientific theories about the age of the earth, transmutation, and natural selection (Robinson 2006; Efron 2007). Community dynamics have also played an important role. For example, in nineteenth-century England, Orthodox leaders maintained a high degree of social integration and persistently challenged Christian stereotypes about Jewish culture. This led many rabbis and lay leaders to position traditional Judaism as accepting of modern science and to minimize potential conflicts between evolution and Judaism (Cantor 2005).

For Orthodox Jews who accepted species' transmutation, evolution was a progressive, goal-directed process, with God as the motive force. The most exuberant champion for evolution among nineteenth-century Orthodox Jews was Naphtali Levy, a Polish born Hebrew and Talmudic scholar, who sent his book *Toledot ha-Adam* (The Generations of Adam) to Darwin in 1876, addressing him as "the Lord, the Prince." Levy argued both that science revealed the truth about Genesis and that rabbinic commentators had anticipated evolutionary ideas. After reading a translation, Darwin expressed that it "had given him more pleasure than he had felt for a long time" (Colp and Kohn 1996; Dodson 2000). For other Orthodox Jews, such as the Italian rabbi Eliyahu Benamozegh and Lithuanian rabbi Abraham Isaac Kook, the Jewish mystical tradition provided a theological framework where the progressive evolution of species could be integrated into a larger vision of spiritual ascent leading toward a messianic future. Some Orthodox rabbis argued that evolution, especially Darwinian evolution, was a lawful, slow process and used this to support their view that Judaism should change slowly, contra the views of Reform Jewish leaders.

At the same time, many Orthodox Jews opposed evolution. They cited the authority of Torah, rabbinic literature, and Jewish theology to argue evolution was materialistic and denied humans an immortal soul. Others argued that *if* evolution were true, it would be consistent with and even strengthen Jewish faith. However, lack of empirical evidence and weak arguments provided good reasons for rejecting evolution, and historical awareness suggested that theories of evolution, like many scientific theories, were transitory and scientists would eventually reject them. In the 1890s, Michael Friedlander, principal of Jews' College in London and translator of Maimonides' *Guide for the Perplexed*, developed an argument later advanced by Menachem Schneerson, the

Many Modern Orthodox Jews have adopted versions of theistic evolution that include natural selection while preserving roles for God's divine action and special revelation of the Torah. While most understood science as irrelevant to the core tenets of Jewish faith – Genesis was about ethics or theology, not science – a few Modern Orthodox scientists have argued that cosmology, geology, and paleontology were keys to a proper interpretation of Genesis. While there have been Modern Orthodox rabbis and scientists skeptical about the veracity of evolution, strong and widespread opposition to evolution has been characteristic of the ultra-Orthodox Jewish community. In the early 1960s, Moshe Feinstein, a preeminent legal scholar, ruled that teachers should tear pages from textbooks that taught heresies regarding creation. While a few ultra-Orthodox scientists have developed arguments for a scientifically grounded young-earth creationism, ultra-Orthodox Jews tend to look to Torah and its traditional commentaries for knowledge about earth history and the history of life. The most prolific and organizationally savvy have been the Lubavitch Hasidim. Beginning in the 1950s, their leader Menachem Mendel Schneerson corresponded with rabbis and scientists to convince them that scientific evidence for evolution was weak and that Judaism required a young-earth creationism. He actively recruited scientists to write and speak and gave his blessing to start the *B'Or Ha'Torah* journal in 1981 and the Torah and Science conferences in 1987, both of which have devoted considerable attention to evolution. Schneerson also influenced the direction of the Torah Science Foundation, founded in 2000, whose leaders adopted a young-earth creationist position that rejected the macroevolution of species (Carmell and Domb 1976; Cantor and Swetlitz 2006; Numbers 2006; Pear 2012).

The strength of ultra-Orthodox opposition to evolution became headline news as a result of the Slifkin affair. Raised in England as a Modern Orthodox Jew, Natan Slifkin moved to Israel, received rabbinic ordination in the ultra-Orthodox world, worked at the Jerusalem Biblical Zoo, and wrote several books (Plate L). His books explore topics in zoology and biology and their relationship to Judaism. He accepts a billion-year-old earth and evolution by natural selection and holds that Talmudic rabbis erred in matters related to natural history and biology (e.g., accepted mythical creatures as real). In 2005 several prominent Israeli and American ultra-Orthodox rabbis issued a ban on the reading, ownership, or distribution of three books by Slifkin. Publishers halted the printing of Slifkin's books, bookstores removed them from their shelves, and Web sites removed Slifkin's essays. Rather than retract his views, Slifkin aggressively defended his view as aligned with rationalist streams within traditional Judaism. Some ultra-Orthodox Jews opposed the ban, but most did so not because they agreed with Slifkin but because they thought the ban unfair and counterproductive. The Slifkin affair reveals an ultra-Orthodox community opposed to evolution but divided about how to manage the confrontation between modern science and Judaism (Slifkin 2001, 2006; Rothenberg 2005; Samuels 2007).

FIGURE 61.2. Michael Friedländer (1833–1910), principal of Jews' College in London, argued that you cannot infer origins from nature and its laws and thus rejected evolution. From M. Friedlander, *Jewish Religion* (London: Shapiro, Vallentine, 1937)

last Lubavitch rebbe (Fig. 61.2). Friedlander argued against scientific certainty about evolution in the same manner as Maimonides had argued against Greek views on eternity: one cannot infer from existing natural laws and processes to their origins. It was perfectly rational to hold that God created each species separately, endowed in ways that lead us to infer (erroneously) that they had originated by evolution (Kaplan 1977; Dubin 1995; Swetlitz 1999; Cherry 2001, 2003; Cantor 2005; Slifkin 2006; Blutinger 2010).

During the past century, Orthodox Jews continued to write about evolution, especially after World War II. The emigration of ultra-Orthodox Jews from Eastern Europe to the open societies of Israel, the United States, and England, and the growing numbers of Orthodox Jews pursuing careers in science, made the relationship of Judaism and science relevant to the life of Orthodox Jews. In addition, Orthodox day schools and modern Orthodox universities, such as Yeshiva University in the United States and Bar-Ilan University in Israel, created the need for curricular decisions about teaching evolution.

GENESIS AND EVOLUTION

The Jewish tradition has a rich history of biblical interpretation, producing diverse ways to interpret the Genesis creation stories. The predominant mode of interpretation has been that Genesis teaches theology or ethics, not science. For most Reform Jews and others who adopted methods of critical biblical study, Genesis is myth or poetry conveying essential teachings, for example, God created the world with a purpose, and humans should protect God's creation. Many Orthodox rabbis held similarly that Genesis taught ethics or theology, and most Modern Orthodox Jews today adopt this approach. This view has historical precedent. For example, the medieval rabbi and philosopher Moses Nahmanides taught that the opening chapter of Genesis embodied the mystical teaching that *yom* refers to kabbalistic *sephirot* and not to any period of time, whether twenty-four hours or one thousand years. While many who adopted this mode of interpretation accepted evolution, some did not. They claimed that evolution or natural selection required commitment to certain ideas such as materialism or purposelessness that contradicted essential teachings of the creation story (Swetlitz 1999; Feit 2006).

Another prominent mode of interpretation holds that Genesis anticipates, references, or teaches what scientists discover about the history of life, including evolution. This concordist approach has strong precedent among medieval philosophers such as Moses Maimonides, who argued that Genesis properly understood references to Aristotelian metaphysics and cosmology. Since Darwin's time, rabbinic sermons, newspaper articles, and letters often pointed to specific words or passages in Genesis that reference an old earth or evolution; for example, *yom* means thousands of years, *yatzar* means development from a preexisting species. In the nineteenth century, Levy's *Toldot ha-Adam* is the best example. Recently, Modern Orthodox physicists Nathan Aviezer and Gerald Schroeder published books bringing contemporary science into Jewish concordism. Aviezer argues that Genesis describes a history of life characterized by a teleological version of punctuated equilibrium. Schroeder argues that Einstein's theories of relativity explain how *both* the seven days of Genesis and the 15 billion years of modern science are literally true. Despite precedents in Jewish tradition and the goal of harmonizing Orthodox Judaism and modern science, the positions advocated by Aviezer and Schroeder remain controversial in the Modern Orthodox world (Cherry 2006; Shatz 2008).

A third mode of interpreting Genesis turns to traditional Jewish commentary for truths about cosmic history and the history of life. Some used this approach to support acceptance of an old earth and evolution. In a well-known rabbinic teaching, God created and destroyed worlds prior to the world described in Genesis. Some Orthodox rabbis took this to mean that Judaism could accept a billion-year-old universe in which evolution could occur, even if punctuated (Shuchat 2005). Most ultra-Orthodox rabbis, however, have looked to traditional Jewish commentary to support a young earth, the fixity of species, and a flood that explains the findings of geology and paleontology. Rabbi Dovid Brown's *Mysteries of the Creation* (1997) offers the most systematic attempt to provide a story of earth and human history grounded in the teachings of Jewish sages from the Talmudic period, which emphasizes catastrophic change, degeneration, and change in nature's laws. In 2005, in response to the intelligent-design debate and the Slifkin affair, Rabbi Chaim Dov Keller, writing in the flagship journal of the ultra-Orthodox Agudath Israel organization, argued that the dominant streams within traditional Jewish commentary interpret *yom* as twenty-four hours and *lemino* as individuals reproducing only from those of the same species, and that these truths are superior to the ever-changing theories of modern evolutionary science (Keller 2006).

NATURAL SELECTION

Darwin's theory of natural selection generated extensive debate about the validity of natural theology based on the argument from design. However, for most Jews this topic received little attention. In the nineteenth century, Jews did not typically turn to organic design as evidence and inspiration for faith in God's wisdom or goodness. Rabbis who did address the topic maintained either that natural selection explained, but did not explain away, divine design or that God's design was at a different level – the harmony of the natural world overall (Swetlitz 1999; Cantor 2005). In the 1960s, the ultra-Orthodox rabbi Avigdor Miller revived the argument from design, which did have precedent in Jewish tradition, and his numerous books and audio tapes have had a broad impact in the ultra-Orthodox world. Jewish advocates of intelligent design have also take up this topic. For both, the intricacies of organic design deny the efficacy of natural selection and prove God's creative power (A. Miller 1995; Robinson 2007).

Jewish discussions about natural selection have focused on key theological issues: providence, teleology, and suffering. Some understood natural selection as opposed to their belief that God guides natural processes or that humanity is the goal of God's creative work. Others thought natural selection compatible with their theology: God explains features of life's evolution that natural selection could not explain; God directs the course of mutations; God's plans involve the broader progress of life but not the details; God's purposes are achieved through the apparently random processes of natural selection (Feit 1990; Sterman 1994; Cherry 2003; Swetlitz 2006). Certain streams of Jewish theology have eased the acceptance of natural selection. Maimonides' view that humanity is *not* the goal of God's creation appeared to accord with the Darwinian view that humans are not the pinnacle of evolution. The tradition of Jewish mysticism derived from Lurianic kabbalah provided many with a theological framework in which to embed evolution by natural selection. The creation story in Lurianic kabbalah involves the withdrawal of God from the world (an act of divine self-restraint) and a shattering of the vessels of creation, which leads to a world full of pain, suffering, and evil in need of redemption through human effort. The restraint on

God's power, the pervasiveness of suffering, and the emphasis on human action were understood to be compatible with a natural world governed by natural selection (Cherry 2011).

Natural selection raised ethical as well as theological issues. Many reform-minded Jews thought Darwin's theory of the evolution of the moral conscience and moral codes was compatible with their view of God's revelation as imminent and progressive (Swetlitz 1999). For more traditional Jews who understood revelation to be a divine-human encounter, evolution by natural selection could be viewed as a prelude to the emergence of the ethical (Soloveitchik 2005). Jews were most concerned about the substance of Darwinian ethics. A few rabbis noticed that Darwin used his theory to explain the altruistic character of moral behavior, and they saw this as confirmation of Jewish ethics. Most, however, argued that natural selection could lead only to egoism and the principle that might makes right, which became reason enough to reject Darwin's theory as a whole. Abraham Treuenfels, a rabbi of the positive-historical school in Germany, a forerunner of the Conservative movement, is a powerful example. While he argued that the theological issues raised by evolution and natural selection could be resolved, the ethical issues could not, and that in itself was grounds for rejecting Darwin's theory of evolution (Treuenfels 1872). Those who were able to separate ethics from natural science, which occurred more and more throughout the twentieth century, could reject Darwinian ethics while accepting natural selection as essential to understanding human emotional and intellectual attributes (Swetlitz 2006).

SOCIAL DARWINISM

For Jews, the social implications of Darwinism involved more than the general application of Darwin's theory to ethics and society. Jews became the objects of social Darwinian thinking. Darwin made a few brief references to Jews in *The Descent of Man* (1871a, 183, 240, 242, 301) but not in relation to racial hierarchy or the struggle for existence. By the end of the century, biologists, physicians, anthropologists, sociologists, and politicians had developed numerous theories that applied natural selection, racial thinking, and eugenics to Jews, both past and present. Best known are theories that viewed history and contemporary society as a racial battle between the superior Aryans and inferior Jews, although the political implications drawn by the authors varied from assimilation to expulsion to extermination. This anti-Semitic social Darwinian biological racism became an important element in the worldview of Adolf Hitler and Nazi ideologues (Efron 1994; Evans 1997; Weikart 2005; Weindling 2010).

During this same period, social Darwinian thinking emerged that portrayed Jews positively and that supported Jewish social and political agendas. Jewish scientists and physicians produced a large body of literature on Jewish racial and national characteristics, and they along with rabbis, writers, and politicians developed theories that explained how persecution, life in the ghetto, and the struggle to survive in a world filled with anti-Semitism had produced those characteristics. Jewish scientists tended to emphasize Lamarckian notions of inheritance, which allowed them to argue that hereditary traits could adapt more quickly to changing circumstances. Ritual practices, especially dietary laws, sexual practices, circumcision, and ritual purity, were translated into eugenic practices that helped to strengthen the hereditary health of Jews and ensure their success in the struggle for existence. In 1910 the *Jewish Chronicle* in Great Britain declared Moses the first and greatest of all eugenicists (Hart 2000, 2007).

Social Darwinist views of history also provided the foundations for competing views about life in the Jewish diaspora. Some, like British folklorist and anthropologist Joseph Jacobs, wanted Jews to integrate into the countries in which they lived and argued that the struggle for survival produced Jews able to survive, thrive, and contribute in the urban, industrial, competitive modern world. Zionists, in contrast, used social Darwinian thinking to support a different political agenda. Many Zionists, including Theodore Herzl, argued that life in the diaspora had led to the degeneration of Jewish racial characteristics. While there were disagreements among Zionists – what about the diaspora was problematic, what were the relative strengths attributed to heredity and environment – all held that a return to Palestine would be, in part, a eugenic project to reinvigorate the Jewish race. And while this interpretation was not dominant among Zionists, it did impact medical and immigration practices in Palestine before the establishment of the state of Israel (Efron 1994; Falk 1998, 2005; Weindling 2005; Passmore 2007).

While social Darwinism had always been controversial, it fell into disfavor among Jews, and others, with the rise of Nazism, World War II, and the Holocaust. These events reinforced trends in Jewish thought that separated ethics and faith from any dependence on science. Most rabbis, Jewish theologians, and Jewish scientists viewed social Darwinism as an illegitimate application of a legitimate biological theory to ethics, economics, and politics. Looking back, they tended to blame philosophers and politicians for this excess, not scientists. In contrast, ultra-Orthodox Jews viewed the Holocaust as evidence that Darwinian evolutionary theory necessarily leads to immoral behavior. Darwinian theory should be rejected because it is a materialist philosophy in the guise of legitimate science (Miller 1962; Swetlitz 2006). More recently, some religious and secular Jews active in the intelligent-design movement have said much the same thing, the most popular venue being the movie *Expelled: No Intelligence Allowed*, co-written and narrated by Ben Stein (Klinghoffer 2008). Many Jews disagree with this position, and the movie prompted the Anti-Defamation League to issue a press release stating that to claim Darwin's theory is key to explaining the Holocaust "trivializes the complex factors that led to the mass extermination of European Jewry" (Anti-Defamation League 2008).

CREATIONISM AND INTELLIGENT DESIGN

Most American Jews, religious and secular, have viewed the creationist and intelligent-design movements as efforts by

Protestant fundamentalists to crush freedom of thought, stifle scientific inquiry, and impose conservative social values in public schools and in the culture at large. Many Jews have opposed the teaching of creationism and intelligent design in public school science classes because of their strong support for the separation of church and state and their desire to maintain a public sphere free of sectarian religious coercion – a secular sphere that has afforded Jews unique opportunities to thrive in American society (Efron 2008).

Starting with the Scopes trial, Jews have been actively involved, as lawyers, judges, and as interested parties to court cases. In 1925, Arthur Garfield Hays was one of three defense lawyers for John Scopes; in 1968, Supreme Court Justice Abe Fortas wrote the majority opinion in *Epperson v. Arkansas*, invalidating an Arkansas statute prohibiting the teaching of evolution in public schools; and in 2004, Jeffrey Selman was lead plaintiff challenging the decision of the Cobb County school district to require a sticker proclaiming "evolution is a theory, not a fact" in science textbooks. Over the years, the American Jewish Committee, American Jewish Congress, Anti-Defamation League, and Reform, Reconstructionist, Conservative, and some Orthodox rabbinical organizations have submitted friend-of-the-court briefs and adopted resolutions, all opposed to teaching creationism or intelligent design in science classrooms. Rabbinic resolutions and sermons consistently support some form of theistic evolution and often hold that scientific explanations are limited or complementary to religious views of creation. But such views should be taught in Jewish day schools and afternoon religious schools, and not in public school science classes (Goldfarb 1981; Cantor and Swetlitz 2006; Robinson 2007; Efron 2008).

While dominant, this standpoint has not been universal. Since the 1940s, ultra-Orthodox Jews have referenced and reviewed creationist literature in books and journals. For most of this time, ideological and institutional factors kept Jewish and Christian creationists apart: the Creation Research Society, for example, required members to accept Jesus Christ as their personal savior; and Orthodox Jews thought Christian creationists misguided because they interpreted scripture literally, without the guidance of authoritative rabbinic commentaries. This changed with the rise of the intelligent-design movement and efforts to create a "big tent" for all opposed to naturalistic theories of evolution, a tent where Jews, too, have found a place. Rabbi Avi Shafran, director of Public Affairs for the ultra-Orthodox Agudath Israel of America, has consistently supported proposals to teach intelligent design in public schools. While ultra-Orthodox Jews attend private Jewish schools, Shafran argued that Jewish interests are best served when no American child is indoctrinated into the "religion of Randomness." In 2005, William Dembski, a leading intelligent-design theorist, spoke at the Conference on Torah and Science, organized by the ultra-Orthodox Lubavitch Hasidim, and participated in a panel on teaching evolution in science classes. A few Modern Orthodox Jews have joined the intelligent-design movement, the most prolific being David Klinghoffer, Senior Fellow at the Discovery Institute. For Dembski, these developments mean the intelligent-design movement is no longer a "Christian thing" but is "going interfaith." However, the extent of support among Orthodox Jews for intelligent design and for teaching intelligent design in the science classroom is unknown (Dembski 2005; Shafran 2005; Numbers 2006).

Challenges to the teaching of evolution in public schools also surfaced recently in Israel, where studies suggest a substantial percentage of Israelis questions the scientific theory of evolution and perceives a conflict between evolution and the biblical story of creation. In February 2010, Gabi Avital, chief scientist for Israel's Ministry of Education, sparked a furor with remarks that questioned the validity of evolutionary theory and global warming. Avital wanted to examine textbooks and curricula and include alternatives to standard scientific theories of evolution. A letter to the education minister from ten leading scientists called for his dismissal, saying his remarks "undermine the standing and importance of science and take us centuries backward, even as the world celebrates the importance of Charles Darwin's discoveries and the great contributions he made to human knowledge and scientific development." A few months later in October 2010, the education minister dismissed Avital. While the reasons for the decision are clouded in speculation, the Avital affair suggests that with the right combination of religion, science, and politics, the controversy about Darwinian theory and religion can be volatile and have real consequences within the Jewish world (NCSE 2010a, 2010b; Dodick, Dayan, and Orion 2010).

HISTORIOGRAPHY

Knowledge of Jewish responses to evolution is in its infancy. We have good studies of how the British and American Jewish communities responded in the nineteenth century and of how several Central and Eastern European rabbis, physicians, and social scientists have engaged evolution. There are also studies of twentieth-century American rabbinic discussions of evolution and of Jewish responses to intelligent design. However, significant gaps remain.

Community-wide studies of Jews in Central and Eastern Europe and Israel would provide us a broader understanding of how they and their religious leaders responded, allowing us to place current individual studies in a broader context. Such studies would also allow us to better understand the history of Jewish creationist and antievolutionary positions, which have been explored in some detail only among post–World War II American Jews. Israel studies could further examine religious, scientific, and ideological interactions in the context of national conflict. Altogether, such studies would provide a better foundation for comparative study among Jewish communities and between Jews and other faiths and a stronger basis for generalizations about Jewish engagement with evolution.

❦ Essay 62 ❦

Religion: Islam

Martin Riexinger

The reception of the theory of evolution by Muslims took place in a setting that differs entirely from the reception by Western Christians and Jews. Because the Islamic world had not participated in the Scientific Revolution, other concepts such as post-Copernican astronomy had become known only slightly earlier (Riexinger 2004, 372–84, 392–410). In many countries, the curricula for religious scholars consisted of the traditional branches of scholarship alone (Riexinger 2009, 246–47). Because the rates of illiteracy remained high in most countries well into the twentieth century, scientific concepts did not find a large audience. And unlike Japan, other East Asian countries that followed Japanese models, and, to a lesser degree, India, the Islamic countries failed to establish institutions of higher education and research capable of producing significant scientific output (United Nations Development Programme 2003).

Furthermore, whereas research on the reception of modern scientific concepts is still at an initial level in much of the Muslim world, for some regions, including North and sub-Saharan Africa and Southeast Asia, even basic studies are still lacking. Hence, this overview on the reception of the theory of evolution, the religious context in which it occurred, the religious responses, and its practical consequences has to be considered preliminary.

ISLAMIC CONCEPTS OF THE CREATION AND THE FLOOD

Occasionally one encounters the apologetic allegation that there is no conflict between Islam and the theory of evolution because the Qurʾān contains no equivalent to the story of creation in Genesis. This allegation is correct with regard to animals, plants, and the time-span of creation. However, the "special creation" of men in various stages according to those in the preparation of pottery is referred to in a number of verses (6:2; 15:26, 28, 33; 23:12; 32:7–9; 37:11; 55:14). In the Qurʾānic reports on the Flood, the exhortative aspect is clearly more important than the account of the details (7:59–64; 11:36–48; 23:23–30), but the story in Genesis is referred to as commonly known. Hence, before the late nineteenth century, it has never been questioned that the Flood was a universal event, although the Qurʾān contains no explicit statement.[1]

Moreover, the Qurʾān is not the only source for religious concepts in Islam. The second major source is the Ḥadīth, sayings and actions attributed to Muḥammad that have been collected in two major and four minor canonical compilations in the mid- to late ninth century.[2] Although the Ḥadīth is primarily important for Islamic law, many traditions are related to beliefs. With regard to creation, many gaps are filled in with reports that often reflect concepts from the Bible or apocryphal Jewish and Christian sources. A particular Islamic concept is that Adam originally measured sixty cubits (Heinen 1982; Schöck 1993). Furthermore, some traditions contain information regarding the time-spans between the prophets, which sum up to a similar time frame to that in the Bible. The creation of Adam figured prominently in popular religious literature, such as the legends of the prophets (*qiṣaṣ al-anbiyāʾ*) and the poems recited on the occasion of *mawlid*, Muḥammad's birthday (Riexinger 2010, 484–85).

Hence the theory of evolution confronted religious preconceptions that contradict it no less than those of traditional Christianity, and they were at least as well entrenched in the collective mind.

FROM THE INITIAL RECEPTION OF THE THEORY OF EVOLUTION TO THE INTERWAR PERIOD

About a decade or two after the publication of the *Origin of Species* the theory of evolution became known in those parts of the Islamic world where the interaction with the West was the most advanced: British India, the Balkan and Western Anatolian provinces of the Ottoman Empire, and the Levant. However, in the case of the latter region, Arab Christians and American missionaries played a central role in its reception and distribution.

In Beirut, American Presbyterians had established the Syrian Protestant College in 1866 (since 1920, the American University Beirut). Although the majority of the teachers objected to Darwin's ideas, some praised his achievement and found an interested audience among some of the students. In 1876 the SPC alumni Shiblī Shumayyil (1850–1917), Yaʿqub Ṣarrūf (1852–1927), and Fāris Nimr (1856–1951) started to publish *al-Muqtaṭaf* (The Digest), a magazine dedicated to science, technology, and the creation of an adequate Arabic vocabulary for both fields. Although the editorial staff was dominated by nominal Christians, the readership was much more diverse, and thus the theory of evolution became known to Muslims in the Arab East and even Muslim scholars in other countries. However, the presentation of the theory of evolution was not based on Darwin's own writings but on a French translation of Ludwig Büchner's *Sechs Vorlesungen über die Darwinsche Theorie* (Jeha 2004; Glaß 2004). The

[1] B. Heller, s.v. "Nūḥ," *Encyclopaedia of Islam*, 2nd ed., viii 108–9.
[2] Burton 1994; J. Robson, s.v. "Ḥadīth," *Encyclopaedia of Islam* iii 23.

FIGURE 62.1. The Young Turks were a group of progressive thinkers at the beginning of the twentieth century who wanted reform of the Ottoman Empire. Their leader was Enver Bey (1881–1922) shown in the middle of this 1919 photograph. Their ideology led them to an enthusiasm for evolutionary theories.

first Arab translation of the *Origin of Species* was prepared by Ismāʿīl Maẓhar, an Egyptian Muslim, between 1928 and 1964 (Glaß 2004, 431–34).

Also in the 1870s, some authors writing in Turkish became aware of the theory of evolution and presented evolutionist concepts in booklets and articles. They too did not read Darwin's own writings but the French translations of Büchner and Haeckel (Demir and Yurtoğlu 2001). Around the turn of the century, the theory of evolution became popular among a group of oppositional intellectuals who became known as "Young Turks" in the West (Fig. 62.1). With an ideological amalgamate derived from ideas of Comte and the German *Vulgärmaterialisten*, authors such as Abdullah Cevdet (1869–1932) challenged the conservative sultan Abdülhamit II (1842–1918, r. 1876–1909), who justified his rule with reference to religious symbols (Hanioğlu 2005; Doğan 2006). An important aspect of this politicized Turkish reception to the theory of evolution via these German sources is the fact that concepts from racialist polygenists that Darwin opposed met with approval (Sami, AH 1296, 15–17).

Whereas the reception of the theory of evolution was concomitant with agnosticism in the Arab East and even an antireligious bent of mind among Turkish intellectuals, the first South Asian Muslims who wrote about the theory of evolution stressed that it was compatible with the basic tenets of Islam. In this respect, the educational reformer and pro-British political leader Sayyid Aḥmad Khān (1817–98) and his detractor, the nationalist publisher Abū l-Kalām Āzād (1888–1958), did not differ (Riexinger 2009, 217–21).

The Islamic opposition to the theory of evolution can also already be traced back to the late nineteenth century. The most important figure in this respect is Ḥusayn al-Jisr al-Ṭarābulusī (1845–1909), a scholar and educator from what is now Lebanon, who promoted the reconciliation of secular and religious learning. However, in 1888 he expressed strong skepticism with regard to evolution in general, and he rejected outright the descent of men in his book *al-Risāla al-ḥāmidiyya* (Tract Dedicated to Abdülhamit II; Ebert 1991). That this book was translated into Ottoman Turkish, Tatar, and Urdu indicates that his negative stance was anything but a marginal position (Riexinger 2009, 225–26).

After the founding of the Turkish Republic in 1923 by Mustafa Kemal (since 1934 Atatürk), an intellectual heir of the Young Turks, the theory of evolution was integrated with the previously mentioned racist concepts into the *Türk Tarihi Tezi* (Turkish History Thesis), which was supposed to demonstrate the foremost role that the Turks in the larger sense played in advancing world civilization. The standard formulation of this concept, *Türk Tarihin Ana Hatları* (The Main Lines of Turkish History), was coauthored by Atatürk's stepdaughter Afet İnan (Laut 2000, 8–11, 48–52, 94–161; Peker, Comert, and Kence 2009, 740). However, conservative religious authors were able to express their objections (Ertuğrul 1929).

In the Arab world, the theory of evolution seems not to have roused much passion. For example, the conservative religious publisher Rashīd Riḍā (1865–1935) did not accept it, but he declared that it was not to be considered a matter of belief and unbelief (Rashīd Riḍā 1930).

In South Asia ʿInāyatullāh Mashriqī (1888–1963), leader of a religious cum political cult inspired by fascist models and considered heretical by most Muslim, included Haeckelian evolutionist concepts into his exegesis of the Qurʾān. In 1941 Abū l-Aʿlā Mawdūdī (1903–79), one of the founding fathers of Islamism, came forth with a scathing critique of the theory of evolution as speculative, cruel, and a denial of God the creator. However, the theory of evolution did not become a major issue for him (Riexinger 2009, 221–30).

RADICAL REJECTION OF THE THEORY OF EVOLUTION

Emergence in Turkey

In the first two and a half decades after World War II, the theory of evolution was attacked in tracts, articles, and commentaries on the Qurʾān by authors from various countries and of different religious orientations (Riexinger 2009, 230–40; 2011, 499–50; 2013). However, all of these statements are rather marginal, and religiously motivated opposition to the theory of evolution did not become a major ideological phenomenon before the 1970s.

The change came about in Turkey and was due to the teachings of a prominent religious figure who had died in 1960. Said Nursi was a Kurdish scholar born in the 1870s. He was educated in traditional *madrasa*s in Eastern Anatolia. He then acquired autodidactically knowledge about natural and social sciences and came forward with a proposal to found a university supposed to close the gap between secular and religious education. However, his religious outlook was rather conservative. In particular, he stressed that all events are directly arranged by God in the framework of an atomistic concept of time and not by an independent natural causality. Although Said Nursi had supported Mustafa Kemal's resistance to the Greeks and British after World War I, he fell afoul of the new regime after the foundation of the republic. He spent the years from 1925 until 1951 either in prison or in banishment. Nevertheless, he would collect a followership to which he dictated his religious ideas, which were published as the *Risale-i Nur Külliyatı* (Collection of Epistles on the Divine Light) after the liberalization of the political system in the 1950s. He does not openly refer to the theory of evolution, but in one of his most famous sermons he denounces the concept of the self-organization of matter as absurd with a parable: How could one imagine that all the pastes and tinctures spilled in a pharmacy through which a storm has blown could bring forth a living creature (Riexinger forthcoming)?

For his followers, called Nurcus (disciples of the Divine Light), the defense of this concept gained importance in the 1970s. They thought it would help them to combat Marxist materialism in a convincing way at a time when the confrontation between the Left and the Right climaxed in Turkey. The first elaborate rejection was presented by Fethullah Gülen (b. 1941), a preacher who addressed students in İzmir, the country's most Westernized city, in various sermons that were recorded on audiocassettes (Gülen 2003). In the following years, several other Nurcus followed him. Some of them now began to adopt the arguments of American, mostly young-earth creationists. Like them, the Nurcus, and some authors coming from other religious movements, did not denounce the theory of evolution as false from the religious point of view. Instead they claimed that it has been scientifically disproved (Riexinger 2013).

After the 1980 coup and especially under the first civilian prime minister, Turgut Özal (1983–89), Islamic creationism received a boost because religious conservatism in general was propagated as an antidote to leftist ideologies. Âdem Tatlı, a botanist and activist of the Nurcu movement, became adviser of the Department of Education. In this position he attacked the theory of evolution by referring to the writings of Henry Morris and Duane Gish. As a result of his efforts, the most offensive aspects of the theory of evolution (natural selection, the descent of men) were deleted from the curricula, whereas the "theory of creation" was introduced as an alternative during this period (Öztürkler 2005; Peker, Comert, and Kence 2009, 741; Riexinger 2013).

During the early and mid-1990s, the debate on the theory of evolution seems to have lost importance in Turkey. However, it reemerged in the end of the decade owing to the activities of Harun Yahya (b. 1956, pen name of Adnan Oktar), an interior architect and hitherto a fringe figure in the Turkish Islamic scene, who propagated, in addition to creationism, anti-Semitic and anti-Masonic conspiracy theories. He realized the opportunities the Internet provided for the propagation of his ideas and had his pamphlets transformed into Web sites with a sophisticated layout.[3] These were translated into other languages. Thus, they became particularly popular among Muslims in the Western diaspora. His activities on the Internet are supplemented by the publication of video-CDs, exhibitions, public lectures, and colorful books, which are often distributed free. In Turkey, his activities met with favor under the government of the religious-conservative Adalet ve Kalkınma Partisi (Party for Justice and Development) of Tayyip Erdoğan since 2002 (Edis 1999; Kence and Sayın 1999; Numbers 2006, 422–26; Shipman 2006; Riexinger 2008, 103–9; Peker, Comert, and Kence 2009, 742–43).[4]

The Arguments of Islamic Creationists

The arguments used by Islamic creationists can be divided between those with a scientific pretense and normative objections. The "scientific" arguments can be summarized as follows (Riexinger forthcoming)

1. The complexity of the living organism cannot be explained without reference to a designer. The organization of the cell, the "Cambrian explosion," and the alleged optimal adaptation of all organisms to particular purposes are the most frequently used examples. Furthermore many authors argue that all life forms are useful for mankind. Some writers claim that the imitation of structures found in living beings for technological purposes (bionics) bears proof for the existence of a designer of the respective models.
2. Species are said to be immutable, because forms deviating from the ideal type would have no chance of survival. Hence, the idea that mutations bring forth new species is derided as nonsensical.
3. The tendency of evolution toward the perfection of life forms is disproved by the persistence of primitive organisms.
4. The theory of evolution provides no explanation for the emergence of life.
5. Arguing against the misrepresentation of Darwinian evolution as saltationism, Islamic creationists reject the possibility of biological change and transition from one species or even higher taxon to another within a generation.
6. In another misrepresentation of the theory of evolution, Lamarckist conceptions are grafted onto Darwin's theory.[5]
7. The alleged principle "survival of the strongest" (instead of the commonly used phrase "survival of the fittest") fails to explain how small animals managed to survive while dinosaurs and mammoths became extinct.
8. The fossil record is said to contradict the theory of evolution. Fossils of "missing links" are usually denounced as forgeries.
9. Actual forgeries such as the Piltdown skull are referred to in order to bolster the claim that the dominance of the theory of evolution results from a conspiracy.
10. Apes and monkeys are exempted from the common praise for the beauty of nature. Instead they are portrayed as exceptionally dumb and vile creatures, in order to ridicule the idea that they could be related to humans.
11. Probability theory disproves the possibility that cells as well as species emerge at random.
12. The Big Bang (interpreted as *creatio ex nihilo*) has disproved the theory of evolution by undermining the concept of pre-eternity of matter, as the alleged precondition for the theory of evolution. This reflects not only how strong Büchner's and Haeckel's appropriation of the theory of evolution has influenced its Muslim opponents but also shows that the theory of evolution is seen as a continuation of Greek philosophy, Islam's main intellectual and ideological challenge in the Middle Ages.
13. Without quantitative methods, evolutionary biology does not deserve the status of science.
14. Because the theory of evolution supposedly cannot be falsified, it is unscientific. This assertion closely parallels similar arguments that American Christian creationists used in attempts to get the exclusive teaching of the theory of evolution in public schools banned.
15. The propagandists of the theory of evolution have to stem a flood of mounting criticism against specific details of their theory. The examples invoked to support this argument are typically borrowed from Christian creationist tracts.

It is remarkable that until the 1990s these arguments were primarily borrowed from American young-earth creationists. However, their claim that earth is just some millennia old was ignored by the Islamic creationists. First, they too considered the Flood a universal event, although recently the idea that it was restricted to Mesopotamia has gained ground. In the 2000s Islamic creationists drew more inspiration from the emerging intelligent-design movements than from young-earth creationists.

---

[3] www.harunyahya.org (Turkish); www.harunyahya.com (English); www.harunyahya.de (German), retrieved on 16 July 2012.

[4] Harun Yahya is not the only one who chose the Internet as a medium for his creationist propaganda. Zaghlūl al-Najjār, an Egyptian geologist, does the same but with far less elaborate presentations and more limited success: http://elnaggarzr.com/en/index.php, retrieved on 16 July 2012.

[5] Gülen (2003, 29): "Why are Muslim and Jewish boys still born uncircumcised?"

On the normative level the Islamic creationists put forth the following arguments:

1. The principle of competition that lies at the heart of the theory of evolution is unethical and can also be refuted with reference to examples of cooperation in nature.
2. There is a consensus among the three monotheist religions with regard to the separate creation of humans and all other species.
3. Some authors claim that the theory of evolution is devoid of aesthetic value. This objection is related to the idea of many conservative and Islamist authors according to which a "dismembered" and "dismembering" modern science has to be replaced by a "centered," holistic concept (Stenberg 1996).
4. The theory of evolution is an atheist belief system originating from the creation myths of the ancient East and classical antiquity according to which the universe has emerged from chaos. Enlightenment materialism is singled out as one further source for the theory of evolution.
5. Darwinism is said to be the basis of racism, capitalism, and Marxism.

The overarching aspect that holds together the "scientific" and the "religious cum ethical" line of argumentation is the correspondence between a harmonious purposeful nature and a harmonious order of society. This aspect is particularly visible in fictional conversion stories where the protagonist discovers that he is not the product of random processes but of God's will. Owing to his insight, he recognizes that life has a purpose, which implies an ethical mandate (Riexinger 2010, 494).

Concepts of Islamic Directed Evolution

Whereas Islamic creationists cling to the concept of immutable species, other Muslim authors do accept the concept of biological change in general or at least insofar humans are not concerned. An example for the latter tendency can be found among Muslims in French-speaking countries who have adopted the arguments of Christian Lamarckists such as the entomologist Pierre Grassé. They insist, however, on the special creation of mankind (Riexinger 2010, 501).

Other authors, who accept the emergence of mankind from prior species do, however, openly reject the theory of natural selection, or they at least pass in silence over this issue. This reflects their attempt to reconcile the theory of evolution with a teleological view of the cosmos. Such an approach has been chosen by authors from a various backgrounds, such as theologians at state-run faculties in Turkey, including Süleyman Ateş and Mehmed Bayrakadar (Riexinger forthcoming) religious thinkers whose ideas are considered heretic by many mainstream Muslims, such as Ghulām Aḥmad Parwez in Pakistan (Riexinger 2009, 240–43); and two engineers, Maḥmūd Shaḥrūr in Syria and Maḥmūd Ṭāhā in the Sudan (Ṭāhā was executed in 1985 by the Islamist regime) (Oevermann 1993; Mahmoud 2007; Riexinger 2010, 500).

Because the respective authors tend to present their ideas as their own original products and do not refer to eventual sources of inspiration, it is difficult to judge whether parallels to similar Christian concepts, for example those of Teilhard de Chardin, are due to borrowing or a result of convergence (Edis 2007, 139).

Most of these reconciliatory approaches seem to be of minor importance. There is, however, one important exception: Iran. Shiite scholars were no less opposed to the theory of evolution than their Sunni counterparts in the early and mid-twentieth century (Arjomand 1998). But at least in Iran the attitude of the Shiite clerics changed in the 1970s because of the efforts of Yādollāh Saḥābī (1905–2002), a geologist who supported the religious opposition to the shah. He wrote two tracts in which he demonstrated the Qur'ānic verses and traditions of the Prophet and the Twelve Imams related to the creation of Adam can be reconciled with the theory of evolution if they are interpreted allegorically.[6] His argumentation was adopted by clerics such as Āyatollāh Meshkīnī-i Ardabīlī (d. 2007), leader of the Council of Experts, the second highest body of control in Iran's political system. Both do, however, avoid discussing natural selection and hence leave space for teleology (Riexinger 2010, 500–1).

The Acceptance of the Theory of Evolution among Muslims

Only recently the acceptance of the theory of evolution in the Muslim world has been investigated in opinion polls. An inquiry conducted among Muslims in Turkey, Pakistan, Kazakhstan, Egypt, Malaysia, and Indonesia in 2007 shows that, with the exception of the former Communist Kazakhstan, the majority of those questioned rejected the theory of evolution. The acceptance was highest in Turkey (22 percent). In Egypt just 8 percent spoke out in favor of the theory. However, a high proportion of the Egyptians but only a few of the Turks, Pakistanis, and Indonesians who were questioned claim to have never thought about it (Hassan 2007). In a similar poll in OECD countries, Turkey turned up as the country with the lowest rate of acceptance of the theory of evolution (23 percent), far behind the United States (39 percent), not to speak of European countries and Japan (Miller, Okamoto, and Scott 2006). However, even in contrast to the United States, the rejection of the theory of evolution is even fairly common among students and teachers of biology (Peker, Comert, and Kence 2009).

Although not much research on the impact of Islamic creationist propaganda has been done, it may be assumed that it contributed to these results. According to a survey among high-school students in an upper- and a lower-middle-class

---

[6] In addition to sayings attributed to Muḥammad, the Shiite Ḥadīth consists also of traditions attributed to the Twelve Imams, that is, Muḥammad's nephew and son-in-law ʿAlī, his two sons Ḥasan and Ḥusayn and the nine descendants of the latter. Shiites regard them Muḥammad's rightful successors and as infallible authorities.

neighborhood in Ankara, those who reject the theory of evolution refer to information they gathered on the Internet or from video-CDs, which means material produced by Harun Yahaya (Öztürkler 2005, 191–92). In Western Europe, no surveys among Muslims in general have been undertaken; however, even among university students, objections seem to be strong (Koning 2006), whereas Muslims in the United States, a community comprising many academics and professionals, are one of the religious groups with the least negative attitude (Pew Research Center 2008).

Concluding Remarks

Because Muslim opponents of the theory of evolution have adopted many auxiliary arguments from Western creationists, it has been argued that Islamic creationism is primarily an ideological import (Peker, Comert, and Kence 2009). This interpretation neglects that the theory of evolution is rejected because of a strong religious motivation stemming from a friction between key concepts of the theory of evolution and inherited religious doctrines, in particular the special creation of men and an anthropocentric teleology of creation. In a given context, Muslim authors adopted those auxiliary arguments easily available to them. These were mostly creationist ones but also, in the French case, "Christian Lamarckism." And even most authors who accept evolution are reluctant to adopt the concept of natural selection as it conflicts with the idea of teleology directed by the benign will of God.

ISLAMIC CREATIONISM: WHY DID IT EMERGE IN TURKEY?

Although the objections to the theory of evolution are massive in the Islamic world, Islamic creationism did not become a notable ideological movement comparable to Protestant creationism in the United States. A certain exception is Turkey, and the positive response of Muslims in Western countries, not only those with a Turkish background, to creationist propaganda is equally noteworthy. As has been pointed out with regard to the role of the theory of evolution for American fundamentalist Protestants, it has become for the "traditional bloc" a *Kulturkampf*, a symbol for a wider spectrum of social developments experienced as negative (Riesebrodt 1993, 96–99).

It seems that the theory of evolution has not been associated with a major ideological current in other Islamic countries, where it is considered one of many un-Islamic belief systems. In Turkey, however it was regarded as a central element of Young Turk ideology and its Kemalist and Marxist heirs. Moreover, a high enrollment until the 1970s in purely secular schools exposed relatively many Turks to the theory of evolution, whereas in other Islamic countries educational policies were much more reluctant to offend religious sensibilities. At least since the 1960s, state theologians, independent religious activists, and free-lance religious authors have also been enrolled in the secular school system, whereas in South Asia most religious scholars are educated in *madrasa*s following traditional curricula.

Essay 63

From Evolution and Medicine to Evolutionary Medicine

Tatjana Buklijas and Peter Gluckman

IN 1991 RANDOLPH NESSE and George C. Williams opened their classic article on "The Dawn of Darwinian Medicine" with the following assertion (Plates LI and LII):

> While evolution by natural selection has long been a foundation for biomedical science, it has recently gained new power to explain many aspects of disease. (Williams and Nesse 1991).

The optimistic statement of the two pioneers of modern "Darwinian" or "evolutionary medicine" raises many questions. Was evolution indeed a foundation for biomedical science? Historians have traditionally located the foundations of modern medicine in clinical disciplines, anatomy, and physiology – or, from the institutional perspective, in the hospital, morgue, and the laboratory. Evolutionary and field studies do not figure in any of the major medical-historical accounts (Bynum 1994; Porter 1999; Cooter and Pickstone 2000). And, if evolutionary studies indeed gained new power to explain disease, why was Nesse and Williams's "manifesto" published in the *Quarterly Review of Biology*, a journal with a traditional emphasis on evolution and, at the time of publication, edited by Williams himself, rather than in a medical journal. Why has evolutionary medicine remained marginalized from the medical community?

Certainly in the two decades following the publication of this article, a new field at the junction of evolutionary biology, molecular and developmental biology, genetics, epidemiology, and clinical medicine has gained in strength, with numerous articles, books, and first advances into medical school curricula (Nesse and Williams 1995; Stearns and Koella 2008; Gluckman, Beedle, and Hanson 2009). The appeal of evolutionary medicine is in providing a coherent framework to organize and explain facts about human biology and disease and in pointing out the importance of ultimate as well as proximal causation in understanding the human condition. But it is still struggling for status and broader acceptance within medicine. Physicians by and large know little about evolution; a survey of U.S. medical school deans several years ago revealed at more than half the medical schools evolutionary biology was not regarded as important knowledge (Nesse and Schiffman 2003). Popular interest in various "evolutionary" hypotheses related to human health – from the "Paleolithic diet" to the fitness advantages of "schizophrenia genes" – is a double-edged sword as the lack of empirical support and the speculative nature of such hypotheses throws an unfavorable light upon the entire field. Injecting rigor into hypothesis formation and broadening the range of methods used to test them is of highest importance, if evolutionary medicine is to gain broad acceptance within biomedical research and clinical community.

This essay first examines the history of the relationship between evolution by natural selection and medicine over the past two centuries. Contrary to the image painted by the historiography of medicine, evolutionary knowledge has indeed informed a wide range of medical disciplines. We end by discussing the current goals, status, and problems of evolutionary medicine.

MEDICINE AND EVOLUTION DURING DARWIN'S LIFETIME

While few physicians nowadays study evolution, most early evolutionary thinking came from doctors, which was a direct consequence of limited educational and career choices for young men interested in natural sciences. Famously, Charles Darwin's own grandfather, the country physician, naturalist, and writer Erasmus, was an evolutionary thinker along the lines of Buffon, who believed in organic transformism (Porter 1989). As this quote from the preface of his *Zoonomia, or the Laws of Organic Life* shows, his medical interests were inextricably linked to the naturalist ones, with both nature and disease, in his view, reducible to basic laws of "organic life": "The purport of the following pages is an endeavour to reduce the facts belonging to animal life into classes, orders, genre and species; and by comparing them with each other to unravel the theory of disease" (E. Darwin 1794). In revolutionary France, Jean Baptiste Lamarck had studied medicine before giving it up to pursue his interest in botany and then take up professorship in invertebrate zoology (Burkhardt 1995). Franz Unger, the Viennese botanist who in the 1840s argued that all plants had arisen from the same ancestor, had trained in medicine in the 1820s (Gliboff 1998). Most famously, Charles Darwin followed in his grandfather's and father's footsteps and enrolled into medical school – only to become "the most famous medical school drop-out" (Desmond and Moore 1991).

These medical backgrounds of evolutionary scholars did not translate evolution into a foundation of medical studies and research. This may be partly explained by contemporary politics: in the 1830s England, only the radical new schools – in the first place the University of London – gave space to the controversial knowledge imported from the politically suspect France, such as Geoffroy St. Hilaire's "unity of composition" that ordered all living beings into a single chain and implicitly challenged the dominant social order (Desmond 1989). As the German states and Habsburg Empire underwent a conservative backlash in the turbulent 1800s, universities stayed away from evolution-related content. But doctors' interest (or lack thereof) in early evolutionary theories was shaped primarily by contemporary medicine. Against the vitalism of the early 1800s, which reached for an elusive *Bildungstrieb* (generative force) to solve the problem of both the beginning of the individual life and the life on earth, medical scientists coming of age in the 1830s and 1840s increasingly limited their inquiry to processes accessible by observation and, increasingly, experiment. They wanted to explain biological phenomena in terms of laws of physics and chemistry and saw the (physiological and biochemical) laboratory as the main site of knowledge production (Cunningham and Williams 1992). Both development and evolution seemed intractable to these efforts.

Medicine's experimental turn of the mid-nineteenth century was, then, probably the reason why Darwin's work – in particular *On the Origin of Species* (1859) and *The Descent of Man* (1871) – produced little obvious, immediate impact on contemporary medicine. Darwin himself did not make it easier with the lack of medical cases in his writings. Certainly, Darwin's medical friends provided him with both a sounding board and useful examples, but most of his evidence came from artificial breeding and naturalist observations. The only major exception is *The Expression of the Emotions in Man and Animals* (Darwin 1872b; Gilman 1979; Browne 1985b). The central argument of the book – that human facial expressions exhibited continuity with the expressions of animals – supported the *Descent*'s thesis that modern humans evolved from "lower" organisms and thus did not have a special status in the nature (Darwin 1871a). Photographs and descriptions of "the insane" provided Darwin with an immediate access to human emotions "as they are liable to the strongest passions and give uncontrolled vent to them" (quoted in Pearn 2010, p. 163; see Fig. 63.1). The book was written in close collaboration with Dr. James Crichton-Browne, superintendent of the West Riding Pauper Lunatic Asylum and a leading Victorian psychiatrist, who provided Darwin not only with research material (photographs and case descriptions) but also with the necessary psychiatric expertise (Pearn 2010).

In his other work, Darwin drew on medical knowledge only very occasionally and nearly always in relation to reproduction and inheritance, concepts at heart of his evolutionary thought, as "variation which is not inherited throws no light on the derivation of species" (Darwin 1868b, 445). *The Descent of Man* drew on a wide range of contemporary knowledge including medicine. In the section on "homological structures," the similarity of diseases and susceptibility to drugs and intoxicating substances between humans and apes was used in support of common descent.

Among Darwin's writings, his discussion of the familial predisposition to diseases in *The Variation of Animals and Plants under Domestication* was probably most directly linked to the concerns of contemporary medicine. He enlisted the examples of hereditary predisposition to gout, insanity, and consumption to support the dominant evidence upon which the concept of natural selection was grounded: domestication and artificial breeding (Darwin 1868b, 451–52). The idea of a heritable predisposition ("diathesis") toward disease

FIGURE 63.1. Along with descriptions and depictions of expressions of emotions in the insane, Darwin also included representations of emotions in "normal" infants, whose expressions were, as they were in the insane, uninhibited and clear. The six photographs on plate I come from two professional photographers, Oscar Gustaf Rejlander of London (figs. 1, 3, 4 and 6) and Adolphe Diedrich Kindermann of Hamburg (figs. 2 and 5), and show the work of facial muscles in young children weeping and crying. From C. Darwin, *Expression of the Emotions* (London: John Murray). These photos are reproduced from the 1890 edition, where the same photos as the original 1872 edition are slightly rearranged.

was not new; indeed it went back to Hippocratic medicine (Ackerknecht 1982). But as "hereditarian" views of the body and disease came to dominate "environmental" ones in the second half of the nineteenth century, Darwin's theory provided a framework for organizing ideas – and expressing worries – concerning the perceived increase in the frequency of certain diseases in modern environments. Some medics, such as the famous clinician Jonathan Hutchinson (1828–1913), went a step further by suggesting that diseases should be seen as species, placed "in natural groups, in connection with their ancestral descent" (Bynum 1983).

Infectious diseases were of greatest interest to medicine in this period. Historians have documented extensively the rise of the germ theory, from Pasteur's 1850s and 1860s experiments that linked microbes to processes ranging from putrefaction and fermentation to plant and animal disease and that decisively rejected the possibility of spontaneous generation, through Joseph Lister's introduction of antisepsis into the surgical theater around 1870, to Robert Koch's quest for specific causes of bacterial diseases from the mid-1870s on, and the early ideas about immunity. Yet the ways in which clinicians and public health experts of that era used Darwin's theory to understand phenomena such as germ specificity, gradual development of the "property of infectiveness," and virulence are still relatively unexplored (Gaudillière and Löwy 2001; Bynum 2002). Sir William Roberts (1830–99), the well-known Manchester physician and medical lecturer with an interest in microbes, explained the apparently puzzling difference in virulence between morphologically indistinguishable bacterial forms with their "capacity for variation" in response to some external cause. His model was taken up by Hubert Airy, a medical inspector and Darwin's correspondent, when explaining the evolved mutual relationship between humans and microbes. It would seem at first, he argued in 1882, that "natural selection" would eliminate disease, but ability to withstand infection was just one of many factors determining human variability. Furthermore, the more virulent germs would attack more people and enhance their own survival. "Natural selection" could explain the demise and extinction of populations that had only recently come into contact with Europeans and, conversely, strong reaction of Europeans to diseases new to this continent. "For man the most serious contingency is that the disease may have a fatal termination: for the microzyme it is not of vital importance that its human nidus should perish; and it would seem that natural selection might find its equilibrium (though continually fluctuating), and be fairly satisfied with so much variation on both sides as should leave man susceptible to infection but not in a fatal degree."[1] Airy's ideas were further developed by Kenneth Millican in his *The Evolution of Morbid Germs* (1883). He argued that the process of natural selection went both ways: not only did the response of human populations change in time but also microbes modified their virulence. The variability of a species increased if one descended the evolutionary scale because the variation emerged more rapidly in organisms with short life-spans and large populations.

In contrast to the received view of "reductionist" bacteriology, evolutionary ideas strongly influenced disciplines studying microbes and parasites. A good example is the development in the 1890s of the notion of "carrier status" or the idea that susceptibility to germs may vary in the population and allow some individuals to spread a germ without becoming sick themselves (Anderson 2004). Yet, by the early twentieth century, much of this influence came to be limited to the fields of parasitology and tropical medicine, which, with the demise of colonial empires, came to play a secondary role in modern medicine.

As medical education underwent radical changes in the 1870s and 1880s, evolutionary science failed to achieve a stable

---

[1] See Airy 1882, cited in Bynum 2002, 62.

FIGURE 63.2. Michael Foster (1836–1907), a student of Huxley and leading physiologist, followed his teacher in thinking that, although evolution is true, its relevance to medicine is minimal and has no place in medical education. This lithograph was published in January 1884 in the *Midland Medical Miscellany and Provincial Medical Journal*. Permission: Wellcome

position in the curriculum. Even its champion, "Darwin's bulldog" Thomas Huxley, who as president of the Royal Society had considerable influence over the content of medical education in this period, believed evolution irrelevant to the problems doctors had to address. Francis Maitland Balfour, the first to teach embryology at Cambridge and also a disciple of the physiologist and Huxley's student Michael Foster, in the 1870s used Haeckel's biogenetic law – the idea that embryos of higher species in their development pass through adult stages of lower species – to explain evolution (Blackman 2006) (Fig. 63.2). At the University of Jena students first heard about evolution in their anatomical and zoological courses, taught by the two most fervent Darwinians in Continental Europe, Karl Gegenbaur and Ernst Haeckel (Nyhart 1995, 151–53). But because the presence of evolution in curriculum depended on research interests of the teacher, it did not achieve stability. Together with embryology, zoology, and other related disciplines, it was exposed to criticism by members of the clinical staff, who argued that it did not provide students with clinically relevant knowledge.

EVOLUTION, HEREDITY, EUGENICS, AND MEDICINE

In the early twentieth century, Darwin's teaching was in crisis. No one doubted evolution, but natural selection as its underlying mechanism was challenged both by neo-Lamarckians and by the new geneticists who argued that significantly new forms appeared suddenly by means of mutation. It has been suggested that the "eclipse of Darwinism" precisely at the time when American medical schools began to systematically introduce science in their curricula sealed the fate of the field for a long time (Nesse and Schiffman 2003). This argument, together with the opposition from the religious groups, may explain evolution's status in the United States, but it does not account for challenges in Europe. These were in some way related to the contemporary political and social turbulences. For the first few decades of the twentieth century, there seems to have been a political divide between the "neo-Lamarckians" – almost all situated on the political left and often Jewish – and the adherents of a hereditarian, arguably "Darwinian" view, on the right (Graham 1977; Slavet 2008).

Darwin and Darwinism have been accused of inspiring a body of knowledge known as eugenics in the English-speaking world and social and racial hygiene in Continental Europe.[2] Eugenics emerged around 1900 when an earlier faith in the positive effects of public health, especially on the mortality and morbidity of the urban poor, was replaced by an increasing pessimism. It was argued that for all the investment in the asylums, hospitals, and laboratory science, mental illness and consumption (tuberculosis) still reigned among the poor, who reproduced at an alarming rate: they bred disease and rebellion within the social order. The concerned middle-class physicians read Darwin's work and argued that modern civilization removed the salutary effects of natural selection, by allowing survival and reproduction of individuals with physical and mental impairment (Bynum 1983). The only way to save humanity, eugenicists argued, was to control reproduction.

The idea of hereditary predisposition to disease preceded Darwin by centuries. But he provided a new account of it. In *The Descent of Man, and Selection in Relation to Sex* (1871), Darwin agreed that, while in primitive societies selection guaranteed the elimination of the weak (in body and mind), in civilized societies these checks were removed by our natural tendency to help the weak (Darwin 1871a; Paul 2003). But the "weakness" was not necessarily hereditary. Darwin believed that immature organisms – in contrast to the adult ones – were susceptible to environmental influences (Hodge 1985; Endersby 2003). Like his cousin Francis Galton (1822–1911), Darwin, a good Victorian, argued that better education would lead to a change in habits.

As long as those in a position of influence subscribed to Lamarckian views and the possibility of changing hereditary traits by modifying the environment, improving living conditions made more sense than controlling breeding. This view began to be supplanted by eugenics around 1900 with the acceptance of August Weissmann's doctrine that "germ cells" and hereditary material are insulated from environmental influences (Paul 1995). The acceptance of eugenics in medicine opened doors to social intervention, based in medical knowledge, which actively promoted advantageous and prevented – by isolation and sterilization – disadvantageous breeding. Literature on the history of eugenics is plentiful and

---

[2] See, for instance, and notoriously, Weikart 2004.

growing; here it is worth recalling that eugenics was not just taught in medical schools in subjects called "social hygiene" or (in Nazi Germany) *Erbbiologie/Erblehre* but also, more subtly, informed medical education across disciplines as textbooks of histology and embryology reveal. A recent article on the infamous "Tuskegee experiment" – the U.S. Public Health Service clinical study of the natural history of untreated syphilis conducted upon poor African American men from 1932 until 1972 – argued that an important reason why this study kept going for so long, well after a reliable treatment was developed and after eugenics was discredited as a scientific paradigm, was the deep influence that early educational experience (in this case at the University of Virginia's medical school) exercised upon the thinking of the study's three key figures (Lombardo and Dorr 2006).

So what role, then, did Darwin's theory of evolution play in eugenics? Darwin never supported intervention into human breeding. Indeed, Darwin's political and social outlook was shaped by the social and political concerns of the 1820s and 1830s, with the antislavery movement playing a crucial role (Desmond and Moore 2009). Imposing a twentieth-century eugenic outlook upon him is anachronistic. Yet Darwin's ambiguous writing combined with readers' confusing his ideas with those of others writing on similar topics – such as Herbert Spencer's "survival of the fittest," first coined in 1864 – may have contributed to these problematic interpretations. Furthermore, while it is true that some scientists who were both "neo-Lamarckian" and politically leftist, such as the controversial Viennese biologist Paul Kammerer, held eugenicist beliefs (Kammerer 1925), nowhere was Darwin received as enthusiastically as in Germany, and nowhere did eugenics – in an extreme form and couched in Darwinian rhetoric – have such a catastrophic impact, with doctors playing a prominent role. But Nazis did not obtain from Darwin a coherent set of ideas: the historian Richard Evans has argued that what they got was a credible language they could use to justify social policies. Darwinism should thus be exculpated of the responsibility for Nazi crimes. Nonetheless, the shadow of these events would hang over studies of humans from an evolutionary perspective.[3]

THE SPECTER OF BIOLOGICAL DETERMINISM, CIRCA 1945 TO 1990

Eugenics began to lose ground as early as the 1930s, and its fall was helped by the discovery of atrocities committed by the Nazi regime – although its demise was much more gradual than commonly believed. Nonetheless, the interest in the evolutionary role and impact of hereditary diseases persisted post 1945. In contrast to earlier views, advocates of the modern synthesis saw Mendelian disorders as well-defined traits that could teach us something about evolutionary pressures in the past. In a 1949 paper, J. B. S. Haldane used the example of thalassaemia major, an inherited deficiency in the production of hemoglobin found among Mediterranean populations exposed to *Plasmodium* parasites, to argue that this mutation increased the fitness of heterozygotes by making erythrocytes resistant to the parasites; nonfatal anaemia associated with the disease was an acceptable cost (Haldane 1949). The idea that diseases are outcomes of balancing selection captured the geneticists' fancy and soon led to speculations concerning provenances of other commonly occurring diseases with a hereditary pattern. In 1964 Julian Huxley and Ernst Mayr proposed the existence of a tentative "Sc gene" to explain the persistent prevalence of schizophrenia at around 1 percent of the population (De Bont 2010).

Others arrived at same questions from different starting points. James Neel, whose work on the effects of radiation in Japan's atomic bomb victims established him as an authority in human genetics, published a programmatic paper in 1958 that warned of the limited value of sophisticated mathematical models for understanding of the actual workings of natural selection upon humans. Instead, he called for (field) studies of selective pressures across diverse human populations (Neel 1958). It was in this paper that Neel first aired the idea that changing dietary patterns – primarily an increase in animal fat and energy intake – in combination with a decrease in physical activity may expose certain selection pressures, famine in the first place, from the human past. The argument came to be known as "thrifty genotype" (Neel 1962). The reception of "Sc gene," "thrifty genotype," and the like was initially mixed – partly owing to the speculative nature of these proposals, but also because of their perceived eugenic message. But the sequencing of the human genome and the related post-1990 advances in genetics/genomics have stirred new interest in these proposals: a quest for "candidate genes" for common diseases showing hereditary or familiar transmission patterns is among the most active fields in genetics.[4]

These proposals, however, had little impact on contemporary clinical medicine. Darwin's ideas about natural selection had much more influence in the fields concerned with infectious diseases and epidemiology, traditionally more open to field studies and ecological approaches (Anderson 2004). A significant application of Darwin's evolutionary theory was Frank Macfarlane Burnet's theory of clonal selection. Burnet – who saw himself as an evolutionary biologist in his own right, a contributor to the modern synthesis – argued that immunity

---

[3] Peter Bowler recently argued that, even if Darwin had never existed, we would have accepted an evolutionary view and possibly a type of social interventionism similar to eugenics. Still, Darwinism, with its emphasis on inheritance at the individual level and especially analogy with artificial selection, may have given rise to a particularly strict form of intervention. See Bowler 2008.

[4] The efforts to locate "schizophrenia" or "thrifty" genes have encountered a range of criticisms, from observations that "candidate genes" explain only a small proportion of the disease risk and ignore the impact of environmental cues during sensitive developmental periods, to the critiques of anthropologists and social scientists who warn of problems potentially arising from this approach (neglect of social determinants of disease or the use of ill-defined ethnic and racial categories to describe risk of disease). See, for example, McDermott 1998; Montoya 2007.

FIGURE 63.3. Frank Macfarlane Burnet (1899–1985), noted Australian virologist, saw selection involved in the functioning of the immune system. This photograph was taken by Bernie Faingold of Denver, photographer renowned for his portraits of famous people such as American presidents. Permission: Wellcome

relies on the selection and multiplication of lymphocytes, the receptor of which fitted with the antigen (Anderson 2004; Park 2006) (Fig. 63.3).

Perhaps the strongest and most obvious exchanges took place within psychiatry and psychology in the 1950s and 1960s, by way of ethology, a discipline studying animal and human behavior that reached its peak in this period. On the basis of his research on birds (jackdaws), the founder of ethology, Konrad Lorenz (1903–89) – who around 1950 became famous for his work on behavioral imprinting – argued that maternal love was essential for healthy infant development and that consequently the child's tie to the mother was a result of an evolved instinctual need (Vicedo 2009). After the Second World War, this argument was taken up by the psychoanalytically inclined U.S. and British psychiatric profession, seeking to boost its scientific credentials by allying with a biological discipline. The emphasis on the (evolutionary) significance of close mother-infant attachment also fed into the contemporary social concerns about the women's work outside the house. Experimental primate research and observational studies followed; while in many cases evidence could not support the "mother love" hypothesis, it remained commonly accepted until the decline of psychoanalytic psychiatry in the 1970s (Vicedo 2010). Later incursions of evolutionary biologists into psychiatry, such as Niko Tinbergen's attempt to explain autism as an outcome of parenting, especially mothering, styles, were met with skepticism or open disapproval (Silverman 2010).

The rejection of evolutionary explanations of disease, or indeed any that favored "nature" over "nurture," in the 1960s and 1970s had to do with social and generational changes in this period. In the early 1960s, Julian Huxley, a leading British eugenicist, could openly state that his interest in a "schizophrenia gene" had an eugenic purpose (Kevles 1985; De Bont 2010). Yet, as the postwar generation came of age and rebelled against established social and medical culture, the implicit biological determinism of evolutionary explanations of disease attracted considerable criticism. Especially controversial were the studies of human behavior and mental abilities, best demonstrated by the controversy surrounding Edward Wilson's *Sociobiology* (1975) (Ruse 1979). Sociobiology was a field that developed in the 1960s, as classical behaviorism waned, to explain social behavior across the animal world by the evolutionary advantages it conferred. It successfully resolved the long-standing paradox of natural selection operating on individuals, while many species exhibit forms of socially beneficial behavior; it showed that, by helping (the related) others, one helped one's own genetic inheritance propagate. As long as sociobiology was focused on animals only, it was not controversial. Yet Wilson's attempt to extend his conclusions to humans met with enormous criticism. The fiercest opponents were two Harvard colleagues, Stephen Jay Gould and Richard Lewontin, who saw Wilson's work as a return into the 1930s eugenics. "I have a strong sense of the historical continuity of biological deterministic arguments at the same time my professional mature arguments have shown me how poorly they are grounded in the nature of the physical world" wrote Lewontin. The questions raised by sociobiology were taken on by other disciplines, particularly evolutionary psychology. Yet here too skepticism rapidly appeared when extreme arguments about the evolution of mental modules in a prehistoric "environment of evolutionary adaptedness" attempting to explain a full range of behaviors were put forward by evolutionary psychologists, notably John Cosmides and Leda Tooby. Given the difficulty of proving hypotheses in this domain (R. C. Richardson 2007), and the clash of cultures surrounding disciplines wishing to explain human behavior, it was perhaps inevitable that the field would be contentious and not accepted as mainstream. Other, less extreme, approaches of integrating psychology with evolution have since emerged (Laland and Brown 2002; Nettle 2009).

DARWINIAN MEDICINE, POST 1991

New interest in the ways in which natural selection shaped the landscape of the human disease emerged in the 1980s. Much of the early work focused on the evolutionary "arms race" between humans and pathogens (Ewald 1987, 1980). It has been argued that this renewed interest arose out of the contemporary concerns around the emergence of AIDS and the rise of antibiotic resistance (Anderson 2004). But the real boost to the study of evolutionary origins of human disease came from an unlikely pair, a young psychiatrist with an interest in evolution, Randolph Nesse, and an older, established evolutionary biologist, George C. Williams, best known for his work on the evolution of senescence and for his arguments undermining group selection. Starting from infections but broadening their outlook to include other human pathological conditions and symptoms – from genetic disorders to mechanical injuries and fevers – in a series of theoretical articles and reviews that summarized the contemporary evolutionary knowledge from the medical perspective, they tackled the question, "Why has evolution left the human bodies vulnerable to disease?" Rather than offering solutions for the "improvement" of humankind, as eugenics once did, the goal of evolutionary (also known as Darwinian) medicine was to provide physicians with a strong framework for organizing facts about the body and disease, and researchers with an innovative outlook that would help them generate new hypotheses. The key ideas were first summarized in a 1991 article (Williams and Nesse 1991) and then further refined and expanded in a book on *Why We Get Sick: The New Science of Darwinian Medicine* (Nesse and Williams 1995). Nesse and Williams were not the first to try to engage physicians in evolutionary science in the posteugenics era: yet earlier attempts received little public recognition (Zampieri 2009). The increase in interest may be explained by the generational shift and a change of medical and social climate in this period. As the Human Genome Project, begun in 1990, promised to provide biomedical researchers a new platform from which they could launch "their assault on disease" (Watson and Cook-Deegan 1990), "nature" once again began to outweigh "nurture" in the consideration of disease causation. In the 2000s, universities have begun to introduce first courses on evolutionary medicine; meetings, conferences, and publications abound; and while the field began in the United States, it has attracted international contributors and audiences (Ellison et al. 2009).

At the same time, the interest in evolutionary medicine is still limited to a relatively small group of supporters. Even in the medical fields traditionally inclined toward (and underpinned by) evolutionary biology such as infectious diseases, and even when discussing phenomena as relevant as the evolution of antibiotic resistance, the very word "evolution" is conspicuously absent from the literature (Antonovics et al. 2007). The strongest interest in evolutionary medicine comes from outside medicine, from anthropology and human biology. Indeed, evolutionary medicine has been a mainstream component of anthropological curricula for more than two decades. While much of that work has been based in solid paleontological and primate research as well as modern biological and anthropological approaches, some health anthropologists have tried to explain a range of human conditions using controversial and untestable hypotheses. At times professional antagonism between the anthropological and the medical community appears reflected in the literature (Trevathan, Smith, and McKenna 2008). This too may have inhibited acceptance of evolutionary thought into the medical framework.

A major problem and challenge for evolutionary medicine is its apparent lack of immediate utility for practically minded clinicians. In response, it has been argued that evolutionary medicine provides an essential perspective in integrating the various domains of human biology. By providing explanations of ultimate causation, it assists in understanding the context in which proximate mechanisms associated with dysfunction lead to symptoms and disease. Because of this, it is argued, evolutionary medicine should be considered a core basic science of medicine; while it does not directly affect decision making, it changes the understanding of both doctor and patient and therefore has considerable value (Gluckman, Beedle, and Hanson 2009). Formal classifications of the pathways by which evolutionary processes impinge on disease risk have been developed (Nesse and Stearns 2008; Gluckman, Beedle, and Hanson 2009; Gluckman et al. 2011). Some important principles have been highlighted including the key point that evolution operates to optimize fitness rather than acting primarily to promote health or longevity: this point alone requires physicians to rethink their understanding of human biology. Other principles include the importance of life history trade-offs, the distinction between proximate and ultimate causation, the importance of multiple forms of heritability, and the genetic inertia that puts humans at health risk when environments such as the nutritional environment change rapidly.

A potentially even more serious challenge arises from within the field. While in the first two decades evolutionary medicine has mostly contented itself with collecting knowledge from established fields such as anthropology, genetics, and clinical medicine, some participants have begun to argue that it needs to develop its own questions, methods, and expertise. Some of the proposed questions and themes have a long history, as we have shown earlier – for instance, coevolution with pathogens, symbionts, and commensals and the ways in which perturbation of one component of the relationship can affect the other species involved. Others are new and include some of the fundamental questions coming from contemporary evo-devo studies. They reflect the current critique of the strongly gene-oriented modern synthesis – for instance, the place of adaptive plasticity in individual development and species evolution; substitution of the genotype-driven model of evolution with an evo-devo model of phenotype evolution; the role of epigenetic inheritance in maintaining traits in a population; and the interaction of biology and culture in determining fitness and, consequently, the implications for the rate and direction of human evolution.

Humans might seem an especially challenging model for evolutionary studies because of their long generation times, few offspring, specific life history, and reproductive patterns, as well as special cultural status limiting the opportunities for intervention. Yet they are also a species the genotype and phenotype of which has been unusually well documented, offering important opportunities to study contemporary evolutionary processes, in particular the ways in which organisms respond to rapidly changing environments (Gluckman and Hanson 2006a, 2006b; Gluckman et al. 2011). Novel ways in which data collected in clinical studies and medical records could be used to track and evaluate the evolution of human traits – including the evolution of disease – have been suggested, and first attempts in this direction have been carried out (Byars et al. 2010; Stearns et al. 2010).

A recent example of how an evolutionary hypothesis may be tested in a clinical environment and how it may contribute knowledge of practical medical and public health significance comes from the area within evolutionary medicine focusing on life history and the role of adaptive plasticity in development and evolution. The basis for the hypothesis was a consistent finding that, at any maternal age and in developed and developing countries alike, the firstborns are born smaller than later-born children. The 1991 U.S. statistics, for instance, showed that second-born weighed at least 500 grams more on average than the firstborn children; firstborn also had a lower proportion of high, and higher proportion of low, birth weight (Cogswell and Yip 1995). Gambian firstborns had significantly depressed birth weights in comparison with infants of higher birth orders (Prentice, Cole, and Whitehead 1987). The phenomenon underlying the limitation of fetal growth is known as "maternal constraint" (Gluckman and Hanson 2004b). The limitation of fetal growth is caused by reduced supply of nutrients to the fetus and is generally thought to be caused by lesser dilatation of uterine arteries in the first pregnancy in comparison with later pregnancies, as the first pregnancy leads to disruption of connective tissue in the vascular wall, allowing greater dilation, and thus greater uterine blood flow, in subsequent pregnancies. While maternal constraint operates in all pregnancies, it is greater in first pregnancies as well as young and short mothers.

Evolutionary biology has provided a convincing explanation for maternal constraint: it has evolved to protect a small female from giving birth to a large infant following mating by a large male. If fetal growth was not constrained, obstructive labor and maternal and fetal death would have ensued. Studies have shown that fetal growth is primarily regulated by the size of the mother and that it may be demonstrated in all monotocous species: the classic experiments were those of Arthur Walton and John Hammond, who studied cross-breeding in horses in the 1930s University of Cambridge School of Agriculture (Walton and Hammond 1938). In human pregnancies involving surrogate mothers, birth size correlates better with the recipient's than donor mother's size (A. A. Brooks et al. 1995). In hominins, an additional reason for constraining fetal growth is the adoption of bipedal posture and the consequent change in the shape of the pelvic canal. Further, as hominins evolved with larger brain sizes relative to other primates, they had to be born at a more altricial stage to allow the large head to exit the pelvis.

It is a reasonable presumption that in the Paleolithic women were mated soon after menarche. While fertility is low in the first year after menarche, within a few years it reaches its maximum. The age of puberty in the Paleolithic is unknown but, on the basis of studies of modern hunter-gatherers where high extrinsic mortality creates a strong selective pressure for accelerated development (Walker et al. 2006), it is likely that it was low (Gluckman and Hanson 2006a). As the female pelvis does not reach its maximal dimensions until at least four years after menarche (Rosenberg and Trevathan 1995), maternal constraint would have been critical in the Paleolithic to match the firstborn fetus to the somatic size of the mother at the time of the pregnancy. A second hypothesis, which is not mutually exclusive, argued that energetically it made sense for the mother to conserve nutrients for subsequent pregnancies in which survival was likely to be higher (Metcalfe and Monaghan 2001). This argument is supported by the finding that short interbirth intervals are linked with small size at birth, reflecting the depleted energetic state of mother (Baker et al. 2009).

Yet the size of the firstborn would not have necessarily been of interest to clinical medicine were it not for their subsequent growth and morbidity patterns. While born small, the firstborn in the course of their first year show rapid catch-up growth and overshoot other infants (Ong et al. 2002). There is an extensive literature linking low (but even extending to normal range) birth weight, catch-up growth, and later metabolic compromise (Hales and Ozanne 2003). The initial evolutionary arguments for this relationship have been summarized as "the thrifty phenotype hypothesis," arguing that the fetus in difficult circumstances trades off growth to survive in utero but might then suffer later adverse consequences (Hales and Barker 1992). More recently, that hypothesis has been extensively refined as predictive adaptive response hypothesis (Gluckman, Hanson, and Spencer 2005). Briefly, this hypothesis posits that the fetus sets its developmental trajectory in accord with the nutritional information it receives from its mother, to optimize its fitness for the environment it anticipates on the basis of the received cues that will be operative later in life. Greater maternal constraint biases the prediction toward a poor nutrition environment (Gluckman and Hanson 2004b). Such a prediction would have had no adverse consequences in a Paleolithic environment, as the predicted environment would have corresponded with the experienced one; yet in the modern nutritional environment, it is inappropriate and mismatched. Obesity arises because the offspring developed with a physiology appropriate for a low-nutrition plane is exposed to a high plane of nutrition, invisible during the fetal period because of the constraint pathways.

Building on these observations and hypotheses, Gluckman and Hanson (2004a) postulated that firstborn children would be more at risk of obesity in middle age. This observation was confirmed in a modern U.K. population for which it was found that at any maternal body weight firstborn children were significantly more obese as adults (between twenty-seven and thirty years old) than children of higher birth order, and that the increase in fat was largely central (Reynolds et al. 2010). The favored explanation of how being born smaller leads to greater visceral obesity relies on epigenetic mechanisms operative in utero affecting intermediary metabolism (Godfrey et al. 2009) and appetite regulatory mechanisms (Vickers et al. 2000).

Different parental attitudes toward infant feeding and thus juvenile weight gain may contribute toward a different weight gain pattern. Either way, even in infancy firstborn children gain weight faster (Stettler et al. 2000). In the current social, clinical, and public health environment, where the family size is falling dramatically and where in China as well as many European countries more than 50 percent of all children are now firstborn, the significance of this phenomenon is evident. While much of the described clinical and laboratory research is ongoing, this example persuasively shows how evolutionary biology, epidemiology, and clinical and laboratory medicine can be jointly recruited to construct and test hypotheses.

Yet, in spite of such promising initiatives, the biggest challenge for evolutionary medicine remains developing a coherent set of approaches to test evolutionary hypotheses. This is a field where "idle Darwinizing," to use Lewontin's famous phrase, is easy, yet it requires increasingly formal evaluation – though it is limited by the generally historical nature of the argument. Nevertheless, there are formal approaches that are possible and are increasingly being followed (Nesse 2011). As the rigor is applied, the discipline will strengthen, and its value to clinical medicine in providing an essential core medical competency may be accepted. Erasmus Darwin's vision may, then, become reality.

Bibliography

Abrams, P. 1983. The theory of limiting similarity. *Annual Review of Ecology and Systematics* 14: 359–76.

Abrard, R. 1928. Contribution à l'étude de l'évolution des Nummulites. *Bulletin de la Société zoologique de France* 28, no. 4: 161–82.

Ackerknecht, E. H. 1982. Diathesis: The word and the concept in medical history. *Bulletin of the History of Medicine* 53, no. 3: 317–25.

Adam, K. 1926. Die Theologie der Krisis. *Hochland* 23: 271–86.

Adams, M., ed. 1994. *The Evolution of Theodosius Dobzhansky: Essays on His Life and Thought in Russia and America*. Princeton: Princeton University Press.

Adriaens, P. R., and A. De Block. 2010. The evolutionary turn in psychiatry: A historical overview. *History of Psychiatry* 21: 131–43.

Agassiz, L. 1840. *Etudes sur les glaciers*. Neuchatel: Jent & Gasmann.

——— 1841. On glaciers and boulders in Switzerland. *Tenth Meeting of the British Association for the Advancement of Science Notices and Abstracts* (Glasgow), 113–14. London: John Murray.

——— 1842. Glaciers, and the evidence of their former existence in Scotland, Ireland, and England. *Proceedings of the Geological Society of London* 3: 327–32.

——— 1850a. Geographical distribution of animals. *Christian Examiner and Religious Miscellany* 48: 181–204.

——— 1850b. The diversity of origin of the human races. *Christian Examiner and Religious Miscellany* 49: 110–45.

——— 1857. *Essay on Classification. Part I of the First Volume of the Contributions to the Natural History of the United States of North America*. Boston: Little, Brown.

Ahlberg, P. E., and J. A. Clack. 2006. Palaeontology: A firm step from water to land. *Nature* 440: 747–49.

Airy, H. 1882. On infection considered from a Darwinian point of view. *Transactions of the Epidemiological Society of London* 4: 246–61.

Alberch, P. 1991. From genes to phenotype: Dynamical systems and evolvability. *Genetica* 84: 5–11.

Allen, C., M. Bekhoff, G. Lauder, eds. 1998. *Nature's Purposes: Analyses of Function and Design in Biology*. Cambridge, Mass.: MIT Press.

Allen, C. E. 1933. Sachs, the last of the botanical epitomists. *Bulletin of the Torrey Botanical Club* 6, no. 5: 341–46.

Allen, G. E. 1969. Hugo de Vries and the reception of "mutation theory." *Journal of the History of Biology* 2, no. 1: 55–87.

——— 1975. The introduction of drosophila into heredity and evolution (1900–10). *Isis* 66: 322–33.

——— 1978. *Thomas Hunt Morgan: The Man and His Science*. Princeton: Princeton University Press.

Allison, A. C. 1954a. Protection afforded by sickle-cell trait against subtertian malarial infection. *British Medical Journal* 1: 290–94.

——— 1954b. The distribution of the sickle-cell trait in East Africa and elsewhere and its apparent relationship to the incidence of subtertian malaria. *Transactions of the Royal Society of Tropical Medical Hygiene* 48: 312–18.

Alter, S. 1999. *Darwinism and the Linguistic Image: Language, Race and Natural Theology in the Nineteenth Century*. Baltimore: Johns Hopkins University Press.

——— 2007a. Race, language and mental evolution in Darwin's *Descent of Man*. *Journal of the History of the Behavioral Sciences* 43, no. 3: 239–55.

——— 2007b. Darwin and the linguists: The coevolution of mind and language, Part I: Problematic friends. *Studies in History and Philosophy of Biological and Biomedical Sciences* 38C: 573–84.

——— 2007c. The advantages of obscurity: Charles Darwin's negative inference from the histories of domestic breeds. *Annals of Science* 64: 235–50.

——— 2008. Darwin and the linguists: The coevolution of mind and language, Part II: The language-mind relationship. *Studies in History and Philosophy of Biological and Biomedical Sciences* 39C: 38–50.

Amigoni, D. 2008. "The luxury of storytelling": Science, literature and cultural contest in Ian McEwan's narrative practice. In *Literature and Science*, ed. S. Ruston, 151–67. Cambridge: D. S. Brewer.

Anderson, L. I., and M. Lowe. 2010. Charles W. Peach and Darwin's barnacles. *Journal of the History of Collections* 22, no. 2: 1–14.

Anderson, W. 2004. Natural histories of infectious disease: Ecological vision in twentieth-century biomedical science. *Osiris* 19: 39–61.

Anonymous. 1859. Darwin's *Origin of Species*. *Saturday Review* 8: 775–76.

——— 1862. Mr. Darwin's orchids. *Saturday Review* 14: 486.

——— 1868. Review of *The Variation of Animals and Plants under Domestication*. *North American Review* 107: 362–68.

——— 1926. Letter to the editor of *Science* from the principal scientific authority of the fundamentalists. *Science* 63: 259.

——— 1943. Prof. K. A. Timiriazev, For.Mem.R.S. (1843–1920). *Nature* 151: 611.

——— 1979. The Christianity Today-Gallup poll: An overview. *Christianity Today*, 21 December: 12–15.

——— 1986. Creationism in NZ "unlikely." *NZ Herald*, 3 July, 14.

——— 2005. The chimpanzee genome. *Nature* 437: 47–108.

Anti-Defamation League. 2008. Anti-evolution Film Misappropriates the Holocaust (29 April 2008). http://www.adl.org/PresRele/HolNa_52/5277_52.htm.

Antonovics, J., J. L. Abbate, C. H. Baker, et al. 2007. Evolution by any other name: Antibiotic resistance and avoidance of the E-word. *PLoS Biology* 5, no. 2: e30.

Appel, T. A. 1987. *The Cuvier-Geoffroy Debate: French Biology in the Decades before Darwin*. New York: Oxford University Press.

Appleton, J. 1975. *The Experience of Landscape*. London: John Wiley.

Arbib, M. 2005. From monkey-like action recognition to human language: An evolutionary framework for neurolinguistics. *Behavioral and Brain Sciences* 28: 105–67.

Ardrey, R. 1961. *African Genesis: A Personal Investigation into the Animal Origins and Nature of Man*. London: Collins.

Argueta Villamar, A. 2009. *El darwinismo en Iberoamérica: Bolivia y México*. Madrid: CSIC.

Arjomand, K. 1998. In defense of the sacred doctrine: Muhammad Husayn al-Shahristānī's refutation of materialism and evolutionary theories of natural history. *Hallesche Beiträge zur Orientwissenschaft* 25: 1–18.

Armstrong, D. F., W. C. Stokoe, and S. E. Wilcox. 1994. Signs of the origin of syntax. *Current Anthropology* 35: 349–68.

Armstrong, D. F., and S. E. Wilcox. 2007. *The Gestural Origin of Language*. New York: Oxford University Press.

Atkinson, Q., and R. D. Gray. 2005. Curious parallels and curious connections: Phylogenetic thinking in biology and historical linguistics. *Systematic Biology* 54: 513–26.

Atran, S. 2002. *In Gods We Trust: The Evolutionary Landscape of Religion*. Oxford: Oxford University Press.

Avdulov, N. P. 1931. Karyo-systematische Untersuchung der Familie Gramineen. *Bulletin of Applied Botany* 44: 1–428.

Avital, E., and E. Jablonka. 2000. *Animal Traditions: Behavioral Inheritance in Evolution*. Cambridge: Cambridge University Press.

Ayala, F. J., ed. 1976. *Molecular Evolution*. Sunderland, Mass.: Sinauer Associates.

Ayres, P. 2008. *The Aliveness of Plants: The Darwins at the Dawn of Plant Science*. London: Pickering & Chatto.

Babcock, E. B. 1924. Species hybrids in *Crepis* and their bearing on evolution. *American Naturalist* 58, no. 657: 296–310.

——— 1947. The Genus Crepis I and II. *University of California Publications, Botany* 21: 22.

Babcock, E. B., and H. M. Hall. 1924. *Hemizonia congesta, a Genetic, Ecologic, and Taxonomic Study of the Hay-Field Tarweeds*. Berkeley: University of California Press.

Babcock, E. B., and G. L. Stebbins Jr. 1938. The American species of crepis: Their interrelationships and distribution as affected by polyploidy. *Carnegie Institution of Washington Publication* no. 504.

Bahr, H. 1894. *Der Antisemitismus: Ein internationales Interview*. Berlin: S. Fischer.

Baker, J., A. M. Hurtado, O. M. Pearson, K. R. Hill, T. Jones, and M. A. Frey. 2009. Developmental plasticity in fat patterning of Ache children in response to variation in interbirth intervals: A preliminary test of the roles of external environment and maternal reproductive strategies. *American Journal of Human Biology* 21, no. 1: 77–83.

Bannister, R. C. 1979. *Social Darwinism: Science and Myth in Anglo-American Social Thought*. Philadelphia: Temple University Press.

——— 2006. Social Darwinism: An American Perennial. http://www.swarthmore.edu/SocSci/rbannis1/SD.htm.

Bapteste, E., and Y. Boucher. 2008. Lateral gene transfer challenges principles of microbial systematics. *Cell* 16: 200–7.

Bapteste, E., Y. Boucher, J. Leigh, and W. F. Doolittle. 2004. Phylogenetic reconstruction and lateral gene transfer. *Trends in Microbiology* 12: 406–11.

Bapteste, E., E. Susko, J. Leigh, D. MacLeod, R. L. Charlebois, and W. F. Doolittle. 2005. Do orthologous gene phylogenies really support tree-thinking? *BMC Evolutionary Biology* 5: 33.

Barkow, J. H., L. Cosmides, and J. Tooby. 1992. *The Adapted Mind: Evolutionary Psychology and the Generation of Culture*. Oxford: Oxford University Press.

Barlow, N. 1945. *Charles Darwin and the Voyage of the* Beagle. London: Pilot Press.

——— 1967. *Darwin and Henslow: The Growth of an Idea*. London: John Murray.

Barnes, J., ed. 1995. *The Complete Works of Aristotle*. Princeton: Princeton University Press.

Barrett, P. H., ed. 1977. *The Collected Papers of Charles Darwin*. 2 vols. Chicago: University of Chicago Press.

Barrett, P. H., P. J. Gautrey, S. Herbert, D. Kohn, and S. Smith, eds. 1987. *Charles Darwin's Notebooks, 1836–1844*. Cambridge: Cambridge University Press.

Barroso, I. 2012. Encode explained: Non-coding but functional. *Nature* 489: 54.

Barth, K. 1960. *Church Dogmatics: The Doctrine of Creation*. Vol. 3. Ed. G. W. Bromiley and T. F. Torrance. London: T&T Clark.

Bartholomew, M. 1975. Huxley's defence of Darwinism. *Annals of Science* 32: 525–35.

Barton, N. H., D. E. G. Briggs, J. A. Eisen, D. B. Goldstein, and N. H. Patel. 2007. *Evolution*. Cold Spring Harbor, N.Y.: Cold Spring Harbor Laboratory Press.

Basalla, G. 1988. *The Evolution of Technology*. Cambridge: Cambridge University Press.

Bateman, A. J. 1948. Intra-sexual selection in Drosophila. *Heredity* 2: 349–68.

Bates, E., D. Thal, and V. Marchman. 1991. Symbols and syntax: A Darwinian approach to language development. In *Biological and Behavioral Determinants of Language Development*, ed. D. M. Rumbaugh, R. R. Schiefelbusch, M. Studdert-Kennedy, and N. Krasnegor, 29–54. Hillsdale, N.J.: Lawrence Erlbaum.

Bates, H. W. 1862. Contributions to an insect fauna of the Amazon Valley. Lepidoptera: Heliconidæ. *Transactions of the Linnean Society of London* 23: 496–566.

——— 1863. *The Naturalist on the River Amazons*. London: John Murray.

——— 1892. *The Naturalist on the River Amazons*. 2nd ed. London: John Murray.

Bateson, B. 1928. *William Bateson, F.R.S., Naturalist: His Essays and Addresses Together with a Short Account of His Life*. Cambridge: Cambridge University Press.

Bateson, G. 1972. *Steps to an Ecology of Mind: Collected Essays in Anthropology, Psychiatry, Evolution, and Epistemology*. Chicago: University of Chicago Press.

Bateson, W. 1894. *Materials for the Study of Variation, Treated with Especial Regard to Discontinuity in the Origin of Species*. London: Macmillan.

——— 1902. *Mendel's Principles of Heredity: A Defence*. Cambridge: Cambridge University Press.

——— 1909. *Mendel's Principles of Heredity*. Cambridge: Cambridge University Press.

——— 1913. *Problems of Genetics*. New Haven: Yale University Press.

——— 1922. Evolutionary faith and modern doubts. *Science* 55: 55–61.

——— 1930. *Mendel's Principles of Heredity*. 4th printing. Cambridge: Cambridge University Press.

Baur, E. 1932. Artumgrenzung und Artbildung in der Gattung *Antirrhinum*, Sektion Antirrinastrum. *Zeitschrift für Inductive Abstammungs- und Vererbungslehre.* 63: 256–302.

Baur, G. 1891. Geography and travel. *American Naturalist* 25, no. 298: 902–7.

——— 1897. Birds of the Galapagos Archipelago: A criticism of Mr. Robert Ridgway's paper. *American Naturalist* 31, no. 369: 777–84.

Bavinck, H. 2004. *Reformed Dogmatics: God and Creation.* Ed. J. Bolt. Trans. J. Vriend. Grand Rapids, Mich.: Baker.

——— 2006. *Reformed Dogmatics: Sin and Salvation in Christ.* Ed. J. Bolt. Trans. J. Vriend. Grand Rapids, Mich.: Baker.

Bearman, P., T. Bianquis, C. E. Bosworth, E. van Donzel, and W. P. Heinrichs, eds. 1960–2007. *The Encyclopaedia of Islam.* 2nd ed. Leiden: Brill.

Beatty, J. 1985. Speaking of species: Darwin's strategy. In *The Darwinian Heritage*, ed. D. Kohn, 265–81. Princeton: Princeton University Press.

——— 1990. Teleology and the relationship between biology and the physical sciences in the nineteenth and twentieth centuries. In *Some Truer Method: Reflections on the Heritage of Newton*, ed. F. Durham and R. D. Purrington. 113–44. New York: Columbia University Press.

Beckner, M. 1959. *The Biological Way of Thought.* New York: Columbia University Press.

Becquemont, D. 2009. *Charles Darwin, 1837–1839: Aux sourcés d'une decouvérté.* Paris: Editions Kime.

Beer, G. 1983. *Darwin's Plots: Evolutionary Narrative in Darwin, George Eliot, and Nineteenth-Century Fiction.* London: Routledge and Kegan Paul.

——— 1985. Darwin's reading and the fictions of development. In *The Darwinian Heritage*, ed. D. Kohn, 543–88. Princeton: Princeton University Press.

——— 1989. Darwin and the growth of language theory. In *Nature Transfigured: Science and Literature, 1700–1900*, ed. J. Christie and S. Shuttleworth, 152–70. Manchester: Manchester University Press.

——— 1996. Introduction: Note on the text. In *On the Origin of Species* by C. Darwin, vii–xxix. Oxford: Oxford University Press.

——— 2009. *Darwin's Plots: Evolutionary Narrative in Darwin, George Eliot and Nineteenth-Century Fiction.* 3rd ed. Cambridge: Cambridge University Press.

Behe, M. J. 1996. *Darwin's Black Box: The Biochemical Challenge to Evolution.* New York: Free Press.

Beilby, J. K., ed. 2002. *Naturalism Defeated? Essays on Plantinga's Evolutionary Argument against Naturalism.* Ithaca, N.Y.: Cornell University Press.

Béjin, A. 1992. Les trois phases de l'évolution du darwinisme social en France. In *Darwinisme et société*, ed. P. Tort, 353–60. Paris: Presses Universitaires de France.

Bell, M. A., M. P. Travis, and D. M. Blouw. 2006. Inferring natural selection in a fossil threespine stickleback. *Paleobiology* 32: 562–77.

Bellon, R. 2003. "The great question in agitation": George Bentham and the origin of species. *Archives of Natural History* 30: 282–97.

——— 2009. Charles Darwin solves the "riddle of the flower," or Why don't historians of biology know about the birds and bees? *History of Science* 47: 373–406.

——— 2011. Inspiration in the harness of daily labor: Darwin, botany and the triumph of evolution, 1859–1868. *Isis* 102: 393–420.

Belt, T. 1874. *The Naturalist in Nicaragua.* London: John Murray.

Bender, B. 1996. *The Descent of Love: Darwin and the Theory of Sexual Selection in American Fiction, 1871–1926.* Philadelphia: University of Pennsylvania Press.

——— 2004. *Evolution and the "Sex Problem": American Narratives during the Eclipse of Darwinism.* Kent, Ohio: Kent State University Press.

Berliner, J. 2008. Jack London's socialistic social Darwinism. *American Literary Realism* 41: 52–78.

Bernard, F. 1895. *Éléments de paleontologie.* Paris: J.-B. Baillière.

Bernstein, S. D. 2001. Ape anxiety: Sensation fiction, evolution, and the genre question. *Journal of Victorian Culture* 6, no. 2: 250–71.

Berry, A., ed. 2002. *Infinite Tropics: An Alfred Russel Wallace Anthology.* London: Verso.

Bickerton, D. 1990. *Language and Species.* Chicago: University of Chicago Press.

——— 2000. How protolanguage became language. In *The Evolutionary Emergence of Language: Social Function and the Origins of Linguistic Form*, ed. C. Knight, M. Studdert-Kennedy, and J. Hurford, 264–84. Cambridge: Cambridge University Press.

——— 2009. *Adam's Tongue: How Humans Made Language, How Language Made Humans.* New York: Hill and Wang.

Blackman, H. J. 2006. Anatomy and embryology in medical education at Cambridge University, 1866–1900. *Medical Education* 40, no. 3: 219–26.

Blackmore, S. 1999. *The Meme Machine.* Oxford: Oxford University Press.

——— 2009. Memetics does provide a useful way of understanding cultural evolution. In *Contemporary Debates in Philosophy of Biology*, ed. Francisco J. Ayala and Robert Arp, 255–72. Hoboken, N. J.: John Wiley & Sons.

Blair, W. F. 1955. Mating call and stage of speciation in the *Microhyla olivacea – M. carolinensis* complex. *Evolution* 9: 469–80.

Blanckaert, C. 1991. "Les bas-fonds de la science française"; Clémence Royer, l'origine de l'homme et le darwinisme social. *Bulletins et mémoires de la Société d'anthropologie de Paris*, n.s., 3, nos. 1–2: 115–30.

Bloch, M. 2000. A well-disposed social anthropologist's problems with memes. In *Darwinizing Culture: The Status of Memetics as a Science*, ed. R. Aunger, 189–203. Oxford: Oxford University Press.

Blouet, B. W. 1987. *Halford Mackinder: A Biography.* College Station: Texas A&M Press.

Blutinger, J. C. 2010. Creatures from before the flood: Reconciling science and Genesis in the pages of a nineteenth-century Hebrew newspaper. *Jewish Social Studies* 16, no. 2: 67–92.

Boakes, R. 1984. *From Darwin to Behaviourism: Psychology and the Minds of Animals.* Cambridge: Cambridge University Press.

Boas, Franz. 1909. The relation of Darwin to anthropology. *Current Anthropology* 42, no. 3: 381–406.

Boitard, P., and Corbié. 1824. *Les pigeons de volière et de colombier.* Paris: Audut.

Bonner, J. T., ed. 1982. *Evolution and Development.* Berlin: Springer Verlag.

Borello, M. E. 2005. The rise, fall and resurrection of group selection. *Endeavour* 29: 43–47.

——— 2010. *Evolutionary Restraints: The Contentious History of Group Selection.* Chicago: University of Chicago Press.

Botha, R. P. 2009. Theoretical underpinnings of inferences about language evolution: The syntax used at Blombos Cave. In *The Cradle of Language*, ed. R. P. Botha and C. Knight, 93–111. Oxford: Oxford University Press.

Boveri, T. 1906. *Die Organismen als historische Wesen.* Würzburg: Königliche Universitätsdruckerei von H. Stürz.

Bowler, P. J. 1976. Malthus, Darwin, and the concept of struggle. *Journal of the History of Ideas* 37: 631–50.

1983. *The Eclipse of Darwinism: Anti-Darwinian Evolution Theories in the Decades around 1900*. Baltimore: Johns Hopkins University Press.

1986. *Theories of Human Evolution, 1844–1944: A Century of Debate*. Baltimore: Johns Hopkins University Press.

1988. *The Non-Darwinian Revolution: Reinterpreting a Historical Myth*. Baltimore: Johns Hopkins University Press.

1996. *Life's Splendid Drama: Evolutionary Biology and the Reconstruction of Life's Ancestry, 1860–1940*. Chicago: University of Chicago Press.

2001. *Reconciling Science and Religion: The Debate in Early-Twentieth-Century Britain*. Chicago: Chicago University Press.

2007. *Monkey Trials and Gorilla Sermons: Evolution and Christianity from Darwin to Intelligent Design*. Cambridge, Mass.: Harvard University Press.

2008. What Darwin disturbed: The biology that might have been. *Isis* 99, no. 3: 560–67.

2009. Geographical distribution in the *Origin of Species*. In *The Cambridge Companion to the "Origin of Species,"* ed. M. Ruse and R. J. Richards, 153–72. Cambridge: Cambridge University Press.

Bowman, R. I. 1961. *Morphological Differentiation and Adaptation in the Galapagos*. Berkeley: University of California Press.

1963. Evolutionary patterns in Darwin's finches. *Occasional Papers of the California Academy of Sciences* 44: 107–40.

Box, J. F. 1978. *R. A. Fisher: The Life of a Scientist*. New York: Wiley.

Boyd, R., and P. J. Richerson. 1985. *Culture and the Evolutionary Process*. Chicago: Chicago University Press.

Boyer, P. 2001. *Religion Explained: The Human Instincts That Fashion Gods, Spirits and Ancestors*. London: William Heinemann.

Brackman, A. C. 1980. *A Delicate Arrangement*. New York: Times Books.

Bradie, M. 1986. Assessing evolutionary epistemology. *Biology and Philosophy* 1: 406–59.

Bradshaw, A. D., and V. B. Smocovitis. 2005. George Ledyard Stebbins. *Biographical Memoirs of Fellows of the Royal Society* 51: 398–408.

Brantlinger, P. 2003. *Dark Vanishings: Discourse on the Extinction of Primitive Races, 1800–1930*. Ithaca, N.Y.: Cornell University Press.

Brewster, D. 1838. Review of Comte's "Cours de Philosophie Positive." *Edinburgh Review* 67: 271–308.

Briggs, D., and S. M. Walters. 1997. *Plant Variation and Evolution*. 3rd ed. Cambridge: Cambridge University Press.

Brinkman, P. D. 2003. Bartholomew James Sulivan's discovery of fossil vertebrates in the Tertiary beds of Patagonia. *Archives of Natural History* 30, no. 1: 56–74.

2010. Charles Darwin's *Beagle* voyage, fossil vertebrate succession, and the gradual birth and death of species. *Journal of the History of Biology* 43, no. 1: 363–99.

Britten, R. J., and E. H. Davidson. 1969. Gene regulation for higher cells – a theory. *Science* 165, no. 891: 349–57.

1971. Repetitive and non-repetitive DNA sequences and a speculation on origins of evolutionary novelty. *Quarterly Review of Biology* 46, no. 2: 111–38.

Broad, C. D. 1944. Critical notice of Julian Huxley's *Evolutionary Ethics*. *Mind* 53: 344–67.

Brockmann, H. J., A. Grafen, and R. Dawkins. 1979. Evolutionarily stable nesting strategy in a digger wasp. *Journal of Theoretical Biology* 77: 473–96.

Bronn, H. 1860. Schlusswort des Übersetzers. In *Über die Enstehung der Arten*, by C. Darwin, 495–520. Stuttgart: Schweizerbart'sche Verlagshandlung.

Brookes, R. 1763a. *The Natural History of Insects, with Their Properties and Uses in Medicine*. Vol. 4. London: Newbery.

1763b. *The Natural History of Waters, Earths, Stones, Fossils and Minerals. With Their Virtues, Properties, and Medicinal Uses; to Which Is Added, the Method in with Linnaeus Has Treated These Subjects*. Vol. 5. London: Newbery.

Brooks, A. A., M. R. Johnson, P. J. Steer, M. E. Pawson, and H. I. Abdalla. 1995. Birth weight: Nature or nurture? *Early Human Development* 42, no. 1: 29–35.

Brooks, J. L. 1984. *Just before the Origin*. New York: Columbia University Press.

Brooks, W. K. 1883. *The Law of Heredity: A Study of the Cause of Variation, and the Origin of Living Things*. Baltimore: John Murphy.

Broom, R. 1931. *The Coming of Man: Was It Accident or Design?* London: Witherby.

Broom, R., and G. W. H. Schepers. 1946. *The South African Fossil Ape-Men: The Australopithecinae*. Pretoria: Transvaal Museum.

Brown, D. 1997. *Mysteries of the Creation*. Southfield, Mich.: Targum Press.

Brown, R. 1833. On the organs and mode of fucundation in Orchideae and Asclepiadeae. *Transactions of the Linnean Society of London* 16: 648–745.

Brown, W. L., and E. O. Wilson. 1956. Character displacement. *Systematic Zoology* 5, no. 2: 49–64.

Browne, J. 1983. *The Secular Ark: Studies in the History of Biogeography*. New Haven: Yale University Press.

1985a. Darwin and the expression of the emotions. In *The Darwinian Heritage*, ed. David Kohn, 307–26. Princeton: Princeton University Press.

1985b. Darwin and the face of madness. In *The Anatomy of Madness: Essays in the History of Psychiatry*, ed. R. Porter, W. F. Bynum, and M. Shepherd, 151–65. London: Tavistock Publications.

1995. *Charles Darwin: A Biography*. Vol. 1, *Voyaging*. New York: Knopf.

2002. *Charles Darwin: A Biography*. Vol. 2, *The Power of Place*. New York: Knopf.

2005a. Commemorating Darwin. *British Journal for the History of Science* 38: 251–74.

2005b. Constructing Darwinism in literary culture. In *Unmapped Countries: Biological Visions in Nineteenth-Century Literature and Culture*, ed. A.-J. Zwierlein, 55–70. London: Anthem Press.

Browne, J., and S. Messenger. 2003. Victorian spectacle: Julia Pastrana, the bearded and hairy female. *Endeavour* 27, no. 4: 155–59.

Buckland, W. 1836. *Geology and Mineralogy Considered with Reference to Natural Theology*. Bridgewater Treatise VI. London: William Pickering.

1842. Evidences of glaciers in Scotland and the north of England. *Proceedings of the Geological Society of London* 3: 333–38, 345–48.

Buckle, H. T. 1857–61. *History of Civilization in England*. 2 vols. London: Parker.

Bulmer, M. G. 2003. *Francis Galton: Pioneer of Heredity and Biometry*. Baltimore: Johns Hopkins University Press.

Bumpus, H. C. 1899. The elimination of the unfit as illustrated by the introduced House Sparrow, *Passer domesticus*. *Biological Lectures, Marine Biological Laboratory, Woods Hole* 11: 209–26.

Burchfield, J. D. 1974. Darwin and the dilemma of geological time. *Isis* 65: 300–21.

1975. *Lord Kelvin and the Age of the Earth*. New York: Science History Publications.

Burkhardt, F. 1988. England and Scotland. The learned societies. In *The Comparative Reception of Darwinism*, ed. T. F. Glick, 32–74. Chicago: University of Chicago Press.

Burkhardt, R. W. 1977. *The Spirit of System: Lamarck and Evolutionary Biology*. Cambridge, Mass.: Harvard University Press.

———. 1983. The development of an evolutionary ethology. In *Evolution from Molecules to Men*, ed. D. Bendall, 429–44. Cambridge: Cambridge University Press.

———. 1995. *The Spirit of System: Lamarck and Evolutionary Biology*. 2nd ed. Cambridge, Mass.: Harvard University Press.

———. 2005. *Patterns of Behavior: Konrad Lorenz, Niko Tinbergen, and the Founding of Ethology*. Chicago: University of Chicago Press.

Burling, R. 2005. *The Talking Ape: How Language Evolved*. Oxford: Oxford University Press.

Burton, J. 1994. *An Introduction to the Hadith*. Edinburgh: University Press.

Bury, J. B. 1920. *The Idea of Progress; An Inquiry into Its Origin and Growth*. London: Macmillan.

Butler, R. J. 1960. Natural belief and the enigma of Hume. *Archiv für die Geschichte der Philosophie* 42: 73–100.

Butler, S. 1887. *Luck, or Cunning, as the Main Means of Organic Modification? An Attempt to Throw Additional Light upon the Late Mr. Charles Darwin's Theory of Natural Selection*. London: Trubner.

Butlin, R. K. 1985. Speciation by reinforcement. In *Orthoptera*, ed. J. Gosálvez, 84–113. Madrid: Fundación Ramón Areces.

Buttemer, W. A. 1992. Differential overnight survival by Bumpus' House Sparrows: An alternative interpretation. *The Condor* 94: 944–54.

Byars, S. G., D. Ewbank, D. R. Govindaraju, and S. C. Stearns. 2010. Natural selection in a contemporary human population. *Proceedings of the National Academy of Sciences* 107: 1787–92.

Bynum, W. F. 1983. Darwin and the doctors: Evolution, diathesis, and germs in 19th-century Britain. *Gesnerus* 40, nos. 1–2: 43–53.

———. 1994. *Science and the Practice of Medicine in the Nineteenth Century*. Cambridge: Cambridge University Press.

———. 2002. The evolution of germs and the evolution of disease: Some British debates, 1870–1900. *History and Philosophy of the Life Sciences* 24, no. 1: 53–68.

Cain, A. J. 1954. *Animal Species and Their Evolution*. London: Hutchison.

———. 1989. The perfection of animals. *Biological Journal of the Linnaean Society* 36: 3–29.

Cain, A. J., and W. B. Provine. 1992. Genes and ecology in history. In *Genes in Ecology*, ed. R. J. Berry, T. J. Crawford, and G. M. Hewitt, 3–28. Oxford: Blackwell.

Cain, A. J., and P. M. Sheppard. 1954. Natural selection in *Cepaea*. *Genetics* 398: 89–116.

Cain, J. 1993. Common problems and cooperative solutions: Organizational activity in evolutionary studies. *Isis* 84, no. 1: 1–25.

———. 2000. Towards a "Greater Degree of Integration": The Society for the Study of Speciation, 1939–1941. *British Journal for the History of Science* 33: 85–108.

———. 2009. Rethinking the synthesis period in evolutionary studies. *Journal of the History of Biology* 42, no. 4: 621–48.

———. 2010. Julian Huxley, general biology, and the London zoo, 1935–1942. *Notes and Records of the Royal Society* 64: 359–78.

Cairns-Smith, A. G. 1986. *Clay Minerals and the Origin of Life*. Cambridge: Cambridge University Press.

———. 1995. *Seven Clues to the Origin of Life*. Cambridge: Cambridge University Press.

Cambridge University. 1831. *Calendar*. Cambridge: J. Smith, printer to the University.

Camp, W. H., and C. L. Gilly. 1943. The structure and the origin of species. *Brittonia* 4: 323–85.

Campbell, B. G., ed. 1972. *Sexual Selection and the Descent of Man, 1871–1971*. Chicago: Aldine.

———, ed. 2006. *Sexual Selection and the Descent of Man: The Darwinian Pivot*. Piscataway, N. J.: Transaction.

Campbell, D. 1974. Evolutionary epistemology. In *The Philosophy of Karl Popper*, ed. P. Schilpp, 413–63. LaSalle, Ill.: Open Court.

Campbell, R. 1996. Can biology make ethics objective? *Biology and Philosophy* 11: 21–31.

Campbell Irons, J. 1896. *Autobiographical Sketch of James Croll LL.D., F.R.S., etc. with a Memoir of His Life and Work*. London: Edward Stanford.

Candolle, A.-P. de. 1819. *Théorie élémentaire de la botanique*. 2nd ed. Paris: Deterville.

———. 1839–40. *Vegetable Organography*. Trans. B. Kingdon. 2 vols. London: Houlston & Stoneman.

———. 1862. Review of *Orchids* by Charles Darwin. *Archives des Sciences Physiques et Naturelles* 15: 173–76.

Cannadine, D. 1999. *The Rise and Fall of Class in Britain*. New York: Columbia University Press.

Canning, G., H. Frere, and G. Ellis. 1798. The loves of the triangles. *Anti-Jacobin*, 16 April, 23 April, and 17 May.

Cannon, W. F. 1961. The impact of uniformitarianism. Two letters from John Herschel to Charles Lyell, 1836–1837. *Proceedings of the American Philosophical Society* 105: 301–14.

Cantor, G. 2005. *Quakers, Jews and Science: Religious Responses to Modernity and the Sciences in Britain, 1650–1900*. Oxford: Oxford University Press.

Cantor, G., and M. Swetlitz, eds. 2006. *Jewish Tradition and the Challenge of Evolution*. Chicago: University of Chicago Press.

Cao, J. 2003. *Zhongguo xueshu sixiang shi suibi* (Essays on Chinese History of Academic Thinking). Beijing: SDX Joint Publishing Company.

Carmell, A., and C. Domb, eds. 1976. *Challenge: Torah Views on Science and Its Problems*. New York: Feldheim Publishers.

Carnegie, A. 2009. *The Autobiography of Andrew Carnegie and the Gospel of Wealth*. Lawrence, Kans.: Digireads Publishing.

Carpenter, G. D. H., and E. B. Ford. 1933. *Mimicry*. London: Methuen.

Carroll, S. 2005. *Endless Forms Most Beautiful: The New Science of Evo-Devo*. New York: Norton.

———. 2008. Evo-Devo and expanding the evolutionary synthesis: A genetic theory of morphological evolution. *Cell* 134: 25–36.

Carroll, S. B., J. K. Grenier, and S. Weatherbee. 2005. *From DNA to Diversity: Molecular Genetics and the Evolution of Animal Design*. Malden, Mass.: Blackwell.

Carruthers, P. 1992. *Human Knowledge and Human Nature*. Oxford: Oxford University Press.

Cartmill, E. A., and R. W. Byrne. 2007. Orangutans modify their gestural signaling according to their audience's comprehension. *Current Biology* 17: 1345–48.

Casebeer, W. D. 2003. *Natural Ethical Facts: Evolution, Connectionism, and Moral Cognition*. Cambridge, Mass.: MIT Press.

Castle, W. E. 1917. Piebald rats and multiple factors. *American Naturalist* 51: 370–75.

Castle, W. E., and J. C. Phillips. 1914. *Piebald Rats and Selection: An Experimental Test of the Effectiveness of Selection and of the Theory of Gamete Purity in Mendelian Crosses*. Washington, D.C.: Carnegie Institution of Washington.

Castree, N. 2009. Charles Darwin and the geographers. *Environment and Planning A* 41: 2293–98.

Cela-Conde, C. J., and F. J. Ayala. 2007. *Human Evolution: Trails from the Past*. Oxford: Oxford University Press.

Chamberlain, H. 1899. *Die Grundlagen des Neuzehnten Jahrhunderts*. 2 vols. Munich: Verlagsanstalt F. Bruckmann.

Chambers, R. 1844. *Vestiges of the Natural History of Creation*. London: Churchill.

⸺ 1846. *Vestiges of the Natural History of Creation*. 5th ed. London: Churchill.

Charlesworth, B. 1994. The genetics of adaptation: lessons from mimicry. *American Naturalist* 144: 839–47.

Chen, J. 1922. Da Er Wen yihou zhi jinhua lun (Evolutionary Theories after Darwin). *Min Duo* (People's Bell) 3, no. 5: 1–24.

Cheney, D. L., and R. M. Seyfarth. 1990. *How Monkeys See the World*. Chicago: University of Chicago Press.

Cherry, S. 2001. Creation, Evolution and Jewish Thought. PhD dissertation, Brandeis University.

⸺ 2003. Three twentieth-century Jewish responses to evolutionary theory. *Aleph* 3: 247–90.

⸺ 2006. Crisis management via biblical interpretation: Fundamentalism, modern orthodoxy, and Genesis. In *Jewish Tradition and the Challenge of Evolution*, ed. G. Cantor and M. Swetlitz, 166–87. Chicago: University of Chicago Press.

⸺ 2011. Judaism, Darwinism, and the typology of suffering. *Zygon* 46, no. 2: 313–34.

Chomsky, N. 1965. *Aspects of the Theory of Syntax*. Cambridge, Mass.: MIT Press.

⸺ 1975a. *Reflections on Language*. New York: Pantheon.

⸺ 1975b. *The Logical Structure of Linguistic Theory*. New York: Plenum Press.

Chorley, R. J., R. P. Beckinsale, and A. J. Dunn. 1973. *The History of the Study of Landforms or the Development of Geomorphology*. Vol. 2, *The Life and Work of William Morris Davis*. London: Methuen.

Chung, P. S. 2009. *Constructing Irregular Theology: Bamboo and Minjung in East Asian Perspective*. Leiden: Brill.

Churchill, L. B. 2010. What is it? Difference, Darwin and the Victorian freak show. In *Darwin in Atlantic Cultures: Evolutionary Visions of Race, Gender and Sexuality*, ed. J. E. Jones and P. B. Sharp, 128–42. New York: Routledge.

Ciccarelli, F., T. Doerks, C. von Mering, C. Creevey, B. Snel, and P. Bork. 2006. Towards automatic reconstruction of a highly resolved tree of life. *Science* 311: 1283–87.

Clark, W. E. Le Gros. 1952. Anatomical studies of fossil Hominoidea from Africa. In *Proceedings of the Pan-African Congress of Prehistory, 1947*, ed. L. S. B. Leakey and S. Cole, 111–15. New York: Philosophical Library.

Clarke, C. 1990. Professor Sir Ronald Fisher, FRS. *British Medical Journal* 301, no. 6766: 1446–48.

Clarke, E., U. H. Reichard, and K. Zuberbühler. 2006. The syntax and meaning of wild gibbon songs. *PLoS One* 1: e73.

Clausen, J. 1951. *Stages in the Evolution of Plant Species*. Ithaca, N.Y.: Cornell University Press.

Clausen, R. E., and T. H. Goodspeed. 1925. Interspecific hybridization in Nicotiana II. A tetraploid *glutinosaTabacum* hybrid: An experimental verification of Winge's hypothesis. *Genetics* 10: 279–84.

Clements, F. E. 1907. Darwin's influence upon plant geography and ecology. *American Naturalist* 43: 143–51.

Clodd, E. 1892. A memoir of the author. In *The Naturalist on the River Amazons*, by H. W. Bates, xvii–lxxxix. London: John Murray.

Cobbe, F. P. 1872. *Darwinism in Morals and Other Essays*. London: Williams and Norgate.

⸺ 1904. *Life of Frances Power Cobbe as Told by Herself*. Ed. B. Atkinson. London: Swan Sonnenschein.

Cockerell, T. D. A. 1897. Physiological species. *Entomological News* 8: 234–36.

Cody, G. D., N. Z. Boctor, J. A. Brandes, T. R. Filley, R. M. Hazen, and H. S. Yoder Jr. 2004. Assaying the catalytic potential of transition metal sulfides for abiotic carbon fixation. *Geochimica et Cosmochimica Acta* 68, no. 10: 2185–96.

Cogswell, M. E., and R. Yip. 1995. The influence of fetal and maternal factors on the distribution of birthweight. *Seminars in Perinatology* 19, no. 3: 222–40.

Cohen, I. B. 1985. Three notes on the reception of Darwin's ideas on natural selection (Henry Baker Tristram, Alfred Newton, Samuel Wilberforce). In *The Darwinian Heritage*, ed. David Kohn, 589–607. Princeton: Princeton University Press.

Cohen, J., F. Briand, and C. Newman. 1990. *Community Food Webs: Data and Theory*. New York: Springer-Verlag.

Cohen, N. W. 1984. The challenges of Darwinism and biblical criticism to American Judaism. *Modern Judaism* 4: 121–57.

Colp, R., and D. Kohn. 1996. "A Real Curiosity": Charles Darwin reflects on a communication from Rabbi Naphtali Levy. *European Legacy* 1: 1716–27.

Conry, Y., 1974. *L'introduction du darwinisme en France au XIXe siècle*. Paris: Vrin.

Cooper, G. 2001. Must there be a balance of nature? *Biology and Philosophy* 16: 481–506.

Cooper, J. M., ed. 1997. *The Complete Works of Plato*. Indianapolis: Hackett.

Cooter, R., and J. Pickstone, eds. 2000. *Medicine in the Twentieth Century*. Amsterdam: Harwood Academic Publishers.

Cornford, F. M. 1932. *Before and After Socrates*. Cambridge: Cambridge University Press.

Corsi, P. 1988. *The Age of Lamarck*. Berkeley: University of California Press.

Cott, H. B. 1940. *Adaptive Colouration in Animals*. London: Methuen and Co.

Coyne J. A., and H. A. Orr. 2004. *Speciation*. Sunderland, Mass.: Sinauer Associates.

Cracraft J. 1989. Speciation and its ontology: The empirical consequences of alternative species concepts for understanding patterns and processes of differentiation. In *Speciation and Its Consequences*, ed. J. A. Endler and D. Otte, 28–59. Sunderland: Sinauer Associates.

Craig, P. 2005. *Centennial History of the Carnegie Institution of Washington*. Vol. 4, *The Department of Plant Biology*. Cambridge: Cambridge University Press.

Crick, F. 1970. Central dogma of molecular biology. *Nature* 227: 561–63.

Crisp, D. J. 1983. Extending Darwin's investigations on the barnacle life-history. *Biological Journal of the Linnean Society* 20: 73–83.

Crockford, C., and C. Boesch. 2003. Context-specific calls in wild chimpanzees, *Pan troglodytes verus*: Analysis of barks. *Animal Behaviour* 55: 115–25.

Croll, J. 1864. On the physical causes of the changes of climate during geological epochs. *The London, Edinburgh, and Dublin Philosophical Magazine and Journal of Science Fourth Series* 28: 121–37.

⸺ 1866. On the eccentricity of the Earth's orbit. *The London, Edinburgh, and Dublin Philosophical Magazine and Journal of Science Fourth Series* 31: 26–28.

⸺ 1867a. On the eccentricity of the Earth's orbit, and its physical relations to the glacial epoch. *The London, Edinburgh, and Dublin*

Philosophical Magazine and Journal of Science Fourth Series 33: 119–31.

———. 1867b. On the change in the obliquity of the ecliptic, its influence on the climate of the polar regions and on the level of the sea. *The London, Edinburgh, and Dublin Philosophical Magazine and Journal of Science Fourth Series* 33: 426–45.

———. 1868. On geological time, and the probable date of the Glacial and the Upper Miocene Period. *The London, Edinburgh, and Dublin Philosophical Magazine and Journal of Science Fourth Series* 35: 363–84.

———. 1875. *Climate and Time in Their Geological Relations: A Theory of Secular Changes of the Earth's Climate.* London: Daldy, Isbister.

Cronin, H. 1991. *The Ant and the Peacock: Altruism and Sexual Selection from Darwin to Today.* Cambridge: Cambridge University Press.

Cunningham, A., and P. Williams, eds. 1992. *The Laboratory Revolution in Medicine.* Cambridge: Cambridge University Press.

Cuvier G. 1817. *Le règne animal distribué d'après son organisation, pour servir de base à l'histoire naturelle des animaux et d'introduction à l'anatomie comparée.* Paris: Déterville.

———. 1830. *Discours sur les revolutions de la surface du globe et sur les changements qu'elles ont produits dans le Règne Animal.* 6th ed. Paris: Edmond D'Ocagne.

Dagan, T., and W. Martin. 2006. The tree of one percent. *Genome Biology* 7, no. 10: 118.

Dagan, T., Y. Artzy-Randrup, and W. Martin. 2008. Modular networks and cumulative impact of lateral transfer in prokaryote genome evolution. *Proceedings of the National Academy of Sciences* 105: 10039–44.

Dalrymple, G. B. 1991. *The Age of the Earth.* Stanford, Calif.: Stanford University Press.

Dalton, R. 2006. Feel it in your bones. *Nature* 440: 1100–1.

Damuth, J., and I. L Heisler. 1988. Alternative formulations of multilevel selection. *Biology and Philosophy* 3, no. 4: 407–30.

Dana, J. D. 1870. *Manual of Geology: Treating the Principles of the Science with Special Reference to American Geological History.* New York: Ivison, Blakeman, Taylor.

Darlington, C. D. 1939. *The Evolution of Genetic Systems.* Cambridge: Cambridge University Press.

———. 1950. Foreword to *On the Origin of Species* by C. Darwin, iv–xx. London: Watts.

———. 1953. *The Facts of Life.* London: George Allen & Unwin, Macmillan.

———. 1980. The evolution of genetic systems: Contributions of cytology to evolutionary theory. In *The Evolutionary Synthesis: Perspectives on the Unification of Biology*, ed. Ernst Mayr and William B. Provine, 70–80. Cambridge, Mass.: Harvard University Press.

Dart, R. 1925. *Austalopithecus africanus*: The man-ape of South Africa. *Nature* 115: 195–99.

———. 1926. Taungs and its significance. *Natural History* 26, no. May: 315–27.

Darwin C. R. 1835. Chiloe Jan^r. 1835 [*Beagle* notes]. DAR 35.328,328a–328j. In *The Complete Work of Charles Darwin Online*, ed. John van Wyhe, 2002. http://darwin-online.org.uk/.

———. 1838. A sketch of the deposits containing extinct mammalia in the neighbourhood of the Plata. *Proceedings of the Geological Society of London* 2: 542–44.

———. 1839a. *Journal of Researches into the Geology and Natural History of the Various Countries Visited by H.M.S.* Beagle, *under the Command of Captain Fitzroy from 1832 to 1836.* London: Colburn.

———. 1839b. *Narrative of the Surveying Voyages of His Majesty's Ships* Adventure *and* Beagle, *between the Years 1826 and 1836, Describing Their Examination of the Southern Shores of South America, and the* Beagle's *Circumnavigation of the Globe.* Vol. 3, *Journal and Remarks, 1832–1836.* London: Henry Colburn.

———. 1840a. On the connexion of certain volcanic phenomena in South America; and on the formation of mountain chains and volcanos, as the effect of the same power by which continents are elevated. *Transactions of the Geological Society of London* 5, 2nd ser., part 3: 601–31.

———. 1840b. On the formation of mould. [Read 1 Nov. 1837.] *Transactions of the Geological Society*, ser. 2, 5, no. 2: 505–9.

———. 1841. Humble-bees. *Gardeners' Chronicle*, 21 August: 550.

———. 1842a. Notes on the effects produced by the ancient glaciers of Caernarvonshire, and on the boulders transported by floating ice. *The London, Edinburgh, and Dublin Philosophical Magazine and Journal of Science Third Series* 21: 180–88.

———. 1842b. On the formation of vegetable mould. *Transactions of the Geological Society of London* 5, 2nd ser., part 3: 505–9.

———. 1842c. *The Structure and Distribution of Coral Reefs. Being the First Part of the Geology of the Voyage of the* Beagle, *under the Command of Capt. Fitzroy, R.N. during the Years 1832 to 1836.* London: Smith, Elder.

———. 1844. *Geological Observations on the Volcanic Islands Visited during the Voyage of H.M.S.* Beagle, *Together with Some Brief Notices on the Geology of Australia and the Cape of Good Hope. Being the Second Part of the Geology of the Voyage of the* Beagle, *under the Command of Capt. FitzRoy, R.N. during the Years 1832–1836.* London: Smith, Elder.

———. 1845. *Journal of Researches into the Natural History and Geology of the Countries Visited during the Voyage Round the World of H.M.S. "Beagle" under the Command of Captain Fitz Roy, R.N. Second Edition, Corrected, with Additions.* London: John Murray.

———. 1846. *Geological Observations on South America. Being the Third Part of the Geology of the Voyage of the* Beagle, *under the Command of Capt. FitzRoy, R.N. during the Years 1832–1836.* London: Smith, Elder.

———. 1851. *A Monograph on the Fossil Lepadidae, or Pedunculated Cirripedes of Great Britain.* London: Palaeontological Society.

———. 1852. *A Monograph on the Sub-class Cirripedia, with Figures of All the Species. The Lepadidae, or Pedunculated Cirripedes.* London: Ray Society.

———. 1854. *A Monograph on the Sub-class Cirripedia, with Figures of All the Species. The Balanidae (or Sessile Cirripedes; the Verrucidae, etc.).* London: Ray Society.

———. 1855. *A Monograph on the Fossil Balanidae and Verrucidae of Great Britain.* London: Palaeontological Society.

———. 1859. *On the Origin of Species by Means of Natural Selection, or the Preservation of Favoured Races in the Struggle for Life.* London: John Murray.

———. 1860a. *On the Origin of Species by Means of Natural Selection, or the Preservation of Favoured Races in the Struggle for Life.* 2nd ed. London: John Murray.

———. 1860b. *Über die Entstehung der Arten im Thier- und Pflanzen Reich durch natürliche Züchtung, oder Erhaltung der vervollkommneten Rassen im Kampfe um's Daseyn.* Trans. H. Bronn. Stuttgart: E. Schweizerbart'sche Verlagshandlung und Druckerei.

———. 1861. *On the Origin of Species by Means of Natural Selection, or the Preservation of Favoured Races in the Struggle for Life.* 3rd ed. London: John Murray.

———. 1862a. On the two forms, or dimorphic condition, in the species of *Primula*, and on their remarkable sexual relations. *Journal of the Proceedings of the Linnean Society (Botany)* 6: 77–96.

———. 1862b. On the three remarkable sexual forms of *Catasetum tridentatum*, an orchid in the possession of the Linnean Society. *Journal of the Proceedings of the Linnean Society (Botany)* 6: 151–57.

1862c. *On the Various Contrivances by Which British and Foreign Orchids Are Fertilised by Insects, and on the Good Effect of Intercrossing*. London: John Murray.

1862d. *De l'origine des espèces, ou Des lois du progrès chez les êtres organizes,* traduit en français sur la 3e édition ... par Mlle Clémence-Auguste Royer, avec une préface et des notes du traducteur. Paris: Guillaumin.

1863. The doctrine of heterogeny and modification of species. *Athenaeum*, no. 25 (April): 554–55.

1864. On the sexual relations of the three forms of *Lythrum salicaria*. *Journal of the Linnean Society of London (Botany)* 8: 169–96.

1865. On the movements and habits of climbing plants. *Journal of the Linnean Society of London (Botany)* 9: 1–118.

1866. *On the Origin of Species by Means of Natural Selection, or the Preservation of Favoured Races in the Struggle for Life*. 4th ed. London: John Murray.

1868a. Arrangement as far as I can make out by comparing the views of various ... [primate family tree sketch]. http://darwin-online.org.uk/content/frameset?viewtype=image&itemID=CUL-DAR80.B91&pageseq=1.

1868b. *The Variation of Animals and Plants under Domestication*. London: John Murray.

1869. *On the Origin of Species by Means of Natural Selection, or the Preservation of Favoured Races in the Struggle for Life*. 5th ed. London: John Murray.

1871a. *The Descent of Man, and Selection in Relation to Sex*. London: John Murray.

1871b. Letters to the editor: "Pangenesis." *Nature* 3: 502–3.

1872a. *The Origin of Species by Means of Natural Selection, or the Preservation of Favoured Races in the Struggle for Life*. 6th ed. London: John Murray.

1872b. *The Expression of the Emotions in Man and Animals*. London: John Murray.

1873. *L'origine des espèces au moyen de la sélection naturelle, ou La lutte pour l'existence dans la nature ... traduit ... sur les 5me et 6me éditions anglaises ... par J.-J. Moulinié*. Paris- Reinwald.

1874. *The Descent of Man, and Selection in Relation to Sex*. 2nd ed. London: John Murray.

1875a. *Insectivorous Plants*. London: John Murray.

1875b. *The Movements and Habits of Climbing Plants*. London: John Murray.

1875c. *The Variation of Animals and Plants under Domestication*. 2nd ed. 2 vols. London: London: John Murray.

1876a. *The Effects of Cross and Self Fertilisation in the Vegetable Kingdom*. London: John Murray.

1876b. *L'origine des espèces au moyen de la sélection naturelle, ou La lutte pour l'existence dans la nature, ... traduit sur la 6e édition anglaise, par Ed. Barbier*. Paris: Reinwald.

1877a. *The Different Forms of Flowers on Plants of the Same Species*. London: John Murray.

1877b. *The Various Contrivances by Which Orchids Are Fertilised by Insects*. 2nd ed. London: John Murray.

1877c. A biographical sketch of an infant. *Mind* 2: 285–94.

1880. *The Power of Movement in Plants*. Assisted by F. Darwin. London: John Murray.

1881. *The Formation of Vegetable Mould, through the Action of Worms*. London: John Murray.

1883. *Prefatory Notice to Fertilisation of Flowers by H. Müller*. London: Macmillan.

1890. *Journal of Researches into the Natural History and Geology of the Countries Visited during the Voyage of H.M.S. "Beagle" round the World, under the Command of Capt. Fitz Roy, R.N.* New ed. London: John Murray.

1909. *The Foundations of the Origin of Species: Two Essays Written in 1842 and 1844*. Ed. Francis Darwin. Cambridge: at the University Press.

1958a. *The Autobiography of Charles Darwin*. Ed. N. Barlow. New York: W. W. Norton.

1958b. *The Autobiography of Charles Darwin 1909–1882. With Original Omissions Restored*. Ed. N. Barlow. London: Collins.

1963. Darwin's ornithological notes. *Bulletin of the British Museum (Natural History) Historical Series* 2: 201–78. Ed. N. Barlow.

1975. *Charles Darwin's Natural Selection, Being the Second Part of His Big Species Book Written from 1856 to 1858*. Ed. Robert C. Stauffer. Cambridge: Cambridge University Press.

1980. The Red Notebook of Charles Darwin. *Bulletin of the British Museum (Natural History) Historical Series* 7. Ed. S. Herbert. London: British Museum (Natural History).

1985–. *The Correspondence of Charles Darwin*. Ed. Frederick Burkhardt et al. Cambridge: Cambridge University Press.

1988. *Charles Darwin's* Beagle *Diary*. Ed. R. Keynes. Cambridge: Cambridge University Press.

1998. *The Expression of the Emotions in Man and Animals*. 3rd ed. Ed. Paul Ekman. London: HarperCollins.

2000. *Charles Darwin's Zoology Notes and Specimen Lists from H.M.S. Beagle*. Ed. R. Keynes. Cambridge: Cambridge University Press.

2002. *Autobiographies*. Ed. M. Neve and S. Messenger. London: Penguin.

2009. *Charles Darwin's Notebooks from the Voyage of the* Beagle. Ed. G. Chancellor and John van Wyhe. Cambridge: Cambridge University Press.

2010. Recollections of the development of my mind and character. In *Charles Darwin: Evolutionary Writings*, ed. J. Secord, 355–425. Oxford: Oxford University Press.

Darwin, C. R., and A. R. Wallace. 1858. On the tendency of species to form varieties; and on the perpetuation of varieties and species by natural means of selection. *Journal of the Proceedings of the Linnean Society of London. Zoology* 3, no. 20 (August): 45–62.

Darwin, E. 1791. *The Botanic Garden, a Poem in Two Parts*. Part 1, *The Economy of Vegetation*. London.

1794. *Zoonomia, or The Laws of Organic Life*. Vol. 1. London: J. Johnson.

1796. *Zoonomia, or The Laws of Organic Life*. Vol. 2. London: J. Johnson.

1803. *The Temple of Nature*. London: J. Johnson.

Darwin, F. 1886. Physiological selection and the origin of species (Letter to editor). *Nature* 34: 407.

ed. 1887. *Life and Letters of Charles Darwin*. 3 vols. London: John Murray.

1892. *Charles Darwin: His Life Told in an Autobiographical Chapter, and in a Selected Series of His Published Letters*. London: John Murray.

1899. The botanical works of Darwin. *Annals of Botany* 13: ix–xix.

Dasmahapatra, K. K., M. Elias, R. I. Hill, J. I. Hoffmann, and J. Mallet. 2010. Mitochondrial DNA barcoding detects some species that are real, and some that are not. *Molecular Ecology Resources* 10: 264–73.

Davidson, E. H. 1990. How embryos work – a comparative view of diverse modes of cell fate specification. *Development* 108, no. 3: 365–89.

2001. *Genomic Regulatory Systems: Development and Evolution*. San Diego: Academic Press.

2006. *The Regulatory Genome: Gene Regulatory Networks in Development and Evolution*. Burlington, Mass.: San Diego, Academic.

Davidson, E. H., and D. H. Erwin. 2006. Gene regulatory networks and the evolution of animal body plans. *Science* 311, no. 5762: 796–800.

2009. An integrated view of precambrian eumetazoan evolution. *Cold Spring Harbor Symposia on Quantitative Biology* 74: 65–80.

Davidson, E. H., K. J. Peterson, et al. 1995. Origin of bilaterian body plans – evolution of developmental regulatory mechanisms. *Science* 270: 1319–25.

Davidson, I. 2003. The archaeology of language origins: States of art. In *Language Evolution: States of the Art*, ed. M. Christiansen and S. Kirby, 140–57. Oxford: Oxford University Press.

2010. The archeology of cognitive evolution. *WIREs Cognitive Science* 1: 214–29.

Davies, R. 2008. *The Darwin Conspiracy*. London: Golden Square books.

Davis, W. M. 1884. Geographic classification. Illustrated by a study of plains, plateaus and their derivatives. *Proceedings of the American Association for the Advancement of Science*: 428–32.

1895. The development of certain English rivers. *Geographical Journal* 5: 127–46.

Dawkins, R. 1976. *The Selfish Gene*. Oxford: Oxford University Press.

1986. *The Blind Watchmaker*. New York: W. W. Norton.

Dawson, G. 2004. The Cornhill Magazine and shilling monthlies in mid-Victorian Britain. In *Science in the Nineteenth-Century Periodical: Reading the Magazine of Nature*, ed. G. Cantor et al., 123–50. Cambridge: Cambridge University Press.

2007. *Darwin, Literature and Victorian Respectability*. Cambridge: Cambridge University Press.

De Beer, G. 1940. *Embryos and Ancestors*. Oxford: Clarendon Press.

1959. Letter of Darwin to G. C. Wallich, 28 March 1882. Some unpublished letters of Charles Darwin. *Notes and Records of the Royal Society of London* 14: 12–66.

De Bont, Raf. 2010. Schizophrenia, evolution and the borders of biology: On Huxley et al.'s 1964 paper in *Nature*. *History of Psychiatry* 21, no. 2: 144–59.

de Chadarevian, S. 1996. Laboratory science versus country-house experiments. The controversy between Julius Sachs and Charles Darwin. *British Journal for the History of Science* 29: 17–41.

De Duve, C. 2002. *Life Evolving*. Oxford: Oxford University Press.

de Vries, H. 1889. *Intracellulare Pangenesis*. Jena: Gustav Fischer.

1900. *Sur la loi de disjonction des hybrids*. Paris: Gauthier-Villars.

1910. *Intracellular Pangenesis*. Trans. C. S. Gager. Chicago: Open Court.

Deacon, T. 1997. *The Symbolic Species: The Co-evolution of Language and the Brain*. New York: Norton.

Dean, D. R. 1980. Graham Island, Charles Lyell, and the craters of elevation controversy. *Isis* 71: 571–88.

Deane-Drummond, C. 2009. *Christ and Evolution: Wonder and Wisdom*. Minneapolis: Fortress Press.

DeAngelis, D., and J. Waterhouse. 1987. Equilibrium and non-equilibrium concepts in ecological models. *Ecological Monographs* 57: 1–21.

Dechaseaux, C. 1934. Principales espèces de Liogryphées liasiques. Valeur stratigraphique et remarques sur quelques formes mutantes. *Bulletin de la Société Géologique de* France 4, no. 5: 201–12.

1936. Pectinidés jurassiques de l'Est du Bassin de Paris. Révision et biogeography. *Annales de paleontology* 25: 1–146.

1939. Megalodon, Pachyerisma, Protodiceras, Diceras, Pterocardium et l'origine des Diceras. *Bulletin de la Société Géologique de France* 5, no. 5: 207–18.

Degler, C. N. 1991. *In Search of Human Nature: The Decline and Revival of Darwinism in American Social Thought*. Oxford: Oxford University Press.

Delio, I. 2008. *Christ in Evolution*. Maryknoll, N.Y.: Orbis Books.

Dembski, W. A. 1995. What every theologian should know about creation, evolution, and design. *Transactions* 3 (May–June): 1–8.

2005. Torah and Science Conference with the Lubavitchers. http://www.uncommondescent.com/intelligent-design/torah-and-science-conference-with-the-lubavitchers/.

Demir, R., and B Yurtoğlu. 2001. Unutulmuş bir Osmanlı düşünürü Hoca Tahsîn Efendî'nin Târîh-i tekvîn-i hilkat adlı eseri ve Haeckelci evrimciliğin Türkiyeye girişi (The History of the Emergence of Creation, the Work of a Forgotten Ottoman Thinker, Hoca Tahsîn Efendi, and the Introduction of Haeckelian Evolutionism into Turkey). *Nüsha* 1: 166–97.

Dennert, E. 1905. *Vom Sterbelager des Darwinismus*. 2nd ed. Halle: Richard Mühlmann.

Dennett, D. 1995. *Darwin's Dangerous Idea: Evolution and the Meaning of Life*. New York: Simon & Schuster.

Depew, D. J., and B. H. Weber. 1995. *Darwinism Evolving: Systems Dynamics and the Genealogy of Natural Selection*. Cambridge, Mass.: MIT Press.

d'Errico, F., and M. Vanhaeren. 2009. Earliest personal ornaments and their significance for the origin of language debate. In *The Cradle of Language*, ed. R. P. Botha and C. Knight, 16–40. Oxford: Oxford University Press.

Derry, M. E. 2003. *Bred for Perfection: Shorthorn Cattle, Collies, and Arabian Horses since 1800*. Baltimore: Johns Hopkins University Press.

Desmond, A. 1982. *Archetypes and Ancestors: Palaeontology in Victorian London, 1850–1875*. London: Blond & Briggs.

1984. Robert E. Grant: The social predicament of a pre-Darwinian transmutationist. *Journal of the History of Biology* 17: 189–223.

1989. *The Politics of Evolution: Morphology, Medicine and Reform in Radical London*. Chicago: University of Chicago Press.

1997. *Huxley: From Devil's Disciple to Evolution's High Priest*. New York: Penguin.

Desmond, A., and J. R. Moore. 1991. *Darwin: The Life of a Tormented Evolutionist*. New York: W. W. Norton.

2009. *Darwin's Sacred Cause: Race, Slavery and the Quest for Human Origins*. London: Allen Lane.

Deutsch, J. 2010. Darwin and barnacles. *Comptes Rendu Biologies* 333, no. 2: 99–106.

Deutscher, P. 2004. The descent of man and the evolution of woman. *Hypatia* 19, no. 2: 35–55.

Dewey, J. 1910. The influence of Darwinism on philosophy. In *The Influence of Darwinism on Philosophy and Other Essays*, 5–12. New York: Holt.

Di Gregorio, M. 1984. *T. H. Huxley's Place in Natural Science*. New Haven: Yale University Press.

1987. Hugh Edwin Strickland (1811–1853) on affinities and analogies, or The case of the missing key. *Ideas and Production* 7: 35–50.

Diamond, J. M., and M. L. Cody. 1975. *Ecology and Evolution of Communities*. Cambridge, Mass.: Harvard University Press.

Dickens, C. 1996. *Great Expectations*. Ed. C. Mitchell. London: Penguin.

1997. *Our Mutual Friend*. Ed. A. Poole. London: Penguin.

Diderot, D. 1943. *Diderot: Interpreter of Nature*. New York: International Publishers.

Digrius, D. M. 2007. Conversations and Contrasting Views an Examination of the Development of Paleobotany in Nineteenth Century Europe, 1804–1895. PhD dissertation, Drew University.

2008. Gregor Mendel. In *Icons of Evolution: An Encyclopedia of People, Evidence, and Controversies*, ed. Brian Regal, 87–113. Westport, Conn.: Greenwood Press.

Dixon, M., and G. Radick. 2009. *Darwin in Ilkley*. Stroud: History Press.

Dixon, T. 2003. *From Passions to Emotions: The Creation of a Secular Psychological Category*. Cambridge: Cambridge University Press.

⸺ 2008. *The Invention of Altruism: Making Modern Meanings in Victorian Britain*. Oxford: Oxford University Press for the British Academy.

Dobzhansky, T. 1937. *Genetics and the Origin of Species*. New York: Columbia University Press.

⸺ 1947. Adaptive changes induced by natural selection in wild populations of *Drosophila*. *Evolution* 1: 1–16.

⸺ 1964. Biology, molecular and organismic. *American Zoologist* 4: 443–50.

⸺ 1973. Nothing in biology makes sense except in the light of evolution. *American Biology Teacher* 35: 125–29.

Dodick, J., A. Dayan, and N. Orion. 2010. Philosophical approaches of religious Jewish science teachers toward the teaching of "controversial" topics in science. *International Journal of Science Education* 32, no. 11: 1521–48.

Dodson, E. O. 2000. Toldot Adam: A little-known chapter in the history of Darwinism. *Perspectives on Science and Christian Faith* 52, no. 1: 47–54.

Domingues, H., M. Romero Sá, and T. F. Glick, eds. 2003. *A recepção do Darwinismo no Brasil*. Rio de Janeiro: Fiocruz.

Domingues, H., M. Romero Sá, M. Angel Puig-Samper, and R. Ruiz. 2009. *Darwinismo, meio ambiente, sociedade*. Rio de Janeiro: MAST.

Donohue, K., ed. 2011. *Darwin's Finches: Readings in the Evolution of a Scientific Paradigm*. Chicago: University of Chicago Press.

Doolittle, W. F. 2000. Uprooting the tree of life. *Scientific American* 282 (February): 90–95.

Doolittle, W. F., and E. Bapteste. 2007. Pattern pluralism and the tree of life hypothesis. *Proc. Acad. Nat. Sci. U.S.A.* 104: 2043–49.

Doolittle, W. F., Y. Boucher, C. L. Nesbø, C. J. Douady, J. O. Andersson, and A. J. Roger. 2003. How big is the iceberg of which organellar genes in nuclear genomes are but the tip? *Philosophical Transactions of the Royal Society of London Series B, Biological Sciences* 358: 39–58.

Doğan, A. 2006. *Osmanlı Aydınları ve Sosyal Darwinizm*. İstanbul: İstanbul Bilgi Üniversitesi Yayınları.

Drčs, M., and J. Mallet. 2002. Host races in plant-feeding insects and their importance in sympatric speciation. *Philosophical Transactions of the Royal Society of London Series B, Biological Sciences* 357: 471–92.

Driesch, H. 1908. *The Science and Philosophy of the Organism*. London: Black.

Drummond, H. 1894. *The Ascent of Man*. London: Hodder and Stoughton.

Dubin, L. 1995. Pe'er ha-Adam of Vittorio Hayyim Castiglioni: An Italian chapter in the history of Jewish response to Darwin. In *The Interaction of Scientific and Jewish Cultures in Modern Times*, ed. Y. Rabkin and I. Robinson, 87–102. Lewiston, N.Y.: Edwin Mellen Press.

Dubinin, N. P., M. A. Heptner, S. J. Bessertnaia, et al. 1934. Eksperimental'nyi analiz ekogenotipov. *Drosophila melanogaster*. *Biologicheskii Zhurnal* 3: 166–216.

Dubinin, N. P., M. A. Heptner, Z. A. Demidova, and L. I. Djachkova. 1936. Geneticheskaia struktura populiatsii i ee dinamika v dikikh naseleniiakh *Drosophila melanogaster*. *Biologicheskii Zhurnal* 5: 939–67.

Dubinin, N. P., and G. G. Tiniakov. 1946. Inversion gradients and natural selection in ecological races of *Drosophila funebris*. *Genetics* 31: 537–45.

Dubow, S. 2006. *A Commonwealth of Knowledge: Science, Sensibility, and White South Africa, 1820–2000*. Oxford: Oxford University Press.

Dunbar, R. I. M. 1993. Coevolution of neocortical size, group size and language in humans. *Behavioral and Brain Sciences* 16: 681–735.

⸺ 1998. *Grooming, Gossip, and the Evolution of Language*. Cambridge, Mass.: Harvard University Press.

Dunlap, T. R. 1999. *Nature and the English Diaspora: Environmental History in the United States, Canada, Australia, and New Zealand*. Cambridge: Cambridge University Press.

Durant, J. 1979. Scientific naturalism and social reform in the thought of Alfred Russel Wallace. *British Journal for the History of Science* 12, no. 40: 31–58.

Durham, W. H. 1991. *Coevolution: Genes, Culture and Human Diversity*. Stanford, Calif.: Stanford University Press.

Dutton, D. 2009. *The Art Instinct: Beauty, Pleasure and Human Evolution*. Oxford: Oxford University Press.

Düsing, C. 1884. Die Regulierung des Geschlechtsverhältnisses bei der Vermehrung der Menschen, Tiere, und Pflanzen. *Jenaische Zeitschrift für Naturwissenschaft 1884* 17: 593–940.

East, E. M. 1932. Further observations on *Lythrum salicaria*. *Genetics* 17: 327–34.

Ebert, J. 1991. *Religion und Reform in der arabischen Provinz: Husayn al-Gisr at-Ṭarābulusī (1845–1909)*. Frankfurt/Main: Lang.

Edis, T. 1999. Cloning creationism in Turkey. *Reports of the National Center for Science Education* 19, no. 6: 30–35.

⸺ 2007. *An Illusion of Harmony: Science and Religion in Islam*. Amherst, N.Y.: Prometheus Books.

Edwards, D. 1999. *The God of Evolution: A Trinitarian Theology*. New York: Paulist.

Efron, J. M. 1994. *Defenders of the Race*. New Haven: Yale University Press.

⸺ 2007. *Judaism and Science*. Westport, Conn.: Greenwood Press.

⸺ 2008. American Jews and intelligent design. http://reilly.nd.edu/assets/65756/rcrefron.pdf.

Egerton, F. N. 1970. Humboldt, Darwin, and population. *Journal of the History of Biology* 3: 325–60.

⸺ 1973. Changing concepts of the balance of nature. *Quarterly Review of Biology* 48: 322–50.

Ekman, P., ed. 1973. *Darwin and Facial Expression: A Century of Research in Review*. New York: Academic.

Elderton, E. M., and K. Pearson. 1915. Further evidence of natural selection in man. *Biometrika* 10: 488–506.

Eldredge, N. 1985. *Unfinished Synthesis: Biological Hierarchies and Modern Evolutionary Thought*. New York: Oxford University Press.

Eldredge, N., and S. J. Gould. 1972. Punctuated equilibrium: An alternative to phyletic gradualism. In *Models in Paleobiology*, ed. T. Schopf, 82–115. San Francisco: Freeman/Cooper.

Eliot, G. 1994. *Middlemarch*. Ed. R. Ashton London: Penguin.

Ellegård, A. [1990] 1958. *Darwin and the General Reader: The Reception of Darwin's Theory of Evolution in the British Periodical Press*. Göteborg: Göteborgs Universitets Årsskrift. Reprint, Chicago: University of Chicago Press.

Ellison, P. T., D. R. Govindaraju, R. M. Nesse, and S. C. Stearns, eds. 2009. *Evolution in Health and Medicine*. Arthur M. Sackler

Colloquia of the National Academy of Sciences. Washington, D.C.: National Academy of Sciences.

Elman, B. A. 2005. *On Their Own Terms: Science in China, 1550–1900*. Cambridge, Mass.: Harvard University Press.

Elton, C. S. 1958. *The Ecology of Invasions by Plants and Animals*. London: Methuen.

ENCODE Project Consortium. 2012. An integrated encyclopedia of DNA elements in the human genome. *Nature* 489: 57–74.

Endersby, J. 2003. Darwin on generation, pangenesis and sexual selection. In *Cambridge Companion to Darwin*, ed. M. J. S. Hodge and G. Radick, 69–91. Cambridge: Cambridge University Press.

Endler, J. 1986. *Natural Selection in the Wild*. Princeton: Princeton University Press.

England, R. 1997. Natural selection before the *Origin*: Public reactions of some naturalists to the Darwin-Wallace papers. *Journal of the History of Biology* 30: 267–90.

Ereshefsky M. 2010a. Darwin's solution to the species problem. *Synthese* 175, no. 3: 405–25.

———. 2010b. Microbiology and the species problem. *Biology and Philosophy* 25: 553–68.

Erskine, F. 1995. The *Origin of Species* and the science of female inferiority. In *Charles Darwin's "The Origin of Species": New Interdisciplinary Essays*, ed. David Amigoni and Jeff Wallace, 95–121. Manchester: Manchester University Press.

Ertuğrul, İ. F. 1928. *Mâddiyûn mezhebinin izmihlâli* (The Dissolution of the Materialist School). Istanbul: Orhaniyye Matbaası.

Erwin, D. H., and E. H. Davidson. 2002. The last common bilaterian ancestor. *Development* 129, no. 13: 3021–32.

———. 2009. The evolution of hierarchical gene regulatory networks. *Nature Reviews Genetics* 10, no. 2: 141–48.

Evans, R. 1997. In search of German social Darwinism. In *Medicine and Modernity*, ed. M. Berg and G. Cocks, 55–79. Cambridge: Cambridge University Press.

Ewald, P. W. 1980. Evolutionary biology and the treatment of signs and symptoms of infectious disease. *Journal of Theoretical Biology* 86, no. 1: 169–76.

———. 1987. Transmission modes and evolution of the parasitism-mutualism continuum. *Annals of the New York Academy of Sciences* 503: 295–306.

Falk, D. 2004. Prelinguistic evolution in early hominins: Whither motherese? *Behavioral and Brain Sciences* 27: 491–503.

Falk, R. 1998. Zionism and the biology of the Jews. *Science in Context* 11: 587–607.

———. 2005. Zionism, race, and eugenics. In *Jewish Tradition and the Challenge of Evolution*, ed. G. Cantor and M. Swetlitz, 137–62. Chicago: University of Chicago Press.

Farber, P. L. 2000. *Finding Order in Nature: The Naturalist Tradition from Linnaeus to E. O. Wilson*. Baltimore: John Hopkins University Press.

Farley, J. 1974. The initial reactions of French biologists to Darwin's *Origin of Species*. *Journal of the History of Biology* 7, no. 2: 275–300.

———. 1977. *The Spontaneous Generation Controversy from Descartes to Oparin*. Baltimore: Johns Hopkins University Press.

Farmer, J. B., and L. Digby. 1912. On the dimensions of chromosomes considered in relation to phylogeny. *Philosophical Transactions of the Royal Society of London* 205: 1–25.

Farrar, F. W. 1865. *Chapters on Language*. London: Longmans, Green.

Faur, J. 1997. The Hebrew species concept and the origin of evolution: R. Benamozegh's response to Darwin. *La Rassegna Mensile Di Israel* 63, no. 3: 43–66.

Fausto-Sterling, A. 1997. Feminism and behavioral evolution: A taxonomy. In *Feminism and Evolutionary Biology: Boundaries, Intersections and Frontiers*, ed. P. A. Gowaty, 42–60. New York: International Thomson Publishing.

Feit, C. 1990. Darwin and drash: The interplay of Torah and biology. *Torah U-Madda Journal* 2: 25–36.

———. 2006. Modern orthodoxy and evolution: The models of Rabbi J. B. Soloveitchik and Rabbi A. I. Kook. In *Jewish Tradition and the Challenge of Evolution*, ed. G. Cantor and M. Swetlitz, 208–24. Chicago: University of Chicago Press.

Felsenstein, J. 1985. Phylogenies and the comparative method. *American Naturalist* 125: 1–15.

———. 2004. *Inferring Phylogenies*. Sunderland: Sinauer Associates.

Fichman, M. 1985. Ideological factors in the dissemination of Darwinism in England, 1860–1900. In *Transformation and Tradition in the Sciences*, ed. E. Mendelsohn, 471–85. Cambridge: Cambridge University Press.

Finnegan, D. 2010. Darwin, dead and buried? *Environment and Planning A* 42: 259–61.

Fisher, J. 1877. The history of landholding in Ireland. *Transactions of the Royal Historical Society* 5: 228–326.

Fisher, R. A. 1915. The evolution of sexual preference. *Eugenics Review* 7, no. 3: 184–92.

———. 1927. Objections to mimicry theory; statistical and genetic. *Transactions of the Royal Entomological Society, London* 75: 269–78.

———. 1930. *The Genetical Theory of Natural Selection*. Oxford: Oxford University Press.

———. 1947. The renaissance of Darwinism. *Listener* 37: 1001.

Fisher, R. A., and E. B. Ford. 1947. The spread of a gene in natural conditions in a colony of the moth *Panaxia dominula* L. *Heredity* 1: 143–74.

Fitch, W. M. 1976. Molecular evolutionary clocks. In *Molecular Evolution*, ed. F. J. Ayala, 160–78. Sunderland, Mass.: Sinauer Associates.

Fitch, W. M., and E. Margoliash. 1967. Construction of phylogenetic trees. *Science* 155: 279–84.

Fitch, W. T. 2009. Musical protolanguage: Darwin's theory of language evolution revisited. http://languagelog.ldc.upenn.edu/nll/?p=1136.

———. 2010. *The Evolution of Language*. Cambridge: Cambridge University Press.

FitzRoy, R. 1839. *Narrative of the Surveying Voyages of His Majesty's Ships* Adventure *and* Beagle, *between the years 1826 and 1836, Describing Their Examination of the Southern Shores of South America, and the* Beagle's *Circumnavigation of the Globe*. Vol. 2. London: Henry Colburn.

Flint, K. 1995. Origins, species and great expectations. In *Charles Darwin's "The Origin of Species": New Interdisciplinary Essays*, ed. D. Amigoni and J. Wallace, 152–73. Manchester: Manchester University Press.

Forbes, P. 2009. *Dazzled and Deceived: Mimicry and Camouflage*. New Haven: Yale University Press.

Ford, E. B. 1931. *Mendelism and Evolution*. New York: John Wiley & Sons.

———. 1964. *Ecological Genetics*. New York: John Wiley & Sons.

Fox, G. E., E. Stackebrandt, R. B. Hespell, et al. 1980. The phylogeny of prokaryotes. *Science* 209: 457–63.

Fracchia, J., and R. C. Lewontin. 1999. Does culture evolve? *History and Theory* 38: 52–78.

Francis, M. 2007. *Herbert Spencer and the Invention of Modern Life*. Stocksfield, U.K.: Acumen.

Fry, I. 2000. *The Emergence of Life on Earth: A Historical and Scientific Overview*. New Brunswick, N.J.: Rutgers University Press.

——— 2006. The origins of research into the origin of life. *Endeavour* 30, no. 1: 24–28.

——— 2009. Philosophical aspects of the origin-of-life problem: The emergence of life and the nature of science. In *Exploring the Origin, Extent, and Future of Life*, ed. C. M. Bertka, 61–79. Cambridge: Cambridge University Press.

——— 2011. The role of natural selection in the origin of life. *Origins of Life and Evolution of the Biosphere* 41(1):3–16: DOI 10.1007/s11084-010-9214-1.

——— 2012. Is science metaphysically neutral? *Studies in History and Philosophy of Biological and Biomedical Sciences* 43: 665–73.

Fulweiler, H. W. 1994. "A dismal swamp": Darwin, design, and evolution in *Our Mutual Friend*. *Nineteenth-Century Literature* 49: 50–74.

Futuyma, D. J. 1986. Reflections on reflections: Ecology and evolutionary biology. *Journal of the History of Biology* 19: 303–12.

Galtier, N., and V. Daubin. 2008. Dealing with incongruence in phylogenetic analyses. *Philosophical Transactions of the Royal Society, Series B: Biological Sciences* 363: 4023–29.

Galton, F. 1871a. Experiments in pangenesis, by breeding from rabbits of a pure variety, into whose circulation blood taken from other varieties had been previously largely transfused. *Proceedings of the Royal Society B: Biological Sciences* 19: 393–410.

——— 1871b. Letters to the editor: Pangenesis. *Nature* 3: 5–6.

——— 1889. *Natural Inheritance*. London: Macmillan.

Gamble, C., and Moutsiou T. 2011. The time revolution of 1859 and the stratification of the primeval mind. *Notes and Records of the Royal Society* 65: 43–63.

Gamble, E. B. 1894. *The Evolution of Woman: An Inquiry into the Dogma of Her Inferiority to Man*. New York: G. P. Putnam's Sons.

Gärtner, C. F. 1849. *Die Bastarderzeugung im Pflanzenreich*. Stuttgart: Schweizerbart.

Gasman, D. 1971. *The Scientific Origins of National Socialism: Social Darwinism in Ernst Haeckel and the German Monist League*. New York: Science History Publications.

——— 1998. *Haeckel's Monism and the Birth of Fascist Ideology*. New York: Peter Lang.

Gates, B. T., ed. 2002. *In Nature's Name: An Anthology of Women's Writing and Illustration, 1780–1930*. Chicago: University of Chicago Press.

Gates, R. R. 1909. The stature of the chromosomes of *Oenothera gigas*, De Vries. *Archive für Zellforschung* 3: 525–52.

Gaudillière, J.-P., and I. Löwy, eds. 2001. *Heredity and Infection: The History of Disease Transmission*. London: Routledge.

Gaudry, A. 1862–67. *Géologie de l'Attique d'après les recherches faites en 1855–56 et en 1860 sous les auspices de l'Académie des sciences*. Paris: F. Savy.

——— 1878. *Les Enchaînements du monde animal dans les temps géologiques*. 3 vols. Paris: Savy.

——— 1896. *Paléontologie philosophique*. Paris: Masson.

Gause, G. F. 1934. *The Struggle for Existence*. Baltimore: Williams & Wilkins.

Gaut, B. S., S. V. Muse, W. D. Clark, and M. T. Clegg. 1992. Relative rates of nucleotide substitution at the *rbc*L locus of monocotyledonous plants. *Journal of Molecular Evolution* 35: 292–303.

Gautier, E. 1880. *Le darwinisme social*. Paris: Derveaux.

Gavrilets, S. 2004. *Fitness Landscapes and the Origin of Species*. Princeton: Princeton University Press.

Gayon, J. 1998. *Darwinism's Struggle for Survival: Heredity and the Hypothesis of Natural Selection*. Cambridge: Cambridge University Press.

——— 2000. History of the concept of allometry. *American Zoologist* 40: 748–58.

——— 2006. Les reconstructions phylogénétiques dans les *Annales de Paléontologie* (1906–1950). Comparaison avec d'autres revues françaises. *Annales de Paléontologie* 92: 223–34.

——— 2009. Pourquoi les paléontologues français ont-ils boudé les phylogénies? Science, philosophie et religion, 1900–1950. In *Au risque de l'existence. Le mythe, la science et l'art*, ed. J. C. Gens and P. Guenancia, 151–77. Dijon: Éditions Universitaires de Dijon.

Gayon, J., and M. Veuille. 2001. The genetics of experimental populations: L'Héritier and Teissier's populations cages, en coll. avec M. Veuille. In *Thinking about Evolution: Historical, Philosophical, and Political Perspectives*, ed. R. Singh, C. Krimbas, D. Paul, and J. Beatty, 77–102. Cambridge: Cambridge University Press.

Geddes, P., and J. A. Thomson. 1889. *The Evolution of Sex*. London: Walter Scott.

Gegenbaur, C. 1870. *Grundzüge der vergleichenden Anatomie*. 2nd ed. Leipzig: Wilhelm Engelmann.

Geikie, A. 1893. Presidential address. *Sixty-Second Meeting of the British Association for the Advancement of Science Report* (Edinburgh), 3–26. London: John Murray.

——— 1909. *Charles Darwin as Geologist*. Cambridge: Cambridge University Press.

Geikie, J. 1872. On changes of climate during the glacial epoch. Concluding paper. *Geological Magazine* 9: 254–65.

Geison, G. L. 1969. The protoplasmic theory of life and the vitalist-mechanist debate. *Isis* 60: 272–92.

Geoffroy Saint-Hilaire, E. 1818. *Philosophie anatomique*. Paris: Mequignon-Marvis.

George, H. [1879] 2005. *Progress and Poverty*. Ed. R. Drake. New York: Robert Schalkenbach Foundation.

Gershenson, S. 1945. Evolutionary studies on the distribution and dynamics of melanism in the hamster *(Cricetus cricetus* L.). I. Distribution of black hamsters in the Ukrainian and Nashkirian Soviet Socialist Republics (U.S.S.R). *Genetics* 30: 207–32.

Gevers, D., F. M. Cohan, J. G. Lawrence, et al. 2005. Re-evaluating prokaryotic species. *Nature Reviews Microbiology* 3: 733–39.

Ghiselin, M. T. 1969. *The Triumph of the Darwinian Method*. Berkeley: University of California Press.

Gibbons, A. 2009. A new kind of ancestor: *Ardipithecus* unveiled. *Science* 326, no. 5949: 36–40.

Gibson, S. A. 2009. Early settler – Darwin the geologist in the Galapagos. *GeoScientist* 19, no. 2: 18–23.

——— 2010. Darwin the geologist in Galapagos: An early insight into sub-volcanic magmatic processes. *Proceedings of the California Academy of Sciences* 61, suppl. 2: 69–88.

Gigerenzer, G. 2000. *Adaptive Thinking: Rationality in the Real World*. Oxford: Oxford University Press.

Gilbert, S., and D. Epel. 2009. *Ecological Developmental Biology*. Sunderland, Mass.: Sinauer.

Gilkey, L. 1970. *Religion and the Scientific Future*. London: S.C.M. Press.

Gillespie, J. 1991. *The Causes of Molecular Evolution*. Oxford: Oxford University Press.

Gilman, S. 1979. Darwin sees the insane. *Journal of the History of Behavioral Sciences* 15, no. 3: 253–62.

Glaß, D. 2004. *Der Muqtaf und seine Öffentlichkeit. Aufklärung, Räsonnement und Meinungsstreit in der frühen arabischen Zeitschriftenkommunikation*. Würzburg: Ergon-Verlag.

Glendenning, J. 2007. *The Evolutionary Imagination in Late-Victorian Novels: An Entangled Bank*. Burlington, Vt.: Ashgate.

Gliboff, S. 1998. Evolution, revolution, and reform in Vienna: Franz Unger's ideas on descent and their post-1848 reception. *Journal of the History of Biology* 31, no. 2: 179–209.

——— 1999. *Gregor Mendel and the Laws of Evolution*. Cambridge: Cambridge University Press.

——— 2008. *H. G. Bronn, Ernst Haeckel, and the Origins of German Darwinism*. Cambridge: MIT Press.

Glick, T. F. 1984. Perspectivas sobre la recepción del darwinismo en el mundo hispano. *Actas, II Congreso de la Sociedad Española de Historia de las Ciencias*, 1:49–64. Zaragoza.

——— 1988. *The Comparative Reception of Darwinism*. Chicago: University of Chicago Press.

——— 2001. The reception of Darwinism in Uruguay. In *The Reception of Darwinism in the Iberian World*, ed. T. F. Glick, M. A. Puig-Samper, and R. Ruiz, 29–52. Dordrecht: Kluwer.

Glick, T. F., and D. Kohn, eds. 1996. *Darwin on Evolution. The Development of the Theory of Natural Selection*. Indianapolis: Hackett.

Glick, T. F., M. A. Puig-Samper, and R. Ruiz, eds. 2001. *The Reception of Darwinism in the Iberian World*. Dordrecht: Kluwer.

Gluckman, P. D., A. S. Beedle, and M. A. Hanson. 2009. *Principles of Evolutionary Medicine*. Oxford: Oxford University Press.

Gluckman, P. D., and M. A. Hanson. 2004a. Living with the past: Evolution, development, and patterns of disease. *Science* 305: 1733–36.

——— 2004b. Maternal constraint of fetal growth and its consequences. *Seminars in Fetal & Neonatal Medicine* 9, no. 5: 419–25.

——— 2006a. Evolution, development and timing of puberty. *Trends in Endocrinology and Metabolism* 17, no. 1: 7–12.

——— 2006b. *Mismatch: Why Our World No Longer Fits Our Bodies*. Oxford: Oxford University Press.

Gluckman, P. D., M. A. Hanson, and H. G. Spencer. 2005. Predictive adaptive responses and human evolution. *Trends in Ecology & Evolution* 20, no. 10: 527–33.

Gluckman, P. D., F. M. Low, T. Buklijas, M. A. Hanson, and A. S. Beedle. 2011. How evolutionary principles improve the understanding of human health and disease. *Evolutionary Applications* 4: 249–63.

Godfrey, K. M., P. D. Gluckman, K. A. Lillycrop, et al. 2009. Epigenetic marks at birth predict childhood body composition at age 9 years. *Journal of Developmental Origins of Health and Disease* 1: S44.

Godfrey-Smith, P. 2009. *Darwinian Populations and Natural Selection*. Oxford: Oxford University Press.

Goethe, J. 1989. *Sämtliche Werke nach Epochen seines Schaffens*. Ed. K. Richter et al. Vol. 12, *Zur Morphologie*. Munich: Carl Hanser Verlag.

Gogarten, J. P., W. F. Doolittle, and J. G. Lawrence. 2002. Prokaryotic evolution in light of gene transfer. *Molecular Biology and Evolution* 19: 2226–38.

Goldfarb, S. J. 1981. American Judaism and the Scopes trial. In *Studies in the American Jewish Experience II*, ed. J. R. Marcus and A. J. Peck, 33–47. Lanham, Md.: University Press of America.

Goldschmidt, R. 1940. *The Material Basis of Evolution*. New Haven: Yale University Press.

Goodrich, E. S. 1912. *The Evolution of Living Organisms*. London: T. C. and E. C. Jack.

Goodwin, B. 1994. *How the Leopard Changed Its Spots*. New York: Charles Scribner's.

Gordon, C. 1939. A method for a direct study of natural selection. *Journal of Experimental Biology* 16: 278–85.

Gotthelf, A. 1999. Darwin on Aristotle. *Journal of the History of Biology* 32: 3–30.

Goudge, T. A. 1961. *The Ascent of Life*. Toronto: University of Toronto Press.

Gould, J. 1839. "Birds," part 3, no. 4. In *The Zoology of the Voyage of H.M.S. Beagle*, ed. C. Darwin, 1–164. London: Smith, Elder.

Gould, S. J. 1977. *Ontogeny and Phylogeny*. Cambridge, Mass.: Belknap Press of Harvard University Press.

——— 1980. Is a new and general theory of evolution emerging? *Paleobiology* 6, no. 1: 119–30.

——— 1982. The importance of trifles. *Natural History* 91, no. 4: 16–23.

——— 1983a. Worm for a century, and all seasons. In *Hen's Teeth and Horse's Toes*, 120–33. New York: W. W. Norton.

——— 1983b. The hardening of the modern synthesis. In *Dimensions of Darwinism: Themes and Counterthemes in Twentieth-Century Evolutionary Theory*, ed. M. Grene, 71–93. Cambridge: Cambridge University Press.

——— 1985. The paradox of the first tier: An agenda for paleobiology. *Paleobiology* 11, no. 1: 2–12.

——— 1989a. Punctuated equilibrium in fact and theory. *Journal of Social and Biological Structures* 12: 117–36.

——— 1989b. *Wonderful life: The Burgess Shale and the Nature of History*. New York: W. W. Norton.

——— 1991. The panda's thumb of technology. In *Bully for Brontosaurus: Reflections on Natural History*, 59–75. New York: Norton.

——— 2002. *The Structure of Evolutionary Theory*. Cambridge, Mass.: Belknap Press of Harvard University Press.

Gould, S. J., and N. Eldredge. 1977. Punctuated equilibria: The tempo and mode of evolution reconsidered. *Paleobiology* 3, no. 2: 115–51.

Gould, S. J., and R. Lewontin. 1979. The Spandrels of San Marco and the panglossian paradigm: A critique of the adaptationist programme. *Proceedings of the Royal Society of London, Series B, Biological Sciences* 205, no. 1161: 581–98.

Gouzoules, S., H. Gouzoules, and P. Marler. 1984. Rhesus monkey (*Macaca mulatta*) screams: Representational signaling in the recruitment of agonistic aid. *Animal Behavior* 32: 182–93.

Grafen, A. 1984. Natural selection, kin selection and group selection. In *Behavioral Ecology: An Evolutionary Approach*, ed. J. Krebs and N. Davies, 62–84. Oxford: Blackwell.

——— 1998. Formal Darwinism, the individual-as-maximizing-agent analogy and bet-hedging. *Proceedings of the Royal Society of London, Series B, Biological Sciences* 266: 799–803.

Graham, L. R. 1977. Science and values: The eugenics movement in Germany and Russia in the 1920s. *American Historical Review* 82, no. 5: 1133–64.

Gramsci, A. 1971. *Selections from the Prison Notebooks*. New York: International Publishers.

Grant, B. R., and P. R. Grant. 1989. *Evolutionary Dynamics of a Natural Population: The Large Cactus Finch of the Galápagos*. Chicago: University of Chicago Press.

Grant, B. S., D. F. Owen, and C. A. Clarke. 1996. Parallel rise and fall of melanic peppered moths in America and Britain. *Journal of Heredity* 87: 351–57.

Grant, K. T., and G. B. Estes. 2009. *Darwin in Galápagos: Footsteps to a New World*. Princeton: Princeton University Press.

Grant, P. R. 1985. *Ecology and Evolution of Darwin's Finches*. Princeton: Princeton University Press.

Grant, P. R., and B. R. Grant. 1997. The rarest of Darwin's finches. *Conservation Biology* 11, no. 1: 119–26.

Grant, P. R., B. R. Grant, K. Petren, and L. F. Keller. 2005. Extinction behind our backs: The possible fate of one of the Darwin's finch species on Isla Floreana, Galápagos. *Biological Conservation* 122: 499–503.

Grant, V. 1966. The selective origin of incompatibility barriers in the plant genus, *Gilia*. *American Naturalist* 100: 99–118.

Gray, A. 1860. Darwin and his reviewers. *Atlantic Monthly* 6: 406–25. Reprinted in Gray 1876, 106–72.

———. 1862a. Fertilization of orchids through the agency of insects. *American Journal of Science and Arts* 34: 420–29.

———. 1862b. Review of *Orchids* by Charles Darwin. *American Journal of Science and Arts* 34: 138–44.

———. 1874. Scientific Worthies III. Charles Robert Darwin. *Nature* 10: 79–81.

———. 1876. *Darwiniana*. New York: D. Appleton.

———. 1887. Darwin's Life and Letters I. *The Nation* 45: 399–402.

———. 1963. Natural selection not inconsistent with natural theology. Reprinted in *Darwiniana: Essays and Reviews Pertaining to Darwinism*, by A. Gray, 72–145. Cambridge, Mass.: Belknap Press of Harvard University Press.

Gray, J. L. 1893. *Letters of Asa Gray*. Boston: Houghton, Mifflin.

Gray, R. D., and F. M. Jordan. 2000. Language trees support the express-train sequence of Austronesian expansion. *Nature* 405: 1052–55.

Green, A. 2003. *Ehyeh: A Kabbalah for Tomorrow*. Woodstock, Vt.: Jewish Lights Publishing.

Green, J. R., and J. Sachs. 1909. *A History of Botany, 1860–1900; Being a Continuation of Sachs "History of Botany, 1530–1860."* Oxford: Clarendon Press.

Greenberg, J. 2007. Why can't biologists read poetry? Ian McEwan's *Enduring Love*. *Twentieth-Century Literature* 53: 93–124.

Greene, J. C. 1959. *The Death of Adam: Evolution and Its Impact on Western Thought*. Ames: Iowa State University Press.

Grene, M., and D. Depew. 2004. *The Philosophy of Biology: An Episodic History*. Cambridge: Cambridge University Press.

Griffiths, P., and R. Gray. 1995. Developmental systems and evolutionary explanation. *Journal of Philosophy* 91: 277–304.

Griffiths, T., and L. Robin. 1997. *Ecology and Empire: Environmental History of Settler Societies*. Edinburgh: Keele University Press.

Grinnell, J. 1917. The niche-relationships of the California thrasher. *The Auk* 34: 427–33.

Gruber, H. E. 1974. *Darwin on Man: A Psychological Study of Scientific Creativity*. With notebook transcriptions by Paul H. Barrett. London: Wildwood House.

Gruber, J. W. 1960. *A Conscience in Conflict: The Life of St. George Jackson Mivart*. New York: Columbia University Press.

Grumett, D. 2005. *Teilhard de Chardin: Theology, Humanity and Cosmos*. Leuven: Peeters.

Guillaume, L. 1927. Révision des Posidonomyes jurassiques. *Bulletin de la Société zoologique de France* 27, no. 4: 217–34.

Gülen, F. 2003. *Yaratılış gerçeği ve evrim*. Istanbul: Nil Yayınları.

Gutleben, C. 2001. *Nostalgic Postmodernism: The Victorian Tradition and the Contemporary British Novel*. Amsterdam: Rodopi.

Haeckel, E. 1862. *Die Radiolarien (Rhizopoda Radiaria). Eine Monographie*. Berlin: Druck und Velag Von Georg Reimer.

———. 1866. *Generelle Morphologie der Organismen*. Berlin.

———. 1868a. *Natürliche Schöpfungsgeschichte*. Berlin: G. Reimer.

———. 1868b. Über die Entstehung und den Stammbaum des Menschengeschlechts. In *Sammlung Gemeinverständlicher Wissenschaftlicher Vorträge*, ed. R. Virchow and F. Holtzendorff, 3: nos. 52 and 53, 109–88. Berlin: Lüderitz'sche Verlagsbuchhandlung, 1868–69.

———. 1869a. Über Entwickelungsgang und Aufgabe der Zoologie. In *Gesammelte populare Vorträge aus dem Gebiete der Entwickelungslehre*, 2:1–24. Bonn: Emil Strauss, 1878–79.

———. 1869b. *Zur Entwickelungsgeschichte der Siphonophoren*. Utrecht: C. Van der Post.

———. 1872. *Die Kalkschwamme*. 3 vols. Berlin: Georg Reimer.

———. 1876. *The History of Creation*. Trans. E. Ray Lankester. 2 vols. London: H. S. King.

———. 1878. *Freie Wissenschaft und freie Lehre*. Stuttgart: E. Schweizerbart'sche Verlagshandlung.

———. 1883. *The History of Creation of the Development, or The Development of the Earth and Its Inhabitants by the Action of Natural Causes*. New York: Appleton.

———. 1902. *The Riddle of the Universe at the Close of the Nineteenth Century*. Trans. J. McCabe. New York: Harper & Brothers.

———. 1907. *Das Menschen-Problem und die Herrentiere von Linné*. Frankfurt: Neuer Frankfurter.

Haffer, J. 1997. "We must lead the way on new paths": The work and correspondence of Hartert, Stresemann, Ernst Mayr – International ornithologists. *Ökologie der Vögel* 19: 1–980.

Hagen, J. B. 1981. Experimental Taxonomy, 1930–1950: The Impact of Cytology, Ecology, and Genetics on Ideas of Biological Classification. PhD dissertation, Oregon State University.

———. 1984. Experimentalists and naturalists in twentieth-century botany: Experimental taxonomy, 1920–1950. *Journal of the History of Biology* 17, no. 2: 249–70.

———. 1992. *An Entangled Bank: The Origins of Ecosystem Ecology*. New Brunswick, N.J.: Rutgers University Press.

———. 2009. Descended from Darwin? George Gaylord Simpson, Morris Goodman, and primate systematics. In *Descended from Darwin: Insights into the History of Evolutionary Studies, 1900–1970*, ed. Joe Cain and Michael Ruse, 93–109. Philadelphia: American Philosophical Society.

Haight, G. S., ed. 1954–78. *The George Eliot Letters*. 9 vols. New Haven: Yale University Press.

Haila, Y. 2002. A conceptual genealogy of fragmentation research: From island biogeography to landscape ecology. *Ecological Applications* 12: 321–34.

Haldane, J. B. S. 1924. A mathematical theory of natural and artificial selection. *Transactions of the Cambridge Philosophical Society* 23: 303–8.

———. [1929] 1967. The origin of life. In *The Origin of Life*, J. D. Bernal, Appendix 2, 242–49. London: Weidenfeld and Nicolson.

———. 1932. *The Causes of Evolution*. London: Longmans.

———. 1949. The rate of mutation of human genes. *Hereditas* 35, no. S1: 267–73.

Hales, C. N., and D. J. Barker. 1992. Type 2 (non-insulin-dependent) diabetes mellitus: The thrifty phenotype hypothesis. *Diabetologia* 35, no. 7: 595–601.

Hales, C. N., and S. E. Ozanne. 2003. The dangerous road of catch-up growth. *Journal of Physiology* 547: 5–10.

Hall, B. H., A. B. Jaffe, and M. Trajtenberg. 2001. The NBER patent citation data file: Lessons, insights and methodological tools. *NBER Working Paper* 8498.

Hamilton, W. D. 1963. The evolution of altruistic behavior. *American Naturalist* 97, no. 896: 354–56.

———. 1964a. The genetical evolution of social behaviour. I. *Journal of Theoretical Biology* 7, no. 1: 1–16.

1964b. The genetical evolution of social behaviour. II. *Journal of Theoretical Biology* 7, no. 1: 17–52.

1972. Altruism and related phenomena, mainly in social insects. *Annual Review of Ecology and Systematics* 3: 193–232.

Hamlin, K. 2009. The birds and the bees: Darwin's evolutionary approach to human sexuality. In *Darwin in Atlantic Cultures: Evolutionary Visions of Race, Gender and Sexuality*, ed. J. E. Jones and P. Sharp, 53–72. New York: Routledge Press.

2011. The "Case of a Beaded Woman": Hypertrichosis and the construction of gender in the Age of Darwin. *American Quarterly* 63, no. 4: 955–81.

Hanioğlu, Ş. 2005. Blueprints for a future society: Late Ottoman materialists on science, religion and art. In *Late Ottoman Society: The Intellectual Legacy*, ed. E. Özdalga, 28–116. London: Routledge.

Hansen, T. F., and S. H. Orzack. 2005. Assessing current adaptation and phylogenetic inertia as explanations of trait evolution: The need for controlled comparisons. *Evolution* 59: 2063–72.

Haraway, D. 1989. *Primate Visions: Gender, Race, and Nature in the World of Modern Science*. New York: Routledge.

Hardin, G. 1960. The competitive exclusion principle. *Science* 131: 1292–97.

Hardy, T. 1967. *Complete Poems: The New Wessex Edition*. Ed. J. Gibson. London: Macmillan.

1998. *Jude the Obscure*. Ed. D. Taylor. London: Penguin.

Harinck, G. 2008. How neo-Calvinists dealt with the modern discrepancy between the Bible and natural sciences. In *Nature and Scripture in the Abrahamic Religions*, ed. J. Van Dermeer and S. Mandelbrote, 2:317–70. Leiden: Brill.

Harker, A. 1909. *The Natural History of Igneous Rocks*. London: Methuen.

Harper, J. L. 1967. A Darwinian approach to plant ecology. *Journal of Ecology* 55: 247–70.

Harris, J. A. 1911. The measurement of natural selection. *Popular Science Monthly* 78: 521–38.

Harris, R., and T. J. Taylor. 1997. *Landmarks in Linguistic Thought, I: The Western Tradition from Socrates to Saussure*. 2nd ed. New York: Routledge.

Hart, M. 2000. *Social Science and the Politics of Modern Jewish Identity*. Palo Alto, Calif.: Stanford University Press.

2007. *The Healthy Jew*. Cambridge: Cambridge University Press.

Harvey, J. 1997. *Almost a Man of Genius: Clémence Royer, Feminism and Nineteenth-Century Science*. New Brunswick, N.J.: Rutgers University Press.

2009. Darwin's angels: The women correspondents of Charles Darwin. *Intellectual History Review* 12, no. 2: 197–210.

Harvey, P. H., and M. D. Pagel. 1991. *The Comparative Method in Evolutionary Biology*. New York: Oxford University Press.

Harwood, J. 1993. *Styles of Scientific Thought: The German Genetics Community, 1900–1933*. Chicago: University of Chicago Press.

Hassan, R. 2007. On being religious: Patterns of religious commitment in Muslim societies. *Muslim World* 97: 437–78.

Haught, J. 2000. *God after Darwin: A Theology of Evolution*. Boulder, Colo.: Westview Press.

2010. *Making Sense of Evolution: Darwin, God, and the Drama of Life*. Louisville: Westminster John Knox Press.

Hauser, M. D., N. Chomsky, and W. T. Fitch. 2002. The faculty of language: What is it, who has it, and how did it evolve? *Science* 298: 1569–79.

Hayden, S., and P. S. White. 2003. Invasion biology: An emerging field of study. *Annals of the Missouri Botanical Garden* 90: 64–66.

Hegel, G. W. F. [1817] 2004. *Philosophy of Nature*. Oxford: Oxford University Press.

Heinen, A. 1982. *Islamic Cosmology: A Study of as-Suyūṭī's al-Hay'a as-sanīya fī l-hay'a as-sunnīya with Critical Edition, Translation and Commentary*. Wiesbaden: Steiner.

Henderson, L. J. [1913] 1970. *The Fitness of the Environment*. Gloucester, Mass.: Peter Smith.

Hendry, A. P., P. Nosil, and L. H. Rieseberg. 2007. The speed of ecological speciation. *Functional Ecology* 21: 455–64.

Henkin, L. 1940. *Darwinism in the English Novel, 1860–1910: The Impact of Evolution on Victorian Fiction*. New York: Russell & Russell.

Henrich, J. 2004. Cultural group selection, coevolutionary processes and large-scale cooperation. *Journal of Economic Behavior and Organization* 53: 3–35.

Henshilwood, C., and B. Dubreuil. 2009. Reading the artifacts: Gleaning language skills from the Middle Stone Age in southern Africa. In *The Cradle of Language*, ed. R. P. and C. Knight Botha, 41–61. Oxford: Oxford University Press.

Henslow, J. S. 1830. On the specific identity of the primrose, oxlip, cowslip, and polyanthus. *Magazine of Natural History* 3: 406–9.

1835. *The Principles of Descriptive and Physiological Botany*. London: Longman, Rees, Orme, Brown, Green, and Longman.

1836. On the requisites necessary for the advance of botany. *Magazine of Zoology and Botany* 1: 113–25.

Herbert, S. 1971. Darwin, Malthus and selection. *Journal of the History of Biology* 4: 209–17.

1974. The place of man in the development of Darwin's theory of transmutation. Part I. *Journal of the History of Biology* 4: 217–58.

1977. The place of man in the development of Darwin's theory of transmutation. Part II. *Journal of the History of Biology* 10: 155–227.

1980. *The Red Notebook of Charles Darwin*. Ithaca, N.Y.: Cornell University Press.

1991. Charles Darwin as a prospective geological author. *British Journal for the History of Science* 24: 159–92.

2005. *Charles Darwin, Geologist*. Ithaca, N.Y.: Cornell University Press.

Herbert, S., S. Gibson, D. Norman, et al. 2009. Into the field again: Re-examining Charles Darwin's 1835 geological work on Isla Santiago (James Island) in the Galápagos archipelago. *Earth Sciences History* 28, no. 1: 1–31.

Herbert, S., and D. B. Norman. 2009. Darwin's geology and perspective on the fossil record. In *Cambridge Companion to the "Origin of Species,"* ed. M. Ruse and R. J. Richards, 129–52. Cambridge: Cambridge University Press.

Herre, E. A. 1987. Optimality, plasticity, and selective regime in fig wasp sex ratios. *Nature* 329: 627–29.

Herschel, J. F. W. 1830. *A Preliminary Discourse on the Study of Natural Philosophy*. London: Longman, Rees, Orme, Brown, Green, and Longman.

Heslop-Harrison, J. 1979. Darwin and the movement of plants: A retrospect. In *Plant Growth*, ed. F. Skoog, 1–26. Berlin and Heidelberg: Springer-Verlag.

Himmelfarb, G. 1959. *Darwin and the Darwinian Revolution*. London: Chatto and Windus.

Hinde, R. A. 1985. Ethology in relation to other disciplines. In *Leaders in the Study of Animal Behavior*, ed. D. Dewsbury, 193–203. London: Associated University Press.

His, W. 1874. *Unsere Körperform und das physiologische Problem ihrer Entstehung*. Leipzig: Vogel.

Hitler, A. [1925] 1943. *Mein Kampf*. Munich: Verlag Franz Eher.

Hodge, C. 1872. *Systematic Theology*. Vol. 2. New York: Charles Scribner.

1874. *What Is Darwinism?* Ed. M. A. Noll and D. N. Livingstone. New York: Scribner, Armstrong and Co.

Hodge, M. J. S. 1985. Darwin as a lifelong generation theorist. In *The Darwinian Heritage*, ed. D. Kohn, 207–43. Princeton: Princeton University Press.

———. 1992. Biology and philosophy (including ideology): A study of Fisher and Wright. In *The Founders of Evolutionary Genetics*, ed. S. Sarkar, 231–93. Dordrecht: Kluwer.

———. 2008. *Before and after Darwin: Origins, Species, Cosmogonies, and Ontologies*. Aldershot: Ashgate.

———. 2009a. Capitalist contexts for Darwinian theory: Land, finance, industry and empire. *Journal of the History of Biology* 42: 399–416.

———. 2009b. The notebook programmes and projects of Darwin's London years. In *The Cambridge Companion to Darwin*, 2nd ed., ed. J. Hodge and G. Radick, 44–72. Cambridge: Cambridge University Press.

———. 2009c. *Darwin Studies: A Theorist and His Theories in Their Contexts*. Aldershot: Ashgate.

———. 2010. Darwin, the Gálapagos and his changing thoughts about species origins: 1835–1837. *Proceedings of the California Academy of Sciences* 61, no. 7: 89–106.

Hodge, M. J. S., and G. Radick, eds. 2009. *The Cambridge Companion to Darwin*. 2nd ed. Cambridge: Cambridge University Press.

Hodgson, G. M. 2004. Social Darwinism in Anglophone academic journals: A contribution to the history of the term. *Journal of Historical Sociology* 47, no. 4: 428–63.

Hoelzel, A. R., J. Hey, M. E. Dahlheim, C. Nicholson, V. Burkanov, and N. Black. 2007. Evolution of population structure in a highly social top predator, the killer whale. *Molecular Biology and Evolution* 24: 1407–15.

Hokstra, H., and J. Coyne. 2007. The locus of evolution: Evo-Devo and the genetics of adaptation. *Evolution* 61: 995–1016.

Hölldobler, Bert, and Edward O. Wilson. 2008. *The Superorganism: The Beauty, Elegance, and Strangeness of Insect Societies*. New York: Norton.

Holmes, J. 2009. *Darwin's bards: British and American poetry in the age of evolution*. Edinburgh: Edinburgh University Press.

Hooker, J. 1862a. Review of *Fertilization of Orchids* by Charles Darwin. *Natural History Review* 2: 371–76.

———. 1862b. Review of *Orchids* by Charles Darwin. *Gardeners' Chronicle* (23 August, 13 September, 27 September): 789–90, 863, 910.

Hooper, J. 2002. *Of Moths and Men: An Evolutionary Tale; Intrigue, Tragedy and the Peppered Moth*. London: Fourth Estate.

Hoquet, T. 2009. *Darwin contre Darwin*. Paris: Le Seuil.

Horwich, P. 1998. *Truth*. 2nd ed. Oxford: Clarendon Press.

Hottes, C. F. 1932. The contributions to botany of Julius von Sachs. *Annals of the Missouri Botanical Garden* 19, no. 1: 15–30.

Hrdy, S. B. 2002. Empathy, polyandry, and the myth of the coy female. In *The Gender of Science*, ed. J. A. Kourany, 171–91. Upper Saddle River, N.J.: Prentice Hall.

Hrdy, S. B., and G. C. Williams. 1983. Behavioral biology and the double standard. In *Social Behavior of Female Vertebrates*, ed. S. L. Wasser, 3–17. New York: Academic Press.

Hu, Z. 2005. *Jingsheng shengwu diaocha suo shigao* (Historical Manuscript of Fan Memorial Institute of Biology). Jinan: Shandong Education Press.

Hubbard, C. E. 1971. William Bertram Turrill, 1890–1961. *Biographical Memoirs of Fellows of the Royal Society* 17: 689–712.

Hubbard, R. 1979. Have only men evolved? In *Women Look at Biology Looking at Women*, ed. M. S. Henifin, B. Fried, and R. Hubbard, 7–35. Boston: G. K. Hall.

Huber, C., and G. Wächtershäuser. 2006. α-Hydroxy and α-amino acids under possible hadean, volcanic origin-of-life conditions. *Science* 314: 630–32.

Huelsenbeck, J. P., and P. Andolfatto. 2007. Inference of population structure under a Dirichlet process model. *Genetics* 175: 1787–802.

Hull, D. L., ed. 1973. *Darwin and His Critics*. Cambridge, Mass.: Harvard University Press.

———. 1974. *The Philosophy of Biological Science*. Englewood Cliffs, N.J.: Prentice-Hall.

———. 1988. *Science as a Process*. Chicago: University of Chicago Press.

Humboldt, A. von. 1814–29. *Personal Narrative of Travels to the Equinoctial Regions of the New Continent, during the Years 1799–1804*. Trans. H. M. Williams, 7 vols. London: Longman.

Hunt, J. 1864. On the Negro's place in nature. *Memoirs of the Anthropological Society* 1: 566–80.

Hunter, T. R. 2009. Rethinking Asa Gray's "Natural Selection not Inconsistent with Natural Theology." Master's thesis, University of Oklahoma.

Huntley, W. 1972. David Hume and Charles Darwin. *Journal of the History of Ideas* 33, no. 3: 457–70.

Hurford, J., M. Studdert-Kennedy, and C. Knight, eds. 1998. *Approaches to the Evolution of Language*. Cambridge: Cambridge University Press.

Hutchinson, G. E. 1961. The paradox of the plankton. *American Naturalist* 95: 137–45.

Huxley, J. S. 1912. *The Individual in the Animal Kingdom*. Cambridge: Cambridge University Press.

———. 1942. *Evolution: The Modern Synthesis*. New York: Harper & Bros.

———. 1943. *Evolutionary Ethics*. Oxford: Oxford University Press.

Huxley, L., ed. 1900. *Life and Letters of Thomas Henry Huxley*. London: Macmillan.

Huxley, T. H. 1857. Lectures on general natural history. Lecture XII: The Cirripedia. *Medical Times & Gazette* 17: 238–40.

———. 1860. Darwin on the origin of species. *Westminster Review* 73: 295–310.

———. 1863a. *Evidences as to Man's Place in Nature*. London: Williams and Norgate.

———. 1863b. *On the Origin of Species, or The Causes of Phenomena of Organic Nature*. New York: D. Appleton.

———. 1869. The anniversary address of the president. *Quarterly Journal of the Geological Society of London* 25: xxviii–xxliii.

———. 1887. On the reception of the "Origin of Species." In *The Life and Letters of Charles Darwin, Including an Autobiographical Chapter, Edited by His Son*, ed. F. Darwin, 2:179–204. London: John Murray.

———. 1893. *Collected Essays: Darwiniana*. London: Macmillan.

———. [1893] 2009. *Evolution and Ethics*. Edited and with an introduction by Michael Ruse. Princeton: Princeton University Press.

———. 1894. *Collected Essays*. Vol. 9, *Evolution and Ethics, and Other Essays*. London: Macmillan.

———. 1896. *Darwiniana: Essays*. New York: D. Appleton.

Ilerbaig, J. 2009. "The view-point of a naturalist": American field zoologists and the evolutionary synthesis, 1900–1945. In *Descended from Darwin: Insights into the History of Evolutionary Studies, 1900–1970*, ed. Joe Cain and Michael Ruse, 23–48. Philadelphia: American Philosophical Society.

Iltis, H. 1932. *Life of Mendel*. Trans. E. Paul and C. Paul. London.

Ingold, T., and K. R. Gibson, eds. 1994. *Tools, Language and Cognition in Human Evolution*. Cambridge: Cambridge University Press.

Innes, S. 2009. The anomalous "Mr Arthrobalanus": Darwin's adaptationist approach to taxonomy. In *A Voyage around the World: Charles Darwin and the Beagle Collections in the University of Cambridge*, ed. A. M. Pearn, 74–76. Cambridge: Cambridge University Press.

International Human Genome Sequencing Consortium. 2001. Initial sequencing and analysis of the human genome. *Nature* 409: 860–921.

Iverach, J. 1894. *Christianity and Evolution*. London: Hodder and Stoughton.

Jablonka, E. 2006. Genes as followers in evolution – a post synthesis synthesis? *Biology and Philosophy* 21: 143–54.

Jablonka, E., and M. Lamb. 1995. *Epigenetic Inheritance and Evolution: The Lamarkian Dimension*. Oxford: Oxford University Press.

——— 2005. *Evolution in Four Dimensions*. Cambridge, Mass.: MIT Press.

James, W. 1880. Great men, great thoughts, and the environment. *Atlantic Monthly* 66: 441–59.

Jann, R. 1997. Revising the descent of woman: Eliza Burt Gamble. In *Natural Eloquence: Women Reinscribe Science*, ed. B. T. Gates and A. B. Shteir, 147–63. Madison: University of Wisconsin Press.

Jeha, S. 2004. *Darwin and the Crisis of 1882 in the Medical Department*. Beirut: American University of Beirut Press.

Jenkin, F. 1867. Review of *The Origin of Species*. *North British Review*: 277–318.

Jensen, J. V. 1988. Return to the Wilberforce-Huxley debate. *British Journal for the History of Science* 21: 161–79.

Johanson, D., and M. Edey. 1981. *Lucy: The Beginnings of Humankind*. New York: Simon & Schuster.

John Paul II. 1997. Address to the Pontifical Academy of Sciences. *Origins, CNS Documentary Service* 9: 24.

Johnson, C. 2007. *Australia's Mammal Extinctions: A 50,000-year History*. Cambridge: Cambridge University Press.

Johnson, D. R. 2010. *Nietzsche's Anti-Darwinism*. Cambridge: Cambridge University Press.

Johnson, D. W., ed. 1909. *Geographical Essays by William Morris Davis*. Boston: Ginn.

Johnson, E. 2008. *Quest for the Living God: Mapping Frontiers in the Theology of God*. New York: Continuum.

Johnson, N. A. 2008. Direct selection for reproductive isolation: The Wallace effect. In *Natural Selection and Beyond: The Intellectual Legacy of Alfred Russel Wallace*, ed. G. Beccaloni and C. H. Smith, 114–24. Oxford: Oxford University Press.

Johnson, P. E. 1991. *Darwin on Trial*. Downers Grove, Ill.: InterVarsity Press.

Jones, G. 2002. Alfred Russel Wallace, Robert Owen and the theory of natural selection. *British Journal for the History of Science* 35: 73–96.

Jones, J. E. 2010. Simians, Negroes, and the "missing link": Evolutionary discourses and transatlantic debates on "The Negro Question." In *Darwin in Atlantic Cultures: Evolutionary Visions of Race, Gender and Sexuality*, ed. J. E. Jones and P. B. Sharp, 191–207. New York: Routledge.

Jones, J. M. 2005. Most Americans engaged in debate about evolution, creation. 13 October. http:/poll./gallup.com.

Jones, J. S., B. H. Leith, and P. Rawlings. 1977. Polymorphism in *Cepaea*: A problem with too many solutions? *Annual Review of Ecology and Systematics* 8: 109–43.

Jordan, D. S. 1922. *The Days of a Man*. Yonkers-on-Hudson: World Book Company.

Jordan, D. S., and V. Kellogg. 1908. *Evolution and Animal Life*. New York: Appleton.

Jordan, K. 1905. Der Gegensatz zwischen geographischer und nichtgeographischer Variation. *Zeitschrift für Wissenschaftliche Zoologie* 83: 151–210.

Joron M. 2003. Mimicry. In *Encyclopedia of Insects*, ed. R. T. Cardé and V. H. Resh, 714–26. New York: Academic Press.

Joyce, G. F., and L. E. Orgel. 2006. Progress toward understanding the origin of the RNA world. In *The RNA World*, 3rd ed., ed. T. R. Cech, J. F. Atkins, and R. F. Gesteland, 23–56. Plainview, N.Y.: Cold Spring Harbor Laboratory Press.

Joyce, R. 2006. *The Evolution of Morality*. Cambridge, Mass.: MIT Press.

Judd, J. W. 1909. Darwin and geology. In *Darwin and Modern Science*, ed. A. C. Seward, 337–84. Cambridge: Cambridge University Press.

——— 1911. Charles Darwin's earliest doubts concerning the immutability of species. *Nature* 88, no. 1292: 8–12.

Kahneman, D., P. Slovic, and A. Tversky. 1982. *Judgment under Uncertainty: Heuristics and Biases*. Cambridge: Cambridge University Press.

Kammerer, P. 1925. *Das Rätsel der Vererbung: Grundlagen der allgemeinen Vererbungslehre*. Berlin: Ullstein.

Kamminga, H. 1982. Life from space: A history of panspermia. *Vistas in Astronomy* 26: 67–86.

Kant, I. [1790] 1951. *Critique of Judgment*. New York: Haffner.

——— 1957. *Kritik der Urteilskraft*. Vol. 5, *Werke*. Ed. W. Weischedel. Wiesbaden: Insel Verlag.

Kaplan, L. 1997. Torah u-Madda in the thought of Rabbi Samson Raphael Hirsch. *Bekhol Derakhekha Daehu* 5: 5–31.

Karpechenko, G. D. 1927. Polyploid hybrids of Raphanus sativus L.x Brassica oleracea, L. *Bulletin of Applied Botany Genetics and Plant Breeding* 17: 305–41.

Kauffman, S. 1993. *The Origins of Order*. Oxford: Oxford University Press.

Kearns, G. 2009. *Geopolitics and Empire: The Legacy of Halford Mackinder*. Oxford: Oxford University Press.

Keith, A. 1923. Man's posture: Its evolution and disorders. *British Medical Journal* 1: 451–54, 499–502, 545–48, 587–90, 624–26, 669–72.

——— 1946. Australopithecinae or Dartians? *Nature* 159: 377.

Keller, C. D. 2006. Evolution vs. intelligent design: A Torah perspective. *Jewish Observer* 39, no. 4: 7–21.

Kellogg, V. L. 1907. *Darwinism To-Day*. New York: Henry Holt.

——— 1918. *Headquarters Nights: A Record of Conversations and Experiences at the Headquarters of the German Army in France and Belgium*. Boston: Atlantic Monthly Press.

Kence, A., and Ü. Sayın. 1999. Islamic scientific creationism: A new challenge in Turkey. *Reports of the National Center for Science Education* 19, no. 6: 18–20, 25–29.

Kennedy, B. A. 2006. *Inventing the Earth: Ideas on Landscape Development since 1740*. Oxford: Blackwell.

Kettlewell, H. B. D. 1955. Selection experiments on industrial melanism in the *Lepidoptera*. *Heredity* 9: 323–42.

Kevles, D. 1985. *In the Name of Eugenics: Genetics and the Uses of Human Heredity*. New York: Alfred Knopf.

Keynes, R. 2001. *Annie's Box: Charles Darwin, His Daughter and Human Evolution*. London: Fourth Estate.

Khaitovich, P., I. Hellmann, W. Enard, et al. 2005. Parallel patterns of evolution in the genomes and transcriptomes of humans and chimpanzees. *Science* 309: 1850–54.

Kimler, W. 1983. One Hundred Years of Mimicry: History of an Evolutionary Exemplar. PhD dissertation, Cornell University.

Kimura, M. 1968. Evolutionary rate at the molecular level. *Nature* 217: 624–26.

———. 1983. *The Neutral Theory of Molecular Evolution*. Cambridge: Cambridge University Press.

King, B. J., ed. 1999. *The Origins of Language*. Santa Fe, N.M.: School of American Research Press.

———. 2004. *The Dynamic Dance: Nonvocal Social Communication in the African Apes*. Cambridge, Mass.: Harvard University Press.

King, J. L., and T. H. Jukes. 1969. Non-Darwinian evolution. *Science* 164: 788–98.

King, M.-C., and A. C. Wilson. 1975. Evolution at two levels in humans and chimpanzees. *Science* 188: 107–15.

Kingsland, S. E. 1991. The battling botanist: Daniel Trembly Macdougal, mutation theory, and the rise of experiment evolutionary biology in America, 1900–1912. *Isis* 82: 479–509.

———. 1995. *Modeling Nature: Episodes in the History of Population Ecology*. Rev. ed. Chicago: University of Chicago Press.

———. 1997. Neo-Darwinism and natural history. In *Science in the Twentieth Century*, ed. John Krige and Dominique Pestre Krige, 417–37. Amsterdam: Harwood Academic.

———. 2005. *The Evolution of American Ecology, 1890–2000*. Baltimore: Johns Hopkins University Press.

Kingsley, C. 1889. *The Water Babies*. London: Macmillan.

Kingsolver, J. G., and S. E. Diamond. 2011, Phenotypic selection in natural populations: What limits directional selection? *American Naturalist* 177: 346–57.

Kirby, S., H. Cornish, and K. Smith. 2008. Cumulative cultural evolution in the laboratory: An experimental approach to the origins of structure in human language. *Proceedings of the National Academy of Sciences* 105: 10681–86.

Kirby, W., and W. Spence. 1815–28. *An Introduction to Entomology, or Elements of the Natural History of Insects*. London: Longman, Hurst, Reece, Orme, and Brown.

Kirch, P. V. 1984. *The Evolution of Polynesian Chiefdoms*. Cambridge: Cambridge University Press.

Kirk, G. S., J. E. Raven, and M. Schofield. 1983. *The Presocratic Philosophers*. Cambridge: Cambridge University Press.

Kitcher, P. 1985. *Vaulting Ambition*. Cambridge, Mass.: MIT Press.

———. 1993. *The Advancement of Science: Science without Legend, Objectivity without Illusions*. Oxford: Oxford University Press.

———. 2011. *The Ethical Project*. Cambridge, Mass.: Harvard University Press.

Kjærgaard, P. C. 2010. The Darwin enterprise: From scientific icon to global product. *History of Science* 48, no. 1: 105–22.

Kleinman, K. 1999. His own synthesis: Corn, Edgar Anderson, and evolutionary theory in the 1940s. *Journal of the History of Biology* 32: 293–320.

Klinghoffer, D. 2008. Don't doubt it: An important historic sidebar. http://www.nationalreview.com/articles/224233/dont-doubt-it/david-klinghoffer.

Klippenstine, D. R., and S. G. Sealy. 2008. Differential ejection of cowbird eggs and non-mimetic eggs by grassland passerines. *Wilson Journal of Ornithology* 120: 667–73.

Knight, C., J. R. Hurford, and M. Studdert-Kennedy, eds. 2000. *The Evolutionary Emergence of Language: Social Function and the Origins of Linguistic Form*. Cambridge: Cambridge University Press.

Knuth, P. 1906–9. *Handbook of Flower Pollination*. Trans. J. R. A. Davis. 3 vols. Oxford: Clarendon Press.

Kohler, R. E. 2006. *All Creatures: Naturalists, Collectors, and Biodiversity, 1850–1950*. Princeton: Princeton University Press.

Kohlstedt, S. G., and M. R. Jorgensen. 1997. "The irrepressible woman question": Women's responses to evolutionary ideology, *Disseminating Darwinism: the Role of Place, Race, Religion and Gender*, ed. R. L. Numbers and J. Stenhouse, 267–93. Cambridge: Cambridge University Press.

Kohn, D. 1980. Theories to work by: Rejected theories, reproduction, and Darwin's path to natural selection. *Studies in the History of Biology* 4: 67–170.

———. 1981. On the origin of the principle of diversity. *Science* 213, no. 4512: 1105–8.

———. 2008. *Darwin's Garden: An Evolutionary Adventure*. Catalogue for an exhibit at the New York Botanical Garden.

———. 2009. Darwin's keystone: The principle of divergence. In *The Cambridge Companion to the "Origin of Species,"* ed. M. Ruse and R. J. Richards, 87–108. Cambridge: Cambridge University Press.

Kohn D., et al. 2005. What Henslow taught Darwin. *Nature* 436: 643–45.

Kolchinsky, E. I. 2008. *Darwinism and Dialectical Materialism in Soviet Russia*. Ed. E.-M. Engels and T. Glick. London: Continuum.

Kölliker, A. 1864. Ueber die Darwin'sche Schöpfungstheorie. *Zeitschrift für wissenschaftliche Zoologie* 14: 174–86.

Kondrashov, A., and M. Shpak. 1998. On the origin of species by means of assortative mating. *Proceedings of the Royal Society of London, Series B, Biological Sciences* 265: 2273–78.

Koning, D. 2006. Anti-evolutionism among Muslim students. *ISIM-Newsletter* 18: 48–49.

Koonin, E. V., Y. I. Wolf, and P. Puigbo. 2009. The phylogenetic forest and the quest for the elusive tree of life. *Cold Spring Harbor Symposium on Quantitative Biology* 74: 205–13.

Kottler, M. J. 1978. Charles Darwin's biological species concept and the theory of geographical speciation. The transmutation notebooks. *Annals of Science* 35: 275–97.

———. 1985. Charles Darwin and Alfred Russel Wallace: Two decades of debate over natural selection. In *The Darwinian Heritage*, ed. D. Kohn, 367–432. Princeton: Princeton University Press.

Kreitman, M. 2000. Methods to detect selection in populations with applications to the human. *Annual Review of Genomics and Human Genetics* 1: 539–59.

Krementsov N. L. 1994. Dobzhansky and Russian entomology: The origin of his ideas on species and speciation. In *The Evolution of Theodosius Dobzhansky*, ed. M. B. Adams, 31–48. Princeton: Princeton University Press.

Krohn, A. D. 1859. Beobachtungen über den Cementapparat und die weiblichen Zeugungsorgane einiger Cirripedien. *Archiv für Naturgeschichte* 25: 355–64.

Kropotkin, P. 2008. *Mutual Aid: A Factor of Evolution*. Charleston, S.C.: Forgotten Books.

Kühn, A. 1955. *Vorlesungen über Entwicklungsphysiologie*. Berlin: Springer.

Kuhn, T. 1970. *The Structure of Scientific Revolutions*. 2nd ed. Chicago: University of Chicago Press.

Kunin, V., L. Goldovsky, N. Darzentas, and C. A. Ouzounis. 2005. The net of life: Reconstructing the microbial phylogenetic network. *Genome Research* 15: 954–59.

Kunte, K. 2009a. The diversity and evolution of Batesian mimicry in Papilio swallowtail butterflies. *Evolution* 63: 2707–16.

———. 2009b. Female-limited mimetic polymorphism: A review of theories and a critique of sexual selection as balancing selection. *Animal Behaviour* 78: 1029–36.

La Vergata, A. 1990. *L'equilibrio e la guerra della natura: Dalla teologia naturale al Darwinismo*. Naples: Morano.

Lack, D. 1940. Evolution of the Galapagos finches. *Nature* 146: 324–27.

⸻ 1945. *The Galapagos Finches (Geospizinae): A Study in Variation*. San Francisco: California Academy of Sciences.

⸻ 1947. *Darwin's Finches: An Essay on the General Biological Theory of Evolution*. Cambridge: Cambridge University Press.

Laland, K. N. 2008. Exploring geneculture interactions: Insights from handedness, sexual selection and niche-construction case studies. *Philosophical Transactions of the Royal Society, Series B, Biological Sciences* 363: 3577–89.

Laland, K. N., and G. R. Brown. 2002. *Sense and Nonsense: Evolutionary Perspectives on Human Behaviour*. Oxford: Oxford University Press.

Laland, K. N., F. J. Odling-Smee, and M. W. Feldman. 2000. Niche construction, biological evolution and cultural change. *Behavioral & Brain Sciences* 23: 131–46.

Lamarck, J. B. 1809. *Philosophie zoologique*. Paris: Dentu.

⸻ 1984. *Zoological Philosophy*. Trans. H. Elliot. Chicago: University of Chicago Press.

Lamotte, M. 1951. Recherches sur la structure génétique des populations naturelles de *Cepaea nemoralis (L.)*. *Bulletin Biologique de la France et de la Belgique (Suppl.)* 35: 1–239.

Lancaster, R. 2003. *The Trouble with Nature: Sex in Science and Popular Culture*. Berkeley: University of California Press.

Lande, R. 1976. Natural selection and random genetic drift in phenotypic evolution. *Evolution* 30: 314–34.

Lane, C. E., and J. M. Archibald. 2008. The eukaryotic tree of life: Endosymbiosis takes its TOL. *Trends in Ecology & Evolution* 23: 268–75.

Largent, M. A. 2008. *Breeding Contempt: The History of Coerced Sterilization in the United States*. New Brunswick, N.J.: Rutgers University Press.

⸻ 2009a. Darwin's analogy between artificial and natural selection in the *Origin of Species*. In *The Cambridge Companion to the "Origin of Species,"* ed. M. Ruse and R. J. Richards, 14–29. Cambridge: Cambridge University Press.

⸻ 2009b. The so-called eclipse of Darwinism. In *Descended from Darwin: Insights into the History of Evolutionary Studies, 1900–1970*, ed. Joe Cain and Michael Ruse, 3–21. Philadelphia: American Philosophical Society.

Larson, E. J. 1997. *Summer for the Gods: The Scopes Trial and America's Continuing Debate over Science and Religion*. New York: Basic Books.

⸻ 2003. *Trial and Error: The American Controversy over Creation and Evolution*. 3rd ed. New York: Oxford University Press.

⸻ 2004. *Evolution: The Remarkable History of a Scientific Theory*. New York: Modern Library.

Larson, S. R., C. M. Culumber, R. N. Schweigert, and N. J. Chatterton. 2010. Species delimitation tests of endemic *Lepidium papilliferum* and identification of other possible evolutionarily significant units in the *Lepidium montanum* complex (Brassicaceae) of western North America. *Conservation Genetics* 11: 57–76.

Laubichler, M. D. 2003. Carl Gegenbaur (1826–1903): Integrating comparative anatomy and embryology. *Journal of Experimental Zoology, Part B: Molecular and Developmental Evolution* 300, no. 1: 23–31.

⸻ 2007. Evolutionary developmental biology. In *The Cambridge Companion to the Philosophy of Biology*, ed. M. Ruse and D. L. Hull, 342–60. Cambridge: Cambridge University Press.

Laubichler, M. D., W. Aird, et al. 2007. The endothelium in history. In *Endothelial Biomedicine*, ed. M. D. Laubichler, W. Aird, et al., 5–22. Cambridge: Cambridge University Press.

Laubichler, M. D., and E. Davidson. 2008. Boveri's long experiment: Sea urchin merogons and the establishment of the role of nuclear chromosomes in development. *Developmental Biology* 314: 1–11.

Laubichler, M. D., and J. Maienschein. 2004. Development. In *The New Dictionary of the History of Ideas*, ed. M. C. Horowitz, 570–74. New York: Charles Scribner's Sons.

⸻ eds. 2007. *From Embryology to Evo Devo*. Cambridge, Mass.: MIT Press.

Laubichler, M. D., and H. J. Rheinberger. 2004. Alfred Kuhn (1885–1968) and developmental evolution. *Journal of Experimental Zoology, Part B: Molecular and Developmental Evolution* 302B: 103–10.

⸻ 2006. August Weismann and theoretical biology. *Biological Theory* 1: 202–5.

Laudan, R. 1987. *From Mineralogy to Geology: The Foundations of a Science, 1650–1830*. Chicago: University of Chicago Press.

Laut, J. P. 2000. *Das Türkische als Ursprache? Sprachwissenschaftliche Theorien in der Zeit des erwachenden türkischen Nationalismus*. Wiesbaden: Harrassowitz.

Lawrence, J. G. 2002. Gene transfer in bacteria: Speciation without species. *Theoretical Population Biology* 61: 449–60.

Lawrence, J. G., and A. Retchless. 2010. The myth of bacterial species and speciation. *Biology and Philosophy* 25: 569–88.

Leakey, L. S. B., P. V. Tobias, and J. R. Napier. 1964. A new species of the genus *Homo* from Olduvia Gorge. *Nature* 202: 7–9.

Leavens, D. A., T. P. Racine, and W. D. Hopkins. 2009. The ontogeny and phylogeny of non-verbal deixis. In *The Prehistory of Language*, ed. R. P. and C. Knight Botha, 142–65. Oxford: Oxford University Press.

Lehninger, A. 1971. *Bioenergetics: The Molecular Basis of Biological Energy Transformations*. Reading, Mass.: Addison-Wesley.

Lennox, J. G. 1992. Teleology. In *Keywords in Evolutionary Biology*, ed. E. F. Keller and E. Lloyd, 334–43. Cambridge, Mass.: Harvard University Press.

⸻ 1993. Darwin was a teleologist. *Biology and Philosophy* 8: 408–21.

⸻ 2001. *Aristotle's Philosophy of Biology*. Cambridge: Cambridge University Press.

⸻ 2010. The Darwin/Gray correspondence, 1857–1869: An intelligent discussion about chance and design. *Perspectives on Science* 18, no. 4: 456–79.

Levin D. A. 1970. Reinforcement of reproductive isolation: Plants versus animals. *American Naturalist* 104: 571–81.

Levine, G. 1988. *Darwin and the Novelists: Patterns of Science in Victorian Fiction*. Cambridge, Mass.: Harvard University Press.

⸻ 2006. *Darwin Loves You: Natural Selection and the Re-enchantment of the World*. Princeton: Princeton University Press.

Levine, L. 1958. Studies on sexual selection in mice. I. Reproductive competition between albino and black-agouti males. *American Naturalist* 92: 21–26.

Levins, R. 1968. *Evolution in Changing Environments*. Princeton: Princeton University Press.

Lewens, T. 2007. *Darwin*. London: Routledge.

Lewes, G. H. 1859–60. *The Physiology of Common Life*. 2 vols. Edinburgh: W. Blackwood.

Lewis, S. M., and C. K. Cratsley. 2008. Flash signal evolution, mate choice, and predation in fireflies. *Annual Review of Entomology* 53: 293–321.

Lewontin, R. C. 1970. The units of selection. *Annual Review of Ecology and Systematics* 1: 1–18.

Lewontin, R. C. 1974. *The Genetic Basis of Evolutionary Change*. New York: Columbia University Press.

Lewontin, R. C., J. A. Moore, W. B. Provine, and B. Wallace, eds. 1981. *Dobzhansky's Genetics of Natural Populations I–XLIII*. New York: Columbia University Press.

Lewontin, R. C., S. Rose, and L. J. Kamin. 1984. *Not in Our Genes: Biology, Ideology and Human Nature*. New York: Pantheon.

L'Héritier, P. 1934. *Génétique et évolution: Analyse de quelques études mathématiques sur la sélection naturelle*. Paris: Hermann, Actualités Scientifiques et Industrielles.

———. 1981. Souvenirs d'un généticien. *Revue de Synthèse* 102: 331-50.

L'Héritier, P., Y. Neefs, and G. Teissier. 1937. Aptérisme des insectes et sélection naturelle. *Comptes Rendus des Séances de l'Academie des sciences*. 204: 907-9.

L'Héritier, P., and G. Teissier. 1934. Une expérience de sélection naturelle, Courbe d'élimination du gène *Bar* dans une population de *Drosophila melanogaster*. *Comptes Rendus des Séances et Mémoires de la Société de Biologie* 117: 1049.

———. 1937a. Elimination des formes mutantes dans les populations de Drosophiles. Cas des Drosophiles bar. *Comptes Rendus des Séances et Mémoires de la Société de Biologie* 124: 880-82.

———. 1937b. Elimination des formes mutantes dans les populations de Drosophiles. Cas des Drosophiles *Ebony*. *Comptes Rendus des Séances et Mémoires de la Société de Biologie* 124: 882-84.

Lieberman, E., J.-B. Michel, J. Jackson, T. Tang, and M. Nowak. 2007. Quantifying the evolutionary dynamics of language. *Nature* 449: 713-16.

Lieberman, P. 1975. *On the Origins of Language*. New York: Macmillan.

———. 1984. *The Biology and Evolution of Language*. Cambridge, Mass.: Harvard University Press.

———. 2006. *Toward an Evolutionary Biology of Language*. Cambridge, Mass.: Harvard University Press.

Liepman, H. P. 1981. The six editions of the "Origin of Species": A comparative study. *Acta Biotheoretica* 30, no. 3: 199-214.

Lightman, B. 2007. *Victorian Popularizers of Science: Designing Nature for New Audiences*. Chicago: University of Chicago Press.

Limoges, C. 1970. *La sélection naturelle: Étude sur la premiere constitution d'un concept (1837-1859)*. Paris: Presses Universitaires de France.

———. 1976. Natural selection, phagocytosis, and preadaptation: Lucien Cuénot, 1886-1901. *Journal of the History of Medicine and Allied Sciences* 31, no. 2: 176-214.

Lindley, J. 1846. *The Vegetable Kingdom*. London: Bradbury and Evans.

Litchfield, H. E., ed. 1904. *Emma Darwin, Wife of Charles Darwin: A Century of Family Letters*. 2 vols. Cambridge: privately printed by Cambridge University Press.

Livingstone, D. N. 1984. *Darwin's Forgotten Defenders*. Grand Rapids, Mich.: Eerdmans.

———. 1987. *Nathaniel Southgate Shaler and the Culture of American Science*. Tuscaloosa: University of Alabama Press.

———. 1992a. Darwinism and Calvinism: The Belfast-Princeton connection. *Isis* 83: 408-28.

———. 1992b. *The Geographical Tradition: Episodes in the History of a Contested Enterprise*. Oxford: Blackwell.

———. 2005. Science, text and space: Thoughts on the geography of reading. *Transactions of the Institute of British Geographers* 30: 391-401.

Locke, J. [1690] 1975. *An Essay concerning Human Understanding*. Ed. P. Nidditch. Oxford: Clarendon Press.

Locke, J. L., and B. Bogin. 2006. Language and life history: A new perspective on the development and evolution of human language. *Behavioral and Brain Sciences* 29, no. 3: 259-80.

Loewenberg, B. J. 1933. The reaction of American scientists to Darwinism. *American Historical Review* 38: 687-701.

Loison, L. 2010. *Qu'est-ce que le néomarckisme? Les biologists français devant l'évolution des espèces*. Paris: Vuibert.

Lombardo, P. A., and G. M. Dorr. 2006. Eugenics, medical education, and the Public Health Service: Another perspective on the Tuskegee syphilis experiment. *Bulletin of the History of Medicine* 80, no. 2: 291-316.

London, J. 1981. *The Call of the Wild, White Fang, and Other Stories*. Ed. A. Sinclair. New York: Penguin.

———. 1992. *The Sea-Wolf*. Ed. J. Sutherland. Oxford: Oxford University Press.

Lorenz, K. [1941] 1982. Kant's doctrine of the a priori in the light of contemporary biology. In *Learning, Development and Culture: Essays in Evolutionary Epistemology*, ed. H. C. Plotkin, 121-43. Chichester: Wiley.

———. 1965. Introduction to *The Expression of the Emotions in Man and Animals* by C. Darwin, ix-xiii. Chicago: University of Chicago Press.

———. 1970. *Studies in Animal and Human Behavior*. Cambridge, Mass.: Harvard University Press.

———. 1974a. Analogy as a source of knowledge. *Science* 185: 229-34.

———. 1974b. Nobel autobiography. http://www.nobelprize.org/nobel_prizes/medicine/laureates/1973/lorenz-autobio.html.

Love, A. C. 2002. Darwin and cirripedia prior to 1846: Exploring the origins of the barnacle research. *Journal of the History of Biology* 35: 251-89.

Love, R. 1983. Darwinism and feminism: The "woman question" in the life and work of Olive Schreiner and Charlotte Perkins Gilman. In *The Wider Domain of Evolutionary Thought*, ed. D. Oldroyd and I. Langham, 113-31. Boston: D. Reidel.

Lovejoy, A. O. 1959. The argument for organic evolution before *The Origin of Species*, 1830-1858. In *Forerunners of Darwin: 1745-1859*, ed. O. Temkin, W. L. Straus, and B. Glass, 356-414. Baltimore: Johns Hopkins University Press.

Lovejoy, O. 1981. The origin of man. *Science* 211: 342-50.

Lovejoy, T. E. 1980. A projection of species losses. In *Global 2000 Report to the President – Entering the 21st Century*, 2:328-31. Washington, D.C.: U.S. Government Printing Office.

———. 1996. Biodiversity: What is it? In *Biodiversity II: Understanding and Protecting Our Biological Resources*, ed. D. E. Wilson, E. O. Wilson, and M. L. Reak-Kudla, 7-14. Washington, D.C.: Joseph Henry Press.

Lowe, P. R. 1936. The finches of the Galapagos in relation to Darwin's conception of species. *Ibis* 6: 310-21.

Lubbock, J. 1867. On the origin of civilization and the primitive condition of man. *Transactions of the Ethnological Society of London*: 1-15.

———. 1870. *Origin of Civilization and the Primitive Condition of Man*. 2nd ed. London: Longmans, Green.

Lucas, P. 1847-50. *Traité philosophique et physiologique de l'hérédité naturelle*. 2 vols. Paris: Baillière.

Lutz, A. M. 1907. A preliminary note on the chromosomes of *Oenothera* lamarckiana and one of its mutants, O. gigas. *Science* 26: 151-52.

Lutz, F. E. 1915. Experiments with Drosophila ampelophila concerning natural selection. *Bulletin of the American Museum of Natural History* 34: 605-24.

Lycett, S. J. 2009. Understanding ancient hominin dispersals using artefactual data: A phylogeographic analysis of Acheulean handaxes. *PLoS One* 4: e7404.

Lyell, C. 1830–33. *Principles of Geology, Being an Attempt to Explain the Former Changes of the Earth's Surface, by Reference to Causes Now in Operation.* London: John Murray.

———. 1842. On the geological evidence of the former existence of glaciers in Forfarshire. *Proceedings of the Geological Society of London* 3: 337–45.

———. 1853. *Principles of Geology, or The Modern Changes of the Earth and Its Inhabitants Considered as Illustrative of Geology.* 9th ed. London: John Murray.

———. 1863. *The Geological Evidences of the Antiquity of Man: With remarks on Theories of the Origin of Species by Variation.* London: Chas. Murray.

———. 1867. *Principles of Geology, or The Modern Changes of the Earth and Its Inhabitants Considered as Illustrative of Geology.* 10th ed. Vol. 1. London: John Murray.

———. 1868. *Principles of Geology, or The Modern Changes of the Earth and Its Inhabitants Considered as Illustrative of Geology.* 10th ed. Vol. 2. London: John Murray.

Lynch, M. 1990. The rate of morphological evolution in mammals from the standpoint of the neutral expectation. *American Naturalist* 136: 727–41.

———. 2007. *The Origins of Genome Architecture.* Sunderland, Mass.: Sinauer.

Ma, J. 1902–2009. Wu jing pian. A translation of the third chapter (Struggle for Existence) of Charles Darwin's *Origin of Species*. In *Jindai kexue zai Zhongguo de chuanbo: wenxian yu shiliao xuanbian* (The Introduction of Modern Science into Late 19th and Early 20th Century China: Selected Works and Documents), ed. Wang Yangzong, 146–53. Jinan: Shandong Education Press.

———. 1920. *Wuzhong yuanshi* (A Translation of Darwin's *Origin of Species*). Shanghai: Zhonghua Book Company.

———. 1985. *Ma Junwu shi zhu* (The Annotated Collected Poems of Ma Junwu). Annot. Tan Xing. Nanning: Guangxi Nationalities Publishing House.

MacArthur, R. H. 1958. Population ecology of some warblers of northeastern coniferous forests. *Ecology* 39: 599–619.

MacArthur, R. H., and E. O. Wilson. 1967. *The Theory of Island Biogeography.* Princeton: Princeton University Press.

Mackinder, H. J. 1904. The geographical pivot of history. *Geographical Journal*, 23: 421–37.

Mackintosh, J. 1991. *Dissertation on the Progress of Ethical Philosophy: Chiefly during the Seventeenth and Eighteenth Centuries.* Bristol: Thoemmes. Facsimile of the 1st ed., Edinburgh: Adam and Charles Black, 1836.

Mahmoud, M. A. 2007. *Quest for Divinity. A Critical Examination of the Thought of Mahmud Muhammad Taha.* Syracuse, N.Y.: University Press.

Majerus, M. E. N. 1998. *Melanism: Evolution in Action.* Oxford: Oxford University Press.

Majumder, P. P. 2004. C. C. Li (1912–2003). His science and his spirit. *Journal of Genetics* 83, no. 1: 101–5.

Mallet J. 1995. A species definition for the Modern Synthesis. *Trends in Ecology and Evolution* 10: 294–99.

———. 2005. Hybridization as an invasion of the genome. *Trends in Ecology and Evolution* 20: 229–37.

———. 2008. Mayr's view of Darwin: Was Darwin wrong about speciation? *Biological Journal of the Linnean Society* 95: 3–16.

———. 2010a. Group selection and the development of the biological species concept. *Philosophical Transactions of the Royal Society, Series B: Biological Sciences* 365: 1853–63.

———. 2010b. Why was Darwin's view of species rejected by 20th century biologists? *Biology and Philosophy* 25: 497–527.

Mallet, J., and M. Joron. 1999. Evolution of diversity in warning color and mimicry: Polymorphisms, shifting balance, and speciation. *Annual Review of Ecology and Systematics* 30: 201–s33.

Malthus, T. 1826. *An Essay on the Principle of Population.* 6th ed. 2 vols. London: John Murray.

Marx, K., and F. Engels. 1975–2005. *Marx/Engels Collected Works.* Moscow: Progress Publishers.

Mather, K. 1943. Polygenic inheritance and natural selection. *Biological Reviews of the Cambridge Philosophical Society* 18: 32–64.

Maturana, H., and F. Varela. 1980. *Autopoeisis and Cognition.* Dordrecht: Kluwer.

May, R. 1974a. *Stability and Complexity in Model Ecosystems.* 2nd ed. Princeton: Princeton University Press.

———. 1974b. Biological populations with non-overlapping generations: Stable points, stable cycles, and chaos. *Science* 186: 645–47.

Maynard Smith, J. 1978. Optimization theory in evolution. *Annual Review of Ecology and Systematics* 9: 31–56.

———. 1982. *Evolution and the Theory of Games.* Cambridge: Cambridge University Press.

———. 1984. Palaeontology at the high table. *Nature* 309: 401–2.

Maynard Smith, J., and E. Szathmáry. 1995. *The Major Transitions in Evolution.* Oxford: Oxford University Press.

Mayr, E. 1942. *Systematics and the Origin of Species from the Viewpoint of a Zoologist.* New York: Columbia University Press.

———. 1951. Taxonomic categories in fossil hominids. *Cold Spring Harbor Symposia on Quantitative Biology*, vol. 15, *Origin and Evolution of Man*, ed. S. Washburn and Th. Dobzhansky, 109–18. Cold Spring Harbor: Biological Laboratory.

———. 1959. Where are we? *Cold Spring Harbor Symposia on Quantitative Biology* 24: 1–14.

———. 1963. *Animal Species and Evolution.* Cambridge, Mass.: Belknap Press of Harvard University Press.

———. 1964. Introduction to Charles Darwin's *On the Origin of Species*, vii–xxvii. Cambridge, Mass.: Harvard University Press.

———. 1980a. G. G. Simpson. In *The Evolutionary Synthesis: Perspectives on the Unification of Biology*, ed. E. Mayr and W. B. Provine, 452–63. Cambridge, Mass.: Harvard University Press.

———. 1980b. Prologue: Some thoughts on the history of the evolutionary synthesis. In *The Evolutionary Synthesis*, ed. Ernst Mayr and William Provine, 1–48. Cambridge, Mass.: Harvard University Press.

———. 1982. *The Growth of Biological Thought. Diversity, Evolution, and Inheritance.* Cambridge, Mass.: Harvard University Press.

———. 1983. How to carry out the adaptationist program? *American Naturalist* 121: 324–34.

Mayr, E., and W. B. Provine. 1980. *The Evolutionary Synthesis: Perspectives on the Unification of Biology.* Cambridge, Mass.: Harvard University Press.

McBrearty, S., and A. S. Brooks. 2000. The revolution that wasn't. *Journal of Human Evolution* 39: 453–63.

McBrearty, S., and N. G. Jablonski. 2005. First fossil chimpanzee. *Nature* 437: 105–8.

McCann, K., A. Hastings, and G. R Huxel. 1998. Weak trophic interactions and the balance of nature. *Nature* 395: 794–98.

McDermott, R. 1998. Ethics, epidemiology and the thrifty gene: Biological determinism as a health hazard. *Social Science & Medicine* 47, no. 9: 1189–95.

McEwan, I. 2005. *Saturday.* London: Jonathan Cape.

McKinney, H. L. 1972. *Wallace and Natural Selection.* New Haven: Yale University Press.

McKirahan, R. 1994. *Philosophy before Socrates.* Indianapolis: Hackett.

McMullin, E. 1985. Introduction: Evolution and creation. In *Evolution and Creation*, ed. E. McMullin, 1–58. Notre Dame: University of Notre Dame Press.

McOuat, G. R. 1996. Species, rules and meaning: The politics of language and the ends of definitions in 19th Century natural history. *Studies in History and Philosophy of Science* 27: 473–519.

Mendel, G. 1966. Experiments on plant hybrids. In *The Origin of Genetics: A Mendel Source Book*, ed. and trans. C. Stern and E. R. Sherwood, 1–55. San Francisco: W. H. Freeman Press.

Mercier, J. 1933. Contribution à l'étude des Métrionhynchnidés (Crodociliens). *Annales de paléontologie* 22: 91–120.

Mesoudi, A., A. Whiten, and K. N. Laland. 2004. Is human cultural evolution Darwinian? Evidence reviewed from the perspective of "The Origin of Species." *Evolution* 58: 1–11.

——— 2006. Towards a unified science of cultural evolution. *Behavioral and Brain Sciences* 29: 329–83.

Metcalfe, N. B., and P. Monaghan. 2001. Compensation for a bad start: Grow now, pay later? *Trends in Ecology & Evolution* 16: 254–60.

Michell, J. 1760. Conjectures concerning the cause, and observations upon the phaenomena of earthquakes; particularly that great earthquake of the first of November, 1755, and whose effects were felt as far as Africa, and more or less throughout almost all Europe. *Philosophical Transactions of the Royal Society* 51: 566–634.

Milam, E. L. 2010. *Looking for a Few Good Males: Female Choice in Evolutionary Biology*. Baltimore: Johns Hopkins Press.

Milinksi, M. 1979. An evolutionarily stable feeding strategy in Sticklebacks. *Zeitschrift für Tierpsychologie* 51: 36–40.

Miller, A. 1962. *Rejoice! O Youth*. New York.

——— 1995. *The Universe Testifies*. New York.

Miller, G. 2000. *The Mating Mind: How Sexual Choice Shaped the Evolution of Human Nature*. New York: Random House.

Miller, J. D., S. Okamoto, and E. C. Scott. 2006. Public acceptance of evolution. *Science* 313: 765–66.

Millican, K. W. 1883. *The Evolution of Morbid Germs: A Contribution to Transcendental Pathology*. London: H. K. Lewis.

Millstein, R. 2007. Hsp90-induced evolution: Adaptationist, neutralist, and developmentalist scenarios. *Biological Theory* 2: 376–86.

——— 2012. Darwin's explanation of races by means of sexual selection. *Studies in History and Philosophy of Biological and Biomedical Sciences* 43: 627–33.

Milne-Edwards, Henri. 1844. Considérations sur quelques principes relatifs à la classification naturelle des animaux, et plus particulièrement sur la distribution méthodique des mammifères. *Annales des Sciences Naturelles (Zoologie)*, 3d ser., 1: 65–99.

Mitchell, S. D. 1995. The superorganism metaphor: Then and now. In *Biology as Society, Society as Biology: Metaphors*, ed. E. Mendelsohn, P. Weingart, and S. Massen, 231–47. Dordrecht: Kluwer.

Mivart, St. G. J. 1871. *The Genesis of Species*. 2nd ed. London: Macmillan.

Molina, G. 1992. Le savant et ses interprètes. In *Darwinisme et société*, ed. P. Tort, 361–86. Paris: Presses Universitaires de France.

Montgomery, G. M. 2009. "Infinite loneliness": The life and times of Miss Congo. *Endeavour* 33, no. 3: 101–5.

Montoya, M. M. 2007. Bioethnic conscription: Genes, race and Mexicana/o ethnicity in diabetes research. *Cultural Anthropology* 22, no. 1: 94–128.

Moody, J. W. T. 1971. The reading of the Darwin and Wallace papers: An historical "non-event." *Society for the Bibliography of Natural History* 5, no. 6: 474–76.

Moore, A. 1889. *Science and the Faith: Essays on Apologetic Subjects*. London: Kegan Paul.

——— [1889] 1890a. The Christian doctrine of God. In *Lux Mundi: A Series of Studies in the Religion of the Incarnation*, ed. C. Gore, 41–81. London: John Murray.

——— 1890b. *Essays Scientific and Philosophical*. London: Kegan Paul.

Moore, D. W. 2005. Most Americans tentative about origin-of-life explanations. http://www.gallup.com/poll/18748/most-americans-tentative-about-originoflife-explanations.aspx.

Moore, G. E. [1903] 1948. *Principia Ethica*. Cambridge: Cambridge University Press.

Moore, J. 1765. *A Treatise on Domestic Pigeons, Comprehending All the Different Species Known in England....* London: C. Barry.

Moore, J. R. 1979. *The Post-Darwinian Controversies*. Cambridge: Cambridge University Press.

——— 1989. Of love and death: Why Darwin gave up Christianity. In *History, Humanity and Evolution*, ed. J. R. Moore, 195–229. Cambridge: Cambridge University Press.

——— 1997. Wallace's Malthusian moment: The common context revisited. In *Victorian Science in Context*, ed. B. Lightman, 290–311. Chicago: University of Chicago Press.

——— 2001. Darwinism gone to seed. *Books and Culture*, March–April: 36–38.

Moorehead, A. 1969. *Darwin and the Beagle*. London: Harper & Row.

Morgan, G. 1998. Emile Zuckerkandl, Linus Pauling, and the molecular evolutionary clock, 1959–1965. *Journal of the History of Biology* 31: 155–78.

Morgan, T. H., A. Sturtevant, H. J. Muller, and C. Bridges. 1915. *The Mechanisms of Mendelian Heredity*. New York: H. Holt.

Morris, H. M., ed. 1974. *Scientific Creationism*. General ed. San Diego: Creation-Life Publishers.

Morris, S. 1978. Darwin and the double standard. *Playboy*, August: 159–60, 208–12.

Morton, A. G. 1981. *History of Botanical Science: An Account of the Development of Botany from Ancient Times to the Present Day*. London: Academic Press.

Morwood, M., and P. Van Oosterzee. 2007. *A New Human: The Startling Discovery and Strange Story of the "Hobbits" of Flores, Indonesia*. New York: Smithsonian Books.

Moss, L. 2004. *What Genes Can't Do*. Cambridge, Mass.: Harvard University Press.

Mullen, L.M., S.N. Vignieri, J.A. Gore and H.E. Hoekstra. 2009. Adaptive basis of geographic variation: Genetic, phenotypic and environmental differentiation among beach mouse populations. *Proceedings of the Royal Society B* 276: 3809–18.

Müller, F. M. 1861. *Lectures on the Science of Language*. London: Longman; 2nd ed., New York: Scribner, Armstrong, 1873.

——— 1873. Lectures on Mr. Darwin's philosophy of language. *Fraser's Magazine* 7: 525–41, 659–78; 8: 1–24.

Müller, G. 2007. Evo-devo: Extending the evolutionary synthesis. *Nature Review of Genetics* 8: 943–49.

Müller, G., and S. Newman, eds. 2003. *Origination of Organismal Form: The Forgotten Cause in Evolutionary Theory*. Cambridge, Mass.: MIT Press.

Müller, H. 1873. *Die Befruchtung der Blumen durch Insekten*. Leipzig: W. Engelmann.

——— 1879. Die wechselbeziehungen zwischen den Blumen und den ihre kreuzung vermittelnden Insekten. In *Handbuch der Botanik*, ed. A. Schenk, 1:1–112. Breslau: E. Trewendt.

——— 1883. *The Fertilisation of Flowers*. Trans. D. W. Thompson. London: Macmillan.

Müller, J. F. T. 1864. *Für Darwin*. Leipzig: Wilhelm Engelmann.

1869. *Facts and Arguments for Darwin*. Trans. W. S. Dallas. London: John Murray.

1879. Ituna und Thyridia. Ein merkwürdiges beispiel von mimicry bei schmetterlingen. *Kosmos* 5: 100–8.

1879. Ituna and Thyridia: A remarkable case of mimicry in butterflies. *Proceedings of the Entomological Society of London*: 20–29.

Müller-Wille, S. 2007. Collection and collation: Theory and practice of Linnaean botany. *Studies in History and Philosophy of Biological and Biomedical Sciences* 38: 541–62.

2009. The dark side of evolution: Caprice, deceit, redundancy. *History and Philosophy of the Life Sciences* 31: 183–200.

Murray J. 1972. *Genetic Diversity and Natural Selection*. Edinburgh: Oliver & Boyd.

Mussgrave, I. 2004. Evolution of the bacterial flagellum. In *Why Intelligent Design Fails*, ed. M. Young and T. Edis, 72–84. New Brunswick, N.J.: Rutgers University Press.

Nadeau, N., and C. Jiggins. 2010. A golden age for evolutionary genetics? Genomic studies of adaptation in natural populations. *Trends in Genetics* 26, no. 11: 484–92.

Nägeli, C. von. 1866. Notes made in preparing his letter to Gregor Mendel, February 25, 1867. In Hugo Iltis, *Life of Mendel*, trans. Eden and Cedar Paul. London: George Allen & Unwin, 1932.

National Center for Science Education. 2010a. Controversy over Evolution in Israel. http://ncse.com/news/2010/02/controversy-over-evolution-israel-005334.

2010b. Gabriel Avital Sacked in Israel. http://ncse.com/news/2010/10/gavriel-avital-sacked-israel-006231.

Neel, J. V. 1958. The study of natural selection in primitive and civilized populations. *Human Biology* 30, no. 1: 43–72.

1962. Diabetes mellitus: A "thrifty" genotype rendered detrimental by "progress"? *American Journal of Human Genetics* 14, no. 4: 353–62.

Nesse, R. M. 2011. Ten questions for evolutionary studies of disease vulnerability. *Evolutionary Applications* 4: 264–77.

Nesse, R. M., and J. D. Schiffman. 2003. Evolutionary biology in the medical school curriculum. *BioScience* 53, no. 6: 585–87.

Nesse, R. M., and S. C. Stearns. 2008. The great opportunity: Evolutionary applications to medicine and public health. *Evolutionary Applications* 1: 28–48.

Nesse, R. M., and G. C. Williams. 1995. *Why We Get Sick: The New Science of Darwinian Medicine*. New York: Time Books.

Nettle, D. 2009. *Evolution and Genetics for Psychology*. Oxford: Oxford University Press.

Newman, W. A. 1987. Evolution of Cirripedes and their major groups. In *Barnacle Biology*, ed. A. J Southward, 3–42. Rotterdam: A. A. Balkema.

1993. Darwin and Cirripedology. *History of Carcinology. Crustacean Issues* 8: 349–434.

Newton, W. C. F., and C. Pellew. 1929. Primula kewensis and its derivatives. *Journal of Genetics* 20: 405–67.

Nicholas, F., and J. Nicholas. 2002. *Charles Darwin in Australia*. Cambridge: Cambridge University Press.

Nietzsche, F. [1882] 1974. *Die fröhliche Wissenschaft (The Gay Science)*. Trans. W. Kaufman. New York: Vintage Books.

Noble, W., and I. Davidson. 1996. *Human Evolution, Language and Mind*. Cambridge: Cambridge University Press.

Noll, F. 1898. Julius von Sachs. A biographical sketch with portrait. *Botanical Gazette* 25, no. 1: 1–12.

Nordenskiöld, E. 1936. *The History of Biology*. 3rd ed. Ed. and trans. L. Eyre. New York: Tudor.

Novoa, A., and A. Levine. 2010. *From Man to Ape: Darwinism in Argentina, 1870–1920*. Chicago: University of Chicago Press.

Numbers, R. L., ed. 1995. *Selected Works of George McCready Price*. In *Creationism in Twentieth-Century America: A Ten-Volume Anthology of Documents, 1903–1961*. New York: Garland Publishing.

1998. *Darwinism comes to America*. Cambridge, Mass.: Harvard University Press.

2003. Science without God: Natural laws and Christian beliefs. In *When Science and Christianity Meet*, ed. D. C. Lindberg and R. L. Numbers, 265–85. Chicago: University of Chicago Press.

2004. Ironic heresy: How young-earth creationists came to embrace rapid microevolution by means of natural selection. In *Darwinian Heresies*, ed. A. J. Lustig, R. J. Richards, and M. Ruse, 84–100. Cambridge: Cambridge University Press.

2006. *The Creationists: From Scientific Creationism to Intelligent Design*. Expanded ed. Cambridge, Mass.: Harvard University Press.

2008. *Prophetess of Health: A Study of Ellen G. White*. 3rd ed. Grand Rapids, Mich.: Eerdmans.

Nyhart, L. K. 1995. *Biology Takes Form: Animal Morphology and the German Universities, 1800–1900*. Chicago: University of Chicago Press.

O'Connor, J. A., F. J. Sulloway, J. Robertson, and S. Kleindorfer. 2010. Philornis downsi parasitisim is the primary cause of nestling mortality in the critically endangered Darwin's medium tree finch (Camarhynchus pauper). *Biodiversity Conservation* 19: 853–66.

Odling-Smee, J., K. N. Laland, and M. W. Feldman. 2003. *Niche Construction*. Princeton: Princeton University Press.

O'Donald, P. 1974. Polymorphisms maintained by sexual selection in monogamous species of birds. *Heredity* 32: 110.

O'Donald, P., N. S. Wedd, and J. W. F. Davis. 1974. Mating preferences and sexual selection in the Arctic Skua. *Heredity* 33: 116.

Oevermann, A. 1993. "Republikanischen Brüder" im Sudan: Eine islamische Reformbewegung im zwanzigsten Jahrhundert. Frankfurt/Main: Lang.

O'Hara, R. J. 1991. Representations of the natural system in the nineteenth century. *Biology and Philosophy* 6: 255–74.

Okasha, S. 2006. *Evolution and the Levels of Selection*. Oxford: Oxford University Press.

Olby, R. C. 1963. Darwin's manuscript of pangenesis. *British Journal for the History of Science* 1: 250–63.

1966. *Origins of Mendelism*. New York: Schocken Books.

1985. *Origins of Mendelism*. 2nd ed. Chicago: University of Chicago Press.

O'Leary, D. 2006. *Roman Catholicism and Modern Science: A History*. New York: Continuum.

2009. From the origin to Humani Generis: Ireland as a case study. In *Darwin and Catholicism: The Past and Present Dynamics of a Cultural Encounter*, ed. L. Caruana, 13–26. New York: Continuum.

Olendzenski, L., and J. P. Gogarten. 2009. Evolution of genes and organisms. *Natural Genetic Engineering and Natural Genome Editing, Annals of the New York Academy of Sciences* 1178: 137–45.

Olenov, J. M., I. S. Kharmac, K. T. Galkovskaja, N. I. Kniazeva, A. D. Lebedeva, and Z. T. Popova. 1937. Natural selection in wild Drosophila melanogaster populations. *Comptes Rendus (Doklady) de l'Académie des Sciences de l'URSS* 25: 97–99.

Ong, K. K., M. Preece, P. M. Emmett, M. L. Ahmed, and D. B. Dunger. 2002. Size at birth and early childhood growth in relation to maternal smoking, parity and infant breast-feeding: Longitudinal birth cohort study and analysis. *Pediatric Research* 52, no. 6: 863–67.

Oparin, A. I. [1924] 1967. The origin of life. In J. D. Bernal, *The Origin of Life*, appendix 1, 199–234. London: Weidenfeld and Nicolson.

[1936] 1953. *The Origin of Life*. Trans. S. Morgulis. New York: Dover Publications.

Oppenheimer, J. 1967. *Essays in the History of Embryology and Biology*. Cambridge, Mass.: MIT Press.

Orel, V. 1996. *Gregor Mendel, the First Geneticist*. Oxford: Oxford University Press.

Orgel, L. E. 1994. The origin of life on the earth. *Scientific American* 271 (October): 53–61.

Ornduff, R. 1984. Darwin's botany. *Taxon* 33: 39–47.

Orr, H. A. 2009. Testing natural selection. *Scientific American* 300 (January): 44–51.

Orr, H. A., and J. A. Coyne. 1992. The genetics of adaptation: a reassessment. *American Naturalist* 140: 725–42.

Orr, J. 1905. *God's Image in Man*. London: Hodder and Stoughton.

Orzack, S. H. 1993. Sex ratio evolution in parasitic wasps. In *Evolution and Diversity of Sex Ratio in Insects and Mites*, ed. Dana Wrensch and Mercedes Ebbert, 477–511. New York: Chapman & Hall.

Orzack, S. H., and P. Forber. 2010. Adaptationism. http://plato.stanford.edu/entries/adaptationism.

Orzack, S. H., E. D. Parker Jr., and J. Gladstone. 1991. The comparative biology of genetic variation for conditional sex ratio adjustment in a parasitic wasp. *Nasonia vitripennis*. *Genetics* 127: 583–99.

Orzack, S. H., and E. Sober. 1994a. Optimality models and the test of adaptationism. *American Naturalist* 143: 361–80.

———. 1994b. How (not) to test an optimality model. *Trends in Ecology and Evolution* 9: 265–67.

———. 2001. *Adaptationism and Optimality*. Cambridge: Cambridge University Press.

Ospovat, D. 1976. The influence of Karl Ernst von Baer's embryology, 1828–1859: A reappraisal in light of Richard Owen's and William B. Carpenter's "paleontological application of 'Von Baer's law.'" *Journal of the History of Biology* 9: 1–28.

———. 1981. *The Development of Darwin's Theory: Natural History, Natural Theology, and Natural Selection, 1838–1859*. Cambridge: Cambridge University Press.

O'Toole, G. B. 1925. *The Case against Evolution*. New York: Macmillan.

Ouattara, K., A. Lemasson, and K. Zuberbühler. 2009. Campbell's monkeys concatenate vocalizations into context-specific call sequences. *Proceedings of the National Academy of Sciences* 106: 22026–31.

Overfield, R. A. 1975. Charles E. Bessey: The impact of the "new" botany on American agriculture, 1880–1910. *Technology and Culture* 16, no. 2: 162–81.

Owen R. 1838. [Notes on the dugong]. *Proceedings of the Zoological Society of London* 6: 28–45.

———. 1840. *Zoology of the Voyage of the H.M.S. Beagle, under the Command of Captain Robert FitzRoy, R.N., during the Years 1832 to 1836. Edited and Superintended by Charles Darwin. Part I: Fossil Mammalia*. London: Smith, Elder.

———. 1849. *On the Nature of Limbs*. London: Voorst.

———. 1859. *On the Classification and Geographical Distribution of the Mammalia, Being the Lecture on Sir Robert Reade's Foundation, Delivered before the University of Cambridge, in the Senate-House, May 10, 1859. To Which Is Added an Appendix "On the Gorilla," and "On the Extinction and Transmutation of Species."* London: John Parker and Son.

———. 1860. Darwin on the origin of species. *Edinburgh Review* 111: 487–532.

———. 1868. *The Anatomy of Vertebrates*. London: Longmans, Green.

Oyama, S., P. Griffiths, and R. Gray, eds. 2001. *Cycles of Contingency: Developmental Systems and Evolution*. Cambridge, Mass.: MIT Press.

Öztürkler, N. 2005. Türkiye'de evrim eğitimin sosyolojik bir değerlendirmesi. PhD dissertation, Ankara Üniversitesi.

Pace, N., G. J. Olson, and C. R. Woese. 1986. Ribosomal rna phylogeny and the primary lines of evolutionary descent. *Cell* 45: 325–26.

Padian, K. 1999. Charles Darwin's view of classification in theory and practice. *Systematic Biology* 48, no. 2: 352–64.

Pagel, M. 2009. Human language as a culturally transmitted replicator. *Nature Reviews Genetics* 10: 405–15.

Pagel, M., and R. Mace. 2004. The cultural wealth of nations. *Nature* 428: 275–78.

Paine, T. 1966. Food web complexity and species diversity. *American Naturalist* 100: 65–75.

Paley, W. [1785] 2002. *The Principles of Moral and Political Philosophy*. Indianapolis, Ind.: Liberty Fund.

———. 1802. *Natural Theology, or Evidences of the Existence and Attributes of the Deity Collected from the Appearances of Nature*. London: R. Faulder.

———. 1809. *Natural Theology, or Evidences of the Existence and Attributes of the Deity, Collected from the Appearances of Nature*. 12th ed. London: J. Faulder.

Palladino, P. 1993. Between craft and science: Plant breeding, Mendelian genetics, and British universities, 1900–1920. *Technology and Culture* 34, no. 2: 300–23.

Pannenberg, W. 1994. *Systematic Theology*. Vol. 2. Edinburgh: T&T Clark.

Parish, W. 1839. *Buenos Ayres and the Provinces of the Rio de la Plata*. London: John Murray.

Park, H. W. 2006. Germs, hosts and the origin of Frank Macfarlane Burnet's concept of "self" and "tolerance," 1936–1949. *Journal of the History of Medicine and Allied Sciences* 61, no. 4: 492–534.

Parker, G. A., and J. Maynard Smith. 1990. Optimality theory in evolutionary biology. *Nature* 348: 27–33.

Passmore, A. 2007. Blut unserer Vater: Evolutionary Theory and Austrian-Jewish Literature of 1900. PhD dissertation, University of Chicago.

Paul VI. 1965. Pastoral Constitution on the Church in the Modern World: *Gaudium et Spes*. Vatican City.

Paul, D. B. 1995. *Controlling Human Heredity: 1865 to the Present*. Atlantic Highlands, N.J.: Humanities Press.

———. 2003. Darwin, social Darwinism and eugenics. In *The Cambridge Companion to Darwin*, ed. J. Hodge and G. Radick, 214–39. Cambridge: Cambridge University Press.

———. 2009. Darwin, social Darwinism and eugenics. In *The Cambridge Companion to Darwin*, 2nd ed., ed. J. Hodge and G. Radick, 219–45. Cambridge: Cambridge University Press.

Pauly, P. J. 1987. *Controlling Life: Jacques Loeb and the Engineering Ideal in Biology*. New York: Oxford University Press.

Pear, R. 2012. And It Was Good? American Modern Orthodox Engagement with Darwinism, 1925–Present. PhD dissertation, Bar-Ilan University.

Pearce, T. 2010. A great complication of circumstances: Darwin and the economy of nature. *Journal of the History of Biology* 43: 493–528.

Pearn, A. M. 2010. "This excellent observer …": The correspondence between Charles Darwin and James Crichton-Browne. *History of Psychiatry* 21, no. 2: 160–75.

Pearson, K. 1892. *The Grammar of Science*. London: Adam and Charles Black.

1894. Contributions to the mathematical theory of evolution. *Philosophical Transactions of the Royal Society of London A* 185: 71–110.

1903. Mathematical contributions to the theory of evolution. XI. On the influence of natural selection on the variability and correlation of organs. *Philosophical Transactions of the Royal Society of London A* 200: 1–66.

Pearson, M. B. 2005. *A. R. Wallace's Malay Archipelago Journals and Notebook*. Privately printed, Linnean Society of London.

Pearson, P. N. 1996. Charles Darwin on the origin and diversity of igneous rocks. *Earth Sciences History* 15, no. 1: 49–67.

Pearson, P. N., and C. J. Nicholas. 2007. Marks of extreme violence: Charles Darwin's geological observations on St Jago (Sao Tiago), Cape Verde islands. *Geological Society of London Special Publications* 287: 239–53.

Peccoud. J., A. Ollivier, M. Plantagenest, and J.-C. Simon. 2009. Adaptive radiation in the pea aphid complex through gradual cessation of gene flow. *Proceedings of the National Academy of Sciences* 106: 7495–500.

Peck, D. 2010. The evolution of Eliza Burt Gamble: Her life, works and influence. http://womeninscience.history.msu.edu/Biography/C-4A-2/eliza-burt-gamble/.

Peckham, M. 1959. *The Origin of Species: A Variorum Text*. Philadelphia: University of Pennsylvania Press.

Pei, W. 1930. Zhongguo yuanren huashi zhi faxian (The Fossil Discovery of Chinese Ape-man). *Kexue* (Science) 14, no. 8: 1127–33.

Peker, D., G. G. Comert, and A. Kence. 2009. Three decades of anti-evolution campaign and its results: Turkish undergraduates' acceptance and understanding of the biological evolution theory. *Science and Education* 19: 739–55.

Penn, D. J. 2003. The evolutionary roots of our environmental problems: Towards a Darwinian ecology. *Quarterly Review of Biology* 73: 275–301.

Pennock, R. T. 1999. *The Tower of Babel: The Evidence against the New Creationism*. Cambridge, Mass.: MIT Press.

Perry, G., et al. 2007. Diet and the evolution of human amylase gene copy number variation. *Nature Genetics* 39: 1256–60.

Peter, I. S., and E. H. Davidson. 2009. Modularity and design principles in the sea urchin embryo gene regulatory network. *FEBS Letters* 583, no. 24: 3948–58.

Petersen W. 1903. Enstehung der Arten durch physiologische Isolierung. *Biologisches Zentralblatt* 23: 468–77.

Pew Research Center. 2008. *U.S. Religious Landscape Survey: Religious Beliefs and Practices; Diverse and Politically Relevant*. Washington, D.C.

Pianka, E. 1970. On r- and K-selection. *American Naturalist* 104: 592–97.

Pierce, C. S. 1992. *The Essential Peirce: Selected Philosophical Writings*. Ed. N. Houser and C. J. W. Kloesel. Vol. 1. Bloomington: Indiana University Press.

Pigliucci, M. 2001. *Phenotypic Plasticity*. Baltimore: Johns Hopkins University Press.

2007. Do we need an extended evolutionary synthesis? *Evolution* 61: 2743–49.

Pigliucci, M., and Kaplan. 2006. *Making Sense of Evolution*. Chicago: University of Chicago Press.

Pigliucci, M., and G. B. Müller. 2010. *Evolution: The Extended Synthesis*. Cambridge, Mass.: MIT Press.

Pimm, S. L. 1998. The forest fragment classic. *Nature* 393: 23–24.

Pinker, S. 1994. *The Language Instinct*. New York: William Morrow.

1997. *How the Mind Works*. New York: Norton.

2010. The cognitive niche: Coevolution of intelligence, sociality, and language. *Proceedings of the National Academy of Sciences* 107: 8993–99.

Pinker, S., and P. Bloom. 1990. Natural language and natural selection. *Behavioral and Brain Sciences* 13: 707–84.

Pinker, S., and R. Jackendoff. 2005. The faculty of language: What's special about it? *Cognition* 95: 201–36.

Pinzón, J. H., and T. C. LaJeunesse. 2010. Species delimitation of common reef corals in the genus Pocillopora using nucleotide sequence phylogenies, population genetics and symbiosis ecology. *Molecular Ecology* 20, no. 2: 311–25.

Plantinga, A. 1983. Reason and belief in God. In *Faith and Rationality*, ed. A. Plantinga and N. Wolterstorff, 18–93. Notre Dame: Notre Dame Press.

1993. *Warrant and Proper Function*. New York: Oxford University Press.

Pollard, K. S., S. R. Salama, N. Lambert, et al. 2006. An RNA gene expressed during cortical development evolved rapidly in humans. *Nature* 443: 167–72.

Popper, K. 1963. Science: Conjectures and refutations. In *Conjectures and Refutations*, 43–87. London: Routledge.

Porter, R. 1977. *The Making of Geology: Earth Science in Britain, 1660–1815*. Cambridge: Cambridge University Press.

1978. Gentlemen and geology: The emergence of a scientific career, 1660–1820. *Historical Journal* 20: 809–36.

1989. Erasmus Darwin: Doctor of evolution? In *History, Humanity and Evolution: Essays for John C. Greene*, ed. J. R. Moore, 39–69. Cambridge: Cambridge University Press.

1999. *The Greatest Benefit to Mankind: A Medical History of Humanity from Antiquity to the Present*. London: Fontana Press.

Poulton, E. B. 1890. *The Colours of Animals*. London: Kegan Paul, Trench, Truebner.

1904. What is a species? *Proceedings of the Entomological Society of London* 1903: lxxvii–cxvi.

1908. *Essays on Evolution, 1889–1907*. Oxford: Clarendon Press.

1913. Mimicry and the inheritance of small variations. *Bedrock* 2: 295–312.

1914. The evolution of mimetic resemblance. *Bedrock* 3: 34–45.

Prairie Starfish Video Productions. 2008. Barnacle Sex Music Video: "Barnacles tell no lies." From the film *Looking for Mr. Good Barnacle*. http://www.youtube.com/watch?v=Nd706ytz_LM&feature=player_embedded.

Prentice, A., T. J. Cole, and R. G. Whitehead. 1987. Impaired growth in infants born to mothers of very high parity. *Human Nutrition – Clinical Nutrition* 41, no. 5: 319–25.

Price, G. M. 1906. *Illogical Geology: The Weakest Point in the Evolution Theory*. Los Angeles: Modern Heretic.

1923. *The New Geology*. Mountain View, Calif.: Pacific Press.

Primer, I. 1964. Erasmus Darwin's temple of nature: Progress, evolution, and the Eleusinian mysteries. *Journal of the History of Ideas* 25: 58–76.

Pritchard, J. K., and Y. Gilad. 2012. Encode explained: Evolution and the code. *Nature* 489: 55.

Pritchard, J. K., M. Stephens, and P. Donnelly. 2000. Inference of population structure using multilocus genotype data. *Genetics* 155: 945–59.

Prodger, P. 2009. *Darwin's Camera: Art and Photography in the Theory of Evolution*. Oxford: Oxford University Press.

Provine, W. B. 1971. *The Origins of Theoretical Population Genetics*. Chicago: University of Chicago Press.

1978. The Role of mathematical population genetics in the evolutionary synthesis of the 1930s and 1940s. *Studies in the History of Biology* 2: 167–92.

1986. *Sewall Wright and Evolutionary Biology*. Chicago: University of Chicago Press.

Pruna, P., and A. G. González. 1989. *Darwinismo y sociedad en Cuba: Siglo XIX*. Madrid: CSIC.

Puig-Samper, M. A., R. Ruiz, and A. Galera, eds. 2002. *Evolucionismo y cultura: Darwinismo en Europa e Iberoamérica*. Madrid: Doce Calles.

Punnett, R. C. 1913. Mendelism, mutation and mimicry. *Bedrock* 2: 146–64.

———. 1915. *Mimicry in Butterflies*. Cambridge: Cambridge University Press.

Pusey, J. R. 1983. *China and Charles Darwin*. Cambridge, Mass.: Council on East Asian Studies, Harvard University.

Qian, T. 1919. Tianyan xinshuo (New Evolutionary Theory). *Kexue* (Science) 4, no. 12: 1209–14.

Quammen, D. 1998. Point of attachment. In *Wild Thoughts from Wild Places*, 226–35. New York: Scribner.

Quatrefages, A. 1870. *Charles Darwin et ses précurseurs français. Étude sur le transformisme*. Paris: Germer Baillière.

Raby, P. 2001. *Alfred Russel Wallace: A Life*. Princeton: Princeton University Press.

Rachootin, S. P. 1985. Owen and Darwin reading a fossil: *Macrauchenia* in a boney light. In *The Darwinian Heritage*, ed. D. Kohn, 155–83. Princeton: Princeton University Press.

Radick, G. 2000. Language, brain function, and human origins in the Victorian debates on evolution. *Studies in History and Philosophy of Biological and Biomedical Sciences* 31C: 55–75.

———. 2002. Darwin on language and selection. *Selection* 3: 7–16.

———. 2007. *The Simian Tongue: The Long Debate about Animal Language*. Chicago: University of Chicago Press.

———. 2008. Race and language in the Darwinian tradition (and what Darwin's language-species parallels have to do with it). *Studies in History and Philosophy of Biological and Biomedical Sciences* 39C: 359–70.

———. 2010a. Darwin's puzzling expression. *Comptes Rendus Biologies* 333: 181–87.

———. 2010b. Did Darwin change his mind about the Fuegians? *Endeavour* 34: 50–54.

Radick, G., and M. Steadman. Forthcoming. Of lice and men: Charles Darwin's debt to the Leeds Museum. Joseph Priestley Lecture, Leeds, 26 November 2009.

Ragan, M. 2009. Trees and networks before and after Darwin. *Biology Direct* 4: 43.

Rahner, K. 1969. *Theological Investigations*. Trans. K. Kruger and B. Kruger. Vol. 6. Baltimore: Helicon.

Rainger, R. 2001. Subtle agents for change: The *Journal of Paleontology*, J. Marvin Weller, and shifting emphases in invertebrate paleontology, 1930–1965. *Journal of Paleontology* 75, no. 6: 1058–64.

Rainger, R., K. Benson, and J. Maienschein, eds. 1988. *The American Development of Biology*. Philadelphia: University of Pennsylvania Press.

Ramm, B. 1954. *The Christian View of Science and Scripture*. Grand Rapids, Mich.: Eerdmans.

———. 1983. *After Fundamentalism: The Future of Evangelical Theology*. San Francisco: Harper & Row.

Ramsay, A. C. 1846. On the denudation of South Wales and the adjacent counties of England. *Memoirs of the Geological Survey of Great Britain, and of the Museum of Economic Geology in London* 1: 297–335.

Ranjard, L., M. G. Anderson, M. J. Rayner, et al. 2010. Bioacoustic distances between the begging calls of brood parasites and their host species: a comparison of metrics and techniques. *Behavioral Ecology and Sociobiology* 64: 1915–26.

Rashīd Ridā, M. 1930. Nazariyyat Dārwīn wal-islām. *Al-Manār* 30: 593–600.

Raup, D. M., S. J. Gould, T. J. M. Schopf, and D. S. Simberloff. 1973. Stochastic models of phylogeny and the evolution of diversity. *Journal of Geology* 81, no. 5: 525–42.

Raven, P. H. 1974. Plant systematics, 1947–1972. *Annals of the Missouri Botanical Garden* 61: 166–78.

———. 1988. Our diminishing tropical forests. In *Biodiversity*, ed. E. O. Wilson, 119–22. Washington, D.C.: National Academy Press.

Reed, H. S. 1942. *A Short History of the Plant Sciences*. Waltham, Mass.: Chronica Botanica.

Reiss, J. 2011. *Not by Design: Retiring Darwin's Watchmaker*. Berkeley: University of California Press.

Reynolds, R. M., C. Osmond, D. I. W. Phillips, and K. M. Godfrey. 2010. Maternal BMI, parity, and pregnancy weight gain: Influences on offspring adiposity in young adulthood. *Journal of Clinical Endocrinology & Metabolism* 95, no. 12: 5365–69.

Reznick, D. N., M. J. Butler, F. H. Rodd, and P. Ross. 1996. Life history evolution in guppies (Poecilia reticulata). 6. Differential mortality as a mechanism for natural selection. *Evolution* 50: 1651–60.

Reznick, D. N., and C. K. Ghalambor. 2005. Selection in nature: Experimental manipulations of natural populations. *Integrative and Comparative Biology* 45: 456–62.

Rhodes, F. H. T. 1991. Darwin's search for a theory of the Earth: Symmetry, simplicity and speculation. *British Journal for the History of Science* 24: 193–229.

Richards, E. 1983. Darwin and the descent of woman. In *The Wider Domain of Evolutionary Thought*, ed. D. Oldroyd and I. Langham, 57–111. Boston: D. Reidel.

———. 1987. A question of property rights: Richard Owen's evolutionism reassessed. *British Journal for the History of Science* 20: 129–71.

———. Forthcoming. Sexing selection: Darwin and the making of sexual selection. Unpublished manuscript, University of Sydney.

Richards, R. A. 2009. Classification in Darwin's *Origin*. In *The Cambridge Companion to the "Origin of Species,"* ed. M. Ruse and R. J. Richards, 173–93. Cambridge: Cambridge University Press.

Richards, R. J. 1987. *Darwin and the Emergence of Evolutionary Theories of Mind and Behavior*. Chicago: University of Chicago Press.

———. 1992. *The Meaning of Evolution: The Morphological Construction and Ideological Reconstruction of Darwin's Theory*. Chicago: University of Chicago Press.

———. 2002a. Race. In *Oxford Companion to the History of Modern Science*, ed. J. Heilbron, 697–98. Oxford: Oxford University Press.

———. 2002b. *The Romantic Conception of Life: Science and Philosophy in the Age of Goethe*. Chicago: University of Chicago Press.

———. 2002c. The linguistic creation of man: Charles Darwin, August Schleicher, Ernst Haeckel, and the missing link in nineteenth-century evolutionary theory. In *Experimenting in Tongues*, ed. M. Dörries, 21–48. Stanford, Calif.: Stanford University Press.

———. 2008. *The Tragic Sense of Life: Ernst Haeckel and the Struggle over Evolutionary Thought*. Chicago: University of Chicago Press.

———. 2009a. Darwin on mind, morals and emotions. In *The Cambridge Companion to Darwin*, 2nd ed., ed. J. Hodge and G. Radick, 96–119. Cambridge: Cambridge University Press.

———. 2009b. The Descent of Man: Review of *Darwin's Sacred Cause: How a Hatred of Slavery Shaped Darwin's Views on Human Evolution*, by Adrian Desmond and James Moore. *American Scientist* 97: 415–17.

———. 2013. *Was Hitler a Darwinian? Disputed Questions in the History of Evolutionary Theory*. Chicago: University of Chicago Press.

Richardson, D. M., and P. Pyšek. 2008. Fifty years of invasion ecology: The legacy of Charles Elton. *Diversity and Distribution* 14: 161–68.

Richardson, J. 2004. *Nietzsche's New Darwinism*. Oxford: Oxford University Press.

Richardson, R. C. 2007. *Evolutionary Psychology as Maladapted Psychology*. Cambridge, Mass.: MIT Press.

Richerson, P., and Boyd R. 2005. *Not by Genes Alone: How Culture Transformed Human Evolution*. Chicago: University of Chicago Press.

Richmond, M. L. 1989. Darwin's study of the Cirripedia. In *Correspondence of Charles Darwin*, ed. S. Smith, F. H. Burkhardt, et al., 4:388–409. Cambridge: Cambridge University Press.

Ridgway, R. 1897. Birds of the Galapagos archipelago. *Proceedings of the United States National Museum* 19, no. 1116: 459–670.

Riedl, R. 1975. *Die Ordnung des Lebendigen: Systembedingungen d. Evolution*. Hamburg: Berlin, Parey.

———. 1983. Evolution and evolutionary knowledge: On the correspondence between cognitive order and nature. In *Concepts and Approaches in Evolutionary Epistemology*, ed. F. Wuketits, 35–50. Dordrecht: D. Reidel.

Riesebrodt, M. 1993. *Pious Passion: The Emergence of Modern Fundamentalism in the United States and Iran*. Berkeley: University of California Press.

Riexinger, M. 2004. *Sanā'ullāh Amritsarī (1868–1948) und die Ahl-i Hadīs im Punjab unter britischer Herrschaft*. Würzburg: Ergon-Verlag.

———. 2008. Propagating Islamic creationism on the internet. *Masaryk University Journal of Law and Technology* 2, no. 2: 99–112.

———. 2009. Responses of South Asian Muslims to the theory of evolution. *Welt Des Islams* 49: 212–47.

———. 2010. Muslim responses to the theory of evolution. In *Handbook of Religion and the Authority of Science*, ed. O. Hammer and J. P. Lewis, 483–509. Leiden: Brill.

———. 2013. Turkey. In *Creationism in Europe*, ed. S. Blancke, H. H. Hjermitslev, and Peter Kjærgaard. Baltimore: John Hopkins University Press.

———. Forthcoming. The emergence of Islamic creationism: Why in Turkey? In *Science, Technology and Entrepreneurship in the Islamic World*, ed. L. Stenberg. Salt Lake City: University of Utah Press.

Rivera, M. C., and J. A. Lake. 2004. The ring of life provides evidence for a genome fusion origin of eukaryotes. *Nature* 431: 34–37.

Roberts, J. H. 1988. *Darwinism and the Divine in America*. Notre Dame: University of Notre Dame Press.

———. 1999. Darwinism, American Protestant thinkers and the puzzle of motivation. In *Disseminating Darwinism: The Role of Place, Race, Religion and Gender*, ed. R. L. Numbers and J. Stenhouse, 145–72. Cambridge: Cambridge University Press.

Roberts, N. 2011. The idea of evolution in geographical thought. In *The Sage Handbook of Geographical Knowledge*, ed. J. A. Agnew and D. N. Livingstone, 441–51. London: Sage.

Robinson, I. 2006. "Practically, I am a fundamentalist": Twentieth-century Orthodox Jews contend with evolution and its implication. In *Jewish Tradition and the Challenge of Evolution*, ed. G. Cantor and M. Swetlitz, 71–88. Chicago: University of Chicago Press.

———. 2007. American Jewish views of evolution and intelligent design. *Modern Judaism* 27: 173–92.

Robson, G. C., and O. W. Richards. 1936. *The Variation of Animals in Nature*. London: Longmans, Green.

Rodriguez-Trelles, F., R. Tarrio, and F. J. Ayala. 2001. Erratic overdispersion of three molecular clocks: GPDH, SOD, and XDH. *Proceedings of the National Academy of Sciences* 98: 11405–10.

———. 2002. A methodological bias toward overestimation of molecular evolutionary time scales. *Proceedings of the National Academy of Sciences* 99: 8112–15.

———. 2006. Rates of molecular evolution. In *Evolutionary Genetics*, ed. C. W. Fox and J. B. Wolf, 119–32. Oxford: Oxford University Press.

Roger, J. 1944. Phylogénie des Céphalopodes Octopodes: Palaeoctopus newboldi (Sowerby, 1846) Woodward. *Bulletin de la Société Géologique de France* 5, no. 5: 83–98.

Rogers, D., and P. Ehrlich. 2008. Natural selection and cultural rates of change. *Proceedings of the National Academy of Sciences* 105: 3416–20.

Rogers, J. A. 1960. Charles Darwin and Russian scientists. *Russian Review* 19, no. 4: 371–83.

———. 1973. The reception of Darwin's *Origin of Species* by Russian scientists. *Isis* 64, no. 4: 484–503.

Romanes, George J. 1883. *Mental Evolution in Animals*. London: Kegan Paul, Trench.

———. 1886. Physiological selection: An additional suggestion on the origin of species. *Journal of the Linnean Society of London (Zoology)* 19: 337–411.

———. 1888. *Mental Evolution in Man: Origin of Human Faculty*. London: Kegan Paul, Trench.

———. 1890. Darwin's latest critics. *Nineteenth Century* 27, May, 831.

Rosenberg, K., and W. Trevathan. 1995. Bipedalism and human birth: The obstetrical dilemma revisited. *Evolutionary Anthropology* 4, no. 5: 161–68.

Rosenzweig, M. L. 1987. Editorial. *Evolutionary Ecology* 1: 1–3.

———. 1992. Species diversity gradients: We know more and less than we thought. *Journal of Mammalogy* 73: 715–30.

Rothenberg, J. 2005. The heresy of Nosson Slifkin. *Moment* 30, no. 5: 37–40, 41–43, 45, 58, 70, 72.

Rowland H. M., J. Mappes, G. D. Ruxton, and M. P. Speed. 2010. Mimicry between unequally defended prey can be parasitic: Evidence for quasi-Batesian mimicry. *Ecology Letters* 13: 1494–1502.

Rudge, D. W. 2005. Did Kettlewell commit fraud? Re-examining the evidence. *Public Understanding of Science* 14, no. 3: 249–68.

Rudwick, M. J. S. 1974. Darwin and Glen Roy: A "great failure" in scientific method? *Studies in History and Philosophy of Science* 5: 97–185.

———. 2008. *Worlds before Adam: The Reconstruction of Geohistory in the Age of Reform*. Chicago: University of Chicago Press.

Rupke, N. A. 1993. Richard Owen's Vertebrate Archetype. *Isis* 8: 231–51.

———. 1994. *Richard Owen: Victorian Naturalist*. New Haven: Yale University Press.

———. 2005. Neither creation nor evolution: The third way in mid-nineteenth century thinking about the origin of species. *Annals of the History and Philosophy of Biology* 10: 143–72.

———. 2009. *Richard Owen: Biology without Darwin*. Chicago: University of Chicago Press.

Ruricola. 1841. Humble-bees. *Gardeners' Chronicle* 30: 485.

Ruse, M. 1973. *The Philosophy of Biology*. London: Hutchinson.

———. 1975a. Darwin's debt to philosophy: An examination of the influence of the philosophical ideas of John F. W. Herschel and William Whewell on the development of Charles Darwin's theory of evolution. *Studies in History and Philosophy of Science* 6: 159–81.

———. 1975b. Charles Darwin and artificial selection. *Journal of the History of Ideas* 36: 339–50.

1979. *Sociobiology: Sense or Nonsense?* Dordrecht: Reidel.

1980. Charles Darwin and group selection. *Annals of Science* 37, no. 6: 615.

1986. *Taking Darwin Seriously: A Naturalistic Approach to Philosophy.* Oxford: Basil Blackwell.

1987. Biological species: Natural kinds, individuals, or what? *British Journal for the Philosophy of Science* 38: 225-42.

1988. Prologue: A philosopher's day in court. In *But Is It Science? The Philosophical Question in the Creation/Evolution Controversy*, ed. M. Ruse, 13-34. Buffalo, N.Y.: Prometheus Books.

1996. *Monad to Man: The Concept of Progress in Evolutionary Biology.* Cambridge, Mass.: Harvard University Press.

1998. *Taking Darwin Seriously: A Naturalistic Approach to Philosophy.* 2nd ed. Amherst, N.Y.: Prometheus Books.

1999a. *The Darwinian Revolution: Science Red in Tooth and Claw.* 2nd ed. Chicago: University of Chicago Press.

1999b. *Mystery of Mysteries: Is Evolution a Social Construction?* Cambridge, Mass.: Harvard University Press.

2000. *Evolution Wars: A Guide to the Debate.* Santa Barbara, Calif.: ABC Clio.

2001. *Can a Darwinian Be a Christian? The Relationship between Science and Religion.* Cambridge: Cambridge University Press.

2003. *Darwin and Design: Does Evolution Have a Purpose?* Cambridge, Mass.: Harvard University Press.

2005. *The Evolution-Creation Struggle.* Cambridge, Mass.: Harvard University Press.

2006. *Darwinism and Its Discontents.* Cambridge: Cambridge University Press.

2008. *Charles Darwin.* Oxford: Blackwell.

2009a. Introduction to *Evolution and Ethics* by Thomas Henry Huxley. Princeton: Princeton University Press.

2009b. *Natural Selection and Heredity* (Philip M. Sheppard). In *Evolution: The First Four Billion Years*, ed. M. Ruse and J. Travis, 755-56. Cambridge, Mass.: Belknap Press of Harvard University Press.

ed. 2009c. *Philosophy after Darwin: Classic and Contemporary Readings.* Princeton: Princeton University Press.

2010. Evolution and progress. In *Biology and Ideology from Descartes to Dawkins*, ed. R. L. Numbers and D. Alexander, 247-75. Chicago: University of Chicago Press.

2012. *The Philosophy of Human Evolution.* Cambridge: Cambridge University Press.

Ruse, M., and J. Travis, eds. 2009. *Evolution: The First Four Billion Years.* Cambridge, Mass.: Belknap Press of Harvard University Press.

Russell, M. J., and W. Martin. 2004. The rocky roots of the acetyl-CoA pathway. *Trends in Biochemical Sciences* 29: 358-63.

Russell, N. 1986. *Like Engendering Like: Heredity and Animal Breeding in Early Modern England.* Cambridge: Cambridge University Press.

Ruxton, G. D., M. P. Speed, and D. J. Kelly. 2004. What, if anything, is the adaptive function of countershading? *Animal Behaviour* 68, no. 3: 445-51.

Rüttimeyer, L. 1868. Referate. *Archiv für Anthropologie* 3: 301-2.

Sabeti, P. C., P. Varilly, B. Fry, et al. 2007. Genome-wide detection and characterization of positive selection in human populations. *Nature* 449: 913-19.

Sachs, J. von. 1887. *Lectures on the Physiology of Plants.* Trans. H. M. Ward. Oxford: Clarendon Press.

1890. *History of Botany (1530–1860).* Trans. I. B. Balfour et al. Oxford: Clarendon Press.

Salmon, M. A. 2000. *The Aurelian Legacy: British Butterflies and Their Collectors.* Berkeley: University of California Press.

Sami, Ş. 1880. *İnsan.* İstanbul: Kütüphane Mihran.

Samuels, B. J. 2007. The Slifkin affair and its American Orthodox Jewish response. Unpublished manuscript, Boston.

Samuelson, N. 2009. *Jewish Faith and Modern Science.* Lanham, Md.: Rowman & Littlefield.

Sapp, J. 2009. *The New Foundations of Evolution: On the Tree of Life.* Oxford: Oxford University Press.

Savage-Rumbaugh, S., S. G. Shanker, and T. J. Taylor. 1998. *Apes, Language and the Human Mind.* New York: Oxford University Press.

Schiebinger, L. 1993. *Nature's Body: Gender in the Making of Modern Science.* Boston: Beacon Press.

Schleicher, A. 1863. *Die Darwinsche Theorie und die Sprachwissenschaft. Offenes Sendschreiben an Herrn Dr. Ernst Haeckel.* Weimer: Hermann Böhlan.

1869. *Darwinism Tested by the Science of Language.* Trans. A. V. W. Bikkers. London: John Camden Hotten.

Schluter, D. 2000. *The Ecology of Adaptive Radiation.* Oxford: Oxford University Press.

Schluter, D., and P. Grant. 1984. Ecological correlates of morphological evolution in Darwin's finch species. *Evolution* 38: 856-69.

Schmalzer, S. 2008. *The People's Peking Man: Popular Science and Human Identity in Twentieth-Century China.* Chicago: University of Chicago Press.

Schneider, L. 2003. *Biology and Revolution in Twentieth-Century China.* Lanham, Md.: Rowman & Littlefield.

Schöck, C. 1993. *Adam im Islam: Ein Beitrag zur Ideengeschichte der Sunna.* Berlin: Schwarz.

Schopf, J. W. 2009. Emergence of precambrian paleobiology: A new field of science. In *The Paleobiological Revolution: Essays on the Growth of Modern Paleontology*, ed. D. Sepkoski and M. Ruse, 89-110. Chicago: University of Chicago Press.

Schwartz, B. I. 1964. *In Search of Wealth and Power: Yen Fu and the West.* Cambridge, Mass.: Harvard University Press.

Sclater, P. L. 1857. On the geographical distribution of the members of the class Aves. *Journal of the Proceedings of the Linnean Society (Zoology)* 2: 130-48.

Scott, E. C. 1996. Creationism, ideology and science. *Annals of the New York Academy of Sciences* 775: 505-22.

Scrope, G. P. 1825. *Considerations of Volcanos: The Probable Causes of Their Phenomena, the Laws Which Determine Their March, the Deposition of Their Products, and Their Connexion with the Present State and Past History of the Globe: Leading to the Establishment of a New Theory of the Earth.* London: W. Phillips.

Sebright, J. S. 1809. *The Art of Improving the Breeds of Domestic Animals. . . .* London: J. Harding.

Secord, J. A. 1981. Nature's fancy: Charles Darwin and the breeding of pigeons. *Isis* 72: 163-86.

1985. Darwin and the breeders: A social history. In *The Darwinian Heritage*, ed. D. Kohn, 519-42. Princeton: Princeton University Press.

1991a. The discovery of a vocation: Darwin's early geology. *British Journal for the History of Science* 24: 133-57.

1991b. Edinburgh Lamarckians: Robert Jameson and Robert E. Grant. *Journal of the History of Biology* 24: 1-18.

2000. *Victorian Sensation: The Secret Authorship, Publication and Extraordinary Reception of Vestiges of the Natural History of Creation.* Chicago: University of Chicago Press.

Sedgwick, A. 1860. Objections to Mr. Darwin's theory of the origin of species. *Spectator* 7: 334-35.

Sedley, D. 2008. *Creationism and Its Critics in Antiquity*. Berkeley and Los Angeles: University of California Press.

Segerdahl, P., W. Fields, and E. S. Savage-Rumbaugh. 2005. *Kanzi's Primal Language: The Cultural Initiation of Primates into Language*. London: Palgrave Macmillan.

Segerstrale, U. 2000. *Defenders of the Truth: The Battle for Science in the Sociobiology Debate and Beyond*. New York: Oxford University Press.

Segré, D., D. Ben-Eli, D. W. Deamer, and D. Lancet. 2001. The lipid world. *Origins of Life and Evolution of the Biosphere* 31: 119–45.

Sepkoski, D. 2009. The "delayed synthesis": Paleobiology in the 1970s. In *Descended from Darwin: Insights into American Evolutionary Studies, 1925–1950*, ed. Joe Cain and Michael Ruse, 179–97. Philadelphia: American Philosophical Society.

———. 2012. *Re-reading the Fossil Record: The Growth of Paleobiology as an Evolutionary Discipline*. Chicago: University of Chicago Press.

Sepkoski, D., and M. Ruse, eds. 2009. *The Paleobiological Revolution: Essays on the Growth of Modern Paleontology*. Chicago: University of Chicago Press.

Seward, A. C. 1909. *Darwin and Modern Science; Essays in Commemoration of the Centenary of the Birth of Charles Darwin and of the Fiftieth Anniversary of the Publication of the "Origin of Species."* Cambridge: Cambridge University Press.

Seyfarth, R. M., D. L. Cheney, and P. Marler. 1980. Monkey responses to three different alarm calls: Evidence of predator classification and semantic communication. *Science* 210: 801–3.

Shafran, A. 2005. Lift up your eyes and see: The "intelligent design" controversy and why it matters. *Jewish Observer* 38, no. 10: 40–43.

Shaler, N. S. 1893. *The Interpretation of Nature*. New York: Houghton Mifflin.

Shanker, S., and B. J. King. 2002. The emergence of a new paradigm in ape language research. *Behavioral and Brain Sciences* 25: 605–26.

Shatz, D. 2008. Is there science in the Bible? An assessment of biblical concordism. *Tradition* 41, no. 2: 198–244.

Shennan, S. 2002. *Genes, Memes and Human History: Darwinian Archeology and Cultural Evolution*. London: Thames & Hudson.

Sheppard, P. M. 1958. *Natural Selection and Heredity*. London: Hutchison.

Shermer, M. 2002. *In Darwin's Shadow*. Oxford: Oxford University Press.

Sherratt, T. N. 2008. The evolution of Müllerian mimicry. *Naturwissenschaften* 95, no. 8: 681–95.

Shetler, S. G. 1967. *The Komarov Botanical Institute: 250 Years of Russian Research*. Washington, D.C.: Smithsonian Institution Press.

Shipman, P. 2006. Turkish creationist movement tours American college campuses. *Reports of the National Center for Science Education* 26, no. 5: 11–14.

Shubin, N. 2008. *Your Inner Fish: A Journey into the 3.5-Billion-Year History of the Human Body*. New York: Pantheon.

Shuchat, R. 2005. Attitudes towards cosmogony and evolution among rabbinic thinkers in the nineteenth and early twentieth centuries. The resurgence of the doctrine of the sabbatical years. *Torah U-Madda Journal* 13: 15–49.

Sidgwick, H. 1876. The theory of evolution in its application to practice. *Mind* 1: 52–67.

Siepielski, A. M., J. D. DiBattista, and S. M. Carlson. 2009. It's about time: The temporal dynamics of phenotypic selection in the wild. *Ecology Letters* 12: 1261–76.

Silverman, C. 2010. "Birdwatching and baby-watching": Niko and Elisabeth Tinbergen's ethological approach to autism. *History of Psychiatry* 21, no. 2: 176–89.

Simberloff, D. 1988. The contribution of population and community biology to conservation science. *Annual Review of Ecology and Systematics* 19: 473–511.

Simberloff, D., D. C. Schmitz, and T. C. Brown. 1997. *Strangers in Paradise: Impact and Management of Nonindigenous Species in Florida*. Washington, D.C.: Island Press.

Simpson, G. G. 1944. *Tempo and Mode in Evolution*. New York: Columbia University Press.

Singh, R. S., J. Xu, and R. J. Kulanthinal. 2012. *Rapidly Evolving Genes and Synthetic Systems*. Oxford: Oxford University Press.

———. 1963. Biology and the nature of science. *Science* 138: 81–88.

Slavet, E. 2008. Freud's 'Lamarckism' and the politics of racial science. *Journal of the History of Biology* 41, no. 1: 37–80.

Slifkin, N. 2001. *The Science of Torah*. Jerusalem: Targum Press.

———. 2006. *The Challenge of Creation*. Ramat Bet Shemesh: Zoo Torah.

Sloan, P. R. 1985. Darwin's invertebrate program, 1826–1836: Preconditions for transformism. In *The Darwinian Heritage*, ed. D. Kohn, 71–120. Princeton: Princeton University Press.

———. 1986. Darwin, vital matter, and the transformism of species. *Journal of the History of Biology* 19: 369–445.

———. 1992. Introductory essay: On the edge of evolution. In *Richard Owen: The Hunterian Lectures in Comparative Anatomy, May–June, 1837*, ed. P. R. Sloan, 3–72. London: Natural History Museum Publications.

———. 2009. Originating species: Darwin on the species problem. In *The Cambridge Companion to the "Origin of Species,"* ed. M. Ruse and R. J. Richards, 67–86. Cambridge: Cambridge University Press.

Slocombe, K. E., S. W. Townsend, and K. Zuberbühler. 2009. Wild chimpanzees (Pan troglodytes schweinfurthii) distinguish between different scream types: Evidence from a playback study. *Animal Cognition* 12: 441–49.

Slocombe, K. E., B. M. Waller, and K. Liebal. 2011. The language void: The need for multimodality in primate communication research. *Animal Behavior* 81: 919–24.

Slocombe, K. E., and K. Zuberbühler. 2005. Functionally referential communication in a chimpanzee. *Current Biology* 15: 1779–84.

———. 2006. Agonistic screams in wild chimpanzees (Pan troglodytes schweinfurthi) vary as a function of social role. *Journal of Comparative Psychology* 119: 66–77.

Smart, B. H. 1839. *Beginnings of a New School of Metaphysics*. Ann Arbor, Mich.: Scholars' Facsimiles & Reprints.

Smith, C. S. 1998. The Alfred Russel Wallace page. http://people.wku.edu/charles.smith/index1.htm.

Smith, J. 2000. Darwin's barnacles, Dickens's *Little Dorrit*, and the social uses of Victorian seaside studies. *Literature Interpretation and Theory* 10, no. 4: 327–47.

———. 2006. *Charles Darwin and Victorian Visual Culture*. Cambridge: Cambridge University Press.

Smith, J. K. A., and A. Yong. 2010. *Science and the Spirit: A Pentecostal Engagement with the Sciences*. Indiana: Indiana University Press.

Smith, K. 2006. Homing in on the genes for humanity: What makes us different from chimps? *Nature* 442: 725.

Smith, R. 1997. *The Fontana History of the Human Sciences*. London: Fontana Press.

Smith, S. 1965. The Darwin collection at Cambridge with one example of its use: Charles Darwin and Cirripedes. *Actes du XIe Congres International d'Histoire des Sciences* 5: 96–100.

Smocovitis, V. B. 1988. Botany and the Evolutionary Synthesis: The Life and Work of G. Ledyard Stebbins. PhD dissertation, Cornell University.

———. 1992. Disciplining botany: A taxonomic problem. *Taxon* 41: 459–70.
———. 1996. *Unifying Biology: The Evolutionary Synthesis and Evolutionary Biology*. Princeton: Princeton University Press.
———. 1997. G. Ledyard Stebbins, Jr. and the evolutionary synthesis (1924–1950). *American Journal of Botany* 84, no. 12: 1625–37.
———. 1999. The 1959 Darwin centennial in America. In *Commemorative Practices in Science*, ed. Clark Elliott and Pnina Abir-Am, *Osiris* 14:274–323. Chicago: University of Chicago Press.
———. 2005. It ain't over til it's over: Rethinking the Darwinian Revolution. *Journal of the History of Biology* 38: 33–49.
———. 2006. Keeping up with Dobzhansky: G. L. Stebbins, plant evolution and the evolutionary synthesis. *History and Philosophy of the Life Sciences* 28: 11–50.
———. 2008. Darwin's botany. In *The Cambridge Companion to the "Origin of Species,"* ed. M. Ruse and R. J. Richards, 216–36. Cambridge: Cambridge University Press.
———. 2009. The "Plant Drosophila": E. B. Babcock, the genus Crepis and the evolution of a genetics research program at Berkeley, 1912–1947. *Historical Studies of the Natural Sciences* 39: 300–55.
Snodgrass, R. E. 1902. The relation of the food to the size and shape of the bill in the Galapagos genus Geospiza. *The Auk* 19, no. 4: 367–81.
Sober, E. 1980. Evolution, population thinking, and essentialism. *Philosophy of Science* 47: 350–83.
———. 1984. *The Nature of Selection: Evolutionary Theory in Philosophical Focus*. Cambridge, Mass.: Bradford Books.
———. 2008. *Evidence and Evolution: The Logic behind the Science*. Cambridge: Cambridge University Press.
———. 2009. Parsimony arguments in science and philosophy: A test case for naturalism. *Proceedings and Addresses of the American Philosophical Association* 83: 117–55.
———. 2010. *Did Darwin Write the Origin Backwards? Philosophical Essays on Darwin's Theory*. Buffalo, N.Y.: Prometheus.
Sober, E., and D. S. Wilson. 1998. *Unto Others: The Evolution and Psychology of Unselfish Behavior*. Cambridge, Mass.: Harvard University Press.
Soloveitchik, J. B. 2005. *The Emergence of Ethical Man*. Jersey City, N.J.: KTAV.
Sommer, M. 2008. History in the gene: Negotiations between molecular and organismal anthropology. *Journal of the History of Biology* 41: 472–528.
Soulé, M. E. 1986. What is conservation biology? *BioScience* 35: 724–34.
Southward, A. J., ed. 1987. *Barnacle Biology*. Rotterdam: A. A. Balkema.
Sparks, R. D. 1926 Congo: A personality. *Field and Stream* 30: 18–20, 72.
Spencer, F. 1990. *Piltdown: A Scientific Forgery*. Oxford: Oxford University Press.
Spencer H. 1852. A theory of population, deduced from the general law of animal fertility. *Westminster Review* 1, no. 468–501.
———. 1855. *Principles of Psychology*. London: Longman, Brown, Green and Longmans.
———. 1857. Progress: Its law and cause. *Westminster Review* 67: 244–67.
———. 1862. *First Principles*. London: Williams and Norgate.
———. 1864. *Principles of Biology*. London: Williams and Norgate.
———. [1884] 1981. *The Man versus the State, with Six Essays on Government, Society and Freedom*. Ed. Eric Mack. Indianapolis: LibertyClassics.
———. 1904. *Autobiography*. London: Williams and Norgate.
———. 2009. *Social Statics Abridged and Revised Together with the Man versus the State*. Charleston, S.C.: BiblioLife.

Sperber, D. 1996. *Explaining Culture: A Naturalistic Approach*. Oxford: Blackwell.
Sperling, S. 1991. Baboons with briefcases: Feminism, functionism and sociobiology in the evolution of primate gender. *Signs* 17, no. 1: 1–27.
Spiegelman, S. 1967. An in vitro analysis of a replicating molecule. *American Scientist* 55: 221–64.
Stam, J. H. 1978. *Inquiries into the Origin of Language: The Fate of a Question*. New York: Harper & Row.
Stamos, D. N. 2006. *Darwin and the Nature of Species*. Albany: State University of New York Press.
Stanford, C. 2003. *Upright: The Evolutionary Key to Becoming Human*. New York: Houghton Mifflin Harcourt.
Stauffer, R. C. 1957. Haeckel, Darwin, and ecology. *Quarterly Review of Biology* 32: 138–44.
———. 1975. *Charles Darwin's Natural Selection; Being the Second Part of His Big Species Book Written from 1856 to 1858*. Cambridge: Cambridge University Press.
Stearns, S. C., S. G. Byars, D. R. Govindaraju, and D. Ewbank. 2010. Measuring selection in contemporary human populations. *Nature Reviews Genetics* 11, no. 9: 611–22.
Stearns, S. C., and J. C. Koella, eds. 2008. *Evolution in Health and Disease*. Oxford: Oxford University Press.
Stebbins, G. L., Jr. 1941. Apomixis in the angiosperms. *Botanical Review* 7: 507–42.
———. 1947. Types of polyploids: Their classification and significance. *Advances in Genetics* 1, no. 403–29.
———. 1950. *Variation and Evolution in Plants*. New York: Columbia University Press.
———. 1968. Ernest Brown Babcock. *Biographical Memoirs of the National Academy of Sciences* 32: 50–66.
———. 1979. Fifty years of plant evolution. In *Topics in Plant Population Biology*, ed. S. Jain, G. B. Johnson, P. Raven, and O. Solbrig, 1–17. New York: Columbia University Press.
———. 1980. Botany and the synthetic theory of evolution. In *The Evolutionary Synthesis*, ed. E. Mayr and W. B. Provine, 139–52. Cambridge, Mass.: Harvard University Press.
Stebbins, R. E. 1988. France. In *The Comparative Reception of Darwinism*, ed. T. F. Glick, 117–63. Chicago: University of Chicago Press.
Stenberg, L. 1996. *The Islamization of Science: Four Muslim Positions Developing an Islamic Modernity*. Lund: Religionshistoriska Avd., Lunds University.
Stenhouse, J. 1999. Darwinism in New Zealand, 1859–1900. In *Disseminating Darwinism: The Role of Place, Race, Religion, and Gender*, ed. R. L. Numbers and J. Stenhouse, 61–89. Cambridge: Cambridge University Press.
Stephens, L. D. 2000. *Science, Race, and Religion in the American South: John Bachman and the Charleston Circle of Naturalists, 1815–1895*. Chapel Hill: University of North Carolina Press.
Sterelny, K. 2009. Darwinian concepts in the philosophy of mind. In *The Cambridge Companion to Darwin*, 2nd ed., ed. J. Hodge and G. Radick, 323–44. Cambridge: Cambridge University Press.
Sterman, B. 1994. Judaism and Darwinian evolution. *Tradition* 29: 48–74.
Stern, C., and E. R. Sherwood, eds. 1966. *The Origin of Genetics*. San Francisco: W. H. Freeman.
Stettler, N., A. M. Tershakovec, B. S. Zemel, et al. 2000. Early risk factors for increased adiposity: A cohort study of African American subjects followed from birth to young adulthood. *American Journal of Clinical Nutrition* 72: 378–83.

Stevens, P. F. 1994. *The Development of Biological Systematics: Antoine-Laurent de Jussieu, Nature, and the Natural System*. New York, Columbia University Press.

Stevens, W. K. 1992. Talks to prevent huge loss of species. *New York Times*, 3 March.

Stevenson, L. 1932. *Darwin among the Poets*. Chicago: University of Chicago Press.

Stiling, R. L. 1991. The Diminishing Deluge: Noah's Flood in Nineteenth-Century American Thought. PhD dissertation, University of Wisconsin at Madison.

Stocking, G. 1968. *Race, Culture, and Evolution: Essays on the History of Anthropology*. Chicago: University of Chicago Press.

——— 1987. *Victorian Anthropology*. New York: Free Press.

Stoddart, D. R. 1966. Darwin's impact on geography. *Annals of the Association of American Geographers* 56: 683–98.

——— 1976. Darwin, Lyell, and the geological significance of coral reefs. *British Journal for the History of Science* 9: 199–218.

Stopes, M. C. 1912. *Botany, or the Modern Study of Plants*. London: T.C. & E.C. Jack.

Stott, R. 2003. *Darwin and the Barnacle*. London: Faber and Faber.

Street, S. 2006. A Darwinian dilemma for realist theories of value. *Philosophical Studies* 127: 109–66.

Strick, J. E. 2000. *Sparks of Life*. Cambridge, Mass.: Harvard University Press.

Strickland, H. E. 1840. Observations upon the affinities and analogies of organized beings. *Magazine of Natural History* 4: 219–26.

Strong, A. H. 1907. *Systematic Theology*. Philadelphia: Griffith & Rowland Press.

Strum, S. C., and L. M. Fedigan, eds. 2000. *Primate Encounters: Models of Science, Gender and Society*. Chicago: University of Chicago Press.

Sulivan, H. N. 1896. *The Life and Letters of the Late Admiral Sir Bartholomew James Sulivan, K.C.B., 1810–1890*. London: John Murray.

Sulloway, F. J. 1979. *Freud, Biologist of the Mind: Beyond the Psychoanalytic Legend*. New York: Basic Books.

——— 1982a. Darwin and his finches: The evolution of a legend. *Journal of the History of Biology* 15, no. 1: 1–53.

——— 1982b. Darwin's conversion: The *Beagle* voyage and its aftermath. *Journal of the History of Biology* 15: 325–96.

——— 1984. Darwin and the Galapagos. *Biological Journal of the Linnaean Society* 21: 29–59.

——— 2009. Tantalising tortoises and the Darwin-Galapagos legend. *Journal of the History of Biology* 42: 3–31.

Sumner, F. B. 1929a. The analysis of a concrete case of intergradation between two subspecies. *Proceedings of the National Academy of Sciences USA* 15: 110–20.

——— 1929b. The analysis of a concrete case of intergradation between two subspecies. II. additional data and interpretations. *Proceedings of the National Academy of Sciences USA* 15: 481–93.

Sundance Channel. 2009. Barnacle (Green Porno, Season 2). http://link.brightcove.com/services/player/bcpid1745093298?bclid=0&bctid=18011211001.

Swarth, H. S. 1929. A new bird family (Geospizidae) from the Galapagos Islands. *Proceedings of the California Academy of Sciences* 18, no. 2: 29–43.

Swetlitz, M. 1999. American Jewish responses to Darwin and evolutionary theory, 1870–1890. In *Disseminating Darwinism: The Role of Place, Race, Religion, and Gender*, ed. R. L. Numbers and J. Stenhouse, 209–46. Cambridge: Cambridge University Press.

——— 2006. Responses to evolution by Reform, Conservative and Reconstructionist rabbis in twentieth-century America. In *Jewish Tradition and the Challenge of Evolution*, ed. G. Cantor and M. Swetlitz, 47–70. Chicago: University of Chicago Press.

Takacs, D. 1996. *The Idea of Biodiversity: Philosophies of Paradise*. Baltimore: John Hopkins University Press.

Tan, J. 1987. *Tan Jiazhen lunwen ji* (Collected Scientific Articles of Tan Jiazhen). Beijing: Science Press.

Tanner, J. E. 2004. Gestural phrases and exchanges by a pair of zoo-living lowland gorillas. *Gesture* 4, no. 1: 1–24.

Tanner, J. E., and R. W. Byrne. 1999. The development of spontaneous gestural communication in a group of zoo-living lowland gorillas. In *The Mentalities of Gorillas and Orangutans: Comparative Perspectives*, ed. R. W. Mitchell, H. L. Miles, and S. T. Parker, 211–39. Cambridge: Cambridge University Press.

Tassy, P. 1998. *L'arbre à remonter le temps*. Paris: Bourgois.

Taub, L. 1993. Evolutionary ideas and "empirical" methods: The analogy between language and species in works by Lyell and Schleicher. *British Journal for the History of Science* 26: 171–93.

Tegetmeier, W. B. 1854. *Profitable Poultry: Their Management in Health and Disease*. London: Darton.

Teilhard de Chardin, P. 1915. Les Carnassiers des phosphorites du Quercy. *Annales de paleontology* 9: 103–92.

——— 1960. *The Divine Milieu*. New York: Harper & Row.

——— 1969. *Christianity and Evolution*. Trans. R. Hague. New York: Harcourt Brace Jovanovich.

——— 1999. *The Human Phenomenon*. Trans. S. Appleton-Weber. Portland, Ore.: Sussex Academic Press.

Tennyson, A. 1973. In Memoriam. In *In Memoriam: An Authoritative Text, Backgrounds and Sources, Criticism*, ed. R. H. Ross, 3–90. New York: Norton.

Theobald, D. L. 2010. A formal test of the theory of universal common ancestry. *Nature* 465: 219–22.

Theunissen, B. 1989. *Eugene Dubois and the Ape-Man from Java: The Story of the First Missing Link and Its Discoverer*. Dordrecht: Kluwer.

——— 2012. Darwin and his pigeons: The analogy between artificial and natural selection revisited. *Journal of the History of Biology* 45: 179–212.

Thompson, J. V. 1830. Memoir IV. On the Cirripedes or barnacles; Demonstrating their deceptive character; the extraordinary metamorphosis they undergo, and the class of animals to which they undisputably belong. In *Zoological Researches and Illustrations, or Natural History of Nondescript or Imperfectly-Known Animals*. Cork. Facsimile reprint, London: Society for the Bibliography of Natural History, 1968.

Thomson, J. A. 1925. *Concerning Evolution*. New Haven: Yale University Press.

Thomson, K. S. 2008. *The Legacy of the Mastodon: The Golden Age of Fossils in America*. New Haven: Yale University Press.

Thomson, W. 1862. Physical considerations regarding the possible age of the sun's heat. *Thirty-First Meeting of the British Association for the Advancement of Science Notices and Abstracts* (Manchester), 27–28. London: John Murray.

——— 1868. On geological time. *Transactions of the Geological Society of Glasgow* 3: 1–28.

——— 1872. Presidential address. *Forty-First Meeting of the British Association for the Advancement of Science Report* (Edinburgh), lxxxiv–cv. London: John Murray.

Tiedemann, F. 1808–14. *Zoologie, zu seinen Vorlesungen Entworfen*. 3 vols. Landshut: Weber.

Tilman, D., S. Kilham, and P. Kilham. 1982. Phytoplankton community ecology: The role of limiting nutrients. *Annual Review of Ecology and Systematics* 13: 349–72.

Tinbergen, N. 1963. On aims and methods in ethology. *Zeitschrift für Tierpsychologie* 20: 410–33.

———. 1965. Behavior and natural selection. In *Ideas in Modern Biology*, ed. J. Moore, 521–42. New York: Natural History Press.

Tocqueville, Alexis de. 2003. *Democracy in America and Two Essays on America*. Trans. G. E. Bevan. New York: Penguin Books.

Todes, D. P. 1989. *Darwin without Malthus: The Struggle for Existence in Russian Evolutionary Thought*. Oxford: Oxford University Press.

Tomasello, M. T. 1995. Language is not an instinct. *Cognitive Development* 10: 131–56.

———. 1999. *The Cultural Origins of Human Cognition*. Cambridge, Mass.: Harvard University Press.

———. 2003. *Constructing a Language*. Cambridge, Mass.: Harvard University Press.

Torrance, T. F. 1981. *Divine and Contingent Order*. Oxford: Oxford University Press.

Travis, J. 1990. Review of *A Model Study in Population Biology*. *Ecology* 71, no. 4: 1631–32.

Treuenfels, A. 1872. *Die Darwin'sche Theorie in ihrem Verhaltniss zur Religion*. Magdeburg: W. Simon.

Trevathan, W. R., E. O. Smith, and J. J. McKenna. 2008. *Evolutionary Medicine and Health: New Perspectives*. Oxford: Oxford University Press.

Trivers, R. L. 1971. The evolution of reciprocal altruism. *Quarterly Review of Biology* 46: 35–57.

Troland, L. T. 1914. The chemical origin and regulation of life. *Monist* 24: 92–133.

Trollope, A. [1861–62] 1981. *The Struggles of Brown, Jones, and Robinson*. New York: Amos Press.

Trow-Smith, R. 1957. *A History of British Livestock Husbandry to 1700*. London: Routledge and Kegan Paul.

———. 1959. *A History of British Livestock Husbandry, 1700–1900*. London: Routledge and Kegan Paul.

Turner, F. M. 1974. *Between Science and Religion: The Reaction to Scientific Naturalism in Late Victorian England*. New Haven: Yale University Press.

Tutt, J. W. 1890. Melanism and melanochroism in British lepidoptera. *Entomologist's Record, and Journal of Variation* 1, no. 3: 49–56.

Tylor, E. B. 1865. *Researches into the Early History of Mankind*. London: John Murray.

Ulanowicz, R. E. 1997. *Ecology, the Ascendant Perspective*. New York: Columbia University Press.

———. 2009. *A Third Window: Natural Life beyond Newton and Darwin*. West Conshohocken, Pa.: Templeton Foundation Press.

United Nations Development Programme. 2003. *Arab Human Development Report, 2003: Creating a Knowledge Society*. New York.

van Frassen, B. 1980. *The Scientific Image*. Oxford: Clarendon Press.

Van Riper, A. B. 1993. *Men among the Mammoths: Victorian Science and the Discovery of Prehistory*. Chicago: University of Chicago Press.

Vandermassen, G., M. Demoor, and J. Braeckman. 2005. Close encounters with a new species: Darwin's clash with the feminists at the end of the nineteenth century. In *Unmapped Countries: Biological Visions in Nineteenth-Century Literature and Culture*, ed. A.-J. Zwierlein, 71–82. London: Anthem.

Vane-Wright, R. I. 1971. The systematics of *Drusillopsis* Oberthür (Satyrinae) and the supposed Amathusiid *Bigaena* van Eecke (Lepidoptera: Nymphalidae), with some observations on Batesian mimicry. *Transactions of the Royal Entomological Society of London* 123: 97–123.

Vardiman, L. 1997. Scientific naturalism as science. *Impact* 293, inserted in *Arts & Facts*: no pagination.

Venter, J. C., M. D. Adams, E. Myers, et al. 2001. The sequence of the human genome. *Science* 291: 1304–51.

Vicedo, M. 2009. The father of ethology and the foster mother of ducks: Konrad Lorenz as expert on motherhood. *Isis* 100, no. 2: 263–91.

———. 2010. The evolution of Harry Harlow: From the nature to the nurture of love. *History of Psychiatry* 21, no. 2: 190–205.

Vickers, M. H., B. H. Breier, W. S. Cutfield, P. L. Hofman, and P. D. Gluckman. 2000. Fetal origins of hyperphagia, obesity, and hypertension and postnatal amplification by hypercaloric nutrition. *American Journal of Physiology* 279: E83–E87.

Vignieri, S. N., J. Larson, and H. E. Hoekstra. 2010. The selective advantage of cryptic coloration in mice. *Evolution* 64: 2153–58.

Virchow, R. 1862. *Vier Reden über Leben und Kranksein*. Berlin: Georg Reimer.

———. 1877. *Die Freiheit der Wissenschaft im modernen Staat*. Berlin: Viegandt, Hempel & Parey.

Viret, J. 1939. Monographie paléontologique de la faune des vertébrés des sables de Montpellier. III. Carnivora Fissipedia. *Travaux du Laboratoire de Géologie de la Faculté des Sciences de Lyon* 37: 1–26.

Vollmer, G. 1983. Mesocosm and objective knowledge: On problems solved by evolutionary epistemology. In *Concepts and Approaches in Evolutionary Epistemology*, ed. F. Wuketits, 69–122. Dordrecht: D. Reidel.

Volterra, V. 1931. *Leçons sur la théorie mathématique de la lutte pour la vie*. Paris: Editions Gauthier-Villars.

von Baer, K. 1828–37. *Entwickelungsgeschichte der Thiere: Beobachtung und Reflexion*. 2 vols. Königsberg: Bornträger.

von Buch, C. L. 1820. Uber die Zussammersetzung der basaltischen Insels und über Erhenbungs-Cratere. *Abhandlungen der Koniglichen Akademie der Wissenschaften, Berlin* 51: 86.

Vorzimmer, P. J. 1965. Darwin's ecology and its influence upon his theory. *Isis* 56: 148–55.

———. 1972. *Charles Darwin: The Years of Controversy; The "Origin of Species" and Its Critics, 1859–1882*. London: University of London Press.

Voss, J. 2009. Monkeys, apes and evolutionary theory: From human descent to King Kong. In *Endless Forms: Charles Darwin, Natural Science and the Visual Arts*, ed. D. Donald and J. Munro, 215–37. New Haven: Yale University Press.

Wächtershäuser, G. 1992. Groundwork for an evolutionary biochemistry: The iron-sulphur world. *Progress in Biophysics and Molecular Biology* 58: 85–201.

Wade, M. J. 1977. An experimental study of group selection. *Evolution* 31, no. 1: 134–53.

Wagner, G. P., C. H. Chiu, et al. 2000. Developmental evolution as a mechanistic science: The inference from developmental mechanisms to evolutionary processes. *American Zoologist* 40: 819–31.

Wagner, G. P., and M. D. Laubichler. 2004. Rupert Riedl and the re-synthesis of evolutionary and developmental biology: Body plans and evolvability. *Journal of Experimental Zoology, Part B: Molecular and Developmental Evolution* 302, no. 1: 92–102.

Wagner, M. 1868. *Darwin'sche Theorie und das Migrationsgesetz der Organismen*. Leipzig: Duncker & Humblot.

———. 1873. *The Darwinian Theory and the Law of the Migration of Organisms*. Trans. J. L. Laird. London: Edward Stanford.

Wake, C. S. 1868. *Chapters on Man*. London: Trübner.

Walker, R. M., M. Gurven, K. Hill, et al. 2006. Growth rates and life histories in twenty-two small-scale societies. *American Journal of Human Biology* 18, no. 3: 295–311.

Wallace, A. R. 1853a. *A Narrative of Travels on the Amazon and Rio Negro*. London: Reeve.

——— 1853b. *Palm Trees*. London: John Van Voorst.

——— 1855. On the law which has regulated the introduction of new species. *Annals and Magazine of Natural History*, 2nd ser., 16: 184–96.

——— 1864. The origin of human races and the antiquity of man deduced from the theory of "natural selection." *Transactions of the Anthropological Society of London* 1: clviii–clxxxvii.

——— 1865. On the phenomena of variation and geographical distribution as illustrated by the Papilionidae of the Malayan region. *Transactions of the Linnean Society of London* 25: 1–27.

——— 1866. The scientific aspect of the supernatural. *English Leader* 2: 59–60, 75–76, 91–93, 107–8, 123–25, 139–40, 156–57, 171–73.

——— 1867. Mimicry, and other protective resemblances among animals. *Westminster and Foreign Quarterly Review* 32: 1–43.

——— 1869a. *The Malay Archipelago*. 2 vols. London: Macmillan.

——— 1869b. Review of *Principles of Geology* by Charles Lyell, 10th ed., and *Elements of Geology*, 6th ed. *Quarterly Review* 126: 359–94.

——— 1870a. *Contributions to the Theory of Natural Selection*. London: Macmillan.

——— 1870b. The measurement of geological time. *Nature* 1: 399–401, 452–55.

——— 1871. A review and criticism of Mr. Darwin's *Descent of Man*. *Academy* 2: 177–83.

——— 1876. *The Geographical Distribution of Animals*. 2 vols. London: Macmillan.

——— 1878. *Tropical Nature, and Other Essays*. London: Macmillan.

——— 1880. *Island Life*. London: Macmillan.

——— 1882. *Land Nationalisation*. London: Trübner.

——— 1886. Physiological selection and the origin of species. *Nature* 34: 467–68.

——— 1889. *Darwinism: An Exposition of the Theory of Natural Selection with Some of Its Applications*. London: Macmillan.

——— 1890. Human selection. *Fortnightly Review* 48: 325–37.

——— 1891. *Natural Selection and Tropical Nature*. London: Macmillan.

——— 1896. Sir Charles Lyell on geological climates and the origin of species. *Quarterly Review* 126 (April): 359–94.

——— 1898. *The Wonderful Century*. London: Sonnenschein.

——— 1900. *Studies Scientific and Social*. 2 vols. London: Macmillan.

——— 1901. *Darwinism*. 3rd ed. London: Macmillan.

——— 1903. *Man's Place in the Universe: A Study of the Results of Scientific Research in Relation to the Unity or Plurality of Worlds*. London: Chapman & Hall.

——— 1905. *My Life: A Record of Events and Opinions*. 2 vols. London: Chapman & Hall.

——— 1910. *The World of Life*. London: Chapman & Hall.

——— 1913a. *Social Environment and Moral Progress*. London: Cassell.

——— 1913b. *The Revolt of Democracy*. London: Cassell.

Walsh, D. 2006. Organisms as natural purposes: The contemporary evolutionary perspective. *Studies in History and Philosophy of the Biological and Biomedical Sciences* 37, no. 4: 771–91.

Walters, S. M., and E. A. Stow. 2001. *Darwin's Mentor: John Stevens Henslow, 1796–1861*. Cambridge: Cambridge University Press.

Walton, A., and J. Hammond. 1938. The maternal effects on growth and conformation in Shire horses–Shetland pony crosses. *Proceedings of the Royal Society B: Biological Sciences* 125: 311–35.

Wang, H. 2004. *Xiandai Zhongguo sixiang de xingqi* (The Rising of Modern Chinese Thoughts). Beijing: SDX Joint Publishing Company.

Wang, Z. 2002. *Jinhua zhuyi zai Zhongguo* (Evolutionism in China). Beijing: Capital Normal University Press.

Warfield, B. B. 1888. Charles Darwin's religious life: A sketch in spiritual biography. *Presbyterian Review* 9: 569–601.

——— 1906. Review of *God's Image in Man*. *Princeton Theological Review* 4: 555–58.

Wasmann, E. 1904. *Die moderne Biologie und die Entwicklungstheorie*. 2nd ed. Freiberg: Herdersche Verlagshandlung.

Waterfield, R., and H. Tredennick, trans. 1990. *Xenophon: Conversations of Socrates*. New York: Penguin.

Waterhouse, G. R. 1843. Observations on the classification of the Mammalia. *Annals and Magazine of Natural History* 4: 399–412.

Waters, C. K. 2003. The arguments in Darwin's *Origin of Species*. In *The Cambridge Companion to Darwin*, ed. J. Hodge and G. Radick, 116–39. Cambridge: Cambridge University Press.

Watkins, T. H. 1992. Review of *The Diversity of Life*. *Washington Post Book World*, 27 September.

Watson, J. D., and R. M. Cook-Deegan. 1990. The Human Genome Project and international health. *Journal of the American Medical Association* 263: 3322–24.

Weber, B. H., and D. J. Depew. 2001. Developmental systems, Darwinian evolution, and the unity of science. In *Cycles of Contingency: Developmental Systems and Evolution*, ed. P. E. Griffiths, R. D. Gray, and S. Oyama, 239–53. Cambridge, Mass.: MIT Press.

———, eds. 2003. *Evolution and Learning: The Baldwin Effect Reconsidered*. Cambridge, Mass.: MIT Press.

Wedgwood, H. 1866. *On the Origin of Language*. London: Trübner.

Weikart, R. 2004. *From Darwin to Hitler: Evolutionary Ethics, Eugenics, and Racism in Germany*. New York: Palgrave Macmillan.

——— 2005. The impact of social Darwinism on anti-Semitic ideology in Germany and Austria, 1860-1945. In *Jewish Tradition and the Challenge of Evolution*, ed. G. Cantor and M. Swetlitz, 93–115. Chicago: University of Chicago Press.

Weindling, P. 2005. The evolution of Jewish identity: Ignaz Zollschan between Jewish and Aryan race theories, 1910–1945. In *Jewish Tradition and the Challenge of Evolution*, ed. G. Cantor and M. Swetlitz, 116–36. Chicago: University of Chicago Press.

——— 2010. Genetics, eugenics, and the Holocaust. In *Biology and Ideology from Descartes to Dawkins*, ed. D. Alexander and R. Numbers, 192–214. Chicago: University of Chicago Press.

Weismann, A. 1889. *Essays upon Heredity and Kindred Biological Problems*. Trans. E. Poulton. Oxford: Clarendon Press.

Weisrock, D. W., R. M. Rasoloarison, I. Fiorentino, et al. 2010. Delimiting species without nuclear monophyly in Madagascar's mouse lemurs. *PloS One* 5: e9883.

Weldon, W. F. R. 1893. On certain correlated variations in Carcinus moenas. *Proceedings of the Royal Society* 54: 318–29.

——— 1895. An attempt to measure the death-rate due to the selective destruction of Carcinus moenas with respect to a particular dimension. *Proceedings of the Royal Society of London* 57: 360–79.

——— 1898. Presidential address to the Zoological Section of the British Association. *Transactions of the British Association*: 887–902.

——— 1902. A first study of natural selection in Clausilia laminata (Montagu). *Biometrika* 1: 109–28.

Wells, W. C. 1818. An account of a female of the white race of mankind, part of whose skin resembles that of a negro; with some observations on the causes of the differences in colour and form between the white and negro races of men. In *Two Essays: One upon*

Single Vision with Two Eyes; The Other on Dew, 425–39. London: Archibald Constable.

West, S. A., A. S. Griffin, and Gardner A. 2007. Social semantics: Altruism, cooperation, mutualism, strong reciprocity and group selection. *Journal of Evolutionary Biology* 20, no. 2: 415–32.

West-Eberhard, M. J. 2003. *Developmental Plasticity and Evolution*. Oxford: Oxford University Press.

Wheeler, W. M. 1911. The ant colony as an organism. *Journal of Morphology* 22: 307–25.

Whewell, W. 1837. *History of the Inductive Sciences*. 3 vols. London: John W. Parker.

White, A. D. 1896. *A History of the Warfare between Science and Theology*. 2 vols. London: Macmillan.

White, G. 1789. *The Natural History and Antiquities of Selborne, in the County of Southampton; with Engravings and an Appendix*. London.

White, P. 2010. Darwin's church. *Studies in Church History* 46: 333–52.

White, T. D., G. G. Wolde, B. Asfaw, et al. 2006. Asa Issie, Aramis and the origin of Australopithecus. *Nature* 440: 883–89.

Wilberforce, S. 1860. Darwin's *Origin of Species*. *Quarterly Review* 108: 225–64.

Wilkinson, J. 1820. *Remarks on the Improvement of Cattle*. Nottingham: H. Barnett.

Williams, G. C. 1966. *Adaptation and Natural Selection: A Critique of Some Current Evolutionary Thought*. Princeton: Princeton University Press.

——— 1988. Huxley's *Evolution and Ethics* in sociobiological perspective. *Zygon* 23: 383–438.

Williams, G. C., and R. M. Nesse. 1991. The dawn of Darwinian medicine. *Quarterly Review of Biology* 66, no. 1: 1–22.

Wilson, D. S. 1975. A theory of group selection. *Proceedings of the National Academy of Sciences* 72: 143–46.

——— 2002. *Darwin's Cathedral: Evolution, Religion, and the Nature of Society*. Chicago: University of Chicago Press.

Wilson, D. S., and E. O. Wilson. 2007. Rethinking the theoretical foundation of sociobiology. *Quarterly Review of Biology* 82: 327–48.

Wilson, E. O. 1975. *Sociobiology: The New Synthesis*. Cambridge, Mass.: Harvard University Press.

——— 1984. *Biophilia: The Human Bond with Other Species*. Cambridge, Mass.: Harvard University Press.

——— 1988. The current state of biological diversity. In *Biodiversity: Papers from the First National Forum on Biodiversity, September 1986*, ed. E. O. Wilson and M. Peters, 3–18. Washington, D.C.: National Academy Press.

——— 1992. *The Diversity of Life*. Cambridge, Mass.: Harvard University Press.

——— 1994. *Naturalist*. New York: Island Press.

——— 1998. *Consilience: The Unity of Knowledge*. New York: Knopf.

Wilson, E. O., and B. Hölldobler. 2005. Eusociality: Origin and consequences. *Proceedings of the National Academy of Sciences* 102: 13367–71.

Wilson, L. G., ed. 1970. *Sir Charles Lyell's Scientific Journals on the Species Question*. New Haven: Yale University Press.

Winge, O. 1917. The chromosomes: Their numbers and general importance. *Comptes Rendu des Travaux du Laboratoire de Carlsberg* 13: 131–275.

Winslow, J. H. 1975. Mr Lumb and Masters Megatherium: An unpublished letter by Charles Darwin from the Falklands. *Journal of Historical Geography* 1, no. 4: 347–60.

Winsor, M. P. 1969. Barnacle larvae in the nineteenth century: A case study in taxonomic theory. *Journal of the History of Medicine and Allied Sciences* 24: 294–309.

——— 1976. *Starfish, Jellyfish, and the Order of Life: Issues in Nineteenth-Century Science*. New Haven: Yale University Press.

——— 1995. The English debate on taxonomy and phylogeny, 1937–1940. *History and Philosophy of the Life Sciences* 17, no. 2: 227–52.

——— 2009. Taxonomy was the foundation of Darwin's evolution. *Taxon* 58: 43–47.

Winter, A. 1998. *Mesmerized*. Chicago: University Press.

Woese, C. R. 1987. Bacterial evolution. *Microbiology and Molecular Biology Reviews* 51, no. 2: 221–71.

——— 1998. The universal ancestor. *Proceedings of the National Academy of Sciences* 95: 6854–59.

——— 2000. Interpreting the universal phylogenetic tree. *Proceedings of the National Academy of Sciences* 97: 8392–96.

Woese, C. R., and G. E. Fox. 1977. The concept of cellular evolution. *Journal of Molecular Evolution* 10, no. 1: 1–6.

Woese, C. R., O. Kandler, and M. L. Wheelis. 1990. Towards a natural system of organisms: Proposal for the domains Archaea, Bacteria, and Eucarya. *Proceedings of the National Academy of Sciences* 87: 4576–79.

Wood, R. J., and V. Orel. 2001. *Genetic Prehistory in Selective Breeding: A Prelude to Mendel*. Oxford: Oxford University Press.

Worster, D. 1994. *Nature's Economy: A History of Ecological Ideas*. Cambridge: Cambridge University Press.

Worthen, M. 2008. The Chalcedon problem: Rousas John Rushdoony and the origins of Christian Reconstructionism. *Church History* 77: 399–437.

Wright, L. 1976. *Teleological Explanations*. Berkeley: University of California Press.

Wright, S. 1931. Evolution in Mendelian populations. *Genetics* 16: 97–159.

——— 1932. The roles of mutation, inbreeding, crossbreeding and selection in evolution. *Proceedings of the Sixth International Congress of Genetics* 1: 356–66.

——— 1977. *Evolution and the Genetics of Natural Populations*. Vol. 3. Chicago: University of Chicago Press.

Wu, D. P., K. Hugenholtz, R. Mavromatis, et al. 2009. A phylogeny-driven genomic encyclopedia of Bacteria and Archaea. *Nature* 462: 1056–60.

Wyhe, J. van, ed. 2002. *The Complete Work of Charles Darwin Online*. http://darwin-online.org.uk/people/van_wyhe.html.

——— 2003. George Combe's *Constitution of Man* and the law of hereditary descent. In *A Cultural History of Heredity II*, 165–74. Max Planck Institut für Wissenschaftsgeschichte, preprint series. Berlin: Max Plank Institut.

——— 2004. *Phrenology and the Origins of Victorian Scientific Naturalism*. Aldershot: Ashgate.

——— ed. 2006. Darwin's "Journal" (1809–1881). http://darwin-online.org.uk/content/frameset?viewtype=side&itemID=CUL-DAR158.1-76&pageseq=1.

——— 2007. Mind the gap: Did Darwin avoid publishing his theory for many years? *Notes and Records of the Royal Society* 67: 177–205.

——— 2009a. *Darwin in Cambridge*. Cambridge: Christ's College.

——— ed. 2009b. *Charles Darwin's Shorter Publications, 1829–1883*. Cambridge: University Press.

——— 2012. Where do Darwin's finches come from?" *Evolutionary Review* 3, no. 1: 185–95.

Wyhe, J. van, and M. J. Pallen. 2012. The "Annie Hypothesis": Did the death of his daughter cause Darwin to "give up Christianity"? *Centaurus* 54: 105–23.

Wynne-Edwards, V. C. 1962. *Animal Dispersion in Relation to Social Behavior*. London: Oliver & Boyd.

———. 1982. Review of *The Genetics of Altruism* by S. A. Boorman and P. R. Levitt. *Social Science and Medicine* 16: 1095–98.

Yahya, H. 2007. *Atlas of Creation*. Vol. 1. Istanbul: Global Publishing.

Yan, F. [1895] 1986. "Yuan qiang" (Whence Strength). In *Yan Fu ji* (Collected Works of Yan Fu), ed. Wang Shi, 1:5–15. Beijing: Zhonghua Book Company.

———. [1898] 1998. *Tianyan lun (The Theory of Evolution). A Translation of T. H. Huxley's Romanes lecture "Evolution and Ethics" and "Prolegomena."* Annot. Feng Junhao. Zhengzhou: Zhongzhou Ancient Books Publishing House.

———. [1903] 1981. *Qunxue yiyan (The Study of Sociology). A Translation of Herbert Spencer's* The Study of Sociology. Beijing: Commercial Company.

Yang, Z., and B. Rannala. 2010. Bayesian species delimitation using multilocus sequence data. *Proceedings of the National Academy of Sciences* 107: 9264–69.

Youatt, W. 1834. *Cattle: Their Breeds, Management, and Diseases*. London: Baldwin and Cradock.

Youmans E. D. L. [1883] 2008. *Herbert Spencer on the Americans and the Americans on Herbert Spencer*. Durham, N.C.: Gadow Press.

Young, R. M. 1985. *Darwin's Metaphor: Nature's Place in Victorian Culture*. Cambridge: Cambridge University Press.

Zahm, J. 1975. *Evolution and Dogma*. Hicksville, N.Y.: Regina Press.

Zampieri, F. 2009. Medicine, evolution, and natural selection: an historical overview. *Quarterly Review of Biology* 84, no. 4: 333–55.

Zelnio, K. 2010. *Ex Omnia Conchis*: Darwin and his beloved barnacles. http://deepseanews.com/2010/02/ex-omnia-conchis-darwin-and-his-beloved-barnacles/.

Zeng, Y.-F., W.-J. Liao, R. J. Petit, and D.-Y. Zhang. 2010. Exploring species limits in two closely related Chinese oaks. *PloS One* 5: e15529.

Zeyl, D., trans. 2000. *Plato's Timaeus*. Indianapolis: Hackett.

Zhang, B., and Z. Wang. 1982. Jinhua lun yu shenchuang lun zai Zhongguo de douzheng (Struggle between Evolutionism and Creationism in China). *Ziran Bianzheng Fa Tongxun* (Journal of Dialectics of Nature) 2: 43–50.

Zhang, J. 2005. *Kexue shetuan zai jindai Zhongguo de mingyun: yi Zhongguo kexue she wei zhongxin* (The Science Association and the Change of Society in Modern China: A Study on the Science Society of China). Jinan: Shandong Education Press.

Zhong, T. [1889] 2009. Zhongxi gezhi zhi xueyi tonglun (Comparison on Gezhi of China and of the West). In *Jindai kexue zai Zhongguo de chuanbo: wenxian yu shiliao xuanbian* (The Introduction of Modern Science into Late 19th and Early 20th Century China: Selected Works and Documents), ed. Wang Yangzong, 1:341–42. Jinan: Shandong Education Press.

Zhou, J. [1937] 2009. Ren ji dongwu zhi biaoqing yizhe bianyan (Translator's preface for The Expression of the Emotions in Man and Animals). In *Jinhua lun de yingxiang li* (The Influence of Evolution), ed. Shen Yongbao and Cai Xingshui, 24–26. Nanchang: Jiangxi Universities Press.

Zhou, Z. [1906] 1961. Gu er ji (Legend of an Orphan). In *Wan Qing wenxue congchao* (Late Qing Literature Series), "*Novel*," ed. A Ying, 4:497–541. Beijing: Zhonghua Book Company.

Zimmer, C. 2006. *Evolution: The Triumph of an Idea*. New York: Harper/Perennial.

———. 2007. Editor's introduction to *The Descent of Man: The Concise Edition*. Ed. C. Zimmer New York: Plume.

———. 2009. *The Tangled Bank: An Introduction to Evolution*. Greenwood Village, Colo.: Roberts.

Zuberbühler, K. 2000. Referential labeling in Diana monkeys. *Animal Behavior* 59, no. 917–27.

Zuberbühler, K., R. Noe, and R. M. Seyfarth. 1997. Diana monkey long-distance calls: Messages for conspecifics and predators. *Animal Behavior* 53: 589–604.

Zuckerkandl, E., and L. Pauling. 1965a. Molecules as documents of evolutionary history. *Journal of Theoretical Biology* 8: 357–66.

———. 1965b. Evolutionary divergence and convergence in proteins. In *Evolving Genes and Proteins*, ed. V. Bryson and H. J. Vogel, 97–166. New York: Academic Press.

Index

Abbot, Francis, 215
Abel, Othenio, 356
abiogenesis, 323
    Darwin and, 324
Abrard, René, 302
Achillea sp, 317
acids, amino, 326, 328, 333, 341, 398–400, 401
action, direct, principle of (Darwin), 177
Adalet ve Kalknma Partisi (Party for Justice and Development), 502
adaptation, 97, 406, 454–56
    chance variation and, 156
    divine design and, 136
    individual organism, 203, 406
    natural selection and, 12, 453
    progress and, 222
    self-destructive, 347
    variation and, 155,
Adaptation and Natural Selection, A Critique of Some Evolutionary Thought (Williams), 157, 351
adaptationism, 333, 408, 409
Adaptive Colouration in Animals (Cott), 143
advantage
    heterozygote, 297, 311
    relative, 156
affinity, 74
    analogy and, in grouping organisms, 82
    defined
        by Strickland, 77
        by Westwood, 77
    elective (Müller), 122
    inextricable web of (Darwin), 440
    keys to, 80
    morphological versus genealogical, 301
    mutual, of organic beings, 100
    relationship and, 78
    resemblance and, 77
    true, 76
Agassiz, Alexander, 228, 362, 367
Agassiz, Louis, 8, 18, 28, 126, 129, 172, 174, 227–28, 259, 477
agnosticism, 16, 211
    Darwin and, 215
Agouti, 57, 62
Agudath Israel of America, 498,

Airy, George, 48
Airy, Hubert, 507
Albatross, 370
Alcippe, 85
All the Year Round, 437
alleles, 275, 283, 297
Allen, Garland, 270
Allmacht (total sufficiency), 233
al-Risāla al-ḥamidiyya (Tract Dedicated to Abdülhamit II, Ṭarābulusī), 501
Alter, Stephen, 422, 426
altruism, 16, 193, 204, 348, 447
    classical problem of, 203, 205, 208, 210
    communal, 350
    egotistic (Wheeler), 349
    natural selection against, 199
    paradox of, 464
    reciprocal (Trivers), 199, 210, 351, 464
    self-sacrificing, 203
Amboseli National Park (Kenya), 423
Ameghino, Florentino, 263
American Scientific Affiliation (ASA), 480
anagenesis, 399
analogy, 8, 33, 37, 83, 187
    artificial selection and, 295
    defined
        by Strickland, 77
        by Westwood, 77
    ecological convergence and, 140
    enduring (Darwin), 93–94
    evolutionary (Kuhn), 456
    of forms, 42
    heuristic function of, 457
    intention and, 77
    Lorenz and, 349
Analogy of Revealed Religion (Butler), 160
anatomy, comparative, 397
Anaxagoras, 35
Anaximander, 33
Anaximenes, 33
ancestor
    African, for humankind, 23, 176, 413, 415
    ape, 66, 185, 223, 240, 251, 413, 415
    brachiating, 418
    common, divergence from, 401
    hominin, 421

last universal common (LUCA), 343–44, 345
    Tarsier, 415
    universal (Woese), 344
Ancestral Law of Inheritance (Galton), 119
Anderson, Edgar, 271, 284, 319, 320
anemia, sickle-cell, 297
angiosperms, 270
Animal Dispersion in Relation to Social Behaviour (Wynne-Edwards), 350–51
Animal Geography (Newbigin), 365
animals
    bisexual, 103
    compound, 202
    culling defective, 90
    culture and, 429
    emotional expression in, 177–78, 183, 506
    environmental adaptions by, 222
    first generations of, 34
    heavily inbred, 91
    highest, 66
    isomorphisms between species of, 40, 42
    limited progenitors of, 476
    lower, 420
    monogamous, 105
    simple, 6
    social, and natural selection, 209, 347, 408
Annales de paleontologie (1906–50), 300–01
Annelides, 82
Answers in Genesis (AiG), 484
anteaters (echidna), 77
Anthropogeographie (Ratzel), 364
anthropology, molecular (Zuckerkandl), 417
Anthropopithecus, 413
Antiquity of Man (Lyell), 223
Anti-Revolutionary Party, 472
anti-Semitism, 241
antithesis, principle of (Darwin), 177
ants, 204, 205
    slave-making, 149
apeiron (indefinite substrate), 33
apes, 178, 223, 423–25
aphasia, 184
Appleton, Jay, 363
Aquinas, Thomas, 37
Arbib, Michael, 422
Archaea (Woese), 341

Archaebacteria (Woese and Fox), 341
Archaeopteryx, 397
archebiosis, 323
archetypes, 36
    Darwin and, 82
    flower, 133
    Geoffroy Saint Hilaire and, 43
    ideal (*Baupläne*), 43
        homology and, 86
    transmutation and, 43
    vertebrate, 83
Ardabīlī, Āyatollāh Meshkīnī-i, 503
Ardipithecus, 418
Ardrey, Robert, 419
Arechavaleta, José de, 261, 263
argument, open question (Moore), 463
Aristotle, 1, 36–37
    Darwin and, 38
Armstrong, David F., 422
Art of Improving the Breeds of Domestic Animals, The (Sebright), 90
Arthrobalanus, Mr, 81–84
Arthur's Seat, Edinburgh, 46
Articulata, 82
artifacts, 434
Association for the Study of Systematics in Relation to General Biology, 290
atavism, 118
Ateş, Süleyman, 503
atheism, 16, 211
Atomism, 33, 153
Atran, Scott, 432
Atticus, 35
Augustine of Hippo, 487
Augustine, Saint, 39, 40
Austen, Jane, 437
Australopithecus afarensis, 417, 418
Australopithecus africanus, 415–16, 417
Autobiography (Darwin, 1887), 95
Aviezer, Nathan, 496
Avital, Gabi, 498
avoidance, 11
Ayala, Francisco, 157
Āzād, Abū l-Kalām, 501

B'Or Ha'Torah (journal), 495
Babbage, Charles, 16, 48
Babcock, Ernest Brown, 271, 272, 317–19
Bacon, Francis, 153, 154, 155, 160
Bagehot, Walter, 365
Bahr, Hermann, 241
Bailey, Liberty Hyde, 229
Bakewell, Robert, 90, 93
Balfour, Francis Maitland, 508
Balfour, John Hutton, 137
bank, entangled, 102, 393, 396
Bapteste, E., 343, 345
Baranetsky, Josip, 265
Barbour, Ian, 475
Barnaby Rudge (Dickens), 439
barnacles (cirripedes), 6, 23, 53, 75, 78, 80–87, 96
    cementing apparatus in, 85,
    cross-fertilization and, 86
    Darwin and, 288
    defined (T. H. Huxley), 85
Barnum, P. T., 449
Barreda, Gabino, 262

Barreto, Tobias, 259
Barth, Karl, 469, 472–73
Basalla, George, 434
Basic Principles of Darwinism, 256
Bastian, Henty Charlton, 323
Bates, Henry Walter, 12, 112, 135, 138, 139–40, 159, 167, 168, 259, 274, 331, 337
Bateson, William, 20, 114, 144, 225, 230, 255, 269, 270, 273–74, 315
bathybius, 261
Baupläne (ideal archetypes), 43
Baur, Erwin, 316
Baur, George, 370–71,
Bavinck, Herman, 470, 471
Bayrakadar, Mehmed, 503
bears, black, 160
Beaton, Donald, 93
Becker, Lydia, 446
Beckner, Morton, 156
Beer, Gillian, 159, 436, 437, 439
bees, 204–05
    bumble, 389
    Melipona domestica, 205
beetles, 72, 167
    ladybird, 256
    myrmecophile, 240
Befruchtung der Blumen durch Insekten (Müller), 138
Beginnings of a New School of Metaphysics (Smart), 184
behavior
    analogies between human and animal, 191
    as adaptation, 464
    combinatorial, 424
    ethical, and biology, 15
    irrational, 432
    social, 346, 348, 349–51
    tradition in, 429
"Behavior and Natural Selection" (Tinbergen), 350
Beijerinck, Martinus, 269
Being, Great Chain of, 41
    God and, 81
Bellamy, Edward, 447
Belloc, Hilaire, 486
Belt, Thomas, 143
Benamozegh, Rabbi Eliyahu, 494
Benedict XVI, Pope, 491
Bentham, George, 135, 137
Bentham, Jeremy, 190, 191
Berg, Lev Semyonovich, 365
Bergmann's rule, 286
Bergson, Henri, 489
Berlinski, David, 241,
Bernard, Claude, 19, 243, 247–48,
Bernard, Félix, 305
Berry, William Basset, 367
Bessey, Charles E., 270
Bhabha, Homi K., 367
Bible, 468, 487, 488
Bickerton, D., 422
Bigelow, Maurice, 229
Bildungstrieb (generative force), 506
Bing Zhi, 250, 255
biochemistry, 324, 397
biodiversity, 393
Bioenergetics: The Molecular Basis of Biological Energy Transformations (Lehninger), 152

biogenetics, 239, 240, 508
    law of (Haeckel), 19, 100, 273
biogeography, 24, 288, 392, 397
"Biographical Sketch of an Infant, A" (Darwin, 1877), 179
Biologus Forum, 475
biology, 15
    Aristotle and, 37
    conservation, 393
    Darwinism beyond, 231–32,
    developmental, 375, 380
    evolutionary, 20, 383
    four-dimensional (Simpson), 356
    molecular, 22, 26, 397–404, 417
        advantages of, 398
        central dogma of (Crick), 410
        multiplicity and, 398
        precision and, 398
        proving evolution, 397
        universality and, 398
    Nazi, 241–42
    plant evolutionary, 314
    reproductive, of flowering plants, 137
    synthetic, 381
    systems, 381
Biology and Creation (Kenyon and Davis), 482
biometrics, 275
Biometry and Biological Statistics (monograph series), 308
Biosystematics, 317, 319
bipedalism, 417–19
Birch, Charles, 393
Black, Davidson, 416
Blackmore, Susan, 432
Blackwell, Antoinette, 447
Bleak House (Dickins), 27
Blombos, 426
blood type, 433
Bloom, P., 422
Blumenbach, Johann Friedrich, 236, 241
Boas, Franz, 180
bobolinks, *Dolichonyx oryzivorus*, 370
Bogin, B., 422
Boitard, Pierre, 120
Bole, S. James, 480
Bonner, John, 380
Bonnier, Gaston, 248
Bopp, Franz, 186
Borel, Emile, 311
Borrello, Mark, 203
Botanic Garden, The (Erasmus Darwin), 46, 47, 48
Botanical Society of America, 266
botany, 11, 20, 22
    Darwin and, 47, 264–72, 313–14
    evolution and, 131–38
    genetics and, 267–70
Botany (Hooker), 252
Botha, R.P., 426
Botocudos Indians, 259
Boucher, Y., 343
Boulle, Marcellin, 300
Boveri, Theodor, 379
Bower, F. O., 265
Bowler, Peter, 348, 363, 472
Bowman, Robert, 373, 374
Boyer, Pascal, 432
Boyle, Robert, 153,

breeding
    in America, 18
    cross-, 67, 92, 118
        heteromorphic, 156
        homomorphic, 156
        inbreeding and selection, 94
        of varieties, 90, 91, 93, 94
    Darwin's perception of, 92
    domestic, 91
    eugenics and, 31, 179
    in-, 67, 90, 91, 94
        Darwin and, 91, 92
        defined, 93
        recombination and, 94
        selection and, 93
    in-and-in, 120
    inter-, 69, 91, 115, 206, 275, 287, 288, 406
    intercrossing, 154
    kinds of, 10
    natural selection and, 88, 94
    out, 120
    purity, 93
    selective, 5, 8, 64, 69, 71
Brefeld, Julius, 265
Breitenbach, Wilhelm, 260
Brewster, David, 16
British Association for the Advancement of Science, 266
Britten, Roy, 380
Britten-Davidson model, 380
Broad, C. D., 463
Broca, Paul, 246
Brongniart, Adolphe Théodore, 267
Bronn, Heinrich Georg, 160, 161, 237, 244
Brooks, William Keith, 229
Broom, Robert, 416, 417
Brown, Arthur I., 373, 480
Brown, Rabbi David, 496
Brown, Robert, 134
Brown, William, 372
Browne, Janet, 48, 137, 448
Brown-Séquard, Charles-Édouard, 121
Bryan, William Jennings, 29, 233, 477
Büchner, Ludwig, 500
Buckland, William, 50, 124, 126
Buckle, Henry T., 117
Buffon, Georges Louis Leclerc, Comte de, 109, 506
Bulletin de la Société géologique de France (1901–50), 301–03
bumblebees, 131
Bumpus, H. C., 332
Burkhardt, Richard J., 349,
Burling, R., 422
Burnet, Frank Macfarlane, 509
Burnett, James (Lord Monboddo), 183
Burns, Robert, 250
Burton, Richard, 14
Buss, David, 108
Butler, Joseph, 160
Butler, Samuel, 27, 218, 222, 224, 440
butterflies, 23, 142, 283
    African swallowtail (*Papilio dardanus*), 338
    Amazon, 135
    brown meadow (*Maniola Jurtina*), 295
    Dismorphia, 140
    Heliconiidae, 337
    Heliconius, 112

Ithomia, 140
Leptalis, 140
mimetic, 339
Pieridae, 338
swallowtail, 338
    Papilionidae, 339
tropical, 337
Byatt, A. S., 442

Cain, Arthur J., 22, 293, 294, 298, 332
Calcagno, Francisco, 261
Call of the Wild, The (London), 441
Cambridge, Octavius Pickard, 217
camouflage, 23, 139–45, 336–39
    defined, 139
    described, 336
    God and, 139
    term, 141
Campbell, Donald, 431, 459–60
campion, white (*Silene alba*), 269
Candolle, Alphonse de, 137, 155, 264
Candolle, Augustin-Pyramus de, 132–33, 136, 264
Cao Juren, 253
Cape Verde Islands, 52, 100
capitalism, laissez-faire, and Spencer, 15
Carlyle, Thomas, 39
Carnegie, Andrew, 197
Carpenter, Frank, 352
Carus, Carl Gustav, 244
Castle, William E., 144, 277
Castro, Cipriano, 262, 263
caterpillars, 336
Catholic Church, 263
Catholicism
    Darwin and, 485–92
    evolution and, 487–88
cattle
    Alderney, 90
    Shorthorn, 90, 91
causation, 53, 70, 150, 290, 348, 356, 379
    backward, 153
    bipedalism and, 419
    efficient (*causa efficiens*, Aristotle), 36
    final (*causa finalis*, Aristotle), 1, 12, 36, 37, 39
        Bacon and, 153
        Cuvier and, 38, 42
        Darwin and, 154, 156
        natural selection and, 38
    formative (*causa formalis*, Aristotle), 36
    geographical, in universal history (Mackinder), 365
    God and, 149
        as first mover, 36, 40
    material (*causa materialis*, Aristotle), 36
    teleological, Aristotle and, 36
Celera Genomics, 403
cell
    eukaryotic, 343
    formation
        by division of existing cells, 122
        free (Schwann), 122
    germ, 508
    versus somatic (Weismann), 349
    proto-, 326
Centaurea, 272

Cevdet, Abdullah, 500
Chadarevian, Soraya de, 264
Chamberlain, Houston Stewart, 242
Chambers, Robert, 8, 23, 27, 44, 140, 220, 437
chance
    craft analogy for, 155
    defined (Gray), 155
    design and, 146–51
    God and, 149, 150–51
    two notions of (Darwin), 155
Chaney, Ralph Works, 321
change
    adaptive, 71
        progressive and, 67
        selection and, 10
    chromosomal
        processes of, 288
    continued, 96
    defined by Lamarck, 41
    domestication by selection and, 95
    mechanisms of (Darwin), 8
    organic, and Erasmus Darwin, 40
    species, and tendency, 163
    upwards from the inorganic, 44
Chapters on Language (Farrars), 183
Chapters on Man (Wake), 186
character, 75–77
    adaptive
        versus essential, 76
        resemblance and, 78
    essential (Owen), 76
    important, 78
    prepotent, 118
characteristics
    acquired, 41, 198
        inheritance of, 41, 98
        pangenesis and, 122
        reversions and, 116–21
    essential (Westwood), 76
    Lamarckian, 431
    sexual, primary and secondary, 105
Chesterton, G. K., 486
Chetverikov, Sergei, 114
Chimpanzee Sequencing and Analysis Consortium, 403
chimpanzees, 429
chloroplasts, 343
Chomsky, Noam, 27, 422–23,
Chonos Archipelago, 81
Christian View of Science and Scripture, The (Ramm), 473, 480
Christianity
    Catholicism, 30
    Protestant, 28
chromatography, 270
chromosomes, 315
    as material carriers of heredity, 316
Chromosomes, The: Their Numbers and General Importance (Winge), 316
Chung, Paul, 475
Church Dogmatics (Barth), 473
Church of England, 212
Cicarrelli, Francesca, D., 344
cladogenesis, 399
Clark, W. E. Le Gros, 416
Clarke, Cyril A., 293, 294, 296, 424

Index

classification, 5
   embryology and, 83
   genealogical, 187
Clausen, Jean, 317
Clayton, Philip, 475
Cleland, Ralph Erskine, 315
Clements, Frederic, 316, 366, 387
Clift, William, 58, 61
clock, molecular, 23, 341, 401, 407, 417
   rheoretical basis for, 401–03
clover, common red (*Trifolium pratense*), 389
Cobb, John B., 475
Cobbe, Frances Power, 192, 212
Cody, Martin, 392
coenobiosis (xenobiosis), 349
coevolution, gene-culture, 433
cognates, 434
cognitivism, 465
colchicine, 288
Cole, Leon, 229
Colenso, John William, 217
Coleoptera, 77
Coleridge, Samuel Taylor, 437
Collection of National News or *The Light Seeker*
   (journal), 252
Collins, Francis, 475
coloration
   aposematic (warning), 141
   Darwin and, 141
   disruptive, 143
   organismal, and pattern, 336
   selection and, 145
Colours of Animals, The (Poulton), 142
Combe, George, 166
community, good of the (Darwin), 203, 346–48
Compendium of Ancient History (Sierra), 262
competition, 104, 179, 193, 199, 209, 253, 295, 371,
   376, 407, 438, 503
   balance of nature and, 387–90
   community structure and, 387–90
   free-market, 197
   human, 471
   interspecific, 386, 388,
   Lack and, 372
   local mate, 24, 105, 334
   Social Darwinism and, 195–97
Complete Works of Darwin Online, The, xii
composition, unity of (Geoffrey Saint
   Hillaire), 506
Comte, Auguste, 247, 258
Concerning Evolution (Thomson), 348
Conchifera, 82
Condillac, Etienne Bonnot de, 183
Congo (mountain gorilla), 449
conjugation, 343
Conklin, Edward Grant, 229
Conry, Yvette, 248
conscience, Darwin and, 461
consilience, 99–102,
Constitution of Man, The (Combe), 166
constraints, developmental, 380
constructionism, niche, 410
Conte, Joseph Le, 228
"Contribution à l'étude de l'évolution des
   Nummulites" (Abrard), 302
contrivance (Darwin), 12, 155

convergence, 140, 191, 415
   mimicry and, 140
Conybeare, William, 50
cooperation, 199–201
Cope, Edward Drinker, 231, 356
Coral Islands, 51–52, 54
Corbié, 120
corn (*Zea mays*), 269, 271
correlation, 161
Correns, Carl, 230, 268, 269, 272, 314, 379
Cosmides, Leda, 464, 510
cosmology
   biology of, 35–36
   evolutionary, 481
cosmos
   defined, 33
   origin of, 32
Cott, Hugh, 143
Covington, Syms, 57
cowbirds, 429
crabs, 275
craft, and theological causation, 36
creation
   continuous, 473
   Darwin's use of, 216
   divine, 331
   ex nihilo, 502
   special, 470
Creation Museum, 484
Creation Research Society (CRS), 481, 498
creationism, 29, 476–84
   creation and, 477–80
   defined, 470
   defined by early Christian theologians, 477
   hyper-, 362
   Islam and, 499–500
   Islamic arguments against, 502–03
   Judaism and, 497–98
   organized, 480
   revival of, 480–82
   rudimentary organs and, 101
   scientific, 481
   Young Earth (YEC), 29, 220
Creed, E. R., 295
Crichton-Browne, Dr. James, 506
Crick, Francis, 299, 409
Crisp, D. J., 86
Critique of Judgment (Kant), 5, 38, 153, 235
Crito (Plato), 34
Croll, James, 126, 128, 129
crows, New Caledonian, 429
Crustacea, 81
crypsis, 336, 339
Cryptophialus minutus, 82,
Ctenomys, 62
cuckoos, 149
Cuénot, Lucien, 248
culture
   evolution of, 27, 428–35
   forms of selection in, 430–32
   gender and, 28
   genetics and, 432–33
   learning and, 429
      processes for, 430
   selection-produced traits and, 27
   transmitting, 429

Cuvier, Georges, 8, 12, 38, 43, 57, 65, 76, 82, 98, 153,
   157, 227, 241, 330, 331
   final causes and, 1
cyanobacteria, 343
Cynips, 288
Cypris, 82
cytoblastemma (Schwann), 122
cytology, 266

Dagan, T., 344
Dale, Langham, 367
Dana, James Dwight, 228
D'Ancona, Umberto, 308
Dante, 37
Darlington, C. D., 271, 315
Darmois, Georges, 311
Dart, Raymond, 415–16, 419
Darwin and Modern Science (ed. Seward), 267
Darwin and the Darwinian Revolution
   (Himmelfarb), 180
"Darwin and the Double Standard" (Morris), 450
*Darwin and the Emergence of Evolutionary Theories
   of Mind and Behavior* (Richards), 181, 348
"Darwin Correspondence Project," xiii
*Darwin on Man: A Psychological Study of Scientific
   Creativity* (Gruber), 181
Darwin on Trial (Johnson), 482
Darwin, Annie (daughter), 214
Darwin, Caroline (sister), 61, 212, 214
Darwin, Charles (paternal uncle), 47
Darwin, Charles Robert
   biographical sketch of, xi–xii, 46, 188–89
   familial advantages of, 3
   influences on, 43
   personal beliefs of, 212–15
   religion and, 212–15
Darwin, Emma Wedgwood (first cousin/wife), xi,
   211, 214
Darwin, Erasmus (paternal grandfather), 4, 20, 40,
   120, 173, 178, 188, 212, 506
   influence of, on Charles Darwin, 46
Darwin, Erasmus Alvey (*Eras*, brother), 3, 16, 47,
   188, 212
Darwin, Francis (son), 91, 134, 137, 138, 160, 211, 265
Darwin, George (son), 11, 112, 206–08, 212
Darwin, Leonard (son), 277
Darwin, Robert Waring (father), xi, 3, 46, 120, 174, 212
Darwin, Susannah Wedgwood (mother), xi, 47,
   188, 212
Darwin: A Reader's Guide (Ghiselin), xiii
Darwinian Theory and the Science of Language
   (Schleicher), 238
Darwinism
   in America, 18
   American anti-, 232–34
   in Argentina, 262–63
   as atheism (Hodge), 29
   in Brazil, 259–61
   challenging, 405–11
   in China, 19, 250–57
   in Cuba, 261
   in France, 22, 300–12
      before 172, 243–49
   genetic, 406
   in Germany, 19, 235–42
   in Great Britain, 17, 22, 218–25

green, 392–95, 396
    in the nineteenth century, 392–95
    in the twentieth century, 395–96
  Huxley and, 218, 222
  in Latin America, 258–63
  Marx and, 19
  materialism and, 222, 233
  as mathematically-formulated progress, 247
  in Mexico, 262
  neo-, 224
  positivist, in Latin America, 258–63
  progress and, 232
  red, 392
  religion and, 18
  in Russia, 19
  Social, 188, 195–201, 231–32, 348, 441
    competition and, 195–97
    creative, 256
    Darwin and, 15
    defined, 179
    eugenics and, 233
    Fisher and, 246
    Gautier and, 247
    in industrialized societies, 197–99
    Judaism and, 497
    Marxism and, 256
    pragmatism and, 232
    Progressivism and, 232
    uses of, after 177, 195
  in South America, 20
  in the United States of America, 226–34
  in Uruguay, 261–62
  in Venezuela, 262
Darwinism (Wallace), 171
Darwinism To-Day (Kellogg), 229, 233
darwinisme social, Le (Gautier), 247
Darwin's Black Box: The Biochemical Challenge to Evolution (Behe), 483
Darwin's Finches (Lack), 371
Darwin's Sacred Cause, Race, Slavery and the Quest for Human Origins (Desmond and Moore), 181
Daubeny, Charles, 136
Daubin, V., 344
Davidson, Eric, 380, 381
Davidson, I., 422, 425,
Davis, Percival, 482
Davis, William Morris, 364,
Dawkins, Richard, 27, 99, 392, 407, 431, 432, 482, 485
"Dawn of Darwinian Medicine, The" (Nesse and Williams), 505
Dawson, Charles, 413
De Genesi ad Litteram (Augustine of Hippo), 487
De Gobineau, Arthur, 259
De Partibus Animalium (Aristotle), 38
de Vries, Hugo, 230, 255, 265, 269, 270–71, 272, 314,
Deacon, T., 422
Dechaseaux, Colette, 301, 303
deism
  Darwin and, 12, 470
  defined, 1, 16
  transmutation and, 40
Deleuze, Gilles, 367
Deluge Geology Society, 480
Dembski, William, 483, 498,
Demiurge, 35, 36

Democritus, 33
Dennert, Eberhard, 240
Dennett, Daniel, 432, 485
Denny, Henry, 173
Department of Energy (DOE), 403
Depew, David, 270
derivation, Owen and, 222
Derrida, Jacques, 367
Descartes, René, 27, 37, 39, 153
descent
  across species, 110
  common-ancestor, 37, 65, 174, 187, 340, 398, 405, 434
  evil passions and, 214
  with modification (Darwin), 82, 86
Descent of Man and Selection in Relation to Sex, The (Darwin, 1871), xi, 10, 14, 16, 103, 104, 110, 137, *173*, 175–77, 208, 216, 420, 452, 508
  central thesis of, 506
  moral sense and, 453
  primary goal of, 412
  second edition, 217
design
  adaptive, 429–30
  intelligent, 30, 33, 35, 241, 482–84
    defined, 482
    Judaism and, 497–98
Desmond, Adrian, 181
determinism
  biological, 509–10
  Spinoza and, 153
Deutsch, J., 86
Deutscher, Penelope, 447
Developmental History of Animals (von Baer), 237
deviation, 117
Dewey, John, 232, 451
Dialogues Concerning Natural Religion (Hume), 38, 214
Diamond, Jared, 332, 392
diathesis (hereditable predisposition), 506
Dickens, Charles, xii, 15, 27, 437–40
Diderot, Denis, 40
Different Forms of Flowers on Plants of the Same Species, The (Darwin, 1877), 137
differentiation, 12
  morphogenesis and, 379
diffusion, 367–367
dimorphism, 135, 154, 156, 260
  mimetic, 339
  sexual, 23, 105, 338
    inheritance and, 106
Discovery Institute, 241
disease, 507
displacement, character, 373
Disquisition about the Final Causes of Natural Things (Boyle), 153
Dissertation on the Progress of Ethical Philosophy, Chiefly during the Seventeenth and Eighteenth Centuries (MacKintosh), 191
distribution, geographical, 100
  Agassiz and, 361
  Darwin and, 362–63
  Hooker and, 363
divergence, 77–78, 367–367
  Darwin and, 10, 75, 78, 98, 102
  defined by Darwin, 208

division of labor analogy with, 98
  Mayr and, 285
  principle of, 169
  racial, 176
  taxonomic, 169
  taxonomy and, 6
diversity, 70, 98, 394, 406
  adaptive, 67
  advantage of, 208, 387
  Darwin's misplaced faith in, 396
  Darwin's threefold, 69
  generating petrological, 54
  racial, 185
Diversity of Life, The (Wilson), 396
Divine Milieu, The (Teilhard de Chardin), 491
Divino Afflante Spiritu (Pius XII), 488
DNA, 22, 299, 397, 398, 407, 417, 434
  point permutations in, 409
Dobzhansky, Theodosius, 108, 111, 157, 255, 279, 284, 290, 307, 317, 319–21, 332, 494
  Darwinian selection and, 21
  modern evolutionary synthesis and, 356, 417
  reinforcement (Wallace effect) and, 114
  static to dynamic evolution and, 289
Dohrn, Anton, 238
Dollo, Louis, 356
dolphins, 76
dominance, de Vries and, 269
Don Pedro, Emperor, 261
Doolittle, W. Ford, 343–44, 345
d'Orbigny, Alcide, 60
Dos partidos en lucha (Two Parties in Battle, Holmberg), 263
Dowdeswell, W. H., 295
Dreiser, Theodore, 232
Driesch, Hans, 238, 239, 349
drift, genetic, 22, 24, 295, 296, 333, 356
  defined, 298
  Wright and, 279, 283, 371
Drosophila, 284, 288
Drummond, Henry, 470
Dubois, Eugene, 413
dugongs, 75
Dunbar, R. I. M., 422
Dunn, L. C., 320
Durkheim, Emil, 248
Düsing, Carl, 331

earth
  age of, 124–26
  continental, 53
  nebular, 68
  oceanic, 54
earthquakes
  at Concepcion (Chile), 50
  volcanoes and, 51
earthworms, 52, 410
Ecological Genetics (Ford), 294
ecology, 266, 288, 397
  behavioral, 335
  community, 393
  Darwinian, 392
  defined, 25
  by Haeckel, 385
  ecosystem, 393
  evolutionary, 383–90, 392

ecology (*cont.*)
    Haeckel and, 25
    invasion, 392, 396
    term, 383
Ecology and Evolution of Communities (Diamond and Cody), 392
ecosystems, 393
Edkins, Joseph, 252
Effects of Cross and Self-Fertilisation in the Vegetable Kingdom, The (Darwin, 1876), 137
Ehrlich, P., 430
eidê (immutable forms)
    evolution and, 35
    Plato and, 35
Ekman, Paul, 181
Elderton, E. M., 332
Eldredge, Niles, 24, 357, 358, 408–09
Eleaticism, 35
Elements of Geology, The (Lyell), 251
elephants, 76
Eliot, George, 440
Elton, Charles, 396
embedding, constitutional (Yarrell), 70
embryology, 25, 80, 378, 397, 413
    homology and, 83
embryos, 100
Emerson, Alfred, 290
Empedocles, 33–34, 37
    Darwin and, 34
empiricism
    knowledge acquisition and, 453
    morality and, 15
En busca del eslabón (Calcagno), 261
Endler, J. A., 332
endosymbiosis, 343
Enduring Love (McEwan), 442
Engels, Friedrich, 19, 251
Enlightenment, 233
Enquiry concerning Human Understanding (Hume), 189
Entstehung des Menschengeschlechts (Haeckel), 184
Entwickelungsgeschichte der Thiere, Der (von Baer), 83
Entwicklung (generation), 378
environment, 231, 259, 278, 316
    mutation and, 311
environment, and Darwin, 163, 391–96
enzyme, 431
Eoanthropus dawsonii (Piltdown Man), 414
Ephrussi, Boris, 307
epistemology, 28,
    reformed, 473
Epistle to the Romans, The (Barth), 472,
Epling, Carl, 319
Epperson vs. Arkansas, 498
equality, 180
equilibria
    developmental physiological (Kühn), 379
    punctuated (Gould), 24, 353, 357–59, 408
equilibrium
    dynamic, 24, 45
    physical, 52
Equus, 289
Erdoğan, Tayyip, 502
Ernst, Adolfo, 262, 263
"Essay" (Darwin, 1844), 5, 70, 71, 83, 91, 103
    delayed publication of, 8

Essay concerning Human Understanding (Locke), 184
Essay on Classification (Agassiz), 8
Essay on the Principle of Population (Malthus), 5, 166
Essays and Reviews (Powell), 29
ethics and morality, 15, 28
    belief and, 216
    Darwin and, 188–94, 195–201
    Descent and, 192–93
    development of Darwin's thoughts on, 189–92
    evolutionary, 28, 461
        empirical, 463
        philosophical, 462
    intelligence and, 208–10
    limits to Darwin's approach to, 193–94
    post-Darwin, 461–67
    religion and, 15
ethnology, 349–51
ethology, 288, 397
    defined, 349
etiology, consequent (Wright), 156
eugenics, 31, 224, 508–09, 511
    Darwinism and, 233
    defined, 179
    Fisher and, 279
    as a God-given duty (Fisher), 280
eukaryotes (complex-celled organisms), 23, 344
Evans, Richard, 509
Evidences as to Man's Place in Nature (Huxley), 175
Evidences of Christianity, The (Paley), 189
Evo-devo, 25, 375–78
evolution, 288
    a priori and, 453
    adaptation and, 222
    adaptative, by natural selection, 410
    adaptionist, 408
    Ancient Greeks and, 1
    as a statistical problem (Weldon), 275
    archetypes and, 18
    as branching tree, 221
    book of Genesis and, 496
    components of (Darwin), 376
    convergent (Jones), 415
    cultural, 428–35
        defined, 428
        Lamarckian, 431
    Darwin and, 4–5, 80, 173–81
    before Darwin, 235–37
    Darwinian and non-Darwinian, 406–07
    Darwinian, as God's evolution (Fisher), 279
    defined, by Wright, 280
    developmental, 163, 375–82, 407
        focus of, 380
        developmental physiological equilibria and, 380
    Enlightenment and, 1
    epistemology of, 451–60
        formal, 459–60
    Erasmus Darwin and, 46
    fossils and, 74
    goal-directed (Broom), 416
    Haeckel and, 413
    human, 26
        human-animal continuity and kinship, 175
        transmutation and, 174
    individual organism, 406

industry and, 18
Islam and, 499–504
Islamic concepts of, 503–04
Judaism and, 493–98
macro-, 356, 409
materialist interpretations of, 489
mega-, 357
micro-, 356
molecular, 417
morphology and, 18
natural selection and, 8, 412
neutral (non-Darwinian) theory of, 26
non-Darwinian, 407
opposition to, 42
origin of life and, 322–29
pace of, 11, 124–30, 275, 293, 374
parallelism in, 224, 416
physiology of (Dobzhansky), 290
post-Darwin, 412–19
as a pre-Darwinian pseudoscience, 12, 20
progress and, 12, 45, 196
quantum (Simpson), 357, 408
religious objections to, 240–41
retrogressive (Huxley), 200
as the selection of small variations in large populations (Fisher), 277
shifting balance theory of (Wright), 279, 280, 281, 283
social, 348
socialism and, 240
subverting religion and science, 4
synthesis period in, 282–92
synthetic experimental, 375, 378–81
testing, 330–35
through natural selection, 163
tiers of (Simpson), 356
transformational, 355
use of the term, in the *Origin*, 163
Evolution (journal), 21, 290, 332
Evolution and Animal Life (Jordan and Kellogg), 349
"Evolution and Ethics" (Huxley), 18, 252
Evolution and the Genetics of Natural Populations (Wright), 307
Evolution in Changing Environment (Levin), 392
Evolution of Living Organisms, The (Goodrich), 275
Evolution of Morbid Germs, The (Millican), 507
Evolution of Woman, The: An Inquiry into the Dogma of her Inferiority to Man (Gamble), 447
Evolution, the Modern Synthesis (J. Huxley), 349
ex nihilo nihil fit, 33
exclusion, competitive (Gause), 386
Excursion, The (Wordsworth), 437
existence
    conditions of, 98, 153
    social, properties of (Wilson), 352
    struggle for, 15, 96, 102, 168, 187, 252, 346
        Darwin and, 203
        Kropotkin and, 200
        Malthus and, 384–85
        variation and, 97
        Yan Fu and, 252
Expelled, No Intelligence Allowed (Stein), 241, 497

experimentation, versus accumulation, 286
explosion, Cambrian, 502
Expression of Emotions in Man and Animals, The (Darwin, 1872), 137, 173, 183, 216, 251, 420, 439, 506
expression, evolutionary history of (Darwin), 177
extinction, 6, 41, 41, 54, 65, 67, 69, 70, 74, 102, 128, 169, 171, 187, 199, 208, 392
    culture of, 262
    Lamarckian lack of explanation for, 406
    transmutation and, 74–75

Falk, D., 422
Fall of Man, The, or the Lives of the Gorillas, A popular Scientific Lecture upon the Darwinian Theory of Development by Sexual Selection, by a Learned Gorilla, 448
fallacy, naturalistic (Moore), 28, 463
Farn, Albert Brydges, 142
Farrar, Frederic W., 183, 217
fauna, harmonic and disharmonic, 370
Ferguson, George, 93
fertilization, 11, 120
    heteromorphic, 135
    homomorphic, 135
Ffinden, George, 217
Fick, Heinrich, 15
Fields, William, 421
finches, 24, 368–74, 385
    Cactornis, 370
    Cactospiza heliobates, 374
    Camarhynchus, 370, 372
    C. pauper, 374
    Certhidea, 370
    C. fusca, 374
    differation by diet, 373
    Fringillidae, 371
    Geospiza, 369, 370, 371, 372
    G. conirostris, 372, 374
    G. difficilis, 384
    G. fortis, 372–74,
    G. fuliginosa, 372, 374
    G. magnirostris, 372
    G. scandens, 372
    Geospizidae, 371
    Platyspiza, 371
fireflies, 338
Fischer, F. E. I., 267
fish, stickleback, 334
Fisher, Joseph, 247
Fisher, Ronald A., 21, 22, 108, 144, 271, 277–81, 283, 294
Fitch, Tecumseh, 402, 422, 425
Fitch, W. M., 422
fitness, inclusive (Hamilton), 203
fittest, survival of the
    Darwin and, xii
    eugenics and, 509
    Huxley and, 200
    Spencer and, 160, 196, 220
    Wallace and, 198
FitzRoy, Captain Robert, 49, 56, 189, 436
FLB, 423
Fleure, Herbert John, 366
Fol, Hermann, 238
Forber, P., 333

force, generative (*Bildungstrieb*), 506
Ford, E. B. (Henry), 145, 279, 283, 293, 294, 295, 296
    Fisher and, 22
Fordyce, John, 215
forms
    analogy of (Kant), 42
    diversity of, 6
    evolution of, 17, 41,
    immutable, 35
    primordial, 476
    progress of, 1
    transitional, 75, 98
    versus functions, 98
Fortas, Justice Abe, 498
Fossil Cirripedia (Darwin 1851, 1855), 80
Fossil Flora of Great Britain (Lindley and Hutton), 266
Fossil Mammalia (Owen), 62
fossils, 4, 18, 40, 44, 353–60, 412–15, 478
    archaeopteryx, 159
    and Darwin, 355
    evolution and, 74
    horse, 51
    intermediate, 397
    marine, 50
    South American, 51
    transitional forms and, 98, 99
    vertebrate, 56
Foster, Michael, 508
Foucault, Michel, 367
Foundations of Comparative Anatomy (Gegenbaur), 238
Foundations of Darwinism, 256
Foundations of the Nineteenth Century (Chamberlain), 242
Fowles, John, 442
Fox, George, 341
Fox, William Darwin (cousin), 212, 217
Franklin, Benjamin, 5, 51
French Lieutenant's Woman, The (Fowles), 442
Freud, Sigmund, 178
Friedlander, Michael, 494
fritillaries, marsh (*Melitaea aurinia*), 294
From Darwin to Hitler (Weikart), 241
fruitflies, 276
    Dobzhansky and, 21
    Drosophila, 22, 255
    D. melanogaster, 308, 317, 332
    D. pseudoobscura, 317, 319
Fryer, John, 251
function, 80
Fundamentals, The (Dennert), 240
Für Darwin (Müller), 259
Fürbringer, Max, 239

Galapagos Archipelago, 5, 24, 51, 100, 368, 370, 371, 384
Galileo, 488, 489
Galtier, N., 344
Galton, Francis, 116, 119, 175, 198, 224, 230, 315, 331, 332, 508
Gamble, Eliza Burt, 443, 447–48, 450
García González, A., 261
Gardeners' Chronicle, 131, 132, 136
Gärtner, Carl von, 118, 120
Gärtner, Joseph, 119

Gasman, Daniel, 241,
Gaudium et spes (Second Vatican Council), 487, 491
Gaudry, Albert, 243, 248, 300
Gause, Georgii F., 308, 311, 371, 386
Gause's Law, 386
Gautier, Emile, 247
Gay Science, The (Nietzsche), 454
Geddes, Patrick, 331, 366
Gegenbaur, Karl, 238, 239, 508
Geikie, Sir Archibald, 53, 126
Geikie, James, 54, 128, 129
gemmules (Darwin), 122
    defined, 122
gender
    Darwin and, 443–50
    racism and, 448
Gene Regulation for Higher Cells: A Theory (Davidson and Britten), 380
gene theory, classical (Morgan), 276
genecology, 316
genera, 74
General Morphology of Organisms (Haeckel), 239, 364
generation, 378–81
    with modification, 168
    sexual and asexual, 67, 119–20
    spontaneous, 66, 67, 322, 323
Genesis Flood (Whitcomb and Morris), 29, 481
Genesis of Species (Mivert), 222
Genetical Theory of Natural Selection, The (Fisher), 108, 144, 271, 277, 283, 307
genetics, 356, 376, 397
    culture and, 432–33
    defined, 11
    ecological, 293–99
        Darwin and, 293–94
        defined (Ford), 293
        Oxford school of, 294
    Mendelian, 314–15
    molecular, 409
    population, 273–81, 283–84, 376, 379
        mathematical models of, 283, 294
    synthesis of, 315–21
Genetics and Evolution: Analysis of Some Mathematical Studies on Natural Selection (L'Héritier), 308
Genetics and Plant Taxonomy (Babcock), 271
Genetics and the Origin of Species (Dobzhansky), 284, 290, 319, 356
genome
    human, 403–04
    and chimpanzee, 403–04
    of Eden (Doolittle), 343
genotype, thrifty, 509, 512
Genus Crepis, The (Babcock), 317
Geoffroy Saint Hilaire, Etienne, 43, 82, 157, 376, 506
Geographical Distribution of Animals, The (Wallace), 171
Geographical Distribution of the Terrestrial Malacological Fauna of the Island of Cuba (Torres), 261
geography, and Darwin, 361–67
Geological Evidences of the Antiquity of Man (Lyell), 187

Geological Recapitulation (Erasmus Darwin), 46
geology, 3, 4
    catastrophists and uniformitarians in, 4
    Darwin and, 46
    flood, 478
    heroic age of, 54
    uniformitarian, 479
Geometer (*Gnophos obscurata*), 142
geometry, 35
geomorphology, 52
George, Henry, 198
Geospiza fuliginosa, 372
germ-plasm, 239
Gershenson, S., 332
Ghalambor, C. K., 332
Ghiselin, Michael T., xiii, 82
Giard, Alfred, 248, 261
gibbons, white-handed, 424
Gilkey, Langdon, 473, 475
Gillespie, J., 332, 333
Gilman, Charlotte Perkins, 443, 447, 450
Gish, Duane, 501
glaciation, continental, 127
Glen Roy, parallel roads of, 52
Glick, T. F., 331
Glossotherium, 62, 62
Gluckman, Sir Peter, 512
God
    Darwin and, 211–17
    materialism and, 485
    mechanical theory of the universe and, 470
    meme, 432
    modification and, 112
God's Image in Man (Orr), 470
Godlewski, Emil, 265
Goebel, Karl, 265
Goeldi, Emilio, 260
Goethe, Johann Wolfgang von, 43, 235, 376
Goette, Alexander, 239, 240
Goldschmidt, Richard, 359, 379
Goodman, Morris, 417,
Goodrich, Edwin S., 275, 277
Göppert, Ernst, 239
Gordon, C., 332,
Gospel of Wealth, The (Carnegie), 197
Gosse, P. H., 137
Goudge, Thomas, 156
Gould, John, 65, 74, 368, 370, 374
Gould, Stephen Jay, 24, 30, 157, 203, 208, 305, 353, 357, 359–60, 408–09, 483, 510
gradualism, 142, 144, 271, 409
    phyletic (Eldridge and Gould), 357
Graham, Billy, 480
Grammar, Universal (Chomsky), 422
Grant, B. S., 296
Grant, Peter and Rosemary, 25, 332, 373–74
Grant, Robert Edmond, 4, 43, 48, 65, 202, 212, 376
    Darwin and, 65, 81
Grassé, Pierre, 503
Gray, Asa, 12, 29, 136, 146, 148, 151, 154, 157, 159, 187, 227, 244, 313, 469, 472
    Darwin and, 8, 18, 154, 169, 215
    design and, 155
Gray, John Edward, 82
Gray, R. D., 434
Great Expectations (Dickens), 437

Green, J. Reynolds, 265
Greenough, George Bellas, 50
Greg, William, 175
Griffin, David Ray, 475
Grinnell, Joseph, 387
growth, correlated, 187
Gruber, Howard E., 181
Guattari, Felix, 367
Guide for the Perplexed (Maimonides), 494
Guillaume, L., 302
Gülen, Fethullah, 501

Haacke, Wilhelm, 230
Habel, Simon, 370
habit
    acquired, and emotional expression, 178
    principle of servicable associated (Darwin), 177
Haeckel, Ernst, 20, 23, 235, 238, 240, 243, 245, 259, 323, 331, 364, 407, 485, 500, 508
    bathybius and, 261
    biogenetics and, 19, 159, 239, 273
    ecology defined by, 385
    embryology and, 413
    human ancestry and, 223
    Nazi ideology and, 241
    professional science and, 20
    racial groups and, 184
    recapitulation and, 348
Haldane, J. B. S., 271, 277, 283, 294, 324–26, 351, 509
Hall, Harvey Monroe, 271, 316, 317
Hallam, Arthur, 27
Ham, Kenneth A., 484
Hamilton, William D., 203, 210, 351, 464
Hamlin, Kimberly, 450
Hammond, J., 512
Hanson, M. A., 512
Hap (Hardy), 440
Hardin, Garrett, 393
Hardy, G. H., 275
Hardy, Thomas, 437, 440
Hardy-Weinberg law, 275–76
Harker, Alfred, 51
Harris, J. A., 332
Harvey, Joy, 445
Harvey, William, 161
Hauser, Marc, 422, 426
Hays, Arthus Garfield, 498
Hefner, Philip, 475
Hegel, Georg Wilhelm Friedrich, 43
Heinricher, Emil, 265
Heller, Edmund, 370, 371
Henderson, L. J., 280
Henshilwood, Christopher, 426
Henslow, John Stevens, 136, 217, 265
    Darwin and, 48, 56, 58, 72, 132, 134
    influence of, on Darwin, 132–34
Heraclitus, 33, 35
Herbert, Sandra, 50, 51
Herbert, William, 118, 120
Herder, Johann Gottfried, 236
heredity, 21, 98, 116–23, 219, 508–09
    blending, 11, 75, 118
    Darwin and, 116–21
    fractional, 118–19
    hard (Weismann), 224
    law of ancestral (Galton), 224

    natural selection and, 11
    original sin and, 471
    predisposition (diathesis) and, 506
    strength of, 117–18
    variant cycle, 328
    variation and, 54
Herland (Gilman), 447
hermaphroditism, 69, 85, 120, 133, 135
herpetology, 288
Herre, Edward Allen, 332
Herschel, John F. W., 4, 48, 49, 95, 133, 155, 184, 221
Hertwig, Oscar and Richard, 238
Herzl, Theodore, 497
heterochrony, 380
heterogenesis, 323
heterozygotes, 275, 297
hierarchy
    biological, shifts in, 203
    Darwinian evolutionary, 179, 185
    Linnaean, 73
    racial, 241, 367–367
Hiesey, William, 317
Higley, L. Allen, 480
Himmelfarb, Gertrude, 180
Hinde, Robert, 351
Hirsch, Emil, 493
His, Wilhelm, 239, 240,
Historia Animalium (Aristotle), 37
"Historical Sketch of the Progress of Opinion on the Origin of Species" (Darwin), 254
History of Botany 1860–1900 (Green), 265
History of Sir Charles Grandison, The (Richardson), 436
History of the Inductive Sciences (Whewell), 133
history, natural, and mathematics, 284–86
Hitler, Adolf, 235
HMS *Beagle*, xi, 3, 5, 49, 56, 189
Hobbes, Thomas, 193
Hodge, Charles, 29, 470
Hodge, Jonathan, 81
Hoekstra, Hopi, 339
Hofstadter, Richard, 348
Holland, Henry, 120
Hölldobler, Bert, 352
Hollings, C. S., 396
Holmberg, Eduardo, 263
homeopathy, 2
hominins, 425–26
Hominoidea, 416
Homo floresiensis, 419
Homo habilis, 417
Homo sapiens, 419
    evolutionary status of, 13
homology, 12, 18, 23, 80, 83, 101, 140
    archetype and, 86
    Aristotle and, 37, 40
    centrality of, to German thinkers, 43
    common descent (Darwin) and, 37
    defined (Owen), 82
    embryological criterion of (Darwin), 84
    embryology and, 83
    Lorenz and, 349
    morphological, 187
    phylogenetic relationships (Darwin) and, 83
    unity of composition and, 83
    vertebrate (Owen), 77
homozygotes, 275

INDEX

Hooker, Joseph Dalton, 10, 136, 172, 220, 264, 265, 288, 313, 323, 363
    Darwin and, 5, 8, 11, 78, 252
    natural selection and, 169
Hooper, Judith, 296
Hoover, Herbert, 233
Hope, Thomas, 47, 49
Hopkins-Stanford expedition, 370
horses, thoroughbred, 93
Hrdy, Sarah Blaffer, 449
Hu Shi, 253
Hubbard, Ruth, 443, 445
Hughes, Charles, 59
Hughes, Ted, 441
Hull, David, 156, 459
Human Genome Project, 26, 511
"Human Progress, Past and Future" (Wallace), 107
"Human Selection" (Wallace), 198
Humani Generis (Pius XII), 487, 489
Humboldt, Alexander von, 49, 49, 50, 166, 227, 361, 363, 393, 436
Humboldt, Wilhelm von, 186
Hume, David, 189, 212, 214
Hunt, James, 241
Hunterian Museum, 58, 61
Huntington, Ellsworth, 366
Hurford, J. M., 422
Hutchinson, G. E., 386
Hutchinson, Jonathan, 507
Hutton, James, 47
Hutton, William, 266
Huxley, Aldous, 8
Huxley, Julian, 108, 143, 145, 224, 284, 289, 319, 321, 349, 372, 463, 509, 510
Huxley, Thomas Henry, xi, 10, 12, 28, 111, 161, 172, 175, 187, 200, 221, 230, 238, 288, 290, 331, 391, 508
    agnosticism and, 215
    barnacles and, 85
    Darwin and, 8, 32, 440
    fossil evidence and, 356
    horticultural colonization and, 253
    hybridism and, 208
    Mivart and, 222
    morphology and, 157
    natural selection and, 31, 111, 144, 463
    Owen and, 13, 223
    religion and, 16
    saltations and, 274, 406
    subordinating religion to science, 220
    Wilberforce and, 18
Hyatt, Alpheus, 228
hybridization, 99, 118, 342–43
    classes (Gärtner), 119
hygiene, social, 241, 509
Hymenoptera, 351
"Hypothesis of Pangenesis" (Darwin, 1865), 121

Ibla, 85,
ice ages, 126–29
Ideas for a Philosophy of the History of Humanity (Herder), 236
identity, community, and material culture, 426
Illogical Geology, The Weakest Point in the Evolution Theory (Price), 478
Illustrations of British Insects (Stephen), 72
Imago Dei, 470, 472

In Memoriam (Tennyson), 27, 44
İnan, Afet, 501
increase, ratio of, 102
Independent Assortment, Law of (Mendel), 268
Individual in the Animal Kingdom, The (J. Huxley), 143, 349
individuality, 202
inductions, consilience of (Whewell), 95
inertia, phylogenetic, 333
Influence of Darwinism on Philosophy, The (Dewey), 451
infrastructure
    gene-first, 327–28
    metabolism first, 327–28
    synthesis and, 282–90
inheritance, 41, 102, 163–64
    acquired characteristics and, 66, 98, 219, 261, 347, 348, 379, 415
    acquired habit and, 178
    agencies affecting, 121
    ancestral law of (Galton), 119
    blending theory of, 293, 385
        adaptation and, 385
    character development and, 106
    at corresponding periods of life, 106
    at corresponding seasons, 106
    cultural, 209, 431, 434
    Darwin and, 376,
    dual-, 428, 433,
    epigenetic, 511
    imitation and, 117
    invention and, 117
    Lamarckian, 163, 219, 222
    laws of, 117
    as limited by sex, 106
    Mendelian, 271, 376
    modification through natural selection and, 362
    natural selection and, 106, 158–64, 430
    non-, 122
    principles of, 106, 117
    processes of, 106
    use-, 452, 453
Innes, John Brodie, 217
Insectivorous Plants (Darwin, 1875), 137
insects, 11, 22
    social, 16, 346–48
    castes and, 347
    sterile castes and, 204–06
Insects and Angels (Byatt), 442
instinct, 99
    Darwin and, 204
    social, 209, 216
instincts, 99
intellectus architypus, 236
International Human Genome Sequencing Consortium, 403
intervention, divine, 5
Introduction of Darwinism in France (Conry), 248
Introduction to Entomology, An, or Elements in the Natural History of Insects (Kirby), 139
Introduction to Population Genetics, An (Li Jingjun), 256
Inuit peoples, 429
Investigations in the Comparative Anatomy of Vertebrates (Gegenbaur), 238
Iris, 271, 288

"Is a New and General Theory of Evolution Emerging?" (Gould), 359
Islam, 30
Island Life (Wallace), 171
islands, continental and oceanic, 370
isolation, 114, 115
    biological, 285
    characterizing, 288
    forms and, 372
    geographical, and Mayr, 285
    physical, 285
    reproductive, 10
    speciation and, 160
isomorphism, 455
Iverach, James, 470, 471

Jacob, François, 312
Jacobs, Joseph, 497
James Island, 51
James, William, 178, 232, 457
Jameson, Robert, 47, 188
Jann, Rosemary, 443, 447
Jeans, James, 277
Jefferson, Thomas, 51
Jenkin, Fleeming, 161, 224, 230
Jenny (orangutan), 420
Jewish Chronicle, 497
Johannsen, Wilhelm, 270
John Paul II, Pope, 30, 487, 489, 491
Johnson, Phillip E., 482
Johnson, William Henry, 449
Jones, Frederick Wood, 415
Jones, Jeanette Eileen, 448
Jordan, David Starr, 228, 232, 349
Jordan, F. M., 434
Journal of Evolutionary Biology, 332
Journal of Researches (Darwin, 1845), 166, 189, 363
Joyce, Richard, 466, 467
Judaism, 30
    evolution and, 493–98
Jude the Obscure (Hardy), 440
Junco, 289

kabbalah, Lurianic, 496
Kammerer, Paul, 509
Kant, Immanuel, 5, 42, 191, 235–36,
    final causes and, 1
Kanzi (bonobo), 421
Kapital, Das (Marx), 19
karyotype, 315
kayaks, 429, 430
Keck, David, 317
Keith, Sir Arthur, 415, 416, 418
Keller, Rabbi Chaim Dov, 496
Kellogg, Vernon, 229, 233, 332, 349
Kemal, Mustafa, 501,
Kendall, May, 443
Kenyon, Dean H., 482
Keplerbund (Protestant naturalists), 240
Kessler, Karl, 367
Kettlewell, Bernard, 22, 142, 293, 294, 296, 332
Khān, Sayyid Ahmad, 501
Kidston, Robert, 267
Kimura, Motoo, 402, 417
King Kong, 422, 448
King, B. J., 422
King, Mary-Claire, 380

559

Kingsley, Reverend Charles, 160, 217, 222, 324
Kingsolver, Barbara, 332
Kinsley, Alfred, 450
Kirby, S., 422
Kirby, William, 139, 204
Kitcher. Philip, 465
Klaatsch, Hermann, 239
Klebs, Georg, 265
Klinghoffer, David, 498
Knight, C. J., 422
Knight, Thomas Andrew, 120, 264
knowledge
    genealogy of, 454
    Heraclitus and, 35
    as human capacity, 452
    logic of (Dewey), 452
    as a state of mind, 451
Koch, Robert, 507
Kohler, Kaufmann, 493
Kohler, Robert, 370
Kohn, David, 169
Kölliker, Alfred, 239
Kölreuter, Joseph, 118, 120
Kook, Rabbi Abraham Isaac, 494
Koonin, E. V., 345
Kosseritz, Carl von, 260
Kovalevsky, Vladiimir, 238
Krohn, August David, 86
Kropotkin, Peter, 200-1, 367
Kühn, Alfred, 379
Kuhn, Thomas, 10, 456
Kulp, J. Lawrence, 480
Kuyper, Abraham, 472

La antigüedad del hombre en el río de la Plata (Antiquity of Man in the Plate Basin River, Ameghino), 263
La raza cósmica (The Cosmic Race, Vasconcelos), 262
labeling, isotope, 270
"Laboratory Science versus Country House Experiments, The Controversy between Julius von Sachs and Charles Darwin" (Chadarevian), 264
Lacerda, João Batista, 260
Lack, David, 351, 370, 371-73, 374, 387
Lamarck, Jean Baptiste de, 4, 6, 23, 41-42, 46, 66, 82, 222, 235, 323, 340, 376, 388, 506
Lamarckism, 98, 347, 406
    Darwin and, 204, 223
    mechanisms of, 163
    neo-, 248
    Social, 348
Lammerts, Walter E., 480, 481
Lamotte, Maxine, 298
Lancaster, Roger, 450
Land Nationalisation (Wallace), 171
Land Nationalisation Society, 198
landbirds, 65
landscape, 363-64
    adaptive (Wright), 21, 278
    morphology (Davis), 364
    phenomenology, 363
Lang, Arnold, 239
language, 14
    animal communication and, 183
    Chomsky and, 27

    communication and, 422
    Darwin and, 182-87
        coevolution with mind, 184-86
        Descent and, 182-83
        heuristics and, 187
        linguistic development and, 187
        origins of, 183-84
        physical aspects of, 183
        racial hierarchy, 186-187
        species analogy, 187
        vocalization and, 183
    Descartes and, 27
    evolution of, post-Darwin, 420-27
    instinct and, 182
    multicomponent, 426
    natural selection and, 452
    origin of (Darwin), 182
Larkin, Philip, 441
Latorre, Lorenzo, 261, 263
law
    doctrine of natural, 167
    hierarchial
        descent and, 166
        natural (Combe), 166
        of mind (Combe), 166
Law of Heredity, The (Brooks), 230
law, natural, Ancient Greeks and, 1
Laws (Plato), 153
Le Dantec, Felix, 248
Leakey, Louis B., 416, 417
Leakey, Mary, 418
Leavens, D., 425
Lebensraum (Living Space, Ratzel), 364
Lehninger, Albert, 152
Leibniz, Gottfried Wilhelm, 38
Leo XIII, Pope, 487,
Leopold, Aldo, 393
"Lessons on the Mathematical Theory of the Struggle for Life" (Volterra), 311
Letter to the Grand Duchess Christina (Galileo), 488
Letters of Darwin, 8
Leucippus, 33
Leukart, Rudolf, 239
Levine, George, 436
Levins, Richard, 392
Levy, Naphtali, 494
Lewes, George Henry, 117
Lewontin, Richard, 157, 354, 510
lexigrams, 421
L'Héritier, Philippe, 306-11
Li Hongzhang, 251
Li Jingjun, 256
Liang Qichao, 253,
Liberman, Mark, 426
lice, 173,
Lieberman, E. J.-B., 431
Lieberman, P., 422
Liepman, H. P., 163
life
    conditions of, 102, 163, 164
        heredity and, 121
    coral of (Darwin), 74-75, 345
    evolution prior to, 326-27
    history of, 23, 54
    net of (Kunin), 345
    organic, 506

    origin of, 23, 322-29
        Darwin and, 323-24
        Oparin-Haldane hypothesis, 324-26
    potato of (Olendzenski and Gogarten), 345
    ring of (Rivera and Lake), 345
    three domains of, 341
    tree of, 6, 16, 23, 41, 64, 66-68, 74, 129, 340-45, 434
        common descent and the, 340
        defined, 341-42
        depicting species propagation, 68
        geographic circumstances and, 67
        as idealization, 344
        natural selection and, 8
        phylogenetic, 341
        purpose, 343
        species and, 344
        universal, 342
        unity of, 398
    web of, 345
Life and Letters of Charles Darwin, 469
Lightman, Bernard, 447
Lillie, Frank, 229
Limnopithecus, 416
Linanthus parrye, 319
Lindley, John, 133, 135, 266
linguistics, 14
links, missing, 412, 413
Linnaeus, Carolus, 5, 38, 74, 78, 80, 82, 241, 316
Linnean Society, 135-135, 137
    ban on discussion of the *Origin of Species*, 135
lion-ant (lion-ant), 384
Lister, Joseph, 507
literature, 436-42
Litopterna, 62
Little Dorrit (Dickins), 87
Living Cirripedia (Darwin 1852, 1854), 80, 81, 82, 84
Locke, J. L., 422
Locke, John, 184
logic, origin of, 454
Loison, Laurent, 248
London, Jack, 27, 232, 441
Lonergan, Bernard, 489-92
loosestrife (*Lythrum*), 137
Lorenz, Konrad, 180, 349-50, 453, 455, 510
Lotka, Alfred J., 308
Lotsy, J. P., 314
Love, Rosaleen, 447
Lovejoy, Owen, 419
Lovejoy, Thomas, 392, 393, 395-96,
Lowe, Percy, 371
Lu Xun, 253
Lubac, Henri de, 491
Lubbock, Sir John, 185, 187, 223, 288
Lucas, Prosper, 117, 119, 121
Lucy (skeleton), 417
Lumb, Edward, 60,
Lutz, F. E., 332
Lwoff, André Michel, 312
Lychnis dioica, 446
Lyell, Charles, 49, 50, 51, 52, 65, 86, 126, 127, 160, 166, 172, 187, 223, 251, 330, 331, 361, 376, 388, 436, 440
    Darwin and, 8, 65, 168
    natural selection and, 169
    religion and, 16

uniformitarianism and, 4
Wallace and, 168
Lysenko, Trofim, 19
Lysenkoism, Soviet, 256, 290

Ma Junwu, 254–55
MacArthur, Robert, 386, 392, 395
Macgillivray, William, 47
Mackinder, Halford John, 365
Mackintosh, James, 189, 190–91
Macleay circles, 75
Macleay, William Sharp, 75, 82
Macrauchenia, 62
macromolecules, 398–99
  informational, 398
Maimonides, Moses, 494, 496,
Maine, Sumner, 259
Malay Archipelago, The (Wallace), 170, 172
Malécot, Gustave, 311
males, complemental (Darwin), 85
Mallet, J., 343
Malthus, Thomas Robert, 5, 45, 96, 153, 166, 251, 367, 376, 437
  Darwin and, 70, 88, 384
mammalogy, 288
Man versus the State (Spencer), 198
Man's Place in Nature (Huxley), 223
Man's Place in the Universe (Wallace), 171
manatees, 75
Mandeville, Bernard, 193
Manual of Geology (Dana), 228
Manual of Mineralogy (Jameson), 47
map, genotype-phenotype, 377
marginalism, 247
Margoliash, E., 402
Maritain, Jacques, 489
Marsh, Frank Lewis, 481
Marsh, Othniel Charles, 231, 356
Marshall, William, 150
Martin Chuzzlewit (Dickens), 439
Martin, W., 344
Martineau, Harriet, 212, 436
Martineau, James, 212
Martínez, Martin C., 262
Marx, Karl, 19, 251, 485
Mashriqī, Ināyatullāh, 501
mastodon, 51, 59, 62,
materialism, 30, 214, 215, 222
Materials for the Study of Variation (Bateson), 273
"Mathematical Contributions to the Theory of Evolution" (series), 332
Mathew, Patrick, 45
Mating Mind, The (Miller), 108
mating, assortive, 387
Matthew, William Diller, 356
maturation, 67
Maurer, Friedrich, 239
Mauss, Marcel, 248
Mawdūdī, Abū l-Aʿlā, 501
Maynard-Smith, John, 108
Mayr, Ernst, xii, 21, 111, 115, 157, 284–86, 289, 290, 315, 320, 321, 332, 356, 376, 417, 509
MaZhar, Ismāʿīl, 500
McCrady, John, 367
McDonald, Roger, 442
McDougall, William, 229

McEwan, Ian, 27, 442
McWhirtier, K. G., 295
Mead, Margaret, 181
medicine, 30, 212
  evolutionary, 509–13
    appeal of, 505
    problems of, 511
Meditations (Descartes), 27, 39
Megalonyx, 62
Mein Kampf (Hitler), 242
melanism, industrial, 22, 295–96,
Meme Machine, The (Blackmore), 432
memetics, 428, 432
  defined (Dawkins), 431
  linguistic, 432
  parasitic, 432
  religious, 432
  theory of (Dawkins), 27, 432
Memorabilia (Xenephon), 34
memory, 192
Mendel, Gregor, 11, 20, 143, 219, 230, 255, 267–70, 314, 385
Mendelism, 230
Mendelism and Evolution (Ford), 294
Mental Evolution in Animals (Romanes), 179
Mental Evolution in Man (Romanes), 179
Mercier, Jacques, 301
mesmerism
  defined, 1
  Wallace and, 167
Messenger, Sharon, 448
Mestre, Arístides, 261
metaethics, 465
metamorphosis, 43
Metaphysics (Aristotle), 36
metaphysics, Darwin and, 68
Methodophile Association (Mexico), 262
mice, Florida beach, 339
Michell, John, 51
microbes, 507
microbiology, 311
microscopy, electron, 270
Middlemarch (Eliot), 440
migration, 283
  law of (Wagner), 160
  with isolation, 67
Miklucho-Maclay, Nikolai, 238
Milinski, M., 334
Mill, John Stuart, 212
Millardet, Pierre-Marie-Alexis, 265
Miller, Geoffrey, 108
Miller, Rabbi Avigdor, 496
Miller, Stanley, 326
Millican, Kenneth, 507
Millstein, R., 333
Milne-Edwards, Henri, 83–84, 244
Milton, John, 436
mimicry, 23, 139–45, 336–39
  aggressive, 338
  Batesian, 140, 337
    pattern diversity in, 338
  crypsis and, 338
  defined, 139, 260, 336
  differential reproduction and, 11
  language and, 182
  Müllerian, 20, 141, 260, 338, 339

Batesian mimicry and, 339
polymorphic female, 339
sex-limited, 338
Mind, Anaxagorean, 36
minimalism, 455
Mirabilis vulgaris x *M. longiflora*, 118
miscegenation, 259, 260
"Missing Link, The" (Jones), 448
mitochondria, 343
Mitten, Annie, 170
Mitten, William, 170
Mivart, St. George, 159, 163, 216, 222, 224, 261, 356, 470, 485
Mixter, Russell L., 480
mockingbirds (*Nesomimus*), 369
Modern Biology and the Theory of Evolution (Wasmann), 240
modification, 128, 130
  descent with (Darwin), 39
  developmental, and genetic mutation, 407
  generation with, 168
  means of, 163
  of an organ (Owen), 76
  selection and, 161
mold, vegetable, 52
Moll, J. W., 265
Mollusca, 61, 82
monad to man, 1
Monism, 33
  panpsychic, 280
monkeys, 177, 183, 423–25
  Campbell's, 424
  Diana, 424
  rhesus macaque, 424
  velvet, 423
Monod, Jacques, 307, 312
Monotremes, 77
Monte Hermoso, 58, 59
Moody Bible Institute, 480
Moore, Aubrey, 469, 471, 472,
Moore, G. E., 28, 463
Moore, James, 166, 181
Moreno, Roberto, 262
Morgan, Lewis Henry, 428
Morgan, Thomas Hunt, 20, 230, 255, 276, 317
morphogenesis, 379
morphology, 101, 157, 273
  defined, 157
  embryology and, 273
  revolt from, 20
Morris, Henry M., 29, 450, 481, 501
Morton, A. G., 270
moths
  Ephestia, 379
  peppered, 142, 288
  scarlet tiger (*Panaxia dominula*), 299
Moulinié, Jean-Jacques, 244
Mr. Darwin's Shooter (McDonald), 442
Müller, Fritz, 20, 141, 259–60, 331
Müller, Hermann, 131, 138
Müller, Johannes, 122
Müller, Max, 15, 185, 452
Müller-Thurgau, Hermann, 265
Muniz, Francisco Janvier, 59
Muqtaṭaf, al- (The Digest), 500
Murchison, Sir Roderick Impey, 50, 266
Murphy, Nancey, 475

Index

Murray, John, 140, 158, 159, 237
Museo Publico de Buenos Aires, 60
Museum d'Histoire Naturelle, Paris, 61
mutation, 283
   de Vries and, 270, 288
   Goldschmidt and, 379
   theory, 314, 321
   theory (de Vries), 315
   variation and, 407
"Mutual Affinities of Organic Beings" (Darwin, 1866), 140
Mutual Aid: A Factor in Evolution (Kropotkin), 201
mutualism, 464
My Life (Wallace), 171
Mylodon, 62
Mysteries of the Creation (Brown), 496

Naegeli, Carl, 314
Naess, Arne, 393
Nagamatsz, A., 265
Nägeli, Carl von, 119
Nahmanides, Moses, 496
Namr, Fāris, 500
Narrative of Travels on the Amazon and Rio Negro, A (Wallace), 167
National Institutes of Health (NIH), 403
National Research Council's Committee on Common Problems of Genetics, Paleontology, and Systematics, 290
National Socialism, and Darwinism, 19
natives, extermination of, 20
nativism
   moral, 461, 464–67
      objectivity and, 466
      undermining morality, 464, 466
Natura non facit saltum, 161
Natural History of Creation, The (Haeckel), 240, 243
Natural History of Religion (Hume), 214
"Natural Selection" (Darwin), 254
Natural Selection and Heredity (Sheppard), 299
Natural Theology, or Evidences of the Existence and Attributes of the Deity (Paley), 148, 150, 153, 189, 215
naturalism, scientific, 222
nature, 160
   as system of stages (Hegel), 43
      balance of, 25, 388, 394
      competition and, 387–90
      defined by Darwin, 160
Nature and Development of the Organic World (Ratzel), 364
Nature Collections: Human Genome, 403
Natürliche Schöpfungsgeschichte (Haeckel), 175, 185, 413
Naturphilosophen, 18, 26, 43, 393
Naturphilosophie, 376, 393
Naudin, Charles, 120
Nauplius, 82
Neefs, Yvette, 311
Neel, James, 509
Neodarwinism (Romanes), 379
Nesse, Randolph, 505, 510
Netto, Ladislao, 260
networks, gene regulatory, 380–81
neurobiology, 397

New Geology, The (Price), 478
Newbigin, Marion, 365
Newell, Norman, 357
Newman, W. A., 85
Newton, Alfred, 170, 337
Newton, Isaac, xii, 5, 95, 250
niche, ecological, 6, 25, 98, 208, 373
   behavioral, 387
   competitive exclusion and, 386
   Darwin and, 385,
   Gause and, 371, 372
   partitioning, 385, 387
      spatial, 385
      temporal, 386
Nicotiana, 271
   tabacum, 316
Niebuhr, Reinhold, 473
Nietzsche, Friedrich, 454
Nina Rodrigues, Raimundo, 260
Ninth Bridgewater Treatise (Babbage), 16
Noble, W., 422
Noll, Fritz, 265
Non-Darwinian Revolution, The: Reinterpreting a Historical Myth (Bowler), 348
Nordenskiöld, Erik, 242
Not by Design, Retiring Darwin's Watchmaker (Reiss), 157
Notebooks (Darwin), 116, 183–84, 437
   contents of, 5, 174
   Notebook A, 68
   Notebook B, 64, 66, 69, 169
   Notebook C, 174, 189
   Notebook D, 68
   Notebook E, 64, 71
   Notebook M, 68, 190, 214, 439, 452, 453
   Notebook N, 71, 174
   Red, 202
Notoungulata, 62
novelty, 80
Nuer peoples, 429
Nuffield Foundation, 22, 294
Nurcus (disciples of the Divine Light), 501
Nursi, Said, 501

Oakley, Kenneth, 414
oceans, barriers of the, 8
Ockham's Razor, 466–67
Of Pandas and People (Kenyon and Davis), 482
Ogle, William, 38
Oken, Lorenz, 18, 43
Olenov, J. M., 332
Oliver Twist (Dickens), 439
On Human Nature (Wilson), 28
"On the Darwinian Theory of Creation" (Kölliker), 239
On the Deathbed of Darwinism (Dennert), 240
On the Developmental History of Siphonophores (Haeckel), 238
On the Interpretation of Nature (Diderot), 40
"On the Law Which Has Regulated the Introduction of New Species" (Wallace), 168
On the Origin of Language (Wedgwood), 183
On the Origin of Species by Means of Natural Selection, or the Preservation of Favoured Races in the Struggle for Life (Darwin, 1859), 16, 68, 132, 295, 313, 330, 361–63, 469

   and religion, 361–63
   and survival of the fittest, 196
   and the good of the community, 203
   chance and, 155
   Chinese editions, 254
   common ancestry and, 340
   creationism and, 476
   Croll and, 128
   crossing and, 91
   diversity and, 75, 78, 393
   editions of, 12, 361–63
   ethics in, 192
   evolution and, 397
   fifth edition, 159, 203
   first edition, xii, 124, 139, 216, 336
   foreign editions, 244
   fourth edition, 125, 159, 208,
   geographic distribution and, 361–63
   glacial epoch and, 127
   good of the community and, 203
   heredity and, 116
   heritable variation and, 91
   human evolution and, 195
   Ilkley edition, 159
   inheritance and, 117
   instinct and, 347
   last edition, 126
   mimicry and, 140
   natural classification and, 86
   niches and, 385
   Paley and, 216
   primary purpose of, 204
   primordial form and, 323, 476
   second edition, 216, 476
   selection and, 88, 103, 104, 139, 154, 354, 361–63, 364, 391, 429, 445, 452
   sixth edition, 129, 159, 216
   social evolution and, 346
   Spanish editions of, 261
   species and, 110, 270, 361–63
   sterility and, 206
   strategy of, 95
   traits and, 297
   translated by Bronn, 237
   tree of life and, 129, 340
   use and disuse in, 293
   variation and, 146, 293
"On the Scope and Methods of Geography" (Mackinder), 365
"On the Tendency of Species to Depart Indefinitely from their Type" (Wallace), 154, 168
On the Various Contrivances by which British and Foreign Orchids are Fertilized by Insects, and on the Good Effects of Intercrossing (Darwin, 1862), 11, 135
ontogeny, recapitulating phylogeny (Haeckel), 19, 239
Oparin, Alexander, 324–26, 328
Oparin-Haldane hypothesis, 324–26
opossums, 77
orchids, 135–36, 154, 155
   Catasetum tridentatum, 135, 136
   Darwin and, 288
   Malaxis, 155
   Primula veris, 156
Orchids (Darwin, 1877), 135–37, 156
organism, defined, by Kant, 42

"Organisms as Historical Beings" (Boveri), 379
organs
    generative, 76
    origination of (Aristotle), 37
Orgel, Leslie, 327, 328
"Origin of Human Races and the Antiquity of Man Deduced from the Theory of Natural Descent, The" (Wallace), 197, 246
"Origin of Isolation, The" (Dobzhansky), 115
origins
    human, 223
    as the mystery of mysteries (Herschel), 4
    problem of, 32
Ornithological Notes (Darwin, 1963), 369
ornithology, 288
Orr, James, 470, 471
orthogenesis (directed variation), 224, 230, 356, 415
Ortiz, Pablo Acosta, 262
Orzack, Stephen, 333, 334
Osborn, Henry Fairfield, 231, 356, 415
Our Corporeal Form and the Physiological Problem of Its Origin (His), 240
Our Mutual Friend (Dickens), 439
Overture, Richard A., 265
Owen, D. F., 296
Owen, Richard, 12, 45, 56, 61, 66, 74, 76, 110, 157, 160, 172, 219, 220, 221
    Darwin and, 4, 43, 82, 134, 324
    Huxley and, 13, 223
oysters (*Gryphaea*), 303
Özal, Turgut, 501

Packard, Alpheus, 228
Paine, Thomas, 389
paleobiology, 24, 353
Paleobiology (journal), 359
paleobotany, 266–67
paleontology, 24, 353–60, 397
    diagrammatic conjecture and, 300–06
    evolution as pattern and, 355
    genetics and, 315–21
    philosophical (Gaudry), 248, 300
    vertebrate, 56–63
paleophytology, 266
Paley, Archdeacon William, 32, 43, 148, 150–51, 153, 189–90, 191, 213, 215, 331,
Palladino, Paolo, 271
Palm Trees (Wallace), 167
Pando, José Manuel, 263
pangenesis (Darwin), 11, 116, 121–23, 164, 314, 385
    defined, 122
    Galton's critique of, 123
    as provisional, 122
Pannenberg, Wolfhart, 473, 475
panspermia, 324
paradigm (Kuhn), defined, 10
Paradise Lost (Milton), 436
parasite (*Balanidae*), 81
parasitism, 349, 432
Parish, Woodbine, 58, 361
Parmenides, 33
Parra, Porfirio, 262
parthenogenesis, 159
Parwez, Ghulām Aḥmad, 503
Pasteur, Louis, 19, 23, 243, 247, 248, 507
Pastoral Constitution on the Church in the Modern World (Second Vatican Council), 491

Pastrana, Julia, 448
pathology, 266
Patterson, J.T., 287
Pauling, Linus, 341, 417
Peacock, George, 48
Peacocke, Arthur, 475
Pearson, Karl, 225, 275, 331, 332,
Pearson, P. N., 54
peas, 22
    Hieracium, 314
    Pisum, 267, 269
    sativum, 314
Peckham, Morse, 158
Peirce, Chatrles S., 232
Pennock,, Robert, 187
Perier, Edward, 248
Peromyscus, 289
Perrier, Edward, 248
Personal Narrative of Travels to the Equinoctial Region of the New Continent (Humboldt), 166, 361, 363, 393, 436
Peters, Ted, 475
Petersen, Wilhelm, 114
Pfeffer, Wilhelm, 265
Phaedo (Plato), 34, 36, 452
Phasmia (Brazilian walking stick insect), 139
pheasant, Argus, 171
phenomenology, landscape, 363
Phenomenon of Man, The (Teilhard de Chardin), 22, 491
phenotype, 382
    defined, 407
    developmental physiological equilibrium and, 380
    extended (Dawkins), 99
Phillips, John, 266
"Philosophical Notes XV–XXIV" (Erasmus Darwin), 46
Philosophical Society, 48
Philosophie Zoologique (Lamarck), 41
philosophy
    moral, Darwinian theses bearing on, 461
    natural, pre-Socratic, 32
Philosophy of the Organism (Driesch), 349
phrenology, 166, 170
    defined, 1
phrenomesmerism, 167
Phylogenetic Method in Taxonomy, The (Hall and Clements), 316
phylogeny, 23
    defined (Haeckel), 159
    universal, 341
"Phylogeny of Octopod Cephalopods" (Rogers), 303
Physical Geography (Herschel), 155
Physics (Aristotle), 36, 153
Physics and Politics, Thoughts on the Application of the Principles of natural Selection and Inheritance to Political Society (Bagehot), 365
physiology, 266, 286
Piaget, Jean, 181
Pickwick Papers (Dickens), 439
pigeons
    fancy, 92–93
    jacobin, 92
    pouter, 92
    rock (*Columbia livea*), 92, 96, 120, 121

runt, 92
tumbler, 92
pigs, guinea, 121
Pinker, Steven, 187, 422, 430
Pithecanthropus alalus, 413
Pithecanthropus erectus, 413
Pittendrigh, Colin, 157
Pius IX, Pope, 240, 485
Pius XII, Pope, 487, 488
"Plan of Creation as Shown in the Animal Kingdom, The" (Agassiz), 227
Planet of the Apes, 448
plankton, paradox of the, 386
Plantinga, Alvin, 28, 454, 473
plants, 5, 11
Plasmodium, 509
plasticity, phenotypic, 410
Plato, 34, 35–36, 153, 452
platypus, 77
pleiotropy, 112, 114
Plethodon, 289
Plinian Society, 189, 212
Pluralism, 333
Plutarch, 35
Politische Geographie (Ratzel), 365
Polkinghorne, John, 475
polymorphism, 295, 296–97,
    balanced, 297
    genetic, 296
    natural, 296
    selected, 296
polyploidy, 270, 316
    allo-, 316
Ponceau, Stephen de, 186
Poor Laws and Paupers Illustrated (Martineau), 436
Popper, Karl, 456, 457–59, 460
poppies (*Papaver somniferum*), 269
population, 45, 96
    cages (*démomètres*, L'Héritier), 307, 308,
    genetics, 273–81, 283–84, 356, 376, 379
    Fisher and Wright and, 283
    mathematical models of, 275, 283, 294, 308–11
pornography, green, 87
Positivism, 247, 305
    defined, 258
    evolutionary (Spencer), 258
    social (Comte), 258
Pouchet, Felix, 323
Poulton, Edward, 141, 144, 145, 274, 296, 332
Powell, Reverend Baden, 29
Power of Movement in Plants, The (Darwin, 1880), 137, 267
Pragmatism, American, 28, 232
Prantl, Karl, 265
predator-avoidance, 11
Preliminary Discourse on the Study of Natural Philosophy (Herschel), 48, 184
Price, George McCready, 478–80, 481
Priestley, Joseph, 188
primroses, evening, 22, 120
    Oenothera lamarckiana, 315, 316
Primula, 133, 135, 154
    kewensis, 316
Principles of Biology (Spencer), 116, 159, 349
Principles of Descriptive and Physiological Botany, The (Henslow), 132

Principles of Geology (Lyell), 4, 49, 65, 166, 436
Principles of Heredity (Mendel), 269
Principles of Moral and Political Philosophy (Paley), 189, 213
Principles of Psychology (Spencer), 223
"Problems of Heredity as a Subject for Horticultural Investigation" (Bateson), 273
Proceedings of the Linnaean Society of London, 8
process, 288–89
Proconsul, 416
production, chance (Darwin), 71
Profitable Poultry (Tegetmeier), 92
progress, 18, 40, 161–63
    defined, in the Age of Enlightenment, 1
    division of labor and, 12
    evolution and, 12, 232
    after the French Revolution, 1
    George and, 198
    organic (Spencer), 45
    Origin of Species and, 13
    social (Spencer), 220
Progress and Poverty (George), 198
"Progress, Its Law and Cause" (Spencer), 15, 196
Progressivism, 232
prokaryotes (simple-celled organisms), 23, 344
"Prolegomena" (Huxley), 252
Proofs and Illustrations of the Attributes of God from the Facts and Laws of the Physical Universe, being the Foundation of Natural and Reveal Religion (Macculloch), 153
propagation, species, and sexual generation, 70
proteins, 324, 326, 341, 380, 397, 398–99, 401
    as enzymes, 327
    switch, 404
proteo-bacteria, alpha, 343
Proteolepas, 86
Protestantism, post-Darwin, 468–75
protozoa, unicellular
    Nummulites, 302
Providence
    defined, 1, 40
    progress and, 40
Providentissimus Deus (Leo XIII), 487
Provine, William B., 308, 321, 332
Pruna, P., 261
psychology, 102
    evolutionary, 464
Puerto San Julian, 59
Punnett, Reginald, 144, 145, 332
Punta Alta, 56, 57, 58, 59
Purkyne, Jan Evangelista, 265
Pusey, James, 253

Qing dynasty (1840–1911), 251–54
Quatrefages, Armand de, 243, 244, 246
Qué es la vida? (What Is Life?, Razetti), 262
Questions about the Breeding of Animals, 91
Qur'ān, and Ḥadīth, 500

races
    Darwinism and, 263
    domestic, natural versus artificial, 69
    slavery and, 173
    species and, 173–77, 178
"Races of Man, The" (Darwin, 1871), 185
Radick, Gregory, 187,
Rahner, Karl, 489, 491

rainforest, Amazonian, 395
Ramm, Bernard, 473, 480
rams, Merino, 91
ranking, 77–78
Ranunculus, 272
Raphanobrassica, 316
rationalism, and knowledge acquisition, 453
rats, 277
    Norway, 429
Ratzel, Friedrich, 364
Raup, David, 359
Raven, Peter, 321, 393, 395, 396
Ray, John, 153
Razetti, Luis, 262
realism
    moral, 467
    scientific, 460
recapitulation, 179, 348
Recollections of the Development of My Mind and Character (Darwin, 2010), 211, 212, 214, 215
recombination, 90
Reed, Howard S., 264
reefs
    coral, and Darwin, 288
    types of (Darwin), 364
Reformed Dogmatics (Bavinck), 470
Reinke, Johannes, 265
Rektorratsrede (Boveri), 379
relations
    complex web of (Darwin), 394, 395
    dominent-recessive, 90
relationship, and consanguity, 77
reliabilism, epistemological, 466
religion, 15, 27, 28
    Darwin and, 16–17, 211–17
    Darwinism and, 18
    morality and, 16
Religion and Science Association, 480
Remak, Robert, 122
Remarks on the Improvement of Cattle (Wilkinson), 90
Renner, Otto, 315
representation, 77
reproduction
    differential, 96, 97
    growth with, 102
Reseda, 133
resemblance, 140
resilience, ecological, 396
"Rethinking the Theoretical Foundation of Sociobiology" (Wilson and Wilson), 352
reticulation, 343, 344
reversion, 93, 120–21
    distant, 118, 121
    gemmules and, 122
Revolt of Democracy, The (Wallace), 171
Reznick, D. N., 332
Rhodes, Frank, 53, 54
ribozymes, 327
Richards, Evelleen, 443, 445
Richards, O. W., 332
Richards, Robert J., 181, 184, 203, 348
Richardson, Samuel, 436
Riḍā, Rashid, 501
Ridgway, Robert, 370–71
Riedl, Rupert, 453
Rimmer, Harry, 480

Rio Carcarvana, 59
Rio Lujan, 59
Rio Negro, 59
Rio Parana, 59
Risale-i Nur Külliyatı (Collection of Epistles of the Divine Right, Nursi), 501
ritualization, and Lorenz, 350
Rivers, Lord, 150
RNA, 327, 328
    micro, 327, 381, 410
    r-, 341
    ribosomal, 399
Roberts, Sir William, 507
Robson, G. C., 332
rock
    igneous, and diversity, 51
    molten (lava), 51
    soup, 54
Rogers, D., 430
Rogers, Jean, 303
roles, sex, and reversal, 105
Rolston, Holmes, 475
Roman, Frédéric, 303
Romanes, George, 160, 171, 179, 186, 239, 379
    Darwin and, 113
Romero, Silvio, 259
roots (root elements), 33
Rosas, General Juan Manuel de, 59, 189
Rosenberg, Alfred, 242
Rousseau, Jean-Jacques, 183
Roux, William, 238
Royer, Clémence, 244, 246, 247
Ruge, Georg, 239
Rural Association (Uruguay), 261
Ruricola (pseudonom), 131
Ruse, Michael, 156, 203, 232, 363, 466–67, 482
Rushdoony, Rousas J., 483
Russell, Bertrand, 28
Russell, Robert J., 475
Rüttimeyer, Ludwig, 239, 240,

Saavedra, Bautista, 263
Sabeti, Pardis, 332
Sachs, Julius von, 21, 137, 264–65, 266, 270, 314
Saḥābī, Yādollāh, 503
Saint-Simon, Henri, 258
saltations, 144, 224, 225, 348, 356, 359, 406
    defined, 22
Salvin, Osbert, 370
Saporta, Gaston de, 267
Sarmiento, Domingo Faustino, 263
Ṣarrūf, Ya'qub, 500
Saturday (McEwan), 442
Savage-Rumbaugh, Sue, 421, 422
Saxifraga, 272
Sc gene, 509, 510
scallops (*Pectinidae*), 301
Scalpellum, 85,
Scelidotherium, 62
Schelling, Friedrich, 18
Schindewolf, Otto, 356
Schlegel, Friedrich, 186
Schleicher, August, 237, 238
Schloss, Jeffrey, 475
Schneerson, Menachem Mendel, 494, 495
Schönborn, Cardinal Christoph, 489

Schopenhauer, Arthur, 178
Schopf, Thomas J. M., 359
Schroeder, Gerald, 496
Schwann, Theodor, 122
Schwartz, B. I., 253
science
    earth, 3, 39
    goals of, 288
    normal (Kuhn), 21
    physical, 39
    professional, 13
    pseudo, 1, 13, 40, 45
Science of Language, The (Müller), 185
"Scientific Aspect of the Supernatural, The" (Wallace), 170
Scientific Creationism (Morris), 481
Sclater, Philip, 171
Scofield Reference Bible, 477
Scopes Monkey Trial (1925), 29, 233
Scott, Dukinfield Henry, 265, 267
Scrope, George Julius Poulett, 50,
seaweed (*Flustra*), 202
Sea-Wolf, The (London), 441
Sebright, Sir John, 5, 89, 90, 120
Sechs Vorlesungen über die Darwinsche Theorie (Buchner), 500
Secord, James, 50, 120
Sedgwick, Adam, 8, 10, 28, 48, 49, 133, 190
    catastrophism and, 4
    Darwin and, 49, 135, 217
Sedgwick, W., 120
Sedley, David, 35
seeds, survival of, in salt water, 8
Segerdahl, Par, 421
selection
    acting at various levels (Thomson), 348
    adaptive change and, 10, 88, 97, 204
    analogy between artificial and natural, 93
    artificial, 5, 8, 88–94, 96, 224, 232, 337
    clonal, 509
    community (Darwin), 199
    Darwin and, 202–10, 376
    efficient causality and, 229
    family, 16, 205, 209
    group, 16, 203, 408
    hybrid sterility and, 206–08
    individual, 16
    individual selectionist (Ruse), 203
    kin (Hamilton), 24, 351, 464
    lasting effects of, 277
    levels of (Darwin), 202–10
    Malthusian sorting and, 71
    methodological, 96
    multilevel, 203
    natural, xi, xii, 12, 45, 64–71, 88–94, 96–98, 102, 141, 187, 295–96, 354, 457, 493, 507
        altruism and, 464
        competition and, 247
        conditions for, 430
        contrasted to sexual, 103, 106
        Darwin and Wallace on, 170, 175
        defined, 5, 97
        divine intervention and, 5
        environment and, 296, 391
        evolutionary change and, 8, 12
        extinction and, 54
        as a false term, 160
        fundamental theorum of (Fisher), 278
        God and, 146, 211, 221
        in Great Britain, 223–25
        group, 21
        hybrid sterility and, 112
        above the individual, 346–48
        inheritance and, 158–64, 430
        Judaism and, 496–97
        mechanistic explanations for, 204
        modification and, 163, 164
        modifications to, 70
        morality and, 15
        in the *Origin of Species*, 154, 160–61
        Orzack and Sober's hypotheses about, 333
        pace of, 11, 97
        personifying nature, 160
        phenotypes and, 430
        populations and, 331
        preservation and, 162
        priority and variety in, 204
        racism and, 247
        reproduction and mutation and, 326
        special creation and, 204,
        Spencer and, 45
        traits and, 330, 331, 332–33
        as ubiquitous, 332–33,
        ubiquity of, 331–34
        unintelligent causes (Hodge) and, 470
        variation and, 11, 96, 160, 161, 430
        as a *vera causa*, 88, 95
        Wallace and, 14, 330
    physiological (Romanes), 113, 160
    Sebright and, 90
    sexual, 8, 97, 103–08, 141, 297, 339
        competition and, 445
        conditions for, 105
        in *Descent of Man*, 14
        design and, 108
        explaining traits not explained by natural selection, 103
        female choice and, 10, 97, 104, 107, 445
        inter-, 104, 108
        intra-, 104, 107
        male combat and, 10, 97, 104
        predictions of, 105
        racial differences and, 175
        sociobiology and, 449
    sexual, and racial differences, 105
    simultaneous double, 106
    stabilizing (Darwin), 358
    tribe-level, 208–10
    unconscious, 96
    undirected variation and, 221
    women and (Wallace), 198
selectionism, genic, 409
Selfish Gene, The (Dawkins), 351, 431
Selman, Jeffrey, 498
Selous, Edmund, 108
Semenov-Tian'-Shanskii, A. P., 114
Semon, Richard, 239
Semple, Ellen Churchill, 366
sense, moral (Darwin), 453
    conditions for development of, 192
Sepkoski, John J., Jr., 24
Seward, Alfred C., 267,
sexism, 28
Shafran, Rabbi Avi, 498
Shaḥrūr, Maḥmūd, 503
Shakespeare, William, 437
Shaler, Nathaniel Southgate, 228, 364
Shanker, S., 422
Shaw, George Bernard, 440
sheep, 96
    New Leicester, 91
Sheppard, Philip M., 22, 293, 294, 296, 298, 299, 332
Shermer, Michael, 172
"Shifting Balance Theory" (Wright), 21
Shumayyil, Shiblī, 500
Sidgwick, Henry, 28, 210, 462
Siepielski, A. M., 332
Sierra Leone Company, 188
Sierra, Justo, 262
sieves (Thomson), 348
Silene, 272
Simpson, George Gaylord, 21, 24, 157, 284, 285, 289, 321, 356–57, 408, 416, 417, 455, 494
    and Dobzhansky, 356
sin, original, 491–92
    heredity and, 471
Sinanthropus pekinensis, 416
Singer, Peter, 393
"Sketch" (Darwin, 1842), 5, 8, 16, 70, 71, 103
Sketches by Boz (Dickens), 439
slavery, 14, 189,
Slifkin, Nathan, 495
Sloan, P. R., 203
Slocombe, K. E., 424,
sloth
    Edentata Xenarthra, 62
    Megatherium, 51, 58, 59, 61, 236
Smart, Benjamin H., 184
Smith, Adam, 183, 189, 376
Smith, Grafton Elliot, 414
Smith, John Maynard, 351, 353
Smith, William, 266
Smocovitis, V. B., 271, 332
snails, 22
    Cepaea nemoralis, 297, 298
snapdragon, common (*Antirrhinum majus*), 316
Snodgrass, Robert E., 370, 371
Snowdonia, 126
Sober, Elliott, 203, 204, 333, 334
Social Darwinism in American Thought (Hofstadter), 348
Social Environment and Moral Progress (Wallace), 171
Social Statics (Spencer), 196
socialism, and Wallace, 15, 16, 198
Société de biologie, 247
societies, cooperative, and Darwin, 15
Society for the Study of Evolution, 290, 356
Society for the Study of Speciation, 290
sociobiology, 23, 346–52
    sexual selection and, 10, 449
Sociobiology, The New Synthesis (Wilson), 24, 28, 352, 464, 510
Socrates, 34–35
Solms-Laubach, Hermann, 267
soul, immortality of (Plato), 34
soup
    hot dilute, of organic polymers (Oparin and Haldane), 326
    lava, 54

South America, 50
sparrows, house, 332
specialisation, 12
speciation, 406
    allopatric (Mayr), 358, 385
    geographical isolation and, 385
    gradual accumulation and, 385
    by hybridization, 270
    isolation and, 114, 160
    Noah's ark and, 481
    rapid, 408
    selection and, 385–87
    sterility and, 206–08
    studies, 288
species, 10, 109–15
    ancestral, 65, 67
    Anderson and, 271
    biological
        Dobzhansky and Mayr and, 320
        Mayr and, 111, 115, 284
    Cuvier and, 330
    Darwin and, 109–15
    de Vries and, 270
    defined
        by Darwin, 109–11
        by Owen, 109, 112
        by physiological isolating mechanisms, 115
        physiologically, 114
    diseases as (Hutchinson), 507
    distinguished from varieties, 110, 111
    doubtful, 111
    formation, and Darwin, 68–70
    genera and, 74
    heritability of variations and, 116
    intersterility of (Buffon), 109
    introgression, 343
    invasive, 396
    Lyell and, 330
    modern synthesis and, 287
    morphological, 112–14
        versus physiological, 112–14
    mutability of, 78
    origin of, 62
    physiological, 112–15
        defined, 112
    physiological (Huxley), 115
    pre-Darwinian concepts of, 340
    reproductive isolation (Mayr) and, 115
    ring, 288
Spence, William, 204
Spencer, Herbert, 19, 20, 24, 196–97, 331
    Darwin and, 440
    equilibrium and, 26
    heredity and, 116, 348
    Jack London and, 441
    laissez-faire capitalism and, 15
    limited view of womanhood and, 444
    Lyell and, 4
    organic progress and, 15, 18, 45
    progressive evolution and, 13, 18, 28, 45, 220, 407, 428
    social Darwinism and, 232
    survival of the fittest and, 12, 160
    Wheeler and, 24
    Wright and, 280
Spiegelman, Sol, 327
Spinoza, Bernard, 153

spiritualism, 170
Spooner, W. C., 267
St. Jago (São Tiago), 50
Stahl, Christian, 265
Stanley, Steven, 359
starfish (*Pisaster ochraceus*), 389
starlings, 429
Stebbins, George Ledyard, 21, 271, 272, 284, 285, 315, 317, 320, 321
Stein, Ben, 497
Stein, Charlotte von, 236
Stenhouse, John, 367
Stephen, James Francis, 72
Sterelny, Kim, 187, 431
sterility
    as a byproduct of reproductive modifications, 208
    defined, 112
    hybrid, 10, 99, 111, 112–14,
        selection and, 112, 206–08
        Wallace and, 112, 114
    partial, 206
    speciation and, 206–08
Stevens, Samuel, 167, 170
Stewart, Dugald, 183, 189, 190
Stoddart, David, 364
Strasburger, Eduard, 238
Street, Sharon, 467
Strickland rules, 77
Strickland, Hugh, 77, 77
Strong, Augustus H., 470, 471, 472
structuralism, process, 406, 411
Structure of Scientific Revolutions, The (Kuhn), 10, 456
"Struggle for Existence" (Darwin), 254
"Struggle for Existence in Human Society, The" (Huxley), 200
Struggles of Brown, Jones, and Robinson, The (Trollope), 438
Stylopids, 77
Sulivan, Lieutenant Bartholomew J., 56, 59
Sulloway, Frank J., 202, 369
Sumner, William Graham, 232, 339
Suñer, Francisco, 263
superorganisms, 348–49
Supplement to Bougainville's Voyage, 40, 160
"Sur la loi des disjunction des hybrids" (de Vries), 269
Swarth, Harry S., 370, 371
Syllabus of Errors, The (Pius IX), 240, 485
symbiosis, 349
sympathy, 193
synthesis
    evolutionary, 282
    expanded (Gould and Eldredge), 408
    extended, 409, 410
    modern evolutionary, 78, 108, 109, 111, 224, 234, 242, 282–92, 321, 349, 353, 405, 406, 409–11, 417, 509, 511
    and Gould, 354
    defined by Mayr, 359
    infrastructure and, 289–90
    new, 417
System of Synthetic Philosophy (Spencer), 196
system, natural, 74
systematics
    genetics and, 315–21

    new, 287
    plant, 316
Systematics and the Origin of Species from the Viewpoint of a Zoologist (Mayr), 290, 320, 372

Ṭāhā, Maḥmūd, 503
Tan Jiazhen, 255
Ṭarābulusī, Ḥusayn al-Jisr al-, 501
Tarde, Gabriel, 259
Tarsier (primate), 415
Tatlı, Âdem, 501
taxon/taxa (species differentiation), 73
taxonomy, 5, 72–79
    adaptionist, 81
    defined, 80
    experimental, 287
    God and, 73, 77
    homology and, 84
    importance of, 72
    Linnaean, 5, 81
    Milne-Edwards's principles of, 84
    plant, 316
    quintarian, 67, 76
        Darwin and, 75, 82
    theological, 77
Taylor, Griffith, 366
Taylor, John James, 212
Tegetmeier, William, 92
Teilhard de Chardin, Pierre, 22, 30, 301, 486, 489–92, 503
teleology, 32, 152–57
    Darwin and, 153–56
    defined, 152, 157
    morphology and, 154
    natural, 153
    post-Darwinian, 156–57
    unnatural, 153
teleonomy (Pittendrigh), 157
Temple of Nature, The (Erasmus Darwin), 437
Tempo and Mode in Evolution (Simpson), 285, 356, 357
tendency, hereditary, 97
Tennyson, Alfred Lord, 27, 44, 383, 391, 437
Ternate essay (Wallace), 168–70
Tess of the D'Urbervilles (Hardy), 440
Tessier, Georges, 306–07, 308–11
Thackeray, William Makepeace, 440
Thales, 33
theism, and first cause, 215
theology
    natural, 153, 215–17, 488
    process, 475
Theory of Chaos, The, 251
Theory of Heavenly Evolution, The (Yan Fu), 252
Theory of Island Biogeography (MacArthur and Wilson), 392
Third Critique (Kant), 235
Third International Conference on Genetics, 315
Thompson, John Vaughn, 82
Thomson, J. Arthur, 331, 348
Thomson, William (Lord Kelvin), 11, 125, 129, 220, 224
Thoreau, Henry, 383
tianyan (evolution), 253
Tianyan lun (The Theory of Heavenly Evolution, Yan Fu), 253, 254

Index

Tiedemann, Friedrich, 237
Tierra del Fuego, 126, 189
Tiktaalik, 397
Tillich, Paul, 473
Timaeus (Plato), 35, 153
time
   geological, 125
   scales, 125
Time Machine, The (Wells), 27
Timiriazev, Kliment Arkeedevich, 267
Tinbergen, Niko, 296, 349, 350, 510
   Wynne-Edwards and, 350
Tiniakov, G. G., 332
"To Build a Fire" (London), 232
Tocqueville, Alexis de, 226
Toledot ha-Adam (The Generations of Adam, Levy), 494, 496
Tomasello, Michael, 422,
Tooby, John, 464, 510
Tooke, James Horne, 183
tools, stone, 426
Torah, 495
Torrance, Thomas F., 473
Torre, Carlos de la, 261
tosca (red earthy clay), 57
Toxodon, 62, 62
Tradescantia, 288
traducianism, 470
Traité philosophique et physiologique de l'hérédité naturelle (Lucas), 117
traits, 208–10, 296, 330–35
   God and, 146
   optimality of, 334
   universal, 283
transduction, 343
transfer, lateral (horizontal) gene, 23, 343
transformation, 343
   saltations and, 144
transformism, 246
transmutation, 4, 40, 166, 168, 174
   archetype-based, 43
   Darwin and, 65
      first endoresement of, 203
   extinction and, 74–75
   fossils and, 62
   Lamarck and, 41
   Lyell and, 65,
"Transmutation of Species" (Darwin), 62
transplantation, reciprocal, 316
Travaux du laboratoire de géologie de la Faculté des sciences de Lyon (1921–43), 306
Treatise on Domestic Pigeons, A, 92
tree
   evolutionary (Lack), 372
   genome, 345
   of one percent, 344
   of organisms, 344
   phylogenetic (Haeckel), 300
   primate family (Darwin), 181
Treuenfels, Rabbi Abraham, 497
tribe, 16
Tristram, Reverend Henry, 337
Trivers, Robert, 108, 180, 351
Troland, Leonard, 326
Trollope, Anthony, 438, 440
Tropical Nature and Other Essays (Wallace), 171
truth, 454–56

   correspondence theory of, 455
   minimalist theory of, 455
Tschermak, Erich von, 230, 268, 269, 272, 314
Turesson, Göte, 316
Turing, Alan, 411
Türk Tarihi Tezi (Turkish History Thesis), 501
Türk Tarihin Ana Hatları (The Main Lines of Turkish History, Inan), 501
Turrill, J. B., 316
Turrill, William Bertram, 272,
Tuskegee experiment, 509
Tutt, J. W., 142, 296
2001, A Space Odyssey, 419
Tylor, Edward B., 185
Tyndall, John, 251
types
   law of the succession of (Darwin), 62
   unity of, 98

Über die Entstehung der Arten (Bronn), 237
ultracentrifugation, 270
Unger, Franz, 122, 506
uniformitarianism, 52, 183
uniformity, 93
unions, trade, and Darwin, 15
Unitarianism, 212
Urey, Harold, 326
utilitarianism, 189
utility, Darwin and, 190

Valen, Leigh Van, 373
van Fraassen, Bas, 459
Vane-Wright, R. I., 339
variability, 90
variation, 11, 54, 80, 96, 102, 288, 376, 406, 431
   adaptive, 69
   as random genetic mutations, 296
   beneficial, 209
   chance, 147, 148
   community good and, 347
   continuous, 161
   Darwin and, 146, 378
   definite (Darwin), 149
   design and, 155
   directed (orthogenesis), evolution and, 224
   fluctuating (Darwin), 149
   God and, 12, 146, 148, 149, 151, 227
   heritable, 88, 90, 91, 116, 297
   as a lottery, 150, 151
   monstrous, 69
   natural selection and, 148
   nature of, 161
   prediction and, 147
   racial, 14, 69
   selection and, 104
   sudden (saltation), 224
   undesigned, 155
Variation and Evolution in Plants (Stebbins), 285, 321
Variation of Animals and Plants under Domestication, The (Darwin, 1868), 88, 93, 116, 121, 134, 149, 155, 164, 314, 337, 506
   Chinese editions of, 255
   sterility and, 206
Vasconcelos, José, 262
Vavilov, Nikolai Ivanovich, 270
Venter, J., 403

"Versuche über Pflanzen-Hybriden" (Mendel), 267
Vestiges of the Natural History of Creation (Chambers), 8, 14, 44, 140, 166, 220, 437
View of the Evidences of Christianity (Paley), 213
Vines, S. H., 265
Virchow, Rudolph, 122, 235, 237, 240, 259, 413
Viret, Jean, 304
Virgularia patagonica, 202
vocalization
   alarm calls, 423–24
   animal communication through, 422
   scream, 424
Vogt, Carl, 184, 244, 444
volcanoes, 49–51
   earthquakes and, 51
Vollmer, Gustav, 455
Voltaire, 212
Volterra, Vito, 308, 311
von Baer, Karl Ernst, 83, 237
von Buch, Christian Leopold, 50
von Ihering, Hermann, 260
Voyage of the Beagle, The (Darwin)
   foreign editions, 244
   popularity of, 8
Voysey, Charles, 215

Wächtershäuser, Günter, 328
Wagner, Moritz, 160, 364
Wake, Charles Staniland, 186
Walden (Thoreau), 383
Wallace, Alfred Russel, 15, 45, 78, 99, 104, 107, 111, 114, 128, 141, 154, 158, 161, 165–72, 175, 197–99, 206–08, 220, 224, 259, 274, 330, 470
   in the Amazon, 167–68
   Bates and, 139, 140, 167
   Darwin and, xi, 8, 14, 96
   early life of, 165–67
   George and, 198
   influence of, 171–72
   Malthus and, 168
   Owen and, 166, 171
   Romanes and, 113
   scientific method of, 14
   sexual dimorphism and, 107
   Spencer and, 171, 197, 198
   in the Tropics, 168–71
Wallace's Line, 170
Walras, Leon, 248
Walton, A. J., 512
warblers
   Mniotiltid, 371
   New England, 386
Ward, Marshall, 265
Warfield, Benjamin Breckinridge, 469, 471
Wasmann, Father Erich, 240
wasps (*Ichneumonidae*), 148, 149, 471
Water Babies, The (Kingsley), 217, 222
Waterhouse, George Robert, 77–78, 82
Waters, Kenneth, 89
Watson, James, 299, 409
Watt, James, 188
Way of All Flesh, The (Butler), 27
Weald, 125
Weald, denudation of the, 99
Weber, Bruce, 270

Index

Wedgwood, Allen (cousin), 213
Wedgwood, Hensleigh (cousin), 183
Wedgwood, Josiah (maternal grandfather), 3, 173, 188
Wedgwood, Josiah (maternal uncle), 52, 212
weed, common (*Crepis*), 271, 288, 317
Weikart, Richard, 241
Weinberg, Wilhelm, 275
Weismann, August, 163, 198, 224, 229, 239, 349, 379, 471, 508
Weldon, W. F. Raphael, 225, 274–75, 332
Wells, H. G., 27
Wells, William, 45
Werner, Abraham, 47
Westwood, John Obadiah, 76
whales, 76, 160
Wheeler, William Morton, 24, 348, 352
Whence Strength (Yan Fu), 252
Whewell, William, 10, 48, 95, 133, 154, 155, 160, 190, 331, 376
Whitcomb, John C., Jr., 29, 481
White Fang (London), 232
White, Ellen G., 478
Why Do We Get Sick? The New Science of Darwinian Medicine (Williams and Neese), 511
Wichura, Max, 119
Wickham, First Lieutenant John C., 56
Wilberforce, Archbishop Samuel, xi, 18, 217, 219, 220
    Darwin and, 134
Wilcox, S. E., 422
Wilder, Burt, 228

Wilkinson, John, 89, 90, 91
Williams, George C., 108, 157, 203, 351, 392, 449, 505, 511
Williamson, John, 266
Williamson, William Crawford, 266–67, 272
    Darwin and, 267
Wilson, Alan, 380
Wilson, David Sloan, 203, 204, 352, 432
Wilson, Edmund B., 20, 230, 510
Wilson, Edward O., 24, 28, 352, 372, 373, 392, 393, 395, 396, 464, 485
Wilson, Woodrow, 233
Winge, Ovjnid, 316
Wisdom of God as Manifest in the Works of Creation, The (Ray), 153
Wise, Rabbi Isaac Mayer, 493
Witham, Henry, 267
Woese, Carl, 341, 344
Women and Economics (Gilman), 447
Wonderful Century, The (Wallace), 171
Woodlanders, The (Hardy), 440
Woodward, A. S., 414
Wordsworth, William, 437
World of Life, The (Wallace), 171
world, six regions of (Wallace), 171
Woronin, Mikhail, 265
Wortley, J. Stuart, 361
Wortman, Julius, 265
Wright, Larry, 156
Wright, Sewall, 21, 144, 277, 278–79, 280–81, 283–84, 308, 371
    Fisher and, 297
Wu Rulun, 253

Wu, D. P., 345
Wücherer, Otto, 260
Wulf, Theodor, 263
Wynne-Edwards, Vero, 24, 203, 350–51
    Wilson and, 352

Xenophon, 34

Yahaya, Haran, 504
Yahya, Harun (Adnan Oktar), 331, 502
Yan Fu, 252–54
Yarrell, William, 69, 118, 120,
Yerkes, Robert Mearns, 449
Youatt, William, 91, 120
Young Turks, 500

Zahavi, Amotz, 108
Zahm, John, 485
Zeno, 33
Zhang Taiyan, 253
Zheng Zuoxin, 250
Zhong Tianwei, 252
Zhou Zuoren, 253
Zittel, K. A. von, 356
zoogony, double (Empedocles), 34
Zoological Philosophy (Lamarck), 323
Zoology of the Beagle (Darwin, 1839), 370
Zoonomia (Erasmus Darwin), 4, 47, 48, 66, 68, 120, 236, 437, 506
zoophytes, 202
Zuberbühler, K., 424,
Zuckerkandl, Emile, 341, 417,
Zur Morphologie (Goethe), 236

576
.8203   Darwin and evolutionary though
Cam   Aurora P.L. APR13
        33164004879032